Engineering Sciences and Fundamentals

Core Programming Topic at the 2012 AIChE Annual Meeting

Pittsburgh, Pennsylvania, USA
28 October - 2 November 2012

Volume 1 of 2

ISBN: 978-1-62276-726-7

Printed from e-media with permission by:

Curran Associates, Inc.
57 Morehouse Lane
Red Hook, NY 12571

Some format issues inherent in the e-media version may also appear in this print version.

Printed by Curran Associates, Inc. (2013)

For permission requests, please contact AIChE
at the address below.

AIChE
3 Park Avenue
New York, NY 10016-5991

Phone: (203) 702-7660
Fax: (203) 775-5177

www.aiche.org

Additional copies of this publication are available from:

Curran Associates, Inc.
57 Morehouse Lane
Red Hook, NY 12571 USA
Phone: 845-758-0400
Fax: 845-758-2634
Email: curran@proceedings.com
Web: www.proceedings.com

TABLE OF CONTENTS

VOLUME 1

VOLUME 2

(7a) Modelling the Fluid Phase Behaviour of Aqueous Mixtures of Multifunctional Alkanolamines and Carbon Dioxide Using Transferable Parameters with the SAFT Approach

Alexandros Chremos, Department of Chemical Engineering, Centre for Process Systems Engineering, Imperial College, London, NJ, United Kingdom, Niall Mac Dowell, Chemical Engineering, Imperial College London, London, United Kingdom, Javier Rodriguez, Imperial College London, London, United Kingdom, Felix Llovell, MATGAS Research Center and ICMAB-CSIC, Bellaterra, Barcelona, Spain, Amparo Galindo, Department of Chemical Engineering, Centre for Process Systems Engineering, Imperial College London, London, United Kingdom, Claire S. Adjiman, Department of Chemical Engineering, Imperial College London,Center for Process Systems Engineering, London, United Kingdom and George Jackson, Centre for Process Systems Engineering, Dept of Chemical Engineering, Imperial College London, London, United Kingdom

In light of growing concern surrounding rising levels of atmospheric carbon dioxide (CO_2), there has been a great deal of renewed interest in alkanolamine-based chemisorption processes as part of the effort to mitigate anthropogenic emissions, particularly from large fixed-point emission sources such as fossil-fuel fired power-stations. The large-scale deployment of alkanolamine-based CO_2 capture technology will entail a significant economic penalty, and the reduction of this cost is a well-recognized imperative. A key element for CO_2 capture is the design of solvents and solvent blends with thermodynamic properties which would allow optimum capture operation processes. To describe and predict the thermodynamic properties of such solvents we have developed molecular models for use with the statistical associating fluid theory (SAFT). Specifically, the purpose is to identify mixtures that are good candidates for CO_2 absorption using transferable models and a group contribution framework within SAFT. In the proposed approach chemical equilibrium is treated with a physical approach, so that reaction products are treated implicitly. Two versions of the SAFT approach will be discussed; the SAFT-VR (SAFT for potentials of variable range) and the SAFT-γ (group contribution) equations of state. In particular, we present the theoretical basis of the applicability of SAFT-VR to carbon dioxide, water and alkanolamine mixtures, and discuss how SAFT-γ opens avenues for an improved thermodynamic prediction. The aqueous solvents examined include monoethanolamine (MEA), diethanolamine (DEA), methyldiethanolamine (MDEA), 2-amino-2-methyl-1-propanol (AMP), and their mixtures with CO_2. An excellent overall predictive capability is demonstrated for the fluid phase behaviour of pure components and binary aqueous mixtures. A good overall description is also obtained for the adsoprtion of CO2 in aqueous mixtures of alkanolamines, particulary for DEA and MDEA. The implicit treatment of the reaction equilibrium is validated by comparing predicted concentrations of reaction speciation products (carbamate and bicarbonate) with available experimental data.

(7b) Molecular Simulations of Amine Functionalized MCM-41 for the Separation of CO2/N2

Santiago Builes, MATGAS Research Center, Bellaterra, Barcelona, Spain and **Lourdes F. Vega**, MATGAS Research Center and Carburos Metálicos, Air Products Group, Barcelona, Spain

Molecular simulations of amine functionalized MCM-41 for the separation of CO_2/N_2

The sorptive behavior of amorphous silica can be enhanced by the use of organic groups. Using certain organic species can increase the selectivity of solid silica for a given compound in a gas mixture. These functionalized materials can be used to capture gases reversibly using adsorption. The effective design of these materials requires a method that can relate the structure of the adsorbent to their performance. This level of understanding can be achieved by using molecular simulations, as they relate the microscopic behavior of the molecules during the adsorption process to the macroscopic behavior of the system, allowing one to tailor materials for specific separation purposes.

We propose a simulation methodology for the design of post-synthesis functionalized silica materials and the calculation of chemisorbed and physisorbed CO_2 using Grand Canonical Monte Carlo molecular simulations. For this purpose realistic model adsorbents are generated using an energy-bias selection scheme for the possible grafting sites. The methodology is robust and can be applied for a large variety of different combinations of solid materials and grafting chains. As an example we evaluate a model MCM-41 grafted with 3-aminopropylsilane for the separation of mixtures of CO_2 and N_2 in the framework of reduction of CO_2 emissions from flue gases. We analyze the effect of physisorption and chemisorption on the separation of CO_2 using grafted amines. The degree of functionalization and the type of bonding of the organic chains to the inorganic adsorbent are studied aiming to optimize the sorption of CO_2 over N_2.

This work was partially financed by the Spanish Government under projects CTQ2008-05370/PPQ, CTQ2011-23255 and CENIT SOST-CO2 (CEN-2008-1027). Additional support from the Catalan Government through 2009SGR-666 and Carburos Metálicos was also provided. S.B. acknowledges a Talent grant from the Comission for Universities and Research of the Generalitat de Catalunya.

(7c) Process Model-Free Analysis of CO2 Capture Systems

John P. O'Connell[1], Nicholas McGregor[2], Jessie Kennedy[2] and Paul M. Mathias[3], (1)Chemical Engineering, University of Virginia, Charlottesville, VA, (2)University of Virginia, Charlottesville, VA, (3)Fluor Corporation, Aliso Viejo, CA

Post-combustion capture of CO2 is typically accomplished by absorption from combustion off-gas into basic aqueous solutions, such as with alkanolamines and ammonia. There are significant energy requirements of regenerating the absorbent and, in the case of ammonia, controlling NH3 emission and absorber cooling. Proposed alternative processes include absorption into ionic liquids and membrane separation. Evaluation of process energy demands is often difficult because varying process conditions are studied and can depend on the process and property models employed.

An approach to comparing energy requirements of different processes can be based solely on steady energy and entropy balances of the streams entering and leaving individual or lumped process units. These two equations, with systematically varied entropy generation in the units, allow solving for two stream variables or heat/work amounts. The analysis provides quantitative descriptions of the effects of irreversibilities as well as can reveal process sections for focused improvement.

This method was previously used to evaluate different sulfur-iodine processes for hydrogen production [1]. In addition to showing reversible and realistic energy requirements, it also uncovered significant differences in property model predictions for the complex mixtures of the S-I system.

The current study uses the process model-free analysis for CO2 absorption systems using Aspen property models. Several cases for NH3 capture with different absorber conditions are compared with MEA systems with similar levels of irreversibility. The energy costs of NH3 recovery and CO2 processing are obtained. Property modeling issues for the CO2 absorption systems have also been found. Suggestions for applying the method to other CO2 processes will be made.

[1]. Process model-free analysis for thermodynamic efficiencies of sulfur–iodine processes for thermochemical water decomposition, J.P. O'Connell, P. Narkprasert, M.B. Gorensek, Int. J. Hydrogen Energy 34 (2009) 4033–4040.

(7d) Electrochemically-Mediated Amine Regeneration for Carbon Dioxide Scrubbing Processes

Michael C. Stern[1], Fritz Simeon[1], Howard Herzog[2] and T. Alan Hatton[1], (1)Chemical Engineering, Massachusetts Institute of Technology, Cambridge, MA, (2)MIT Energy Initiative, Massachusetts Institute of Technology, Cambridge, MA

Electrochemical-mediated amine regeneration (EMAR) is an alternative strategy for facilitating CO_2 desorption and regeneration of loaded amine solutions exiting the absorber in CO_2 scrubbing processes. The traditional methods used is either a thermal-swing regeneration or pressure-swing regeneration. The strong affinity for amine molecules to form aqueous complexes with cupric ions is exploited in the EMAR process to release the CO_2 from the amines isothermally through competitive binding. Electrochemical systems offer several advantages over thermal systems including significantly lower lean loadings, less required internal reconfiguration of the turbine systems, and more flexible operation that allows power to be redirected back to the grid in the event of a spike in energy demand.

Development of an efficient EMAR process requires both investigation into the chemistry and system design to address the thermodynamic, kinetic, and transport phenomena. Selection of the correct amine molecule to achieve desirable thermodynamic and kinetics properties requires careful consideration of the CO_2 reaction equilibrium and the stability constant of the copper complex formation. While strong copper binding is required to elicit the release of the CO_2, amines that complex too strongly to the cupric ions will greatly hinder the electrochemical kinetics of the amine/copper reactions. The kinetics of these reactions in the presence of CO_2, which have not been previously investigated, will be discussed.

A variety of voltammetric techniques will be used to assess the electrochemical kinetics for several different amine species in the presence and absence of CO
electrode setup to a bench scale system capable of running at several amperes. The results will be used to show the feasibility of large scale systems through kinetic models.

(7e) Dynamics Study and Gas Solubility in Phase Change Ionic Liquid for CO2 Capture

Hao Wu and Edward J. Maginn, Department of Chemical and Biomolecular Engineering, University of Notre Dame, Notre Dame, IN

Phase change ionic liquids (PCIL) undergo a solid to liquid phase transition when they react with CO_2. The heat of fusion during the phase change can be utilized to reduce the energy requirements for CO_2 capture. Here we present the results of a molecular simulation study in which the water solubility, dynamics and liquid structure of PCILs are computed and used to gain an understanding of these liquids.

Water solubility in PCILs is important because in the regeneration step, the separation of water from the IL solution requires additional thermal energy to vaporize the water. Water solubility is determined in molecular simulation by calculating the solvation free energy. These are related to the Henry's Law constants, which are then compared with experimental data. The enthalpy and entropy of solvation are also obtained from the simulations. A hydrogen bond analysis is performed and the structure of the liquid is studied in an effort to explain the solubility calculations. A trend of hydrophobicity in a series of PCILs is predicted.

It has also been found that the viscosity of aprotic heterocyclic anion-based PCILs does not increase due to CO_2 complexation, unlike conventional amines. Molecular dynamics simulations are performed to investigate the relationship between dynamic properties and liquid structure of PCILs. In this work, a prediction is made on dynamics and by analogy the viscosity of these ionic liquids by simulation and experiments. It is found that, like the experimental results, the dynamics are unaffected by reaction with CO_2 when aprotic heterocyclic anion-based PCILs are used.

(7f) Reversible Ionic Liquids for CO2 Capture: A Thermodynamic Study

Amy L. Rohan[1,2], Jackson R. Switzer[1,2], Kyle M. Flack[2,3], Emily C. Nixon[2,3], Amber C. Rumple[2,3], Elizabeth J. Biddinger[1,2], Manish Talreja[1,2], Pamela Pollet[2,3], Charles A. Eckert[1,2,3] and Charles L. Liotta[1,2,3], (1)School of Chemical and Biomolecular Engineering, Georgia Institute of Technology, Atlanta, GA, (2)Specialty Separations Center, Georgia Institute of Technology, Atlanta, GA, (3)School of Chemistry and Biochemistry, Georgia Institute of Technology, Atlanta, GA

We have developed reversible ionic liquids (RevILs) for CO_2 capture that are capable of dual mode absorption: chemical and physical. Chemical absorption takes place when the silylamine-based solvents chemically react with CO_2 forming an ammonium-carbamate ion pair (RevIL). Physical absorption is then observed due to the solubility of CO_2 in the RevIL. The chemically and physically absorbed CO_2 may be released by heating the CO_2-swollen RevIL to temperatures below 100°C.

CO_2 capture with reversible ionic liquids is governed by the thermodynamics of the process. The reaction equilibrium constant, K, of the silylamine and CO_2 dictates the extent of reaction at capture conditions and reversal temperatures. The additional CO_2 that is captured through physical absorption is determined by the solubility of the CO_2 in the RevIL (Henry's constant). During regeneration (release of CO_2) the RevIL is heated to the reversal temperature. The energy required to complete the reversal is structure-specific and depends on the heat capacity of the RevIL, the regeneration temperature and the enthalpy of reaction.

We have identified structural changes that may be made to the RevIL to increase CO_2 capture capacity and reduce the energy required for reversal.

(7g) Thermodynamic Model for CO_2 Absorption by Phase-Change Ionic Liquids

Luke D. Simoni, Joan F. Brennecke and Mark A. Stadtherr, Department of Chemical and Biomolecular Engineering, University of Notre Dame, Notre Dame, IN

Towards the goal of realizing a cost-efficient CO_2 capture system that would be effective for point sources, like post-combustion flue gas from coal-fired power plants, novel compounds and methods are needed that can absorb or adsorb CO_2, and subsequently release it for sequestration, with little parasitic energy penalty to the overall operation. Phase-change ionic liquids (PCILs) are a newly discovered class of materials; they are solid salts that, upon binding with CO_2, undergo a phase change to liquid. This change is readily reversible, releasing the CO_2 for storage and regenerating the solid PCIL. As CO_2 absorbents, PCILs are attractive because: 1) The heat of fusion provides a significant part of the energy needed to release the CO_2 from the absorbent, thus reducing the parasitic energy penalty, and 2) The CO_2 uptake is high (nearly one mole of CO_2 per mole of PCIL, at typical post-combustion flue gas conditions).

For purposes of preliminary design and optimization of CO_2 capture systems based on PCILs, as well as to assist in designing the most effective PCIL molecules, we have developed an absorption isotherm model that captures the basic chemical and physical principles and the underlying thermodynamics of the CO_2 absorption process. This model allows us to relate key operating conditions for the absorber and stripper to basic physical properties such as the enthalpies of CO_2 absorption (chemical and physical), the PCIL enthalpy of fusion and the PCIL melting temperature. In this presentation, we will describe the development of this model and the use of experimental data for parameter estimation. Variations of the model, based on differing assumption about the underlying phase behavior, will also be discussed. We show that this model provides a good prediction of CO_2 uptake by PCILs, and can be used to determine realistic values of physical parameters describing the system.

(30a) Molecular Simulation of Phase Equilibria and Self-Assembly: Progress and Challenges

Dr. Athanassios Z. Panagiotopoulos, Department of Chemical and Biological Engineering, Princeton University, Princeton, NJ

This lecture provides a review of molecular simulation methods for modeling phase equilibria and self-assembly, topics greatly influenced by Prof. Keith Gubbins. The Gibbs ensemble is a direct approach proposed over twenty years ago that is suitable for calculations of moderate accuracy. Combined with Kofke's Gibbs-Duhem integration, it allows the relatively routine calculation of phase diagrams for given intermolecular potentials. Alternatively, histogram reweighting Monte Carlo methods provide good accuracy near critical points and provide paths for overcoming hysteresis and long time scales inherent in self-assembly of surfactants and polymers. In recent years, with the advent of fast, scalable, open-source packages for molecular dynamics calculations the time scales over which self-assembly and microphase separation can be studied have been extended to μs, even for realistic potential models with explicit solvent. However, there is still a significant need for development of coarse-grained models that correctly capture structure and thermodynamics in order to extend the time and length scales of systems that can be modeled.

In-Pore Superhigh Pressure Effect on Solid Phase Transition and Organic Crystal Synthesis

Koki Urita,[1] Toshihiko Fujimori,[2] Kozue Abe,[3] Yuichi Shiga,[3] Tsutomu Itoh[2], Yoshiyuki Hattori,[4] Tomonori Ohba[3], Takayoshi Arai[3], Kenji Hata[5,] Masako Yudasaka,[5] Sumio Iijima,[5,6] Isamu Moriguchi,[1] Hirofumi Kanoh[3] and Katsumi Kaneko[2*]

[1]Department of Engineering, Nagasaki University, 1-14 Bunkyo-machi, Nagasaki 852-8521, Japan
[2] Research Center for Exotic Nanocarbons, Nagano, 380-8553, Japan
[3]Department of Chemistry, Graduate School of Science, Chiba, 263-8522, Japan
[4]Faculty of Textile Science and Technology, Shinshu University, Tuneda, Ueda 386-8567 Japan,
[5]National Institute of Advanced Industrial Science and Technology, Ibaraki 305-8561, Japan
[6]Department of Physics, Meijo University, Tempaku, Nagoya 468-8502, Japan
*Corresponding author, e-mail kkaneko@shinshu-u.ac.jp, fax 81-26-269-5737

Confinement of KI nanocrystals in nanoscale pores of single wall carbon nanohorn (SWCNH) and activated carbon fiber (ACF) below 0.1 MPa induces the formation of high pressure phase solid, although compression under 1.9 GP is necessary for formation of the high pressure phase. Also formation of sulfur-atomic chain crystal having metallic nature is found in the single wall carbon nanotube (SWCNT) and double wall carbon nanotube (DWCNT) below 0.1 MPa, although mettalic sulfur needs compression by more than 90 Gpa in the bulk phase. This remarkably strong compression effect of carbon nanospaces was applied to organic synthesis. Organic synthetic reaction, which occurs only above high pressure of 1GPa without nanopore systems, efficiently progresses in a glass flask under an atmospheric pressure.

Introduction

The solid nanospaces can offer intensive interaction potential fields for molecules. We clearly showed that the gas-phase disproportionation reaction of $(NO)_2$ to N_2O and NO_2 above 20 MPa occurs efficiently in the slit-shaped carbon nanospaces.[1] We named the effect of carbon nanospaces a quasi-high-pressure effect. The quasi-high-pressure effect of the slit-shaped nanopore spaces was also evidenced in the electrochemical reduction of CO_2 to CO by Fujishima et al..[2] The confinement of CH_4 molecules in the tube spaces of single wall carbon nanohorn (SWCNH) elevates the boiling temperature according to the temperature dependence of the rotational vibration spectra.[3] SWCNH has more defective structure than SWCNT; the defective structure has been more characterized with DFT-assisted surface enhance Raman scattering spectroscopy[4,5]. TEM and synchrotron X-ray diffraction studies showed that the confinement of KI in the tube spaces of SWCNHs induced formation of the high pressure solid phase requesting the compression by more than 1.9 GPa.[6] Gubbins et al gave the theoretical possibility for the quasi-pressure reaction.[7] Recently they have revisited this high pressure effect with the relevance to the experimental results.[8,9] Then the dependence of the superhigh pressure effect on the nanoscale pore geometry should be studied. Here the superhigh pressure effect by tube- and slit-shaped nanoscale pore spaces is compared with each other. Superhigh pressure organic synthesis is one of current subjects in organic chemistry and thereby an epoch making simple route for superhigh pressure synthesis should be invented. We have evidenced that the carbon nanospaces could induce superhigh pressure organic synthetic reaction, producing the bulk amount of the target organic product. Here we selected superhigh pressure reaction of 1,2-epoxy cyclohexane (ECH) into *trans*-cyclohexanediol (CHD) which occurs at 303 K above 1 GPa[10] and Cu-pthalocyanine synthesis which needs the compression of 0.9 GPa[11], as the target reaction.

Experimental

Confinement of KI nanocrystals in the slit-shaped pores of activated carbon fiber (ACF)s of 1.2 nm in width was carried out by sealing of ACFs and KI in the glass tube at 1073 K for 8 hrs after pretreatment at 423 K in vacu. The synchrotron X-ray diffraction(XRD) measurements were performed at beam-line

BL02B2, SPring-8, where the wavelength of the incident x-ray was 1.002(9) Å. The atomic structures of KI in pores of ACF were directly observed by a TEM (JEOL: JEM-2010) with an accelerating voltage of 200 kV.

ECH was dissolved in water and acetone solution in the presence (and absence) of activated carbon fiber (ACF) or single wall carbon nanotube (SWCNT) and the mixed solution was refluxed at 333 K for 24 h in a glass flask at an ambient pressure. The nanopore structure of ACF and SWCNT was determined by N_2 adsorption at 77 K. The products were filtrated and dried in *vacuo*. They were characterized by TD-DTA, FT-IR, TEM, NMR, and DART-MS. The FT-IR was measured using a KBr disk.

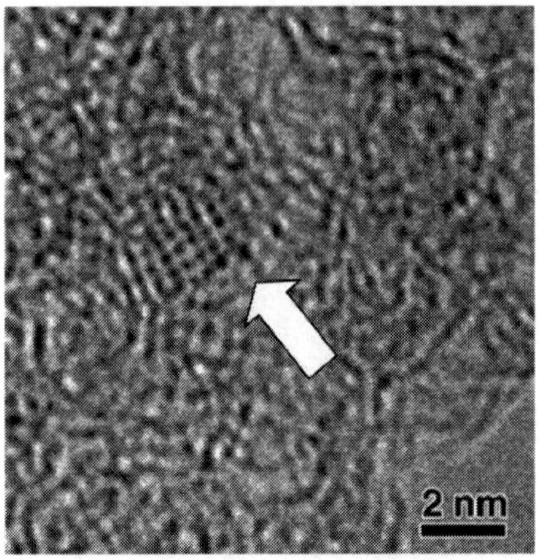

Figure 1. TEM image of KI in the slit pore space of ACF. The arrow Indicates the nanocrystal.

Results and discussion

Figure1 shows the TEM picture of a nano KI crystal confined in the slit-shaped pore space. The crystallite of less than 2 nm in size has a plate-like structure, agreeing with the slit-like pore. The lattice distances are 0. 39 nm and 0.37 nm, being different from those of the crystal lattice constants of the ambient pressure phase (a = b = c = 0.7061 nm). However, these distances are also different from those of high pressure phase observed in SWCNH(a = b = 0.35 nm, c = 0.304 nm or 0.386 nm). Synchrotron XRD had additional weak peaks which could not assign to KI crystals of the ambient pressure phase. These should be associated with new structure formed due to the superhigh pressure effect. The KI nanocrystals of high pressure phase was unstable in the slit-shaped pores, because the accessibility of water vapor in the atmosphere was easier than the tube space on exposing the KI-doped ACF to the air.

Sulfur was introduced in the tube spaces of SWCNT and DWCNT and then the S structure was determined by TEM and synchrotron XRD. We found a single chain crystal of sulfur atoms in those nanotube spaces, which give sharp XRD peaks indicating the complete 1D crystal.[12]

Figure 2 shows FT-IR spectra of the products on ACF, CHD, and ACF, indicating that CHD is produced. DART-MS also clearly evidenced the formation of CHD. The formation of CHD was also ascertained in the presence of single wall carbon nanotube. The CHD products were obtained in the bulk crystals; the yields of CHD were 40 % for ACF and 24 % for SWCNT, respectively. CHD was not produced in the absence of nanopore systems.[13]

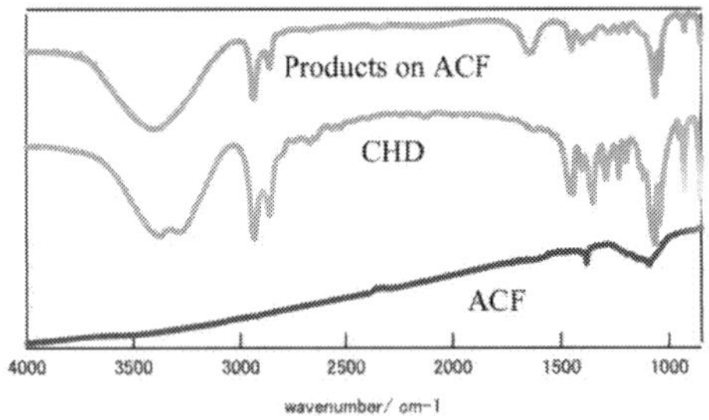

Fig. 2. FT-IR spectra of products on ACF, CHD, and ACF.

Conclusions

Confinement of materials in nanoscale pore spaces near ambient pressure induces high pressure phase transition and synthetic reaction which needs the high pressure of more than 1 GPa.

Acknowledgment.

This work was supported by Grant-in-Aid for Scientific Research (A) (No. 21241026). K. K., T. I., and T. F. were supported by Exotic Nanocarbons, Japan Regional Innovation Strategy Program by the Excellence, Japan Science and Technology Agency.

References

[1] J.Imai, M. Souma, S. Ozeki, T.Suzuki and K. Kaneko, *J. Phys. Chem.* 95(1991)9955.

[2]T. Yamanoto, D. A. Tryk K.Hashimoto and A. Fujishima, *J. Electrochem. Soc.* 147 (2000) 3393.

[3] S. Hashimoto, T. Fujimori, H. Tanaka, K. Urita, T. Ohba, H. Kanoh, T. Itoh, M. Asai, H. Sakamoto, S. Niimura, M. Endo, F. Rodoriguez-Reinoso and K. Kaneko, *J. Amer. Chem. Soc.* 133 (2011) 2022.

[4] T. Fujimori, K. Urita, T. Ohba, H. Kanoh, K. Kaneko, *J. Am. Chem. Soc.* ***132***, 6764–6767 (2010).

[5] T. Fujimori, K. Urita, D. Tomanek, T. Ohba, I. Moriguchi, M. Endo, K. Kaneko, *J. Chem. Phys.* **136**, 064505-1 064505-5 (2012).

[6] K. Urita, Y. Shiga, T. Fujimori, Y. Hattori, H, Kanoh, T. Ohba, H. Tanaka, M. Yudasaka, S. Iijima, I, Moriguchi, F. Okino, M. Endo and K. Kaneko, *J. Amer. Chem. Soc.* 133(2011) 10344.
[7] E. E. Santiso, A. M. Gorge, K. E. Gubbins and M. B. Nardlli, *J. Chem. Phys.* 125 (2006) 084711-1.
[8] Y. Long, J. C. Palmer; B. Coasne, M. Sliwinska-Bartkowiak; K. E. Gubbins, *Phys. Chem. Chem. Phys.* 13 (2011) 17163.
[9] Y. Long, J. C. Palmer; B. Coasne, M. Sliwinska-Bartkowiak; K. E. Gubbins, *Micro. Meso. Mater.* 154 (2012)19.
[10] H. Kotsuki, M. Kataoka, and H. Nishizawa, *Tetrahedron Lett.* 34 (1993) 4031.
[11] S. J. Edmondson, J. S. Hill, N. S. Isaacs, P. C. H. Mitchell, *J. Chem. Soc. Dalton Trans.* 1990, 1115.
[12] T. Fujimori et al, To be submitted.
[13] K. Abe et al, To be submitted.

(30c) Isotropic-Polar Phase Transition in An Amphiphilic Fluid

Martin Schoen, Stranski-Laboratory for Physical and Theoretical Chemistry, Technical University, Berlin, Berlin, Germany; Department of Chemical and Biomolecular Engineering, North Carolina State University, Raleigh, NC

We present Monte Carlo simulations of the isotropic-polar (IP) phase transition in an amphiphilic fluid carried out in the isothermal-isobaric ensemble. Our model consists of Lennard-Jones spheres where the attractive part of the potential is modified by an orientation-dependent function. This function gives rise to an angle dependence of the intermolecular attractions corresponding to that characteristic of point dipoles. Our data show a substantial system-size dependence of the dipolar order parameter. We analyze the system-size dependence in terms of the order-parameter distribution and a cumulant involving its first and second moments. The order parameter, its distribution, and susceptibility observe the scaling behavior characteristic of the classical 3D-Heisenberg universality class. Because of this scaling behavior and because all cumulants have a common intersection irrespective of system size we conclude that the IP phase transition is continuous. Considering pressures $1.3 \leq P \leq 3.0$ we demonstrate that a line of continuous phase transition exists which is analogous to the Curie line in systems exhibiting a ferroelectric transition. Our results are can be explained semi-quantitatively by a simple mean-field theory adapted from the theory of IP phase transitions in fluids in which molecules carry an electromagnetic point dipole.

(30d) Insight Into Phase Behavior and Microstructure of Complex Fluids – in Honor of Keith Gubbins' 75th Birthday

Walter G. Chapman, Bennett D. Marshall II, Kai Gong, Deepti Ballal and Kenneth R. Cox, Chemical and Biomolecular Engineering, Rice University, Houston, TX

To design and optimize processes and products in the energy, chemicals, and materials industries, engineers need computational tools to predict bulk and interfacial equilibrium properties for complex fluids in terms of temperature, pressure, composition, and molecular geometry. Systems that have been particularly challenging to model include associating or polar molecules as well as polymers and amphiphiles. Applications range from flow assurance and enhanced oil recovery in the energy industry to paints and coatings and advanced materials.

Advances in statistical mechanics have led to successful molecular models for bulk fluids such as the statistical associating fluid theory (SAFT). SAFT enables engineers to predict the effects of molecular weight, polydispersity, polarity, association, and compressibility on the bulk phase behavior of mixtures containing solvents, monomers, and polymers. Extensions using density functional theory show great promise in modeling interfacial properties and molecular structure, including the behavior of fluids at hydrophilic and hydrophobic surfaces, segregation in polymer blends, self assembly of surfactants and block copolymers, and polymer depletion forces in polymer/colloid systems.

We briefly review the physical basis for the SAFT approach and the extension to interfacial systems. Some recent extensions and applications of the SAFT approach for bulk and interfacial systems are presented. These points are illustrated through comparisons with molecular simulation as well as phase equilibria data for monomers, solvents, and copolymer systems.

(30e) Phase Transitions of Liquids Confined in Nano-Pores; The Novel Ice Structures

Malgorzata Sliwinska-Bartkowiak, Institute of Physics, Adam Mickiewicz University, Poznan, Poland

Melting of fluids confined in nano-porous materials has been extensively studied using both experiments and simulations . This effort is relevant to the fundamental understanding of the behavior of nano-dimensional fluids and solids, and the influence of surface forces on such materials. Nano-porous materials, such as activated carbons, silicas, etc., play a prominent role in chemical processing, particularly in separations and as catalysts and catalyst supports. Fluids confined in such porous materials possess many novel properties that can form the basis of future nanotechnologies. Recent studies for pores of simple geometry have shown a rich phase behavior associated with melting in confined systems. The melting temperature may be lowered or raised relative to the bulk melting point, depending on the nature of the adsorbate and the porous material. Much of the apparently complex phase behavior is a result of competition between the fluid-wall and fluid-fluid intermolecular interactions. Of particular interest have been the effects of confinement on phase transitions, on the structure of water and ice in nano-pores , and on dynamical properties of water in one-dimensional pores considered as a model of water channels.

We report experimental results on the structure and melting behavior of ice confined in multi-walled carbon nanotubes (MWCN) and ordered mesoporous carbon CMK-3, which is the carbon replica of SBA-15 silica template. The silica template has cylindrical mesopores with micropores connecting the walls of neighboring mesopores. The structure of the CMK-3 consists of carbon rods connected by smaller side-branches, with quasi-cylindrical mesopores of average pore size 4.5 nm and micropores of 0.6 nm. Neutron diffraction, dielectrics relaxation and differential scanning calorimetry have been used to determine the structure of the confined ice and the solid - liquid transition temperature. The results are compared with the behavior of water in MWCN of inner diameters of 2.4 nm and 4 nm studied by the same methods. For water in CMK-3 we observe the existence of cubic ice, tetragonal : ice VIII and ice IX nanocrystal forms, while we find cubic ice in the case of MWCN and CMK-3 mesopores. These crystal forms, which occur in bulk water only at temperatures below 180 K in the case of cubic ice, and at pressures of hundreds or thousands of MPa in the case of ice VIII and IX, are stabilized by the confinement.

(30f) Nucleation From Solution: How Slow Diffusion Changes Everything

Baron Peters, Chemical Engineering, University of California Santa Barbara, Santa Barbara, CA

We show that a previously un-noted coupling exists between slow diffusion transport and barrier crossing in nucleation. The effect is modeled using the diffusion equation, an Ostwald-Freundlich boundary condition, and a mass balance linking nucleus size to flux across the nucleus-solution interface. The model retains the classical nucleation theory as a special case because of the common theoretical foundations behind classical nucleation theory and the Ostwald-Freundlich equation. However, the model also shows that certain types of concentration profiles can drive a classically pre-critical nucleus over the nucleation barrier. We identify the separatrix as a function of both nucleus size and characteristics of the local concentration field. Our analysis suggests that local concentration fluctuations "stochastically gate" nucleation events, analogous to stochastic gating in the theory of chemical reactions. Our analysis provides new insight on the mechanisms of two-step nucleation processes. Finally, our analysis provides a starting point for further statistical field theory analyses of local concentration fluctuations and their effects on nucleation rates. Time permitting, we will also show how slow diffusion on the free energy landscape can influence polymorph selection.

B. Peters, *J. Chem. Phys.* 135, 044107 (**2011**)

B. Peters, *J. Chem. Phys.* 131, 224103, (**2009**)

(30g) Molecular Modeling of Crystallization of Flexible Organic Molecules

Erik E. Santiso[1], Andrei V. Kazantsev[1], Manolis Vasileiadis[1], Claire S. Adjiman[1], Constantinos C. Pantelides[1] and Bernhardt L. Trout[2], (1)Department of Chemical Engineering, Imperial College London,Center for Process Systems Engineering, London, United Kingdom, (2)Massachusetts Institute of Technology, Cambridge, MA

Crystallization is an extremely important unit operation the food, chemical, and pharmaceutical industries, among others. However, understanding the molecular-level details of nucleation and polymorph selection in molecular crystals remains a challenging problem. On the one hand, the stochastic nature of nucleation, and the fact that is usually involves only a few molecules, make it very hard to probe it experimentally. On the other hand, the time scales involved are well beyond the reach of traditional molecular simulation methods.

In this work, we present some recently developed methods for modeling the nucleation of molecular crystals, and for predicting possible crystalline polymorphs. Nucleation is modeled by constructing minimum free energy paths in a space of order parameters constructed to be sensitive to the structure of a particular polymorph. These minimum free energy paths are constructed using the string method in collective variables, and free energy barriers are then estimated using milestoning with Voronoi tesselations. Finally, we discuss how to construct plausible crystal structures for unknown polymorphs by using a global optimization method coupled with an *ab initio* potential.

(31a) Biomaterial-Cell Interface: Implications for Drug Delivery

Samir Mitragotri, Chemical Engineering, University of California, Santa Barbara, Santa Barbara, CA

Polymeric nanoparticles have found application in varied fields including drug delivery and medical imaging. The behavior of therapeutic nanoparticles in the body is determined by their interactions with cells including macrophages, endothelial cells and cancer cells. Such interactions determine their therapeutic performance including circulation half-life, drug release rates and toxicity. My talk will focus on recent advances in engineering of nanoparticles to control their biological interactions. In particular, I will focus on how physical attributes of particles including shape and deformability can be controlled to improve fundamental understanding of cell-material interactions and therapeutic performance of carriers. We have devised methods to generate particles of several distinct morphologies and studied their impact on key processes in drug delivery including phagocytosis, circulation, adhesion of vascular walls, and targeting. Based on this understanding, we have designed novel particles that demonstrate enhanced targeting.

Our studies demonstrate that particle morphology provides a new dimension in engineering of polymeric carriers and opens up new opportunities in drug delivery. In addition to morphology, we also demonstrate that controlling mechanical properties of carriers also offers unique opportunities. Specifically, we have synthesized flexible particles made from proteins that mimic the physical and functional properties of body's own circulating cells such as red blood cells and platelets. Particles that mimic the size, shape and flexibility of natural circulating cells offer advantages that are typically lacking in conventional spherical polymeric particles.

The motivation to use physical properties of nanoparticles to control biological function is provided by the biology itself. In nature, numerous examples can be found where physical aspects, such as shape, mechanical properties and compartmentalization are crucial to biological function. Physical attributes such as size, shape and mechanical properties form essential building blocks of biology. This realization forms the basis of the new paradigm in design of nanoparticles.

(31b) Single Molecule Tracking At Wet Interfaces

Daniel K. Schwartz, Chemical and Biological Engineering, University of Colorado at Boulder, Boulder, CO

Interactions between molecules and surfaces lead to complex and highly-varied interfacial behavior, where heterogeneity may arise from spatial variation of the surface/interface itself or from molecular configurations (i.e. conformation, orientation, aggregation state, etc.). The direct observation of adsorption, interfacial diffusion, and desorption of individual fluorescent molecules permits the characterization of heterogeneous interfacial behavior in ways that are inaccessible to traditional ensemble-averaged methods. Moreover, spectral information can be used to simultaneously track molecular configuration. Single-molecule tracking experiments have traditionally been limited by small sample sizes (e.g. a few hundred molecules), leading to poor statistical significance and a lack of sensitivity to rare populations. However, new advances in high-throughput tracking methods now enable hundreds of thousands of molecules to be followed in a given experiment. This approach has recently been used to characterize heterogeneous molecule-surface interactions including: multiple modes of diffusion and desorption associated with both internal and external molecular configuration, Arrhenius activated interfacial transport, spatially dependent interactions, and others.

(31c) Engineering of Self-Propelling Particles That Move in Programmed Trajectories and Haul Cargo

Orlin Velev, Chemical and Biomolecular Engineering, North Carolina State University, Raleigh, NC

The fabrication of complex self-motile particles is an emerging topic of fascinating potential, yet few principles of particle propulsion have been studied and few concrete applications exist to date. We will report the progress in addressing the challenges of designing physical mechanisms of moving such particles, and identifying areas of their practical use. First we will discuss how external uniform AC electrical fields can be used as means of providing energy, propelling and controlling the motion of Janus metallo-dielectric spheres and miniature semiconductor diodes. The asymmetry of these particles makes them move by particle-localized AC and DC electrohydrodynamic propulsion. We will demonstrate how the diode particles can be steered by controlling the symmetry of the AC field and will hypothesize how an additional level of complexity can be engineered to turn them into prototypes of self-propelling microdevices. In the second part of the talk we will present a new class of gel-based self-propelling particles moving by Marangoni effect in a programmed oscillatory mode. These gel boats floating on the water surface are driven by a hydrogel reservoir releasing an ethanol flux that is periodically disrupted by the bulk flows around the particles. The pulse interval and the distance propelled in a pulse by these gel floaters were interpreted based on the rate of ethanol mass-transfer. The model allowed us to design particles that perform "dances" of repeatable periodic sequences of forward/backward, or rotary steps, and travel in complex pre-programmed trajectories on the liquid surface. We will also present new types of functionalized self-propelling floaters driven by surfactant release that can find applications as mixers and cargo microcarriers, and may serve as a new platform for environmental remediation.

(39e) Thinking Outside Our Box: On Being a Fluid Mechanical "Missionary"

Todd M. Squires, Department of Chemical Engineering, University of California at Santa Barbara, Santa Barbara, CA

This abstract is meant to be provocative.

The fluid mechanics community has seen incredible advances in the last century: a deep, physical understanding of a wide variety of fluid phenomena; the mathematical formulation of fluid mechanical problems (and particularly the value of non-dimensionalization); asymptotic analysis for accurate approximations (and the physical intuition that arises from such approximations); and powerful computers and algorithms that enable direct numerical simulations that would have been inconceivable a few decades ago.

That said, I fear that we in the fluid mechanics community have evolved so far, and in a direction that many of our colleagues in Chemical Engineering and other disciplines have not, that the wealth of expertise and intuition we have developed does not have the impact on the broader community that it should. In addition to framing the broader discussion in this session about "New Directions" for fluid mechanics, I will briefly discuss some work we did that was part research, part academic "outreach". A large community has been working to develop bio-sensors; some come from ChE, but many are in Chemistry, Mechanical or Electrical Engineering, Physics, Materials, Biology, Bioengineering, in other disciplines and in industry. With few exceptions, researchers in these disciplines are not trained to think about the convective, diffusive, and reactive transport of solute in flowing solutions, and yet these transport issues are central to the operation, performance, and interpretation of these sensors. A controversy had emerged regarding the practicality (or even possibility) of nanowires as sensors for extremely dilute (e.g. femtomolar) solutions. The question was a quantitative one, albeit with a large parameter space, and yet the debate was largely qualitative. Published COMSOL computations for one system would be used to justify assertions about a completely different system, and so on.

We constructed a simple model system that captured the key features of many of these sensor systems, and fleshed out the variety of behaviors and parameter regimes that arise. We then set ourselves the challenge of trying to develop and convey a fundamental understanding of these transport processes *without* requiring partial differential equations, much less matched asymptotic equations. Our central goal was to try to help a wide variety of researchers, with an equally wide variety of backgrounds, to learn how to think about transport in these systems.

I will argue that our field would benefit from more of this kind of "missionary" work – the need is out there, and yet the gap between how we understand solve problems, and how other fields are trained to do so, is severe enough that necessary expertise does not find its way to the communities it could benefit. The more we interact with other fields and disciplines, the more impact we will have. Doing so will additionally expose us as researchers to interesting challenges faced by other disciplines.

(39a) What Makes a Flow Interesting to Industrial and Academic Researchers?

Patrick T. Spicer, Microstructured Fluids Group, Procter and Gamble Co., West Chester, OH

A variety of motivations drive research in academia and industry, but it's safe to say that activity often occurs in areas that are either funded, interesting, or novel. While it's usually clear when work is funded, interest and novelty in a given area is often a matter of perspective that can vary significantly from industry to academia. Research topics with enormous economic impact in industry can sometimes be ignored by academics because the topics seem trivial or the impact of the work is not obvious. Industry often does a poor job communicating its needs because of its short-term focus, its broad interests, and a lack of experience working with academics. In this talk I will draw on examples of novel flows and industrial unanswered questions in the area of complex fluid flows and contrast them with my perception of the current trends in the same field for academia. The intent is to communicate needs for the field that may be unknown outside of industry, but also to suggest ways of better communicating research motivations in both directions.

(39b) Complex Fluid Flows in Chemical Product Development

Christopher Tucker and Suraj Deshmukh, The Dow Chemical Company, Midland, MI

Many Chemical Products that we use everyday are prepared, sold or used as highly complex formulations which may contain liquids and dissolved or dispersed solids. These are typically a complex medley of flow modifiers, surface active agents, fillers, polymers, colorants and solvents. This concoction of materials leads to a complex flow behavior under various conditions, many of which are not well understood. Some examples of these products include paints & coatings, adhesives, slurries and pastes, thermosetting or film forming polymers and consumer products such as laundry detergents, skin lotions and shampoos. Flow behavior can be a key determining factor of the product characteristics such as particle size and distribution in dispersed polymer systems or other emulsion products. Flow properties can also be critical in the application performance and consumer perceptions of products such as paints, lotions, etc.

In this talk, we will use practical examples of these ubiquitous products to describe some of these complex flows. We will describe how rheological characterization of these systems is fraught with complications of high normal stresses, high inertia and slip behavior while surface flow behavior characterization is confounded by multi-component dynamic wetting, time-dependent behavior and fluid instabilities. We will describe some of our efforts in using a scientific and statistical approach in understanding this behavior as well as strategies used to modify flow properties to meet those needs. We will also endeavor to outline a few of the open-ended flow problems that remain important but not well understood.

(39c) Fluid-Like Properties of Powders: Rheology in the Intermediate Flow Regime

James N. Michaels, Pharmaceutical Materials Laboratory, Merck & Co Inc., West Point, PA

Manufacture of solid dosage forms is a microcosm of powder technology, involving synthesis and processing of fine powders through a variety of particle formation, communition, blending, compaction, and coating processes. Common to these operations is the bulk flow of powders in a regime in which particles experience a range of interactions from collisions to more enduring frictional contacts with neighboring particles. This is termed the "intermediate" flow regime, and it has only recently begun to be studied with rigor. Quantitative models for industrially-relevant powder flows do not exist, and for this reason the design and scaling of these unit operations is largely empirical. To begin to address this gap, NSF funded a powder flow collaboratory, lead by the International Fine Particle Research Institute (IFPRI), to bring together a small group of experimentalists and modelers in order to explore the state of development of powder flow models. In the collaboratory, the results of two IFPRI-funded flow experiments, a particle-level study of two-dimensional hopper flows and a continuum-level study of Couette flows were provided to the modelers in order to test the fidelity of their models. The results of this effort revealed key gaps, both experimental and theoretical, in our understanding of and ability to model dense, active powder flows. I will present in this talk the key findings and a set of industrially relevant results achieved during the NSF funded program.

(39d) Fluid Flow Challenges to Meet for the Growing Demands for Energy

Shaffiq Jaffer, Total American Services, Boston, MA

Global population is expected to reach 9 billion by the year 2050 putting tremendous pressure on delivering energy to meet the expected demands. An integral part of this solution will continue to be the use of fossil fuels, conventional and unconventional. Due to various issues, geopolitical to technical costs to new technology requirements, it is expected that the production rate of oil will plateau around 100 Mbbpd. However, this assumes tremendous improvements in the ability to recover oil in existing an new oil reserves. In large part, the technical challenges for this recovery improvements are governed by transport phenomena and physical chemistry. From the fluid flow in low permeable reservoirs with nano scale pores with surface heterogeneity, to the challenge of oil extraction from unconventional resources like oil sands in multiphase processes.

The challenges facing society is to get to sustainable / renewable energy. The industry is still very small today and is projected to be still a small fraction percentage overall energy used in 2030 but this is largely due to technical challenges, high cost and the need for some significant breakthrough technologies / solutions. Within the renewable energy space, fluid flow challenges are predominantly related to solution processing, coating processes, separation processes and catalytic processes. As our ability to control features at the nano scale improves, the interplay between transport process and physical chemistry becomes even more important for renewables technology.

This talk will aim to provide an overview of future energy demand in the face of climate challenges and help place in context some of the engineering challenges we face and where fluid flow related problems are or will be present.

(57a) Mode Specific Elastic Constants and the Effects of Curvature On the Binding of Lipid Chain Anchors to DPPC/DOPC/Cholesterol Model Lipid Bilayers

Mark J. Uline, Chemical Engineering, U. South Carolina, Columbia, SC and Igal Szleifer, Biomedical Engineering, Northwestern University, Evanston, IL

Using a theoretical model of a bilayer membrane containing cholesterol, dipalmitoyl-phophatidylcholine (DPPC), and dioleoylphosphatidylcholine (DOPC) that qualitatively reproduces phase diagrams of giant unilaminar vesicles (GUVs) of the same three components [R. Elliott, I. Szleifer, and M. Schick, Phys. Rev. Lett., **96**, p.098101 (2006)], we calculate the bending and saddle-splay constants for gel, liquid-ordered (lo), and liquid-disordered (ld) phases. By proper expansion of the free energy, the molecular theory enables us to determine the effects of the mode of membrane bending deformation on the value of the elastic constants for different phases. In particular, we refer to the ability of the molecules to arrange the composition between the two monolayers upon deformation. The effect of "blocked" vs "free" exchange of lipids across the two monolayers on the values of the bending constant is as high as 50 k_bT in the ld phase to as high as 200 k_bT in the lo phase. These results show that one must strongly consider the mode of deformation in determining the mechanical properties of lipid bilayers, as has been shown in previous work by several authors.

We also calculate how the curvature of lipid vesicles determines the amount of binding of molecules with lipid tail anchors. By explicitly determining the chemical potential difference of species across a curved bilayer under different modes of deformation in both liquid-ordered and liquid-disordered phases, we are able to calculate the equilibrium binding concentrations of lipid tail anchors as a function of membrane curvature, concentration of lipids, and solution environment. Our results are in excellent agreement with recent experiments [N. S. Hatzakis *et. al.*, *Nature Chem. Bio.*, **5**, 835, 2009].

(57b) Simulating Water Sorption On Protein Matrices

Jeremy C. Palmer, Department of Chemical and Biological Engineering, Princeton University, New Jersey , NJ and Pablo G. Debenedetti, Chemical and Biological Engineering, Princeton University, Princeton, NJ

Understanding the role that hydrating water plays in governing the structure, dynamics, and biological activity of proteins is crucial to identifying key mechanisms for maintenance and repair of biological substrates, mitigating the effects of extreme environmental conditions, and developing strategies for long-term preservation of proteins. Although a vast number of experimental studies demonstrate strong correlations between solvent nature and protein behavior, identification of specific mechanisms responsible for protein-solvent coupling remains a formidable challenge. A notable example is the putative protein dynamical transition[1] (PDT), which is often associated with abrupt changes in the magnitudes of atomic fluctuations at ~220 K as measured using a variety of techniques, including quasi-elastic neutron scattering[1,2], Mossbauer spectroscopy[3], and terahertz dielectric response[4]. Reducing the amount of adsorbed water hydrating the protein system suppresses the PDT[5], shifting it to higher temperatures, until the signatures vanish completely under extreme dehydration conditions[2,5]. However, despite great efforts to characterize the PDT, a consensus has not been reached regarding its mechanistic origins[6]. This is in part due to the limited ability of available empirical techniques to fully resolve molecular structure and motions of protein systems at the nanoscale[6]. Our aim is to use advanced molecular simulation techniques to investigate the structural and dynamic changes that occur in protein systems upon rehydration at a resolution that is beyond the current capabilities of empirical methods. Our recent efforts have resulted in the development of a new simulation protocol to compute water sorption isotherms (hydration level vs. relative humidity) for protein systems, allowing for the first time, for a rigorous connection to be made between the simulated state conditions and those imposed during real experiments. Using our new approach, we have calculated water sorption isotherms for crystals and powders of Ubiquitin and Lysozyme. We find that the simulated isotherms and enthalpies of hydration are in good qualitative agreement with experimental measurements, suggesting that our approach provides a reasonable description of the hydration thermodynamics. We also show that the dynamical properties of the proteins are altered during the hydration process, and that they exhibit glassy behavior at low hydration levels and gradually recover internal flexibility as water content is increased.

1. W. Doster; S. Cusack; W. Petry, **Nature**, 337, 754, 1989.

2. S. Magazu; F. Migliardo; A. Benedetto, **J. Phys. Chem. B.,**115, 7736, 2011.

3. D. Matyushov, **J. Phys. Chem. B**., 115, 10715, 2011.

4. A. Markelz; J. Knab; J. Chen; **Chem. Phys. Lett.,** 442, 413, 2007.

5. V. Kurkal; R. Daniel; J. Finney; M. Tehei, R. Dunn, J. Smith; **Biophys. J.,** 89, 1282, 2005.

6. T. Becker; J. Hayward; J. Finney; R. Daniel; J. Smith; **Biophys. J**., 87, 1436, 2004.

(57c) Role of Solvent Additives in the Hydrophobic Hydration and Association of Methane Molecules

Dheeraj Tomar, Chemical Engineering, Johns Hopkins University, Baltimore, MD and Dilip Asthagiri, Johns Hopkins University, Baltimore, MD

In the hydrophobic hydration of molecules such as methane, often the picture is one of a clathrate-like cage forming around the hydrophobic solute. Here we evaluate this picture in the context of additives such as tetrahydrofuran that are known to encourage gas-hydrate formation. Indeed, the ability of THF as a facile promoter of methane-hydrate formation has made THF-induced enclathration of much utility in studying gas-hydrates. To explore the molecular role of such additives, we calculate the hydration free energy and potential of mean force between two methane molecules in pure water and in aqueous solutions containing various additives such as THF, methanol, urea, tetramethylamine n-oxide (TMAO), and mixtures of urea and TMAO. All the calculations are founded on a newly developed framework of obtaining thermodynamic properties of hydration based on regularizing the solute-system binding energy distribution. This approach gives a transparent accounting of the contributions from the solute-medium interactions within a defined local domain (here approximately the size of a hydrate cage), the packing contributions accounting for evacuating the local domain in a solution but without the solute, and long-range interactions of the solute with the material outside the local domain. Results from this analysis will be discussed.

(57d) Structural Flexibility of the α-Helical Domain in a G-Protein-Coupled-Receptor-Gs Protein Complex

Harish Vashisth, Chemistry and Biophysics, University of Michigan, Ann Arbor, MI and Charles L. Brooks III, Department of Chemistry and Biophysics Program, University of Michigan, Ann Arbor, MI

G-protein-coupled receptors (GPCRs) are membrane proteins mediating cellular responses to hormones and neurotransmitters, which couple binding of small extracellular diffusible ligands to conformational changes and activation of cognate heterotrimeric G-proteins. Prototypical GPCRs have a conserved structural topology of seven transmembrane helices, which undergo conformational reorganization on binding to small molecules, and allow for the docking of G-proteins. Recent crystal structure (PDB 3SN6) of a G-protein-bound neurotransmitter GPCR, β_2-adrenergic-receptor, has provided structural bases for the active receptor-G-protein complex. This structure has revealed that the α-helical domain of the G_α-subunit of the hetereotrimeric G-protein is significantly displaced in comparison to its nucleotide-bound form. Furthermore, single-particle electron-microscopy analysis (EM) of this complex has shown that the α-helical domain is relatively flexible in the absence of nucleotide. However, the exact mechanistic details of this large-scale conformational change remain unclear. We first study this conformational change in the G_α-subunit of the G-protein using temperature-accelerated molecular dynamics (TAMD) simulations, and molecular dynamics flexible-fitting (MDFF). We find that the TAMD-generated closed-like conformations of the G_α-subunit could be relatively easily fitted into simulated low-resolution EM-maps of its nucleotide-bound form, and the placement of entire complex is consistent with experimental EM maps. Additionally, we attempt to understand the thermodynamic basis of this conformational change in the active complex. We therefore further refine the "activation-pathway" generated by TAMD to a minimum-free energy path (MFEP) using string method in collective variables (CVs) with and without nucleotide. Preliminary results of this thermodynamic characterization will also be presented.

(57e) Hydrophobic Association with Amphiphilic and Superhydrophilic Cosolutes

Qing Shao and Shaoyi Jiang, Chemical Engineering, University of Washington, Seattle, WA

Hydrophobic association in aqueous solution plays an important role in biological and chemical processes such as protein folding, protein-protein aggregation and enzyme-substrate association. This critical association can be affected by the environmental factors such as temperature, pressure and co-solutes. A recent experiment (Keefe A. and Jiang S., *Nature Chemistry*, 2012, 4, 59-63) showed that the poly-betaine and poly-(ethylene glycol) conjugated proteins have different substrate affinities. This observation implicates differing influences of carboxybetaine (CB) and ethylene glycol (EG) on hydrophobic association. Zwitterionic CB is superhydrophilic and EG is amphiphilic. In this work, we explored the effects of CB and EG to the hydrophobic association. We selected the association of two methane molecules as a model and the potential of mean force of this association with CB and EG of various concentrations were investigated. The structure and dynamics of the aqueous solution were also investigated to explore the relation between the variation of hydrophobic association and the change of aqueous environment. The simulation results offer fundamental information at the molecular level for the rational design of related biomaterials.

(57f) Force Fluctuations Play a Key Role in Biomolecular Kinetics

Elena F. Koslover, Biophysics Program, Stanford University, Stanford, CA and Andrew J. Spakowitz, Chemical Engineering, Biophysics, Stanford University, Stanford, CA

A wide array of biological processes occur at rates that vary significantly with force. Instantaneous molecular forces fluctuate due to thermal noise and active processes, leading to concomitant fluctuations in biomolecular rate constants. While such fluctuations are generally ignored in the current prevailing view of force-dependent enzyme kinetics, we employ a widely-applicable analytic method to demonstrate that force fluctuations can have a dramatic impact on effective rates.

A critical question in cellular biology is how the proteins responsible for reading and replicating the genome process packaged DNA. We address one aspect of this problem by considering the kinetics of RNA polymerase as it transcribes through DNA wrapped around nucleosomes, the lowest level of the chromatin packing hierarchy. We develop a simple mechanical model for the forces experienced by RNAP as it approaches the nucleosome and consider the pausing behavior of the polymerase under these fluctuating forces. Incorporating fluctuations in the amount of downstream DNA unwrapped from the nucleosome yields predictions for the pausing timescales that differ by an order of magnitude from those under the average force alone.

We use our model to illustrate the broad range of behaviors that can arise in biomolecular processes that are susceptible to force fluctuations. The fluctuation time scale, which varies significantly for in vivo biomolecular processes, yields qualitatively different results for overall rates and dramatically alters the force regime of relevance to the transition. For our example system, polymerase pausing rates are dominated by transient excursions to high forces. These calculations thus motivate the need for single-molecule measurements of enzymatic rates under forces far above the in vivo average. Our results demonstrate that accounting for force fluctuations is essential to developing a quantitative understanding of enzyme kinetics in the dynamic environment of a living cell.

(57g) Toward Self-Assembly of Nonviral Gene Delivery Complexes by Pegylated-Poly(Lysine) and siRNA Molecules Via Molecular Dynamics Simulations

Iris W. Fu[1], Cade B. Markegard[1] and Hung D. Nguyen[2], (1)Chemical Engineering and Materials Science, University of California, Irvine, Irvine, CA, (2)Department of Chemical Engineering and Materials Science, University of California, Irvine, Irvine, CA

The design of self-assembling molecules by anionic nucleic acids and cationic polymers is of great interest in nonviral gene delivery. However, the kinetic mechanisms by which cationic polymers encapsulate anionic nucleic acids in forming polyplexes are not fully elucidated. Moreover, the complexation behavior of such polyplexes is highly dependent on several major characteristics such as molecular weight, charge density, structure of the polymer, ionic strength of buffers, and etc. Our studies examine the molecular-level kinetic mechanisms of self-assembly by PEGylated-poly(Lysine) and siRNA molecules to form polyplexes via molecular dynamics simulations using multiscale models. We also investigate the role of external biological stimuli (e.g. pH, salt concentration, PEGylated-poly(Lysine)/SiRNA concentrations, or temperature) that affect the stability and size distribution of the assembled complexes. The findings of this research will guide experimentalists in improving siRNA delivery by understanding the interactions present between anionic siRNA and cationic PEGylated-poly(Lysine) molecules and controlling the size of encapsulated complexes for gene therapy.

(57h) Macromolecular Crowding Effects On Coupled Folding and Binding

Apratim Bhattacharya, Chemical Engineering, Lehigh University, Bethlehem, PA, Young C. Kim, Naval Research Laboratory and Jeetain Mittal, Department of Chemical Engineering, Lehigh University, Bethlehem, PA

The ability of proteins to carry out various cellular functions is usually tied to their ability to adopt specific three-dimensional structure. However, it is now recognized that many proteins actually perform their function without adopting a single stable structure. These intrinsically disordered proteins (IDPs) are highly prevalent in biology, involved in signaling and regulation, and extensively implicated in human diseases. The mechanism of binding of IDPs is also intriguing. Two mechanisms have been advanced: (i) conformational selection: a pre-equilibrium scenario where the bound structure of the IDP is present in the unbound population, and the target binds to this pre-formed structure and (ii) coupled binding and folding: an induced fit scenario where the IDP only assumes its bound structure after committing to binding.

Owing to the extremely crowded cellular milieu, it becomes an imperative issue to consider the mechanism of IDP binding in the presence of crowders. We have performed Replica Exchange Monte Carlo (REMC) simulations to study binding of a well studied protein complex - pKID/KIX. Under bulk conditions, the unstructured pKID was found to bind KIX via induced fit in earlier studies. We find that macromolecular crowding can significantly impact protein binding thermodynamics as well as kinetics. Moreover, it can change the mechanism of binding from induced fit to conformational selection by stabilizing the folded state in solution.

(79a) Colloidal Hydrodynamics: The Good, the Bad and the Ugly

John F. Brady, California Institute of Technology / Division of Chemistry and Chemical Engineering, Pasadena, CA; Chemical Engineering and Mechanical Engineering, California Institute of Technology, Pasadena, CA

What do corn starch, swimming spermatozoa, DNA and self-assembling nanoparticles have in common? They are all (or can be modeled as) 'particles' dispersed in a continuum suspending fluid where hydrodynamic interactions compete with thermal (Brownian) and interparticle forces to set structure and determine properties. These systems are 'soft' as compared to molecular systems largely because their number density is much less and their time scales much longer than atomic or molecular systems (both scaling with the colloidal to atomic size cubed). In this talk I will describe the common framework for modeling these diverse systems and the essential features that any hydrodynamic modeling must incorporate in order to capture the correct behavior (the Good). Actually computing the hydrodynamics in an accurate and efficient manner is the real challenge (the Bad) and I will describe past successes and current efforts. The Ugly? Well, that may be in the eye of the beholder.

(79b) Advective Symmetry Breaking in Phoretic Migration of a Colloidal Particle

Aditya S. Khair, Chemical Engineering, Carnegie Mellon University, Pittsburgh, PA

The phoretic migration of a colloidal particle is animated by an imposed gradient of a scalar field: e.g. neutral solute gradients cause diffusiophoresis, and temperature gradients drive thermophoresis. It is usually assumed that the scalar field evolves solely via diffusion, owing to its large diffusivity and the modest colloidal particle velocity. (That is, the Peclet number for scalar-field transport is assumed to be zero). This leads to Morrison's celebrated result that the translational phoretic velocity of a colloid is independent of its size, shape, or orientation with respect to the imposed gradient (moreover, the colloid does not rotate). However, intuitively, as the colloid moves it sets up a fluid flow that advects the same scalar field that instigated its motion. In this talk, using asymptotic expansions, we explore the first effects of advection on the phoretic motion of colloidal particles (i.e. at the experimentally relevant conditions of small but finite Peclet number). In particular, we show that advection leads to symmetry breaking in the phoretic motion of fore-aft asymmetric particles, wherein the particle velocity depends on the direction of the imposed gradient. Moreover, we demonstrate that advection drives phoretic rotation of non-spherical colloids, which ultimately align with the imposed gradient.

(79c) Diffusiophoretic Motion of a Colloid At Nano-Scale Due to a Surface Chemical Reaction

Nima Sharifi-Mood, Chemical Engineering, Department of Chemical Engineering, Levich institute, New York, NY, Joel Koplik, Department of Physics, Levich Institute, City College of New York and Graduate Center,City University of New York, New York, NY and Charles Maldarelli, Levich Institute and Chemical Engineering, The City College of New York, New York, NY

Diffusiophoresis is a chemo-mechanical transduction mechanism for propelling colloid particles in a fluid in which the driving force for the motion is due to the intermolecular interactions between solute molecules surrounding the particle and the colloid itself. When solutes are asymmetrically distributed around the particle, the net interaction exerted on the colloid due to solute and solvent is unbalanced, and the particle is propelled. In the self (active) diffusiophoretic motion the concentration gradient required for the motion is provided by the colloid particle itself. In this case, for instance an asymmetric distribution of a catalyst on the particle surface brings about the concentration gradient for the motion. Prior theoretical studies of the mechanism from the coarse grained continuum approach have shown the coupled mass transfer and hydrodynamic motion in the limit in which the net van der Waals forces between all species cause a slip of the fluid on the surface of the colloid. Having assumed the length scale L of the intermolecular interaction (typically of order 1-100 nm) to be much smaller than the colloid radius, the diffusiophoretic velocity (active and passive) is to leading order independent of a (particle radius).

In this presentation, we first provide the detailed hydrodynamic analysis of stokes flow regime in order to obtain the terminal diffusiophoretic (active and passive) velocity of the colloid particle with an arbitrary shape. The mass transfer analysis is undertaken in the limit in which the fluxes generated by diffusion and the intermolecular forces are larger than the convective flux (small P'eclet number) and the reaction is slow relative to the diffusion of the solutes (small Damk"{o}hler number). Numerical solutions of the solutal mass transfer equations using a finite element technique are obtained and coupled to an integral analytical solution for doffusiophoretic velocity, to provide numerical solutions for the terminal diffusiophoretic velocity. Asymptotic calculations are also undertaken for the concentration field and velocity as L/a is small. Two interactions are examined, an exponential interaction and a long range attractive van der Waals attraction which is computed by pairwise additivity and formulated to include the attraction of the solvent with the colloid. For each interaction, the velocity decreases as the colloid radius decreases with the interaction parameters constant. For small L/a, the exponential potential decreases with an order one correction in L/a while this correction is logarithmic for the van der Waals potential.

Secondly, the diffusiophoretic motion of a colloid was considered from an atomistic scale utilizing molecular dynamics simulation in order to obtain a better insight in to how the interactions between different species in the system bring about the motion at nanoscale. These simulations are undertaken with the Lennard-Jones potential interaction between all species in the system in a priodic box. The atomistic simulation results are compared with our coarse grained continuum theory predictions in the region of overlap in order to probe the validation of continuum theory.

(79d) Electrohydrodynamic Interactions of Colloidal Particles Under Quincke Rotation

Debasish Das, Mechanical Science and Engineering, University of Illinois, Urbana-Champaign, Urbana, IL and David Saintillan, Department of Mechanical Science and Engineering, University of Illinois at Urbana-Champaign, Urbana, IL

Quincke rotation, a well-observed phenomenon in particle suspensions, denotes the spontaneous rotation of dielectric particles immersed in a slightly dielectric liquid when subjected to a high enough DC electric field. It occurs when the charge relaxation time of the particles is greater than that of the fluid medium, causing the particles to become polarized in a direction opposite to that of the electric field and therefore giving rise to an unstable equilibrium position. When slightly perturbed, the particles start to rotate, and if the applied electric field exceeds a critical value this perturbation does not decay and the particle rotation reaches a steady state with a constant angular velocity obtained by balancing the viscous torque with the electric torque due to the induced dipole. Motivated by applications in particle suspensions, where Quincke rotation can lead to a decrease in effective shear viscosity, we use a combination of numerical simulations and asymptotic theory to study the effect of electrohydrodynamic interactions between particles under Quincke rotation. We study the prototypical case of two equally sized spheres carrying no net charge and interacting with each other both electrically and hydrodynamically. We use the classic method of reflections to capture far-field interactions, and solve a coupled system of time-dependent ordinary differential equations for the dipole moments, angular velocities, and positions of the two spheres capturing interactions up to order $O(R^{-3})$ (where R is the separation distance between the sphere centers). We find that interactions between particles cause a decrease in angular velocity with respect to the single sphere case, as well as an increase in the critical electric field strength required for spontaneous rotation to occur.

(79e) High-Throughput Microrheology in a Microfluidic Device

Kelly M. Schultz, Department of Chemical and Biological Engineering, University of Colorado at Boulder, Howard Hughes Medical Institute, Boulder, CO and Eric M. Furst, Department of Chemical and Biomolecular Engineering, University of Delaware, Newark, DE

Rheological characterization of hydrogels enables the design and engineering of these scaffolds for therapeutic applications, such as wound healing and tissue regeneration. Due to growing complexity of hydrogels in efforts to recapitulate the extracellular matrix, materials have become synthetically laborious and expensive. Therefore, during the characterization of this material an ideal method would maximize the number of samples measured while minimizing the amount of material used and measurement acquisition time. To meet this end, we have developed a technique that combines sample preparation within a microfluidic device with multiple particle tracking (MPT) microrheological measurements, named μ^2rheology. In μ^2rheology, 50 – 100 droplets are created within a microfluidic device varying the composition of each droplet using linear pumping programs. MPT measurements are taken of stationary droplets sealed within the device. MPT is a passive microrheological technique that measures the Brownian motion of fluorescently labeled probe particle embedded within a material. The particle trajectories are directly related to rheological properties, such as creep compliance and viscosity, using the Generalized Stokes-Einstein Relation. MPT is chosen as the measurements technique due to the inherently small samples size required for accurate measurements (4-40 μL), short acquisition time and ability to determine sample homogeneity. In this work we validate μ^2rheology by measuring droplets from a serial dilution of glycerine. Viscosity measurements from μ^2rheology agree with both tabulated data and individually prepared MPT measurements. The utility of the technique is illustrated with measurements of the high molecular weight heparin (HMWH) overlap concentration and creation of a gelation state diagram of a HMWH-PEG hydrogelator, which identifies the composition where a gel will form. Both measurements illustrate the fine gradient in composition achieved by microfluidic sample preparation and precise rheological measurements using MPT. Overall, our methods can be used to thoroughly screen the material rheological properties, yielding a high information density, but requiring only small volumetric amounts.

(79f) Large Amplitude Oscillatory (micro-)Rheology

James W. Swan, Department of Chemical and Biomolecular Engineering, University of Delaware, Newark, DE, Roseanna N. Zia, Mechanical Engineering, Princeton University and John F. Brady, California Institute of Technology / Division of Chemistry and Chemical Engineering, Pasadena, CA

A microstructural understanding of the origin of the large-amplitude rheological response of common (e.g. colloidal dispersions and polymer melts) and exotic (e.g. slug mucus) materials, as well as the behavior of various constitutive models when forced in the same fashion, is an on-going challenge. We have conducted a detailed study of the response to large amplitude oscillatory flows (LAOF) from a micro mechanical perspective -- the microrheology of colloidal dispersions. The suspension microstructure, satisfying the Smoluchowski equation, is decomposed into temporal normal modes which are coupled together by the imposed flow. The solution for these coupled structure functions provides considerable insight into the physical processes involved in large amplitude flows, in particular, the competition between hydrodynamic and thermal forces. We compare the linear and non-linear response for suspensions with and without hydrodynamic interactions and contrast colloidal probes driven by an imposed force and an imposed velocity. We also show how the microrheological behavior connects to macrorheological LAOS.

(79g) Modeling of Nanoparticle Motion in an Incompressible Newtonian Fluid: A Comparison Between Fluctuating Hydrodynamics and Generalized Langevin Approaches

Balakrishnan Uma, UPenn, Philadelphia, PA, Portonovo S. Ayyaswamy, Mechanical Engineering and Applied Mechanics, University of Pennsylvania, Philadelphia, PA, David M. Eckmann, School of Medicine, University of Pennsylvania, Philadelphia, PA and Ravi Radhakrishnan, Department of Bioengineering, University of Pennsylvania, Philadelphia, PA

In order to obtain an accurate description of nanocarrier dynamics in the bloodstream and near wall interactions, we have developed a formalism where both the Brownian motion as well as the hydrodynamic interactions are important. We employ a direct numerical simulation adopting an arbitrary Lagrangian-Eulerian based finite element method to simulate the Brownian motion of nanocarrier in an incompressible Newtonian fluid. Thermal fluctuations are implemented using three different approaches: (1) fluctuating hydrodynamics approach; (2) generalized Langevin approach; (3) a hybrid approach combining (1) and (2) to incorporate memory effects. At thermal equilibrium, the numerical predictions for temperature of the nanoparticle, velocity distribution of the particle, decay of the velocity autocorrelation function, diffusivity of the particle and particle-wall interactions in the presence and absence of adhesive interactions are computed for various flow conditions. We show that the hybrid approach simultaneously preserves the canonical distribution as well as the hydrodynamic correlations.

References:

1. Nanoparticle Brownian motion and hydrodynamic interactions in the presence of flow fields, B. Uma, T. N. Swaminathan, R. Radhakrishnan, D. M. Eckmann, P. S. Ayyaswamy, Physics of Fluids, 2011, 23, 073602. DOI: 10.1063/1.3611026.

2. Generalized Langevin dynamics of a nanoparticle using a finite element approach: Thermostating with correlated noise, B. Uma, T. N. Swaminathan, P. S. Ayyaswamy, D. M. Eckmann, R. Radhakrishnan, J. Chem. Phys., 2011, 135, 114104. DOI: 10.1063/1.3635776; Erratum, 136, 019901, 2012.

3. A hybrid formalism combining fluctuating hydrodynamics and generalized Langevin dynamics for the simulation of nanoparticle thermal motion in an incompressible fluid medium, B. Uma, D. M. Eckmann, P. S. Ayyaswamy, R. Radhakrishnan, Molecular Physics, 2012, in press. (DOI: 10.1080/00268976.2012.663510)

4. Hybrid scheme combining fluctuating hydrodynamics and generalized Langevin dynamics for nearly neutrally buoyant particle in stationary medium, B. Uma, D. M. Eckmann, R. Radhakrishnan, P. S. Ayyaswamy, ASME Journal of Heat Transfer, 2012, in press.

(79h) Nanoparticle Transport in the Pore Space of a Packed Column and Investigation of Adsorption Effects with the Use of Lattice Boltzmann Simulations

Ngoc Hong Pham, Chemical, Biological and Materials Engineering, The University of Oklahoma, Norman, OK, Daniel P. Swatske, School of Chemical, Biological, and Materials Engineering, The University of Oklahoma, Norman, OK, Daniel E. Resasco, School of Chemical, Biological and Materials Engineering and Carbon Nanotube Technology Center, University of Oklahoma, Norman, OK, Jeffrey Harwell, Chemical Engineering, University of Oklahoma, Norman, OK, Bor-Jier Shiau, School of Petroleum and Geological Engineering, University of Oklahoma, Norman, OK and Dimitrios V. Papavassiliou, School of Chemical Biological and Materials Engineering, The University of Oklahoma, Norman, OK

Engineered nanoparticles have been found to be potentially useful for enhanced oil recovery processes. By injecting these particles into the well, it is expected that the particles will pass through the pore throats and go deeply into the reservoir rock to reflect information about the reservoir. Thus, injected nanoparticles must satisfy some crucial pre-requisites, such as: long-term stability, no aggregate generation, and minimal retention by the rock. In this study, the motion of nanoparticles inside the pore space of a crushed Berea sandstone column is explored numerically by using the Lattice Boltzmann Method (LBM) in conjunction with the Lagrangian Scalar Tracking (LST) algorithm [1], which is utilized to track the trajectories of individual nanoparticels in the flow field generated by the LBM. The crushed Berea column is modeled to be a column of 150μm-diameter spheres packed in a random manner. Particles lost due to size exclusion are neglected in our simulation, since the characteristic length of a qualified nanoparticle is usually one order of magnitude smaller than the characteristic dimension of the consolidated reservoir rock pores. Therefore, the main factor hindering the particle motion is particle adsorption on the pore surfaces, which is modeled as a pseudo-first order adsorption process. The numerical approach is validated with experimental results and then the simulations are used to provide quantitative results on the effects of porosity, tortuosity, nanoparticle diffusivity and adsorption kinetics on the transport behavior of the nanoparticles.

ACKNOWLEDGEMENTS

The financial support of the Advanced Energy Consortium (AEC BEG08-022) and the computational support of XSEDE (CTS090017) are acknowledged.

REFERENCES

1. Voronov, R.S., VanGordon, S., Sikavitsas, V.I., and D.V. Papavassiliou, "Efficient Lagrangian scalar tracking method for reactive local mass transport simulation through porous media," *Int. J. for Numerical Methods in Fluids*, **67**, 501-517, 2011

(79i) The Dynamics of Adhesion Between a Pair of Vesicles

L. Gary Leal and Johann Walter, Department of Chemical Engineering, University of California, Santa Barbara, Santa Barbara, CA

An important problem to assess the stability of vesicle suspensions is to establish the conditions in which vesicles with weakly attractive bilayers adhere to each other in the presence of thermal agitation or flow. When they adhere, this may lead to the growth of multi-vesicle "particles" that are large enough to generate creaming or sedimentation instabilities, even in a suspension of submicron vesicles. We study the dynamics of a pair of interacting unilamellar vesicles when a large non-adsorbing polymer in the suspending fluid causes such a depletion attraction force between them. The mechanical behavior of the vesicle membrane is modeled carefully by taking into account its resistance to macroscopic curvature changes as well as its ability to increase its surface area, both by pulling out thermal undulations and by increasing the headgroup area. The problem is studied numerically by coupling a boundary integral formulation of the Stokes equations with a finite element model for the mechanics of the membranes, while the non-hydrodynamic interactions are treated within the framework of the Derjaguin approximation. The numerical results are generally consistent with an earlier theoretical study (Ramachandran et. al. *Langmuir* (2011)), but find stronger adhesion energies than previously predicted in the case of weakly-adhering vesicles. The deformability of the vesicles is shown to enhance the interaction energy. Surface area increases account for this effect, while resistance to curvature changes is found to play a negligible role for vesicles larger than a few hundred nanometers. The dynamics of the phenomenon are also studied, in particular by subjecting the pair of vesicles to a weak, time-dependent flow.

(92a) Integrated Experimental and Computational Studies of Energy-Relevant Interfaces

Peter T. Cummings, Center of Nanophase Materials Sciences, Oak Ridge National Laboratory, Oak Ridge, TN

Interfaces are ubiquitous in energy-relevant systems. Examples include the subterranean mineral-fluid interfaces that govern carbon sequestration, the electrode-fluid interfaces in batteries and supercapacitors, and the fluid-solid interfaces at which heterogeneous catalysis takes place. For many years, we have studied mineral-fluid interfaces with a combination of molecular (e.g., X-ray reflectivity, quasi-elastic neutron scattering and neutron spin echo) and bulk (e.g., titration) experimental probes closely integrated with molecular dynamics simulations using fluid-solid forcefields derived from *ab initio* calculations. Recently, as part of the activities of the Fluid Interface Reactions Structure and Transport (FIRST) Engineering Frontier Research Center (EFRC), we have extended this approach to the study of interfaces encountered in batteries, supercapacitors, and heterogeneous catalysis. The FIRST Center performs fundamental research on fluid-solid interfaces based on the premise that the next generation of electrical storage devices with superior performance will require a fundamental knowledge of the nanoscale architecture of the interface, the effect of nanotexture on interfacial properties, and the structural and dynamic changes that occur during charge and discharge cycles. In this presentation, we will provide an overview of our research on the molecular-level modeling and experimental characterization of energy-relevant interfaces.

Interfacial Properties of Gas Hydrates in Energy Applications

Carolyn A. Koh, Zachary Aman, E. Dendy Sloan, Amadeu K. Sum
Center for Hydrate Research, Chemical & Biological Engineering Dept.
Colorado School of Mines, Golden, CO 80401

Gas hydrates are solid inclusion compounds that form when water and gas come into contact at high pressures and low temperatures. Understanding the interfacial properties of gas hydrates is important in a number of different energy applications, including the transportation of hydrocarbons in subsea oil and gas pipelines, containing an oil/gas well blowout, energy recovery (with carbon dioxide injection) from natural hydrated sediments, as well as energy storage in clathrate hydrate materials. This paper will present an overview of the state-of-the-art of hydrate science and engineering with specific emphasis on the interfacial properties of gas hydrate systems. In the case of hydrocarbon transportation, gas hydrate formation can lead to blockage of the subsea pipelines. A key strategy to preventing pipeline blockages due to hydrates is to control the interfacial properties of the hydrates by minimizing hydrate-hydrate particle cohesion to maintain a transportable dispersion of hydrates in the hydrocarbon phase, and/or minimizing particle-surface adhesion and resulting deposition. This strategy can be also applied to reduce the effects of hydrate formation in containing an oil/gas blowout/spill, in which the resulting plume of oil and gas can be well within the hydrate stability region, exhibiting analogous mechanistic behavior to that in the pipeline in terms of hydrate particle/surface interactions and deposition on the containment device.

(92c) CO_2/N_2 Separation in Metal Organic Framework Materials

Perla B. Balbuena, Department of Chemical Engineering, Texas A&M University, College Station, TX

CO_2/N_2 Separation in Metal Organic Framework Materials

Perla B. Balbuena

Department of Chemical Engineering, Texas A&M University,

College Station, TX 77843

In recent years, metal-organic frameworks (MOFs), a class of nanoporous materials, have received substantial attention due to their special adsorption properties for gas storage and separation. The separation of CO_2 from N_2 present in flue gas streams is especially significant due to the problems associated with CO_2 pollution of the atmosphere. Since gas separation is a combination of thermodynamic and kinetic processes, both adsorption and diffusion properties need to be taken into account. A huge challenge is that CO_2 and N_2 have similar kinetic diameters (CO_2 3.30 Å, N_2 3.64 Å), making the separation based purely on size very difficult. On the other hand, the modular nature of the MOFs makes them attractive for introducing specific functional groups which may enhance adsorption selectivity. Additional challenges are due to the chemical and thermal stability of the framework materials and the presence of impurities in flue gas such as water, O_2, NO_x, and SO_x. Here we review our recent computational efforts in understanding the behavior of MOF materials for gas separations. In particular we will refer to our results using ab initio and density functional theory as well as Grand Canonical Monte Carlo and classical molecular dynamics methods. We will address important practical aspects of MOFs such as the effect of the post-synthesis activation processes, as well as their thermal and chemical stability. In addition, we will introduce a new concept based on shape selectivity for the design of effective MOF materials.

(92d) From the Fundamentals of Porous Carbons Poromechanics to Applications for CO2 Sequestration and Gas-Shale

Roland J.-M. Pellenq[1], Laurent Brochard[1], Matthieu Van Damme[2] and Franz-Josef Ulm[1], (1)UMI MSE2, MIT-CNRS, 1-374, MIT, Cambridge, MA, (2)Multiscale Materials, ENPC- Navier Institute, Champs-sur-Marne - 77455 Marne- la- Vallée cedex 2, France

Coal is made of a microporous carbonaceous solid matrix and exhibits a high adsorption capacity. Gas-shale are in their matured state very close to coal. Adsorption of fluids in microporous carbons is known to induce dimensional changes. Understanding of adsorption-induced strain is critical for the industrial development of Enhanced Coal Bed Methane recovery (ECBM) and for gas-shale production. In this work, we present a general poro-mechanics modeling valid for microporous solids under adsorption and the application of this modeling to the specific case of carbon dioxide (CO2) and methane (CH4) adsorption in coal and kerogen.

We derived extended poro-mechanics constitutive equations valid whatever the size and morphology of pores i.e. valid for disordered microporous carbons such as coal. We then performed molecular simulations of adsorption of CH4 adsorption in a flexible molecular model of disordered microporous carbon representative of a kerogen/coal matrix. According to these molecular simulations, the adsorbed amount depends almost linearly on the strain of the solid. Therefore, we proposed a simplified poro-mechanics modeling in which adsorption in coal is assumed to depend linearly on the strain. We then applied this poro-mechanics modeling of coal (i) to estimate the so called 'differential swelling' induced by ECBM in coal beds at different geological depths (ii) to determine the thermodynamics of

(92e) Combining Water, Argon and Nitrogen Adsorption for an Advanced Characterization of Ordered Nanoporous Carbons

Matthias Thommes[1], Juergen Morell[2], Katie Cychosz[1] and Michael Froeba[3], (1)Quantachrome Instruments, Boynton Beach, FL, (2)Institute of Inorganic and Applied Chemistry,, University of Hamburg, Germany , Hamburg, Germany, (3)(2) Institute of Inorganic and Applied Chemistry, , University of Hamburg, Hamburg, Germany

During recent years, major progress has been achieved in the understanding of the adsorption and phase behavior in ordered micro- and mesoporous materials with simple pore geometries, which has led to major advances in the structural characterization by physical adsorption. However, the effects of geometrical and chemical heterogeneities of the pore walls on the sorption, phase, and wetting behaviors of fluids in nanoporous materials such as nanoporous carbons are still under investigation. Within this context, the potential use of water as a probe for surface chemistry and pore structure characterization of nanoporous carbon materials has led to a lot of interest. The use of water adsorption is attractive because it can be performed at room temperature, water has a very small kinetic diameter which allows it to enter pores even smaller than the ones accessible to carbon dioxide, argon, nitrogen and it is sensitive to surface chemistry. On the other hand, the interpretation of water adsorption data is not straightforward, mainly because details of the water adsorption isotherms are strongly affected by both pore structure and surface chemistry and the underlying mechanism of water adsorption in nanoporous carbons is still under investigation. Also, the majority of the water adsorption studies reported so far have been performed on fairly disordered porous carbon materials (e.g. activated carbon fibers, activated carbon [1-3]. Hence, in order to further assess the potential of water adsorption for textural and surface characterization, we have performed a systematic, physical adsorption characterization study on ordered micro/mesoporous carbons (CMK-1, CMK-3, CMK-8) by combining N2 (77.4K), Ar (87K) and water adsorption (over a temperature range from 293 – 313 K). An accurate micro/mesopore size analysis is obtained from the nitrogen and argon adsorption data by applying dedicated DFT methods, i.e. NLDFT (non-local density functional theory) and novel QSDFT (quenched sold density functional theory) which takes into account the effects of surface heterogeneity/roughness on nitrogen and argon adsorption [4].

The analysis of our water adsorption data obtained for these ordered nanocarbons in combination with nitrogen and argon adsorption (coupled with QSDFT pore size analysis) sheds light into the mechanisms of water adsorption and clearly demonstrates the potential of water as a powerful, complimentary adsorptive for the physical adsorption characterization of nanoporous carbons.

[1] T. Ohba, H. Kanoh, K. Kaneko, J. Phys. Chem B 108, 14964 (2004)

[2] D.D.Do and H.D. Do, Carbon 38, 767 (2000

[3]Lodewyckx, P., Raymundo-Piñero, E., Wullens, H., Vaclavikova, M.,

Béguin, F.:In: CD Proceedingsof the International Carbon Conference, Nagano, Japan(2008)

Lodewyckx, P., Vansant, E.F.. Carbon 37, 1647–1649

(1999)

[4] Thommes M, Morlay C, Ahmad R, Joly J.P, Adsorption. 17 (2011) 633

[5]G.Y. Gor, M. Thommes, K.A. Cychosz, A.V. Neimark, Carbon 50 2012, 1583-1590. ;A.V Neimark, P.I Ravikovitch, Y. Lin, M. Thommes, Carbon, 47, 1617 (2009)

(92f) Modeling the Self-Assembly of Ordered Nanoporous Silica Materials

Lin Jin[1], Scott M. Auerbach[2] and **Peter A. Monson**[1], (1)Chemical Engineering, University of Massachusetts, Amherst, MA, (2)Chemistry and Chemical Engineering, Univ. of Massachusetts, Amherst, MA

Porous materials are ubiquitous in modern technology. Traditional applications include catalysis and adsorption separations, while more recent applications include membrane separations, biosensors, drug delivery and low-k dielectrics for microelectronics. The development of new porous materials with controlled pore structure that can be tailored for specific applications is a very active area of materials research. Understanding how such materials form and how to control the pore geometries will help facilitate advanced applications of these materials. Molecular modeling of syntheses of porous materials is challenging because large system sizes are needed to have representative sampling of the processes, and long time scales are required to observe the formation of ordered structures. In this presentation we describe the development and application of models of porous material assembly, focusing on silica materials. We show how quite simple models based on the assembly of corner-sharing silica tetrahedra can describe the self-assembly of a wide range of ordered nanoporous materials, including ordered mesoporous silicas and all-silica zeolite frameworks.

(92g) Understanding the Ion Exchange Properties in Zeolites From Monte Carlo Simulations

Alain H. Fuchs, CNRS, Paris, France

We have adapted a grand ensemble Monte Carlo simulation method to directly compute, for the first time to our knowledge, univalent cation exchange isotherms in zeolites. The computed isotherms for the exchange of sodium in NaY faujasite by lithium, potassium, rubidium, and cesium ions, respectively, are in good agreement with the experimental ones. They display the three main types of behavior observed in zeolites, namely, a monotonous evolution of selectivity throughout the exchange process (Li+), a selectivity reversal (K+), and an incomplete exchange (Rb+ and Cs+). The initial stage of the cation exchange is shown to be dominated by the hydration energy of the cations in the external aqueous solution. The final part of the process is often dominated by the cationframework and cationcation interactions. A crossover between these two regimes explains the frequently observed reversal of selectivity phenomenon. The incomplete exchange observed in the case of Rb+ and Cs+ is shown to correspond to a blocked state of the system for highest accessible composition of the aqueous solution. This stable state is shown not to be linked to an inability of the cesium cations to cross the six-ring window in order to penetrate into the smallest cages.

(93a) Shape and Stability of Liquids That Partially Wet Solids

Paul H. Steen, Chemical and Biomolecular Engineering, Cornell University, Ithaca, NY and Joshua B. Bostwick, Theoretical & Applied Mechanics, Cornell University, Ithaca, NY

The shape at rest of a liquid that partially wets a solid depends on the contact-angle at the three-phase common line. Moreover, in the absence of other forces, the relative affinities of the triplet of materials (liquid, solid and gas) predicts the entire shape of the liquid at equilibrium and its stability depends on the mobility of the contact-line, as is well-known. What seems not as well-appreciated is that, for mobile contact-lines, fully three-dimensional shapes can be *more stable* than axisymmetric shapes. For example, a mobile axisymmetric sessile drop on a planar substrate can be unstable to a non-axisymmetric shape. These observations have a number of fundamental implications for the inviscid spreading behavior of drops and rivulets on solids. We shall illustrate these in this talk.

(93b) Droplet Dynamics in a Steady Electric Field Via 3D Spectral Boundary Elements

Ye Yao, Kevin Beussman and Yechun Wang, Mechanical Engineering, North Dakota State University, Fargo, ND

The development of an efficient and accurate numerical scheme for electrically actuated droplet motion is always a challenging problem. A three-dimensional spectral boundary element method has been developed to investigate the dynamics of a neutrally buoyant droplet in another immiscible fluid subjected to a steady electric field. Both the dispersed fluid and the continuous phase are assumed leaky dielectric. The droplet is not charged initially. Good agreement has been found by comparing with analytical solutions and experimental results for droplets in uniform electric fields. Benefited from the fully three-dimensional algorithm that we recently developed, the droplet deformation and migration induced by non-uniform electric fields have been investigated. We computationally predict the droplet behavior under the influence of physical properties of the system: resistivities, permitivities and viscosities, as well as the electric capillary number. The numerical scheme developed by this study and computational results will provide foundation for the computational investigation of dielectrophoresis in digital microfluidics in which an electric field is applied to actively manipulate droplet motion.

(93c) Electrostatically Driven Particle Adsorption On Polyelectrolyte or Chemically Functionalized Substrates

Kunal Savaji, Chemical Engineering, City College of New York, New York, NY and Alexander Couzis, The City College of New York, New York, NY

Layer-by-layer electrostatic assembly using polyelectrolytes is a versatile approach for obtaining nanoparticle (NP) multilayers. The outcome is dependent on pH and ionic strength that affects both the magnitude of charge on the polyelectrolyte chain and the NP surface, as well as conformation of the polyelectrolyte chains. Motivated by improved packing of NPs and better understanding of the electrostatic adsorption process, we have characterized the adsorption of different nanoparticles (e.g. silver, silica, manganese dioxide) on functionalized surfaces. We utilized poly (diallyldimethylammonium chloride) (PDDA) as the polyelectrolyte that binds both to the substrate and the adsorbate particles electrostatically (physisorption), and comparatively also (3-aminopropyl) trimethoxysilane (APS) that will bind on oxide-covered substrates (e.g. silicon wafers) by covalent chemical bonds. APS possess an ionizable amino group that will attract oppositely charged NPs. The two systems (PDDA and APS) provide useful insight into (ir) reversibility of the NP adsorption. NP desorption has to be separated from the desorption of the PDDA or APS molecules themselves. Our choice of different NP in turn allows accounting for the effect of particle physico-chemical properties and particle size and morphology on the NP packing in the film.

To gain insight into the actual mechanism of the adsorption process, efforts are made to ascertain the rate determining transport step and hence propose a model describing the kinetics of the electrostatically driven adsorption process. The experimental saturation NP packing density and the kinetics of the adsorption on the substrate were compared with Random Sequential Adsorption model. Parameters influencing the process are thought to be the ionic strength of the suspension fluid, pH of the suspension that would govern the charge density on both the interfaces, particle size, temperature of adsorption and the bulk concentration of the particles. To account for these parameters kinetic runs were carried out with different substrates and particles as described above and we investigate if the Random Sequential Model can satisfactorily describe the adsorption process in cognizance with the experimental results taking into account the parameters mentioned above.

We seek to apply our expertise of NP electrostatic assembly to nanostructured battery cathode fabrication of MnO_2 and nickel particles for the Zn/MnO_2 battery.

(93d) Lattice Boltzmann Simulations of a Single n-Butanol Drop Rising in Water

Alexandra E. Komrakova, Department of Chemical and Materials Engineering, University of Alberta , Edmonton, AB, Canada and Jos Derksen, Department of Chemical & Materials Engineering, University of Alberta, Edmonton, AB, Canada

Emulsions, suspensions, and dispersions of immiscible liquids are widely encountered throughout engineering processes and everyday life. Knowledge of dispersed systems characteristics (apparent viscosity, rheology, terminal drop velocity, mass transfer rate) is required for safe and efficient operation of equipment dealing with these systems. The nature of liquid-liquid flows where millions of droplets interact with each other, break up and merge, and form swarms, makes the determination of these characteristics complicated. To get fundamental understanding of complex dispersed system behavior, simplified problems are studied experimentally and numerically: single drop motion and breakup under shear flow or gravity, head-on drop collision and coalescence. The findings of these investigations can be scaled on the real systems and also utilized in engineering models and correlations describing the dispersions.

In the present study the motion of an organic *n*-butanol drop in water under the influence of gravity was simulated using a diffuse interface free energy lattice Boltzmann method (LBM). Simulation results are useful for liquid-liquid extraction processes, since an important design parameter - terminal drop velocity - was determined. A pure two-liquid system without mass transfer between phases was considered. To reduce the computational cost while studying the long-term behavior of a moving interface, all calculations were carried out in a moving reference frame. A wide range of drop diameters that covered the flow conditions relevant to extraction process was examined. For each drop diameter the evolution of drop velocity in time, the terminal rise velocity and drop's shape were determined. The results were compared to the experimental and numerical results obtained with other numerical techniques and to semi-empirical correlations: the deviation of simulated terminal velocity from other results is within 5% for smaller drops and below 15% for large oscillating drops. The capability of the method to capture the drop shape deformation especially in oscillating regime was also demonstrated. As an example of computational results, the *n*-butanol drop deformation with velocity streamlines at different time steps is shown in the figure below.

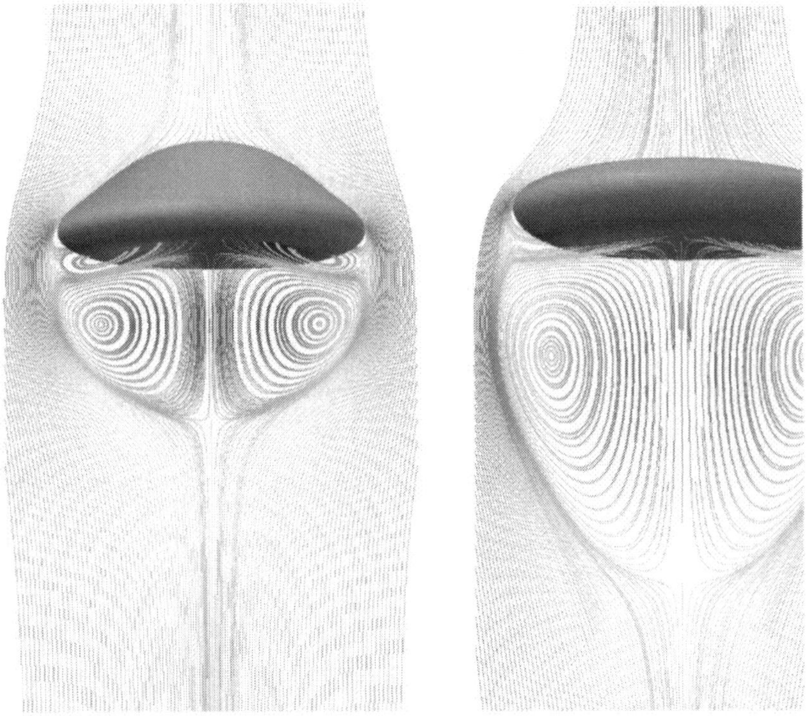

(93e) CFD Modelling of Bubble Column in Heterogeneous Regime:1D and 2D

Xiaoping Guan, Department of Chemical and Biological Engineering , Zhejiang University, Hangzhou, China

Bubble column has been widely applied in many chemical industrial processes, including oxidation, hydrogenation, carbonylation, and Fischer-Tropsch synthesis etc. Depending on states of gas dispersion, two kinds of flow patterns in bubble column were classified: homogenous and heterogeneous. The homogeneous regime occurred at relatively low superficial gas velocities and with uniformly sized bubbles. The heterogeneous regime operated at high superficial gas velocities and characterized by a steep radial profiles of gas holdup and liquid velocity. In this paper, a CFD model was developed for the calculations of gas holdup and liquid axial velocity profile in bubble column in the heterogeneous regime. Two key lateral forces was considered in the model: The lateral lift force and the turbulent diffusion force. In 1D simulation, the balance of these two forces determined gas holdup profile, which was used to compute axial liquid velocity profile. CFD and experimental results agreed very well.In 2D simulation, a new method was developed to deterimine the bubble diameter ,which was used to caculate drag force. Also the two lateral forces were considered. Again, simulation results agreed very well with experimental ones.

(93f) CFD Approach for the Design of Microcapsules Based On the Formation of Drops Using Viscous Non-Newtonian Fluids Through Fan Jet Nozzles

Jose Nogareda[1], Mariano Martín[1], Eva M. Martín del Valle[2] and Miguel A. Galán[2], (1)Chemical Engineering, University of Salamanca, Salamanca, Spain, (2)Department of Chemical Engineering, University of Salamanca, Salamanca, Spain

Currently a number of applications such as cosmetics, pharmaceutical or food industry rely on the use of microcapsules to bring the costumer the appropriate feeling, release the drug at the proper rate or transport it across the body, provide good taste or texture to the food, the proper essence etc. New applications require a better controlled design of the particles that involves the use of high viscosity non Newtonian liquids which must be handled at low temperatures so that the microcapsules and the content of them survive the process. The design of these microcapsules is a tedious and expensive process and a systematic approach is presented in order to reduce the design time and effort.

In the present paper we use a CFD modeling approach for the design of the microcapsules being able to predict drop size formation for a large number of experimental conditions involving flow rates and liquid properties. Close attention is paid to the non-Newtonian behavior of the solution which, in the end, is responsible for the droplet generation and size. Thus special emphasis is placed upon the experimental characterization of the liquid phase, mainly its rheology, and how the shear stress generated by the geometry and the air flow focusing the liquid through the nozzle results in a particle size. The experimental set up consists of a device develop within the group[1] were the liquid is fed from a pressurized tank into a 350 mm diameter nozzle and forced to exit an orifice just below the nozzle by regulating the pressure in the tank. Air is also fed into the nozzle chamber and its pressure can also be controlled. The generated thread which exits the nozzle orifice breaks into droplets of specific size which are gelated in a solution of barium chloride (2% wt) placed downstream. Both air and liquid pressures are regulated by a control panel which controls a pneumatic cabinet. The cabinet is fed of compressed air from a pressurized bottle. The obtained particle solution is then filtered and washed with two volumes of distilled water and particles are collected in a flask with distilled water. After that, particles are placed into an ultrasound sonicator and sonicated for 40 minutes. Then microcapsules are ready for particle analysis. We simultaneously record the thread/cone dispersion and the droplet formation using high speed video techniques.

We first develop an empirical model for predicting the Sauter mean diameter as function of the pressure drop between the liquid and the gas flows and the apparent viscosity of the liquid so as to calibrate the experimental device and understand the phenomenon as well as to establish the range of the operating variables. Next, based on the experimental experience we developed a Finite Element based model using the commercial package COMSOL, involving the creation of the complex geometry, which is capable of reproducing the mean drop size generated by the equipment, in the range of 20 – 60 mm. In this sense we are a step closer to define the operating conditions and the properties of the liquid phase in order to design the proper microcapsules.

References

(1) Cervero, JM; Nogareda, J;del Valle, EMM; Galan, MA; Chem. Eng. J. 174. 2-3 699-708

(97a) Applied Filtered Density Function

Peyman Givi, Mechanical Engineering and Materials Science, University of Pittsburgh, Pittsburgh, PA

The optimal means of capturing the detailed, unsteady physics of turbulent reacting flows, as it is now widely believed, is via large eddy simulation (LES). The challenge in LES is accurate and consistent modeling of the subgrid scale (SGS) quantities. The filtered density function (FDF) methodology; including its mass weighted form, the filtered mass density function (FMDF), has proven particularly effective for this purpose. The FDF is essentially the probability density function (PDF) of the SGS quantities. Therefore, it provides all the statistical information pertaining to these quantities.

In its stand-alone form, the FDF must account for the joint statistics of all the relevant physical variables. The most sophisticated FDF closure available to-date is the frequency-velocity-scalar FMDF (FVS-FMDF), and a simpler version (VS-FMDF) which does not include the SGS frequency. Hydrodynamic closure in incompressible, non-reacting flows has been successfully achieved via the velocity-FDF (V-FDF) and the one which has been utilized the most only considers the scalar field (S-FDF and S-FMDF). This is the most elementary form of FDF when it was first introduced. Since then, the FDF has experienced widespread usage, and is now regarded as one of the most effective and popular means of LES worldwide. Some of the most noticeable contributions in FDF by others are in its basic implementation, fine-tuning of its sub-closures, and its validation via laboratory experiments.

In this chapter, a review is presented of modern advances in FDF modeling and simulation. These are primarily the development of new sub-closures in FDF transport, and construction of efficient solvers for this transport. With these developments, it is now possible to conduct LES of turbulent reactive flows in complex configurations. Examples are provided of recent FDF-based simulations with some speculation of the type of complex flows to be simulated within the near future.

(97b) Emerging Concepts in the Computational Modeling of Mixing and Combustion

Venkat Raman, Department of Aerospace Engineering & Engineering Mechanics, University of Texas at Austin, Austin, TX

The last decade has seen an explosive growth in computational power, which has transformed the modeling of turbulent reacting flows. With the availability of vast computational resources, model-based design of turbulent reactors is now seen as a feasible goal. In this talk, progress in the modeling complex multiscale processes will be first discussed. In particular, the evolution of large eddy simulation (LES) methodology as a predictive computational tool for reactive flows will be presented. To enable computational design, it is important to build confidence in the models. In the second part of the talk, the integration of experiments and modeling to quantify uncertainties will be discussed. An important aspect of this integration is the design of experiments that provide the most information for building model confidence. For this purpose, a new generation of computational tools based on information theory is being developed. A few of these emerging concepts including Bayesian analyses of chemical kinetics and adjoint analyses of reacting flows will be presented.

(97c) Advanced Laser Diagnostics Measurements for Reactive Mixing Flows

Robert Lucht, Mechanical Engineering, Purdue University, West Lafayette, IN

The development and application of advanced laser diagnostic techniques for reactive mixing flows are described. Techniques to be discussed include high-data-rate planar imaging techniques such as OH planar laser-induced fluorescence (PLIF). We discuss the acquisition of OH PLIF images at data rates of 5 kHz. Planar imaging techniques for measuring the extent of molecular mixing in nonreacting flows are also discussed. Coherent anti-Stokes Raman scattering measurements of reactive mixing processes are also discussed. Dual-pump CARS is used for the simultaneous measurement of temperature and species concentrations. Recent measurements of concentration and temperature fields using dual-pump H_2/N_2 CARS for a reactive jet in crossflow in a high-pressure combustion test rig are discussed. We also discuss the use of ultrafast laser systems for measurements in reactive mixing flows. We have demonstrated the use of ultrafast laser systems for 5 kHz temperature measurements in flames using chirped-probe-pulse femtosecond CARS. We also present the results of recent femtosecond CARS measurements in turbulent nonpremixed jet flames, and discuss the potential for concentration measurements using this technique.

(97d) Self-Assembling Process of Flash Nanoprecipitation in a Multi-Inlet Vortex Mixer

Ying Liu, Chemical Engineering, University of Illinois at Chicago, Chicago, IL, Robert Prud'homme, Chemical Engineering, Princeton University, Princeton, NJ and Rodney O. Fox, Chemical & Biological Engineering, Iowa State University, Ames, IA

We present an experimental and numerical study of selfassembled polymeric nanoparticles in the process of flash nanoprecipitation using a multi-inlet vortex mixer (MIVM). The four-stream MIVM allows control of both the supersaturation and the final solvent quality by varying stream velocities. The design also enables the separation of reactive components prior to mixing. It is critical for the nanoprecipitation process to have a short mixing time, so that the solvent replacement starts homogeneously in the reactor. The properties of the nanoparticles depend on the competitive kinetics of polymer aggregation and organic solute nucleation and growth. We report the existence of a threshold Reynolds number (of 1600) over which nanoparticle sizes become independent of mixing. A similar value of the threshold Reynolds number is confirmed by numerical simulation along with experimental study of competitive reactions (the socalled "Bourne reactions") in the MIVM. Flow patterns in the MIVM at various Reynolds number are microscopically visualized by mixing iron nitrate ($Fe(NO_3)_3$) and potassium thiocyanate (KSCN) to precipitate $Fe(SCN)_x^{(3-x)}$.

(97e) Clathrate Hydrate Formation in Emulsions Subjected to Shear Flow

Genti Zylyftari[1], Prasad U Karanjkar[2], Amit Ahuja[2] and **Jeffrey F Morris**[3], (1)Levich Institute, City College of New York, New York, NY, (2)Chemical Engineering, Levich Institute, City College of New York, New York, NY, (3)Levich Institute and Chemical Engineering, Levich Institute, City College of New York, New York, NY

Clathrate hydrate formation is a phase transition from fluid to solid associated with forming cages of host (water) molecules around a guest molecule, typically of hydrocarbon form, with examples including methane, propane, and cyclopentane. The last of these, cyclopentane, forms hydrates at atmospheric pressure and temperatures above the freezing point of water; these features make it rather convenient for study of the influence of the hydrate-forming reaction on the properties of an emulsion containing hydrate formers in the organic phase; we study cyclopentane hydrate formation in a water-in-oil emulsion stabilized by surfactants, focusing on sodium mooleate (an oil soluble surfactant). Our interest is in the rheology of the mixtures, and how bulk flow effects this reaction process.

Hydrate formation is relatively unusual in the reaction generating a solid phase at a fluid-fluid boundary, and this spatial localization of the reaction will be a focus in our considerations here. In particular, it wil be shown that the formation at an interface results in a pronounced influence of surface-active agents upon the reaction rate and the resulting morphology of the reactant crystals. We will describe novel hollow conical crystals and their relationship to the surfactant loading; additionally, the influence on morphology of the cystalline stucture by the surfactant will be analyzed, and correlated to observations of the evolution of rheological properties as the reaction to form hydrate progresses. An interesting feature in this system is the role of flow-induced stresses. We observe that the reaction can be shutdown by mass transfer limitations, as the hydrate shell formed at the interface isolates the water from the organic phase. As a consequence, the role of flow is not simple mixing, but also involves fracturing of shells of hydrate around water drops allowing surface wetting of existing hydrate crystals. Reaction rate jumps suddenly when this phenomenon occurs, as evidenced by the rapid release of energy associated with the exothermic hydrate formation.

(97f) Reactive Infiltration Instabilities in Fractured and Porous Rocks

Anthony J.C. Ladd, Department of Chemical Engineering, University of Florida, Gainesville, FL and Piotr Szymczak, Department of Physics, University of Warsaw, Warsaw, Poland

Reactive infiltration instabilities occur in a wide range of geophysical and geotechnical systems. The simplest such instability occurs when fluid flows between two soluble plates, which is an idealized model of fractured limestone. Even when the initial aperture is uniform at the nanoscale, an instability in the reaction front develops leading to the formation of pronounced solutional channels or "wormholes". We have suggested that this instability may explain the onset of large underground caves systems, by allowing a much deeper penetration of reactant than is possible by uniform opening of the fracture [1].

Reactive infiltration in porous media has been investigated experimentally[2], numerically[3], and theoretically [4-6]. The predictions of the linear stability analysis vary considerably depending on the underlying model assumptions. In this talk I will outline how these different theoretical approaches can be viewed as limiting cases of a more general theory. Our results span the range of geologically feasible Peclet and Damkohler numbers [7-8]. In particular we show that the convective limit [5,6] is singular, with a wavelength selection and growth rate that is very sensitive to small amounts of diffusion. We also show that the opposite limit of diffusion-dominated dissolution [4] spans a much smaller parameter range than has been supposed.

In this talk I will briefly summarize some of our most important results and discuss their relevance to geophysical systems.

[1] Szymczak, P. & Ladd, A. J. C. 2011. The initial stages of cave formation: Beyond the

one-dimensional paradigm. Earth Planet. Sci. Lett. 301, 424-432.

[2] Fredd, C. N. & Fogler, H. S. 1998. Influence of transport and reaction on wormhole formation

in porous media. AIChE J. 44, 1933-1949.

[3] Hanna, R. B. & Rajaram, H. 1998. Influence of aperture variability on dissolutional growth

of fissures in karst formations. Water Resourc. Res. 34, 2843-2853.

[4] Chadam, D., Hoff, D., Merino, E., Ortoleva, P. & Sen, A. 1986. Reactive infiltration

instabilities. J. Appl. Math. 36, 207-221.

[5] Sherwood, J. D. 1987. Stability of a plane reaction front in a porous medium. Chem. Eng. Sci.

42, 1823-1829.

[6] Hinch, E. J. & Bhatt, B. S. 1990. Stability of an acid front moving through porous rock. J.

Fluid Mech. 212, 279-288.
[7] Szymczak, P. & Ladd, A. J. C. 2012. Reactive-infiltration instabilities in rocks. Fracture dissolution. J. Fluid Mech. To be Published.

[8] Szymczak, P. & Ladd, A. J. C. 2012. Reactive-infiltration instabilities in rocks. Dissolution of a porous matrix. In Preparation.

(97g) Reducer Performance of Chemical Looping Reactors: Mixing and Reaction Effects

L. - S. Fan, Liang Zeng, Andrew Tong and Qiang Zhou, William G. Lowrie Department of Chemical and Biomolecular Engineering, The Ohio State University, Columbus, OH

With the pressing need for clean, efficient, and cost-effective energy sources, the chemical looping strategy has evolved as a promising alternative to the traditional carbonaceous fuel conversion process. The syngas chemical looping (SCL) process co-generates electricity and hydrogen with in-situ CO_2 capture from coal derived syngas through the cyclic reduction and oxidation of an iron oxide based oxygen carrier. The SCL process features countercurrent gas-solid moving-bed reactor that converts syngas to CO_2 while reducing the oxygen carriers. The countercurrent reducer in the SCL process at the Ohio State University has been operated for a combined time of more than 300 hours in a 2.5 kW_{th} bench scale moving-bed reactor and a 25 kW_{th} subpilot scale integrated system. A syngas conversion in excess of 99.5% and an oxygen carrier conversion of nearly 50% have been obtained. The thermodynamic analysis indicates that the countercurrent moving-bed reducer offers better gas and solids conversion behavior, compared to the fluidized-bed reducer. In this study, effects of the mixing and the reaction in the reducer of a chemical looping reactor system operated in the two distinct flow patterns represented by these two types of reactors, i.e., countercurrent moving-bed and the fluidized-bed, are described in terms of experimental results and reactor modeling. Further, the importance of the thermodynamic properties of the oxygen carrier in the reducer performance is highlighted in this study.

(117a) Invited: Theoretical Modeling of Clathrin Self-Assembly On an Undulating Membrane

Andrew J. Spakowitz, Chemical Engineering, Biophysics, Stanford University, Stanford, CA

Clathrin proteins stabilize the budding and curvature of cell membranes that are characteristic of endocytosis, the process by which cells internalize cargo. These elastic, three-legged, pinwheel-like triskelia are also capable of aggregating into a diverse array of two- and three-dimensional nanostructures. Observations of clathrin aggregation on membranes in vivo show the proteins forming two distinct structures, depending on membrane conditions. Specifically, large, crystalline “plaques” are observed predominantly where the underlying membrane is adhered to a rigid substrate, while short-lived, high-curvature “pits” are found on all membranes. Such findings indicate a pathway through which intercellular transport is modulated by the surrounding environment. We develop a simplified physical model of clathrin self-assembly to address the physical mechanisms that underlie the observed experimental behavior. In this model, the proteins bind to one another to form honeycomb lattices (6-membered rings) on flat surfaces, and insert non-6-membered defects and void spaces into this lattice depending on the elastic properties and binding affinities of the subunits. Our model also predicts the lattice’s response to membrane indentations, such as those occurring in the initial stages of endocytosis. When a spherical indentation is imposed on a crystalline lattice, the structure tears apart and gives way to a void space around the indentation. We observe the opposite effect when an indentation is imposed on a fluid clathrin phase, as the proteins aggregate into a localized crystalline patch. These conditions mimic those found often in cells undergoing endocytosis. During this process, the bulk of the membrane is freely fluctuating, while a localized patch wraps intercellular cargo. Our results indicate that the physical structure of the clathrin subunit is ideal for facilitating cell uptake, and the modulation of forces felt by the membrane plays a critical role in this process.

(117b) Evaluating Photosynthetic Antenna Complex Assemblies for Efficient Solar Energy Utilization

Sándor Á. Kovács, William P. Bricker, Jing Jiang and Cynthia S. Lo, Energy, Environmental and Chemical Engineering, Washington University in St. Louis, St. Louis, MO

Photosynthesis is a robust and effective mechanism for harnessing solar energy in natural organisms and a closer look reveals photosynthetic systems possess inherently efficient energy capture and energy transfer mechanisms. When coupled with engineered materials, these biohybrid systems may provide a low-cost source of clean and renewable power. In this work, protein-surface docking and molecular dynamics free energy methods are used to probe the structure and stability of the biomolecular interface in such a device. We compare the relative free energies of assembly of two photosynthetic trimeric proteins: FMO from green sulfur bacteria and and PCP from dinoflagellates. We then consider the interaction of these proteins with various biomaterial surfaces, including silica and gold, and compare free engergies of binding. In addition to the native amino acid sequences, the effects of point mutations deisgned to mimic engineered peptide sequences with higher surface affinities are also considered. The degree of denaturation of these biohybrid assemblies is assessed, which leads to implications for retention of photosynthetic activity and ultimately long-term stability. These results predict how such complexes may be assembled and patterned in nanostructured biohybrid devices for efficient solar energy utilization.

(117c) Investigating Repressor Binding in the Tetracycline Operon by Molecular Dynamics Simulations

Dan S. Bolintineanu[1], Victor Vivcharuk[1], Katherine Volzing[1] and Yiannis Kaznessis[2], (1)Chemical Engineering and Materials Science, University of Minnesota, Minneapolis, MN, (2)Department of Chemical Engineering and Materials Science, University of Minnesota, Minneapolis, MN

The tetracycline operon is an important gene network component commonly used in synthetic biology applications. At the heart of the tet operon system is the highly specific binding of the tet repressor protein (TetR) to its cognate DNA sequence (tetO). We employ molecular dynamics (MD) simulations coupled with the free energy perturbation method to investigate the binding affinity of TetR to different tetO mutants. We also carry out *in vivo* tests in *E. coli* for a series of promoters based on these mutants, and obtain reasonable agreement between experimental green fluorescent protein (GFP) repression levels and binding free energy differences computed from molecular simulations. In all cases, the wild type tetO sequence yields the strongest TetR binding, which is observed both experimentally in terms of GFP levels and in simulation in terms of free energy changes. Two of the four tetO mutants we tested yield relatively strong binding, although somewhat weaker than the wild type, whereas the other two mutants tend to be significantly weaker. The clustering and relative ranking of this subset of tetO mutants is generally consistent between our own experimental data, previous experiments with different systems and the free energy changes computed from our simulations. Overall, this work offers detailed quantitative insights into an important synthetic biological system and demonstrates the potential of molecular simulations to quantitatively predict biologically relevant behavior.

(117d) Self-Assembly of Complex DNA Architectures Via Multiscale Simulations for Nanotechnology and Alterative Energy

Cade B. Markegard[1], Iris W. Fu[1] and Hung D. Nguyen[2], (1)Chemical Engineering and Materials Science, University of California, Irvine, Irvine, CA, (2)Department of Chemical Engineering and Materials Science, University of California, Irvine, Irvine, CA

Elucidation of the molecular-level mechanisms by which DNA units are self-assembled nanostructures might provide new informative tools in the engineering of novel biomaterials. Our studies examine the role of DNA sequences and the environmental conditions on the morphology and mechanical properties of DNA assemblies by performing molecular dynamics simulations. We use a newly-developed coarse-grained model that bridges the gap between detailed atomistic models and extremely simplified models by capturing geometric and energetic details yet it is sufficiently simple to allow simulations of the spontaneous self-assembly of many DNA units simultaneously. We first examine the dependence of persistence length and melting temperature of as a function of ionic strength, composition and chain length. Our results from single-molecule simulations agree qualitatively with experimental data. To examine the kinetic mechanisms involved in DNA self-assembly, we perform constant-temperature simulations to observe the whole process of DNA assembly starting from random configurations of relatively large DNA systems. We also perform replica-exchange simulations to delineate a phase diagram characterizing different types of structure exhibited for each sequence as a function of the condition being examined. The findings of this research will guide experimentalists to identify systems of novel biomaterials with advantageous morphological properties.

(117e) Predicting the Nature of the Protein Corona Surrounding Nanoparticles

Lauren Ridge and Carol K. Hall, Chemical and Biomolecular Engineering, North Carolina State University, Raleigh, NC

Our knowledge of the risks of exposure to nanoparticles is essentially in its infancy since the current *in vivo* and *in vitro* studies of "nanotoxicology" span only a minute subset of the many different types of nanoparticles being made today. A key step in analyzing the risks posed by nanoparticles is to determine the composition of the corona of plasma proteins adsorbed onto the surface of nanoparticles. The nature of the protein corona is expected to have a marked influence on the fate of nanoparticles in the body.

The goal of this research is to develop an algorithm to predict the affinities of proteins for a given nanoparticle based on their specific properties, and hence predict which proteins are most likely to form a corona around that nanoparticle. Discontinuous molecular dynamics simulations with the coarse-grained force field PRIME20 were used to calculate the adsorption of polyalanine (polyA) and polytryptophan (polyW) onto a hydrophobic nanoparticle. In single-protein systems, a decrease in temperature led to an increase in adsorption for all concentrations and peptide lengths. However, the effect of increased protein concentration on adsorption was not as straightforward, as short chains (A_{16} and W_{16}) had a direct relationship between adsorption and protein concentration, while longer chains had an inverse relationship between adsorption and protein concentration. On the whole, polyA was more likely to adsorb than polyW under identical conditions. For polyA, the degree of adsorption appears to affect the deformation of chains adsorbed on the nanoparticle surface. For short chains, A_{16}, as the density of the protein corona increases, so does the likelihood that an adsorbed protein will deform and lose its α-helical character, but not for the longer A_{30}. Competitive adsorption between polyA and polyW was also simulated. When W_{30} competes with short A_{16}, the adsorption of polyW depends directly on concentration, and when W_{30} competes with longer A_{30}, the adsorption of polyW depends inversely on concentration, trends that are similar to those found in the single-protein simulations.

(117f) Thermodynamics of Misfolding of Polyglutamine Peptides

Aaron M. Fluitt and Juan J. de Pablo, Department of Chemical and Biological Engineering, University of Wisconsin-Madison, Madison, WI

Huntington's disease is one of nine inherited neurodegenerative diseases associated with the expansion of a polyglutamine repeat in a disease-specific protein. In each case, disease symptoms appear only if the polyglutamine expansion exceeds a certain threshold length; in Huntington's disease, the threshold is approximately 36 residues. A hallmark of the expanded polyglutamine diseases is the formation of amyloid fibers rich in aggregated polyglutamine in certain regions of the brain. It has long been thought that the expansion of polyglutamine beyond the threshold length enables the formation of a metastable, misfolded conformer that nucleates aggregation. In this work, we calculate the free energies of different conformers of polyglutamine of lengths spanning the pathological threshold. We employ atomistic molecular dynamics simulations in explicit water to furnish results suitable for direct comparison to *in vitro* spectroscopic experiments. The results provide insight into the nature of the length dependence of polyglutamine misfolding and aggregation.

(117g) Charge Effects On the Fibril-Forming Peptide KTVIIE: A Two-Dimensional Replica Exchange Simulation Study

Joo-hyun Jeon, Carolyn E. Mills and M. Scott Shell, Chemical Engineering, University of California Santa Barbara, Santa Barbara, CA

The assembly of peptides into ordered nanostructures is increasingly recognized as both a bioengineering tool for generating new materials as well as a critical aspect of aggregation processes that underlie neurological diseases such as Alzheimer's, Parkinson's and Huntington's. A major problem is understanding how extremely subtle sequence changes can lead to profound and often unexpected differences in self-assembly behavior. To better delineate the complex interplay of different microscopic driving forces in such cases, we develop a novel methodology to quantify and compare the propensity of different peptide sequences to form small oligomers during early self-assembly stages. This Umbrella Sampling Replica Exchange Molecular Dynamics (UREMD) method performs a REMD simulation along peptide association reaction coordinates using umbrella restraints.

With this method, we study a set of sequence-similar peptides that differ in net charge: K+TVIIE-, K+TVIIE, and +K+TVIIE. Interestingly, experiments show that only the monovalent peptide, K+TVIIE, forms fibrils, while the others do not. We examine dimer, trimer, and tetramer formation processes of these peptides, and compute high-accuracy potential of mean force (PMF) association curves. The PMFs recapitulate a higher stability and equilibrium constant of the fibril-forming peptide, similar to experiment, but reveal that entropic contributions to association free energies can play a surprisingly significant role. The simulations also show behavior reminiscent of experimental aggregate polymorphism, revealed in multiple stable conformational states and association pathways. Our results suggest that sequence changes can have significant effects on self-assembly through not only direct peptide-peptide interactions but conformational entropies and degeneracies as well.

We also use this method to study the self-assembly of a diphenylalanine peptide that was recently found to form nanotubes in aqueous solution. To address the potential role of electrostatic interactions in this assembly process, we compare UREMD results of a zwitterionic diphenylalanine dimer with that of a modified non-charged diphenylalanine dimer.

(117h) Regularizing Binding Energies and the Thermodynamics of Hydration --- an Approach to Study the Hydration Free Energy of a Protein From All-Atom Simulations

Dilip Asthagiri, Johns Hopkins University, Baltimore, MD, Valery Weber, IBM, Zurich, Switzerland and Dheeraj Tomar, Chemical Engineering, Johns Hopkins University, Baltimore, MD

The interaction of a hydrated solute with the solvent spans a wide range of energies. Interactions with the solvent adjacent to the solute are typically strong (relative to thermal energy), and, in the case of ionic solutes, the strength of these interactions can rival covalent bonds. Farther away from the solute-solvent interface, the interaction weakens and the distribution of interaction energies approaches a Gaussian. Traditionally, in computing the thermodynamic properties of hydration, the problem of disparate energy scales is dealt by alchemically transforming the solute from a non-interacting solute to a fully interacting solute. But at the scale of a globular protein such approaches can prove challenging. Here, guided by the quasichemical organization of the potential distribution theorem, we present an approach that readily applies to systems ranging in complexity from those described by first principles (*ab initio*) potentials, a challenge that has thus far remained formidable, and to study the hydration thermodynamics of a protein, potentially at a level of resolution that is now routine for small solutes in water. The latter possibility can help address fundamental questions regarding the thermodynamics that govern the structure and stability of proteins. In this presentation, we will show results of our approach for the hydration of water, of a peptide model in different osmolyte mixtures, and finally for the protein cytochrome C. The potential utility of this approach for studying ion and osmolyte-specific effects in protein solution thermodynamics will be indicated.

(117i) Rapid Calculation of Protein Pk_a Values Using Rosetta

Krishna Praneeth Kilambi and Jeffrey J. Gray, Chemical and Biomolecular Engineering, Johns Hopkins University, Baltimore, MD

We developed a Rosetta-based Monte Carlo method to calculate pK_a values of protein residues that commonly exhibit variable protonation states (Asp, Glu, Lys, His and Tyr). We tested the technique by calculating pK_a values for 270 residues from 34 proteins. The standard Rosetta score function, which is independent of any environmental conditions, failed to capture pK_a shifts. After incorporating a Coulomb electrostatic potential and optimizing the solvation reference energies for pK_a calculations, a method allowing side-chain flexibility achieved a root-mean-square deviation (RMSD) of 0.83 from experimental values (0.68 after discounting 11 predictions with an error over 2 pH units). Additional degrees of side-chain conformational freedom for the proximal residues facilitated capture of charge-charge interactions in a few cases and achieved an overall RMSD of 0.85 pH units. Adding backbone flexibility increased overall RMSD to 0.94 pH units but improved relative pK_a predictions for proximal catalytic residues. The method also captures large pK_a shifts of lysine and some glutamate point mutations in *staphylococcal* nuclease. Thus, a simple and fast method based on the Rosetta score function and limited conformational sampling produces pK_a values that will be useful when rapid estimation is essential, as in docking, design and folding.

(118a) Exergy Analysis of Modular Organic Rankine Cycle (ORC) Power Plants

Maciej Lukawski[1,2,3], Pall Valdimarsson[4,5] and Jefferson W. Tester[1,2,3], (1)School of Chemical and Biomolecular Engineering, Cornell University, Ithaca, NY, (2)Atkinson Center for a Sustainable Future, Cornell University, Ithaca, NY, (3)Cornell Energy Institute, Cornell University, Ithaca, NY, (4)Gas and Process Division, Atlas Copco, Cologne, Germany, (5)School of Science and Engineering, Reykjavik University, Reykjavik, Iceland

This research investigates thermodynamic and economic feasibility of using modular Organic Rankine Cycle (ORC) power plants for generation of electricity from geothermal fluids or waste heat coming from engine exhaust gases or cement plants. Analysis of market demand for ORC power plants is also performed. Thermodynamic characteristics of different heat sources are analyzed and a set of design conditions maximizing ORC unit performance for the most promising applications is determined. A thermodynamic model of ORC power plant using recuperator and wet mechanical-draft cooling tower was developed. Four hydrocarbon working fluids were selected for detailed modeling to determine their thermodynamic performance for a year-round operation. For a heat source at 150 °C, isopentane provides the best thermodynamic performance. Isopentane and n-pentane were proved to be unable to fully utilize low outdoor temperatures, which resulted in a significant decrease in annual energy output. Use of n-butane reduces the size and cost of heat exchangers when compared to isobutene, but not enough to compensate for the decrease in net power output.

Thermoeconomic optimization is the main feature of the design procedure. This method is based on exergy analysis and allowed us to find the optimal power plant configuration and size of the heat exchangers. Traditional approaches to design and optimization of power plants focus on maximizing fuel utilization efficiency. Our approach is different and is centered on cost-optimal design of each component to minimize the levelized cost of electricity (LCOE) generated by the power plant.

As the first stage of thermoeconomic optimization, a detailed economic analysis is made for four different applications: geothermal plants using water from conventional hydrothermal or former oil and gas reservoirs, as well as waste heat recovery plants coupled with diesel engines or clinker coolers in cement plants. Case-specific parameters, such as interest rates, project lifetime, load factor or type of steel used for heat exchangers were specified for each application. Costs of single components are described by non-linear equations in which a key characteristic parameter that scales with component cost is used as a variable. Exergy flow rates in each of the systems are calculated and costs are assigned to these streams in order to identify cost-ineffective processes.

Analysis has shown that, depending on the type of application, the levelized cost of electricity ranges from 5.07 to 7.55 U.S. ¢(2008)/kWh, making modular ORC power plants thermodynamically and economically feasible. While the majority of commercially available ORC waste heat recovery units do not incorporate recuperators, the cost-optimal design for each application considered in this study included this component. In geothermal power plants, use of regenerative heat exchange lowers the utilization efficiency, if based on incoming exergy of heat source. However, recuperator also reduces the size of preheater and evaporator made out of costly stainless steel and lowers LCOE. In waste heat recovery applications the LCOE is somewhat lower than for the best design without the regenerator. In the same way as in the geothermal unit, the regenerator increases the thermal efficiency of the cycle. In waste heat applications, decreased efficiency of heat recovery has a limited impact.

Finally, sensitivity analysis shows the impact of heat source characteristics and macroeconomic variables on levelized cost of power. As expected, economic performance of geothermal units is highly sensitive to temperature and flow rate of geothermal water.

(118b) Modeling Water Injected Gas Turbines - Performance and Emissions

Mohammed Shafi Syed, Kerry M. Dooley and F.Carl Knopf, Chemical Engineering, Louisiana State University, Baton Rouge, LA

Natural gas-fired cogeneration systems are commonly used for large-scale industrial energy production – both electricity generation and heat recovery. Industrial cogeneration currently represents about 8% of the U.S. total electricity generation capacity. Plans call for cogeneration to increase to 20% of the generation capacity by the end of 2030 (ORNL/TM- 2008/224). Industrial cogeneration systems attain both high thermal efficiency and low emissions. The attainment of low emissions from natural gas fired turbines, in particular low NO_xemissions, is of considerable environmental importance especially as coal becomes a less favorable fuel source.

Our current project addresses emissions and performance modeling of the 20 MW natural gas-fired cogeneration system located at Louisiana State University. Water injection is used to help lower emissions. Data reconciliation and gross error detection are performed to adjust measured variables and determine efficiency. A continuous emission monitoring system (CEMS) has been recently installed to measure both the O_2 and NO_x concentrations in the exhaust; CO is also measured. These concentrations have been used to refine a kinetics model, based on GRI-Mech 3.0, in order to predict NO_x, CO and O_2 concentrations leaving the system. The kinetics model is used within a chemical reactor network consisting of PSRs and PFRs to represent the turbine combustion and dilution zones. Changes in measured emission species combined with the reactor network simulation are used to help indicate the onset of mechanical problems in the combustion zone.

(118c) Injection of Carbon Dioxide and Nitrogen Into Methane Hydrate Reservoirs: Binary Hydrateressim Simulations

Nagasree Garapati, Chemical Engineering, West Virginia University, morgantown, WV; National Energy Technology Laboratory, Morgantown, WV and Brian Anderson, Chemical Engineering, National Energy Technology Laboratory/West Virginia University, Morgantown, WV

Natural gas hydrates, non-stoichiometric crystalline inclusion compounds are likely to contain more carbon than in all other fossil fuel reserves combined. Different techniques that are currently being proposed for production of CH_4 from the gas hydrates include depressurization, thermal stimulation and injection of inhibitors. These methods involve dissociation of gas hydrate and the release of a significant amount of water that may cause geomechanical stresses on the reservoir that could lead to subsidence. A newly proposed method is to replace CH_4 in the hydrate by the injection of pure CO_2 or CO_2+N_2 mixtures, which serves dual purpose of long-term storage of a greenhouse gas (CO_2) and the production of natural gas, while maintaining the hydrate structure *in situ*.

Field-scale experiments and the equipment required for production tests on hydrate deposits are very expensive. Therefore, reservoir simulations can be used to predict the production potential of gas hydrates. HydrateResSim (HRS) is the only open-source code available for public through National Energy Technology Laboratories (NETL), written in FORTRAN. The current version of HRS can only simulate pure methane hydrates and the relationship between equilibrium temperature and pressure is given by a simple, 1-D regression expression parameterized by Moridis (2003). In this work the code has been modified to simulate mixed hydrates including CH_4-CO_2 and CH_4-N_2 in order to understand simultaneous CH_4 production rate as well as CO_2 sequestration rate in naturally-occurring hydrate reservoirs. The original code allows distribution of heat and up to 3 components (H_2O, CH_4, and inhibitors) whereas the modified code, can allow distribution of heat and up to 4 components (H_2O, CH_4, CO_2/N_2 and inhibitors) between four possible phases (gas, aqueous, ice, and hydrate). A major component to the new code is the implementation of two-component equilibrium surface. The equilibrium surface obtained using a cell potential method is incorporated in to the code in tabular form and a bi-linear interpolation is used to interpolate data at given conditions. The phase equilibrium is given as input to the code in two data files, $T_{eq}=f(P,y_{CH}\ )$ & $P_{eq}=g(T,y_{CH}\ )$ where T is temperature (°C), P is pressure (MPa) and y_{CH} is CH_4 composition in gas phase. Simulations have been performed to analyze the behavior of the hydrate reservoir by injecting pure CO_2 and N_2.

A methane hydrate reservoir is considered with coexisting hydrate and aqueous phases and at the boundary CO_2 gas is injected and the reservoir behavior is studied. It is found that CO_2 hydrate was formed due to the available free water phase which caused dramatic reduction in permeability of the reservoir to the injection of fluid. Therefore it is necessary to remove the free water within the vicinity of the injection well in a reservoir so as to avoid the excessive hydrate formation and blockage close to the well. In order to remove the free water before injection of CO_2 gas the reservoir can be flooded with N_2 gas. The reservoir behavior with injection of N_2 is also studied and it is observed that there is an ice formation near the injection well which can block the well. The solution could be injection of N_2+CO_2 mixture and to study the exchange process in the reservoir by injection of mixed gas. Therefore HydrateResSim has been modified to allow for the simulation of a ternary gas system, CH_4-N_2-CO_2.

(118d) Determination of Binary Parameters of Liquid-Liquid Equilibrium for Biodiesel + Glycerol + Ethanol Systems Using Local Composition Models

Nathan Sombra[1], Frederico do Carmo[1], Rílvia Santiago-Aguiar[2] and **Hosiberto de SantxAna**[1], (1)Chemical Engineering Department, Universidade Federal do Ceará, Fortaleza, Brazil, (2)Chemical Engineering Department, Universidade Federal do Ceará, Fprtaleza, Brazil

In the last decades, biodiesel has been considered an alternative fuel for diesel engines, due its characteristics, e.g., non-toxicity, renewability, ignition properties, and almost zero sulfur content. During biodiesel production a liquid-liquid equilibrium (LLE) among biodiesel, glycerol, and alcohol may occur. For this reason, the knowledge of LLE data is essential for design, unit operations processes, and process modeling. A set of data on LLE (biodiesel + glycerol + ethanol), taken from different sources of the literature and also from experimental measurements determined by our research group was used in this work. A total number of 69 tie-lines have been listed. These data were modeled by using two local composition models (NRTL and UNIQUAC equations). Binary parameters for biodiesel + glycerol + ethanol systems have been obtained from these data. Additionally. a set of new binary parameters for glycerol-ethanol has been obtained for biodiesel + glycerol + ethanol systems. Ternary LLE diagrams were constructed and it could be observed a satisfactory behavior by comparing experimental and calculated compositions and tie-lines, with a global root mean square deviation of 1.13 % and 1.66 % for NRTL and UNIQUAC models, respectively. Consequently, biodiesel + glycerol + ethanol systems can be represented by using both models tested.

(118e) Regenerable MgO-Based Sorbents for CO2 Removal in IGCC Processes

Shahin Zarghami, Chemical and Biological Engineering , illinois institute of technology, Chicago, IL and Javad Abbasian, Chemical and Biological Engineering Department, Illinois Institute of Technology, Chicago, IL

Integrated Gasification-Combined Cycle (IGCC) processes are among the leading contenders for power generation conversion in the 21st century because of their significantly higher efficiencies and potential environmental advantages compared to pulverized coal combustion (PC) processes. The higher efficiency in IGCC processes reduces the emissions of CO_2 per unit of power generated by 30%. Furthermore, the higher concentration of CO_2 and high pressure prevailing in the pre-combustion fuel gas, leads to additional advantages IGCC, if carbon capture and sequestration technology in implemented in power generation processes. In IGCC processes equipped with high temperature particulate removal and hot/warm temperature sulfur removal subsystems, to avoid efficiency losses, it is desirable to remove CO_2 in the temperature range of 300° to 500°C, which makes regenerable MgO-based sorbents ideal for such operations. In this temperature range, CO_2 removal results in shifting of water-gas-shift (WGS) reaction toward significant reduction in carbon monoxide (CO) concentration, generating additional hydrogen in the process, which is considered as a higher value product. However, highly durable, reactive, and attrition resistant sorbents are required for such application.

This paper presents the results obtained with a highly reactive and attrition resistant regenerable MgO-based sorbent which can simultaneously remove carbon dioxide and enhance hydrogen production in a single reactor. A brief overview on sorbent preparation technique will be presented and the effects of various variables such as calcination temperature, calcination time, impregnation method, and drying methods on the physical properties of the sorbents are discussed. The effects of physical and chemical properties of the sorbents on their CO_2 reactivity and absorption capacity are also presented and discussed.

(118f) Hollow Fiber-Supported Designer Ionic Liquid Sponges for Post-Combustion CO2 Scrubbing

Jong Suk Lee[1], Patrick Hillsheim[2], Dongkun Huang[1], Ryan P. Lively[3], Kyung Hee Oh[1], Sheng Dai[2] and William J. Koros[1], (1)Chemical & Biomolecular Engineering, Georgia Institute of Technology, Atlanta, GA, (2)Chemical Sciences Division, Oak Ridge National Laboratory, Oak Ridge, TN, (3)Algenol Biofuels, Bonita Springs, FL

A proof of concept study for a new type of carbon capture system is considered for post-combustion CO2 capture based on porous hollow fiber sorbents with ionic liquids sorbed in the cell walls of the fiber. This study proves that delicate morphological features in the open-celled porous wall can be maintained during the infusion process. Mixtures of task specific ionic liquid (i.e. [BMIM][Tf2N]) and superbase (i.e. DBU) were loaded into polyamide-imide (PAI) fibers by a so-called two-step non-solvent infusion protocol. In the protocol, methanol carries ionic liquids into the pore cell walls of hollow fibers and then hexane carries superbase to create an efficient CO2 sorbent. Our ionic liquid/superbase impregnation technique overcomes a serious increase in mass transfer resistance upon reaction with CO2, thereby allowing its large scale utilization for post-combustion CO2 capture. The investigation on the effect of different pore former additives (different molecular weights of polyvinylpyrrolidone, lithium nitrate, and their mixtures) suggested a potential solution to produce highly interconnected open cell pore structures of PAI hollow fibers. Lastly, a lumen side barrier layer was successfully formed on the bore side of neat PAI fibers by using a mixture of Neoprene® with crosslinking agents (TSR-633) via a post-treatment process. The lumen layer will enable heat removal from the fiber sorbents during their application in rapid thermal swing cycling processes.

(118g) Prediction of CO_2 Adsorption Properties in Zeolites Using Force Fields Derived From Periodic Dispersion-Corrected DFT Calculations

Hanjun Fang[1], Preeti Kamakoti[2], Ji Zang[1], Stephen Cundy[2], Charanjit Paur[2], Peter I. Ravikovitch[2] and David S. Sholl[1], (1)School of Chemical & Biomolecular Engineering, Georgia Institute of Technology, Atlanta, GA, (2)ExxonMobil Research and Engineering, Annandale, NJ

We demonstrate a new approach to develop transferable force fields describing molecular adsorption in zeolites by combining dispersion-corrected Density Functional Theory (DFT) calculations and classical atomistic simulations. This approach is illustrated with the adsorption of CO_2 in zeolites. Multiple dispersion-corrected DFT methods were tested for describing CO_2 adsorption in Na-ferrierite. The DFT-D2 approach was found to give the best agreement with high level quantum chemistry results and experimental data. A classical force field for CO_2 adsorption in siliceous zeolites was then developed based on hundreds of DFT-D2 calculations that probed the full range of accessible volume in purely siliceous chabazite (Si-CHA) via random sampling. We independently performed experiments with Si-CHA measuring CO_2 isotherms and heats of adsorption by microcalorimetry. Excellent agreement was obtained between adsorption isotherms predicted with our first-principles-derived force field and our experiments. The transferability of this force field was examined using available adsorption isotherms for CO_2 in siliceous MFI and DDR zeolites, again with reasonably good agreement between calculated and experimental results. The methods demonstrated by these calculations will be broadly applicable in using molecular simulations to predict properties of adsorbed molecules in zeolites and other nanoporous materials.

(139a) Engineering Metabolic Modules for Electrofuels Production in Proteobacteria

Curt R. Fischer, JT Sauls, John Sexton, Reshma Shetty and Jason Kelly, Ginkgo Bioworks, Boston, MA

Electrofuels as a concept seeks to produce liquid transportation fuels from carbon dioxide and renewable electricity. Ginkgo Bioworks's approach to electrofuels is to use formic acid salts as a vector of electricial energy for biology-based fuels production. Unlike other vectors such as hydrogen, formate is water-soluble and easy to handle safely, which simplifies process scale up. We have designed and are building a set of metabolic modules aimed at (i) converting the electrons carried by formate into central metabolic intermediates such as NAD(P)H, (ii) converting the carbon carried by formate into central metabolic intermediates such as pyruvate, and (iii) producing liquid, water-immiscible hydrocarbons from pyruvate. Together these modules enable the biosynthesis of liquid fuels from formate salts.

Recently we have also begun to implement these modules in alternate host organisms (alphaproteobacteria instead of *E. coli*). These hosts offer a number of potential advantages over *E. coli* but require the fabrication and characterization of synthetic biology "parts" including promoters, plasmid replication and selection elements, etc. that will also be discussed.

(139b) Metabolic Engineering of Clostridium Ljungdahlii for Production of Fuels and Other Chemicals From Carbon Dioxide Via Microbial Electrosynthesis or with Carbon Monoxide or Hydrogen As the Electron Donor

Toshiyuki Ueki, Ching Leang, Kelly P. Nevin and Derek R. Lovley, Department of Microbiology, University of Massachusetts, Amherst, MA

Clostridium ljungdahlii is an acetogenic microorganism that has the natural ability to use hydrogen as the electron donor to reduce carbon dioxide as its terminal electron acceptor with the production of acetate that is excreted from the cell. Carbon monoxide can also serve as an electron donor for acetate production. Under some conditions ethanol and butanediol are produced in addition to acetate, but usually in lesser quantities. Recent studies in our laboratory have also demonstrated that *C. ljungdahlii* can directly accept electrons from a negatively poised electrode in order to reduce carbon dioxide to acetate. The ability of microorganisms like *C. ljungdahlii* to convert carbon dioxide to organic products with electrons derived from an electrode as the driving force is known as microbial electrosynthesis. Carbon dioxide is reduced to organic products in *C. ljungdahlii* via the Wood-Ljungdahl pathway. Acetyl-CoA is a key intermediate in the Wood-Ljungdahl pathway, which offers the possibility of genetically modifying *C. ljungdahlii* to make a diversity of products more valuable than acetate. Techniques for genetic manipulation of *C. ljungdahlii* were developed which permitted deleting genes from, or inserting genes into, the chromosome. Initial gene manipulation studies focused on the mechanisms for hydrogen uptake and identified a single hydrogenase gene that was required for growth on hydrogen. This hydrogenase-deficient strain continued to reduce carbon dioxide on negatively poised electrodes, demonstrating that hydrogen was not an intermediate in electron transfer from electrodes to *C. ljungdahlii*. Introducing the appropriate genes and/or deleting genes for unwanted pathways made it possible to divert metabolism in *C. ljungdahlii* away from acetate and ethanol production and toward the production of more desirable products. A strain for the production of the transportation fuel butanol was developed, as were strains that can produce other multi-carbon organic commodities from carbon dioxide with an electrode or hydrogen serving as the energy source. Furthermore, these strains could ferment sugars to desired products. These results demonstrate that *C. ljungdahlii* is a genetically tractable chassis organism for the production of fuels and other commodities from a diversity of renewable energy sources.

(139c) Bioelectrosynthetic Reduction of Carbon Dioxide to Biofuel Precursors Using a Mixed Microbial Consortium Isolated From Bovine Rumen

Ryan Hammonds[1], Kristi Whitehead[1], Stephen Creager[2], Jung Oh[3] and J. Michael Henson[1], (1)Biological Sciences, Clemson University, Clemson, SC, (2)Clemson University, Anderson, SC, (3)Chemistry, Clemson University, Clemson, SC

Microbially-catalyzed reduction of carbon dioxide to higher-molecular weight compounds through consumption of electrical current offers prospects for electrochemical production of renewable liquid transportation fuels. Recently, pure cultures of chemoautotrophic microorganisms were reported to produce electrofuel precursors, such as acetate, from carbon dioxide and electrical current. A consortium consisting of several different microbial populations acting as a stable community might offer an energetically favorable approach for production of these electrofuels. In order to evaluate this concept, bovine rumen contents were enriched under strict anaerobic conditions for autotrophic, acetogenic microbes for several generations using hydrogen and carbon dioxide. A stable acetogenic consortium was obtained and transferred to bioelectrosynthesis cells supplied with electric current from a potentiostatic controller. Current consumption and biofuel precursor production were measured. Initial electron recovery as biofuel precursors exceeded 80% efficiency. To further improve rates and efficiency of biofuel precursor production, a combinatorial approach to evaluate growth conditions, active microbial community structure, and reactor geometry is in progress.

(139d) Electroautotrophic Synthesis of Acetate and Methane

Christopher W. Marshall[1], Edward V. LaBelle[1], Erin J. Fichot[2], Daniel E. Ross[2], R. Sean Norman[2] and **Harold D. May**[1], (1)Marine Biomedicine & Environmental Science Center, Medical University of South Carolina, Charleston, SC, (2)Environmental Health Sciences, University of South Carolina, Columbia, SC

Electroautotrophic synthesis of chemicals from CO_2 offers an attractive way of producing fuels that would be sustainable, domestic and environmentally benign, particularly in relation to climate change. Autotrophic rates of acetate and methane production by microbes with hydrogen gas as an electron donor have been demonstrated in the multiple kg m^{-3} h^{-1} range. However, this approach requires a source of hydrogen. Generation of hydrogen by electrolysis, or direct delivery of electrons through an electrode, could serve as the source of hydrogen if sustainable energy, such as solar or wind, is used to drive the production of electrical energy. The overall method becomes even more attractive if the use of rare earth elements is avoided and only microbial catalysts are employed. This approach is especially advantageous if the microbes operate within the electrolysis/electrochemical cell. Using a community of microorganisms selected from a brewery waste we have produced high amounts of methane and acetate from reductive electrical current in electrochemical cells with only microbial catalysts. The data thus far indicate that the generation of both products is dependent upon the biological catalysts associated with the electrodes, and the production rates observed for both products exceed any published rates as of the time of this submission. The electromethane could readily be dropped into existing natural gas infrastructure and used as a transportation fuel. Acetate is a commodity chemical but is also a building block that may be used to produce other fuels. We are attempting to convert that acetate into alcohols (ethanol and butanol) with communities of microbes selected within the electrochemical cells. The rates of production, electrochemical data associated with the catalysis, reactor design, microbial catalyst identification and community characteristics (examined metagenomically) will be discussed.

Integrated Electromicrobial Conversion of CO_2 to Higher Alcohols

Han Li,[1,2] Paul H. Opgenorth,[3] David G. Wernick,[1] Steve Rogers,[4] Tung-Yun Wu,[1]
Wendy Higashide,[4] Peter Malati,[5] Yi-Xin Huo,[4] Kwang Myung Cho,[4] James C. Liao[1,2,3,6]*

A man-made photovoltaic device is relatively efficient in converting sunlight to electricity, but the electrical energy generated is difficult to store. The biological photosystems, on the other hand, are limited by the intrinsic design and biomaterials available, for which no near-term improvements are in sight (*1*). One way to circumvent both problems is to link man-made solar cells to biological CO_2 fixation and fuel production. Theoretically, H_2 generated by solar electricity can drive CO_2 fixation in lithoautotrophic microorganisms engineered to synthesize high energy-density liquid fuels. However, the low solubility, low mass transfer rate, and safety issues of H_2 in microbial cultures limit the efficiency and scalability of such processes.

Compared with H_2, formic acid is a favorable energy carrier. Electrochemical production of formic acid from CO_2 and H_2O can achieve relatively high efficiency (*2*, *3*). Formate is highly soluble and is readily converted to CO_2 and NADH in the cells, providing a safe replacement for H_2. However, the high solubility of formate increases the cost of separation. Accumulated formate will decompose at the anode, decreasing the yield of the process (*2*). Therefore, simultaneous electrochemical formate production and biological formate conversion to higher alcohols is desirable (Fig. 1A). Unfortunately, introduction of electricity to microbial cultures may impede cell growth.

As such, an integrated process for reduction of CO_2 to liquid fuel powered by electricity requires (i) metabolic engineering of a lithoautotrophic organism to produce liquid fuels, (ii) electrochemical generation of formate from CO_2 in fermentation medium, and (iii) enabling microbes to withstand electricity. In this work, we chose *Ralstonia eutropha* H16 as the production host and isobutanol and 3-methyl-1-butanol (3MB) as the target fuels, which can be used in the internal combustion engines. We introduced the set of genes previously reported (*4*, *5*) for isobutanol and 3MB production into *R. eutropha* H16 (fig. S1, A to E). We also disrupted polyhydroxybutyrate synthesis (fig. S1, A and B) and used the heterologous isobutanol and 3MB production pathway as the new metabolic sink for carbon and reducing equivalents. In a pH-coupled formic acid feeding fermentor, the engineered strain LH74D produced fuels with the final titer of over 1.4 g/l (~846 mg/l isobutanol and ~570 mg/l 3MB) (fig S2B).

Next, we used an In foil cathode and a Pt anode in the culture medium bubbled with air containing 15% CO_2 to produce formate electrochemically (Fig. 1A). However, growth of *Ralstonia* was inhibited when electric current was introduced. Studies using *Escherichia coli* showed transient inhibition of electric current on cell growth (fig. S3). When the current stopped, cell growth resumed, suggesting that unstable compounds such as reactive oxygen or nitrogen species might be responsible for the growth inhibition.

To test this hypothesis, we constructed three *Ralstonia* strains with different reporter plasmids to detect the presence of reactive oxygen and nitrogen species. The plasmids each contained a *lacZ* gene driven by a promoter of *Ralstonia katG*, *sodC*, or *norA* gene, genes which have been shown to be induced by hydrogen peroxide (H_2O_2), superoxide free radicals (O_2^{-}), and nitric oxide (NO), respectively (*6*–*8*). When the plasmid-bearing strains were exposed to electrolysis, expression of β-galactosidase from *sodC* and *norA* promoters were induced but not from the *katG* promoter (Fig. 1B). These results suggested that O_2^- and NO trigger a stress response in *Ralstonia* cells and inhibit growth. To circumvent this toxicity problem, a porous ceramic cup was used to shield the anode (Fig. 1A). This inexpensive shield provides a tortuous diffusion path for chemicals. Therefore, the reactive compounds produced by the anode may be quenched before reaching the cells growing outside the cup. Using this approach, healthy growth of *Ralstonia* strain LH74D and production of over 140 mg/l biofuels were achieved with the electricity and CO_2 as the sole source of energy and carbon, respectively (Fig. 1C).

This integrated process to convert CO_2 to liquid fuels does not depend on biological "light reactions." Electricity generated from photovoltaic cells, wind turbines, or off-peak grid power sources can be used to drive CO_2 fixation and fuel production. Thus, this process provides a way to increase photosynthetic efficiency by coupling man-made photoelectric generation device with biological CO_2 fixation and fuel production capability. The approach demonstrated here can also be applied to produce other compounds, thus opening the possibility of electricity-driven bioconversion of CO_2 to a variety of chemicals. Furthermore, because formate is a by-product of chemical dehydration and conversion of biomass (*9*), transformation of formate into liquid fuel will also play an important role in the biomass refinery process.

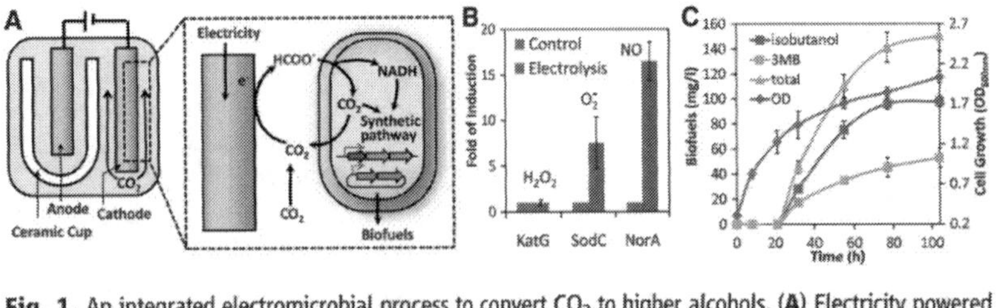

Fig. 1. An integrated electromicrobial process to convert CO_2 to higher alcohols. **(A)** Electricity powered the electrochemical CO_2 reduction on the cathode to produce formate, which is converted to isobutanol and 3MB by the engineered *R. eutropha*. **(B)** *R. eutropha* strains harboring a reporter gene driven by promoters that respond to reactive oxygen and nitrogen species showed increased reporter activity upon electrolytic exposure. **(C)** Engineered strain *R. eutropha* LH74D (supporting online material) showed healthy growth and produced over 140 mg/l biofuels in the integrated electromicrobial reactor with electricity and CO_2 as the sole energy and carbon sources, respectively. Error bars indicate SD (n = 3).

References and Notes

1. R. E. Blankenship *et al.*, *Science* **332**, 805 (2011).
2. K. S. Udupa, G. S. Subramanian, H. V. K. Udupa, *Electrochim. Acta* **16**, 1593 (1971).
3. S. Ikeda, T. Takagi, K. Ito, *Bull. Chem. Soc. Jpn.* **60**, 2517 (1987).
4. S. Atsumi, T. Hanai, J. C. Liao, *Nature* **451**, 86 (2008).
5. S. Atsumi *et al.*, *Appl. Microbiol. Biotechnol.* **85**, 651 (2010).
6. D. Touati, *J. Bacteriol.* **170**, 2511 (1988).
7. C. Michán, M. Manchado, G. Dorado, C. Pueyo, *J. Bacteriol.* **181**, 2759 (1999).
8. A. Pohlmann, R. Cramm, K. Schmelz, B. Friedrich, *Mol. Microbiol.* **38**, 626 (2000).
9. H. Mehdi *et al.*, *Top. Catal.* **48**, 49 (2008).

Acknowledgments: This work is supported by the Electrofuel program of Advanced Research Projects Agency–Energy (ARPA-E).

Supporting Online Material

www.sciencemag.org/cgi/content/full/335/6076/1596/DC1
Materials and Methods
Figs. S1 to S4
Tables S1 and S2
References (*10*–*22*)

8 December 2011; accepted 14 February 2012
10.1126/science.1217643

[1]Department of Chemical and Biomolecular Engineering, University of California, Los Angeles, CA 90095, USA. [2]The Molecular Biology Institute, UCLA, Los Angeles, CA 90095, USA. [3]Department of Chemistry and Biochemistry, UCLA, Los Angeles, CA 90095, USA. [4]Easel Biotechnologies, Los Angeles, CA 90024, USA. [5]Department of Materials Science and Engineering, UCLA, Los Angeles, CA 90095, USA. [6]Institute of Genomics and Proteomics, UCLA, Los Angeles, CA 90095, USA.

*To whom correspondence should be addressed. E-mail: liaoj@seas.ucla.edu

Rhodobacter as a platform for autotrophic fuel production

Nymul Khan[1], Alex Rajangam[1], Eric Nybo[2], Joe Chappell[2], Wayne Curtis[1]

[1] *Department of Chemical Engineering, Pennsylvania State University, University Park, PA*

[2] *Department of Plant and Soil Sciences, University of Kentucky, Lexington, KY*

Overview

Our overall ARPA-E project goal seeks to produce a renewable hydrocarbon feedstock by the genetic engineering of a C_{34} branched hyrdrocarbon of Botrococcus braunii into *Rhodobacter capsulatus*. We have shown the ability of this genetically engineered autotrophic organism to produce C_{30+} isoprenes from H_2, O_2 and CO_2. My specific component of this project has been to examine fundamantal kinetics of autotrophic growth towards the production of bioreactactor systems and multiplexed autotrophic growth capabilities for screening the genetically engineered bacteria.

Introduction

Liquid fuels are an essential part of today's economy. The transportation sector is the largest consumer of liquid fossil fuels (13.6 million bbl/day in 2011, www.eia.gov/forecasts/aeo/MT_liquidfuels.cfm) and the second largest source of CO_2 emission (1.9 billion metric tons in 2008, http://www.eia.gov/oiaf/1605/ggrpt/carbon.html) after power plants. Coal and natural gas account for 67% of US electricity generation and together generate the major portion of the CO_2 (about 2.3 billion tons in 2008, http://www.eia.gov/oiaf/1605/ggrpt/carbon.html). So, besides being an unsustainable source of energy, consumption of fossil fuels has high environmental impact due to the generation of CO_2. Furthermore, it is estimated that the demand for liquid fuels for transportation sector is likely to increase during the next 20-30 years (Figure 1).

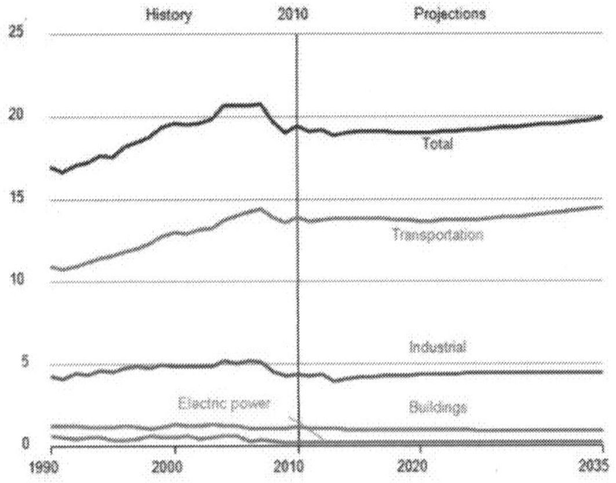

Figure 1: Consumption of petroleum and other liquids by sector, 1990-2005 (million bbl/day). http://www.eia.gov/forecasts/aeo/MT_liquidfuels.cfm#uses

In comparing liquid fuel products to other energy storage forms, it is apparent that a liquid hydrocarbon fuel is a much more efficient energy carrier by weight and volume than thermal, mechanical, battery and hydrogen energy storage. It is probably not possible to replace liquid fuels for many current end uses. Therefore, there is an urgent need to develop a renewable source of liquid fuels. Our proposed process is to take the distributed energy of solar radiation and store it in a liquid fuel molecule that can be easily used in the existing infrastructure of refining, storage and distribution. Figure 2 shows a schematic of the process. The electricity generated by photovoltaics will be used to split water to produce H_2 and O_2. These gases plus the CO_2 from power plants or other point sources will be used as the major substrates to grow an autotrophic bacterium (*Rhodobacter capsulatus*), genetically engineered to produce a liquid fuel, botryococcene.

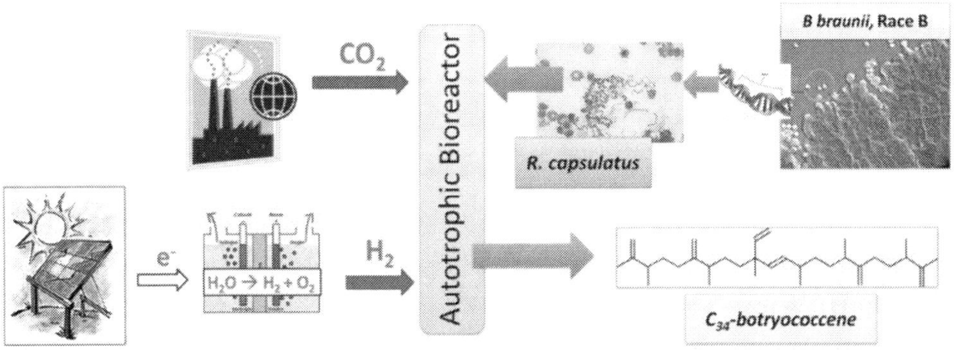

Figure 2: Schematic of the proposed biofuel production process in the context of supporting technologies of photovoltaics (P.V.) and carbon capture.

Botryococcenes are a group of C_{34}-branched hydrocarbons naturally produced by the micro-algae *Botryococcus braunii* Race B. These oxygen-free molecules are extremely energy dense (43.8 KJ/g as compared to ethanol at 29.7 KJ/g). As a hydrophobic long chain hydrocarbon, it easily separates as a distinct layer from water and has characteristics very similar to crude oil (Hillen, Pollard, Wake, & White, 1982). Hillen and co-workers also showed that botryococcene can be hydrocracked with a commercial catalyst to produce 67% gasoline, 15% aviation, 15% diesel and 3% residual oil. The octane rating of the gasoline fraction was much higher than natural crude and very low nitrogen and sulfur levels to affect the catalyst operation. As such, it can be a drop-in feedstock to the current petroleum industry. In fact, *B. braunii* oils appear to have contributed as much as 1.4% to the total hydrocarbons in oil shales (McKirdy & Cox, 1986).

Non-photosynthetic growth of organisms on CO_2 has been observed for a long time (Kelly, 1992) and has been proposed as a platform for the production of single cell protein, including an interface to H_2 production by electrolysis (Aragno & Schlegel, 1992). *Rhodobacter capsulatus* is a purple non-sulfur bacterium which has an extremely versatile metabolic capacity for growth under variety of growth conditions including aerobic heterotrophic, anaerobic photoheterotrophic and dark autotrophic using H_2 as the electron donor, O_2 as the electron source and CO_2 as the carbon source (Madigan & Gest, 1979). It has been chosen as the organism for this work for both its diverse metabolic capacity and our previous experience of working with this organism for membrane protein expression.

An autotrophic process is quite unique in its requirement for multiple gas phase substrates. While traditional aerobic fermentation processes are already limited by gas-liquid mass transfer, it is not difficult to imagine that that will be the limiting factor in this process as well. To make a rough comparison, while it takes ~0.7 L O_2 to grow 1 g of *E. coli*, it would take ~12.5 L of gas mix (in stoichiometric composition) to grow 1 g of *R. capsulatus* (based on yield of *R. capsulatus* on the different gases, Madigan & Gest 1979). This poses some unique restrictions on the bioreactor and genetic engineering of the organism. For example, from a preliminary calculation using the yield of the bacteria on the different gas species, the average cost of electricity and gasoline it was immediately clear that the growth-associated fuel production is not a feasible scenario for this process. Figure 3 shows a comparison of just the electricity cost per gallon of fuel corresponding to the fraction of fuel accumulated per unit biomass. Even at a very high fraction of fuel accumulation the cost is exceedingly high (and this does not include the operating or other downstream processing costs). So, the

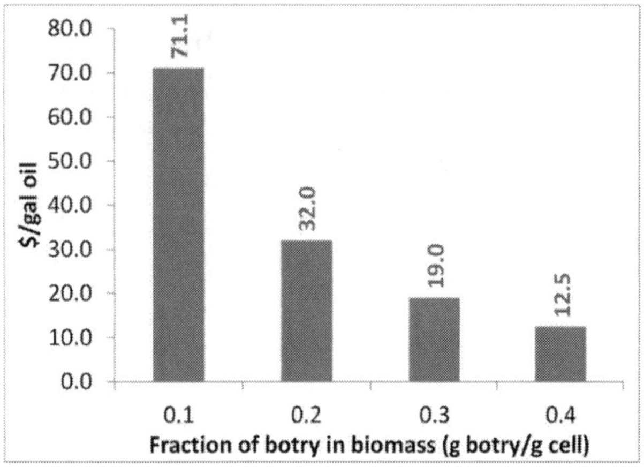

Figure 3: Cost of electricity per gallon of fuel at different fuel to biomass ratio.

fuel production must be growth-dissociated. The parameter that determines the quantity of gas that would be consumed under very low or no-growth condition is the **maintenance coefficient** (and the true yield) of the bacteria. While the maintenance coefficients of bacteria growing heterotrophically are quite abundant in the literature, those for autotrophic bacteria (like *R capsulatus*) are not very common. In this paper, we present our recent work related to the measurement of this parameter for *R capsulatus*.

True yield and maintenance coefficient

Maintenance coefficient of an organism is the amount of energy it needs to stay alive and maintain itself without cell division. On the other hand, true yield of an organism is the amount of substrates it would consume for growth only and not for maintenance. Both of these parameters are very important not only for accurately assessing the baseline economic feasibility of the proposed process but also indicate energetic efficiency of the organism. The yield and maintenance coefficient of *R capsulatus* on H_2, the most expensive substrate, is not available in the literature. In fact, values of these parameters are generally very scarce for any organism under autotrophic conditions.

In principle, maintenance coefficient and true yield can be calculated if the yield on the substrate is known as a function of the growth rate. (Pirt, 1965) derived the following equation as the inter-relationship among specific growth rate (μ) and yield (Y_{XS}) and true yield (Y_{XS}^T) and maintenance coefficient (m_S).

$$\frac{1}{Y_{XS}} = m_S \frac{1}{\mu} + \frac{1}{Y_{XS}^T} \qquad (1)$$

Y_{XS} and μ are both experimentally measurable parameters. Therefore plotting $\frac{1}{Y_{XS}}$ against $\frac{1}{\mu}$ a straight line can be obtained whose slope and intercept would be equal to the m_S and $\frac{1}{Y_{XS}^T}$ respectively. This has been traditionally used in the determination of true yield and maintenance coefficient. However, in a batch system the growth rate varies continuously, therefore an accurate measure of instantaneous growth rate and yield are not possible. At the same time, the culture goes through physiological changes as it progresses through the different growth stages and therefore the yields obtained this way may not refer to the same condition. So, true yield and maintenance are measured in continuous systems, in which the growth rate of an organism can be fixed at a particular value by fixing the dilution rate (inverse of residence time in a reactor).

Cell balance on a continuous reactor:

$$V\frac{dX}{dt} = FX_0 - FX \quad (2)$$

where V = volume of reator, $\frac{dX}{dt}$ = growth rate, F = flow rate, X_0 and X = inlet and outlet cell concentration respectively. With 0 inlet cell concentration at steady state,

$$\frac{1}{X}\frac{dX}{dt} = \frac{F}{V} \quad (3)$$

From the definitions of specific growth rate, $\mu = \frac{1}{X}\frac{dX}{dt}$ and dilution rate, $D = \frac{F}{V}$, we have $\mu = D$ in a steady state continuous flow reactor.

Multiple steady states are established by varying the dilution rate and yield values are measured at these steady states. The corresponding yield and dilution rates are plotted on a $1/Y$ vs. $1/D$ graph to get the true yield and maintenance coefficient. However, this method is very time consuming and produces a limited number of data points. These parameters can alternatively be measured in deceleration-stat or acceleration-stat (D-stat or A-stat) (Barbosa et al., 2005) rather than a true chemostat. Instead of going from one dilution rate to the next in a step and waiting for a steady state to be established, the dilution rate is varied smoothly from one to the next. If the rate is slow enough for the cells to adapt physiologically, the culture goes through a succession of pseudo steady-states and the yield and dilution rate in this period is used to calculate the maintenance and true yield. The D-stat method is followed in our determination of the yield and maintenance coefficient of *R capsulatus*.

Materials and methods

The experimental setup and schematic of the autotrophic bioreactor is shown below (Fig. 4). H_2, O_2 and CO_2 gases are mixed via three mass flow controllers (Brooks Instruments, CA) and delivered to the reactor after going through a humidifying column. The outlet gas passes through a drying column, then through a mass flow controller (Omega Engineering) that measures the flow rate and a GC that measures the composition. The composition data is collected once every hour. For safety concerns about using explosive gas mix (H_2+O_2), the reactor used in this work is a 250 ml media bottle. The liquid flow rate into and out of the reactor is controlled via a highly precise peristaltic pump (Watson-Marlow), whose rotation can be controlled to within 0.01 rpm by a 0-5 V analog input. This is very critical in maintaining an accurate and consistent flow rate and smoothly varying it for D-stat. The actual flow rate of the inlet media is calculated by periodically recording the weight of the media bottle which sits on a balance and fitting appropriate regression lines through them dynamics. The culture is recirculated via a second peristaltic pump for online optical density measurement. Samples are taken periodically from the outlet for offline OD and dry weight measurements. The media used for growth is based on a previous work in the literature (Madigan & Gest, 1979) where the autotrophic growth of *R capsulatus* was first demonstrated. A pH probe continuously measures the pH of the culture. As discussed before, the culture

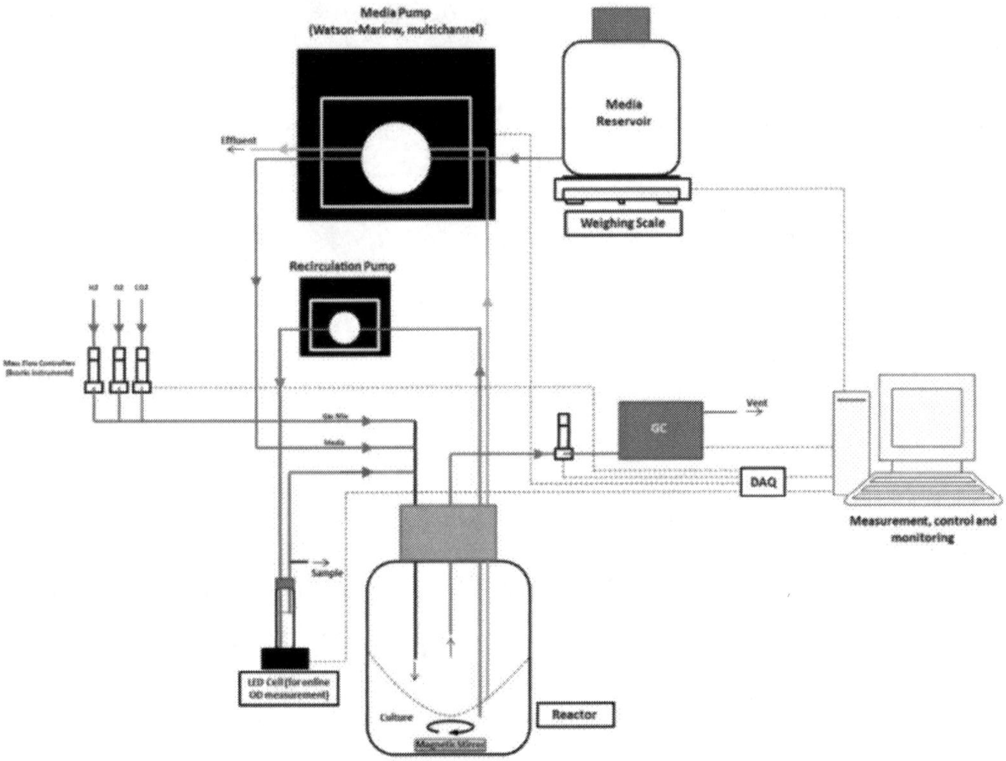

Figure 4: Schematic of the experimental setup.

is allowed to come to a steady state at a particular dilution rate and then the dilution rate is decreased slowly. The deceleration rate is based on previous literature work on *R capsulatus* (Hoekema, Douma, Janssen, Tramper, & Wijffels, 2006) and experience with running the experiment.

Another critical aspect of the experiment is the measurement of gas flow rate and composition, which is much more difficult to do accurately than liquid phase because of strong temperature and pressure dependence. Mass flow meters, although capable of measuring flow rate of a single gas very accurately, are not very accurate for a mixture of gases and at low flow rates. A soap-bubble meter was therefore placed at the end of the line to calibrate the outlet flow based on manual measurements. This was checked against the flow rate based on composition correction of the MFM and they compared very well. Furthermore, a state-of-art GC was used to analyze these gases continuously in a single pass for composition measurement.

Results and discussion

The experimental data is shown from Fig. 5 where the dilution rate was decreased from about 0.08/hr to about 0.06/hr at a rate of 0.094/day^2. This deceleration rate was based conservatively on a previous work of (Hoekema et al., 2006) where they measured the yield and maintenance of *R capsulatus* in the photoheterotrophic growth mode, producing H_2. Optical density plot (Fig 5A) shows that the expected pattern of variation is followed as the dilution rate is varied. Fig 5C shows the precision of media flow variation. The final calculated results are plotted in Fig 5D. The true yield and maintenance coefficient values are shown below (Table 1).

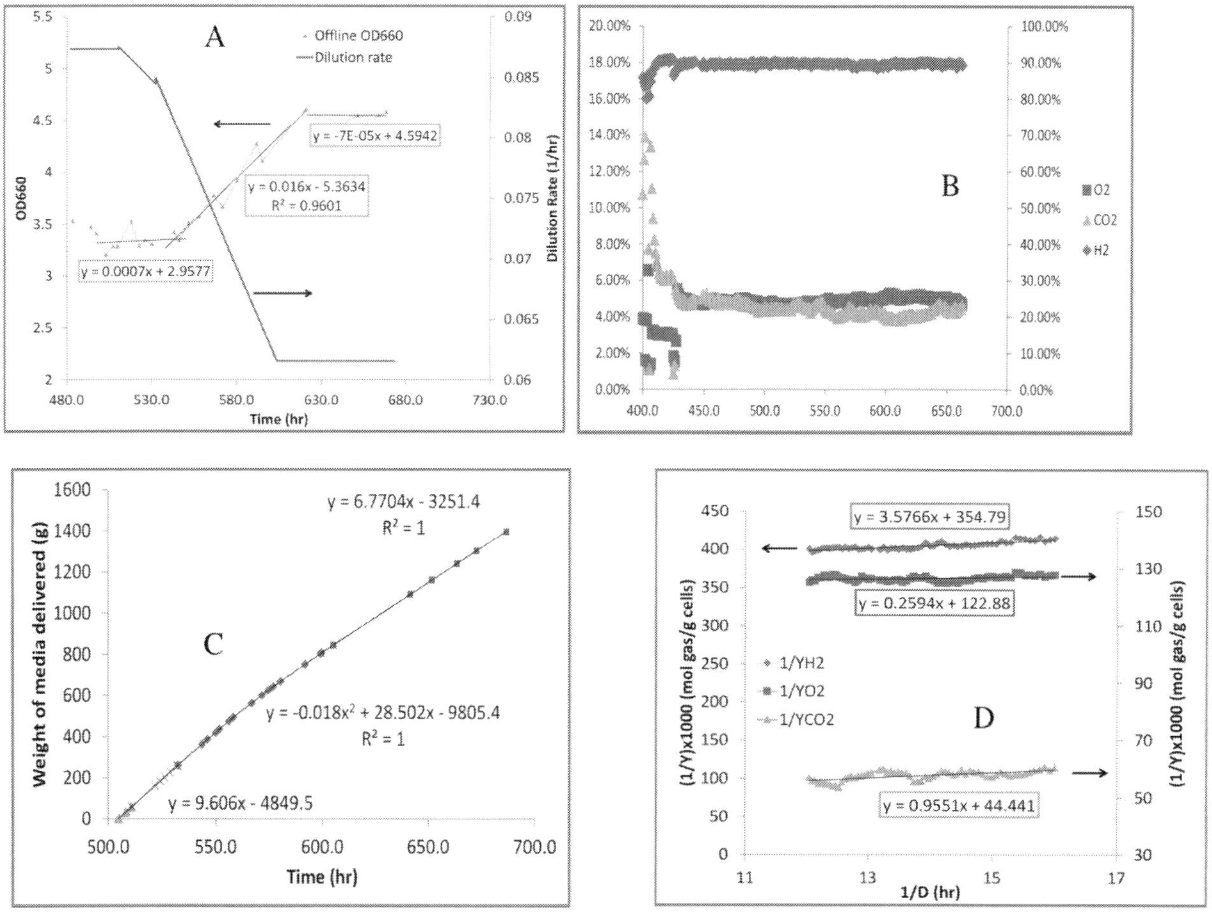

Figure 5: Experimental data for *R. capsulatus* true yield and maintenance coefficient determination. (A) Ramp down of dilution rate (red) and the corresponding OD increase (green), (B) gas phase composition, (C) media bottle weight to measure flow rate, (D) 1/Y vs. 1/D.

These values compare quite well with the values for *R eutropha* reported in the literature. Table 2 lists the true yields and maintenance coefficients of *R eutropha* and a couple of methanogens. Siegel & Ollis (1984) noted that the maintenance coefficient of *R eutropha* is a strong function of dissolved O_2 concentration. As observed by other researchers (Howells, 1982), O_2 can be both inhibitory and limiting for the growth of a bacteria at different concentrations and is strongly strain dependent. Under continuous culture conditions, the steady state in between inhibitory and limiting concentrations is unstable and tends to move either toward washout at higher concentration or towards O_2 limitation (Siegel & Ollis, 1984). We can see from the values that the maintenance coefficient increases significantly under H_2-limited conditions. The explanation offered by the authors is that the NAD-reducing hydrogenase in *R eutropha* is deactivated by O_2 and becomes an additional metabolic burden to replace the deactivated hydrogenases at high O_2 concentration, hence the increase in maintenance cost. High O_2 concentration is also found to be inhibitory to *R capsulatus* (Madigan & Gest, 1979; Siefert & Pfennig, 1979). These facts lead us to believe that this is a general phenomenon in lithotrophic CO_2 fixation and give us an idea of the possible operating regime for such a process. Future studies will examine the effect of these parameters for *R eutrophoa* as well.

Table 1: Experimental values of true yield and maintenance coefficient.

	True yield, Y_{XS} (g cells/mol gas)	Maintenance coefficient, m_S (mol gas/g cells.hr)
H_2	2.82	3.58×10^{-3}
O_2	8.13	2.60×10^{-4}
CO_2	22.52	9.55×10^{-4}

The yield and maintenance coefficient of *M. bryantii* were determined in batch growth experiments. The yield was calculated from q_{max} (maximum specific substrate utilization rate) and μ_{max} (maximum specific growth rate), which were determined under non-limiting H_2 concentration. Also, they treated the maintenance coefficient as "endogenous metabolism" i.e. oxidation of biomass to support cell maintenance, and determined the maintenance rate, b, from a biomass decay experiment where there was no H_2 present. This is clearly different from a true maintenance coefficient where the biomass is not really oxidized to provide for maintenance but the actual substrate is consumed by active cell machinery to maintain itself. Since they have not exactly provided the maintenance coefficient, we have calculated the value from yield (Y) and maintenance rate (b). Again, it is not really compatible since they were calculated under two different physiological conditions. However, it gives us something for comparison. The values for *M arboriphilus* were obtained under H_2-limited continuous culture conditions (Morii, Koga, & Nagai, 1987). The true yield and maintenance values were not given directly in the paper but the dilution rates and the substrate utilization rates were used to calculate the values using method outlined above.

Table 2: True yield and maintenance coefficient of some organisms on H_2.

Organism	True yield (g cells/mol H_2)	Maintenance coefficient (mol H_2/g cells.hr)	Experimental conditions	Reference
Ralstonia eutropha	4.1	0.001	O_2 limited continuous culture	(Siegel & Ollis, 1984)
	4.5	0.025	H_2 limited continuous culture	
Methanobacterium bryantii M.o.H	0.325	0.0115	Non-limited batch culture	(Karadagli & Rittmann, 2005)
Methanobrevibacter arboriphilus A2	0.40	0.0106	H_2 limited continuous culture	(Morii et al., 1987)

The values for these methanogens are included for comparison. It is readily apparent that the yield and maintenance are both much worse for these organisms. It is quite likely because the methanogenic process is anaerobic and the yields are expected to be lower than an aerobic process. It needs to be borne in mind that these organisms are actually producing a product that is much more reduced than biomass and that is also a reason for low yield and high maintenance. The energetic analysis of a fuel producing process therefore must include the metabolic effect of producing the fuel molecule.

Economic analysis

The preliminary economic assessment of the electricity cost per gallon of fuel has been made for the non-growth scenario for a comparison against the growth based scenario. For calculations, literature values for *R eutropha* and the experimental values shown in the previous section for *R capsulatus* were used. Cost of H_2 and CO_2 are taken to be \$3.82/kg and \$87/ton respectively. The H_2 cost is based on the current solar electricity prices but as solar energy becomes more widespread this is expected go down and affect the economics correspondingly.

Baseline scenario. Growth based. Shown in Figure 3.

Proposed scenario. At very low growth rates maintenance energy would be the main form of energy consumption. At this point, we lack the important parameter of fuel production rate by the bacteria to make an accurate estimate. A work-around to provide a quantitative perspective is to use a similar value for another organism from literature. *Chlorella vulgaris* which produces lipid under severely growth-limited conditions was chosen as a reasonable basis. From (Tuerk, 2011) this value is estimated to be 0.001256 g lipid/g cells.hr. This value was used as a baseline fuel production value and varied until 0.1 g borty/g cells.hr to observe the effect on cost of fuel (Fig 6). (Residence time in the reactor was a parameter in this calculation which was set to 10 hr. The cell retention was set to 0.9). The high fuel cost calculated for low production rates demonstrates that even in a non-growth associated production scenario maintenance energy requirements play a significant role in determining the cost of fuel production.

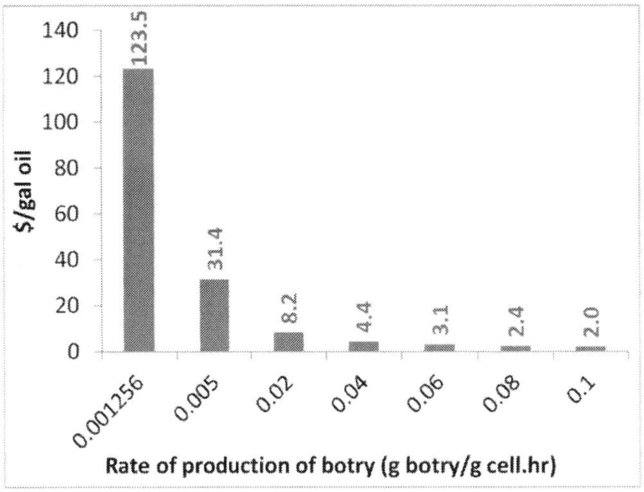

Figure 6: Cost of electricity per gallon of fuel under non-growth associated fuel production.

Future work on this will include studying the effect of various parameters (like O2 tension, nitrogen source etc.) on the maintenance coefficient of *R capsulatus*. Similar studies will be done on *R eutropha* for an unbiased comparison of the two organisms.

References

Aragno, M., & Schlegel, H. G. (1992). The chemolithotrophic prokaryotes. In A. Balows, H. G. Truper, M. Dworkin, W. Harder, & K. Schleifer (Eds.), *The Prokaryotes, 2nd Edition. A Handbook on the Biology of Bacteria: Ecophysiology, Isolation, Identification, Applications.* (pp. 332–343). New York: Springer-Verlag.

Barbosa, M. J., Zijffers, J. W., Nisworo, A., Vaes, W., van Schoonhoven, J., & Wijffels, R. H. (2005). Optimization of biomass, vitamins, and carotenoid yield on light energy in a flat-panel reactor using the A-stat technique. *Biotechnology and bioengineering*, *89*(2), 233–42. doi:10.1002/bit.20346

Hillen, L. W., Pollard, G., Wake, L. V., & White, N. (1982). Hydrocracking of the oils of Botryococcus braunii to transport fuels. *Biotechnology and bioengineering*, *24*(1), 193–205. doi:10.1002/bit.260240116

Hoekema, S., Douma, R. D., Janssen, M., Tramper, J., & Wijffels, R. H. (2006). Controlling light-use by Rhodobacter capsulatus continuous cultures in a flat-panel photobioreactor. *Biotechnology and bioengineering*, *95*(4), 613–626. doi:10.1002/bit

Howells, E. R. (1982). Single-cell protein and related technology. *Chemistry Industry*, (15), 508–511.

Karadagli, F., & Rittmann, B. E. (2005). Kinetic characterization of Methanobacterium bryantii MoH. *Environmental science & technology*, *39*(13), 4900–4905.

Kelly, D. P. (1992). The chemolithotrophic prokaryotes. In A. Balows, H. G. Truper, M. Dworkin, W. Harder, & K. Schleifer (Eds.), *The Prokaryotes, 2nd Edition. A Handbook on the Biology of Bacteria: Ecophysiology, Isolation, Identification, Applications.* (pp. 332–343). New York: Springer-Verlag.

Madigan, M. T., & Gest, H. (1979). Growth of the photosynthetic bacterium Rhodopseudomonas capsulata chemoautotrophically in darkness with H2 as the energy source. *Journal of bacteriology*, *137*(1), 524–30. Retrieved from http://www.pubmedcentral.nih.gov/articlerender.fcgi?artid=218479&tool=pmcentrez&rendertype=abstract

McKirdy, D., & Cox, R. (1986). Botryococcane in a new class of Australian non-marine crude oils. *Nature*, *320*. Retrieved from http://www.nature.com/nature/journal/v320/n6057/abs/320057a0.html

Morii, H., Koga, Y., & Nagai, S. (1987). Energetic analysis of the growth of Methanobrevibacter arboriphilus A2 in hydrogen-limited continuous cultures. *Biotechnology and bioengineering*, *29*(3), 310–5. doi:10.1002/bit.260290304

Pirt, S. J. (1965). The maintenance energy of bacteria in growing cultures, *163*(991), 224–231.

Siefert, E., & Pfennig, N. (1979). Chemoautotrophic growth of Rhodopseudomonas species with hydrogen and chemotrophic utilization of methanol and formate. *Archives of Microbiology*, *122*(2), 177–182. Retrieved from http://www.springerlink.com/index/V323851512727030.pdf

Siegel, R. S., & Ollis, D. F. (1984). Kinetics of growth of the hydrogen-oxidizing bacterium Alcaligenes eutrophus (ATCC 17707) in chemostat culture. *Biotechnology and bioengineering*, *26*(7), 764–70. doi:10.1002/bit.260260721

Tuerk, A. L. (2011). *An assessment of photosynthetic biofuels and electrofuels technologies under rate-limited conditions.* Pennsylvania State University.

(142aa) Viscoelastic Properties of Entangled Polymer Melts From Particle Rheology Simulations

Mir Karim[1], Tsutomu Indei[2], Jay D. Schieber[3] and Rajesh Khare[1], (1)Department of Chemical Engineering, Texas Tech University, Lubbock, TX, (2)Department of Chemical and Biological Engineering and Center for Molecular Study of Condensed Soft Matter, Illinois Institute of Technology, Chicago, IL, (3)Illinois Institute of Technology, Chicago, IL

Particle microrheology has become a well-established technique for the experimental determination of micro-scale viscoelastic properties of complex fluids. Particle motion at the nanoscale and its relation to the medium viscoelastic properties is of interest for several applications in medical and materials sciences. The viscoelastic properties at these length and time scales are strongly influenced by the specific chemical interactions in the system. Molecular simulations can explicitly account for these chemical interactions within the system to accurately capture the dynamics at time and length scales that are much lower than those that can be accessed in microrheology.

Recently we showed that[1] the particle motion in a melt of short polymer chains as obtained by molecular simulations can be analyzed by using a treatment which considers it as stochastic dynamics in a continuum medium to yield the storage (G') and loss (G'') moduli of the medium. Specifically, both passive and active particle rheology approaches were used to determine the frequency dependence of G' and G'' for a polymer melt represented by short, unentangled bead-spring chains of length 20. An important contribution of that work was the demonstration that particle and medium inertia need to be explicitly accounted for in such an analysis of particle motion at the nanoscale. In this work, the particle rheology simulation approach is extended to a system consisting of long, entangled polymer chains. Simulation results are used to investigate the rich physics in this system that follows from the interplay of the length scales represented by the particle size, chain tube diameter and the chain size.

1. M. Karim, S. C. Kohale, T. Indei, J. D. Schieber, and R. Khare, submitted.

(142ac) Simulation Guided Design and Experimental Validation of Microfluidic Devices for Stretching DNA Using Electrophoresis and Pressure-Driven Flow

Chiou-De Huang, Chemical Engineering, National Taiwan University, Taipei, Taiwan, Cheng-Han Li, Department of Chemical Engineering, National Taiwan University, Taipei, Taiwan and Chih-Chen Hsieh, Chemical Engineering Department, National Taiwan University, Taipei, Taiwan

We have used computational fluid dynamics and Brownian dynamics simulations to design a microfluidic device for stretching DNA. The device is an implementation of a new conformation pre-conditioning strategy. The key of the strategy is to "pre-stretch" DNA in the direction orthogonal to its final stretching direction. This microfluidic device is predicted by the simulations to yield substantial improvement on DNA stretch at given Deborah number and contraction geometry. To test our design, we have performed DNA stretching experiments using either electrophoresis or pressure-driven flow as the stretching force. The pre-conditioning strategy is found working only for the case using the pressure-driven flow. We have analyzed the evolution of DNA conformation, the maximum DNA extension and its distribution. The simulations and experimental results are in reasonably good agreement for DNA driven by pressure-driven flow but are qualitatively different for DNA driven by electrophoresis. We will discuss the causes of the discrepancy between simulations and experiments.

(142ad) Microfluidic Generation of Ultra-Thin Oil Film On Inertia Bubble for the Enhancement of Microextraction At High Phase Ratio

Kai Wang, Kang Qin, Wanlu Zhang and Guangsheng Luo, State Key Lab of Chemical Engineering, Department of Chemical Engineering, Tsinghua University, Beijing, China

Liquid-liquid extraction process plays very important role in chemical engineering processes. During recent years, microfluidic technologies have become effective tools to realize on-line microextraction or give fundamental study of mass transfer process at micrometer scale. For the mass transfer enhancement of liquid-liquid extraction process in microchannel at high phase ratio ($Q_C/Q_D > 10$), inertia gas phase is added to the liquid-liquid system to form gas-in-oil-in-water styled microflow system in this work. The gas phase is injected in the dispersed oil by a dual-coflowing microfluidic device to help increase the phase ratio of dispersed phase and prepare ultra-thin oil films on inertia bubbles with thicknesses ranging from 0.3 μm to 4.1 μm, which can significantly reduce the mass transfer distance in the oil extractant. The mass transfer enhancing effect of this gas-in-oil-in-water styled microflow system is characterized by extracting Sudan Red IV from SDS-PVA-aqueous solution to PDMS-alkane solution and the results indicated the extraction efficiency of this gas-in-oil-in-water system is tens of times higher than the common oil-in-water system.

(142ae) Fractionation Dynamics of Soft and Stiff Particles in Micro-Flow

Yeng-Long Chen, Institute of Physics, Academia Sinica, Taipei, Taiwan

Recent studies have demonstrated the fractionation of different cell types (healthy and diseased) by cell mechanical properties in microfluidic channels [1,2,3]. We investigate the fractionation mechanism of soft and stiff particles in microfluidic flow with the lattice Boltzmann-Brownian dynamics (LB-BD) method. For soft particles in microflow, it is found that lower particle volume fraction and higher shear rate enhances particle migration towards the microchannel center, due to particle deformation induced near-wall hydrodynamic field. For stiff particles, particle migration is found at non-dilute particle volume fractions, where flow inertia around the particle induces migration to an off-center position in the channel. In a mixture of soft and stiff particles, contributions from shear-induced particle deformation, flow inertia and particle volume fraction are systematically investigated. It is found that fractionation may be achieved with higher efficiency at higher soft particle volume fractions.

[1] H.W. Hou et al., Lab Chip 10, 2605 (2010).

[2] S.C. Hur et al., Appl. Phys. Lett. 99, 044101 (2011).

[3] E. J. Lim et al., Lab Chip, DOI: 10.1039/c2lc21100a (2012).

(142af) Effect of Inertia On Particle Trajectories Around a Spherical Obstacle

Sumedh R. Risbud, Mingxiang Luo, Joelle Frechette and German M. Drazer, Chemical and Biomolecular Engineering, Johns Hopkins University, Baltimore, MD

We investigate inertia effects on the motion of a spherical particle past a fixed spherical obstacle of the same radius under the action of a constant force in a quiescent ambient fluid, using lattice Boltzmann simulations. The obstacle is located at the origin of coordinates, while the particle starts its motion far upstream to the obstacle along x-axis with an initial offset $y^{-\infty} = b_{in}$. We use the distance of closet approach of the particle from the obstacle and the far downstream offset $y^{\infty} = b_{out}$ above x-axis, as functions of the initial offset, as indicator variables to probe inertia effects on the trajectories. The computational method allows us to isolate and study the individual effects of particle and fluid inertia, quantified by Stokes and Reynolds numbers, respectively. We observe that, (a) as expected the particle trajectories are asymmetric, (b) the particle attains a smaller distance of closest approach from the obstacle due tothe presence of particle and/or fluid inertia, (c) the far downstream offset b_{out} is smaller than the far upstream one b_{in} when particle inertia is significant. Fluid inertia alone, on the other hand, does not significantly affect the behaviour of the far downstream offset for $Re<2.5$.

SINGLE IMAGE TURBULENCE ANALYSIS FOR DRAG REDUCTION AFFECTED BY FLEXIBILITY OF POLYMERS

Dr. Ruri Hidema, Kobe University, Kobe, JAPAN
Prof. Hiroshi Suzuki, Kobe University, Kobe, JAPAN
Dr. Yoshiyuki Komoda, Kobe University, Kobe, JAPAN
Prof. Hidemitsu Furukawa, Yamagata University, Yonezawa, JAPAN

Abstract

An experimental study has been performed in order to investigate the relationship between the extensional viscosity of polymers and the turbulent drag reduction. Polyethyleneoxide as flexible polymer and hydroxypropyl cellulose as rigid rod-like polymer were added to the two-dimensional turbulent flow that was visualized by the interference pattern of a flowing soap film and analyzed by an originally developed single-image analysis. The power spectra of interference images were obtained, which is related to the water layer fluctuations in turbulence. The power spectra show a scaling behavior and the power components give the information of drag reduction. It was found from these results that the energy transfer mechanisms are different in streamwise and normal directions. In the normal direction, the energy transfer is prohibited by the orientation of polymers, while the energy transfer in the streamwise direction is prohibited by extensional viscosity of polymers. In order to get the precise information of frequency fluctuation in the streamwise direction that is related to the extensional viscosity, a wavelet-coefficient of interference images in the streamwise direction was calculated. The characteristic correlation scales were observed as a periodic structure in wavelet-coefficients, which was changed by the concentration of polymers. These scales may relate to the energy transfer in turbulence.

Introduction

The addition of very small amounts of flexible polymer to a water flow reduces drag in a turbulent flow, which is known well as drag reduction phenomena, and used in many industries to improve the energy efficiency. Although this effect has been known for more than half a century, the physical mechanism that causes this drag reduction has been still unclarified [1,2].

The first important work on the drag reduction was made by Lumley [3]. He argued the increase of the apparent viscosity in the flow outside of the viscous sublayer is the key for drag reduction, which is due to the molecular extension of polymers. The important experiments performed by Virk and Wagger [4] also suggested that the polymers are only effective when they are in an elongated shape, like a rod or an extended thread depending on the initial conditions of the polymer [5]. The extension of polymers occurs only under the rapid strain rates, and gives the anisotropic effects in the fluids. On the other hand, de Gennes and Tabor [7,8] suggested another mechanism to explain drag reduction with polymers. They mentioned that the coil-stretch transition doesn't occur in turbulent flows with a randomly fluctuating strain rate. In addition, if the polymers are moderately stretched in the turbulence, apparent viscosity mainly extensional viscosity could not change measurably. In their opinion, the major effect arises only when the elastic energy stored by the partially stretched polymers becomes comparable to the turbulent energy.

However, the both analysis have the shortcomings to explain which a strain works for stretching polymers, that is, shear strain or extensional strain. Smith *et al.* showed that DNA molecules were effectively stretched by extensional flow [5]. Indeed, extensional viscosity of polymers can reach much higher compared to the intrinsic viscosity that is related to the shear viscosity (Fig. 1(a)). Thus, Hidema *et al.* [9] mentioned that an abrupt increase of the extensional viscosity depends on the extensional strain rate will be the key for the turbulent drag reduction. However, the effect of polymers on the turbulent drag reduction is different between those by flexible polymer and by rod-like rigid polymers, since the flexible polymer is stretched abruptly under the extensional strain while the rod-like rigid polymers is gradually oriented under the extensional strain, which was confirmed by dynamic right scattering and the flow birefringence [9, 10]. These differences change the effective increase of extensional viscosity (Fig. 1(b, c)). In our previous study, we focused on how this difference of extension and orientation make anisotropic effects on the turbulent drag reduction, which is related to energy transfer in turbulence related to the drag reduction. It was proposed that the energy transfer in the normal direction and that in the streamwise direction were different. In the normal direction, the orientation of polymers prohibits the energy transfer, while the extensional viscosity prohibits the energy transfer in the streamwise direction. In this study, we calculate wavelet coefficients of the water layer fluctuations in the streamwise direction in order to observe how the spatial frequency is affected by polymers, since this direction is affected by extensional viscosity.

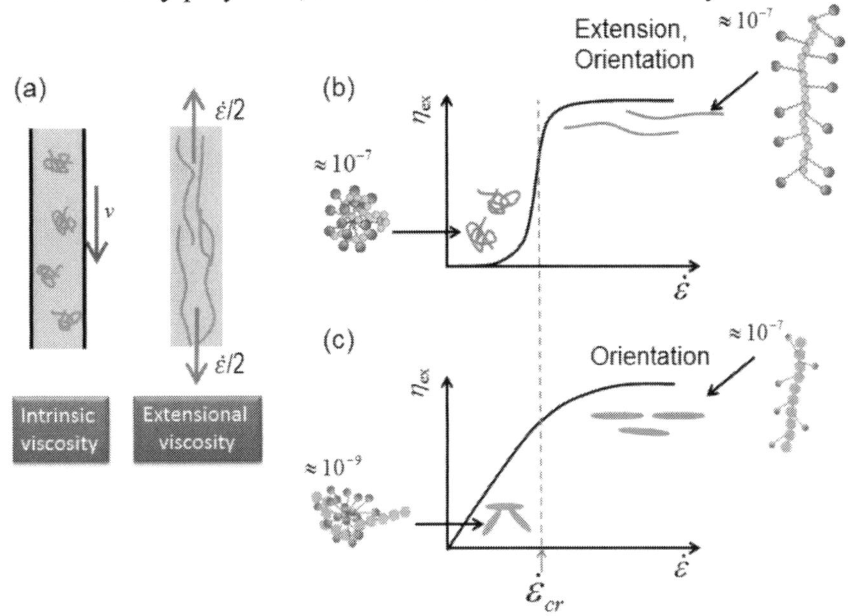

Figure 1. The extension of polymers causes extensional viscosity (a). Flexible and rod-like rigid polymers show the different increment of an extensional viscosity (b, c).

In order to observe the turbulent flow affected by stretched polymers, two-dimensional turbulence made by flowing soap films is used (Fig.2). As shown in Fig. 2(c), both sides of the flowing soap films are free surfaces and the water layer is sandwiched by surfactants. The thickness of water layer is much larger than vanishingly small surfactant molecules, and also the surface area is infinity compared to the thickness. Additionally, surface tension is negligible due to the surfactant in the surface of flowing soap films. Therefore, soap film is considered as a 2D water layer flow. Soap films reflect illumination light at the front and the back surfaces, which make interference patterns of the film. Since interferences of the illumination lights are affected by the thickness of the water layer, the interference patterns have information of the dynamics of the water layer as 2D flows. When a grid is inserted to the flow, extensional strain occurs at the grid, which causes polymer stretching or polymer orientation. Thus, in the case when polymers are added to the flow, the 2D grid turbulence is affected by the polymers. Flows were visualized and analyzed by single-image processing of the interference patterns. This visualization and image processing are proposed by our previous work, and named as Film Interference Flow Imaging method (FIFI) [15]. The extensional strain applied at FIFI is roughly calculated by Eq.(1).

$$S(t) = S_0 \exp(-\dot{\varepsilon}t) \tag{1}$$

,where S_0 [m^2] is the cross-section area before deformation, $S(t)$ [m^2] is the cross-section area after deformation, t [s] is the time required for deformation and $\dot{\varepsilon}$ [s^{-1}] is the extensional rate. In our case, $\dot{\varepsilon}$ is calculated as 351s^{-1} by the mean velocity, width of the flow, diameter of the grids and thickness of the film.

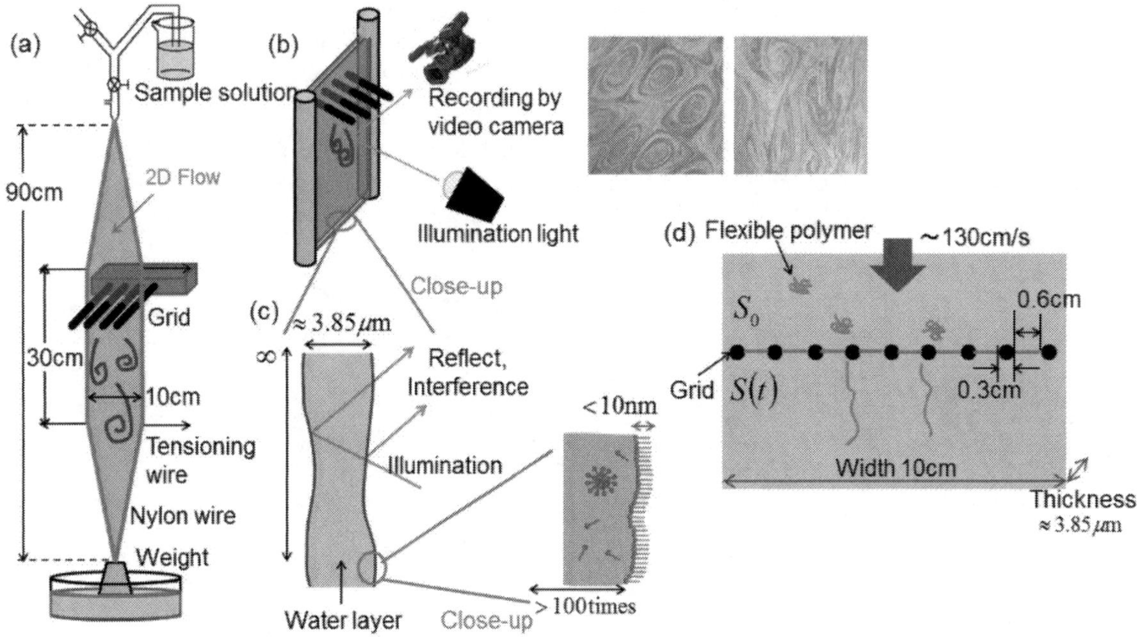

Figure 2. The experimental set up of flowing soap films. (a) is the whole image, (b) is the close up figure of the flowing soap film channel, that is 2D flow, (c) shows the cross section of a 2D flow. (d) explains how the extensional strain occurs in this flow around the grid.

Experimental

Materials

Soap solutions contained sodium dodecylbenzenesulfonate (SDBS) as a surfactant at the concentration of 2 wt%. As the flexible polymers, polyethyleneoxide (PEO, molecular weight of 3.5x10^6) was used at the concentration of 0.25, 0.5, 0.75, 1.0, 1.5 and 2.0x10^{-3} wt%. As the rod-like rigid polymers, hydroxypropyl cellulose (HPC, molecular weight is greater than 1.0x10^6) was used at the concentration of 0.01, 0.02 and 0.05 wt%. The overlap concentration of PEO was about 0.012 wt% and that of HPC was roughly 0.15 wt%.

Turbulence Visualization by Flowing Soap Films

The experiments were carried out in an apparatus shown in Fig. 2(a). The thickness of water layer $h(t)$ [m] was about 3.85 μm with the mean velocity $V(t)$ [m/s] of 130 cm/s when the flow flux

$Q(t)$ [m^3/s] was 0.5 ml/s, since there is the relationship like $h(t) = Q(t)/V(t)W$ where W [m] is the width of the soap film channel. The interference images of soap films were recorded with a digital video camera (Panasonic TM700) at the data acquisition area which was located at 20 cm behind the grid. The shutter speed of the video camera was 1/3000 s. A time interval in a series of images was adjusted to 1/60 s. Each of the frames acquired by the camera was converted into RGB form files with a spatial resolution of 640x360 pixels.

Single-Image Processing by Film Interference Flow Imaging

The interference images were analyzed by a 2D-FFT and a wavelet transformation [12].

A 2D-FFT software calculated the power spectrum of the interference images. Since the RGB pixel intensity of the images is related to the thickness of water layer, the power spectrum of the image shows the dynamics of 2D flows. Here, the power spectrum $<I^2(k_x, k_y)>$ was calculated by the pixel intensity G with a hamming window, where the k_x [m^{-1}] and k_y [m^{-1}] are the spatial frequency that are streamwise and normal directions in an interference image. For measurements presented here, the data acquisition area that was clipped from the video frames has 256x256 pixels, which correspond to 2.56x2.56 cm^2, thus, the k_x and k_y range from 1/2.56 to 1/0.02 cm^{-1}, that is, 0.391 to 5.00 cm^{-1} as the frequency. $<I^2(k_x, k_y)>$ of the pixel intensity G characterizes the strength of the thickness fluctuation of water layer on spatial frequency, k_x and k_y. Here the power spectrum in both frequencies are plotted to discuss energy transfer of turbulence in streamwise and normal directions.

The wavelet analysis is applied for analyzing the spatial fluctuation of the water layer [13]. A streamwise line data of RGB pixel intensity of the image is used for the analysis. The data of about 6000 pixels corresponding to 65 cm is clipped by connecting consecutive about 30 images. 30 images were taken in 0.5 seconds since the time interval in a series of images was 1/60s. A mother wavelet, $\varphi(X)$ on a variable X, the "Mexican hat" wavelet is adopted as follows.

$$\varphi(X) = (1 - X^2)e^{-\frac{X^2}{2}} \tag{2}$$

With this, the wavelet transform coefficient, $W(a, b)$ is defined for space variation of streamwise water layer fluctuation of intensity G, $G(x)$, as follows.

$$W(a,b) = \frac{1}{\sqrt{a}}\int \varphi\left(\frac{x-b}{a}\right)G(x)dx \tag{3}$$

where x [m] is the length scale, and a [m] and b [m] are wavelet scale and a sequence scale parameters, respectively. In the present study, the wavelet coefficient is normalized with the maximum wavelet value, W_{max}, at each measuring point.

Results and Discussion

Figure 3 shows the examples of interference patterns of 2D grid turbulence. Turbulence was generated at the comb in the flowing soap films, and was convected in the observation area. The interference patterns of 2D turbulence visualize the thickness of water layer fluctuations, which is considered as a passive scalar. The characteristics of 2D turbulence are an inverse energy cascade from small scales to large scales and an enstrophy cascade from large scales to small scales. The inverse energy cascade is seen through the interference patterns of 2D turbulence since the apparent size of the eddies in Fig.3(a) is larger than the injection scale. By adding of polymers, the eddies in the interference patterns clearly become long in streamwise direction and narrow in normal direction. This feature is especially seen in the case of the addition of PEO as shown in Fig.3(b) – (d). This indicates that the polymer inhibits inverse energy transfers in normal direction. In the case of the addition of HPC, the effect can be slightly observed at much higher concentration compared with that in the case of PEO. Such phenomena are precisely analyzed by FIFI as shown in following paragraphs.

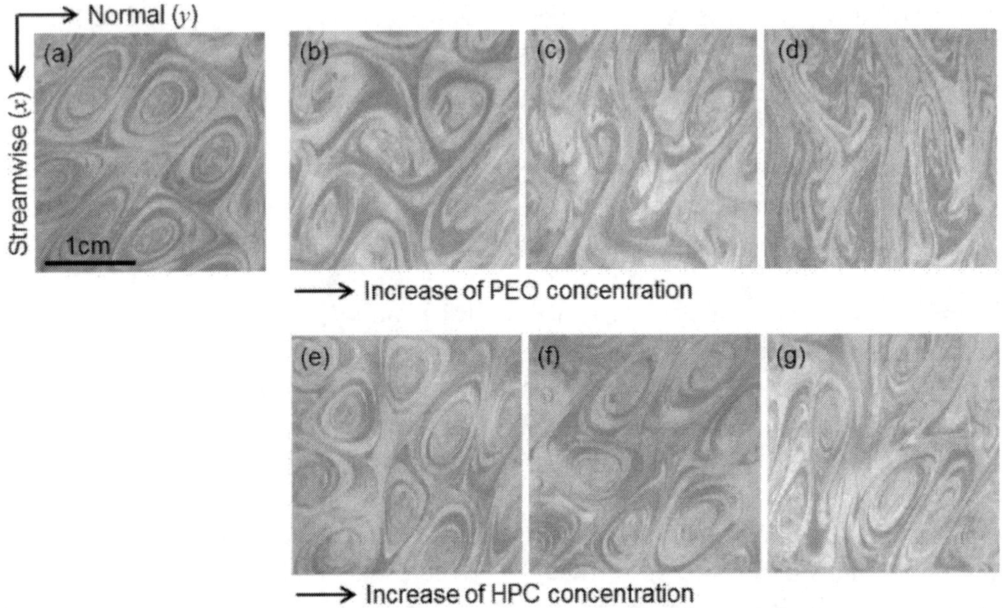

Figure 3. Interference images of the turbulence in flowing soap films. The images are taken at SDBS 2wt% solution (a), + PEO 0.5x10^{-3}wt% solution (b), + PEO 1.0x10^{-3}wt% solution (c), + PEO 2.0x10^{-3}wt% (d), + HPC 0.01wt% solution (e), + HPC 0.02wt% solution (f) and + HPC 0.05wt% solution (g) for a mean velocity is about 130 cm/s. The black bars indicate 1 cm.

The example of the power spectra $<I^2(k_x, k_y)>$ of the interference pattern is shown in Fig.4. Figure 4(a) shows the two-dimensional representation of $< I^2(k_x, k_y)>$, where the darkness of the image display the intensity. Two arrows in the image show the direction that corresponds to the streamwise and normal directions in the interference patterns. k_x [m-^{1}] and k_y [m^{-1}] are the spacial frequencies which have the information of water layer fluctuations directed in streamwise and normal directions. $<I^2 (k_x, 0)>$ is shown in Fig. 4(b) as an example, which shows the scaling behaviors. From them, it is found that the power component almost fits with the -5/3 that is theoretically predicted for a 2D turbulence. The variation of the power components is shown in Fig.4(c) and (d). Figure 4(c) shows the variation when the PEO is added. In this case, the value was changed from -5/3 to -1 in both streamwise and normal directions. At lower concentrations of PEO, the variation rate of the power component is small but the rate increases rapidly from 0.5 to 0.75 x 10^{-3} wt%. Figure 4(d) shows the variation when the HPC is added. In this case, the value was changed gradually from -5/3 to -1 in normal directions. However the variation of the scaling exponent in the streamwise direction was small and has absence of a pattern in the case of HPC. As mentioned above, in 2D turbulence, there are two cascades. One is an inverse energy cascade and the other is an enstrophy cascade. The inverse energy cascade is seen through the vortex mergers in the interference patterns. The scaling exponent detects the prohibition of the inverse energy cascade due to polymers, that is, the variation from -5/3 to -1. Thus, it is considered that PEO prohibit the inverse energy cascade in both directions, while HPC prohibits the inverse energy cascade only in the normal direction. This difference is due to the different mechanism of energy transfer in turbulence when polymers are added. In the normal direction, the orientation of polymers prohibits the energy transfer. Thus it is happen in both PEO and

HPC cases. On the other hand, the extensional viscosity prohibits the energy transfer in the streamwise direction. The extensional viscosity of HPC is lower compared to PEO, thus HPC is less effective in the streamwise direction.

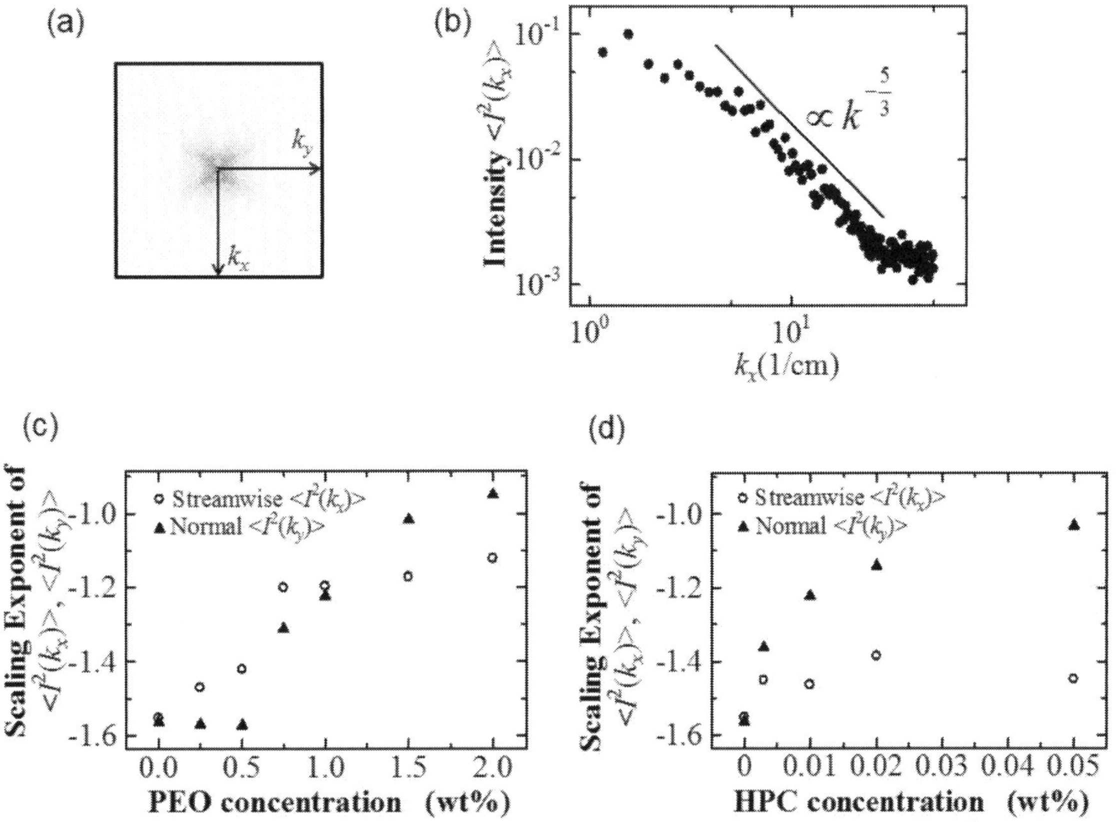

Figure 4. (a) The two-dimensional representation of the power spectrum, $< I^2(k_x, k_y)>$ of the interference image corresponds to the Fig.3(a). The k_x and the k_y directions in the image represent the spatial frequencies in the streamwise and the normal directions. (b) is the example of the scaling forms of the power spectrum. (c) and (d) are the power component of the power spectrum in each direction at PEO (a) and HPC (b) additives.

Since the streamwise direction is affected by extensional viscosity, wavelet transform is applied for analyzing the spatial fluctuation. Figure 5(a)-(c) shows the wavelet transform results on the pixel intensity in the streamwise direction G(x). In the figure, a dark region corresponds to a minus region of wavelet coefficients indicated by an attached color bar. In the case of the polymer-free solution, a value of the wavelet coefficient is random in any wavelet scales. When turbulent drag reduction occurred by adding PEO, periodic fluctuations appeared in some scales. On the other hand, a periodic structure does not appeared in the case of HPC. In order to see how the periodic structures are changed by polymer concentrations, that is, effectiveness of drag reduction, the root mean square values of wavelet coefficients, W_{rms}, normalized by the maximum value of wavelet coefficients, W_{max}, are plotted to the wavelet scale (Fig. 5(d, e)). In the case of the polymer-free solution, only a broad peak is observed, which may relate to a vortex size in 2D turbulence. This peak is enhanced by adding PEO $0.25x10^{-3}$ wt%. The first peak is relatively decreases with the PEO concentration, and the second peak appears. The wavelet scales of the first peak is slightly shifted to the smaller scale, and the wavelet scales of the second peak is shifted to the larger scale (Fig. 5(d)). In the case of HPC, the first and the second peaks are vaguely separated. From these results, it is suggested that an original vortices in 2D turbulence disappears when turbulent drag reduction occurs, and another periodic structure appears in turbulence. This is related to the energy transfer in the streamwise direction, which is dominated by the extensional viscosity.

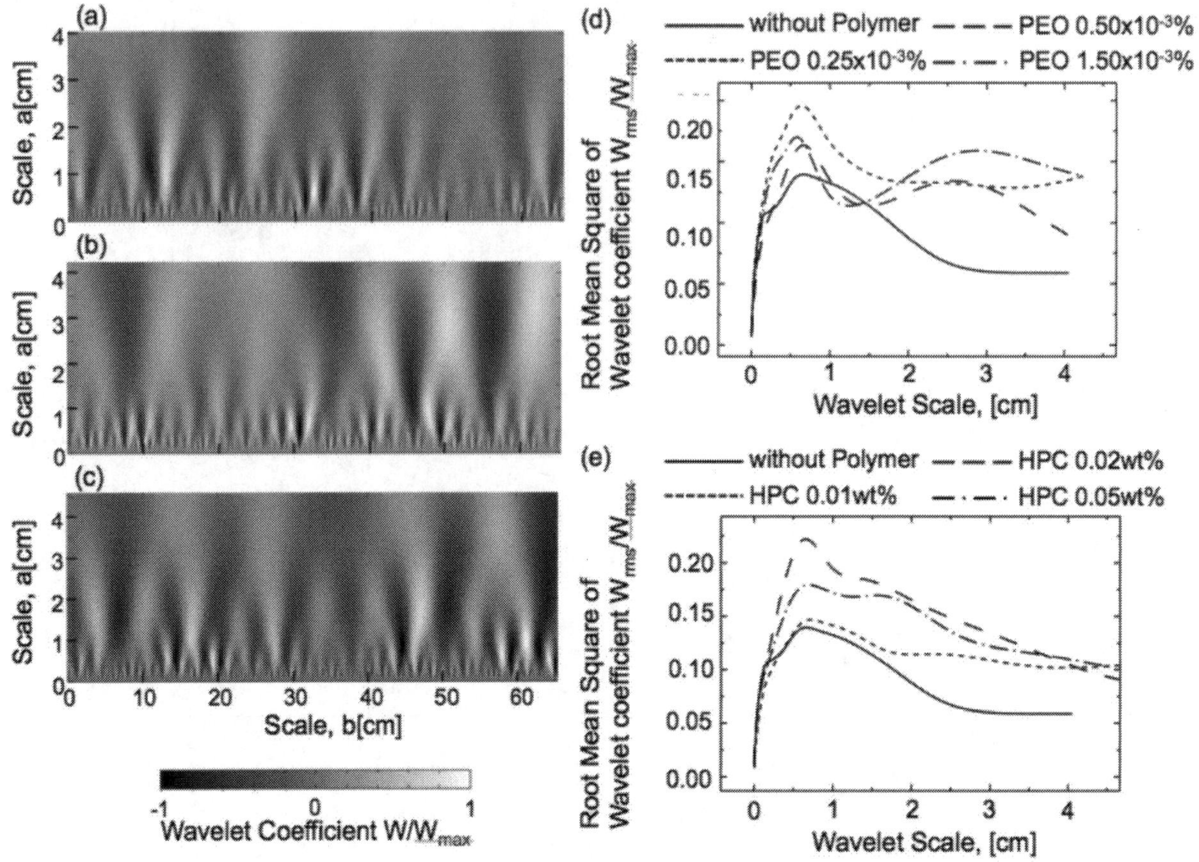

Figure 5. Wavelet-coefficient contours for interference images of the SDBS 2wt% solution (a), + PEO 2.0×10^{-3} wt% solution (b) and + HPC 0.05 wt% solution (c). (d) and (e) are the root mean square value of wavelet coefficient for the PEO added solution and the HPC added solution.

Conclusion

An experimental study has been performed in order to investigate how the extensional viscosity of polymers affects on turbulent drag reduction. PEO as a flexible polymer and HPC as a rod-like rigid polymer were added to the 2D turbulence, and drag reduction effects were visualized and analyzed by image processing. From these results, the following conclusions were obtained.

1. The power spectra of interference images were obtained, which is related to the water layer fluctuations in turbulence. The scaling behavior of the power spectra show give the information of energy transfer in turbulence. In the normal direction, the power components were changed from -5/3 to -1 by adding polymers. This variation indicates the prohibition of the inverse energy transfers in 2D turbulence. In the streamwise direction, the value was changed to -1 by adding PEO, while the value was merely changed by adding HPC. From these results, it is suggested that the mechanism of drag reduction in turbulence is different in each direction. In the normal direction, the orientation of polymers prohibits the energy transfer. On the other hand, the

extensional viscosity prohibits the energy transfer in the streamwise direction. The extensional viscosity of HPC is lower than that of PEO, thus HPC is less effective in the streamwise direction.
2. Wavelet coefficients were obtained by calculating the pixel intensity in the streamwise direction. When turbulent drag reduction occurrs, periodic fluctuations appears in some scales. The periodic structure is changed by the concentration of PEO. An original structure of turbulence disappears by the drag reduction, and new structures appear. This is related to the increase of extensional viscosity in the streamwise direction, which changes the energy transfer in turbulence.

Acknowledgement

This study is supported in part by the Grant-in-Aid for Young Scientists(B) (Project No.: 24760129) and the Grant-in-Aid for Scientific Research(B) (Project No.: 24360319) from the Japan Society for the Promotion of Science (JSPS).

Reference

1. Sreenivasan, K.R. and White, C.M. (2000), "The onset of drag reduction by dilute polymer additives, and the maximum drag reduction asymptote", *J. Fluid Mech.*, 409, pp.149-164
2. den Toonder, J.M., Hulsen, M.A., Kuiken, G.D.C. and Nieuwstadt, F.T.M (1997), "Drag reduction by polymer additives in a turbulent pipe flow:numerical and laboratory experiments", *J. Fluid Mech.*, 337, pp.193-231
3. Lumley, J. (1973), "Drag reduction in turbulent flow by polymer additives", *J. Polymer Sci.*, 7, pp.263-270.
4. Virk, P.S. and Wagger, D.L. (1990), IUTAM Symp., pp.201-213. Springer.
5. Smith, D.E. and Chu, S. (1998), "Response of Flexible Polymers to a Sudden Elongational Flow", *Science*, 281, pp.1335-1340.
6. Wei, T. and Willmarth, W.W. (1992), "Modifying turbulent structure with drag-reducing polymer additives in turbulent channel flow", *J. Fluid Mech.*, 245, pp.619-641.
7. de Gennes, P.G., Physica A (1986), "Towards a scaling theory of drag reduction", 140A, pp.9-25.
8. Tabor, M. and de Gennes, P.G. (1986), "A cascade theory of drag reduction", *Europhys. Lett.*, 2, pp.519-522.
9. Hidema, R. and Furukawa, H. (2012), "Development of Film Interference Flow Imaging Method (FIFI) Studying Polymer Stretching Effects on Thin Liquid Layer", *e-J. Surf. Sci. Nanotech*. 10, pp.335-340.
10. Hidema, R., Ushiki, H. and Furukawa, H. (2011), "Polymer Effects on Turbulence in Flowing Soap Films Studied with Film Interference Flow Imaging Method", *J. Solid Mech. Mater. Eng.* 5, pp.838-848.
11. Kellay, H. and Goldburg, W. I. (2002), "Two-dimensional turbulence: a review of some recent experiments", *Rep. Prog. Phys.*, 65, pp.845-894.
12. Hidema, R., Yatabe, Z., Shoji, M., Hashimoto, C., Pansu, R., Sagarzasu, G. and Ushiki, H. (2010), "Image analysis of thickness in flowing soap films. I: effects of polymer", *Exp. Fluids.*, 49, pp.725-732.
13. Suzuki, H., Nguyen, H. P., Nakayama, T., Usui, H. (2005), "Development characteristics of fluctuating velocity field of drag-reducing surfactant solution flow in a duct", *Rheol. Acta.*, 44, pp.457-464.

(142ah) Rivulet Flow of Liquid in Inclined Pipelines

Subbarao Duvvuri, Department of Chemical Engineering, Universiti Teknologi PETRONAS, Tronoh, Malaysia and Shuhaimi Mahadzir, Chemical Engineering, Universiti Teknologi PETRONAS, Tronoh, Malaysia

Packed towers are widely used in process industries for carrying out unit operations such as distillation, absorption, water cooling and catalytic gas-liquid reactions in which flowing gas needs to be brought into intimate contact with liquid flowing on the particles in the form of rivulets or films. Wetting efficiency, defined as the ratio of gas-liquid contact area to the particle surface area, directly affects the mass transfer rates. It has been observed that wetting efficiency can be less than one depending on the equipment variables (such as type of liquid distributor, particle shape, size, materials of construction, cocurrent or counter current flows) as well as operating variables (such as liquid and gas flow rates). Hydrodynamics of packed beds is generally analyzed by considering the bed to consist of parallel tortuous tubes through which gas and liquid may flow either co- or counter-currently. To develop an understanding of the wetting efficiency in packed towers, liquid flow through a single inclined tube by gravity in presence of gas flowing in co- or counter current flow is investigated. Width of liquid rivulet flowing by gravity is measured as a function of liquid and gas flow rates. Effect of pipe inclination , pipe diameter, liquid viscosity and surface tension on rivulet width are also investigated. The observations are analyzed through the concept of friction factor.

(142ai) Constant Elongational Stress Measurements of LD-PE and PS Using a Filament Stretching Rheometer

Nicolas Alvarez[1], Jose M. R. Marin[1], Qian Huang[1], Henrik K. Rasmussen[2] and Ole Hassager[1], (1)Department of Chemical and Biochemical Engineering, Technical University of Denmark, Kgs. Lyngby, Denmark, (2)Department of Mechanical Engineering, Technical University of Denmark, Kgs. Lyngby, Denmark

Traditionally, extensional rheological material parameters are characterized using uniaxial elongational deformation of polymer melts under constant strain rate. Elongational experiments performed under constant stress, i.e. elongational creep experiments, are considerably less reported, since few elongational rheometers are capable of applying a constant stress deformation. The filament stretching rheometer (FSR), which historically controls the strain rate and measures the stress on a cylindrical polymer sample (filament) undergoing uniaxial deformation at the midplane, is adapted to run in constant stress mode using an active control scheme. The details of the active control scheme for both controlled stress and controlled strain rate experiments are outlined. Constant stress experiments for both LDPE and PS are discussed and comparisons are made to the linear viscoelastic envelope (LVE) and previously reported constant strain rate experiments using the FSR for the same materials. We evaluate the viability of extensional creep experiments to measure the steady-state extensional viscosity of polymer melts.

(142aj) Mixing in Sessile Drops Merging On a Surface

Ying Zhang[1], Samuel Oberdick[1], Stephen Garoff[1] and **Shelley L. Anna**[2], (1)Physics, Carnegie Mellon University, Pittsburgh, PA, (2)Chemical Engineering and Mechanical Engineering, Carnegie Mellon University, Pittsburgh, PA

We investigate the mixing of two sessile drops that merge on a surface. The drops consist of low viscosity glycerol-water mixtures deposited on a silicone elastomer surface with contact angle near 90°. We observe the shape of the drops and the location of their intersection by placing a fluorescent dye in one drop and using a laser light sheet to image a plane perpendicular to the surface. The initial healing of the meniscus bridge between the merging drops, and the damping of capillary waves appearing on their surfaces occur on timescales comparable to the inertio-capillary relaxation time. However, the interface between the two fluids remains sharp, broadening diffusively over several minutes. The shape of the merged drops and the boundary between them also continues to evolve on a timescale of minutes. This later motion is controlled by a gravity current established by the density difference between the liquids in the two initial drops. The gravity current leads to the eventual stratification of the liquids to form two separate layers within the stationary shape of the merged drop. We characterize the long timescale flow arising from the gravity current as a function of the drop sizes and the density and viscosity differences between the two merging fluids. We use dimensional analysis to aid in elucidating the essential physics.

(142ak) High Throughput Cell Separation Device for the Enrichment of Circulating Tumor Cells

Eric Lin, Hyeun Joong Yoon, Yang Wang, Trinh M. Pham and Sunitha Nagrath, Department of Chemical Engineering, University of Michigan, Ann Arbor, MI

Isolation and capture of circulating tumor cells (CTC) from the peripheral blood of cancer patients can greatly contribute to the cancer diagnosis, biomarker discovery and the track of therapeutics. Various antibody-based techniques have been developed to isolate, capture, and enumerate CTCs. However, the main disadvantage of antibody-based capture is the difficulty in retrieving live cells for downstream analysis, presumed tumor marker expression and their levels. Moreover, the time-consuming process, including biochemical functionalization, preparation and sample processing, as well as the low throughput hinder its utilization in clinic. Here we present a high throughput cell separation device for the enrichment of CTCs. By utilizing the balance between various hydrodynamic forces, this label-free process is promising for the separation of CTCs based on the size difference of CTCs and other blood cells. This microfluidic device contains complex PDMS channel, allowing cells to be efficiently separated at high throughput while ensuring great recovery and purity of CTCs.

(142al) Slow Growth of the Rayleigh-Plateau Instability in Aqueous Two Phase Systems

Iwona Ziemecka, Volkert van Steijn, Ger J. M. Koper, Jan H. van Esch and Michiel Kreutzer, Chemical Engineering, Delft University of Technology, Delft, Netherlands

Microfluidic uses of aqueous two-phase systems (ATPS) are limited due to the difficulty in making monodisperse droplets from aqueous polymer solutions that are immiscible. This paper teaches how to generate such *water-in-water* microdroplets from a thread based on a study of the Rayleigh-Plateau instability. Careful vibration-free experiments with controlled actuation of the flow allowed direct measurement of the growth rate of this instability. In our experiments, we used the well-known aqueous two phase system of dextran and polyethylene glycol solutions. We report growth rates of the order of 1/s for this system, which is more than an order of magnitude slower than the growth rates in an analogous experiment with two immiscible Newtonian fluids with viscosities and interfacial tension that closely matched the ATPS experiment. Viscoelastic effects and adhesion to the walls were excluded as explanations for the observed behavior. The results are remarkable, because all current theory suggests that such dilute polymer solutions should break up faster, not slower, than the analogous Newtonian case.

The effect of the Non-Newtonian blood flow rheology on the flow through the arterial network

Alex Apostolidis & Antony Beris
Department of Chemical and Biomolecular Engineering
University of Delaware
Newark, DE, 19713
USA

1. Introduction

Blood is a complex fluid with non-Newtonian characteristics. It has a shear-thinning behavior and exhibits a yield stress (viscoplasticity) with potential history effects (thixotropy) [1]. In order to understand the complex rheological behavior of this fluid, one needs to consider its composition. Blood is a suspension in plasma of formed elements such as platelets, leucocytes and erythrocytes. Of these cells, by far the greatest proportion (~98%) consists of erythrocytes, which also occupy 40-50% of the blood volume. An important parameter when studying blood rheology is the hematocrit (Hct), which defines the volumetric fraction of the RBCs in the blood. To a significant extent, the rheology of blood is determined by the behavior of RBCs at different shear rates.

Two major effects arise from the presence of red blood cells: rouleaux formation and cell alignment [2]. At sufficiently low shear rates (< 1 sec^{-1}) these cells aggregate into column structures called rouleaux. These structures can develop into more complicated ones at even lower shear rates, justifying the presence of a yield stress. However, this process is reversible as stresses greater than 0.1 dyn/cm^2 cause rouleaux disaggregation which largely explains the shear-thinning character of blood [3]. Furthermore, an increase in the shear rate causes the deformation of the dispersed RBCs and the alignment of their long axes with the streamlines. This is another factor contributing to the shear-thinning behavior.

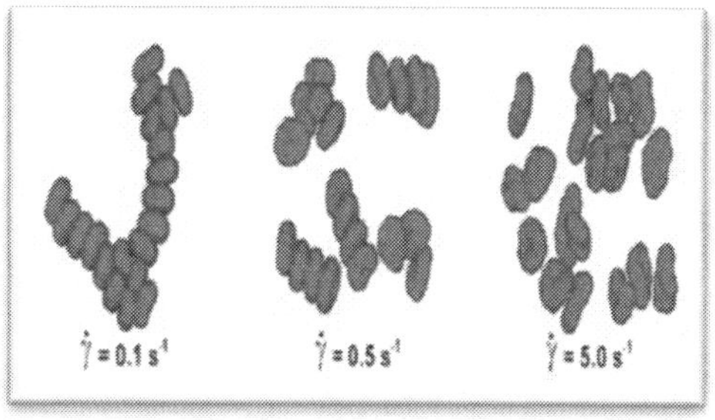

Figure 1: Visualization of aggregation at different shear rates. Hct=10%. [4]

Plasma plays a key role in the unraveling of the aforementioned mechanisms. In particular, the fibrinogen concentration in plasma is known to be strongly connected to the existence of yield stress [3,5]. The physical interpretation of this is that fibrinogen, by bridging adjacent RBCs, allows for the erythrocyte aggregation and rouleaux formation to take place. Thus, even though plasma in a Newtonian fluid, the non-Newtonian rheology of blood is dominated by the interactions of red cells with this plasma protein.

A crucial parameter when modeling blood flow is the vessel (or tube) diameter. For diameters of 30μm-500μm the tube hematocrit, Hct, and the apparent viscosity of blood decrease as the diameter decreases [3]. These dependencies are identified as the Fahraeus and the Fahraeus-Linqvist effect, respectively, and they rise to due to the migration of RBCs away from the wall [6]. In moderately small tubes (40-500μm) the cell-free region at the wall is only a small fraction of the cross-sectional area. However, as the diameter further decreases (<20μm) the fraction becomes substantial and a two phase model is required to describe the system. Barbee and Cokelet [6] have been able to predict the pressure-flow data of blood in narrow tubes (down to 29μm) from their macro-rheological properties determined in large tubes, provided that the proper tube hematocrit is used. Our analysis is valid for diameters down to 29μm, where the continuum model still holds [6].

2. Methodology

We have focused our analysis on simple, steady-state, Poiseuille flow. Steady flows give insight into the importance of non-Newtonian rheology, without facing the mathematical complications that accompany the rheology of non-Newtonian fluids. Moreover, the pulsatile flow, the examination of which is part of our future work, can be decomposed into a steady and a zero-flow oscillatory component. Thus, even in the transient analysis of the problem a determination of the steady component is still required.

A large number of constitutive equations have been proposed to describe the rheology of blood. Our model is based on the Casson equation:

$$\sqrt{\tau} = \sqrt{\tau_y} + \sqrt{\mu\dot{\gamma}} \text{ for } \tau \geq \tau_y, \quad (1)$$

$$\dot{\gamma} = 0 \text{ for } \tau \leq \tau_y$$

where τ is the shear stress; τ_y is the yield stress; μ is the viscosity and $\dot{\gamma}$ is the shear rate.

Most of the experimental research on blood rheology is conducted with the use of capillary viscometers which yield the shear stress versus *pseudo*-shear rate data (see Figure 2).

Due to the difficulty in reaching very low *pseudo*-shear rates, as well as the unavoidable experimental noise in the capillary at low shear, the capillary viscometer is not considered an appropriate instrument for obtaining the yield stress of blood [8]. A more accurate estimate can be obtained by using the Couette viscometer. Being able to convert capillary viscometry data into a material property function in the form of a shear stress versus shear rate curve can provide important rheological information on blood.

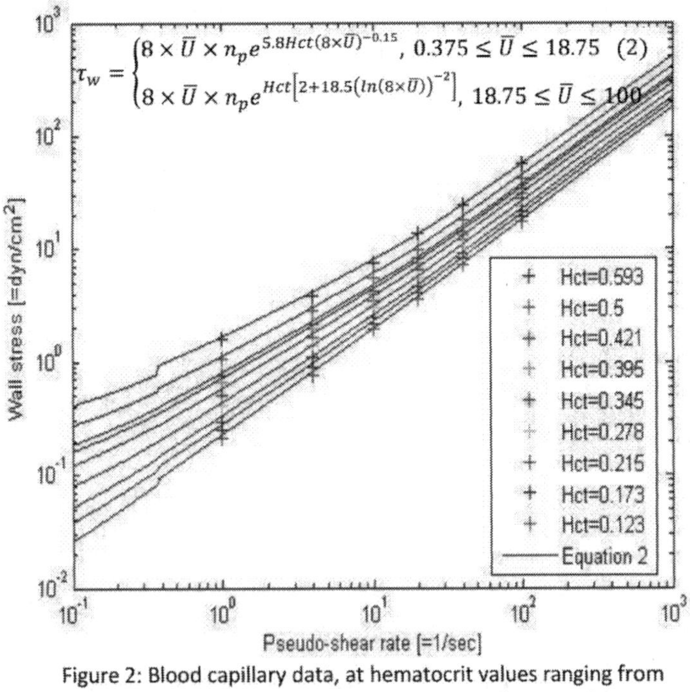

Figure 2: Blood capillary data, at hematocrit values ranging from 0.123 to 0.593. The solid curves are computed from Eq. (2). Diameter=811μm, Temperature=23°C. [18]

We have derived a formula that allows for the conversion of capillary data to Couette:

$$\dot{\gamma}(\tau_w) = \tau_w \frac{d(\bar{U})}{d\tau_w} + 3(\bar{U}) \quad (3)$$

where τ_w is the wall shear stress and $\bar{U}$ is the pseudo-shear rate in the capillary.

The conversion formula is central to our analysis as it can be used to test whether blood can be approximated and modeled as a simple fluid. In addition, through the use of Eq. (3), we test the possibility of describing blood as a generalized Newtonian fluid.

3. Results

Using the capillary data of Figure 2 along with the conversion formula (Eq. (3)) we have developed the corresponding Casson plots ($\sqrt{\tau}$ vs $\sqrt{\dot{\gamma}}$) for these blood samples. Figure 3 depicts the capillary-converted Casson plots for three of the Hct values of Barbee's data. Same behavior is noted for the rest of the Hct values (data not shown).

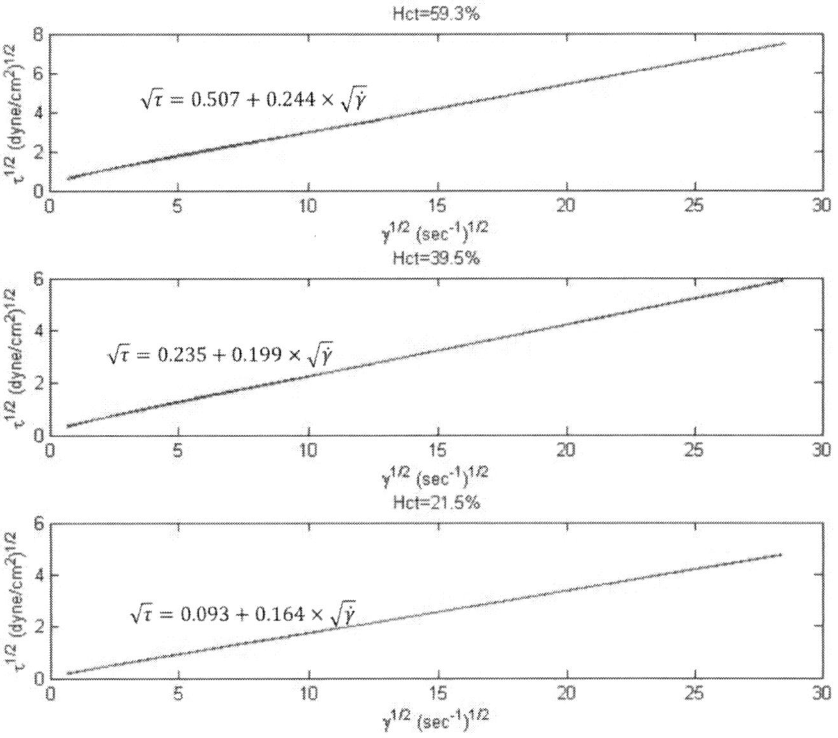

Figure 3: Casson plots ($\sqrt{\tau}$ vs $\sqrt{\dot{\gamma}}$) for the Hct values of Figure 1. The blue line represents the Couette flow data as computed from Eq. (2) and Eq. (3), for an arbitrary, input vector of pseudo-shear rate data, $\overline{U}$. The red line is a linear fit. The fitted equations are displayed on each plot.

The Casson plots show a linear correlation between the square root of shear stress ($\sqrt{\tau}$) and that of shear rate ($\sqrt{\dot{\gamma}}$), therefore proving that the Casson model is very well representing the data. From each fitted straight line we extract the apparent viscosity and the yield stress of blood. We have developed parametric equations that describe the dependence of these macro-rheological properties on the hematocrit, Hct. The effects of temperature and fibrinogen concentration on the plasma viscosity and the yield stress, respectively, are also included.

$$\tau_0 = \{2.2839 \times [(Hct - 0.04)^2]^2 + 0.1329 \times (Hct - 0.04)^2 + 0.002187\} \times \left(\frac{C_f}{0.3}\right)^2 \quad (4), \qquad 0.15gm\% \le C_f \le 0.35gm\%$$

$$\mu = (0.075 \times Hct^2 + 0.02616 \times Hct + 0.17788) \times \exp\left(-\frac{2893.137}{T_0} + \frac{2893.137}{T}\right) \quad (5),$$

where C_f = fibrinogen concentration in g% and T_0 = reference temperature = 296.2K.

The predictions of these equations are compared to experimental data, both in a Couette and a Poiseuille type of flow (see Figure 4 and Figure 5).

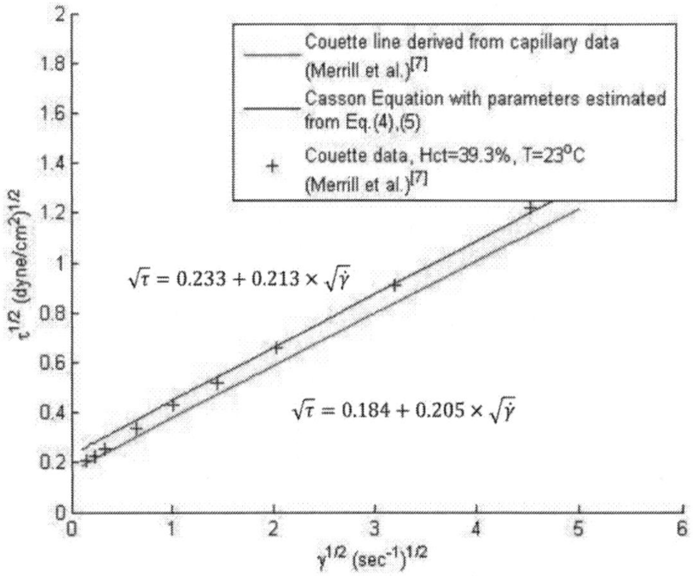

Figure 4: Comparison of experimental Couette data with the Casson model. The parameters of the Casson equation are predicted from Eq. (4) and (5). The red line represents the predicted Couette behavior based on capillary data of the same blood sample (capillary data not shown). Eq. (2) was used to convert the fitted line of Merrill's capillary data to Couette.

Based on capillary data of blood with a hematocrit value of 39.3% [7] (data not shown), and through the use of Eq. (2), we have derived the "Couette line" (red) of Figure 4. The experimental data points of Figure 4 correspond to the same blood sample when tested in a Couette viscometer. From a rheological point of view, the ability to describe the Couette data, even with low accuracy, with a function that originates from Poiseuille data means that the fluid can be considered simple. Therefore, despite the complex nature of blood, we can assume that the local stress is a function only of local properties.

The predictions of the equations that we have proposed are also included in Figure 4 (blue line). The viscosity prediction (slope on Casson plot) compares well with the data, while an offset (~25%) is apparent in the case of yield stress (y axis intercept). This could be attributed to a potential difference in the fibrinogen concentration between the blood sample upon which our model is based (Barbee's data) and the one used by Merrill et al. Even though the fibrinogen content of these samples is not known, we have indications that support our hypothesis. Using a c_f value of 0.218 gm% in Eq. (4) reduces the error of the yield stress prediction to less than 3%. This improvement is noted for both experimental data sets in the work of Merrill et al (39.3% Hct and 20% Hct, the latter case not shown).

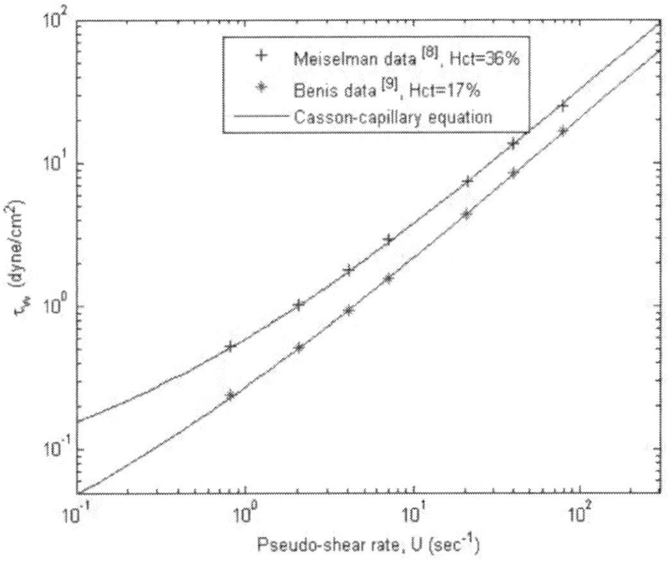

Figure 5: Comparing the Casson model against capillary data.

Finally, we have tested our model against capillary data of different hematocrits (Figure 5). The Casson model for capillary flow is in excellent agreement with the experimental data of Meiselman and Benis.

4. Conclusions

A proper interpretation of rheological data for blood has been offered. In our analysis we carefully distinguish between the processing of Poiseuille and Couette data, a subject that has been sometimes confused in literature. We propose a formula that allows for the interpretation of capillary data as Couette. Inherent to that formula (Eq.3) is the assumption of a simple fluid. We have shown that the simple fluid assumption holds for blood. We have demonstrated that blood can be described as a generalized Newtonian fluid and that the Casson model is very well representing the data. A new set of parametric equations has been suggested for the prediction of the apparent viscosity and yield stress, accounting for factors other than the hematocrit, such as the fibrinogen concentration and temperature. Our analysis is valid down to vessel diameters of ~30μm.

5. References

[1]. Owens RG, A new microstructure-based constitutive model for human blood. *J Non-Newtonian Fluid Mech*; **140**:57-70, 2006

[2]. Either, CR and Simmons, CA, Introductory Biomechanics; From Cells to Organisms. New York, NY: Cambridge University Press, 2007

[3]. Merrill, EW Rheology of Blood, *Physiological Reviews*. 49:863-888, 1969

[4]. Fedosov, DA, *et al.*, Predicting human blood viscosity in silico, *PNAS*, **108**: 11772-11777, 2011

[5]. Merrill, EW, Gilliland, ER, LEE, TS and Salzman EW, Blood rheology: Effect of fibrinogen deduced by addition. *Circulation Res.* **18**:437-446, 1966

[6]. Cokelet, GR Blood rheology interpreted through the flow properties of the red cell. In: *Microcirculation*, edited by J. Grayson and W. Zingg. New York: Plenum, **1**: 9-32, 1976.

[7]. Merrill, EW, Benis, AM, Gilliland, ER *et al*, Pressure-flow relations of human blood in hollow fibers at low flow rates. *J. Appl. Physiol.*; 20(5):954-967, 1965

[8]. Meiselman, HJ, Some Physical and Rheological Properties of human blood, Sc. D. Thesis, M.I.T., Cambridge, Mass., 1965

[9]. Benis, AM, The flow of human blood through models of the microcirculation, Sc. D. Thesis, M.I.T., Cambridge, Mass., 1964

(142an) A Continuous Label-Free Cell Separation Device Combining Inertial Microfluidics and Dielectrophoresis

Tae Hyun Kim[1], Hyeun Joong Yoon[2], Eric Lin[2], Saarika Ketkar[2] and Sunitha Nagrath[2], (1)Department of Electrical Engineering and Computer Science, University of Michigan, Ann Arbor, MI, (2)Department of Chemical Engineering, University of Michigan, Ann Arbor, MI

Separating cells from whole blood samples is an essential preparatory step in cell biology research and in many medical assays. Various microscale separation techniques have evolved using biochemical labels resulting in isolation of cells in specific methods. However, in spite of its high sensitivity, these methods might not be quite desirable since it may alter the cellular characteristics, hinders post-capture processing (release, enumeration, and manipulation etc.), and limited by its low throughput. Here we have fabricated a high throughput, label-free microfluidic device to separate cancer cells utilizing its electrical polarizability and hydrodynamic properties for identification. Size based hydrodynamic focusing was achieved with spiral shaped microchannels followed by dielectrophoresis (DEP) to successfully isolate MCF-7 breast cancer cells from blood samples. In addition, dielectrophoretic characteristics and behavior of the cell in response to DEP forces also helped prevent cell clogging at the outlet ensuring a more precise focused separation with enhanced reproducibility.

(142ao) Detailed Simulations of Lagrangian Particle Transport in Turbulent Reacting Flows

Guangyuan Sun[1], David O. Lignell[1] and John C. Hewson[2], (1)Chemical Engineering, Brigham Young University, Provo, UT, (2)Sandia National Laboratories, Albuquerque, NM

An understanding of particle dynamics in turbulent flows is important to many biological, environmental and industrial fields. We are performing fundamental investigation into dispersive transport and time-temperature histories of Lagrangian particles in turbulent reacting flows. Of specific focus is the defeat of biological agents using explosively dispersed gas through thermal or chemical reaction. It is of great significance to predict the statistics of the interaction of biological aerosol with the reacting flow, particularly the full particle temperature probability density function (PDF). A great challenge in modeling of particle motions in computational fluid dynamics is the accurate prediction of fine-scale aerosol-fluid interactions. Direct numerical simulation (DNS) studies provide a detailed description of two-phase flows, but they are very expensive computationally, even at moderate Reynolds numbers and particle densities. Reynolds-averaged Navier-Stokes (RANS) and large eddy simulation (LES) approaches are computationally efficient, but cannot resolve fine structures and require modeling of unknown subgrid statistics. A computationally affordable stochastic modeling approach, one-dimensional turbulence (ODT), is a proven method that captures the full range of length and time scales (in one-dimension), and provides detailed statistical information of fine-scale turbulent-particle mixing and transport.

Limited results of particle transport in ODT have been previously reported in non-reacting flow. Here, we present the implementation and validation of a Lagrangian particle model into an existing reacting ODT code. Results of particle transport in three flow configurations--channel flow, homogeneous isotropic turbulence, and jet flames--are presented. The work investigates the functional dependence of the statistics of particle-flow interactions. A parametric study is performed varying temperature profiles, Reynolds numbers, and particle Stokes numbers. Particle temperature histories and PDFs are presented. The simulations also allow the analysis of time scale and the sensitivity of initial and boundary conditions, and other physical parameters. Flow statistics of the ODT particle model are compared to both experimental measurements and DNS data. The successful validation of Lagrangian particle transport in ODT has the potential for much more efficient investigation of fine-scale turbulence-particle interactions and access to a wider Reynolds number range than available using DNS.

(142ap) Separation and Isolation of Biomolecules Using Motion of Magnetic Spheres in Microfluidic Channels

Anubhav Tripathi, School of Engineering, Brown University, Providence, RI, Kenneth Morabito, Biomedical Engineering, Brown University, Providence, RI and Jinkee Lee, School of Mechanical Engineering, Sungkyunkwan University, Suwon, South Korea

The use of microfluidic devices and magnetic beads for applications in biotechnology has been extensively explored over the last decade. This work presents a simple technique for isolation and separation of biomolecules using magnetic bead movement on a microchip with zero-flow conditions. The microchip consisted of two well reservoirs (W1 and W2) connected via a tapered microchannel. The bead motion was first analyzed on the microchip and a permanent magnet was moved alongside the microchip, dragging the beads at equivalent speed through the microchannel between the two wells. It was found that greater than 95 % bead transfer from W1 to W2 occurred at an optimal velocity of 0.7 mm/s. Prior to testing our microchip with enzymatic reactions, we characterized our enzyme, alkaline phosphotase (AP), and substrate, 6-8-difluoro-4-methylumbelliferyl phosphate (DiFMUP), with an off-chip control. AP was used to dephosphorylate the non-fluorescent DiFMUP to fluorescent 6,8-difluoro-7-hydroxy-4-methylcoumarin (DiFMU). Next, three assays were performed using our microchip: non-specific adsorption of DiFMUP, isolation of AP, and separation of AP. Biotinylated AP was immobilized on streptavidin coated magnetic beads for the isolation and separation assays. Our non-specific adsorption assay indicated that we were able to transfer our beads with less than 0.002% carryover of DiFMUP. Our isolation assay indicated that we were able to efficiently isolate biotinylated AP with our streptavidin conjugated magnetic beads. 100% of DiFMUP was converted to DiFMU within 15 minutes. Our separation assay elucidated the binding capacity of our beads, wherein leftover unbound AP converted 100% of DiFMUP within 10 minutes and samples with less than full bead capacity (i.e. all AP was transferred) did not convert any of the substrate. The K_m value for the isolation assay was approximately 16% greater than our off-chip control indicating that the immobilization of AP onto the beads has a minor inhibitory effect on the reaction kinetics of the enzyme.

(142aq) Effect of Viscoelasticity On the Collective Behavior of Swimming Microorganisms

Yaser Bozorgi and **Patrick T. Underhill**, The Isermann Department of Chemical and Biological Engineering, Rensselaer Polytechnic Institute, Troy, NY

Hydrodynamic interactions of swimming microorganisms can lead to coordinated behaviors of large groups. Previous theoretical work has shown that in mean-field theories the interactions give rise to a linear instability of the uniform isotropic state if the organisms push themselves forward. However, that work is done with the organisms suspended in a Newtonian fluid, while many organisms exist in a non-Newtonian medium. Using a mean-field theory and the Oldroyd-B constitutive equation, we show how viscoelasticity of the suspending fluid alters the hydrodynamic interactions and therefore the ability of the group to coordinate. We quantify the ability to coordinate by the initial growth rate of a small disturbance from the uniform isotropic state. For small wavenumbers the response is qualitatively similar to a Newtonian fluid but the Deborah number determines the effective viscosity of the suspension. At higher wavenumber, the response of the fluid to small amplitude oscillatory shear flow leads to a maximal growth rate at a particular wavelength. This is in stark contrast to the Newtonian result.

(142ar) Modelling Particulate Flow During Composites Manufacturing

Staffan Lundstrom, Engineering Sceince and Mathematics, Lulea University of Technology, Lulea, Sweden

Modelling particulate flow during composites manufacturing

T. Staffan Lundstrom and Vilnis Frishfelds

Fluid Mechanics

Department of Engineering Sciences and Mathematics

Lulea University of Technology

Lulea 971 87, Sweden;

E-mail: staffan.lundstrom@ltu.se

During Liquid Composite Moulding (LCM) processes like Resin Transfer Moulding and Vacuum Infusion a liquid resin is forced to impregnate a preform consisting of fabrics. The fibre forming the fabric are often gathered in bundles resulting in a dual-scale porosity, < 10 μm inside the fibre bundles and > 100 μm between the bundles. The detailed geometry of such fabrics is important for the permeability of the preform but is also vital for applications when the resin is doped with particles to create multifunctional composites. The added particles can, for example, enhance the fire resistance, toughen the material, introduce electrical conductivity and shielding properties to the final composite. In order to achieve satisfactory properties of these functional materials, it is vital to have a known spatial distribution of the functionality throughout the material and it is of great interest to develop methods to control the distribution of particles during manufacturing. One way to obtain a desired particle distribution is to control the flow during impregnation.

In the present study the motion of a liquid resin doped with solid particles is numerically modelled as it moves perpendicular to a couple fiber bundles. The system is discretisised with modified Voronoi diagrams and the dissipation rate of energy is minimised with respect to the stream function. When testing different sizes of the particles, it is shown that there is a qualitative agreement between model and experiments previously performed. The capability of the model is furthermore demonstrated by studying the filtration of particles at fully saturated conditions for different size and position distributions of the particles as well as the geometry of the fibre bundles, see Figure 1.

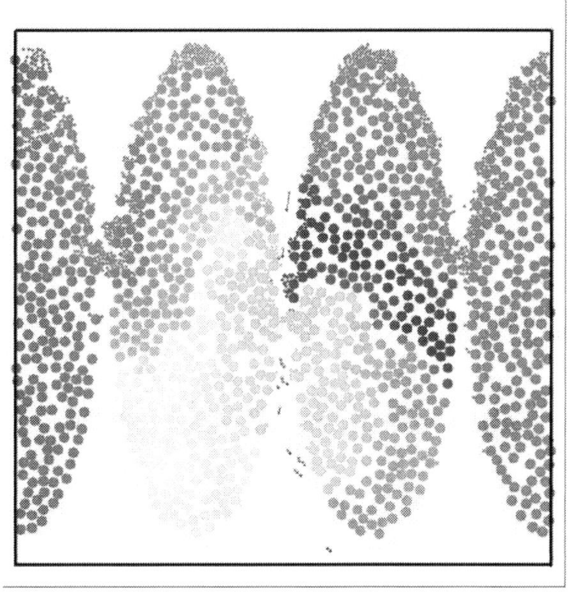

Figure 1. A case when there initially are no particles within the fibre bundles. The large circles are the fibres and the small the particles. The colours denote the stream function. At this moment in time most of the flow is in the middle most interbundle channel.

(142as) Microfluidic Characterization of Immiscible Oil-Seawater-Silicate Multiphase Flow

Chuntian Hu and Ryan L. Hartman, Department of Chemical and Biological Engineering, The University of Alabama, Tuscaloosa, AL

The phase behavior of solid-liquid-liquid flow is a problem of considerable societal importance, including applications in the production of pharmaceuticals and fine chemicals, petroleum and natural gas production, and water sustainability. As an example, the Deep Water Horizon catastrophe is regarded as the worst ecological disaster in the history of the United States. Remediation immediately following the disaster involved spraying dispersants in the deep sea, at the reservoir head, and on the ocean surface. Understanding the complex interactions of immiscible phases in such an emergency is necessary to minimize the environmental impact. Microfluidic devices were exploited to develop such understanding, and as a result new insight on the stability of oil-seawater multiphase laminar flow in the presence of colloidal aggregates has been revealed. The characterization of such phase behavior has the potential to impact multiphase microscale flows that involve surfactant chemistry. The need to develop, from first principles, controlled heterogeneous operations in microfluidic systems is growing.

(142at) Formation of Droplets in Microfluidic T-Junction At Small Capillary Number

Piotr M. Korczyk[1], Volkert van Steijn[2] and Piotr Garstecki[1], (1)Institute of Physical Chemistry, Polish Academy of Sciences, Warsaw, Poland, (2)Department of Chemical Engineering, Delft University of Technology, Delft, Netherlands

Microfluidic devices can generate monodisperse drops with precise control over their size and contents. The mechanism for drop formation in cross-junctions and T-junctions is well established. A drop grows into the main channel and obstructs the flow of the continuous phase, which then squeezes the interface of the forming drop until it breaks and a drop releases into the main channel. Drops formed by this *squeezing mechanism* have a volume that linearly scales with the ratio of flow rates, that is, for a fixed value of the capillary number (Ca). However, for non-fixed values of Ca, drop size also depends on the capillary number, because the fraction of the continuous phase flowing around forming drops is Ca-dependent. Current models for drop size prediction either ignore the flow of the continuous phase around forming drops completely or set it to a fixed value independent of the capillary number.

Here, we present a theoretical model that overcomes these limitations by taking into account these flows and their dependence on the capillary number. We found good quantitative agreement between our model and our experiments, which show that the fraction of the continuous phase flowing around forming drops significantly increases for decreasing Ca. To further validate our model, we formed partially-wetting drops, which do not allow the continuous phase to flow around them. As expected, their dynamics can be predicted based on continuity arguments only, in sharp contrast to non-wetting drops.

(142au) Nanoliter Droplet Microfluidic Viscometer with Additive-Free Operation

Eric Livak-Dahl, Jaesung Lee and Mark A. Burns, Chemical Engineering, University of Michigan, Ann Arbor, MI

Measurement of the viscosity of a sample solution is an important analytic technique for a variety of applications including medical diagnosis, pharmaceutical development, and industrial processing. The use of droplet-based (for example, water-in-oil) "digital" microfluidics for viscosity measurements allows nanoliter-scale sample volumes to be used, much smaller than those in either standard macro-scale rheometers or single-phase microfluidic viscometers. By observing the flow rate of a sample plug driven by a controlled pressure through a drastic constriction, we achieve accurate and precise measurement of the plug viscosity without addition of labels or tracer particles. Sample plugs in our device geometry had a volume of ~30 nL, and measurements had an average error of 7.2% with an average relative standard deviation of 2.7%. We tested glycerol-based samples with viscosities as high as 101 mPa*s, with the only limitation on samples being that their viscosity should be higher than that of the continuous oil phase.

(142av) Numerical Modelling of Transitional and Turbulent Flow Through a Pore Doublet Model

Gunnar Hellstrom, Engineering Sciences and Mathematics, Lulea University of Technology, Lulea, Sweden and Staffan Lundstrom, Engineering Sceince and Mathematics, Lulea University of Technology, Lulea, Sweden

Numerical modelling of transitional and turbulent flow through a pore doublet model

J. Gunnar I. Hellstrom, Shervin Khayamyan, T. Staffan Lundstrom and

L. Hakan Gustavsson

Fluid Mechanics

Department of Engineering Sciences and Mathematics

Lulea University of Technology

Lulea 971 87, Sweden;

E-mail: staffan.lundstrom@ltu.se

Transitional and turbulent flow through a pore doublet model is studied numerically to mimic the interconnectivity of pore flow. The model consists of two parallel pipes with different diameters branched via a Y-shaped splitter from a (wider) header pipe. Recently performed experiments on this system have shown that transition in the system starts by turbulent slugs in the laminar flow at pipe Reynolds number larger than 2000. This is also confirmed by a peak of the skewness value. Frequency analysis furthermore shows that the pipes communicate at low frequency, but that higher frequencies are filtered out possibly due to the experimental set-up. The implication of the experimental results as to porous media flow is that flow is redistributed when the flow becomes turbulent in the larger pores.

In the present work simulations of the flow through a virtual model of the pore doublet model are performed with a number of different models for the transitional and turbulent flow. The simulations are verified and the results from the simulations are compared to the experiments and conclusions are presented.

(142aw) Kinetic Simulation of an Annular Reactor Using CFD

Ardson do S. Vianna Jr. and José Carlos G. Peres, Chemical Engineering, University of São Paulo, São Paulo, Brazil

The objective of this work is to use the advanced oxidation processes (AOP) to remove organic materials in waste water. An annular reactor has been used to carry out this process, and its fluid dynamics was well understood. However, the reaction mechanism of AOP encompasses very large kinetic constants, such as 10^{10}L/mol/s. It can be seen rapid variation in the species concentration. In the present work, the simulation of reactions in an annular reactor was developed using CFD commercial code, the ANSYS 13.0. At first, the simple reactions with low kinetic constant were considered. Then, multiple reactions with low and high constants were simulated. Finally, a pseudo first order kinetic of phenol degradation was studied. A sequence of calculations was carried out using successive refinement of several meshes, and it can be observed mesh convergence. The k-ω turbulence model well represents the fluid dynamics of the reactor. The geometry directly influences the profiles of species concentration. The degradation of phenol can be simulated using relaxation factors of 10^{12} kg/m^3/s at least. The satisfactory results indicate that the strategy may be applied to simulate reactions using CFD tool.

(142ax) Stability of a Two-Phase Column Bioreactor for Algae Growth

Justin Smith, Department of Chemical Engineering, The University of Tulsa, Tulsa, OK, Daniel W. Crunkleton, Chemical Engineering, University of Tulsa, Tulsa, OK and Selen Cremaschi, Department of Chemical Engineering, University of Tulsa, Tulsa, OK

Stability of a Two-Phase Column Bioreactor for Algae Growth

Justin Smith, Daniel Crunkleton, Selen Cremaschi

Department of Chemical Engineering, University of Tulsa, 800 South Tucker Drive, Tulsa, OK 74104, USA

We perform a linear stability analysis of algae growth in a two-phase air-water bubble column. To accomplish this, the 2D, Cartesian, Navier-Stokes equations for fluid flow are coupled using the Boussinesq approximation with the reactive concentration balance for algae growth. A system of seven equations is obtained and a normal mode perturbation of the horizontal and vertical components of the velocity of both the continuous and dispersed phases, the gas volume fraction, pressure, and algae concentration is performed. The equations are linearized by setting any product of multiple perturbations equal to zero under the assumption that as the perturbation size is small, the size of a product of multiple perturbations is negligible. A stability matrix is then constructed, the determinant of which gives a dispersion relation with roots that can be analyzed for stability as a function of several values of key process parameters, including initial algae concentration, and reaction rate, defined here as algae growth rate. By varying the size of both the horizontal and vertical wave vector components of the perturbation, the stability of the bioreactor is determined by observing changes to the amplitude and sign of the real parts of the roots. The stability characteristics of the system indicate that for base case force parameters in a system with drag and virtual mass forces, bubble induced turbulence, and bubble pressure forces, algae growth is most stable with small values for algae growth rate and initial algae concentration and becomes less stable as the initial algae concentration is increased. The asymptotic behavior of the current study compares favorably with previous linear stability results that considered the two-phase system in the absence of a reactive concentration balance.

(142ay) Hybrid Field Regulation of Long DNA Configurations and Dynamics in Flows

Fangfang Ren, kartik Kumar Rajagopalan, Yingbo Zu and **Shengnian Wang**, Chemical Engineering/Institute for Micromanufacturing, Louisiana Tech University, Ruston, LA

Polymer flows are important to a variety of chemical, polymer, and petroleum practices. Many existing non-Newtonian effects are still not well understood, partially due to the unclear individual macromolecule dynamics. The introduction of direct visualization and manipulation of single long DNA molecule provides a wealth of physical insights in complex flow study. However, the diversity and stochastical tumbling of macromolecules causes large fluctuations of their configurations when measuring individual molecule dynamics, which impedes its wide adoption in polymer flow research. Here we report a hybrid field microfluidics (HFM) approach to overcome these issues. In HFM, a desired electric bias is imposed in appropriate moments and locations to effectively regulate the concentration and molecular configurations of DNA molecules as well as their dynamics in flows. With lambda DNA as the model molecules, we systematically studied the trapping, the chain relaxation, the coil-stretch dynamics, and the accumulation of macromolecules under different flow conditions for both Newtonian and non-Newtonian fluids. Our observations are further compared to other pioneer work and further correlated with bulk phenomena to establish connections with macromolecular dynamics and interactions at microscopic level.

(142az) Response of Endothelial Cells to Stagnation Point Flows

Maggie A. Ostrowski[1], Ngan F. Huang[2], Gerald G. Fuller[1] and Alexander R. Dunn[1], (1)Chemical Engineering, Stanford University, Stanford, CA, (2)Department of Cardiovascular Medicine, Stanford University, Stanford, CA

Human blood vessels are lined with a single cell layer called the endothelium. This mechano-sensitive layer senses and responds to the hydrodynamic forces of the blood imposed upon it by the pumping action of the heart. Under conditions of steady shear stress, endothelial cells orient parallel to the direction of blood flow and provide a healthy, tight barrier that protects against vessel diseases such as atherosclerosis.

Vascular disease occurs preferentially in locations where vessels branch or bifurcate. This geometry results in unsteady flow patterns referred to as "disturbed flow." Specifically, stagnation points resulting from flow recirculation, combined with impinging and upwelling flows, are thought to cause an unhealthy endothelial cell layer that is prone atherosclerosis. Although the subject of intense interest, the underlying molecular mechanisms connecting disturbed flow to vascular disease remain incompletely understood.

To investigate the behavior of endothelial cells exposed to disturbed flow, we built a novel submerged impinging jet device that produces a stagnation point flow, and that allows us to observe the behavior of endothelial cells in this flow environment on the hours/days timescale. We observe dramatic changes in cell migration, orientation, and crowding as a function of Reynolds number. Our observations also reveal a remarkable ability of endothelial cells to sense and respond to spatial gradients in fluid shear. We discuss these results in the context of human physiology and disease.

(142b) Chiral Rheology of Condensed DPPC Domains

Kyuhan Kim[1], SiYoung Q. Choi[2], Joseph A. Zasadzinski[3] and Todd M. Squires[2], (1)Chemical Engineering, Univerisity of California, Santa Barbara, Santa Barbara, CA, (2)Chemical Engineering, University of California, Santa Barbara, Santa Barbara, CA, (3)Department of Chemical Engineering and Materials Science, University of Minnesota, Minneapolis, MN

We present the asymmetric responses of condensed phase dipalmitoylphosphatidylcholine (DPPC) monolayers due to clockwise and counterclockwise deformations. The phospholipid, especially DPPC, forms condensed grainy liquid crystal monolayer, whose individual grains are chiral: due to tilt of tail group at the interface and a chiral point within the molecular structure, bio-derived DPPC, R enantiomer, grains curl in counterclockwise direction while S-DPPC , made by only synthetically, curls in clockwise direction. By applying constant stress to the interlocked condensed domains of R DPPC at condensed phase, we show that the strain responses are asymmetric: deformations are easier in clockwise direction than counterclockwise direction. In addition, we manipulate the degree of chirality of DPPC grains by adding racemic DPPC which contains no chiral properties, and we show that this asymmetry of strains decreases as adding more racemic DPPC to R DPPC. We conclude that this chirality of DPPC domains leads to a chiral viscoelastic responses.

(142ba) Augmentation of Mass Transport of Solutes in Flowing Suspensions Due to Secondary Currents Driven by Second Normal Stress Differences

Arun Ramachandran, Department of Chemical Engineering and Applied Chemistry, University of Toronto, Toronto, ON, Canada

It has been long known that the diffusivity of solutes in suspensions of both rigid and deformable particles is enhanced by shear-induced self diffusion[1,2]. The augmentation of the solutal diffusivity, which is defined as D_{eff}/D_0-1, scales as $g(\Phi)Ua^2/BD_0$, where D_{eff} is the observed diffusivity, D_0 is the molecular diffusivity, B is a characteristic dimension of the cross-section, a is the particle radius and $g$is a monotonically increasing function of the volume fraction Φ.

For suspensions of rigid particles flowing through conduits of non-axisymmetric cross-sections, it was recently demonstrated[3] that the flow is not unidirectional; the main flow along the axis is accompanied by a secondary flow within the cross-section driven by second normal stress differences (N_2). The magnitude of the secondary currents scales as $k\delta(\Phi)U$, where k is a constant that depends on the degree of non-axisymmetry of the geometry, and δ is proportional to N_2, which is an increasing function of Φ. Notably, it is independent of the particle radius (for non-Brownian suspensions only).

In this work, we demonstrate that N_2-driven secondary currents represent a new, convective mechanism for mass transport in flowing suspensions. The characteristic Peclet number for the secondary current convection is $k\delta\ UB/(D_0+gUa^2/B)$. It can be seen that when the particle size is small enough to render the enhancement of diffusivity by shear-induced self-diffusion weak, the Peclet number can still be large, since the magnitude of the secondary currents does not depend on particle size. Therefore, in this limit, secondary convection, and not the traditionally-considered shear-induced self-diffusion, could be the dominant mechanism for enhancement of mass transfer. For large flow velocities, when shear-induced self-diffusion is the dominant diffusive mechanism for mass transfer, the Peclet number reduces to $k\delta B^2/\ ga^2$, independent of the suspension velocity. We find that for moderately concentrated suspensions, the Peclet number in this limit can be much larger than 1, because of the large value of the aspect ratio $B/\ a$ in most practical suspension flows. Thus, secondary convection can provide additional enhancement of mass transfer over that due to shear-induced self-diffusion, and the relative augmentation can be by as much as a factor of 4 for some realistic examples. However, it must be stated that since the streamlines associated with secondary currents are closed, the enhancement of the mass transfer rate by convection asymptotes to a constant for high Peclet numbers.

The practical relevance of this new mechanism of mass transfer is in the transport of large molecular weight solute molecules in blood flow. Such molecules are associated with small diffusivities, and are associated with significant concentration gradients within blood vessels (higher mass transfer resistances) during their transport from blood into tissues. Our model and calculations will permit improved modeling of the mass transfer of such solutes.

References

1. A. L. Zydney and C. K. Colton, Augmented solute transport in the shear flow of a concentrated suspension, *Phys. Hyd.* **10**, 77–96 (1988).

2. W. Cha and R. L. Beissinger, Augmented solute transport to surfaces in sheared suspensions. *J. Colloid Interface Sci.* **178**, 1–9 (1996).

3. A. Ramachandran and D. T. Leighton, The influence of secondary flows induced by normal stress differences on the

shear-induced migration of particles in concentrated suspensions, *J. Fluid Mech.* **603**, 207-243 (2008).

Effect of Non-Cohesive Particulates on Yield Stress of Waste Sludge Simulants

Ibraheem R. Muhammad and John P. Kizito
North Carolina A&T State University, Department of Mechanical Engineering
Greensboro, NC

Abstract

Nuclear waste material often exhibit non-Newtonian behavior, so to achieve efficient processing of the material, the rheology of the waste material must be understood. The goal of the current study was to determine the effect of coarse, non-cohesive particles on a mixture exhibiting a yield stress. A simple, inexpensive, and easy to run experimental approach, was used to estimate the yield stress based on the definition that yield stress is the critical shear stress that must be applied before the material begins to flow. In the current study, experimental tests were run using a capillary tube rheometer and a water manometer. The simulants used for testing included kaolin clay/water mixtures and kaolin/sand/water mixtures. The kaolin particles (D = 1.36 μm) represented the cohesive particles and the sand particles (100 μm < D < 800 μm) represented the non-cohesive particles. The yield stress was estimated using a non-intrusive, visual observation technique. The yield stress increased exponentially with increased weight % kaolin. The yield stress decreased in kaolin/sand mixtures as the amount of sand was increased. The results of the technique are compared to yield stress data using shear rate and shear stress data. Results were also compared to previous researchers and it was shown that the data was in good agreement. The technique used can be applied as an initial means of measurement for yield stress in not only nuclear waste facilities, but other processes in which yield stress needs to be calculated as well.

Introduction

Yield stress is an important parameter in processes involving materials that behave as non-Newtonian fluids. Non-Newtonian fluids are found widespread in nature. In nuclear waste remediation, the sludge can behave as a non-Newtonian fluid, exhibiting a yield stress, which must be overcome before the fluid begins to flow. It is important to know the yield strength of the solids material in off-bottom suspension processes, so that enough hydrodynamic force is provided be the mixer [1].

Though it's a highly debated topic, yield stress can be simply viewed as the amount of stress required for a flow to begin to flow. Yield stress can be measured indirectly by extrapolating the shear stress-shear rate data to the zero shear rate or it can be measured directly [2-4]. Yield stress can be measured using a number of different rheometers, such as capillary, concentric cylinder, wide gap rotational, cone-and-plate, or parallel plate [5]. The pressure driven, capillary tube rheometers are advantageous compared to other rheometers because they are easily constructed and operated, while providing some of the most accurate viscosity results [3].

In mixtures containing both fine and coarse particles it is expected that the particles in both ranges have an important role in determining the yield stress. Particle size is important as the surface area increases with decreasing particle size, allowing more surface area for forces to act upon. So as the particle size decreases, the yield stress increases [6]. The fine particles are deemed cohesive because of forces such as van der Waals, electrostatic forces, etc. The effect of coarser, non-cohesive, particles on cohesion in mixtures and ultimately, the yield stress, must be understood [7].

Sand/ kaolin/water mixtures can give an idea of the effect of non-cohesive particles on cohesive mixtures. It is important to be able to accurately measure the yield stress of the mixtures so that precise data can be obtained in studies where it is being used as a simulant. The purpose of this initial study is to provide a simple technique for determination of the yield stress, which can be used as a means of first measurement of yield stress for future studies. The actual yield stress results can be used to give some general knowledge of the behavior of yield stress fluids containing both cohesive and non-cohesive particles.

Experimental Methods and Materials

Experimental tests have been run using different mixtures of EPK kaolin clay (Edgar Minerals) and MICROSIL CGS silica-quartz (US Silica) in water. The kaolin clay is used as the fine, cohesive particles. The kaolin clay has an average particle size of 1.36 μm and a specific surface area of 28.52 m^2/g. The sand particles are used as the coarser, non-cohesive particles. The particle size of the sand is given as a histogram, shown in Figure 1.

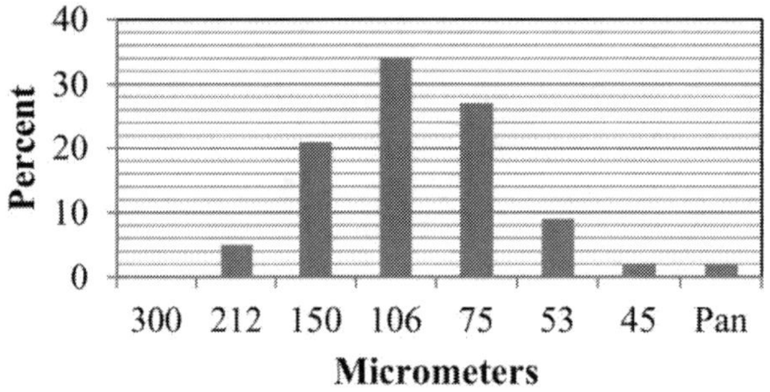

Figure 1: Histogram of particle size of sand used for studies.

Tests were run where the weight % of the total solids, φ_T, is varied from 50 to 60 %. Since the total solids are varied with different amounts of kaolin and sand, it is mathematically defined as

$$\varphi_T = \varphi_K + \varphi_S \quad (1)$$

where φ_K is the weight percent of kaolin and φ_S is the weight percent of sand. The total weight % solids were varied from 50 to 60%. Within those total solids content, the clay-sand weight % was varied from 50 to 100% for each component. Though tested, some of the results are not presented due to limitations in the equipment, which is discussed later.

A capillary tube rheometer was constructed to measure the yield stress of the mixtures. Figure 2 shows a schematic of the apparatus used for the tests. The apparatus includes a PVC reservoir (H = 11.7 mm and D = 10.2 mm) that is pressurized by compressed air. The capillary tube is made of brass and is connected to the bottom of the reservoir. The aspect ratio (L/R) of the tubes ranged from 9.625 - 90. The pressure drop across the capillary tube is measured using a 3.05 m (10") water, U-tube manometer. The maximum pressure that can be measured using the manometer is 29.84 kPa. The maximum shear stress that can be measured is about 783 Pa. The water manometer is advantageous due to its ease of construction, lost cost, and sensitivity.

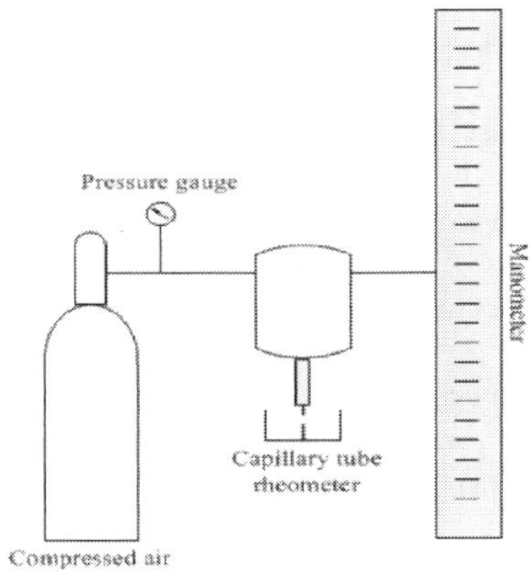

Figure 2: Schematic of capillary tube rheometer apparatus.

A simple, fast, and easily constructed visual technique is used to measure the yield stress, kaolin clay, sand, and water mixtures are created by carefully measuring the appropriate amount of each component and manually mixing. Mixing takes place for about 15 minutes to ensure consistency throughout the mixture and then it is allowed to age for a period 2.5 to 3 hours. The mixture is transferred to the reservoir where it is filled to the half-way mark. The mixture is slightly mixed for 2 minutes while in the reservoir in attempt to eliminate any air pockets. The compressed air is increased until the mixture starts to flow out of the capillary tube. The pressure is then stopped and the material extruding from the tube is cleared. This ensures that more accurate measurement of initial flow is taken as the fluid is filled to the very tip of the tube. The compressed air is then increased very slowly and the moment at which incipient motion occurs is recorded. The height difference in the manometer is measured and the yield stress is calculated. Each trial run is repeated 4 times.

The validity of the visual technique is studied by calculating yield stress using a Brookfield LV viscometer. Also, studies are run by following the same procedure as above but calculating the volumetric flow rate of material exiting the capillary as a function of pressure drop.

The equations for flow in a capillary is derived assuming fully developed, steady, isothermal, laminar flow with no flow in the r and Θ direction. It is assumed there is no slip at the walls and the fluid is incompressible with viscosity independent of pressure. The maximum shear stress occurs at the wall and can be calculated as

$$\tau_w = \frac{\Delta P R}{2L} \tag{2}$$

$$\Delta P = \rho g h \tag{3}$$

where ΔP is the pressure drop across capillary tube, R is the radius of the capillary tube, L is the length of the capillary tube, ρ is the density of water, g is the gravity acceleration constant, and h is the height difference in the manometer. The apparent wall shear rate can be expressed as

$$\dot{\gamma}_{aw} = \frac{4Q}{\pi R^3} \tag{4}$$

Several models exists to model non-Newtonian fluids, including the Bingham, Casson, and Power –law [3]. Some fluids can be modeled using more than one model, which would provide different yield stress values. Thus, the yield stress can be looked at as not being an absolute quantity [2]. The power law model is used to model the mixtures in the current study. For power law fluids flowing through a pipe, the shear stress can be expressed as

$$\tau = K\dot{\gamma}^{n} \tag{5}$$

where K is the consistency coefficient and n is power law index (i.e. n < 1 for shear thinning and n > 1 for shear thickening). The shear rate at the wall can then be expressed as

$$\dot{\gamma}_{w} = \frac{4Q}{\pi R^{3}}\left[\frac{3}{4}+\frac{1}{n}\right] = \dot{\gamma}_{aw}\left[\frac{3}{4}+\frac{1}{n}\right] \tag{6}$$

Thus, the viscosity can be calculated as

$$\eta = \frac{\tau_{w}}{\dot{\gamma}_{w}} \tag{7}$$

Manometer Calibration

Though calibration is not needed using manometers, some testing was done to ensure the system was able to measure pressure accurately. The pressure calculated by the height difference in the manometer was tested against the readings of a pressure gauge. Figure 3 displays percent of error between the manometer and pressure gauge readings. The highest error was only 1.9%. This is due to human error, which results from inaccurate measuring of the meniscus in the tubes.

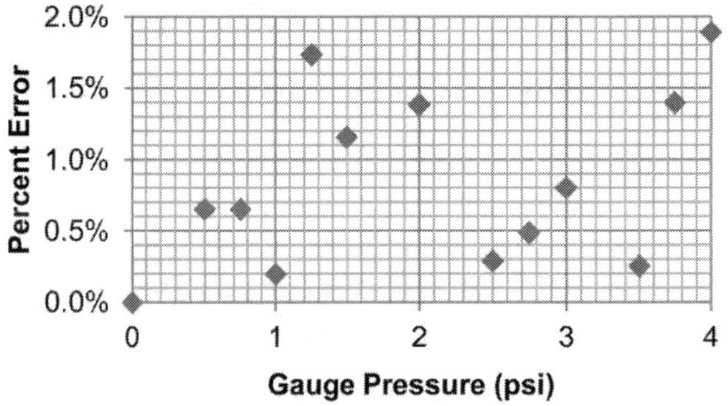

Figure 3: Percent error of readings from manometer compared to pressure gauge.

Results

Comparisons of Technique and Brookfield Viscometer Results

An initial test was run testing the shear stress versus shear rate for a mixture of 50/50 mixture of water and clay. Tests were run using a Brookfield LV Viscometer and the capillary tube rheometer. The results of the Brookfield viscometer are shown in Figure 4. The results were modeled using the power law model where the consistency coefficient was found to be 127.95 Pa·s and the power law index was found to be 0.1563. Since n < 1, the model represents a fluid with shear thinning behavior.

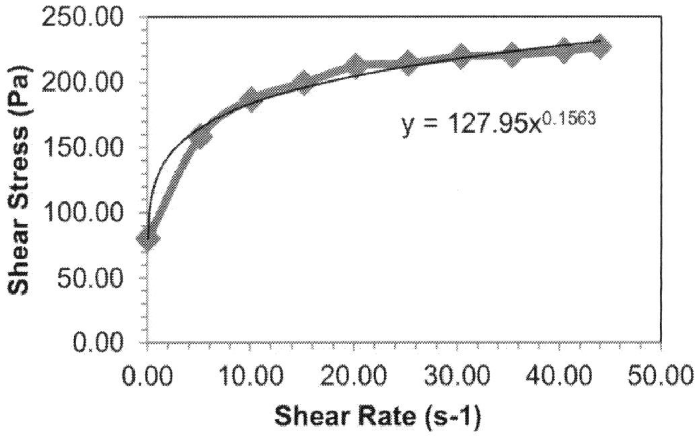

Figure 4: Shear stress vs. shear rate for 50/50 mixture of kaolin/water using Brookfield viscometer.

Figure 5 combines the results of shear stress versus shear rate for the Brookfield viscometer and the capillary tube rheometer using the power law model. The Brookfield viscometer is used for low shear rate regions and the capillary tube rheometer is used for high shear rate. The consistency coefficient was found to be 139.8 and the power law index was found to be 0.1028. These values differ from the results of the Brookfield viscometer, as expected.

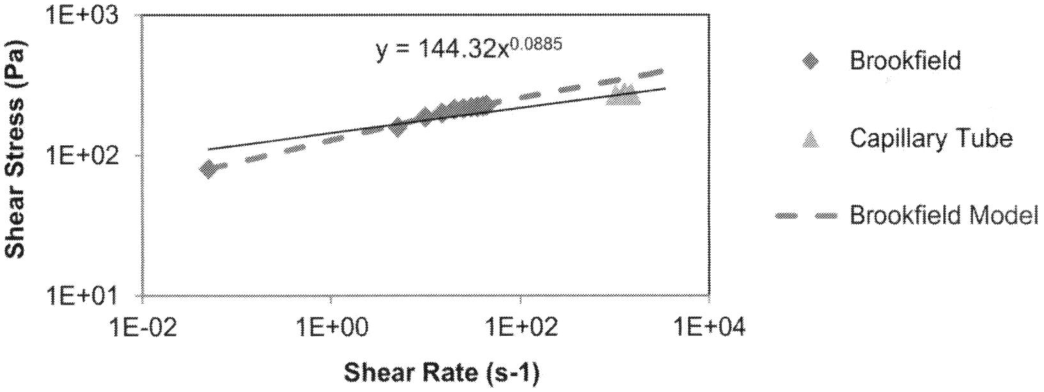

Figure 5: Compiled data for shear stress vs. shear rate for 50/50 mixture using Brookfield viscometer and capillary rheometer.

The yield stress for the mixture found from the Brookfield viscometer was about 78 Pa. This value is slightly lower than value found for the mixture using the capillary rheometer, which was found to be 92.35 Pa. The two values are reasonably close. It was not expected that the two values would be the same since the methods were different. However, the values are reasonable enough to allow for the method of yield stress determination in this study to be used further with some confidence.

Kaolin/Water Mixtures

Tests were completed for mixtures containing 50 to 60% total solids by weight of kaolin clay. Figure 6 shows the results when the solid portion of the mixture was comprised completely of the kaolin. There was an exponential increase in yield stress with an increasing amount of solids concentration. The data was fitted with an exponential equation with an R^2 value of 0.9965. The equation is shown in Equation (8). The data point for the 60 % solids was not included because the yield stress was higher than the maximum shear that can be measured using the rheometer, 783 Pa.

$$\tau_y = 9e-05 \exp[27.639 \cdot \Phi_k] \tag{8}$$

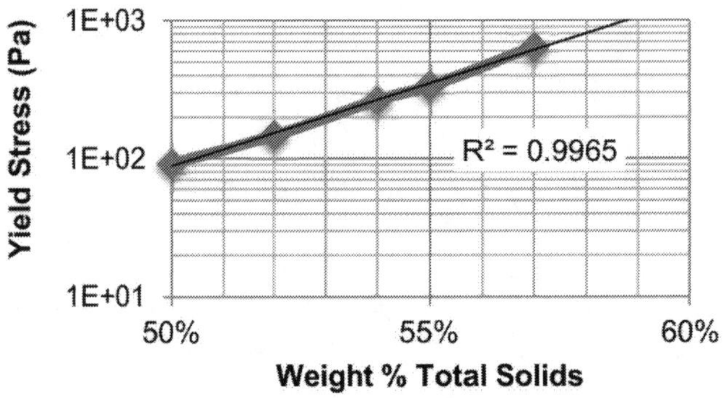

Figure 6: Yield stress for different wt % total solids of kaolin.

Some authors, including Burns et al. [8] , Gauglitz et al. [9], and Rassat et al. [10] have studied the shear strength of kaolin/water mixtures for different concentrations of kaolin. The results from previous authors and those from the current study are shown in Figure 7. The yield stress data and that of the shear strength data from the previous studies can be deemed analogous as they defined the shear strength as the static yield stress [11]. The static yield stress can be viewed as the property of a material that differentiates it from being a solid and liquid [12]. All of the studies used EPK kaolin clay. Comparing the sets of data shows that the yield stress results vary greatly. However, results of the current study fall in between those of previous studies.

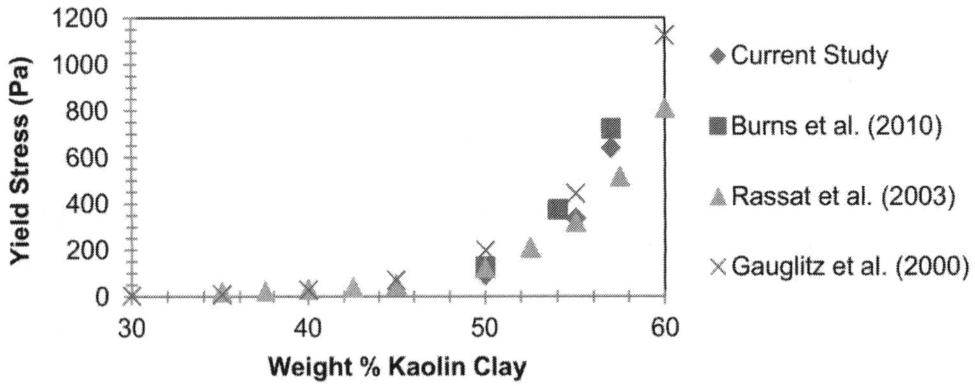

Figure 7: Comparison of yield stress results from current study and that of previous studies.

Kaolin/Sand/Water Mixtures

Tests were run for different concentrations of total solids, φ_T, in which the fraction of kaolin, φ_K, and fraction of sand, φ_S, was varied as well. Figure 8 shows the results for the kaolin/sand mixtures. As the kaolin concentration is increased, the yield stress is increased. As the sand concentration increased, the yield stress decreased. For each concentration of total solids, there is a limit of larger particles, or sand, which can be added before the mixture loses its yield strength completely. Such a limit can be deemed the flowability limit. It was expected that the yield strength of the fine particles would decrease as larger particles were added to the mixture. The interparticle force decreased as the larger particles were added because the spacing between the particles increased.

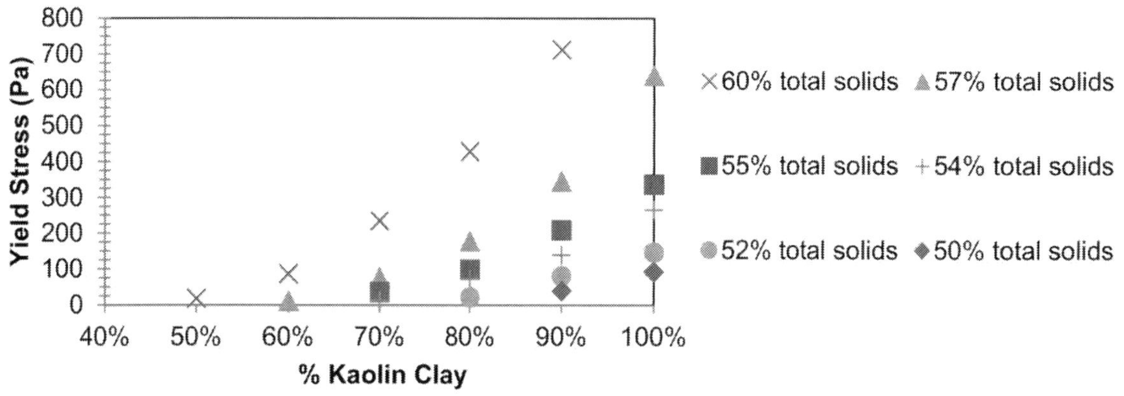

Figure 8: Yield stress of mixtures of 50 – 60 wt% total solids in which the solids were composed of different ratios of kaolin/sand.

Equations were formed to calculate yield stress as a function of total solids, with different percentages of kaolin and sand. Equation (9) displays the yield stress as a function of φ_T for a 90/10% mixture of kaolin/sand.

$$\tau_{y,90/10\%} = 2e-05 \exp\left[28.932 \cdot \Phi_T\right] \quad (9)$$

The equations for kaolin/sand mixtures of 80/20% and 70/30% are shown in equations (10) and (11), respectively.

$$\tau_{y,80/20\%} = 2e-07 \exp\left[37.75 \cdot \Phi_T\right] \quad (10)$$

$$\tau_{y,70/30\%} = 5e-10 \exp\left[44.836 \cdot \Phi_T\right] \quad (11)$$

The current study is mainly focused on providing a simple method of providing yield stress measurements of materials can be determined. Such a technique can be very useful as a means of initial, inexpensive measurement tool for materials in which little rheological information is known. The technique should mainly be used for materials that do not flow unless supplied with a pressure force. When setting up such a system, comparisons should be made with other rheometer data, or some calibration with a fluid of known viscosity.

Conclusion

The current study uses a simple, inexpensive technique as a means of initial measurement of yield stress of kaolin/water and kaolin/sand/water mixtures. Yield stress values were found for highly concentrated mixtures ranging from 50 to 60 wt % total solids. Results were able to be modeled using the constitutive relations of the power law model. Results for kaolin/water mixtures were compared to that of previous researchers and found to be reasonable. The yield stress results from the current study should not at all be used as means of an absolute quantity, though they can be used for comparison.

The technique used for the current study was found to be a sound method for determining yield stress. It can be used for initial measurements of yield stress. For practical purposes, the technique would be useful when trying to determine the force needed to move, dislodge, or suspend yield stress materials (i.e. suspension of sludge materials using jet mixers). Compared to other methods of calculating the yield stress, the method of the current study is much simpler, inexpensive, and less time consuming.

References

1. Powell, M.R., Onishi, Y., and Shekarriz, R., *Research on jet mixing of settled sludges in nuclear waste tanks at Hanford and other DOE sites: A historical perspective*. 1997, Pacific Northwest Laboratory: Richland, Washington. p. 90.
2. Nguyen, Q.D. and D.V. Boger, *Measuring the Flow Properties of Yield Stress Fluids*. Annual Review of Fluid Mechanics, 1992. **24**(1): p. 47-88.
3. Macosko, C.W., *Rheology : principles, measurements, and applications*. Advances in interfacial engineering series. 1994, New York [etc.]: Wiley-VCH.
4. Barnes, H., *The yield stress -- a review or "panta rei" -- everything flows?* Journal of Non-Newtonian Fluid Mechanics, 1999. **81**: p. 133-178.
5. Richardson, R.P.C.a.J.F., *Non Newtonian Flow in the Process Industries*. 1999, Woburn, MA: Butterworth-Heinemann.
6. TRASK, P.D., *EFFECT OF GRAIN SIZE ON STRENGTH OF MIXTURES OF CLAY, SAND, AND WATER*. Geological Society of America Bulletin, 1959. **70**(5): p. 569-580.
7. Wells, B.E., Enderlin, C.W., Gauglitz, P. A., and Peterson, R.A., *Assessment of Jet Erosion for Potential Post-Retrieval K-Basin Settled Sludge*. 2009, Pacific Northwest National Laboratory: Richland, WA.
8. Burns, C.A., P.A. Gauglitz, and R.L. Russell, *Shear Strength Correlations for Kaolin/Water Slurries: A Comparison of Recent Measurements with Historical Data*. 2010, Pacific Northwest National Laboratory: Richland, WA.
9. Gauglitz, P.A., M.M. Denn, and W.R. Rossen, *Mechanics of Bubbles in Sludges and Slurries*. 1999, Pacific Northwest National Laboratory: Richland, WA.
10. Rassat, S.D., et al., *Physical and Liquid Chemical Simulant Formulations for Transuranic Waste in Hanford Single-Shell Tanks*. 2003: Richland, Washington.
11. Poloski, A.P., et al., *Non-Newtonian Slurry Simulant Development and Selection for Pulse Jet Mixer Testing*. 2004: Battelle Memorial Institute, Pacific Northwest Division.
12. Xu, N. and C.S. O'Hern, *Measurements of the yield stress in frictionless granular systems*. Physical Review E, 2006. **73**(6): p. 061303.

(142bc) On the Persistence Length Dependence of Single Molecule Dynamics in Shear Flow

Amit Kushwaha, Chemical Engineering, Stanford University, Stanford, CA and **Eric Shaqfeh**, Mechanical and Chemical Engineering, Stanford University, Stanford, CA

The dynamics of single-molecules in shear flow has been an active area of research due to the advancements in the field of fluorescence microscopy, which provides a means to directly visualize polymer configurations. Several researchers have studied the dynamics of free and tethered polymers in shear flow using theory, experiments, and numerical simulations (Smith et al (1999), Hur et al. (2000), Schroeder et al (2005), Doyle et al. (2000)). Most of these studies were performed for flexible polymers with a large number of degrees of freedom—i.e. lambda-DNA with approximately 150 Kuhn lengths—and, all studies displayed complex dynamic conformational behavior. For example, the tumbling of a lambda-DNA in free shear flow shows rather complicated dynamics in which different parts of the DNA can undergo independent tumbling events simultaneously. On the other end of the spectrum, a stiff rod in free shear flow goes through a distinguishable tumbling motion, allowing one to precisely measure the mean tumbling times which can be directly compared to theory or simulations. Although considerable research has been devoted to understanding the dynamics of a flexible polymer in shear flow, rather few studies have considered how the shear dynamics of a polymer is affected as one changes the flexibility moving from the rod-like to the flexible limit.

Recently, Guihua et al. (2011) examined the extensional behavior of a series of tethered DNA chains with different lengths (17.3 μm, 7.8 μm, and 4.2 μm) under shear flow. Although even the smallest of these DNA chains lies in the flexible range with 40 Kuhn-lengths, the fractional extensional data for three chain lengths was found to depend intrinsically on the number of Kuhn steps as an independent parameter. More recently, Harasim et al. (personal communication) have visualized the dynamics of semiflexible actin filaments having persistence length 320 times that of lambda-DNA. They reported the tumbling times for different lengths of actin filaments with varying flow strengths, and also proposed an analytical model. In this talk, we use Brownian dynamics to simulate the experimental data from both Guihua et al. (2011) and Harasim et al., spanning the flexible to the stiff polymer regime. We use these simulations to examine the stochastic nature of tumbling dynamics and relate it to the mean fractional extension in free shear flow. In addition, we also study tethered single molecules in shear flow and analyze the effect of stiffness on the scaling behavior of this different dynamical system. We ultimately compare the physical mechanisms of tumbling and length fluctuations between the free and tethered polymers systems over the broad range of stiffness.

(142bd) Controllable Microfluidic Production of Gas-in-Oil-in-Water Emulsions for Hollow Microspheres with Thin Polymer Shells

Jianhong Xu[1], Ran Chen[2], Pengfei Dong[2], Yundong Wang[3] and Guangsheng Luo[2], (1)Department of Chemical Engineering, The State Key Lab of Chemical Engineering, Beijing, China, (2)Department of Chemical Engineering, The State Key Laboratory of Chemical Engineering, Beijing, China, (3)Department of Chemical Engineering, The State Key Laboratory of Chemical Engineering, Beijing, China

Controllable Microfluidic production of Gas-in-Oil-in-Water emulsions for hollow microspheres with thin polymer shells

Jian-Hong Xu,* Ran Chen, Peng-Fei Dong, Yun-Dong Wang, and Guang-Sheng Luo

(The State Key Laboratory of Chemical Engineering, Department of Chemical Engineering, Tsinghua University, Beijing, 100084, China)

* Corresponding author. Tel.: +861062783870; Fax: +861062770304.

E-mail address: xujianhong@tsinghua.edu.cn

Encapsulation of microbubbles can be very useful. It helps eliminate the bubble dissolution and coarsening by separating the inner gas compartment from the outer fluid, ensuring the various potential applications of microbubbles in lightweight materials, aerated food products, waste water treatment3, ultrasound contrast agents and pharmaceuticals. Moreover, such a method can also be applied in the preparation of hollow particles, which are of great interest in areas such as energy-storage materials, catalyst supports and drug delivery carriers. Traditional ultrasonic approach provides an effective way in the encapsulation of the microbubbles, but it lacks control over the size and size distribution of the particles and the shell thickness. Here we introduce a simple and novel capillary microfluidic device by injecting a smaller capillary into another capillary and making sure that the tapered orifices align where the single-step emulsification carries on (Fig. 1). This simple microfluidic device ensures the flowing independency of the inner and middle phases and finally eliminates the defects. In this communication, we mainly introduce the application of such microfluidic device in the controllable fabrication of hollow microspheres with thin shells using G/L/L emulsion as templates.

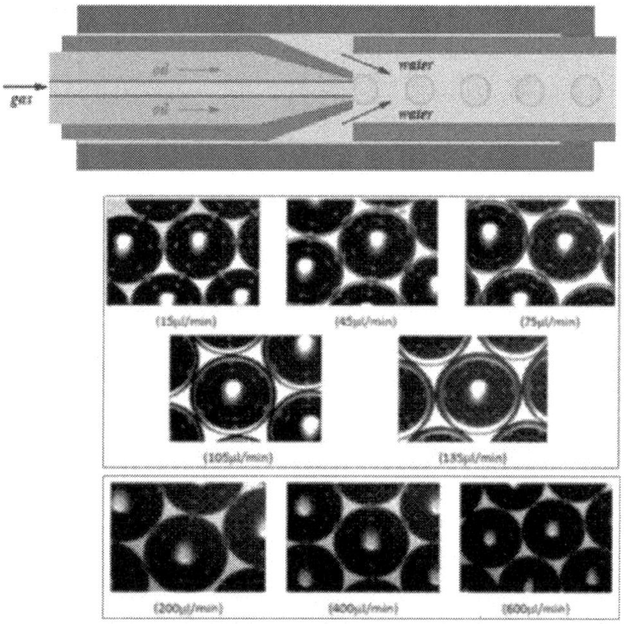

Fig.1 The sketch of the microfluidic capillary device and the microscope images of G/O/W droplets with different thickness of the oil layer, synthesized by tuning the flow rates of the middle phase.

We successfully produced Gas-in-Oil-in-Water double emulsions with high gas volume fraction using the simple and novel capillary microfluidic device. The size of the encapsulated microbubbles and the thickness of the oil layer can be easily controlled by adjusting the flow rates of oil and water phases. Such emulsions were used as the template for preparation of hollow core-shell microspheres with thin shells, as shown in Fig. 2.

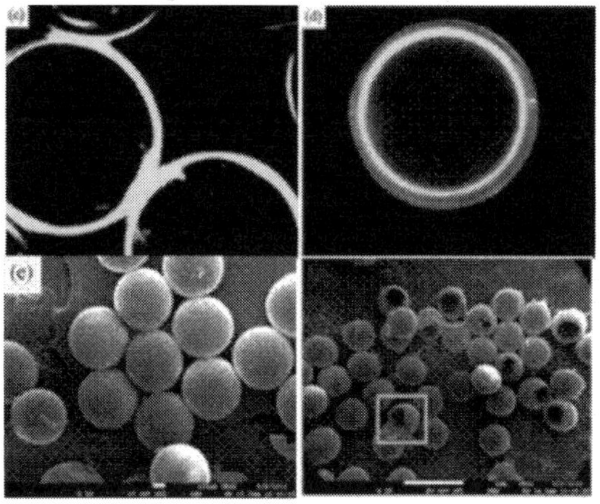

Fig. 2 The LSCM and SEM photograph of the hollow microspheres after polymerization.

The hollow core-shell microspheres may find many applications in areas such as industrial catalysis, pharmaceuticals and delivery of cosmetics and nutrients. We believe that the approach we have demonstrated here is promising because it is suitable for the preparation of G/L/L emulsions with controllable structures, and it is a purely flow-controlled method, which is very simple to adjust the size and structure of G/L/L emulsions and hollow microspheres. Additionally, our approach is a one-step technology to realize microbubble encapsulation and hollow microspheres preparation. In the future work, we will try to explore the applications of the G/L/L emulsions in the area of the enhancement of Liquid/Liquid extraction, gas-liquid-liquid mass transfer and heterogeneous reactions.

Key words: G/L/L emulsion; microfluidic; hollow microsphere

Acknowledgement

The authors gratefully acknowledge the supports of the National Natural Science Foundation of China (21036002, 21136006), A Foundation for the Author of National Excellent Doctoral Dissertation of PR China (FANEDD 201053) and National Basic Research Program of China (2012CBA01203).

(142be) Dynamics of Nanoparticle Dispersions in Model Confined Media

Kai He, Department of Chemical and Biomolecular Engineering , University of Houston, Houston, TX, Jacinta C. Conrad, Department of Chemical and Biomolecular Engineering, University of Houston, Houston, TX, Ramanan Krishnamoorti, Chemical & Biomolecular Engineering, University of Houston, Houston, TX and Scott Retterer, Cnms, Oak Ridge National Laboratory, Oak Ridge, TN

Understanding the dynamics of nanoparticle dispersions, especially in confined media, have significant technological and environmental implications. In this study, we investigated the dynamics of nanoparticle dispersions in micro-patterned channels that serve as a model for confined media using fluorescence differential dynamic microscopy (f-DDM). We first confirmed that the free-diffusion dynamics of dilute aqueous dispersions of fluorescently labeled polystyrene nanoparticles, ranging in diameter from 100 to 400 nm, measured using f-DDM in unconfined media are consistent with those obtained using dynamic light scattering. We then examined the dynamics of nanoparticles in micro-patterned channels, which contained rectangular arrays of cylindrical posts of diameter 500 nm separated by a spacing ranging from 400 nm to 10 microns. Using f-DDM we showed that the diffusion coefficients of the nanoparticles decrease as the particles are increasingly confined by the cylindrical posts and surprisingly exhibit a much stronger dependence on the post volume fraction than that expected by simple tortuosity arguments.

(142bf) The Role of Mucous Viscosity and Esophageal Strictures On Bolus Flow in Diseased Esophagi

Hedieh Saffari, Chemical Engineering, University of Utah, Salt Lake City, UT, Mike Hansen, Chemical Engineering , university of Utah, salt lake city, UT, Kathryn Peterson, Department of Internal Medicine, Division of Gastroenterology, University of Utah, Salt Lake City, UT, Gerald Gleich, Department of Dermatology, University of Utah, Salt Lake City, UT and Leonard F. Pease III, Chemical Engineering and Pharmaceutics & Pharmaceutical Chemistry, University of Utah, Salt Lake City, UT

Hydrodynamic flow through the gastrointestinal tract is significantly perturbed in diseased states. For example, in eosinophil esophagitis (EoE), an allergic condition of the esophagus putatively caused by invasion of white blood cells called eosinophils, the diameter of the esophagus narrows dramatically in a feature called a stricture and mucin genes are upregulated affecting the viscosity of the thin mucous layer lining the esophageal wall. No existing model of esophageal peristalsis accounts for these changes and their impact on wall stresses, which may affect patient sensation or pain.

Here we model the flow of a swallowed bolus in normal and diseased esophagi by solving the non-dimensionalized axisymmetric Navier-Stokes equations for both the bolus and mucous layers. We calculated wall stresses and mean flow rates, as well as their dependencies upon esophageal stricture length ratios and mucous-to-bolus viscosity ratios. We find that sharp strictures generate large increases in both normal and shear stresses in the esophagus wall, and that a more viscous mucous layer reduces these stresses. These results have significant implications for understanding EoE symptoms (e.g. patient pain) and the role of mucous in peristaltic transport throughout the entire GI tract.

(142bg) Drop Circulation and Liquid-Liquid Extraction in Hele Shaw Flow

Christine Cardinal Roberts[1], Rekha R. Rao[2], Lindsey Gloe[2] and Martin B. Nemer[2], (1)Thermal &Fluid Experimental Sciences, Sandia National Laboratories, Albuquerque, NM, (2)Sandia National Laboratories, Albuquerque, NM

Understanding interfacial mass transport on a droplet scale is essential for modeling liquid-liquid extraction processes. A model system is investigated whereby neodymium is extracted from buffered aqueous droplets to a continuous phase of dodecane, tributyl phosphate and thenoyltrifluoroacetone. A thin flow-focusing channel is used to generate monodisperse aqueous phase droplets of varying size, with drop diameter to channel height ratios ranging from 0.5 to 2. Liquid streamlines within the droplets are inferred by particle tracking, revealing a fountain flow recirculation pattern. Larger droplets that are more confined in the channel move with a slower velocity relative to the continuous fluid and have a stronger recirculation pattern in the plane of the microfluidic chip. Knowledge of the recirculation pattern is then applied to understand species mass transfer from the droplet fluid. Quantitative mass transfer measurements are obtained using high speed imagery and spectrophotometry for various flow rates and droplet sizes.

This research is supported by the Laboratory Directed Research and Development program at Sandia National Laboratories. Sandia National Laboratories is a multi-program laboratory operated by Sandia Corporation, a wholly owned subsidiary of Lockheed Martin Corporation, for the U.S. Department of Energy's National Nuclear Security Administration under contract DE-AC04-94AL85000.

(142bh) Modeling of Cryoprotectant Transport Across Cell Membranes in a Micromixer

Thomas Scherr[1], Shelby Pursley[2], William Monroe[2] and Krishnaswamy Nandakumar[1], (1)Cain Dept. of Chemical Engineering, Louisiana State University, Baton Rouge, LA, (2)Biological and Agricultural Engineering, Louisiana State University, Baton Rouge, LA

Cryopreservation of reproductive cells offers scientists a cost-effective alternative to maintaining a live strain of a genetically modified species. Prior to cryopreservation, it is desirable to remove intracellular water to prevent ice crystal formation. This is done by loading a cryoprotectant, such as glycerol, into the cell. Microfluidic devices present a useful platform for this process because of their high-throughput nature, reproducible results, and small requisite sample volumes. Despite these advantages, there have been very few studies investigating the use of a microfluidic device for such purposes.

In the present study, we numerically model the transport phenomena of glycerol across a zebrafish sperm cell membrane in a microchannel: namely, the cells and cryoprotectant moving throughout the micromixer, and the cryoprotectant transport across the sperm cell membrane. The former is modeled using the finite volume method (in the computational fluid dynamics solver FLUENT) along with the discrete particle method to solve the three-dimensional Navier-Stokes equations, species continuity equation, and Newton's Second Law. The latter is simulated in MATLAB using the previous simulations' data as inputs to the Kedem-Katchalsky equations for solute transport across a membrane. Our simulations idealize sperm cells as spherical particles, and we use the fluid mixture of glycerol and water. In experimental studies, data is usually collected at the end of the cryoprotectant loading process; the use of numerical methods allows us to simultaneously probe every cell as it moves throughout the micromixer.

Throughout the simulations, we focus on each cell's internal cryoprotectant concentration and the cell's size. As the magnitude of cryoprotectant influx is different from the magnitude of water efflux, the cell's size will change as a result (large enough changes being a cause of cell death during cryopreservation). We characterize the effect of inlet flow rate, ranging from 0.05 μL/min to 2 μL/min. Varying inlet flow rates will change the mixing performance of the device, which in turn affects the loading of cryoprotectant. The two processes occur on different time scales, so if cells do not have a sufficient residence time, they do not reach equilibrium with the micromixer environment. We also investigate the effect of inlet concentration over the range of 4.5 M glycerol to 9 M glycerol; more concentrated loading requires longer cellular residence times and cellular sizes show varying levels of the "shrink-swell" response as the inlet glycerol concentration changes. These results lay the foundation for an optimal microfluidic cryopreservation protocol.

(142bi) Analysis of Groove Micromixers: Shape Optimization Study

Mranal Jain, Tyler Nesmith and Krishnaswamy Nandakumar, Cain Dept. of Chemical Engineering, Louisiana State University, Baton Rouge, LA

Micromixing is a key step in realizing fast analysis time in many bio-chemical, biological and detection applications of Lab-on-a-chip (LOC) devices. The conventional T-mixer design requires longer channel lengths and times to achieve complete mixing owing to its dependence on transverse diffusion. The performance of a homogeneous T-mixer can be enhanced significantly by the stimulation of secondary/ transverse flows in the microchannel. The groove based micromixers (pressure or electrokinetic) generate helical flows within the microchannel to augment the mixing performance [1, 2]. These micromixers are extensively studied with respect to geometric parameters such as groove width, groove spacing, channel height etc. However, the effect of groove shape is not systematically studied with respect to mixing performance. Previous studies have focused on two or three different groove shapes, typically involving slanted grooves, asymmetric herringbone grooves and their variations. In this computational study, we analyze the effect of groove shape on micromixing performance and search for the optimal groove shape. The groove shape is parametrically represented by Bézier curves which could take any shape within a constrained plane. The control points of Bézier curve are chosen as optimization parameters to identify the optimal groove shape which maximizes the mixing for given operating conditions. The optimal groove shape problem is studied for both pressure driven and electrokinetically driven flows in microchannel. With the assumption of thin electric double layer (EDL), electrokinetic flow is modeled using Helmholtz-Smoluchowski slip condition within a three dimensional finite-element model. The resulting optimal design generates the most favorable transverse flow structure to provide optimal mixing performance. Various parametric studies are carried out to compare the optimal groove structure with other common groove types (staggered, herringbone etc.) micromixers.

1. Stroock AD, Dertinger SKW, Ajdari A, Mezic I, Stone HA, Whitesides GM. Chaotic mixer for microchannels. Science. 2002 JAN 25;295(5555):647-51.

2. Johnson TJ, Ross D, Locascio LE. Rapid microfluidic mixing. Anal Chem. 2002 JAN 1;74(1):45-51.

(142bj) CFD Simulation of Non-Isothermal Flows of Viscoelastic Fluids in 4:1 Contractions

Florian Habla, Alexander Woitalka, Andreas Obermeier and Olaf Hinrichsen, Catalysis Research Center and Chemistry Department, Technische Universität München, München, Germany

Introduction

Non-isothermal viscoelastic fluid flows can be found in a wide range of industrial applications such as injection molding, polymer blending and extrusion processing. It is of course of major importance to understand and predict such type of flows. Fortunately, numerical simulation evolved as an important tool to guide the engineer in the task of developing and improving such processes. However, there is still a gap between the engineer's needs and the current state-of-the-art, especially in complex flows such as is the case for non-isothermal viscoelastic fluids flows.

We therefore focus our research on developing a comprehensive CFD method to simulate such type of flows. In this work, focus is set on the thermo-rheological modeling and the numerical method. Furthermore, we apply the developed model to simulation of non-isothermal viscoelastic contraction flows.

Theory

In viscous fluids, mechanical energy is fully dissipated as heat, whereas in solids the mechanical energy is stored and can be released into mechanical energy again [1]. The behavior is known as pure entropy elasticity and energy elasticity, respectively. Viscoelastic fluids behave in between those of fluids and solids. As a result, viscoelastic fluids do show both of the aforementioned behaviors, which severely complicates the thermo-rheological modeling. One of the present approaches uses an a-priori defined split factor [2], which is being adopted in this work.

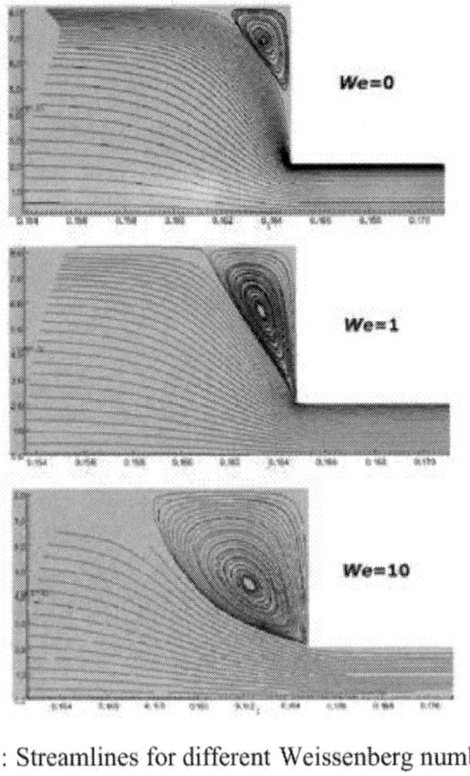

Fig. 1: Streamlines for different Weissenberg numbers in an axisymmetric contraction.

The Oldroyd-B model is used to model viscoelasticity [3]. Temperature dependence of the viscosity and the relaxation time is modeled with the Williams-Landel-Ferry model (WLF) [4].

The finite volume method is used for discretization. Special attention has to be paid to the convection scheme, which has a major influence on the accuracy and stability of the numerical method. We therefore evaluated several state-of-the-art differencing schemes in a stagnation point flow. The DEVSS technique is used to explicitly stabilize the solution method by introducing an elliptic operator into the momentum equation [5].

Results and discussion

Contraction flows are extensively studied in literature. Especially for viscoelastic fluid flows, this is an important test case, aiming at the stability of numerical algorithms due to the singularity at the reentrant corner. The failure in obtaining converged solutions at high Weissenberg numbers is probably the central problem in literature, which is known as the High-Weissenberg-Number-Problem (HWNP).

In this work, we were able to obtain converged solutions up to Weissenberg numbers of We = 10 for an axisymmetric contraction, see Fig. 1. As expected, the recirculating flow length increases with increasing Weissenberg numbers and the vortex area becoming more bulgy, see also Fig. 1. Initially the vortex is solely at the salient corner, but from We = 1 the vortex connects with the re-entrant corner.

When imposing a temperature jump at the walls in the area of the contraction, we find the vortex length to increase with increasing temperature jump, see Fig. 2. Herein, pure entropy elasticity was assumed.

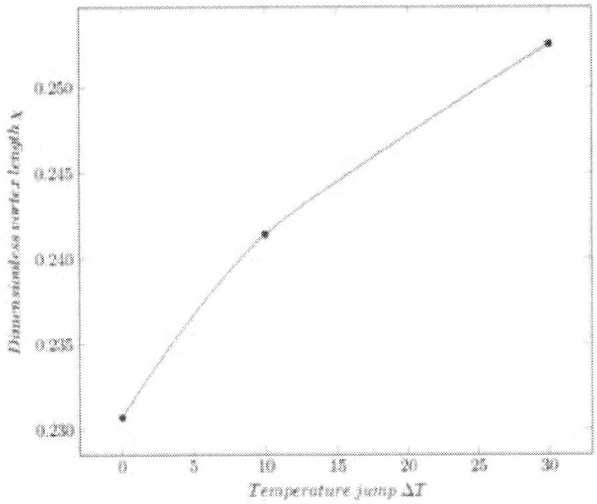

Fig. 2: Dimensionless vortex length as a function of the temperature jump at the contraction for We = 1.

Conclusions

In this work we developed a general finite volume model capable of simulating non-isothermal viscoelastic fluid flows. The model was used to simulate axisymmetric 4:1 contraction flows and results were obtained up to a Weissenberg number of We = 10. The recirculating flow length was found to increase both with increasing Weissenberg number and temperature jump.

References

[1] G. W. M. Peters, F. P. T. Baaijens, Modelling of non-isothermal viscoelastic flows, J. Non-Newtonian Fluid Mech. 68 (1997) 205-224.

[2] A. Wachs, J.-R. Clermont, Non-isothermal viscoelastic flow computations in an axisymmetric contraction at high

Weissenberg numbers by a finite volume method, J. Non-Newtonian Fluid Mech. 95 (2000) 147-184.

[3] R. B. Bird, O. Hassager, Dynamics of Polymeric Liquids, Vol. 1, second ed., Wiley Interscience, 1987.

[4] J. D. Ferry, Viscoelastic Properties of Polymers, Wiley, New York, 1980.

[5] R. Guénette, M. Fortin, A new mixed finite element method for computing viscoelastic flows, J. Non-Newtonian Fluid Mech. 60 (1995) 27-52.

(142bk) Particle Dynamics and Scaling in Inertial Microfluidic Flows

Amy Reece and John Oakey, Chemical & Petroleum Engineering, University of Wyoming, Laramie, WY

Microfluidic inertial focusing has enabled the development of miniaturized flow cytometers, size-selective sorting devices and other high-throughput particle screening tools by reliably positioning particles to well-defined lateral and longitudinal locations. The equilibrium positions of inertially focused particles has been extensively studied and the constitutive phenomena empirically characterized in straight channels of square and rectangular cross section. In order to design higher throughput devices with a broader range of applications, a better understanding of particle dynamics during focusing is required. This talk will focus on the elucidation of scaling relationships that determine relative equilibrium positions of various sized particles with different concentrations in channels of varying cross-sectional geometry, including non-traditional cross-sections. We have investigated the competing inertial forces that contribute to lateral focusing and determine the threshold at which Stokes' flow begins to influence lateral focusing. Experiments to quantify the dynamics of lateral particle migration have been used to quantify the shear and wall-induced inertial lift forces, which is used to verify our scaling and also inform alternate designs for optimal focusing. Finally, we demonstrate new high-throughput particle analysis applications that are enabled by an improved understanding of particle dynamics during inertial focusing.

(142bl) The Microdroplet Generation in Supercritical CO2 by a High Pressure Microfluidic Device

Kang Qin, Kai Wang, Tao Wang and Guangsheng Luo, State Key Lab of Chemical Engineering, Department of Chemical Engineering, Tsinghua University, Beijing, China

The microdroplet generation in supercritical CO_2 by a high pressure microfluidic device *

Kang Qin[1], Kai Wang[2], Tao Wang[3], Guangsheng Luo

The State Key Laboratory of Chemical Engineering, Department of Chemical Engineering, Tsinghua University, Beijing 100084, China

The biphasic system constituted with supercritical carbon dioxide ($ScCO_2$) and another liquid phase, has some unique characters in the chemical reaction and separation processes, as well as its strong ability to couple reaction and separation in one integrated process. To increase the mass transfer rate between the $ScCO_2$ phase and the liquid phase, the microdispersion processes of several $ScCO_2$/liquid systems are investigated in this study with a high pressure T-junction microfluidic device. The visible microfluidic device with an online high-speed camera imaging system can withstand a pressure of 25MPa and the generations of water and ionic liquid droplets in $ScCO_2$ are experimentally studied. The effects of pressure, temperature and dispersion method on the two-phase flow pattern and the dispersion size of droplets are discussed and a model equation is established for the microdispersion process in $ScCO_2$/liquid systems. Comparisons between droplet generation processes in $ScCO_2$, liquid CO_2 and gaseous CO_2 are also made for understanding the characters of microdispersed $ScCO_2$/liquid systems. The differences of microdispersion of unsaturated and saturated liquid with CO_2 were observed.

Key words: supercritical carbon dioxide; biphasic system; droplet generation; microfluidic; T-junction microchannel

---

* Supported by National Science Foundation of China (21076111)

1 Presenter, qin-k10@mails.tsinghua.edu.cn

2 Corresponding author, kaiwang@tsinghua.edu.cn

3 Corresponding author, taowang@tsinghua.edu.cn

(142bm) Hydrogel-Filled Capillaries As in-Plane Barriers in Microfluidic Devices

Andrea Kadilak[1], Amanda Card[1], Thomas Kelly[1] and Leslie M. Shor[2,3], (1)Department of Chemical, Materials, and Biomolecular Engineering, University of Connecticut, Storrs, CT, (2)Department of Chemical, Materials and Biomolecular Engineering, University of Connecticut, Storrs, CT, (3)Center for Environmental Sciences and Engineering, University of Connecticut, Storrs, CT

Hydrogels are highly cross-linked networks of hydrophilic polymers able to absorb large quantities of water, but still retain their structure at highly hydrated states. Biopolymer hydrogels including alginate, agar, and agarose are widely used in biomedical and molecular biology applications for biological separations and as a matrix for cell support. Hydrogels have also been incorporated into microfluidic devices to develop time-dependent chemical gradients for cellular behavior studies. This research introduces a novel configuration of agarose-filled capillaries within a microfluidic device that can be used for two purposes: (i) to establish stable, reproducible, in-plane chemical gradients for diverse microfluidic applications; and (ii) to create in-plane migration barriers for cells such as microorganisms. We report protocols for preparing devices with gel confined to capillary barriers. A high-throughput optical method was used to quantify diffusivity of dyes and antibiotics through hydrogels and optical microscopy was used to characterize bacterial and protozoan motility through agarose-filled capillaries.

(142bn) A Peg-DA Microfluidic Device for the Study of Cellular Chemotaxis

Mahama A. Traore and Bahareh Behkam, Mechanical Engineering, Virginia Tech, Blacksburg, VA

A PEG-DA MICROFLUIDIC DEVICE FOR THE STUDY OF

CELLULAR CHEMOTAXIS

Mahama A. Traore[1], Bahareh Behkam[1, 2]

1. Mechanical Engineering Department, Virginia Tech, Blacksburg, VA
2. School of Biomedical Engineering and Sciences, Virginia Tech, Blacksburg, VA

Corresponding Author: Bahareh Behkam, Email: Behkam@vt.edu

<>INTRODUCTION

Microfluidic devices have been extensively used for the study of both prokaryotic and eukaryotic cells under well-defined chemical and mechanical conditions. In many cases, the establishment of steady-state (temporally invariant) spatially varying chemical gradients within which cells can be studied over extended periods of time and in absence of any fluid flow is of outmost importance.

In this work, we have microfabricated a microfluidic device made of the biocompatible and photopolymerizable polyethylene glycol diacrylate (PEG-DA). We have demonstrated that PEG-DA of different molecular weight can be used in order to achieve customizable chemical gradients. This microfluidic device is composed of three channels fabricated using photolithography methods and operates under the principle of diffusion as shown in Fig 1.

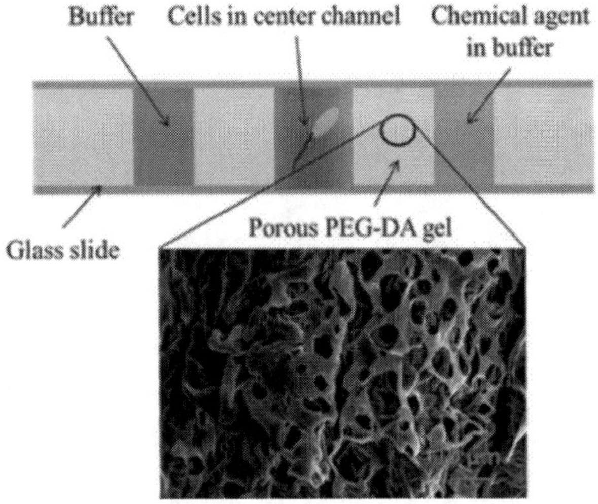

Figure 1. Schematic showing a cross-section of microdevice and SEM image of a 2000 Da molecular weight porous gel

A solution of a chemical agent of choice (e.g. casamino acids) in a buffer (e.g. PBS) is flowed in one of the outer channels while just the buffer solution is flowed in the other outer channel. Cells (i.e. subject of study) reside in the buffer solution in the center channel. The chemical agent diffuses through the porous gel wall to the center channel,

generating a well-defined spatially varying concentration field. The gel wall thickness and porosity can be designed to accommodate for the required chemical diffusion rate. COMSOL multiphysics simulations, diffusion coefficient measurements by optical spectrophotometry and diffusion characterization using fluorescence microscopy were employed to develop custom-devices with desired gel wall characteristics to determine the threshold chemical gradient at which *Escherichia coli* (*E. coli*) strain RP437 begin to exhibit a chemotactic behavior towards casamino acids. <>
<>METHODOLOGY

Fabrication of microfluidic device

The microfluidic device is fabricated using a two-step photolithography process (Fig 2). Briefly, the glass surface was first rendered hydrophilic using oxygen plasma for duration of 5 minutes and then functionalized using a TPM (3-(Trichlorosilyl)propyl methacrylate) solution. PEG-DA gels of a desired molecular weight (700 Da, 2000 Da, 6000 Da or 10000 Da) were exposed to UV light (EFOS Ultracure, UV spot lamp, Mississauga, ON, Canada) at a wavelength of 365nm for 20 s through a photomask with the chosen pattern.

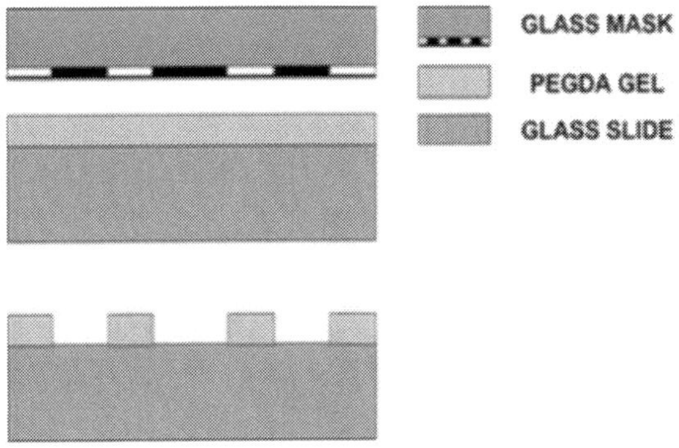

Figure 2. Process flow fabrication of patterned PEG-DA gel in microfluidic device

Figure 3. Image of a 700 Da MW PEG-DA microfluidic device. For demonstration purposes, red and green fluids are flown in the two outer channels.

The fabricated gel pattern is assembled with the other layers of the device; Figure 3 shows an assembled microfluidic chemotaxis assay device.

Diffusion properties characterization

The diffusion coefficients of the desired chemical agent through the fabricated gels were determined using a Franz diffusion cell[1] as seen in Fig. 4. Known concentration of a chemical agent in buffer present in the donor compartment diffuses over time across the porous PEG-DA membrane and into the acceptor chamber. Samples from the acceptor compartment are collected over time and their absorbance was measured using a UV spectrophotometer. For chemical agents that cannot be detected by spectrophotometry, the chemical gradient in the center channel of the microfluidic device was quantified using a different method[2]. A fluorescent dye solution of known concentration with molecular weight close to the molecular weight of the chemical agent of interest was flown in the outer channel and diffused through the PEG-DA porous walls to the center channel. Fluorescence microscopy images were used to find the chemical concentration profiles within the center channel and thus the diffusion coefficient values. A picture of the fluorescence profile within the center channel can be seen in Fig 4.

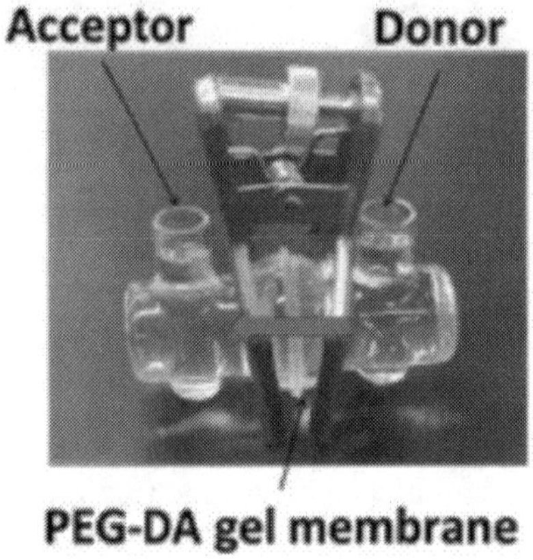

Figure 4. Franz diffusion cell used for diffusion coefficient measurements.

Figure 5. Fluorescent gradient profile across microfluidic channel using a 700 Da molecular weight gel

RESULTS The diffusion coefficient of casamino acid through PEG-DA gels of four different molecular weights (700 Da, 2000 Da, 6000 Da, and 10000 Da) were characterized using a Franz diffusion cell as described above. The diffusion coefficient of the gels was calculated using the mean first passage time measurements[1] method; the results are summarized in Table 1.

Table 1. Diffusion coefficient of casamino acid through PEG-DA gel as a function of the gel molecular weight.

PEG-DA gel (Da)	700	2000	6000	10000
Diffusion coefficient ($10^{-6} cm^2/s$)	1.5±0.5	2.66±0.53	3.61±0.9	5.19±0.98

COMSOL simulations were run to determine the chemical gradient profile in the center channel where the cells will be inserted. The chemical gradient profile in the center channel depends on the dimensions of the device, the molecular weight of the gel used, and the initial chemical concentration in the source channel. In these simulations, all other parameters were kept the same except the molecular weight of the PEG-DA gel used to fabricate the device.

The initial chemical concentration of the chemical diffusing from the source channel was 3.84×10^{-4} M, and the chemical concentration in the buffer channel was maintained at zero. The chemical concentration gradient in the center channel after 1 hour is shown in Fig. 5. As expected the more porous gels have a higher chemical gradient slope. This can be attributed to the fact that the chemical in the source channel diffuses at a faster rate to the center channel, thus creating a steeper linear gradient profile.

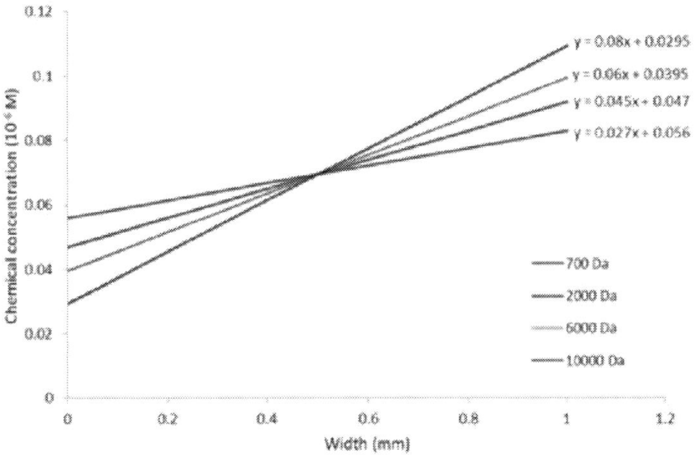

Figure 6. Chemical concentration across center channel of microfluidic device

The chemical gradient values obtained above were verified experimentally using a fluorescein solution of identical initial concentration in the source channel. Fluorescent images of the center channel of the device were taken and analyzed using Image J (free software provided by the NIH). Fluorescein gradients within the center channel were then determined and compared within 20% of the chemical concentration gradients obtained through simulations. Table 2 demonstrates the results from the simulations and the experiments. The differences between the experimental and theoretical values can be attributed to the fact that the molecular weight of fluorescein (332 Da) used in the fluorescent quantification of the diffusion coefficient is different from the molecular weight of the casamino acids (250 Da).

Table 2. Comparison between experimental and simulation chemical gradients in the center channel.

Fluorescein gradient in center channel (10^{-4}M/mm)				
	700 Da	2000 Da	6000 Da	10000 Da
Simulations	0.027	0.045	0.060	0.080
Experiments	0.024	0.051	0.071	0.097
Difference (%)	11.1	13.3	18.3	21.2

A chemotaxis sensitivity assay was conducted using *E. coli* bacteria strain RP437 (fluorescent bacteria) towards casamino acids. The molecular weight of the gel used in this experiment was 700Da. The lowest molecular weight gel was chosen for this experiment because of the low chemical gradient it generates in the center channel. *E. coli* bacteria positively responded to the chemical attractant present in the source channel. Fig 6. shows the *E. coli* bacteria response towards casamino acids in the center channel.

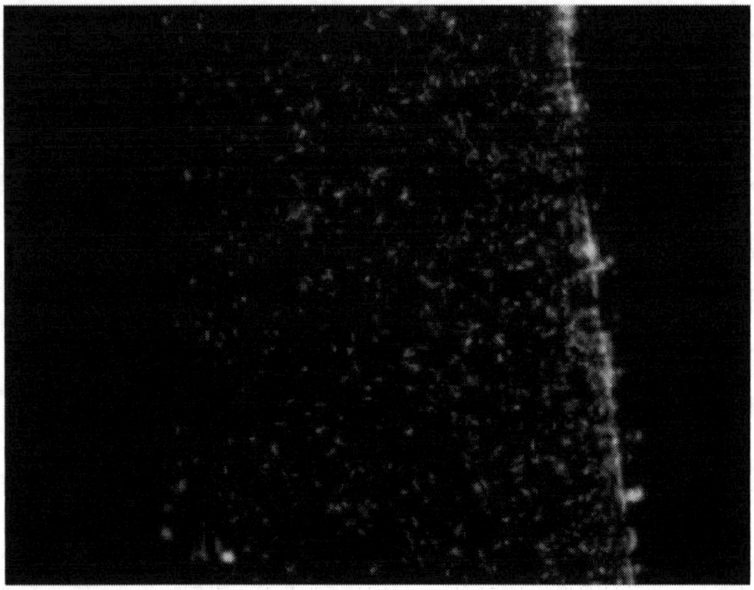

Figure 7. Response of *E. coli* bacteria strain RP437 in center channel to a gradient of casamino acids.

DISCUSSION

The fabrication of a microfluidic device with different gradient profiles can be achieved using different molecular weight gels. Bacterial chemotaxis studies usually require smaller chemical gradients to be established in the center channel. The use of the 700 Da molecular weight PEGDA gel can accommodate for the chemotaxis study of bacteria. The ability to control the chemical gradient in the center channel can help determine the threshold chemical gradient needed to provoke the chemotactic behavior of a strain of bacteria as well as the saturation chemical concentration.

CONCLUSIONS

In this work, multiple microfluidic devices were fabricated using a photopolymerizable hydrogel PEG-DA at different molecular weights. Different gel structures lead to significantly different diffusion coefficients and therefore distinctly different chemical gradient field within the main channel. The capability of realizing well-defined and customizable steady-state chemical gradient fields in absence of fluid flow opens many opportunities for the study of various behaviors of bacteria and mammalian cells within a controlled environment. REFERENCES

1. Wenquan Liang *et al.* Journal Chem Phys, 2006, 125, 0447071-0447077.

2. Mingming Wu *et al.* Lab on a Chip, 2007, 7, 763-769.

(142bo) Diamagnetic Particle Trapping in Ferromicrofluidics

James Wilbanks[1], Garrett Kiessling[1], Zeng Jian[1], Chen Chen[2] and Xiangchun Xuan[1], (1)Mechanical Engineering, Clemson University, Clemson, SC, (2)Clemson University, Clemson, SC

Concentrating particles to a detectable level is often necessary and critical in many areas such as environment monitoring, food safety, and water quality control. We present in this talk a magnetic method for trapping and concentrating particles in a continuous ferrofluid flow through a straight rectangular microchannel. A pair of attracting permanent magnets is embedded into the PDMS slab on the two sides of the microchannel to produce strong magnetic field gradients in the flow direction. As they are less magnetizable than the ferrofluid, diamagnetic particles are repelled away from the high magnetic field region and trapped in front of the magnets where the flow velocity is counterbalanced by the magnetophoretic particle velocity. We experimentally study how the effects of flow rate, location of each magnet, and particle size on the particle trapping performance by directly visualizing the particle behavior. We also develop an analytical model to simulate the particle transport and trapping in ferrofluid flows under the experimental conditions.

(142bp) Migration Due to Deformation, Buoyancy and Inertia in Confined Shear Flows

David T. Leighton Jr., University of Notre Dame, Notre Dame, IN and **Seemit Praharaj**, Chemical and Biomolecular Engineering, University of Notre Dame, Notre Dame, IN

It has long been known that suspensions of deformable particles (e.g. blood) will marginate and segregate in shear flows [1]. For example, normal flexible red blood cells will marginate towards the center of a capillary [2] or arteriole displacing stiffer leukocytes or malarially infected red blood cells towards the walls. Such deformation induced migration has both physiological consequences and has been used as the basis for microfluidic separations in the laboratory. Despite this, quantitative experimental study of fundamental properties such as the rate of migration of cells in confined geometries is lacking.

In this study we focus on a very simple problem: particle migration due to buoyancy, deformation and inertia in simple shear flow such as that which occurs in the parallel plate geometry. Through a balance between the sedimentation velocity and the migration mechanisms the equilibrium position of particles at infinite dilution can be calculated using the models of Chan & Leal [3] and Ho & Leal [4]. Thus, measurement of both the sedimentation velocity and the equilibrium position of the particles yields information about the migration process.

In our work we have shown that the equilibrium position of 10μm diameter polystyrene spheres in a 100μm gap is solely function of Re^2_p/Re_s, where $Re_p = \gamma a^2/\nu$ is the particle Reynolds number, and $Re_s = U_s a/\nu$ the sedimentation Reynolds number. The observed positions closely correspond to that predicted from the model of Ho & Leal [4] for inertial migration with no adjustable parameters. Red blood cells (unlike rigid polystyrene spheres) experience migration due to both deformation and inertia. The ratio of the two drift mechanisms is characterized by the ratio, $k_d[(6\pi\mu^2_0 a)/(\sigma h^2 \rho_0)]F(z^*)/G(z^*)$, where μ_0 is the viscosity, a is the particle radius, σ is the effective surface tension, h is the gap width, ρ_0 is the fluid density, $F(z^*)$ and $G(z^*)$ are dimensionless functions of z^* and k_d is an O(1) function of the viscosity ratio. For red blood cells in water for this gap size the two migration mechanisms are of comparable magnitude. In a more viscous suspending medium, however, migration is dominated by deformability allowing independent investigation of the two migration mechanisms. In particular, the equilibrium position of the cells relative to gravity provides a direct measure of the particle stresslet - the particle contribution to the bulk stress of a dilute suspension of deformable particles that is not readily measurable in other ways. In our experiments we have studied the behavior of heparinized human blood in isotonic sucrose solutions, with 10% dextran added to increase the viscosity to 0.07 P. These experiments showed $(N_1-N_2)/\gamma\mu\varphi$ increased from 0.04 to 0.25 as γμ increased from 5 to 30 dynes/cm². While an erythrocyte is not a drop, the equilibrium positions in the gap were found to reasonably coincide with that predicted for drops by the model of Chan & Leal [3] with an effective surface tension of 0.47 dynes/cm. Further experiments have shown that a careful balance of buoyancy, inertia, and deformation may allow the use of the parallel plate geometry to both probe migration properties and result in an effective separation mechanism.

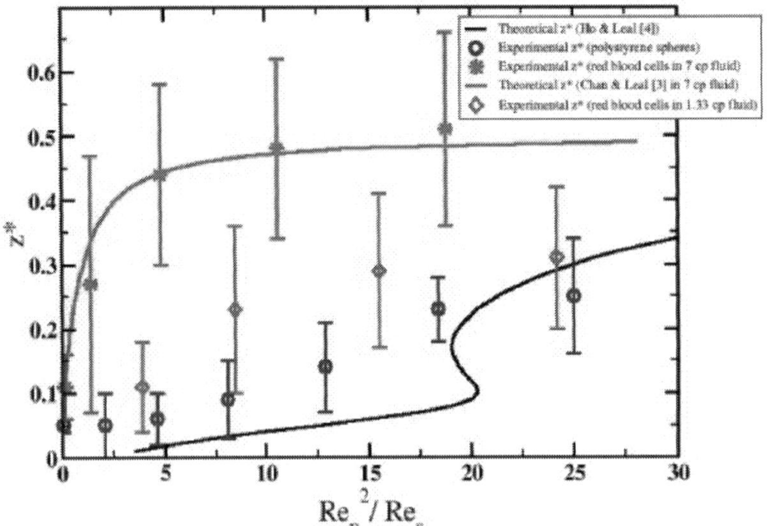

Figure 1: Variation in the vertical equilibrium position as a function of Re_p^2/Re_s for O polystyrene spheres suspended in water, ◊ red blood cells suspended in isotonic sucrose solution (1.33 cP) and * red blood cells suspended in viscous isotonic sucrose solution (7 cP). Theoretical curves are obtained by balancing

buoyancy with inertia (Ho & Leal [4]) or deformation induced lift (Chan & Leal [3]) with an effective surface tension of 0.47 dynes/cm.

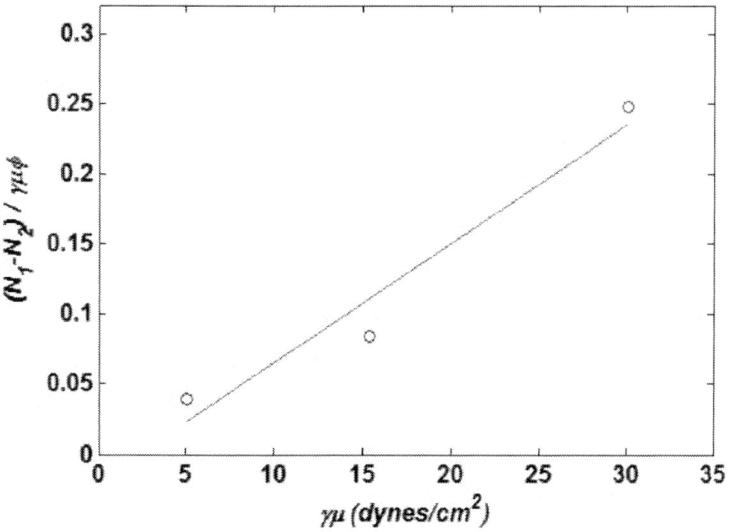

Figure 2: Variation of dimensionless particle stresslet with γμ (dynes/cm²)

References

[1] R. Fahraeus and T. Lindqvist. The viscosity of blood in narrow capillary tubes. Am. J. Physiol., 96:562-568, 1931.

[2] P. Olla. Simplified model for red cell dynamics in small blood vessels. Phys. Rev. Lett., 82(2):453-456, 1999.

[3] P.C. Chan and L.G. Leal. The motion of a deformable drop in a second-order fluid, J. Fluid Mech., 92(1):131-170, 1979.

[4] B. P. Ho and L. G. Leal. Inertial migration of rigid spheres in two-dimensional unidirectional flows. J. Fluid Mech., 65(2):365-400, 1974.

See more of this Session: Fluid Mechanics Poster Session

See more of this Group/Topical: Engineering Sciences and Fundamentals

(142bq) Hydrodynamics and Dry Pressure Drop in Structured Packing

Anup Sunkarwar, Chemical engineering, Indian institude of technology roorkee, roorkee, India and **Dr Vimal Kumar**, Chemical engineering, IIT Roorkee, Roorkee, India

Hydrodynamics and Dry Pressure Drop in Structured Packing

Anup Sunkarwar, Vimal Kumar

Department of Chemical Engineering, Indian Institute of Technology Roorkee

Email: anup.14023@gmail.com, vksinfch@iitr.ernet.in

In recent years supercritical fluids have been used in the extraction of natural raw materials in food, pharmaceutical, beverages industries, etc. (Krukonis, 1994; Gamse, 2005). The liquid phase supercritical fluid extraction (SFE) has been carried out in packed columns with structured packing due to their relatively high surface area, free volume, high mass transfer efficiency and high column capacity. As compared to the unstructured packing, structured packing has high cost and low capacities at high flow rates. However, with an accurate design columns with structured packing can be design with low cost and high separation efficiency.

It is difficult to assess the efficiency and influence of shape and geometry in structured packing at severe conditions (at high temperature and pressure) during the process. Computational fluid dynamics (CFD) tools can be very helpful in predicting hydrodynamic behavior and pressure drop under severe conditions also reduces number of experiments to predict the efficiency of the process. An extensive literature is available where hydrodynamics and pressure drop is studied using computational fluid dynamic (Hodson et al., 1997; Von Scala et al., 1999; Petre et al., 2003a and 2003b). Fernandes et al. (2008) computationally estimated the dry pressure drop in the Sulzer EX structured gauze packing considering two approaches: the simpler model with two corrugated packing sheets with periodic boundary conditions and the complex model with thirteen packing sheets representing one-third of the actual packing element, tightly inserted into a cylindrical tube. They compared the pressure drop results with experimental data for supercritical carbon dioxide of Meyer (1998), Olano (1995) and Stockfleth and Brunner (1999). They further extended the dry pressure drop work to wet pressure drop calculations using computational fluid dynamic tool in a whole Sulzer EX packing element (Fernandes et al., 2009).

In the present work the hydrodynamics and dry pressure drop numerically simulated in structured packing (Flexipac 1Y, Sulz packing) using commercial computational fluid dynamic software, FLUENT 6.3, ANSYS. Also the hydrodynamics and pressure drop is studied in a newly proposed zigzag structured design considering supercritical fluid extraction using CO_2. The zigzag pattern resulted into centrifugal forces and hence better interaction between the fluid elements, which may increase the efficiency of the extraction process.

It is observed that the present simulated predictions using realizable k-ε model considering pressure gradient effects is in good agreement with the results reported in the literature. The prediction were analysed in terms of resistance factor (Ψ) versus Reynolds number for the comparison of different models and literature results. The realizable k-ε model considering pressure gradient effects model is used to study the pressure drop inside the newly developed geometry. The new geometry with the same computational domain of two sheet packing, but with the changed geometric parameter with the zigzag orientation is developed. The effect of the geometric packing parameter variation on the dry pressure drop friction term inside the newly developed structured packing is studied. It has been observed that the zigzag pattern structured packing resulted into small pressure drop as compared to the Sulz EX packing. Also the wall effects of the column with the zigzag patterned geometry are reduced, which is one of the reasons that the pressure drop is less as compared to the other structured packings. The flow is developed only in the inner channels for the sulz packing while it is developed in the entire computational domain of the zigzag packing leading to the less pressure drop as compared to the sulz packing. Fluid element interaction between different fluid elements is more in the zigzag packing as compared to sulz EX packing due to the zigzag orientation where the flow experiences maximum directional changes. Therefore the zigzag patterned packing have less pressure drop and better mixing performance. Further work is proposed for the development of entire column and to carry out the simulations for CO_2absorption process.

References

Fernandes, J., Lisboa, P.F., Simoes, P.C., Mota, J.P.B., Saatdjian, E., 2009. Application of CFD in the study of supercritical fluid extraction with structured packing: dry pressure drop calculations, J. Supercrit. Fluids 50, 61–68.

Fernandes, J., Simoes, P.C., Mota, J.P.B., Saatdjian, E., 2008. Application of CFD in the study of supercritical fluid extraction with structured packing: dry pressure drop calculations, J. Supercrit. Fluids, 47, 17–24.

Gamse, T., Industrial applications and current trends in supercritical fluid technologies,

Hemijska Industrija 59 (9/10) (2005) 207–212.

Hodson, J.S., Fletcher, J.R., Porter, K.E., 1997. Fluid mechanical studies of structured distillation packings, in: Institution of Chemical Engineers Symposium Series, AIChE, New York, 142, 999–1007.

KatapakM®catalyst supports, in: International Symposium on multifunctional reactors, Chem. Eng. Sci. 54(10), 1375–1381.

Krukonis, V., G. Brunner, G., Perrut, M., 1994. Industrial operations with supercritical fluids: current process and perspectives for the future, in: G. Brunner, M. Perrut, (Eds.), Institut National Polytechnique de Lorraine, Proceedings of the 3rd International Symposium on Supercritical Fluids, Tome 1, 1994.

Lemmon, E.W., McLinden, M.O., Friend, D.G., 2005. Thermophysical properties of fluid systems, in: P.J. Linstrom, W.G. Mallard (Eds.), NIST Chemistry WebBook, NIST Standard Reference Database Number 69, National Institute of Standards and Technology, Gaithersburg, MD, 2005, http://webbook.nist.gov.

Meyer, J.-T., 1998. Druckverlust und Flutpunkte im Hochdruckgegenstrom-kolonnen

mitüberkritischen Kohlendioxid, Ph.D. Dissertation, Technische Universität Hamburg-Harburg, Germany.

Olano, S., Nagura, S., Kosuge, H., Asano, K., 1995. Mass transfer in binary and ternary distillation by a packed column with structured packing, J. Chem. Eng. Japan, 28.

Petre, C.F., Larachi, F., Iliuta, I. Grandjean, B.P.A., 2003a. Pressure drop through structured

Petre, C.F., Larachi, F., Iliuta, I. Grandjean, B.P.A., 2003b. Tailoring the pressure drop of structured packings through CFD simulations, Chem. Eng. Process. 42, 535–541.

Stockfleth, R. Brunner, G., 1999. Hydrodynamics of a packed countercurrent column for the gas extraction, Ind. Eng. Chem. 38, 4000–4006.

Stockfleth, R., Brunner, G., 2001. Holdup, pressure drop, and flooding in packed countercurrent columns for the gas extraction, Ind. Eng. Chem. 40, 347–356.

vonScala, C., Wehrli, M., Gaiser, G., 1999. Heat transfer measurements and simulation of packings: breakdown into the contributing mechanisms by CFD modeling, Chem. Eng. Sci. 58, 163–177.

(142br) Changes in Fiber Shape and Rheology of Cellulosic Suspensions During Enzymatic Hydrolysis

Emilio J. Tozzi[1], David M. Lavenson[2], Nardrapee Karuna[3], Tina Jeoh[3], Michael J. McCarthy[4] and Robert L. Powell[5], (1)Department of Chemical Engineering and Materials Science, University of California, Davis, Davis, CA, (2)Chemical Engineering and Materials Science, University of California Davis, Davis, CA, (3)Biological and Agricultural Engineering, University of California, Davis, Davis, CA, (4)Biological and Agricultural Engineering, University of Calfiornia Davis, Davis, CA, (5)Chemical Engineering and Material Science, University of California-Davis, Davis, CA

The initial stage of cellulose hydrolysis, or liquefaction, is of particular importance in bioprocesses that use microorganisms to produce chemicals or fuels. A rapid liquefaction stage is desirable to reduce mixing energy input and reactor volume. During this process large changes in rheology occur that are relevant for economical scale up of the process. Slurry rheology depends not only on solids concentration but also on fiber shape, both of which change during hydrolysis. The evolution of the particle length-to-width ratio is believed to be important, but has not been extensively characterized in the literature. To probe the effect of concentration and fiber shape on rheology during hydrolysis we combine fiber length and width measurements with rapid rheology measurements based on magnetic resonance imaging. Two fiber types were used that differed in the initial average length. For both fiber types rapid changes in rheology were observed during the first ten minutes of hydrolysis. The yield stresses measured displayed good correlation with crowding numbers.

(142bs) Direct Numerical Simulation of Elastic Turbulence in Taylor-Couette Flow of Dilute Polymeric Solutions

Nansheng Liu, Materials Research and Innovation Laboratory, Department of Chemical and Biomolecular Engineering, University of Tennessee, Knoxville, Knoxville, TN; Department of Modern Mechanics, University of Science and Technology of China, Hefei, China and B. Khomami, Chemical and Biomolecular engineering, Sustainble Energy and Education Research Center (SEERC), The University of Tennessee, Knoxville, TN

Direct Numerical Simulation (DNS) of elastic turbulence has posed tremendous challenges to researchers engaged in developing first principles models and simulations that can accurately and robustly predict complex spatio-temporal behavior of viscoelastic flows. To this end, we have focused our attention in performing DNS of elastic turbulence in a prototypical curvilinear flow, namely, the Taylor-Couette (TC) flow. Our computations with a prototypical constitutive equation for dilute polymeric solutions, namely, the FENE-P model are capable of reproducing the essential features of the experimentally observed elastic turbulence in TC flow of this class of fluids, i.e., randomly fluctuating fluid motion excited in a broad range of spatial and temporal scales. Moreover, the experimentally measured Power Spectral Density (PSD) of the radial velocity fluctuations with two continuous regions of power-law decay, namely, -1.1 and -2.2 at low and high frequencies is accurately captured. In addition, careful examination of the probability density functions (PDFs) of the velocity and polymer stress fluctuations show that highly localized "turbulent-like" flow structures in the wall region are mainly responsible for the stochastic characteristics of the flow.

(142bu) Correlation Between Radial Filtration and Normal Stress Distribution in Highly Concentrated Suspensions Undergoing Constant-Force Squeeze Flow

Mohsen Nikkhoo[1], Allyson Hofman[2] and Francis A. Gadala-Maria[1], (1)Department of Chemical Engineering, University of South Carolina, Columbia, SC, (2)Calvin College, Grand Rapids, MI

Drastic changes in the normal stress distribution in highly concentrated suspensions of non-colloidal spheres in highly-viscous Newtonian fluids undergoing constant-force squeeze flow are explained in terms of the migration of the liquid phase away from the center of the sample. The normal stress is measured using pressure sensitive films, and the solid volume fraction is measured by taking samples from different locations in the sample, dissolving the liquid, and then drying and weighing the solids. Increases in the solid volume fraction near the center cause the normal stress to increase by an order of magnitude in that region and to sharply decrease beyond that.

Changes in the volume fraction of solids due the liquid phase filtration depend on the viscosity of the suspending fluid, the size of the particles, and the initial volume fraction of solids. Under some conditions, the volume fraction of solids remains essentially constant during the squeeze test indicating that liquid phase filtration does not take place to any significant degree; however, under other conditions, the volume fraction of solids increases throughout the sample as the squeezing proceeds and liquid is expelled in preference to the solids. In these latter cases the concentration increases are largest toward the center of the samples. Criteria for the occurrence of the liquid phase migration in suspensions undergoing squeeze flow are presented and discussed.

Coalescence Kinetics and Sintering in Polyester Particles

Suresh Ahuja, Xerox Corporation, Webster, NY

Introduction:

Kinetics of coalescence and sintering in polymer particles and their composites have been studied for several decades for their importance to fundamental research, modeling and industrial applications. It has found its use in several polymer processes, such as rotational molding, powder coating, laser induced sintering and cold compression molding. Polymer powder particles, when in contact with each other at elevated temperatures, tend to decrease their total surface area by coalescence. This process, called sintering, is usually accompanied by a decrease in the total volume of the particulate bed. The driving force is surface tension and the main factor opposing particle coalescence is the resistance to flow, viscosity. The sintering process proceeds in two stages, first by developing interfaces and bridges between adjacent particles, followed by a stage in which the density increases by elimination of the inter-particle cavities.

Frenkel (1,7) modeled coalescence of two isolated spherical drops as being viscous (Newtonian) coalescence, which was later corrected by Eshelby (2).As shown in Fig.1, a_0 is the initial particle radius, x is the neck radius, y is the distance from the particle center to the contact plane, and Θ is the coalescence angle, between the axial plane of symmetry and the line from the particle center to the neck radius. They derived their model by performing a mechanical energy balance, which resulted in the work of surface tension equaling the work done by stresses.

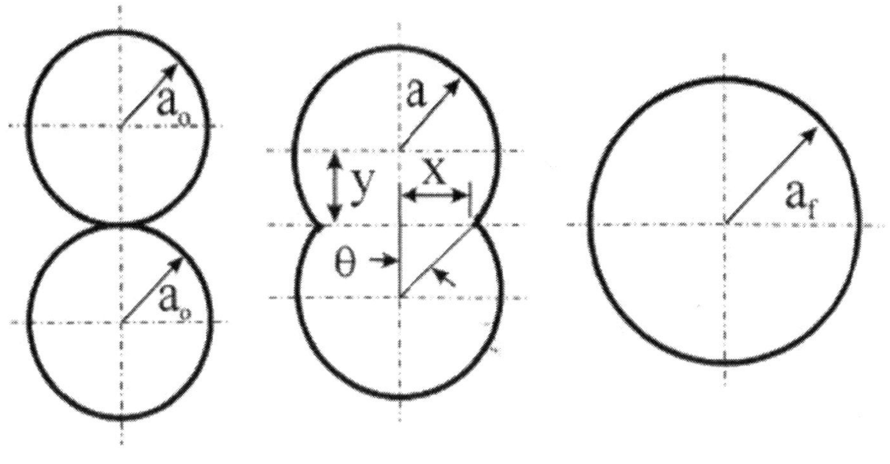

Fig.1 Coalescence of two spherical particles

$$\frac{x}{\alpha}=\left(\frac{\Gamma t}{\eta\alpha}\right)^{1/2}$$

---[1]

where x, α, Γ, t, η are the neck radius formed between the particles, the radius of particles, material surface tension, time and material viscosity. The model is based on the balance of the work of surface tension and the viscous dissipation.

Following Scherer's (3) work, densification can be calculated to be

$$Kt = 4\left[\frac{3}{4\pi}\right]^{1/3} p^{1/3}\left[1-\left(\frac{\rho_o}{\rho}\right)^{1/3}\right]$$

--[2]

where the parameter K, which can be termed as "Scherer's densification coefficient," is estimated from

$$K = (\sigma\, n^{1/3}/\eta)$$

--[3]

The densification coefficient depends on p, the number of pores per particle and n, the number of pores per unit volume both can be estimated from bulk density of the packing and particle size.

Newtonian Model of Coalescence:

Pokluda et al. (4) applied the conservation of mass with constant density to account for the time dependence of the particle radius,

$$a(t) = a_o\left(\frac{4}{[1+\cos\theta(t)]^2[2-\cos\theta(t)]}\right)^{1/3},$$

---[4]

where Θ is referred to as the coalescence angle, x/a = sinΘ, as shown in Fig. 1. This time-through the definition for the surface area of the two spheres:

$$S = 4\pi a^2(1 + \cos\theta).$$
---[5]

Differential equation of the following form can be obtained

$$\frac{d\theta}{dt} = \frac{\Gamma}{\eta a_o} \frac{2^{-5/3}\cos\theta\sin\theta}{K_1^2[2-\cos\theta]^{5/3}[1+\cos\theta]^{4/3}},$$
--- [6]-

Where Γ is the surface tension, η is the viscosity, and K_1 arises from the definition of the extension rate,

$$\dot{\varepsilon} = -\frac{1}{2}\frac{\partial v_y}{\partial y} = K_1 \frac{d\theta}{dt},$$
---[7]

$$K_1 = \frac{\tan\theta}{2} - \frac{\sin\theta}{6}\left[\frac{2(2-\cos\theta)+(1+\cos\theta)}{(1+\cos\theta)(2-\cos\theta)}\right].$$
------------------------------[8]

There are difficulties in applying Frenkel model to polymer coalescence. Particle radius is assumed to be constant. Also, the viscosity is assumed to be Newtonian. Polymers unless below entanglement are shear thinning that is viscosity decreases with increase in shear rate. Therefore, Frenkel Model over-predicted the coalescence rate. Next concept in modeling is to consider coalescence of polymer particles where polymer was viscoelastic and followed the Upper Convected Maxwell (UCM) constitutive model.

Steady-state UCM Model of Coalescence

Bellehumeur et al. (5) assumed that that the flow field is homogeneous, and that at any instant the stresses are at steady state at any instant during the coalescence process (though the stresses may change with the extension rate throughout the process).

$$8\left(\lambda K_1 \frac{d\theta}{dt}\right)^2 + \left(-2\lambda K_1 + \frac{a_o \eta}{\Gamma}\frac{K_1^2}{K_2}\right)\frac{d\theta}{dt} - 1 = 0,$$
--------------------------------- [9]

where K_1 in their calculations is the small angle form of Equation[8]

$$K_2 = \frac{2^{-5/3} \cos\theta \sin\theta}{(1+\cos\theta)^{4/3}(2-\cos\theta)^{5/3}}.$$

---[10]

Using increasingly large values for the relaxation time that were not based on experimental data, they found that the steady-state UCM coalescence model was in close agreement with their experimental data for materials with low Deborah numbers (De).

Equation (9) predicts that the coalescence rate decreases with an increase in the relaxation time. Even though two propylene ethylene copolymers coalesced at a slower rate than predicted by the Newtonian model but coalescence times for PTFE and several acrylic resins, also viscoelastic were significantly shorter than what was predicted by the Newtonian model. The relaxation times were adjusted to produce better agreement with data, and these values were unrealistically large.

Transient UCM model

Scribeben et al.(6) extended to the case of time-dependent material functions using the Upper Convected Maxwell (UCM) model. A numerical scheme was developed to solve the ordinary differential equations (ODE) for the stresses, which are embedded in the ODE based on the mechanical energy balance. Initial conditions required to solve the set of non-linear ODEs were obtained from visualization experiments of the coalescing drops as the theory for elastic contact gave unrealistically high values of the initial neck radius. The transient model predicted that coalescence was accelerated by increasing the relaxation time, the opposite relationship of what was predicted by the steadystate UCM formulation and was capable of quantitatively predicting the experimental coalescence rates at times when viscoelasticity was important.

An equation in time is obtained as

$$\frac{a_o 2^{2/3} K_1}{3\Gamma}\left(\tau_{xx} - \tau_{yy}\right)\frac{(1+\cos\theta)^{4/3}(2-\cos\theta)^{5/3}}{\cos\theta \sin\theta} - 1 = 0.$$

------------------- [11]-

For the transient single-mode UCM model, two additional ordinary differential equations are used to solve for the stresses. These are

$$\tau_{xx} = e^{-\left(\frac{1}{\lambda}-2\dot{\varepsilon}\right)t}\left(\frac{2\eta\dot{\varepsilon}}{(1-2\lambda\dot{\varepsilon})}e^{\left(\frac{1}{\lambda}-2\dot{\varepsilon}\right)t} + C_{xx}\right),$$

.---[12]

$$\tau_{yy} = e^{-\left(\frac{1}{\lambda}+4\dot{\varepsilon}\right)t}\left(\frac{-4\eta\dot{\varepsilon}}{(1+4\lambda\dot{\varepsilon})}e^{\left(\frac{1}{\lambda}+4\dot{\varepsilon}\right)t} + C_{yy}\right),$$ --[13]

where the integration constants Cxx and Cyy are determined from the expressions:

$$C_{xx} = \left[\tau_{xx}(t_n) - \frac{2\eta\dot{\varepsilon}}{(1-2\lambda\dot{\varepsilon})}\right]e^{\left(\frac{1}{\lambda}-2\dot{\varepsilon}\right)t_n},$$

--[14]

$$C_{yy} = \left[\tau_{yy}(t_n) + \frac{4\eta\dot{\varepsilon}}{(1+4\lambda\dot{\varepsilon})}\right]e^{\left(\frac{1}{\lambda}+4\dot{\varepsilon}\right)t_n}.$$

-- [15]-

The choice of the initial condition may be a major issue in accurately predicting coalescence of two polymeric drops. It is applicable on isothermal viscous coalescence and the experimentally observed value is used throughout the calculations. Theoretical models were compared to experiments (7,8)

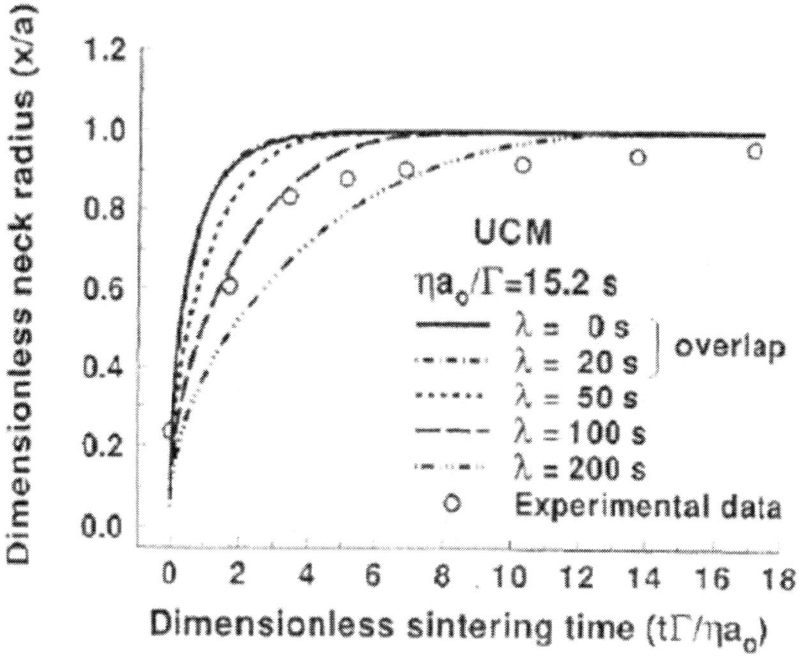

Fig 2: Modified – Frenkel model predictions using the upper convected Maxwell (UCM) model and experimental data

Experimental and Materials

Styrene butadiene of different molecular weights and crosslinked polyesters were made and characterized for viscosity and viscoelasticity using ARES rheometer.. From elastic and viscous modulus, relaxation modulus as function of time was used to get λ relaxation time.

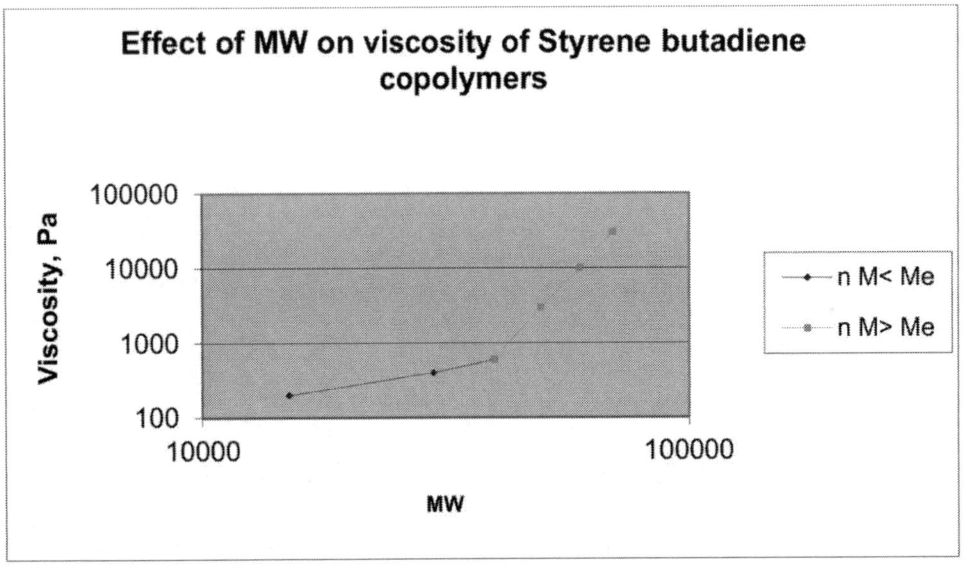

Fig.3: Dependence of viscosity on molecular weight MW of styrene butadiene polymers below and above entanglement.

An application to sintering and coalescence lies in digital imaging (9).. A digital imaging particle is made up of binder polymer of a certain Tg and molecular weight. Viscosity of toner is dependent on molecular weight of polymer which in turn is dependent on temperature and shear rate. As the molecular weight increases, polymer chains become entangled and viscosity becomes shear rare dependent.

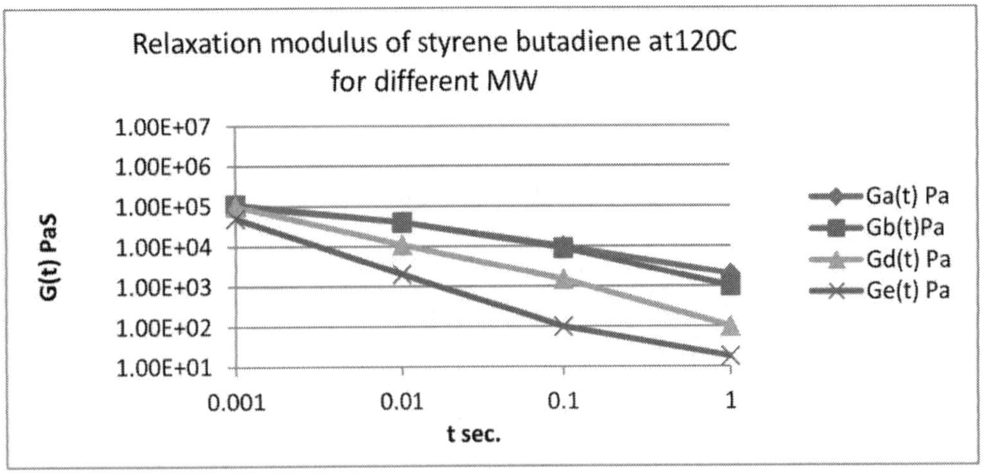

Fig. 4 Relaxation modulus as function of time at different molecular weight of styrene butadiene

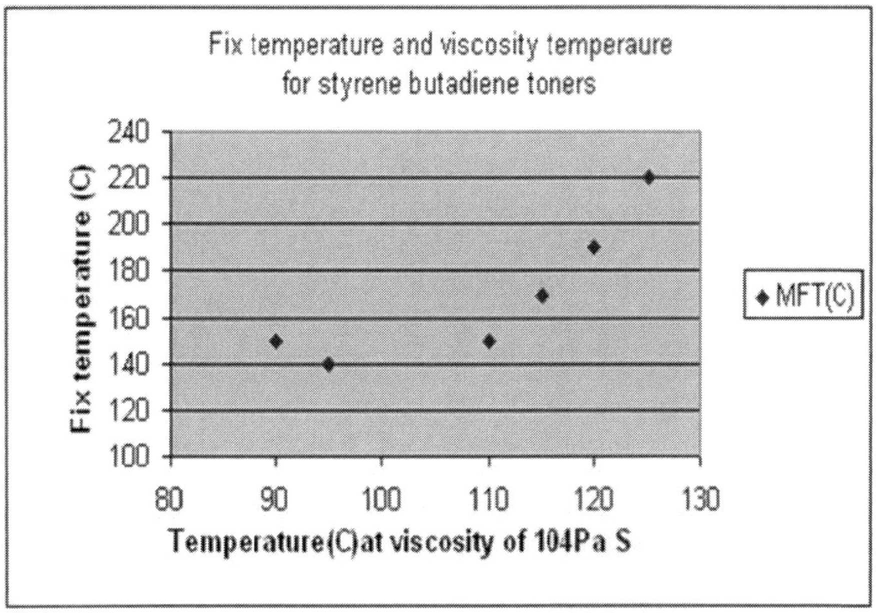

Fig.5 Fix temperature (temperature required to have adequate adhesion to paper) as function of temperature for 10,000P viscosity

Image fix is a function of viscosity of toner at the fusing temperature and brittleness of the image as it solidifies on the media. If a toner has low molecular weight, it would have low viscosity at the fusing temperature. It would coalesce fast but the image is likely to be brittle and break. At low molecular weights, lower critical stress due to lower cohesive strength dominates over gain in faster coalescence due to lower viscosity. If molecular weights are high, toner coalesce is inadequate and image adhesion to the media would be poor.

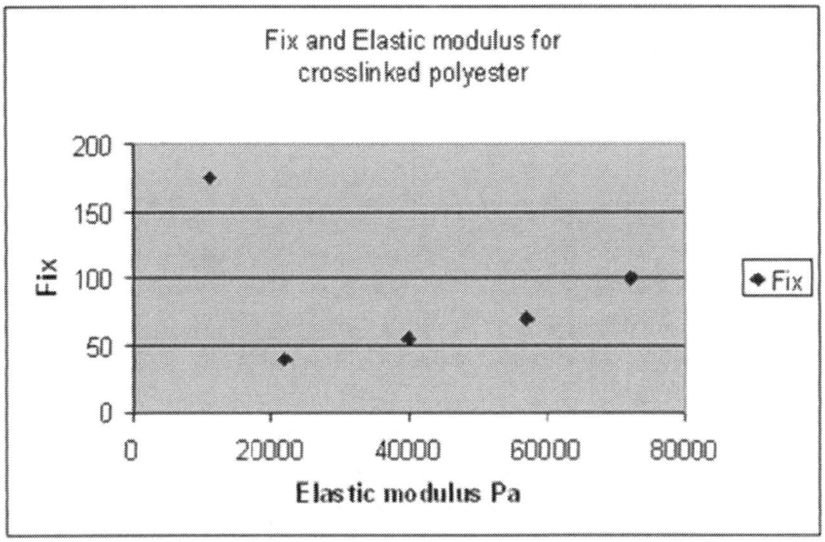

Fig. 6 Image Fix and Elastic modulus of Crosslinked polyesters

Initially as the molecular weight of the styrene butadiene polymers increases, adhesion of the polymer toner (digital image), image fix to the substrate paper surface increases and temperature required to achieve that adhesion decreases. Viscosity of the sintered/coalesced liquid is low compared with the modulus of the solidified image. Temperature required to achieve a fixed viscosity would decrease with increase in molecular weight MW (Fig.5). Beyond a certain molecular weight MW, adhesion to the paper surface at a certain time is more dependent on relaxation modulus and relaxation time of the sintered/coalesced digital image than by the increase in solid modulus of the solidified image. Higher temperature is required to get to the same fix or adhesion to the substrate. Similar behavior is observed in cross-linked polyesters as shown in Fig.6.

Conclusion:

Kinetics of sintering and coalescence models are presented which are of importance in wide variety of applications from rotational molding to digital imaging. In digital imaging, sintering and coalescence of particle is essential and evaluation of digital images are strongly dependent on relaxation behavior.

References

(1) Frenkel J., J. Phy., 9, (1945) 385-391
(2) Eshelby, J.D.. J.Metals,1Metals Transactions(1949),185, p.806.
(3) G. W. Scherer, J. Am. Ceram. Soc., (1991)**74**, 1523
(4) Pokluda O., Bellehumeur C.T., Vlachopoulos J., (1997), AIChE J., 43, 3253-3256.
(5) Bellehumeur, C.T., M. Kontopoulou and J. Vlachopoulos (1998),. Rheol. Acta, 37, 270-278.
(6) Scribben, E., D. Baird and P. Wapperom (2006). Rheol. Acta, 45, 825-839.
(7) Kontopoulou, M. and J. Vlachopoulos (2001). Polym. Eng. Sci., 41, 155-169.
(8) Muller, J.D. , M. Bousmina and A. Maazouz (2008). Macromol., 41, 2096-2103
(9)Ahuja,,S.(2012)," Toner Rheology and Digital Imaging", Short Course at NIP(Imaging Science and Technology) Meeting

(142bw) Force Driven Separation of Particles and Drops by Deterministic Lateral Displacement

German Drazer, Chemical and Biomolecular Engineering, Johns Hopkins University, Baltimore, MD

We present the separation of solid particles and drops in force-driven deterministic lateral displacement (f-DLD), a promising high-throughput continuous separation method in microfluidics. We perform both macroscopic experiments in scaled up systems as well as microfluidic experiments. The solid particles or drops are driven through a square array of cylindrical obstacles and, for some forcing orientations, they migrate in different directions. These experiments demonstrate the separation capabilities —and provide constraints for the design— of f-DLD for particles and drops of multiple sizes. We show that in all cases the trajectories of the particles follow selected locking directions in the lattice and present a simple collision model, based on particle-obstacle repulsive interactions, that accurately describes the trajectories of the particles for all forcing directions. Finally, we also demonstrate that a single line of cylindrical obstacles rotated with respect to the driving force is capable of performing binary separations. Using scaled up models proved to be a succesful approach to capture the behavior of microfluidic systems and test the deterministic limit.

(142bx) Development of a Viscoelastic Two-Phase CFD Model in Openfoam

Laura Dietsche[1], Florian Habla[2] and Olaf Hinrichsen[2], (1)Core R&D - Engineering Sciences, The Dow Chemical Company, Midland, MI, (2)Catalysis Research Center and Chemistry Department, Technische Universität München, München, Germany

When polymers with different rheologies are blended together, there is often a need to break up large polymer droplets into smaller droplets. It is desirable to be able to explore the effects of the polymer rheology, equipment design, and operating conditions on droplet break-up efficiency using computational fluid dynamics (CFD). Previous CFD studies looked at polymer droplets flowing past a series of wires using shear thinning viscosity models and two-phase volume-of-fluid (VOF) methodology. The droplet break-up predictions were in qualitative agreement with experimental results. However, it was felt that the predictions would be improved if viscoelastic effects were also included. Since there are presently no commercial CFD codes that can perform this type of simulation a project was initiated to develop a new solver which handles viscoelastic two-phase flows. The first generation of this solver combined the VOF method and the recently added viscoelastic capabilities in OpenFOAM[R] to simulate free-surface flows of viscoelastic materials using various constitutive equations, including Oldroyd-B, FENE-CR, Phan-Thien-Tanner, and Giesekus models. The numerical algorithm is validated using two well-known viscoelastic free-surface effects, namely the Die Swell effect and the Weissenberg (Rod Climbing) effect, as wells as transient and steady-state deformation of single drops subject to shear and planar elongational flows. Application of this model to polymer droplet break-up is also explored.

(142bz) Rapid Mixing in Microchannel Due to Flow Instability Induced by Flexible Wall

M.K.S. Verma, Department of Chemical Engineering, Indian Institute of Science, Bangalore, India and V. Kumaran, Chemical Engineering, Indian Institute of Science, Bangalore, India

An important limitation in microfluidic applications is the slow mixing due to the laminar nature of the flow. Mixing takes place by molecular diffusion, and the time for mixing across a distance L scales as (L^2/D), which is approximately 10 to 1000 seconds when we use a length in the range (100 micron-1 mm), and with typical diffusion coefficient of 10^{-9} m^2s^{-1}. Here, we report the development of a fast and homogenized microfluidic mixer, where cross-stream mixing is induced by a flow instability due to the flexibility of one of the wall of a channel of rectangular cross-section. The flexible wall of microfluidic channel is prepared with varying range of shear modulus (17–550 kPa) of PDMS gel and the pressure difference between the inlet and outlet is used to drive fluid flow. It is found that the laminar flow becomes unstable, and there is a transition to a chaotic flow with large cross-stream fluctuations. This results in decrease of mixing times by five orders of magnitude than those in a laminar flow. The mixing is characterized by two methods: a macroscopic dye-stream image analysis technique based on images of the channel, and a microscopic ionic conductivity of the fluid at the outlets. The results of both procedures show very good mixing of the fluid streams within milliseconds after the transition. The Reynolds number for transition decreases as the shear modulus of the wall is decreased, and it is about 200 for the softest gels used here. Wall oscillations are also detected at the onset of mixing using a laser scattering technique where a laser beam is scattered off microbeads embedded in the flexible wall. Even though there is near complete mixing, the increase in dissipation is modest, and the pressure drop is only about 50% higher than that for a laminar flow in the same channel. These experiments show that use of soft walls is practically feasible and a low-cost method to induce instability and generate fast and effective mixing in microfluidic applications. This is also advantageous over conventional passive methods which typically involve large increase in pressure gradients and complicated fabrication, and also over active methods since there are no moving parts.

(142c) Porous Media Model and Collective Behavior of Colloidal Particles Trapped At a fluidic Interface

Eric S.G. Shaqfeh, Chemical Engineering and Mechanical Engineering, Stanford University, Stanford, CA, Shenghan Yan, Chemical Engineering, Stanford University, Stanford, CA and Gerald Fuller, Stanford University, Stanford University, Stanford, CA

It is well known that charged colloidal particles may form an effective two di- mensional suspension at a fluidic interface. Research toward the understanding of the dynamics and collective behavior of these suspended particles is at the core of engineering Pickering emulsions, which have a broad range of practical applications. Many experiments have been performed where two particle laden interfaces have been brought into close contact in a controlled manner and various observations, including particle "bridging", have been made in an attempt to understand the stabilization mechanism of interfacial particles in a Pickering emulsion. One of the most interest- ing observations is the tendency for the particles on one interface to "evacuate" and those on the other interface to "aggregate" during the close approach of the surfaces.

In this work, we propose to understand the mechanism of particle evacuation- aggregation via a combined experimental and theoretical approach. First, we per- formed real-time experiments where two particle-laden water-decane interfaces were brought into contact. Many phenomena including particle evacuation-aggregation and bridging were observed. We then developed a Brownian dynamics simulation of the evacuation-aggregation including the important relevant interparticle interactions that we presumed were important in describing the phenomena. In order to do so, we had to answer three questions. First, what are the relevant aspects of the charged particle interaction within the same interface? Second, what is the charge interaction across the two approaching interfaces? Third, what are the flow effects, including the flow between the two interfaces during approach, on the particle motion and how can we model such a flow? Toward this goal, we have incorporated both reasonable electric inter-particle interactions from available literature studies and flow interac- tions via a porous media model that relates the particle velocity to the local surface coverage through the effective permeability of a porous media. Thus the flow effects are captured in a mean field sense. The BD simulations were able to capture the evac- uation/aggregation qualitatively and, in most instances, quantitatively. In particular the diameter of the evacuated area decreases with increasing surface coverage in both simulations and experiments, and we will describe the physical mechanisms leading to this behavior by analyzing the particle force balance in the BD simulations.

(142ca) Brownain Dynamics Simulation of Non-Newtonian Bead-Spring Microstructure Based On Kinetic Theory

Kyung-Hyo Kim, Chemical Engineering, University of Illinois at Chicago, Chicago, IL

The progression of disease is oftentimes accelerated by the changes in the mechanical behavior of cells. The rheological material properties of biological cells such as red blood cells are important factors in the occurrence and onset development of diseases. A better fundamental understanding of flow and its rheological properties will help rational approach to prevention of serious conditions and its costs, and development of new treatments as well.

The empirical data of the human blood, for example, has non-Newtonian behavior and its elastic properties have been measured. The viscosity drops as the shear rate increases instead of remaining constant as in the Newtonian fluids. Often times the pure mathematical Casson model has been used to model the behavior of the blood since it gives the characteristic of power-law slope of the viscosity similar to that of experimental data. However, the model neither predicts the time-dependent behavior nor fit all the range of shear rate empirical data where the viscosity tends to have nearly constant viscosity at the low and high shear limits. More importantly, the Casson model gives us no insight into the dynamics of cells in complex flows.

The Brownian Dynamic simulation will allow predicting the movement of biological cell in methodological way with the consideration of all the forces. Instead of using an empirical constitutive equation, we can use kinetic theory where the stress tensor depends on the configuration of the microstructure of the fluid. The microstructure of cell is constructed with spherical beads and each bead is subject to be influenced by a variety of forces. In general, the equations of motion for each bead of the microstructure can be represented by the sum of all the forces acting on the bead. The main forces under consideration are a hydrodynamic force in the form of a Stokes' law drag, the Brownian force, intermolecular forces, and constraint forces. The hydrodynamic forces tend to distort the microstructure as oppose to intermolecular forces which tend to restore the microstructure to its original shape. The Brownian forces randomize the orientation of the microstructure. In equilibrium systems, the expression configurational distribution function can be directly obtained by equilibrium statistical mechanics. For non-equilibrium system, we can derive a second-order partial differential equation for the configuration-space distribution function, so called the diffusion equation, by combining the force balances on the beads with the equation of continuity in configuration space. The system can be and with or without hydrodynamic interaction.

The model can have degrees of complexity in order to have good estimate of the cell configuration. For such model, the potential forces on the beads are related by tension in the spring and the connection between beads and springs. Thus, the equation of motion which results in the stochastic differential equation (SDE) of motion depends on the configuration of microstructure. Once the displacement of beads in the microstructure is calculated from the stepwise integration (Euler scheme solution) of SDE, geometrical holonomic constraints can be applied to adjust positions of the beads so that the biological cell conserves its overall size throughout the motion of flow while allowing the shape to change. This adjustment is made for each time step by Lagrange multiplier to enforce the constraint.

There are various approaches to apply constraints to the system. It can be done at different level; Fokker-Planck equation, SDEs, or numerical integration scheme. Although it should be theoretically equivalent, there can be some deviation in these approaches depending on how strictly the constraint is satisfied since they are approximation schemes. The numerical integration scheme in Cartesian coordinates is used here to solve SDE. The time-dependent Lagrange multipliers that are associated with constraints are determined to satisfy at each time step. The Lagrange multipliers method is used for optimization to find the maximum or minimum of a function subject to constraints. The Lagrange formulation has advantage of giving the magnitude of correction while the derivative of constraints that are set gives direction.

A discrete model of a triangular bead-spring ring with a holonomic constraint is considered as suspension of deformable microstructure suspended in a dilute solution of Newtonian fluid. The microstructures are represented by bead-spring model where hydrodynamic resistant site referred as "beads" with inter-bead potentials referred as "springs." The model is designed to capture the most essential features of cells. This is a ring structure with three identical beads, each connected to two adjacent springs. A simple linear Hookean spring is used to give flexibility to deform. The intermolecular force acting on each bead through two adjacent springs has the relation between the potential forces on the beads and the tension in the spring. The polymer kinetic theory is used with the hypothesis that the constraints are yet preserved while microstructure deforms. Without constraint, this model becomes a Rouse model with ring closure effect.

Constraints introduce difficulty that the positions of the beads are no longer independent since they are connected by the equations of the constraint. To test the method of multiple constraints, a three bead-spring model with area and sum of length square constraints was tested. The sum of length square constraint keeps the total length of the connected springs so that we give restriction to deformation yet still have some degree of flexibility. The rheological material properties were obtained for different type of flow conditions. Averaged rheological material properties of the fluid can be calculated with the components of stress tensor. The calculation of the material properties that characteristics the fluid depend on the type of flow. The model is tested under shear and shear free flow. Kramers-Kirkwood form of stress tensor is used with stochastic approach with actual bead velocity instead of the momentum-space-averaged velocity of the bead.

The bead-spring-ring model with an area constraint was constructed to represent the microstructures of cells and the results clearly show the expected shear-thinning effect for viscosity. Neither of the Rouse nor the Zimm model result gives a decrease of viscosity with increasing shear rate. The result of Kramers chain, a bead-rod model, with multiple constraints shows the shear thinning effect. Therefore, a sample ring closure model with constraint without hydrodynamic interaction is constructed.

The result of in steady shear flow with two-constraint model was superposed to that of one-constraint for comparison. An assumption was made that the friction tensor is isotropic.

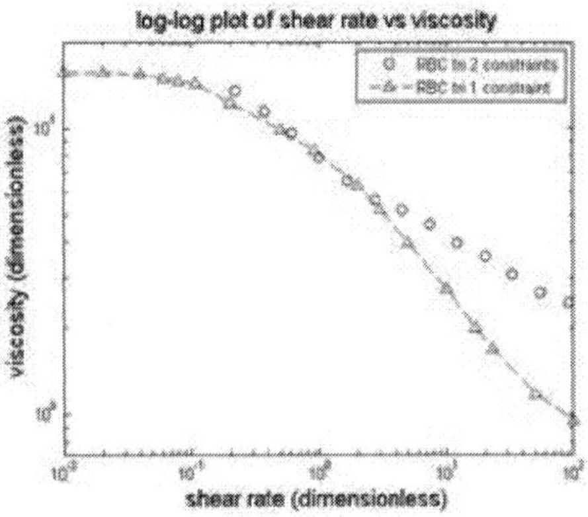

Figure 1. A comparison of result of two-constraint method against one-constraint of a triangle model. A log-log plot of viscosity with respect to shear rate for a steady shear flow.

The result with area and sum of length square constraint shows that the slope drops in the mid region of shear rate. This is due to the elimination of degree of freedom that the model does not have flexibility to change its shape compare to one constraint. The following graph shows the comparison between one-constraint method and two-constraint method for elongational flow with same area size.

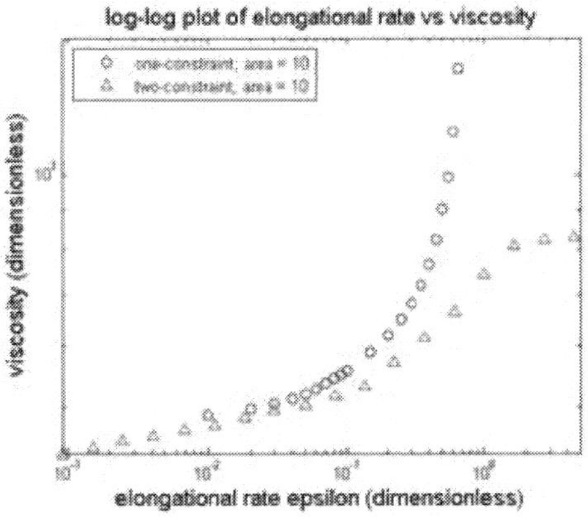

Figure 2. A comparison of result of two-constraint method against one-constraint of a triangle model. A log-log plot of viscosity with respect to elongational rate for a steady elongational flow.

Unlike the one-constraint result where the viscosity increases exponentially as the value of elongation rate increases in the log-log plot, the rate of increase in viscosity plateaus after certain point. This effect is shown in other polymer fluids. In the on constraint model where the constraint is only an area, the model stretches infinitely which is unlikely in reality.

Further study on more complicated geometric modeling of microstructures can be done with hydrodynamic interaction. Setting up the constraint equation so that the constraint is differentiable by the bead position is inevitable to use the constraint method with Lagrange multiplier.

(142cb) Understanding DNA in Nanoconfinement Through Monte Carlo Simulations and Scaling Theory

Liang Dai, BioSym, Singapore-MIT Alliance for Research and Technology, Singapore, 117543, Singapore, Johan van der Maarel, Department of Physics, National University of Singapore, Singapore, Singapore and Patrick S. Doyle, Chemical Engineering, Massachusetts Institute of Technology, Cambridge, MA

DNA behavior in nanoconfinement is a burgeoning research area. Nanodevices with well-defined geometries are particularly useful in DNA separations and single molecule genomic assays. In pursuit to better control DNA in these nanodevices, it is crucial to systematically understand DNA behavior under various degrees of confinement. DNA behavior in moderate and strong confinement regimes have been well understood from the blob theory and deflection theory, respectively. However, the transition regime between those two regimes is unclear. In particular, some contradictory results from experiments exist in literatures. We perform Monte Carlo simulations of DNA in slits and tubes, spanning from very weak to very strong confinement, with special focus on the transition regime. We find that the competition of three lengths (persistence length, chain width, slit height / tube diameter) results in some sub-regimes in addition to the Odijk regime (strong confinement) and the classic de Gennes regime (moderate confinement).

(142cc) A Rivulet Flow Model for Wetting Efficiency in Packed Beds

Subbarao Duvvuri, Department of Chemical Engineering, Universiti Teknologi PETRONAS, Tronoh, Malaysia and Shuhaimi Mahadzir, Chemical Engineering, Universiti Teknologi PETRONAS, Tronoh, Malaysia

Packed towers are widely used in process industries in unit operations such as distillation, absorption, water cooling and catalytic gas-liquid reactions in which flowing gas needs to be brought into intimate contact with liquid flowing on the particles in the form of rivulets or films. The width of rivulets increases with increase in liquid flow rate and eventually the liquid film spreads all over the surface of the particles. Wetting efficiency, defined as the ratio of gas-liquid contact area to the particle surface area, directly affects the mass transfer rates in packed towers. It has been observed that wetting efficiency can be less than one depending on the operating variables (such as pressure, temperature, gas and liquid flow rates), system variable (such as density, viscosity, surface tension) and equipment variables (such as type of liquid distributor, particle shape, size and materials of construction). Rivulet flow on an inclined plane surface is analyzed through the concept of friction factor to develop a correlation for the width of rivulets as a function of liquid flow rate. This is extended to estimate wetting efficiency in packed towers. Literature data are used to validate the model equations. Inertial forces dominate over the viscous and surface tension forces in determining the wetting efficiency in packed towers.

(142cd) Investigation of the Motion of Small Particles in Microfluidic Flows

Minh Vo, Chemical, Biological and Materials Engineering, The University of Oklahoma, Norman, OK and Dimitrios V. Papavassiliou, School of Chemical Biological and Materials Engineering, The University of Oklahoma, Norman, OK

The motion of micro- and nano-particles in microfluidic flows is important in several practical cases, such as in MEMS applications and in the transport of fine particles in the pore space of porous media. The goal of this work is to calculate the trajectory of individual particles in microflows by combining the Lattice Boltzmann method (LBM) with a Lagrangian particle tracking algorithm. The former method provides the velocity profile within the flow field and the latter utilizes this velocity to determine the velocity and the position of particles via force balance equations [1].

Initially, particles are released into a micro-slit with no slip boundary condition. When moving in the flow, the particles experience forces that include gravity, drag, buoyancy, and lift. Depending on the size of the particles, Brownian motion is also important. The paper will present cases where each specified force becomes dominant at low Reynolds number flow and with different particle shapes (both spherical and non-spherical particles). The diameter of particles investigated is in the range of nanometers to micrometers, and the size of the microchannel is between 5 and 10 micrometers. In the particle tracking algorithm, the particles travel in each time step due to convection, diffusion (Brownian motion) and due to the effect of total force. Convective displacement can be calculated by using the velocity obtained from LBM. Displacement due to Brownian motion is calculated by the application of random jumps that follow a normal distribution with a zero mean and a standard deviation that depends on the Schmidt number of the particle, according to Einstein's theory for stochastic motion of particles. In this manner, particles with Schmidt number ranging from 100 to 10,000 are investigated. Lastly, displacement of particles because of resultant force is determined by Newton's law of motion. Additionally, the interaction between the particles and the wall can be accounted by defining a Morse-like potential in regions very close to the wall [2]. In this manner, the deposition rate of the particles can be calculated, and the relative effects of the hydrodynamics forces, the Brownian motion and the particle-surface interactions can be quantified. As expected, the effect of the Brownian motion dominates the particle trajectories for nanoscale particles, while hydrodynamic forces play an increasingly important role when the particle shape becomes cylindrical with high aspect ratio.

ACKNOWLEDGEMENTS

The financial support of the Advanced Energy Consortium (AEC BEG08-022) and the computational support of XSEDE (CTS090017) are acknowledged.

REFERENCES

1. Voronov, R.S., VanGordon, S., Sikavitsas, V.I., and D.V. Papavassiliou, "Efficient Lagrangian scalar tracking method for reactive local mass transport simulation through porous media," *Int. J. for Numerical Methods in Fluids,* **67**, 501-517, 2011
2. Kolmakov G., Ravanur, R., Tangirala, R., Emrick, T., Russell, T.R., Crossby A.J., and A.C. Balazs, "Using nanoparticle-filled capsules for site-specific healing of damaged substrates: Creating a repair and go system," *ACS Nano,* **4**(2), 1115-1122, 2010.

(142ce) Nonlinear Rheology of Colloid-Polymer Composites

Tianying Jiang, Chemical and Biomolecular Engineering, University of Illinois Urbana-Champaign, Urbana, IL and Charles F. Zukoski, Department of Chemical & Biomolecular Engineering, University of Illinois at Urbana Champaign, Urbana, IL

Colloid-polymer composites have high potential application in improving mechanical properties of materials as an important class of complex fluid. Rules of controlling macroscopic flow properties of colloid-polymer composites are not well understood because of limited developed knowledge in the microscopic view of colloid interaction and polymer dynamics. Even less systematic study has been carried out to investigate the effect of particle size, interaction, concentration with varied polymer dynamics on nonlinear response of the composites under shear both theoretically and experimentally. Here we study the rheology of silica particles with varied size suspended in unentangled polymer melts by applying both continuous shear and dynamic shear. Before studying the dense suspension in which our particular interests are, low particle concentration condition is studied to understand the basic physics of particles interaction in dilute condition. When moving to high concentration system within the region of glass and weak gel formation where the difference is observed with polymer molecular weight tuned, significant difference is observed in the flow properties of the enhanced polymer melts. Specially, shear thickening behavior, nonlinear rheology and yielding behavior are searched to understand roles of particle concentration, size and polymer molecular weight in terms of confined particle and polymer dynamics.

Linear Viscoelastic Properties of Salt-tolerant HPAM Solutions with Ultrahigh Molecular Weight

Xiaoming Xiao[1], 1. School of Chemical Engineering and Technology, Tianjin university, Tianjin, China
Luhong Zhang[1,2], 1. School of Chemical Engineering and Technology, Tianjin university, Tianjin, China
2. National Engineering Research Center of Distillation Technology, Tianjin, China

Abstract

In this paper, the linear viscoelastic properties of salt-tolerant partially hydrolyzed polyacrylamide (HPAM) solutions with ultrahigh molecular weight are presented. By means of frequency sweep using Haake Rheostress 150 rheometer, the dynamic measurements of HPAM solutions with ultrahigh molecular weight are investigated comprehensively, and the concentration of solutions range from 3000 mg·L-1 to 7000 mg·L-1. The results show that the HPAM solutions are linear-viscoelastic when the strain is $0<\gamma<50\%$ at the angular velocity of 1 rad/s and all the HPAM solutions show shear thinning flow behavior in the oscillatory frequency range of 0.01-10 Hz. Secondly, different from the HPAM solutions with low molecular weight and low concentration, the tested solutions show a greater elastic modulus than viscous modulus throughout the whole frequency range investigated, plus, according to a three-zone model for viscoelastic liquids, the dynamic rheological behaviors of all the HPAM solutions are assigned to plateau zone. Finally, the elastic modulus and viscous modulus increase exponentially with the increase of HPAM solution concentration.

Key words

salt-tolerant partially hydrolyzed polyacrylamide; oscillatory frequency; complex viscosity; elastic modulus; viscous modulus

Introduction

At present, polymer flooding has been successfully applied to enhance crude oil recovery in China [1-4]. Recent investigations revealed that, compared with low-molecular weight HPAM solutions, ultrahigh-molecular weight HPAM solutions enjoy a better performance of high viscosity and greater residual resistance factor. Accordingly, when the same polymer injection amount is applied using higher molecular weight HPAM solutions, the oil recovery rate increases [5-7]. The rapid development of polymer flooding and its auxiliary equipment technologies [8-11] has called for the application of salt-tolerant, high concentration HPAM of ultrahigh molecular weight solutions in preparing the mother polymer liquor. HPAM solutions with low molecular weights and low concentrations have been widely studied with respect to their viscoelasticities and influencing factors [12, 13]. Luhong Zhang [14] investigated the steady shear behaviors of HPAM solutions at the molecular weight of 2.5×10^7 g·mol^{-1}, and found that HPAM solutions had the properties of viscoelasticity and shear-thinning. However, there has been little research on the dynamic viscoelasticities of salt-tolerant, high concentration HPAM solutions of unltrahigh molecular in China and abroad thus far.

Using dynamic oscillatory measurements, this paper studied the linear viscoelastic properties of the salt-tolerant HPAM solutions with molecular weight of 2.5×10^7 g·mol^{-1} and 3.5×10^7 g·mol^{-1} at the concentrations of 3000 mg·L^{-1}, 5000 mg·L^{-1}, 6000 mg·L^{-1}and 7000 mg·L^{-1} at a temperature of 15°C. The effects of oscillatory frequency, HPAM solution concentration and molecular weight on the complex viscosity (η*), elastic modulus (G') and viscous modulus (G") were investigated. Results from this paper about the rheological properties of the salt-tolerant HPAM solutions are important in handling such solutions in terms of mixing, pumping, storage, and transportation processes [15].

Experiment

Materials

The salt-tolerant partially hydrolyzed polyacrylamide with relative average molecular weights of 2.5×10^7 g·mol^{-1} and 3.5×10^7 g·mol^{-1} was provided by the Daqing oilfields. The polyacrylamide was 25% ~30% hydrolyzed and contained 90% of solid content, as specified by the supplier. The chemical structure of the ultrahigh molecular weight HPAM is shown in Figure 1. The samples were dried in vacuum for 24 hours prior to use.

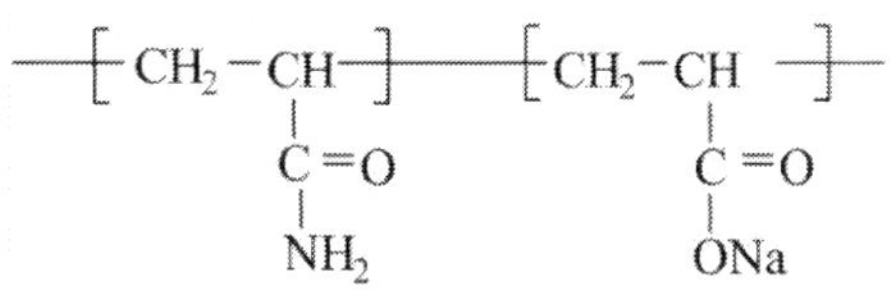

Figure 1. The chemical structure of HPAM with ultra high molecular weight

Preparation of samples

Mixtures of polymer powder and 300mL distilled water at room temperature (about 13°C) were stirred gently using a Laboratory Mixer (Cole-Parmer Instrument Company, Modd50002-35). The concentrations of HPAM solutions were 3000 mg·L^{-1}, 5000 mg·L^{-1}, 6000 mg·L^{-1}, and 7000 mg·L^{-1}. After the samples were prepared, they were stored for 24 hours to eliminate entrapped air-bubbles to obtain homogeneous solutions. In order to avoid possible errors, the procedure and time span for each sample were the same.

Rheological measurements

Oscillatory rheological experiments were performed on the Haake RS150 RheoStress instrument operating in a strain-controlled mode at 15°C. A stainless steel cone and plate measure system with a diameter of 17.5±0.01mm and a cone angle of 1° was used. The range of the rheometer torque was from 0.5×10^{-4} to 150mNm and the strain resolution was better than 6×10^{-7} rad. Temperature control was conducted using a HAAKE C25 controller, the range of which was 0-120°C with accuracy of ±0.01°C. The gap between the cone and plate was 0.052mm. A controlled variable speed lift was used to fasten the cone and plate (Auto Gap) to ensure the highest accuracy for tests. Before measuring, all the samples were kept between the cone and the plates at the same temperature for 20 min.

For all samples, an amplitude sweep was conducted at the angular velocity of 1rad/s to determine the linear viscoelastic range. Frequency sweep tests were performed using the frequencies between 0.01Hz and 10 Hz, with the strain being changed to maintain a constant angular displacement as the frequencies were swept. Finally a time sweep at 1Hz was carried out on the HPAM solution at a concentration of 6000 mg·L^{-1}.

The experimental dynamic rheological data (elastic modulus, G', viscous modulus, G'', and complex viscosity, η*, complex modulus G^*) were obtained directly from the Haake rheology advantage data analysis software RheoWin Pro2.91, and were also used for data evaluation and analysis. The deviation was within 5% between duplicate runs.

Results and discussions

The effect of oscillatory frequency on rheological properties

The viscoelastic behaviors of polymer solutions are usually independent of strain amplitudes when the strain amplitude is smaller than a critical value [16-18], the region under which is usually called the linear viscoelastic (LVE) region. Amplitude sweep is always the first step in any oscillatory rheometry experiment to ensure operations are within the LVE region [18-20].

The amplitude sweep results showed that all the tested samples in this study lied in the LVE region when the strain was lower than 50%. In the experiments of the frequency sweep on HPAM solutions, a strain of 5% was chosen to avoid possible splash of the samples out of the gap between the cone and plate. One can observe from Figure 2 (a), (b) the relationship between η^* of HPAM solutions and the oscillation frequency. The HPAM solution concentrations ranged from 3000 mg·L^{-1} to 7000 mg·L^{-1}, and the oscillation frequency from 0.01 to 10Hz. It was found that the complex viscosity decreased as the oscillatory frequency increased. The fact that η^* decreased almost linearly as the frequency increased indicated that all HPAM solutions with tested concentrations exhibited shear-thinning flow behavior. The molecular structure of the polymer was studied using the oscillatory measurements. The increase of frequency changed the ultrahigh-molecular configuration of the polymer, resulting in the divergence of the long-chain polymer molecular from its equilibrium configuration. This in turn segregated the entanglement between molecules and thus allowed relatively easy movement between the macromolecules. As a result, the complex viscosity decreased.

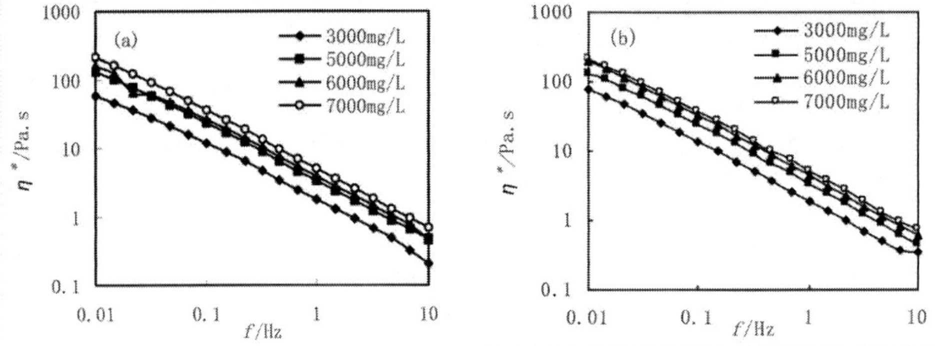

Figure 2. The relationship between complex viscosity η^* and oscillatory frequency for the HPAM solution with ultra high molecular weight (a) 2.5×10^7 (b) 3.5×10^7

Figure 3 (a), (b) illustrated the changes in elastic modulus (G') and viscous modulus (G") of HPAM solutions of different concentrations as a function of frequency. The elastic modulus climbed upward as the oscillatory frequency increased, whereas the viscous modulus was almost independent of the frequency. It can be explained that, the elastic property was related to the continuous changing of the polymer molecular conformation, while the viscosity was generated by the relative movements between the molecular chains. The long chains of the ultrahigh-molecular weight HPAM increased the number of molecular configurations and thus improved the flexibility of the ultrahigh-molecular chains. As a result, the high flexibility allowed the elastic deformation to be stored in the ultrahigh-molecular system of the HPAM solutions when exposed to a force. Hence, the elastic modulus of the solution was dominant.

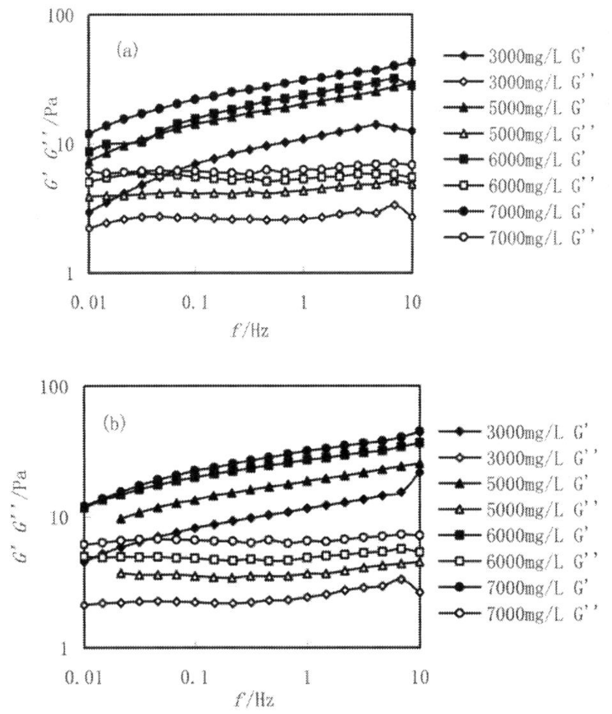

Figure 3. The relationship between dynamic moduli and oscillatory frequency for the HPAM solution with ultra high molecular weight (a) 2.5×10^7 (b) 3.5×10^7

Compared with the low-molecular weight, low concentration HPAM solutions[10,11,13], solutions with ultrahigh-molecular weight of 2.5×10^7 g·mol^{-1} and 3.5×10^7 g·mol^{-1} appeared to have no crossover points for G' and G" within the frequency range investigated, and the elastic modulus was always higher than the viscous modulus $G' > G''$. At 6.81Hz, the elastic modulus was 3.95 to 5.70 times higher than the viscous modulus with the molecular weight of 2.5×10^7 g·mol^{-1}, and the figure was 4.64～6.02 times with the molecular weight of 3.5×10^7 g·mol^{-1}. These phenomena indicated that the elastic behavior of the HPAM solutions was predominant over the viscous behavior.

In Figure 4[21, 22], the G' and G'' of a viscoelastic liquid was plotted logarithmically as a function of frequency based on the relative magnitude of G' and G''. The figure consisted of three zones: terminal, plateau and transition. According to this three-zone model, the tested HPAM solutions in this paper fell under the category of the plateau zone within the oscillatory frequencies investigated, indicating that the ultrahigh-molecular chains in the HPAM solutions had become entangled. Therefore, the molecular interaction was strengthened and the HPAM solutions behaved as a solid-like material.

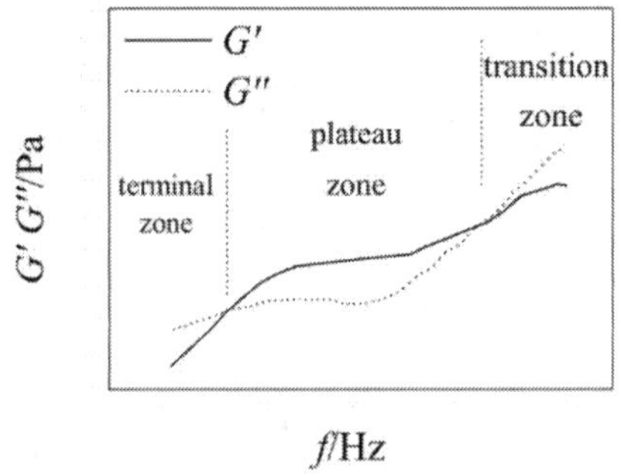

Figure 4. Illustration of three-zone model for viscoelastic liquid [18, 19]

The effect of concentration on rheological properties

Figure 5 shows the effects of different HPAM concentrations on the dynamic properties at 1 Hz. Both G' and G'' increased significantly along with the increase of the HPAM concentrations, and the logarithmic values of G' and G'' exhibited a linear function of the HPAM concentrations. The relationship of the elastic modulus and the viscous modulus with HPAM concentration can be expressed as the following power law:

$$G'=G_1 exp(a_1 c) \quad (1)$$

$$G''=G_2 exp(a_2 c) \quad (2)$$

where G_1, G_2 represents either the elastic modulus or the viscous modulus, a_1 and a_2 are constants, and c is the concentration of the salt-tolerant HPAM solutions. Table 1 shows the parameters G_1, G_2 and a_1, a_2 of eqs.(1) and (2) for solutions with molecular weights of 2.5×10^7 g·mol^{-1} and 3.5×10^7 g·mol^{-1}. R_1^2 and R_2^2 in the table are the correlation coefficients for the regression of the equations (1) and (2).

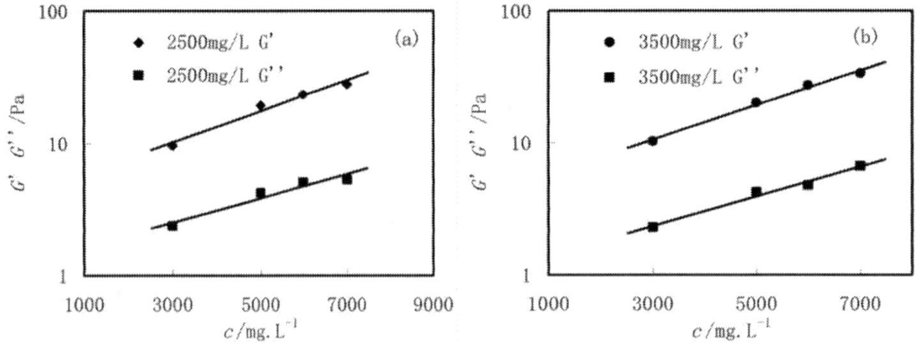

Figure 5. Concentration dependence of dynamic moduli for different HPAM solutions at 1 Hz (a) 2.5×10^7 g·mol^{-1} (b) 3.5×10^7 g·mol^{-1}

Table 1. Parameters of Eq. (1) for HPAM solutions

HPAM solutions with defferent molecular weight/ (g·mol^{-1})	G_1	a_1	R_1^2	G_2	a_2	R_2^2
2.5×10^7	4.27	0.0003	0.97	1.34	0.0002	0.94
3.5×10^7	4.49	0.0003	0.99	1.08	0.0003	0.98

The effect of molecular weight on rheological properties

Figure 6 contrasts the curves of the elastic modulus and viscous modulus of the HPAM solutions with molecular weights of 2.5×10^7 g·mol^{-1} and 3.5×10^7 g·mol^{-1}. The concentration of the solutions was 7000 mg·L^{-1} and different frequencies were used in the experiments. It was found that the HPAM solution with molecular weight of 3.5×10^7 g·mol^{-1} had higher elastic modulus and viscous modulus than that with molecular weigh of 2.5×10^7 g·mol^{-1}. The reason was that the entanglement and the molecular interactions became stronger when molecular weights were higher, making it more difficult for the polymer molecules in the HPAM solutions to move and thus increasing the viscous modulus. In addition, since the molecular chains tended to be longer for higher molecular weight solutions, the solution with molecular weight of 3.5×10^7 g·mol^{-1} had more ultrahigh-molecular configurations of internal rotation and its molecular chains became more flexible, which enhanced the elasticity of the HPAM solution. It was interesting to note that the differences of elastic modulus and viscous modulus between the solutions with molecular weights of 2.5×10^7 g·mol^{-1} and 3.5×10^7 g·mol^{-1} were not significant. This was because both HPAM solutions with molecular of 2.5×10^7 g·mol^{-1} and 3.5×10^7 g·mol^{-1} were within the ultrahigh molecular range. Similar behaviors could be observed with HPAM solutions at different concentrations, for instance, 3000 mg·L^{-1}, 5000 mg·L^{-1}, 6000 mg·L^{-1}. (Not shown)

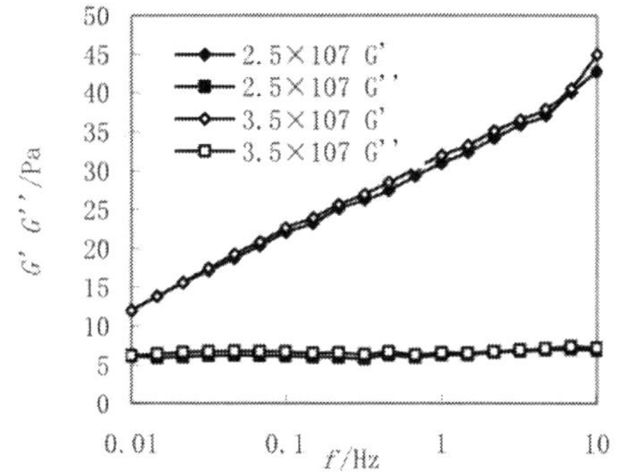

Figure 6. The comparisons between dynamic moduli and oscillatory frequency for the HPAM solution with molecular weight of 2.5×10^7 and 3.5×10^7

Conclusions

At the strain amplitude of 5%, all the investigated HPAM solutions showed shear thinning flow behavior at the oscillatory frequencies of 0.01~10 Hz. The elastic modulus increased with an increase in frequency while the viscous modulus changed very little. Unlike HAPM solutions with low molecular weights and low concentrations, the tested HPAM solutions had no crossover points between elastic modulus and viscous modulus and the elastic behavior of the solutions was predominant over the viscous behavior within the frequency range investigated. Based on a three-zone model for viscoelastic liquids, the dynamic rheological behaviors of all the HPAM solutions were assigned to the plateau zone. The elastic modulus and viscous modulus exhibited an exponential increase with the increase of the HPAM solution concentrations. The dynamic moduli of the HPAM solution with molecular weight of 3.5×10^7 g·mol^{-1} were a little higher than those of the solution with molecular weight of 2.5×10^7 g·mol^{-1}.

References

1. Fang DB, Guo RW, Ha RH (2006), "Acrylamide polymers", Beijing,China:

Chemical Industry Press.
2. Zhang Y, Wu X, Guo S, Dong K, Yang L (2006), "Performance Properties and Displacement Efficiency of High Concentration Solution of Superhigh Molar Mass HPAM Polymer", *Oilfield Chemistry*, December, pp. 345-348.
3. Moritis G. (1996), "New Technology, Improved Economics. Boost EOR Hopes", *Oil and Gas Journal*, April, pp. 39-61.
4. Taylor KC, Nasr-El-Din HA (1998), "Water-soluble hydrophobically associating polymers for improved oil recovery: A literature review", *Journal of Petroleum Science and Engineering*, September, pp. 265-280.
5. Yang F, Yang X, Wang D, Li J, Liu Y (2003), "A new enhanced oil recovery by high-concentration polymer flooding", *Journal of Daqing Petroleum Institute*, December, pp. 24-26.
6. Li J, Yang F, Yang X (2005), "Effects of timing of high-concentration polymer flooding on oil displacement efficiency", *Journal of Daqing Petroleum Institute*, October, pp. 16-18.
7. Liao G, Niu J, Shao Z, Chen P (2004), "Application and experience of industrialized polymer flooding in Daqing oilfield", *Petroleum Geology & Oilfield Development in Daqing, February*, pp. 48-51.
8. Hou J, Liu Z, Zhang S, Yue X, Yang J (2005), "The role of viscoelasticity of alkali/surfactant/polymer solutions in enhanced oil recovery", *Journal of Petroleum Science and Engineering*, June, pp. 219-235.
9. Perrin CL, Tardy PMJ, Sorbie KS, Crawshaw JC (2007), "Experimental and modeling study of Newtonian and non-Newtonian fluid flow in pore network micromodels", *Journal of Colloid and Interface Science*, March, pp. 542-550.
10. Zhang L, Yue X, (2007), "Mechanism for viscoelastic polymer solution percolating through porous media", *Journal of Hydrodynamics*, April, pp. 241-248.
11. Balhoff MT, Thompson KE (2006), "A macroscopic model for shear-thinning flow in packed beds based on network modeling", *Chemical Engineering Science*, January, pp. 698-719.
12. Ait-Kadi A, Carreau PJ (1987), "Rheological Properties of Partially Hydrolyzed Polyacrylamide Solutions", *Journal of Rheology*, October, pp. 537-561.
13. Lewandowska K. (2006), "Comparative studies of rheological properties of polyacrylamide and partially hydrolyzed polyacrylamide solutions", *Journal of Applied Polymer Science*, February, pp. 2235-2241.
14. Zhang L, Zhang D, Jiang B. (2006), "The rheological behavior of salt tolerant polyacrylamide solutions", *Chemical Engineering & Technology*, March, pp. 395-400.
15. Ghannam MT, Esmail MN. (1998), "Rheological properties of aqueous polyacrylamide solutions", *Journal of Applied Polymer Science*, August, pp. 1587-1597.
16. Mancini F, Montanari L, Peressini D, Fantozzi P (2002), "Influence of Alginate Concentration and Molecular Weight on Functional Properties of Mayonnaise", *LWT - Food Science and Technology*, September, pp. 517-525.
17. Li WH, Lynam C, Chen J, Liu B, Zhang XZ, Wallace GG. (2007), "Magnetorheology of single-walled nanotube dispersions", *Materials Letters*, June, pp. 3116-3118.
18. Phan-Thien N, Safari-Ardi M. J. (1998), "Linear viscoelastic properties of flour-water doughs at different water concentrations", *Journal of Non-Newtonian Fluid Mechanics*, January, pp. 137-150.
19. Walberer JA, McHugh AJ. (2001), "The linear viscoelastic behavior of highly filled polydimethylsiloxane measured in shear and compression", *Journal of Rheology*,

January, pp. 187-201.

20. Robertson CG., Roland CM, Paulo C, Puskas JE. (2001), "Linear viscoelastic properties of hyperbranched polyisobutylene", *Journal of Rheology*, May, pp. 759-772.
21. Hsieh TT, Tiu C, Simon GP, Wu RY. (1999), "Rheology and miscibility of thermotropic liquid crystalline polymer blends", *Journal of Non-Newtonian Fluid Mechanics*, September, pp. 15-35.
22. Lai G, Li Y, Li G. (2008), "Effect of concentration and temperature on the rheological behavior of collagen solution", *International Journal of Biological Macromolecules*, April, pp. 285-291.

TURBULENT HEAT TRANSFER IN A MICROFIN TUBE WITH TWISTED TAPE INSERT

Smith Eiamsa-ard and Vichan Kongkaitpaiboon
Mahanakorn University of Technology, Bangkok, Thailand
Petpices Eiamsa-ard and Monsak Pimsarn
King Mongkut's Institute of Technology Ladkrabang, Ladkrabang, Thailand

Biography

Smith Eiamsa-ard is an Associate Professor of Mechanical Engineering at Mahanakorn University of Technology, He obtained his D.Eng. from King Mongkut's Institute of Technology Ladkrabang
Vichan Kongkaitpaiboon is an Assistant Professor of Mechanical Engineering at Mahanakorn University of Technology, He obtained his D.Eng. from Mahanakorn University of Technology
Petpices Eiamsa-ard is D.Eng. candidate at King Mongkut's Institute of Technology Ladkrabang
Monsak Pimsarn is an Assistant Professor of Mechanical Engineering at King Mongkut's Institute of Technology Ladkrabang, He obtained his PhD. from University of Connecticut

Abstract

The purpose of the present study is to investigate experimentally the heat transfer, turbulent flow friction and thermal performance factor characteristics in a microfin tube with a twisted tape insert. In the experiment, heat transfer augmentation is expected from both the turbulence of flow near the tube wall produced by fin surface and the swirling flow generated by the twisted tape. Effect of the twisted tape at different twist ratios of y/W = 3, 4 and 5 on the heat transfer enhancement characteristics is also reported. In the experimental set-up, cold water was passed through the uniform heat flux tube in a range of Reynolds number between 5000 and 17000. The experimental results obtained are compared with those from plain tubes. The results show that the thermal performance from using the microfin tube fitted with twisted tape is considerably higher than that from the microfin tube alone. The Nusselt numbers and friction factors are found to be, respectively, 3.0, 4.0 and 5.0 times over the microfin tube alone for the microfin tube combined with twisted tape. In addition, the results demonstrate that as the twist ratio (y/W) decreases, the twisted tape will give better heat transfer enhancement. The microfin tube combined with twisted tape provides higher thermal performance factor than those the microfin tube alone around 172%, 142% and 124%, respectively, for y/W = 3, 4 and 5. The results were correlated in the form of Nusselt number as a function of Reynolds number, Prandtl number and twist ratio.

Keywords: heat transfer, twisted tape, microfin tube, turbulent flow, swirl flow, heat exchanger

Introduction

The technology of enhanced heat transfer has received serious attention over past decades; heat transfer augmentation techniques find application mainly in the design of more compact heat exchanger at various industries, especially refrigeration, automotive and chemical process industries. The enhancement of heat transfer has become an important factor in achieving these goals and has captured the interest of many researchers. Active and passive methods have been used to improve heat transfer in chemical reactors, heat exchanger and in flow systems for several decades. The principle of the passive technique involves in either surface treatment, such as coated surface, rough surface and extended surface or flow manipulation such as swirl flow and putting additive into the flow. Active techniques, which require an extra external power source, include mechanical aids, surface vibration, fluid vibration, electrostatic fields, injection or suction of fluid and jet impingement. One of the most favorable passive techniques is twisted tape inserts because they are inexpensive and can be easily employed to the existing system. The previous studied found that the twisted tape insert can help to generate the swirling flow and stronger turbulence in tubes. This causes a thinner boundary layer and longer resident time of the flow leading to an increase in the heat transfer coefficient. However, an increase in pressure drop is the penalty of the twisted tape technique.

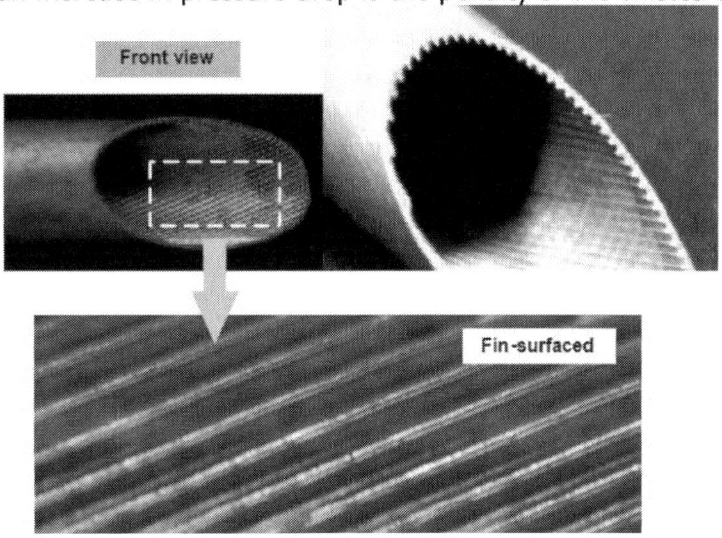

Fig. 1. Photograph of microfin tube in different location views.

The modified inner tube wall have been also widely employed in thermal engineering applications by introducing the turbulence flow near the tube wall such as microfin tube, helical tube corrugated tube and groove tube, etc. The effect of turbulence flow and boundary layer disruption is to enhance the heat transfer coefficient and momentum transfer. The turbulence flow with high turbulence intensity can improve convection of the tube wall by increasing the effective axial Reynolds number, decreasing the flow cross-section area, and increasing the mean velocity and temperature gradient. It can help to produce the higher heat fluxes and momentum transfer due to the large effective driving potential force but also higher pressure drop.

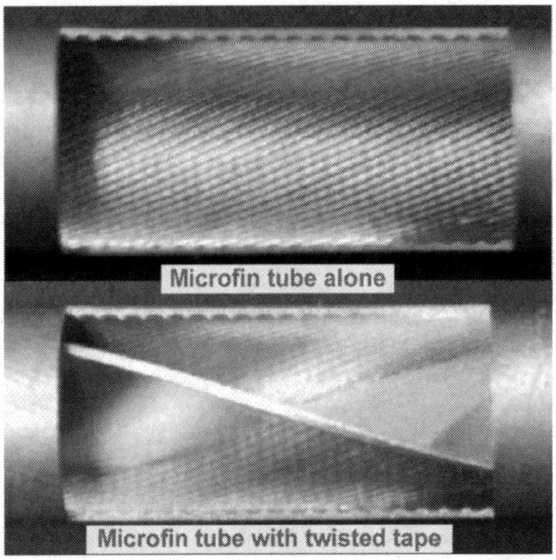

Fig. 2. Photograph of microfin tube with/without twisted tape insert in counter-flow arrangement.

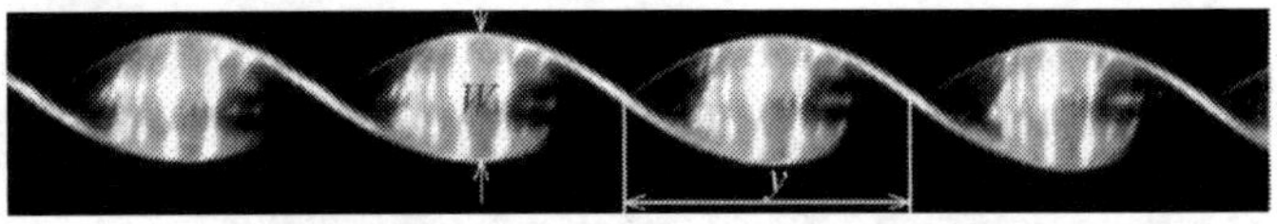

Fig. 3. Photograph of twisted tape.

Yong-Zhang and Mao-Cheng [1] performed heat transfer and pressure drop characteristics of a tube fitted with edgefold twisted tape. They found that the tube fitted with ETT provided higher heat transfer rate and friction factor than that typical twisted tape (TT) due to the higher tangential and asymmetrical velocities. Jafari Nasr and Khalaj [2] showed the thermohydraulic characteristics in a corrugated tube fitted with a twisted tape which predicted by the artificial neural network (ANN) approach. Murugesan *et al.* [3] reported the heat transfer and friction factor characteristics of a circular tube fitted with plain twisted tapes (PTT) and trapezoidal-cut twisted tapes (TTT). Al-Fahed *et al.* [4] conducted the heat transfer characteristics in tube fitted with tight-fit tapes (typical twisted tapes) and loose-fit tapes (modified twisted tapes) at various twisted ratios. Wang *et al.* Promvonge [5] examined the effect of the regularly spaced short-length twisted tape (effects of twist ratio and rotated angle) on turbulent heat transfer using computational fluid dynamic (CFD) technique. They found that the tape with larger rotated angle provided higher heat transfer and flow friction. Chang *et al.* [6] presented the heat transfer and friction factor and thermal performance factor characteristics in a heat exchanger tube fitted with serrated twisted tape in turbulent regimes. They observed that heat transfer attributed to the serrated twisted tape. The effects of twist ratio, perforation hole diameter ratio and spacing between holes (pitch) ratio were also reported. Apart from the modified twisted tapes mentioned above, other forms of twisted tapes such as serrated twisted tape (Eiamsa-ard and Promvonge [8]), alternate clockwise and counter-clockwise twisted-tapes (Wongcharee and Eiamsa-ard [9]): right-left helical tape (Sivashanmugam [10]: Eiamsa-ard *et al.* [11]).

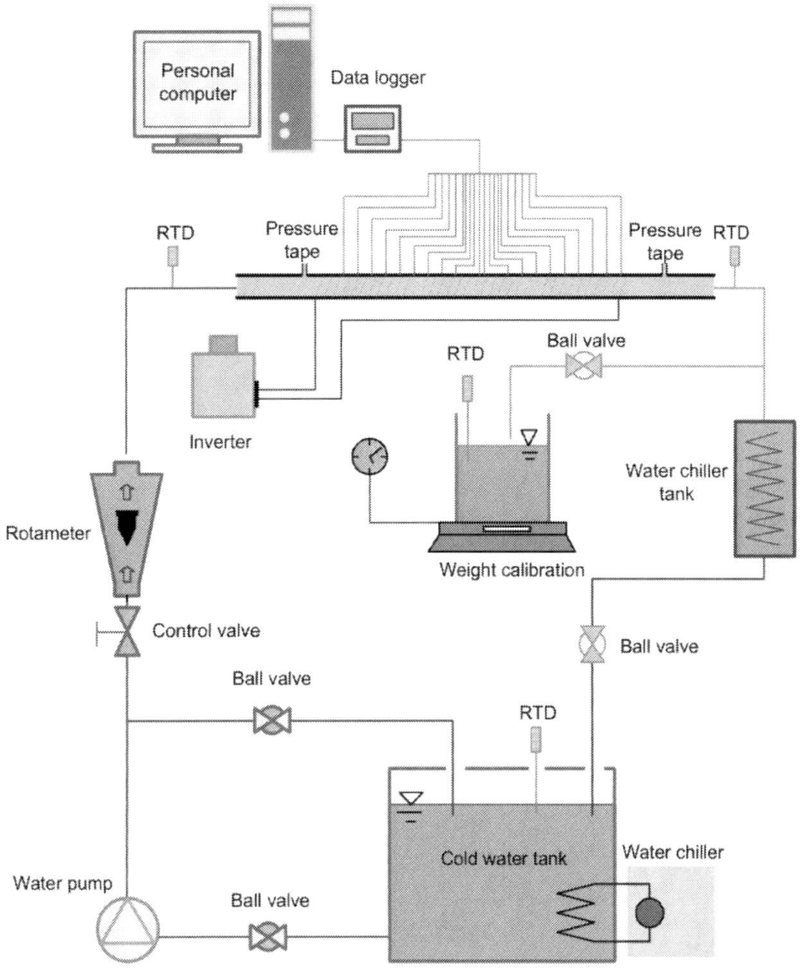

Fig. 4. Schematic diagram of the experimental heat transfer set-up.

In the present study, apart from improvement by swirl flows stemming from the centrifugal forces occurring from a twisted tape insert, the goal is to increase the heat transfer further with the passive method by incorporating a microfin tube. In this paper the influences of (1) microfin tube alone and (2) combination of the microfin tube and the twisted tape, on heat transfer and friction factor characteristics are experimentally investigated. In the experiments, twisted tapes were inserted into the microfin tube at three different twist ratios of y/W = 3, 4 and 5, respectively. All of the experiments are carried out at the same inlet conditions with the Reynolds number, based on the tube diameter, in a range of 5000 to 17000. In addition, results of the thermal performance factor are also described in this paper.

Microfin tube fitted with twisted tape

In the experiments, a microfin tube was made of copper with mean diameter of 8.88 mm and length of 600 mm as seen in Fig. 1. The tube was wound with electrical SWG Nichrome heating wire. The terminals of the Nicrome wire were connected to the varaic transformer.

The electrical output power was controlled via a variac transformer. Fifteen K-type thermocouples were tapped along the tube wall for monitoring the local temperatures of the surface tube wall. The heating tube and thermocouple were covered with insulation to minimize heat loss to surrounding. The heat transfer and pressure drop experiments were conducted individually. The heat transfer experiment was conducted under a uniform heat flux. During the heat transfer test, the test tube was heated by continually winding flexible electrical wire. On the other hand, the pressure drop (friction) test was performed under an isothermal condition without turning on the heater. In the experiments, the twisted tapes were installed into the microfin tube at three different twist ratios of y/W = 3, 4 and 5 as shown in Fig. 2. Details of the twisted tape are also presented in Fig. 3.

Experimental apparatus and procedure

The test loop consists of a water pump, data logger, pressure transmitter, thermocouple/RTD, rotameter, and heat transfer test section is depicted in Fig. 4. In the apparatus setting above, the water was pumped to the overhead water tank then discharged through the rotameter to measure the volumetric flow rate and directed to the heat transfer test section. For each test run, it was necessary to record the data of temperature, volumetric flow rate and pressure drop of the bulk water at steady state conditions in which the inlet water temperature was maintained constant at 25°C. During each test, local wall temperatures, inlet and outlet water temperatures, pressure drop across the test section and flow velocity were measured. The isothermal friction factor was calculated from pressure drop measured by manometer at room temperature. In the experiment, the pressure taps were located around 30 mm upstream and 20 mm downstream of the test section. The Reynolds number of cold water was varied from 5000 to 17,000. Uncertainties of measurements are estimated based on ANSI/ASME. The uncertainties of axial velocity, pressure and temperature are found to be within ±7%, ±5% and ±0.5%, respectively. In addition, the uncertainties of non-dimensional parameters are ±5% for Reynolds number, ±10% for Nusselt number and ±12% for friction.

Data reduction

The steady state convective heat transfer rate is assumed to be equal to the heat transfer to the working fluid (water) in the test tube as expressed in Eq. (1):

$$Q_{water} = Q_{conv} \tag{1}$$

The heat supplied by electrical heater plates in the test tube is found to be 3 to 7% higher than the heat absorbed by water at thermal equilibrium. This is due to convection and radiation heat losses from the test tube to surroundings. Thus, only the heat transfer rate absorbed by fluid is taken for internal convective heat transfer coefficient calculation. The convection heat transfer from the test tube can be written by

$$Q_{conv} = hA(\tilde{T}_s - T_b) \tag{2}$$

where

$$T_b = (T_o + T_i)/2 \tag{3}$$

$$\tilde{T}_s = \Sigma T_s / 12 \tag{4}$$

where the average surface temperature ($\tilde{T}_s$) is calculated from 12 data of the local surface temperatures (T_s) lined between the inlet and the exit of the test tube. The average heat transfer coefficient (h) and the average Nusselt number (Nu) are then estimated as follows:

$$h = \dot{m}C_{p,water}(T_o - T_i)/A(\tilde{T}_s - T_b) \tag{5}$$

$$Nu = hD/k \tag{6}$$

The Reynolds number is given by

$$Re = UD/\nu \tag{7}$$

Friction factor (f) can be written as

$$f = \frac{\Delta P}{(L/D)\rho U^2/2} \tag{8}$$

in which U is mean water velocity in the tube. All of thermo-physical properties of the water are determined based on the overall bulk water temperature.

For a constant pumping power

$$(\dot{V}\Delta P)_p = (\dot{V}\Delta P)_s \tag{9}$$

and the relationship between friction and Reynolds number can be expressed as:

$$(f\,Re^3)_p = (f\,Re^3)_s$$

$$Re_p = Re_s(f/f_p)^{1/3} \tag{10}$$

As preliminary design guidance to the selection of a technique, the thermal performance factor can be evaluated based on the power consumption per unit mass of fluid. The thermal performance factor criteria, η is defined as the ratio between the heat transfer coefficient for the tube with heat transfer promoter (h) and the value for the plain tube (h_p) at identical pumping power.

$$\eta = \left.\frac{h}{h_p}\right|_{pp} = \left.\frac{Nu}{Nu_p}\right|_{pp} = \left(\frac{Nu}{Nu_p}\right)\left(\frac{f_p}{f}\right)^{-1/3} \tag{11}$$

Experimental results

Effect of microfin tube

Effect of using the microfin tube on heat transfer characteristics is shown in Fig. 5. It is found that the microfin tube gives higher heat transfer rate than the plain tube. The maximum Nusselt number increases at about 38% when compared with those from the plain tube. This can be attributed to better mixing of fluids between the wall using fin-surfaced wall causing the turbulence flow and pressure gradient in the radial direction. The boundary layer along the tube wall would be redeveloped along the fin wall resulting in more heat flow through the fluid. Furthermore, the turbulence enhances the turbulence fluctuations, which leads to even better convection heat transfer. Thus, the higher the Reynolds number is, the greater the Nusselt number. The relationship between the friction factor and Reynolds number for using the microfin tube is presented in Fig. 6. In the figure, it is worth noting that friction factor of the microfin tube is considerably higher than that of the plain tube. This is because of higher surface area and the dissipation of dynamic pressure of the fluid at high viscosity loss near the tube wall. The friction factor is found to reduce gradually for increasing Reynolds number. In addition, it is found that the thermal performance factor of the microfin tube decreases with increasing Reynolds number. The microfin tube alone yields the maximum thermal performance factor at 1.12. This may be explained by the fact that good contacts of the core fluid with the tube wall from turbulence between the fin elements.

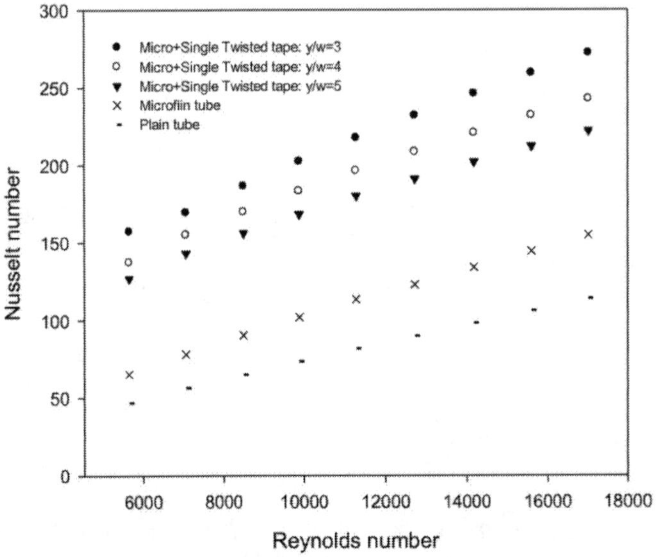

Fig. 5. Nusselt number versus Reynolds number.

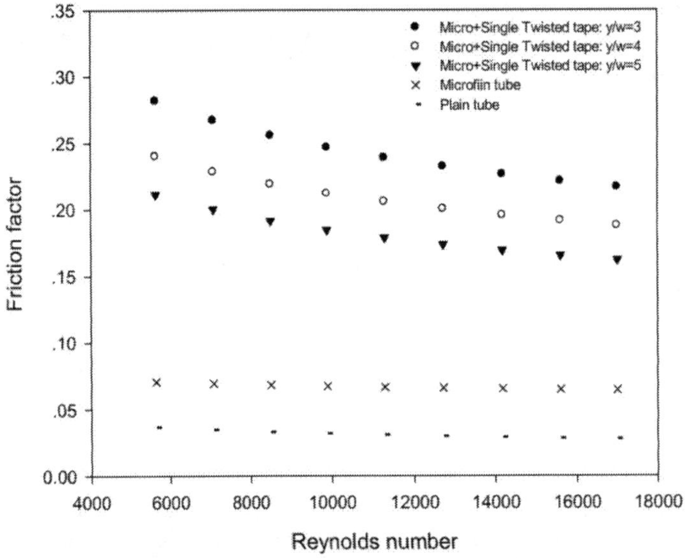

Fig. 6. Friction factor versus Reynolds number.

Effect of combined microfin tube and twisted tape insert

The results of the measurements with using the enhancement devices are also shown in Fig. 5. In the figure, it can be seen that use of the microfin tube with the twisted tape yields a higher heat transfer rate than that of the microfin tube alone. The increase in Nusselt number can be explained by two mechanisms: firstly, the surface of the microfin tube may be act as extended heat transfer surfaces and secondly, it is due to strong turbulence/swirl flow created by the twisted tape. In general, the average heat transfer rate for microfin tube with the twisted tape is found to be 63 to 98% better than that for the microfin tube alone. The corresponding increase in the mean Nusselt number of microfin tube with the twisted tape is about 124 to 172% over the plain tube.

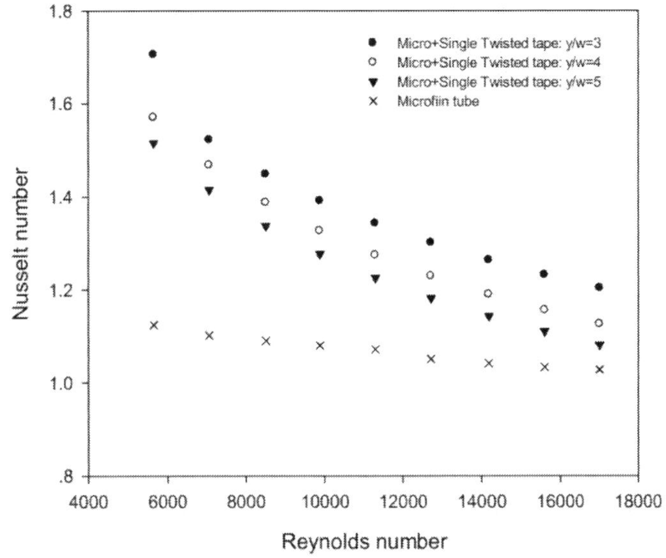

Fig. 7. Thermal performance factor versus Reynolds number.

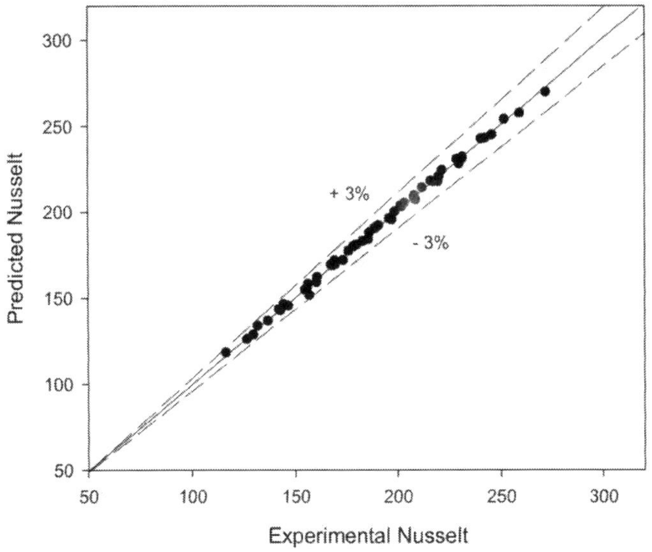

Fig. 8. Comparison of Nusselt numbers predicted by correlations with measured data.

It is apparent that the use of the microfin tube along with the twisted tape also provides higher friction factor than that of the microfin tube alone due to larger contact surface areas and flow disturbance at the core as shown in Fig. 6. Besides, the presence of the microfin tube together with twisted tape reduces flow areas, resulting in a high speed rotating flow. This leads to the substantial pressure loss action of the fluid higher than one without the twisted tape. For the microfin tube with the twisted tape, the increase in friction factor is found to be around 171 to 262% above one without the tape. It is reveal that the mean friction factors of using microfin tube fitted with twisted tape are found to be about 3.6, 3.1 and 2.7 times for the y/W = 3.0, 4.0 and 5.0, respectively, over the microfin tube alone.

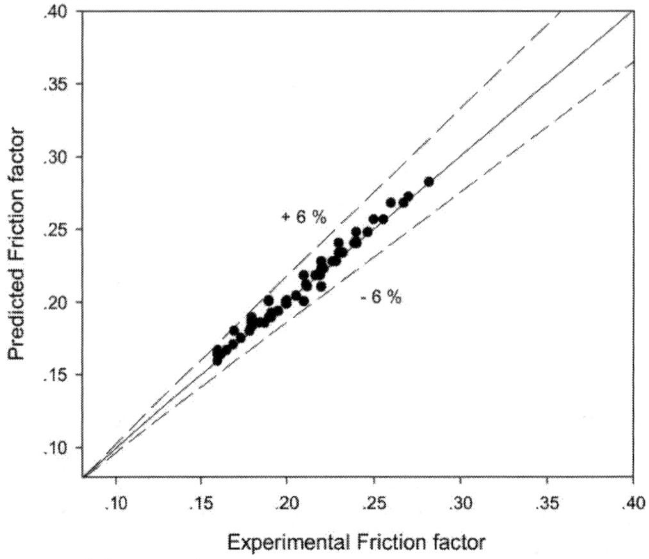

Fig. 9. Comparison of friction factors predicted by correlations with measured data.

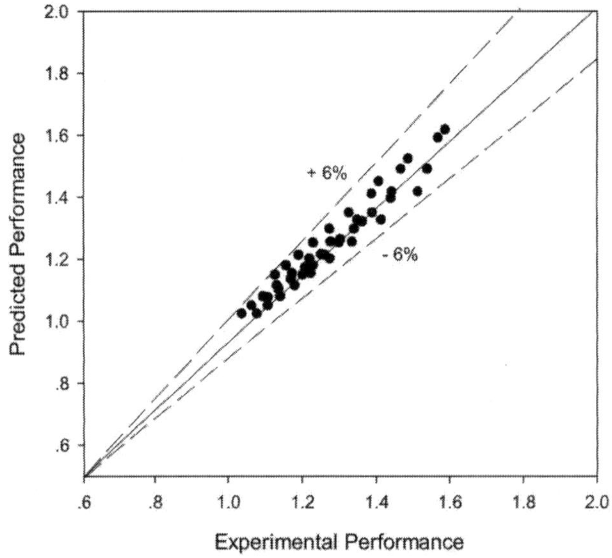

Fig. 10. Comparison of thermal performances predicted by correlations with measured data.

The influence of microfin tube with a twisted tape insert on thermal performance factor is presented in Fig. 7. It can be observed in the figure that use of the microfin tube fitted with the twisted tape leads to higher thermal performance than those of the microfin tube alone. Thermal performance varied between 1.17 and 1.52, 1.10 and 1.40, and 1.05 and 1.35 for the microfin tube with twisted tape at y/W = 3, 4 and 5, respectively. The thermal performance tends to reduce with the increase of Reynolds number. The microfin tube with twisted tape shows a faster decrease in the thermal performance than the others for increasing Reynolds number. Close examination reveals that the thermal performances for all enhancement devices are higher than unity. This suggests that the thermal performance with/without the twisted tape is not feasible in terms of energy saving, especially at higher Reynolds numbers.

Figures 8 to 10 show comparisons between the present experimental data and the present predictions by the correlations below. In the figures, results of the present correlations reasonably agree well within ±3%, ±6% and ±6%, in comparison with experimental data for the Nusselt number, friction factor and thermal performance factor. The present results are correlated with the Nusselt number, friction factor and thermal performance factor as seen below.

Correlation for microfin tube alone:

$$\mathrm{Nu} = 0.048\,\mathrm{Re}^{0.76}\,\mathrm{Pr}^{0.4} \tag{12}$$

$$f = 3.14\,\mathrm{Re}^{-0.23} \tag{13}$$

$$\eta = 1.26\,\mathrm{Re}^{-0.05} \tag{14}$$

Correlation for microfin tube with twisted tape:

$$\mathrm{Nu} = 2.11\,\mathrm{Re}^{0.467}\,\mathrm{Pr}^{0.4}(y/W)^{-0.36} \tag{15}$$

$$f = 3.97\,\mathrm{Re}^{-0.23}(y/W)^{-0.56} \tag{16}$$

$$\eta = 24.50\,\mathrm{Re}^{-0.29}(y/W)^{-0.17} \tag{17}$$

Conclusions

An experimental study has been conducted to investigate heat transfer and friction loss behaviors in a microfin tube fitted with a twisted tape. The experimental result shows that apart from the friction factor, the heat transfer rate can be substantially improved by using microfin tube fitted with a twisted tape compares to those the microfin tube alone. This is due to better mixing behavior between two streams from using the twisted tape: swirling flow around the tape and the turbulence flow along the fined wall. The maximum increase in heat transfer rate and friction factor are found to be about 2.4 and 4 times the microfinn tube for the flow range studied.

Acknowledgements

The authors would like to gratefully acknowledge the Thailand Research Fund (TRF) and Thailand Toray Science Foundation was founded (TTSF) for the financial support of this research.

References

1. Yong-Zhang, C. and Mao-Cheng, T. (2010), "Three-Dimensional Numerical Simulation of Thermalhydraulic Performance of a Circular Tube with Edgefold-Twisted-Tape Inserts," *Journal of Hydrodynamics*, 22, pp. 662-670.
2. Jafari Nasr, M.R. and Khalaj, A.H. (2010), "Heat Transfer Coefficient and Friction Factor Prediction of Corrugated Tubes Combined with Twisted Tape Inserts using Artificial Neural Network," *Heat Transfer Engineering*, 31, pp. 59-69.
3. Murugesan, P., Mayilsamy, K. and Suresh, S. (2011), "Heat Transfer in Tubes Fitted with Trapezoidal-Cut and Plain Twisted Tape Inserts," *Chemical Engineering Communications*, 198, pp. 886-904.
4. Al-Fahed S., Chamra L.M. and Chakroun W. (1998), "Pressure Drop and Heat Transfer Comparison for Both Microfin Tube and Twisted Tape Inserts in Laminar Flow," *Experimental Thermal and Fluid Science*, 18, pp. 323-333.
5. Wang, Y., Hou, M., Deng, X., Li L., Huang, C., Huang, H., Zhang, G., Chen, C. and Huang, W. (2011), "Configuration Optimization of Regularly Spaced Short-Length Twisted Tape in a Circular Tube to Enhance Turbulent Heat Transfer using CFD Modeling," *Applied Thermal Engineering*, 31, pp. 1141-1149.
6. Chang S.W., Jan Y.J. and Liou J.S. (2007), "Turbulent Heat Transfer and Pressure Drop in Tube fitted with Serrated Twisted Tape," *International Journal of Thermal Sciences*, 46, pp. 506-518.
8. Eiamsa-ard, S. and Promvonge, P., (2010), "Thermal Characteristics in Round Tube Fitted with Serrated Twisted Tape," *Applied Thermal Engineering*, 30, pp. 1673-1682.
9. Eiamsa-ard, S., Wongcharee, K., Eiamsa-ard, P. and Thianpong, C. (2010), "Thermohydraulic Investigation of Turbulent Flow Through a Round Tube Equipped with Twisted Tape Consisting of Centre Wings and Alternate-Axes," *Experimental Thermal and Fluid Science*, 34, pp. 1151-1161.
10. Sivashanmugam, P. (2008), "Modeling of Heat Transfer Augmentation in a Circular Tube Fitted with Right-Left Helical Inserts in Laminar Flow using CFD," *Chemical Product and Process Modeling*, 3, A23.
11. Wongcharee, K. and Eiamsa-ard, S. (2011), "Friction and Heat Transfer Characteristics of Laminar Swirl Flow Through a Round Tube Inserted with Alternate Clockwise and Counter-Clockwise Twisted-Tapes," *International Communications in Heat and Mass Transfer*, 38, pp. 348-352.

(142ch) Transition in a Dense Granular Flow

V. Kumaran and S. Maheshwari, Chemical Engineering, Indian Institute of Science, Bangalore, India

The effect of base roughness on the flow down an inclined plane is studied using the discrete element method (DEM), in which the interaction force between particles is modeled using a spring-dashpot model in which the force due to deformation consists of a spring force proportional to the relative displacement of the surface of contact between the both parallel and perpendicular to the surfaces, and a damping force proportional to the relative velocity. The simulations are carried out using the open source code, LAMMPS (Large-scale Atomic Molecular Massively Parallel Simulator). The base consists of a random configuration of frozen particles, and the base roughness is varied by varying the ratio of the diameters of the frozen and moving particles. As the base particle diameter is decreased, a discontinuous change is observed in both the relative arrangement of particles and in the flow dynamics at a critical base particle diameter. Above the transition particle diameter, the relative particle arrangement is disordered, the flow is less dense (volume fraction 0.59 or less), whereas below the transition diameter, the particles flow in the form of layers sliding past each other, with hexagonal ordering within the layers, and the flow is more dense (volume fraction in the range 0.62-0.64). In both regimes, it is found that Bagnold law is valid, and the stress is proportional to the square of the strain rate. However, the Bagnold coefficients (ratio of stress and square of strain rate) are higher, by more than an order of magnitude, in the disordered state in comparison to the ordered state. The physical basis for the transition will be discussed using both theories for flow instabilities in homogeneous fluids, as well as the instability of a layered flowing structure.

(142ci) Microfluidic Production of Food Emulsions Containing Anisotropic-Shaped Fat Particles

Jihye Kim, Chenical Engineering, Texas Tech University, Lubbock, TX and Siva A. Vanapalli, Chemical Engineering, Texas Tech University, Lubbock, TX

Shape of particles such as ice crystals and fat particles contribute significantly to the rheology of food suspensions. For example, at the same volume fraction, a suspension of ellipsoids can generate larger shear viscosity that a suspension of spherical particles. Thus, designing particles of anisotropic shape and using them as food additives offers a promising route to control the rheology of food suspensions. However, methods to produce anisotropic food particles are scarce. Here, we report the development of microfluidic techniques to generate ellipsoidal fat particles.

We assemble cross-flow microfluidic devices, using commercially available parts and fittings. We use palm oil (Sans Trans 35) as the dispersed phase and 2% Tween 20 as the continuous phase. We preheat both the continuous and dispersed phases and inject them using syringe pumps into the cross-flow device. Oil droplets produced at the cross-junction are carried downstream. We control the temperature of the downstream section to induce crystallization in the flowing droplets. Droplet generation is viewed using optical microscopy and image analysis is conducted to quantify droplet sizes.

We find that confinement of the droplets in the microchannel together with partial-solidification of the droplets enables optimal production of ellipsoidal fat particles. The control parameters including local temperature, residence time, and flow rates of the incoming fluids sensitively affected the ability to continuously produce fat particles. The drop speed required for the production of ellipsoidal particles was lower than that for spherical fat particles. In addition, higher aspect ratio fat particles were obtained at lower flow rates of the carrier fluid.

(142cj) Continuum/Molecular Hybridization Model for Small Scale Systems

Sesha Hari Vemuri[1], Robert L. Smith[2], Pil Seung Chung[1], Lorenz T. Biegler[1] and Myung S. Jhon[1], (1)Chemical Engineering, Carnegie Mellon University, Pittsburgh, PA, (2)Dept. of Chemical Engineering, Carnegie Mellon University, Pittsburgh, PA

Considerable effort is being paid to improve the performance and the efficiency of micro/nano electro-mechanical systems (M/NEMS), where nanoscale/microscale information transfer plays a critical role. The hierarchical nature of these systems consists of complicated nano-scale multiphysical phenomena between various sub-component interfaces with macro/mesoscale fluid flow. The continuum based theory breaks down for these systems as the confinement size is comparable to or smaller than the molecular/pseudo-particle mean free path. On the other hand, although atomistic/molecular dynamics (MD) simulation methodologies have the capabilities of calculating these phenomena, it is extremely difficult to extend simulations to real systems due to the memory requirement and computational time limitations.

We have been successful in the past in developing two separate theories to describe the multi-scale/multi-physical phenomena: (i) a novel mesoscale lattice Boltzmann method (LBM), which has the capabilities of describing transient phenomena with parallel computation facilitating the handling of complex geometries. The LBM scheme was developed for analyzing the dynamics for high Knudsen number compressible flows and heat transfer [1-3], and it was found to accurately capture the characteristic velocity and temperature slip at the interface in the nano-scale by introducing mixed boundary conditions. (ii) In the molecular scale, we developed a coarse-grained molecular dynamics model (CGMD) (with the molecular parameters obtained from ab-initio calculations) which can accurately describe the nanoscale systems with particle and solid surface coupling [4, 5]. In this study, we will hybridize the above two parallel approaches on continuum and molecular levels, and will thoroughly examine the critical fluid–fluid and fluid–solid interfaces within the multiscale framework [6].

This will be achieved by generating a buffer zone at the component interface where both CGMD and LBM solutions are valid. Here, the mean particle velocities obtained from CGMD provide boundary conditions for the LBM solution at one side of the overlap region and the constrained dynamics algorithm forces the instantaneous mean particle velocity to equate the continuum solution at the other boundary. The average continuum velocity in each cell is obtained by averaging the velocities of all CGMD particles within a differential volume that is centered on the point of interest. The equations of motion in the CGMD are also modified to match the external force obtained from the LBM via proper scaling.

To validate our hybrid model, we examine our results of the test case for simple fluid flow in a confined geometry and will extend our analysis towards the complex multi-physical phenomena in N/MEMS applications.

[1] W. T. Kim, M. S. Jhon, Y. Zhou, I. Staroselsky, and H. Chen, "Nanoscale air bearing modeling via lattice Boltzmann method," J. Appl. Phys. **97**, 10P304 (2005)

[2] H.M Kim, D. Kim, W. T. Kim, P.S. Chung, and M. S. Jhon, "Langmuir slip model for air bearing simulation using the lattice boltzmann method," *IEEE Trans. Magn*., **43**, 2244 (2007)

[3] S.S. Ghai, W.T. Kim, R.A. Escobar, C.H. Amon, and M.S. Jhon, "A novel heat transfer model and its application to information storage systems," J. Appl. Phys., 97, 10P703 (2005)

[4] S. H. Vemuri, P.S. Chung, R.L. Smith, L.T. Biegler, and M.S. Jhon, "Perfluoropolyether Lubricant Interactions with Novel Graphene Overcoat via Coarse-grained Molecular Dynamics," *IEEE Trans. Magn*., (accepted)

[5] P.S. Chung and M.S. Jhon, "Molecular spreading of pure and binary mixture PFPE nano films on carbon-

overcoated disks," *IEEE Trans. Magn.*, **46**, 2405 (2010)

[6] P.S. Chung, M.S. Jhon, and L.T. Biegler, "The holistic strategy in multiscale modeling," *Adv. Chem. Eng.*, **40,** 59 (2011)

(142ck) Mass Transfer Characteristic of Gas-Liquid Segmented Flow in T-Junction Microchannels

Jing Tan, The State Key Laboratory of Chemical Engineering, Beijing, China, Jianhong Xu, Department of Chemical Engineering, The State Key Lab of Chemical Engineering, Beijing, China, Yangcheng Lu, Chemical Engineering, Tsinghua University, Beijing, China and Guangsheng Luo, Department of Chemical Engineering, The State Key Laboratory of Chemical Engineering, Beijing, China

In this paper, the mass transfer characteristic of gas-liquid segmented flow in T-junction microchannels was systematically investigated, in which the mass transfer in both the formation and flowing stage was distinguished and studied respectively. Gas-liquid segmented flow was realized in T-junction microchannles with different contacting angles and an in situ method was developed to measure the mass transfer characteristic in the formation stage. The result showed that the slug formation time was in the range of 0.2-0.4 s and the mass transfer during the formation stage could contribute around 30-40% of the total transferred solute. The overall mass transfer coefficient during the slug formation stage was in the range of 1.4-5.5 * 10^{-4} m/s. The comparison between the experimental results and the calculated data by diffusion model indicated that the recirculation flow in the gas slug significantly intensified the mass transfer with the enhancement factor in the range of 5-15. A correlation equation was developed to predict the enhancement factor, considering both the influences of the surface renewal and the moving speed of the gas phase. The calculated data was in good coincidence with the experimental results. On the other hand, a systematical study on the mass transfer performance of segmented gas-liquid flows during flowing stage was conducted and the perpendicular shearing method was applied to generate the gas-liquid microflow. An in situ measuring method was developed, with which the overall mass transfer coefficient in the flowing stage could be determined by measuring the change of gas slug volume versus flowing distance. Meanwhile, the overall mass transfer coefficient in both straight and curve shaped microchannels were determined and compared. The influences of both the gas and liquid flow rates, the gas concentration and the curve radii of main channel on the overall mass transfer coefficient were investigated. The overall mass transfer coefficient in both straight and curve shaped microchannels respectively ranged in 1-3* 10^{-4} m/s and 2-7 * 10^{-4} m/s, indicating the mass transfer in segmented gas-liquid flows can be enhanced by changing the main channel shape. Semi-empirical equations have been developed to correlate overall mass transfer coefficient in straight and curve shaped microchannels and calculated results show that the equations can predict the experimental values well.

(142cl) Capsule Motion in Constricted Microchannels

Sun-Young Park and Panagiotis Dimitrakopoulos, Chemical and Biomolecular Engineering, University of Maryland, College Park, MD

The study of the interfacial dynamics of artificial or physiological capsules (i.e. membrane-enclosed fluid volumes) in Stokes flows has seen an increased interest during the last few decades due to their numerous industrial, biological and physiological applications.

In this study, we investigate the motion of capsules in a microfluidic channel with a rectangular constriction by employing our three-dimensional Interracial Spectral Boundary Element method for elastic capsules. In particular, we study strain-hardening capsules with comparable shearing and area-dilatation resistance.

In our talk we will present the effects of capsule size and flow rate on the capsule dynamics in the confined solid geometries. As the channel s flow rate increases or the capsule size increases (i.e. higher hydrodynamic forces), the capsule deformation increases. In addition, as the capsule size increases, its velocity decreases while the additional pressure difference increases significantly, owing to the smaller lubrication gap between the capsule surface and the solid walls.

(142cm) Bridge Dynamics for Liquid Micro-Deposition: Computations and Experiments

Yechun Wang, Artur Lutfurakhmanov and Iskander Akhatov, Mechanical Engineering, North Dakota State University, Fargo, ND

The dyanmics of liquid bridge is an important problem which occurs in a variety of physical and technological processes, including ink-jet printing, microlithography, coating processes etc. In this study, we focus on the behavior of liquid bridge formed between a capillary and a substract with a direction application in capillary based droplet deposition, an important component of microlithography.

The experimental process of the capillary based liquid deposition can be divided into two steps. First, a static liquid bridge between the capillary and the substrate is formed when the liquid meniscus on the capillary tip touches the substrate. Second, the capillary retraction away from the substrate causes the bridge elongation and rupture, leaving a micro/nano-droplet on the substrate.

A three-dimentional spectral boundary element method is applied to describe the bridge dynamics while the capillary is retracted away from the substrate. Modeling results are compared with experimental data. The influences of the liquid properties, pressure, and retraction speed will be discussed and analyzed. The motion of the three-phase contact line as well as the relation between the contact line speed and the contact angle will be discussed based on experimental observations and numerical simulations.

(142cn) Numerical Simulations of Dilute Sedimenting Suspensions At Finite-Particle Reynolds Numers

Radompon Sungkorn, Chemical and Materials Engineering, University of Alberta, Edmonton, AB, Canada and Jos Derksen, Department of Chemical & Materials Engineering, University of Alberta, Edmonton, AB, Canada

Sedimenting suspensions can be found in various natural and man-made processes. Examples include the sedimentation of dust in the atmosphere, the centrifugation of proteins, and the deposit of contaminants in waste water. Attempts have been made by researchers to understand the fundamental properties of sedimenting suspensions using numerical simulations. However, a numerical method that can correctly reproduce experimental observations is not yet achieved.

We present an alternative numerical method for simulations of suspension flows with application to sedimenting suspensions at finite-particle Reynolds numbers. The locally averaged conservation equations for the fluid phase are discretized using an extended lattice-Boltzmann scheme. A Lagrangian particle tracking model is used to track the trajectories of individual particles. The present method enables us to simulate large-scale suspension systems with reasonable computational expenses.

The present method is able to capture the main features of sedimenting suspensions observed in experiments, *e.g.*, swirl structures, helical structures, and saddle points. At low-particle Reynolds numbers, the simulated particle velocity fluctuation amplitudes closely follow the scaling derived from the experimental data in the literature. At finite-particle Reynolds numbers, our simulation results suggest a transition in the fluctuation amplitudes at a particle Reynolds number around 0.08 (Fig. 1(a)) in accordance with the work of Yin & Koch [1] who used a surface-resolved numerical simulation method. We numerically demonstrate that there exists a range of domain sizes in which the fluctuation amplitudes have a strong domain size dependence, beyond which the fluctuation amplitudes saturate (Fig. 1(b)). This is the first time that the dependency of the fluctuation amplitudes on the domain size is numerically reproduced. Furthermore, we found that the magnitude and domain size dependence of the fluctuation amplitudes at finite-particle Reynolds numbers are well-represented by introducing new fluctuation amplitudes scaling functions and characteristic scaling function in the correlation derived by Segre et al. [2].

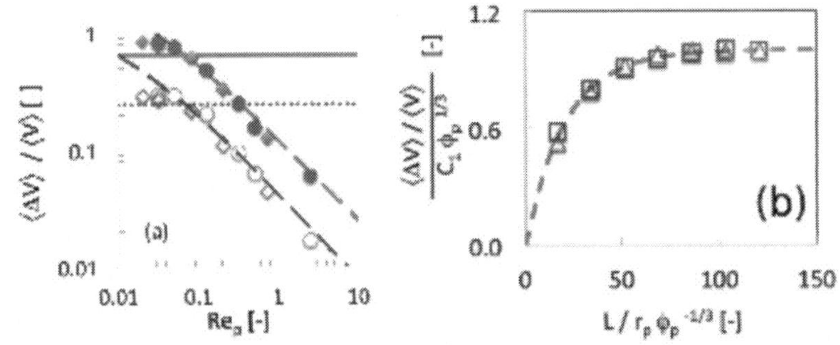

Figure 1. (a) Long-term average fluctuation amplitudes ratio as a function of the particle Reynolds number in the direction parallel (filled symbols) and perpendicular (empty symbols) to gravity. The solid and dotted lines are the calculation from the empirical correlation presented in the literature and our correlation, respectively. (b) Long-term average fluctuation amplitudes in the direction parallel (squares) and perpendicular (triangles) to gravity versus the domain size. Dotted line is the calculation from our correlation.

References

[1] Yin X., Koch D. L. Velocity fluctuations and hydrodynamic diffusion in finite-Reynolds number sedimenting

suspensions. Phys. Fluids. 2008;20:043305.

[2] Segre P. N., Herbolzheimer E., Chaikin P. M. Long-range correlations in sedimentation. Phys. Rev. Lett. 1997;79:2574-2577.

(142co) Strong Electrokinetic Flows about a Charged Colloidal Particle

John Y. Liow and Aditya S. Khair, Chemical Engineering, Carnegie Mellon University, Pittsburgh, PA

The motion of a charged micron-sized particle instigated by a hydrodynamic flow or electric field is of fundamental interest in colloid science. Notably, the majority of previous work in this area has been restricted to the "linear-response" regime wherein the ionic (Debye) cloud that screens a charged particle in an electrolyte solution is only slightly distorted by the "weak" applied flow or field. In this poster, we instead focus on "strong" flows or electric fields, wherein the Debye cloud can be highly distorted. Specifically, we consider two paradigmatic problems, via numerical solution of the nonlinear electrokinetic equations: (i) a charged particle held fixed in a uniform fluid flow (the strong flow regime being relevant to ultra-centifugation and streaming flows in microfluidics); and (ii) the electrophoretic motion of a sphere in an imposed electric field (the strong field regime being relevant to electrophoresis in non-polar solvents). In case (i) we determine the force-velocity relationship for a wide range of imposed flow strengths and Debye layer thicknesses. In case (ii) we compute the electrophoretic mobility as a function of the applied field strength. In both cases, we observe marked departures from the respective classic linear-reponse predictions, primarily as a result of the strongly deformed Debye layer in the high flow/field regime.

(142cp) Microscale Transport with Application to Sensing Stability in Non-Aqueous Suspensions

Nichole Asermely[1], Shelley L. Anna[2] and Lynn Walker[1], (1)Chemical Engineering, Carnegie Mellon University, Pittsburgh, PA, (2)Chemical Engineering and Mechanical Engineering, Carnegie Mellon University, Pittsburgh, PA

Microscale Transport with Application to Sensing Stability in Non-aqueous Suspensions

Nichole M. Asermely[1], Shelley L. Anna[1,2] and Lynn M. Walker[1]

Center for Complex Fluids Engineering

[1]Department of Chemical Engineering, [2]Department of Mechanical Engineering

Carnegie Mellon University, Pittsburgh, Pennsylvania 15213, USA

Colloid stability in non-aqueous media is an important aspect of numerous commercial products such as coatings, fuels, and pesticides. While colloid science in aqueous systems has been intensively studied and well characterized, the interparticle forces which occur in non-aqueous systems and how these connect to macroscopic stability are less understood. More important is that tools to characterize colloidal interactions and stability in non-aqueous media are not as well developed. Microfluidics offers the capability to rapidly vary experimental parameters, such as solvent quality or stabilizer composition, over a wide range in a single experiment. The development of microfluidic platforms for high throughput analysis of colloidal stability will require improved understanding of microscale transport in low Reynolds number flow. We are investigating the microscale transport issues associated with two different approaches to sensing the initial stages of colloidal aggregation for characterization of colloidal stability in non-aqueous systems. The transport and fluid mechanics aspects in characterization of colloidal stability will be presented and analyzed.

(142cq) Hollow Polymer Three-Dimensional Micro-Lattice Heat Exchangers

Christopher S. Roper, Kevin J. Maloney and Alan J. Jacobsen, HRL Laboratories, LLC, Malibu, CA

Title: Hollow Polymer Three-Dimensional Micro-lattice Heat Exchangers

Authors: C. Roper, K. Maloney, and A. Jacobsen

Abstract:

Compact heat exchangers with micro-scale flow paths offer higher overall heat transfer coefficients and higher surface area to volume ratios compared to traditional heat exchangers [1]. Polymer heat exchangers offer superior corrosion resistance and lower weight compared to metallic heat exchangers of similar size and geometry [2]. Compact polymer heat exchangers with micro-scale flow paths combine these benefits [3]; however, the inherent 2-D nature of each fabrication step used for traditional polymer microfluidics impedes production of compact polymer heat exchangers with sufficiently large overall size for many applications.

The recent development of a scalable 3-D micro-lattice material formed via a self-propagating photopolymer waveguide process [4] has enabled a new class of multifunctional mini- and micro-heat exchangers [5,6]. In the fabrication process, controlled feature sizes can be selected between 10 microns and 10 mm. Additionally, micro-lattice panels tens of mm thick, tens of cm long, and tens of cm wide can be fabricated with an exposure time of less than one minute. Utilizing a solid polymer micro-lattice as a sacrificial scaffold enables the fabrication of hollow micro-lattice heat exchangers. To date, the fabrication and testing of hollow micro-lattice heat exchangers constructed of nickel and copper have been reported [5,6].

In this presentation, we will for the first time discuss the design, fabrication, and testing of *polymer* wall hollow micro-lattice heat exchangers, thus overcoming previous limitations of compact polymer heat exchangers with micro-scale flow paths. Briefly, to form these heat exchangers, parylene is conformally coated onto a solid polymer micro-lattice. The solid polymer is then selectively removed, leaving behind a three-dimensionally ordered, interconnected network of hollow tubes (Figure 1). This tubular network separates two fluidically isolated, but interpenetrating volumes which have a high interfacial area to volume ratio (50 – 10,000 m^2/m^3 depending on the micro-lattice architecture). Headers formed in the fabrication process allow parallel distribution of flow inside each hollow tubular lattice member. Hollow polymer heat exchangers with internal tubular lattice member diameters ranging from 200 to 800 microns have been fabricated. Testing results of these hollow polymer heat exchangers will be presented.

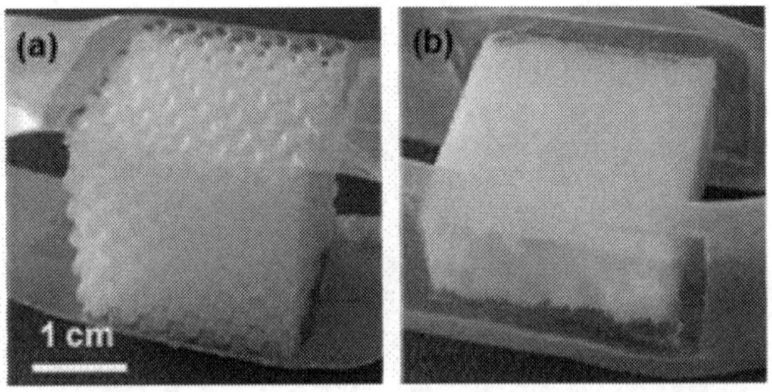

Figure 1: Polymer hollow micro-lattice heat exchanger cores with lattice member internal diameters of (a) 800 microns and (b) 200 microns.

References:

[1] Kays and London: *Compact Heat Exchangers*, 3rd Edition, Krieger Publishing Company, 1984.

[2] T'Joena, Park, Wang, Sommers, Han, Jacobi: *A review on polymer heat exchangers for HVAC&R applications*, International Journal of Refrigeration, 32 (2009) 763-779.

[3] Harris, Kelly, Wang, McCandless, Motakef: *Fabrication, Modeling, and Testing of Micro-Cross-Flow Heat Exchangers*, Journal of Microelectromechanical Systems, vol. 11 no. 6 (2002) 726-735.

[4] Jacobsen, Barvosa-Carter, Nutt: *Micro-scale Truss Structures formed from Self-Propagating Photopolymer Waveguides*, Advanced Materials, 19 (2007) 3892-3896.

[5] Roper, Kolodziejska, Maloney, Fink, Schaedler, and Jacobsen: *Recent Progress and Applications of Micro-Lattice Materials Formed Via a Self-Propagating Photopolymer Waveguide Process,* Technologies for Future Micro-Nano Manufacturing Workshop, (2011) 251-253.

[6] Maloney, Fink, Schaedler, Kolodziejska, Jacobsen, and Roper: *Multifunctional Heat Exchangers Derived from Three-Dimensional Micro-Lattice Structures,* International Journal of Heat and Mass Transfer, 55 (2012) 2486-2493.

(142cr) Effect of Strain, Shape, and Surface Roughness On the Rheology of Colloidal Gels

Lilian C. Hsiao, Department of Chemical Engineering, University of Michigan, Ann Arbor, MI and Michael J. Solomon, Chemical Engineering, University of Michigan, Ann Arbor, Ann Arbor, MI

Rigid, load-bearing structures have long been known as critical factors in influencing the shearing and bending properties of granular materials and fiber networks. Recently, our exploration of isostaticity in colloidal gels showed that microstructural rigidity plays an important role in their nonlinear elastic rheology. Here, we synthesize spherical and spheroidal poly(methy methacrylate) colloids with different degrees of surface roughness and incorporate depletion attractions to create weak colloidal gels. We impose simple shear flows of various strains on the gels and use a confocal laser scanning microscope to characterize the 3D structural change after deformation. This is done with a UV light-activated photopolymer, which allows particle configurations to be locked in place rapidly (< 0.6 s) after yielding. Concurrent non-linear oscillatory and stress relaxation measurements are used to determine the effect of roughness and shape on the nonlinear elasticity of the gels. In particular, surface roughness offers a potentially powerful tool for refining and controlling the rheology of colloidal gels with attractive pair potentials.

(142cs) Dropwise Condensation On a Cold Surface: Drop Distributions and Surface Functionalization

Ashley Macner, Susan Daniel and Paul Steen, Chemical and Biomolecular Engineering, Cornell University, Ithaca, NY

In order to design effective surfaces for large-scale dropwise condensation, an understanding of how the surface functionalization affects drop growth and coalescence is needed. The long term technological goal is to be able to propose a set of design conditions to NASA to help them achieve maximum heat transfer rates under microgravity conditions in their management of waste heat generated from electronics and the cabin environment. We are specifically interested in how the shape of a drop size distribution is tied to the surface functionalization.
Transient dropwise condensation from a vapor phase onto a cold and chemically treated surface was studied. The experimental setup consisted of a glass slide sitting on top of an aluminum plate which rested on a glass Pyrex container filled with boiling water. The aluminum plate had a rectangular shaped hole cut into it which exposed the glass slide to the vapor. Water was condensed onto the underside of the glass slide which had been chemically treated to yield a uniform contact-angle. Vapor deposition with dodecyltrichlorosilane and heptadecafluorotrichlorosilane yielded contact-angles ranging from neutral wetting to hydrophobic. The temperature of the vapor and the condensing surface were not controlled over the course of the experiment which lasted for 50 minutes. Hence, the vapor cooled and the glass slide heated up over the course of the experiment. The drops were imaged from above and statistics related to their size and population density were collected.
The process of condensation occurs by the three concurrent processes of nucleation, growth, and coalescence of drops. The growth of a drop can be divided into two categories: that by vapor diffusion onto the liquid-gas interface and that by surface "diffusion" where smaller drops merge with the perimeter of an existing drop. Coalescence is when two drops of comparable size grow large enough to touch and merge into a single drop.
For a typical condensation experiment, the progression of a generation of drops advances through two stages: an increase in drop density as a result of nucleation and a decrease in drop density as a result of macroscopic coalescence events. How quickly the transition from one stage to the next occurs is dependent upon the characteristics of the surface. It was found that for increasing contact-angle, the time needed to reach the maximum drop density (i.e. transition point) increases. In addition, the rate of decrease in the drop density increases as the contact-angle increases.

The transition point can also be determined by observing the time evolution of drop size distributions (number of drops versus drop radius). Initially, the distribution starts off as exponential as the first generation of drops comes into view. The distribution then transitions to a gamma distribution at the transition point. When the drop coalescence events are large enough to regenerate large areas of the surface for subsequent nucleation events, a bimodal distribution is observed. For long times, an exponential distribution (with a longer tail) is again observed. Because the experiment is transient in nature, the shape of the distribution can be used to predict the number of drop generations and their stage of development.

(142ct) An Exact Solution for the Capillary Driven Transport in Porous Media in the Presence of Sudden Expansion

Eric M. Benner, Chemical and Nuclear Engineering, University of New Mexico, Albuquerque, NM and Dimiter N. Petsev, Chemical and Nuclear Engineering, The University of New Mexico, Albuquerque, NM

Theoretical analysis of the capillary driven transport of liquid in porous media is given for special cases of sudden expansion. Elliptical and oblate spheroid coordinate transforms where used to find transient analytic solutions for two and three-dimensional expansions (Figures 1 and 2). The fluid flow is simplified to potential functions, which reduces the equations to single dimensional, time dependent problems. The asymptotic expression of the two-dimensional case is compared with the circular result of Hyvaluoma. In three dimensions a spherical expansion is derived which is compared to the asymptotic solution of the oblate spheroid. In both two and three dimensions, the expressions for expansion near the entrance are compared to the one-dimensional Lucas-Washburn equation and are shown to deviate from the classical result due to the sudden expansion. The system in the two-dimensional case additionally includes the time dependent effects of a constant evaporation rate, allowing a steady state to be reached. Two characteristic time-scales, one related to fluid capillary pressure and viscous resistance and the other to the evaporation rate, give a dimensionless evaporation number which shows the transport mechanism that dominates the time-dependent flow.

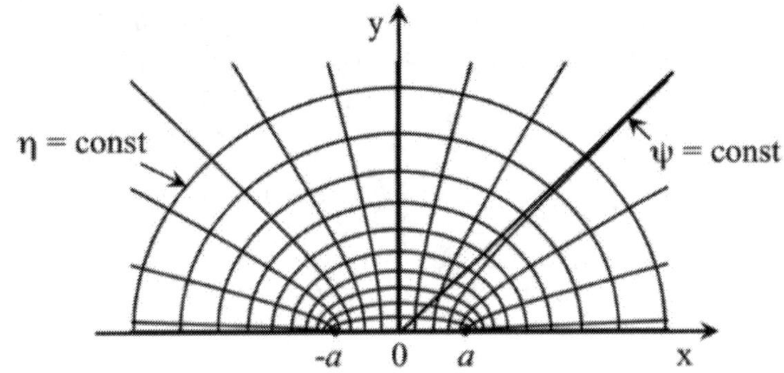

Figure 1: Elliptic coordinates, used to describe the flow in an expanding 2D porous domain. The points $-a$ and a are the foci of the ellipses.

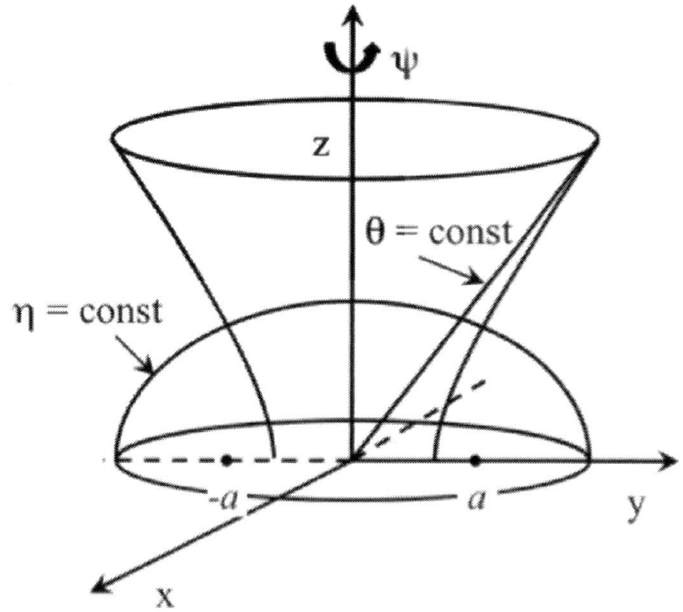

Figure 2: Oblate spheroid coordinates, used to describe the flow in an expanding 3D porous domain. The points $-a$ and a are the foci of the oblate surface corresponding to eta. Only systems with axial symmetry are considered.

(142cu) Coarse-Graining Solvent Interactions in Bead-Spring Chain Models

Rangarajan Radhakrishnan and Patrick T. Underhill, The Isermann Department of Chemical and Biological Engineering, Rensselaer Polytechnic Institute, Troy, NY

Bead-spring chain models have been successfully used to model dilute solutions of long flexible polymers. Solvent polymer interactions need to be incorporated accurately in simulations of bead-spring chains to capture the dynamics of these long molecules. We have developed two different methods to study solvent effects in good and poor solvents. In our first method we try to capture the solvent effects of a molecule in a good solvent by means of a dumbbell model. Using this model, we analyze the implications of solvent effects in elongational flow. We find that the coil-stretch transition and its hysteresis can be significantly affected by solvent effects. In our second method, we capture the solvent effects of a molecule in a poor solvent. Using a model with short ranged attractions and repulsions we have found that shear flow enhaces the formation of globules. We have analyzed the enhancement in terms of how flow pushes the polymer over the free energy barriers to forming globules.

(142cv) Suspension Properties At Finite Reynolds Number

Hamed Haddadi, Chemical Engineering, The City College of New York , New York, NY and Jeffrey F Morris, Levich Institute and Chemical Engineering, Levich Institute, City College of New York, New York, NY

The role of particle scale inertia in suspensions of neutrally buoyant monodisperse spherical particles is examined. Using numerical simulation via Lattice Boltzmann Method (LBM), trajectories and velocities of particles are calculated in simple shear flow. Based on relative pair velocities obtained for a range of Reynolds number and volume fractions, the variation of probability density for the vector, **r**, separating the center of the two spheres is calculated for a moving material point in **r** space. Shear induced self-diffusivities computed from the slope of the mean-square particle displacement and by integrating the corresponding velocity autocorrelation are also presented. To study the bulk stress of suspensions, the portraits of first and second normal stress difference are demonstrated in shear plane by sampling pair space, and are used along with the simulated microstructure of suspensions to explain the behavior of normal stress differences in suspensions at various levels of volume fraction and inertia.

(142cw) Rheology of Lamellar Mesophases

V Kumaran, Chemical Engineering, Indian Institute of science, Bangalore, India

Lamellar fluids typically contain alternating water/surfactant or water/oil/surfactant layers in equilibrium. Under shear, one would expect the layers to align with
unit normal velocity gradient direction. In this alignment, the fluid should have viscosity only a few times that of water by the inverse sum rule for the viscosity. However, real lamellar fluids have viscosities that are several orders of magnitudes higher than water. For this reason, lamellar mesophases are used in products which require very high viscosity, such as hair conditioners and butter substitutes. The reason for these high viscosity is a puzzle.

The modeling of lamellar mesophases is complicated due to the two-way coupling between structure and rheology. A perfectly aligned lamellar phase, for example, exhibits fluid like behaviour when the normal to the lamellae is along the shear or vorticity direction, but has a solid like resistance to flow when the normal is along the flow direction. In addition, even though a perfect defect free stack of layers is the final equilibrium state, real samples are rarely defect free due to kinetic constraints. The lamellar spacing is typically small compared to macroscopic scales (the distance between layers in lyotropic liquid crystalline phases is usually a few hundred Angstroms and a macroscopic sample contains 10^4-10^6 lamellae), and so a flowing lamellar mesophase cannot be modeled using a microscopic description. It is necessary to use different simulation techniques (molecular, mesoscale, macroscale) for accurately capturing the rheology of lamellar phases.

First, we present a multiscale modeling methodology to link molecular and mesoscale simulations. A mesoscale model based on a concentration field (phase field) description is then used to examine the rheology of a lamellar phase under a linear shear flow. For sufficiently large system sizes, the final steady state is not a perfectly aligned state, but rather a disordered state where there is a dynamical balance between the annealing of defects under shear and
the spontaneous creation of defects. There are two different defect creation mechanisms observed in our simulations, an undulation mechanism along the extension axis and a compressional mechanism along the compression axis. Order parameters are used to quantify the extent of disorder in the system, and these are found to be linked to the rheology.

(142cx) Single Molecule Polymer Dynamic Measurements in an Automated Hydrodynamic Trap

Christopher Brockman and Charles M. Schroeder, Chemical and Biomolecular Engineering, University of Illinois at Urbana-Champaign, Urbana, IL

For nearly two decades, fluorescently-labeled double stranded DNA (dsDNA) has been the model system for studying single molecule polymer dynamics in non-equilibrium conditions. Single polymer molecule studies utilizing double stranded DNA (dsDNA) have provided a treasure trove of information regarding polymer dynamics; however, dsDNA is a semi-flexible polymer with markedly different local molecular properties and a large persistence length ($l \approx 53$ nm) compared to flexible polymer chains, such as polystyrene ($l \approx 0.7$ nm) and ssDNA ($l \approx 0.6$ nm). Recently, we developed a biochemical synthesis platform for producing long strands of fluorescently-labeled single stranded DNA (ssDNA). In addition, our lab has also developed a microfluidic-based, automated hydrodynamic trap to confine individual particles in free solution at the stagnation point in planar extensional flow. In this work, we extend the use of the automated hydrodynamic trap to study chain dynamics of both dsDNA and ssDNA in extensional flow.

As a proof of principle, we trapped single molecules of dsDNA for up to 10 minutes enabling a host of experiments, including repeated relaxation time measurement an individual molecule and chain behavior for both randomly varying and oscillating strain rates. We then used the automated trap to measure ssDNA relaxation for chains with varying molecular weights and compared the results to our dsDNA measurements to elucidate the differences between relaxation time scaling for "real" and "ideal" chains. Overall, we seek a molecular-based understanding of the non-equilibrium dynamics of flexible polymer chains, which is crucial for control in processing and molecular self-assembly.

1. C. Brockman, S. Kim, F. Latinwo, C. M. Schroeder, "Direct Observation of Single Flexible Polymers using Single Stranded DNA", Soft Matter, 7, 8005-8012 (2011).

2. M. Tanyeri, E. M. Johnson-Chavarria and C. M. Schroeder, "Hydrodynamic Trap for Single Particles and Cells", Applied Physics Letters, 96, 224101 (2010).

3. Johnson-Chavarria E.M., Tanyeri M., Schroeder C.M., "A Microfluidic-based Hydrodynamic Trap for Single Particles", Journal of Visualized Experiments, 47, (2011).

(142cy) Spin-up Flow Experiments with Brownian Ferrofluids and the Existence of Couple Stresses

Isaac Torres-Diaz and Carlos Rinaldi, Chemical Engineering, University of Puerto Rico at Mayaguez, Mayaguez, PR

Ferrofluids are colloidal suspensions of magnetic nanoparticles with interesting fluid mechanics due to the presence of magnetic body couples and non-symmetric viscous stresses. Ferrofluid flow in a cylindrical geometry under the influence of a uniform rotating magnetic field, spin-up flow, has received much attention in the ferrohydrodynamics field since it could prove or disprove the existence of couple stresses in these fluids. Previous experiments in the spin-up geometry show qualitative agreement between the magnitude and direction of the velocity profile in the bulk and profiles predicted by the spin diffusion theory. It has been demonstrated that non-uniformities in the rotating magnetic field do not drive these flows. However, previous experiments considered ferrofluids consisting of magnetite nanoparticles at high concentrations, wherein the relaxation time could not be experimentally determined and where only a minor fraction of the particles relax by the Brownian mechanism. Additionally, measurable flows were obtained only at high magnetic fields, whereas the solution available for the ferrohydrodynamics equations was applicable for small magnetic fields. Here we consider water based ferrofluids with cobalt ferrite nanoparticles at low concentrations (< 0.01 v/v) which display the Brownian relaxation mechanism to test the ferrohydrodynamics equations, elucidate the existence of couple stress, and determine the value of the spin viscosity in these flows. We used an ultrasound technique to measure bulk velocity profiles in the spin-up geometry, varying the magnetic field intensity and frequency generated by a two pole stator winding. Additionally, torque measurements in a cylindrical geometry were considered. Results show that measured velocity profiles had rigid-body like behavior in the bulk, and no dependence in the axial direction. The magnitude of the velocity reached by these dilute Brownian ferrofluids (< 0.01 v/v), is in the same order as reported velocities using magnetite more concentrated ferrofluids (~ 0.1 v/v). Experimental velocity profiles are in quantitative agreement with the predictions of the spin diffusion theory, with a value of the spin viscosity of $\sim 10^{-8}$ kg m/s, two orders of magnitude larger than the value calculated with magnetite ferrofluids, and twelve orders of magnitude larger than estimated using dimensional arguments valid in the limit of infinite dilution. Additionally, the magnitudes of experimental torques are in the same order of magnitude as theoretical predictions. These results present further and conclusive evidence of the existence of couple stresses in ferrofluids and their role in driving the phenomenon of spin-up flow.

COMPUTATIONAL MODELING OF A MICROCHANNEL HEAT EXCHANGER: PRESSURE, VELOCITY, AND TEMPERATURE PROFILES

Javier A. Narvaez (a), Robert Wilkens (a), Hugh Thornburg (b), and Chad Hunter (b)

(a) University of Dayton
(b) Wright-Patterson Air Force Base

Abstract ID# 270641 (password: 494371)

The design of smaller but more efficient heat exchangers is a problem of increasing importance in thermal energy management. In order to transfer more energy per unit volume, heat exchangers tend to have micro-sized contact regions between hot sources and cold streams. Due to the intricate geometries of these micro-sized heat exchangers, modeling them for pressure, velocity, and temperature profiles is a challenging but critical problem in computational fluid dynamics. In this paper a Lytron CP20 cold plate heat exchanger under constant heat flux is modeled using Star-CCM+. Water was used in this simulation. Pressure, velocity, and bulk and wall temperature profiles are presented and compared with experimental results. Regions of especially high wall temperature are highlighted and discussed. Critical details on the way modeling was performed and on the limitations of the software used are also discussed.

Submitter's E-mail Address: jnarvaez0410@yahoo.com
Keywords: Computational Methods, Heat Transfer and Transport Phenomenon
Session Selection:
Fluid Mechanics Poster Session: Monday, October 29, 2012: 3:15 PM-5:45 PM
Slot: 3:

(142d) CFD Approach for the Analysis of NON-Newtonian Polymeric Microparticles Production

Cristina Rodriguez Rivero, Eva M. Martín del Valle and Miguel A. Galán, Department of Chemical Engineering, University of Salamanca, Salamanca, Spain

We have developed a method to produce uniformly sized biocompatible microparticles from highly viscous non-Newtonian polymers. It is based on the growth of Rayleigh instabilities[1], assisted by controlled vibration, in polymer fluid jets, which leads to the breakup of the jet into droplets. These are later stabilized giving rise to the solid microparticles. The system can process higher ranges of viscosity than conventional techniques, which results in a higher mechanical resistance of the produced particles. The technique has been characterised in a previous work, where a semiempirical model for predicting particle size was also proposed[2].

In the present work a modelling approach, performed through Computational Fluid Dynamics (CFD) simulations (Ansys-Fluent ®), is assessed as a mean to validate the system behaviour. The simulations cover and aim to describe the jet breakup and the formation of the droplets under different conditions of superimposed vibrations. Simulation results and comparison with experiments and predictions based on the semiempirical model are discussed.

Close attention is paid to the nature of the used materials since most biocompatible polymers display non-Newtonian fluid behaviour to some degree. This makes the study of the system more difficult, in which non-Newtonian parameters play a key role and clearly influence the way the jet evolves. Thus, CFD has been also used as a way to test which non-Newtonian effects influence the system most. In addition, it also enables to determine viscoelastic parameters (such as relaxation and retardation times) which have not been possible to be obtained through experimentation. With this aim some constitutive equations were implemented and included to the standard equations already gathered by the commercial software via User-Defined-Functions (UDF). This programming tool is also used to establish the inlet boundary conditions as oscillating displacements.

This systematic approach is also useful in planning large scale systems and testing how changes in the materials and geometry of the nozzles affect the system. It is also of great convenience for the theoretical analyses in progress.

1. Rayleigh L. On the capillary phenomena of jets. *Proc. R. Soc. London* 1879;29.

2. Rodríguez-Rivero C, Del Valle EMM, Galán MA. Development of a new technique to generate microcapsules from the breakup of non-Newtonian highly viscous fluid jets. *AIChE Journal.* 2011;57(12):3436-3447.

(142db) Single Chain Dynamics of Submicron Sized Semiflexible and Flexible Biological Polyelectrolytes in Flow Fields

Jeong Yong Lee[1], Myung-Suk Chun[2] and Hyun Wook Jung[1], (1)Department of Chemical and Biological Engineering, Korea University, Seoul, South Korea, (2)Complex Fluids Laboratory, National Agenda Res. Division, Korea Institute of Science and Technology (KIST), Seoul, South Korea

Polyelectrolytes are characterized by charged soft matters bearing ionic groups and accompanying counter-ions, such as water-soluble biomolecules, DNA, and polysaccharides. This implies complication arising from Debye screening effect and high sensitivity to external fields. We investigated conformation and translational diffusion of the single chain of submicron sized biological polyelectrolyte xanthan in solvents by employing coarse-grained Brownian dynamics simulations [1,2]. Our coarse-grained model goes beyond other simulations, which dealt with mainly the long chain of DNA [3,4], as they do not consider simultaneously both the long-range electrostatic and hydrodynamic interactions between pairs of beads. Based on order-disorder transition of polysaccharide backbone, semiflexible and flexible chains can be prepared by native (double strand) and denatured (single strand) xanthan, respectively. Simulation parameters are obtained from the viscometric method of rheology data on each chain. Contrary to conformational properties in uniform flows, the influences by flow strength as well as flow type are evident in both simple shear and extensional-like flows, exhibiting a sigmoidal transition of radius of gyration. Transition to a higher plateau and independence of electrostatic screening can be encountered earlier with increasing flow strength, as a special feature in extensional-like flow. The translational self-diffusion increases with increasing either the flow strength or the electrostatic screening in uniform and simple shear flows. In order to bridge theoretical and experimental approaches, we will present typical results of the single molecule tracking performed on fluorescein-labeled xanthan by applying epi-fluorescence microscopy under controlled flow fields. Present study on conformational dynamics of length and time scales in external flow fields can play an important role in a challenge of design to develop well-suited microfluidic devices. (This work was supported by the Basic Research Fund (20100021979) and the Converging Research Center (2009-0082136) through the National Research Foundation of Korea.)

1. J. Jeon, M.-S. Chun, Structure of flexible and semiflexible polyelectrolyte chains in confined spaces of slit micro/nanochannels, J. Chem. Phys. 126, 154904 (2007).
2. M.-S. Chun, C. Kim, D.E. Lee, Conformation and translational diffusion of a xanthan polyelectrolyte chain: Brownian dynamics simulation and single molecule tracking, Phys. Rev. E 79, 051919 (2009).
3. Y.L. Chen, M.D. Graham, J.J. de Pablo, G.C. Randall, M. Gupta, P.S. Doyle, Phys. Rev. E 70, 060901R (2004).
4. C.M. Schroeder, R.E. Teixeira, E.S.G. Shaqfeh, S. Chu, The dynamics of DNA in the flow-gradient plane of steady shear flow: observations and simulations, Macromolecules 38, 1967 (2005).

(142dc) Mechanism of Flow Induced Segregation in Blood and Other Multicomponent Suspensions

Amit Kumar and Michael D. Graham, Chemical and Biological Engineering, University of Wisconsin-Madison, Madison, WI

Blood is a multicomponent mixture comprising mostly of red-blood-cells (RBCs) along with trace amounts of other components like leukocytes, platelets, and circulating tumor cells (in the case of cancer). Under physiological flow conditions both the leukocytes and the platelets segregate near the walls of the blood vessel, a phenomenon commonly known as margination, while the RBCs tend to migrate away from the walls. The key physical differences between RBCs, leukocytes, and platelets are their relative size and rigidity: the leukocytes are larger than RBCs, while the platelets are smaller; both are considerably stiffer than RBCs. However, how these differences in properties lead to the observed segregation behavior is poorly understood. In this work we focus on the model system of a fluid-filled elastic capsule mixture in which individual components differ in size and rigidity. Using detailed boundary integral simulations we delineate the effect of both of these key properties on the flow induced segregation behavior and relate these to the observations of leukocyte and platelet margination in blood flow. To gain a mechanistic understanding of these results, we introduce a novel hydrodynamic-Monte Carlo simulation technique, which incorporates two of the main ingredients of flow dynamics in confined suspensions: the wall induced migration and pair collisions. In particular, the model clarifies the important role played by heterogeneous collisions, i.e. collisions between two different species in a mixture, in the observed segregation behavior. The model is also shown to reproduce the results of the detailed boundary integral simulations, while requiring only a fraction of the latter's computational cost. The insights and tools presented in this poster will be helpful, e.g., in designing drug delivery particles for optimal vascular targeting and for designing microfluidic devices for separating/enriching trace components of blood.

(142dd) Probing the Microstructures of Adhesive Hard-Spheres Under Large Amplitude Oscillatory Shear (LAOS) Deformation with Time-Resolved Small-Angle Neutron Scattering (tOR-SANS)

Jung Min Kim[1], A. Kate Gurnon[2], Aaron P. R. Eberle[3], Lionel Porcar[4] and Norman J. Wagner[2], (1)Department of Chemical and Biomolecular Engineering , University of Delaware, Newark, DE, (2)Department of Chemical and Biomolecular Engineering, University of Delaware, Newark, DE, (3)Advanced Characterization, ExxonMobil Research and Engineering Company, Annandale, NJ, (4)Large Scale Structures Group, Institut Laue-Langevin, Grenoble, France

Large amplitude oscillatory shear (LAOS) is becoming a popular method to study the nonlinear response of complex fluids and polymers. Many reports attempt to rationalize the LAOS response in terms of proposed microstructural changes. However, interpretation of the measurements is largely qualitative. To address this, we have developed new experimental methods to directly measure the microstructure during LAOS. The yielding and flow of colloidal gels and glasses is of significant academic and industrial interest. Here, we synthesize a model adhesive hard-spheres (AHS) system comprised of octadecyl-grafted silica in tetradecane, which thermoreversibly gels upon cooling [1]. Rheology and light scattering are used to determine the point of dynamic percolation (gelation and glass formation) as a function of particle concentration and temperature. Steady shear flow and LAOS rheological experiments are performed and the microstructure is probed simultaneously via time-resolved small-angle neutron scattering (tOR-SANS) in the 1-3 plane. With decreasing temperature the dispersion gels and the strength of the attraction is extracted by modeling the structure facture determined by SANS. Under strong shear, the systems near and in the gelled state exhibit highly anisotropic 2-D SANS patterns that show a "butterfly" structure consistent with previous reports [2, 3]. This corresponds to the gel microstructure stretching along the flow direction. Under LAOS deformation the butterfly patterns are observed to oscillate. Analysis of the patterns reveals how the oscillatory shear flow organizes the microstructure. Identification of a structural order parameter enables constructing the first structural-Lissajous plots for this class of materials, which presents the microstructure as a function of the instantaneous strain and strain-rate of the shear deformation. A detailed analysis of the third harmonic using a recently developed theory applicable for the onset of nonlinearity, the so-called medium amplitude oscillatory shear, or MAOS regime is performed. As the model system can be varied from near hard sphere behavior to a strong gel, this analysis provides insight in the relative importance of hydrodynamic and thermodynamic forces to the microstructure and rheology of colloidal dispersions. This new experimental technique and analysis method provides a new pathway to link rheology and microstructure under time-dependent deformation and provides valuable new insights into suspension micromechanics.

1. Eberle, A.P.R., N.J. Wagner, and R. Castaneda-Priego, *Dynamical arrest transition in nanoparticle dispersions with short-range interactions*. Physical Review Letters, 2011. **106**(10).

2. Hoekstra, H., et al., *Multi length scale analysis of the microstructure in sticky sphere dispersions during shear flow*. Langmuir, 2005. **21**(24): p. 11017-11025.

3. Thareja, P., et al., *Shear-induced phase separation (SIPS) with shear banding in solutions of cationic surfactant and salt*. Journal of Rheology, 2011. **55**(6): p. 1375-1397.

(142de) Effect of Polymer Additives On Blood Flow in the Microcirculation: Simulations and Theory

Kushal Sinha[1], Rafael Henriquez[2], Amit Kumar[3] and Michael D. Graham[3], (1)Department of Chemical and Biological Engineering, University of Wisconsin Madison, Madison, WI, (2)Chemical Engineering, University of Wisconsin, Madison, WI, (3)Chemical and Biological Engineering, University of Wisconsin-Madison, Madison, WI

In recent experimental studies, it has been shown that introducing drag reducing polymers (DRP) to blood in nanomolar concentration has many beneficial physiological effects, including the prevention of lethality from hemorrhagic shock in animals. In micro-channel experiments, with suspensions of red blood cells (RBCs), DRPs have been shown to reduce cell free layer thickness and affect the orientation of the cells. The underlying mechanisms for these phenomena are still not understood. In the present study, we investigate the dynamics of RBCs in Newtonian fluids and in dilute solutions of DRP in microcirculation. Cell membranes are modeled with energy penalty for area changes and bending. Polymer molecules are modeled as bead-spring chains with finitely extensible nonlinearly elastic springs. Three dimensional simulations are performed with a novel Stokes flow formulation of the immersed boundary method for the capsules, combined with Brownian dynamics for the polymer molecules. We present our study for a range of microchannel diameter, from arterioles and venules which are 30-75 μm in internal diameter to capillaries which are few microns in diameter. For larger channel, we report migration of RBCs, cell free layer thickness, apparent viscosity, pressure drop and wall shear stress in microcirculation with and without DRPs. In capillary scale channel, we report the shape transitions and dynamics of RBC and the effect of elasticity and bending rigidity of membrane and the effect of adding polymers.

(142df) A Second-Order Accurate Immersed Boundary-Lattice Boltzmann Method for Particle-Laden Flows

Qiang Zhou, William G. Lowrie Department of Chemical and Biomolecular Engineering, The Ohio State University, Columbus, OH and L. S. Fan, Chemical and Biomolecular Engineering, The Ohio State University, Columbus, OH

A new and efficient direct numerical method with second-order accuracy is presented for fully resolved simulations of incompressible viscous flows laden with rigid particles. The method combines the state-of-the-art immersed boundary method (IBM), the multi-direct forcing method and the lattice Boltzmann method (LBM). Previously the combination of IBM and LBM could only achieve first-order accuracy though LBM is a second order method. The IBM was recently improved based on the traditional solver of incompressible Navier-Stokes equations. First, the multi-direct forcing method is adopted in the improved IBM to better approximate the no-slip/no-penetration (ns/np) condition on the surface of the particles. Second, a slight retraction of the Lagrangian grid from the surface towards the interior of the particles with a fraction of the Eulerian grid spacing helps increase the accuracy of the method. These two improvements are adopted in the present IB-LBM. The method is further improved by: 1) an over-relaxation technique in the procedure of multi-direct forcing method; and 2) an implementation of the classical fourth order Runge-Kutta scheme in the coupled fluid-particle interaction. The over-relaxation technique is demonstrated to yield higher orders of convergence when the retraction distance is fixed. The use of the classical fourth order Runge-Kutta scheme helps the overall IB-LBM achieve the second order accuracy and provides more accurate predictions of the translational and rotational motion of the particle. A novel finding of this study is the demonstration that the retraction allows a super-convergence of the method, which is around fourth order, with a proper range of the retraction distance. The new method has been validated by several benchmark applications and promises to be a very efficient and high-fidelity technique for fluid-particle interaction problems. More challenging problems with larger numbers of particles will be tested in the near future.

(142dg) Pair-Particle Dynamics in Sheared Colloidal Suspensions: Simulation and Theory

Ehssan Nazockdast, Chemical Engineering, Levich Institute, CCNY, New York, NY and Jeffrey F Morris, Levich Institute and Chemical Engineering, Levich Institute, City College of New York, New York, NY

We study the dynamics and microstructure of Brownian suspensions averaged to pair-particle level as a function of pair separation vector $\mathbf{r}$. This is carried out by sampling of configurations obtained from Accelerate Stokesian Dynamics simulation. The particles interact only through Brownian forces and hydrodynamic interactions with the Newtonian fluid. The dimensionless parameters defining the structure are the volume fraction, Φ, and Pectlet number, Pe, which is the ratio of hydrodynamic to Brownian forces. We report the average pair velocity and pair trajectory as well as fluctuating velocity field and dispersions around the average pair trajectory for a range of Pe and Φ. This work is motivated by the theory we have developed for prediction of microstructure of colloidal suspensions in simple-shear flow. The method determines the steady pair distribution function $g(\mathbf{r})$ from a solution of the Smoluchowski equation reduced to pair level. The theory is based on number of assumption allowing for formulation of the pair relative convective and diffusive flux associated with hydrodynamic interactions of "bath" particles with the pair. The sampling results are compared against the predicted values for further insight and examination of the validity of the involved approximations. Finally the time-dependent microstructure and rheology of colloidal suspensions is studied using the developed theory in flow conditions such as startup and oscillatory shear flows.

(142e) Development and Characterization of A Microcapsules Production Technology, Based On the Formation of Drops From Viscous Liquids Trough Fan Jet Nozzles

Jose M. Cervero, Eva M. Martín del Valle and Miguel Ángel Galán, Department of Chemical Engineering, University of Salamanca, Salamanca, Spain

The aim of the present work is to present a new technique to produce monodispersed microcapsules ranging from 30 to 300 micrometers, from the controlled breakup of a high viscous non-Newtonian fluid as raw material, using a fan jet nozzle.

The system employed allows working in a wide range of viscosities, higher than conventional techniques, producing a jet which is disrupted by a coaxial rapid air current. The system allows to work in a wide range of viscosities (up to 10000 mPa·s), and also has more control in the monodispersion of the droplet size distribution.

Behavior studies show the influence of liquid and air flow rate and liquid viscosity.

A semiempirical model has been developed using wave propagation mechanisms to explain the thread disruption phenomena. These model equations predict the particle size as a function of dimensionless groups which govern the main acting forces in the process (We, Oh and liquid to air mass flow ratio). The predicted and experimental particle sizes were compared obtaining a good level of accuracy in the prediction.

Numerical and Experimental Study on Liquid Film Flows on Packing Elements in Absorbers

Yoshiyuki Iso, Mariko Kato, Jian Huang, Shinsuke Matsuno

Heat & Fluid Dynamics Department, Research Laboratory, IHI Corporation
1, Shin-Nakahara-Cho, Isogo-Ku, Yokohama 235-8501, Japan
Email: yoshiyuki_iso@ihi.co.jp

ABSTRACT

To reduce global warming and greenhouse gas emissions, it is important to develop Carbon Capture and Storage (CCS) technology, especially for coal fired power plants. Post-Combustion CO_2 Capture (PCC) technology is one of the greatest potential technologies for reducing CO_2 emissions, according to its economy and efficiency. Gas-liquid interfacial flows (wetting flows), such as the flue gas and the liquid solvent, are applied into the absorber process for PCC in order to capture CO_2. Efficient control of liquid solvent flows by using packing elements in packed columns can increase the gas-liquid interfacial area and the mass transfer rate. For the optimum design of packing elements in these kinds of processes, it is very important to predict the detailed behavior of gas-liquid interfacial flows in packing elements, such that the mass transfer between the gas and liquid may be enhanced and the flooding and channeling of the liquid flow may be prevented.

The present study focuses on detailed descriptions of the transition phenomena between the film flow and the rivulet flow, as well as how such phenomena are affected by wall surface texture treatments. This study develops a three-dimensional numerical prediction technique using Computational Fluid Dynamics (CFD) with the Volume of Fluid (VOF) model as well as a lab-scale experimental testing technique. Through the comparison of two geometry cases (smooth wall and wavy wall), the numerical and experimental results show quantitatively that surface texture treatments can help to prevent the liquid channeling and can increase the wetted area. Additionally, the simulation results agree well with the experimental results at points such as the gas-liquid interfacial surface shape and the wetted area.

1. INTRODUCTION

Global warming and greenhouse gas emission reduction have become a major issue in the world. It is necessary to develop Carbon Capture and Storage (CCS) technology for coal fired power plants in order to reduce greenhouse gas emissions. According to these backgrounds, Post-Combustion CO_2 Capture (PCC) technology has the great potential as the economical and efficient technology for reducing CO_2 emissions, because PCC can be applied into both retro-fit and new construction of coal fired power plants as well as can control the CO_2 capture rate.

Gas-liquid interfacial flows (wetting flows), such as the flue gas and liquid solvent, are applied into CO_2 absorbers for PCC in order to capture CO_2 from the flue gas into the liquid solvent. Efficient control of liquid solvent flows by using packing elements in packed columns is important to increase the gas-liquid interfacial area and the mass transfer rate between the gas and the liquid. In typical packed columns, liquid is fed from the top of the packing element, and then the liquid film flow is formed on packing element surfaces. Simultaneously, gas is fed from the bottom of the packing element and makes contact counter-currently flows with the liquid film. There are two types of packing elements: structured packing and random packing. Structured packing elements can increase the gas-liquid interfacial area efficiency and the throughput of gas flow capacity more than random packing elements.

Control of liquid film flows by using packing elements is one of the key design factors in packed columns. In particular, the channeling flow of liquid (e.g. liquid film break-up) significantly reduces the gas-liquid interfacial area. To prevent this phenomenon, it is very important to predict the detailed behavior of liquid film flows for the design and development of packing elements. Therefore, the present study focuses on gas-liquid interfacial flows on an inclined wall, which is a model simplified from a typical structured packing element in absorbers.

There were numerous previous theoretical and experimental studies on typical liquid falling film flows (Nusselt, 1916[1]; Phan and Narain, 2007[2]; Wang, 2009[3]). With certain geometrical features on the surface of the wall, flow properties will change significantly and that is the focus of the present study. Several experimental studies have been conducted for specific packed columns in order to measure the mass transfer performance and to develop the empirical modeling (Bravo et al., 1985[4] & 1992[5]; Spiegel and Meier, 2003[6]; Murrieta et al., 2004[7]; Sidi-Boumedine and Raynal, 2005[8]; Ataki et al., 2006[9], [10]; Chen et al., 2007[11]; Raynal et al., 2009[12]). In most cases, however, these experimental studies were carried out by using air-water flows instead of the actual fluids and by using smaller or simpler experimental models than industrial ones. Thus, the fluid properties effects and scale effects on the hydrodynamics and mass transfer are not yet well known.

Recently, Computational Fluid Dynamics (CFD) simulations have become a useful alternative to experiments, owing to the fact that multiphase flow measurements in systems with complex geometries (such as packing elements) are very difficult and highly expensive, especially at the industrial large scale with actual flow conditions. In recent years, several studies can be found in the literature, which employed the CFD simulations of gas-liquid two-phase flows in packed columns (Spiegel and Meier, 2003[6]; Valluri et al., 2005[13]; Ataki et al., 2006[9], [10]; Chen et al., 2007[11]; Raynal et al., 2007[14] & 2009[12]; Kenig, 2008[15]; Ludwig and Dziak, 2009[16]; Lan et al., 2010[17]). In these studies, two-dimensional or three-dimensional CFD simulations were used with the Volume of Fluid (VOF) model, which is a surface tracking technique with the Eulerian mesh. Several useful results were obtained including the liquid thickness hold up, gas pressure drop, validation of CFD and so on.

Additionally, several studies exist on not only the gas-liquid two-phase film flows on an inclined wall but also the gas-liquid-liquid three-phase ones (Hoffmann et al., 2005[18] & 2006[19]; Repke et al., 2007[20]; Xu et al., 2012[21]). Both the CFD and experiment have been carried out to predict the liquid film break-up, rivulet and droplet formation. These studies have shown that rivulet flows are influenced strongly by three dimensional effects and therefore cannot be predicted by analytical approaches and two-dimensional CFD simulations.

Despite these previous efforts, the detailed descriptions of the transition phenomena between the film flow and rivulet flow, as well as how it is affected by the wall surface texture treatment and liquid flow rate, is still lacking. The present study attempts to close this gap by focusing on the effects of the change of liquid flow rate and the wall surface texture treatments on the characteristics of wetting flows which have the drastic flow transition between the film flow and rivulet flow. The study may shed some insights on the potential ways to prevent the channeling and flooding of liquid flow and to enhance the mass transfer performance.

In what follows, this study develops a three-dimensional numerical prediction technique using CFD with the Volume of Fluid (VOF) model (Iso and Chen, 2011[22]) as well as a lab-scale experimental testing technique. We investigate that the effects of wall surface texture treatments on the liquid film flow through the comparison of two geometry cases (smooth wall and wavy wall).

2. NUMERICAL AND EXPERIMENTAL METHOD

2.1 Flow Configuration

The outline of the flow configuration of the interfacial flow on the inclined wall is shown in **Fig. 1**. The geometry and flow conditions are set up according to the existing studies for wetting flows on the inclined wall (Hoffmann et al., 2005[18] & 2006[19]; Repke et al., 2007[20]; Xu et al., 2012[21]). The flow configuration is a model simplified from a typical commercial structured packing element in absorbers. The dimension of the inclined stainless steel wall plate is $60\ \mathrm{mm} \times 50\ \mathrm{mm}$ (the first dimension is the length of flow field, and the second dimension is the width of one), which is held by the supports on the left and right sides of same material. The inclined angle of the wall plate (α) is 60 degree with respect to the horizontal ground, which is common used inclinations of commercial structured packing elements.

Air and water are used as the gas-liquid two-phase fluids at the normal pressure (1 atm) and room temperature (25 °C). For air, its density ρ_g is $1.185\ \mathrm{kg/m^3}$, viscosity μ_g is $1.84\times 10^{-5}\ \mathrm{Pa\cdot s}$; for water, its density ρ_l is $997\ \mathrm{kg/m^3}$, viscosity μ_l is $8.9\times 10^{-4}\ \mathrm{Pa\cdot s}$, surface tension σ is $0.072\ \mathrm{N/m}$. The static contact angle for air and water on the stainless steel (θ) is 70 degree (standard deviation: 4.5 degree), which is obtained by our measurement using the DMs-400 equipment system manufactured by Kyowa Interface Science Co.,LTD.

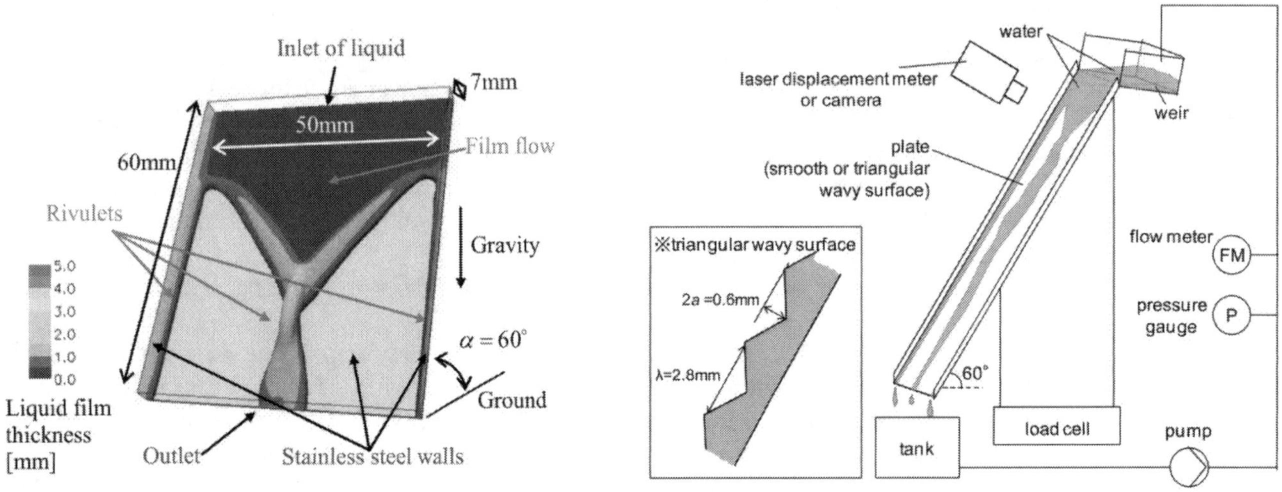

(a) Computational region and a liquid flow image (b) Experimental setup

Figure 1 Outline of the interfacial flow on the inclined wall

2.2 Numerical Method

The computational region and grids of two geometry cases (smooth wall and wavy wall) are shown in **Fig. 2**. For the wavy plate used in this paper, the amplitude a and length λ of the wavy surface texture are selected based on a commercial packing element (Raynal et al., 2007[14] & 2009[12]). In this case, the amplitude of the wavy surface texture is of the same order (several hundred μm order) as the liquid film thickness.

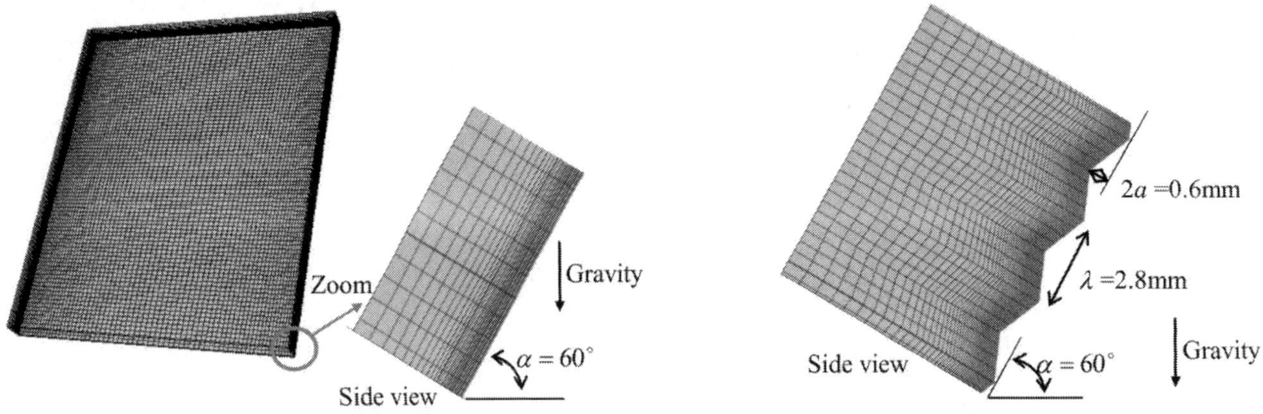

(a) Smooth wall (b) Details of the wavy surface texture

Figure 2 Computational grids of two geometry cases ((a) smooth wall and (b) wavy wall)

In this study, CFD simulations are carried out with the commercial code FLUENT (Version 6.3), ANSYS Inc., as a platform. The three-dimensional and transient model for gas-liquid two-phase interfacial flow simulations by a Volume of Fluid (VOF) model are used to predict the wetting flows, because the existing studies have shown that rivulet flows are influenced strongly by three dimensional effects and therefore cannot be predicted by analytical approaches and two-dimensional CFD simulations (Hoffmann et al., 2005[18] & 2006[19]; Ataki et al., 2006[9], [10]; Repke et al., 2007[20]). We use the CFD models in this study, which has been validated in terms of the interfacial surface shape, the transition point between the film flow and rivulet flow, and the liquid film thickness on an inclined plate with a smooth surface (Iso and Chen, 2011[22]).

The CFD models are based on the universal transport equations of extensive magnitudes in differential volume of the fluid. In case of hydrodynamics modeling, two conservation equations of the mass and momentum are solved numerically. The mass conservation leads to the continuity equation, and the momentum conservation gives the Navier-Stokes equation, as follows:

$$\frac{\partial \rho}{\partial t}+\nabla \cdot(\rho \cdot \mathbf{v})=0 \tag{1}$$

$$\frac{\partial(\rho \cdot \mathbf{v})}{\partial t}+\nabla \cdot(\rho \mathbf{v} \cdot \mathbf{v})=-\nabla p+\nabla \cdot\left[\mu\left(\nabla \mathbf{v}+\nabla \mathbf{v}^{\mathrm{T}}\right)\right]+\rho \mathbf{g}+\mathbf{S}_{m} \tag{2}$$

where v and p are the velocity vector and static pressure of the fluid, respectively. $\boldsymbol{S}_m$ is the source term vector expressing the momentum sources such as the surface tensions, the external forces and so on, which are described later in this section.

In this study, the set of these two equations is applied into the single field shared among all phases, which is called the shared-field approach, by using the Eulerian method. Turbulence is not taken into account since the range of the Reynolds number of the liquid flow indicates laminar (Reynolds number of the liquid flow in this study: $\mathrm{Re}_{l\mathrm{N}} \leq 300$, where $\mathrm{Re}_{l\mathrm{N}}$ is defined later in this paper).

Additionally, using the VOF model, the interfacial surface tracking technique is applied to the fixed Eulerian mesh. The set of the mass and momentum equations can be accomplished with the relation describing the interfacial surface position. The interfacial surface position is not explicitly tracked but is represented by the volume fraction of each fluid. The volume fractions are evolved with an advection equation. This relation is described by:

$$\frac{\partial(\rho_k f_k)}{\partial t}+\nabla \cdot(\rho_k f_k \cdot \mathbf{v}_k)=S_{f_k} \tag{3}$$

where f_k and v_k are the volume fraction and velocity vector of the phase k, respectively. S_f is the source term expressing the mass sources such as the mass transfer between phases, supplied mass and so on. In this case, S_f is zero based on the assumption that the mass transfer between gas and liquid can be negligible.

In this VOF model, the volume fractions are assumed to be continuous functions of space and time, and the sum of volume fractions is equal to unity. If $f_k = 0$, the calculation cell is empty of the phase k. When $f_k = 1$, the calculation cell is filled with the phase k. For $0 < f_k < 1$, the calculation cell contains the interfacial surface between the phase k and one or more other phases. The set of the mass and momentum equations is dependent on the volume fractions of all phases through the local fluid properties such as the density, viscosity, etc. The local fluid properties in the calculation cell are calculated by using a linear interpolation as the volume fraction average.

The effect of surface tension along the interface between each pair of phases is implemented with the continuous surface force (CSF) model proposed by Brackbill et al. (1992)[23]. The surface tension can be considered in terms of the pressure jump across the interfacial surface. In this CSF model, the addition of surface tension to the VOF model is reflected into a source term $\boldsymbol{S}_m$ in the momentum equation. Additionally, in order to consider the effects of wall adhesion, contact angles between the phases and walls are imposed as a wall boundary condition. The wall adhesion model proposed by Brackbill et al. (1992)[23] is used. The contact angle (θ) derived from static data is implemented by setting the angle of the gradient of the volume fraction at the wall boundary condition. The local curvature of the interfacial surface in the calculation cell next to the wall is used to adjust the body force term in the surface tension calculation.

The three-dimensional and transient flow is simulated in this study. The SIMPLE algorithm is used to deal with the coupling computations between the flow and pressure fields. A body force weighted discretization scheme is used for the pressure correction equation, a QUICK discretization scheme is used for the momentum equation, and a modified HRIC discretization scheme is applied to the transport equation of the volume fraction. The time step size of the transient simulation is chosen as 2.5×10^{-5} sec. The total actual time of the transient simulation needed to reach to a stable flow (quasi-steady convergent solution) is around several seconds in the present cases.

The liquid inlet condition is assumed as a uniform film flow defined with a constant velocity value perpendicular to the inlet boundary, which is consistent with the experimental liquid feeding conditions (where an overflowing weir is used in order to form the uniform film flow). The liquid inlet dimension is fixed as 50 mm × 0.4 mm in all cases (the first dimension is the width of liquid inlet, and the second dimension is the thickness of one). Here, the thickness of the liquid inlet is estimated by using Nusselt theory (1916)[1] at the averaged liquid flow rate in all cases for full film flows in this simulation. The liquid flow rate is varied in this paper as $1.1 \times 10^{-6}\,\mathrm{m^3/s} \leq Q_{l\,in} \leq 1.4 \times 10^{-5}\,\mathrm{m^3/s}$

($0.056\text{m/s} \le v_{l in} \le 0.67\text{m/s}$, $25 \le \mathrm{Re}_{l\mathrm{N}} \le 300$). The main plate and both side supports are implemented as no-slip walls with the given contact angle (θ). The other boundaries, which are upward, downward outlet and top boundaries, are set to the pressure outlet conditions by using the defined static pressures (the gauge static pressure at all pressure outlets is chosen as zero). Isothermal flow conditions are considered in this simulation.

Additionally, as can be seen in **Fig. 2**, the computational grid is made very fine for ensuring an accurate resolution of the liquid flow; the grid density and convergence have been validated by using several numerical tests, in which the doubled grid density showed the same results as the grid density used in the present paper. The two types of grids are used for the smooth wall geometry (the number of cells is approximately 1.7×10^5 cells) and wavy wall geometry (the number of cells is approximately 5.5×10^5 cells) in this paper. Non-uniform grids are used in the simulations, such that the grid density inside the liquid film and near the interfacial surface region is finer than that in the region adjacent to surrounding gas.

2.3 Experimental Method

The experimental setup is shown in **Fig. 1 (b)**. The dimension of the inclined stainless steel wall plate (SUS304 2B, surface roughness 3.2 μm) is 250 mm × 50 mm (the first dimension is the length of flow field, and the second dimension is the width of one), which is held by the supports on the left and right sides of same material. The inclined angle of the plate (α) is 60 degree with respect to the horizontal ground. The liquid distributor, which is an overflowing weir, is used to deliver a uniform liquid film thickness at the liquid inlet. The liquid flow rate is varied in this paper as $1.1 \times 10^{-6}\,\text{m}^3/\text{s} \le Q_{l in} \le 1.4 \times 10^{-5}\,\text{m}^3/\text{s}$.

Experimental data are measured by using a camera and optical techniques at the normal pressure (1 atm) and room temperature (11~14 °C). The digital camera (Nikon D200) is set vertically to the inclined plate to take pictures of the liquid flow behavior. The wetted area of liquid on the inclined plate is measured by using an image analysis system to calculate a wetted area from a picture. The measurement region on the inclined pale is 60 mm × 50 mm (the first dimension is the length of flow field, and the second dimension is the width of one) to compare experimental data with numerical data.

3. DIMENSIONLESS GROUPS FOR INTERFACIAL FLOWS

In order to apply the framework to the present interfacial flows, in this section, the dimensionless groups are defined by using the suitable representative quantities. Consider the traditional expression of the falling film thickness introduced by Nusselt (1916)[1], i.e. the well-known Nusselt theory, in which the balance of only two forces between gravitational force and friction force on a fluid element in the film is considered by assuming the steady continuous liquid film flow to be formed uniformly on the plate. Using these classical works of Nusselt theory, both the falling film thickness and the velocity distribution can be calculated.

In the case of the falling liquid film flow on the inclined plate, the liquid film thickness δ_N is expressed as follows:

$$\delta_\mathrm{N} = \left[\frac{3\mu_l \cdot Q_l}{\left(\rho_l - \rho_g\right) \cdot g \sin\alpha \cdot w} \right]^{1/3} \tag{4}$$

where Q_l is the volumetric flow rate of liquid. w is the width of the liquid film on the plate, which is assumed to be equal to the plate width as the full film flow develops completely on the whole plate area. Thus, the dimensionless groups for the present interfacial flows can be defined as:

$$\mathrm{Re}_{l\mathrm{N}} = \frac{\rho_l \cdot V_{l\mathrm{N}} \cdot \delta_\mathrm{N}}{\mu_l} \tag{5}$$

$$\mathrm{Fr}_{l\mathrm{N}} = \frac{V_{l\mathrm{N}}^{\ 2}}{g \cdot \delta_\mathrm{N}} \tag{6}$$

$$\mathrm{We}_{l\mathrm{N}} = \frac{\rho_l \cdot V_{l\mathrm{N}}^{\ 2} \cdot \delta_\mathrm{N}}{\sigma} \tag{7}$$

where $V_{l\,N}$ is the averaged liquid film velocity according to:

$$V_{l\,N} = \frac{Q_l}{\delta_N \cdot w} \tag{8}$$

In what follows, the present results are discussed based on these dimensionless groups. In particular, the Weber number $We_{l\,N}$ is used for considering the balance between inertia force and surface tension, which is an important parameter for the flow transition phenomenon.

4. RESULTS AND DISCUSSION

Wall surface treatments, such as small texture or perforations, are often used in the several commercial industrial packing elements in order to increase the wetted area and to prevent the liquid channeling (Valluri et al., 2005[13]; Ataki et al., 2006[9], [10]; Raynal et al., 2007[14] & 2009[12]). In this section, the effects of wall surface texture treatments on the flow transition of wetting flows are investigated.

The numerical and experimental results of the interfacial surface shape between gas and liquid are shown in **Fig. 3** and **Fig. 4**, for different Weber numbers ($0.04 \le We_{l\,N} \le 1.44$, $25 \le Re_{l\,N} \le 230$). The view in **Fig. 3** and **Fig. 4** is parallel to the inclined wall. Numerical results are shown by using iso-surfaces which are defined by the specific volume fraction of liquid, f_l=0.5. Two types of flow patterns are observed in **Fig. 3** and **Fig. 4**: the full film flow which is developed on the entire plate, and the channeling flow for which both of the film flow and rivulet flows are formed on the plate. The decrease of the wetted area by the channeling flow at the low Weber number $We_{l\,N}$ occurs due to the effect of surface tension (mainly exhibited as cohesive force between fluid particles at the interfacial surface region), which tends to minimize the interfacial surface area against the inertia force.

The liquid flow patterns on the wall surface texture in **Fig. 4** are also quite different than ones on the smooth wall in **Fig. 3** at the same condition of Weber number $We_{l\,N}$. For example, the channeling flow occurs in the case of the smooth wall at $We_{l\,N} \approx 1.0$ in **Fig. 3**, yet at the same Weber number condition the full film flow is retained on the full plate with the assistance of the texture treatments in **Fig. 4**.

Figure 5 shows the calculated wetted area A_w / A_t obtained by numerical and experimental results as a function of the Weber number ($0.04 \le We_{l\,N} \le 2.24$, $25 \le Re_{l\,N} \le 300$). It is apparent quantitatively that the wall surface texture treatment gives larger wetted area than the counterpart on the smooth wall (under the present flow condition).

Thus, the present results show that the surface texture treatments can indeed help to prevent liquid channeling and can increase the wetted area. We believe that the main reason for the occurrence of increasing the wetted area on the wavy wall is that the liquid film break-up is inhibited due to the spreading of the liquid flow in spanwise direction by the wavy wall geometry. **Figure 6** shows the comparison of liquid spanwise velocity on the gas-liquid interfacial surface obtained by the numerical simulation. As seen in **Fig. 6**, the liquid spanwise velocity is diffused due to the liquid spreading in spanwise direction by the wavy wall geometry as compared with the smooth wall. Especially, in **Fig. 6**, this diffusion effect by the wavy wall is observed noticeably in the region closely both side walls, in which liquid flow is affected by the meniscus force.

Additionally, the simulation results, on both of smooth and wavy walls, agree well with the experimental results at points such as the gas-liquid interfacial surface shape and the wetted area, as seen in **Fig. 3, Fig. 4** and **Fig. 5**. These validations demonstrate that the present simulation, which uses the VOF model, is capable of predicting gas-liquid interfacial flows on an inclined plate with a high accuracy in a reasonable calculation time period.

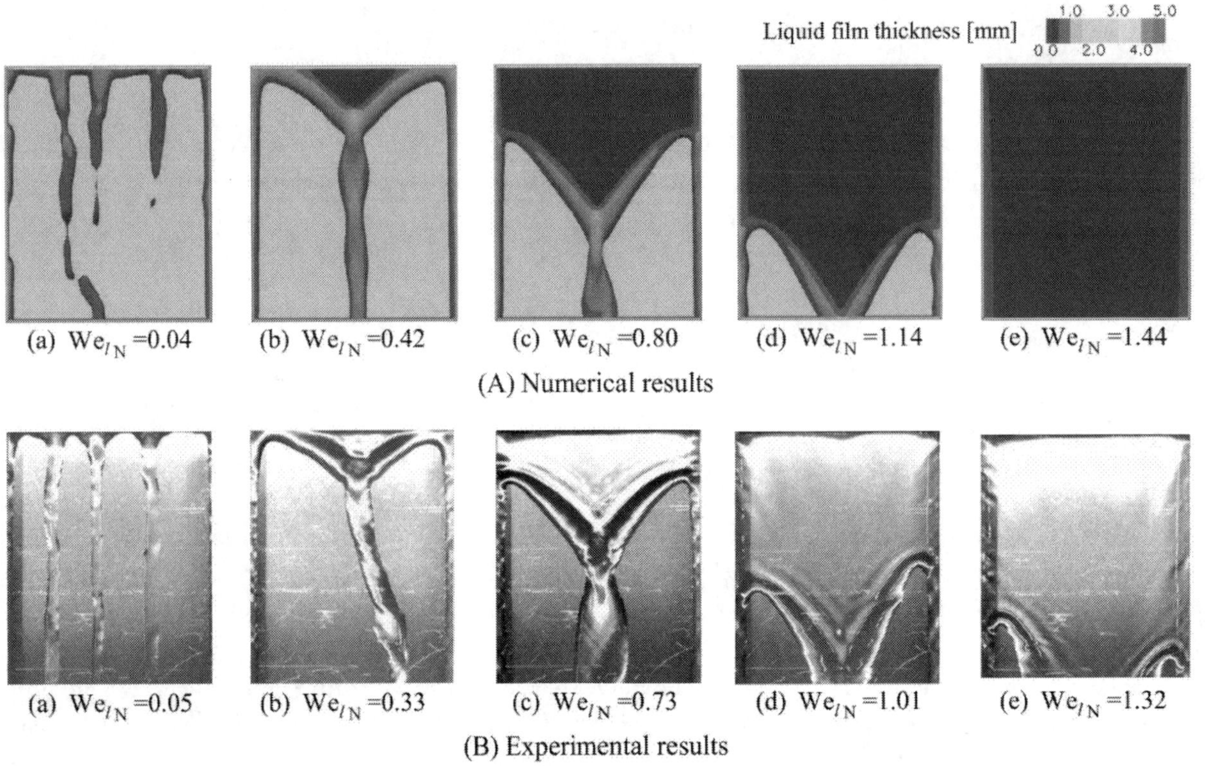

Figure 3 Instantaneous liquid flow patterns on the smooth wall while the liquid flow rate is increasing

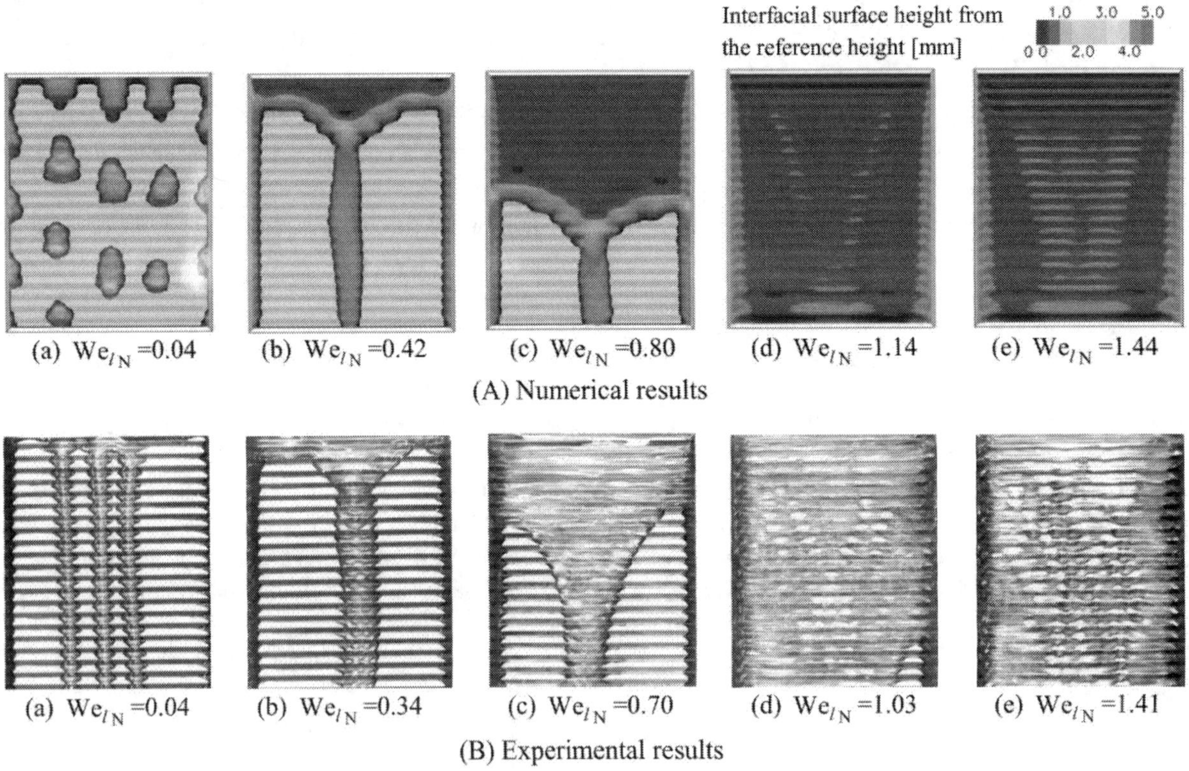

Figure 4 Instantaneous liquid flow patterns on the wavy wall while the liquid flow rate is increasing

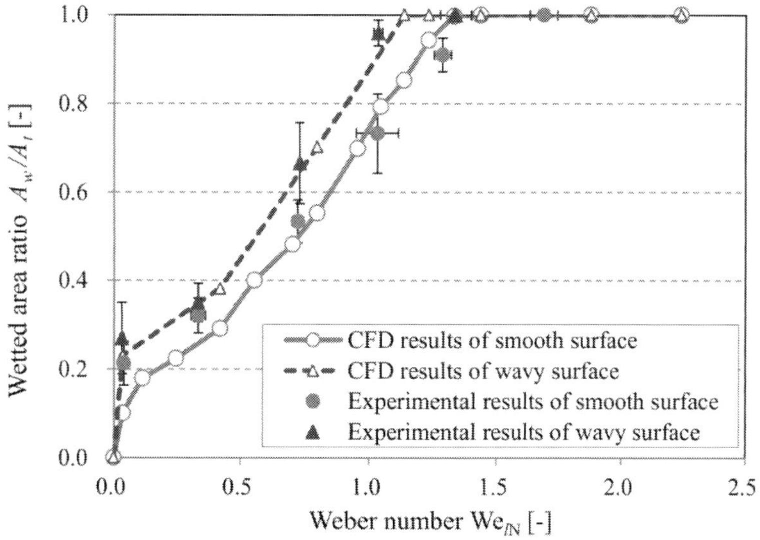

Figure 5 Comparison of wetted area ratio A_w/A_t between the smooth surface and the wavy surface while the liquid flow rate is increasing

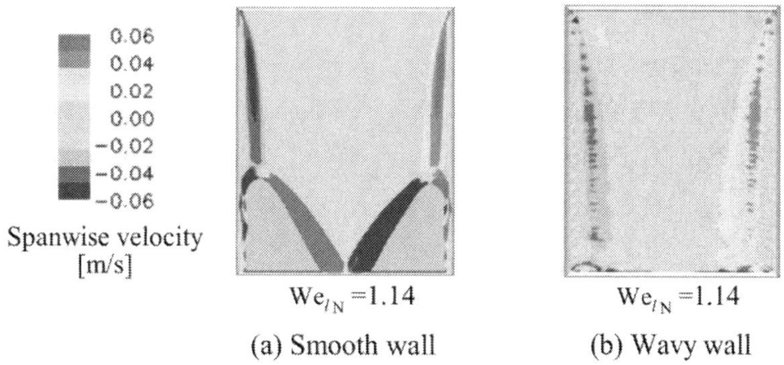

Figure 6 Comparison of liquid spanwise velocity on the gas-liquid interfacial surface

5. CONCLUSIONS

In this study, the effects of the liquid flow rate and the wall surface texture treatments on the characteristics of interfacial flows (wetting flow) are investigated by using the three-dimensional gas-liquid two-phase interfacial flow simulation with the VOF model as well as a lab-scale experimental testing technique. The following main conclusions are drawn:

(1) With a typical texture geometry used in a commercial industrial packing element, the present numerical and experimental results show that the surface texture treatments can help to prevent liquid channeling and can increase the wetted area quantitatively by comparing between the two geometry cases (smooth wall and wavy wall). The main reason for the occurrence of increasing the wetted area on the wavy wall is that the liquid film break-up is inhibited due to the spreading of the liquid flow in spanwise direction by the wavy wall geometry.

(2) The applicability and reliability of the present numerical simulation are validated via the comparison with experimental results. The simulation results, on both of smooth and wavy walls, agree well with the experimental results at points such as the gas-liquid interfacial surface shape and the wetted area. These validations demonstrate that the present simulation, which uses the VOF model, is capable of predicting gas-liquid interfacial flows on an inclined plate.

On the next step, using this prediction technique, we will systematically investigate the effects of surface texture treatments of packing elements on the gas-liquid interfacial flow. We will also apply these numerical and experimental prediction techniques into gas-liquid interfacial flows on packing elements with more complicated geometries under similar conditions of the amine-based CO_2 capture process in order to develop the advance design of packing elements, which have the high absorption performance and the low gas pressure loss.

NOMENCLATURE

v	velocity [ms^{-1}]
t	time [s]
p	static pressure [Pa]
g	gravitational acceleration [ms^{-2}]
S	source term, Sm [Nm^{-3}] or Sf [$kgm^{-3}s^{-1}$]
f	volume fraction [-]
V	representative velocity [ms^{-1}]
L	representative length [m]
Re	Reynolds number [-]
Fr	Froude number [-]
We	Weber number [-]
Q	volumetric flow rate [m^3s^{-1}]
w	width [m]
A	area [m^2]
a	amplitude of the surface texture [m]

Greek letters

α	inclined angle of the wall plate [degree]
ρ	density [kgm^{-3}]
μ	viscosity [Pas]
σ	surface tension [Nm^{-1}]
θ	contact angle for air and water on the wall [degree]
δ	liquid film thickness [m]
λ	length of the surface texture [m]

Subscripts

l	liquid phase
g	gas phase
in	inlet
m	momentum
k	phase k
f	volume fraction
N	Nusselt theory
w	wetted surface
t	total surface of the plate

REFERENCES

[1] Nusselt, W., 1916, "Die oberflachenkondensation des wasserdampfes," Zeitschr. Ver. Deut. Ing., 60, pp. 541-546 and pp. 569-575.

[2] Phan, L. and Narain, A., 2007, "Nonlinear stability of the classical Nusselt problem of film condensation and wave effects," Trans. of the ASME, J. of Applied Mechanics, 74, pp. 279-290.

[3] Wang, X., He, M., Fan, H. and Zhang, Y., 2009, "Measurement of falling film thickness around a horizontal tube using laser-induced fluorescence technique," J. of Physics, Conf. Series, 147, 012039.

[4] Bravo, J. L., Rocha, J. A. and Fair, J. R., 1985, "Mass transfer in gauze packings," Hydrocarbon Processing, 64, 1, pp. 91-95.

[5] Bravo, J. L., Rocha, J. A. and Fair, J. R., 1992, "A comprehensive model for the performance of columns containing structured packings," Inst. Chem. Eng. Symp. Series, 128, pp. A439-A457.

[6] Spiegel, L. and Meier, W., 2003, "Distillation columns with structured packings in the next decade," Trans. IChemE, 81, Part A, pp. 39-47.

[7] Murrieta, C. R., Seibert, A. F., Fair, J. R. and Rocha-U, J. A., 2004, "Liquid-side mass-transfer resistance of structured packings," Ind. Eng. Chem. Res., 43, pp. 7113-7120.

[8] Sidi-Boumedine, R. and Raynal, L., 2005, "Influence of the viscosity on the liquid hold-up in trickle-bed reactors with structured packings," Catalysis Today, 105, pp. 673–679.

[9] Ataki, A., Kolb, P., Buhlman, U. and Bart, H. J., 2006, "Wetting performance and pressure drop of structured packings: CFD and experiment," IChemE, Symposium Series, 152, pp. 534-543.

[10] Ataki, A., and Bart, H. J., 2006, "Experimental and CFD simulation study for the wetting of structured packing element with liquid," Chemical Eng. and Tech., 29, 3, pp. 336-347.

[11] Chen, J., Liu, C., Li, Y., Huang, Y., Yuan, X. and Yu, G., 2007, "Experimental investigation of single-phase flow in structured packing by LDV," Chin. J. Chem. Eng., 15, 6, pp. 821-827.

[12] Raynal, L., Rayana, F. B. and Royon-Lebeaud, A., 2009, "Use of CFD for CO2 absorbers optimum design: from local scale to large industrial scale," Energy Procedia, 1, pp. 917-924.

[13] Valluri, P., Matar, O. K., Hewitt, G. F. and Mendes, M. A., 2005, "Thin film flow over structured packings at moderate Reynolds numbers," Chemical Eng. Science, 60, pp. 1965-1975.

[14] Raynal, L. and Royon-Lebeaud, A., 2007, "A multi-scale approach for CFD calculations of gas-liquid flow within large size column equipped with structured packing," Chemical Eng. Science, 62, pp. 7196-7204.

[15] Kenig, E. Y., 2008, "Complementary modeling of fluid separation processes," Chemical Engineering Research and Design, 86, pp. 1059-1072.

[16] Ludwig, W. and Dziak, J., 2009, "CFD modelling of a laminar film flow," Chemical and Process Eng., 30, pp. 417-430.

[17] Lan, H., Wegener, J. L., Armaly, B. F. and Drallmeier, J. A., 2010, "Developing laminar gravity-driven thin liquid film flow down an inclined plane," Trans. of the ASME, J. of Fluids Engineering, 132, 8, 081301, pp.1-8.

[18] Hoffmann, A., Ausner, I., Repke, J. U. and Wozny, G., 2005, "Fluid dynamics in multiphase distillation processes in packed towers," Computers and Chemical Eng., 29, pp. 1433-1437.

[19] Hoffmann, A., Ausner, I., Repke, J. U. and Wozny, G., 2006, "Detailed investigation of multiphase flow behaviour on inclined plates," Trans. IChemE, Chemical Eng. Research and Design, 84, A2, pp. 147-154.

[20] Repke, J. U., Ausner, I., Paschke, S. Hoffmann, A. and Wozny, G., 2007, "On the track to understanding three phases in one tower," Trans. IChemE, Chemical Eng. Research and Design, 85, A1, pp. 50-58.

[21] Xu, Y., Yuan, J., Repke, J. U. and Wozny, G., 2012, "CFD study on liquid flow behavior on inclined flat plate focusing on effect of flow rate," Engineering Applications of Computational Fluid Mechanics, 6, 2, pp. 186-194.

[22] Iso, Y. and Chen, X., 2011, "Flow Transition Behavior of the Wetting Flow between the Film Flow and Rivulet Flow on an Inclined Wall," Trans. of the ASME, J. of Fluid Engineering, 133, 9, 091101, pp. 1-7.

[23] Brackbill, J. U., Kothe, D. B. and Zemach, C., 1992, "A continuum method for modeling surface tension," J. of Comp. Physics, 100, pp. 335-354.

Bubble Growth and Expulsions in Natural Circulation Boiling Loop

Arnab Karmakar and Swapan Paruya
Chemical Engineering Dept., NIT Durgapur, India

Abstract

The present experimental study reports bubble growth, expulsion time and expulsion period (cycle time) under the varied conditions of heater power and inlet subcooling in a natural circulation boiling loop. Wider variations of the expulsion time and expulsion periods have been observed at lower heat flux and higher inlet subcooling with the help of high-speed video. The analysis reveals that the expulsions are coupled with the bubble growth phenomenon, and the associated flow oscillations are highly complex depending on inlet subcooling and heater power.

Introduction

Natural circulation boiling loop (NCBL) containing steam-water mixture has extensive applications in boiling water reactors (BWRs), simplified boiling water reactors (SBWRs), nuclear steam generators, thermal drum-type boilers and all the thermosyphon reboiler units of chemical, refining and petrochemical plants, etc. due to its excellent heat removal properties without the application of pump. During startup operation, NCBL experiences various nonlinear flow instabilities[1]. One of the major instabilities is geysering phenomenon induced by wall-superheating at boiling incipience. The mechanism and dynamic characteristics of the geysering have been studied by many investigators[2-7]. The instability originates from complex interactions among - (1) nucleation sites, bubble through microlayer (thin layer of superheated liquid between bubble and wall) and wall-bubble interface, (2) bubbles themselves (coalescence and collision) and (3) bubbles and bulk liquid. These interactions involve the exchange of heat, mass and momentum. However, a cycle of geyseing is completed in three consecutive stages - (1) refilling through condensation of vapor (increasing static head) to stop boiling, (2) incubation to initiate bubble formation, and (3) expulsion of liquid and vapor as a result of self-evaporation induced by a decrease in static head. The expulsion is usually identified by the surge of loop flow rate in NCBL.

Efforts have been made in the recent past to identify geysering to be periodic or chaotic. As described above, the phenomenon is highly complex. It is therefore always challenging to investigate the same. Here we present an experiment based on picture motion browser (PMB) with the help of Sony Handy Cam (HRD-XR160, 24 frames/sec, 3.3 mega pixels) to visualize the bubble growth and the associated expulsion phenomenon in geysering.

Visualization of Bubble Growth and Flow Pattern Transition

Description of Natural Circulation Facility

Our NCBL mainly consists of six parts — heated section, adiabatic riser, condenser, down-comer, upper and lower plenum. The rectangular loop was fabricated of SS 316. Eight resistance temperature detectors (RTDs) made of SS-310 with the range of 0–400°C measure fluid temperature at various points of the loop. Three pressure sensors are installed at heater inlet, heater outlet and the top of the riser. The differential pressure transmitter (DPT) made of SS-310 with the range of 0–10 bar measures pressure drop over the length of 0.6 m in the heated section. The electromagnetic flow meter (EFM) with 20 mm polytetrafluoro ethylene lining and the range of 0-1.8 m^3/hr is placed at the mid of the lower plenum to measure the loop flow. For uniform and safe heating six segmental heater-coils (Kanthal super: $MoSi_2$) are wound around the aluminum tube of the heater length. The aluminum tube is co-axially placed outside the heated pipe. In the annular section between the process pipe and the aluminum pipe, the heat is transferred by radiation and convection. An accurate control of heater power has been made by a thyristor power controller in PID mode. All signals from sensors come to SCADA panel. After comparing the signals with the set values the controllers send manipulating signal from SCADA panel to the valve actuator for cold-leg temperature control, the thyristor panel for heater power control and the solenoid valve for pressure control. The data-logger software (DAQ) in the computer records the data from all sensors via SCADA panel @ 0.5 Hz. The controller signals are stored online in the computer using controller software. The controllers are tuned online. It is required to mention that the two-phase valve controlling the loop resistance is fully open in all experiments.

Bubble Growth and Flow Pattern Transition

When violent vapor generation (vapor jet) starts in the heated section at a heater power Q and inlet subcooling ΔT_{sub} (=T_{sat}-T_{in}), the flow oscillations become severe. With PMB of the recorded video of bubble growth through the watch glass (diameter 25 mm) just above the heater section, we have analyzed the flow pattern transitions and the bubble shape in high-speed boiling flow. At the start of the expulsion stage, developing bubbly flow is observed at the top of the heater section. If the developing bubbly flow is not enough to transfer the heat from wall to bulk liquid, the wall superheat ΔT_{sat} (=T_{wall}-T_{sat}) increases. It expedites the bubble growth and bubbly flow changes to distorted vapor-slugs flow with high velocity. Consequently the expulsion phenomenon becomes vigorous. The slugs have numerous structures with complex geometry.

At the lower Q and high ΔT_{sub}, expulsions occur at irregular and low frequency. At the higher Q and lower ΔT_{sub}, expulsions occur at more regular and higher frequency compared to that of previous conditions. During rapid expulsion, both bubbles and liquid flow upward with very high velocity. The form drag and the skin drag of the vapor-bubbles in contact with bulk liquid increases with their size. However, the vapor-jet pulls the liquid up in the channel during the expulsions. Figure 1 shows flow-patterns transition captured at Q=5 kW and ΔT_{sub} = 30°C. The slug-bubbles were noticed to form with the frequencies of 20–23 sec^{-1} in the duration of 2–3 sec in an expulsion. The bubble velocity in the developed bubbly flow was evaluated to be 0.31 m/sec. The developing bubbly flow is shown in Figure 1(a). Just after 0.2 sec the developing spherical bubbles coalesce to form the developed bubbly flow, which is shown in Figures 1(b-c). With the initiation of fully developed bubbly flow, the violent vapor expulsion starts, and the flow regime quickly changes to the slug one as a consequence of the coalescence of the developed spherical bubbles shown in Figures 1(d-e). During the violent expulsion of vapor and liquid, big slug-bubbles form due to the coalescence of small ones and then get condensed to small ones in the riser section due to the significant increase of interfacial heat transfer area (refer Figures 1(f-p)). After some time, the expulsions stop. At the time instants of 1.28, 2.40, 2.60 and 3.0 sec, the slug-bubbles are observed to be big and glazy ones. At the intermediate time, the less glazy and smaller slug bubbles form. At 2.64 sec and 3.04 sec, the bubble collapse due to condensation is observed.

At higher ΔT_{sub} and lower Q vapor–mushrooms are observed to form. The vapor–mushrooms are trailed by vapor stems at their bottom. A liquid macrolayer exists in the region between wall and vapor stem (bellow the mushroom), whereas a thin liquid film called microlayer forms between the vapor stem and the vapor–mushrooms. These macrolayer and microlayer participate in determining boiling heat transfer characteristics. In the subcooled boiling flow, the vapor-mushrooms and vapor stems are also reported in the literature[8-9]. Figure 2 displays flow-patterns transition captured at Q=4 kW and ΔT_{sub} = 30 °C. Figures 2(a-d) show the gradual growth of the large slugs from small slugs or bubbles. Despite of higher buoyancy force, the large slugs will have less upward-velocity compared to small slug bubbles due to the dominance of wall resistance and drag force of the larger ones in contact with liquid. The large slugs sweep away large amount of heat from the wall. If the heater power is lower, the big slug bubbles do not form frequently. The smaller slug bubbles form just after the rapid expulsion of the larger slug bubbles. The smaller ones coalesce with the larger ones to form a mushroom-like shape shown in Figure 2(e). Figures 2(f-h) show the different size and shape of the vapor-mushrooms at different time of the expulsion. The different size distributions of bubbles in turn produce shear stress fluctuation in the vapor-liquid interface and create the distortion of the interface. As a result, the vapor–liquid interface becomes more complex in case of lower heater power. Consequently heat transfer characteristics in flow boiling become a highly complex phenomenon.

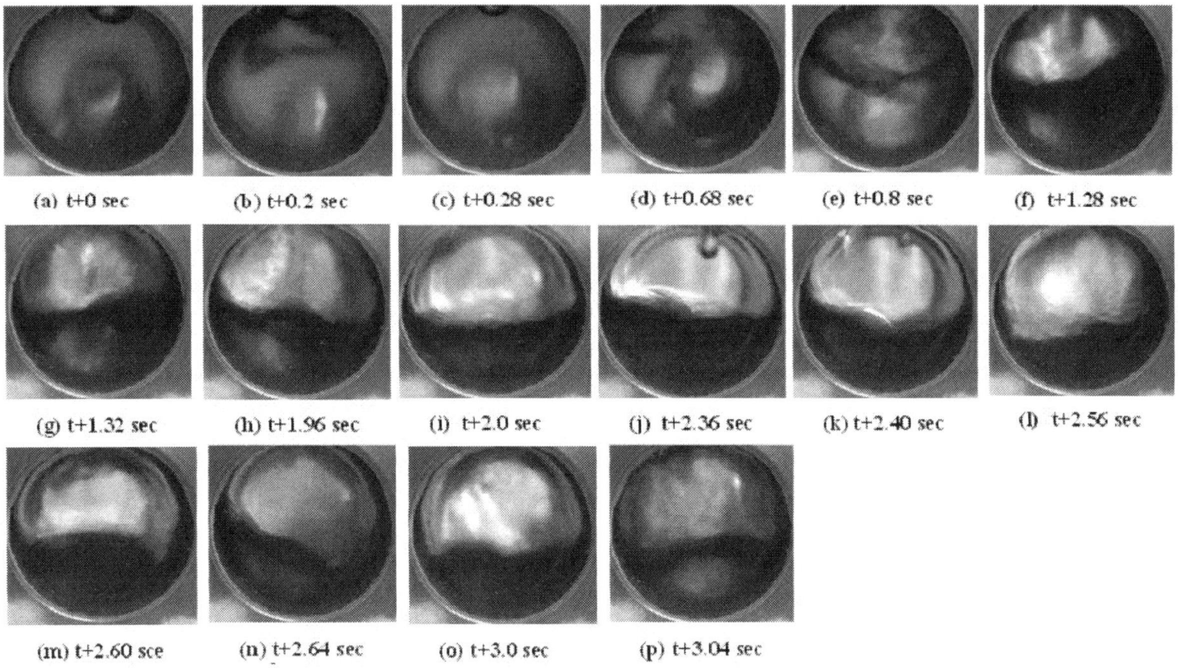

Figure 1. Flow-patterns transition captured at Q=5 kW and ΔT_{sub}= 30 °C

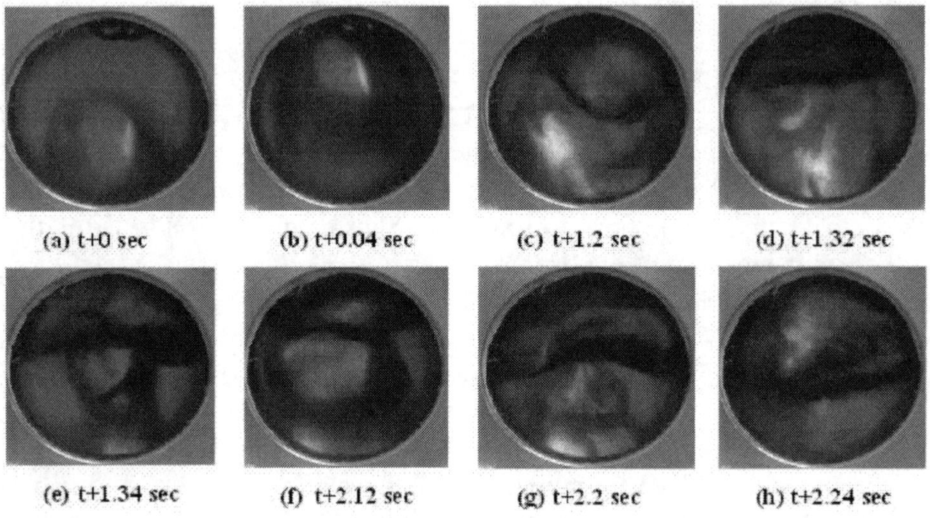

Figure 2. Flow-patterns transition captured at Q=4 kW and ΔT_{sub}= 30 °C.

Figure 3 present the flow-patterns transition captured Q=4 kW and ΔT_{sub}=50°C. In this case, the bubbles grow in a more irregular time intervals with the wider size distribution. Another thing is noticed that more mushroom-like bubbles are observed at higher ΔT_{sub} compared to those at lower ΔT_{sub}. At higher ΔT_{sub}, the strong temperature gradient across the radial distance of channel imposes a higher degree of nonlinearity in the boiling phenomenon. As the size distribution of bubble widens with the inlet subcooling, the collisions among the bubbles of different size in the upward bubble-flow increase. So formations of the mushroom-like assemblies of complex structure are very frequent due to the distortion of the vapor–liquid interface.

Figure 4 shows the expulsion time and cycle time in consecutive 24 cycles at different Q and ΔT_{sub}. At Q=5 kW and ΔT_{sub} = 30°C, the expulsion time varied in the range of 3.92–5.8 sec, whereas the cycle time in the range of 11.68–15.42 sec. The points are close to each other compared to lower heater power. At Q=4 kW and ΔT_{sub} = 30°C, the expulsion time and cycle time vary widely in the range of 3.48–11.8 sec and 9.44–31.84 sec, respectively. For an increase of ΔT_{sub} to 50 °C at Q=4 kW, the variations are in the range of 5.24–31.44 sec and 23.04–73.52 sec, respectively. One may easily conclude that at high ΔT_{sub} and low Q, the expulsion time and cycle time are larger and have wider distribution, and both expulsion time and frequency are highly irregular.

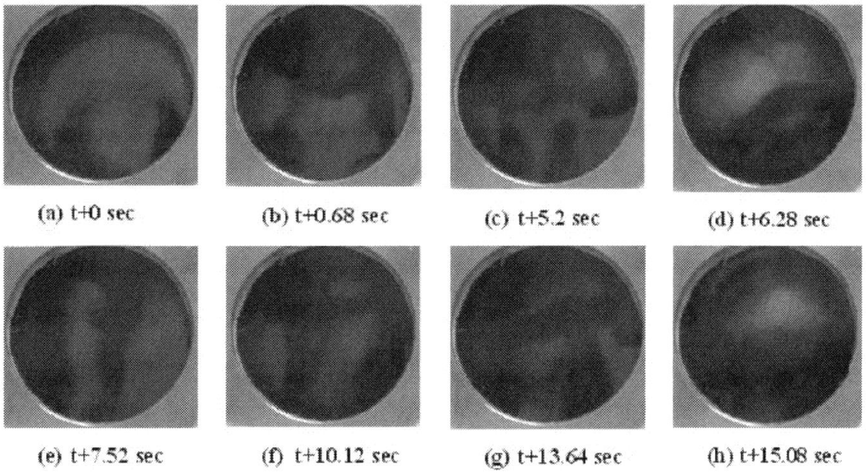

Figure 3. Flow-patterns transition captured at Q=4 kW and ΔT_{sub}= 50 °C.

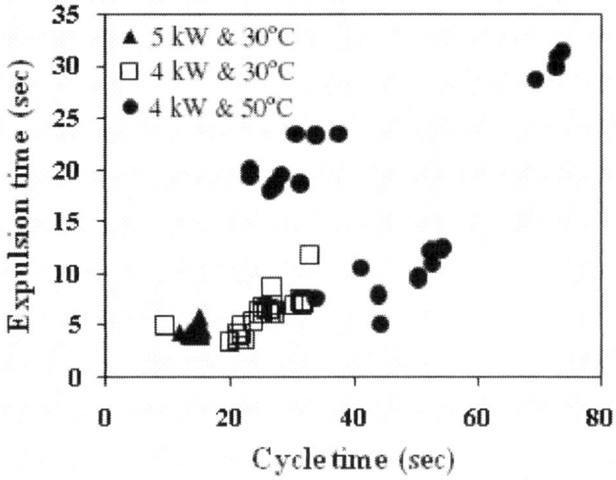

Figure 4. Expulsion time vs. cycle time at different Q and ΔT_{sub}

The phenomena may be explained by the fact that with increase in heat flux, both bubble waiting time (t_w) and bubble departure time (t_d) reduce due to the rapid growth of the bubble, which in turn induce the high inertia effect on the bubble to depart away from the heated surface. Increased wall superheat at higher heat flux causes the rate of heat and mass transfer in the bubble from micro-layer region to increase; as a result the growth rate of the bubble increases. Figure 4 displays substantial reduction of both expulsion time and cycle time as Q is increased from 4 kW to 5 kW. Here, short expulsion time is due to increase of frequency $\left(\frac{1}{t_w+t_d}\right)$ of bubble release as t_w and t_d reduce at high Q. Due to rapid bubble growth and heat transfer rate at Q=5 kW, short expulsion time is required to sweep away the heat. At higher ΔT_{sub}, the bubble frequency decreases as a result of increased t_w and t_d. Figure 4 shows that the expulsion time and cycle time are much higher at ΔT_{sub}=50 °C than those at ΔT_{sub}=30 °C. After the bubbles depart from heated surface, the subcooled liquid fills the region, and time required to generate next bubble from the surface will increase as the extent of subcooling increases; thus the waiting time of bubble increases. The bubble-departure time increases as a result of low growth rate of bubble at high subcooling due to extended bubble condensation.

Interestingly, the expulsions have been indicated by high loop flow rate W_{loop}. The time series of W_{loop} at different Q and ΔT_{sub} are shown in Figure 5. The peak flow rates are the signature of the expulsions at different time intervals. In Figure 5(a), the expulsion stage and the refilling stage are indicated by vertical dotted line a, b, c and d. In the region ab, developing bubbly flow observed, where loop flow rates start to increase. The region bc corresponds to the violent jet-flow in the heater section (expulsion), where the very high flow rates were observed. In the region cd, the loop flow rates decrease suddenly due to the collapse of bubbles in the riser of NCBL. The cold liquid the riser enters the heater section just after violent expulsion. Reverse flow or flow termination occurs and a non-boiling condition exists in the heated section. After the region cd, the incubation starts and nucleate boiling is initially observed in the heated channel. This stage is indicated by the flat line in the time series just after the region cd. At the low ΔT_{sub} and high Q, the frequency of flow oscillations increases and becomes more regular as evident in Figure 5(c).

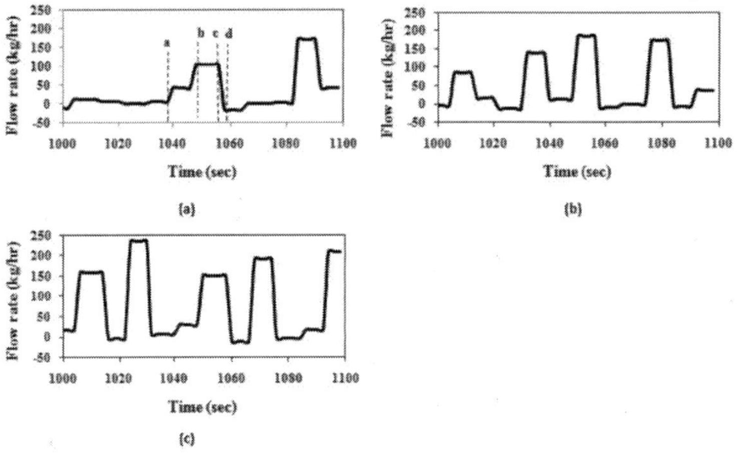

Figure 5. Time series of loop flow rate at (a) Q=4 kW and ΔT_{sub}=50 °C, (b) Q=4 kW and ΔT_{sub}=30 °C, (a) Q=5 kW and ΔT_{sub}=30 °C.

Conclusions

The present experimental investigations suggest that the expulsion phenomenon is highly complex and chaotic, and influenced by bubble growth and condensation. Mainly two major shapes of bubbles have been identified based on PMB. They are vapor-slugs at high Q and low ΔT_{sub}, and vapor-mushrooms at low Q and high ΔT_{sub}. The chaotic phenomenon depends on Q and ΔT_{sub}. At the lower Q and higher ΔT_{sub}, the expulsion time and cycle time have been observed to be larger and have wider distribution; both expulsion time and frequency are highly irregular. The resulting flow oscillations become more chaotic and encounter reverse flow.

Acknowledgement

The authors thankfully acknowledge the financial support of Department of Science and Technology, New Delhi, India under SERC scheme (Sanction # SR/S3/CE/089/2009).

References

1. Goswami N, Paruya S. Advances on the research on nonlinear phenomena in boiling natural circulation loop. *Prog. Nucl. Energy* 2011:53;673–697.
2. Griffith P. Geysering in liquid-filled lines. *ASME Paper 62-HT-39*, 1962; vol. 1.
3. Chexal VK, Bergles AE. 1973. Two phase instabilities in a low pressure natural circulation loop. In: *Proceedings of the 13th National Heat Transfer Conference of AIChE Sym.*1973:69;37–45.
4. Aritomi M, Chiang JH, Mori M. Geysering in Parallel Boiling Channels. *Nucl. Eng.Des.* 1993:14;111.
5. Jiang SY, Yao MH, Bo JH, Wu SR. Experimental simulation study on startup of the 5 MW nuclear heating reactor. *Nucl. Eng. Des.* 1995;158:111–123.
6. Baars A, Delgado A. Non-linear Effects in a Natural Circulation Evaporator: Geysering Coupled with Manometer Oscillations. *Heat and Mass Transfer* 2006:43;427.
7. Paruya S, Saha AK, Bhattacharya P. Validations of thermohydraulic models for geysering in a natural circulation loop using an impedance needle probe. *Ind. Eng. Chem. Res.* 2009;48:2020–2033.
8. Gaertner RE. Photographic Study of Nucleate Pool Boiling on a Horizontal Surface. *J. Heat Transfer* 1965:87;17–29.
9. Lay JH and Dhir VK. Shape of a Vapour Stem During Nucleate Boiling of Saturation Liquids. *J. Heat Transfer* 1995:117;394–401.

(142h) Numerical Investigation of the Effect of Surfactants On the Dynamics of the Organic Drops in Water Column

Abhijit Rao, Cain Department of Chemical Engineering, Louisiana State University, Baton Rouge, LA, Rupesh K. Reddy, Cain Department of Chemical Engineering, Lousiana State University, Baton Rouge, LA and Krishnaswamy Nandakumar, Cain Dept. of Chemical Engineering, Louisiana State University, Baton Rouge, LA

Chemical dispersants are widely used to counter the adverse effects of oil spills on the environment. Surfactants are the active ingredients in dispersants which cause reduction in interfacial tension resulting in the production of smaller oil droplets. Surfactant distributes between the bulk phase (water) and the interface to achieve equilibrium between the two phases.

In this study we present a numerical model which captures the effect of surfactant on the dynamics of the hydrocarbon droplets. In the preliminary results we show how the variation in concentration of surfactant affects the terminal settling velocities of Carbon tetrachloride droplets in water. The system under consideration consists of 0.02 ml (d_p = 3.4 mm) CCl_4droplets, containing 0. 2M of acetic acid, settling in an extraction column of internal diameter 3cm and length 70cm, containing water with surfactant TritonX-100. Numerical simulations are carried out at different concentrations of surfactant in the bulk phase. As the concentration of surfactant in continuous phase remains constant in each case, a uniform interfacial tension prevails across the droplet surface. The relation between the bulk surfactant concentration and interfacial tension is established using Langmuir-Szyszkowski equation. The results are compared with those obtained in the experiments conducted by Chen et al. ("Adsorption Behavior of surfactants and mass transfer in single drop extraction", January 2000,Vol 46. No1,AICHE Journal, pp.160-168). The mass transfer coefficients for transfer of acetic acid from droplet to continuous phase are taken from the data provided in the experiment.

We also present a model which is applicable in scenarios of continuous spray of surfactant on the swarm of hydrocarbon drops. In this case, the concentration of surfactant in the continuous phase is not constant. A convection-diffusion equation is included in the model to capture the transport of surfactant from continuous phase to dispersed phase and depending on the surfactant concentration at the interface, the interfacial tension is calculated. The results are extended to the system with droplets resulting from the breakup of a liquid jet. Applications of this model can be found in understanding the deep water oil spill scenarios.

(142j) Instability in the Presence of Evaporation for Binary Liquids

Ranga Narayanan, Chemical Engineering Department, University of Florida, Gainesville, FL and Erdem Uguz, Chemical Engineering Department, University Of Florida, Gainesville, FL

Instability in the presence of evaporation for binary liquids

R. Narayanan and K.E. Uguz

University of Florida, Gainesville, FL. 32611

There are several applications in chemical engineering in which convection phenomena play a very important role, ranging from crystal growth and the coating of films to the casting of alloys and the space processing of glasses and semiconductors and in most of those applications involve mixtures of fluids. Common to, many of these problems is the occurrence of an interface such as a fluid-fluid or fluid-solid interface and when there is an interface there often occurs a form of interfacial instability wherein a sudden change in interfacial pattern occurs as system control variable crosses a critical value.

In this study, evaporative convection accompanies the classical Rayleigh and Marangoni convection phenomena. Having two or more fluids introduce solutal effect into the system such as concentration dependence of density and concentration dependence of surface tension. Due to these new effects binary mixtures behave different than pure ones.

Pure evaporative convection, in the absence of gravity and surface tension, can occur if the system is bounded by impermeable walls and there is a temperature gradient at the interface. The closed system guarantees zero total evaporation rate. Note that in the closed system the liquid which evaporates at the trough has to condense at the crest. As a result of evaporation there will be an upwards flow at troughs and a downward flow at the crest as shown in Fig 1 ,which is an indication of convection.

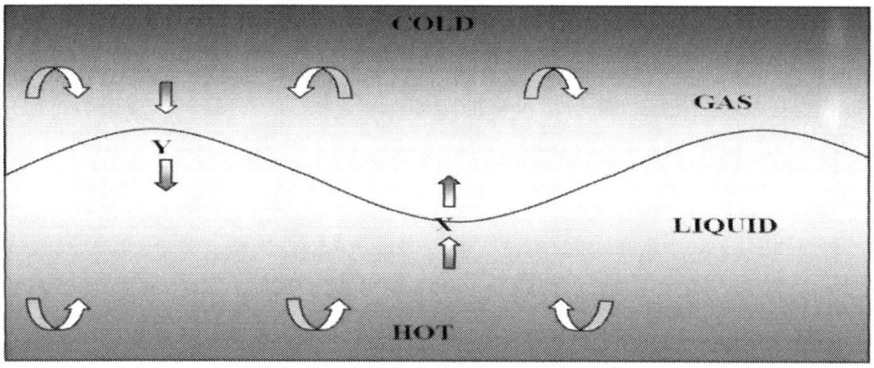

In this work the effect of mass fraction, vapor dynamics and heating arrangement (heating from below or above) are investigated. Calculations are done for a deflecting interface both in the presence and absence of gravity.

The results show that the deeper the liquid layer, the more unstable the system becomes. However, a similar statement does not hold for the vapor layer height. For the vapor layer height, there is a trade off between the buoyancy-driven convection and the stabilizing effect of vapor flow of phase change even when buoyancy is ignored. Thus there is a turning point of stability for the vapor height where the destabilizing effect of the buoyancy overcomes the stabilizing role of vapor flow arising from phase change itself.

(142k) Modeling of Multi-Layered Spin Coating Process

Pankaj Doshi and Akash Arora, National Chemical Laboratory, Pune, India

Spin coating is the most commonly used method in the industries to make single layer and multi-layered thin films of desired thickness and functionality. These thin films finds applications in device fabrication ranging from integrated chips in microelectronics industry to organic solar cells. In general, multilayer structure of thin films provides hardness, toughness and multifunctional character to the material for desired applications. In this paper, a lubrication theory is used to derive the evolution equations governing the flow of two layered film on a spinning disc. The set of two nonlinear evolution equations is solved using a numerical technique based on Galerkin Finite Element Method (G/FEM). With the help of numerical simulations, the effect of disc speed and physical properties of the liquids like ratios of viscosity, surface tension and density, on the uniformity of the film so form are studied. The ultimate goal of this research is to elucidate the interplay of liquid layers in the formation of layered films, and find the optimum parameter space for a given application.

(142l) Inverse Approach for Turbulent Models Using Optimized Lattice Boltzmann Method

Sesha Hari Vemuri[1], Lorenz T. Biegler[1] and Myung S. Jhon[1,2], (1)Chemical Engineering, Carnegie Mellon University, Pittsburgh, PA, (2)School of Advanced Material Science and Engineering, Sungkyunkwan University, Suwon, South Korea

Turbulent flow plays an important role in many areas of applied science and engineering. The simulation of turbulent flow is a challenging subject, and it is still remained as an unsolved problem due to the difficulties associated with accurately analyzing chaotic fluid motion at high Reynolds number. Here, we utilize the lattice Boltzmann method (LBM) [1] as an alternative methodology in resolving issues in computational fluid dynamics (CFD) where the governing equations can be derived through the Chapman-Enskog multi-scale expansion via the Boltzmann transport equation (BTE). LBM, a descritized form of BTE, is a powerful tool for fluid flow simulations as well as multiscale modeling methodology due its advantages of algorithmic simplicity, transient nature, and ease in the treatment of complex geometrical boundaries, in addition to its amenability to parallel computing platforms. Furthermore, the methodology of LBM is "rule based," and one may be able to find appropriate physical models based on experimental findings as shown in this study via what is termed an inverse approach.

We developed an extensive single-phase flow analysis tool based on LBM with non-reflecting boundary conditions to avoid spurious disturbances at the boundaries with the Smargorinsky large eddy simulation (LES) model for turbulent flows [2]. Our model describes most of the essential characteristics of the complex flows without losing the critical competitive edge of computational resource efficiency. In this study, we exploit the abovementioned rule-based characteristics to obtain an accurate LBM model, by minimizing the disparity between our numerical results and the experimental results of high Reynolds number flows for various geometries (square cylinder with rounded and square corners at various orientations towards the incoming flow) by a novel inverse modeling approach. This will result in an accurate performance comparison of a broad range of turbulence models via an "optimized" set of multiple relaxation time parameters, as well as the capability to adjust the model automatically when the experimental data is inconsistent with simulation. This preliminary test case will help us to observe the robustness of LBM, unlike conventional CFD approaches which have no easy implementation for this scenario. Furthermore, we will extend our model towards complicated flow systems consisting of multi-physical phenomena, which will lead to deeper insight into the area of turbulence simulation..

[1] S. Chen and G.D. Doolen, Annu. Rev. Fluid Mech., **30**, 329 (1998).

[2] D. Kim, H. M. Kim, M. S. Jhon, S. J. Vinay III, and J. Buchanan, Chin. Phys. Lett., **25**, 1964, (2008)

(142m) Modeling Precipitation Reactions in Turbulent Flow with Qmom Incorporated Into LES

Alex W. Abboud, Sean T. Smith, Tony Saad and Jeremy Thornock, Chemical Engineering, Institute for Clean and Secure Energy, University of Utah, Salt Lake City, UT

For precipitation in a liquid turbulent pipe flow, a population balance in large-eddy simulation will be presented. The moments of the population balance will be used to track the size distribution of the particles as it evolves in the flow field. The flow field will be simulated in a computational fluid dynamics program using large-eddy simulation. A quadrature method of moments solver has been incorporated into the LES code to solve the population balance of several competing precipitates. The chemistry for the problem has a timescale separation, such that aqueous species react quickly enough to reach equilibrium, while the precipitates form on the same timescale as the flow. In order to avoid the computational expense of the non-linear equilibrium solve, the chemistry has been pre-tabulated as a function of the mixture fraction and the extent of reaction to be used as a look-up table. To provide coupling between the overlapping mixing and precipitation time scales a model for the scalar variance of the mixture fraction has been incorporated using a simplified subgrid scale mixing model based on the multi-environment approach.

(142n) Diffusion Driven Flows in Multi-Ionic Systems

Abhishek Kar[1], Tso-Yi Chiang[1] and Darrell Velegol[2], (1)Chemical Engineering, The Pennsylvania State University, University Park, PA, (2)Chemical Engineering, Penn State University, University Park, PA

Diffusioosmotic flows occur in gradients of salt concentration, by a much-studied process. The essential physics is that the ions in the salt have different diffusion coefficients, and in order to maintain electroneutrality, a spontaneous electric field arises in the system. This electric field drives a diffusioosmotic flow at charged surfaces, and diffusiophoresis of charged particles. A common approach is to impose an external salt gradient on a system, and then to follow the transport of particles as they move by diffusiophoresis. In this talk, we do not impose salt gradients, but rather, allow the particles themselves to self-generate the gradients in ionic strength. Using mineral particles, such as calcite (CaCO3) or gypsum (CaSO4), we enable flows to be generated and tracer particles to be moved. To better study the process, and delineate it from any convective or density driven flow mechanism, we consider the case where salt gradients generated from divalent salts in the presence of monovalent ions can generate flows in vertical systems of 'a reservoir and a sink' system. We study this in the presence of monovalent ionic species (e.g. KCl and NaCl) and generate an experimental and theoretical model showing diffusion driven flows in such cases. We show that the velocity of these tracer particles can be related to essential parameters of the systems, such as the Ksp of the minerals, the diffusion coefficients of the ions resulting from the dissolved mineral, the zeta potentials of the particles and surfaces, and even the change of zeta potentials resulting from the dissolution of minerals. These data, and our interpretations of them, will be presented in this talk.

(142o) Density-Functional Theory and Small Angle X-Ray Scattering Experiments for the Structure Factor of Solvent-Free Nanoparticle-Organic Hybrid Materials

Hsiu-Yu Yu, Samanvaya Srivastava, Lynden A. Archer and Donald L. Koch, School of Chemical and Biomolecular Engineering, Cornell University, Ithaca, NY

Nanoparticle-organic hybrid materials (NOHMs) contain inorganic nanocores surface functionalized by oligomeric chains in the absence of unattached solvent molecules [1]. In this talk, we present a density-functional theory and small angle X-ray scattering (SAXS) experiments to characterize the static structure factor of NOHMs. The structure factor of NOHMs is qualitatively different from that of hard-sphere suspensions. In monodisperse NOHMs, the fact that each particle along with its tethered oligomers occupies the same volume implies that the static structure factor approaches zero at zero wave number. This behavior has been predicted by a density-functional theory [2] and confirmed by molecular dynamics simulations [3]. Our experimental results show that the structure factor in NOHMs with small core particle volume fraction is much smaller than that in hard spheres. However, we observe a nonzero structure factor at small wave number and this might be attributed to unavoidable polydispersity.

Here, we synthesize NOHMs that consist of silica nanoparticles surface-grafted with polyisoprene corona. Some systems are synthesized to be approximately monodisperse while others consist of blended bidisperse mixtures. The density-functional theory considers a bidisperse suspension of oligomer-tethered particles with different core radii and/or oligomer grafting densities. These variations yield variations in the tethered fluid volume per particle and distort the bulk structure at large length scales such that the structure factor exhibits a nonzero value at zero wave number. The theory predicts a much larger effect of bidispersity in grafting density than bidispersity in core radius on the structure factor at zero wave number. The bidispersity of core radius with a fixed grafting density has a relatively small effect on the structure factor at small wave number because the volume of the particle grows as a^3 and the volume of the tethered fluid grows at a somewhat similar rate, as the $O(a^2)$ surface area, where $a$is the particle radius. Our results also suggest that scattering experiments could be used to evaluate the uniformity of the oligomer grafting achieved in the synthesis procedures.

[1] P. Agarwal, H. Qi, and L. A. Archer, *Nano Lett.* **10**, 111 (2010).

[2] H.-Y. Yu and D. L. Koch, *Langmuir* **26**, 16801 (2010).

[3] A. Chremos, A. Z. Panagiotopoulos, H.-Y. Yu, and D. L. Koch, *J. Chem. Phys.* **135**, 114901 (2011).

(142p) The Evolution of Kinetic Energy in Actively Forced Confined Mixing Layers

Wei Zhao, University of South Carolina, Columbia, SC and Guiren Wang, Mechanical Engineering, University of South Carolina, Columbia, SC

The evolution of kinetic energy in actively forced confined mixing layers

Wei Zhao; Guiren Wang

Abstract The receptivity of flow in the nozzle section of confined mixing layer is investigated at different forcing frequencies and intensities. The results are compared to the velocity fluctuations in the mixing chamber adjacent to the trailing edge.

It's found the largest receptivity (at 5.3 Hz) in nozzle section doesn't necessarily lead to highest velocity fluctuations. At 5.3 Hz, the velocity fluctuations decrease fast in the mixing chamber along flow direction. This is contrary to the unforced case. Besides, the role of mean vertical velocity in momentum transfer and energy transport is still undetermined. Hence, the evolution of kinetic energy in confined mixing layer is investigated. As the Reynolds number we applied is only around 3000, close to the trailing edge, the homogeneous and isotropic turbulence state cannot be achieved. Therefore, axisymmetric approximation is adopted instead to simplify the dissipation terms.

Experimental results indicate, compared to the unforced case, the dissipation terms in forced flow (5.3 Hz) exhibit small difference. Hence, the fast decreasing of the kinetic energy cannot be the result of energy dissipation. Further investigations show that the total production term in energy equation becomes highly negative, when the flow is forced. Its magnitude is 25 times larger than the total dissipation term.. In the process, the term $-v'^2(dV/dy)$ has the most significant contribution to the total production term, where v'^2, V and y are vertical velocity fluctuation, average vertical velocity and vertical position respectively. It implies the existence of V may inhibit the development of turbulence and the mixing. This is questionable, because intuitively the varied V in vertical direction should cause the stretching of spanwise vortices and results in earlier break-down which should be propitious to the mixing process. Further study is needed to understand the energy transport phenomena in this complicated flow.

ANALYSIS OF THE PARTICLE VISCOSITY AND DRAG COEFFICIENT IN DILUTE GAS-SOLID FLOWS IN DUCTS

Rodrigo K. Decker, Jonathan Utzig, Henry F. Meier
Regional University of Blumenau – FURB, Department of Chemical Engineering, São Paulo St., 89030-000, Blumenau – Brazil.

Introduction

In the last decade the use of Computational Fluid Dynamic (CFD) models to describe the behavior of pneumatic gas-solid flows in industrial process has received a lot of attention in the literature. Furthermore, the use of commercial software has encouraged a lot of companies to investigate their processes by means of numerical simulation based on numerical models. However, depending on the constitutive equations applied in the CFD simulations for these analysis, some problems may happen and an uncorrected answer may be obtained. One common problem in very dilute gas-solid flows is the particle clusters formation. These clusters are formed in dilute flows mainly due to changes in the main flow direction caused by the presence of curves [1,2,3,4].

One way to investigate these phenomena can be addressed to the drag coefficient model between the gas and particulate phases which should be able to predict particle cluster formation [5, 6]. In this sense, the objective of this study is to identify the behavior of different drag models, since basic models that do not recognize clusters behavior until more accurate ones that considers not just the particle itself but also the behavior of clusters when they are formed. The results of these models are compared in terms of mean particle axial velocity profiles with experimental data obtained with a two dimensional Particle Image Velocimetry (2D-PIV) system, and in terms of particle volume fraction profiles for different radial positions along the vertical duct. Furthermore, the influence of the particle viscosity in the mean particle velocity profile is also investigated first of all neglecting its value and then considering a value equal to the gas phase.

Methodology

Experimental Setup

In the test facility showed in Figure 1, a Pitot tube (A) is connected to a differential pressure transmitter in order to measure the inlet mean velocity of the gas phase. This velocity is kept constant and controlled by the frequency converter of the radial ventilator located at the end of the test facility, which is also responsible for maintaining the entire system under negative pressure. After the Pitot tube, particles are loaded into the system by a conveyor screw (B) that is also controlled by a frequency converter, responsible for controlling the mass load ratio. The measurements are obtained in the Plexiglas tubes (C) that are followed by the cyclone separator (D), which is responsible for separating the solid phase from the gas phase. The big particles are collected in the hopper of the cyclone and the small particles are dragged out through the cyclone outlet into the filter bag (E). There are also pressure sensors in all devices of the test facility. All measured data are controlled by a data acquisition system in a computer and the data are treated and recorded in known time steps.

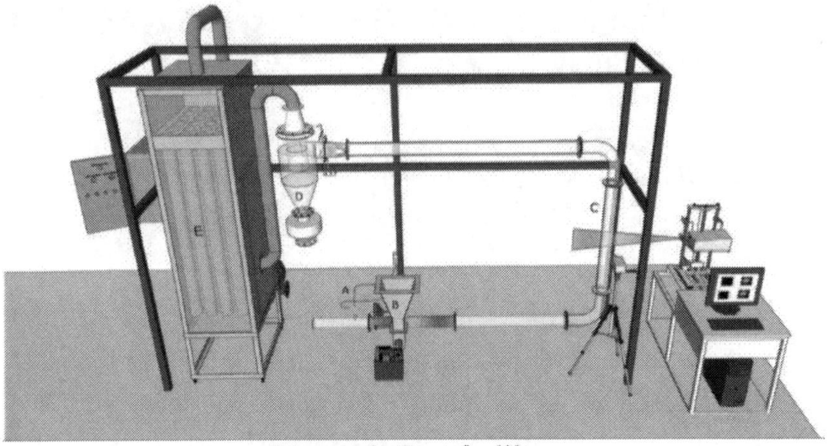

Figure 1. Test facility.

Operational and Geometric Conditions

In the test facility experiments were developed for an initial mean gas phase velocity of 5 m/s. A mass load ratio of 39 g/m^3 and a Reynolds number for the gas phase of $2.8x10^4$ were also applied in the experiments. The FCC (fluid catalytic cracking) catalyst particles employed in this study are in the solid phase. These particles have a density of 1845 kg/m^3 and a mean particle diameter, obtained by a Malvern particle size analyzer (MASTERSIZER), of 56.7μm. The PIV measurements were taken in two different Plexiglas ducts as shown in Figure 2. The vertical duct section has a length of 1180 mm and the second horizontal duct section, 2100mm, both with a diameter of 100mm. In each subsection, velocity fields and axial velocity profiles were obtained for the following axial positions:

Region 1: from 147 to 214 mm;
Region 2: from 545 to 614 mm;
Region 3: from 995 to 1059 mm;
Region 4: from 140 to 270 mm;
Region 5: from 475 to 613 mm;
Region 6: from 1218 to 1367 mm.

The reference point (0 mm) for regions 1, 2 and 3 is the beginning of the vertical section and for regions 4, 5 and 6, the beginning of the 2nd horizontal section.

After calibration of the PIV system, 200 frames were taken with a time step of 80μs.

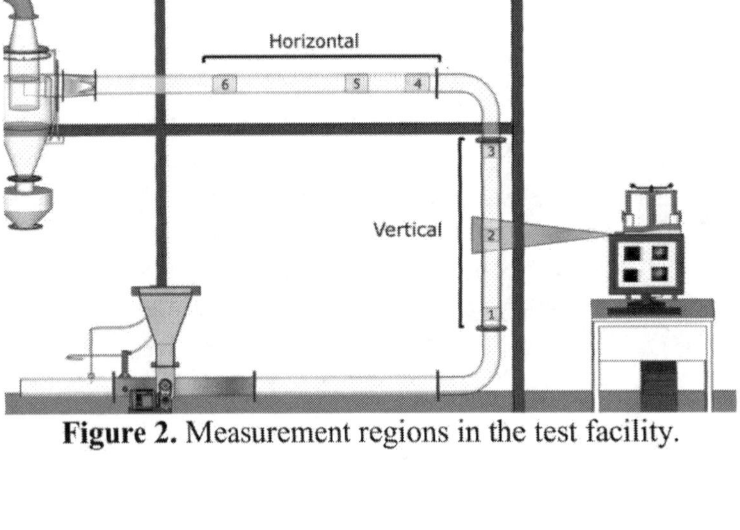

Figure 2. Measurement regions in the test facility.

Mathematical Modeling

Conservation Equations

The mathematical model applied in the numerical studies is based on two principles of fluid dynamics: mass and momentum conservation for the gas and the solid phases using an Eulerian-Eulerian (EE) approach.

The following hypotheses are formulated:

- phases are interpenetrable;
- spherical particles have a mean diameter (two-phase model);
- turbulence is dealt with statistically where the Reynolds-Averaged-Navier-Stokes equations are obtained after the Reynolds decomposition and application of time averaging.

The mass conservation equations for both phases, gas and solid, are respectively expressed as follows:

$$\frac{\partial}{\partial t}\left(f_g \rho_g\right)+\nabla .\left(f_g \rho_g v_g\right)=0 \tag{1}$$

$$\frac{\partial}{\partial t}\left(f_s \rho_s\right)+\nabla .\left(f_s \rho_s v_s\right)=0 \tag{2}$$

The momentum equations for the gas and the solid phases are:

$$\frac{\partial}{\partial t}\left(f_g \rho_g \mathbf{v}_g\right)+\nabla .\left(f_g \rho_g \mathbf{v}_g \mathbf{v}_g\right)=-f_g \nabla .\left(\mathbf{T}_g^{ef}\right)+f_g \rho_g \mathbf{g}-f_g \nabla p_g+\left(\mathbf{F}_{res}\right)_{g-s} \tag{3}$$

$$\frac{\partial}{\partial t}\left(f_s \rho_s \mathbf{v}_s\right)+\nabla .\left(f_s \rho_s \mathbf{v}_s \mathbf{v}_s\right)=-f_s \nabla \cdot\left(\mathbf{T}_s^{(t)}\right)+f_s \rho_s \mathbf{g}\text{-}f_s \nabla p_s-\left(\mathbf{F}_{res}\right)_{g-s} \tag{4}$$

where the subscripts g and s are the gas phase and the solid phase respectively, ρ is the density of each phase, **v** is the velocity vector of each phase, f is the volume fraction of each phase, $\mathbf{T}_g^{ef}$ is the effective tensor that acts on the real fluid, $\mathbf{T}_s^{(t)}$ is the turbulent tensor of the solid phase, **g** is the gravity, p is the pressure in each phase and $\left(\mathbf{F}_{res}\right)_{g-s}$ is the gas-solid interaction term.

Turbulence Modeling

The effective tensor for the gas phase is similar to the Newton model using the Boussinesq hypothesis, which is directly proportional to the deformation rate, expressed by

$$\mathbf{T}_g^{ef}=-2\mu_g^{ef}\mathbf{D}_g \tag{5}$$

where μ^{ef} is the effective viscosity and $\mathbf{D}_g$ is the deformation rate.

The deformation rate tensor is expressed by Eq. 6:

$$\mathbf{D}_g=\frac{1}{2}\left[\nabla \mathbf{v}_g+\left(\nabla \mathbf{v}_g\right)^T\right] \tag{6}$$

The effective viscosity of the gas phase is a result of the interaction of the molecular and turbulent viscosity:

$$\mu_g^{ef}=\mu_g+\mu_g^{(t)} \tag{7}$$

where $\mu_g^{(t)}$ is the turbulent viscosity.

The turbulent viscosity of the gas phase is obtained by an isotropic turbulence model (standard k-ε model) with two differential equations: one for the kinetic turbulence energy and the other for its dissipation rate, as follows:

$$\frac{\partial}{\partial t}\left(f_g \rho_g k_g\right)+\nabla.\left(f_g \rho_g \mathbf{v}_g k_g\right)=f_g \nabla.\left(\frac{\mu^{ef}}{\sigma^k}\nabla k_g\right)+f_g\left(G_g-\rho_g \varepsilon_g\right) \tag{8}$$

$$\frac{\partial}{\partial t}\left(f_g \rho_g k_g\right)+\nabla.\left(f_g \rho_g \mathbf{v}_g k_g\right)=f_g \nabla.\left(\frac{\mu^{ef}}{\sigma^k}\nabla k_g\right)+f_g\left(G_g-\rho_g \varepsilon_g\right) \tag{9}$$

with

$$G_g=-\mathbf{T}_g^{(t)}:\nabla \mathbf{v}_g \tag{10}$$

Consequently,

$$\mu_g^{(t)}=C_\mu \rho_g \frac{k_g^2}{\varepsilon_g} \tag{11}$$

For the development of an inviscid model, the momentum equations (Eq. 4) can be rewritten not considering the stress and pressure terms of the solid phase, as follows:

$$\frac{\partial}{\partial t}\left(f_s \rho_s \mathbf{v}_s\right)+\nabla.\left(f_s \rho_s \mathbf{v}_s \mathbf{v}_s\right)=f_s \rho_s \mathbf{g}-\left(\mathbf{F}_{res}\right)_{g-s} \tag{12}$$

Constitutive Equations

The resistive force between the gas and solid phases can be modeled by the equation:

$$\left(\mathbf{F}_{res}\right)_{g-s}=\beta_w\left(\mathbf{v}_g-\mathbf{v}_s\right) \tag{13}$$

where β_w is the interface coefficient between the gas and the solid phases and can be calculated according to the Syamlal and O'Brien model [7]and according to the Gidaspow model [8].

Gidaspow model:

For the dilute flow (fg > 0.8), the interface coefficient can be modeled according to Wen- Yu correlation [9] and the EMMS theory [6]. This last one takes into account the effect of cluster formation. All these models are associated with the drag coefficient, Cd_{g-s}, between the gas and the solid phases:

$$\beta_w=w\beta_{g-s}, \tag{14}$$

with:

$$\beta_{g-s}=\frac{3}{4}Cd_{g-s}\frac{f_g \rho_g\left|\mathbf{v}_g-\mathbf{v}_s\right| f_s}{d_s} \tag{15}$$

Wen Yu's Correlation:

$$w=f_g^{-2.7} \tag{16}$$

EMMS (Energy Minimization Multi-Scale) Theory:

For $0.74 \leq f_g \leq 0.82$,

$$w=-0.5760+\frac{0.0214}{4(f_g-0.7463)^2+0.0044} \tag{17}$$

For $0.82 \leq f_g \leq 0.97$,

$$w = -0.0101 + \frac{0.0038}{4(f_g - 0.7789)^2 + 0.0040} \quad . \tag{18}$$

For $f_g > 0.97$,

$$w = -31.8295 + 32.8295 f_g \tag{19}$$

For a concentrated flow (f_g <0.8) the Ergun equation [10] is employed and it is expressed as

$$\beta_w = 150\mu_g \left(\frac{f_s}{f_g d_s} \right)^2 + 1.75 \frac{\rho_g f_s |\mathbf{v}_g - \mathbf{v}_s|}{d_s} \tag{20}$$

There are several empirical correlations to calculate the drag coefficient between the gas and solid phases, $Cd_{g\text{-}s}$, for different flow regimes. These correlations are basically characterized by the Reynolds number of the solid phase, as follows [11]:

- Stokes regime, where $0 < Rep \leq 1$:

$$Cd_{g-s} = \frac{24}{\mathrm{Re}_p} \tag{21}$$

- Viscous regime, where $1 < Rep \leq 1000$:

$$Cd_{g-s} = \frac{24}{\mathrm{Re}_p}\left[1 + 0.15\,\mathrm{Re}_p^{\,0.687}\right] \quad . \tag{22}$$

- Inertial regime, where $Rep > 1000$

$$Cd_{g-s} = 0.44 \quad \ldots \tag{23}$$

Syamlal and O'Brien Model:

$$\beta_w = \frac{3}{4} \frac{f_g f_s \rho_g}{v_{r,s}^2 d_p} C_D \left(\frac{\mathrm{Re}_p}{v_{r,s}} \right) |\mathbf{v}_g - \mathbf{v}_s| \quad . \tag{24}$$

where

$$C_D = \left(0.63 + \frac{4.8}{\sqrt{\mathrm{Re}_p / v_{r,s}}} \right) \tag{25}$$

and

$$v_{r,s} = 0.5(A - 0.06\,\mathrm{Re}_p + \sqrt{(0.06\,\mathrm{Re}_p)^2 + 0.12\,\mathrm{Re}_p(2B - A) + A^2} \quad \ldots \tag{26}$$

with

$$A = f_g^{4.14},\ B = 0.8 f_g^{1.28} \text{ for } f_g \leq 0.85 \quad . \tag{27}$$

or

$$A = f_g^{4.14},\ B = f_g^{2.65} \text{ for } f_g > 0.85 \tag{28}$$

The particle Reynolds number is defined as

$$\mathrm{Re}_p = \frac{\rho_g f_g |\mathbf{v}_g - \mathbf{v}_s| d_s}{\mu_g} \quad \ldots \tag{29}$$

Initial and Boundary Conditions

The boundary conditions for the numerical experiment are described as follows:

Inlet: initial velocity and mass load ratio are known and all other flow properties are considered to be uniform;
Outlet: pressure is known and constant;
Wall: condition for the particle phase is free-slip and condition for the gas phase is no-slip.

Numerical Grid and Characteristics

Figure 3 contains a lateral and a frontal view of the numerical grid applied in the simulations, with an 165000 hexahedrical volumes . The applied solution in this case is also a 3-D and transient model and the software used for these developments was ANSYS FLUENT 12. After this analysis, the numerical grid with 165000 hexahedrical volumes was defined for further simulations, comparing the different drag models between the phases.

The convergence criteria was 10^{-6} for the residual values and 10^{-3} for the other variables. The time step applied was 10^{-3}s with 5s of real time. Second-order interpolation schemes were applied for the momentum equations and first-order interpolation schemes, for the other variables.

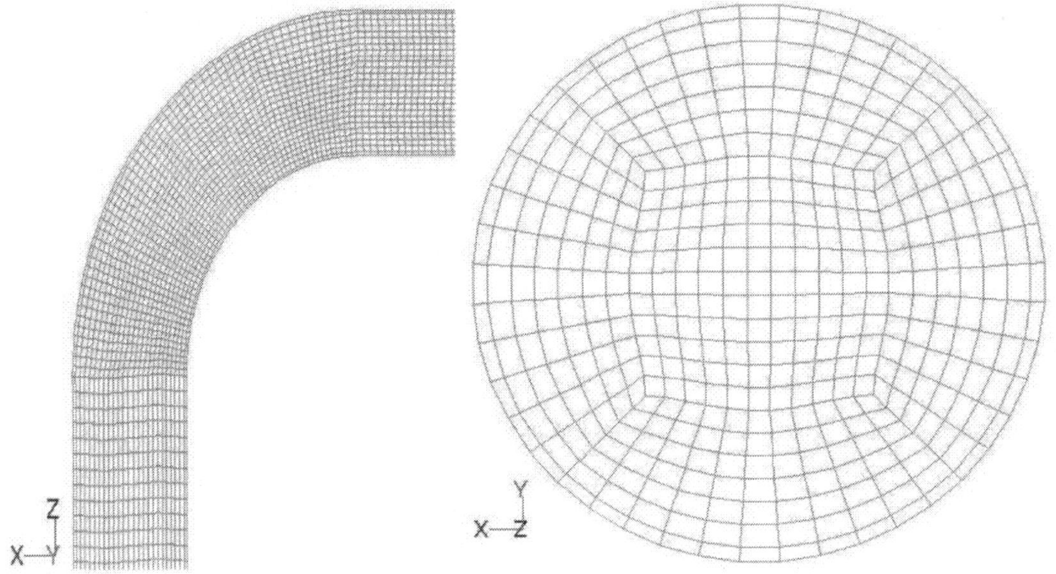

Figure 3. Mesh concentration in the curve region (a) Lateral view (b) Frontal view.

Results and Discussions

The results presented in Figure 4 and 5 represents the efforts to investigate the influence of different drag models in relation to cluster formation and consequently its influence on the particle velocity profiles for dilute gas-solid flows. In both figures all profiles have a similar and good qualitative behavior in relation to the experimental data at the beginning of the vertical section (L/D=1.8) as well as at the end of the vertical (L/D = 10.3) and horizontal (L/D = 13) sections. The biggest difference between all profiles was observed in Figure 4, in a region located immediately after the curve, responsible for the change of the main flow direction. In Figure 4a it is possible to observe that the mean particle velocity results obtained with the EMMS theory are better represented mainly in the radial positions between -50mm and -10mm, when comparing with the experimental data. Furthermore, the results obtained with the EMMS theory showed that this methodology was able to describe the behavior of the particle velocity in regions in the radial profile where numerical problems appear (20-40mm). These problems were just observed with the application of Fluent CFD software from ANSYS and it is directly related to the particle volume fraction. In this case the software considers that the volume fraction values obtained between 20mm and 40mm (Figure 4b) in the radial position are so small, that no numerical value is attributed for the particle velocity. As a consequence, a descontinuity is formed in the particle velocity profile. The EMMS theory, in this case, do not recognized such problem because particle were aggregated in small clusters (Figure 4b), avoiding the minimal volume fraction value in which such problems happen.

Another important result obtained in this study is related to the influence of the particle viscosity. In all developed simulation no difference in the particle velocity profile was obtained changing the particle viscosity value from zero to the value of the gas phase viscosity. That probably occurred due to the very dilute characteristics of the gas-solid flow investigated in this study.

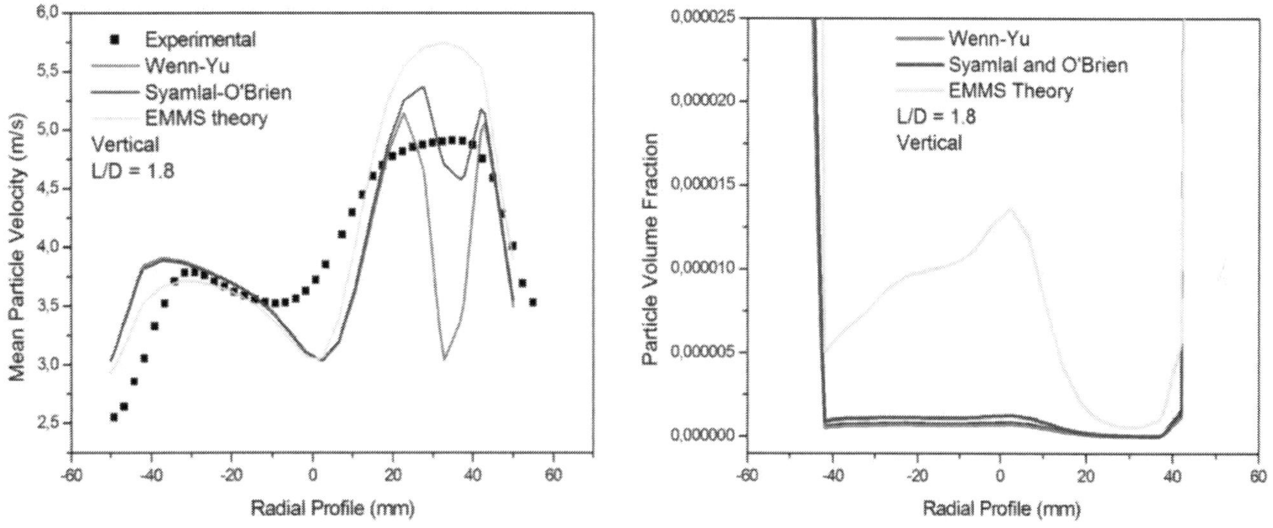

Figure 4. a) Mean particle axial velocity ; b) Particle volume fraction (L/D = 1.8).

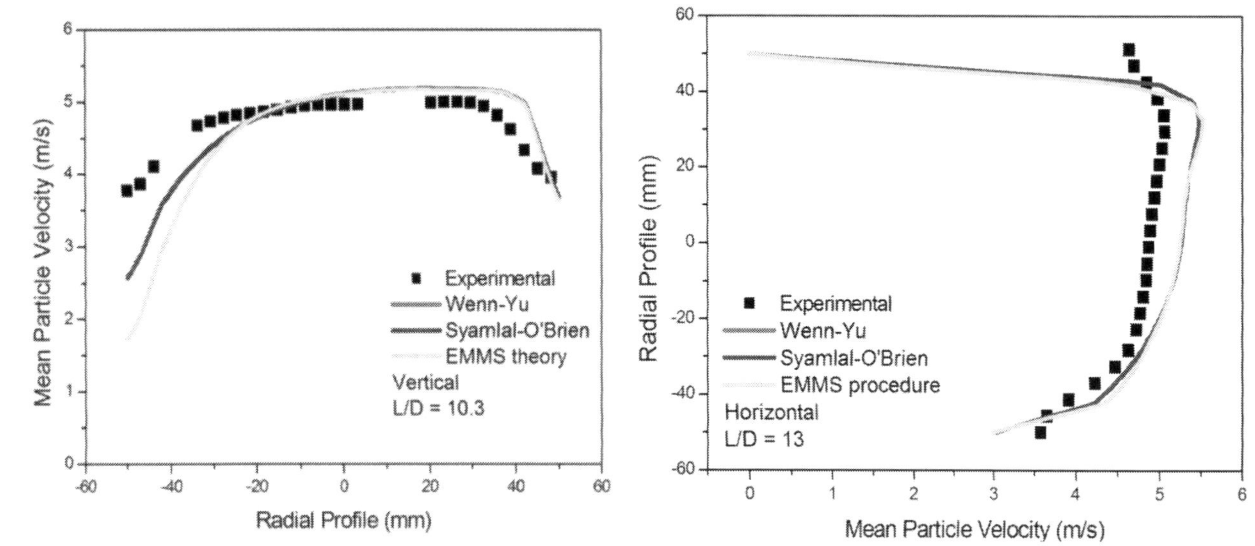

Figure 5. Mean particle axial velocity in the developed flow region of the duct sections.

Conclusions

The main conclusions can be obtained in relation to the EMMS theory, which showed to aggregate particles into clusters even though in very dilute flow conditions. However, due to the dilute characteristics of the flow, no significative change was observed in the particle mean velocity profile when comparing the different drag coefficient models. The consideration of an a inviscid model, which do not considers the stress and pressure terms of the solid phase, presented good qualitative results when comparing with experimental data.

It was also able to conclude during the numerical experiments that the different values of particle viscosity do not change the final results of the mean particle velocity profiles. This observation is addressed to the dilute flow conditions used during the simulations.

The next steps of this study, is to investigate the influence of these models in more concentrated particle flows.

Acknowledgments

The authors are grateful to PETROBRAS, cooperation term 0050.0045549.08.4, for the financial support that guaranteed the development of this work.

References

1. Akilli, H.; Levy, E. K.; Sahin, B., (2001). Gas-solid behavior in a horizontal pipe after a 90° vertical-to-horizontal elbow, Powder Technology, 116, 43-52.
2. Kuan, B.; Yang, W.; Schwarz, M.P.(2007). Dilute gas–solid two-phase flows in a curved 90° duct bend: CFD simulation with experimental validation, Chemical Engineering Science, 62, 2068 – 2088.
3. Yang, W.; Kuan, B., (2006). Experimental investigation of dilute turbulent particulate flowinside a curved 90° bend, Chemical Engineering Science, 61, 3593 – 3601.
4. Yilmaz, A.; Levy, E. K., (2001). Formation and dispersion of ropes in pneumatic conveying, Powder Technology,114, 168–185.
5. Benzarti, A.; Mhiri, H.; Bournot, H., (2012). Drag models for simulation gas-solid flow in the bubbling fluidized bed of FCC particles, World Academy of Science, Engineering and Technology, 61.
6. Yang, N.; Wang, W.; Ge, W.; Li, J., (2003). CFD simulation of concurrent gas-solid flow in circulating fluidized beds with structure-dependent drag coefficient, Chemical Engineering Journal, 96, 71-80.
7. Syamlal, M.; O'Brien, T. J., (1989). Computer simulation of bubbles in a fluidized bed, AIChE Symposium Series, 85, 22-31.
8. Gidaspow, D., (1994). Multiphase flow fluidization: continuum kinetic theory description, in Academic Press, Boston.
9. Wen, C. Y.;Yu, Y. H., (1966), Mechanics of fluidization, Engineering Progress Symposium Series, 62, 100-111.
10. Ergun, S., (1952). Fluid flow through packed columns, Chem Eng Progr, 48, 89-94.
11. Shiller, L.; Neuman, A. Z., (1933). A drag coefficient correlation, Verein Deutschen Ingenieure Zeitung, 77, 318-320.

(142r) Simulation of Reaction and Mixing Processes in Turbulent Aqueous CaCO3 Precipitation

Derek D. Harris, Chemical Engineering, BYU, Provo, UT and David O. Lignell, Chemical Engineering, Brigham Young University, Provo, UT

Simulation of reaction and mixing processes in turbulent aqueous $CaCO_3$ precipitation

Derek D. Harris, David O. Lignell

Abstract – AIChE 2012 Annual Meeting

Particulate systems in turbulent flow are computationally expensive to model accurately due to nonlinear interactions between multiphase mixing and reaction processes. In aqueous carbonate precipitation, surface energies, activity coefficients, and the particle size distribution are needed to accurately calculate the rates and timescales of nucleation and growth processes. Crystals formed often have different polymorphic structures which exhibit different equilibria with the liquid phase. Liquid Schmidt numbers are 500 times larger than those found in typical gaseous systems, necessitating a high grid density to resolve the turbulent fluctuations. One-dimensional turbulence (ODT) has proven to be a useful tool in resolving turbulent mixing fluctuations and is used to show the effects of mixing on the precipitation of polymorphic particles. These fluctuations are important to resolve because of how non-linear the nucleation and growth rates are with super saturation.

A kinetic model for carbonate formation has been devised. Model results are presented of mixing between two adjacent aqueous streams of $NaCO_3$ and $CaCl_2$. Flow and particle statistics are recovered using multiple realizations of the stochastic mixing process in ODT. The quadrature method of moments (QMOM) is used to model the solid phase particle balance equation for each of three anhydrous $CaCO_3$ polymorphs and a common amorphous phase. The non-ideal aspects of this system are taken into account using a Debye-Huckel-type model.

A parametric study is presented by varying the temperature of a single stream, the Reynolds number, and the NaCl concentration. These parameters have a strong effect on the timescales of the nucleation and growth kinetics, and the diffusive, and turbulent mixing processes. The interaction between the transport and reaction processes, and their associated timescales is quantified. The results of these detailed simulations will aid in model development for complex flow configurations (using e.g., LES and RANS simulations) and uncertainty quantification in crystal systems.

(142s) Hydrodynamic Behavior of Carbon Nanotubes and Characterization of Length Distributions

Carlos Silvera Batista[1], Jeffrey A. Fagan[2], Constantine Khripin[1] and Ming Zheng[1], (1)Polymer Division, National Institute of Standards and Technology, Gaithersburg, MD, (2)Polymers Division, National Institute of Standards and Technology, Gaithersburg, MD

Single-walled carbon nanotubes (SWCNTs) are 1D, cylindrical, structures of carbon with long persistence lengths and consistent diameters. In this talk, I will discuss the use of doubly sorted SWCNTs (by buoyancy and length), which are effectively colloidal rods, to explore experimentally the effectiveness of theoretical approximations for the hydrodynamic drag of a freely rotating rod. The objective of this work is establish and validate the use of Analytical Ultracentrifugation (AUC) as a technique to measure the length distribution of rodlike colloidal particles including SWCNT dispersions. This is necessary for real applications of nanotube dispersions, as the transport, optical, and thermal properties, as well as the toxicity of SWCNTs have all been demonstrated to depend on the length. To date however, length distributions of suspensions have most commonly been determined using Atomic Force Microscopy (AFM). Unfortunately for this purpose, AFM is time consuming, tedious, does not measure the distribution in situ, and can be biased by both the method of sample preparation as well as the image analysis. AUC would be thus advantageous because it allows the direct measurement of ensembles as they exist in the liquid phase. In this talk, I will present measurements and analysis of SWCNT samples with narrow distributions in length, diameter and buoyancy as measured through AUC and compare them against independent measurements conducted with AFM. Using this data, the validity of hydrodynamic theory for this application is verified.

(142t) Dynamics of Single Semiflexible Biopolymers: Equilibrium Properties and Dynamics in External Flows

Harishankar Manikantan and David Saintillan, Department of Mechanical Science and Engineering, University of Illinois at Urbana-Champaign, Urbana, IL

Recent advances in experimental capabilities using microfluidics has led to a renewed interest in the dynamics of biopolymers in external flow fields. We present a detailed and efficient simulation method for solutions of semiflexible polymers using slender-body theory in Stokes flow. We consider inextensible polymers of length of the order of the persistence length, and develop an algorithm that incorporates Euler-Bernoulli elasticity as well as Brownian forces. The model is shown to work excellently in reproducing established equilibrium properties of single free polymers. Furthermore, we report agreement with previously observed scaling laws with respect to filament rigidity. We then focus on flow fields commonly observed in microfluidic devices – particularly linear shear flows and flows near hyperbolic stagnation points. In shear flows, Brownian forces are seen to dislodge polymers from an otherwise stable axis leading to a characteristic tumbling motion – a phenomenon that has been previously observed experimentally. The frequency of tumbling is seen to vary with flow strength according to previous scaling predictions. In the hyperbolic flow geometry, if the filament is aligned with the extensional axis, fluctuations due to Brownian forces are suppressed due to tension in the filament. A theory is presented to predict this suppression, and numerical results agree with the proposed theory. When aligned with the compressional axis, a competition between elasticity and the line tension induced in the filament due to flow causes it to buckle in a way analogous to Euler buckling of beams, taking on higher mode shapes with increasing flow strength. Using a linear stability analysis, we are able to predict the critical flow strength associated with each of these mode shapes in the non-Brownian case and supplement it with numerical simulations. In the presence of Brownian forces, the thresholds for the so called stretch-coil transition are rounded by thermal fluctuations – however, numerical simulations show that these instability thresholds are broadly consistent with non-Brownian predictions.

(142u) Equilibrium and Non-Equilibrium Dynamics and Structure of Soft Sphere Fluids

Yajun Ding, James F. Gilchrist and Jeetain Mittal, Department of Chemical Engineering, Lehigh University, Bethlehem, PA

We perform equilibrium and non-equilibrium molecular dynamics to study transport properties and structure of soft sphere fluids. The soft spheres are modeled using inverse power potential ($v(r) = \varepsilon(\sigma/r)^{-n}$) with n varied from 4 to 36. For equilibrium simulation, diffusion coefficient is computed based on mean squared displacement and viscosity is calculated according to the Green Kubo relation. We scaled the density with the freezing density for systems with different softness. This scaling law effectively collapses diffusion coefficient and viscosity from n = 6 up to 36, with n = 4 slightly deviating from our scaled curve. For non-equilibrium simulation, SLLOD algorithm is applied to simulate the simple shear flow. Systems with different softness are studied at the same scaled density. Shear viscosity and two-body translational structural order parameter are examined. The strong correlation between shear viscosity and translational structure order is found. Also this kind relation is dependent on the softness of spheres interaction. A close investigation of cumulative order integral illustrates the relation between shearing, softness and structure ordering.

(142v) Non-Singular Boundary Integral Methods for Fluid Mechanics

Evert Klaseboer, Institute of High Performance Computing, Singapore, Singapore, Qiang Sun, Mechanical Engineering, National University of Singapore, Singapore, Singapore and D.Y.C. Chan, Particulate Fluids Processing Centre, The University of Melbourne, Parkville, Australia

A formulation of the boundary integral method that is applicable to the Laplace equation for the potential problem, the Helmholtz equation, and equations associated with Stokes flow and linear elastic deformations have been developed [1]. The approach is based on removing the usual weakly singular integral and the Cauchy principal value integral in the boundary integral method by subtracting the solution of a specially constructed related problem. We demonstrate the details of our approach using the potential problem from which it is easy to see how the method can be extended to the more complicated cases of Stokes flow and linearly elastic deformations.

This non-singular boundary integral formulation simplifies numerical implementations of this popular technique. Validation of the approach is obtained by comparing numerical results for problems in Stokes flow for which analytic results are known [2, 3]. The flexibility of this method in colloidal hydrodynamics is demonstrated with examples where there may be a mix of boundary conditions based on the surface velocity or the surface traction.

[1] E. Klaseboer, Q. Sun and D. Y. C. Chan *Non-singular boundary integral methods for fluid mechanics applications* J Fluid Mech **696** (2012) 468-478.

[2] E. Klaseboer, R. Manica, D. Y. C. Chan and B. C. Khoo *BEM simulations of potential flow with viscous effects as applied to a rising bubble* Engg Analysis Boundary Elements **35** (2011) 489-494.

[3] Q. Sun, E. Klaseboer, B. C. Khoo and D. Y. C. Chan *Implementation of a non-singular boundary integral method for Stokes flow* Engg Analysis Boundary Elements (to be published 2012)

Stereo-PIV Measurements of a Single Phase Flow in a Pipe Separator

Eyitayo A. AFOLABI and J.G.M LEE

Newcastle University, Newcastle upon Tyne, UK.

e.a.afolabi@ncl.ac.uk

Abstract

This paper focuses on investigating the hydrodynamic behaviour of single phase flow in a three-phase pipe separator. Using the Stereoscopic-Particle Image Velocimetry technique, the velocity distributions of swirling flow were measured, which can mostly reflect the turbulent flow properties in a pipe separator.

Introduction

In the past, the multiphase separation technology used in the oil and gas industry has been based on conventional vessel-type separators which are expensive, heavy and bulky in size. Nowadays, compact separators are widely used as an effective and economical alternative to conventional separators especially in offshore platforms in oil and gas production operations. The compact separator is simpler to operate, more lightweight, has neither moving nor internal parts, requires less floor space, and involves lower capital and operational costs. A cyclone is a device that spins a continuous phase stream to remove entrained dispersed phases by centrifugal force. It operates at velocities where the flow is found to be typically turbulent and often quite complex. This is due to the fact that the cyclone operates with strong swirl intensity together with a flow reversal, flow separation and three-dimensional boundary layers with strong streamline curvature.

Presently, the cylindrical cyclone or pipe separator has potential application as a free water knockout system in equipment for the upstream oil and gas production. This includes down-hole, surface (onshore and offshore) and subsea separation. Other applications include, use in flare gas scrubbers, slug catchers and portable well testing equipment. Preliminary studies have shown that the three phase pipe separator is an effective device to perform the partial separation of air-oil-water mixtures at moderate velocities (Vazquez, 2001). However, the cylindrical cyclone performs as a mixer rather than separator at high velocities. A lack of understanding of the complex multiphase flow behaviour inside the three-phase cylindrical cyclone prevents complete confidence in its design. Detailed investigation of the hydrodynamic flow behaviour will allow the correct prediction of separation performance, which is necessary for improved design in all applications. The objective of this work is to investigate the hydrodynamic behaviour of a single phase flow within the three phase pipe separator using the Stereoscopic Particle Image Velocimetry (PIV) technique.

As shown in Figure 1, the three phase pipe separator consists of a tangential inlet inclined at an angle of 27° to the vertical cylindrical body and an oil finder which is an inner concentric pipe extended through the bottom. The air-oil-water mixture enters through the inclined inlet designed to promote the pre-separation of the gas-liquid mixture. The tangential inlet with reduced area produces a swirling motion in the vertical cylindrical pipe. The gas flows upwards to the gas outlet and leaves the single system compact separator. The liquid-liquid mixture moves to the lower section of the vertical pipe. As a result of differences in density, the centrifugal effect segregates the oil-water mixture, thereby concentrating the oil at the center of the pipe whereas the water moves towards the wall region. The oil rich core formed at the center flows through the oil finder and the water rich fraction flows to the annulus between the pipe wall and the oil finder, leaving the single stage three phase separator through the water-rich outlet. Erdal (2001) used tangential, horizontal and inclined inlets to generate strongly non- axisymmetric flow within a Gas-Liquid Cylindrical Cyclone (GLCC). He ran a single phase glycerine and water flow through a vertical cylindrical pipe. Tangential and axial velocity profiles with associated velocity fluctuations were measured using a Laser Doppler Velocimetry (LDV) system.

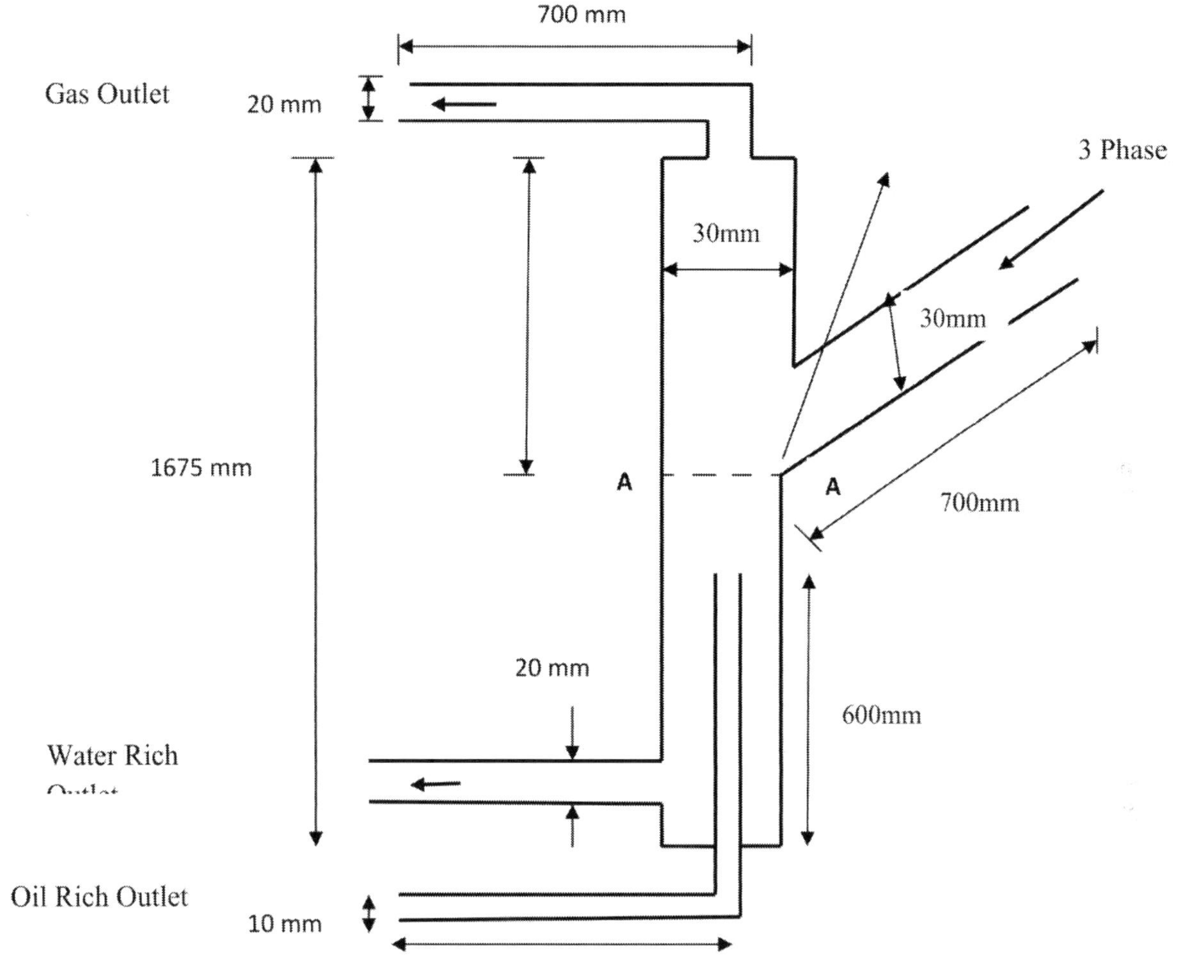

Figure 1: Single Stage Three-Phase Cylindrical Cyclone System

The axial velocity contour map showed that the reversed axial velocity region has a helical (spiral) shape. The axial velocity measurements showed downward flow near the wall and small upward flow near the center of the tube. He found that the tangential velocity measurements showed that free vortex occurs near the wall region and forced vortex at the center of the cyclone. Turbulent kinetic energy is relatively high near the vortex center and in the region near the inlet. The experimental results were then used to develop a new correlation for vortex center wavelengths in order to improve existing correlations for axial and tangential velocities (Erdal, 2001).

Stereoscopic Particle Image Velocimetry (SPIV) is a new non-intrusive visualization experimental technique which is ideally suited to determining the whole field three-dimensional fluid velocity in a pipe separator. Stereoscopic PIV allows the determination of mean velocity and detailed turbulence quantities such as Reynolds stresses. One of the first attempts to measure the velocities in a cylindrical cyclone using Stereoscopic PIV was presented by Liu et al (2006). They investigated the swirling flow structure in a gas cyclone by measuring the instantaneous whole field tangential, axial and radial velocities of air flow. The time-averaged tangential velocity profile showed that the tangential velocity of the gas generated an inner quasi-forced vortex and outer quasi-free vortex. The axial velocity profile revealed an inner upward flow and outer downward flow. Furthermore, they observed a reverse flow at the inner core of the forced vortex around the cyclone axis. This flow reversal is generated by the motion of precessing vortex core (PVC), which is usually associated with vortex breakdown and insurgence of reverse flow(Liu et al., 2006). An extensive study has been conducted by Vasquez (2001) on the performance of this three phase cylindrical cyclone and concluded that it worked effectively as a free water knock-out device for mixtures with high water content and low oil volume. The work of Vasquez (2001) did not provide the hydrodynamic flow behaviour of the complex, swirling and turbulent flow in the three-phase pipe separator. This is the main focus that this research work is going to address. Previous study by Erdal (2001) was limited to a LDV measurement of a single phase flow that passed through a cyclone with a single outlet. However, this research work is aimed at providing SPIV measurement of single flow passing through three different outlets of a pipe separator.

Methodology

Experimental Facility and Flow Loop:

The three phase flow facility used in this study is based on one of the geometries developed and patented for multiphase flow separation by the Separation Technology Project of the University of Tulsa, USA. A 30mm ID laboratory prototype of a three phase pipe separator was fabricated to investigate the hydrodynamic behaviour of the turbulent flow and separation efficiency. The separator test section was constructed using a transparent perspex tube. As shown in Fig. 1, the main body of the separator is a 30-mm ID vertical cylinder with a height of 1675 mm. The inlet is a 30-mm ID pipe, inclined at an angle of 27 degree from the horizontal and connected tangentially at 585 mm from the top of the vertical main body. The inlet configuration was designed to be gradually reduced to 25% of the cross-sectional area of the inlet tube. The gas outlet is a 20-mm ID pipe attached to the top of the vertical main pipe. The water outlet is a 20-mm ID pipe connected at 185-mm from the bottom of the vertical main pipe. The oil outlet is a 10mm ID pipe, 700-mm long inserted into the main vertical pipe through its base. A seal was used to keep the oil outlet located in a concentric position.

S-PIV Set-up:

The stereo-PIV system used for this investigation was manufactured by TSI Inc and loaned from the EPSRC engineering instruments pool. A CFR-200 double pulsed Nd: YAG laser system designed by Big Sky Laser was used as the light source to illuminate the tracer particles in the flow under investigation with a 200mJ per pulsed laser, and each light pulse had duration of 9ns at a wavelength of 532nm. As shown in Figure 2, the laser beam passed through a TSI Model 610015 light arm, which helped in aligning the beam to the arm's optical axes so that the beam was not clipped but transmitted properly. Cylindrical and spherical lenses were adjusted to achieve a lightsheet of 1 mm thickness, which was used to illuminate the flow in the test section under investigation. Two TSI Model 630059 POWERVIEW TM Plus 4MP PIV cameras with CCD sensors were installed on the Scheimpflug mounts so as to satisfy the stereoscopic camera condition. The two cameras looked at angle of $+45^{\circ}$ and -45° to the light sheet. By tilting the image sensor plane and the lens principle to the Scheimpflug condition, the plane best of focus could be found so that it was aligned with the lightsheet (Prasad, 2000c). These cameras were positioned at both sides of the light sheet and then connected to a 64 bit frame grabber to capture and digitize images and then communicate with a computer. A TSI 610035 laser pulse synchronizer was used to synchronize the image capturing and laser pulses. A water prism of 45 degrees was constructed and moved to the test section of the experimental rig in order to minimize optical distortion arising from refraction through the perspex wall. The prism was filled with water and optical rails were fitted to the frame supporting the experimental rig in order to align the cameras with the laser beam. The single phase water flow was seeded with silver coated hollow glass spheres marketed by TSI Inc (10089-SLV). The density of these particles is 1.65 g/cc with a nominal mean diameter of 14μm. They were mixed with water in the storage tank and allowed to be uniformly dispersed in a few complete circulations of the flow loop. This seeding material has very good fluid and imaging properties for the current experiment (Melling, 1997).

Two hundred pairs of images were recorded by the acquisition system at each measuring location. The software INSIGHT 3G from TSI Inc was used to controlled the hardware while capturing, retrieving and processing raw images in order to gain data on velocity vector fields. In order to evaluate the recorded images via cross-correlation, the complete image is divided into many interrogation windows. The correlation function operates on the intensities inside each interrogation window. The window overlap is the overlap among neighbouring interrogation windows, and together with interrogation window size determine the grid size of a vector field. In this work, a multi-pass processing scheme was used; consisting of two passes with interrogation windows of 64 x 64 pixels and 50% overlap. Then another pass was applied with an interrogation area of 32 x32 pixels and 50% overlap. Erroneous vectors were removed and replaced by a global velocity filter and local 3 x 3 median filters. The outliers were finally interpolated to fill gaps in the vector fields.

Camera Calibration

A single plane calibration target populated by a Cartesian grid of 2mm white marker dots with a 3-mm cross at the centre on a black background was cut to size to fit into the test section. The calibration target was moved using a micrometer in seven steps of 0.5mm and images of the target were recorded and analysed by perspective calibration method of INSIGHT 3G software of TSI Inc. The calibration image analysis produced a set of calibration points used to create a calibration mapping function, which was then used to combine 2-D PIV vector fields in order to obtain the three-dimensional vector field.

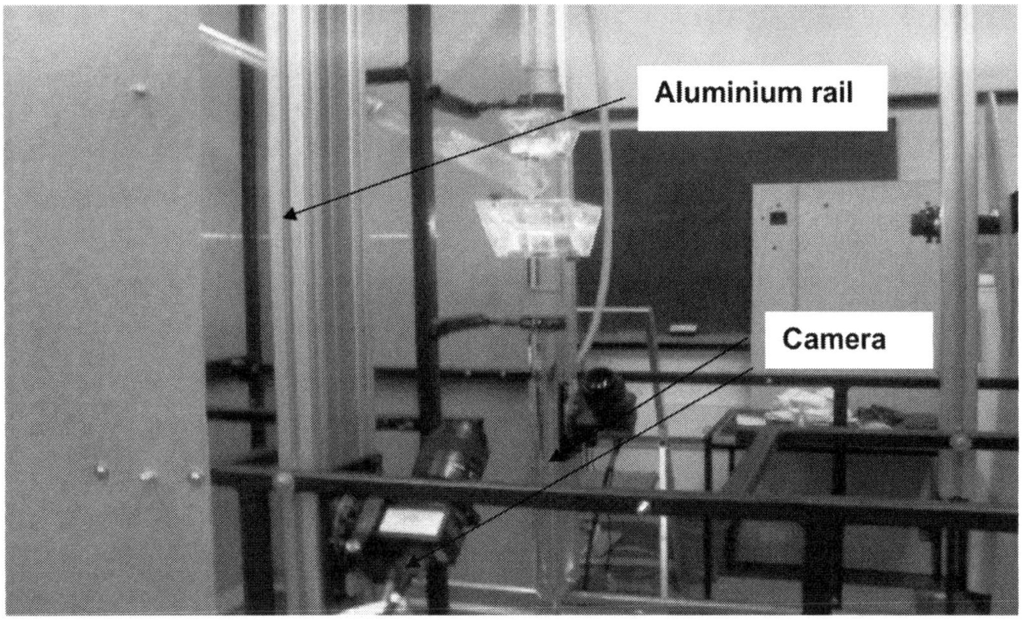

Figure 2: S-PIV Experimental Set-up

Results and discussion

The single phase flow experiment was run with a water flow rate of $196 cm^3/s$. The outlets were restricted with rubber bungs such that the percentage of water as a fraction of the inlet mass flow was 60% through the air outlet, 33% through the water-rich outlet and the balance through the oil-rich outlet. The experimental data presented were extracted along the y=0 line at three axial positions Z= Z= -395 mm,-75 mm and 295 mm of the pipe separator, as shown in Figure 3.

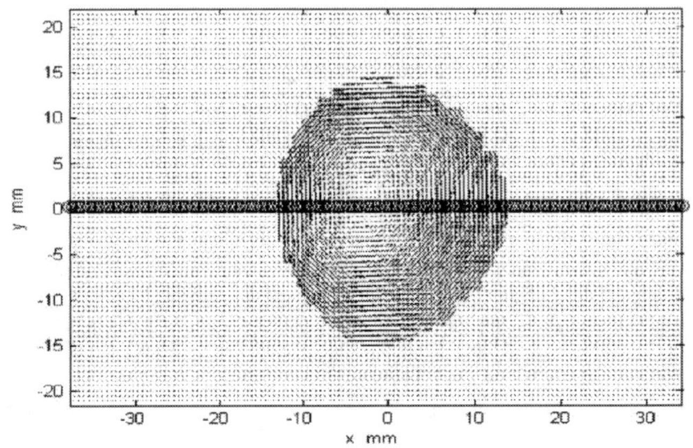

Figure 3: Experimental Data: Position of y=0 sections

Tangential Velocity

Figure 4 shows the comparison of the mean tangential velocities of the water flow at the Z=-0.395 mm,-0.075 mm and 0.295 mm axial positions. At all axial positions, it can be observed that the tangential velocity increases moving away from the center of the tube before reaching a maximum and then dropping close to the wall due to wall friction. At Z=295mm, the maximum tangential velocity occurs at a radius of 7.5mm. However, at Z=-395mm the maximum velocity occurs at a radius of 10mm and at Z= -75mm, maximum velocity is observed at a radius of 12.5mm. By identifying this location, the tangential velocity distribution can be described as a continuous flow stream with an outer and inner region.

The inner region is characterized as a forced vortex where tangential velocity increases directly with radius. However, the outer region is characterized as a free vortex and the rate of rotation is greatest at the center and then decreases progressively (Kelsall, 1952). This can be described by the relationship:

$$v_\theta = Cr^n \quad (1)$$

where v_θ is the tangential velocity, C is a constant of proportionality, and r is the radial coordinate of the vortex. For a forced vortex, n= 1 and a free vortex, n=-1. The observations made are similar to the reports of earlier researchers in the literature (Leeuwner and Eksteen, 2008; Slack and Wraith, 1997). At Z=-75mm, the forced vortex extends over 75% of the surface area of the separator's plane and decreases to 67% and 50% at Z= -395mm and 295mm respectively. However, the free vortex occupies 50% of the surface area of the separator's plane at Z=295mm and decreases to 33.3% and 25% at the Z= -395mm and -75mm axial positions respectively. This means that the surface area occupied by the forced vortex decreases moving away from the inlet section. However, the free vortex increases when moving away from Z=-395mm.

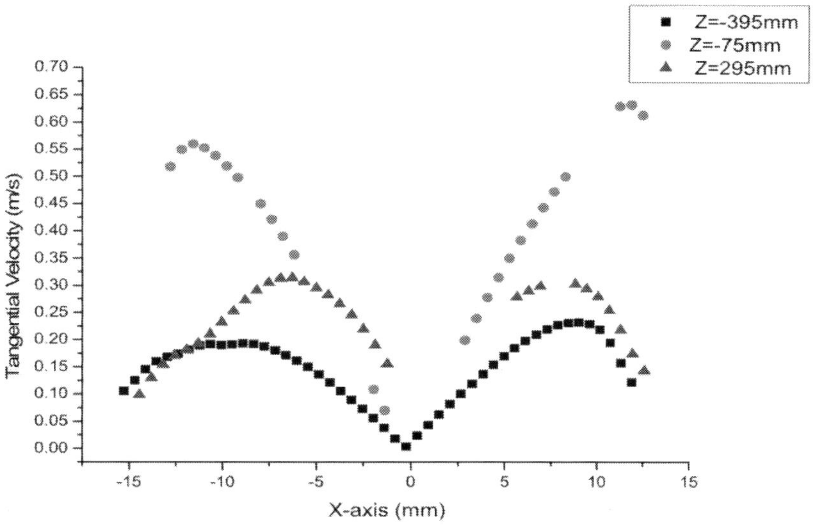

Figure 4: Comparison between the Tangential Velocity Profiles at Three Axial Positions in the Cyclone

The comparison of the tangential velocity profiles at the three axial positions shown in Figure 4 reveals that the magnitude of mean tangential velocity decreases as we move away from the inlet section towards the outlets. The reason for this is mainly the decay in swirl intensity as the flow moves away from the inlet. Therefore, as we move away from the inlet section, the tangential velocity gradually decreases and relatively less strong centrifugal fields are generated. Tangential velocity magnitude is a measure of the centrifugal force in the vortex region and this helps to improve the separation efficiency of the pipe separator. It is generally accepted that tangential velocities are directly proportional to separation efficiencies. Therefore, most of the fluid separation occurs immediately at the region where the inlet section joins the vertical axis of the cyclone.

Axial Velocity

Figure 5 shows the comparison between the axial velocity profiles at the three different axial positions in the separator. In agreement with other researchers such as Peng, et al (2002) and Monredon et al (1992), the positive axial velocity profile is hereby referred to as region of flow in an upward direction with respect to the rotation axis. A negative axial velocity profile will be taken as a region of flow in a downward direction from the rotation axis.

Furthermore, at Z=-395mm and -75mm, the positive values of axial velocity decrease with an increase in radial distance. It reaches zero at some distance away from the center of the tube. However, close to the separator wall, the axial velocity reaches a minimum and then increases. This is due to higher friction at the separator wall. Axial velocity profiles at Z=-395 and -75mm reveal a combination of downward and upward flow patterns at the wall and center of the tube respectively. However, at Z=295mm there is no negative axial velocity at any x-axis coordinate.

As can be seen in Figure 5, a large amount of water flows in an upward direction at the center of the tube and in a downward direction near to the wall at Z=-75mm. For example, the maximum positive and negative axial velocities are 0.175m/s at Z=-75 mm. The maximum positive axial velocity among the three axial positions is 0.175m/s at Z=-75mm. At Z=-395mm, a large downward flow is observed close to the wall together with small amounts of upward flow at the center of the tube. The location of the maximum axial velocity on the x-axis is at the center of the tube at Z=-75mm, but off center at Z= -395mm. Similar observations of axial velocity flow patterns were made by Dlamini et al (2005) and Leeuwner and Eksteen (2008).

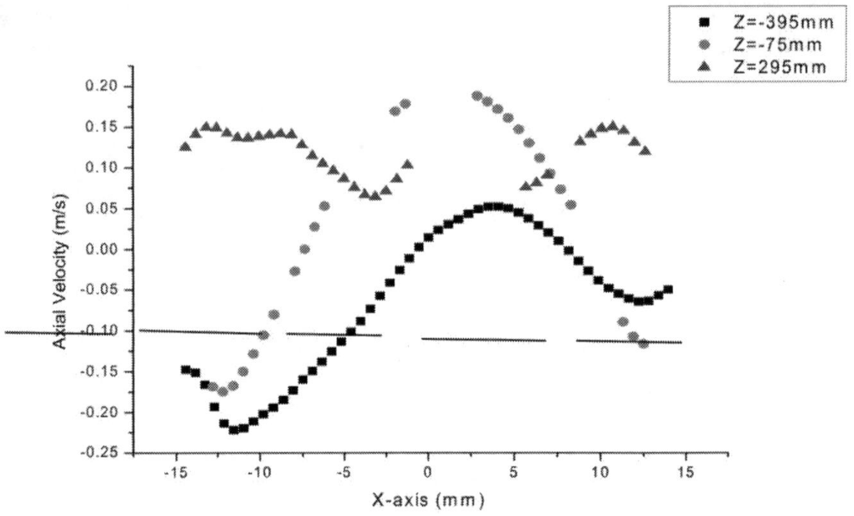

Figure 5: Comparison between the Axial Velocity Profiles at three axial positions in the Cyclone

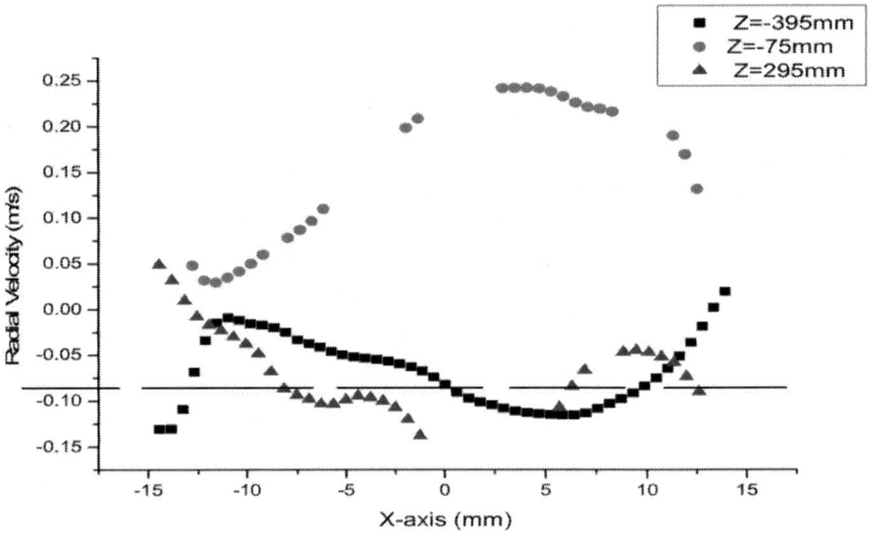

Fig. 6: Comparison between the Radial Velocity Profiles at Three Axial Positions in the Cyclone

Radial Velocity

Figure 6 shows a graph of the comparison between the mean radial velocity profiles at the three axial positions in the cyclone. In agreement with other researchers such as Kelsall (1952) and Peng et-al (2007), radial velocity flow is said to be directed inward when the radial velocity is negative and outward when its value is positive. As can be seen in Figure 6, radial flow is found to be inwards as we move away from the center of the tube at Z=−395mm. However, close to the wall at the positive value of the x-axis, outward radial velocity begins to develop. At Z= −75mm, radial velocity is observed to be directed outwards at all values of the x-axis. The radial velocity profile at Z= 295mm shows inward flow as we move away from the center of the tube. However, close to the wall at a negative value of the x-axis, outward flow is observed. In Figure 6, it can be observed that the magnitude of radial velocity decreases as the flow moves towards the outlets.

Conclusions

The stereoscopic PIV measurements were taken to investigate the single phase, water flow fields at three different axial positions within the pipe separator. Tangential velocity measurements showed that a forced vortex occurs at the center of the tube, with a free vortex near the wall region. The location of zero tangential velocity was observed to be off the axis of the pipe separator and the shape of the tangential velocity profile found to be independent of axial positions within the pipe separator. The tangential velocity profiles indicate that the magnitude of the mean tangential velocity decreases as we move away from the inlet section towards the outlets. The axial velocity measurements had shown a downward flow near the wall of the pipe separator and an upward flow near the center of the tube. There is no general pattern for the radial velocity within the pipe separator. At the inlet section, the radial velocity was found to be directed outward. However, away from the inlet region, a pattern of both outwards and inwards flows was observed.

Acknowledgment

This research has been supported by the PTDF, Nigeria and EPSRC (TSI Inc.)

References

Erdal, F. M. (2001) ''Local Measurements and Computational Fluid Dynamic Simulations in a Gas Liquid Cylindrical Cyclone Separator,''. thesis. Ph.D. dissertation, The University of Tulsa.

Kelsall, D. F. (1952) 'A study of the motion of solid particles in a hydraulic cyclone', Trans. Inst. Chem. Eng, 30, pp. 87- 108.

Leeuwner, M. J. and Eksteen, J. J. (2008) 'Computational fluid dynamic modelling of two phase flow in a hydrocyclone', Journal of The South African Institute of Mining and Metallurgy, 108, (4), pp. 231-236.

Liu, Z., Jiao, J., Zheng, Y., Zhang, Q. and Jia, L. (2006) 'Investigation of turbulence characteristics in a gas cyclone by stereoscopic PIV', AIChE Journal, 52, (12), pp. 4150-4160.

Melling, A. (1997) 'Tracer particles and seeding for particle image velocimetry', Measurement Science and Technology, 8, (12), pp. 1406-1416.

Monredon, T. C., Hsieh, K. T. and Rajamani, R. K. (1992) 'Fluid flow model of the hydrocyclone: an investigation of device dimensions', International Journal of Mineral Processing, 35, (1-2), pp. 65-83.

Peng, W., Hoffmann, A. C., Boot, P. J. A. J., Udding, A., Dries, H. W. A., Ekker, A. and Kater, J. (2002) 'Flow pattern in reverse-flow centrifugal separators', Powder Technology, 127, (3), pp. 212-222.

Peng, W., Hoffmann, A. C., Dries, H., Regelink, M. and Foo, K. K. (2007) 'Reverse-flow centrifugal separators in parallel: Performance and flow pattern', AIChE Journal, 53, (3), pp. 589-597.

Prasad, A. K. (2000a) 'Particle Image Velocimetry', current Science, 79, (1), pp. 51-60.

Prasad, A. K. (2000b) 'Stereoscopic Particle Image Velocimetry', Experiemnets in Fluids, 29, pp. 103 116.

Prasad, A. K. (2000c) 'Stereoscopic particle image velocimetry', Experiments in Fluids, 29, (2), pp. 103-116.

Prasad, A. K. and Jensen, K. (1995) 'Scheimpflug Stereocamera for Particle Image Velocimetry in Liquid Flows.', Applied Optics, 34, (30), pp. 7092 - 7099.

Raffel, M., Willert, C. and Kompenhans, J. (1998) Particle Image Velocimetry: A Practical Approach. London: Springer.

Slack, M. D. and Wraith, A. E. (1997) 'Modelling the velocity distribution in a hydrocyclone', 4th International Colloquium on Process Simulation, 11–13 June, Espoo Finland, pp. 65–83.

TSI Inc. (2006) Model 640050 Stereoscopic PIV Camera System. :

Vazquez, C. O. (2001) Multiphase flow separation in Liquid-Liquid cylindrical cyclone and Gas-Liquid -Liquid cylindrical cyclone compact separators thesis. The University of Tulsa.

(142x) Intrachain Hydrodynamic Interactions for DNA in Slits

Jeremy J. Jones[1], Johan van der Maarel[2] and Patrick S. Doyle[1], (1)Chemical Engineering, Massachusetts Institute of Technology, Cambridge, MA, (2)Department of Physics, National University of Singapore, Singapore, Singapore

Advancements in nanofabrication techniques have given rise to a new generation of micro and nanodevices for use in biological assays. These devices are also ideal platforms to study fundamental issues regarding DNA polymer conformations and dynamics in confinement. Over the past decade there has been extensive work by our group and others in examining DNA in slits. Despite this body of work, there is still not agreement about concepts such as "complete" screening of hydrodynamics or whether one can assume blobs in a DNA chain are non-draining. Here we use an intrachain density-density correlation function to extract information about the dynamic correlations at different length and time scales within single DNA chains. Using this technique we are able to measure long-time correlated intrachain dynamics that would otherwise be masked by translational diffusion of the DNA center of mass. This affords us opportunity to explicitly examine the effects of hydrodynamics on DNA segmental motion across different length scales. Our results resolve outstanding issues raised in prior studies regarding intrachain hydrodynamic interactions and their effects on chain dynamics.

(142y) Effect of Attraction Strength On Microchannel Flow of Colloid-Polymer Mixtures

Rahul Pandey, Chemical and Biomolecular Engineering, University of Houston, Houston, TX and Jacinta C. Conrad, Department of Chemical and Biomolecular Engineering, University of Houston, Houston, TX

Flows of dense colloidal suspensions through very thin geometries appear frequently in technological applications, such as inkjet printing and direct write assembly, crude oil recovery, and drug delivery. While the flow properties of hard-sphere suspensions in micro-scale geometries have been studied extensively, the effects of inter-particle attraction on the confined flow properties are not well understood. In this study, we investigated the flow properties of colloidal suspensions at a volume fraction of =0.15 with varying strengths of attraction using bulk rheology and confocal microscopy. We used a slightly charged model system of poly(methylmethacrylate) particles that are suspended in a refractive-index and density-matched solvent mixture of decahydronaphthalene and cyclohexylbromide, and induced a controlled short-range depletion attraction between the particles by adding non-absorbing linear polystyrene. As the strength of attraction was increased, the steady shear viscosity increased for all values of the shear rate. Fits of the Herschel-Bulkley model to the shear stress-shear rate data showed that the strongly attractive suspension yielded at higher stresses, consistent with the formation of an interconnected network of particles. We directly imaged the flow of these suspensions in microchannels using confocal microscopy. Increasing the strength of attraction between the particles resulted in changes in the shape of the flow velocity profile that reflected the emergence of wall slip and rotational motion of clusters of particles. In addition, the formation of strong bonds between particles suppressed two mechanisms for densification present in the weakly-attractive suspension, shear-induced migration and consolidation.

(142z) Harness the Structure and Rheology Relationship of Hydrogel-Biopolymer Mixture to Design Biomimetic Lubricants

Raymond Seekell III, Chemical and Biomolecular Engineering, University of Notre Dame, Notre Dame, IN; Chemical Engineering, Villanova Univesity, Villanova, PA and Yingxi Elaine Zhu, Department of Chemical and Biomolecular Engineering, University of Notre Dame, Notre Dame, IN

One superlubricant in the human body is the synovial fluid. It is a complex fluid comprised of a high molecular weight polysaccharide, hyaluronic acid (HA), and a globule protein, lubricin. It encompasses aspects of a boundary layer lubricant as well as a hydrodynamic lubricant. HA creates a highly viscouselastic network able to adsorbs large amounts of energy; Lubricin prevents surface contact by means of a "ball-bearing" effect. One of the debilitating effects of Osteoarthritis is a decrease in the elasticity of synovial fluid. A common remedy is to inject the joint with aqueous solutions of HA. However, this is a short-term solution. This research explores the possibly enhancing these HA solutions through the addition of micron sized hydrogel particles based on poly(N-isopropylacrylamide) (PNIPAM). Various concentration of PNIPAM hydrogel particles and HA were explored using bulk measurements on a cone-and-plate rheometer. This presentation will show that suspensions of HA and PNIPAM hydrogel particles can be successfully tune to have comparable rheological properties to synovial fluid.

(224i) Vector Separation of Particles Using an Array of Slanted Ramps

Jorge A. Bernate, Sumedh R. Risbud and German Drazer, Chemical and Biomolecular Engineering, Johns Hopkins University, Baltimore, MD

The separation of the different components of complex suspensions with high selectivity is a crucial step in many applications. Bernate and Drazer (PRL 2012) showed that an array of slanted grooves and a uniform force field can be used to induce vector separation, in which different species migrate in different directions in a planar microfluidic device. In the deterministic limit, the extent of deflection of a given particle depends on: i) the magnitude of the hydrodynamic drag advecting the particle relative to retarding force due to the external field that the particle experiences in the ramps leading out of the grooves, and ii) the width of the ramps. The stronger the response of a given particle to the external field, the more it will get deflected surmounting the ramp. Here, we present experiments in a microfluidic platform exploiting the same mechanism and show the versatility and high selectivity of the device. First, we separate beads based on their size and density simply using gravity. We then separate magnetic particles from non-magnetic ones in the presence of a magnetic field, and show that subpopulations of cells magnetically labeled can be similarly depleted resulting in the enrichment of unlabeled cells.

(145a) New Directions for Entropic Forces in Hard Particle Fluids

Sharon C. Glotzer, Chemical Engineering , University of Michigan, Ann Arbor, MI

Entropic forces in hard particle fluids can order particles into organized assemblies. Liquid crystal, plastic (rotator) crystal, quasicrystal, and both simple and complex crystal phases have all been reported for hard shapes that interact solely through excluded volume interactions. We present the results of extensive Monte Carlo simulations of nearly 200 different fluids of hard shapes and their thermodynamic assemblies. We show how particle shape and local measures of order in the dense fluid can be used to predict the category of assembly. We discuss the concept of entropic "bonding" and demonstrate the directional nature of entropic forces in fluids of non-spherical hard particles. We show how these concepts lead naturally to predictive design rules for colloidal and other matter.

In collaboration with: K. Ahmed, L. Boehling, E. Chen, P. Damasceno, M. Engel, T. Grubb, A. Haji Akbari, E. Irrgang, A.S. Keys, D. Klotsa, J.A. Millan, R. Newman, S. Nola, D. Ortiz, B. Schultz, R. Smith, & G. van Anders

This work supported by the U.S. Department of Defense Army Research Office, Air Force Office of Scientific Research, and Assistant Secretary of Defense for Research and Engineering.

(145b) A Pseudo-Quantum Description of Classical Vacancy Diffusion in Crystals

Giovanni Ciccotti, Dipartimento di Fisica, Università di Roma , Roma, Italy, Simone Meloni, School of Physics, University College Dublin, Ireland, Pierre-Antoine Geslin, Laboratoire d'Etudes des Microstructures (LEM), CNRS-ONERA, France and Eric Vanden-Eijnden, Courant Insitute of Mathematical Sciences, New York University, New York, NY

We introduce an observable field to describe the dynamics of a single vacancy in a crystal. This field is the probability density $\rho(x \mid r)$ to find the vacancy pseudo-particle at the point x when the atoms are in a configuration r. We considered the case of a classical and a "quantum" pseudo-particle, to which correspond a classical (Boltzman-like) and a ground state "quantum" density, respectively. These densities are obtained assuming that the vacancy pseudo-particle is subject to the potential energy field generated by the atoms in the sample. In our description, the fictitious inverse temperature β^* and the coefficient α of the quantum kinetic energy term are tunable parameters that make the classical/"quantum" density localized in the regions of relevant minima of the potential energy field. We show that the classical density is unsatisfactory. In fact, at finite temperature this density results multimodal (see Fig. 1/A) and is, therefore, difficult to use it to identify the vacancy position. On the contrary, the "quantum" density is well behaving: localized in the relevant minima of the potential and smooth (Fig. 1/B). We use the "quantum" $\rho(x \mid r)$ in rare event simulations to investigate vacancy diffusion paths in a 2D crystal.

Further progresses to study vacancy related phenomena at high temperature, where vacancy formation and annihilation must be taken into account, could come from "quantum" statistics.

In conclusion, we have shown how a ground state quantum description of a vacancy allows to identify the dynamics of point defects in solids.

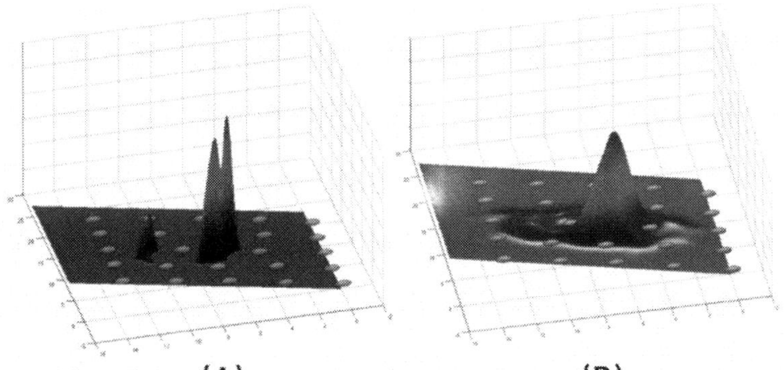

(A) (B)

Figure 1: Classical (A) and "quantum" (B) $\rho(x \mid r)$ in a configuration along a local vacancy migration path at $T = 2.5\ \varepsilon/k_B$.

(145c) 2-Component Van Der Waals EoS and the Second Osmotic Virial Coefficient

Benjamin Widom, Chemistry, Cornell U., Ithaca, NY

The second osmotic virial coefficient of a solute in a solvent is a measure of the effective solvent-mediated solute-solute interactions.

Normally the properties of a two-component mixture are known from an equation of state in the form p=p(d1,d2,T) where d1 and d2 are the densities of the two components, but this contains the osmotic virial coefficients only implicitly. To obtain the osmotic coefficients requires transforming the equation of state from p=p(d1,d2,T) to z2=z2(z1,d2,T), where z1 and z2 are the activities of the solvent and solute, respectively. Then by the McMillan-Mayer solution theory, the second osmotic virial coefficient is related to and may be identified from the coefficient of the second-order term in the expansion of z2 in powers of d2 at fixed z1 and T. It is shown here how that transformation of the equation of state may be effected. The process is illustrated with the two-component van der Waals equation of state applied to obtain the second osmotic virial coefficient of propane in water. The resulting virial coefficient is compared with that obtained by integrating the solute-solute pair correlation function in the limit of infinite dilution as found by molecular dynamics computer simulations.

(145d) Dissipative Particle Dynamics: Extensions and Emerging Applications

John K. Brennan[1], Joshua D. Moore[1] and Martin Lisal[2], (1)Weapons and Materials Research Directorate, U.S. Army Research Laboratory, Aberdeen Proving Ground, MD, (2)E. Hala Laboratory of Thermodynamics, Academy of Sciences of the Czech Republic, Prague, Czech Republic

Allowing for simulation studies at spatial-temporal scales that are inaccessible to molecular methods, the dissipative particle dynamics method (DPD) [1,2] has become a robust tool for studying soft matter, including polymers, surfactants and colloids. Since its development over 20 years ago, the fundamentals of the DPD method are now fairly well-established [3,4], as are many technical subtleties [e.g., 5], with frequently-emerging improvements [e.g., 6]. Extensions of the DPD method that impose energy conservation have also been developed, allowing for internal energy exchange between particles [7,8]. These factors led to many DPD simulation studies that provided insight into a variety of soft matter phenomena. However, DPD is now being applied in other areas such as biophysics [9] and detonation physics [10,11]. Furthermore, it is expected that DPD will play an ever-increasing role in multiscale modeling approaches via bridging of the atomistic and continuum scales.

In this talk, we will present the status of one such emerging DPD application, namely, the study of the thermo-mechanical response of nanocomposites. A number of simulation challenges exist for these materials, including the development of coarse-grain models and DPD methods that can capture the following known thermo-mechanical responses: (i) phase transitions; (ii) structural rearrangements and mechanical deformation (e.g., elastic-plastic deformation); and (iii) chemical reactions. We will present the status of our ongoing project, including progress on coarse-grain model development [11], implementation of improved integrator algorithms for various DPD methods [12], the development of a constant-pressure, constant-enthalpy DPD method [12], and the development of a DPD framework that can mimic chemical reactivity (DPD-RX).

This effort is supported by the Institute for Multi-Scale Reactive Modeling of Insensitive Munitions (MSRM), which is a multi-team effort led by the U.S. Army Research Laboratory and the U.S. Army Armament Research, Development and Engineering Center, involving various other national laboratories and academic groups totaling over 20 scientists. ML acknowledges funding through a Cooperative Agreement with the U.S. Army Research Laboratory.

References:

[1] Hoogerbrugge, P.J., and J.M.V.A. Koelman, Europhys. Lett., 19, 155 (1992).

[2] Koelman, J.M.V.A., and P.J. Hoogerbrugge, Europhys. Lett., 21, 363 (1993).

[3] Español, P., and P.B. Warren, Europhys. Lett., 30, 191 (1995).

[4] Groot, R.D., and P.B. Warren, J. Chem. Phys., 107, 4423 (1997).

[5] Groot, R.D., J. Chem. Phys., 118, 11265 (2003).

[6] Groot, R.D., and S.D. Stoyanov, Phys. Rev. E, 78, 051403 (2008).

[7] Bonet Avalos, J., and A.D. Mackie, Europhys. Lett., 40, 141 (1997).

[8] Español, P., Europhys. Lett., 40, 631 (1997).

[9] Venturoli, M., M.M. Sperotto, M. Kranenburg and B. Smit, Phys. Rep., 437, 1 (2006).

[10] Stoltz, G., Europhys. Lett., 76, 849 (2006).

[11] Moore, J.D., S. Izvekov, M. Lísal and J.K. Brennan, Proceedings of the 17th Shock Compression of Condensed Matter, Chicago, IL (2011).

[12] Lísal, M., J.K. Brennan and J. Bonet Avalos, J. Chem. Phys., 135, 204105 (2011).

(145e) Property Prediction with Analogous Series: A Continuum View of Discrete Molecular Entities

Brian K. Peterson, Process Modeling, ExxonMobil Research and Engineering, Annandale, NJ

Property Prediction with Analogous Series: A Continuum View of Discrete Molecular Entities

Brian K. Peterson

ExxonMobil Research & Engineering

Depending on the data available and other constraints, the best choice of a method for predicting the properties of materials might include the use of theory, molecular simulation, quantum chemistry, various correlative methods, or combinations of several of these. When faced with sparse data and limited time, the data-based correlative method of Analogous Series offers surprising ease and accuracy if data of a certain structure is available. The method is introduced by several examples and its principles are explained with reference to a hypothetical continuum of molecular species which is intuitive for those with experience in statistical mechanical theory and molecular simulations. We demonstrate that given small amounts of data on two families of materials related to the material of interest, the method can be used to predict a wide variety of thermophysical and other properties and is especially useful for detecting outliers and anomalous data. We have so far applied the method to predicting critical and boiling temperatures, infinite dilution activity coefficients, GC retention times, closed-cup flash points, heat capacities, adsorption in zeolites, and kinetic rate constants.

(145f) SAFT-γ Coarse Grained Models for the Molecular Simulation of Complex Fluids with a Top-Down Methodology

Carlos Avendano[1], Thomas Lafitte[2], Vassilis Papaionannou[3], Olga Lobanova[3], Claire Adjiman[4], Amparo Galindo[3], Erich A. Muller[3] and **George Jackson**[3], (1)Cornell University, Ithaca, NY, (2)Chemical and Biological Engineering, Princeton University, Princeton, NJ, (3)Chemical Engineering, Imperial College London, London, United Kingdom, (4)Centre for Process Systems Engineering, Imperial College London, London, United Kingdom

An application of the "top-down" concept for the development of accurate coarse-grained intermolecular potentials of complex fluids is presented. With the common "bottom-up" procedure, coarse-grained models are constructed from a suitable simplification of a full detailed atomistic representation, and minor refinements to the intermolecular parameters are made by comparison with limited experimental where necessary. By contrast in the top-down approach, a molecular-based equation of state is used to obtain an effective coarse-grained intermolecular potential that reproduces the macroscopic experimental thermophysical properties over a wide range of conditions. In order to demonstrate our procedure, coarse-grained models for carbon dioxide (CO_2), light green-house gases, n-alkanes, alkylbenzene water and alkylpolyoxyethylene non-ionic surfactants are obtained from a recent implementation of the statistical associating fluid theory of variable range (SAFT-VR) employing a Mie (generalised Lennard- Jones) potential. For example, the parameters of a single-site Mie model of CO_2 and a coarse-grained Mie-site model of the linear alkanes are estimated by optimising the equation of state's description of the experimental vapour-pressure and saturated liquid density data. Our SAFT-γ coarse-grained models are used in Monte Carlo molecular simulation to assess the adequacy of the description of the fluid phase behaviour and properties which were not used to develop the potential model such as the enthalpy of vaporisation, interfacial tension, density profiles, supercritical densities and second-derivative thermodynamic properties. The accuracy of the CG SAFT-γ models are found to be of similar quality to that of more sophisticated all-atom or united atom intermolecular potentials. The models for the aqueous solutions of alkylpolyoxyethylene surfactants provide a good description of these complex micellar solutions at a fraction of the computational cost of the more detailed models.

(145g) Thermodynamic and Kinetic Models of the Emergence of Biological Homochirality

Pablo G. Debenedetti, Department Chemical and Biological Engineering, Princeton University, Princeton, NJ, Harold W. Hatch, Chemical and Biological Engineering, Princeton University, Princeton, NJ, Frank H. Stillinger, Department of Chemistry, Princeton University, Princeton, NJ and Francesco Ricci, Princeton University, Princeton, NJ

Chiral asymmetry choices exhibited by molecules that are present in living organisms constitute a scientifically challenging set of observations. Such geometric preferences favoring one enantiomer over its mirror image are obvious in the observed structures of amino acids, sugars, and the biopolymers that they form. These facts automatically generate fundamental questions about how those chiral asymmetries arose spontaneously in the terrestrial biosphere. We have formulated thermodynamic and kinetic models of chiral amplification that provide molecular-level insight into possible scenarios for the emergence of chiral imbalance in a prebiotic and presumably racemic world.

In the thermodynamic case, we model the phase behavior of a ternary mixture composed of two enantiomeric forms of a chiral molecule and a non-chiral liquid solvent. The mean-field solution of the model allows the calculation of a ternary phase diagram, a prominent feature of which is the existence of two symmetric triple points involving coexistence of a liquid phase enriched in one of the enantiomers, a racemic crystal, and an enantiopure crystal. Over broad ranges of initial composition, including liquid mixtures containing almost equal amounts of the two enantiomers, thermodynamic equilibrium results in liquid-phase chiral amplification, in agreement with experimental observations.

The kinetic model involves an auto-catalytic reaction leading to the formation of a chiral compound, inhibition, and molecular diffusion. Numerical solution of the model via kinetic Monte Carlo allows the identification of two types of behavior. In one, the system evolves towards a mixture containing equal amounts of the two chiral enantiomers. In the symmetry-broken regime, the system evolves spontaneously towards large excess of one or the other chiral enantiomers.

(147a) The Mobilities of Micro- and Nano-Particles At Interfaces of Nematic Liquid Crystals

Dan Abras[1], Gaurav Pranami[1] and **Nicholas L. Abbott**[2], (1)Chemical and Biological Engineering, University of Wisconsin-Madison, Madison, WI, (2)Department of Chemical and Biological Engineering, University of Wisconsin-Madison, Madison, WI

This presentation will describe the use of single-particle tracking to measure the diffusion coefficients of chemically functionalized micro- and nano-particles at interfaces between aqueous phases and a nematic liquid crystal (LC). For hydrophobic particles with diameters of 2.3 ± 0.2 μm that homeotropically anchoring the LC, we measured anisotropic diffusion, qualitatively consistent with the influence of nematic ordering of the LC at the interface on the local rheological environment. Analysis of the magnitudes of the diffusion coefficients reveal that the ordering of the LC about the microparticles is perturbed in the interfacial environment relative to the bulk, leading to low drag on the microparticles at the aqueous-nematic interface. In contrast, for hydrophobic nanoparticles (diameters of 141 ± 11 nm) at the LC-aqueous interface, almost indistinguishable diffusion coefficients were measured at the interface and in bulk LC when the displacements of the nanoparticles in the two environments were in the same directions relative to the far-field director of the LC. These results and others to be described in this presentation reveal fundamental differences to exist between the interfacial mobilities of micro- and nanoparticles at LC-aqueous interfaces, and that a relative insensitivity to interfacial environment appears to be a property of the smaller particles studied in our experiments. These findings are discussed in light of past studies of the diffusion of particles at either isotropic liquid interfaces or in bulk LCs.

(147b) Stokes-Einstein and Desorption-Mediated Diffusion of Protein Molecules At the Oil-Water Interface

Indira Sriram, Robert Walder and Daniel K. Schwartz, Chemical and Biological Engineering, University of Colorado at Boulder, Boulder, CO

We directly observe the diffusion of individual bovine serum albumin molecules at the oil-water interface over more than two orders of magnitude of oil viscosity using total internal reflection fluorescence microscopy. Our system consists of silicone oils, and fluorescently labeled proteins. At low oil viscosities, on the order of 100 cP, the molecular mobility is dominated by the expected Stokes-Einstein mode of interfacial diffusion as the effective diffusion coefficient is inversely proportional to both the protein hydrodynamic radius and the oil viscosity. However, at higher oil viscosities, we observe a different diffusive regime, with anomalously large effective diffusion coefficients that are relatively insensitive to the oil viscosity and are similar to the measured diffusion coefficient of the same protein at a hydrophobic solid-liquid interface. This transition occurs at approximately 2000 cP. We propose that at high oil viscosity, interfacial diffusion is dominated by desorption-mediated diffusion. This has been previously observed at many solid-liquid interfaces.

(147c) Interfacial Transport of Polymeric Nanomedicine in the Proximity of Cellulose Surfaces

Mustafa Akbulut and Ming Zhang, Chemical Engineering, Texas A&M University, College Station, TX

The increased production and commercial use of nanoparticulate drug delivery systems combined with a lack of regulation to govern their disposal may result in their introduction to soils and ultimately into groundwater systems. To better understand how such particles interact with environmentally significant interfaces, we study the adsorption, desorption, and removal behavior of polyethylene glycol-based nanoparticulate drug delivery systems on and from cellulose, which is the most common organic compound on Earth. In this presentation, I am going to talk about the dynamics of adsorption for such nanoparticles. Our studies indicated that such an adsorption process is only partially reversible, and most of the adsorbate particles do not desorb from the cellulose surface even upon rinsing with a large amount of water. The rate constant of adsorption decreases with increasing particle size. Furthermore, hydrodynamic forces acting parallel to the surfaces are found to be of great importance in the context of particle dynamics near the cellulose surface, and ultimately responsible for the removal of some fraction of particles via rolling or sliding. As the particle size increases, the removal rates of the particles increases for a given hydrodynamical condition.

(147d) The Impact of External Shear Forces On the Transport of Cell Membrane Species in Heterogeneous Lipid Bilayers

Chih-Yun Hsia, Chemical and Biomolecular Engineering, Cornell University , Ithaca, NY and Susan Daniel, Chemical and Biomolecular Engineering, Cornell University, Ithaca, NY

The impact of external shear forces on the transport of cell membrane species in heterogeneous lipid bilayers

Many biological processes in the cell are controlled by interactions between biomolecules residing in the cell membrane. Interactions between these species can be controlled and facilitated by the local membrane composition and the affinity of the species for a particular membrane type. The cell membrane has distinct lipid micro-domains of co-existing phases, called lipid raft and liquid-disordered phases, and this feature has been suggested to play a key role in regulating several cellular activities, as some proteins have been shown to exhibit different activity levels depending on specific lipid interactions.

However, studying dynamic changes in biomolecule partitioning between different lipid phases and identifying stimuli that may change affinity for a particular phase has been a significant challenge for researchers. Either the experimental approaches used in vivo result in too many artifactual consequences (i.e. detergent contamination, multivalent labeling enrichment), or they are limited by in vitro methods that can only assess equilibrium conditions.

Our group recently developed a new in vitro approach to study kinetics of partitioning using patterned heterogeneous supported lipid bilayers (SLB) inside a microfluidic device. The patterned heterogeneous SLBs system is a native-like membrane environment that removes the drawbacks of the previous approaches since it allows membrane species to retain their native structure, fluidity, and functionality. Using this approach, we are able to study how external shear forces impact biomolecule partitioning between heterogeneous lipid phases and facilitate interactions among membrane-bound species.

In this study, we developed an analytical model that we validated with experiments using fluorescent tracers to report on the velocity of biomolecules within the supported bilayers. This model helps us to determine the velocity profile in the interfacial regions between two phases. With this information, we are able to characterize the transport behavior of membrane-bound species subjected to external shear stress to determine the influence of shear forces on the partitioning process in heterogeneous membranes. We suggest that this platform can be used for studying heterogeneous cell membrane environments that is subjected to shear, such as in epithelial tissue. Such studies might contribute to our understanding of biological function in the vasculature and disease processes that occur there, such as atherosclerosis.

(147e) Autophobing On Liquid Subphases Via Interfacial Transport of Amphiphilic Molecules

Ramankur Sharma[1], Roomi Kalita[1], Timothy Corcoran[2], Ellen Peterson[3], Todd M. Przybycien[4], Robert D. Tilton[4] and Stephen Garoff[5], (1)Department of Chemical Engineering, Carnegie Mellon University, Pittsburgh, PA, (2)Department of Medicine, University of Pittsburgh, Pittsburgh, PA, (3)Department of Mathematical Sciences, Carnegie Mellon University, Pittsburgh, PA, (4)Department of Chemical Engineering and Department of Biomedical Engineering, Carnegie Mellon University, Pittsburgh, PA, (5)Department of Physics, Carnegie Mellon University, Pittsburgh, PA

Autophobing is a phenomenon in which a low surface tension drop of pure amphiphilic molecules or a solution of amphiphiles fails to completely wet a high energy substrate and instead forms a lens. Considerable research has examined autophobing on solid substrates, where questions still remain about the balance of surface tensions in the final state and what molecular structures produce those surface tensions. Autophobing has also been observed on liquid subphases. Autophobing on liquids has been attributed to molecular transport through one of the bulk phases rather than to interfacial transport. Motivated by Marangoni driven spreading on liquid subphases, we investigate the mechanism of autophobing on liquid subphases via interfacial transport of amphiphiles. We deposit drops of a pure amphiphilic compound (oleic acid) or of a surfactant solution (dodecanol in hexadecane) on water or glycerol-water mixture subphases. The low solubility and vapor pressure of the deposited amphiphiles make it highly unlikely that bulk phase transport of the molecules can cause the drop to autophobe on the short timescales observed here. We measure surface tensions relevant to the spreading both before and after drop deposition, determine if the amphiphiles have moved from the drop onto the surrounding subphase surface outside the lens, and estimate the packing of those molecules on the surface. In all cases, we observe the rapid escape of amphiphiles across the drop contact line promptly after drop deposition and formation of a monolayer that is close to collapse on the subphase outside the drop. The monolayer formation decreases the subphase-air surface tension external to the drop and reverses the sign of the spreading coefficient from positive (before the monolayer forms) to negative, forcing the drop to retract into a lens with an increased contact angle. Measurement of all three surface tensions that exist at the end of drop spreading reveals that it is the reduction in the surface tension of the subphase outside the drop that is solely responsible for inhibiting the spreading of the drop and initiating its retraction. Further, we capture the dynamic retraction of an oleic acid drop on a glycerol-water mixture subphase by simultaneously tracking the movement of the drop contact line and the capillary ridge that advances ahead of the drop contact line. We clearly see that the drop does not spread completely to a thin wetting film with zero degree contact angle despite the initially positive spreading coefficient. Rather it first spreads only to a finite extent and then starts retracting once the amphiphilic molecules have oriented themselves to form a dense monolayer outside the drop. In addition, we also develop a mathematical model to predict the diameter of the lens given all the surface tensions, densities and the deposited drop volume. We compare the experimentally measured lens diameters with the mathematical model and find the values to be in good agreement.

(147f) Dissolution Processing of Nanostructured Polymers

Mohammad Ghasemi, Marina Tsianou and **Paschalis Alexandridis**, Department of Chemical and Biological Engineering, University at Buffalo - The State University of New York (SUNY), Buffalo, NY

Nanoscale and mesoscale order in polymers, manifested by chain crystallinity and organization in spherulites or fibers, significantly affects the dissolution of nanostructured polymers, and constrains their subsequent physical or chemical processing. Cellulosic biomass provides a nice example of a composite material that is recalcitrant to solvents. The efficient utilization of cellulose as a starting material for the synthesis of high value-added functional polymers and chemicals and also for biofuel production provides the motivation for the current project. We consider here the transport phenomena governing the dissolution of solid cellulose, e.g., solvent penetration, transformation from crystalline to amorphous domains, specimen swelling, and polymer chain untangling, and analyze the effect of various parameters on the kinetics of dissolution.

Nanoscale and mesoscale order in polymers, manifested by chain crystallinity and organization in spherulites or fibers, significantly affects the dissolution of nanostructured polymers, and constrains their subsequent physical or chemical processing. Cellulosic biomass provides a nice example of a composite material that is recalcitrant to solvents. The efficient utilization of cellulose as a starting material for the synthesis of high value-added functional polymers and chemicals and also for biofuel production provides the motivation for the current project. We consider here the transport phenomena governing the dissolution of solid cellulose, e.g., solvent penetration, transformation from crystalline to amorphous domains, specimen swelling, and polymer chain untangling, and analyze the effect of various parameters on the kinetics of dissolution.

(147g) Multicomponent Effects in the Diffusion of Hydrophobic Solutes in Microemulsions

Ronald J. Phillips[1], Stephanie R. Dungan[2] and Wyatt J. Musnicki[1], (1)Chemical Engineering and Materials Science, U.C. Davis, Davis, CA, (2)Food Science & Technology, Chemical Engineering and Materials Science, University of California, Davis, Davis, CA

A major motivation for the use of microemulsions in food and medical applications is to alter the rate at which valuable solutes are delivered or released. In microemulsions, solute transport is often characterized with a single, effective diffusion coefficient. In this presentation, we show how this approach breaks down when concentrations are not dilute. Holographic interferometry was used to monitor transport of hydrophobic solutes within microemulsions. Transport is driven by concentration differences across the sample, either in the solute (at constant [surfactant]), or in the surfactant (at constant [solute]). Especially for more hydrophobic solutes, the transport kinetics cannot be interpreted by straightforward application of Fick's second law. Multicomponent interaction effects must be taken into account to interpret qualitative changes in the interference fringes, even at particle concentrations as low as 5%. Furthermore, the off-diagonal terms in the diffusion matrix are not generally small relative to the diagonal terms. By incorporating multicomponent effects into the analysis of the interference fringes, their importance can be demonstrated and measured.

(151a) A Hybrid Fluctuating Hydrodynamics and Molecular Dynamics Approach for Simulation of Biomolecular Systems

Jhih-Wei Chu, Chemical Engineering, University of California, Berkeley, Berkeley, CA and Nikolaos Voulgarakis, Chemical Engineering, University of California Berkeley, Berkeley, CA

A Hybrid Fluctuating Hydrodynamics and Molecular Dynamics Approach for Simulation of Biomolecular Systems

To go beyond the accessible time- and length-scales of atomistic simulation while retaining the essential forces for representing biomolecular systems, such as solvation, hydrophobicity, interfacial phenomena, and hydrodynamics interactions, we develop a hybrid fluctuating hydrodynamics and molecular dynamics (hybrid FHD/MD) simulation method. The hybrid FHD/MD method is also multiphysics in nature as field and particulate variables are hybridized in the simulation model. We will discuss the devised approaches of mixing particles and fields to achieve this hybridization. The ability to represent important phenomena such as the Stoke's law, hydrodynamic relaxation, solvation free energy, and hydrodynamic collapse at the same time by the simulation model will also be illustrated. The inter-domain coupling between enzyme protomers in protein dynamics will be employed as an application example to illustrate how this method can be used to obtain insights and dynamics information that are difficult to acquire otherwise.

(151b) A New Multiscale Algorithm and Its Application to Coarse-Grained Peptide Models for Self-Assembly

Scott Carmichael, Department of Chemical Engineering, University of California Santa Barbara, Santa Barbara, CA, M. Scott Shell, Chemical Engineering, University of California Santa Barbara, Santa Barbara, CA and Rafael Modestino Prato, Chemical Engineering, University of California, Santa Barbara, Santa Barbara, CA

A New Multiscale Algorithm and Its Application to Coarse-Grained Peptide Models for Self-Assembly

Scott Carmichael, Rafael Prato, and M. Scott Shell

Department of Chemical Engineering, University of California, Santa Barbara

Email: carmichael@engr.ucsb.edu shell@engr.ucsb.edu

Peptide aggregation plays a role in a number of neurodegenerative diseases, such as Alzheimer's, Huntington's, and Parkinson's. Here we aim to develop an accurate molecular-scale picture of the aggregation process using a multi-scale computational approach. Recently, Shell [1] developed a coarse-graining methodology that is based on a thermodynamic quantity called the relative entropy; a measure of how different two molecular ensembles behave. By minimizing the relative entropy between a coarse-grained (CG) system and an all-atom (AA) system, an optimized coarse-grain model can be obtained.
Here we develop a corresponding numerical strategy for optimizing CG models of arbitrary peptide sequences, and apply it to a model polyalanine molecule (Ala)15. Specifically, we optimize the CG force-field parameters that control the behavior of one, two, and three bead per amino acid models of (Ala)15. Simulations of the optimized models peptide are performed, and histograms of various structural correlations (e.g., bond length, bond angle, etc.) are extracted and compared to an AA system. We obtain excellent agreement in these distribution functions for all of the optimized force-fields, indicating the CG models are able to accurately reproduce much of the behavior of the original AA system. Moreover, the models yield single-molecule folding curves and free energy surfaces whose accuracy improves with the level of model detail.
We subsequently use these CG models to simulate the self-assembly of large-scale peptide systems that are too computationally demanding to study using conventional AA simulations. Simulations containing many CG (Ala)15 peptides form a β-sheet structure with excellent agreement with the experimentally-observed [2], structure typical of alanine rich peptides.We use these large-scale simulations to provide a detailed picture of the self-assembly mechanism and the effects of chain length and concentration.

[1] Shell, M. S. J. Chem. Phys. 2008, 129, 144108-7.
[2] Forood, et al., Biochem. and Biophys. Research Comm. 1995, 211, 7-13.

(151c) A Novel, Coarse-Grain Model for Chemically-Specific, Protein-Surface Interactions: Validation and Application

Shuai Wei, Brandon Loong and Thomas A. Knotts IV, Department of Chemical Engineering, Brigham Young University, Provo, UT

Protein-surface interactions are key phenomena in a variety of technologies, but a complete, fundamental understanding of the underlying biophysics involved remains elusive. In recent years, coarse-grain protein and surface models have been the preferred tools to examine such systems due to resolution limitations of experimental techniques and computational demands of all-atom simulations. While many insights into the topic have been gained with this approach, studies to date have suffered from a major limitation: the surface has been modeling in only a rudimentary way even by coarse-graining standards. Specifically, the vast majority of all coarse-grain simulations have used either a hard surface or a short-range, repulsive surface which has interacted with all sites in the protein the same way regardless of the chemistry of the surface or the identity of the residues in the protein.

Moving the research to the next level requires a better model for the system that takes into account sequence-specific, residue-level interactions between the protein and the surface. We have spent the last several years developing such a model. The first version of the model showed promise, but we encountered several challenges that required taking a slightly different approach. This new formalism maintains the computational efficiency characteristic of all coarse-grain representations, but takes into account the hydrophobic nature of both the surface and the individual residues comprising the protein.

This presentation will outline the latest version of the model. The limitations of previous approaches will be explained first followed by a description of the current functional form of the model. The details of the parameterization against experimental adsorption data for small peptides will then be explained. This will be followed by results from simulations used to validate the transferability of the model to larger, more-relevant proteins such as lysozyme and alpha-amylase. The agreement between the simulation and experimental data for these cases is remarkable and is obtained without reparameterization. The presentation will end with a side-by-side comparison of folding mechanisms for various proteins produced using the old and the new models. Taken as a whole, the results offer significant hope that the rational design of protein-surface technologies is possible.

(151d) Stability of Heterochromatin Condensation Due to Cooperative Binding

Peter J. Mulligan, Chemical Engineering, Stanford University, Stanford, CA, Elena F. Koslover, Biophysics Program, Stanford University, Stanford, CA and Andrew J. Spakowitz, Chemical Engineering, Biophysics, Stanford University, Stanford, CA

Gene regulation in eukaryotic cells requires the segregation of silenced genomic regions into densely packed heterochromatin, leaving the active genes in euchromatin regions more accessible. We develop a physical model that explains how such genomic states can be stably created and inherited. Epigenetic marks such as methylation at histone 3 lysine-9 (H3K9) enable chromatin condensation via bridging interactions from heterochromatin protein 1 (HP1). The underlying physical behavior can be mapped to a random-field Ising-like model with complex connectivity that accounts for the underlying chromatin structure. We use replica field theory to calculate HP1 binding for various low energy chromatin fiber structures. Each H3K9 site represents a binding site, with cooperative interactions when sites are within a contact distance. Energetic parameters are based on mononucleosome binding data. We observe phase segregation of HP1 into methylated regions, with a threshold of methylation above which we see high accumulation of HP1. This matches experimental observations that repression of heterochromatin genes occurs even with changing methylation levels. While larger differences in HP1 binding energy to H3K9me3 versus H3K9 might be expected to enhance HP1 binding in methylated regions, the differences in the dissociation constants has been found to be small. We find that smaller differences actually enhance the stability of the system to gains or losses in methylation marks and allow the cell to modulate HP1 concentration to decondense fibers for replication. Finally, we demonstrate that cooperative binding of HP1 is the key driver for HP1 accumulation into heterochromatin regions. These results match experimental observations showing strong HP1 accumulation both in vitro and in vivo. Our work presents a clear physical pathway by which epigenetic marks can control gene expression through chromatin condensation. These marks allow for behavior that is robust to methylation fluctuations, even with the addition of new histones post-replication.

(151e) Investigation of the Proton Translocation Mechanism in Lactose Permease of E. Coli by a Hybrid QM/MM Approach

Pushkar Y. Pendse, Department of Chemical and Biomolecular Engineering, University of Maryland, College Park, College Park, MD and Jeffery Klauda, Department of Chemical and Biomolecular Engineering, University of Maryland, College Park, MD

In the biological cells, molecular traffic in and out of the cell is mainly controlled by the membrane transport proteins. The membrane transport proteins play critical roles in human physiology and their malfunctions are responsible for a range of human diseases. The Major Facilitator Superfamily (MFS) is one of the largest families of membrane transport proteins whose members are found in almost all types of organisms and are very diverse in terms of substrate transport. We are studying Lactose Permease (LacY) of *E.coli* as a structural and functional model for the MFS proteins. LacY, a symporter, couples the energetically downhill transport of protons with the energetically uphill transport of sugar molecules but the mechanism of this energy coupling is unknown. A hybrid quantum mechanics/molecular mechanics (QM/MM) protocol is used to study the mechanism of proton translocation. Self-consistent-charge-density-functional-tight-binding (SCC-DFTB) is used for the QM calculations and the CHARMM MM force field is used to describe the MM region, while the generalized solvent boundary potential (GSBP) method is used to describe electrostatic interactions far away from the active site. The transition between the QM and MM regions is treated using the generalized hybrid orbital (GHO) approach.

It has been established that protonation of LacY is a precursor to sugar binding but the details of the proton binding site are unclear. An extensive mutational analysis indicated that residues Tyr236, Asp240, Arg302, Lys319, His322, and Glu325, which form a tight hydrogen bond/salt bridge network at the center of the molecule, are involved in proton binding but no single amino acid could be identified as an individual proton acceptor (Smirnova et al., *Biochemistry*, 2009). Therefore it was postulated that a water molecule coordinated within this tight network acts as a proton acceptor. We are investigating the role of structural water molecule/s in proton binding. Multiple independent simulations are carried out with only the structural water in the QM region to determine the stability of protonated water and its form. To study whether any of the residues in the surrounding hydrogen bond network can act as a proton acceptor, several independent simulations are carried out with each of these residues included in the QM region in addition to the protonated water. The energetics of proton transfer from the protonated water to each residue are calculated using the potential of mean force (PMF) approach. The results from these proton binding studies will be used to further probe the transport of proton from its binding site to Glu269, which is an irreplaceable residue believed to be involved in both sugar binding as well as proton translocation. The results from this study will help not only in understanding the overall mechanism of LacY but also in gaining useful insights into the proton translocation by other MFS proteins such as a multidrug efflux protein EmrD.

(151f) A Next-Generation Coarse Grain DNA Model for Engineering and Biology

Daniel M. Hinckley, **Gordon S. Freeman** and Juan J. de Pablo, Department of Chemical and Biological Engineering, University of Wisconsin-Madison, Madison, WI

We present a new version of the 3-Site-Per-Nucleotide (3SPN) coarse grain model for simulation of DNA. By incorporating anisotropic potentials between bead sites, we are able to remove Go-type interactions, thus freeing DNA to explore non-canonical configurations such as hairpins and holliday junctions. We demonstrate that the resulting model displays the correct trends of sequence-dependent mechanical properties. These include the effect of sequence on the persistence length of double-stranded DNA. We use metadynamics to show unambiguously that the model reproduces experimental thermodynamic trends, including the dependence of melting temperature on oligomer and salt concentration. Furthermore, we use Forward Flux Sampling (FFS) to calculate DNA hybridization rate constants and find that these calculated rate constants are in close agreement with values from experiment. The ability to capture the structural, thermodynamic, and kinetic behavior of DNA makes this new model appropriate to any coarse grain simulation where structural resolution of DNA is desired, including both biological and nanoengineering applications.

(151g) Computational Studies On the Effects of DNA Methylation On Its Structure and Dynamics

Shyno Mathew and **Vanessa Ortiz**, Chemical Engineering, Columbia University, New York, NY

DNA methylation occurs in CpG dinucleotides (C and G connected by a phosphodiester bond) when a methyl group is added at the carbon 5 position of the C base ring. In higher eukaryotes, DNA methylation has an important role in regulating gene expression. It is one of the most common epigenetic events taking place in the mammalian genome, and approximately 1% of human DNA is methylated. It is known that when cytosine bases are methylated, the gene is silenced, but the precise role of methylated DNA in this process is still debated, and the mechanism by which methylation can affect gene expression remains poorly understood.

One hypothesis is that the methylation state of a DNA segment can alter the positions of nucleosomes, which in turn, determine gene expression. Several studies point to a strong link between CpG methylation and nucleosome positioning, but these effects might be indirect because methylation influences the binding of other factors, which could in turn trigger nucleosome repositioning. Because methylation chemically modifies DNA, the methylation status of a DNA sequence could influence its flexibility and, thus, its affinity to the nucleosome. Recent experimental studies suggest that CpG methylation can decrease the ability of DNA to bend into the major groove at the methylated CpG step, and can thereby influence nucleosome positioning.

To test the idea that methylation can affect the ability of DNA to bend, we have performed a molecular dynamics study of short DNA segments using an all atom force field. DNA conformations were characterized by calculating the six rigid body base-pair step parameters, the minor groove width and the hybridization state. These properties were measured and compared for double stranded DNA segments in both the unmethylated and methylated states. In addition, all atom trajectories for the unmethylated and methylated cases were analyzed to extract a new set of parameters for a coarse grain model for DNA that can now be used to study the effects of methylation in larger DNA systems such as nucleosomes.

(151h) A Molecular Model for the Thermodynamic Stability of Hepatitis B Virus

Jehoon Kim, Chemical and Environmental Engineering, University of California, Riverside, Riverside, CA, Roya Zandi, Department of Physics and astronomy, University of California, Riverside, Riverside, CA and Jianzhong Wu, Chemical and Environmental Engineering, UC Riverside, Riverside, CA

We establish an efficient and appropriate thermodynamic model for the stability of a virus nucleocapsid particle. For the Hepatitis B virus (HBV), an effective coarse-grained model is constructed, which includes all pertinent molecular components (i.e., capsid proteins and their flexible domains, nucleic acids, and ions in the electrolyte background) and solvent-mediated interactions: the hydrophobic association of capsid subunits, the electrostatic interactions, the entropy, the excluded volume, and the macromolecular crowding effects. Based on such an explicit framework, we investigate the stability of the viral particle with a focus on physiological cellular conditions. We consider the free energy of the empty capsid formation and the interaction free energy of pre-genomic RNA and C-terminal domain (CTD) of the capsid proteins. In comparison with experimental investigations for HBV, our model presents appropriate equilibrium values for the capsid assembly and supports the balanced electrostatic interaction hypothesis for the capsid stability and genome encapsidation. Moreover, this model presents a generic theoretical approach for determining the optimal genome length of viral particles.

(151i) DNA Base Dimers Are Stabilized by Hydrogen Bonding Interactions Including Non-Watson-Crick Pairing near Graphite Surfaces

Akshaya Shankar[1], Anand Jagota[1] and Jeetain Mittal[2], (1)Chemical Engineering, Lehigh University, Bethlehem, PA, (2)Department of Chemical Engineering, Lehigh University, Bethlehem, PA

Single and double stranded DNA are increasingly being paired with surfaces and nanoparticles for numerous applications. Unlike the majority of DNA structures in bulk that are stabilized by canonical Watson-Crick pairing between Adenine-Thymine and Guanine-Cytosine, those adsorbed on surfaces are often stabilized by non-canonical base pairing, quartet formation, and base-surface stacking. Not much is known about these kinds of interactions. To build an understanding of the role of non-Watson-Crick pairing on DNA behavior near surfaces, one requires basic information on DNA base pair stacking and hydrogen bonding interactions.
All-atom molecular simulations of deoxyribonucleosides in two cases - in bulk water and strongly adsorbed on a graphite surface, are conducted to study the relative strengths of stacking and hydrogen bond interactions for each of the ten possible combinations of base pairs. The key information obtained from these simulations is the free energy as a function of distance between two bases in a pair. We find that stacking interactions exert the dominant influence on the stability of DNA base pairs in bulk water as expected.
The strength of stability for these stacking interactions is found to decrease in the order, Gua-Gua > Ade-Gua > Ade-Ade > Gua-Thy > Gua-Cyt > Ade-Thy > Ade-Cyt > Thy-Thy > Cyt-Thy > Cyt-Cyt.
On the other hand, mutual interactions of surface adsorbed base pairs are stabilized mostly by hydrogen bonding interactions in the order, Gua-Cyt > Ade-Gua > Ade-Thy > Ade-Ade > Cyt-Thy > Gua-Gua > Cyt-Cyt > Ade-Cyt > Thy-Thy > Gua-Thy.
Interestingly, several non-Watson-Crick base pairings, that are commonly ignored, have similar stabilization free energies due to inter-base hydrogen bonding as Watson-Crick pairs. This clearly highlights the importance of non-Watson-Crick base pairing in the development of secondary structures of oligonucleotides near surfaces.

(151j) Simulations of Microtubule Forces Derived From Cytoskeletal-Bound Motors

Anthony J.C. Ladd[1], Gaurav Misra[2,3], Jun Wu[1], Robert J. Russell[1], Nandini Shekhar[1], Tanmay Lele[1] and Richard Dickinson[1], (1)Department of Chemical Engineering, University of Florida, Gainesville, FL, (2)Laboratory for Rehabilitation Neuroscience, University of Florida, (3)Department of Bioengineering, Stanford University, Stanford, CA

Microtubules are key components of force transmission in cells, and play a role in cell locomotion, transport, and mitosis. Experiments in Lele's group[1] have shown that microtubules severed by laser ablation do not straighten, as would be expected from the large bending moments along their lengths. Instead, segments near newly created minus ends typically increased in curvature following severing, while segments near new microtubule plus ends depolymerize before any observable change in shape. However, in dynein-inhibited cells, segments near the cut straightened rapidly following severing. These observations suggest that microtubules are subject to significant tangential forces, and that lateral motion of the microtubule is opposed by a large effective friction rather than elastic forces. To help interpret the experimental results we have developed a numerical model for intracellular microtubule mechanics, accounting for dynein-generated forces on the microtubules.

At the length scales of mammalian cells, microtubules behave as semi-flexible filaments and can be coarse-grained using the Kirchoff theory for elastic rods. We have supplemented the Kirchoff model[2] with the stochastic growth and collapse of microtubules[3] (the dynamic instability), and by a model for dynein generated forces[1]. Numerical simulations of the buckling of a single microtubule can explain both the enhanced buckling at the minus end of a severed microtubule and the apparently frozen shape of the plus end. I will present simulations of the dynamics of individual microtubules that explain how motor forces result in the localization of short-wavelength buckles near the cell periphery. Our results suggests that microtubule shapes *in vivo* reflect a *dynamic* force balance, where bending moments are opposed by dynein-motor forces, including an effective friction from the stochastic binding and unbinding of the motors. Simulations of the motion of the centrosome are consistent with a mechanism for centrosome centering driven by pulling forces exerted by dynein motors. The simulations also explain how tension on the centrosome can be reconciled with buckled filaments near the cell periphery. I will present results of ongoing calculations that aim to discover if discrete distributions of molecular motors can explain experimentally observed fluctuations in filament shapes.

[1] J. Wu et al., *Molec. Biol. Cell*, 22:4834-4841, 2011.

[2] A. J. C. Ladd and G. Misra, *J. Chem. Phys*. 130:124909, 2009.

[3] T. Mitchison and M. Kirschner *Nature* 312: 237-42, 1984.

Fabrication of Monodisperse Silica Microspheres by a sol-emulsion-gel method: Synthesis and characterization

Cong Gao[a,b], Xiaobo Qi[b], Man Zhang[c], Ruizhuang Yang[b], Weixing Huang[a*], Bo Li[b], Zhanwen Zhang[b]*

[a] *School of Chemical Engineering, Sichuan University, Chengdu, Sichuan 610065, China*

[b] *Research Center of Laser Fusion, China Academy of Engineering Physics, Mianyang, Sichuan 621900, China*

[c] *Institute of Computer Application, China Academy of Engineering Physics, Mianyang, Sichuan621900, China*

* Corresponding authors: xbqi@caep.ac.cn (X. Qi), hwx@scu.edu.cn (W. Huang).

Abstract: To fabricate monodispersed silica microspheres, a sol-emulsion-gel process combined with a T-shaped junction microfluidic technology was developed.. The microfluidic system was made up of two T-shaped microchannels, one of which was used as a milimixer in which partial hydrolyzed TEOS sol mixed with dilute ammonia solution, while the other was the droplet generator in which emulsion droplets formed. The sol emulsions were transported into heat medium to fabricate gel microspheres. The gel microspheres were characterized by optical and scanning electron microscopy, Fourier transform infrared spectroscopy (FTIR), thermal gravimetric analysis (TGA) and x-ray diffraction (XRD) analysis. Micrographs of the microspheres illustrate the uniform diameter and good sphericity. SEM of the microspheres reveal the nanoporous structure. The result show that monodispersed silica gel microspheres can be obtained based on the uniform reactions in the monosized sol droplets. The size of the silica microspheres can be controlled by adjusting the flow rates of the continuous phase and dispersed phase. Monodisperse silica microspheres with diameters range between 398.65μm to 479.31μm were fabricated. FTIR measurements show the existence of residual C-H and Si-OH groups in the gel microspheres. These groups were then oxidized and removed by calcination. A weight loss of 7.96% during calcination was detected by the TGA measurement. XRD analysis reveal that the microspheres stayed in a amorphous state after calcination. The droplet formation process have been stablized by using the milimixer, which benefits the formation of monosized droplets. The present method is also characterized by no-surfactant. The microfluidic device in the present experiment can be extended to synthesize other particles by sol-emulsion-gel process .

Key words: monodispered microspheres, sol-emulsion-gel, silica microspheres, microfluidics

1. Introduction

Because of the advantages such as biocompatibility, chemical inertness, excellent heat resistance, spherical shape, and controlled texture, silica microspheres have become increasingly important for a variety of applications in pharmaceutical[1], hydrogen storage [2, 3] ,electronic packaging[4, 5], chromatographic columns[6] etc. These applications require high purity powders. Hydrolysis of tetraethyl orthosilicate (TEOS) is a versatile technique to synthesize pure silica[7, 8].

Methods such as solution growth[1, 9] , spray dring[10], emulsion technique[11-14] have been developed to synthesize silica microspheres. Spray drying technique is suitable for bulk production, but hard to control the quality of the product. Sol-emulsion-gel route is popular to synthesize silica microspheres because it is well suited to control shape,density, amount and surface properties of spheres[11]. L.M.Yang et al prepared micrometer-sized hard silica spheres in water-in-oil(W/O) emulsion[12]. With the sol-emulsion-gel method, hollow silica microspheres can also be synthesized[11, 13, 14].

Monodispersed silica spheres are very promising because they can easily be packed into existing reactors, columns, fixed and fluidized beds, which favors efficiency improving. However, few work focuses on monodisperse silica microspheres and the reported conventional routines are not suitable for fabricating monodisperse silica microspheres. A recent developing technology termed microfluidic technology is an ideal method for obtaining uniform size droplets[15-17]. Microfluidic presents numerous advantages over conventional batch synthesis methods. Its high potentialities stem from the continuous production of highly monodisperse liquid reactors and the ability to manipulate and functionalize each one independently "on line". It paves the way to manufacture monodisperse particles. Masatoshi [16], Gao[3] reported a combination of sol-gel chemistry and millifluidic/microfluidic to fabricate engineering materials.

In the present paper, a new method for monodisperse silica microspheres is developed by combing the sol-gel chemistry with microfluidic technique.

2. Materials and experimental procedure

2.1 Materials

In the present experiment, silica gel microspheres were synthesized from tetraethyl orthosilicate (TEOS) as a source of silica. During the process of pre-hydrolysis, hydrochloric acid(HCl,36wt.%)was used as catalyst, ethanol as cosolvent for TEOS and deionized water(DW).Ammonia solution(NH_3,25~28%) was diluted by ethanol and DW. Silicon oil was used as the continuous fluid to strip the sol solution into droplets in the process of microfluidic generation of sol droplets. Petroleum ether was employed to wash silicon oil from the gel microspheres. All chemical reagents were analytically pure.

2.2 Preparation of sol solution

A flow chart of producing silica gel microspheres is shown in Fig.1. The sol solution was prepared by the following procedure: 30ml of TEOS, 7ml deionized water were sequentially added into 30ml of ethanol at 60° C. Then,0.35ml of hydrochloric acid(36wt%) were added drop wise to the mixture at 60° C by stirring at a rate of 1200 rpm for 6h, for partially-hydrolyzing TEOS. On the other hand, deionized water, ethanol and ammonia solution were mixed at a volume ratio of 200:300:1 to prepare dilute ammonia solution. The dilute ammonia solution and the precursor solution were then stored separately for next stage at room temperature in sealed bottles.

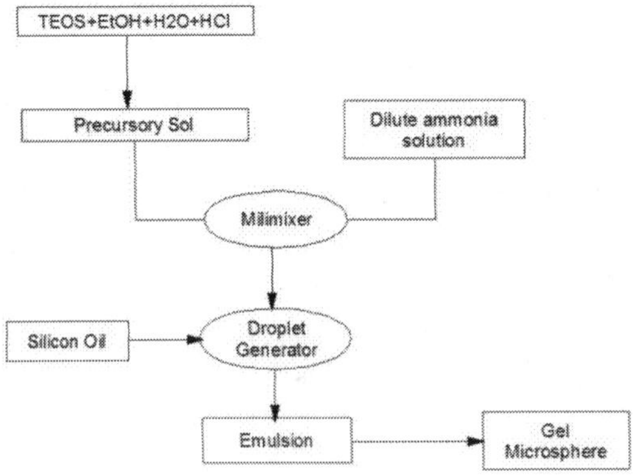

Fig.1 Flow chart for fabrication silica microspheres by sol-emulsion-gel process

2.3 Formation of gel microspheres

The gel microspheres fabrication set-up is shown schematically in Fig.2. Two PMMA chips with T-shaped microchannels were employed, one of which was used as a milimixer, the other was droplet generator. The two chips are connected by a FEP microtube (outer diameter:1.6mm, inner diameter:0.5mm).

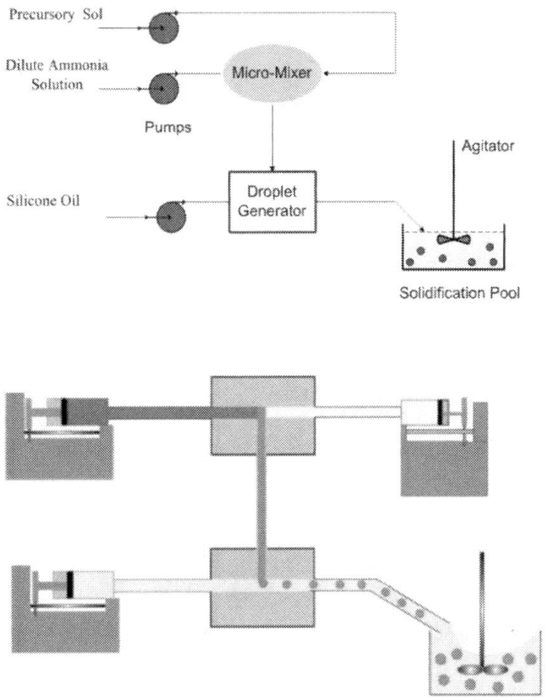

Fig.2 Schematic of the droplet generating set-up

As soon as the sol derived from partial hydrolyzation is mixed with the dilute ammonia solution, complete hydrolyzation, condensation and gelation take place to form the gel. In a base condition, the sol converts into gel very quickly. The gelation time for the present experiment counting from the mixing of the two precursory fluid(the sol and the dilute ammonia solution) is around 2 minutes. For a stable droplet formation process, a milimixer was employed. The sol was injected into the microchannel by a syringe pump. The dilute ammonia solution was injected from opposite direction by syringe pump. The two fluids were injected at the same flow rate for a mixing volume ratio at 1:1. The two fluids collided at the T-junction, mixed and flowed perpendicularly into the droplet generator. The sol and the dilute ammonia solution mixed to form a homogenous sol solution by both molecular dispersion and macro mixing caused by fluid flow. At the junction of the generator, the sol solution was then stripped into droplets by the perpendicular silicon oil flow. The W/O(water in oil) emulsion was delivered into silicon oil, which was agitated at 90 rpm under constant temperature (60°C) in a Teflon beaker. Hydrolyzation, condensation and gelation took place very quickly and gel microspheres formed. After 24 h to age, the gel microspheres were filtered and washed by petroleum ether.

2.4 Characterization

The gel microspheres were characterized by optical and scanning electron microscopy, Fourier transform infrared spectroscopy, XRD and TGA analysis. The shape, diameter and monodispersity of the microspheres were measured by optical microscopy. The microstructure was revealed by SEM. The average diameter ($\overline{D}$) and the monodispersity index (C_V) of the sol droplets and the dried gel microspheres were determined by measuring the diameters of 50 microspheres. $\overline{D}$ is the arithmetic mean diameter of the measured microspheres. C_V is defined by the following equation $C_V=\delta/\overline{D}\times 100\%$, where δ is the diameter standard deviation of the measured microspheres

3. Results and discussion

Fig.3 shows the optical images of the silica microspheres without heat treatment. It is clear that these microspheres are uniform in size and have good sphericity. These microspheres were fabricated at variable flow rates. At the same continuous flow rate (15ml/min), as the flow rates of sol and ammonia solution increased from 0.8ml/min to 1.2ml/min, silica microspheres diameter increased from 398.65μm to 479.31μm.The monodisperse index Cv is around 3.0, which indicates that the microspheres have excellent monodispersity.

a b c

Fig.3 Optical Micrographs of the silica microspheres(scale bar represents 200μm,Qc-flow rate of the oil phase.Qd1,Qd2 flow rates of the sol and dilute ammonia solution
a.Qc=15ml/min,Qd1=Qd2=0.8ml/min, $\overline{D}$ =398.65μm,Cv=2.77,
b.Qc=15ml/min,Qd1=Qd2=1ml/min,$\overline{D}$=445.42μm,Cv=2.85,
c.Qc=15ml/min,Qd1=Qd2=1.2ml/min,$\overline{D}$=479.31μm,Cv=2.56)

In order to examine the thermal property of the silica microspheres, TGA measurements were conducted in air from 40 °C to 1400 °C at a heating-up rate of 10 °C /min.The TGA profile of the silica microspheres is shown in Fig.4. Four weight loss regions are distinguishable on the curve. The first weight loss of 1.35% up to 100 °C can be attributed to the elimination of small molecular such as water , petroleum ether and ethanol. The second weigh loss of 1.6% in the region of 100~300 °C can be ascribed to crystallization water and the combustion of residual petroleum ether. The third region is between 300~800 °C with weight loss around 4.36%, which can be attributed to oxidization of residual –CH groups. A weight loss of 0.65% can be detected during the fourth region between800~1250 °C, which may be due to volatilization of carbon element. According to the TGA measurement, it can be concluded that almost all the impurities can be removed at 800 °C, which is further illustrated by FTIR results in the coming part.

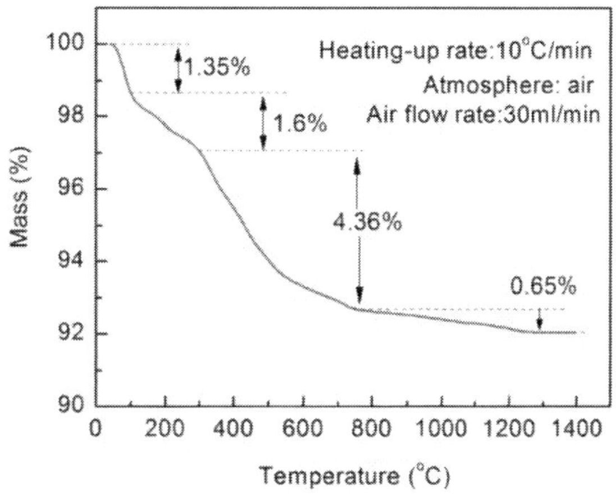

Fig.4 TGA measurement of the silica microspheres

According to TGA measurement, almost all the impurities can be removed by calcinations at 800 °C. To further examine the composition changes, infrared spectra ranging from 450 to 4000cm^{-1} of the microsphers with and without calcinations were tested by FTIR as shown in Fig.5. Peaks at 2967,1400 cm^{-1} are assigned to the bending and stretching vibrations of C-H groups, which were oxidized and removed by calcinations in air. Peaks at 970 cm^{-1} is due to stretching vibration of Si-OH group which was also oxidized and eliminated. The absorption peaks at 3443,1640 cm^{-1} can be assigned to hydroxyl group of physical absorbed water. Bending vibrations of Si-O-Si result in the peaks at 460 cm^{-1}, while the asymmetric vibration of Si-O-Si is corresponding to peaks at 1088cm^{-1}. From the above results, pure silica microspheres can be obtained by calcinations at 800 °C.

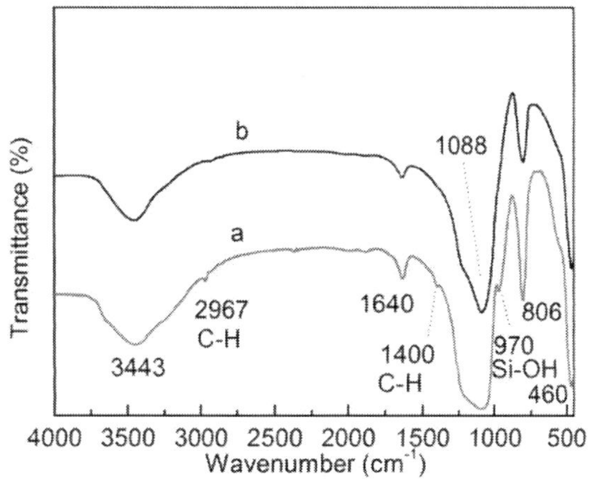

Fig.5 FTIR spectra of the microspheres (KBr disc, a. for silica microspheres without calcinations, b. for silica microspheres calcined at 800 °C for 2h)

X-ray diffraction measurements were conducted to examine the phase state of the silica microspheres. Fig.6 compares the X-ray diffraction pattern of the silica microspheres kept at 800 °C for 2h with that of those without heat treatment. It shows that the silica microspheres are amorphous.

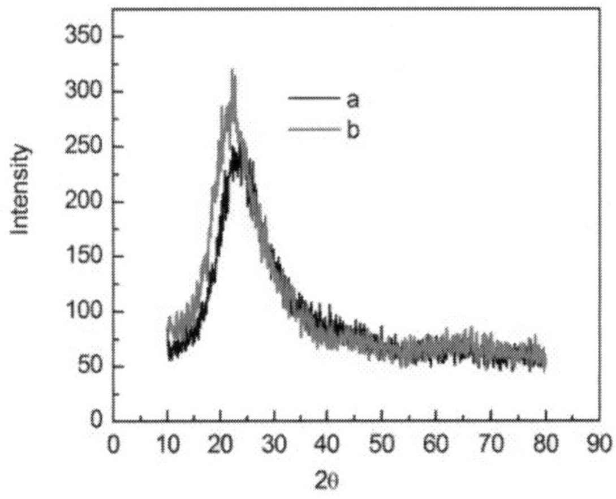

Fig.6 XRD of the microspheres: a.without calcination, b. kept at 800°C for 2h

Fig.7 shows the SEM of the surface of silica microspheres. The surface of the microspheres look smooth and dense at a low magnification. The higher magnification micrographs reveal the nanoporous structure. The nano pores shrinked due to heat treatment, which is illustrated by comparison of Fig.4a and b. Heat treatment densified the microspheres.

a. without heattreatment

b. kept at 800 ^{o}C for 2h

Fig.7 SEM of the surface of silica microspheres

4. Conclusion

To fabricate monodisperse silica microspheres, the sol-emulsion-gel process combined with microfluidic technology was developed. The experimental results show:

Monodisperse silica microspheres can be obtained by the device in the present paper. Silica microspheres with diameters ranging from 398.65μm to 479.31μm were fabricated. The monodispersity index is less than 3. The spherical shape and uniform diameters are illustrated by optical micrographs. SEM reveals that the smooth microspheres are nanoporous. The TGA and FTIR measurements indicate that impurities can be removed by calcination at temperature of 800 ^{o}C, thus very pure silica microspheres can be fabricated. XRD profiles illustrated the silica microspheres stayed in amorphous state after calcinations.

As a result, amorphous silica microspheres with good monodispersity can be fabricated with the method in the present paper. Because the production can be scaled up by paralleling a large number of microfluidic devices, mass production of monodisperse silica microspheres can be achieved. In the future study, improvement of the microfluidic device will be conducted in order to enlarge the diameter ranges of silica microspheres.

Acknowledgement

Special appreciation must go to Yiyang Liu, Mengsheng Yang, Bin Tang and YongJin for their assistance in sample analysis.

References

1. A. Yu, Y. Wang, E. Barlow, F. Caruso. *Mesoporous Silica Particles as Templates for Preparing Enzyme-Loaded Biocompatible Microcapsules.* Advanced Materials, 2005. 17(14): p. 1737-1741.
2. Li Du, Shijun Liao, Quanbing Liu et al. Shan Ji. *Porous grape-like spherical silica with hydrogen storage capability, synthesized using neutral dual surfactants as templates.* International Journal of Hydrogen Energy, 2009. 34(9): p. 3810-3815.
3. Cong Gao, Xiaobo Qi, Zhanwen Zhang, Sufen Chen, Bo Li, Weixing Huang. *Fabrication of monodisperse precursor gel microspheres for hollow glass microspheres by combining the sol-microemulsion-gel process with a T-shaped microfluidic device.* International Journal of Hydrogen Energy, 2011. 36(16): p. 9758-9766.
4. Ali Serpengüzel and Abdullah Demir. *Silicon microspheres for near-IR communication applications.* Semiconductor Science and Technology, 2008. 23(6): p. 064009.
5. Chao Xu, Katarzyna Wygladacz, Yu Qin, Robert Retter, Michael Bell, Eric Bakker. *Microsphere optical ion sensors based on doped silica gel templates.* Analytica chimica acta, 2005. 537(1): p. 135-143.
6. Albert J.De Vries, Madeleine LePage, Raymond Beau, Claude L.Guillemin. *Evaluation of porous silica beads as a new packing material for chromatographic columns. Application in gel permeation chromatography.* Analytical Chemistry, 1967. 39(8): p. 935-939.
7. C.J.Brinker and G.W. Scherer.*Sol-gel science: the physics and chemistry of sol-gel processing.* 1990: Academic Pr.
8. Sakka, S. *Handbook of sol-gel science and technology: processing, characterization, and*

applications. Vol. 1. 2005: Springer.

9. Jenny Ho, Wei Zhu, Huanting Wang, Gareth M.Forde. *Mesoporous silica spheres from colloids.* Journal of colloid and interface science, 2007. 308(2): p. 374-380.
10. M.Todea, R.V.F. Turcu, B.Frentiu et al. *Amorphous and nanostructured silica and aluminosilicate spray-dried microspheres.* Journal of Molecular Structure, 2011. 1000(1): p. 62-68.
11. R.K.Singh,A. Garg, R.Bandyopadhyaya et al. *Density fractionated hollow silica microspheres with high-yield by non-polymeric sol–gel/emulsion route.* Colloids and Surfaces A: Physicochemical and Engineering Aspects, 2007. 310(1): p. 39-45.
12. L.M.Yang, Y.J.Wang, Y.W.Sun et al. *Synthesis of micrometer-sized hard silica spheres with uniform mesopore size and textural pores.* Journal of colloid and interface science, 2006. 299(2): p. 823-830.
13. Bo Peng, Min Chen, Shuxue Zhou et al. *Fabrication of hollow silica spheres using droplet templates derived from a miniemulsion technique.* Journal of colloid and interface science, 2008. 321(1): p. 67-73.
14. Wen Pan,Junwei Ye, Guiling Ning et al. *A novel synthesis of micrometer silica hollow sphere.* Materials Research Bulletin, 2009. 44(2): p. 280-283.
15. W.Engl, R. Backov, and P. Panizza, *Controlled production of emulsions and particles by milli-and microfluidic techniques.* Current Opinion in Colloid & Interface Science, 2008. 13(4): p. 206-216.
16. Masatoshi Tachibana, Wilfrid Engl, Pascal Panizza et al. *Combining sol–gel chemistry and millifluidic toward engineering microporous silica ceramic final sizes and shapes: An Integrative Chemistry approach.* Chemical Engineering and Processing: Process Intensification, 2008. 47(8): p. 1317-1322.
17. Rhutesh K. Shah, Ho Cheung Shum, Amy C. Rowat et al. *Designer emulsions using microfluidics.* Materials Today, 2008. 11(4): p. 18-27.

(158b) Manipulating Size-Selective Dispersability of Gold Nanoparticles in Gas Expanded Liquid Systems Using Ligand/Solvent Steric Effects

Pranav S. Vengsarkar[1], Jennifer N. Duggan[2] and Christopher B. Roberts[1], (1)Department of Chemical Engineering, Auburn University, Auburn, AL, (2)Chemical Engineering, Auburn University, Auburn University, AL

Nanotechnology is a field which involves the study of the conversion of two and three dimensional assemblies of molecular scale building blocks, into nanomaterials/nanoparticles. It is an important topic in modern research due to its multidisciplinary applicability and future potential in many underdeveloped applications. Nanoparticles of most materials have unique size-dependent physicochemical properties due to quantum effects and a high surface area to volume ratio. However, the current methods used to produce monodisperse nanoparticles have several operational shortcomings such as the use of expensive equipment, large quantities of solvents and low throughputs. The synthesis procedures used to generate monodisperse nanoparticles usually necessitate then use of expensive reagents, high temperatures and can result into nanoparticles having non-uniform morphologies. A technique to size-selectively precipitate and fractionate ligand-stabilized gold and silver nanoparticle dispersions in organic solvents has been developed in our laboratory which utilizes the pressure tunable physicochemical properties of CO_2 Gas-eXpanded Liquids (GXLs). This size-selective fractionation technique is based on the controlled reduction of the solvent strength through increases in the concentration of CO_2 (a known nonsolvent for aliphatic stabilizing ligands) via pressurization. These changes in solvent strength affect the subtle balance between the osmotic repulsive forces and the Van der Waals forces of attraction between differently sized nanoparticles necessary to maintain a stable dispersion. Through modest changes in CO_2pressure, increasingly smaller sized nanoparticles can be controllably precipitated from the dispersion.

The aim of this particular study is to investigate how the steric nature of a solvent and ligand affects the ligand-solvent interaction and thus the size dependent dispersability of gold nanoparticles in a GXL system. The nanoparticles precipitate from GXLs due in part to collapsing of the ligands under worsening solvent conditions. It is hypothesized that bulky, branched ligands should collapse to a lesser extent in an alkyl solvent due to less favorable solvent-ligand interactions when compared to straight-chain ligands and result in lower precipitation pressures. It is attempted in this study to increase the efficiency and effectiveness of this GXL process by combining branched ligands and branched solvents. Gold nanoparticles with a broad size distribution were synthesized using the popular Brust-Schiffrin method with various branched alkyl ligands, dispersed in alkyl solvents of varying structure, and then precipitated using the GXL technique. n-Hexane was chosen as the control solvent and the effect of several of its isomers on the precipitation and fractionation process will be discussed. The precipitation characteristics of ligand-stabilized gold nanoparticles were quantified by measuring the intensity of the surface plasmon resonance band for the nanoparticle dispersion using UV-vis spectroscopy at various levels of CO_2 pressurization. The characterization of these nanoparticles was performed using transmission electron microscopy (TEM) to analyze their size distribution and hence judge the efficacy of the size-selective fractionation. The pressures necessary to induce precipitation, the pressure range over which the nanoparticles precipitate, and the fractionation efficiency from each of these solvents provides new physicochemical insights that help further our fundamental understanding of the GXL precipitation phenomenon.

(158c) Salt-Assisted Analysis of Chitosan/Tripolyphosphate Micro- and Nanogel Formation

Yan Huang and **Yakov Lapitsky**, Department of Chemical and Environmental Engineering, University of Toledo, Toledo, OH

Biobased micro- and nanogels ranging from tens of nanometers to microns in diameter can be prepared by ionically crosslinking chitosan with tripolyphosphate (TPP). These soft materials attract widespread interest as potential drug and gene carriers. To this end, much empirical work has been done to characterize the effect of process and formulation parameters on their size distribution, swelling and stability. Despite these efforts, however, the mechanism of their formation remains poorly understood. This lack of understanding stems (at least in part) from the very rapid kinetics of chitosan/TPP gelation, which make the micro- and nanogel formation process very difficult to study. To address this, we used stopped-flow and dynamic light scattering (DLS), isothermal titration calorimetry (ITC) and transmission electron microscopy (TEM) to dissect the mechanism of ionotropic chitosan/TPP gelation in the presence of monovalent salt (NaCl). This revealed that the micro- and nanogel formation rates are extremely sensitive to TPP and NaCl concentrations, and scale approximately with $[TPP]^5$ and $[NaCl]^{-5}$. By slowing down their formation using high NaCl and low TPP concentrations, the temporal evolution in micro- and nanogel structure could be probed by DLS and TEM. The ionotropic gelation was shown to occur in two stages: first, the formation of small primary nanoparticles (average diameter ~ 30 nm); then, their secondary aggregation into larger micro- and nanogels. Furthermore, by slowing down the ionotropic gelation process, we were able to dramatically enhance the colloidal stability during chitosan/TPP mixing, and produce particles with more-uniform size distributions.

(158d) Understanding the Effects of Synthesis and Treatment On Network Stability and Surface Chemistry of Silica

Chris S. Ewing[1], J. Karl Johnson[2], Joseph J. McCarthy[1] and Götz Veser[3], (1)Chemical and Petroleum Engineering, University of Pittsburgh, Pittsburgh, PA, (2)Dept. of Chemical and Petroleum Engineering, University of Pittsburgh, Pittsburgh, PA, (3)Department of Chemical and Petroleum Engineering, University of Pittsburgh, Pittsburgh, PA

Porous silica nanoparticles have many unique features enabling a wide range of applications including drug delivery, imaging and catalysis. One of the most attractive features of silica for nano-applications is the ease of preparation and compatibility of the synthesis precursors with a variety of materials. It has recently been shown that freshly synthesized solid colloidal silica can be made porous by simple hydrothermal treatment resulting in etching, and thus porosity, with some control over final porosity based on the duration of treatment. The complex structure of silica, however, poses some challenges to the extension of this method to various systems. For example, experimental observations suggest that the mechanism of silica etching in water occurs through the breaking of siloxanes (Si-O-Si), and that the stability of these siloxanes is drastically altered by heat treatment in air or in vacuo. Furthermore, it has been shown that the surface of silica calcined in air at high temperatures could be fully rehydroxylated, without etching of the structure, by boiling in water. In order to fully exploit the potential of silica materials, a thorough understanding of both structural stability and surface chemistry is necessary.

The present study investigated the effects of synthesis, pretreatment and hydrothermal treatment on the porosity and surface chemistry of the silica nanoparticles with the overall goal of controlling these material properties. Silica nanoparticles were synthesized over a wide size range (6 – 500nm) and characterized via BET, TEM, and surface silanol quantification via probe molecule titration and UV-Vis spectroscopy. Calcination in air at 500 °C increased structural stability, making the silica resistant to etching in water as reflected in constant surface areas through multiple cycles of calcination and hydrothermal treatment. An interesting correlation between surface silanol concentration and apparent surface area, measured by both nitrogen and argon physisorption, was observed. Dehydroxylation of the sample resulted in a decrease of measured surface area, proportional to the reduction of surface silanol concentration. The effect of surface silanols on gas adsorption is currently being investigated via Monte Carlo computational studies.

<>

(158e) Ionic Liquid Mediated Assembly of CaCO3–Chitosan Nanoboxes

Mustafa Akbulut, Chemical Engineering, Texas A&M University, College Station, TX and Zhiping Luo, Texas A&M, College Stataion, TX

Nanoscale objects are the focus of a great deal of attention not only as critical components in the emergence of cellular life but also as small-scale materials with advanced functionalities and properties that can be isolated or assimilated into numerous applications. In particular, hollow nanostructures have recently received increasing attention owing to their emerging applications in catalysis, separation processes, drug delivery, and energy storage and conversion devices. In this presentation, an innovative approach using ionic liquid (1-butyl-3-methyl-imidazolium chloride) mediated auto-templating assembly of $CaCO_3$ and chitosan to produce well-defined hollow inorganic–organic nanoboxes and nanoframes will be described. With this method, by varying experimental conditions, size and shell-thicknesses of hollow nanostructures can be adjusted from 200 to 400 nm and 15 to 75 nm, respectively. Nanostructure shells had a degradation rate of 1-5 nm/hr in water at 25 °C. This feature is very beneficial for drug delivery applications, and can allow precise programming of the release kinetics of the therapeutic agent within nanoboxes by tuning the wall thickness of nanoboxes.

(158f) Synthesis of Nanoparticles Through the Use of Reversible Ionic Liquids

Steven R. Saunders[1,2], Amy L. Rohan[1,2], Emily C. Nixon[2,3], Elizabeth J. Biddinger[1,2], Pamela Pollet[2,3], Charles L. Liotta[1,2,3] and Charles A. Eckert[1,2,3], (1)School of Chemical and Biomolecular Engineering, Georgia Institute of Technology, Atlanta, GA, (2)Specialty Separations Center, Georgia Institute of Technology, Atlanta, GA, (3)School of Chemistry and Biochemistry, Georgia Institute of Technology, Atlanta, GA

Many methods of nanoparticle synthesis are cost, energy and time intensive and methods that have the potential to overcome these hurdles garner wide interest. A facile nanoparticle synthesis technique with the potential to produce zero waste is possible through the use of reversible ionic liquids (ionic liquids that are capable of switching their ionic nature on or off through the use of an external stimulus). The RevILs discussed herein are formed when CO_2 is introduced to a silylamine (e.g., 3-(aminopropyl) triethylsilane) to form an ammonium-carbamate ion pair. These RevILs can be returned to their original molecular (i.e., non-ionic) form when heated or sparged with an inert gas. Through adjustments in the structure of the silylamine and the degree of conversion of the molecular form to the ionic form the properties (e.g., viscosity, density, polarity, etc.) of the solvent medium can be drastically tuned. Nanoparticles with relatively narrow size distribution and an average size less than 10 nm can be produced using these RevILs in a process akin to a reverse-micelle synthesis with the ability to recycle the RevIL. Methods of nanoparticle synthesis and deposition onto a support will be discussed.

(158g) Copolymer-Coated Stabililization of Superparamagnetic Iron Oxide Nanoclusters At High Ionic Strength and High Temperature Designed by Combinatorial Materials Chemistry

Hitesh G. Bagaria[1], KiYoul Yoon[1], Bethany Neilson[2], Victoria Cheng[1], JaeHo Lee[1], Andrew Worthen[1], Christopher W. Bielawski[2], Steven Bryant[3], Chun Huh[3] and Keith P. Johnston[1], (1)Chemical Engineering, The University of Texas, Austin, TX, (2)Chemistry, The University of Texas, Austin, TX, (3)Petroleum and Geosystems Engineering, The University of Texas, Austin, TX

Electrosteric stabilization of copolymer-coated nanoparticles is uncommon for high monovalent salinity (upto 5 M NaCl), and rarely reported for concentrated divalent cations. However, high mono and divalent concentrations are routinely encountered in subsurface applications including electromagnetic imaging and oil recovery. Stabilization of nanoparticles at high calcium salinity, especially at high temperatures (up to 100 °C), is challenging because of the high counterion association of calcium with anionic polymer stabilizers. With the goal of identifying a suitable polymer for NP stabilization in such high calcium high temperature conditions, we conducted a combinatorial study by coating various compositions of acrylic/sulfonic acid copolymers on platform citrate-coated iron oxide (IO) nanoclusters. Hundreds of experiments were conducted by judicious choice of coating parameters - pH, concentration and type of electrolytes, temperature and polymer concentration. Based on these results, a select class of sulfonic acid copolymers was found to provide nanoparticle stability in API brine (8% NaCl + 2% $CaCl_2$ w/v) for at least 3 weeks at 90 °C. A fundamental explanation is provided to describe electrosteric stabilization at these unusually harsh conditions. This enhanced stability will enable the study of NPs for various applications including oil and gas exploration and production.

(173a) A Molecular View of Structure and Dynamics in Nanocomposite Glasses At Rest and Under Active Deformation

Juan J. de Pablo, Department of Chemical and Biological Engineering, University of Wisconsin-Madison, Madison, WI

There is considerable interest in understanding the behavior of polymeric glasses and polymer nanocomposites at multiple length scales, particularly under active deformation. Over the past two years, we have used a combination of Monte Carlo and molecular dynamics simulations to examine the local and global thermophysical properties of several coarse grain models of polymeric and nanocomposite glasses both at rest and under constant stress or constant strain rate deformation. Our simulations and recent experimental studies have demonstrated that an applied stress can lead to a decrease in the segmental relaxation times by several orders of magnitude. Many questions regarding the origins of these behaviors remain unanswered. Our simulations of an entangled polymer glass and a nanocomposite glass suggest that nanoparticles impart mechanical reinforcement onto the polymer, requiring larger stresses to achieve the same deformation. In both materials, the dynamics correlate well with the instantaneous strain rate, although there are situations, having to do with the amount of strain, where this correlation breaks down. Overall, simulated trends for the dynamics during creep agree very well with recent experiments pure polymer glasses. For large deformations, we also examine how nanoparticles of different size and shape influence the cavitation and fracture of polymeric glasses. We find that a distinct correlation exists between the local mechanical properties of the material at rest and the positions were voids are nucleated and grown into crazes, thereby providing a connection between structure and failure. Finally, results will also be presented for the structure, mechanical properties, and deformation of ultra-stable glasses prepared by vapor deposition. Such glasses exhibit unique thermophysical properties that differ considerably from those of ordinary glasses, and the origins and implications in the context of stability and aging will be discussed.

(173b) Localization Model of Relaxation in Glass-Forming Liquids

David S. Simmons[1], Marcus T. Cicerone[2] and Jack F. Douglas[2], (1)Deparment of Polymer Engineering, University of Akron, Akron, OH, (2)Polymers Division, Material Measurement Laboratory, National Institute of Standards and Technology, Gaithersburg, MD

By definition, solidifying glass-forming materials exhibit upon cooling a precipitous drop in the viscous and structural relaxation rates that govern deformation and flow. This solidification transition also reflects a particle localization transition, in which the material's constituent particles cease to freely explore space and become locally entrapped. A fundamental understanding of deformation and flow in glass-forming materials would thus seem to require an understanding of the relationship between relaxation and particle localization, and achieving such an understanding has accordingly been a longstanding goal of soft matter physics.

In this talk, we present a new model for structural relaxation in glass-forming materials that accurately relates the alpha structural relaxation time to an experimentally accessible measure of localization – the Debye-Waller factor – in a variety of simulated and experimental glass-formers. This "localization model" follows from both free volume theory and elastically activated transport models of relaxation. It suggests that the growth in structural relaxation time is indeed directly related to particle localization, albeit in a way that depends both upon the anisotropy and anharmonicity of particle localization and upon the onset temperature of glass formation. Finally, the emphasizes the existence of a well-defined congested fluid regime, intermediate between a solid and liquid, that initiates with a localization onset condition at a temperature well above the glass transition.

(173c) Viscoelastic Bulk Modulus of Polystyrene and Nanocomposites

Ran Tao, Texas Tech University, Lubbock, TX and **Sindee L. Simon**, Chemical Engineering, Texas Tech University, Lubbock, TX

The relationship between the viscoelastic bulk and shear moduli has been debated for the last fifty years since Leaderman's suggestion that the two responses arise from different molecular mechanisms. Using a custom-built pressurizable dilatometer, we have examined the pressure-volume-temperature (PVT) behavior and the time-dependent bulk modulus of several polymers, including polystyrene and star polystyrene in our laboratory. Our results indicate that although there is quantitative agreement between the slopes of the retardation spectra for the bulk and shear responses at short times, the long-time chain mechanisms available to the shear response are not available to the bulk response. This latter point, that chain modes are unavailable to the bulk response, was also recognized by Markovitz and by Berry and Plazek. In fact, much of the apparent controversy concerning the origin of the bulk and shear responses may arise from this detail. However, the short-time retardation spectra for the shear response is significantly higher than that for the bulk, a finding that is, as yet, unexplained. The PVT and pressure relaxation experiments are currently being extended to polystyrene nanocomposites containing silica nanoparticles.

(173d) A Simple Model for the Deformation-Induced Relaxation of Glassy Polymers

Ronald G. Larson, University of Michigan, Ann Arbor, MI, Suzanne Fielding, University of Durham, Durham, United Kingdom and Michael Cates, University of Edinburgh, Edinburgh, United Kingdom

Strain hardening has long been an observed feature of polymer glasses in extension; explanations to date have often been phenomenological. Recent experiments by Ediger and coworkers (Lee et al. *Science* 323, 231, 2009) have found strain hardening to be accompanied by a striking and unexplained dip in the segmental relaxation time, i.e., a non-monotonicity in the segmental relaxation time under constant load. Here we explain such behavior and develop a simple constitutive equation for polymer glasses by combining a minimal model flow-induced liquefaction of a glass with a description of the stress carried by strained polymers. Under constant load, liquefaction of segmental motion permits strong flow that creates polymer-borne stress. This slows the deformation enough for the segmental modes to re-vitrify, causing strain hardening. In this way, the observed non-monotonicity in the segmental relaxation modes is produced, both on loading and on unloading. We compare predictions of this theory with measurements of segmental relaxation time during loading by the Ediger group, and show semi-quantitative agreement of our theory with these data. We also address the elastic recoil and change in segmental relaxation time that occur during unloading of stress.

(173e) What Is the Origin of "Strain Hardening"?

Shi-Qing Wang[1], Shiwang Cheng[2] and Panpan Lin[2], (1)Department of Polymer Science, University of Akron, Akron, OH, (2)University of Akron, Akron, OH

In uniaxial compress, polymer glasses show a sequence of events as a function of strain in the order of elastic deformation, yielding, plastic flow and "strain hardening". Plastic flow corresponds to the regime where the stress is observed to drop and level off. The so called strain hardening occurs when further straining beyond the plastic flow regime produces rising stress. Since the stress level and the corresponding "modulus" G_R is 100 times the elastic plateau modulus measured in the melt state, the origin of stresses has become a subject of intensive investigation. Modeling this strain hardening behavior using the formula of the classical rubber elasticity theory is clearly fruitless and misleading as remarked by Kramer [1]. But in polymer physics only two types of force are known to be prevalent: viscous forces as seen in stead shear flow of entangled polymeric liquids and elastic forces of entropic origin as seen in vulcanized rubbers. Can we explain the strain hardening in terms of a combined of these forces?

[1] Kramer, E. J. J Polym Sci Part B: Polym Phys 2005, 43, 3369–3371.

(173f) Does Graphene Increase Tg of Polymer Nanocomposites?

Ken-Hsuan Liao, Chemical Engineering and Materials Science, University of Minnesota in Twin Cities, Minneapolis, MN, Steven Maslo, Chemical Engineering, University of Minnesota, Minneapolis, MN, Shingo Kobayashi, Chemical and Materials Engineering, University of Minnesota, Minneapolis, MN, Hyunwoo Kim, Dow Chemical Company, Midland, MI, Ahmed Abdala, Chemical Engineering Program, the Petroleum Institute, Abu Dhabi, United Arab Emirates and Christopher W. Macosko, Department of Chemical Engineering and Materials Science, University of Minnesota, Minneapolis, MN

Does Graphene Increase T_g of Polymer Nanocomposites?

Ken-Hsuan Liao[1], Steven Maslo[1], Shingo Kobayashi[1], Hyunwoo Kim[2], Ahmed Abdala[3], Chris Macosko[1]

[1]Chemical Engineering & Materials Sciences, University of Minnesota, Minneapolis MN; [2]The Dow Chemical Co., Midland MI; [3]Chemical Engineering, The Petroleum Institute, Abu Dhabi UAE

Graphene recently has attracted large interest because of its 2-D structure, outstanding electrical and mechanical properties. Due to the high surface area and aspect ratio, graphene is capable to improve polymers' mechanical properties, gas impermeability and electrical conductivities. However, there is considerable disagreement as to whether the glass transition temperature (T_g) changes with incorporation of graphene in polymers. We review the literature and find that solvent or melt blending of graphene with typical thermoplastics does not result in a change in polymer T_g. However, aqueous-soluble polymer which can form strong hydrogen bonds with graphene derived from graphene oxide (GO) can show a significant T_g increase. Also blending polymers with chemically modified graphene or *in situ* polymerization of monomers in the presence of GO derived graphene, are capable to provide strong covalent bonds between matrix polymers and fillers, which cause T_g increase. We illustrate this with new experiments on PMMA.

(173g) Dynamics of Glass-Forming Liquids: Is the Finite Temperature Divergence Real?

Jing Zhao and Gregory B. McKenna, Chemical Engineering, Texas Tech University, Lubbock, TX

There is significant interest in the apparent divergence of the dynamics of glass-forming liquids at temperatures above absolute zero. One considered signature of the complex liquid is the so-called Vogel-Fulcher-Tammann (VFT) or the Williams, Landel, Ferry (WLF) super-Arrhenius behavior of the dynamics or the viscosities. Yet, because it is extremely difficult to make measurements in the equilibrium state far below the glass transition temperature range, the actual continuation of this apparent VFT or WLF divergence is little tested. Here we examine the problem by making measurements on a poly(vinyl acetate) (PVAc) polymer in the vicinity of the glass transition using both mechanical measurements and dielectric measurements. By aging the samples for long times we are able to establish a "plateau" in the aging behavior that we associate with the nominal equilibrium response. Three findings of considerable interest are made. First, the equilibrium dynamics for the dielectric response as much as 16 K below the nominal T_g are found to deviate from the VFT extrapolation and tend towards an Arrhenius behavior, albeit one with very high activation energy. On the other hand, the mechanical response only deviates mildly from the VFT extrapolation and description of these results as consistent with an Arrhenius type of temperature dependence would be problematic. Finally, the dielectric and mechanical responses also age into equilibrium very differently with the dielectric response coming to a stationary response as much as an order of magnitude faster than the mechanical response. The significance of such non-divergence of time scales as temperature is lowered becomes important in the evaluation of models of the glass transition itself. Models in which divergence is predicted are precluded by the present dielectric findings, viz., simple free volume models and the configurational entropy based models from the middle of the 20th century as well as more modern models such as that of Lubchenko and Wolynes [1]. On the other hand, modern treatments such as that from Elmatad et al [2], from Mauro et al [3] or from Dyre [4] should be further investigated. Furthermore, the present results also suggest the need for more concerted experiments in which very long time equilibrium is achieved in order to establish the detailed nature of the deviations from the VFT/WLF-types of response.

[1] V. Lubchenko, and P. Wolynes, Annual Review of Physical Chemistry, 58, 235-266, 2007.

[2] Y. Elmatad, D. Chandler, and J. Garrahan, J. Phys. Chem. B, 113, 5563–5567, 2009.

[3] J. Mauro, Y. Yue, A. Ellison, P. Gupta, and D. Allan, PNAS, 19780–19784, 2009.

[4] J. Dyre, Reviews of Modern Physics, 78, 953-972, 2006.

(179a) Elastic Induced Colloidal Interactions in Liquid Crystals

Marcello Cavallaro Jr.[1], Mohamed Amine Gharbi[2], Daniel Beller[3], Lu Yao[1], Randall Kamien[3], Shu Yang[3] and Kathleen J. Stebe[1], (1)Chemical & Biomolecular Engineering, University of Pennsylvania, Philadelphia, PA, (2)University of Pennsylvania, Philadelphia, PA, (3)University of Pennsylvania

In this work, we exploit well defined boundary conditions to control the formation of defects in nematic liquid crystal (NLC) 5CB. When supported between micropost arrays, disclination lines form and bear the signature of the micropost shape owing to the elasticity of the medium. The existence of these disclination lines can be tuned due to the thermotropic nature of the material simply by varying temperature thereby providing an element of control not present in typical isotropic fluids. When colloids are introduced to a nematic liquid crystal, elastic distortions arise that have an associated Frank free energy cost. This excess energy is reduced by interacting with the micropost induced disclination lines and subsequent colloid-micropost attractions arise. We consider both isotropic particles (spheres) and more complex anisotropic particles. Further, we observe these effects 5CB-air interfaces as well as in bulk 5CB. This work is experimental and supplemented with analysis to support the observed behavior.

(179b) Structure and Chemistry Dependence of Pollen Adhesion to Surfaces

Ismael J. Gomez, Haisheng Lin and J. Carson Meredith, Chemical and Biomolecular Engineering, Georgia Institute of Technology, Atlanta, GA

Nature provides remarkable examples of evolutionary-optimized microscale particles with structures and/or chemistries tailored for effective adhesion to a variety of surfaces. Pollen displays a remarkable range of ornamentations consisting of combinations of spines and grooves composed of sporopollenin, a crosslinked organic material rich in fatty acids. In addition, some species of pollen have a viscous coating known as pollenkitt, which appears to be related to dispersion mechanisms, e.g., insect versus wind-pollenation. It is not clear, however, how the structural features and pollenkitt contribute quantitatively to pollen adhesion to surfaces in air. This question has significance not only in plant biology, but also in epidemiology of asthma and allergies, and utilization of pollen as biotemplates for advanced materials. We report here the measurement of the effect of pollenkitt and geometry on adhesion of pollen from 5 plant species on 5 model surfaces. For pollen grains with intact pollenkitt, the adhesion forces varied with the hydrophilicity of the counter surface, and both capillary and van der Waals forces contributed to adhesion. In contrast, the interaction of pollen grains without pollenkitt was independent of counter surface type and was due primarily to van der Waals forces. Pollens exhibiting a nano-sized reticulate (grooved) structure (olive, poplar) had lower adhesion than those with echinate (spiked) structures (ragweed, dandelion, sunflower). In addition, increasing spike size and aspect ratio led to an increase in adhesion (from low to high: ragweed, dandelion, sunflower), which could be explained by increased contact area of the larger spikes.

(179c) Engineering Inorganic Nanoparticle and Organic Nanorod Hybrid Materials

Li Li, Sunxi Wang and Guangzhao Mao, Chemical Engineering and Materials Science, Wayne State University, Detroit, MI

Molecular self-assembly is an important bottom-up approach in nanotechnology of engineered materials in nanoscale. Incorporating both organic and inorganic components can combine and sometimes enhance the functions belonging to different material groups. Here we propose a universal strategy to fabricate inorganic/organic thin film with particle/rod structure based on seed-mediated nucleation. The inorganic nanoparticles used are thiolate-protected CdSe, CdS, and Au nanoparticles as well as electrodeposited Au nanoparticles. The organic crystalline materials are *n*-carboxylic acids, tetrathiafulvalene charge transfer salt, and Krogmann salt. Solution-based spin coating and electrochemical methods are both successful in making the particle/rod structure. We use cyclic voltammetry (CV), atomic force microscopy (AFM), and scanning electron microscopy (SEM) to characterize the thin film structure. Inorganic nanoparticles serve as nucleating seeds to initiate the nucleation of organic crystals and confine their growth morphology. The resulting particle/rod structure is crucial in understanding the underling mechanism such as seed-mediated nucleation and engineering of nanomaterials.

(179d) Colloidal Particle Halos in Ternary Particle System

Shunxi Ji, Chemical Engineering, Virginia Tech, Blacksburg, VA and John Y. Walz, Chemial Engineering, Virginia Tech, Blacksburg, VA

Colloidal particle halos are a concept in which small particles are trapped in a shallow energy well close to the surface of larger particles. The object of this work is to experimentally determine if halos of submicron particles around larger, micron-sized particles could be produced by the presence of nanoparticles. Specifically, the nanoparticles should produce attractive depletion forces between the micro- and submicro-particles of sufficient magnitude to create a halo of the submicroparticles. Colloidal probe atomic force microscopy (CP-AFM) was used to measure the force profile between a single silica microparticle and silica plate in a solution containing both submicroparticles and nanoparticles, also of silica. The experiments were performed at pH 4.5, such that all surfaces were negatively charged. The measured force profiles were compared to those obtained by using the solution of submicroparticles only or nanoparticles only, and also with a force-balance model that predicted the equilibrium force profile.

(179e) Separation of Particles Using Depletion Forces in a Packed Column

Francisco Guzman, Chemical Engineering, Virginia Tech, Blacksburg, VA and John Y. Walz, Chemial Engineering, Virginia Tech, Blacksburg, VA

Attractive depletion forces were used to enhance the separation of micron-sized colloidal particles flowing through a packed bed. The dispersion consisted of a binary mixture of negatively charged polystyrene sulfonate (PS) particles (diameter of 0.82 and 1.5 micron) in a solution of negatively-charged, 7 nm diameter silica nanoparticles, and the column packing was 0.5 mm diameter silica beads. The experiments consisted of a short-pulse injection of the PS particles into the nanoparticle dispersion at the inlet of the packed column. The nanoparticles created an attractive depletion force between PS particles and the column packing which reduced the rate at which the microparticles passed through the column. Because the magnitude of the depletion force scales with the particle size, the impact of the force on the rate of transport varied with size. The degree of separation was quantified by analyzing particle breakthrough curves and measuring the concentration of particles in the collected effluent. Control experiments without any nanoparticles were done using a carrier fluid with an equivalent ionic strength and pH as the solution of nanoparticles. These experiments illustrate a novel application of depletion forces for particle separations.

(179f) Experimental and Theoretical Investigation of the Depletion Force Generated by CTAB Micelles

Greg K. James, Chemical Engineering, Virginia Tech, Blacksburg, VA and John Y. Walz, Chemial Engineering, Virginia Tech, Blacksburg, VA

The technique of colloidal probe atomic force microscopy was used to measure the equilibrium depletion force between a relatively large (30 μm diameter) spherical silica particle and a flat silica plate in solutions of cetyltrimethylammonium bromide (CTAB) surfactant. Previous work has indicated that the observed depletion force from CTAB micelles is smaller than that predicted by hard-sphere models. The objective of this work was to establish if CTAB micelles could be modeled using the well established equations for charged, hard-sphere depletants, or if corrections to these approaches are needed to account for the unique physical properties of micelles. Experiments were performed in solutions of CTAB below and sufficiently above the critical micelle concentration that not only equilibrium depletion forces but also long-range structural forces could be discerned. The force profiles were then compared to models that have been found to work well in systems using charged nanospheres as the depletants.

(179g) Stabilization of Silica Nanofluids At Neutral pH and Evaluate the Effect of Coating Concentration

Saba Lotfizadeh[1], Dan Reilly[2], Hassan A. AlJama[3] and Themis Matsoukas[1], (1)Department of Chemical Engineering, The Pennsylvania State University, University Park, PA, (2)Pennsylvania State University, University park, PA, (3)Pennsylvania State university, University park, PA

The presence of nano-sized particles in thermal fluid is known to enhance the thermal properties of the suspension, such as thermal conductivity and thermal diffusivity, relative to the base fluid. Significant enhancements have been reported even at low volume fractions of nanoparticles. This enhancement is important becasue it would lead to decrease the size and capital costs of heat transfer equipement and more efficient use of energy. Commercial nanofluids are now available in the market but a practical problem that limits their utilitiy is poor colloidal stability, which results in aggregation, increased viscosity precipitation, and loss of thermal enhacement. Our approach is to attach covaelently a self-assembled monolayer (SAM) of suitable chains onto the particle surface to (a) passivate the surface of the particles and (b) to improve stability via ionization of surface groups on the SAMs themeselves. We have tested and show results for the following aminosilanes: 3-aminopropyltriethoxysilane (APES), aminopropyldimethylethoxysilane (APMS), and N-[3-(trimethoxysilyl)propyl]ethylenediamine (TMPE) were used to modify the surface of silica colloidal nanoparticles. We provide colloidal characterization for each SAM, report optimum coating conditions, and evaluate the effect of each SAM concentration on the stability of the nanofluid at different pH values. Since TMPE shows to be the best candidate for the coating process, thermal stability test was also done for it.

(179h) Effects of Interfacial Characteristics On Impact Sensitivity of PBX

Jung Seoh Shim, High Energy, Agency for Defense Development, Daejeon, South Korea, Jung Soo Park, high energy, Agency for Defense Development, Deajeon, South Korea and Hyoun Soo Kim, high energy , Agency for Defense Development, Deajeon, South Korea

A Plastic Bonded Explosive (PBX's) is mainly composed of nitramine explosive and polymer binder. A great number of serious applications of PBX require the good adhesion between nitramine crystals and the binder, which depends on the surface characteristics of a filler and binder. In the pursuit of the better design to achieve the enhanced adhesion, profound knowledge of the surface and interfacial characteristics of explosive and binder should be exploited. In this work, the influences of physicochemical properties between RDX and binders such as interfacial tension(γ_{SL}), latent heat(Q_m), and density(ρ) on impact sensitivity of PBX were investigated. The major contribution factor to impact sensitiveness of PBX was the interfacial tension, compact to other surface properties.

References

1. H. Hengjian, D. Haishan, Z. Ming, C. Kemei and Z. Yufen, "A Quasi-Theoretic Method for the Selection of Desensitizers in the Designs of Explosive Formulations", 2003 Int'l Autumn Seminar on Propellants, Explosive and Pyrotechnics, Vol. V. pp.54(2003)

2. O. Planinsek, R. Pisek, A Trojak, S. Srcic, "The utilization of surface free-energy parameters for the selection of a suitable binder in fluidized bed granulation", Int'l Journal of pharmaceutics, Vol.207, Issues 1-2, pp(77-87(2000)

(179i) The Nonlinear Viscoelastic Characteristics of Polymer Binder On Pbxs

Hyoun-Soo Kim, High Explosive Team, Agency for Defense Development, Daejon, South Korea, Jung Seob Shim, High Energy, Agency for Defense Development, Daejeon, South Korea and Jung Min Lee, high energyt, Agency for Defense Development, Daejeon, South Korea

Plastic bonded explosive(PBX) is mainly composed of the nitramine explosive and the polymer binder. PBX is characterized by high velocity and pressure of detonation, low vulnerability and good thermal stability. Many important applications of PBX require the good mechanical properties and the good adhesion between explosives and the binder.

Time dependent compressive properities of composite exploseves have been determined by stress-strain curves obtained at different strain rates and temperatures. Both the small deformation and ultimate property can be representes by master curves, which are functions only of the experimental time scale, along with a temperature function. The mechanical properites of adhesion between the binder and explosive particle has been investigated by determining RDS. The interface was observed throught SEM.

It is supposed that this phenomenon is derived from a strong interaction force between binder molecules and a strong adhesion force between the binder and explosive particles. This results will predict the interaction between explosive powder and polymer binder from viscoelastic properties.

(179j) Corrosion Modeling of Magnesium Alloys Using Internal State Variable

Chris Walton, Holly J. Martin, M.F. Horstemeyer and P.T. Wang, Center for Advanced Vehicular Systems, Mississippi State University, Mississippi State, MS

Magnesium, a lightweight metallic alloy, is currently being investigated for use within the automotive industry because of its low density, excellent castability, easy machinability, and high mechanical stiffness [1,2]. When exposed to saltwater, magnesium has a high corrosion rate, relegating its current use to unexposed locations within the vehicle [3-5]. By altering the microstructure with the addition of various elements, including aluminum, zinc, manganese, and rare earth elements, the corrosion rate of magnesium can be reduced [6-8]. In addition to the alteration of the microstructure, the method of magnesium formation can also affect the corrosion rate. The presence of an as-cast skin, which consists of very small grains formed during alloy cooling, improves the corrosion resistance of the magnesium alloy ten-fold [7,9]. While casting is commonly used to form parts of a vehicle, such as control arms or engine cradles, extruded metal is also used. Therefore, in addition to understanding the additive effect of elements to corrosion resistance, the method of magnesium alloy formation must be included. A detailed understanding of magnesium's response to corrosion condition, through the modeling of the pit nucleation and growth, is needed.

In the Center for Advanced Vehicular Systems (CAVS) at Mississippi State University, a complete corrosion model based upon internal variable theory is introduced that captures the effects of general, pitting, and intergranular corrosion in a multiplicative decomposition manner. General corrosion is signified by volume loss of the material. Pitting corrosion is used as localized corrosion for the sake of our theoretical model's paradigm and implementation. Lastly, the definition of intergranular corrosion in our context is localized corrosion that occurs at the grain boundaries caused by precipitates and segregation leading to the formation of microgalvanic cells. The total corrosion damage is the addition of the general, pitting, and intergranular corrosion. A series of magnesium alloys are used to demonstrate the relative importance of the different mechanisms developed for this model.

[1] A. Jambor, M. Beyer, Mater. Des. 18 (1997) 203-209.

[2] F.H. Froes, D. Eliezer, E.L. Aghion, JOM. 50 (1998) 30-34.

[3] BA Shaw, Corrosion Resistance of Magnesium Alloys, in: L.J. Korb, ASM (Eds.), ASM Handbook, Vol. 13A: Corrosion, Ninth Ed., ASM International Handbook Committee, Metals Park, 2003, 692.

[4] G.L. Makar, J. Kruger, Int. Mater. Rev. 38 (3) (1993) 138-153.

[5] G. Song, A. Atrens, Adv. Eng. Mater. 5 (12) (2003) 837-858.

[6] O. Lunder, T.K. Aune, K. Nisancioglu, NACE Corros. 43 (1987) 291-295.

[7] G. Song, A. Atrens, M. Dargusch, Corros. Sci. 41 (1999) 249-273.

[8] R. Ambat, N.N. Aung, W. Zhou, Corros. Sci. 42 (2000) 1433-1455.

[9] G. Song, Adv. Eng. Mater., 7 (2005) 563-586.

(179k) Engineering of Pickering Emulsion Templated Colloidosomes and Fundamentals of Emulsification with Nano-Particles

Hongzhi Wang[1], Virendra Singh[2] and Sven Behrens*[1], (1)Chemical & Biomolecular Engineering, Georgia Institute of Technology, Atlanta, GA, (2)School of Chemical & Biomolecular Engineering, Georgia Institute of Technology, Atlanta, GA

Colloidosomes are semipermeable microcapsules with external shell formed by colloidal particles, providing promising technique for encapsulations with controllable permeability, mechanical properties and bio-compatibility. Our work explores double Pickering emulsion, also known as particle stabilized emulsions, as precursors to synthesize composite colloidosomes for triggered and sustainable release control with envisioned applications in flexographic printing and biomedical aspects. To generate such colloidosomes, a prerequisite is the controlled formation and stabilization of Pickering emulsions. Unfortunately, our current fundamental understanding of Pickering emulsions is still far from satisfactory and often fails to predict reliably the stability and the type of Pickering emulsions. We therefore focus on understanding the fundamentals of Pickering emulsions by researching emulsifications with polymeric nano-particles. Our study suggests two widely neglected electrostatic effects in Pickering emulsions, one is the effect of image charge repulsion on the emulsion formation, and the second is the effect of dipole-interaction on the preferred emulsion type. The obtained new insights will benefit not only our design of colloidosomes but realms relevant to Pickering emulsions

(179l) Rational Surfactant Design for Hydrofluoroolefins (HFO-1234yf and HFO-1234ze(E)) , the Next Generation Refrigerant/Propellants: A Molecular Dynamics Approach

Lin Yang, Department of Chemical Engineering and Materials Science, Wayne State University, Detroit, MI, Sandro R. P. da Rocha, Wayne State University, Detroit, MI and Cláudio Dariva, Núcleo de Estudos em Sistemas Coloidas – NUESC, Universidade Tiradentes, Aracaju, Brazil

Hydrofluoroalkane (HFA) refrigerants/propellants have a relatively high global warming potential (GWP), at 1430 for HFA-134a and 3200 for HFA-227. With the advent of a series of global warming-related regulations, new alternatives to the traditionally used HFAs are thus being sought. Hydrofluoroolefins (HFOs), with their zero ozone depleting potential, and low GWP - 6 for HFO-1234ze and 4 for HFO-1234yf, are promising next generation working fluids in a wide range of applications, from foams, to medical aerosols, to refrigerants. The design of HFO-philic species and surface active agents for HFO-based systems is thus of great relevance.

In this work we use atomistic molecular dynamics simulations to probe the microstructure of HFO interfaces, including HFO-air (vacuum) and HFO-water, and to study the activity of surfactants at those interfaces. We investigate a series of non-ionic surfactants, with alkyl- [representative of the Brij surfactant class ($CH_3(CH_2)_n(OCH_2CH_2)_nOH$)],and propylene oxide-based HFO-philic moieties. The surface tensions for both HFOs are seen to match the experimental results within a wide range of temperature (270-300 K). The interfacial tensions for HFO-1234ze(E)-water and HFO-1234yf-water are 40.6±1.0 and 44.4±1.0 respectively. With a surfactant coverage of 112Å^2/surfactant the interfacial tension undergo a drop of 12.0 and 11.3 respectively for HFO-1234ze(E)-water and HFO-1234yf-water. No experimental results are available yet for comparison. The surfactant results will be discussed at varying surface coverage, and a detailed analysis of the structure of the surfactant-modified interfaces will be provided.

(179m) Effect of Colloidal Surface Roughness On Colloid Force Measurements

Yi Zou and Guangzhao Mao, Chemical Engineering and Materials Science, Wayne State University, Detroit, MI

Interactions between colloidal particles greatly affect dispersions. While several measurement techniques such as rheology and light scattering offer indirect methods for probing the dispersion interaction, AFM colloidal force measurements provide the data about surfactant-colloid, and colloid-colloid interactions. In this study, heat treatment process is optimized to achieve minimum surface roughness, which is less than 1nm. This enables us to focus on physiochemical property and illustrated how geometry affects colloids dispersion. Interactions between PS colloids are measured in absence of dispersant and in presence of three different dispersants, *i.e.* pentaethylene glycol monododecyl ether (C12E5), poly(ethylene oxide)$_{129}$–poly(propylene oxide)$_{56}$–poly(ethylene oxide)$_{129}$ (Pluronic® F108), and one acrylic anionic surfactant Joncryl® 60. It is found that in absence of surfactant, adhesion increases significantly with increase of surface roughness. The presence of surfactants greatly reduces colloidal adhesion, however, colloidal surface roughness impacts colloidal interactions and therefore the surfactant performance as pigment dispersants. For example, at low concentration (0.2%), only Pluronic F108 could completely eliminate adhesion between rough PS colloids while all three dispersant show little difference for smooth PS colloids. One of the main reasons is that adsorption layer thickness of F108 is larger than others. The significance of this study directly illustrates the relationship between surfactant property – adsorption physics – colloids interaction. This also will be helpful in discovering new materials for commercial dispersant.

(179n) Polystyrene Nanoparticle Interactions with Albumin: Changes in Particle Characteristics and Interparticle Forces

Amir M. Farnoud, Chemical and Biochemical Engineering, University of Iowa, Iowa City, IA and Jennifer Fiegel, Pharmaceutics and Translational Theraputics; Chemical and Biochemical Engineering, The University of Iowa, Iowa City, IA

The biological activity of nanoparticles for pulmonary drug delivery applications is dependent on their interactions with lung proteins. Such interactions can change nanoparticle physicochemical properties and ultimately affect their in-vivo fate. However, studies of such interactions are complicated due to the wide range of proteins in the lungs and potential competition behavior in protein adsorption onto nanoparticle surfaces. Thus, mechanistic information from simpler systems is needed before a more complex system with multiple proteins can be tackled. The current study focuses on interactions of albumin, the most common protein in the conducting airways with negatively charged carboxyl modified polystyrene nanoparticles (200 nm).

Nanoparticle interactions with proteins were studied by monitoring size and charge of nanoparticles in suspension after incubation with albumin for different time periods. A significant reduction in particle charge along with a slight increase in particle size was observed after incubation. The changes in particle properties were time dependent with the maximum size and minimum charge occurring at the longest interaction time (2 hours). The DLVO model was used to predict the interparticle forces between nanoparticles in suspension. The model was modified to predict the forces after adsorption of one layer of albumin on nanoparticle surfaces. Based on the model estimations, a significant drop in aggregation energy barrier will occur once the nanoparticles are completely covered with a layer of albumin. These results suggest that nanoparticle surfaces are partially covered by albumin upon incubation.

(179o) Spontaneous Gathering and Unraveling of DNA On Patterned Lipid Bilayer

Chin-An Lee, Chemical Engineering, National Taiwan University, Taipei, Taiwan and Chih-Chen Hsieh, Chemical Engineering Department, National Taiwan University, Taipei, Taiwan

We study the dynamic and conformational behavior of DNA absorbed on cationic supported lipid bilayer on a patterned glass surface. The pattern consists of periodic shallow trenches with the width of 10 microns and the depth of 300 nm. The trenches are separated by 10 microns. When DNA adsorbed on the lipid bilayer in a trench, it first adopts a 2-dimensional self-avoiding conformation and diffuses on the bilayer until some of its segments reach the trench wall. DNA then spontaneously gathers and unravels along the root of the trench wall within few seconds. For lambda DNA, it easily extends to more than 50% in our typical setup. This phenomenon is controlled by the competition between the electrostatic interaction and conformational entropy of DNA. We examine the influence of DNA molecular weight and the charge density of lipid bilayer on the DNA behavior, and develop a simple model to describe our observation.

(179p) Effects of Radioactivity On Surface Interaction Forces and Aggregation of Aerosols

Costas Tsouris[1], Joanna McFarlane[1], Yong-ha Kim[2], Hyojin Kweon[2], Sotira Yiacoumi[2] and Ida Lee[3], (1)Energy and Transportation Science Division, Oak Ridge National Laboratory, Oak Ridge, TN, (2)School of Civil and Environmental Engineering, Georgia Institute of Technology, Atlanta, GA, (3)Department of Electrical Engineering and Computer Science, The University of Tennessee at Knoxville, Knoxville, TN

Explosion of radiological devices is expected to generate radioactive aerosols, which contribute to the transport of radioactivity in the environment. Self-charging of radioactive aerosols is an important characteristic since it affects growth and transport of aerosols in a radioactive plume. Current tests and models of radioactive plumes do not take into consideration the different behavior of radioactive particles, which can become electrostatically charged due to the decay process. The main objective of this study is to investigate effects of self-charging of radioactive aerosols on surface charge, surface forces, and the aggregation rate. Scanning surface potential microscopy (SSPM) is used to measure the surface potential of radioactive surfaces as a function of the decay rate. Atomic force microscopy (AFM) is employed to quantify the adhesion force and force-distance profile between a particle and a planar substrate and to compare measurements with and without the involvement of radioactivity. The mean charge number of radioactive aerosols caused by self-charging is estimated by an approximate model considering particle size and decay rate. Self-charging effects on the collision efficiency of particles in the Brownian regime is calculated as a function of van der Waals, electrostatic, and viscous forces. Results can be incorporated in transport models to accurately predict aerosol transport after nuclear accidents or explosions of radiological devices.

(179q) Removal of Model Fats From Cotton Fabrics with Fabric Abrasion Using a Tribometer

Ruben Mercade-Prieto, Chemical Engineering, University of Birmingham, Birmingham, United Kingdom, Serafim Bakalis, Department of Chemical Engineering, University of Birmingham, Birmingham, United Kingdom and Carlos Amador, Procter and Gamble Technical Centre, Newcastle upon Tyne , United Kingdom

The detergency of fabrics with oily stains using surfactants has been extensively researched in the past. In recent years, significant efforts have been made to reduce the large consumption of energy during laundry by lowering the washing temperature. With time, the washing temperature in washing machines will approach that commonly used during hand washing: that of the tap water. These energy saving measures bring an old challenge: at 'cold' temperatures (eg. <30 °C) many fats solidify, limiting the detergency of surfactants. This and other problems, persistent in the old-fashioned hand washing in 'cold' waters used everywhere around the world, are being tackled by the industry using different strategies, for example by developing new enzymes efficient at 'cold' temperatures.

The present study considers the non-chemical action involved during laundry cleaning, the mechanical effects that occurs with the hydrodynamic flow, during flexing and when abrading the fabrics. Mechanical action, abrasion in particular, has been found to be efficient in the removal of water insoluble stains, eg. carbon black with mineral oil (Lee *et al. Fibers and Polymers* (2008) 9:101). The understanding of this common sense observation is very scarce, particularly due to the lack of systematic and controlled experiments.

Here, it is reported the first methodology to systematically study the effect of abrasion between two fabrics on the removal of fats. The fats selected are those relevant for 'cold' washing: have melting points around room temperature. Well controlled abrasion conditions were possible by using a tribometer (mini Traction Machine, PCS Instruments, UK), using a rotating silicone disk covered with a soiled fabric and a static silicone ball covered with non-soiled fabric (inset Figure 1). The silicone disk was rotated at 10 mm/s for 12 min (following the washing time suggested in the ASTM D4265-98), resulting in about 57 rotations, while the load of the static ball was typically kept at 1 N. The contact diameter between the ball and the disc is about 2 mm, resulting in an estimated contact pressure of ~0.3 MPa. Soiled samples were prepared using commercial white cotton fabrics fully immersed in the given fat with an optic lipophilic dye (Sudan Black B) in the liquid form (ie. at a temperature higher than the melting temperature) for 5 min. Excess fat on the fabric was absorbed with tissue paper under pressure, and the samples where left overnight in a fridge or at room temperature. The amount of fats in the samples was quantified by solubilising the fats in propanol followed by UV quantification of the lipophilic dye. About 40-60% of the weight of the stained fabrics is fat.

Tribology experiments performed at different temperatures (10-45°C) while using dionized water as the 'cleaning' solution shows the remarkable influence of the melting temperature in pure fats stains. In fats that easily crystallize, such as octadecane (mp ~28°C), little fat is removed using water and abrasion does not improve cleaning. Above the melting temperature, the quantity of octadecane in the abraded and non-abraded parts of the fabric is reduced significantly (about to half the initial fat amount), but in the abraded part more fats is removed. Similar behaviour was obtained using hexadecane (mp ~16°C) and heptadecane (mp ~21°C), although abrasion was found more effective in hexadecane bellow the melting temperature provably due to incomplete crystallization on the fabric. Hence, fabric abrasion below the melting point in these pure fats has a limited effect, regardless of the actual temperature. Remarkably, although abrasion significantly reduces the fat left in the abraded parts of the fabric above the melting point, this improvement is found constant also with temperature. Hence, temperatures higher than the melting temperature of the fats do not improve the removal of the abraded and non-abraded fabrics in water.

The use of a chemically different model fat, undecanoic acid (mp ~26°C), shows a similar temperature dependence than the three previous alkanes, however abrasion is much less effective. Only at temperatures around the melting point does abrasion significantly improve the removal of the stains.

The abrasion effect observed in pure fats differs significantly with that found in a complex fat mixture that is lard, representing a more real stain. In lard, abrasion was found to have a significant effect at all temperatures, except at highest temperature tested (~45°C) when not much lard is left in the fabric (Figure 1), higher than the melting point.

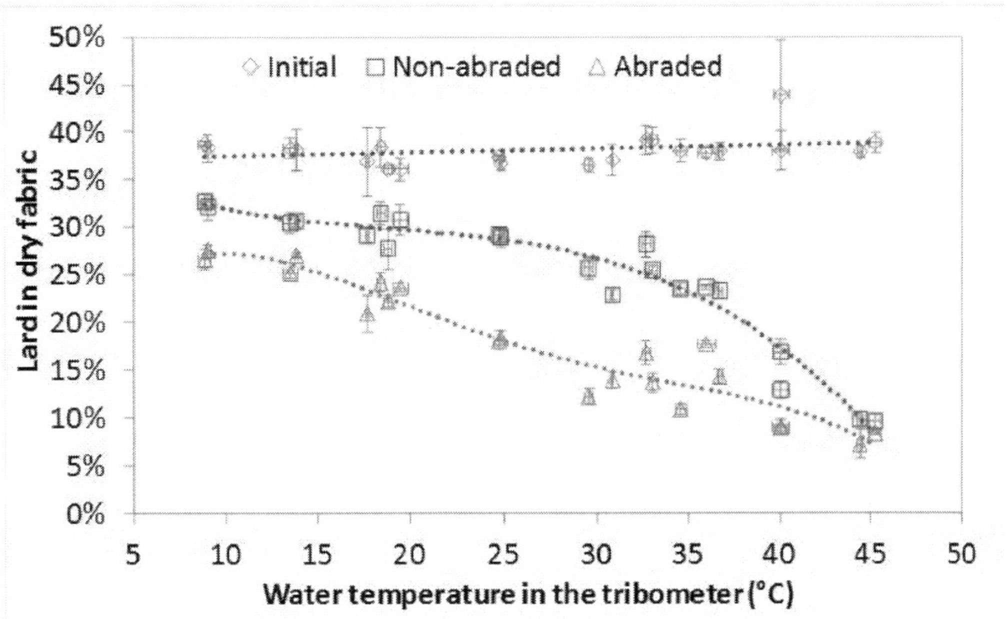

Figure 1. Effect of abrasion in cotton fabrics stained with lard kept at ~22°C for 1 day, using water as 'cleaning' solution. Abrasion was performed in a tribometer at 10 mm/s with a load of 1 N for 12 min, following a previous soaking step of 12 min at the desired temperature. 'Initial' data points represent the amount of lard in the dry stained fabric. Inset photograph shows the location of the 'Non-abraded' and the 'Abraded' sampling areas. Vertical error bars are the standard deviation of 3 sampling areas; horizontal error bars represent the standard deviation of the water temperature in the tribometer cell. Dotted lines are a guide to the eye.

Tribology experiments on lard stained cotton fabrics using 0.5 g/L linear alkylbenzene sulfonate (LAS) solutions at different temperatures shows that this simple surfactant solution had a negglegible effect on removal of lard compared to water. In order to simulate a commercial laundry detergent, the standard detergent IEC 60456 A* was used at a concentration of 4 g/L. The detergent solution provides marginal improvements at temperatures <26°C, but lard removal is greatly enhanced at higher temperatures. Hence, using abrasion similar low remaining lard concentrations can be achieved in water or in detergent solutions but at much lower temperatures. Or in other words, if the water temperature is reduced for energy conservation as discussed previously, fabric abrasion can offset the lower washing efficiency.

In pure fats the efficiency of the detergent solution was more salient at temperatures above the melting points, where the amount of fat remaining in the fabric in abrading and non-abrading conditions was reduced compared to water only (except for hexadecane where the difference was not significant). Abrasion still significantly reduces the alkanes left at temperatures higher than the melting point, but it had a negligible effect for undecanoic acid.

The use of a tribometer allows studying parameters relevant in abrasion in a controlled manner. Experiments using lard stains show that the load had no effect of the removal between the 0.3-1 N tested. Kinetic experiments in water show two distinct behaviours. On one hand in pure fats (below or above the melting point) as well as lard above the melting point (44°C) the effect of abrasion changes little between 1 min and 1 h. On the other hand, for lard below 40°C the amount of lard in the abraded areas diminishes with time, while in the non-abraded remains constant.

The effect of abrasion in lard is suggested to be related to the spreadability of lard, which depends on the solid content fraction, as for instance lard is observed to be accumulated outside the abraded area particularly at low temperatures. In liquid or in crystalline fat stains, whatever abrasion can do is done very quickly (5 revolutions or less).

The mechanistic effects of abrasion on the removal of fats were further studied using fluorescency microscopy (Leica MacroFluo). Cotton fabrics were stained as before but using Pyrromethene 546 ($l_{abs,max}$ ~495 nm, Exciton, USA) as a fluorescent lipophilic dye. These samples were then tested under the tribometer as above. Quantification of the fat removal in the abraded areas using fluorescence microscopy is comparable to the solubilisation method. The abraded areas show damage in the fabric, as observed from the larger number of superficial loose fibers. Interfabric abrasion works best to remove the superficial fat stains, whereas is less effective when the stains are far from the fabric surface, for example in the contact region between yarns. Abrasion also improves fat removal even when the water temperature is above the melting point of the stain (eg. >40°C for Lard, Fig. 1). Microscopy shows that the stain concentration in the abraded yarns is lower, see Fig. 2, provably some liquid fats have been squeezed out. Moreover, in the non-abraded fabric free liquid fat droplets are found trapped between yarns and loose fibres (Fig. 2). Abrasion facilitates the detachment of these droplets improving fat removal.

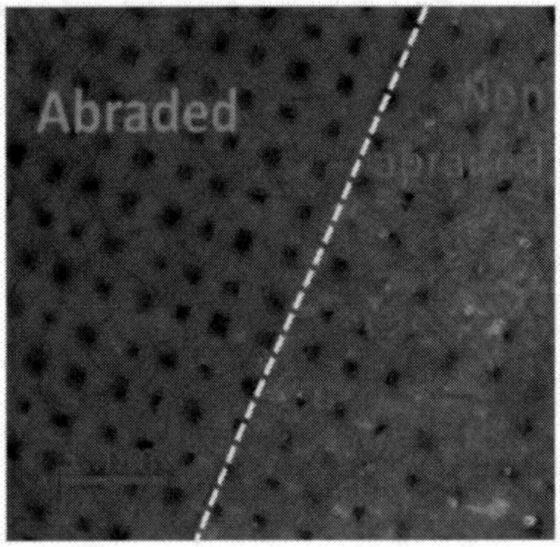

Figure 2. Fluorescence microscopy of a cotton sample soiled with lard and a lipophilic fluorescent dye after a tribology experiment in water at 44°C. Dotted line shows the approximate locus of the abraded (left) and non-abaraded (right) regions. Note the higher fluorescence in the non-abraded yarns, as well as the entrapped liquid lard drops.

(179r) Micron-Size Particle Deposition Scenarios by Reflection Interference Contrast Microscopy

Jose C. Contreras-Naranjo and Victor M. Ugaz, Artie McFerrin Department of Chemical Engineering, Texas A&M University, College Station, TX

Different deposition scenarios are expected when micron-size particles are deposited on a surface depending upon environmental conditions (relative humidity, temperature), deposition medium (air, liquid), and time effects. For instance, when using a liquid as deposition medium, a liquid meniscus can form between the particles and the substrate at the end of the drying process; depending on the conditions, the meniscus can dry out completely or partially, and the particles might undergo deformation due to capillary forces, in addition to regular adhesion forces, creating a variety of deposition scenarios. Reflection Interference Contrast Microscopy (RICM) offers a unique and convenient view of the deposition phenomenon as the minimum separation distance between particle and substrate, contact area, and particle contour can be accurately quantified when looking at the sample from below using monochromatic light. This information collected from hundreds of individual particles is used to describe the microscopic effects of the aforementioned variables on particle deposition which represents valuable information for resuspension and adhesion models.

(179s) The Use of Insulin Fragments to Elucidate the Kinetic Differences Between Fibrillation of Insulin From Different Species

Daniel Forciniti, Cbe, Missouri University of Science and Technology, Rolla, MO and Paulina Morales, Chemical Engineering, Missouri University of Science and Technology, Rolla, MO

Amyloid fibrils are ordered aggregates of peptides or proteins that are associated with many diseases (e.g. type 2 diabetes mellitus, Alzheimer's disease, Parkinson's disease, etc). Amyloid diseases are apparently unconnected; however, intermolecular secondary structure (mostly β-sheets) is present in all amyloid aggregates. One protein that forms amyloid fibrils is insulin. We have found that bovine insulin forms amyloid fibrils faster than human insulin. In addition, we have found that the lag time for fibril formation depends on the species and on the presence of a solid/liquid interface.

Some of the results may be explained by considering the (minute) differences in the polypeptide chains of both proteins. Bovine insulin differs from human insulin in three of its amino acids; alanine substitutes threonine twice, and valine substitutes isoleucine once. It is known that Thr, Val and Ile are "beta sheet" forming amino acids but Ala is an alpha helix forming one. Moreover, aromatic amino-acids, like Thr, are found to have the highest amyloidogenic propensity.

We decided to explore those differences further. We synthesized the following peptides: a) Leu-Ser-Cys-Ile-Ser-Thr-Cys-Cys-Gla-Glu and b) Leu-Ser-Cys-Val-Ser-Ala-Cys-Cys-Gla-Glu. These segments include the regions where the differences between bovine and humane insulin are found. Because of the limited solubility of these peptides a Lys chain was added to the C-termini. The peptides were exposed to conditions known to induce the formation of fibrils by the hormone (pH ~ 2 and 60 C) in the absence and in the presence of solid/liquid interfaces. The solid/liquid interfaces were created by including liposomes of various degrees of fluidity. The peptides did not form fibrils at low concentrations in bulk but they did at higher concentrations. The liposomes seem to inhibit the formation of fibrils. Differences in the kinetics of fibril formation were observed.

(179t) Influence of the Spin Rotational Brownian Motion On the Magneto-Rheological Effect In a Rod-Like Hematite Particle Suspension In a Simple Shear Flow

Akira Satoh, Machine Intelligence & System Engineering, Akita Prefectural University, Akita, Japan and Yasuhiro Sakuda, Faculty of System Science and Technology, Akita Prefectural University, Yuri-Honjo, Japan

In order to clarify the behavior of a dilute suspension composed of rod-like hematite particles, we have derived the basic equation of the orientational distribution function for a simple shear flow. The behavior is strongly dependent on the direction of a magnetic field, and in the present study a magnetic field is assumed to be applied in the shear flow direction. The equation has been solved numerically in order to investigate mainly the influence of the spin rotational Brownian motion on the orientational distribution, the magneto-rheological effect and the characteristics of magnetization. A prolate spheroidal hematite particle has an important characteristic that it is magnetized in a direction normal to the particle axis. Hence, the orientational distribution function depends not only on the particle direction but also on the magnetic moment direction. This implies that the spin rotational Brownian motion about the particle axis has to be included in the basic equation in addition to the ordinary rotational Brownian motion. The main results obtained here are summarized as follows. The present results are in good agreement with those of the theory without that motion in that the position and the height of a peak of the orientational distribution and also the whole distribution shape agree significantly well with each other. Hence, it is seen that the spin rotational Brownian motion does not significantly influence the orientational distribution. In contrast, the spin rotational Brownian motion has a quantitative effect on the viscosity, although the dependence of the viscosity on the magnetic field strength is in qualitatively good agreement with that of the theory without the spin rotational Brownian motion. Moreover, the magneto-rheological effect can be obtained more significantly for a larger aspect ratio or for a more prolate hematite particle. Since the magnetization has a strong relationship with the orientation of the magnetic moment, the effect of the spin rotational Brownian motion appears in this characteristic more significantly.

(179u) Wetting Transition of Water-Alcohol Mixture On Smooth and Pillared Hydrophobic Surfaces: A Molecular Dynamic Study

Sandip Khan, Department of Chemical Engineering, Indian Institute of Technology Kanpur, Kanpur, India and Jayant K. Singh, Chemical Engineering, Indian Institute of Technology, Kanpur, Kanpur, India

The mixture of ethanol and water plays a different role in wetting transition as the hydroxyl group of ethanol molecules can participate in the hydrogen bond network of water molecules and the ethyl group is comparatively small and hydrophobic in nature. It has been found that the ethanol molecules accumulate in the three phase contact line and the vapor-liquid interface on the smooth graphite surface and the contact angle of the droplet is found to decrease with increase in ethanol concentration[1]. Therefore, the interfacial tension across the contact line or line tension must play an important role in wetting of ethanol solution. In this study using molecular dynamics simulations, we rigorously calculate the line tension of water-ethanol solution on smooth graphite surface and system size effect on the contact angle of water-ethanol solution as a function of ethanol concentration.

The wetting behavior of ethanol solution on rough surfaces is quite different from that on smooth surface. Ethanol molecules are found to accumulate in the grooves rather than on the top of the pillar which is in contrast to the result of water-urea mixture on rough graphite surface[2]. Interestingly, we observe that as long as pillar spacing is comparable with the size of ethanol molecules, the ethanol molecules always prefer to stay in the grooves and hence enhance the wetting on the rough surfaces. We extensively study the wetting behavior of water-ethanol solution on rough graphite surfaces at different ethanol concentrations. We carefully examine the different wetting states as a function of surface fraction and pillar height. The transition from Cassie to Wenzel state is also found for certain surface fraction and pillar height with increase in ethanol concentration. For such pillared surfaces, the ethanol-water solution is found in the partial Wenzel state at lower ethanol concentration whereas in higher concentration it is in the fully Wenzel state which is good agreement with the results recently conducted in experimentally[3].

[1] M. Lundgren, N.L. Allan, and T. Cosgrove, Langmuir **18**, 10462 (2002).

[2] T. Koishi, K. Yasuoka, X.C. Zeng, and S. Fujikawa, Faraday Discuss. **146**, 185 (2010).

[3] J.B. Boreyko, C.H. Baker, C.R. Poley, and C.-H. Chen, Langmuir **27**, 7502 (2011).

(179v) Immiscible Phase Mixing and Reaction in Taylor-Couette-Poiseuille Flow

Mohammad Aljishi, Chemical and Biomolecular Engineering, Cornell University, Ithaca, NY and Yong L. Joo, School of Chemical & Biomolecular Engineering, Cornell University, Ithaca, NY

Taylor-Couette flow allows for examining hydrodynamic instabilities while stimulating vortex motion, offering a highly active interface for mass transfer, and phase mixing. Mixing between immiscible hydrocarbon-water mixtures in Taylor-Couette-Poiseuille flow was examined by assessing the effects of varying the Taylor number, axial flow rate, fluid viscosity, surface tension and volume fraction on the range of distinct flow patterns that can be identified. The various flow regimes were investigated using flow visualization and laser sheet illumination techniques.

Experiments were also carried out to study the transport and reaction during water- hydrocarbon mixing with varying operational parameters and several reactor geometries, with differing gap widths and aspect ratios. Within the resolution of the experiments, it was found that certain flow transitions were subcritical, exhibiting hysteresis, while others remained stable. Specifically, quasi-static acceleration rates of the inner cylinder were found to be critical in generating specific flow patterns at their corresponding Taylor numbers.

(179w) Gold Nanoparticle Films with Varying Surface Roughness Obtained by A Gas-Expanded Liquid Deposition Process

Jie Zhong[1], Christopher B. Roberts[1] and William R. Ashurst[2], (1)Department of Chemical Engineering, Auburn University, Auburn, AL, (2)Deaprtment of Chemical Engineering, Auburn University, Auburn, AL

Alkanethiol-stabilized gold nanoparticles (AuNPs) were deposited via a gas-expanded liquid (GXL) technique utilizing CO_2-expanded hexane onto different substrates. The surfaces of the substrates were modified by coating silicon (100) substrates with various organic self-assembled monolayers (SAMs). The films were characterized and the rms surface roughness values were obtained to determine the nanoparticle film deposition and growth characteristics. Based on the critical surface tension of the SAMs, the film growth indicates a shift between layer-by-layer deposition and island formation, which also results in a change in the surface roughness. Moreover, the surface roughness is influenced by the concentration of the nanoparticle solution, the average size, the size distribution of AuNPs and the nature of the AuNP surface chemistry. The manner in which the surface roughness was affected by each of the influences was investigated and possible explanations for the results under different conditions were developed. By varying the surface roughness, the real contact surface area of the substrate will be changed as well. Therefore, the potential exists to engineer nanoparticle films to provide reduction in adhesion that occurs between surfaces due to interfacial forces of MEMS devices, and these aspects will be discussed in detail.

(179x) An Investigation of Surface Modification On Nanostructured Polymer-Metal Interfaces

Wei Wang, Shaowen Ji and Ilsoon Lee, Department of Chemical Engineering and Materials Science, Michigan State University, East Lansing, MI

Interfaces between heterogeneous materials are of researchers' interest because merits of different materials can be possibly combined with the thorough knowledge of them. The complexity of this research usually exists due to their inherent lack of affinity to each other, although it can be overcame by surface modifications. The interface between polymer and metal is one of them. A variety of different types of interactions could exist at the interface between polymer and metal, depending on the modification strategies. To name a few, hydrophobic and hydrophilic interactions, electrostatic forces, hydrogen bonds are all possible interactions. With better understanding of the mechanism behind these interactions, some toxic and corrosive pretreatment before coating can be improved or even eliminated. In our study, we will investigate and make use of these interactions to assemble metal films on to nanostructured polymer particles and 3D foams. The metal coating on to nanostructured polymer particles with different sizes, different types of polymer particles with different shapes and also 3D cellular foam materials will be presented.

(181a) Predicting the Properties of CO2-Natural Gas Mixtures for CCSS Projects

Adel Elsharkawy, Petroleum Engineering, Kuwait University, Kuwait, Kuwait

There is mounting evidence that rapid global warming is caused by CO2 emission into atmosphere. There is a consensus among nations that CO2 emission to the atmosphere must be controlled via creating economically viable technologies for capturing, separation and long term storage of CO2 (CCSS). One of many possible proposed sequestrations is the injection of CO2 into depleted oil and gas fields. However for planning such CO2 sequestration projects, accurate estimation of CO2-natural gas properties is required.

In this study a large data bank containing physical properties of various mixtures of CO2-natural gas properties at various temperatures and pressures has been collected. The data bank has been used to assess the accuracy of the industry standard methods as well as the recently published ones. The study indicated that these methods have series errors when they applied at high pressures, high temperatures, or at high content of CO2.

A new method is proposed in this study to calculate compressibility factor and density of CO2-natural gas mixtures. It is capable of estimating the CO2- natural gas mixtures properties at various temperatures and pressures and for any level of CO2 content. The new method is useful for designing pipeline, compression and storage of CO2 in CCSS projects.

(181aa) Generalized Non-Linear QSPR Models for Surface Tension

K. M. Yerramsetty, B. J. Neely and K. A. M. Gasem, School of Chemical Engineering, Oklahoma State University, Stillwater, OK

Surface tension (ST) is of importance for many processes and phenomena, such as mass and heat transfer, gas injection displacement and flow through porous media. At low values, ST is the dominant fluid property which determines relative permeabilities and residual liquid saturations in gas condensate systems. The correlations available in the literature for predicting ST currently have a limited range of applicability or poor suitability for generalization. Further, most literature models exhibit large errors near the critical point of the fluid, and rely on experimentally measured physical properties as input for their predictions. Therefore, there exists a need for generalized models capable of providing *a priori* predictions for compounds that have no relevant experimental data or for compounds that are yet to be synthesized.

Previously, researchers at Oklahoma State University (OSU) have developed a unified framework for correlating saturation properties. This scaled-variable-reduced-coordinates (SVRC) model is based on the corresponding states theory (CST) and scaling-law behavior and in general is capable of representing saturation properties within their experimental uncertainties. The modeling approach involves the use of quantitative structure-property relationship (QSPR) methodology to generalize the model parameters of the developed SVRC model. Specifically, we use SVRC to develop the behavior model, and QSPR to generalize the SVRC model parameters. This approach, in the past, has proven to be more effective than the typical efforts to develop generalized models directly using QSPR techniques. In this work, we extend the SVRC model to correlate ST values and generalize the model parameters using structure-property modeling. A database of 2,417 data points involving 180 fluids was used in the development of this model. When applied to an external dataset containing 922 data points involving 71 fluids, the ST model predictions were, on average, within 2.6% of the reported values.

(181ab) Excess Free Energy of the Reactive Systems Ethanol + Acetic Acid, I-Butanol + Acetic Acid and 3-Methyl-1-Butanol + Acetic Acid At 323 K

Jairo A. Duran[1], Yohana Paez[2], Alejandro Lopez Gomez[3], Alvaro Orjuela[1] and Gerardo Rodriguez[4],
(1)Department of Chemical and Environmental Engineering, Universidad Nacional de Colombia, Bogota, Colombia,
(2)Department of Chemical and Environmental Engineering, Universidad Nacional de Colombia, Bogotá, Colombia,
(3)Departament of Chemical and Environmental Engineering, Universidad Nacional de Colombia, Bogota, Colombia,
(4)Departamento de Ingeniería Química y Ambiental, Universidad Nacional de Colombia, Bogotá, Colombia

Excess Free Energy of the Reactive Systems ethanol + acetic acid, *i*-butanol + acetic acid and 3-methyl-1-butanol + acetic acid at 323 K.

Jairo Duran, Yohana Páez, Alejandro López, Alvaro Orjuela

Grupo de Procesos Químicos y Bioquímicos. Department of Chemical and Environmental Engineering. Universidad Nacional de Colombia. Bogotá. Colombia

Reliable thermodynamic data is important for evaluation and design of chemical processes especially in systems where biobased chemicals are involved. For example phase equilibrium predictions obtained from generalized methods (e.g. UNIFAC) tend to fail with polar or branched structures common in biobased molecules. Among the variety of biobased chemicals, esters play an important role in the current chemical industry; therefore, experimental evaluation on the thermodynamics of such materials is required . Esterification systems have been studied widely and in most cases, thermodynamic information has been evaluated more than once. However, reactive systems, i.e., alcohol + carboxylic acid show problems related with reaction products (ester and water) which can change phase equilibrium behavior of the mixture. Several authors have pointed the importance of carrying out phase equilibrium measurements under mild conditions where reaction extent is negligible.

In this work, isothermal VLE of the binary systems containing ethanol, *i*-butanol and 3-methyl-1-butanol with acetic acid have been measured at 323 K. A hermetic and jacketed-glass cell (100 mL) equipped with a magnetic stirrer was used in experiments. Degasified mixture of acid + alcohol was charged into the cell continuously stirring at 500 rpm and the temperature inside the cell was maintained at 323 K ± 0.1 K by using heating fluid circulation from an isothermal bath. Using a turbo pump (Pfeiffer Vacuum GmbH, Germany), pressure was reduced in the chamber below 0.05 kPa ± 0.03 kPa. Equilibrium pressure in the cell was measured with an absolute capacitance manometer (Baratron® model 625 D, MKS Instruments, US) with an accuracy of ± 0.03 kPa. Procedure was repeated more than once in each point in order to ensure a correct pressure reading. Samples were analyzed by gas chromatography. Acetate fractional yield was not greater than 0.7% (mole basis).

Activity coefficients for the liquid phase and compositions of the vapor phase were calculated taking into account the dimerization of acetic acid in the vapor phase. Additionally, Gibbs – Duhem equation was used to evaluate excess free energy in the systems. All systems showed negative deviations from the ideality. The Herington test of thermodynamic consistency was performed to evaluate data reliability. The effect of excess volume with total pressure was neglected in thermodynamic consistency calculations. Three data sets were consistent using the criteria of Herington and Kojima *at al.* Equilibrium data were correlated with three parameter NRTL model. Maximum deviation of 0.8 kPa was found. Negative values of G^E showed the strong interactions in liquid phase due the formation of hydrogen bonds. This effect is stronger in ethanol + acetic acid than 3-methyl-1-butanol + acetic acid system.

Keywords: Esterification, vapour – liquid equilibrium, excess free energy, thermodynamic consistency.

(181ac) Simulation and Experimental Benchmarking of the Thermophysical Properties of N-Functionalized Alkylimidazoles

C. Heath Turner[1], Zhongtao Zhang[1], Alexander Cooper[2], Matthew S. Shannon[3] and Jason E. Bara[3], (1)Chemical and Biological Engineering, University of Alabama, Tuscaloosa, AL, (2)The University of Virginia, Charlottesville, VA, (3)Chemical & Biological Engineering, University of Alabama, Tuscaloosa, AL

Molecular simulations are used to probe the thermophysical properties of a series of *N*-functionalized alkylimidazoles, ranging from *N*-methylimidazole to *N*-heptylimidazole. These compounds have been previously synthesized, and their solvation properties have been shown to be potentially useful for CO_2 capture from industrial sources. We report first-principles calculations, which are used to fit electrostatic charges to the molecular models, which are then implemented in a series of molecular dynamics simulations. Over a range of different temperatures, we benchmark the simulated densities and heat capacities against experimental measurements. Also, we predict the Henry's constants for CO_2 absorption and probe the solvents' structures using molecular simulation techniques, such as fractional free volume analysis and void distributions. The simulations demonstrate excellent agreement when comparing density values, and the predictions of the heat capacities and Henry's constants for CO_2 absorption are in very good qualitative agreement with the experiments. Overall, we find that our simulations are able to closely reproduce the experimental benchmarks and add additional insight into the molecular structure of these fluids, with respect to their observed solvent properties.

HYDROCARBON MIXTURES: PHASE BEHAVIOR, DENSITY, AND MODELING WITH PC-SAFT AND A VOLUME-TRANSLATED, CUBIC EQUATION OF STATE.

Babatunde A. Bamgbade[a,b,]*, Yue Wu [a,b], Mark A. McHugh[a,b], Hseen Baled[a,c], Robert M. Enick[a,c], Ward Burgess[a], Deepak Tapriyal[a], and Bryan D. Morreale[a]

[a] National Energy Technology Laboratory, Office of Research and Development, Department of Energy, Pittsburgh, PA (USA)
[b] Department of Chemical and Life Science Engineering, Virginia Commonwealth University, Richmond, VA 23284 (USA)
[c] Department of Chemical and Petroleum Engineering, University of Pittsburgh, Pittsburgh, PA (USA)
* Corresponding author; E-mail: bamgbadeba@vcu.edu; Phone: 571-315-2199

Abstract

Density and phase behavior information of normal, branched, cyclic, and aromatic hydrocarbons and their mixtures at different pressures and temperatures are needed for both industrial applications and modeling studies. In the petrochemical industry, density data on hydrocarbon mixtures with CO_2 and/or water at pressures in excess of 240 MPa and temperatures of 533 K are used for the design and optimization of processes related to the production of oil, condensate and natural gas from ultra-deep wells, and the sequestration of CO_2in deep saline aquifers. In modeling studies, high-pressure and high-temperature hydrocarbon density data for pure-component hydrocarbons and their mixtures are used to test different equations of state (EOS). The present study reports experimental high-pressure, high-temperature phase behavior and density data for pure component and binary hydrocarbon mixtures. The density data are modeled with the Peng-Robinson EOS (PR EOS), a modified Soave-Redlich-Kwong EOS that incorporates a high temperature, high pressure volume-translation (HTHP VT-SRK EOS), and the Perturbed Chain Statistical Associating Fluid Theory (PC-SAFT EOS).

Introductions

Modeling fluid flow in petroleum reservoirs and production wells requires reliable density and viscosity predictions of the brine and the hydrocarbon phases. This study focuses on the accurate prediction of hydrocarbon density at extreme conditions associated with ultra-deep formations with pressures and temperatures of up to 276 MPa and 533 K, respectively. We present the modeling capability of the HTHP VT-SRK EOS and the HTHP PC-SAFT EOS for hydrocarbon systems. SAFT based EOS first developed by Chapman and coworkers in the 1990s [1-4], are known to have superior capability in describing the physics of fluid systems. The

SAFT equation represents the Helmholtz free energy as a sum of terms accounting for hard sphere repulsion, dispersion interactions, chain connectivity, and association between segments.

Theory

Volume-Translated Cubic EOS

Cubic EOSs provide accurate predictions for vapor molar volumes but are not capable of predicting accurate liquid volumes. The correction widely employed to overcome this shortcoming of cubic EOS is to shift or translate the predicted liquid molar volume by incorporating the deviations between the predicted molar volume and the experimental volume, into the cubic EOS. This deviation, known as the volume-translation, greatly improves liquid density predictions in the sub-critical region, but is insignificant in the gas phase due to the large value of the vapor molar volume as compared to the liquid phase.

The volume translation parameter c is the difference between the predicted liquid molar volume and the experimental molar volume, which is shown here only for the SRK EOS.

$$P = \frac{RT}{v + c - b} - \frac{a_c \alpha}{(v + c)(v + c + b)} \tag{1}$$

Special emphasis in this study is placed on densities in the HTHP region of 7 - 275 MPa and 278 - 533 K. Averaged c values are obtained by fitting the SRK EOS to density isotherms. Baled et al. [5] show that pressure has very little effect on c, and the variation of c with temperature is adequately correlated with equation 2.

$$c = \Delta v = v_{EoS} - v_{\exp} = A + B \cdot T_r \tag{2}$$

The parameters A and B are represented by the following correlation obtained from fitting density isotherms for 17 different hydrocarbons [5]. Values for the parameters of Equation 3 for the SRK EOS are listed in Table 1.

$$A, B = f(M, \omega) = k_0 + k_1 \exp\left(\frac{-1}{k_2\, M\, \omega}\right) + k_3 \exp\left(\frac{-1}{k_4\, M\, \omega}\right) + k_5 \exp\left(\frac{-1}{k_6\, M\, \omega}\right) \tag{3}$$

Table 1. Parameters A and B needed to calculate the volume translation for the SRK EOS [5].

Constants	***A*** **(cm³/mol)**	***B*** **(cm³/mol)**
k_0 (cm³/mol)	0.2300	-0.3471
k_1 (cm³/mol	46.843	-29.748
k_2	0.0571	0.0644
k_3	23161	-347.04
k_4	0.0003	0.0010
k_5	267.40	-88.547
k_6	0.0053	0.0048

The HTHP VT-SRK EOS can be extended to mixtures using the following mixing rule [6-8]:

$$c_m = \sum_i x_i\, c_i \tag{4}$$

where x_i and c_i are the mole fraction and the volume translation term of pure component *i* in the mixture, respectively. The HTHP VT-SRK EOS has been tested on binary mixtures using the conventional van der Waals mixing rules for the parameters.

$$a_m = \sum_i \sum_j x_i\, x_j\, (a_i a_j)^{0.5} (1 - k_{ij}) \tag{5}$$

$$b_m = \sum_i x_i b_i \tag{6}$$

where k_{ij} is the binary interaction parameter between molecules i and j. For alkane-alkane pairs, k_{ij} can be set to zero if experimental VLE data are unavailable. For other pairs, especially those including non-hydrocarbons, k_{ij} should be different from zero and must be determined [9].

Representative results for HTHP VT-SRK EOS

Figure 1 shows an improved performance of the HTHP VT-SRK EOS (a) over the SRK EOS (b), in modeling *PVT* properties of pentane in the vapor phase, liquid phase, and the vapor-liquid saturation curve. [10-14] For the 325.85, 423.05, and 520.45 K isotherms, the calculated; mean absolute percent deviation (MAPD) of the model from experimental data is 1.11% with a standard deviation (SD) of 0.74%. The HTHP VT-SRK EOS does an excellent job of representing the high pressure density data for pentane. The performance of the HTHP VT-SRK EOS is exactly the same as the SRK EOS in the vapor region since the volume translation is not significant in this region. In all other regions the HTHP VT-SRK EOS shows a significant improvement over the SRK EOS, however neither model accurately predicts the dew point line. Panel (c) shows the capability of both models to predict experimental binary mixture of hexane and hexadecane at an hexane mole fraction of 0.4 [15]. The HTHP VT-SRK EOS clearly shows a significantly superior capability to model this binary mixture density data.

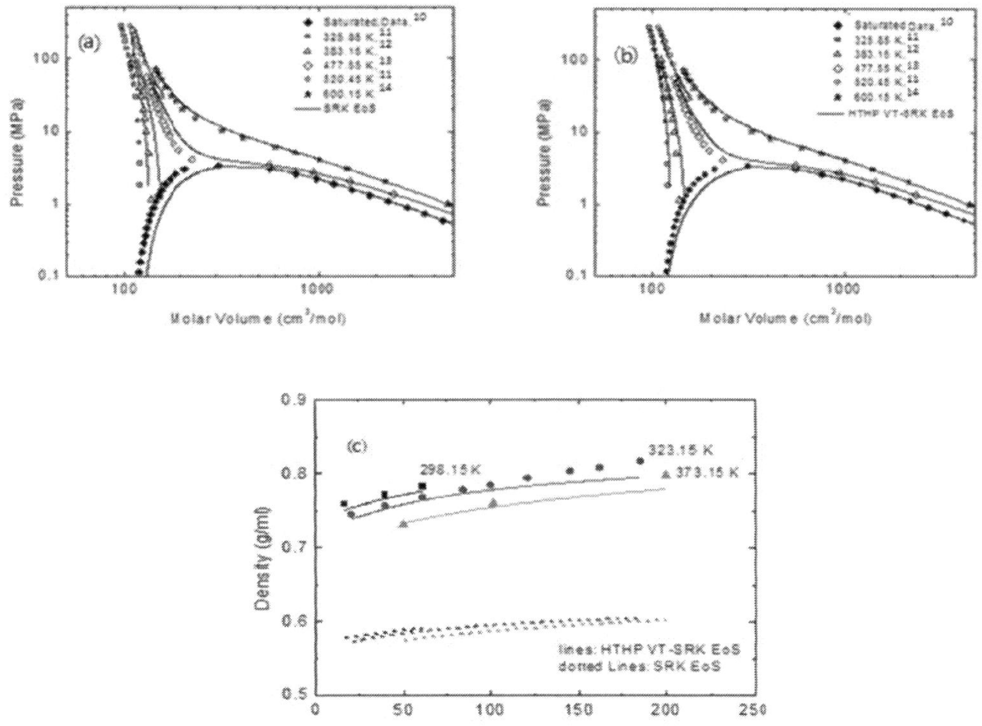

Figure 1. Comparison of the SRK EOS (a) and HTHP VT-SRK EOS (b) for modeling *PVT* data for pentane [10-14]. Panel () shows the comparison of the two models for density prediction a binary mixture of hexane and hexadecane at hexane mole fraction of 0.4 [15].

PC-SAFT EOS

The PC-SAFT EOS is a modified version of the original SAFT EOS proposed by Chapman *et al.*,[1] where the reference fluid is a system of hard chains, instead of hard spheres. The three pure-component PC-SAFT parameters (m, the number of hard spheres in a molecule; σ, the temperature-independent hard core diameter of a sphere; ε/k_B, the attractive energy of dispersion divided by the Boltzmann constant) are typically obtained by fitting sub-critical *PVT* data. [16] Here a new set of pure-component PC-SAFT parameters are obtained by fitting the equation to HTHP density data. Subcritical density data are not used to obtain these new parameters. A downhill simplex method is used to determine the best values for the three pure-component PC-SAFT parameters by minimizing the value of the MAPD between the experimental density and the PC-SAFT prediction.

The PC-SAFT EOS for hydrocarbon systems that do not exhibit chemical association, such as hydrogen bonding, is given as a sum of Helmholtz free energies.

$$\tilde{a}^{res} = \tilde{a}^{hc} + \tilde{a}^{disp} \tag{7}$$

where $\tilde{a}^{hc}$ and $\tilde{a}^{disp}$ are the reduced hard chain reference and dispersion terms, respectively. For a multicomponent fluid,

$$\tilde{a}^{hc} = m\tilde{a}^{hs} + \tilde{a}^{chain} = m\tilde{a}^{hs} + \sum_i x_i(1-m_i)\ln\left(g_{ii}^{hs}\right) \tag{8}$$

$$m = \sum_i x_i m_i \tag{9}$$

$$\tilde{a}^{hs} = \frac{1}{\xi_0}\left[\frac{(\xi_2)^3 + 3\xi_1\xi_2\xi_3 - 3\xi_1\xi_2(\xi_3)^2}{\xi_3(1-\xi_3)^2} - \left[\xi_0 - \frac{(\xi_2)^3}{(\xi_3)^2}\right]\ln(1-\xi_3)\right] \tag{10}$$

$$\xi_k = \left(\frac{\pi\rho_{num}}{6}\right)\sum_i x_i m_i (d_i)^k; k = 0,1,2,3 \tag{11}$$

$$d_i = \sigma_i\left[1 - 0.12\exp(-3\varepsilon_i / k_B T)\right] \tag{12}$$

$$g_{ii}^{hs} = \frac{1}{1-\xi_3} + \frac{3d_i}{2}\frac{\xi_2}{(1-\xi_3)^2} + 2\left[\frac{d_i}{2}\right]^2\frac{\xi_2^2}{(1-\xi_3)^3} \tag{13}$$

In these equations g_{ii}^{hs} represents the hard-sphere fluid radial distribution function. The terms m_i, σ_i, and ε_i/k_B represent, for component i, the pure-component PC-SAFT parameters, which are the segment number, temperature-independent segment diameter and the energy parameter respectively, while d_i represents the hard-sphere diameter at a given temperature. The variable ρ_{num} is the total number density of molecules (Å^3), while $\xi_0, \xi_1, \xi_2,$ and ξ_3 are functions of the average pseudo-molecular size. The dispersion energy is calculated for a reference fluid comprised of hard chains and it is given in equation 13, where PC-SAFT applies the second-order perturbation theory of Barker and Henderson [17,18] to a hard chain reference fluid.

$$\tilde{a}^{disp} = -\pi\rho_{num}\left[2I_1(\eta,m)\sum_i\sum_j x_i x_j m_i m_j \sigma_{ij}^3 \frac{\varepsilon_{ij}}{k_B T} + mI_2(\eta,m)C\sum_i\sum_j x_i x_j m_i m_j \sigma_{ij}^3\left(\frac{\varepsilon_{ij}}{k_B T}\right)^2\right] \quad (14)$$

$$\sigma_{ij} = 0.5(\sigma_i + \sigma_j) \quad (15)$$

$$\varepsilon_{ij} = (1-k_{ij})\sqrt{\varepsilon_i \varepsilon_j} \quad (16)$$

The terms I_1 and I_2 in equation 13 represent integrals approximated as power series in density [16], while the combining rules given in equations 14 and 15 are used to calculate the size and energy parameters σ_{ij} and ε_{ij} for a pair of unlike segments. The binary interaction parameter k_{ij} is used to correct the segment-segment interactions for these unlike segments. C is given as

$$C = \left(1 + m\frac{8\eta - 2\eta^2}{(1-\eta)^4} + (1-m)\frac{20\eta - 27\eta^2 + 12\eta^3 - 2\eta^4}{[(1-\eta)(2-\eta)]^2}\right)^{-1} \quad (17)$$

$$\eta = \frac{\pi}{6}\rho_{num}\sum_i x_i m_i d_i^3 \quad (18)$$

Density calculations with PC-SAFT are based on the method of Gross and Sadowski [16], in which the density at a given system pressure is determined by adjusting the value of the reduced density, η, until the calculated pressure equals the experimental pressure. Pure component HTHP density values are used for fitting the PC-SAFT EOS from which the pure component parameters, σ, ε/k_B, and m, are obtained.

Representative results for PC-SAFT EOS

Figure 2 shows the performance of the HTHP PC-SAFT EOS in modeling the experimental density data of pentane. The figure also shows the performance of the model with Gross and Sadowski's PC-SAFT (G-S PC-SAFT) parameters. It is clear that the HTHP PC-SAFT does a better job, having a MAPD of 0.78% and SD of 0.60%, as compared to the G-S PC-SAFT model that produced a MAPD of 1.33% and SD% of 1.19%. Figure 3 shows the performance of both the GS PC-SAFT and the HTHP PC-SAFT to model the experimental binary mixture density of hexane and hexadecane at an hexane mole fraction of 0.4 [15].

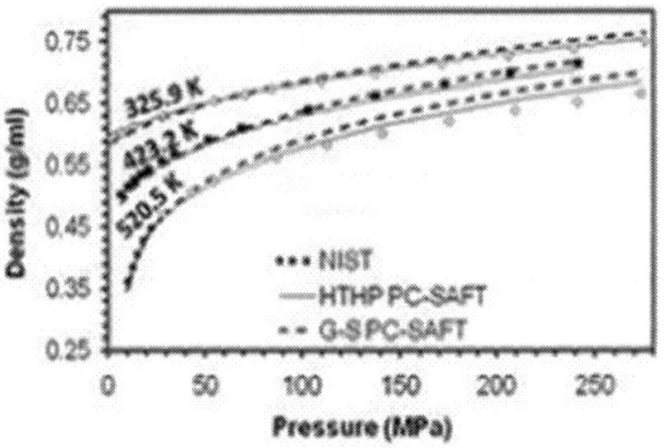

Figure 2. Comparison of the HTHP PC-SAFT EOS and the G-S PC-SAFT EOS with the experimental density data of pentane, ○ [11], ■ [19]. Note, the points labeled NIST are from smoothed correlation and do not represent actual experimental data points [20].

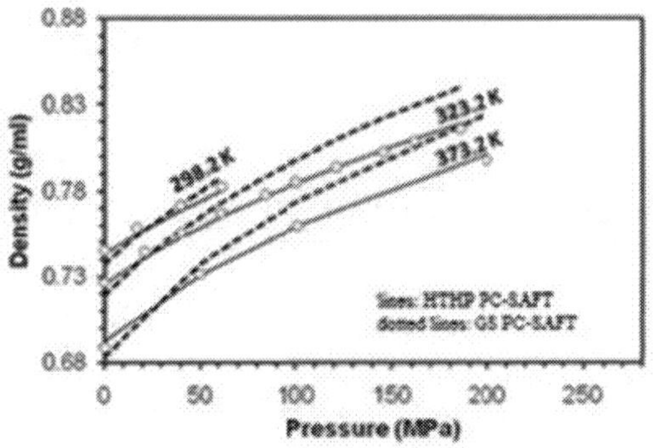

Figure 3. Comparison of the HTHP PC-SAFT and GS PC-SAFT to model experimental density of a binary mixture of hexane and hexadecane at hexane mole fraction of 0.4 [15].

Conclusion

The VT-SRK EOS and the PC-SAFT EOS are correlated to pure component, single-phase HTHP density data, up to 275 MPa and 533 K, to calculate new parameter sets. These two models outperform other EOS models in the extreme temperature and pressure conditions expected for ultra-deep well production of oils, condensate, and natural gas. Further examples will be provided to demonstrate the performance of both the VT-SRK and the PC-SAFT equations of state for modeling high pressure density data for hydrocarbons.

Acknowledgments

This technical effort was performed in support of the National Energy Technology Laboratory's Office of Research and Development support of the Strategic Center for Natural Gas and Oil under RES contract DE-FE0004000.

References

1. Chapman WG, Gubbins KE, Jackson G, and Radosz M. New reference equation of state for associating liquids. Ind. Eng. Chem. Res., Aug 1990;29:1709-1721.
2. Chapman WG, Jackson G, and Gubbins KE. Phase-equilibria of associating fluids chain molecules with multiple bonding sites. Mol. Phys., Dec 10 1988;65:1057-1079.
3. Huang SH, and Radosz M. Equation of state for small, large, polydisperse, and associating molecules. Ind. Eng. Chem. Res., Nov 1990;29:2284-2294.
4. Huang SH, and Radosz M. Equation of state for small, large, polydisperse, and associating molecules - extension to fluid mixtures. Ind. Eng. Chem. Res., Aug 1991;30:1994-2005.
5. Baled H, Enick RM, Wu Y, McHugh MA, Burgess W, Tapriyal D, and Morreale BD. Prediction of hydrocarbon densities at extreme conditions using volume-translated srk and pr equations of state fit to high temperature, high pressure pvt data. Fluid Phase Equilib., 2012;317:65-76.
6. de Sant'Ana HB, Ungerer P, and de Hemptinne JC. Evaluation of an improved volume translation for the prediction of hydrocarbon volumetric properties. Fluid Phase Equilib., Jan 25 1999;154:193-204.
7. Ji WR, and Lempe DA. Density improvement of the srk equation of state. Fluid Phase Equilib., Apr 15 1997;130:49-63.
8. Wang LS, and Gmehling J. Improvement of the srk equation of state for representing volumetric properties of petroleum fluids using dortmund data bank. Chem. Eng. Sci., Sep 1999;54:3885-3892.
9. Soave G, Gamba S, and Pellegrini LA. Srk equation of state: Predicting binary interaction parameters of hydrocarbons and related compounds. Fluid Phase Equilibria 2010;299:285-293.
10. Vagaftik NB. Tables on the thermophysical properties of liquids and gases. Washington: Hemisphere; 1975.
11. Liu K, Wu Y, McHugh MA, Baled H, Enick RM, and Morreale BD. Equation of state modeling of high-pressure, high-temperature hydrocarbon density data. J. Supercrit. Fluids, Dec 2010;55:701-711.
12. Audonnet F, and Pádua AAH. Simultaneous measurement of density and viscosity of n-pentane from 298 to 383 k and up to 100 mpa using a vibrating-wire instrument. Fluid Phase Equilib., 2001;181:147-161.
13. Lee AL, and Ellingto.Rt. Viscosity of n-pentane. J. Chem. Eng. Data, 1965 1965;10:101-&.
14. Das TR, Reed CO, and Eubank PT. Pvt surface and thermodynamic properties of n-pentane. J. Chem. Eng. Data, 1977 1977;22:3-9.
15. Dymond JH, Young KJ, and Isdale JD. P, ϱ, t behaviour for n-hexane + n-hexadecane in the range 298 to 373 k and 0.1 to 500 mpa. J. Chem. Thermdyn., 1979;11:887-895.
16. Gross J, and Sadowski G. Perturbed-chain saft: An equation of state based on a perturbation theory for chain molecules. Ind. Eng. Chem. Res., Feb 2001;40:1244-1260.
17. Barker JA, and Henderson.D. Perturbation theory and equation of state for fluids - square-well potential. J. Chem. Phys., 1967 1967;47:2856-2861.
18. Barker JA, and Henderson.D. Perturbation theory and equation of state for fluids .2. A successful theory of liquids. J. Chem. Phys., 1967 1967;47:4714-4721.
19. Byun HS, DiNoia TP, and McHugh MA. High-pressure densities of ethane, pentane, pentane-d(12), 25.5 wt % ethane in pentane-d(12), 2.4 wt % deuterated poly(ethylene-co-butene) (peb) in ethane, 5.3 wt % hydrogenated peb in pentane, 5.1 wt % hydrogenated peb in pentane-d(12), and 4.9 wt % hydrogenated peb in pentane-d(12)+23.1 wt % ethane. J. Chem. Eng. Data, Sep-Oct 2000;45:810-814.
20. Nist chemistry webbook. 2011.

(181ae) On the Self-Assembly of Skin Lipids Using Coarse-Grained Molecular Dynamics Simulation

Shan Guo[1] and Clare McCabe[1,2], (1)Chemical and Biomolecular Engineering, Vanderbilt University, Nashville, TN, (2)Chemistry, Vanderbilt Univerisity, Nashville, TN

It has been shown that the skin's ability to act as a barrier to harmful agents entering the human body and to reduce water loss from the body is related to the structure of the lipids between the corneocytes in the outmost layer of skin, also known as the stratum corneum (SC). To obtain a molecular-level understanding of how the SC lipids (namely ceramides, cholesterol and free fatty acids) self-assemble into bilayers, and other structures, we have developed coarse-grained models for these lipids that allow self-assembly to be studied on time scales accessible to molecular dynamics simulation. In particular, we will present results for mixed lipid systems involving ceramides 2 and 3, fatty acids and cholesterol. The accuracy of the coarse-grained models is validated by comparing the self-assembled structures against those obtained from simulations of pre-assembled atomistic lipid bilayers and experimental work where possible. The effect of changing the cholesterol content of the lipid systems and unsaturation in lipid tails on the self-assembly process will be discussed.

(181af) Modeling Thermodynamic Behavior of Non-Ideal Mixed Surfactant Systems During Isothermal Titration Calorimetry Experiments

Suvid Joshi[1], Barbara L. Knutson[2] and Stephen E. Rankin[2], (1)Department of Chemical and Materials engineering, University of Kentucky, Lexington, KY, (2)Chemical and Materials Engineering, University of Kentucky, Lexington, KY

Mixed micellar systems provide a large number of advantages over single surfactant micelles due to synergistic effects that sometimes arise due to favorable energetics of mixing between the surfactants. Because of this synergy, mixed surfactants find many applications in both industry and the laboratory. Here we present a study of thermodynamics in a mixed micellar cationic / nonionic surfactant system. Cetyltrimethylammonium bromide (CTAB), the cationic surfactant, is selected because it is commonly used in synthesis of templated silica materials. N-octyl-β-D-glucopyranoside (C8G1) is a non-ionic detergent which is biodegradable and also potentially provides ability to create imprinted sites for glucose in a silica material. Apart from our long-term interest in using such mixtures of cationic and carbohydrate-based surfactants to synthesize materials with selective sites for carbohydrate adsorption, this system serves as a model non-ideal mixed surfactant system which presents challenges in developing a comprehensive model of the thermodynamics of micellization.

Here, the heat evolved during isothermal titration calorimetry (ITC) of CTAB / C8G1 systems will be measured and analyzed using a model that accounts for nonideal mixing. The objective is to not only determine the non-ideal interaction parameter (β) for the system but also to develop a model able to predict the entire shape of the heat vs. concentration curve in an ITC experiment. The model described to do this is based on incorporating a regular solution theory activity coefficient model into Clint's pseudo-phase separation model of micellization in mixed surfactant system. This model calculates the change in surfactant monomer concentrations with respect to the total surfactant concentration (and thus the change in micelle composition with respect to concentration). Consistent with the regular solution theory approach, the cmc of the system is calculated based on Rubingh's equation. By rigorously accounting for the heats associated with partitioning of surfactants between the bulk solution and micelle phases after each injection in the ITC experiment, quantities including individual component CMCs, enthalpies of demicellization of pure surfactants and the interaction parameter (b) are determined. The model also allows the design of experiments for rapid determination of these paratmers. This work would thus help in efficiently selecting surfactant mixtures best suited for synthesis of selective adsorbent materials based on mixing behavior of the constituents.

(181ag) High-Pressure Vapor-Liquid Equilibria for the Carbon Dioxide – Heptane – Dodecane

María F. Torres-Mondragón, Departamento de Ingeniería Química Petrolera, Instituto Politecnico Nacional - E.S.I.Q.I.E., México, D. F., Mexico, Giovanni J. Villaseñor-Orozco, Departamento de Ingeniería Química Petrolera, Instituto Politécnico Nacional - E.S.I.Q.I.E., Mexico, D. F., Mexico, **Octavio Elizalde-Solis**, Departamento de Ingeniería Química Petrolera, INSTITUTO POLITECNICO NACIONAL-ESIQIE, México, D. F., Mexico and Luis A. Galicia-Luna, INSTITUTO POLITECNICO NACIONAL-ESIQIE, Mexico, Mexico

Experimental vapor-liquid equilibrium (VLE) measurements are reported for the ternary system carbon dioxide – heptane – dodecane at temperatures of (313, 343 and 373) K. The operating principle of the apparatus where VLE measurements were carried out is based on the static-analytic method; its main advantage is the on-line sampling via a ROLSI sampler-injector, a titanium alloy equilibrium cell and a gas chromatograph. Equilibrium ratios indicate a good internal consistency of the obtained data with an uncertainty less than 0.008 MPa, 0.02 K and 1% for the mole fraction composition in both phases. Separation factors obtained from experimental xi, yi mole fractions, indicate that heptane is feasible to be separated over dodecane at the three different temperatures. VLE was predicted using the Peng-Robinson Equation of State with the classical and Wong-Sandler mixing rules. Binary interaction parameters were obtained by correlating the VLE of the concerning binary systems reported in the literature.

Suppression of Coffee-Ring Deposition by Controlled Hydrophobic Interactions

Sulki An, Ji-Ho Yoon, Byeong-Soo KIM, Jeong-Woo Park and Youngrok Seo, Energy & Resources Engineering, Korea Maritime University, Busan, South Korea

Introduction

When a drop of liquid dries with a pinned edge on solid surface, it shows characteristic ring-like deposit at the edge. The uniform deposition of particles from solutions is a crucial factor in printing and coating processes and future technology such as inkjet printing, DNA chip and self-assembly. This phenomenon is known as the coffee-ring effect, which is mainly due to the unbalanced capillary flow during the drying process because of droplet evaporation, and Marangoni flow. Marangoni flow is the mass transfer along an interface between two fluids due to surface tension gradient. This fluid system has many parameters that size(particle, droplet), shape, and concentration of the solid constituents, the temperature. In this study, we provide a useful guide to control the deposit pattern of particles by changing the solvent properties with a small amount of hydrophobic and/or hydrophilic chemicals. The polystyrene particles with uniform diameters of 0.8, 1.3, 2, and 3 μm were used as suspended particles for the drying process. We found that several organic chemicals such as 3-methyl-1-butanol, tert-butyl methyl ether, cyclopentane, 1,4-dioxane, tetrahydrofuran, and 1-propanol change the solvent chemistry, leading to the suppression of coffee-ring deposition of particles.

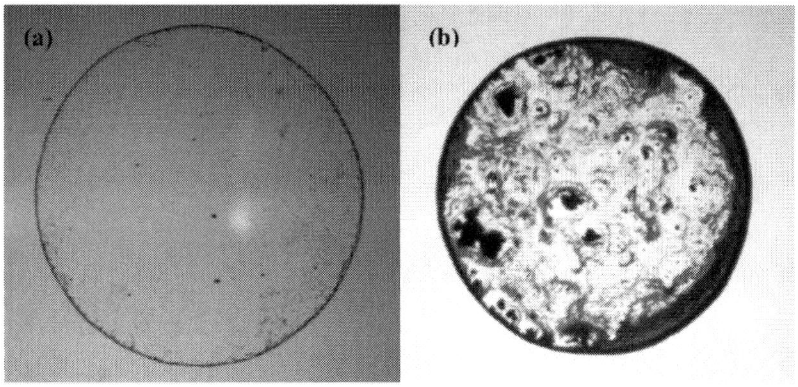

Figure 1 . Set of deposition from dried water drops (a) containing 0.05 % polymer particle(d=2 μm) (b) containing 0.1% polymer particle(d=2 μm)

Table 1. Chemical formula and typical structure of organic chemicals

3-Methyl-1-butanol	tert-Butyl methyl ether	Cyclopentane	1,4-dioxane	Tetrahydrofuran	1-Propanol
$C_5H_{12}O$	$C_5H_{12}O$	C_5H_{10}	$C_4H_8O_2$	C_4H_8O	C_3H_8O
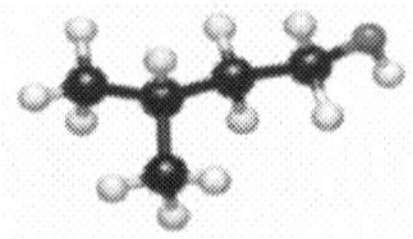	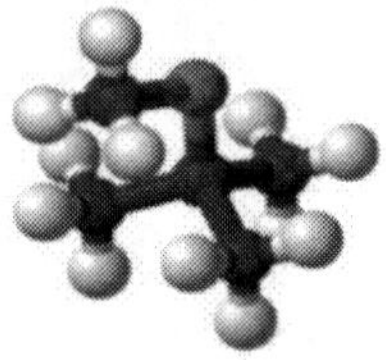	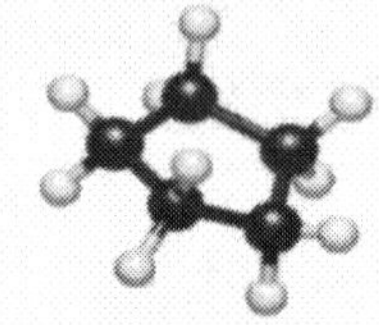	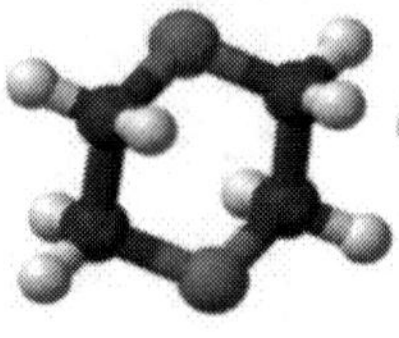		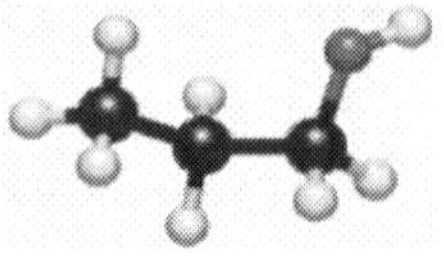

References

1. Hua Hu, and Ronald G. Larson (2006), "Marangoni Effect Reverses Coffee-Ring Depositions," *J. Phys. Chem. B*, March 23, pp. 7090-7094
2. Tim Still, Peter J. Yunker, and Arjun G. Yodh (2012), "Surfactant-Induced Marangoni Eddies Alter the Coffee-Rings of Evaporating Colloidal Drops," *Langmuir*, February 27, pp. 4984-4988

Loop Heat Pipes Exploiting Nano-porous Wicks
- Development and Application of Materials Sustaining Fluid under Tension

Among many different technologies for transferring heat, the heat pipe (HP) – closed-circuit systems in which a working fluid transfers its latent heat as it cycles between an evaporator and a condenser by capillary pumping force – is one of the most efficient systems known today and has offered an attractive alternative in an array of applications[1]. However, two main limitations have constrained the application of conventional HPs: 1) the maximum capillary head developed by the micro-porous wicks of ~1 bar that limits the maximum heat loads, the length, and their orientation with respect to gravity or accelerations.; and 2) the saturation state in the condenser region creates an unbounded liquid-vapor interface that causes undesired transients and additional resistance to heat transfer.

The transpiration process in vascular plants operates by a mechanism similar to the recirculation of liquid through the wick of HPs[2]. Unlike in HPs, though, the water pressure in plants is able to reach large negative values (down to -100 bars) to overcome gravity, viscous drag in the liquid path, and the reduced chemical potential of water in sub-saturated soils. The secret behind the amazing abilities of plants is a the structure of plants cell walls, which serve as nano-porous membranes, allowing the water pressure to be less than zero (under tension). Inspired by plants, our research is focused on the development of a nano-porous loop heat pipe (NPLHP) exploiting nano-porous wicks, including a theoretical model of the coupled transport and thermodynamic phenomena and an experimental platform with which to test design concepts for NPLHPs.

After presenting the design considerations of a NPLHP, we will focus on a central experimental challenge in its development: nano-porous membranes that allow for the pure liquids to come to metastable equilibrium at negative pressures with sub-saturated vapors. We will discuss several membrane structures developed in our laboratory and our experimental methods for characterizing them. We will then turn to an intriguing open question about water under tension: several experimental studies focusing on cavitation thresholds (the pressure at which the thermodynamically metastable liquid turns into vapor by nucleation) of water under tension have been reported, yet there is no consensus[3]. Many experiments using various methods have found the stability limit of water at room temperature is between -20 ~ -30MPa[4-6], while experiments in quartz inclusions appear to reach a stability limit of -140 MPa (with the use of an extrapolated equation of state). We will describe how measurements based our distinct experimental approach compare with those in the literature and discuss the implications for our understanding of the phase diagram of water and on the practical limitations on the use of water at negative pressure in technologies such as NPLHPs.

1 Faghri, A. *Heat Pipe Science and Technology*. (Taylor and Francis, 1995).

2 Nobel, P. S. *Physicochemical and Environmental Plant Physiology*. 2nd edn, (Academic Press, 1999).

3 Caupin, F. & Herbert, E. Cavitation in water: a review. *Comptes Rendus Physique* **7**, 1000-1017 (2006).

4 Briggs, L. J. Limiting Negative Pressure of Water. *Journal of Applied Physics* **21**, 721-722 (1950).

5 Wheeler, T. D. & Stroock, A. D. Stability Limit of Liquid Water in Metastable Equilibrium with Subsaturated Vapors. *Langmuir* **25**, 7609-7622, doi:10.1021/la9002725 (2009).

6 Davitt, K., Arvengas, A. & Caupin, F. Water at the cavitation limit: Density of the metastable liquid and size of the critical bubble. *EPL* **90**, 16002 (2010).

(181aj) Thermodynamic Derived Properties for Carbon Dioxide - Ethanol - Decane Mixtures At High Pressures

Héctor S. Zamora-López, Laboratorio de Termodinámica - SEPI, Instituto Politécnico Nacional - E.S.I.Q.I.E., Mexico, D. F., Mexico, Luis A. Galicia-Luna, INSTITUTO POLITECNICO NACIONAL-ESIQIE, Mexico, Mexico, Octavio Elizalde-Solis, Departamento de Ingeniería Química Petrolera, INSTITUTO POLITECNICO NACIONAL-ESIQIE, México, D. F., Mexico and Irma P. Hernández-Rosales, Área de Ciencias Básicas e Ingeniería, Universidad Autónoma de Nayarit, Tepic, Nayarit, Mexico

Thermodynamic derived properties (isobaric expansivity, isothermal compressibility, thermal pressure coefficient and internal pressure) are reported in this work for the ternary system carbon dioxide (1) + ethanol (2) + decane (3) at five different compositions in the intervals of (313 – 363) K and (1 – 20) MPa. Compositions for the ternary system were prepared by varying the ethanol / decane relation and trying to keep constant the presence of carbon dioxide at about 0.2 mole fraction. These were {x2 ethanol - x3 decane - (1 - x2 - x3) carbon dioxide} x2 = 0.0657, 0.1986, 0.4087, 0.6042, 0.7109. Derived properties were obtained from an empirical correlation of 5 adjustable parameters based on density data. The experimental apparatus is based on the static – synthetic method coupled to a commercial vibrating tube densitometer from Anton Paar. Experimental densities are reported within an uncertainty of 0.2 $kg \cdot m^{-3}$; these were obtained from vibrating periods at each temperature and pressure using water and nitrogen as reference fluids.

Pressure-Induced Reversible and Irreversible Structural Transitions of a Hydrogen-Bonded Organic Framework

Byeong-Soo Kim, Jeong-Woo Park, Sulki An, Youngrok Seo, Ji-Ho Yoon*
Department of Energy & Resources Engineering, Korea Maritime University, Busan 606-791, Korea

Abstract

Organic clathrate can be used as the potential host materials for gas storage, because they often form stable inclusion compounds with a variety of guest species over a wide range of temperature and pressure conditions. In this study, we investigated the pressure-dependent behavior of hydroquinone (HQ) clathrate compounds using a diamond anvil cell. The guest-free and N_2-loaded HQ clathrates were synthesized by the gas-phase reaction, followed by a controlled heating process to remove CO_2 guest molecules. The pressure-driven structural transformation of HQ clathrate compounds was observed by a combination of synchrotron X-ray powder diffraction and Raman spectroscopy up to 10 GPa. The N_2-loaded HQ clathrate was transformed into a new high-pressure phase at ~ 4.5 GPa. Upon returning pressure, the new structural phase was reversibly recovered to the original b-form HQ clathrate at ambient condition. In contrast, the guest-free HQ clathrate transformed into the a-form HQ at ~ 0.5 GPa and remained the a-form HQ upon decreasing pressure to ambient condition. This indicates that the pressure-induced structural transition of the guest-free HQ clathrate is completely irreversible. This result provides a useful knowledge on the effect of guest molecules on the stability of structural integrity of organic clathrate compounds.

Clathrates are crystalline inclusion compounds formed by the physically stable interaction between host species and relatively small guest molecules occupied in the cages of the host framework.[1-4] Clathrate compounds have been recognized as energy storage or gas separation / sequestration media because they store a large amount of gases per unit volume of the solid phase. Hydroquinone (HQ), phenol, and Dianin's compounds are well-known as organic clathrates, which are an extensively studied class of inclusion compounds. In particular, β-form HQ clathrate is a simple representative of this family and comprises a hydrogen-bonded organic framework (HOF), which entraps relatively volatile guest molecules.[2] The HOFs have an inherent flexibility that allows them to conform to guest molecules with different sizes and shapes such as Ar, Xe, SO_2, CH_4, and methanol. It also represents potential host channel structures defined by the hexagonal entrance of hydrogen-bonded host (HQ) molecules. The flexible hexagonal entrance allows the migration of guests into the channel cages and the guest release from the HOFs via the dynamic pore-widening process, leading to controlled, selective, and reversible storage of small molecules.[5]

An important clue on how the guest–host interaction in inclusion compounds would occur under high pressure has been shown in clathrate hydrates.[6-8] These studies represent that the guest molecules such as CH_4 and H_2 play an important role in the structural transition of the clathrate hydrates. It is clear that the small guests encaged in the clathrate hydrates lead to changes of the hydrogen bonding nature of the host molecules under high pressure. On the other hand, at ambient pressure, the β-form HQ clathrate reveals the structural transformation to α-form HQ at the same temperature of 380 K for CO_2, CH_4, and methanol guests.[9-11] This indicates that the combined effect of guest–host interactions, the size and shape of guest molecules, and the flexibility of host frameworks may affect the structural transition of clathrate compounds. A recent study shows that the β-form HQ clathrate with CH_4 guests undergoes the structural transition at around 5 GPa.[12] However, the effect of guest molecules on the stability of structural integrity of the HOF has remained unclear. We report here that, under high pressure, the structural transformation of organic clathrates and thereby the stability of the HOF can strongly be affected by guest molecules encapsulated in the cages.

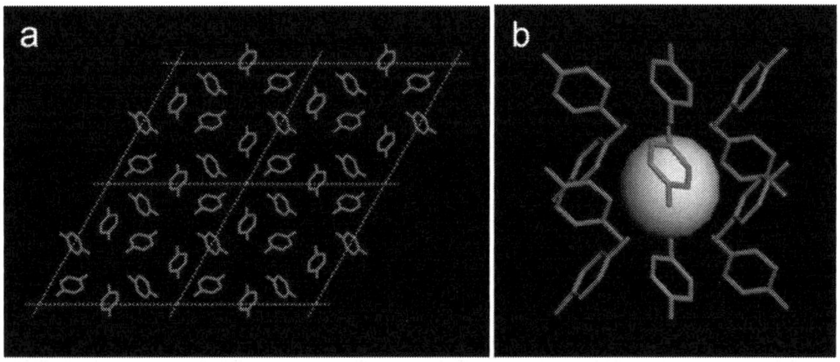

Figure 1. Views of β-form HQ clathrate structure: (a) Capped-stick representation viewed along the (001) direction. Dotted lines define a unit cell. (b) Cage structure formed with twelve HQ molecules, viewed perpendicular to the c-axis. The space volume of the cage for guests is represented by yellow sphere. Hydrogen atoms are omitted for clarity.

HQ exists in three different crystalline forms: α, β, and γ.[1] The rhombohedral α-form HQ is stable at ambient conditions and transforms to the β-form HQ clathrate when small guest molecules are incorporated into the ordered HQ lattice structure. It is also known that the γ-form HQ forms by sublimation or rapid evaporation of HQ solution in ether. Figure 1 shows the crystalline structure of the β-form HQ clathrate and the cage structure. The unit cell consists of 9 HQ molecules and 3 cages. Therefore, the molar ratio of HQ molecules to cages for β-form HQ clathrate is 3:1, which enables a high guest-molecule loading capacity. In this study, the N_2-loaded HQ clathrate was synthesized by the gas-phase reaction (see Supporting Information). The chemical analysis indicates that 67% of the cages of the HQ clathrate are filled with N_2 molecules, which results in a chemical formula of $0.67N_2{\cdot}3HQ$. The guest-free HQ clathrate was also prepared by the gas-phase synthesis of CO_2-loaded HQ clathrate, followed by controlled heating process to remove CO_2 guests.[13] Raman spectroscopy provides information on the crystal structure of the formed HQ clathrate and the dynamics of guests occupying the cages. The two split Raman bands at 1160 cm^{-1} are characteristics of α-form HQ, whereas the spectrum of the β-form HQ clathrates features a single peak (Figure 2a). The Raman peak at 2322 cm^{-1} represents the presence of N_2 molecules encaged in the β-form HQ clathrate framework.[14] The X-ray diffraction (XRD) patterns and solid-state ^{13}C NMR results also show a clear difference between α-form HQ and β-form HQ clathrates and a close structural similarity of N_2-loaded and guest-free β-form HQ clathrates (Figures 2b and 2c). This can be attributed to the different dimensionality in the hydrogen bonding networks between α-form HQ and β-form HQ clathrate.

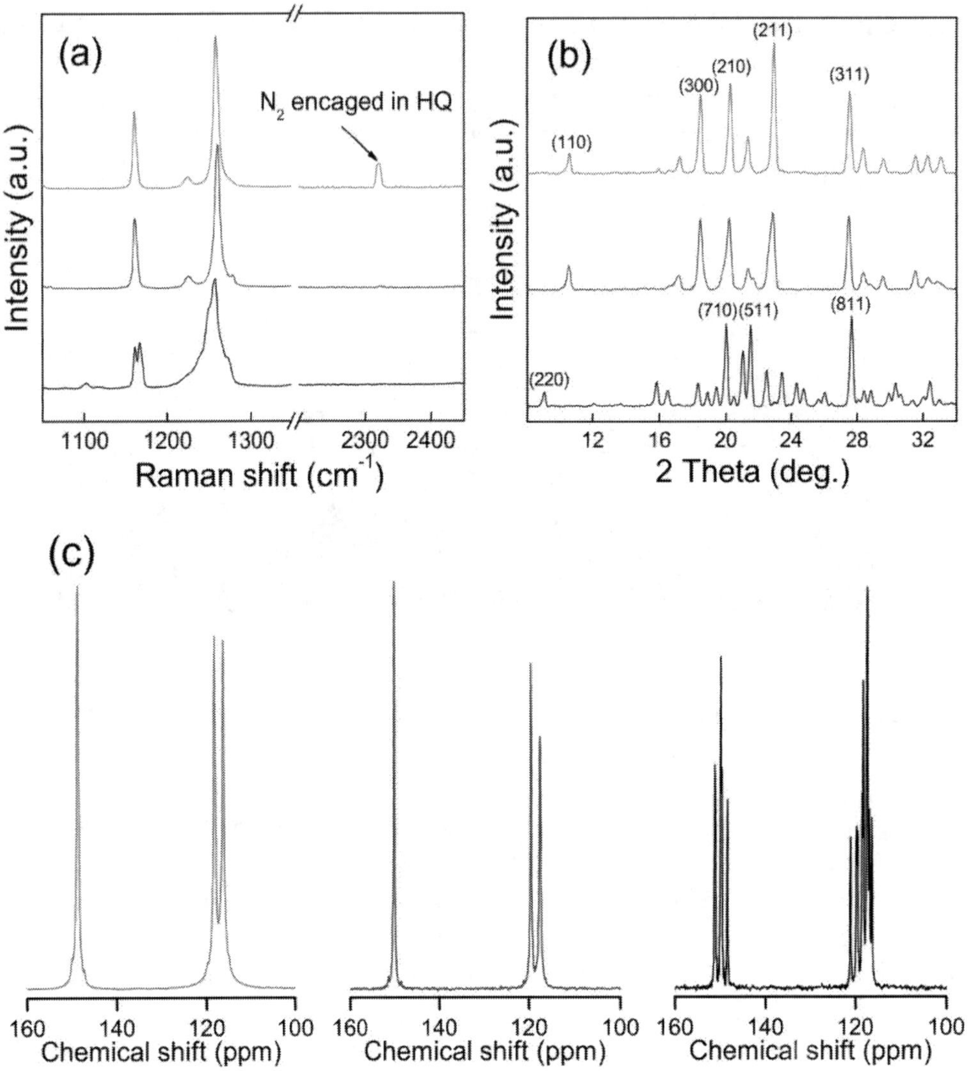

Figure 2. (a) Raman spectra, (b) X-ray diffraction patterns, and solid-sate ^{13}C NMR results of HQ compounds: (red) N_2-loaded β-form HQ clathrate. (blue) Guest-free β-form HQ clathrate. (black) α-form HQ.

Pressure-induced changes in the Raman spectra of N_2-loaded HQ clathrate are shown in Figure 3 in the frequency ranges of 1100–1250 and 2300–2450 cm^{-1}. All the C–O stretching vibrational modes (1160 cm^{-1}) of HQ (host) molecules and the N–N stretching vibrational modes (2322 cm^{-1}) of N_2 (guest) molecules exhibit substantial blue shift with increasing pressure. In addition, the Raman bands show progressive broadening and thereby decrease in intensity as pressure increases. This is the characteristic signature from the compression of the unit cell volume including the special contraction of the cage structure.

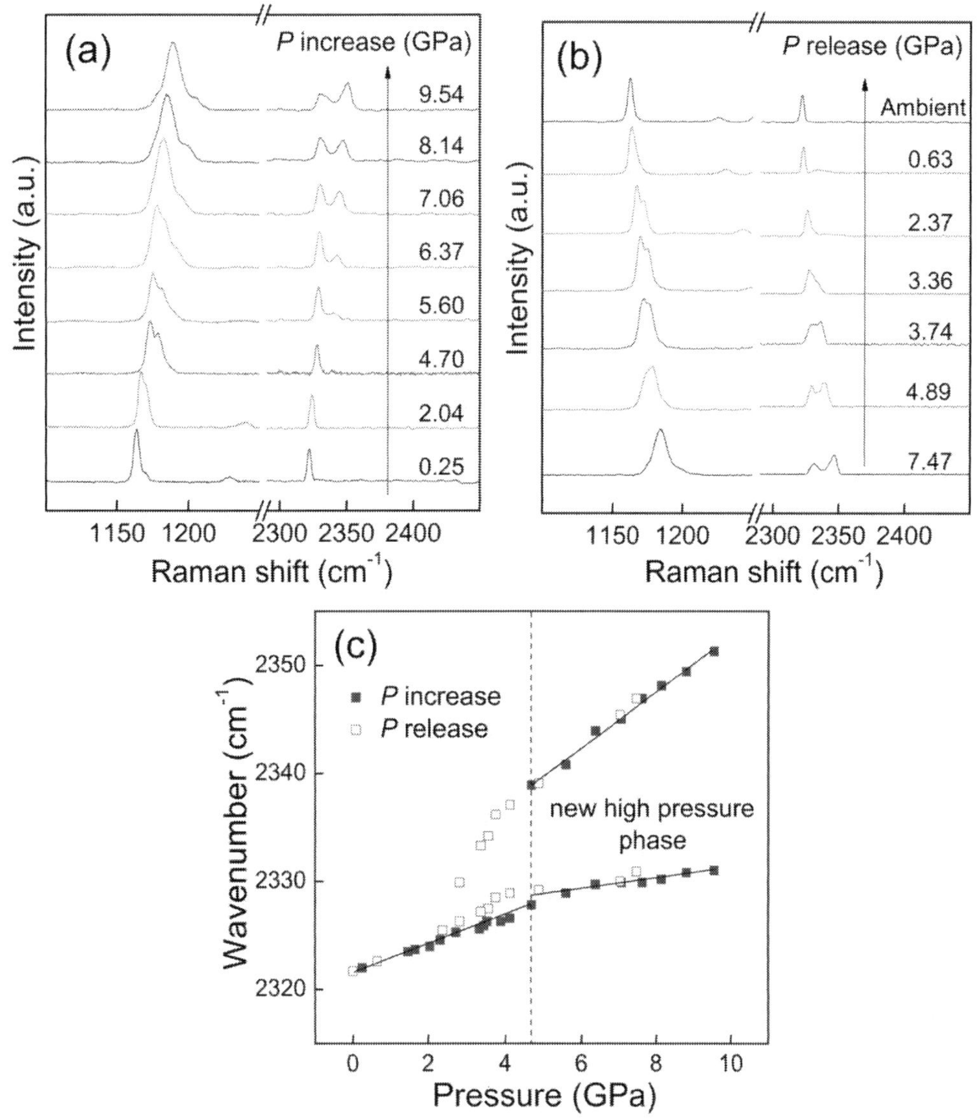

Figure 3. Pressure dependence of Raman spectra of N_2-loaded HQ clathrate (a) with increasing pressure and (b) upon releasing pressure. (c) Pressure dependence of Raman frequency for the symmetric N–N stretching mode of N_2 molecules in β-form HQ and new high pressure phase. Solid lines are the guide for the eyes.

Abrupt changes in the Raman spectra are observed at near 4.7 GPa (Figures 3 and S1), signaling the formation of a new high pressure phase. In particular, a new Raman peak appears in the higher frequency side to the N–N stretching vibrational mode of N_2 molecules encaged in the HQ clathrate framework. With further increasing pressure to 9.54 GPa, the intensity of the higher frequency peak gradually increases, whereas the intensity of the lower peak (for N_2 trapped in the cages) decreases. Furthermore, the higher frequency peak shows a remarkable increase in its frequency versus pressure, while the lower peak shows a progressive increase in frequency with pressure (Figure 3c). The frequency difference between two peaks is $\Delta v = 11.1\ cm^{-1}$ at 4.7 GPa and becomes larger with increasing pressure, leading to $\Delta v = 20.3\ cm^{-1}$ at 9.54 GPa. The pressure-induced shift of the Raman frequency peaks varies linearly with increasing pressure up to 10 GPa. We note that the pressure dependence of the higher frequency peak is very close to that of the N–N stretching vibrational mode (v2) of the N_2 solid phase.[15,16] From these results, we conjecture that the HOF of N_2-loaded β-form HQ clathrate exhibits its structural transition at near 4.7 GPa, and thereby the N_2 guests are removed from the cages of the HOF and transform to the N_2 solid phase. With further increasing pressure, the amount of the N_2 solid phase gradually increases, as shown in Figure 3a. It is interesting to note that, at ambient temperature, the δ–β phase transition of the N_2 solid occurs at 4.67 GPa.[17,18] This implies that the structural transformation of organic clathrate compounds might be associated with the phase transition of guest molecules encapsulated in the cages of the HOF. As a result, the higher frequency Raman peaks observed at pressures higher than 5 GPa can be assigned to the N–N stretching vibron of the N_2 solid δ–phase.

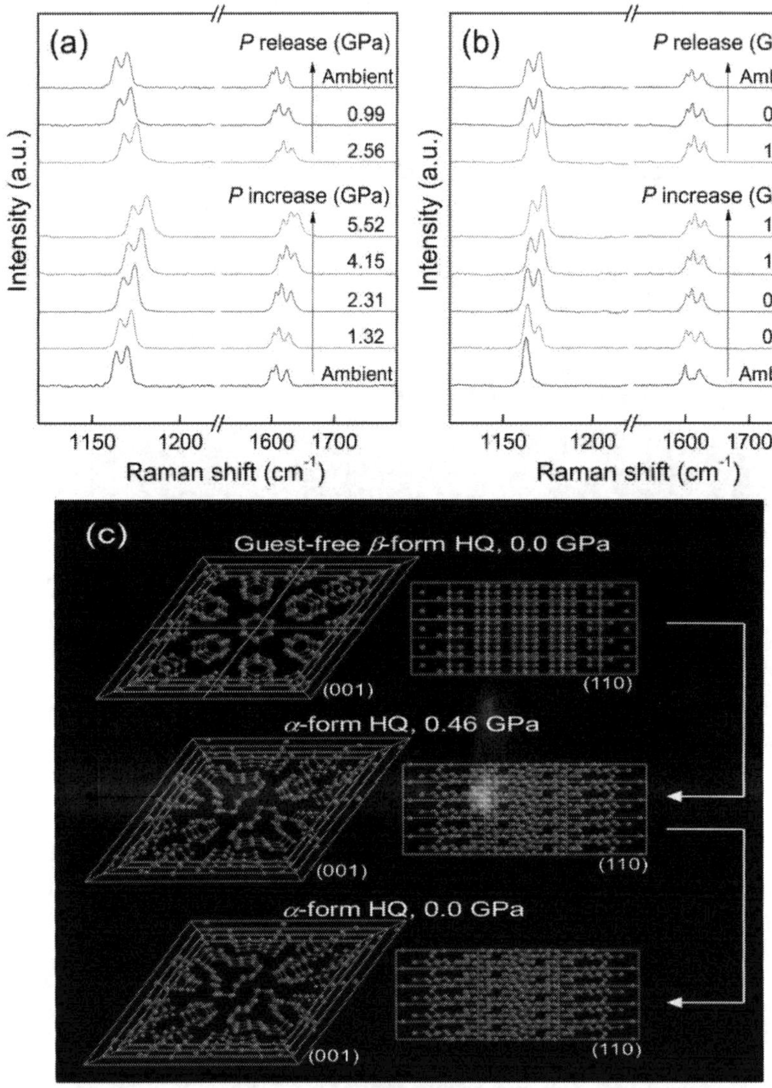

Figure 4. Pressure dependence of Raman spectra of (a) α-form HQ and (b) guest-free β–form HQ. (c) Hydrogen bonding networks of guest-free β–form HQ and α-form HQ, viewed perspectively along the (001) and (110) directions. Oxygen atoms of HQ molecules are shown as red spheres. Thick red sticks between oxygen atoms represent hydrogen bonding. White lines define a unit cell. Carbon and hydrogen atoms of HQ molecules are omitted for clarity.

The intensity of the Raman peak for the N_2 solid phase gradually decreases with decreasing pressure, and the discontinuous change in the Raman frequency is also observed at 4.7 GPa (Figures 3b and 3c), indicating the structural transition. With further decreasing pressure to 2.37 GPa, the Raman peak of the N_2 solid β–phase disappears, which is most likely due to enclathration of all N_2 molecules into the cages of the HOF. It should also be noted that, at ambient temperature, the melting of the N_2 solid β–phase to fluid N_2 occurs at 2.4 GPa.[17] Upon releasing pressure to ambient condition, the crystal structure is recovered to the original β-form HQ clathrate, indicating that the pressure-induced structural transition with N2 guests is completely reversible.

Pressure-dependent changes in the Raman spectra of guest-free β-form HQ clathrate are shown in Figure 4 and compared with α-form HQ in the spectral region of 1100–1250 and 1500–1800 cm^{-1} (Figure 4). From our previous studies on CH_4- and CO_2-loaded HQ clathrates,10,14 the relative intensity ratio of three unresolved peaks at around ~1600 cm^{-1} has been used as an indicator to determine complete conversion of α-form HQ to the β-form HQ clathrates. As shown in Figure 4a, there is no indication of structural transition for α-form HQ, as pressure increases up to 5.52 GPa. However, the guest-free β-form HQ clathrate partially transforms to α-form HQ at 0.24 GPa (Figure 4b). With further increasing pressure to 0.46 GPa, the crystal structure shows complete transformation to α-form HQ. Another important feature to be noted is that, upon releasing pressure to ambient condition, the crystal product remains the α-form HQ structure. This reveals that the pressure-induced structural transition of the guest-free β-form HQ clathrate is completely irreversible and the HOF of α-form HQ is more stable and dense at ambient condition (Figure 4c).

In summary, we presented here the pressure-induced structural changes of N_2-loaded and guest-free HQ clathrates. Even though they have the same crystalline structure of β-form HQ, the phase transition behavior under high pressure is completely different. The N_2-loaded β-form HQ clathrate shows the reversible structural transition to a new high pressure phase, whereas the guest-free β-form HQ clathrate undergoes the irreversible phase transition to nonporous α-form HQ. These results suggest that guest molecules encaged in the HOFs play an important role in the structural transitions under high pressure. Reorganizations of the HOF and modifications of the cage geometry point toward novel properties of clathrate compounds under pressure.

ACKNOWLEDGMENT

This research was supported by Basic Science Research Program (2010-0007026) and Nuclear R&D Program (M2AM06-2008-03931) through the National Research Foundation of Korea (NRF) funded by the Ministry of Education, Science and Technology.

References

1. Palin, D. E.; Powell, H. M. (1945), "Hydrogen Bond Linking of Quinol Molecules," *Nature*, 156, pp. 334-335.
2. Atwood, J. L.; Davies, J. E. D.; MacNicol, D. D. (1984), "Inclusion Compounds," *Academic Press*: FL.
3. van der Waals, J. H.; Platteeuw, J. C. (1959), "Clathrate solutions," *Adv. Chem. Phys.*, 2, pp. 1-57.
4. Palin, D. E.; Powell, H. M. (1947), "The Structure of Molecular Compounds. Part III. Crystal Structure of Addition Complexes of Quinol with Certain Volatile Compounds," J. Chem. Soc., pp. 208-221.

5. Lee, Y.-J.; Han, K.W.; Jang, J.S.; Jeon, T.-I.; Park, J.; Kawamura, T.; Yamamoto, Y.; Sugahara, T.; Vogt, T.; Lee, J.-W.; Lee, Y.; Yoon, J.-H. (2011), "Selective CO2 Trapping in Guest-Free Hydroquinone Clathrate Prepared by Gas-Phase Synthesis," *ChemPhysChem.*, 12, pp. 1056-1059.
6. Loveday, J. S.; Nelmes, R. J.; Guthrie, M.; Belmonte, S. A.; Allan, D. R.; Klug, D. D.; Tse, J. S.; Handa, Y. P. (2001), "Stable Methane Hydrate above 2 GPa and The Source of Titan's Atmospheric Methane," *Nature.*, 410, pp. 661-663.
7. Chou, I.-M.; Sharma, A.; Burruss, R. C.; Shu, J.; Mao, H.-K.; Hemley, R. J.; Goncharov, A. F.; Stern, L. A.; Kirby, S. H. (2000), "Transformations In Methane Hydrates," *Proc. Natl. Acad. Sci. U.S.A.,* 97, pp. 13484-13487.
8. Mao, W. L.; Mao, H. K.; Goncharov, A. F.; Struzhkin, V. V.; Guo, Q.; Hu, J.; Shu, J.; Hemley, R. J.; Somayazulu, M.; Zhao, Y. (2002), "Hydrogen Clusters in Clathrate Hydrate," *Science.*, 297, pp. 2247-2249.
9. Jang, J. S.; Jeon, T.-I.; Lee, Y.-J.; Yoon, J.-H.; Lee, Y. (2009), "Characterization of α-Hydroquinone and β-Hydroquinone Clathrates by THz time-domain Spectroscopy," *Chem. Phys. Lett.*, 468, pp. 37-41.
10. Lee, J.-W.; Lee, Y.; Takeya, S.; Kawamura, T.; Yamamoto, Y.; Lee, Y.-J.; Yoon, J.-H. (2010), "Gas-Phase Synthesis and Characterization of CH4-Loaded Hydroquinone Clathrates," *J. Phys. Chem. B.*, 114, pp. 3254-3258.
11. Nam, B.-U.; Kim, B.-S.; Lee, H.-H.; Yoon, J.-H. (2012), "Structural Transformation and Guest Dynamics of Methanol-loaded Hydroquinone Clathrate Observed by Temperature-dependent Raman Spectroscopy," *J. Phys. Chem. A.*, 116, pp. 2435-2438.
12. Lee, Y.; Lee, J.-W.; Lee, H.-H.; Lee, D.-R.; Kao, C.-C.; Kawamura, T.; Yamamoto, Y.; Yoon, J.-H. (2009), "High Pressure Investigation of α-form and CH4-loaded β-form of Hydroquinone Compounds," *J. Chem. Phys.*, 130, pp. 124511-124516.
13. Lee, Y.-J.; Han, K. W.; Jang, J. S.; Jeon, T.-I.; Park, J.; Kawamura, T.; Yamamoto, Y.; Sugahara, T.; Vogt, T.; Lee, J.-W.; Lee, Y.; Yoon, J.-H. (2011), "Selective CO2 Trapping in Guest-Free Hydroquinone Clathrate Prepared by Gas-Phase Synthesis," *ChemPhysChem.*, 12, pp. 1056-1059.
14. Lee, J.-W.; Choi, K. J.; Lee, Y.; Yoon, J.-H. (2012), "Spectroscopic Identification and Conversion Rate of Gaseous Guest-loaded Hydroquinone Clathrates," *Chem. Phys. Lett.*, 528, pp. 34-38.
15. Schiferl, D.; Buchsbaum, S.; Mills, R. L. (1985), "Phase Transitions in Nitrogen Observed by Raman Spectroscopy from 0.4 to 27.4 GPa at 15 K," *J. Phys. Chem.*, 89, pp. 2324-2330.
16. McCluskey, M. D.; Hsu, L.; Wang, L.; Haller, E. E. (1996), "Infrared Absorption of Solid Nitrogen at High Pressures," *Phys. Rev. B.*, 54, pp. 8962-8964.
17. Vos, W. L.; Schouten, J. A. (1989), "Improved Phase Diagram of Nitrogen Up to 85 Kbar," *J. Chem. Phys.*, 91, pp. 6302-6305.
18. Belak, J.; Lesar, R.; Etters, R. D. (1990), "Calculated Thermodynamic Properties and Phase Transitions of Solid N2 at Temperatures $0 \leq T \leq 300$ K and Pressures $0 \leq P \leq 100$ GPa," *J. Chem. Phys.*, 92, pp. 5430-5442.

(181al) Density Measurements of Tetrahydrofuran + Water Solutions At Atmospheric Pressure

Jose Domenzain-Gonzalez[1], Alfredo Pimentel-Rodas[1], Luis A. Galicia-Luna[1], Dominique Richon[2] and Octavio Elizalde-Solis[3], (1)INSTITUTO POLITECNICO NACIONAL-ESIQIE, Mexico, Mexico, (2)CEP/TEP - Centre Énergétique et Procédés, Mines Paris-Tech, Fontainebleau, France, (3)Departamento de Ingeniería Química Petrolera, INSTITUTO POLITECNICO NACIONAL-ESIQIE, México, D. F., Mexico

In this work, we report experimental density data for the binary mixtures of tetrahydrofuran + water at atmospheric pressure (0.785 mbar and 1 bar) and various temperatures. These data are compared with data reported in the literature. Obtained densities were determined using a commercial vibrating-tube densimeter from Anton Paar (model: DMA HPM). Experimental methodology was verified by measuring densities of heptane and the maximum deviation obtained was 0.01%. The mixtures were prepared in a flask under vacuum by the method of successive weighings. Excess molar volumes for the tetrahydrofuran + water system have been calculated using the experimental densities and correlated with the Redlich–Kister equation. Optimized parameters for this equation have been adjusted on experimental results. The partial molar volumes and partial excess molar volumes at infinite dilution have also been calculated for each component. The tendency of the data to higher water concentration is due to the effect called negative partial molar expansibility.

(181am) CO2 + H2O + 1-Propanol Semiclathrate Hydrate Phase Equilibrium

Ma Fernanda Sánchez-Mora[1], Luis A. Galicia-Luna[1] and Amir H. Mohammadi[2], (1)INSTITUTO POLITECNICO NACIONAL-ESIQIE, Mexico, Mexico, (2)CEP/TEP - Centre Énergétique et Procédés, Mines Paris-Tech, Fontainebleau, France

Clathrate hydrates are ice-like crystalline compounds wich form through a combination of water "host" and suitably sized "guest" molecules, tipically under low temperatures and elevated pressures. In clathrate hydrates, water molecules form hydrogen-bonded cavity structures encapsulating the guest molecules. Gas hydrates could cause operational problems in the chemical industry. To inhibit the formation of gas hydrates, glycols and alcohols are normally used. The knowledge about gas hydrate phases equilibrium in the presence of an inhibitor is of interest to calculate the limit of hydrate formation. In this work, we first describe the detail of an experiment for measuring the dissociation conditions of CO_2 + 1-propanol (5 and 10 %wt) + H_2O clathrate hydrates; then, we report the dissociation conditions at temperatures between (273 and 280) K and pressures up to 5 MPa, which were measured by an isochoric pressure-search method. The reported experimental values are compared with some selected experimental data from the literature and the agreement is found to be acceptable.

(181an) Vapor–Liquid Equilibrium for the Binary System Limonene + Propane

Ricardo García-Morales, Lab. Termodinamica, SEPI, INSTITUTO POLITECNICO NACIONAL-ESIQIE, Mexico, D.F., Mexico, Luis A. Galicia-Luna, INSTITUTO POLITECNICO NACIONAL-ESIQIE, Mexico, Mexico and Octavio Elizalde-Solis, Departamento de Ingeniería Química Petrolera, INSTITUTO POLITECNICO NACIONAL-ESIQIE, México, D. F., Mexico

Abstract

The increasing interest on alternative separation processes has made the supercritical fluid extraction an affordable method due to its several advantages. However, the lack of good experimental data makes really hard to know the phase behavior of studied systems and develop these new processes.

In this work, vapor-liquid equilibrium for the binary system limonene + propane was measured at three different temperatures (323, 348 and 373) K using a static-analytic method coupled to a mobile sampler-injector. The ROLSI sampler has the advantage of taking a small enough sample to avoid perturbations of the equilibrium conditions. Within pressure steady state, samples of the vapor or liquid phase were sent to the gas chromatograph to determine its composition using the above mentioned sampler and the analytical and static equipment are connected on-line with a thermo-regulated tubing. This work is the next step of a systematic study for the limonene extraction by using supercritical fluids.

(181ao) A New Equipment for Hydrate Formation: Measurements of Hydrate Phase Equilibria for Nitrogen + Water

Hugo I. Pérez-López, Laboratorio de Termodinámica, SEPI-ESIQIE, Instituto Politécnico Nacional, Mexico D.F., Mexico and Luis A. Galicia-Luna, INSTITUTO POLITECNICO NACIONAL-ESIQIE, Mexico, Mexico

Gas hydrates, or clathrate hydrates are, ice-like solid crystalline compounds, which are former through combination of water and small guest molecules, like nitrogen, under low temperatures and elevated pressures. Gas hydrates formation phenomenon has been used for developing separation processes. In this work we describe the details for experimental equipment developed for measuring the dissociation conditions of hydrates. The equipment is suitable to operate at temperatures ranging from (243 to 423) K and pressures up to 40 MPa. The present methodology is based on the isochoric search method (search for equilibrium pressure). Experimental dissociating conditions for nitrogen + water are reported up to 35 MPa and in the temperature interval of (272 and 280) K in order to validate the experimental methodology. The dissociation data generated in this study together with literature data are compared with prediction of thermodynamic model. Good agreement between the data indicates the reliability of the experimental technique used in this study.

(181ap) Volumetric Properties for the Carbon Dioxide + Heptane + Dodecane

Luis Camacho-Camacho[1], Luis A. Galicia-Luna[1] and Octavio Elizalde-Solis[2], (1)INSTITUTO POLITECNICO NACIONAL-ESIQIE, Mexico, Mexico, (2)Departamento de Ingeniería Química Petrolera, INSTITUTO POLITECNICO NACIONAL-ESIQIE, México, D. F., Mexico

This work reports (P, r, T, xi) measurements in the compressed fluid phase for the ternary system carbon dioxide – heptane – dodecane in the (313 to 363) K temperature range and at pressures up to 25 MPa. The obtained data were measured by means of a static-synthetic method based apparatus which couples a feeding cell with an Anton Paar vibrating tube densimeter reported in previous papers. Measurements were performed with an estimated uncertainty, to be less than ±0.01 MPa for pressure, ±0.03 K for temperature and ±0.001 for . Obtained data were correlated with the five-parameter Modified Toscani Swarcz equation to represent the behavior of this mixture. Derived thermodynamic properties (isobaric expansivity, isothermal compressibility, thermal pressure coefficient and internal pressure) were also calculated by using the parameters of the equation above mentioned. PrT properties for this ternary system have not been reported in open literature before this work. Experimental data obtained form part of a major study about vapor-liquid equilibria, densities and solubility of n-alkanes with or in carbon dioxide.

(181aq) Compressed Liquid Densities for Carminic Acid + Ethanol in the Temperature Range of (313 to 363) K

Rubén P. Mendo-Sánchez, Laboratorio de Termodinámica, SEPI-ESIQIE, Instituto Politécnico Nacional, Mexico D.F., Mexico, Luis A. Galicia-Luna, INSTITUTO POLITECNICO NACIONAL-ESIQIE, Mexico, Mexico and Octavio Elizalde-Solis, Departamento de Ingeniería Química Petrolera, INSTITUTO POLITECNICO NACIONAL-ESIQIE, México, D. F., Mexico

Thermodynamic properties obtained through experimentation are the fundamental basis for the development of models used to represent and predict the behavior of fluids. Volumetric properties are important in chemical engineering for the design, simulation and optimization of separation processes. This work presents density measurements for the binary mixture (carminic acid + ethanol) at temperatures from (313 to 363) K and pressures from (10 to 200) bar using a commercial vibrating tube densimeter (VTD) from Anton Paar calibrated with water and nitrogen as reference fluids. The VTD was connected through a tubing line to a loading cell in order to perform density measurements for the studied system. A needle valve allows the homogeneous fluid contained in the loading cell to flow and fill the VTD. The method used was the static synthetic. In order to observe directly the system behavior, the experimental vibrating period, temperature and pressure are registered and plotted using a data acquisition program (Agilent, VEE 7.0). The multimeter, temperature indicator and evaluation unit are communicated with the computer through RS-232/Bluetooth and a multi-parallel RS-232/USB serial interface adapter (Model: 2403, Sealevel) Experimental data were correlated with 6-parameter equation.

(181ar) Application of the CPA Equation of State to Ionic Liquids and Their Mixtures

Filipa M. Maia[1], Ioannis Tsivintzelis[2], Oscar Rodríguez[1], Georgios Kontogeorgis[2] and **Eugenia A. Macedo**[1],
(1)Departamento de Engenharia Química, Faculdade de Engenharia da Universidade do Porto, Porto, Portugal,
(2)Chemical and Biochemical Engineering (CERE), Technical University of Denmark, Kgs. Lyngby, Denmark

Ionic liquids (ILs) are a class of solvents with certain characteristics that make them very different than traditional molecular solvents. They are defined as ionic compounds with a melting temperature lower than 100 °C and they are typically constituted by bulky organic cations (like imidazolium, pyridinium, pyrrolidinium, phosphonium or ammonium) and organic or inorganic anions (for example, bis(trifluoromethylsulfonyl)imide, tetrafluoroborate or halides). Some of their attractive properties are very low vapour pressure, high chemical and thermal stability, selectivity and solvation capability, and easiness of recovery [1]. The typical cations that constitute ILs allow different substituents, like alkyl chains of different lengths or even with other functional groups. This leads to a huge number of different possible cations and anions, which can originate an enormous variety of ILs. Additionally, it is possible to adjust some of the properties of ILs (like melting point, density, viscosity or hydrophobicity) by making simple changes in the structure of its ions. For this reason, ILs are regarded as 'designer solvents' which can be specifically tailored for each application. Consequently, there has been a growing interest in ILs for applications in different fields such as catalysis, analytical chemistry, electrochemistry, separation processes or biotechnology.

With such a vast number of different ILs available, extensive experimental work becomes time-consuming and expensive. Therefore, it would be of great value to have predictive tools for the calculation of certain properties of ILs as well as their mixtures with other components. In this work, the Cubic Plus Association (CPA) equation of state [2] has been applied to model the properties of two ILs and their phase behavior in mixtures with CO_2 and water. The two ILs are 1-ethyl-3-methylimidazolium bis(trifluoromethylsulfonyl)imide ([C_2mim][NTf_2]) and 1-butyl-3-methylimidazolium bis(trifluoromethylsulfonyl)imide ([C_4mim][NTf_2]). Several sets of pure compound parameters were estimated for both ILs by fitting the model results to experimental vapor pressure and liquid density data. Associating schemes 2B and 4C were considered for the ILs. According to the results obtained, the model is able to accurately describe the properties of the pure ILs. After this, the sets of pure compound parameters were used to model the vapor-liquid equilibria (VLE) of the two ILs with CO_2. A good description of the experimental VLE data was achieved for both ILs using one temperature-independent binary interaction parameter. The average absolute deviations (AAD) obtained were around 8-13% for [C_2mim][NTf_2] and 12% for [C_4mim][NTf_2]. Similar results were obtained for the modeling of the VLE using both associating schemes considered for the ILs. Finally, the liquid-liquid equilibria (LLE) of the two ILs with water were also modeled. For this kind of phase equilibrium, the model does not provide equally satisfactory results. For the LLE with water and considering associating scheme 4C for the ILs, CPA cannot accurately describe the phase behavior. However, with the optimum sets of pure compound parameters for associating scheme 2B and a temperature-independent binary interaction parameter, good results are obtained, with AAD of 7.3 and 42.4% for the IL-rich phase (for [C_2mim][NTf_2] and [C_4mim][NTf_2], respectively) and 15.5 and 8.6% for the water-rich phase. Additional work is ongoing for the application of the CPA equation of state to other ILs from the same family as well as from other families.

References:

[1] Welton, T. (1999) Chem. Rev. 99, 2071-2084.

[2] Kontogeorgis, G.M., Voutsas, E.C., Yakoumis, I.V., Tassios, D.P. (1996) Ind. Eng. Chem. Res. 35, 4310-4318.

(181as) Vapor-Liquid Equilibria of the Binary Systems (N,N,N,N'-Tetramethylethylenediamine + Water), (Tetramethylpropylenediamine + Water) or (Diethylenetriamine + Water) At Several Temperatures

Zahida Bouzina[1], Amina Negadi[1], Ilham Mokbel[2], Jacques Jose[2] and Latifa Negadi[1], (1)Laboratoire de Thermodynamique Appliquée et Modélisation Moléculaire, Université de Tlemcen, Tlemcen, Algeria, (2)Laboratoire des Sciences Analytiques-UMR 5280, UCB-Lyon1, Villeurbanne, France

Amines are an important class of compounds used in different fields of industry, they represent a particularly interesting family of molecules for the purpose of testing group-contribution models and analyze of the intramolecular effects, especially the proximity effect.

The present paper is part of a research program on thermodynamic properties of aqueous solutions of amines. The purpose of this work is to investigate vapor-liquid equilibria (VLE) of (N,N,N',N' tetramethylethylenediamine (TMEDA + water), (tetramethylpropylenediamine (TMPDA) + water), or (diethylenetriamine (DETA) + wáter) binary mixtures with a view to use the results to determine the interaction parameters for predictive group contribution methods. A survey of the literature shows that there is no data available in the literature.

The vapor pressures of (TMEDA + water), (TMPDA + water), or (DETA + water) and of all the pure components were measured by means of two static devices at temperatures between 273 and 363 K. The data were correlated with the Antoine equation. From these data, excess Gibbs functions were calculated for several constant temperatures and fitted to a fourth-order Redlich-Kister equation using Barker's metho.

The NRTL and UNIQUAC models have also been used and good results were obtained in the prediction of the total pressure.

(181at) Investigation of the Thermodynamic Properties of (Furfuryl Alcohol + Toluene), (Furfuryl Alcohol + Ethanol), or + (Furfural + Toluene) Binary Mixtures At Several Temperatures

Latifa Bendiaf[1], Amina Negadi[1], Ilham Mokbel[2], Jacques Jose[2] and Latifa Negadi[1], (1)Laboratoire de Thermodynamique Appliquée et Modélisation Moléculaire, Université de Tlemcen, Tlemcen, Algeria, (2)Laboratoire des Sciences Analytiques-UMR 5280, UCB-Lyon1, Villeurbanne, France

Vapor-liquid equilibrium (VLE) data are essential for the design of separation processes and equipment as well as for the extension of thermodynamic models. The present paper is part of a research program concerning the investigation of the thermodynamic properties of biofuels containing systems.

The vapor-liquid equilibria data for the (furfuryl alcohol + toluene), (furfuryl + ethanol), or (furfuryl + toluene) binary mixtures measured using a static device at temperatures between 263 and 363 K are reported.

The data were correlated with the Antoine equation. From these data excess Gibbs functions (G^E) were calculated for several constant temperatures and fitted to a fourth-order Redlich-Kister equation using the Barker's method. The three binary systems exhibit positive deviations in G^E for all investigated temperatures over the whole composition range.

The NRTL, UNIQUAC and UNIFAC models have also been used and good results were obtained in the prediction of the total pressure and thermodynamic excess properties.

(181au) Using an Integration Technique to Trace Coexistence Curves: Applications to Multi-Component Phase Equilibrium

Suma Magadi, The University of Akron, Akron, OH and **Donald P. Visco Jr.**, University of Akron, Akron, OH

Process simulation is an important tool in the design, development, analysis and optimization of technical processes. The simulation work, performed on computers, is mainly applied to chemical plants, chemical and petroleum engineering processes. Thermodynamic computations in process simulation, both in the field of physical properties and of phase equilibria, consumes about 60 to 80% of the computer time of which phase equilibrium calculations take up about 75% of the time. Phase equilibrium calculations mostly make use of conventional root finding algorithms. Hence, the development of faster and more efficient algorithms for thermodynamic computations becomes highly essential. To this end, a technique known as the Integrate method has been developed that involves the differential form of the equilibrium constraints. These relationships are solved using numerical integration techniques and have been shown to produce results that both take less time (than the root-finding technique) and are of useful accuracy. In the current study, we evaluate the method for use in VLE and LLE for a variety of equations of state, from van der Waals to SAFT.

Self-Consistency Analysis of Physical Property and Molecular Descriptor Databases Using a Variety of Prediction Techniques - ID# 255431

Mordechai Shacham, Chem. Eng. Dept., Ben-Gurion University, Beer-Sheva, Israel

Michael Elly, Chem. Eng. Dept., Ben-Gurion University, Beer-Sheva, Israel

Inga Paster, Chem. Eng. Dept., Ben-Gurion University, Beer-Sheva, Israel

Neima Brauner, School of Engineering, Tel-Aviv University, Tel-Aviv, Israel

Abstract

The dominant descriptor version of the targeted QSPR method (TQSPR1, Shacham and Brauner, Chem. Eng. Sci., 66, 2606, 2011) is used to predict 16 constant properties for close to 1800 compounds in order to assess the range of applicability of the prediction technique and to analyze the consistency of a property and molecular descriptor database.

It is demonstrated that the TQSPR1 method can model the properties within the data uncertainty level for most groups of compounds included in the database. Common causes of poor prediction accuracy were identified as: 1. inconsistencies in the 3D molecular structure of the compounds; 2. use of molecular descriptors outside their range of applicability; 3. mixing compounds at different phases for properties defined at a standard state and 4. use of descriptors whose asymptotic behavior do not match that of the modeled property for long range extrapolation to high carbon number compounds.

The results of this study enable improving the consistency of the physical property and descriptor databases for increasing the robustness of QSPRs that can be derived for a variety of properties and compounds.

Introduction

Pure-compound property data are widely used in process design, simulation and optimization, environmental impact assessment, hazard and operability analysis and additional diverse areas as chemistry and chemical engineering. Presently pure compound property databases are the (such as the DIPPR database, Rowley et al. 2010) serve as the main sources of property data. For providing the maximal benefits to the users, these databases typically contain both experimental and predicted data. Both types of data are associated with certain levels of uncertainty. For experimental data the challenge is the selection of the "best" value from several reported values, while for predicted values, only an average "prediction error" is reported, which is based on a particular training set of compounds and may not be applicable to the particular compound considered. New, more accurate prediction techniques are being continuously developed that enable replacing older less accurate predicted values by more accurate ones.

However, periodical screening of the database for identifying data points that need to be replaced represent a great challenge, because of the huge amount of data involved. To make this task easier, a system that can screen all the data in the database and flag potentially erroneous or low precision data is required. Once such a point has been identified, various state of the art property prediction techniques, such as the multi-level group contribution method of Marrero and Gani, 2001, asymptotic behavior correlations (ABCs, Marano and Holder, 1997), Targeted Quantitative Structure Property Relationships (TQSPRs, Brauner et al., 2006) and the Reference Series method (Shacham et al., 2012) can be used to determine whether the particular value needs to be and can be replaced by a more accurate one.

The QSPR type property prediction methods require molecular descriptors for representing the structure of the molecule. In recent years computer programs that can calculate several thousands of descriptors have emerged. Checking the accuracy and consistency of the molecular descriptors, and the correctness of the associated molecular structure files (that often provided in MOL format), represent additional major challenges. Flagging potentially erroneous property data points can also help in identifying incorrect MOL files.

In the system we have developed, the single (dominant) descriptor version of the TQSPR method (Shacham et al., 2007) is used for the initial screening of the database by selecting in turn every compound (*target* compound) in the database and predicting all its available constant properties. If the difference between the recommended database value and the predicted value is considerably higher than the uncertainty value assigned to the property in the data base, the data point is flagged as potentially erroneous. The flagged data points require additional analysis, as there can be various sources to the large differences.

In the following some of the results of the consistency analysis of the physical property and molecular descriptor databases are reported.

The Dominant Descriptor Targeted QSPR Method (TQSPR1)

The Dominant Descriptor Targeted QSPR (TQSPR1) method was described in detail by Shacham and Brauner (2011). Only a brief review of the method is given here. The first stage of the method involves identification of a training set structurally related to the target compound. The similarity between the *target* compound (the compound for which the property need to be predicted) and a potential *predictive* compound is measured by the partial correlation coefficient between the vector of the molecular descriptors of the target compound and that of a potential predictive compound. The training set is established by selecting the p predictive compounds (usually $p = 10$) with the highest correlation with the target compound for which experimental property values y_i are available.

The selected training set is used for the development of a TQSPR1 model for a particular property of the target compound. A linear structure-property relation is assumed of the form:

$$\mathbf{y} = \beta_0 + \beta_1 \boldsymbol{\zeta}_D + \boldsymbol{\varepsilon} \qquad (1)$$

In this equation $\mathbf{y}$ is a p-dimensional vector of the respective property values, $\boldsymbol{\zeta}_D$ is a p-dimensional vector of the (dominant) molecular descriptor (to be selected via a stepwise regression algorithm), β_0, β_1 are the corresponding model parameters to be estimated, and $\boldsymbol{\varepsilon}$ is a p-dimensional vector of random errors.

To identify the dominant descriptor, we examine the partial correlation coefficient between the vector of (target) property value of the compounds included in the training set (*i.e.*, $\mathbf{y}$) and the vector of descriptor values for these compounds, for all descriptors available in the database. These correlation coefficients will be further called *descriptor-property* (*D-P*) correlation coefficients for the particular property. The dominant descriptor, ζ_D, is the descriptor that has the *highest D-P correlation coefficient*. The TQSPR1 so-obtained can be subsequently employed for calculating estimated property values for the target compound and for other compounds in the similarity group that do not have measured data, by applying Eq. 2:

$$\tilde{y}_t = \beta_0 + \beta_1 \zeta_{Dt} \qquad (2)$$

In this equation $\tilde{y}_t$ is the estimated unknown property value of the targeted compound and ζ_{Dt} is the corresponding dominant molecular descriptor value.

Measures of the consistency of the predicted property values

There are several measures associated with the training set, which can be used for assessing the consistency of the predicted target property values for the members of the training set and also for the target compound. To represent the quality of the fit of the TQSPR1 for the members of the training set, a *Training Set Average Error* (*TSAE*) is used:

$$TSAE(\%) = \frac{100}{p}\sum_{i=1}^{p}\left|y_i - (\beta_0 + \beta_1\zeta_i)\right|/y_i \qquad (3)$$

The advantage of *TSAE* over statistical measures (such as variance or standard deviation) is that it can be compared with the uncertainty values of the reported data (U_i) or the maximal uncertainty value (U_{max}). A consistent set of property values for the training set should yield $TSAE << U_{max}$. If no reported property value (y_t) is available for the target compound the *TSAE* can be used as an estimate for the prediction error (δy). Otherwise, the difference between the reported and predicted values is used as an estimate for the prediction error for the target compound:

$$\delta y(\%) = 100(y_t - \tilde{y}_t)/y_t \qquad (4)$$

If $\delta y < U_t$ (where U_t is the uncertainty of the target property data) the predicted property value for the target compound is consistent with the property values of the members of the training set. Detection of inconsistency in predicted values requires investigation of its source, since the reason is often the existence of erroneous or inconsistent property and/or descriptor values in the database.

The range of *variation of the property value within the training set*, *TSRV*, may have an effect when converting absolute errors to relative errors. TSRV is defined by

$$TSRV(\%) = 100\left|\frac{(y_{max} - y_{min})}{y_{avg}}\right| \; ; \; y_{avg} = \frac{1}{p}\sum_{i=1}^{p} y_i \qquad (5)$$

where y_{avg} is the average of the property value for the members of the training set

Methodology

The database used in this study was described in detail by Shacham and Brauner, 2011. It contains physical property data for 1798 compounds. Included in this data base are numerical values and data uncertainty for 34 properties (critical properties, normal melting and boiling temperatures, heat of formation, flammability limits etc.). All the property data is from the DIPPR database (Rowley et al., 2010).

The database contains 3224 molecular descriptors generated by the Dragon, version 5.5. software (DRAGON is copyrighted by TALETE srl, http://www.talete.mi.it) from minimized 3D molecular structures that were obtained from Rowley, 2010.

A Visual Basic program that uses the TQSPR1 method was developed. The program can operate in batch mode and attempts to predict all the properties for all the compounds in the data base. Most of the computations reported were carried out using this program. For the examples presented here, some additional details were computed with a special version of the

SROV (MATLAB) program of Shacham and Brauner (2003), which was revised in order to fit the needs of the TQSPR1 algorithm.

For identification of the training sets, a subset of 294 selected descriptors mostly for the 1-D from the "atom centered fragment" and "functional group count" categories were used. This subset was selected based on its ability to identify compounds belonging to the target compound's homologous series (if available) and discriminate between compounds according to the number and location of branches and double bonds (Paster, 2012).

Consistency analysis for normal boiling temperature of n-propyl formate and its training set

The details of the consistency analysis for a target compound and its training set for a particular target property is demonstrated by developing a TQSPR1 model for normal boiling temperature (T_b) of n-butyl formate (number of carbon atoms, $n_C = 4$) as target compound. The target compound belongs to the *n*-formate homologous series.

Data regarding the molecular structure of the target compound and the first 10 similar compounds that were identified as training set members are shown in Table 1. The first three members of the training set are the immediate neighbors of the target compound in the *n*-formate series, however there are also compounds in the training set that does not belong to these series: isobutyl and sec-butyl formate, ethyl and *n*-propyl propionate and n-propyl acetate. All the compounds comprise of the same functional groups (CH_3, CH_2 and COO), however there are differences in the location of the various groups. The compounds included contain between 3 to 8 carbon atoms. There are five compounds in the training set for which $n_C = 5$ which is the closest to the target compounds' $n_C = 4$.

For prediction of T_b for the target compound, property data for the members of the training set are needed. The T_b values and their uncertainties (from the DIPPR database), for the target compound and the members of the training set, are shown in Table 2. The uncertainties for all but one compound are <1%, for compound No. 6 it is <3%.

From among the 3224 descriptors, the one that has the highest correlation with T_b is the *Xu* index. This descriptor belongs to the 2D "topological" descriptors category and it is calculated from the adjacency and distance matrices of the molecule. Figure 1 shows the plot of the T_b data of the training set members versus *Xu*. The T_b values are aligned along a straight line: $T_b = 235.9526+19.4903Xu$, with a correlation coefficient of $R^2 = 1.0$. Using this TQSPR and the descriptor value $Xu = 5.969$ for the target compound yields $T_{b,target} = 352.29$ K, which is different from the DIPPR recommended value (353.97 K) by $\delta y = 0.47$ %. Introducing the training set's δy values from Table 2 into Eq. 4 yields: $TSAE = 0.447$. Thus, in this case the T_b values of the target compound and the training set members are consistent within the data uncertainty level and the *TSAE* provides a very good estimate for the prediction error of the target compound's property value.

Consistency analysis for sixteen properties of *n*-propyl formate and its training set

Analysis similar to the analysis shown in the previous section for T_b was carried out for 15 additional properties. The selected 16 constant properties included in this study from the DIPPR database are listed in Table 3. Included in the Table are the symbols of the properties (as defined by DIPPR), short descriptions of the properties and their units. Out of the 34 properties, 18 were excluded for several reasons. Some were excluded because they are categorized as "defined", meaning that they can be calculated from other properties and/or from the molecular structure (for example critical compressibility factor which is calculated from critical temperature, pressure and volume or molecular weight which can be calculated from the atomic weight of the individual atoms). Excluded are also properties for which most of the values in the database are predicted and/or of very high uncertainty).

The TQSPR1 method was used to predict the 16 properties shown in Table 3 for the target compound, *n*-propyl formate. The results of the prediction are shown in Table 4.

The following information is presented for each property predicted: the maximal uncertainty of the reported property values in the training set: U_{max}, *TSRV* (Eq. 5), the parameters (β_0, β_1) and the (dominant) descriptor used in the TQSPR1 model, *TSAE* (Eq. 3), the pertinent descriptor value for the target compound, the value of the target property value and the uncertainty as reported by DIPPR, the TQSPR prediction for the property value, and the δy (Eq. 4)

The same training set (shown in Table 1) was used for all the properties, except for *AIT*. In case of AIT, four members of the training set had to be replaced by compounds less similar to the target because of the lack of their *AIT* values in the database.

Observe that $TSAE < U_{max}$ for all the properties included in Table 1, implying that these property values are consistent within the data uncertainty level for the members of the training set. As for the predicted property values, for most properties $\delta y < U_t$. In the two cases where δy is marginally larger than U_t (*FP*, *TC*) the higher U_{max} value can explain the difference. Thus, for the 16 properties included in the study the predicted values are consistent with the training set data and the prediction errors are within the experimental error limits.

The *TSRV* (Eq. 5, 4th column in Table 4) needs to be considered when comparing *TSAE* and δy values for various properties. These relative error indicators can be scaled to the range of variation of the property value within the training set (by dividing them by TSRV).

High prediction error caused by inconsistency in the 3D molecular structure files

Table 5 shows a summary of results of the prediction of selected properties for selected compounds. The results shown in this table will be used for demonstrating some common pitfalls and reasons for excessive prediction errors.

The first case involves the prediction of the critical volume (V_C) for 1-decanol. The QSPR in this case includes the descriptor *Ds*. This descriptor belongs to the 3-D WHIM descriptors category and it represents " D total accessibility index / weighted by atomic electrotopological states". Comparing the *TSAE* value (*TSAE* = 5.73 %) with *Umax* (*Umax* = 25 %) shows that the TQSPR1 is consistent with the training set's V_C data within experimental error level. However, for the target compound, 1-decanol the prediction error is excessive (δy = 117 %). The detailed data and results of the prediction, for this case are shown in Table 6. Observe that all the training set members belong to the 1-alkanol homologous series. In Fig. 2 the V_C data for the members of the training set and the target compound are plotted versus the descriptor *Ds.* All the members of the training set (except for 1-tetradecanol and 1-eicosanol) are aligned nicely along a straight line. For 1-tetradecanol and 1-eicosanol their larger distance from the straight line can be explained by their higher level of uncertainty (U_i < 25 %). However the V_c data for 1-decanol is clearly an outlier in this plot. Plot of the descriptor *Ds* vs. n_C indicates that the *Ds* descriptor value is inconsistent with the descriptor values of the other members of the 1-alkanol series.

Graphical representation of the 3-D molecular structure (MOL) files of 1-decanol and its two immediate neighbors in the homologous series (1-nonanol and 1-undecanol, Fig. 3) reveals the cause of the inconsistency of the *Ds* values. The reason is that the $-CH_2OH$ group in the 1-

decanol structure file is rotated relatively to its position in the structure files of the other members of the 1-alkanol series. Indeed, upon modifying the 1-alkanol structure as to match the configuration of the other members of the 1-alkanol series, the resulting descriptor value is *Ds* = 0.229. Introducing this value into the TQSPR1 shown in Table 5 yields V_C = 0.541 m^3/kmol, where the prediction error is δy = 16 %. While such an error is still considered as relatively large, its magnitude can be attributed to the *Umax* value (= 25 %) of the training set data.

The second case of this category involves the prediction of the liquid molar volume (*LVOL*) for *n*-hexatriacontane. The QSPR in this case includes the descriptor *Mor22v*. This descriptor belongs to the "3D-MoRSE descriptors" group and it represents "3D-MoRSE - signal 22 / weighted by atomic van der Waals volumes". The training set in this case consists of members of the n-alkane homologous series. Comparing the *TSAE* value (*TSAE* = 0.09 %) with *Umax* (*Umax* = 5 %) shows that the QSPR represents excellently the training set data. As in the previous case, for the target compound, n-hexatriacontane the prediction error is excessive (δy = 37 %). Analysis of the results indicates that the *Mor22v* descriptor value is inconsistent with the descriptor values of the other members of the n-alkane series. Comparing the molecular structure file of *n*-hexatriacontane with the files of the rest of the n-alkane group shows that the *n*-hexatriacontane structure is rotated (with respect to the direction of the –H atoms in the -CH_2 group) with respect to the other compounds of the series. Revising the *n*-hexatriacontane structure to match the structure of the rest of the n-alkane series yields *Mor22v* value of 2.11 (instead of 1.241). Introducing this descriptor value into the TQSPR shown in Table 5 yields the estimated *LVOL* value of 0.652 m^3/kmol with δy = -0.23 %. Thus, in this case a set of *Mor22v* descriptors, which is obtained by using consistent molecular structures files of the training set and the target compound, yields a very accurate *LVOL* prediction for the target compound.

We have carried out some studies (Paster et al., 2009 and 2010) on the effects of the 3-D molecular structure variation on the calculated values of the 3-D descriptors and have found that some groups of descriptors are more sensitive to variations of the 3D structure than the others. The present study shows, however, that the dangerous situation arises when the 3D structure of the target compound is inconsistent with those of the training set members. Thus, it is very important to obtain the 3D optimized structure of the target compound using the same tools and algorithms that were used for the rest of the compounds of the database.

High prediction error caused by phase change within the training set and/or the target compound

Several properties which were included in this study are measured at a standard state of *T* = 298.15 K and *P* = 100 kPa. These include *ENT*, *HCOM*, *HFOR*, *HSTD*, *RI* and *SSTD* (see Table 3). The properties *ENT* and *HFOR* are defined for the (ideal) gas phase, while *HCOM*, *HSTD*, *RI* and *SSTD* correspond to the substance actual phase at the standard state. The property *LVOL* is defined as liquid molar volume at 298.15 K, but it is actually measured at T_b if T_C < 298.15 K, and at the triple point temperature if the latter exceeds 298.15 K. Thus, when attempting to predict *HCOM*, *HSTD*, *RI*, *SSTD* and *LVOL*, the inclusion of compounds in the training set which are in a different phase than the target compound at the standard state may have a considerable impact on the prediction accuracy.

In the 3rd row of Table 5 a summary of results of the prediction of *SSTD* for n-tridecane is shown. The QSPR in this case includes the descriptor *EEig13d*. This descriptor belongs to the 2-D "edge adjacency indices" category and it represents the " eigenvalue 13 from edge adj.

matrix weighted by dipole moments". Comparing the *TSAE* value (*TSAE* = 6.88 %) with *Umax* (*Umax* = 3 %) shows that the QSPR does not represent adequately the training sets *SSTD* data. Also, for the target compound, *n*-tridecane the prediction error is larger than *Ut* and *Umax* (*Ut* = 1%, *Umax* =3% while δy = 3.9 %). The detailed data and results of the prediction, for this case are shown in Table 7. Observe that all the training set members belong to the *n*-alkane homologous series. In the range of $12 \leq n_C \leq 17$ (up to *n* -heptadecane) the *n*-alkanes are in liquid phase at 298.15 K. For $n_C \geq 18$ the *n*-alkanes are in solid phase at this temperature.

Figure 5 depicts the *SSTD* data for the members of the training set and the target compound versus the descriptor *EEig13d.* There are clearly three separate groups of data points and the data cannot be represented by a straight line as indicated by the very low correlation coefficient value of R^2 = 0.24. One of the reasons for that is the sharp "step" change in the *SSTD* value when moving from the liquid phase to solid phase compounds in standard state. The liquid and solid phase compounds clearly belong to two separate populations as shown in Fig. 6 where *SSTD* is plotted versus n_C.

Another problem is that the value of the descriptor *EEig13d* is zero for the compounds n-dodecane (n_C = 12) and n-tridecane (n_C = 13). These zero values are inconsistent with the *EEig13d* descriptor values for $n_C \geq 14$ and they are generated by the Dragon program to indicate that this descriptor is undefined for $n_C \leq 13$. To prevent such errors, which are caused by replacing undefined descriptor values by "pseudo" zeros, undefined descriptor values should be clearly identified. Descriptors that are undefined for some of the training set members or for the target compound must be removed from consideration while selecting the dominant descriptor.

To find a satisfactory TQSPR1 for representing the *SSTD* of *n*-tridecane, a training set which includes only compounds that are liquid at the standard state should be selected. Indeed using the five compounds with T_m <298.15 K as a training set, can yield very accurate TQSPR1, based on a variety of dominant descriptors that are identified as highly correlated with the training set *SSTD* data. One option is to use n_C as the dominant descriptor. In this case the QSPR: *SSTD* = 103280 + 32285.1351 * n_C, with *TSAE* = 0.018 and δy = 0.022 are obtained, indicating excellent representations of both the training set and the target compound.

In Fig. 7 the *HSTD* data for the members of the training set and the target compound are plotted versus n_C. As in the case of *SSTD* there is also a "step" change in the *HSTD* values when moving from the liquid to the solid phase compounds. However in this case the change is much more moderate. Consequently, the QSPR developed for this property using all ten members of the training set and the *Mor15p* descriptor (row 4 in Table 5 and Fig. 8) yields predictions within the data uncertainty limits for the training set (*TSAE* = 1.45 % while *Umax* = 3%) and the target compound (δy = -0.18% and *Ut* = 1 %). For the properties *LVOL*, *HCOM* and RI there are slight changes in the trend of the data when moving from the liquid to the solid phase region. The small change can be detected in the residual plot only, but still TQSPR1s can be developed, which are consistent with the property data for the training set and the target compound within the data uncertainty level.

It can be concluded that a training set which consists of compounds that are not in the same phase as that of the target compound at the standard state may result in erroneous predictions. Comparing the *TSAE* and *Umax* values may assist the detection of such situations in most cases.

High prediction error caused by long range extrapolation

From the 1-alkene series the compound with the highest n_C for which property data are available in the 1-alkene series is 1-triacontene with n_C= 30. The training set for predicting properties of 1-triacontene is comprised of ten 1-alkenes in the range of $11 \leq n_C \leq 20$, as no data for 1-alkenes in the range of $20 < n_C < 30$ are available .

In the 7th row of Table 5 the summary of the results for the prediction of *LVOL* for 1-triacontene is shown. The QSPR in this case includes the descriptor *EEig10x*. This descriptor belongs to the 2-D "edge adjacency indices" category and it represents the "eigenvalue 10 from edge adjacency matrix weighted by edge degrees ". Comparing the *TSAE* value (*TSAE* = 0.99 %) with *Umax* (*Umax* = 1 %) shows that the QSPR is consistent with the property data within uncertainty level. However, for the target compound the prediction error is excessive (δy = 21.4 % while Ut = 10%).

Paster et al. (2011) have shown that if property prediction involves extrapolation from the training set to higher carbon numbers the behavior of the descriptor at $n_C \rightarrow \infty$ should match the asymptotic behavior of the modeled property. They have further noted that unlike T_C, T_m etc. that converge to constant values at $n_C \rightarrow \infty$, the liquid molar volume is additive in nature with a monotonic incremental change with increasing n_C and it does not converge to a constant value. We have found that the *EEig10x* descriptor converges to the value of 3.0 for the *n*-alkane series at $n_C \rightarrow \infty$ (the value of *EEig10x* = 2.992 was obtained for n_C =320). Thus, while this descriptor can represent a training set (with a narrow range of n_C values) within experimental error limits, the TQSPR1 is not applicable for a long range extrapolation because of its inappropriate asymptotic behavior. This is the cause of the large prediction error for 1-triacontene. Consequently, a descriptor which is collinear with the property in the training set range and its asymptotic behavior matches that of *LVOL* should be selected for this prediction.

In the 8th row of Table 5 the summary of the results of prediction of *LVOL* of 1-triacontane with a QSPR based on the *RHyDp* descriptor is shown. The *RHyDp* descriptor belongs to the topological descriptors category, it represents a "reciprocal hyper-distance path index" and its asymptotic behavior is characterized by monotonic incremental change with increasing n_C, just like *LVOL*. This QSPR is highly consistent with the training sets data (*TSAE* = 0.06) and its prediction error (δy = 1.67 %) is well below the data uncertainty.

Conclusions

This study involved the prediction of 16 constant properties for close to 1800 compounds using the "dominant descriptor" version of the Targeted QSPR method, in order to determine the range of applicability and potential sources of high prediction errors. It was shown that in a typical case, such as prediction of properties of *n*-propyl formate, using a training set that contains alkyl formates, propionates and acetates it is possible to model the properties so that the QSPR predictions are consistent with the property data within data uncertainty level for members of the training set. The predicted values for the target compound are consistent with the training set data and the prediction errors are within the experimental error limits.

One potential source of large prediction errors was identified as erroneous and/or inconsistent 3D molecular structure for the target compound and/or some of the members of the training set. To prevent this inconsistency it is very important to use the same computational tools and algorithms to obtain the optimized 3D structure for all the potential predictive compounds in the database as well as for the target compounds. Otherwise, the use of 3D descriptors in QSPRs should better be avoided.

The use of molecular descriptors outside their range of definition can also result in inconsistencies and large prediction errors. To prevent such errors, the validity range of the various descriptors needs to be clearly identified, and descriptors that are undefined for some of the training set members and/or the target compound must be removed from consideration while selecting the dominant descriptor.

Properties that are defined at a standard state are dependent also on phase of the compound at that state. In this study we have found that mixing liquid and solid phase compounds (in the training set and the target compound) leads for very large prediction error in predicting absolute entropy at standard state, while the effects on the enthalpy of formation, enthalpy of combustion and liquid molar volume are more moderate.

It was also demonstrated that when the prediction involves long range extrapolation to high carbon numbers, the asymptotic behavior of the selected dominant descriptor must match the asymptotic behavior of the modeled property and be highly correlated with the training set property data.

This study enables improving the consistency of the physical property and descriptor databases for increasing the robustness of QSPRs which can be obtained for a variety of properties and compounds.

References

1. Brauner, N; Stateva, R. P.; Cholakov, G. St.; Shacham, M. (2006). Structurally "Targeted" Quantitative Structure-Property Relationship Method for Property Prediction. *Ind.Eng.Chem.Res.*;45:8430-8437.
2. Marano, J.J.; Holder, G.D. (1997a). General Equations for Correlating the Thermo-physical Properties of n-Paraffins, n-Olefins and other Homologous Series. 2. Asymptotic Behavior Correlations for PVT Properties. *Ind.Eng.Chem.Res.*;36:1895.
3. Marrero, J.; Gani, R. (2001). Group-contribution based estimation of pure component properties. *Fluid Phase Equilibrium.*;183-184:183-208.
4. Paster I.; Shacham M.; Brauner N. (2009). Investigation of the Relationships between Molecular Structure, Molecular Descriptors and Physical Properties. *Ind. Eng. Chem. Res.*; 48:9723–9734.
5. Paster I.; Tovarovski G.; Shacham M.; Brauner N. (2010). Combining Statistical and Physical Considerations in Deriving Targeted QSPRs Using Very Large Molecular Descriptor Databases. Pp. 61 - 66 in S. Pierucci and G. Buzzi Ferraris (Eds), 20th European Symposium on Computer Aided Process Engineering – ESCAPE 20, June 6- 9, Ischia, Naples, Italy
6. Paster I.; Shacham M.; Brauner N. (2011). Adjustable QSPRs for Prediction of Properties of Long-chain Substances. *AIChE J*; 57(2):423–433.
7. Rowley, R.L.; Wilding, W.V.; Oscarson, J.L.; Yang, Y.; Zundel, N.A. (2010). DIPPR Data Compilation of Pure Chemical Properties Design Institute for Physical Properties, (http//www.aiche.org/dippr), Brigham Young University Provo Utah.
8. Rowley, R. L. (2010). Personal communications.
9. Shacham, M.; Brauner N. (2003). The SROV Program for Data Analysis and Regression Model Identification. *Computers chem. Eng.*;27(5):701-714.
10. Shacham, M.; Kahrs, O.; St Cholakov, G.; Stateva, R.; Marquardt, W.; Brauner, N. (2007). The Role of the Dominant Descriptor in Targeted Quantitative Structure Property Relationships. *Chem. Eng. Sci.*; 62(22):6222-6233.
11. Shacham M.; Brauner N. (2011). Analysis and Refinement of the Training set in Predicting a Variety of Constant Pure Compound Properties by the Targeted QSPR Method". *Chem. Eng. Sci.*;66: 2606–2615 .
12. Shacham M.; Paster I.; Brauner N. (2012). Property Prediction and Consistency Analysis by a Reference Series Method. *AIChE J*, published online, DOI: 10.1002/aic.13813.

Table 1.
Molecular structure for the target compound: *n*-propyl formate and its selected training set members

Name	Structural Formula	No. of C atoms
n-butyl formate	$HCOO(CH_2)_3CH_3$	5
ethyl formate	$HCOOCH_2CH_3$	3
n-pentyl formate	$HCOO(CH_2)_4CH_3$	6
isobutyl formate	$HCOOCH_2CH(CH_3)_2$	5
n-propyl acetate	$CH_3COO(CH_2)_2CH_3$	5
n-hexyl formate	$HCOO(CH_2)_5CH_3$	7
sec-butyl formate	$HCOOCH(CH_3)C_2H_5$	5
ethyl propionate	$CH_3CH_2COOCH_2CH_3$	5
n-propyl propionate	$CH_3CH_2COOCH_2CH_2CH_3$	6
n-heptyl formate	$HCOO(CH_2)_6CH_3$	8
***n*-propyl formate**	$\mathbf{HCOO(CH_2)_2CH_3}$	**4**

Table 2
Training set, target property data and predicted values for modeling of T_b with a TQSPR for *n*-propyl formate

	DIPPR Data		TQSPR1 Prediction		Value of *Xu*
Name	*NBP* (K)	*U* (%)	*NBP (K)*	δy (%)	Descriptor
n-butyl formate	379.25	<1	378.17	0.28	7.297
ethyl formate	327.46	<1	325.47	0.61	4.593
n-pentyl formate	405.45	<1	403.26	0.54	8.584
isobutyl formate	371.22	<1	373.48	0.61	7.056
n-propyl acetate	374.65	<1	373.48	0.31	7.056
n-hexyl formate	428.65	<3	427.64	0.24	9.835
sec-butyl formate	366.55	<1	371.33	1.30	6.946
ethyl propionate	372.25	<1	371.33	0.25	6.946
n-propyl propionate	395.65	<1	396.84	0.30	8.255
n-heptyl formate	451.25	<1	451.38	0.03	11.053
***n*-propyl formate**	**353.97**	**<1**	**352.29**	**0.47**	**5.969**

Table 3
Constant properties included in this study from the DIPPR database

No.	Symbol	Property description	Units
1	AIT	Auto Ignition Temperature	K
2	ENT	Absolute Entropy of Ideal Gas at 298.15 K and 100000 Pa	J/kmol*K
3	FP	Flash Point	K
4	HCOM	Net Enthalpy of Combustion Standard State (298.15 K)	J/kmol
5	HFOR	Enthalpy of Formation of Ideal gas at 298.15 K and 100000 Pa	J/kmol
6	HFUS	Enthalpy of Fusion at Melting Point	J/kmol
7	HSTD	Enthalpy of Formation in Standard State at 298.15 K and 100000 Pa	J/kmol
8	HSUB	Heat of Sublimation at the triple point	J/kmol
9	LVOL	Liquid Molar Volume at 298.15 K	m^3/kmol
10	MP	Melting Point (1 atm)	K
11	NBP	Normal Boiling Point (1 atm)	K
12	PC	Critical Pressure	Pa
13	RI	Refractive Index at 298.15 K	--
14	SSTD	Absolute Entropy in Standard State at 298.15 K and 100000 Pa	J/kmol*K
15	TC	Critical Temperature	K
16	VC	Critical Volume	m^3/kmol

Table 6
Training set and target property data and predicted values for modeling of V_C with a TQSPR for 1-decanol

			DIPPR Data		**TQSPR1 Prediction**		**Value of *Ds***
Comp. No.	**Name**	n_C	*Vc (m^3/kmol)*	*U (%)*	*Vc (m^3/kmol)*	**δy (%)**	**Descriptor**
1	1-undecanol	11	0.715	<3	0.68481	4.22	0.227
2	1-tridecanol	13	0.86	<5	0.82838	3.68	0.225
3	1-tetradecanol	14	0.802	<25	0.90016	12.24	0.224
4	1-pentadecanol	15	1.01	<5	0.97194	3.77	0.223
5	1-hexadecanol	16	1.08	<6	1.0437	3.36	0.222
6	1-heptadecanol	17	1.16	<7	1.1155	3.84	0.221
7	1-octadecanol	18	1.23	<8	1.1873	3.47	0.22
8	1-nonadecanol	19	1.31	<10	1.2591	3.89	0.219
9	1-eicosanol	20	1.12	<25	1.2591	12.42	0.219
10	1-nonanol	9	0.576	<3	0.61302	6.43	0.228
Target	**1-decanol**	**10**	**0.645**	**<3**	**1.4026**	**117.46**	**0.217**

Table 7
Training set and target property data and predicted values for modeling of *SSTD* with a TQSPR for n-tridecane

			DIPPR Data		TQSPR1 Prediction		EEig13D	
Comp. No.	**Name**	**n_C**	***SSTD (J/kmol*K)***	***U* (%)**	***SSTD (J/kmol*K)***	**δy (%)**	**descriptor**	**T_m (K)**
1	*n*-tetradecane	14	555430	<1	589847.7	6.20	-1.95	279.01
2	*n*-dodecane	12	490660	<1	502471.3	2.41	0	263.57
3	*n*-pentadecane	15	587520	<1	584336.3	0.54	-1.827	283.07
4	*n*-hexadecane	16	619650	<1	576987.7	6.88	-1.663	291.31
5	*n*-heptadecane	17	652240	<1	568698.2	12.81	-1.478	***295.13***
6	*n*-octadecane	18	480200	<1	560094.9	16.64	-1.286	***301.31***
7	*n*-nonadecane	19	503550	<3	551491.7	9.52	-1.094	305.04
8	*n*-eicosane	20	526900	<3	543157.4	3.09	-0.908	309.58
9	*n*-heneicosane	21	550250	<3	535226.3	2.73	-0.731	313.35
10	*n*-docosane	22	573610	<3	527698.5	8.00	-0.563	317.15
Target	***n*-tridecane**	**13**	**522870**	**<1**	**502471.3**	**3.90**	**0**	**267.76**

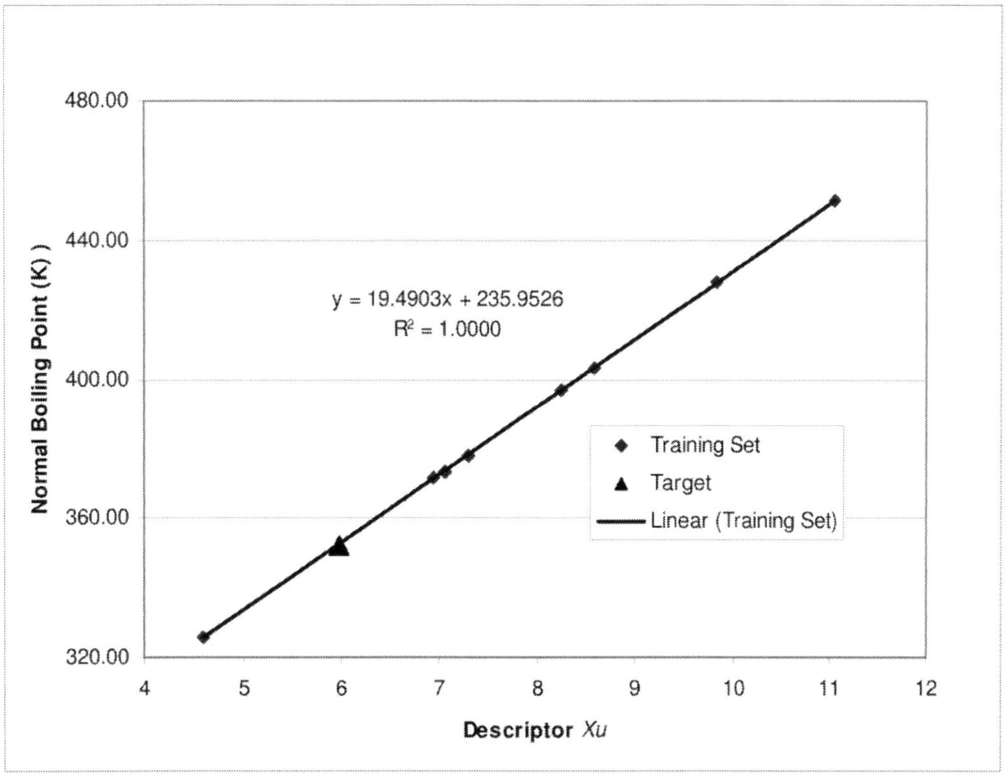

Fig. 1. Plot of T_b of *n*-propyl formate and its training set versus the descriptor *Xu*

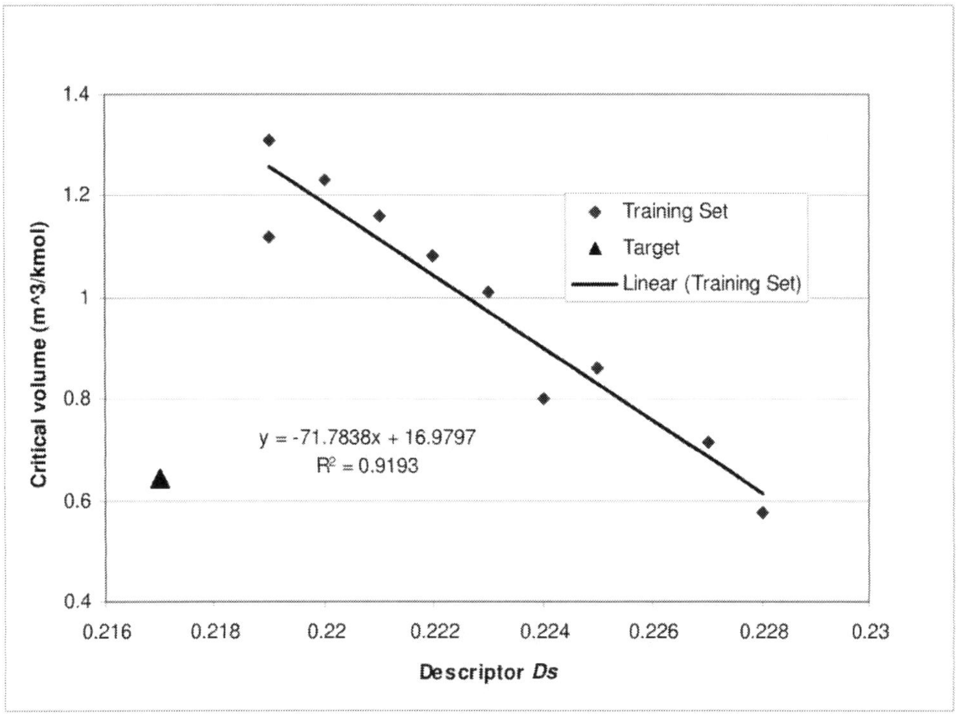

Fig. 2. Plot of V_c of 1-decanol and its training set versus the descriptor *Ds*

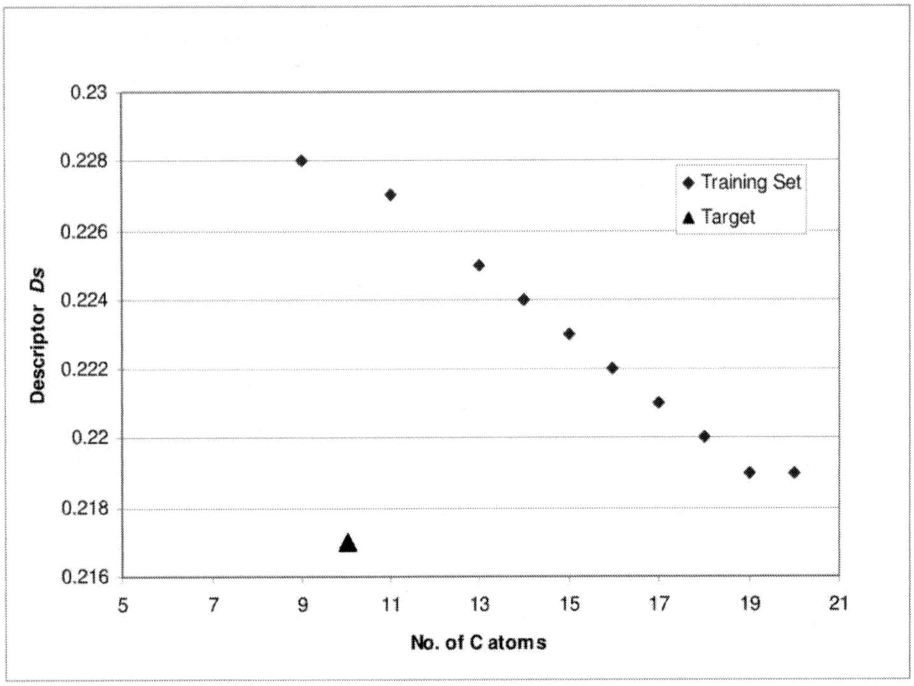

Fig. 3. Plot of the descriptor *Ds* of 1-decanol and its training set versus n_C.

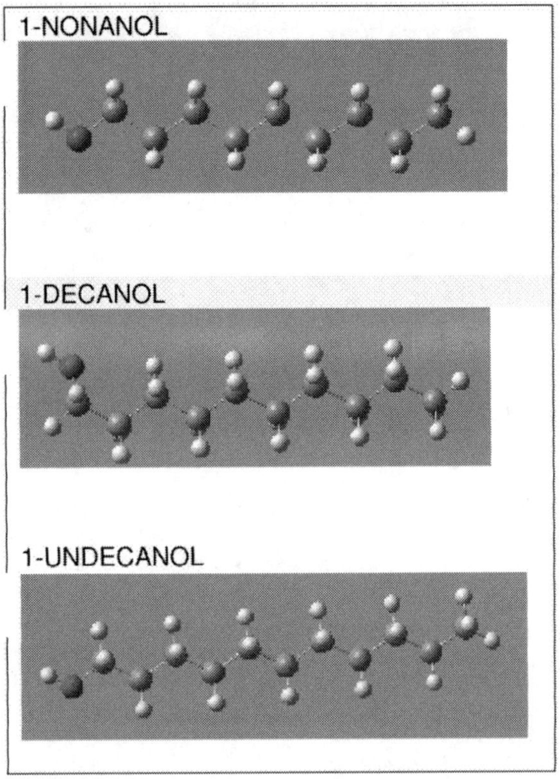

Fig. 4. The 3-D structure (MOL) files of 1-decanol (target compound) and its two immediate neighbors in the homologous series

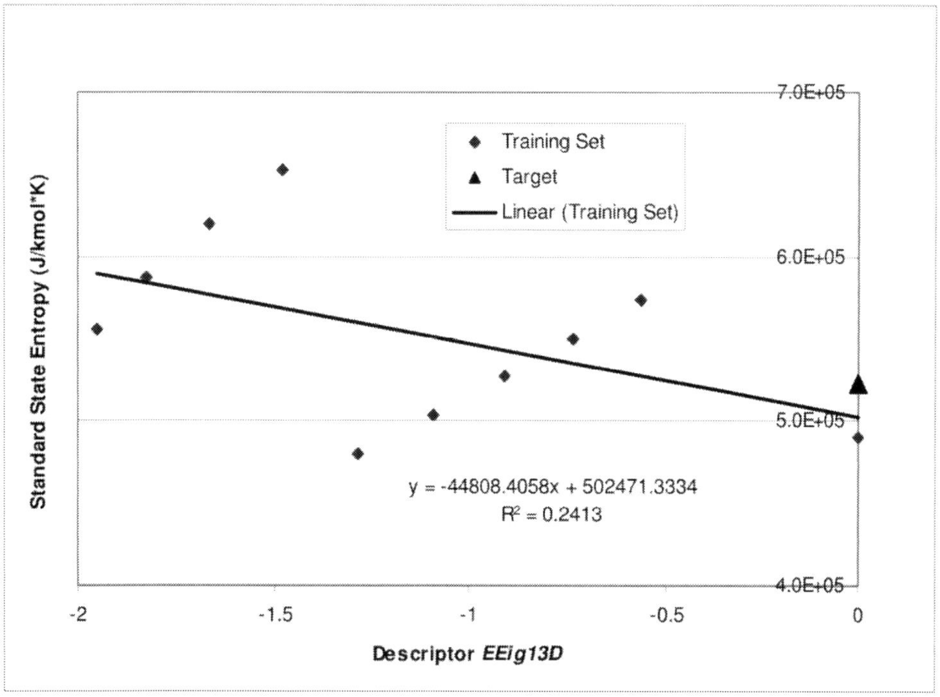

Fig. 5. Plot of *SSTD* of *n*-tridecane and its training set versus the descriptor *EEig13D*

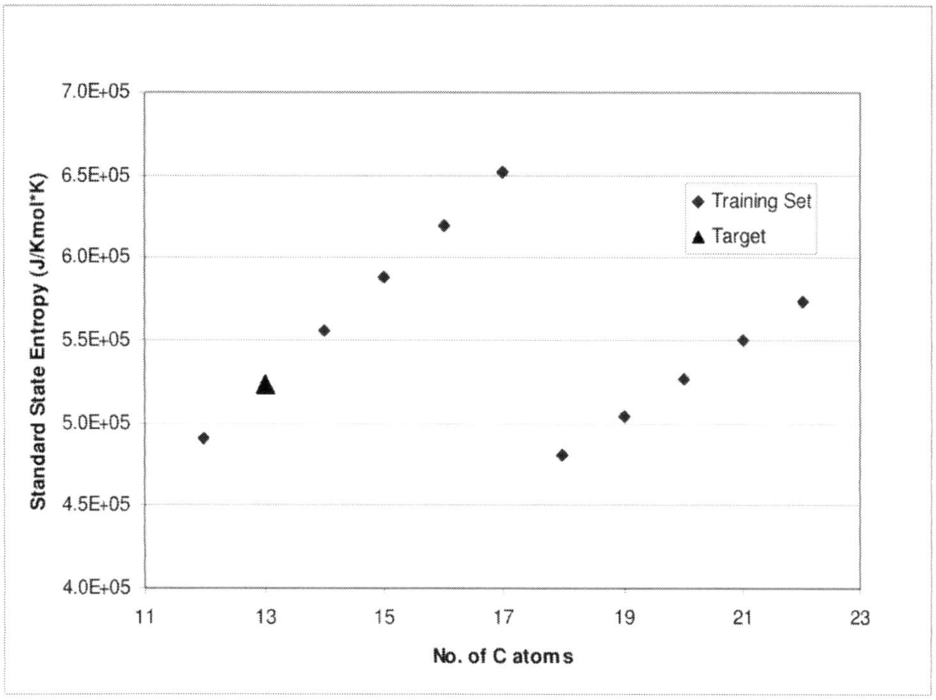

Fig. 6. Plot of *SSTD* of *n*-tridecane and its training set versus n_C

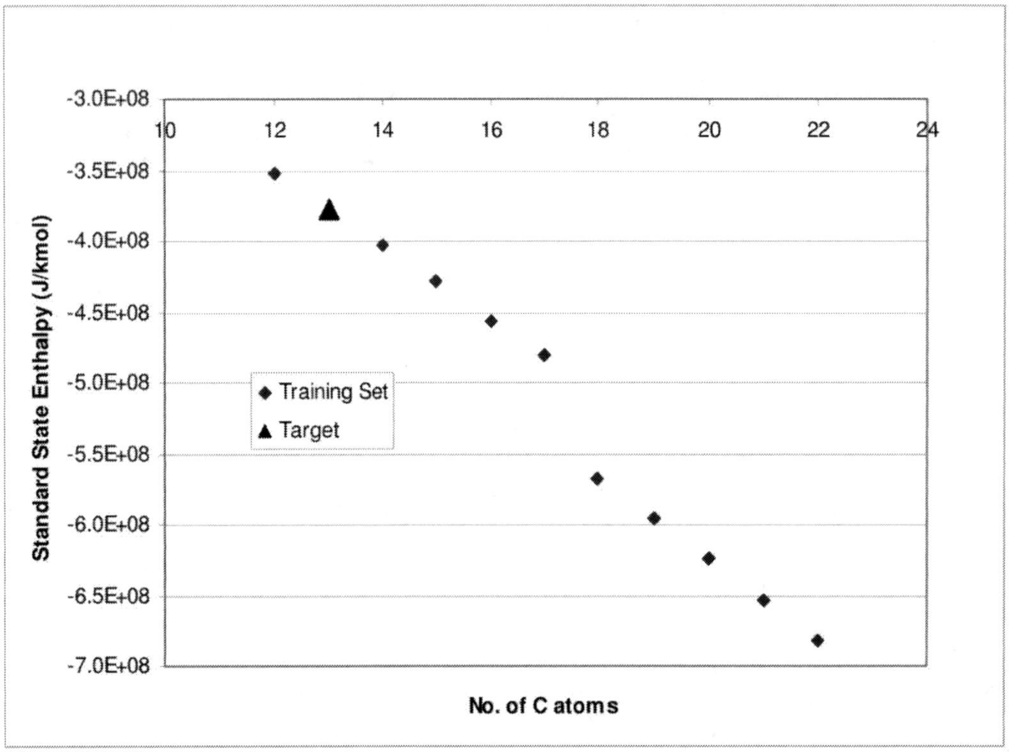

Fig. 7. Plot of *HSTD* of *n*-tridecane and its training set versus n_C.

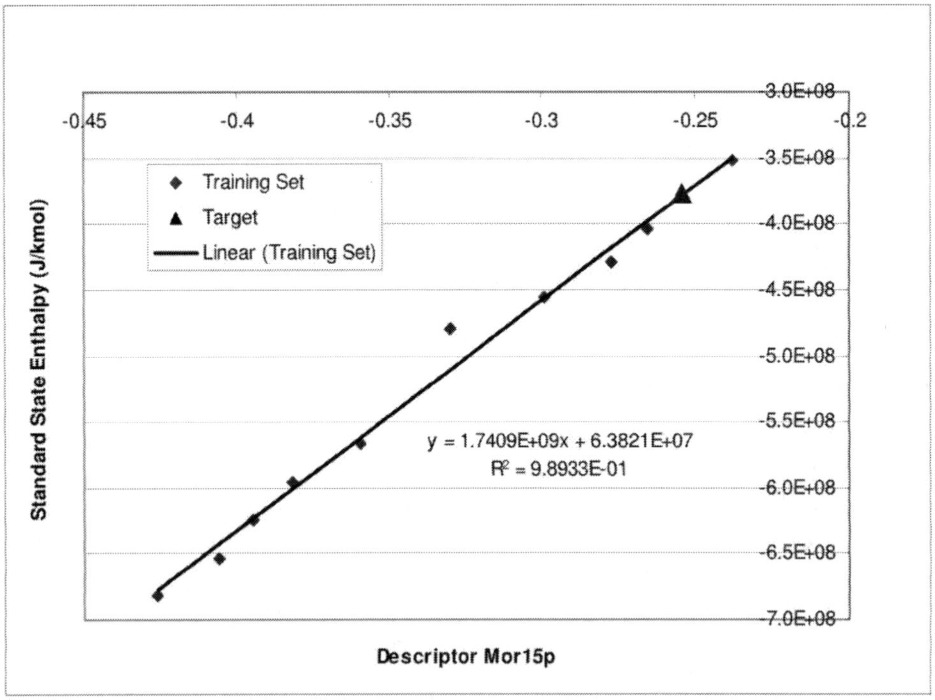

Fig. 8. Plot of *HSTD* of *n*-tridecane and its training set versus the descriptor *Mor15p*

Table 4
Prediction of constant properties of *n*-propyl formate with the TQSPR1 method

Property		Training Set		TQSPR1 model				Descriptor	DIPPR Property[1]		Predicted Property[1]	
No.	Symbol	*Umax*	TSRV	β_0	β_1	DD	TSAE	Value	Value	*Ut*	Value	δy
1	AIT	25	21.84	878.8999	-818.3105	R4p	2.58	0.24	728.15	-	682.5	6.27
2	ENT	5	46.45	153415.1122	19777.9619	CID	0.23	10.839	3.678E+05	3	3.678E+05	3.14E-03
3	FP	10	25.43	226.9051	-23.9229	Mor05e	0.95	-1.572	270	1	264.51	2.03
4	HCOM	3	-101.03	742962384.8	-615998813.2	VRA1	0.41	4.609	-2.041E+09	3	-2.096E+09	-2.70
5	HFOR	3	-22.48	-152635152.4	-142419766.4	piPC02	0.97	1.792	-4.076E+08	3	-4.079E+08	-6.17E-02
6	HFUS	25	115.89	5781669.457	-2695094.046	Mor03u	4.18	-2.36	1.320E+07	10	1.214E+07	8.01
7	HSTD	3	-24.81	124726829.3	-6336963.415	ZM1V	0.82	90	-4.452E+08	10	-4.456E+08	-8.98E-02
8	HSUB	10	48.58	42424139.54	1670761.232	RDF025u	0.99	11.337	6.040E+07	10	6.137E+07	1.60
9	LVOL	1	67.92	0.000049094	0.0051122	IAC	0.30	19.303	0.097942	1	0.09873	0.80
10	MP	5	36.85	195.9889	81.1295	Mor22e	3.76	-0.199	180.25	1	179.84	0.23
11	NBP	3	31.97	235.9527	19.4903	Xu	0.45	5.969	353.97	1	352.29	0.47
12	PC	10	64.79	3708568.004	-1057090.248	EEig04d	1.40	-0.247	4.020E+06	3	3.970E+06	1.25
13	RI	0.2	3.78	1.4953	-0.19938	Xt	0.09	0.601	1.375	-	1.3754	3.20E-02
14	SSTD	10	53.91	103492.0836	34485.5067	VRp1	0.41	4.657	265900	5	2.641E+05	0.68
15	TC	5	21.21	428.0744	7.5182	piID	0.33	13.832	538	1	532.07	1.10
16	VC	25	72.46	-0.3336	0.035199	Ss	0.93	17.5	0.285	5	0.28238	0.92

[1]Values for the target compound

Table 5
Prediction of selected constant properties for selected compounds by the TQSPR1 method

									Property Value for Target Compound				
		Property	Training Set		TQSPR1 model				Descriptor	DIPPR Property		Predicted Property	
No.	Compound	Symbol	Umax	TSRV	β_0	β_1	DD	TSAE	Value	Value	Ut	Value	δy
1	1-decanol	VC	25	113.7984	16.9797	-71.7838	Ds	5.73	0.217	0.645	3	1.4026	117.4645
2	n-hexatriacontane	LVOL	5	27.283	0.059774	0.28075	Mor22v	0.09	1.241	0.65069	1	0.40818	37.2699
3	n-tridecane	SSTD	3	32.903	502471.3334	-44808.4058	EEig13d	6.88	0	522870	1	502471.3334	3.9013
4	n-tridecane	HSTD	3	-87.3574	63820939.76	1740938839	Mor15p	1.47	-0.254	-3.78E+08	1	-3.78E+08	-0.18203
5	n-tridecane	LVOL	1	69.4015	0.069398	0.00079767	ISIZ	0.18	219.66	0.24563	1	0.24461	0.41458
6	n-tridecane	HCOM	3	-74.5586	-158416112	11500536037	Mor29p	0.07	-0.697	-8122900000	1	-8174289730	-0.63265
7	1-triacontene	LVOL	1	27.9235	0.2708	0.072119	EEig10x	0.99	2.061	0.53383	10	0.41944	21.4288
8	1-triacontene	LVOL	1	27.9235	0.065716	0.0088287	RHyDp	0.06	52.01	0.53383	10	0.5249	1.673

(181c) Effect of Charge Asymmetry and Charge Screening On Structure of Superlattices Formed by Oppositely Charged Colloidal Particles

Ganeshprasad Pavaskar[1], Siddharth Sharma[1] and **Sudeep Punnathanam**[2], (1)Indian Institute of Science, Bangalore, India, (2)Chemical Engineering, Indian Institute of Science, Bangalore, India

Colloidal suspensions made up of oppositely charged particles have been shown to self-assemble into substitutionally ordered superlattices. For a given colloidal suspension, the structure of the superlattice formed from self-assembly depends on its composition, charges on the particles, and charge screening. In this study we have computed the pressure-composition phase diagrams of colloidal suspensions made up of binary mixtures of equal sized and oppositely charged particles interacting via hard core Yukawa potential for varying values of charge screening and charge asymmetry. The systems are studied under conditions where the thermal energy is equal or greater in magnitude to the contact energy of the particles and the Debye screening length is smaller than the size of the particles. Our studies show that charge asymmetry has a significant effect on the ability of colloidal suspensions to form substitutionally ordered superlattices. Slight deviations of the charges from the stoichiometric ratio are found to drastically reduce the thermodynamic stability of substitutionally ordered superlattices. These studies also show that for equal-sized particles, there is an optimum amount of charge screening that favors the formation of substitutionally ordered superlattices.

(181d) Vapor-Liquid Equilibria of Ternary Systems Containing 1-Butanol (or 2-Butanol), + 1-Hexene and + Methylbenzene At 313.15 K

Chahrazed Ghezouali[1], Aouicha Belabbaci[1], Rosa M. Villamañan[2], Maria C. Martin[2], Amina Negadi[1], Miguel A. Villamañan[2] and Latifa Negadi[1], (1)Laboratoire de Thermodynamique Appliquée et Modélisation Moléculaire, Université de Tlemcen, Tlemcen, Algeria, (2)Grupo de Termodinámica y Calibración (TERMOCAL), Departamento Ingeniería Energética y Fluido mecánica, E.T.S. de Ingenieros Industriales, Universidad de Valladolid, Valladolid, Spain

Ethers and alcohols are usually added to gasolines in order to improve their combustion and reduce pollution. Phase equilibrium data of oxygenated mixtures are important for predicting the vapor-phase composition that would be in equilibrium with hydrocarbon mixtures. In addition to the well-established industrial uses of experimental data for completely miscible (ethers or alcohols + hydrocarbons) systems, there is a general scientific interest in using such experimental data in combination with theories or mathematical models to improve our understanding of molecular interactions in non-ideal liquid systems.

The present work is part of a research program on vapor liquid-equilibria of binary and ternary mixtures containing 1-butanol or 2-butanol with several hydrocarbons.

In this work experimental isothermal P-x-y data for the ternary systems (1-butanol + 1-hexene + methylbenzene) and (2-butanol + 1-hexene + methylbenzene) at 313.15 K are reported.

A static VLE apparatus, consisting of an isothermal total pressure cell, has been employed for measuring the vapor-liquid equilibrium of ternary systems. The apparatus and measuring technique are based on that by Van Ness and co-workers.

Data reduction by Barker's method provides correlations for GE, using the Wohl expansion; the Wilson, NRTL, and UNIQUAC models have also been applied successfully. Good results have been obtained for all. The investigated mixtures exhibit a positive deviation from ideality.

(181e) Estimating the Compressibility Factor for CO2-Natural Gas Mixtures

Adel Elsharkawy, Petroleum Engineering, Kuwait University, Kuwait, Kuwait

There is a consensus among nations that CO2 emission to the atmosphere must be controlled via creating economically viable technologies for capturing, separation and long term storage of CO2 (CCSS). One of many possible proposed sequestrations is the injection of CO2 into depleted oil and gas fields. However for planning such CO2 sequestration projects, accurate estimation of CO2-natural gas properties is required.

In this study a large data bank containing physical properties of various mixtures of CO2-natural gas properties at various temperatures and pressures has been collected. The data bank has been used to assess the accuracy of the industry standard methods in calculating the compressibility factor. The study indicated that these methods have series errors when they applied at high pressures, high temperatures, or at high content of CO2.

A new method is proposed in this study to calculate compressibility factor and density of CO2-natural gas mixtures. It is capable of estimating the CO2- natural gas mixtures properties at various temperatures and pressures and for any level of CO2 content

(181f) Energy and Diffusivity Landscapes From Smoluchowski Analyses and Shape Matching of Colloidal Crystallization Dynamics

Daniel J. Beltran-Villegas and Michael A. Bevan, Chemical and Biomolecular Engineering, Johns Hopkins University, Baltimore, MD

Understanding concentrated colloidal dynamics in the presence of different pairwsie interactions and external fields provides a basis to predict the temporal evolution of colloidal microstructures in colloidal crystallization. However, a theory of concentrated colloidal dynamics does not yet exist that rigorously includes both statistical mechanical (free energy changes) and fluid mechanical (hydrodynamic interactions) contributions associated with changing microscopic configurations. In this talk, we report the development of an order-parameter based Smoluchowski Equation (SE) model that accurately captures the dynamic evolution of initially disordered colloidal fluid configurations into colloidal crystals in the presence of either tunable depletion or electric field induced interactions. We first identify appropriate order parameters to monitor colloidal crystallization. These order parameters are computed by means of novel shape matching techniques. With this foundation, we apply the SE analysis to extract free energy and "diffusivity landscapes" from the BD and SD simulations to understand the role of multi-body hydrodynamic interactions in colloidal crystallization dynamics. Finally, we demonstrate the use of these landscapes in the SE model to estimate key dynamic information, such as mean passage times between microscopic states and crystallization success ratios. These results provide a basis to control colloidal crystal assembly processes via open- and closed- loop control to produce perfect single colloidal crystals.

(181g) Solid-Liquid Equillibria of Five Binaries Composed of in-Process Products in Hydrodeoxygenation Reaction of Triglyceride of Fatty Acid

Koh-hei Ohya, College of Industrial Technology, Nihon University, Narashino, Japan, Taka-aki Hoshina, College of Industrial Technology, Nihon Unversity, Narashino, Japan, Tomoya Tsuji, College of Indusrial Technology, Nihon University, Narashino, Japan, Kouji Maeda, School of Mechanical and Systems Engineering, University of Hyogo, Himeji, Japan, Hidetoshi Kuramochi, Center for Material Recycle and Waste Management Research, National Institute for Environmental Studies, Tsukuba, Japan and Masahiro Osako, Center for Material Recycle and Waste Management Research, National Institute for Environmental Studies, Tsukuba, Japan

In recent Japan, a new method has been proposed for synthesis of biodiesel fuel (BDF). It is quitedifferent from conventional ones, that is, fats and oils will be perfectly converted to propane, and some blanched alkanes, having 16-18 carbon atoms in a molecule, by hydrodeoxygenation reaction at high temperatures and the pressures. In our previous study, we consider the utilization of so called trap grease. The trap grease is discharged as free-floating grease from waste water reservoir of restaurant, food courts, food industry and so on. It was mainly composed of triglyceride of fatty acids, and free fatty acid. The conventional hydrodesulfurization catalysts were also effective for the hydrodeoxygenation reaction of these compounds. Then, the triglyceride, composed of unsaturated/saturated fatty acids, are finally converted to alkanes via some in-process products, unsaturated fatty acid, saturated fatty acid, and alcohols. However, the melting point of these products have not monotonic tendencies. For example, the melting point of unsaturated fatty acid are lower than that of saturated. On the contrary, that of saturated fatty acid higher than that of alcohol. That of alcohol generally lower than that of alkane. The melting point, and solid-liquid equilibrium (SLE) are important for design of the flow type reactor. So, in this study, SLE were measured for the five binaries, composed of in process products, oleic acid + triolein, stearic acid +oleic acid, heptadecanol + stearic acid, octadecanol + stearic acid, and octadecane + octadecanol under the atmospheric pressure in the temperature range from 284.64 to 341.59 K under the atmospheric pressure. The apparatus is based on direct measuring the cooling temperature, and the detail description was reported elsewhere. The reliability has been already ensured for measuring SLE of benzene + cyclohexane. As a result, oleic acid + stearic acid showed a complete miscibility. The others were simple eutectic system, and their eutectic points. The all SLE were well correlated with Schroder - van Laar equation applying Redrich-Kister activity coefficient model.

(181h) Fundamentals Under the Phase Transition of Binary Mixtures

Yiping Tang, Honeywell Process Solutions, London, ON, Canada

The VdWs type equation of state (EOS) has been used for decades in chemical physics. In phase transition region, however, the EOS has a smooth VdWs loop, and cannot show the real phase transition with a discontinuity and gap between two phases. The VdWs loop is generally believed unphysical and has to be corrected by Maxwell equal-area rule. In this presentation, we show a new theory to study phase behavior by semicanonical (sGC) ensemble for binary mixtures at the given temperature and pressure. The sGC ensemble is one advance of grand canonical ensemble for studying the phase behavior of pure fluids recently. In the sGC ensemble, subsystems are merged together successively and the molecules within fluctuate in larger and larger scopes. The corresponding Gibbs free energy is transformed until being stable. The final Gibbs free energy yields the real phase behavior, including the discontinuity at phase transition. The process of the transform leads to the new sGC method to calculate phase equilibria with composition fluctuations. The sGC method needs no initialization and the solution is physically warranted, regardless of the type and number of phases as well as models. Several typical examples of phase diagram calculations including VLE, LLE, SLE and their crossovers will be present. The sGC transform suggests that the VdWs loop is physically justified and composition fluctuations with the macroscopic magnitude bring the discontinuity of phase transition.

(181i) Hydrogen Solubility in Triolein and Oleic Acid for Estimation of Hydrodeoxygenation Reaction Rate At High Temperature and Pressure

Tomoya Tsuji, College of Indusrial Technology, Nihon University, Narashino, Japan, Koh-hei Ohya, College of Industrial Technology, Nihon University, Narashino, Japan, Taka-aki Hoshina, College of Industrial Technology, Nihon Unversity, Narashino, Japan, Kouji Maeda, School of Mechanical and Systems Engineering, University of Hyogo, Himeji, Japan, Hidetoshi Kuramochi, Center for Material Recycle and Waste Management Research, National Institute for Environmental Studies, Tsukuba, Japan and Masahiro Osako, Center for Material Recycle and Waste Management Research, National Institute for Environmental Studies, Tsukuba, Japan

Now, in Japan, hydrodeoxygenation reaction of fats has been paid much attention as a new method to produce biodiesel fuel (BDF) of the next generation. According to the method, triglyceride of fatty acid may be converted to propane, and some blanched alkanes by use of a desulfurization reactor in conventional petroleum refinery. In our previous study, we considered the utilization of trap grease as a carbon resources. The trap grease is free-floating grease in waste water reservoir of restaurant, food courts, food industry and so on, and mainly composed of triglyceride of fatty acids, and free fatty acid. Moreover, we have shown that the conventional hydrodesulfurization catalysts were effective for the hydrodeoxygenation reaction at 573.2 K. Though the phase equilibrium and the reaction rate are essential for designing the processes, there is few available data. So, in this study, a new apparatus was designed for a measurement of hydrogen solubility in triolein, and oleic acid up to 473.2 K. The apparatus was based on a recirculation method and the cell volume about 500 cm^3. A cylinder, with the volume about 150 cm^3, was connected with cell, and the inlet and the outlet of the cylinder have VCR assemblies. So, the liquid sample was picked up repeatedly. After weighing mass of the picked up sample, the dissolved hydrogen volume was measured by a flow meter. The experimental data of hydrogen solubility showed a tendency like Henry's low up to 15 MPa and 473.2 K. The temperature dependence of Henry constant was correlated with a conventional relation. The next, by use of the extrapolated Henry constant at 573.2 K, the binary parameters, in mixing rule of Peng-Robinson (PR) equation of state, were determined. Finally the phase diagram and the reaction rate were evaluated by PR eq. at the desired reaction temperature, 573.2 K.

(181j) Present Status of Asog Parameters for Process Design

Katsumi Tochigi[1], Hiromasa Taguchi[2], Masaki Tateishi[1], Hiroyuki Matsuda[1] and Kiyofumi Kurihara[1],
(1)Department of Materials and Applied Chemistry, Nihon University, Tokyo, Japan, (2)PreFEED Corporation, Tokyo, Japan

Activity coefficients are fundamental to process development and equipment design in the chemical industries. ASOG group-contribution method is one predictive method for activity coefficients. Some thermophysical properties can be evaluated using the activity coefficients.

This article shows the present status of ASOG parameters, and reviews with the prediction of the following thermophysical properties and application;

1.Vapor-liquid equilibria

a. Low-pressure vapor-liquid equilibria

b. High-pressure vapor-liquid equilibria (PRASOG)

c. Infinite-dilution activity coefficients

d. Water activities in sugar food and polyols

e. Vapor-liquid equilibria for systems containing electrolytes

f. Vapor-liquid equilibria for systems containing polymers

2. Solubilities

a. Liquid-liquid equilibria (ASOG-LLE)

b. Gas solubilities

c. Solid-liquid equilibria

d. Solid-vapor equilibria (PRASOG)

3. Excess enthalpies and excess entropies (ASOG-HE)

4. Liquid kinematic viscosities (ASOG-VIS)

5. Application

a. Solvent selection for extractive distillation and liquid-liquid extraction

b. Distillation calculation

c. Thermodynamic efficiency

d. Solvent effect on reaction rate

Some thermophysical properties can be calculated using CD-ROM.

PREDICTION OF THERMODYNAMIC PROPERTIES FOR STRUCTURAL ISOMERS BY USE OF THE LATTICE CLUSTER THEORY EQUATION OF STATE FOR PURE COMPOUNDS

Kai Langenbach, TU Berlin, Berlin, Germany
Sabine Enders, TU Berlin, Berlin, Germany

Introduction

Many industrial processes and products use or contain polymers. Typical examples for this are plastic packaging [1], catalysis [2–6] , super absorber [7,8] and chemical engineering [9–12]. In order to process mixtures containing polymers, knowledge of their thermodynamic behavior is essential. While experimental methods [e.g. 13] can gain inside into this knowledge, the huge number of possible compounds in polymer mixtures due to polydispersity with respect to molar mass as well as chemical composition and short chain branching demands accessibility by means of theoretical methods. These theories should be able to account for molar mass and short chain branching.

In the past years physically based equations of state (EOS) have emerged as powerful tools for this purpose. One example for a group of this kind of EOS is the SAFT-family (Tan *et al.* [14] give a comprehensive review), which focuses on the interactions between alike and different functional groups of the molecules under consideration. While these theories principally allow the calculation of thermodynamic behavior for different isomers of the same chemical composition, a new set of parameters is needed in order to do that. Also excess free enthalpy models are often used, with the shortcoming that they cannot model effects of pressure on the thermodynamic properties. However, this dependence is of crucial importance for many applications, such as the polymerization of poly-ethylene (PE), which is performed at elevated pressures in the magnitude of 10 to 100 MPa. Also the phase behavior of PE depends strongly on the short chain branching and on the molar mass distribution. This was measured by de Loos *et al.* [15] exemplary for the cloud-point pressures of linear and branched PE in ethylene.

A known method to incorporate architecture directly into the EOS is the Lattice Cluster Theory (LCT) by Freed and co-workers [16–20] that uses a Mayer [21] like cluster expansion in the inverse coordination number $1/z$ and the reduced interaction energy ε/kT – with k being the Boltzmann constant and T being the temperature – in order to introduce molecular structure into the partition function and hence, into the free energy of the system under consideration. The architecture is represented by a set of combinatorial numbers calculable directly from molecular structure [22]. By introducing void lattice cells [18] the theory is made applicable for compressible fluids. The general expression for the free energy is lengthy, complex and potentially numerically unstable, since it involves the addition and subtraction of very many terms of equal size, which often poses numerical risks due to machine accuracy. Hence, LCT needs to be simplified in order to use it properly [23].

In order to test the predictive ability of compressible LCT and to allow the fitting of pure compound parameters Langenbach and Enders [24] developed the LCT equation of state (LCT-EOS) for pure components with the greater goal in mind to develop a useable multi-component theory. This contribution is concerned with the further presentation of this EOS. New calculation examples are shown and the theory and results are discussed. It is possible for several saturated alkane isomers with the same number of carbon atoms as their linear counterparts to predict their difference in vapor pressure from their molecular architecture and a fit to the experimental data of the n-alkane only [24].

Theoretical Background

LCT-EOS is based on a diagrammatic cluster expansion on a not deformable Bravais lattice with z nearest neighbors for each lattice site [18]. This lattice is populated with nM segments of interconnected occupied lattice sites, with M being the number of segments of one molecule, and n_v empty lattice sites resulting in a total of N_l sites on the lattice and a volume V of

$$V = n_v v_{cell} + nMv_{cell} \tag{1}$$

where $v_{cell} = \sigma^3 N_{AV}$ is the volume of a lattice cell with σ being the size of a cubic lattice cell and N_{AV} being the Avogadro constant. The occupied segments may interact with their nearest neighbor occupied segments with the nearest neighbor interaction energy ε, but voids interact neither with voids nor with occupied segments. The afore mentioned connectivity between species' segments is ensured through the expansion procedure [18] of the canonical partition function Z resulting in a dependence of the set of connectivity numbers or combinatorial structure numbers N_g with $g \in \{1;2;3;\perp;+;\perp';1,1;1,2;2,2\}$ in the thermodynamic limit. These are directly calculable from the molecular architecture of the species under consideration by using graph theoretical procedures [24]. For a complete derivation of the LCT please refer to Foreman and Freed [25]. Here only the results for pure compounds are given. After taking the appropriate limit and restricting the multi-component version of Dudowicz and Freed [18] to one species, the free energy per lattice site reads [24]

$$-\frac{F_{comp}^{LCT}}{N_l kT} = \frac{S_{0,comp}}{N_l k} - \frac{E_{1,0,comp}}{N_l kT} - \frac{E_{2,0,comp}}{N_l kT} \tag{2}$$

where F_{comp}^{LCT} is the free energy of a compressible one component system $S_{0,comp}$ is the entropic contribution to the free energy, $E_{1,0,comp}$ is the first order energetic contribution to the free energy and $E_{2,0,comp}$ is the second order energy contribution to the free energy. These contributions can be calculated as a series in the void segment fraction

$$\phi_v = \frac{n_v}{n_v + nM} \tag{3}$$

and together with a mean field term [24] as

$$\frac{S_{0,comp}}{N_l k} = -\phi_v \ln\phi_v - \frac{(1-\phi_v)}{M}\left(\ln(1-\phi_v) + \ln\frac{2z}{M} - M\ln z + M - 1\right) + \sum_{p=0}^{4} Q_p^{(s)}\phi_v^p \tag{4}$$

$$-\frac{E_{1,0,comp}}{N_l kT} = \frac{\varepsilon}{kT}\sum_{p=0}^{5} Q_p^{(\varepsilon)}\phi_v^p \tag{5}$$

$$-\frac{E_{2,0,comp}}{N_l kT} = \left(\frac{\varepsilon}{kT}\right)^2 \sum_{p=2}^{6} Q_p^{(\varepsilon^2)}\phi_v^p \tag{6}$$

where the $Q_p^{(s)}$, $Q_p^{(\varepsilon)}$ and $Q_p^{(\varepsilon^2)}$ are structure coefficients that are functions of the N_g and M only and calculable with the formulas of Langenbach and Enders [24]. In order to calculate phase equilibria the EOS and the chemical potential need to be derived. The EOS is derived by differentiating the free energy with respect to volume using Eq. (1) [24].

$$P = -\left(\frac{\partial F_{comp}^{LCT}}{\partial V}\right)_{T,n} = -\frac{1}{v_{cell}}\left(\frac{\partial F_{comp}^{LCT}}{\partial n_v}\right)_{T,n}$$

$$= \frac{kT}{v_{cell}}\left(\begin{array}{l} -\ln(\phi_v) - (1-\phi_v)\left(1+\frac{1}{M}\right) + Q_0^{(s)} + \sum_{p=1}^{4}\left(p-(p-1)\phi_v\right)Q_p^{(s)}\phi_v^{p-1} \\ +\frac{\varepsilon}{kT}\left(Q_0^{(\varepsilon)} + \sum_{p=1}^{5}\left(p-(p-1)\phi_v\right)Q_p^{(\varepsilon)}\phi_v^{p-1}\right) + \left(\frac{\varepsilon}{kT}\right)^2\left(\sum_{p=2}^{6}\left(p-(p-1)\phi_v\right)Q_p^{(\varepsilon^2)}\phi_v^{p-1}\right) \end{array}\right) \tag{7}$$

The chemical potential per lattice site $\bar{\bar{\mu}}$ needs to be derived using a differentiation with respect to segment numbers nM at constant volume. From Eq. (1) it follows that

$$\left(\frac{\partial n_v}{\partial nM}\right)_{T,V} = -1 \tag{8}$$

and by that Eq. (9) follows.

$$\frac{\bar{\bar{\mu}}}{RT} = -\ln(\phi_v) + \frac{1}{M}\ln(1-\phi_v) + \frac{1}{M} - 1 + \sum_{p=1}^{4} pQ_p^{(s)}\phi_v^{p-1} + \frac{\varepsilon}{kT}\sum_{p=1}^{5} pQ_p^{(\varepsilon)}\phi_v^{p-1} + \left(\frac{\varepsilon}{kT}\right)^2\sum_{p=2}^{6} pQ_p^{(\varepsilon^2)}\phi_v^{p-1} \tag{9}$$

In reference [24] this was derived somewhat different, but the obtained chemical potentials differ only by a constant value, *i.e.* the standard state was chosen a bit different. By equating the pressure and segmental chemical potential in two phases the vapor-liquid equilibrium may be calculated, if the combinatorial structure numbers N_g are known. To calculate these numbers with the help of graph theory the united atom graph of the respective molecule are chosen [24]. For the case of alkanes, a suitable choice of united atom groups are the CH_x groups with $x \in [0,4]$.

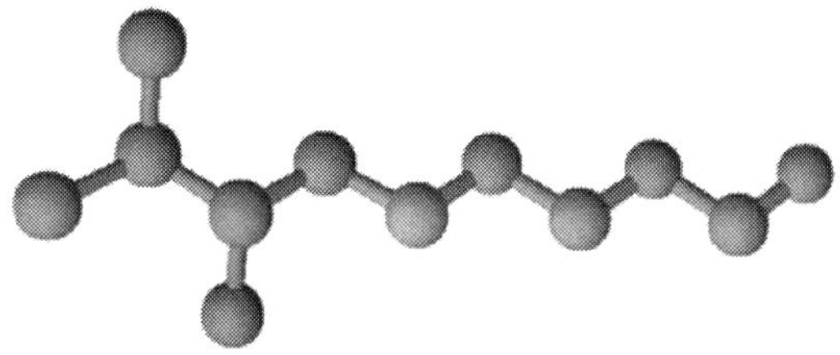

Figure 1: United atom structure of 2,3-dimethyldecane

Table 1: Combinatorial number for species used in this contribution

compound	M	N_1	N_2	N_3	$N_\perp$	$N_{\perp'}$	N_+	$N_{1,1}$	$N_{1,2}$	$N_{2,2}$
n-heptane	7	6	5	4	0	0	0	10	12	3
2,3-dimethylpentane [24]	7	6	7	6	2	5	0	8	10	2
n-dodecane	12	11	10	9	0	0	0	45	72	28
2,3-dimethyldecane	12	11	12	11	2	5	0	43	80	36

An example for the united atom structure of 2,3-dimethyldecane is depicted in Figure 1. The combinatorial numbers calculated with the procedures of Langenbach and Enders **[24]** for the species used in this contribution are listed in Table 1.

Results and Discussion

To test the predictive capability of LCT-EOS Langenbach and Enders [24] fitted parameter for several alkanes for two distinct coordination numbers and for 3 different kinds of parameterizations with the conclusion that $z=6$ is the optimal choice for predicting the pressure difference between isomers and that 2 parameter suffice two describe the vapor-liquid equilibrium of n-alkanes with carbon numbers larger than six in good agreement with experimental data. Also a temperature dependence of the interaction energy

$$\varepsilon = \varepsilon_0 + \varepsilon_0 \varepsilon_1 T^{-1} \qquad (10)$$

may improve calculations for lower temperatures, if only high temperature data is available for the fit [24]. All calculations are performed with the respective set of parameters from Table 2. There are three sets of parameters for the heptanes in order to show and discuss the influence of parameterization.

Table 2: Parameters for the compounds used in the calculations

compounds	σ **[nm]**	ε_0/k **[K]**	ε_1/k **[K]**	z
heptane [24]	32.3313	262.674	-	6
heptane [24]	31.8963	109.13	-	12
heptane [24]	32.4655	235.986	34.4882	6
dodecane	32.0897	266.675	4.70693	6

Figure 2 depicts three calculated vapor pressure curves in comparison with experimental data [26,27]. The solid line is calculated with $z=6$ and two parameters, while the dashed line is calculated with $z=12$ and two parameters and the dot-dashed line is calculated with $z=6$ and three parameters. The parameter sets were fit using the P-v-T-Data of the experimental points shown as circles. The agreement is reasonable, but not perfect, especially with the data points of Nichols *et al.*, but the two curves are only distinguishable in the region closer to the critical point. Since it is known, that the structural influence is best captured with $z=6$ [24], this will be the coordination number used in Figures 3 to 4. The calculated vapor pressure for three parameters is in better agreement with the experimental data, as was expected. Thus the three parameter version is used for the calculations shown in Figures 3 to 4.

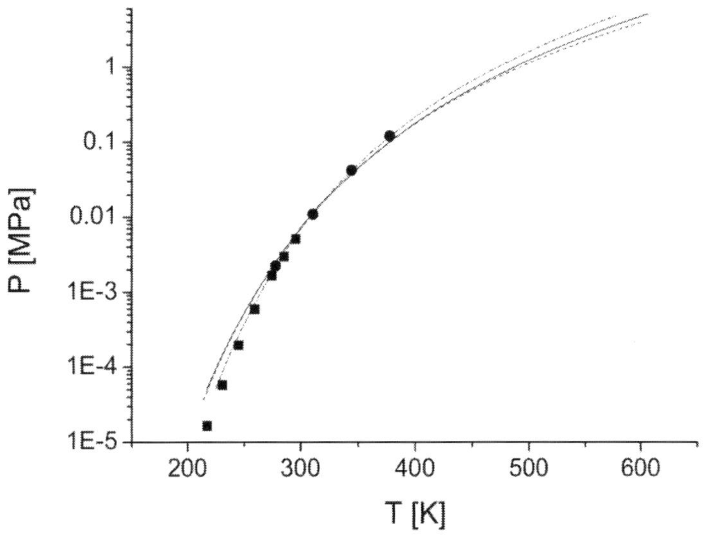

Figure 2: Vapor pressure of n-heptane. The symbols are experimental data (circles [26] are used in the fit, whereas squares [27] are not), the dashed line is the calculated vapor pressure with coordination number $z = 12$ with two parameters, the solid line is the calculated vapor pressure with coordination number $z = 6$ and two parameters and the dot-dashed line is calculated with $z = 6$ and three parameters.

In Figure 3 previous results [24] are shown for two isomers of heptane for which experimental data [26–28] was available. In this case the three-parameter version of LCT-EOS was used with coordination number $z = 6$. The parameter were fit to the four experimental P-v-T data sets of n-heptane [26] depicted as filled squares. With these parameters, together with the *a priori* known combinatorial numbers of Table 1, the vapor pressure curve of 2,3-dimethylpentane is predicted [24]. This prediction was discussed by Langenbach and Enders [24] and is shown here only for completeness.

Figure 4 shows the same kind of prediction for two dodecane isomers. Again parameters were only fit to experimental data of the pure n-alkane [29]. The calculated vapor pressure curve for n-dodecane is in excellent agreement with experimental data [29], except for the low temperature regime, where the high temperature series expansion of Dudowicz and Freed [18] is expected to fail. The prediction of the vapour pressure of 2,3-dimethyldecane is not perfect yet, but comes very close to the experimental data [30].

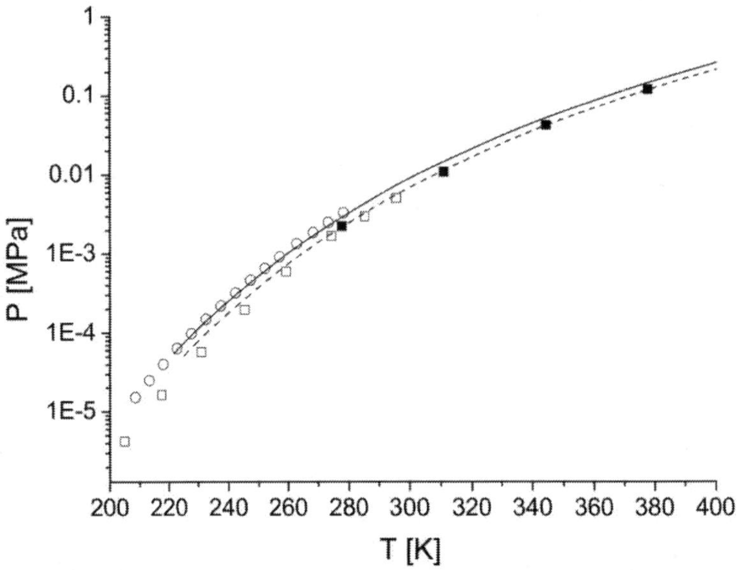

Figure 3: Calculated [24] vapor pressure of n-heptane and 2,3-dimethylpentane compared to experimental data [26–28]. Squares are vapor pressures of n-heptane and circles are vapor pressures of 2,3-dimethylpentane. Only filled squares were used in the fit [24]. The broken line is the calculated [24] vapor pressure of n-heptane and the solid line is the predicted [24] vapor pressure of 2,3-dimethylpentane. Calculations were done with the 3- parameter version of LCT-EOS.

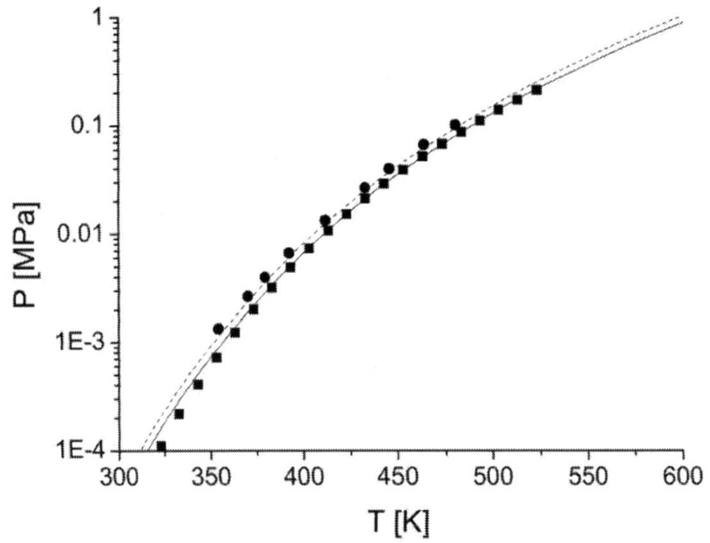

Figure 4: Calculated vapor pressure of n-dodecane and 2,3-dimethyldecane compared to experimental data **[29,30]**. Squares are vapor pressures of n-dodecane and circles are vapor pressures of 2,3-dimethyldecane. The solid line is the vapor pressure calculated with LCT-EOS (3-parameter) and the broken line is the vapor pressure prediction for LCT-EOS with the same parameters.

Summary and Outlook

The accessibility of structural influence on thermodynamic properties is of vital importance in the field of polymer science, since polymer species mostly have polydispersity with respect to molar mass as well as short-chain branching. The Lattice Cluster Theory of Freed and co-workers [16–20] provides a method to assess this influence by means of a theoretical description of the local correlation effects arising from molecular architecture. In order to use this theory, simplifications have to be introduced, to avoid possible numerical pitfalls from the heavily used summations of equally sized numbers. One such simplification, the LCT-EOS [24], is discussed here, which is important to fit pure compound parameter of *e.g.* solvents like linear or branched alkanes.

With LCT-EOS vapor-liquid equilibria of pure linear alkanes were fitted using different methods for parameterization. The parameterization with coordination number $z = 6$ is best suited to investigate the influence of structure on this kind of equilibrium [24]. Here we show, that both the parameterization with $z = 6$ and the one with $z = 12$ are only distinct in the near critical region, where mean field theories are known to be little accurate, because they do not account for critical fluctuations [31]. For $z = 6$ and parameters fit to the respective linear alkane predictions are shown for the vapor pressures of their isomers. For the investigated substances the predictions for the 2,3-dimethyl-isomers were in very good agreement with experimental data.

The LCT-EOS has proven to be a powerful tool for the investigation of structural influence on thermodynamic behavior based on theoretical considerations. The next step, in order to generate a tractable version of the multi-component LCT-EOS, is to minimize the number of summations necessary for calculating the thermodynamic properties of arbitrary mixtures. Results for two components are currently in preparation and will be published soon [32].

References

[1] O.G. Piringer, A.L. Baner, Plastic Packaging: Interactions with Food and Pharmaceuticals, Wiley-VCH, 2008.

[2] L. Tuchbreiter, S. Mecking, Macromol. Chem. Phys. 208 (2007) 1688–1693.

[3] G. Zhang, X. Fan, J. Kong, Y. Liu, Y. Xie, Prog. Chem. 20 (2008) 326–338.

[4] J.-D. Marty, E. Martinez-Aripe, A.-F. Mingotaud, C. Mingotaud, J. Colloid Interface Sci. 326 (2008) 51–54.

[5] K. Tahara, H. Shimakoshi, A. Tanaka, Y. Hisaeda, Dalton Trans. 39 (2010) 3035–3042.

[6] M. Cypryk, P. Pospiech, K. Strzelec, K. Wąsikowska, J.W. Sobczak, J. Mol. Catal. A: Chem. 319 (2010) 30–38.

[7] T. Tanaka, Sci. Am. 244 (1981) 124–136, 138.

[8] A. Pourjavadi, F. Zeidabadi, S. Barzegar, J. Appl. Polym. Sci. 118 (2010) 2015–2023.

[9] M. Seiler, Fluid Phase Equilibr. 241 (2006) 155–174.

[10] M.A. Rehim, H.M. Fahmy, Z.E. Mohamed, M.H. Abo-Shosha, N.A. Ibrahim, Pigm. Resin Technol. 39 (2010) 149–155.

[11] Y. Chen, P. Peng, Z. Guo, J. Yu, M. Zhan, J. Appl. Polym. Sci. 115 (2010) 3687–3696.

[12] M. Irfan, M. Seiler, Ind. Eng. Chem. Res. 49 (2010) 1169–1196.

[13] T.W. De Loos, W. Poot, G.A.M. Diepen, Macromolecules 16 (1983) 111–117.

[14] S.P. Tan, H. Adidharma, M. Radosz, Ind. Eng. Chem. Res. 47 (2008) 8063–8082.

[15] T.W. de Loos, W. Poot, R.N. Lichtenthaler, J. Supercrit. Fluids 8 (1995) 282–286.

[16] K.F. Freed, J. Phys. A: Math. Gen. 18 (1985) 871–877.
[17] K.F. Freed, M.G. Bawendi, J. Phys. Chem. 93 (1989) 2194–2203.
[18] J. Dudowicz, K.F. Freed, Macromolecules 24 (1991) 5076–5095.
[19] J. Dudowicz, M.S. Freed, K.F. Freed, Macromolecules 24 (1991) 5096–5111.
[20] J. Dudowicz, K.F. Freed, Macromolecules 24 (1991) 5112–5123.
[21] J.E. Mayer, M.G. Mayer, Statistical Mechanics, Third Printing, John Wiley & Sons Inc, 1940.
[22] S. Enders, K. Langenbach, P. Schrader, T. Zeiner, Polymers 4 (2012) 72–115.
[23] C. Browarzik, D. Browarzik, S. Enders, Fluid Phase Equilibria 328 (2012) 49–60.
[24] K. Langenbach, S. Enders, Fluid Phase Equilibria 331 (2012) 58–79.
[25] K.W. Foreman, K.F. Freed, in:, I. Prigogine, S.A. Rice (Eds.), Advances in Chemical Physics, John Wiley & Sons, Inc., Hoboken, 2007, pp. 335–390.
[26] G.F. Carruth, R. Kobayashi, J. Chem. Eng. Data 18 (1973) 115–126.
[27] W.B. Nichols, H.H. Reamer, B.H. Sage, Ind. Eng. Chem. 47 (1955) 2219–2221.
[28] A.G. Osborn, D.R. Douslin, J. Chem. Eng. Data 19 (1974) 114–117.
[29] N.B. Vargaftik, Tables on the thermophysical properties of liquids and gases : in normal and dissociated states, Hemisphere Pub. Corp. ; distributed by Halsted Press, Washington; New York, 1975.
[30] E. Terres, L. Brinkmann, D. Fischer, D. Huellstrung, W. Lorz, G. Weisbrod, Brennstoff-Chem. 40 (1959) 279–280.
[31] M.E. Fisher, Rev. Mod. Phys. 46 (1974) 597–616.
[32] K. Langenbach, S. Enders, C. Browarzik, D. Browarzik, J. Chem. Thermodyn. in preparation (2012).

(181l) Micelle/Water Partition Coefficients Using COSMO-RS: Conformational Study

Elke Sponsel, Liudmila Mokrushina and **Wolfgang Arlt**, Chair of Separation Science and Technology, University of Erlangen-Nuremberg, Erlangen, Germany

In the field of chemical engineering and life sciences, the prediction of phase equilibria in complex multicomponent systems is of growing interest. One of the most popular and effective methods here is the COSMO-RS model, which allows for the predictions based only on the chemical structure of the system components. As shown by our group, the results depend strongly on the choice of conformations. For small molecules, sets of minimum energy conformations from the common conformational analysis made in vacuum give adequate results. Large flexible molecules however may have a huge number of conformations and the prediction results vary significantly for different conformations. Therefore, the reliable conformations of large flexible molecules, which afford reproducible results and lead to stable *a-priori* predictions on phase equilibria, should be identified.

In this work, the partitioning of small solutes (alkanes, alcohols, etc.) in the micellar solution of the non-ionic surfactant Triton X-100 (TX100) in water is studied as an example. Micelles hereby are treated as a macroscopic phase being in equilibrium with the aqueous surrounding (pseudo-phase approach), so that the partitioning of a solute between these coexisting pseudo-phases is determined by the thermodynamic equilibrium and thus the micelle/water partition coefficient (K^{MW}) can be calculated based on the solute limiting activity coefficients in the two pseudo-phases. For any small solute molecule with the limited number of conformations, the same set of the minimum energy conformations was used in all calculations. However, both pseudo-phases also contain large flexible amphiphilic molecules, for which adequate conformations have to be found. The conformations of TX100 were generated using two methods: the force-field-based energy-minimization method (hyperchem HC) and the condensed phase molecular dynamics (MD, Gromacs). The HC conformational search was carried out in vacuum, while the MD simulations were made in water and in octanol (the model solvent for micellar pseudo-phase) to account for the solvent effects. The conformations obtained by both methods were then used to calculate K^{MW} of the solutes using COSMO-RS and their influence on the latter was studied.

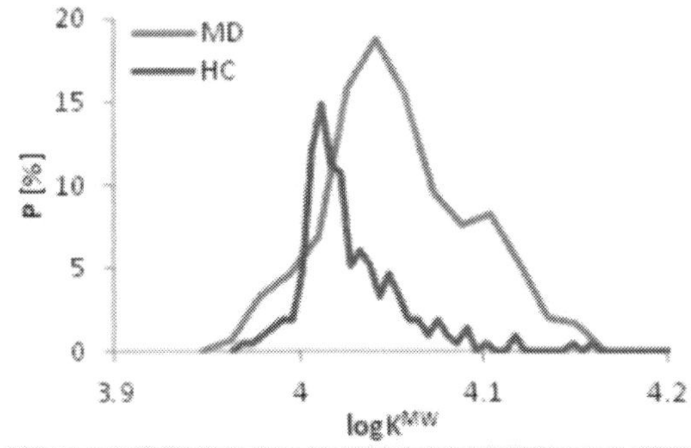

The figure shows the probability distribution of the logarithmic K^{MW} of octanol calculated by COSMO-RS for different single conformations. The maxima observed in the distributions indicate that some of the conformations occur more often than the others and thus will lead to the more reproducible values of K^{MW}, i.e., are reliable for K^{MW}. However, using this method to identify reliable conformations, great calculation efforts have to be made. Thus, it would be advantageous to be able to select such conformations directly from the conformational space obtained from MD or HC, even before performing the time-consuming DFT/COSMO geometry optimization calculations. Such a selection should be based on physical parameters. In this study, we consider two of them, the radius of gyration and conformation specific energy as well as their combination. It has been shown that if the selection is made from the 20% most probable conformations of any of criteria, the outliers in the probability distributions of K^{MW} are avoided. If both criteria are considered simultaneously, a single surfactant conformation can be identified that leads to the most probable result for K^{MW}. Thus, the probability-based methodology leads to reliable and stable *a-priori* predictions of solute partitioning in micellar systems as well as of phase equilibria in general.

We thank DFG (MO 2199/1-1) for the financial support.

(181m) Lattice Model Based Upon Information Theory

Thomas Wallek, Martin Pfleger and Andreas Pfennig, Institute of Chemical Engineering and Environmental Technology, Graz University of Technology, Graz, Austria

Lattice Models

Lattice theories are widely-used approaches for modeling condensed-phase behaviour, allowing for relatively straightforward realizing conceptions on intermolecular energies. Such models can conveniently be verified by molecular simulation. While the predominant number of lattice models, like those in the broadest sense following Guggenheim's work, is based upon classical statistical thermodynamics, the present contribution describes application of an information-theory based approach, subsumed under the 'method of discrete modeling'.

The Method of Discrete Modeling

Discrete Modeling is based upon straightforward formulation of internal energy U and entropy S of the system as functions of distribution variables, i.e. frequencies of discrete states of the lattice sites. These distribution variables are directly related to the molecule numbers N_i of the system components i by summation over all states. S is derived from Shannon's approach for information, interpreting 'probability' as the relative frequency of a discrete state of a lattice site among all possible states. Since both U and S are formulated as homogeneous functions of the distribution variables, Euler's homogeneous function theorem can be utilised for establishing a set of equations that permits the formulation of system properties like U, S and free enthalpy G, as explicit functions of molecule numbers N_i. From these functions, activity coefficients and other excess properties can easily be derived for model evaluation. The method has successfully been applied to the ideal gas model by derivation of the equation of state, heat capacity as well as the Maxwell-Boltzmann distribution of energies [1]-[2], while the present contribution focuses on application of discrete modeling to condensed phases [3].

Application and Analysis

Application of the method is illustrated by a simple example lattice model with focus on demonstrating both analogies and differences between conventionally used statistical thermodynamics and discrete modeling as well as identifying the potential of discrete modeling for effective implementation of conceptions on intermolecular energies.

References

[1] Pfleger M., 'A Statistical Approach for Deducing Equations of State', oral presentation, 8th European Congress of Chemical Engineering together with ProcessNet-Annual Meeting, Berlin, 2011

[2] Pfleger M., 'Applying Discrete Modeling to Ideal Gas and Real Gases', oral presentation, 20th International Congress of Chemical and Process Engineering, Prague, 2012

[3] Wallek T., 'Thermodynamic Modeling Based Upon Information Theory', oral presentation, 20th International Congress of Chemical and Process Engineering, Prague, 2012

(181n) Deep Eutectic Solvents: An Experimental and Molecular Simulation Study

Sasha L. Perkins, Paul Painter and Coray M. Colina, Department of Materials Science and Engineering, Pennsylvania State University, University Park, PA

Ionic liquids (ILs) have received prominent interest due to their low melting points, negligible vapor pressure, non-flammability, and unusual solvent properties. Because of these interesting properties, ILs are being used for different applications such as separations, fuel cells, lubricants, novel solvents, and for other materials engineering applications. Deep eutectic solvents (DES), considered ILs analogues, show promise for many engineering applications, since many of them are readily available and inexpensive compared to some of the typical ILs[1].

In this work, a combined experimental and molecular simulations approach is used to characterize hydrogen bonding and ionic interactions in order to understand the interactions of ILs with mineral surfaces and their role in hydrocarbon/sand separation processes. Specific attention is focused on choline chloride-based eutectic mixtures, *i.e* ions in choline chloride and a complex agent (i.e. urea, malonic acid, etc) because they are relatively inexpensive and environmentally benign.

Experimental IR spectra are reported for choline chloride-urea mixtures at different molar ratios and for the pure components. Ab-initio and atomistic molecular dynamics simulations has also been performed on these systems. The geometries of the systems studied were first optimized by using gas phase ab-initio calculations in Gaussian for individual ions and molecules. Atomistic simulations were performed for 1 to 2 choline chloride to urea mixture ratios using the General Amber Force Field, GAFF, functional form. Atomic partial charges obtained from the ab-initio simulations where transferred to the force field. Comparison of the experimental and simulated IR spectra of the choline chloride/urea system in a 1:2 ratio, showed interesting interactions around the NH_2coupling vibrations that are seen by the reduction of three peaks to two peaks when relating solid urea to the complex mixture. Finally, a hydrogen bond analysis between the moieties shows that a complex structure forms due to the presence of stable hydrogen bonds. The influence of these hydrogen bonds on stability is discussed, as well as their influence on the phase behavior of this deep-eutectic IL system.

1. Abbott, A.P., Capper, G., Davies, D.L., Rasheed, R.K. & Tambyrajah, V. Chem Commun. 2003, 70-71

(181o) Unraveling the Mechanisms of Crystal Nucleation and Growth At the Nanoscale

Caroline Desgranges and Jerome Delhommelle, Department of Chemistry, University of North Dakota, Grand Forks, ND

We discuss results recently obtained on crystal nucleation and growth in metals and alloys as well as in semiconductors. In the case of metals, we apply our approach to the formation of nanocrystallites in single-components systems as well as in binary systems. More specifically, we show how our approach accounts for the formation of gold nanoparticles exhibiting twinned FCC crystals as observed in experiments. We also unravel the interplay between the composition and the crystalline structure during the nucleation and growth of Pd-Ni nanoparticles. In the case of semiconductors, we show how the nucleation of a crystalline nucleus of silicon is preceded by the formation of a nanosized droplet of a metastable low density liquid precursor. We also show that the formation of the crystalline nucleus subsequently occurs at the interface between the two liquid phases. This result suggests that theories, like the classical nucleation theory, and empirical rules, like Ostwald's rule, should be modified to account for the role of liquid polymorphs in the nucleation process.

(181p) Modeling of the Surface Tension of Pure Components and Mixtures Using the Density Gradient Theory and a Theoretically Derived Influence Parameter Correlation

Bianca Breure and **Cor J. Peters**, Chemical Engineering, Petroleum Institute, Abu Dhabi, United Arab Emirates

Interfacial tension is a very important thermophysical property for many technological processes, including industrial separations, (enhanced) oil recovery, and specialty products such as paints and detergents. Since the determination of interfacial properties by experimental techniques is time-consuming and expensive, reliable methods to predict these properties are highly desired. This work focuses on the application of the density gradient theory for the prediction of surface tensions of pure components and mixtures with hydrocarbons. With the density gradient theory surface tensions can be predicted when the Helmholtz free energy of the homogeneous fluid is known as well as an influence parameter, which determines the importance of the contribution of density gradients in the interface to the total Helmholtz energy of the system. The PC-SAFT equation of state is used to calculate the Helmholtz energy and other thermodynamic properties of the system. Because of its good predictive capabilities PC-SAFT has been widely applied for phase equilibria calculations of pure components and mixtures, including complex systems such as solutions with polymers and petroleum mixtures. The influence parameter has a molecular definition, but is often calculated from experimental data or semi-empirical expressions. In this work a correlation for the influence parameter is developed using its theoretical definition as starting point. With the developed correlation influence parameters can be calculated for a component or pseudo-component, provided that its PC-SAFT pure component parameters are known and its phase behavior is well-predicted. It will be shown that the developed influence parameter correlation in combination with the density gradient theory and the PC-SAFT equation of state gives good predictions of surface tensions for pure components as well as binary and ternary mixtures with hydrocarbons.

(181r) Comparison of Vapor-Liquid and Liquid-Liquid Equilibrium Prediction From Combination of PRSV and Cosmosac Through Different Mixing Rules

Chieh-Ming Hsieh and Shiang-Tai Lin, Department of Chemical Engineering, National Taiwan University, Taipei, Taiwan

The fluid phase equilibrium behaviors of chemicals are important in the design, development, optimization, and control of chemical processes. In our previous study, a new revision of the COSMO-SAC model is proposed to provide better accuracy in predictions of both vapor-liquid equilibrium (VLE) and liquid-liquid equilibrium (LLE). However, the COSMO-SAC model, an excess Gibbs energy (G^{ex}) model, is not sufficient to directly apply to describe pressure and volume effects of fluid phase equilibrium. A simple way is to combine the COMO-SAC model and PRSV equation of state (EOS) through a G^{ex}-based mixing rule. In this study, the performance of previously published mixing rules, such as MHV1 and WS mixing rule, are investigated. The effects of different mixing rules are studied systematically to find out strengths and weaknesses of the models. Furthermore, the previously developed PR+COSMOSAC equation of state is also included in this work. This study is based on a large set of VLE (426 binary) and LLE (278 binary) systems covering a wide range of functional groups, temperature (from 208 to 553 K), and pressure (up to 100 MPa) to understand the range of applicability of the present approach. The MHV1 provides the best accuracy in VLE prediction among these three mixing rules, but is insensitive to pressure in the LLE prediction. The PR+COSMOSAC equation of state gives the best description of pressure effects on the LLE immiscible gap.

(181s) Equilibrium Conditions of Hydrate-Forming Pickering Emulsions

Christopher Hassan, Chemical Engineering, CCNY, New York, NY and Jae W. Lee, Department of Chemical Engineering, The City College of New York, New York, NY

Cyclopentane (CP), water, and hydrophobicized silica nanoparticles, when homogenized, form stable water-in-oil (W/O) emulsions. However, there are no available data regarding the phase equilibrium conditions of CP + H_2O hydrates in the presence of nano-particles, which is crucial to developing models with which to accurately predict the formation of hydrates in deep-sea oil pipelines if combined with high-pressure gas hydrate formers (methane, ethane, propane, etc.) such as what occurred during the BP oil spill. In this presentation, the dissociation temperatures of CP + H_2O hydrates at atmospheric pressure (101.325 kPa) will be determined for particle concentrations ranging from (0.5 to 7.5) weight percent of cyclopentane using a Micro-DSC. Emulsion droplet size distribution analyses were performed using an optical microscope. Also, the combined surface-active agent and particle system will be used to understand the equilibrium change from the bulk phase CP hydrates as a model system for hydrate formation in crude oil and gas pipelines.

(181t) Viscosity Modeling of Highly Asymmetric Mixtures

Aimee Ruiz-Llamas, ESIQIE-Departamento de Ingeniería Química, Instituto Politécnico Nacional, México, D.F., Mexico and Ricardo Macias-Salinas, ESIQIE-Departamento de Ingeniería Química, Instituto Politécnico Nacional, Mexico, D.F., Mexico

In the present work, a segment-based Eyring-UNIQUAC viscosity model has been developed for the accurate estimation of the dynamic viscosity of highly asymmetric mixtures including polymer solutions. The original viscosity modeling framework based on the Eyring theory was slightly modified to provide a more practical theory-based model for the viscosity of liquid mixtures containing complex components that cannot be typically described by the component-based approach. For the excess Gibbs energy of activation, a segment-based solution model (UNIQUAC) was used to overall represent the non-ideal viscous effects typically encountered in asymmetric mixtures where the size and shape of the constituent molecules greatly differ. The resulting model was successfully validated during the representation of experimental dynamic viscosities of various highly asymmetric mixtures containing mainly hydrocarbons as well as polymer solutions within a wide ranges of temperature, polymer molar mass and composition. The present segment-based viscosity model should be useful and promising for modeling the liquid viscosity in a number of asymmetric mixtures such as petroleum, petrochemicals and polymer solutions.

(181u) Real Pure Component Systems in the View of Discrete Modeling

Martin Pfleger, Thomas Wallek and Andreas Pfennig, Institute of Chemical Engineering and Environmental Technology, Graz University of Technology, Graz, Austria

Introduction

Discrete modeling is a kind of thermodynamic modeling introducing discrete models for the state functions entropy, internal energy, particle numbers, and system volume. The key item of the method is the usage of Shannon's formulation of information [1] $H = -\Sigma\, p_i \log p_i$, where the p_i are the probabilities of a particle to reside in the system state i. H can be interpreted as a measure of entropy. In combination with Gibbs' thermodynamics this leads to an explicit expression of the entropy given by internal energy, system volume and particle numbers $S=S(U,V,N)$, that can be used e.g. to deduce thermic and caloric equations of state.

The Case of Ideal Gas

The case of the ideal gas illustrates the most important step of the method: the definition of the system's state, i.e. what variables define the mechanical state of a particle, and how they have to be used in order to achieve the correct expression for the discrete entropy. With this expression only few steps are necessary to obtain the ideal-gas law. But additionally the analysis of the discrete model of the internal energy yields the Maxwell-Boltzmann distribution of energies and the heat capacity for the ideal gas - an impressive proof of the method.

Treating the Real Gas

With the understanding of discrete modeling of the system's state in the case of the ideal gas it becomes now clear how the interaction forces between real gas particles have to be incorporated in the discrete model not only for the internal energy, but also for the entropy in form of an expression for the potential energy. It will be shown, how a relatively simple approach of the potential energy yields the Van der Waals equation. Based on this result the modeling of more sophisticated approaches will be discussed.

References

[1] Shannon, C.E., (1948), A mathematical theory of communication, Bell Syst. Tech. J. 27, 379, 623.

(181v) Vapor-Liquid Equilibria of (Water + Glycerol), (Water + 1,3-Propanediol), or (Ethanol + Glycerol) Binary Mixtures At Several Temperatures: Measurements and Modeling

Manel Zaoui-Djelloul Daouadji[1], Amina Negadi[1], Ilham Mokbel[2], Jacques Jose[2] and Latifa Negadi[1], (1)Laboratoire de Thermodynamique Appliquée et Modélisation Moléculaire, Université de Tlemcen, Tlemcen, Algeria, (2)Laboratoire des Sciences Analytiques-UMR 5280, UCB-Lyon1, Villeurbanne, France

Petroleum is the main energy source utilized in the world, but its availability is limited and the search for new renewable energy sources is of major interest. Biofuels, such as ethanol and biodiesel, are among the most promising sources for the substitution of fossil fuels. Biodiesel can replace petroleum diesel, as it is produced from animal fats and vegetable oils, which generate about 10% (w/w) glycerol as the main byproduct. The excess glycerol generated may become an environmental problem, since it cannot be disposed of in the environment. One of the possible applications is its use as carbon and energy source for microbial growth in industrial microbiology. Glycerol bioconversion in valuable chemicals, such as 1,3-propanediol, dihydroxyacetone, ethanol, succinate, ...

The present paper is part of a research program concerning the thermodynamic properties of biofuels containing systems. In this presentation, the vapor-liquid equilibria of (water + glycerol), (water + 1,3-propanediol), and (ethanol + glycerol) binary mixtures measured using a static device at temperatures between 263 and 363 K are reported.

The data were correlated with the Antoine equation. From these data excess Gibbs functions (G^E) were calculated for several constant temperatures and fitted to a fourth-order Redlich-Kister equation using the Barker's method. The three binary systems exhibit positive deviations in G^E for all investigated temperatures over the whole composition range.

The NRTL, UNIQUAC and UNIFAC model have also been used and good results were obtained in the prediction of the total pressure and thermodynamic excess properties.

(181w) Modeling Ternary Liquid-Liquid Equilibria of Polylactides-Water-1,4-Dioxane with PC-SAFT Eos for TIPS Scaffold Fabrication

Giovanni Cocchi, Dipartimento di Ingegneria Chimica, Mineraria e delle Tecnologie Ambientali, University of Bologna, Bologna, Italy, Maria Grazia De Angelis, Dicma, University of Bologna, Bologna, Italy, Gabriele Sadowski, Department of Biochemical and Chemical Engineering, Laboratory of Thermodynamics, Technische Universität Dortmund, Dortmund, Germany and Ferruccio Doghieri, Department of Chemical Engineering, University of Bologna, Bologna, Italy

Poly(L-Lactic Acid) or PLLA and Poly(D-Lactic Acid) or PDLA and their copolymers are biocompatible and biodegradable polymers that could be used for fabrication of porous scaffolds suitable for tissue engineering applications. Among the many fabrication procedure listed in literature, phase separation methods seems to be adequate for producing porous structures with pore size distribution compatible with tissue growth [1,2]. It is believed that the pore size distribution and pore topology can be tailored by adjusting the thermodynamic and kinetic parameters of the process. Thermally Induced Phase Separation processing of polylactides can be exploited by controlled cooling of solution of polylactides in 1,4-dioxane, that is a good solvent of the polymer, containing a given amount of water, that acts as a non-solvent. Upon cooling the system will demix and two phases will appear, namely a polymer lean solution and a polymer rich solution, that is going to give rise to the solid phase of interest. Although in the case of optically pure PLLA it is know that crystallization plays a major role in the formation of the solid phase, knowledge of the liquid-liquid equilibria of the ternary solution is fundamental in any modelling or optimization effort of the whole TIPS fabrication process. Liquid-liquid equilibria of ternary systems comprising a macromolecular component is a challenging task for every kind of thermodynamic model. Moreover the presence of water and its peculiar role of non solvent, that upon cooling promotes phase separation, calls for the use of a model that can endow the underlying physics of hydrogen bonding.

Actually, as shown by Mannella et al. [3], modelling the phase behaviour of the binary mixture made by 1,4-dioxane and water is a task that can't be accomplished by several commonly used activity coefficient models. Moreover it seems that even the modelling capabilities of Equation of State based on statistical mechanics theory, like the Sanchez Lacombe EOS, are deeply challenged by this mixture. The Perturbated Chain – Statistical Associating Fluid Theory (PC-SAFT) Equation of State is a model based on a well defined picture of the intermolecular interactions that has been shown to be able to predict pure substances properties and phase equilibria of water, polar substances, macromolecular species and components of biological interests [4,5,6,7]. In the present work we discuss the modelling the pure component properties of 1,4-Dioxane as well as those of the polylactides, the binary Vapor-Liquid Equilibria of 1,4-Dioxane –Water and the ternary systems PDLA-1,4-Dioxane-Water and PLLA-1,4-Dioxane-Water with PC-SAFT EOS. Comparisons with several sets of experimental data from the literature [2,8,9,10] show that PC-SAFT EOS performs very well respect to this very challenging systems, correctly depicting the underlying physics.

References:

[1] C. Schugens, V. Maquet, C. Grandfils, R. Jerome, P. Teyssie, "Polylactide Macroporous Biodegradable Implants for Cell Transplantation. II. Preparation of Polylactide Foams by Liquid-Liquid Phase Separation", Journal of Biomedical Materials Research 30,(4):449-461, (1996).

[2] T. Tanaka, D.R. Lloyd, "Formation of Poly(L-lactic acid) Microfiltration Membranes Via Thermally Induced Phase Separation", Journal of Membrane Science, 238:65-73, (2004).

[3] G.A. Mannella, V. La Carrubba, V. Brucato, "On the calculation of free energy of mixing for aqueous polymer solutions with group-contribution model", Fluid Phase Equilibria, 299:222-228, (2010).

[4] J. Gross, G. Sadowski, "Perturbed-Chain SAFT: an Equation of State Based on a Perturbation Theory of Chain Molecules", Industrial & Engineering Chemistry Research, 41:1084-1093, (2002).

[5] J. Gross, G. Sadowski, "Application of the Perturbed-Chain SAFT Equation of State to Associating Systems", Industrial & Engineering Chemistry Research, 41:5510-5515, (2002).

[6] M. Kleiner, "Thermodynamic Modeling of Complex Systems: Polar and Associating Fluids and Mixtures", PhD Thesis, TU Dortmund, 2008.
[7] L.F. Cameretti, "Modeling of Thermodynamic Properties in Biological Solutions", PhD Thesis, TU Dortmund,2008.
[8] T.Takamuku, A. Yamaguchi,M.Tabata,N. Nishi,K. Yoshida, H. Wakita,T. Yamaguchi, "Structure and dynamics of 1,4-dioxane-water binary solutions studied by X-ray diffraction, mass spectrometry, and NMR relaxation", Journal of Molecular Liquids, 83:163-177(1999).
[9] Dechema Chemistry Data series vol.I, part. 1a: Vapor Liquid equilibrium data collection, p. 288-303.
[10] P. Van De Witte, P.I. Dijkstra, J.W.A. Van Den Berg, J. Feijen, "Phase Behavior of Polylactides in Solvent-Nonsolvent Mixtures", Journal of Polymer Science: Part B: Polymer Physics, 34:2553-2568 (1996).

(181x) Free Energy Landscape for Lower Critical Solution Temperature Transition of Thermoresponsive Poly(N-isopropylacrylamide)

Hongbo Du, Chemical Engineering, University of Arkansas, Fayetteville, AR and Xianghong Qian, Ralph E Martin Department fo Chemical Engineering, University of Arkansas, Fayetteville, AR

Thermo-responsive polymers exhibit hydrophobic−hydrophilic phase transition at their lower critical solution temperatures (LCST). For poly(N-isopropylacrylamide) (PNIPAM), the LCST phase transition is due mainly to the presence of a hydrophilic amide group on the side chain and a hydrophobic isopropyl group in each NIPAM unit as well as a hydrophobic C-C backbone. However, the driving force for the LCST transition, particularly at the molecular level remains controversial and unresolved. Here classical molecular dynamics simulations coupled with metadynamics technique was used to investigate the free energies associated with the folding and unfolding processes (i.e. hydrophobic-hydrophilic transition) of PNIPAM in water at the transition temperature of 305 K. In addition, the role of intermolecular and intramolecular hydrogen bonding interactions on the LCST transitions was also elucidated. Our results indicate that LCST is mainly entropically driven and that unfolding process has a higher barrier and is energetically less favored compared to the folding process at the transition temperature.

(181y) Binary SLE of Lipidic Ionic Liquid Systems

John R. McCabe[1], A. Shay Thigpen[1], Ali AL-Hashem[1], Arsalan Mirjafari[2], Kevin N. West[1] and James H. Davis Jr.[1,2], (1)Chemical & Biomolecular Engineering, University of South Alabama, Mobile, AL, (2)Chemistry, University of South Alabama, Mobile, AL

Lipidic ionic liquids are a class of ionic liquids characterized by the inclusion of large, nonpolar groups in the structure of the ionic liquids, typically including an aspect of structural asymmetry to decrease the melting point. These species were first described by our group in 2009 and since we have expanded the original class of compounds, 1-methyl-3-alkylimidazolium bistriflimides with unsaturations in the alkyl chain, to include long chains with branches, ether groups and thioethers as well as pyridinium and ammonium-based lipidic ionic liquids. We have focused much of our recent efforts on the synthesis and characterization of species containing long chains with thioether groups, synthetically obtained through the use of thiol-ene "click chemistry."

In addition to utilizing these species as solvents for separations and reactions, we are interested in the thermophysical properties and phase behavior of mixtures of lipidic ionic liquids with other ionic liquids and with molecular species. The knowledge from these studies will be valuable in understanding where lipidic ionic liquid may be of practical use as additives in other ionic liquids and also in understanding how they may be used for selective extraction of natural lipids from biological substrates. In this work we describe the solid/liquid equilibrium of mixture of these ionic liquids with each other and with molecular species.

(181z) Calculation of Vapor-Liquid Coexistence of Grafted Nanoparticles

Siladitya Mukherjee[1], Christopher R. Iacovella[1] and Peter T. Cummings[1,2], (1)Chemical and Biomolecular Engineering, Vanderbilt University, Nashville, TN, (2)Center of Nanophase Materials Sciences, Oak Ridge National Laboratory, Oak Ridge, TN

It is well known that the aggregation/dispersion of nanoparticles can have a significant impact on the properties of the system. One approach to control the aggregation is to graft polymers to the surface of the nanoparticles. The resulting behavior is therefore dictated by contributions from both the nanoparticle and polymer interactions and significant changes to the properties can be achieved by tuning these interactions. In this work we use molecular dynamics (MD) simulations to calculate the vapor-liquid coexistence of grafted nanospheres under a wide range of conditions. We use a coarse-grained model, generic enough to be applicable to a wide range of systems, but with parameters informed by recent studies of the interactions between silica nanoparticles [1] and the interactions between alkane chains known to accurately reproduce phase equilibria [2]. NVT MD simulations are performed using the GPU-enabled HOOMD-Blue [3-4] package, where vapor-liquid equilibrium is calculated using the quench dynamics technique. Vapor-Liquid coexistence is examined as a function of nanosphere interaction strengths, grafting density and grafting length. We find that the critical temperature increases with nanoparticle-nanoparticle interaction strength relative to a pure polymer system. We also observe that the critical temperature is relatively unaffected by polymer length for low grafting densities.

References:

1. C.K. Lee and C.C. Hua, J. Chem. Phys. 132, 224904 (2010)

2. M.G. Martin and J.I. Siepmann, J. Phys. Chem. B 102 (1998)

3. J. A. Anderson, C. D. Lorenz, and A. Travesset, Journal of Computational Physics 227(10): 5342-5359

4. http://codeblue.umich.edu/hoomd-blue

(188b) Nonspherical Janus Particles At an Oil-Water Interface

Bum Jun Park[1], Chang-Hyung Choi[2], Chang-Soo Lee[2] and **Daeyeon Lee**[1], (1)Chemical and Biomolecular Engineering, University of Pennsylvania, Philadelphia, PA, (2)Department of Chemical Engineering, Chungnam National University, Daejeon, South Korea

We study the configurations of nonspherical Janus particles such as ellipsoids, dumbbells, and cylinders at an oil-water interface. To determine the equilibrium configurations, we numerically calculate and minimize the attachment energy of each Janus particle as a function of the orientation angle as well as the vertical displacement with respect to the interface. We find that the equilibrium orientation of nonspherical Janus particles strongly depends on the particle characteristics, such as their size, shape, aspect ratio, surface wettability, and the location of the Janus boundary. In general, Janus particles with a high aspect ratio or a small difference in wettability between the two sides tend to have tilted orientations. Under certain circumstances, Janus ellipsoids and cylinder have meta-stable configuration as indicated by the presence of secondary minima in the attachment energy profile. The size of Janus particles plays an important role in determining the distribution of orientation angles of these particles at the fluid-fluid interface. Janus dumbbells, on the other hand, possess primary energy minimum only, indicating that their orientation at the fluid interface would be homogenous. Our study provides a detailed guidance for designing and synthesizing the non-spherical Janus particles, especially for their applications in multiphase fluid mixture stabilization.

(188c) Self-Assembly of Micro-Rods in Nematic Liquid Crystal Controlled by Boundary Conditions

Mohamed Amine Gharbi[1], Gaoxiang Wu[2], Marcello Cavallaro Jr[2], Daniel Beller[2], Randall Kamien[2], Kathleen J. Stebe[2] and Shu Yang[2], (1)University of Pennsylvania, Philadelphia, PA, (2)University of Pennsylvania

While the physics of spherical colloids confined in a nematic slab has been thoroughly investigated (see, e.g. [1-3]), interactions between anisotropic particles are less well studied [4-6]. Here, we report the behavior of micro-rods confined in a nematic cell. The anchoring conditions on the particles and on the bounding walls of the cell are well defined. Depending on the type of anchoring, cylinders spontaneously form 1D structures, chains parallel to the director, and 2D ordered structures, crystals with hexagonal symmetry. These latter structures are not observed for isotropic particles. By application of an external field, the anchoring conditions on the bounding walls can be dynamically controlled, and transitions between the different structures can be generated. Recent progress in understanding the particle self-assembly process is presented.

[1] P. Poulin and D. Weitz, *Physical Review E*, **57**, 626 (1998).

[2] M. Škarabot and *el.*, *Physical Review E*, **77**, 031705 (2008).

[3] I. Musevic and *el.*, *Science*, **313**, 954 (2006).

[4] D. Andrienko and *el.*, *Physical Review E*, **69**, 021706 (2004).

[5] F. R. Hung, *Physical Review E*, **79**, 021705 (2009).

[6] U. Tkalek and *el.*, *Soft Matter*, **4**, 2402 (2009).

(188d) Electric-Field Assisted Assembly of Colloidal Dimers

Fuduo Ma, Sijia Wang and **Ning Wu**, Chemical Engineering, Colorado School of Mines, Golden, CO

Anisotropic colloids could possess anisotropic properties in geometry, chemical composition or surface functionality. As such, they can recognize each other and assemble via directional interactions, potentially making large numbers of crystalline arrays and meso-structures that isotropic spheres cannot form. A colloidal dimer is an ideal candidate for studying the impacts of specific anisotropic property on the assembly of anisotropic particles, because versatile synthetic methods are available to create and tune the anisotropic properties in geometric, chemical composition, surface functionality, and physical properties. In this talk, we will present our experimental and theoretical explorations on the assembly of colloidal dimers with anisotropies in both geometric size and surface charge. Surprisingly rich patterns and crystalline arrays have been observed under the influence of AC electric fields. Our modeling demonstrates that the competition and balance between the double layer interaction, the induced dipole-dipole interaction, the electro-hydrodynamic flow, and the image forces determine different phases and packing paths between colloidal dimers. All those patterns are strongly dependent on the frequency of electric fields, which makes this field-driven assembly strategy particularly attractive. Potential applications in the field of display will also be discussed.

(188e) Near Field Capillary Interactions Between Microparticles with Undulated Contact Lines

Lu Yao, Lorenzo Botto, Marcello Cavallaro Jr., Blake Bleier, Valeria Garbin and Kathleen J. Stebe, Chemical & Biomolecular Engineering, University of Pennsylvania, Philadelphia, PA

Microparticles at fluid interfaces interact by capillarity. Far from contact, particles are known to attract and interact as quadrupoles in the interface. In the near field, geometric details dominate interactions. In practice, particles or particle agglomerates can have highly irregular surfaces that can create undulated interfaces in the near field. The interactions owing to these undulations may be repulsive or attractive (J. Lucassen, Colloids Surfaces, 65 (1992), pp. 131–137). Here we present a fundamental study between particles with undulated interfaces in the near field.

We have designed and fabricated microparticles with corrugated edges in the form of sine waves. When placed at interfaces, the corrugations pin the contact line, creating undulations in the interface with wavelengths and amplitudes dictated by the particle shape. Far from contact on unbounded interfaces, the particles interact like the expected capillary quadrupoles. In the near field, interactions depend on the details of the near-field undulations. Identical microparticles assemble with their corrugated edges in phase. Those with the same wavelengths and amplitude but out of phase corrugations shift to align their features. Microparticles with different wavelengths but the same amplitude rotate weakly to form oblique assemblies. To study the interactions of aligned undulated features, we confine the microparticles to an interface in a channel that is slightly wider than the dimension of the microparticles. For non-identical particles, we report an equilibrium separation distance due to near-field capillary repulsion.

(188f) Additive-Mediated Electrochemical Synthesis of Anisotropic Copper Crystals for Methanol Oxidation

Rajesh Venkatasubramanian, Chemical & Biomolecular Engineering, Tulane University, New Orleans, LA and Noshir S. Pesika, Chemical and Biomolecular Engineering, Tulane University, New Orleans, LA

Anisotropic metallic micro- and nanocrystals synthesis and characterization, either in suspension or localized on a surface is an area of immense interest due to its potential use in many applications ranging from catalysis, microelectronics, biosensors and solar cells. In this work, a simple and versatile electrochemical approach to synthesize anisotropic copper crystals in presence of shape directing additives has been proposed. SEM, TEM and XRD characterization confirm the anisotropic morphology and crystal structure of synthesized copper particles. Br- ions in the additive adsorbs preferentially to selective crystal faces on copper particles thereby altering the relative crystal growth kinetics leading to anisotropic morphology. Electro-oxidation of methanol is chosen as a model experiment to characterize the superior electro-catalytic properties of anisotropic copper crystals. Our future work will be focused on investigating the feasibility of using such non-noble metal micro- and nanostructures as a suitable alternative for anode interface in direct methanol fuel cell applications.

(188g) Magnetic Janus Particles: Preparation, Self-Assembly, Simulation, and Application

Sangyeul Hwang[1], Trung D. Nguyen[2], Jaewon Yoon[3], Srijanani Bhaskar[3], Sharon C. Glotzer[4] and Joerg Lahann[5], (1)Chemical Engineering Department, University of Michigan, Ann Arbor, MI, (2)Department of Chemical Engineering, University of Michigan, Ann Arbor, MI, (3)Macromolecular Science and Engineering, University of Michigan, Ann Arbor, MI, (4)Chemical Engineering , University of Michigan, Ann Arbor, MI, (5)Department of Chemical Engineering, Macromolecular Science and Engineering, Biomedical Engineering, and Materials Science and Engineering, University of Michigan, Ann Arbor, MI

We report the preparation of different shaped and sized magnetic Janus particles (MJPs) using electrohydrodynamic (EHD) co-jetting. Spherical MJPs with 20 μm sizes in diameter are directly prepared by the EHD co-jetting. The MJPs are composed of poly(lactic-co-glycolic acid) (PLGA) in both compartments, where one compartment has magnetite (28nm in diameter) and the other has not. The EHD co-jetting also produces bicompartmental PLGA fibers with much thicker diameters (up to 450 μm in a diameter), and subsequent cryosectioning of the fibers using a microtome fabricates cylinder-shaped MJPs. A newly developed method to change a particle shape from cylinder to sphere is introduced, allowing us to easily manufacture large-sized spherical MJPs.

This study also demonstrates a simple approach to harness the rotational freedom of the MJPs using a magnetic field gradient. Two-dimensionally ordered assemblies of a high concentration of the MJPs are also directed within a boundary system. Various assembled structures having from low to high particle density including hexagonally close-packed one are achieved by changing the field gradient strength. By altering the field direction from a perpendicular to a parallel orientation, the assembled MJPs orthogonally rotated to simultaneously reconfigure their assembled structures. All assemblies are reversible and applicable to the opposite orientation. Furthermore, a simulation developed in this study, which is well matched with the experimental results, reveals that the functions of the gradient field strength and the induced dipole interactions of the particles determine the degree of alignment and the particle packing density in the assembled structures.

Finally, the large-sized spherical MJPs, where the magnetite compartment has a coloring additive (carbon black) and the none-magnetic compartment contains a yellow-colored dye (Yellow 14), are utilized to demonstrate a color switchable device. The simple magnet operation assembles the MJPs to show a black color or yellow color at a time. This process induced by the permanent magnet is energy efficient and reproducible.

Effect of Particle Size and Anisotropy On Yield Stress and Dynamic Modulus in Latex

Suresh Ahuja, Xerox Corporation, Webster, NY

Introduction

Colloidal dispersions such as latex are found in many traditional applications, such as, paints, inks, cosmetics, pharmaceuticals or foods. Study of colloids has been useful in tissue engineering, photonic crystals, inkjet technology, ceramics, microfluidics and digital imaging. The control of particle size and its morphology are critical in producing high quality latex for applications in digital images, coatings, adhesives, impact modifiers, and medical diagnostics. Understanding the development of non-equilibrium morphologies where the possible number of particle structures is large is a challenge. Semi-batch emulsion polymerization in the presence of seed particles that eliminates the particle nucleation stage and with the regulated concentration of monomer in the reaction system represents an effective tool to investigate the particle growth mechanisms. Monomers used in emulsion polymerization are butadiene, styrene, acrylonitrile, acrylate ester and methacrylate ester monomers, vinyl acetate, and vinyl chloride. An effective stabilizer such as ionic and non-ionic surfactants and protective colloid which can be physically adsorbed or chemically put onto the particle surface, is required to prevent the interactive latex particles from coagulation.

Colloidal Stability

Colloidal stability can be achieved via the electrostatic stabilization mechanism [1], the steric stabilization mechanism [2,3] or both. Most of the commercial latex products are manufactured by semibatch or continuous reaction systems due to the very exothermic nature of free radical polymerization and rather limited heat transfer capacity in large-scale reactors. One major difference among the above polymerization processes is the residence time distribution of the growing particles within the reactor. The Smith–Ewart theory predicts that the number of latex particles nucleated per unit volume of water is proportional to the surfactant concentration and initiator concentration

The studies on diffusion-controlled nucleation and growth of particle nuclei and the molecular diffusion of monomer from the aqueous phase to the particle–water interface are two examples in investigating polymerization mechanisms and kinetics. Phase behavior, dynamics and flow of colloidal dispersions are strongly controlled by particle volume fraction. Progressive crowding eventually leads to a vitrification when particles are trapped within a virtual cage provided by nearest neighbors. Long range particle motion slows down and the low shear viscosity diverges when a critical concentration is approached,

When the stabilizer layer is either relatively thin with respect to the particle radius or relatively rigid, the viscosity curves still coincide with those of Brownian hard spheres, provided the stabilizer layer is included in the particle volume. The effect of the stabilizer can then be expressed by means of an effective volume fraction which can be calculated from the core volume fraction of the particles themselves, the particle radius and the thickness of the stabilizer layer[4]. As shown in Fig.1, when the stabilizer layer becomes thicker and/or softer, the previous procedure is not adequate anymore and results in overestimation of the viscosity, especially at higher volume fractions

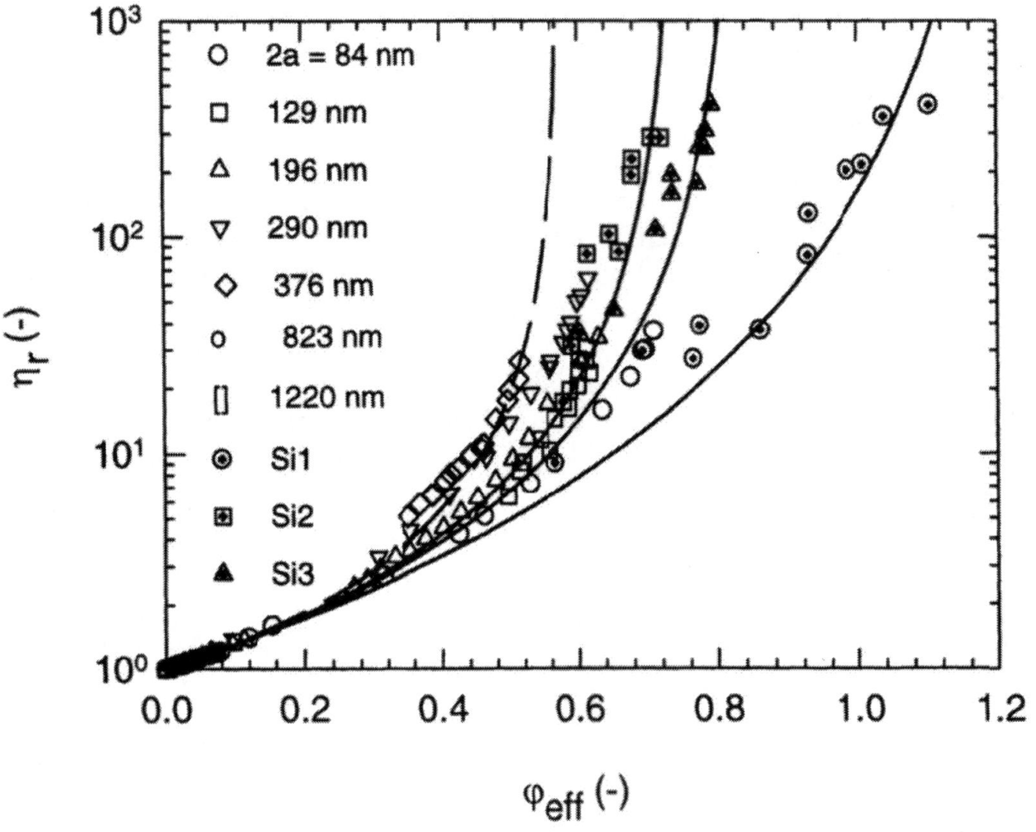

Fig1. Limiting high shear viscosities for various sterically stabilized PMMA and silica dispersions

Oscillatory measurements, in particular on concentrated colloidal dispersions, can provide detailed information about the material under consideration. In such experiments the viscous and elastic contributions can be separated into the loss modulus and storage modulus . In Brownian hard spheres the Brownian motion is responsible for the elastic part, whereas the high frequency limit of loss modulus measures the pure hydrodynamic contribution in the equilibrium structure. In concentrated, sterically stabilized, systems the stabilizer layers overlap and elastic modulus measures the interaction forces between layers on neighboring particles during the oscillatory motion. The effect of anisotropy and the effect of microstructure induced by anisotropy on dynamical transitions such as gelation and the glass transition are also of broad interest in soft matter physics. However, there are few experimental data available to test recent simulations and theoretical studies of the interaction of colloidal anisotropy, microstructure, and slow dynamics [5].

Plateau Modulus

In a study of PS-microgel particles suspended in an isorefractive organic solvent allowing for light scattering experiments and an aqueous polymer latex dispersion with short range repulsive interactions based on a commercial polymer latex system. Both systems essentially behave like hard spheres, their zero-shear viscosity diverges at a volume fraction of 0.58, linear viscoelastic behavior is well-described by the mode coupling theory and the absolute values of the plateau moduli are close to those reported for other hard sphere systems[6].

Fig.2 displays the plateau modulus and the longest relaxation time from oscillatory shear experiments as a function of particle volume fraction.

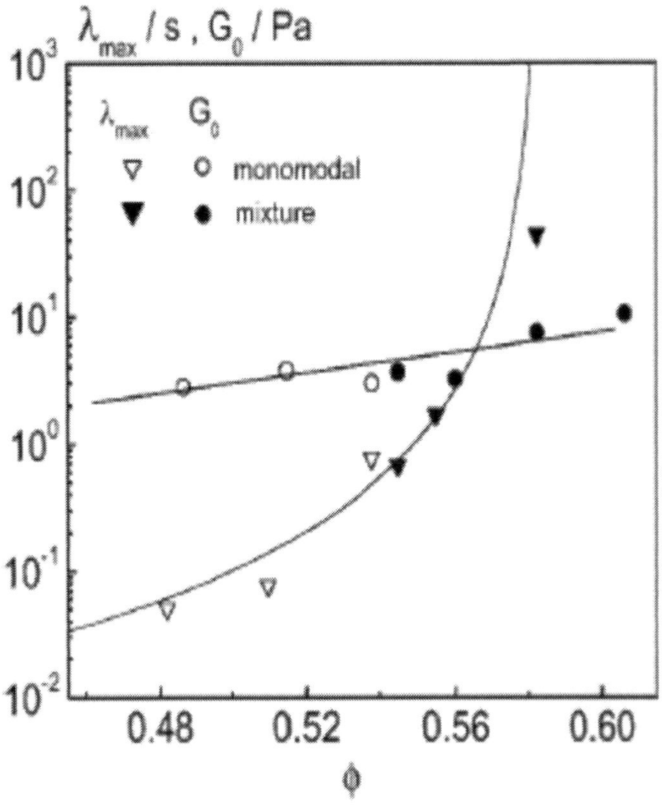

Fig 2 Maximum relaxation time, lmax (triangles), and plateau modulus G0 (circles) obtained from the frequency dependence of linear viscoelastic modulus and for mono- and bimodal PS microgel. The straight line is to guide the eye with respect to the G0 data and the curve is a fit of the functional form of Equation to the maximum relaxation data with solid fraction of 0.58.

$$\eta_0/\eta_S = A(1 - \phi/\phi_g)^{-\gamma}$$

--(1)

Method, Material s and Discussion

Acrylic latexes prepared by emulsion polymerization were subjected to steady shear and sinusoidal dynamic deformation. Viscosity of 0.6ϕ was measured at different shear rates. At the lowest volume fractions the suspensions show shear rate independent behavior. As the particle-particle separation distance is large compared to the range of steric inter-particle forces, only the Brownian diffusion dominates. The viscosity remains constant with shear rate as the particles can diffuse freely in the suspension. As the volume fraction increases the inter-particle distance becomes comparable to the particle radius and the hydrodynamic and surface interactions dominate. The viscosity increases with increasing volume fraction and changes from one that is independent of shear rate.to shear thinning. The yield stress of latexes obtained by polynomial fits in shear stress shear rate curves is related to the additional force required to break the bonds between particles or polymer chains that hold the network together before flow can begin. In the attractive suspension, there are attractive surface forces between the particles. Only when the applied stress exceeds a critical amount, the yield stress, does the system flow.

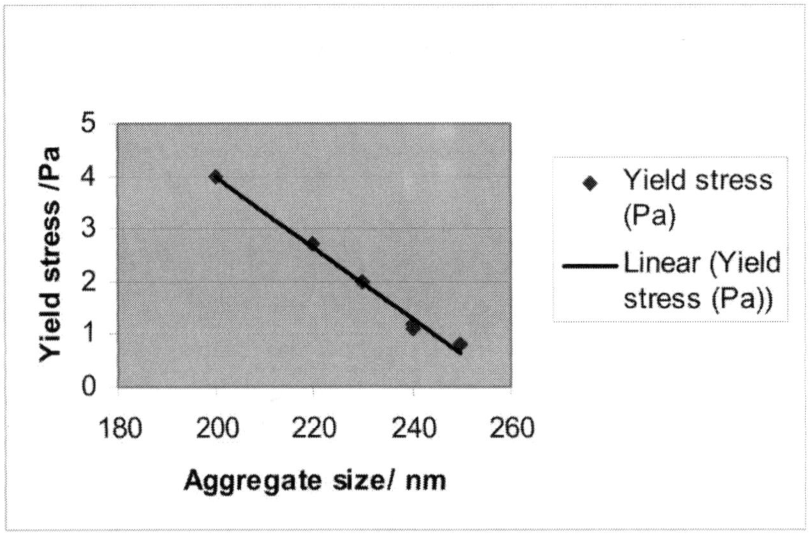

Figure 3 dependence of yield stress on aggregate size

Acrylic latex differed in anisotropy. Some were potato shaped and some spherical. Both modulus and yield stress were found to be higher for potato shaped aggregates compared to spherical aggregates for the same solid concentration. This effect was attributed to anisotropy.

Conclusion

Colloidal dispersions like latex depending on solid concentration, particle (aggregate) size and anisotropy(rod, disc) of aggregates form plateau modulus and give yield stress.

References

[1] Y. A. Vlasov, N. Yao and D. J. Norris, Adv. Mater., 1999, 11, 165–

169.

[2] J.W. Bender, N.J. Wagner, J. Colloid Interf. Sci. 172 (1995) 171.

[3] R.A. Lionberger, W.B. Russel, J. Rheol. 41 (1997) 399.

[4] G. Biebaut, Ph.D. Thesis, Katholieke Universiteit Leuven, Leuven,Belgium, 1999.

[5] Pfleiderer, P.; Milinkovic, K.; Schilling, T. Europhys. Lett.(2008), 84, 16003

[6] Willenbacher, N; Vesaratchanon, J.S., Thorwarthb, O., and Bartschb, E., Soft Matter, (2011), 7, 5777

(200a) Imaging Colloidal Dynamics with Digital Holographic Microscopy

Vinothan N. Manoharan, School of Engineering and Applied Sciences and Department of Physics, Harvard University, Cambridge, MA

Although colloidal dynamics in concentrated suspensions have been studied extensively with optical and confocal microscopy, very few real-space experiments have explored the dynamics of interacting particles in dilute suspensions. In these systems, standard microscopy techniques often prove too slow to image the rapidly diffusing particles. We use an interferometric technique called digital holographic microscopy to reveal the dynamics and structure of colloidal suspensions with millisecond time resolution and nanometer-scale spatial precision in all three dimensions. These experiments have revealed unexpected aspects of seemingly simple phenomena, such as the diffusion of particles in clusters or on emulsion droplets, or the binding of a particle to a liquid interface. In the latter case, we find that the dynamics are surprisingly slow: particles may take weeks or even months to relax to equilibrium at oil-water interfaces, a consequence of the hindered motion of the contact line near the solid substrate. These studies may yield some insights into the mechanisms and failure modes of colloidal self-assembly.

(200b) Dynamics of a Janus Droplet in An External Flow

Ubaldo Cordova, Chemical Engineering, University of Puerto Rico at Mayagüez, Mayaguez, PR, Misael Diaz, Department of Chemical Engineering, University of Puerto Rico at Mayagüez, Mayagüez, PR, Andrey Ivantsov, Institute of Continuous Media Mechanics UB RAS, Perm, Russia and Sergey Shklyaev, Department of Chemistry, The Pennsylvania State University, University Park, PA; Department of Chemical Engineering, University of Puerto Rico–Mayagüez, Mayagüez, PR

Janus droplets are of crucial importance in the design of multifunctional materials. In particular, creation, control, and polymerization of Janus droplets are frequently used in construction of Janus particles needed for many technological applications. In this study, the dynamics of the simplest compound droplet—a Janus droplet that consists of two fluids occupying equal hemispherical domains, such that the internal interface (inside the droplet) remains flat—is considered at low Reynolds number. The droplet is placed into a space-uniform external flow. Thus, analytical solution in Lamb's series can be constructed. It is shown that fluid flow can be represented as a superposition of that for two limiting cases: (i) the external flow perpendicular to the internal interface and (ii) the flow parallel to the interface. In case (i) there is no torque applied to the droplet, therefore, it materializes a stable orientation of the Janus droplet (the fluid of smaller viscosity has to be at the upper stream for the stable configuration). It is important to note that even for equal viscosities of internal fluids the problem is not reduced to the Hadamard-Rybczynski problem, because the internal interface prohibits the motion inherent to the simple droplet. In contrast, for case (ii) there is the above-mentioned reduction; for small difference in viscosities of internal fluids the overall torque is proportional to this difference. The motion of the droplet under shear flow is also studied. This shear leads to additional torque, which changes a stable configuration of a torque-free Janus droplet in such a way that the drop's axis of symmetry is tilted to the external fluid. This, in turn, generates a component of the force (and, hence, a drop velocity) normal to the external flow. These analytical results are in a good agreement with numerical computations by Fluent. The latter method becomes especially efficient for more realistic Janus droplets with curved interfaces.

(200c) Brownian Motion of Janus and Patchy Particle Swimmers

Ilona Kretzschmar, Chemical Engineering, The City College of New York, The City University of New York, New York, NY

The ability to control and drive particle motion at the micro- and nanometer scale has fascinated researchers across the world for several decades and has recently received renewed attention due to the advent of Janus particle research. A Janus particle is a particle with two chemically or physically distinct halves. If one of the two halves is chosen such that it is reactive toward a solute molecule present in the solution surrounding the Janus particle, an asymmetric concentration gradient of product molecules is formed. The asymmetric concentration gradient results in an uneven body force experienced by the Janus particle leading to the directed, ballistic motion of the particle, a process also called diffusiophoresis. Diffusiophoretic velocities of a few microns per second have been reported. Unfortunately, the presence of Brownian rotation leads to a randomization of the motion at long time steps and prohibits the use of this motion for directional transport.

Over the past few years, we have studied the behavior of patchy particles, i.e., particles with reactive material patches ranging from 11 to 50% of the particle surface. As expected, we find a reduced velocity for particles with smaller patches due to the less pronounced concentration gradient. We find that the velocity is particle radius independent when 2.4 and 5 μm particles are studied in good agreement with predictions by Golastanian. Surprisingly, we find that the velocity of the patchy particles does not decay as fast as the decrease in patch sizes would suggest, but rather that patchy particles exhibit an enhanced diffusiophoretic velocity. Careful analysis of the experimental data reveals that the patch asymmetry is the reason for the enhanced velocities and a potential explanation involving hindered Brownian rotation of these particles near solid surfaces will be presented that could potentially enable the directed motion of Janus and Patchy particles over much longer distances.

(200d) Electrostatic Charge Shuttles As Chemically Powered Colloidal Motors

Aaron M. Drews[1], Hee Young Lee[2] and Kyle J. M. Bishop[1], (1)Chemical Engineering, Penn State University, University Park, PA, (2)Penn State University, University Park, PA

As nanotechnology progresses from the synthesis of diverse structures to the design of functional systems, there is an increasing demand for methods by which to power and propel nanoscale machines. The sustained operation of such machines requires a steady input of energy which must be delivered remotely (chemical fuels) and converted efficiently into useful mechanical motions. Despite impressive demonstrations of chemically powered, nanoscale propulsion, existing systems remain limited by poor efficiencies (typically ~1E-09) which prohibits their operation in low fuel environments (*e.g.*, biological media). Here, we describe a new strategy by which electrochemical processes are coupled to mechanical motion through electrostatic forces acting on colloidal "charge shuttles". With the use of high-speed optical microscopy and electrical measurements we show that this approach is *ca.* one billion times more efficient than existing nanomotors. Moreover, careful electrode design and surface modification *via* self-assembled monolayers affords low-voltage operation (~1V), thereby allowing power to be sourced electrochemically. Our experimental results are explained by a simple model which accounts for particle charging at an electrode and its movement in the applied field, and further indicates that the underlying forces scale favorably down to molecular dimensions. In closing, we demonstrate rectification of the system's oscillatory behavior for linear actuation or directed propulsion, and highlight the intriguing dynamics of multi-particle systems.

(200e) The Structure of Agglomerates Consisting of Polydisperse Nanoparticles

Max Eggersdorfer and Sotiris E. Pratsinis, Particle Technology Laboratory, Swiss Federal Institute of Technology (ETH Zurich), Zurich, Switzerland

The Structure of Agglomerates Consisting of Polydisperse Nanoparticles

M.L. Eggersdorfer and S.E. Pratsinis

Particle Technology Laboratory, Institute of Process Engineering, Department of Mechanical and Process Engineering, ETH Zurich, Sonneggstrasse 3, CH-8092 Zürich, Switzerland.

Ph. +41 (0) 44 632 31 80; Fax. +41 (0) 44 632 15 95

Agglomeration is encountered in many natural or industrial processes, like growth of aerosol particles in the atmosphere, material synthesis by aerosol processes or even flocculation of minerals in oceans or colloidal particles. These particles collide by different mechanisms and stick together forming irregular or fractal-like agglomerates. Typically, the structure of these agglomerates is characterized with the fractal dimension, D_f, and pre-exponential factor, k_n, of simulated agglomerates of monodisperse primary particles (PP) for ballistic or diffusion-limited particle-cluster and cluster-cluster collision mechanisms. This concept has served well a wide spectrum of aerosol and colloidal particles, in particular by coagulation. In fact, a number of characterization techniques and process design concepts have been developed capitalizing on these D_f values to extract other particle properties (e.g. collision diameter, primary particle size) and design reactors for manufacturing such particles. What might have been overlooked in characterization and simulations of such particles is that the above D_f values have been developed for agglomerates of monodisperse primary particles. For coagulating aerosols and colloids, however, this needs to be carefully examined as Brownian coagulation-driven particle formation leads to polydisperse particles. Figure 1 shows a TEM image of a fractal-like zirconia nanoparticle made by scalable flame spray pyrolysis having a primary particle size distribution with geometric standard deviation of $s_g \approx 1.58$.

Here, the effect of PP polydispersity on D_f and k_n is investigated with agglomerates consisting of 16 – 1024 PP with closely controlled size distribution (s_g = 1-3). Figure 2 shows a snapshot of a numerically generated diffusion-limited cluster-cluster agglomerate (DLCA) with s_g = 1.58. These simulations are in excellent agreement with the classic structure (D_f and k_n) of agglomerates consisting of monodisperse PPs made by four different collision mechanisms as well as with agglomerates of bi-, tri-disperse [1] and normally distributed PPs [2]. Broadening the PP size distribution of agglomerates decreases monotonically their D_f and for sufficiently broad PP distributions ($s_g > 2.5$) the D_f reaches about 1.5 and k_n about 1 regardless of collision mechanism [3]. Furthermore with increasing PP polydispersity, the corresponding projected area exponent, D_a, and pre-exponential factor, k_a, decrease monotonically from their standard values for agglomerates with monodisperse PPs. So the PP polydispersity determines for $s_g > 2.5$ the agglomerate structure rather than the collision mechanism. So D_f as well as D_a can be an indication for PP polydispersity in mass–mobility and light scattering measurements, if the dominant agglomeration mechanism is known, like diffusion-limited and/or ballistic cluster-cluster coagulation in aerosols or colloidal systems.

[1] Bushell, G. and Amal, R. J. Colloid Interface Sci. 205 (1998) 459.

[2] Tence, M., Chevalier, J. P. and Jullien, R. J. Phys. 47 (1986) 1989.

[3] M.L. Eggersdorfer, S.E. Pratsinis, Aerosol Sci. Technol. 46 (2012) 347.

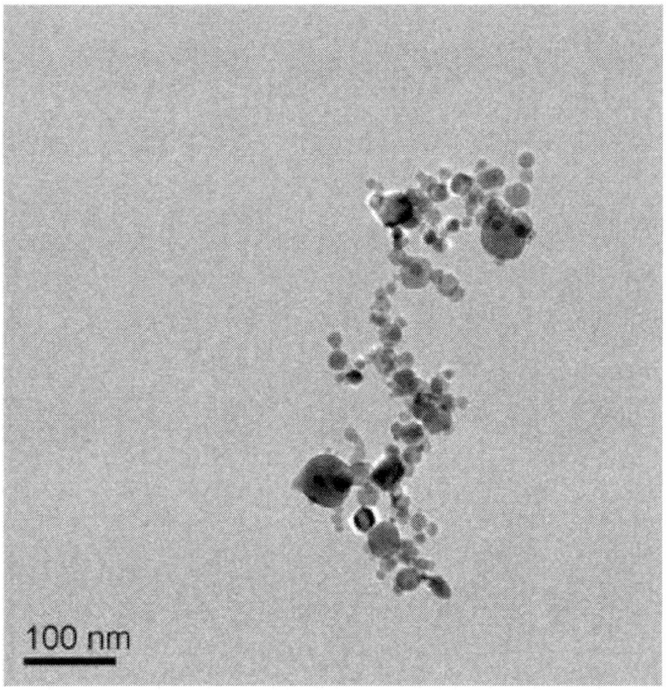

Figure 1: TEM image of a size-selected zirconia agglomerate with a mobility diameter, d_m = 110 nm, and a primary particle size distribution with geometric standard deviation of $s_g \approx 1.58$.

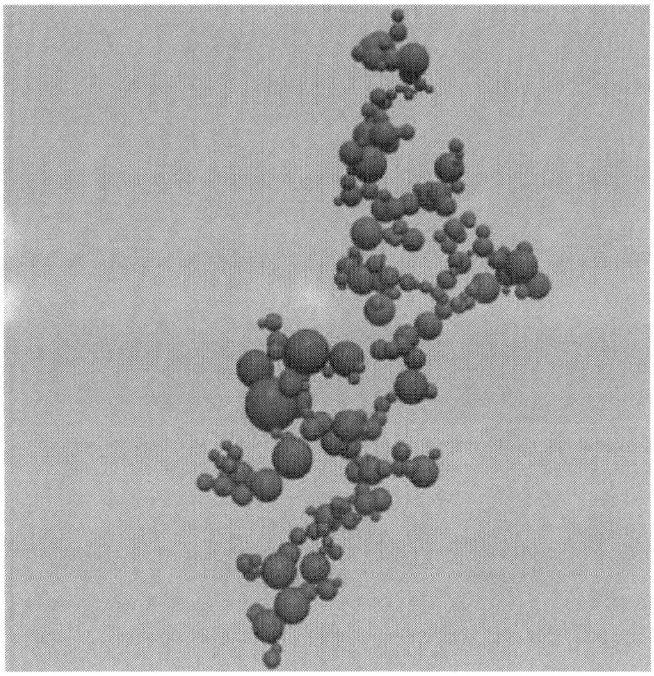

Figure 2: Snapshot of a diffusion-limited cluster-cluster agglomerate (DLCA) consisting of 256 primary particles with $s_g = 1.58$.

(200f) Breakup Dynamics of Colloidal Clusters in Shear Flow

Yogesh M. Harshe, Department of Chemistry and Applied Biosciences, Institute for chemical and bioengineering, ETH Zurich, Zurich, Switzerland and **Marco Lattuada**, Department of Chemistry and Applied Biosciences , Institute for Chemical and Bioengineering, ETH Zurich, Zurich, Switzerland

In this work, we present a broad spectrum computational study of the breakage dynamics of colloidal clusters in the presence of shear flow. We have formulated a model, which uses Stokesian dynamics to estimate the hydrodynamic interactions among the particles in a cluster, DLVO theory to describe the normal interparticle interactions, and tangential interactions through discrete element method to account for contact forces. Fractal clusters composed of monodisperse spherical particles were generated using different Monte Carlo methods, covering a wide range of cluster masses and fractal dimensions. The study is divided in to two parts, one related to first breakage events in colloidal clusters exposed to shear forces, while the second part focuses on their long term dynamics. The time required to reach the first breakage event of a cluster was used to compute a characteristic breakage rate constant for various flow magnitudes under simple shear conditions. The breakage rate constant showed a power law dependence on the product of the aggregate size and the applied stress, with values of the corresponding exponents depending only on the aggregate fractal dimension and the type of flow field, whereas the prefactor of the power law relation also depends on the size of the primary particles comprising a cluster. The long-term dynamics was quantified by monitoring the evolution of the average clusters size, fractal dimension, average particle coordination number and number of fragments. When the time was rescaled by the characteristic time for the first breakage event, the dynamics all clusters, independent of mass and fractal dimension, showed a universal behavior. The results of these simulations provide new insights on the complex interplay between cluster morphology and hydrodynamic and interparticles interactions in sheared dispersions of clusters.

References:

Harshe Y.M. and Lattuada M., Langmuir 2012, 28, 283–292

Harshe Y.M. and Lattuada M., Soft Matter, submitted

(200g) The Role of Shear Flow in Heteroaggregation Behavior of Hydrogel Microparticles and Nanoparticles

Kapil V. Deshpande, Chemical Engineering, Rutgers University, Piscataway, NJ and Nina C. Shapley, Chemical and Biochemical Engineering, Rutgers, the State University of New Jersey, Piscataway, NJ

Microparticles coated with oppositely charged nanoparticles are very versatile compared to their uncoated counterparts. Nanoparticle coating is useful for improving microparticle stability and flowability. These coated particles can also be used for pH dependent tunable release, among other applications. In this work, we examine how a system of oppositely charged hydrogel microparticles and nanoparticles interacts in a shear flow environment. We aim to determine the effect that shearing has on the heteroaggregation phase behavior of the interacting particles.

The experimental system that we used to model this interaction consists of cationic green fluorescently tagged chitosan nanoparticles and anionic untagged alginate microparticles in water. The interactions are characterized using fluorescence spectroscopy to quantify the interaction taking place. Optical microscopy is used to confirm the results. Particle heteroaggregation behavior is studied at various levels of shear for different fixed concentration ratios of alginate and chitosan particles to determine the changes in the phase diagram.

(200h) Elastic Recoil of Flexible Colloidal Chains

Julie E. Byrom, Chemical & Biomolecular Engineering, Rice University, Houston, TX and **Sibani Lisa Biswal**, Chemical and Biomolecular Engineering, Rice University, Houston, TX

Elasticity describes the ability of a material to return to its original shape after being deformed by an external force. Flexible elastomers can be stretched or compressed with minimal force. Extensible elastomers can be stretched significantly before rupturing. We have developed a classic bead-spring-bead model chain to visualize elastic recoil using paramagnetic particles and DNA. The bending rigidity of these chains can be tuned by adjusting the properties of the DNA. Semiflexible filaments, where the length, L, is comparable to the persistence length, Lp, have been designed. The driving force for elastic recoil in a rubber-like elastomer is the increased entropy of the relaxed state relative to the stretched state. We expect that this is the same mechanism controlling elastic recoil in DNA-linked chains. Our semiflexible chains can be compressed with a magnetic field and the elastic recovery can be visualized. In the absence of an external magnetic field, the chain is in its relaxed state and has higher entropy because there are many more ways of arranging than a stretched polymer chain. The effect of an external force is to decrease the entropy, which is recovered when the magnetic field is removed and the chain recoils to its relaxed state. Understanding how elasticity couples to the hydrodynamics of these chains, resulting in unique chain conformations will be discussed.

(200i) Brownian Motion of Stiff Filaments in Confined Media

Nikta Fakhri[1,2], Frederick C. MacKintosh[3], Brahim Lounis[4], Laurent Cognet[4] and Matteo Pasquali[2], (1)Third Institute of Physics-Biophysics, University of Göttingen, Göttingen, Germany, (2)Chemical and Biomolecular Engineering, Rice University, Houston, TX, (3)Department of Physics and Astronomy, Vrije Universiteit, Amsterdam, Netherlands, (4)Laboratoire Photonique, Numérique et Nanosciences - LP2N, Université de Bordeaux, Institut d'Optique Graduate School & CNRS, Talence 33405, France

The thermal motion of stiff filaments in a crowded environment underlies the behavior of such disparate systems as polymer materials, nanocomposites, and the cell cytoskeleton. Despite decades of theoretical studies, the fundamental dynamics of such systems remains a mystery. Using near-infrared video microscopy, we study the thermal diffusion of individual single-walled carbon nanotubes (SWNTs) confined in porous agarose networks. SWNTs are the ideal system to study confined dynamics of stiff filaments. SWNTs are slender (typical diameters 0.7-1.2 nm), sufficiently long to be visualized by optical microscopy (3-15 μm), stiff (20-150 μm, scaling with d^3), and share many dynamical characteristics with polymers. Surprisingly, we find that even a small bending flexibility strongly enhances their motion: the rotational diffusion constant is proportional to the filament bending compliance and is independent of the network porosity. This study establishes conclusively the dynamics of individual stiff filaments in crowded environments, elucidating the fundamental physics of how backbone flexibility affects mobility and diffusion. In addition to their relevance for the fundamental physics of stiff polymers, our results can be exploited in biological and technological contexts.

(201a) Enhanced Sampling of Peptide Adsorption and 2D Self-Assembly with Parallel Tempering Metadynamics [Invited Talk]

Jim Pfaendtner and Michael Deighan, Chemical Engineering, University of Washington, Seattle, WA

Efficient sampling of the interaction of surfaces with peptides and proteins continues to be a pressing challenge. Classical molecular dynamics (MD) simulations are frequently insufficient for this task as these systems can become trapped in minima that prevent equilibration of relevant probability distributions. Monte Carlo also suffers from low efficiency due to well-known challenges with explicit solvation. In this talk we will present a strategy for efficient simulation of peptide adsorption and 2D self-assembly on surfaces using a new algorithm: parallel tempering metadynamics in the well-tempered ensemble (PTMetaD-WTE). This algorithm allows for biasing of a few relevant collective variables (CVs) while using the parallel tempering scheme (PT) to achieve a random walk through temperature space and overcome enthalpic barriers due to "hidden CVs". Importantly, the algorithm helps overcome well-known scaling challenges that have plagued PT in explicitly solvated systems by coupling PTMetaD with the well-tempered ensemble (WTE) of Bonomi and Parrinello.

First we will describe the behavior of model systems based on the alpha and beta LK peptides interacting with a number of different model surfaces including self-assembled monolayers (SAMs) [methyl terminated, carboxyl terminated] and graphene. The adsorption behavior is compared to experimental measurements from sum frequency generation (SFG) and surface plasmon resonance (SPR). Following this, we will show preliminary results from simulation of the self-assembly of a monolayer of LK peptides in an explicitly solvated environment.

(201b) Lipid Self Assembly by Statistical Temperature Monte Carlo and Molecular Dynamics

Lili Gai[1], Katie A. Maerzke[1], Christopher R. Iacovella[1], Clare McCabe[1] and Peter T. Cummings[1,2], (1)Chemical and Biomolecular Engineering, Vanderbilt University, Nashville, TN, (2)Center of Nanophase Materials Sciences, Oak Ridge National Laboratory, Oak Ridge, TN

The recently developed statistical temperature Monte Carlo (STMC) and statistical temperature molecular dynamics (STMD) [1] methods have been applied to study the self-assembly process of amphiphilic molecules (bilayer and micelle formation). STMC and STMD are extensions of the Wang-Landau (WL) method [2], from which a flat energy distribution is also attained by updating the statistical temperature rather than the density of states (as is done in the WL method). However, this updating of the intensive quantity temperature provides a continuum description for the complete density of states without imposing limitations on the size of the energy bins.

As a test of the ability of the STMC and STMD methods to study self assembly, two systems have been studied: a lattice based implicit solvent 3-segment model and an off-lattice explicit-solvent 3-segment model. In the lattice model, varying sizes of energy bins are simulated with the STMC method. A bilayer structure is found to form at low temperatures, with phase transitions to clusters as temperature increases. The results are in good agreement with phase diagrams calculated by traditional Metropolis and the original WL methods. Additionally by allowing the use of larger energy bins, the simulation converges faster by the STMC method than by the original WL method.

For the off-lattice model, a systematic study of the thermodynamic properties will be presented, as well as the effect of the strength of molecular interactions (hydrophobic and hydrophilic) on the structures formed discussed. The STMD method makes well use of the WL sampling idea with molecular dynamics move, which is applicable to search global minima for complex systems, where effective STMC moves are not practical.

1. J. Kim, J. E. Straub and T. Keyes, "Statistical-Temperature Monte Carlo and Molecular Dynamics Algorithms," *Phys. Rev. Lett.* **97** (2006) 050601

2. F. Wang and D. P. Landau, "Efficient, Multiple-Range Random Walk Algorithm to Calculate the Density of States," *Phys. Rev. Lett.* **86** (2001) 2050–2053.

(201c) Programmable Self-Assembly of Nanoparticle Clusters Using DNA-Mediated Interactions

Jonathan D. Halverson and Alexei V. Tkachenko, Center for Functional Nanomaterials, Brookhaven National Laboratory, Upton, NY

Nanoparticles have novel optical, electronic, magnetic and chemical properties due to their small size. Envisaged applications such as nanoplasmonics require the nanoparticles to be precisely arranged into structures with a specific overall shape. Approaches to bring about the self-assembly of micrometer size particles such as patchy regions or complementary shapes are not directly applicable to nanopartices. However, nanoparticles coated with ssDNA can form effective bonds with other particles through the hybridization of complementary strands. We have conducted computer simulations using a coarse-grained model to study the self-assembly of pre-programmed finite clusters from an initially homogeneous solution of these particles. Our designs produce the target cluster in nearly perfect yield and free of errors. In addition, by controlling the kinetic pathway the clusters are produced rapidly.

Research carried out in whole at the Center for Functional Nanomaterials, Brookhaven National Laboratory, which is supported by the U.S. Department of Energy, Office of Basic Energy Sciences, under Contract No. DE-AC02-98CH10886.

(201d) Simulation of Amyloid Beta in the Presence of Aggregation Inhibitors

David C. Latshaw II and Carol K. Hall, Chemical and Biomolecular Engineering, North Carolina State University, Raleigh, NC

In Alzheimer's Disease (AD), the amyloid β (Abeta) peptide aggregates and forms oligomer and fibril structures, causing a decline in thought processes and memory, ultimately resulting in dementia and death. Compounds that inhibit the formation of these structures are being examined by both academic and industrial researchers. Understanding the mechanisms of aggregation inhibition on a molecular level could provide insight into the characteristics that would be most effective for disrupting the aggregation process, further aiding drug design. To examine the effect of inhibitors on protein aggregation, discontinuous molecular dynamics simulations (DMD) combined with an intermediate resolution protein model developed in the Hall group, PRIME20, were applied to a peptide/inhibitor system. The system contained 48 Abeta(16-22) and a variable number of model inhibitors, represented here by simple hydrophobic spheres. The effects of varying the number, size, and interaction strength of the spheres were examined. Preliminary results indicate that aggregation is promoted when the number of hydrophobic spheres of fixed size is increased, leading to an increase in effective concentration. When the size of the hydrophobic spheres at fixed inhibitor concentration is increased, aggregation is decreased because the increased mass and interaction range decreases peptide mobility. The effect of increasing the interaction strength between the spheres and peptides depends on temperature. Peptide aggregation is promoted until a critical interaction strength above which aggregation is hindered.

(201e) Nanocubes and Nucleosomes: One Story of Assembly and Another of Disassembly [Invited]

Gaurav Arya, Irina V. Dobrovolskaia, Bo Gao and Andrea Tao, Department of Nanoengineering, University of California, San Diego, La Jolla, CA

I will discuss the role of computational methods in providing new insights into the assembly and disassembly of nanoscale systems. First, I will discuss Monte Carlo simulations of polymer-grafted Silver nanoparticles that helped design strategies to assemble them, within a polymer thin film, into 1D plasmonic strings possessing tunable orientation and electromagnetic properties [1]. We demonstrate the critical role played by the grafted polymer chain length in modulating the relative strengths of van der Waals attraction and steric repulsion and thereby dictating interparticle orientations. Second, I will discuss the development of a mesoscale model of the nucleosome and its Brownian dynamics simulations that helped elucidate the dynamics of force-induced unraveling of DNA from histone octamers [2]. We demonstrate the role of non-uniform DNA/histone interactions in stabilizing the second turn of DNA relative to the first turn and reveal the complex flipping and rocking motions of the octamer during unraveling.

[1] B. Gao, G. Arya, & A. R. Tao, "Self-orienting nanocubes for the assembly of plasmonic nanojunctions," *Nat. Nano.*, in press (2012).

[2] I. V. Dobrovolskaia & G. Arya, "Dynamics of forced nucleosome unraveling and role of nonuniform histone-DNA interactions," submitted (2012).

(201f) Phase Behavior of Thermodynamically Small Assemblies of Colloidal Nanoparticles

Ray M. Sehgal[1], Daniel J. Beltran-Villegas[2], Michael A. Bevan[2], Dimitrios Maroudas[3] and David Ford[1], (1)Chemical Engineering, University of Massachusetts Amherst, Amherst, MA, (2)Chemical and Biomolecular Engineering, Johns Hopkins University, Baltimore, MD, (3)Chemical Engineering, University of Massachusetts Amherst, Amhest, MA

Self and directed assembly into highly ordered configurations of small clusters of colloidal nanoparticles is an area of great technological and scientific interest; of particular technological importance is the assembly of such colloidal nanoparticle clusters into crystalline phases. These thermodynamically small systems range from ~10 to ~100 particles in size. The phase behavior of these small systems is inherently different from the typical thermodynamic behavior of bulk phases. Specifically, such colloidal assemblies exhibit coexistence of phases or configurations over a broad range of physical conditions. In this presentation, we report results of a systematic investigation of the phase behavior of colloidal systems, which interact via electrostatic repulsion and depletion attraction; this interparticle potential can be derived experimentally and used for accurate statistical mechanical analyses of the fundamental thermodynamics and kinetics of the colloidal clusters.

In order to study the coexisting configurations present in clusters of colloidal nanoparticles, we conducted Monte Carlo (MC) simulations based on experimentally derived interparticle potentials and employed windowed Monte Carlo-umbrella sampling (MC-US) to generate free-energy landscapes (FELs). We constructed these FELs over a wide range of interparticle interaction strength and cluster size and obtained a comprehensive picture regarding the possible stable configurations of such colloidal assemblies at equilibrium, as well as the phase-transition behavior observed between them. The FELs were generated with respect to the proper number and type of order parameters required to provide an adequate description of the underlying phase-transition dynamics; the determination of the dimensionality of the order-parameter space and the choice of order parameters for the analysis are described and discussed in detail. By analyzing these FELs, we can predict the conditions for the formation of stable crystalline nuclei for the assembly of crystalline phases of colloidal nanoparticles. The predicted conditions for critical crystalline cluster nucleation are used to mark the onset of crystallization for clusters of colloidal nanoparticles. This FEL analysis yields phase-diagram information, which can describe not only the bulk-like phase behavior (i.e., coexistence at a single point) but also the complex phase behavior arising from the system smallness inherent to these colloidal clusters.

(201g) Disentangle Model Differences and Fluctuation Effects in DPD Simulations of Diblock Copolymers

Qiang (David) Wang, **Paramvir Sandhu** and Delian Yang, Department of Chemical and Biological Engineering, Colorado State University, Fort Collins, CO

Disentangle Model Differences and Fluctuation Effects in DPD Simulations of Diblock Copolymers

Paramvir Sandhu, Delian Yang,and Qiang (David) Wang

Department of Chemical and Biological Engineering, Colorado State University, Fort Collins, CO

In the widely used dissipative particle dynamics (DPD) simulations[1], polymers are commonly modeled as discrete Gaussian chains interacting with soft, finite-range repulsions. In the original DPD simulations of microphase separation of diblock copolymer melts by Groot and Madden[2], the simulation results were compared and found to be consistent with the phase diagram for the "standard model" of continuous Gaussian chains with Dirac δ-function interactions obtained from self-consistent field (SCF) calculations. Since SCF theory is a mean-field theory neglecting system fluctuations/correlations while DPD simulations fully incorporate such effects, the model differences are mixed with the fluctuation/correlation effects in their comparison. Here we report the SCF phase diagram for exactly the same model system as used in DPD simulations. Comparing our phase diagram with that for the standard model highlights the effects of chain discretization and finite-range interactions, while comparing our phase diagram with DPD simulation results unambiguously (without any parameter-fitting) reveal the effects of system fluctuations/correlations neglected in the SCF theory.

[1] P. J. Hoogerbrugge and J. M. V. A. Koelman, **Europhys. Lett. 19**, 155 (1992); R. D. Groot and P. B. Warren, **J. Chem. Phys. 107**, 4423 (1997).

[2] R. D. Groot and T. J. Madden, **J. Chem. Phys. 108**, 8713 (1998).

Keywords: Dissipative particle dynamics simulations, self-consistent field calculations, diblock copolymers, fluctuations.

Session to submit: 01A08 Computational Studies of Self-Assembly

(213a) Understanding Hydrophobic Interactions On the Nano-Scale Using Beta-Peptide Oligomers

Claribel Acevedo[1], Derek Ma[1], Samuel H. Gellman[2] and **Nicholas L. Abbott**[3], (1)Chemical and Biological Engineering, University of Wisconsin-Madison, Madison, WI, (2)Department of Chemistry, University of Wisconsin-Madison, Madision, WI, (3)Department of Chemical and Biological Engineering, University of Wisconsin-Madison, Madison, WI

Helical oligomers of beta-peptides represent a particularly promising type of organic nanostructure for investigations of intermolecular forces because (i) the helical secondary structure can be designed to be very stable and because (ii) control of the beta-amino acid sequence can lead to precise patterning of chemical functional groups over the surfaces of the helices. This presentation will describe the use of force spectroscopy to quantify the interactions of single beta-peptide oligomers, each of which display stable and well-defined three-dimensional chemical nanopatterns, with hydrophobic surfaces. Whereas many prior reports of single molecule force measurements of oligo-alpha-peptides and macromolecules exist – the secondary and/or tertiary structures of these species are not preserved during their interactions at interfaces, and thus the three-dimensional chemical patterns that underlie previously reported force measurements are generally not known. By using beta-peptide oligomers that display the same chemical functional groups in stable and distinct spatial nano-patterns, we have demonstrated that it is possible to relate changes in measured forces to changes in three-dimensional chemical nano-patterns. Overall, the results to be described in this presentation will show how beta-peptide oligomers can be used to study intermolecular interactions that arise from precisely defined chemical nano-patterns. A particular focus of the talk will be directed to understanding hydrophobic interactions, thus providing insights into the mechanisms through which changes in chemical patterns presented by organic nanoscopic objects can affect their self-assembly behavior in aqueous environments.

(213b) Particle-Polymer Interaction in Hydrogels

Paul Takhistov, School Env.Biolog.Sciences, Rutgers, The State University of New Jersey, New Brunswick, NJ and Phong Huynh, Rutgers University, New Brunswick, NJ

The presence active pharmaceutical ingredient (API) in biopolymer network impacts the film formation and plays role in control release. One of the most important factors that affect the formation of polymer matrix is the interaction of polymer – surfactant – API particles. The film formation of several biopolymers loaded API has been investigated. Dynamic light scattering technique is applied to analyze particle size of API particles and their zetapotential in different film precursor solutions. SEM is used to reveal the arrangement of API particle in the film matrix. The interaction of polymer – polymer and polymer – surfactant in these mixtures is analyzed by surface tension and rheology measurements. A sensitive gravimetric system is exploited to detail the drying kinetic of those considered mixtures. Our results show that, the surfactant plays a role in adsorption behavior of polymer on API particles when API is loaded in a polymeric film precursor solution while polymer-polymer interaction is key factor in the formation of binary polymeric network

(213c) Surface Chemistry and Van Der Waals Forces of the Air-Water and Oil-Water Interface: Homo- and Hetero- Interactions Between Drops and Bubbles

Rico Tabor, Hannah Lockie, Chu Wu, D.Y.C. Chan, Franz Grieser and **Raymond R. Dagastine**, the Particulate Fluids Processing Centre, The University of Melbourne, Parkville, Australia

Understanding the interactions between droplets of oil and gas bubbles is central to tuning the desired properties of foodstuffs such as ice cream and mousse, cosmetics and pharmaceuticals, and in mineral flotation and separation. Fundamentally, the interactions between dissimilar fluid interfaces (drops and bubbles) or fluid interface with a solid are interesting problems. Dissimilar fluid interfaces may charge and deform to different extents. In addition, both dissimilar fluid interfaces and fluid interfaces interactions with solids can be intrinsically stable based on material choice not surface chemistry. We have used the atomic force microscope (AFM) to analyse collisions between bubbles and oil droplets - of around 100 microns diameter - in surfactant-free aqueous solutions [1] or a between a bubble and flat surfaces[2,3]. Homo- (bubble-bubble and drop-drop), hetero- (bubble- drop, bubble-solid) interactions were examined to show the role of colloidal and hydrodynamic forces that, together with interfacial deformations, dictate the stability in these systems.

Through carefully chosen experiments and quantitative modelling, it was possible determine important properties such as surface potentials and the Hamaker functions for the oil drops, air bubbles and surfaces. Between like pairs of drops or bubbles, electrical double-layer forces were predictably repulsive, and Van der Waals forces were attractive. However, it was found that within a small pH-range, double-layer forces between dissimilar bodies could be rendered attractive. Similarly, material choice allowed the sign of the Van der Waals force to be selected to be either positive or negative, and ionic strength also has a surprising role via screening when Van der Waals forces are small. Collisions at high speed show that local deformation of the bodies is controlled by their internal Laplace pressures, defining the profile of liquid film trapped between them.

These results give new insight into how coalescence and interactions can be controlled for fluid handling in microfluidics, as well as dispersion and formulation science. Material choice, pH and ionic strength can be used to enhance or avoid coalescence in hetero-bodied (dissimilar materials) systems, suggesting new, facile pathways for self-assembly at to add functionality to colloidal systems.

(1) Tabor, R. F.; Manica, R.; Chan, D. Y. C.; Grieser, F.; Dagastine, R. R., Repulsive van der waals forces in soft matter: Why bubbles do not stick to walls. *Physical Review Letters* **2011,** 106, 064501.

(2) Tabor, R. F.; Morfa, A. J.; Grieser, F.; Chan, D. Y. C.; Dagastine, R. R., Effect of gold oxide in measurements of colloidal force. *Langmuir* **2011,** 27, 6026-6030.

(3) Tabor, R. F.; Wu, C.; Lockie, H.; Manica, R.; Chan, D. Y. C.; Grieser, F.; Dagastine, R. R., Homo- and hetero-interactions between air bubbles and oil droplets measured by atomic force microscopy. *Soft Matter* **2011,** 7, 8977-8983.

(213d) Application of the SAFT-VR Density Functional Theory to the Prediction of Interfacial Properties of Mixtures of Relevance to Reservoir Engineering

Felix Llovell[1], Niall Mac Dowell[2], Felipe J. Blas[3], Amparo Galindo[4] and George Jackson[4], (1)MATGAS Research Center, Bellaterra, Barcelona, Spain, (2)Chemical Engineering, Imperial College London, London, United Kingdom, (3)Departamento de Física Aplicada, University of Huelva, Huelva, Spain, (4)Centre for Process Systems Engineering, Dept of Chemical Engineering, Imperial College London, London, United Kingdom

Interfacial phenomena of complex mixtures play a crucial role in a number of industrial applications. In particular, they are of key importance in oil-reservoir engineering. The exploitation of oil reserves involves several complicated stages of extraction, where the presence of vapour-liquid and liquid-liquid equilibria have a marked effect on the operating conditions. The optimization of these conditions not only depends on the bulk phase behaviour, but also on the interfacial properties such as the interfacial tension between the phases, the interface profiles, the interfacial thickness, and the surface adsorption (the relative enrichment of one compound over another).

The SAFT-VR DFT formalism, which has recently been extended to mixtures [F. Llovell et al., J. Chem. Phys. 133, 024704 (2010)], is applied to carefully selected binary mixtures that are representative of the types of system relevant to enhanced oil recovery and carbon dioxide sequestration processes. The approach is based on the statistical associating fluid theory for attractive potentials of variable range (SAFT-VR) and includes a specific DFT treatment, where the short range interactions are treated at the local level and the long-range interactions are treated with an appropriate perturbative contribution. The correlations between the fluid particles are taken into account with an average correlation function following the ideas of Toxvaerd [S. Toxvaerd, J. Chem. Phys. 64, 2863 (1976)]. Binary mixtures of short and long chain n-alkanes, carbon dioxide + n-alkanes, water + carbon dioxide and water + n-alkanes are investigated over wide ranges of temperature and pressure. Molecular parameters for each pure compound are estimated from the bulk vapour-liquid equilibrium data alone, and the unlike binary dispersive energy parameter is adjusted to provide the best overall description of the phase equilibrium properties of the mixtures. No additional parameters are required in the implementation of the SAFT-VR DFT theory. Predicted vapour-liquid and liquid-liquid interfacial tensions are found to be in quantitative agreement with experimental data in most cases. Density profiles which are characterised by noteworthy adsorption features are also presented. The fraction of free hydrogen bonds in the neighbourhood of the interface is also reported for the aqueous mixtures studied.

(213e) Local Equilibrium of the Gibbs Interface in Two-Phase Systems

Thierry Savin, Marco Schweizer and Hans Christian Öttinger, ETH Zürich, Zürich, Switzerland

The classical approach to equilibrium interfacial thermodynamics in phase coexistence describes the interface as a separate two-dimensional thermodynamic system. Excess densities then characterize the autonomous interface. Gibbs' original formulation relies however on the uniformity of intensive variables throughout the system. Therefore, it cannot be directly applied to nonequilibrium situations such as evaporation/condensation processes or heterogeneous catalysis, where jumps in temperature and chemical potential across the interface can occur. We present a conceptually clear formulation of local equilibrium for interfaces. In particular, we gain new insights by expressing, in terms of gauge transformations, the ambiguity in locating the interface. The gauge invariance of thermodynamic properties appears to be equivalent to conditions for jumps of bulk densities across the interface, and we are notably able to rigorously generalize the Clapeyron relationships to nonequilibrium situations. We support our theoretical predictions by performing stringent tests with nonequilibrium molecular dynamics simulations of a coexisting liquid/vapor Lennard-Jones fluid.

(213f) Shear Free and Blotless Cryo-TEM Imaging: A New Method for Probing Early Evolution of Nanostructures

Jinkee Lee[1], Amitesh Saha[2], Jens Rieger[3], Arijit Bose[2] and Anubhav Tripathi[4], (1)School of Mechanical Engineering, Sungkyunkwan University, Suwon, South Korea, (2)Department of Chemical Engineering, University of Rhode Island, Kingston, RI, (3)BASF, Ludwigshafen, Germany, (4)School of Engineering, Brown University, Providence, RI

Cryogenic transmission electron microscopy (cryo-TEM) is a powerful method to image native state morphologies of nanoscale soft and hard objects suspended in solvents. Sample preparation is a critical step towards producing images at length and time scales of interest. We demonstrate a nearly shear-free sample thinning method which simultaneously allows imaging of evolving nanostructures at sub-second time scales. This device breaks the trade-off between high shear and short time scales typical in current cryo-TEM sample preparation methods. We demonstrate the low-shear feature of the new method by imaging worm-like micelles, showing an interconnected network, in contrast to the traditional sample preparation method which shows aligned micelles at similar time points. The time resolution of this method is demonstrated by imaging morphologies of calcium carbonate (formed through the reaction of calcium chloride with sodium carbonate) at sub-second time scales, capturing its evolution from an amorphous to a crystalline state. The impact of hyperbranched polyglycerol additives on the amorphous to crystalline transition in calcium carbonate at short times is examined. Early images at low shear provide unique fundamental insights into mechanisms of nanostructure evolution, thus offering a new paradigm for research in materials sciences, soft matter and biological sciences.

(213g) Role of Draining Channels in Hydrodynamic Interactions

Rohini Gupta and **Joelle Frechette**, Chemical and Biomolecular Engineering, Johns Hopkins University, Baltimore, MD

The adhesion and locomotion mechanisms employed by tree frogs for flooded conditions offer the ultimate solution for the need of strong, reversible, *reusable, tunable*, and *water tolerant* adhesives. The presence of structured toe-pads (hexagonal epithelial cells separated by channels) has been proposed to facilitate drainage of fluid and reduce hydrodynamic repulsion during approach. These channels may remain closed during retraction for hydrodynamic adhesion and open for pull-off, both of which require active/passive control of toe-pad deformation.

We present results from our investigation of the role of draining channels on normal hydrodynamic interactions in the absence of elastohydrodynamics (rigid surfaces). The surface force apparatus is employed to measure hydrodynamic forces between a smooth silver film and a structured SU-8 surface driven towards or away from each other at a constant drive velocity. The structured SU-8 comprises of hcp array of cylindrical posts. The hydrodynamic forces are analyzed within the framework of Reynolds' continuum approach in the lubrication limit for smooth surfaces in cross-cylinder geometry.

Reduction in repulsion upon approach with respect to that predicted by Reynolds' approach in the lubrication limit is analyzed using a scaling argument based on the geometric parameters associated with the structures (diameter of posts, channel width and depth, and surface coverage), and the range of hydrodynamic interactions, which depends on the fluid viscosity and drive velocity. Viscosity-dominated drainage of fluid in the presence of a network of channels intrinsically results in competition between drainage from in contact region and drainage through the network of channels. The scaling argument can, then, be used to predict regimes of preferential drainage of fluid through network of channels vs drainage of fluid from in contact region for a single post or the cross-cylinders as a whole as the surface separation changes. These predicted regimes are compared with those observed experimentally for different combinations of diameter of posts, channel width and depth, and surface coverage as a function of drive velocity to highlight the universal nature of the scaling argument employed.

(218a) Using Computational Methods At Boeing to Solve Emerging Material Issues

Kady Gregersen[1], Andrea R. Browning[2], Samuel Tucker[3] and Stephen Christensen[2], (1)The Boeing Company, seattle, WA, (2)The Boeing Company, Seattle, WA, (3)The Boeing Company

Boeing has been utilizing numerical modeling methods such as FEM and CFD for many years. More recently Boeing has begun incorporating atomistic and quantum level simulations for standalone simulations and as inputs in higher level modeling. Computational methods are being used to support emerging issues with existing material systems as well as provide a greater fundamental understanding of their behavior. The development of new material systems and the capability to provide support for existing materials demand a versatile approach that quickly delivers innovative solutions at a reduced cost.

Currently, atomistic MD is used to predict the thermal and mechanical responses of potential polymer candidates for use as sealants, coatings, and matrices in fiber-reinforced composites. Molecular dynamics is also used to study transport properties of materials such as diffusivity and viscosity, which can be evaluated with equilibrium MD using Green-Kubo relationships. Finally, quantum methods are used to predict materials' resistance to UV degradation and to study the reaction mechanisms of epoxy-amine reactions for new polymer development.

The need for these methods is fast becoming evident with an ever greater demand for reduced development cycles and material solutions. Methods are being developed that are defining a critical direction for the discovery of high performance materials to meet the needs of advanced aircraft structures. Materials discovery augmented with computational methods can yield significant reduction in development cycle time and cost.

(218b) Separation of Ethanol and Water: Molecular-Level Insights On Extraction and Adsorption Systems

J. Ilja Siepmann[1,2], Samuel J. Keasler[1] and Peng Bai[2], (1)Chemistry, University of Minnesota, Minneapolis, MN, (2)Chemical Engineering and Materials Science, University of Minnesota, Minneapolis, MN

The high energy cost of producing ethanol from fermentation broths is one of the biggest obstacles to the economical use of bio-renewable fuels for transportation and other applications. The separation of ethanol from a concentrated aqueous solution, usually by distillation, is one of the most energy intensive steps, motivating the development of more efficient alternatives, including liquid-liquid solvent extraction and adsorption in zeolites. Coupled-decoupled configurational-bias Monte Carlo (CBMC) simulations in the isobaric-isothermal version of the Gibbs ensemble (GE) using the TraPPE (transferable potentials for phase equilibria) force field have been performed to screen a series of long-chain alcohols and diols for their ethanol capacity and for their selectivity for ethanol over water. The distribution of hydrogen-bonded aggregates has been analyzed and used to identify several key features of these aggregates that promote high selectivity or capacity, as well as characteristics of the extraction solvent that enhance the formation of these favorable types of aggregates. The CBMC-GE methodology is also applied to investigate the selective adsorption of ethanol in zeolites from aqueous solutions over a wide range of composition and two temperatures.

(218c) Solvation Free Energies and 1-Octanol/Water Partition Coefficients of Nitroaromatic Compounds Using Molecular Simulation Techniques

Alauddin Ahmed and Stanley I. Sandler, Department of Chemical and Biomolecular Engineering, University of Delaware, Newark, DE

Nitroaromatic compounds (NACs) have long been used as explosives, solvents, herbicides, insecticides, dyes and polyurethane foams in industry. These compounds are environmental contaminants and xenobiotics, and are largely distributed water mediated transportation. The extent of solvation of the NACs affects the level of contamination in soil, sediment, and ground water. The properly model environmental impact of these nitroaromatic compounds, the solubility and most importantly the solvation free energies in water and in 1-octanol leading to the octanol-water partition coefficients are key components. We have calculated solvation free energies of a set of nitroaromatic compounds that includes multifunctional nitro groups using expanded ensemble molecular dynamics simulation algorithm. TIP3P and SPC-Fw water models have been used as the explicit solvent models in our simulations, and it has been found that the choice of the water model has a significant effect on the Gibbs solvation free energy simulations in water. GAFF, CGenFF, OPLS-AA and TraPPE-UA force fields parameters have been used to model the nitroaromatic compounds either with the pre-optimized partial charges for atom types, with Chelpg charges obtained from quantum calculations or with AM1-BCC charges estimated from a semi-empirical method. The GAFF and CGenFF force field parameters with Chelpg charges are found to reproduce experimental data relatively well with TIP3P water model. OPLS-AA force field performs equally well with the GAFF/CGenFF force fields, even with the pre-optimized partial charges assigned with atom types. The performance of different force fields were analyzed based on the number of nitro groups and it was found that most of the force fields studied here are somewhat limited in their ability to accurately predict the solvation behavior of NACs with multiple nitro groups.

(218d) Molecular Behavior of Water Confined in TiO2 Nano-Slits Using Classical, Reactive and Ab Initio Molecular Dynamics

Ming-Jie Wei[1] and Peter T. Cummings[1,2], (1)Department of Chemical and Biomolecular Engineering, Vanderbilt University, Nashville, TN, (2)Center of Nanophase Materials Sciences, Oak Ridge National Laboratory, Oak Ridge, TN

Titanium dioxide (TiO_2) has drawn great attention from scientists and engineers in the past decade because it is expected to play an important role in helping solve many serious environmental and pollution challenges. Essentially all applications of TiO_2 involve contact between water or aqueous solution and TiO_2 surfaces. Mineral-aqueous electrolyte solution interfaces are of common occurrence in natural and industrial environments. In the interfacial region, the solution properties and the crystal structure both deviate from the characteristics of the bulk phases, and an electrical double layer (EDL) is formed in the interfacial region by absorption of ionic species and countercharges from the aqueous solution in the vicinity of the interface. The EDL exerts a profound influence on the stability and transport properties of colloids, the mobilities of trace elements in aqueous solutions in contact with solid phases, nutrient bioavailability and bacterial attachment, corrosion, crystal nucleation and growth, as well as abundant natural and industrial processes. Thus, for many applications, it is necessary to understand the dynamics of water and electrolytes in this interfacial region.

Since the dynamics of water and electrolytes confined in nanopores are very different from bulk, these dynamic properties need to be investigated directly. Molecular dynamics (MD) simulations have been used to observe the dynamics of fluid confined in nanopores. Our previous work[1-5], using classical MD simulation, could not describe reactions in the interfacial region. In this work, we report MD simulations using a reactive force field (ReaxFF) and *ab initio* MD (CPMD) to compare with our previous work, with the goal of obtaining more accurate dynamics for water confined in nano-slit pores, and to understand the role of bond vibration and bond breaking (reactivity) in the dynamics.

References:

(1) Předota, M.; Bandura, A. V.; Cummings, P. T.; Kubicki, J. D.; Wesolowski, D. J.; Chialvo, A. A.; Machesky, M. L. *The Journal of Physical Chemistry B* **2004**, *108*, 12049.

(2) Mamontov, E.; Vlcek, L.; Wesolowski, D. J.; Cummings, P. T.; Wang, W.; Anovitz, L. M.; Rosenqvist, J.; Brown, C. M.; Garcia Sakai, V. *The Journal of Physical Chemistry C* **2007**, *111*, 4328.

(3) Mamontov, E.; Wesolowski, D. J.; Vlcek, L.; Cummings, P. T.; Rosenqvist, J.; Wang, W.; Cole, D. R. *The Journal of Physical Chemistry C* **2008**, *112*, 12334.

(4) Mamontov, E.; Vlcek, L.; Wesolowski, D. J.; Cummings, P. T.; Rosenqvist, J.; Wang, W.; Cole, D. R.; Anovitz, L. M.; Gasparovic, G. *Physical Review E* **2009**, *79*, 051504.

(5) Wei, M.-J.; Zhou, J.; Lu, X.; Zhu, Y.; Liu, W.; Lu, L.; Zhang, L. *Fluid Phase Equilibria* **2011**, *302*, 316.

(218e) Fick Diffusion Coefficients In Ternary Liquids From Equilibrium Molecular Dynamics

Xin Liu[1], Erin McGarrity[2], Ana Martin Calvo[3], Sofia Calero[4], Andre Bardow[1] and Thijs JH Vlugt[5], (1)RWTH Aachen University, Aachen, Germany, (2)Delft University of Technology, Delft, Netherlands, (3)University Pablo de Olavide, Seville, Spain, (4)Department of Physical, Chemical, and Natural Systems, University Pablo de Olavide, Seville, Spain, (5)Process & Energy, Delft University of Technology, Delft, Netherlands

An approach for computing ternary Fick diffusivities directly from equilibrium Molecular Dynamics (MD) simulations is presented and demonstrated for a liquid mixture of chloroform-acetone-methanol. In our approach, Fick diffusivities are obtained from Maxwell-Stefan (MS) diffusivities and the so-called matrix of thermodynamic factors. MS diffusivities describe the friction between different molecular species and can be directly computed from equilibrium MD simulations. The thermodynamic factor is the concentration derivative of the activity coefficient describing the deviation from ideal mixing behaviour. It is important to note that all mutual diffusion experiments measure Fick diffusion coefficients, while previous equilibrium molecular dynamics simulation did only provide MS diffusivities. The required thermodynamic factor to convert MS into Fick diffusivities and vice versa, however, has been difficult to extract from both simulations and experiments leaving a gap between theory and applications. Here, we extend an efficient computing scheme for the thermodynamic factor recently proposed by the authors and co-workers [1] for ternary complex mixtures. The novel computing scheme extracts the thermodynamic factor from density fluctuations in small subsystems embedded in a larger simulation box. The computed thermodynamic factor is in excellent agreement with experimental VLE data as well as calculations using the COSMO-SAC method. Our approach provides an efficient route to obtain multicomponent diffusion coefficients in liquids based on a consistent molecular picture and therefore bridges the gap between theory and experiment.

[1] X. Liu, S.K. Schnell, J-M Simon, D. Bedeaux, S. Kjelstrup, A. Bardow, T.J.H. Vlugt, J. Phys. Chem. B, 2011, 115, 12921-12929.

(218f) Using Molecular Simulations to Determine the Effective Permeabilities of Mixed Matrix Membranes of Polydimethylsiloxane with Increased Zeolite Loadings of Silicalite-1

Paul F. Harten, Sustainable Technology Division, U.S. Environmental Protection Agency, Cincinnati, OH

In this research, the effective permeabilities of a mixed matrix membrane of polydimethylsiloxane and silicalite-1 to ethanol and water are of principle interest; with the ultimate interest being how much of the zeolite should be loaded into the polymeric membrane to achieve the most functional membrane separation of ethanol from water. This efficient membrane separation of ethanol from water in a fermentation broth can significantly reduce the energy cost associated with biofuel production.

The effective permeability of a mixed matrix membrane to a penetrant is approximated using Maxwell's equation for electrical conductivity through heterogeneous media. For this approximation, it is necessary to have the permeability of both components of the mixed matrix membrane and their volume fractions.

The permeability of each membrane component can be expressed as the product of a diffusivity coefficient and a solubility coefficient. Diffusivity coefficients are determined using molecular dynamics and examining how the mean squared displacement of an ethanol or water atom within the bulk material changes with the length of the time interval. The solubility coefficients are calculated by analyzing the absorption of ethanol and water atoms by the bulk material using the Widom insertion method within Grand Canonical Monte Carlo simulations. Finally, the results of these approximations are compared with available experimental data.

(218g) Ethane/Ethylene Separation by Simulated Moving Bed: From Molecular Simulations to Process Design

Miguel A. Granato, Department of Chemical Engineering, LSRE - Laboratory of Separation and Reaction Engineering - Associate Laboratory LSRE/LCM, Faculdade de Engenharia - Universidade do Porto, Porto, Portugal, Vanessa F. D. Martins, Chemical Engineering, LSRE - Laboratory of Separation and Reaction Engineering – Associate Laboratory LSRE/LCM, Faculdade de Engenharia - Universidade do Porto, Porto, Portugal, João C. dos Santos, Chemical Engineering, LSRE - Laboratory of Separation and Reaction Engineering - Associate Laboratory LSRE/LCM, Faculdade de Engenharia - Universidade do Porto, Porto, Portugal, Miguel Jorge, Departamento de Engenharia Química, Universidade do Porto, Porto, Portugal and Alirio E. Rodrigues, Laboratory of Separation and Reaction Engineering, University of Porto - Faculty of Engineering, Porto, Portugal

Ethylene (ethene) is the major worldwide industrial feedstock. Separation of paraffin/olefin mixtures, in particular ethane/ethylene and propane/propylene, involves huge distillation columns operated at a very high reflux ratio, one of the most energy consuming processes in chemical industry. Alternative technologies are Simulated Moving Bed (SMB) and Vacuum Swing Adsorption (VSA). In the perspective of ethane/ethylene separation by SMB, the choice of a suitable desorbent is a key issue. This study encompasses molecular simulation calculations for determination of adsorption parameters, whose results will be then used in a mathematical model for evaluating the performance of an SMB unit. Propane is proposed as a desorbent for separation of the mixture by data-driven simulations, in which calculations are made to predict properties with little or no input from experiment. This work is entirely done in silico, by using available force field parameters for the molecular simulations part, and reliable mathematical models for the SMB part. Experimental data are solely used for comparison with the molecular simulation results, which are subsequently expanded to calculate adsorption properties for separating the mixtures, without further experimental work.

(224a) Hydrodynamic Instabilities of Chemotactic Bacteria

Donald L. Koch, Chemical and Biomolecular Engineering, Cornell University, Ithaca, NY and T.V. Kasyap, School of Chemical and Biomolecular Engineering, Cornell University, Ithaca, NY

Over the past decade, experimental studies of swimming micro-organisms and numerical simulations of hydrodynamically interacting self-propelled particles have shown that such living fluids exhibit collective dynamical motions on length scales large compared with the size of an individual swimmer even in the absence of any hydrodynamic or chemical gradients. A continuum theory showed that living fluids consisting of pushers such as bacteria are subject to a hydrodynamic instability that involves the coupling of the shear motion caused by the active stress produced by the swimmers and their orientation distribution. Recent experiments, however, suggest that even more vigorous convective motions may arise in bacterial suspensions subject to chemical gradients. In this talk we discuss a new mechanism of instability that arises from the coupling of the bacteria concentration and the competition between chemotactic driven and the convective motions produced by an active stress. The stability of a suspension of chemotactic bacteria confined in an infinitely long channel and subjected to a stationary, linear chemo-attractant gradient is investigated. While swimming, individual bacteria exert a force dipole on the fluid which at the continuum level leads to a stress depending upon the bacterial orientation and number density fields. The presence of the attractant gradient causes bacteria to tumble less frequently when swimming along the gradient leading to a mean orientation and a non-zero chemotactic drift velocity. A balance of the fluxes due to chemotaxis and the random run-tumble motion of bacteria yields an exponentially varying number density profile across the channel in the base state. The associated bacterial stress field is also exponentially varying and is normal. This spatially non-uniform base state is unstable to fluctuations in the bacterial concentration exceeds a critical value determined by a Peclet number defined as Pe = U_0 H/D where U_0 is the chemotactic velocity, D the run-and-tumble diffusivity and H the channel thickness. The instability resulting from the coupling between the active stress driven fluid flow and the bacterial concentration manifests as rectangular convection patterns driven by spanwise bacterial clusters.

(224b) Occlusive Flow of a Red Blood Cell

Thierry Savin, ETH Zürich, Zürich, Switzerland and L. Mahadevan, Harvard University, Cambridge, MA

When red blood cells (RBCs) move through narrow capillaries in the microcirculation, they deform as they flow. In pathophysiological processes such as sickle cell disease and severe malaria, RBC motion and flow are severely restricted. This can lead to vaso-occlusive crises with dramatic consequences. To understand the dynamics close to the threshold of occlusion, we used a combination of experiment and theory to study the motion of a single swollen RBC through a narrow glass capillary of varying inner diameter. By tracking the motion of the squeezed cell as it is driven by a controlled pressure drop, we measure the RBC velocity as a function of the pressure gradient as well as the local capillary diameter, and find the velocity of the cell varies with pressure head as a power-law that depends upon the lubrication gap between the cell and the capillary wall, consistent with a simple elasto-hydrodynamic model. Our study highlights the role of lateral confinement in the nearly arrested pressure-driven flow of soft confined objects and allows us to quantify the increase in the effective viscosity associated with this regime.

(224c) The Continuous Chromatographic Separation of Molecules/Particles Using Optical Electric Fields

Nicolas Alvarez[1], Claus Jeppesen[2], Kresten Yvind[2], Iwao Teraoka[3], N. Asger Mortensen[2] and Ole Hassager[1], (1)Department of Chemical and Biochemical Engineering, Technical University of Denmark, Kgs. Lyngby, Denmark, (2)Department of Photonics Engineering, Technical University of Denmark, Kgs. Lyngby, Denmark, (3)Chemical and Biological Science, Polytechnic Institute of New York University, Brooklyn, NY

We introduce a new field-flow fractionation technique, whereby molecules are separated based on their differential interaction with optical electric fields, i.e. electric fields with frequencies in the visible and near-infrared range. When solute particles are introduced to electric field gradients they are polarized based on their size and refractive index. The resulting attractive or repulsive polarization potential influences a solute particle's lateral position with respect to the gradient in the electric field. By coupling well described optical electric fields with a non-uniform flow, we demonstrate how such potentials can be used to continuously separate nanometer scale solute particles in a flowing two dimensional microchannel. Comparisons are made between theoretical axially uniform optical fields and numerical simulations considering optical electric fields with finite width. We discuss the impact of the governing dimensionless groups on the resolution of separation and the dispersion of solute molecules after separation. An experimental apparatus is introduced and comparisons are made between theoretical and experimental results.

(224d) Micro and Nanoparticle Trapping and Manipulation with Fluid Flow

Melikhan Tanyeri and **Charles M. Schroeder**, Chemical and Biomolecular Engineering, University of Illinois at Urbana-Champaign, Urbana, IL

The ability to trap and control the motion of single particles has led to key advances in science and engineering. Particle trapping is a pervasive technology that underlies nanoscience and biological analysis. In this presentation, I will introduce two new microfluidic methods based on the sole action of hydrodynamic forces for the confinement and manipulation of micro- and nanoscale particles in free-solution. Both trapping methods utilize an automated feedback-control mechanism for active control of fluid flow within microchannels using an integrated on-chip metering valve. The first method is the "**hydrodynamic trap**" where a target particle is confined at a microchannel junction by actively controlling the stagnation point of a planar extensional flow field This method enables trapping, 2-D manipulation, stretching and sorting of objects ranging from single molecules to individual cells using the sole action of fluid flow. The second method is the "**microfluidic Wheatstone bridge**" which facilitates rapid sampling of particles, cells and droplets in continuous flow for monitoring and analysis. This dynamic sampling method allows for "on-demand" trapping of multiple particles at the "bridge" channel without stopping or diverting the main sample stream. Currently, we are pushing the limits of these methods for confining small nanoparticles (< 50 nm) and large macromolecules. We anticipate that these microfluidic-based techniques will enable new scientific studies in the fields of catalysis, colloidal science, polymer dynamics, single molecule studies, systems biology, and cellular mechanics.

1. M. Tanyeri, E. M. Johnson-Chavarria and C. M. Schroeder, Applied Physics Letters (2010) **96**, 224101.

2. M. Tanyeri, M. Ranka, N. Sittipolkul and C. M. Schroeder, Lab on a Chip (2011) **11**, 1786.

(224e) Mapping Chaotic Flow States in Microscale Rayleigh-Bénard Convection to Identify Regimes of Accelerated DNA Replication

Aashish Priye and Victor M. Ugaz, Artie McFerrin Department of Chemical Engineering, Texas A&M University, College Station, TX

We have previously shown that chaotic advection in micro-scale flow geometries can be harnessed to greatly enhance the rate of thermally activated biochemical reactions (e.g., the polymerase chain reaction (PCR)). Here we present a 3D analysis of coupled flow and reaction that provides a time-resolved mapping of the DNA replication process as it unfolds. Chemical reaction kinetics are incorporated into the 3D flow model, enabling us to obtain a measure of global DNA replication rate within the entire reactor volume, expressed in terms of a characteristic doubling time. Mass action kinetics were assumed for the denaturing, annealing, and extension steps. Annealing and extension were modeled as bimolecular reactions to capture the contribution of the primers and dNTPs. Reversibility of denaturing and annealing reactions was also taken into account. Reaction models of varying complexity were compared via addition of side reactions. individual stream traces were examined for different geometries, expressed in terms of their aspect ratio (height/diameter, *h/d*). The residence time of a fluid element in a particular reaction zone and the number of times that fluid elements pass through the zones were determined for 300 randomly chosen stream traces yielding a distribution of residence times and frequencies of ingress in each of the reaction zones. At larger values of *h/d* we find that individual fluid elements on average spend more time in each temperature zone. But the number of times a fluid element passes through each zone is much larger in wider geometries (lower *h/d*), owing to the chaotic characteristics of the flow field. Since our experiments show much faster DNA replication at low *h/d*, the number of excursions into each reaction zone appears to play a more important role than the transit time through each temperature range. Finally a reaction map was constructed to determine the cylindrical cell geometries and other parameters that should be used to enhance the efficiency of microscale Rayleigh Bénard convection PCR. These results highlight the importance of the flow field's 3D characteristics.

(224f) Electrokinetically Driven Micro-Turbulence in Microfluidics with Re in the Order of 1

Guiren Wang[1], Fang Yang[2] and Wei Zhao[1], (1)University of South Carolina, Columbia, SC, (2)Department of Mechanical Engineering & Biomedical Engineering Program, University of South Carolina, Columbia, SC

We report an electrokinetically driven micro-turbulent flow phenomenon in a microchannel, where the Reynolds number (Re) based on bulk flow and channel width is in the order of 10. As we know the critical Re in a channel flow is about 2000 and a pressure driven flow with Re in the order of 10 is definitely laminar, hence, the fluid mixing is very flow. Although there have been electrokinetical micromixers published that can increase mixing, the flows are believed not to be turbulence, but chaotic. Here we design a novel electrokinetically forced pressure driven flow with Re in the order of 10 to achieve turbulence like flow. To confirm that the flow is turbulence, Laser Induced Fluorescence Photobleaching Anemometer (LIFPA) is used to measure the corresponding flow velocity features. LIFPA enables ultrahigh temporal and spatial resolution measurement. We report the measured flow features: high dissipation and diffusion rate, multiscale continuous power spectra and 3-D flow. The results open a new view of turbulence at microscale.

(224g) Mixing Enhancement in Three-Dimensional Helical Microchannels

Shahab Shojaei-Zadeh, Rutgers, The State University of New Jersey, Piscataway, NJ and **Michael C. Fechtmann**, Mechanical and Aerospace Engineering, Rutgers, The State University of New Jersey, Piscataway, NJ

We present experimental and numerical results on mixing performance for the flow of two miscible liquids in a three-dimensional helical microchannel. A novel yet straightforward method is developed to fabricate 3-D helical microchannels with different pitch length in a polydimethylsiloxane (PDMS) mold. This geometry provides a constant Dean number flow, due to the constant radius of curvature of the helix, which differentiate it from the existing planar spiral microchannels. The mixing performance is characterized, using digital image processing techniques, based on the mixing length which is defined as the stream-wise distance in the microchannel where the mixing has reached 90% of the full intensity. Experimental results show a minimum mixing length vs. Reynolds number as a common feature for all the microchannels of similar diameter and different pitch length. Increasing the Reynolds number beyond this minimum mixing length adversely influences the mixing performance by increasing the measured mixing length due possibly to the decrease in the residence time of the two streams in the microchannel. Numerical simulations are presented to support the experimental observations.

(224h) Two Phase Microfluidics with Molten Polymers

Shu-che Peng[1], Shailesh Nagarkar[1], Justin Lowen[1], Alex Boardman[1], Steven D. Hudson[2], Kalman Migler[2], Doyoung Moon[3] and **Sachin Velankar**[4], (1)University of Pittsburgh, (2)Polymers Division, National Institute of Standards and Technology, Gaithersburg, MD, (3)National Institute of Standards and Technology, (4)University of Pittsburgh, Pittsburgh, PA

Drop-based methods for fabricating complex particles in microfluidic devices have been studied heavily over the past several years. Typically, a two-phase oil/water flow is set up in a microfluidic device under conditions where capillary breakup leads to controlled generation of drops. Subsequent solidification of the drops yields particles. These approaches suffer from material limitations: they rely on fluids that can be rapidly solidified, often by UV crosslinking. We seek to overcome these material limitations, specifically we aim to unite microfluidics with polymer processing to develop a process that can fabricate polymer particles out of virtually any thermoplastic polymer.

Handling molten polymers in microfluidic devices requires entirely new concepts in device design. A new high-temperature microfluidic platform based on micromachined metal foils has been developed. The flow of the molten polymer is driven by gas pressure rather than positive displacement pumps. This talk will show that – despite the high viscosity and low interfacial tension of molten polymers – capillary breakup is still a viable route to form drops in microfluidic devices. However the mode of drop breakup is quite different in polymeric systems as compared to oil/water systems. Furthermore, the parameter space (i.e. pressure ranges) in which drops can be generated is relatively narrow, and drop sizes are highly sensitive to the operating conditions. A simple "circuit" model is developed which successfully explains these latter observations and points to operating guidelines for using pressure-driven flow in two-phase microfluidics. In summary, the vast developments in drop-based microfluidics can be transplanted to molten polymers, thus permitting the fabriction of complex particles out of a variety of thermoplastic polymers.

(224i) Vector Separation of Suspended Particles Using an Array of Slanted Open Cavities

Jorge A. Bernate[1], Chengxun Liu[2], Liesbet Lagae[2], Konstantinos Konstantopoulos[1] and German Drazer[1],
(1)Chemical and Biomolecular Engineering, Johns Hopkins University, Baltimore, MD, (2)IMEC, Leuven, Belgium

We present a microfluidic device that takes advantage of the flow characteristics in a channel in which the bottom surface is patterned with slanted ridges that create open cavities, to passively and continuously fractionate suspended particles based on their size and settling velocity. We exploit the flow that is created along the cavities to induce the vector separation of suspended particles with dimensions comparable to the height of the ridges and that move in the vicinity of the patterned surface. The width and height of the ridges are much smaller than those of the main channel to reduce the effect of recirculation in the bulk. We shall show that two regimes can be distinguished depending on the settling velocity of the particles compared to their average velocity across the cavities imposed by the fluid flow. When these velocities are similar, heavier particles settle deeper into the open cavities and exhibit larger displacements along these cavities. On the other hand, if sedimentation is negligible, smaller particles are advected by the flow deeper into the cavities and therefore deflect more than larger particles. First, we illustrate the flow characteristics by numerically solving the particle-free flow. We then present experiments separating silica and polystyrene beads of the same and different size. We also show the potential of this method to separate cells showing that different blood components can be deflected to a different extent. Finally, we discuss the potential versatility of this platform using force fields other than gravity, e.g., electric, dielectrophoretic, and magnetic, to manipulate the settling velocity of the particles. In particular, we show the enrichment of rare blood cells by depleting magnetically labeled leukocytes.

(235a) Comparative Dynamics of Soft Colloidal Glass, Low Molecular Glass Former, and High Molecular Glass Former

H. Henning Winter, Division of Chemical, Bioengineering, Environmental and Transport Systems, National Science Foundation, Arlington, VA

After many years of modeling the liquid-to-solid transition for physical gels and soft glasses, the difference between these two material classes is still in dispute. This is where experimental rheology can help, since pronounced rheological differences between gels and glasses were found in the *immediate approach of the liquid-to-solid transition from the liquid side*. Three very different glass formers of known linear viscoelasticity were chosen to exemplify the dynamics near the glass transition in contrast to gelation. The first system is a colloidal suspension of repulsive particles with increasing particle loading (collaboration with the Helmholtz Institute in Berlin, Germany; M. Ballauff and M. Siebenbuerger); the low molecular glass formers were studied in the group of G. McKenna, Texas Tech in Lubbock, Texas. The polymeric glass formers were polystyrene and polybutadiene. The alpha relaxation time of these systems follows the BSW relaxation pattern, but superposition master curves are not possible in a global way.

(235b) Carbon Nanotubes As Mechanical Probes of Equilibrium and Non-Equilibrium Biopolymer Networks

Nikta Fakhri[1], Matteo Pasquali[2], Frederick C. MacKintosh[3] and Christoph Schmidt[1], (1)Third Institute of Physics-Biophysics, University of Göttingen, Göttingen, Germany, (2)Chemical and Biomolecular Engineering, Rice University, Houston, TX, (3)Department of Physics and Astronomy, Vrije Universiteit, Amsterdam, Netherlands

A biological cell is a soft material with complex mechanical properties arising from a composite and highly dynamic intracellular protein polymer network (consisting of actin, microtubules and intermediate filaments). Equilibrium soft materials show strain fluctuations due to random thermal stresses; living biological materials exhibit, in addition, non-equilibrium, internal stress fluctuations generated by active force generators (e.g. motor proteins). Understanding the dynamic properties of such heterogeneous and structurally complex networks requires probes that span length-scales from nanometers to micrometers.

We use single-walled carbon nanotubes (SWNTs) as multi-scale micro-probes. SWNTs are nanometer-diameter hollow carbon filaments with micrometer lengths and a tunable bending stiffness. Their persistence length varies between 20 – 100 microns. Therefore they show significant thermal fluctuations on the micron scale.

We study the motion of individual SWNTs in reconstituted actin (equilibrium) and actin-myosin (non-equilibrium) networks by near infrared fluorescence microscopy. At long times, SWNTs reptate in the networks. At short times SWNTs can sample the spectrum of local stresses in both equilibrium and non-equilibrium networks. We can calculate complex shear moduli from recorded fluctuations and observed power-law scaling in equilibrium actin networks. In the non-equilibrium networks we observed strong local shape fluctuations reflecting the activity of the molecular motors.

(235c) Modeling the Response of "Dual-Crosslinked" Functionalized Nanoparticles to Mechanical Deformation

Balaji Iyer[1], Isaac Salib[1], Victor. Y. Yashin[1], German Kolmakov[2], Krzysztof Matyjaszewski[3] and Anna Balazs[1], (1)Chemical & Petroleum Engineering, University of Pittsburgh, Pittsburgh, PA, (2)Physics, New York City College of Technology, Brooklyn, NY, (3)Department of Chemistry, Carnegie Mellon University, Pittsburgh, PA

We investigate the role of stretchable polymeric arms and force-dependent kinetics of dual cross-links on the mechanical response of a network formed by polymer-grafted nanoparticles (PGN). Each of the individual particles of the network is composed of a rigid core and a corona of grafted polymeric arms with functionalized end groups that can form stable and labile bonds. The presence of both the labile and stable bonds leads to a "dual-crosslinked" system. The overlap of the coronas of adjacent particles enables interaction between functionalized end groups that lead to the formation of the nanoparticle network. Interaction between polymer-coated colloids is combined with the Bell model for rupture and formation of bonds to model the interaction of array of particles. The mechanical response of the network is captured through the stretching and the force-dependent kinetics of the dual cross-linked polymer arms in the network. We show that the presence of stretchable arms allows for rearrangements leading to increases in the strength and ductility of the PGN network. We also show that the force-dependent kinetics of labile bonds modifies the initial mechanical response, while the stable bonds determine the ductility of the network.

(235d) Swelling-Induced Folding of Crosslinked Polymer Films

Sachin Velankar, University of Pittsburgh, Pittsburgh, PA, Victoria Lai, Dept. of Chemical Engineering, University of Pittsburgh, Pittsburgh, PA, Derek Breid, University of Pittsburgh, PA, Andrew Flowers, University of Pittsburgh and Richard Vaia, Wright Patterson Air Force Base

When a crosslinked polymer film attached to a rigid substrate is swollen with a solvent, the swelling induces an in-plane compressive stress in the film. If this stress is sufficiently large, a creasing pattern is known to be developed on the free surface. Here we show that the same experiment can also result in a completely different kind of instability: if the film is weakly attached to the substrate, the swelling-induced compressive stress nucleates buckle delamination of the film from the substrate. Surprisingly, the buckles do not have a sinusoidal profile, instead, the film near the delamination buckles slides towards the buckles causing growth of sharp folds of high aspect ratio. The folds persist even after the solvent evaporates. We characterize the dependence of such fold formation on the film thickness and swellability of the film.

We also show that the geometry of folding depends on the size of the region of the film exposed to solvent. A very small region of exposure (realized by placing a small drop of solvent on the film) does not induce delamination. Remarkably, with moderate sized drops, the delamination and folding occurs around the perimeter of the drop, thus culminating in a corral with tall walls. We quantify the parameters (drop volume, film thickness) which demarcate the transitions between no fold formation, corral formation, and multiple fold formation.

Finally this talk will discuss the implications of this research to fabrication of high-aspect ratio features on a surface.

(235e) Polymer Gels for Defense Applications

Randy A. Mrozek[1], Yelena R. Sliozberg[1], Jan W. Andzelm[1], Roza Mahmoodian[2], Z. Ilke Kalcioglu[2], Simona Socrate[3], Krystyn J. Van Vliet[2] and Joseph Lenhart[1], (1)Macromolecular Science and Technology Branch, US Army Research Laboratory, Aberdeen Proving Ground, MD, (2)Materials Science and Engineering, MIT, Cambridge, MA, (3)MIT, Cambridge, MA

This talk will present research spanning a broad program in polymer gels for defense applications. A polymer gel is a physically or chemically cross-linked polymer that is highly swollen by solvent. The gel properties can be tuned by varying the polymer chemistry, solvent type, polymer-solvent architecture and molecular weight, and solvent loading. In addition, small molecule additives and fillers can be incorporated into the gel formulation to enhance the properties further. This tunability offers the potential for gel implementation in an array of Army related technologies ranging from combat casualty care and tissue surrogates, to robotics and electronics devices. While potentially versatile materials, several obstacles hinder widespread deployment of gel-based technologies including: 1) limited operational temperature windows and material lifetimes; 2) poor toughness and durability; 3) unstable performance in other harsh environmental conditions (ie. extreme mechanical, electrical, chemical, and radiation environments); and 4) limited multi-functional capability. This talk will summarize our progress in these areas, and demonstrate how multifunctional soft polymer composites can be realized through the judicious selection of materials and processing approaches. The relevance of these materials will primarily be discussed in terms of tissue simulants but they also represent a broad platform for various technologies including robotics, smart clothing, armor, sensors, energy storage, battlefield medicine, etc.

(235f) Nanoparticle Gels in the Molecular Limit

Nikola Dudukovic, Department of Chemical and Biomolecular Engineering, University of Illinois at Urbana-Champaign, Urbana, IL and Charles F. Zukoski, Department of Chemical & Biomolecular Engineering, University of Illinois at Urbana Champaign, Urbana, IL

Anisotropic or patchy interactions between nanoparticles are expected to result in low volume fraction stiff gels. Here we explore gel formation in suspensions of ~1.5 nm particles which form at extremely low concentrations (<0.001 wt%). One specific type involves hydrophobic peptide based molecules – Fmoc-diphenylalanine (Fmoc-FF). We find that when water is added to a DMSO/Fmoc-FF solution, gels are formed along a line of increasing Fmoc-FF concentration with decreasing water concentration. Rigid (G' > 10^5 Pa) space filling gels are observed at Fmoc-FF concentrations as low as 0.1 wt%. Gelation in these systems is associated with formation of fibrils that branch to fill space. Here we report on the kinetics of gelation and demonstrate that these gels are reversible in the sense that they can be disrupted mechanically and rebuild strength over time. We attempt to understand the gelation process as arising from increasing strength of attraction between Fmoc-FF molecules with increasing water concentration. We find no indication that the system is approaching a spinodal for fluid-fluid phase separation as the gel point is reached or that the average diffusivity of species in solution decreases steadily at the gel point is approached, suggesting that gelation is associated with crossing a solubility boundary as opposed to the slow growth of structures as the strength of interparticle attraction grows. We present studies of gelation kinetics, mechanics of the resulting gels and changes in kinetics of relaxation of density fluctuations as the gel point is approached.

(235g) Thermoresponsiveness of Integrated Poly(N-isopropylacrylamide) Hydrogels with Ultra-Thin Silicon

Prithwish Chatterjee, Yuping Pan, Eric Stevens, Hanqing Jiang and Lenore L. Dai, School for Engineering of Matter,Transport and Energy, Arizona State University, Tempe, AZ

Novel materials based on various environmentally sensitive polymers hold multiple important applications. However, the functionality of these materials alone is often limited, and thus the integration of these, with other functional materials like Silicon, is strongly desired. Here we demonstrate the capability of integrating thermoresponsive poly(N-isopropylacrylamide) (PNIPAAm) hydrogels with Silicon nanoribbons. This will enable the stiff silicon ribbons to become adaptive and drivable by the soft environmentally sensitive substrate, such as becoming mechanically stretched and compressed when inducing a temperature change. Furthermore, we investigate how advanced lithographic techniques can be used to generate patterned deformation on the above integrated structures. We also explore bi-layer hydrogel structures formed by the integration of different types of polymers of PNIPAAm. These structures have been determined to achieve tunable curvature under the influence of different stimuli. The integration of soft materials with traditional semiconductor materials could have interesting implications for the development of novel “smart” devices.

(236a) Preprogrammed Noncontact Patterning of Polymer Films

Joshua Katzenstein, Dustin W. Janes, Julia D. Cushen, Nikhil Hira, Dana McGuffin and Christopher J. Ellison, Department of Chemical Engineering, University of Texas at Austin, Austin, TX

Many contact and non-contact methods have been developed over the years for the formation of three-dimensional topography in polymer films. Some of these techniques include electrohydrodynamic patterning, where pattern formation relies on an electric field applied to a dielectric polymer film and lithographically induced self-assembly, which exploits close proximity of a liquid polymer film to a patterned template. In these methods, the patterned template is placed a controlled distance away from the polymer film throughout the entire topography development process, which inherently limits throughput. These techniques are often slow, sometimes requiring hours for the pattern to fully develop with polymer melts. Alternatively, contact methods, such as micro-contact printing or imprint lithography, utilize 'stamps' or templates to directly transfer a relief pattern into a polymer film. Here we describe a new non-contact strategy for high-speed patterning of arbitrary shapes in polymer films. In this approach, the topographical pattern can be preprogrammed and stored in a smooth solid film, and later revealed without use of a wet or dry etch step, unlike traditional photoresist methods. Micro- and nano-scale features created by this method could be used for facile creation of soft-lithographic stamps, microfluidic devices or roll-to-roll manufacturing of textured surfaces for anti-biofouling coatings.

(236b) Patterning Polymer and Metal Microstructures On Curved Substrates Via Initiated Chemical Vapor Deposition (iCVD)

Christy D. Petruczok and Karen K. Gleason, Chemical Engineering, Massachusetts Institute of Technology, Cambridge, MA

Fabrication of polymer microstructures is an essential step in the production of microelectronic, optical, and biomedical devices. The ability to pattern features on curved substrates is of particular interest, as it enables the fabrication of devices with unique or improved properties. A patterning technique for curved substrates has been developed using Initiated Chemical Vapor Deposition (iCVD). In this process, vapor-phase monomer(s) and thermally-labile initiator pass through an array of heated filaments. The initiator decomposes into radicals; these radicals and the monomer(s) adsorb onto a cooled substrate, and simultaneous free radical polymerization and thin film formation occur. Since surface tension and de-wetting effects are absent, iCVD thin films conform to the geometry of the underlying substrate. Additionally, the benign reaction conditions inherent to this low-energy process allow for complete retention of polymer functionality.

A thin film of poly(4-vinylpyridine) (P4VP) is deposited on substrates using iCVD. After deposition, the photosensitivity of the film is enhanced via functionalization with 10,12-tricosadiynoic acid (TDA). The functionalized film is exposed at 254 nm, inducing 1,4 addition photopolymerization of the TDA. Unexposed regions remain soluble and are removed upon developing. Incorporation and polymerization of TDA have been verified using FTIR, Raman, and UV-visible spectroscopy. Functionalization improves the UV sensitivity of the film by a factor of 10, and the resulting value (71 ± 10 mJ/cm^2) is comparable to that of commercial products. Processing time is on the order of 1 h and multiple substrates can be prepared simultaneously. Thicknesses of patterned polymer films range from 40-200 nm, and features on the order of 1 μm have been obtained. Polymer microstructures have been patterned on glass rods with diameters of 2 to 6 mm, then used as masks to pattern metal features. The vapor-phase processing used in this method also facilitates the fabrication of bifunctional polymeric surfaces; since no solvents are used, underlying polymer layers cannot be dissolved as new layers are formed. Surfaces consisting of alternating regions of poly(1H,1H,2H,2H-perfluorodecyl acrylate) and functionalized P4VP have been patterned; successful fabrication was confirmed by comparing the static contact angles of water droplets measured in each region of the patterned substrate.

(236c) Poly(styrene)-b-Poly(acrylic acid) Block Copolymers: Phase Separation Behavior and Directed Self Assembly

Jing Cheng[1], Richard Lawson[2], Wei-Ming Yeh[2], Nathan Jarnagin[1], Andrew Peters[3], Laren M. Tolbert[4] and Clifford Henderson[5], (1)Georgia Institute of Technology, Atlanta, GA, (2)Georgia Institute of Technology, (3)Chemical Engineering, Georgia Institute of Technology, Atlanta, GA, (4)School of Chemistry and Biochemistry, Georgia Institute of Technology, Atlanta, GA, (5)School of Chemical & Biomolecular Engineering, Georgia Institute of Technology, Atlanta, GA

Directed self-assembly (DSA) of block copolymers is a promising technology for extending the patterning capability of current lithographic exposure tools. For example, production of sub-40 nm pitch features using 193 nm exposure technologies is conceivably possible using DSA methods without relying on time consuming, challenging, and expensive multiple patterning schemes. Significant recent work has focused on demonstration of the ability to produce large areas of regular grating structures with low numbers of defects using phase separation-based self-assembly of poly(styrene)-b-poly(methyl methacrylate) copolymers (PS-b-PMMA). While these recent results are promising and have shown the ability to form patterns with pitches approaching 20 nm using DSA, the ability to advance to even smaller pitches will be dependent upon the ability to develop new block copolymers with higher χ values and the associated alignment and block removal processes required to achieve successful DSA with these new materials. This paper reports on work focused on identifying higher χ block copolymers and their associated DSA processes for sub-20 nm pitch patterning. Specifically, this paper will review our work focused on poly(styrene)-b-poly(acrylic acid) materials (PS-b-PAA). We will review the synthesis and characterization of the block copolymers themselves, their phase separation behavior, and their performance in DSA. It will be shown that poly(styrene)-b-poly(acrylic acid) copolymers (PS-b-PAA) are one promising material for achieving substantially smaller pitch patterns than those possible with PS-b-PMMA while still utilizing simple hydrocarbon polymers. Characterization of the χ value for PS-b-PAA places it at approximately 0.18, which is roughly 4.5 times greater than the χ for PS-b-PMMA ($\chi_{PS\text{-}b\text{-}PMMA} \sim 0.04$). One particular advantage of PS-b-PAA polymers is that it is anticipated that much of the learning that has been done with the PS-b-PMMA system, such as development of highly selective plasma etch-based block removal procedures, can be directly leveraged or transferred to the PS-b-PAA system. This paper will specifically review the results of both solvent vapor annealing and thermal annealing processes and their differences in achieving DSA on topographic and chemoepitaxial guiding layers.

(236d) A Study of the Air-Water Interfacial Properties of Biodegradable Polyesters and Their Block Copolymers with Poly(ethylene glycol): Toward Rational Design of a Polymeric Lung Surfactant

Hae-Woong Park and **You-Yeon Won**, School of Chemical Engineering, Purdue University, West Lafayette, IN

It has been reported that the surface pressure-area isotherm of poly(D,L-lactic acid-*ran*-glycolic acid) (PLGA) at the air-water interface exhibits several interesting features: (1) a plateau at intermediate compression levels, (2) a sharp rise in surface pressure upon further compression, and (3) a marked surface pressure-area hysteresis during compression-expansion cycles. In order to investigate the molecular origin of this behavior, we conducted an extensive set of surface pressure and AFM imaging measurements with PLGA materials having several different molecular weights and also a poly(D,L-lactic acid-*ran*-glycolic acid-*ran*-caprolactone) (PLGACL) material (in which the caprolactone monomers were incorporated as a plasticizing component). The results suggest that: (i) the plateau in the surface pressure-area isotherm of PLGA (or PLGACL) occurs due to the formation (and collapse) of a continuous water-free monolayer of the polymer under continuous compression; (ii) the PLGA monolayer becomes significantly resistant to compression at high compression because at that condition the collapsed domains become large enough to become glassy (such behavior was not observed in the non-glassy PLGACL sample); and (iii) the isotherm hysteresis is due to a deformation (i.e., coarsening) of the collapsed domains that occurs under high compression conditions. We also investigated the monolayer properties of PEG-PLGA and PEG-PLGACL diblock copolymers. The results demonstrate that the tendency of PLGA (or PLGACL) to spread on water allows the polymer to be used as an anchoring block to form a smooth biodegradable monolayer of block copolymers at the air-water interface. These diblock copolymer monolayers exhibit protein resistance.

(236e) Fundamentals of Diblock Copolymer Phase Separation and Directed Self-Assembly Processes: Detailed Analysis of Lamellae Formation

Andrew Peters, Chemical Engineering, Georgia Institute of Technology, Atlanta, GA, Richard Lawson, Georgia Institute of Technology, Peter J. Ludovice, School of Chemical and Biomolecular Engineering, Georgia Institute of Technology, Atlanta, GA and Clifford Henderson, School of Chemical & Biomolecular Engineering, Georgia Institute of Technology, Atlanta, GA

Directed self assembly (DSA) of block copolymers, which relies on the use of surface features to guide the phase separation of a block copolymer into structures with long range order and precise registration, is a very promising technique for producing sub-30 nm pitch regular patterns. These patterns could be used as an enhancement to current lithographic techniques for manufacturing integrated circuit features below the current minimum pitch of ~80nm provided by leading edge 193nm lithography tools. In order to speed the development of such techniques and to understand the behavior and limitations of such processes, computer simulation of such methods are proving invaluable. In this work, the number of atoms required to simulate a useful size of polymer during the DSA process has been reduced by coarse graining the atomistic polymer into a mesoscale model form based on Kuhn segment beads to represent the polymer chain. Such mesoscale models have been developed and carefully parameterized to reproduce the behavior of potentially useful DSA block copolymers (e.g. PS-b-PMMA). These mesoscale models have been implemented into MD simulations using GPU-based computing based on a modified form of the HOOMD-Blue software package. These simulations have been used to probe several interesting questions related to block copolymer phase separation and DSA processes. First, the scaling of the pitch of phase separated patterns with the polymer degree of polymerization and chi value will be analyzed. This scaling is an important issue for those designing polymers to achieve selected target pitches for integrated circuit manufacturing. Comparisons of simulation results to a large amount of experimentally determined pitch data which indicate that the simulations are able to quantitatively predict and explain the observed behavior quite well will be shown. While it is commonly known that the order-disorder transition (ODT) in bulk diblock copolymer phase separation occurs at a XN value of ~10.4, the nature of the effective ODT in DSA processes utilizing thin diblock polymer films on topographical or chemical guiding substrate layers is not well understood. This paper will discuss the concept of an effective ODT for such thin film DSA processes and will show how this ODT depends on a a variety of factors including film thickness, nature of the guiding layer, and diblock copolymer properties. Finally, the effect of diblock copolymer polydispersity on the ODT and the resulting pattern quality (e.g. edge roughness behavior) will be discussed.

(236f) Image Formation in Thin Films of Polymeric Resists

Gila Stein, Dept. of Chemical and Biomolecular Engineering, University of Houston, Houston, TX and Ginusha Perera, Chemical and Biomolecular Engineering, University of Houston, Houston, TX

Next-generation nanopatterning requires imaging materials that achieve sub-20 nm resolution in ultrathin films. Current manufacturing practices are based on projection lithography with chemically-amplified (CA) resists; however, such ``top-down'' lithographic processes are approaching their intrinsic resolution limits, so alternative techniques like ``bottom-up'' block copolymer (BCP) self-assembly are increasingly attractive. The physicochemical parameters that control imaging in top-down and bottom-up lithography are very different, but we find that interactions at the free surface and substrate interfaces will control the three-dimensional image in both systems. To examine BCP self-assembly, we cast thin films of lamellar poly(styrene-b-methyl methacrylate) (PS-PMMA) copolymers on epitaxial templates. The three-dimensional domain shape is measured with variable-incident-angle small-angle X-ray scattering (SAXS). We find significant deformations in domain shape near the free surface and substrate interfaces. This behavior is attributed to the different surface tensions of PS and PMMA, as well as different affinities of PS and PMMA for the underlying epitaxial template. To study imaging in CA resists, we consider a model resist system of poly(4-hydroxystyrene-co-tertbutylacrylate) loaded with triphenylsulfonium photoacid generator (acid catalyst). Thin films of CA resist are nanopatterned with electron-beam lithography, baked to generate the chemical image, and the reaction products are measured with SAXS. We find that image resolution varies with depth into the film, where the reaction front is significantly broader near the film surface. Such behavior is consistent with a surface excess of acid catalyst, depth-dependent catalyst diffusion rates, or both of these factors. These studies demonstrate that the types of interactions at each interface can dominate a nanopatterning process based on polymer thin films.

(236g) Surface Segregation of Well-Defined Comb Polymers

Boxi Liu[1], Roderic P. Quirk[1], Mark D. Foster[1], Renfeng Hu[2] and David T. Wu[3], (1)Department of Polymer Science, The University of Akron, Akron, OH, (2)Chemical Engineering Department, Colorado School Of Mines, Golden, CO, (3)Chemical Engineering and Chemistry Departments, Colorado School Of Mines, Golden, CO

Interfacial segregation phenomenon in polymer blends of different chain architectures may be employed to control interfacial properties. It has been theoretically predicted by Wu *et al.*[1] that a long-chain, branched polymer blended with its linear analog will be preferentially segregated to the interfaces of the blend film. Among the long-chain branching architectures, comb polymers are promising candidates for achieving substantial surface segregation for tailoring surface properties, since their high degree of branching provides a significant driving force for surface segregation when chain ends prefer the surface. We prepared comb polystyrenes with well-defined architectural details by living anionic polymerization via the "grafting-through" approach. Neutron reflectivity (NR) and secondary ion mass spectrometry (SIMS) analyses reveal that the comb polymers that are still miscible with linear analogs in the bulk segregate so strongly to the surface that the surface concentration of combs is nearly 100 vol% under certain conditions. The level of surface segregation is affected by factors including comb architecture, bulk concentration, and chain end chemistry. Surprisingly, thermal annealing is not necessary to achieve strong surface segregation for these comb polymers. Wu's linear response theory suggests that a modest surface attraction on the order of kT for chain ends is sufficient to generate surface composition profiles that match the experimentally measured surface composition and segregation length scale.

Acknowledgements: NSF support (CBET 0730692 & 0731319) is gratefully acknowledged.

Reference:

1 Wu, D. T.; Fredrickson, G. H. *Macromolecules* **1996**, *29*, 7919-7930.

(240a) Effect of Supercritical Solvents On the Synthesis of Higher Alcohols Over a Cu-Co Based Catalyst

Rui Xu[1], Charlotte Stewart[1], Sihe Zhang[1] and Christopher B. Roberts[2], (1)Department of Chemical Engineering, Auburn University, Auburn University, AL, (2)Department of Chemical Engineering, Auburn University, Auburn, AL

There is sustained interest in the synthesis of higher alcohols from synthesis gas (syngas) due to the potential for the production of high octane value fuels and fuel additives. Higher alcohols are also valuable in view of their applicability as either raw materials or solvents for producing various chemicals. The abundance of diverse carbonaceous sources for producing syngas, including coal, natural gas and biomass, makes the catalytic synthesis of higher alcohols from syngas appealing as an alternative to petroleum resources. However, due to the highly exothermic nature of the alcohol synthesis reactions and the underlying kinetics for these reactions, this procedure suffers from poor productivity and selectivity towards higher alcohols. These limitations have prevented the development of a large scale and economically feasible process.

Supercritical fluids (SCFs) have unique properties, especially density depended properties, which are intermediate between those of a liquid and a gas. The density of a SCF, while lower than that of a liquid, is considerably higher than that of a gas. The diffusivity, while lower than that of a gas, is significantly higher than that of a liquid. These characteristics of a SCF provide several incentives for considering SCF reaction media in fixed bed higher alcohol synthesis, such as enhanced heat transfer relative to gas phase operation, improved accessibility to catalytic active sites and increased thermal uniformity.

Higher alcohol synthesis from syngas involves an overall reduction in the number of moles (i.e. volume reduction) and is highly exothermic. As such, the high pressure, dense phase solvent conditions of a supercritical reaction medium affords the opportunity to enhance the reactions via Le Chatelier's principle and to improve heat transfer relative to gas phase operation due to the improved heat transfer capacity of the SCF. Higher selectivity towards $C_{2+}OH$ can be expected when applying a properly selected SCF medium to the process.

In this work, 0.5 wt% K promoted $CuCo/ZnO/Al_2O_3$ was investigated as a catalyst for the synthesis of higher alcohols from syngas. The catalyst was prepared by co-precipitation methods and tested in a fixed bed reactor using a supercritical fluid as the reaction medium. Several types of supercritical solvents were tested, including polar and non-polar hydrocarbons. In addition, argon was utilized as an inert balancing gas for the comparison of gas phase and SCF phase operations. The results of experiments performed when using supercritical hexanes as the reaction media for this catalytic system illustrate that the presence of the SCF medium improves the heat transfer from the catalyst bed as evidenced by a significant reduction in the formation of CH_4. Furthermore, the results of these catalytic investigations demonstrate that the presence of the supercritical medium has a significant effect on the selectivity and the productivity towards higher alcohols.

(240b) Multiphysics Approach to Developing a New Heavy Oil Upgrading Technology

Michael T. Timko[1], Yuko Kida[2], Pushkaraj Patwardhan[2], Ayten Ates[3], Robin Edwards[2], Caleb Class[2], Guang Wu[2], Sadegh Dabiri[4], Ali M. Al-Somali[5], Ki-Hyouk Choi[5], Jefferson W. Tester[6], Ahmed Ghoniem[2] and William H. Green Jr.[7], (1)Energy Initiative, MIT, Cambridge, MA, (2)MIT, Cambridge, MA, (3)Cumhuriyet University, Sivas, Turkey, (4)Notre Dame University, South Bend, IN, (5)Research & Development Center, Saudi Aramco, Dhahran, Saudi Arabia, (6)Department of Chemical and Biomolecular Engineering, Cornell University, Ithaca, NY, (7)Department of Chemical Engineering, Massachusetts Insititute of Technology, Cambridge, MA

Continuous increases in the sulfur content and average molecular weight of remaining petroleum reserves are straining refinery capacity using current technologies. New technologies are required for cleaner, energy efficient utilization of sulfur rich heavy oils. Treating heavy oils and bitumen in supercritical water has been known since the 1980s to reduce their molecular weight distributions and sulfur contents. Crucially, supercritical water treatment does not promote coke formation, does not require catalysts, and reduces heavy oil asphaltene content. However, commercialization has been held back by contradictory literature data and a lack of fundamental understanding of the chemical reactivity, thermodynamic phase behavior, and transport phenomena at supercritical conditions. This paper will present an overview of an on-going research effort to provide a deeper fundamental engineering understanding of the relevant physical and chemical processes. Experimental work has been performed on model systems to measure global rate constants, close material balances, and identify decomposition products. Catalysts have been deployed to help understand chemical mechanisms and evaluate the potential of catalytic enhancement of the supercritical water process. Computational work has utilized these experimental data to construct extensive reaction networks that elucidate the role of water, the water gas shift reaction, and primary reaction pathways. Transport and thermodynamic simulations provide understanding of the relevant heat, mass, and momentum time scales and how they interact with the complex, multi-component, near-critical thermodynamic phase behavior and relevant chemical time scales.

Transesterification of Bio-oils in a Binary Supercritical Mixture of Methanol and Carbon Dioxide: Parametric Effects on Yield and Selectivity

Aaron Gonzales, Jared Bouquet, Ryan Tschannen, and Sunggyu Lee
Ohio University, Athens, OH

Abstract

Biodiesel is an oxygenated fuel created from renewable bio-oils and is usable in unmodified diesel engines. The conventional process for biodiesel production proceeds through transesterification of triglycerides using an alcohol, most often methanol, and a small amount of base to catalyze the reaction. This work illustrates a novel route for continuous biodiesel production that proceeds non-catalytically in a mixture of carbon dioxide and methanol at elevated temperatures and pressures. The process utilizes carbon dioxide's ability to act as a solvent at high pressures and temperatures to reduce the amount of methanol necessary for the transesterification to proceed non-catalytically. The experiments were conducted in a 0.1 L Haynes® 282 alloy tubular reactor with a fixed molar ratio of 28:1 methanol to triglyceride and a carbon dioxide flow rate of 3 SLPM. The space time varied from 30 seconds to 90 seconds, the reactor pressure from 1500 psig to 4500 psig, and the temperature ranged from 240°C to 320°C.

Introduction

Diesel is used heavily as a transportation fuel in the United States. The principal source of diesel fuel consumed in the United States is petroleum oil; however, with depleting easily extracted petroleum reserves and the political nature of petroleum, costs associated with petroleum-based diesel continue to increase. One way to reduce the need for petroleum crude is to use a bio-based alternative such as biodiesel.

Biodiesel is the term used to describe diesel like substances produced from plant or animal oils that are chemically reacted with alcohols. One conventional method for biodiesel production is through the batch transesterification of triglycerides using an alcohol, most often methanol, in the presence of small quantities of base to catalyze the reaction. This produces three alkyl ester molecules which can be used in conventional diesel engines with no modification.

This work proposes a novel method to produce biodiesel continuously in a binary solvent mixture of carbon dioxide and methanol, and without the presence of additional catalyst. It relies on supercritical carbon dioxide's ability to act as a solvent for organic species in order to reduce the amount of methanol needed to produce the biodiesel.

Materials and Methods

Carbon dioxide has been widely utilized in its supercritical phase due to its mild critical temperature and pressure, 31.1°C and 7.38 MPa respectively, and its high affinity for organic compounds. It was chosen to act as an inert solvent in order to reduce the methanol needed for the transesterification. These experiments were performed with Research Grade Carbon Dioxide supplied from Praxair.

Methanol is the alcohol chosen for the transesterification. The critical temperature and pressure of methanol are higher than those of carbon dioxide, 239.5°C and 8.1 MPa respectively, so reducing the amount of methanol needed to solvate the triglycerides reduces the energy input. For these experiments, histological grade methanol was purchased from Fisher Scientific.

The vegetable oil used in these experiments is Canola Oil #500 from Columbus Vegetable oils.

The test apparatus is a modification of the Ohio University, Sustainable Energy and Advanced Materials Lab's Multi-Fuel Reformer (MFR). The original application of the apparatus is continuous supercritical water reformation of hydrocarbons to produce synthesis gas; however, the setup of the process unit makes it ideal to do the preliminary work on this method. Figure 1 shows a simplified PFD of the MFR system. The reactor is a 0.1 L Haynes® 282® alloy tubular reactor with separate gas and liquid feeds to the reactor. Pressure is controlled using a control valve. The gaseous products leave the unit and proceed to a wet test meter to determine the flow rate of the exiting gas while the liquid is collected for sampling via GC/MS. Liquid samples are analyzed using GC/MS to determine whether the transesterification reaction takes place or not at the various conditions.

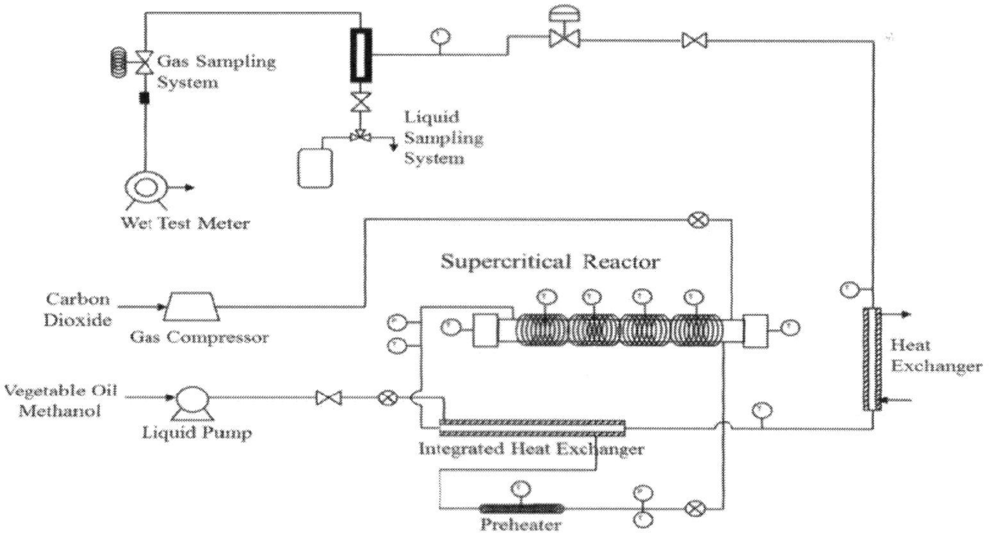

Figure 1. An Experimental System for Supercritical Biodiesel Production

Various temperatures, pressures, and space times were examined to assess the process efficacy of the binary supercritical fluid system as well as to determine the effects of process conditions on the transesterification.

A Study on Formaldehyde and Acetaldehyde Reactions in Supercritical Water

Ryan Tschannen[1], Aaron Gonzales[1], Byung Gwon Lee[2], and Sunggyu Lee[1]

1: Ohio University, Athens, OH
2: Korea Institute of Science and Technology, Seoul, Korea

Abstract

Over-abundance of glycerin from biodiesel production has led to numerous studies on ways to convert and utilize glycerin. One technique employed is to decompose glycerin in a supercritical water medium. This can result in valuable intermediate chemicals or complete gasification of glycerin into syngas. To effectively use this process, a complete and comprehensive understanding of the reaction pathway and kinetics of the decomposition of glycerin is desired. Two of the intermediate compounds produced during glycerin decomposition in supercritical water are formaldehyde and acetaldehyde. This study was carried out to elucidate the various mechanistic reactions of formaldehyde and acetaldehyde in supercritical water. Experiments were conducted non-catalytically in a continuous, 0.1 L Haynes® 282® alloy reactor at 22.4 MPa. Temperatures of 500°C to 700°C were investigated at a feed water-to-carbon molar ratio of 5:1 and a space time of 45 seconds.

Background

Glycerin is a byproduct of producing biodiesel from triglycerides. As the production and demand of biodiesel grows, the need to find uses for glycerin becomes more significant. A large amount of research has focused its efforts into finding potential uses for glycerin. One solution is to process the glycerin using supercritical water. At near-critical conditions, glycerin can undergo various decomposition reactions to form a variety of useful chemicals such as acrolein, 3-hydroxylpropanal, hydroxyacetone, acetaldehyde, and formaldehyde.[1-3] At temperatures further above the critical point of water, glycerin and its intermediates can be completely converted to permanent gases such as hydrogen, carbon monoxide, and methane.[3,4] These gases can be used in processes that require synthesis gas as a feed stock or further processed and purified for fuel cell applications.

Supercritical water refers to water at or above both the critical pressure (22.08 MPa) and critical temperature (374°C) of water. At these conditions, supercritical water is a highly energetic reaction medium. It allows compounds, such as glycerin, to undergo decomposition, dehydration, and free radical reactions without the aid of a catalyst.[2] Glycerin has three primary decomposition pathways in supercritical water and they are outlined in Figure 1.[3] The first path involves glycerin dehydrating into hydroxyacetone and then decomposing into formaldehyde and acetaldehyde. In the second path, glycerin dehydrates into 3-hydroxypropanal, which then decomposes into formaldehyde and acetaldehyde. The third path involves glycerin dehydrating into 3-hydroxypropanal and then dehydrating again to form acrolein. Of these three pathways, two of them involve acetaldehyde and formaldehyde as intermediates. It is believed that these compounds then undergo further decomposition to form gaseous products.[3,4]

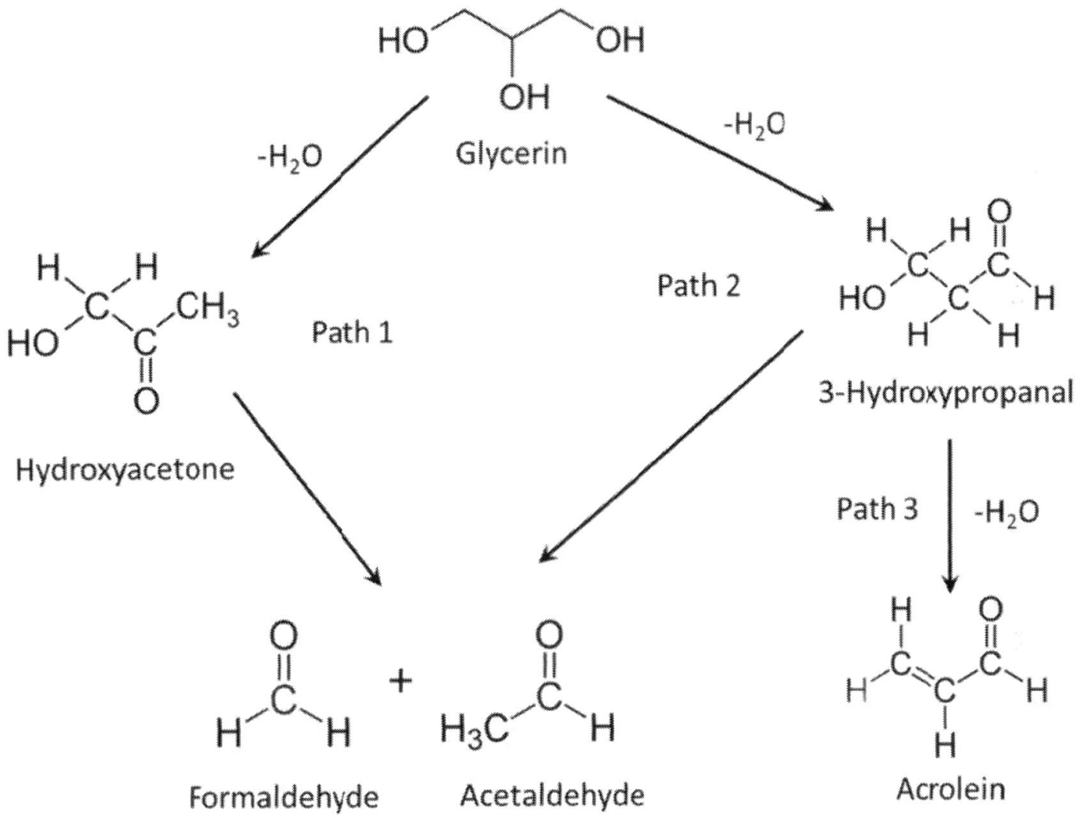

Figure 1. Glycerin Decomposition Pathways

Our lab has done previous studies on the decomposition pathway of glycerin in supercritical water. To better understand the decomposition of glycerin and obtain useful kinetics for the many reactions present, more studies need to be performed. This study attempts to elucidate the gases and liquid intermediates produced from two intermediate products of glycerin decomposition, viz., formaldehyde and acetaldehyde.

Apparatus and Procedure

Experiments were performed on a custom-built process unit capable of continuously reforming a variety of compounds using supercritical water. The process unit consists of a liquid feed system, integrated heat exchanger, preheat system, 0.1 L Haynes® 282® Alloy reactor, water-cooled heat exchanger, pneumatic control valve, sample collection systems, and data acquisition devices. Figure 2 shows a process flow diagram of the process unit. Effluent gas was analyzed using a wet test meter to measure volumetric flow rate and a Thermo Scientific Trace GC Ultra gas chromatograph with a TCD to measure gas composition. The Trace GC Ultra was calibrated to quantify hydrogen, nitrogen, carbon monoxide, methane, carbon dioxide, ethene, ethane, propene, and propane. Liquid samples were analyzed using an Agilent 5975C MSD Series gas chromatograph mass spectrometer.

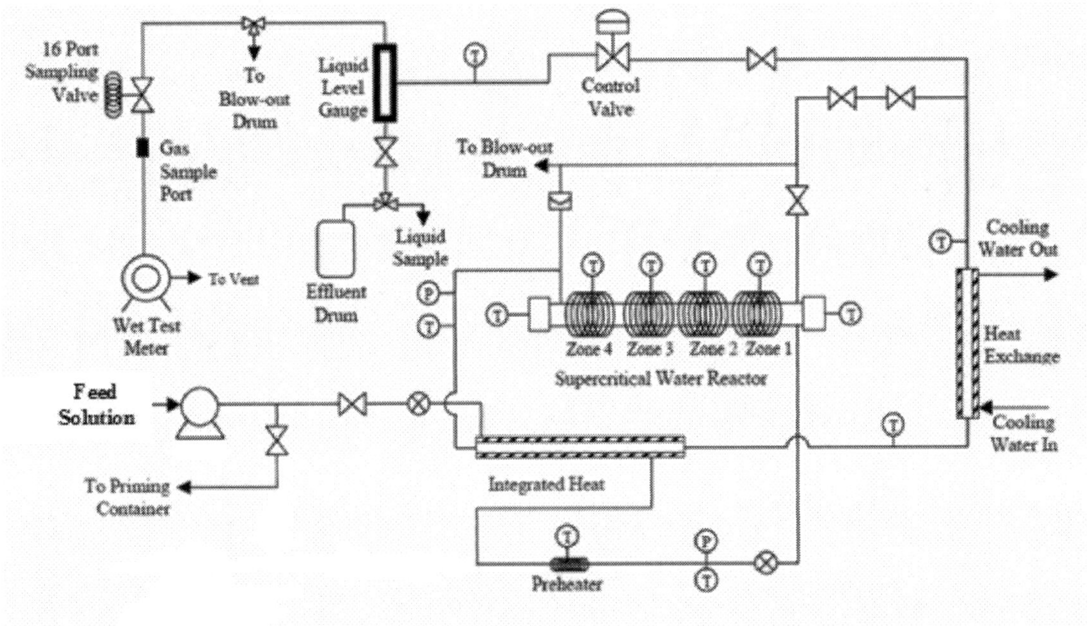

Figure 2. Supercritical Water System Process Flow Diagram

Feed solutions were prepared using deionized water, 37 wt% formaldehyde solution from Fisher Scientific, and 35 wt% acetaldehyde solution from Sigma Aldrich. The formaldehyde and acetaldehyde solutions were diluted to their molar feed concentrations of 5 moles of water per 1 mole of carbon using deionized water. This water to carbon ratio is similar to a previous study using a crude glycerin solution.[4] Reactor space times were calculated using the Peng-Robinson equation of state and the reactor entrance conditions.

The process unit was brought up to the experimental temperature and pressure using deionized water as a feed. After the reactor achieved the desired conditions, the feed was switched to the formaldehyde solution or the acetaldehyde solution. Steady state was achieved when the gas composition of three consecutive gas samples were within a 0.1 mol% difference. Once the system reached steady state, the system was switched back to deionized water and the reactor heaters were turned off.

Acknowledgements

This research work is partially supported by a research funding from the Korea Institute of Science and Technology (KIST).

References

1. Guo, Simao; Guo, Liejin; Cao, Changqing; Yin, Jiarong; Lu, Youjun; Zhang, Ximin (2012), "Hydrogen Production from Glycerol by Supercritical Water Gasification in a Continuous Flow Tubular Reactor," International Journal of Hydrogen Energy, Vol. 37, 5559-5568.

2. Buhler, W.; Dinjus, E.; Ederer, H. J.; Kruse, A.; Mas, C. (2002), "Ionic Reactions and Pyrolysis of Glycerol as Competing Reaction Pathways in Near- and Supercritical Water," Journal of Supercritical Fluids, Vol. 22, 37-53.

3. Bouquet, J. S. (2012), "The Effect of Supercritical Water on Crude Glycerin Solution," PhD Dissertation, Missouri University of Science and Technology, Rolla MO.

4. Bouquet, J. S.; Tschannen, R. E.; Gonzales, A. C.; Lee, S. "The Effects of Carbon-to-Oxygen Ratio upon Supercritical Water Reformation of a Crude Glycerin Solution for Hydrogen Production", Paper #682g, presented at Symposium Session on Alternative Fuels I, at the 2011 AIChE Annual Meeting, Minneapolis, MN, October 20, 2011.

(240e) Investigation of Ni Catalysts for Biomass Deoxygenation

Jacob G. Dickinson and Phillip E. Savage, Chemical Engineering, University of Michigan, Ann Arbor, Ann Arbor, MI

Phenols and heterocyclic oxygen containing compounds are often found in the reaction product when high water content biomass (e.g. microalgae) is processed hydrothermally. Hydrothermal catalytic deoxygenation of these molecules may provide significant processing advantages over conventional hydrotreating because upon removal of oxygen, the molecules lose their solubility in water. This may simplify the oil/water separation after crude oil processing.

Literature has shown that platinum catalysts are effective for deoxygenation reactions in supercritical water, but the high cost of platinum may make such processing not economically feasible. Therefore, we studied the deoxygenation of o-cresol in supercritical water over a variety of nickel catalysts. In particular, we examined the hydrothermal stability and activity maintenance of these catalysts. We also examined the effects of catalyst synthesis procedures, residence time, and hydrogen concentration of the reaction products. Preliminary results show that $Ni/SiO_2Al_2O_3$ and Ni/C are effective catalysts for the deoxygenation of benzofuran and phenols.

(240f) Experimental Analysis of the Partial Oxidation of Hexadecane in a Hydrothermal Tubular Reactor

Yousef Alshammari, Imperial College London, London, United Kingdom and **Klaus Hellgardt**, Chemical Engineering, Imperial College London, London, United Kingdom

Experimental Analysis of the Partial Oxidation of Hexadecane in a Hydrothermal Tubular Reactor

Yousef M. Alshammari, and Klaus Hellgardt

Department of Chemical Engineering, South Kensington Campus, Imperial College London, SW7 2AZ, UK

Abstract

Steam reforming of hydrocarbons is a major process for supplying hydrogen and syngas for various energy and chemical applications. One novel approach for hydrogen production is the conversion of hydrocarbons underground under downhole hydrothermal conditions. Hydrothermal systems may provide unique processing properties including enhanced heat and mass transfer, faster reactions rates, in addition to the environmental benefits of using water as a green solvent [1]. We previously reported the thermodynamic analysis [2,3] and noncatalytic experimental investigation of hydrothermal reforming of hexadecane, a heavy hydrocarbon model, under nonoxidative conditions [4]. This work reports the first parametric investigation of the partial oxidation of hexadecane in hydrothermal tubular flow reactor system shown in Figure 1. The effects of residence time, temperature, and oxygen to carbon ratio on the conversion of hexadecane and product distribution are being investigated with the aim of developing a kinetic model and understanding the gasification reactions of hydrocarbons under downhole conditions. Compared with our previous work on nonoxidative reforming of hexadecane, we report the effect of introducing oxygen into the reactor system to initiate an adiabatic reforming reaction raising the initial reservoir temperature, as shown in Figure 2, which increases the overall gasification efficiency.

Figure 1. Schematic of the hydrothermal tubular flow reactor system

Figure 1. Adiabatic temperature profile from partial oxidation of hexadecane for different mass concentrations of H_2O_2

References

[1] GUO, Y., WANG, S. Z., XU, D. H., GONG, Y. M., MA, H. H. & TANG, X. Y. 2010. Review of catalytic supercritical water gasification for hydrogen production from biomass. *Renewable and Sustainable Energy Reviews*, 14,334-343

[2] ALSHAMMARI, Y. M. & HELLGARDT, K. 2012. Thermodynamic analysis of hydrogen production via hydrothermal gasification of hexadecane. *International Journal of Hydrogen Energy*, 37,5656-5664.

[3] ALSHAMMARI, Y. M. & HELLGARDT, K. Reservoir-Based Hydrothermal Reforming of Hydrocarbons. *AIChE Annual Meeting 2010*, Salt Lake City, UT, USA, 2010, (363e).

[4] ALSHAMMARI, Y. M. & HELLGARDT, K. Continuous-Flow Reforming of Hydrocarbons Underground: Kinetic Analysis. *AIChE Annual Meeting 2011*, Minneapolis, MN, USA, 2011, (713f).

(240g) Oil Upgrading and Desulfurization by Supercritical Water

Mohammed Albahar[1], **Emad Shafei**[1], Abdulaziz Ghabbani[2], Mohammad Garhoush[1], Joo-Hyeong Lee[1], Ali H. Alshareef[1], Ashok K. Punetha[1], Ali M. Al-Somali[1] and Ki-Hyouk Choi[1], (1)Research & Development Center, Saudi Aramco, Dhahran, Saudi Arabia, (2)Saudi Aramco, Dhahran, Saudi Arabia

In spite of many available upgrading technologies such as coking and hydrocracking processes, various nonconventional approaches have been studied to make a breakthrough to manage imminent challenge on refining industry. Supercritical water, as one of promising nonconventional technologies, has been explored to have firm proof-of-concept for upgrading petroleum feedstocks by many research groups. Major issues related to the proof-of-concept are the role of supercritical water in the reaction, reaction mechanisms, extent of desulfurization and upgrading, and probable corrosion by reactant and product. In spite of many reports on the issues, there was no clear evidence for the involvement of hydrogen in water toward upgrading reaction. Hydrogen is required to maintain performance in desulfurization and upgrading of petroleum without generating solid coke. This paper will present experimental results on the source of hydrogen in thermal cracking of hydrocarbons. Participation of water in the cracking reaction of paraffin and sulfur compound was proven by analysis of cracking product. Also, recent progress of supercritical water technology application to petroleum desulfurization and upgrading will be summarized.

(251a) The Performance of Imidazolium Ionic Liquids and Their Mixtures in Inhibiting Methane Hydrate Formation

Anthony R. Richard, Department of Chemical and Petroleum Engineering, University of Wyoming, Laramie, WY and Hertanto Adidharma, Department of Chemical & Petroelum Engineering, University of Wyoming, Laramie, WY

The performance of imidazolium ionic liquids, a novel ionic liquid inhibitor for gas hydrate, in inhibiting methane hydrate at low and high ionic liquid concentrations is investigated in a pressure range of 100–300 bar. The effects of different substituents and functional groups of ionic liquid on hydrate dissociation conditions are further explored. Experiments on methane hydrate dissociation conditions in the presence of mixed ionic liquid and conventional inhibitors, such as sodium chloride and monoethylene glycol (MEG), as well as mixtures containing two ionic liquids are also performed to investigate any possible synergistic effects. It is observed that single component solutions of ionic liquid demonstrate a progressive increase in inhibition effect with increasing concentration, which may surpass the effectiveness of MEG at high concentrations. Ionic liquid substituents and functional groups that are effective in inhibiting hydrate formation will be discussed. The thermodynamic inhibition performance of ionic liquid mixtures, which shows synergistic effects, will also be presented.

(251b) Understanding the Stability of Mixed Hydrates Containing Propane, Ethane, and Methane Under Deep Water Conditions

Srinath Velaga and Brian Anderson, Chemical Engineering, National Energy Technology Laboratory/West Virginia University, Morgantown, WV

Methane hydrates are globally distributed in sediments along the continental margins and potentially contain more energy than all fossil fuel reserves. However, methane is also a potential greenhouse gas which could play a major role in global climate change. Understanding the stability of gas hydrates can help us to understand their role in the climate change. Three main factors affect the stability of hydrates: Temperature (*T*), Pressure (*P*) and composition. Hydrates become unstable when they are exposed to pressures and temperatures outside the hydrate stability zone (HSZ) in a process commonly called dissociation. However, hydrates can also become unstable even when the pressure and temperature are within the HSZ but the concentrations of the hydrate forming gases are below their fully-saturated levels in the surrounding water phase. This process can be described as dissolution. In situ observations of marine outcrops of gas hydrates indicate that these hydrates exposed to surrounding seawater are more stable than predicted using diffusion-controlled models based on the surrounding methane saturations. Naturally-occurring hydrates may not be simple structure I methane hydrates and may contain higher hydrocarbons like ethane, propane etc., which forms structure II (sII) hydrates. Therefore, these mixed hydrates may act to stabilize the hydrates. In this work, the dissolution of sII hydrates in the presence of water has been studied using molecular dynamics (MD) simulations to understand if and how the presence of ethane and propane may stabilize the hydrate. Lattice constants for sII hydrates were calculated and compared to experimental values to validate the OPLS potentials used for the hydrocarbon guest molecules. The effect of higher hydrocarbons, such as ethane and propane, on the stability of gas hydrate was studied by changing the composition in the hydrate phase keeping the methane composition constant in the large cage and small cages. Also, the effect of methane composition was also studied by changing the methane composition in large and small cages.

(251c) Cubic- and PC-SAFT-Based Thermodynamic Models and a Free Volume Viscosity Model for Hydrocarbons At Pressures to 276 Mpa and Temperatures to 533 K

Ward A. Burgess[1], Deepak Tapriyal[1,2], Bryan D. Morreale[3], Hseen Baled[1,4], Robert M. Enick[4], Babatunde Bamgbade[1,5], Yue Wu[1,5] and Mark A McHugh[1,5], (1)National Energy Technology Laboratory (NETL), Office of Research and Development, Department of Energy, Pittsburgh, PA, (2)URS, Pittsburgh, PA, (3)Office of Research and Development, U.S. Department of Energy, National Energy Technology Laboratory, Pittsburgh, PA, (4)Department of Chemical and Petroleum Engineering, University of Pittsburgh, Pittsburgh, PA, (5)Department of Chemical and Life Science Engineering, Virginia Commonwealth University, Richmond, VA

Because of their simplicity and the relative ease of predicting pure-component parameters, cubic equations of state (EoSs) are often used in industrial settings to model the density and phase equilibrium for pure hydrocarbons and their mixtures. However, cubic EoSs typically provide unreliable phase density predictions at the high-temperature, high-pressure (HTHP) conditions associated with ultradeep petroleum reservoirs. The PC-SAFT equation of state gives density predictions that are substantially more accurate than those provided by the cubics. A modified set of pure-component PC-SAFT parameters fit to experimental data obtained at HTHP conditions (that is, temperatures from ambient to 533 K and pressures from 6.9 to 276 MPa) gives density predictions that agree with reference data to better than ±1%. However, a hurdle to the more widespread use of this equation is the difficulty in obtaining values for the pure-component PC-SAFT parameters m, σ, and ε/k. A reliable means of predicting said parameters is needed because reference density data are not always available at HTHP conditions. Therefore, in this study, a group-contribution (GC) method for predicting HTHP PC-SAFT parameters is presented, with the overall parameter values a function of contributions from the various functional groups and aliphatic, aromatic, and naphthenic carbons comprising the molecule. In this manner, the GC HTHP PC-SAFT parameters distinguish between isomers, with density values predicted to within ±1% in the HTHP region.

In addition, a free volume-based theory is adapted to HTHP conditions to calculate solution viscosity. This three-parameter model also requires the density at a given pressure and temperature. Free volume theory viscosity predictions when cubic EoS, SAFT-based EoS, and experimental data were used to obtain density inputs were compared. When coupled to density predictions provided by the PC-SAFT equation of state, viscosity is predicted to within better than ±2% for pure components and better than ±5% for mixtures of components for which the molecular size of the components vary by no more than 50%.

(251d) A New Volume Translated Peng-Robinson Equation of State for Saturated and Single-Phase Liquid Densities

Agelia Abudour, Sayeed A. Mohammad, Robert L. Robinson Jr. and Khaled A. M. Gasem, School of Chemical Engineering, Oklahoma State University, Stillwater, OK

A new volume translation method is presented for the Peng-Robinson equation-of-state. In a departure for earlier studies in the literature, the volume translation function in this study was developed for both saturated and single-phase liquid density predictions. The method includes only one fluid-specific parameter, which has been generalized in terms of molecular properties such as critical compressibility factor, acentric factor and dipole moment. The method was tested with a large database of highly accurate data for liquid densities. The database contained 65 pure fluids involving several classes of chemical compounds that varied strongly in terms of their molecular size, shape, chain-length, asymmetry and polarity. The model was developed based on these 65 fluids and an additional 20 fluids were used for validating the generalized model.

Results indicate that the volume translation method is capable of precise representations of the saturated liquid densities, yielding an overall average absolute percentage deviation (%AAD) of 0.6 for 65 fluids involving more than 12,000 data points. Generalization of the model was also performed, which provided predictions with %AAD of 0.8 for the same database. Further, the validation of the model was also conducted by utilizing data for liquid densities of 20 fluids not used in the model development. The generalized method yielded liquid densities with a %AAD of 0.96, which is comparable to the deviations obtained for the fluids used in model development.

The volume translation approach was extended to predict compressed liquid densities in the single-phase region. Of the fluids tested, the generalized model provided predictions with a %AAD of 1.8. Thus, the volume translation approach developed in this study appears capable of reliable and accurate predictions of both saturated and compressed liquid densities for diverse classes of molecules.

(251e) A Single Pore Model for Displacement of Heavy Oil with Carbon Dioxide

Truynh Quoc My Duy Tran[1], Mohammed Ahmad[2], Parthasakha Neogi[2] and Baojun Bai[3], (1)Department of Chemical and Biological Engineering, Missouri University of Science and Technology, Rolla, MO, (2)Chemical and Biological Engineering, Missouri University of Science and Technology, Rolla, MO, (3)Department of Geological Science and Engineering, Missouri University of Science and Technology, Rolla, MO

When a gas displaces a viscous liquid in a pore, a thin liquid film is entrained as the gas finger moves forward. The fluid mechanics of the problem (Bretherton problem), particularly those at low capillary numbers and where the entrained thin film needs disjoining pressure, are known in form of entrained film thickness as a function of capillary pressure. To this problem, mass transfer in the form of dissolution of CO2 in oil under high pressure has been added with accompanying changes in density, viscosity and diffusivity (of CO2) in the liquid phase. The physical property changes are taken from existing data. The results, obtained using FLUENT and pertinent to CO2 flooding of heavy oils, will be discussed. They include the effect on profile shapes, capillary numbers, the thickness of the oil films left behind as well as the net mass transfer rates. They also include the deficiencies of FLUENT for this problem, where the time restrictions confine it to larger capillary numbers. However, the capillary numbers are not too large and the method can, in principle, be extended to the low capillary numbers. At higher (but still low) capillary numbers, the results are relevant to bubble columns.

(251f) Viscosity of Several Hydrocarbons At Temperatures to 533 K and Pressures to 276 Mpa

Hseen Baled[1,2], Robert M. Enick[2,3], Ward A. Burgess[1], Deepak Tapriyal[1,4], Bryan Morreale[1], Yee Soong[1], Babatunde Bamgbade[1,5], Yue Wu[1,5] and Mark McHugh[1,5], (1)National Energy Technology Laboratory (NETL), Office of Research and Development, Department of Energy, Pittsburgh, PA, (2)Department of Chemical and Petroleum Engineering, University of Pittsburgh, Pittsburgh, PA, (3)National Energy Technology Laboratory (NETL), Office of Research and Development, Department of Energy, Pittsburgh, PA, (4)URS, Pittsburgh, PA, (5)Department of Chemical and Life Science Engineering, Virginia Commonwealth University, Richmond, VA

Modeling fluid flow in petroleum reservoirs and production wells requires reliable predictions of the phase behavior, density, and viscosity of brine and hydrocarbon mixtures. This study focuses on the accurate prediction of hydrocarbon viscosity at the high-temperature, high-pressure (HTHP) conditions associated with ultradeep formations. Because reservoir compositional simulators typically model crude oil, condensate, and natural gas as a mixture of components, this study focuses on model hydrocarbons. Viscosity experiments are performed with a high-temperature, high-pressure, windowed, rolling ball viscometer constructed from Inconel 718 and using Inconel 718 balls. Pyrex 7740 tubes and balls can also be used with this viscometer for the measurement of extremely low-viscosity fluids, such as methane and propane. The viscometer has been calibrated with n-octane at different temperatures and pressures to 276 MPa. Experimental viscosity results for n-pentane, n-decane, n-hexadecane, and n-octadecane are reported at pressures to ~ 276 MPa and temperatures to ~ 533 K. The frictional theory of viscosity (F-theory) and the free volume theory of viscosity (FV theory) are used to predict the viscosity. F-theory under-predicts the viscosity by as much as 20% at pressures near 276 MPa, but this problem is greatly diminished by including a correction term which is a function of system temperature, boiling temperature, critical temperature, critical pressure, and repulsive pressure. For n-alkanes, viscosity predictions from the corrected F-theory are comparable with those obtained from FV theory. When coupled to density predictions made using an accurate equation of state (EoS), FV theory typically gives mean absolute percentage deviations (MAPDs) from experimental viscosity values of less than 3%. Three EoSs are considered for density predictions: the perturbed-chain statistical associating fluid theory (PC-SAFT) EoS, the HTHP volume translated Peng-Robinson (HTHP-VT-PR) EoS, and the HTHP volume translated Soave-Redlich-Kwong (HTHP-VT-SRK) EoS. Coupling the PC-SAFT densities to FV theory results in the most accurate viscosity predictions at HTHP conditions.

(251g) A Modified Capillary Viscometer: Dynamic Viscosity Measurements for Heptane up to 30 Mpa and 353 K

Alfredo Pimentel-Rodas, Luis A. Galicia-Luna and Jose J. Castro-Arellano, INSTITUTO POLITECNICO NACIONAL-ESIQIE, Mexico, Mexico

Abstract

The knowledge of the viscosity values play an important place on the development, design and redesign of several processes. It is affordable to have data within a good accuracy including uncertainty analysis to use it on a future scaling industrial process. In this work, results of dynamic viscosities measurements for heptane are presented. The viscosity have been measured with a modified capillary viscometer designed and built to operate at pressures up to 40 MPa and temperatures up to 473.15 K. The viscosity was measured at pressures up to 30 MPa and temperatures up to 353 K. The total uncertainties of the dynamic viscosity measurements were estimated to be less than 1%. Experimental improvements are carrying out in order to have uncertainty in viscosity less than 0.5 %. Measured viscosities were compared with predictions and correlations previously reported in the literature, where the differences between the literature and the measurements reported here are less than 0.5%.

(268a) Thermodynamic and Kinetic Constraints On Viral Packaging [Invited]

Michael F. Hagan, Physics, Brandeis University, Waltham, MA

During the replication of many viruses, hundreds to thousands of protein subunits assemble around the viral nucleic acid to form a protein shell called a capsid. Capsid formation is a multicomponent, out-of-equilibrium assembly process for which kinetic effects and thermodynamic constraints compete to determine the outcome. Success or failure hinges crucially on the interactions among the assembling components, and most viruses have evolved their structural components to drive highly efficient assembly. In this presentation, we will describe computer simulations that represent the assembly of a capsid around a nucleic acid or a synthetic polyelectrolyte. We will investigate how the length of nucleic acid which can be packaged during a dynamical assembly process depends on solution conditions. We will then compare the results to equilibrium calculations to determine the extent to which kinetics or if thermodynamics is a limiting factor. Finally, we will describe how one virus may be assembling a metastable complex primed for release of its nucleic acid cargo.

(268b) Thermodynamics of Protein-Mediated Self Assembly On Cell Membranes

Natesan Ramakrishnan, UPenn, Philadelphia, PA and Ravi Radhakrishnan, Department of Bioengineering, University of Pennsylvania, Philadelphia, PA

In eukaryotic cells, the internalization of extracellular cargo into the cytoplasm via the endocytosis machinery is an important regulatory process required for a large number of essential cellular functions, including nutrient uptake, cell-cell communication, and modulation of cell-membrane composition. Endocytosis is orchestrated by a variety of proteins. These proteins are implicated in membrane deformation/bending, cargo recognition and vesicle scission. While the involvement of these proteins have been established and their roles in membrane deformation, cargo recognition, and vesicle scission have been identified, current conceptual understanding falls short of a mechanistic description of the cooperativity and the bioenergetics of the underlying vesicle formation and growth which we address here using theoretical models based on an elastic continuum representation for the membrane and coarse-grained representations for the proteins. We present a quantitative model for describing how cell-membrane topologies are actively mediated and manipulated by intracellular protein assemblies. We formulate a minimal model by restricting our focus to a few proteins in the clathrin coat assembly: clathrin, epsin and AP-2, and BAR-domains and their role in the stabilization of various budding morphologies on the cell membrane. We describe the energetics of (topologically-invariant) deformation in planar membranes as well as in membranes with intrinsic curvature by the Helfrich Hamiltonian coupled with orderparameters associated with protein arrangements. Our approach is versatile in describing membrane geometries in both small and large deformation limits and for arbitray shapes and thermal undulations of the membrane. Our results yield membrane energies as well as entropy changes. A rich variety of membrane phase behavior is obtained by varying the extent and degree of induced curvature and the concentration of epsins and BAR domains on the membrane.

References

1. Minimal Mesoscale Model for Protein-Mediated Vesiculation in Clathrin-Dependent Endocytosis, N.J. Agrawal, J. Nukpezah, R. Radhakrishnan, PLoS: Computational Biology, 6(9) e1000926, 2010. doi:10.1371/journal.pcbi.1000926. Pubmed ID: 20838575.

2. Systems Biology and Physical Biology of Clathrin-Mediated Endocytosis: An Integrative Experimental and Theoretical Perspective, V. Ramanan, N. J. Agrawal, J. Liu, S. Engles, R. Toy, R. Radhakrishnan, Integrative Biology (RSC Journal), 2011, 3(8), 803-815. DOI: 10.1039/c1ib00036e. Pubmed ID: 21792431.

3. Mesoscale Modeling and Simulations of Spatial Partitioning of Curvature Inducing Proteins under the Influence of Mean Curvature Fields in Bilayer Membranes, J. Liu, R. Tourdot, V. Ramanan, N. J. Agrawal, R. Radhakrishnan, Molecular Physics, 2012, in press. (DOI:10.1080/00268976.2012.664661)

4. N. Ramakrishnan, R. Radhakrishnan, to be published.

(268c) Monolayer and Electrode Geometry Effects On the Formation, Structure, and Conductance of Molecular Junctions

William R. French[1], Christopher R. Iacovella[1] and Peter T. Cummings[2], (1)Chemical and Biomolecular Engineering, Vanderbilt University, Nashville, TN, (2)Center of Nanophase Materials Sciences, Oak Ridge National Laboratory, Oak Ridge, TN

The construction of basic circuit elements (*e.g.*, diodes, transistors, switches) from molecular building blocks constitutes a potential route to smaller, faster, and cheaper electronics. One method for wiring individual molecules between metallic electrodes is the mechanically controllable break junction (MCBJ) technique, in which molecules self-assemble between two fractured nanowire tips. Here, we report atomistically detailed simulations of the MCBJ technique for forming molecular junctions composed of benzene-1,4-dithiolate (BDT) chemically attached between two Au nanotips. The simulations utilize a hybrid molecular dynamics-Monte Carlo (MD-MC) approach and account for the influence of important environmental factors (*e.g.*, monolayer effects) on the behavior of the junctions.[1] Using this tool, we demonstrate the impact of monolayer density and tip geometry on the likelihood of molecular junction formation, and on the bonding geometry and tilt angle of bridged molecules. Since the primary property measured in experiment is conductance, we next calculate the low-bias conductance of the structures resulting from the MD-MC simulations. We employ a combined density functional theory and Green's function approach to calculate conductance; self-interaction corrections[2] are included to accurately describe the energy-level lineup between BDT and the Au leads. Using this approach we are able to systematically study structure-conductance relationships in Au-BDT-Au junctions, helping to further inform and interpret experimental results.

[1] W. R. French, C. R. Iacovella and P. T. Cummings, "Large-Scale Atomistic Simulations of Environmental Effects on the Formation and Properties of Molecular Junctions", *ACS Nano*, **2012**, 6 (3), 2779-2789.

[2] C. Toher and S. Sanvito, "Effects of Self-Interaction Corrections on the Transport Properties of Phenyl-Based Molecular Junctions", *Phys. Rev. B*, **2008**, 77, 155402.

(268d) Inverse Design of Pairwise Interactions for Self-Assembly of Low-Coordinated Lattice Structures

Avni Jain, The Department of Chemical Engineering, The University of Texas at Austin, Austin, TX, Jeffrey R. Errington, Chemical and Biological Engineering, University at Buffalo, The State University of New York, Buffalo, NY and Thomas M. Truskett, Department of Chemical Engineering and Institute for Theoretical Chemistry, The University of Texas at Austin, Austin, TX

Engineering the effective interactions between suspended nanoparticles or colloids to promote their assembly into specific lattice structures is an important and rapidly growing area of soft-matter research. Low-coordinated (e.g., diamond) crystals are often chosen to be the "target" lattices in such studies due to their potential as novel structures for photonic band gap materials. Computer simulations have shown that particles with anisotropic ("patchy") interactions can spontaneously form such equilibrium structures [1-5]. However, whether simpler isotropic interactions can also be designed to drive self-assembly into low-coordinated lattice structures – with perhaps more favorable kinetics than systems with anisotropic interactions— is a long-standing open question. In this study, we use inverse statistical mechanical methods to derive simple, repulsive pair potentials that display diamond and other low-coordinated ground states [6,7]. We characterize aspects of the equilibrium and dynamic behavior of these model systems and discuss their experimental realizability in polymeric colloidal materials.

[1] Z.L Zhang, A.S Keys, T. Chen, S.C Glotzer, 'Self-Assembly of patchy particles into diamond structures through molecular mimicry', *Langmuir*, 21, pp. 11547-11551 (2005)

[2] Z.L Zhang, S.C Glotzer, 'Self Assembly of patchy particles', *Nano Letters,* 204, pp. 1407-1413 (2004)

[3] F. Romano, F.Sciortino, 'Colloidal self-assembly: Patchy from bottom up', *Nature Materials,* 10, 171-173 (2011)

[4] F. Romano, E. Sanz, F.Sciortino, 'Phase Diagram of a tetrahedral patchy particle model for different interaction ranges', *Journal of Chemical Physics*, 132, 184501 (2010)

[5] E. Bianchi, R. Blaak, C.N Likos, 'Patchy Colloids: state of the art and perspectives', *Physical Chemistry Chemical Physics*, 13, pp.6397-6410 (2011)

[6] S. Torquato, 'Inverse Optimisation techniques for targeted self-assembly', *Soft Matter,* 5, pp. 1157-1173 (2009)

[7] E. Marcotte, F.H Stillinger, S.Torquato, 'Unusual ground states via monotonic convex pair potentials', *Journal of Chemical Physics,* 134, 164105 (2011)

(268e) Packings and Self-Assembly of a Family of Hard Truncated Triangular Bipyramids

Amir Haji-Akbari, Department of Chemical and Biological Engineering, Princeton University, Princeton, NJ, Elizabeth R. Chen, Chemical Engineering, University of Michigan, Ann Arbor, MI, Michael Engel, Department of Chemical Engineering, University of Michigan, Ann Arbor, MI and Sharon C. Glotzer, Chemical Engineering , University of Michigan, Ann Arbor, MI

For decades, hard-core interactions have been extensively used as models of colloidal systems with short-range repulsions. In recent years, theoretical and computational studies of hard particles have received further attention due to a flurry of activity in manufacturing colloidal particles of varying shapes. One remarkable feature of hard particles is their ability to assemble themselves into complicated structures. For instance, hard tetrahedra [1] and hard triangular bipyramids [2] have been observed to form a dodecagonal quasicrystal. A recent study of hard truncated tetrahedra has revealed the formation of several atomic crystal isostructures such as diamond, beta-tin and high pressure lithium [3]. All these studies underscore how geometric anisotropy of a particular building block can give rise to complicated structures with potentially interesting optical, electrical and/or mechanical properties. In this work, we consider a family of hard truncated triangular bipyramids and observe eight distinct families of packings in this system. We also establish that the formation of the quasicrystal is robust in the system and can be observed for truncations as high as 45 per cent.

1. A. Haji-Akbari, M. Engel, A. S. Keys, X. Y. Zheng, R. Petschek, P. Palffy-Muhoray, S. C. Glotzer, *Disordered, quasicrystalline and crystalline phase of densely-packed tetrahedra*, **Nature** 2009 462: 773-777.

2. A. Haji-Akbari, M. Engel, S. C. Glotzer, *Degenerate Quasicrystal of Hard Triangular Bipyramids*, **Phys. Rev. Lett.** 2011 107: 215702.

3. P. F. Damasceno, M. Engel, S. C. Glotzer, *Crystalline Assemblies and Densest Packings of a Family of Truncated Tetrahedra and the Role of Directional Entropic Forces*, **ACS Nano** 2012 6(1), 609-614.

(268f) Self-Assembly of Grafted Nanoparticles in Thin Films

Thomas Lafitte[1], Sanat K. Kumar[2] and Athanassios Z. Panagiotopoulos[1], (1)Chemical and Biological Engineering, Princeton University, Princeton, NJ, (2)Department of Chemical Engineering, Columbia University, New York, NY

We present a computational study of the structural properties of free-standing thin films of solvent-free polymer grafted nanoparticles, investigated via molecular dynamics simulations. A recent study [1] of the bulk demonstrated that polymer grafted nanoparticles in solvent free condition displayed structural behaviour which deviate significantly from that of simple liquids. We focus here on the possibility to obtain stable 2-dimensional pattern formation of spherically grafted nanoparticles when suspended as a free-standing film. Morphological phase diagrams in terms of particle-particle and particle-polymer attractions are systematically studied for nanoparticles with different grafted chain length. We observe formation of defined structures ranging from dispersed particles, finite stripes to percolating networks in 2 dimensions. Although the formation of stripes for this system is generally understood to be linked to the grafted chain deformation upon aggregation of the particle cores, resulting in an effective "dipolar" interparticle potential, we find that this anisotropic self-assembly can be systematically reproduced by coarse-graining the interactions into 2-dimensional isotropic potentials obtained by a simple inverse-Boltzmann procedure. This suggests that the main driving force for the anisotropic self-assembly for these uniformly grafted nanoparticle is due to the presence of a short-range attraction and long-range repulsion.

[1] Chremos, A.; Panagiotopoulos, A. Z., *Phys. Rev. Lett.* 2011, 107, 105503

(272a) Turbulent Dissolution, Drag Reduction and Degradation of Pipe Wall-Injected Polyox

Preetinder S. Virk, Department of Chemical Engineering, Massachusetts Institute of Technology, Cambridge, MA

Turbulent dissolution, drag reduction and degradation of wall-injected Polyox P309 solutions was studied by simultaneous axial pressure profile and flow visualization measurements at Reynolds numbers from 15000 to 150000 in a transparent test pipe of ID 16 mm and L/D 430. Friction factors for solvent, deionized water, adhered to the Prandtl-Karman law for fully-developed turbulent flow in smooth pipes. The P309 Polyox, of MW ~ $8x10^6$ with N_{bb} ~ $0.5x10^6$ backbone chain links, was injected as a concentrated solution C_{inj} = ei 500 or 1000 wppm through 6 equi-spaced circumferential ports at friction-normalized injection velocities $V_{inj}^+ \sim 1 << $ pipeline mean velocity $V^+ \sim 25$. Downstream polymer concentrations C_{line} varied from 0.2 to 20 wppm, exhibiting eventual fractional drag reductions DR_{line} from 0.10 to the asymptotic maximum drag reduction MDR ~ 0.80 possible at the present Re. Polymer dissolution into the flow was assessed from the initial increase in fractional drag reduction towards the eventual downstream DR_{line}, which provided a characteristic development distance $(L/D)_{dr}$ and also from flash photography which provided a characteristic distance $(L/D)_{pv}$ where red-dyed strands of the injected solution disappeared. Finally, at the lowest C_{line} and highest Re, drag reduction was observed to increase, peak, and then decline with increasing downstream distance. Since the peak drag reduction corresponded roughly to the disappearance of the red-dyed strands, that is, to full polymer dissolution, the subsequent downstream decline in drag reduction could be ascribed to polyox degradation and used to estimate its kinetics.

(272b) Characteristics of Flow in a Packed Pebble-Bed Reactor

Rahman Abdulmohsin and Muthanna Al-Dahhan, Chemical and Biological Engineering, Missouri University of Science and Technology, Rolla, MO

In general, moving bed where the solid particles are contacted with gas phase have found applications in industry such as two-phase flow catalytic hydro-processing of heavy oil and the pebble bed reactors (PBRs) for the 4th generation nuclear energy. Despite these recent interests, there is still lack of understanding of the complex gas flow structure and mixing pattern. Therefore, in this work, the complex flow structure of the gas phase which is used for cooling has been investigated experimentally in a 0.3 m diameter cold-flow randomly packed pebble-bed unit. An advanced gaseous tracer technique was developed to measure the residence time distributions (RTDs) of the gas phase. The dispersion and the degree of axial mixing of the gas phase have been quantified in terms of axial dispersion coefficients and Peclet numbers, respectively, over a wide range of Reynolds numbers which cover both laminar and turbulent flow conditions. The non-ideal flow behavior of the gas phase in packed pebble-bed has been described using one-dimensional axial dispersion model (ADM). Experiments results indicate that the axial dispersion coefficient increases as gas flow velocity increased. Also, it was found that the flow pattern of the gas phase deviates from ideal plug-flow reactor (PFR) model depending on the gas flow rate.

(272c) Droplet Behavior in a Taylor Vortex

Jian Qiao, Chemical and Biomolecular Engineering, National University of Singapore, Singapore 117576, Singapore and **Chi-Hwa Wang**, Department of Chemical and Biomolecular Engineering, National University of Singapore, Singapore, Singapore

Typical applications of Taylor-Couette device include chemical reaction, filtration, extraction and cell cultivation, in which the dispersed phase may appear in the form of droplets sometimes, and therefore the understanding of the droplet behavior in the vortex flow is important for optimizing the performance of this type of devices. In this study, the droplet of aqueous phase was introduced into a Taylor-Couette device with an oleic continues phase. An aspect ratio (Γ) of 6 and a radius ratio (η) of 0.67 for the Taylor-Couette device were applied. The droplet motion was captured by a high speed camera. It was found that the droplet behavior is mainly determined by the Reynolds number of the Taylor-Couette flow and the interfacial tension between the two fluids. The droplet with high surface tension displays a dancing behavior in low Reynolds number region and breaks into small droplet in high Reynolds number region. However, the droplet with low surface tension is only stable in high Reynolds number regime and breaks after the Reynolds number is reduced.

(272d) Bubble Behavior in Ionic Liquid [Bmim]BF4

Xin Zhang, Haifeng Dong, Xiangping Zhang and Suojiang Zhang, Institute of Process Engineering, Chinese Academy of Sciences, Beijing, China

Ionic liquids (ILs) have been widely considered as potential green substitutes for traditional organic solvents in a variety of chemical processes. However, knowledge related to the hydrodynamics of ionic liquids especially the bubble behavior is still very limited compared with the conventional molecular solvents. In this work, a photographic image analysis method was established to study the bubble behavior in ionic liquid [bmim]BF_4 by using high speed image pick-up system, operated at 500 frames per second with a resolution of 1280×1024 pixels. The bubble behavior, such as bubble size distribution, average gas holdup, Sauter mean diameter and interfacial area at different gas superficial velocities, liquid temperatures as well as axis positions of the bubble column were investigated. The experimental results showed that the bubble diameter in [bmim]BF_4 increased as gas superficial velocity and axis position were raised but decreased with increasing temperature. The total gas holdup in [bmim]BF_4 increased at increasing temperature due to the decrease of liquid viscosity. The total gas holdup also increased at increasing gas superficial velocity due to the increase of gas momentum which favored the formation of small bubbles. A new correlation based on the experimental data was proposed for the prediction of bubble Sauter diameter in [bmim]BF_4. The values predicted by the new correlation are in good agreement with the experimental results with a relative error less than 10%.

References

[1] Krishna, R., Ellenberger, J., 1996. Gas holdup in bubble column reactors operating in the churn-turbulent flow regime. AIChE Journal 42, 2627-2634.

[2] Dong, H., Wang, X., Liu, L., Zhang, X., Zhang, S., 2010. The rise and deformation of a single bubble in ionic liquids. Chemical Engineering Science 65, 3240-3248.

(272e) Molecular Dynamics Simulations of Diffusion and Clustering Along Critical Isotherms of Medium-Chain n-Alkanes

J. Wambui Mutoru[1], Wendell Smith[2], Corey S. O'Hern[3] and Abbas Firoozabadi[1], (1)Department of Chemical and Environmental Engineering, Yale University, New Haven, CT, (2)Department of Physics, Yale University , New Haven, CT, (3)Department of Mechanical Engineering and Materials Science; Department of Physics, Yale University , New Haven, CT

Understanding the transport properties of n-alkanes and other molecular fluids in the critical region is important for a number of industrial and natural systems. It is well-established that the Fickian diffusion coefficient vanishes at the critical point in binary fluids. However, there are strongly conflicting results for the behavior of the self diffusion coefficient D_s near the critical point in single-component systems. Further, there is no molecular-scale understanding of the behavior of D_s in single-component molecular fluids in the critical regime. Thus, we perform extensive molecular dynamics simulations of single-component fluids composed of medium-chain n-alkanes (n-pentane, n-decane and n-dodecane) using anisotropic united-atom potentials. For each system, we quantify the self-diffusion coefficient, cluster sizes and lifetimes along critical isotherms. We show that D_s decreases and the average size of molecular clusters increases as a function of density along critical isotherms, but D_s remains finite at the critical point. This work confirms that the self-diffusion coefficient is finite at the critical point in single-component molecular fluids.

(272f) Surface Stabilization of Heteroepitaxial Thin Films Induced by the Combined Action of Electric and Thermal Fields

Georgios I. Sfyris, Dwaipayan Dasgupta and Dimitrios Maroudas, Chemical Engineering, University of Massachusetts Amherst, Amhest, MA

Stressed elastic solids are known to undergo surface morphological instabilities; inhibiting such instabilities is important in the fabrication of devices. It is well known, for example, that the competition between elastic strain energy and surface energy can destabilize the surface of a stressed elastic solid through the Asaro-Tiller or Grinfeld (ATG) instability. For device fabrication purposes, heteroepitaxial film/substrate systems are of particular interest. Typically, these consist of a coherently strained thin film grown epitaxially on a substrate of a different material; due to lattice mismatch between the two constituent materials, the film is subjected to misfit strain, which can destabilize the planar film surface according to the so-called Stranski-Krastanow (SK) morphological instability that leads to the formation of islands on the film surface. In recent studies, we have shown that, in such a material system, the application of a properly directed and sufficiently strong electric field can inhibit the SK instability and maintain the planar state of the film surface morphology. Furthermore, we have shown that the requirements for the strength of the externally applied electric field can be minimized by the use of substrate engineering techniques, among which most notable is the use of finite-thickness compliant substrates.

In this presentation, we explore the surface morphological response of a coherently strained epitaxial thin film under the simultaneous action of the applied electric field and a temperature gradient; such thermal fields are common in device fabrication and service. We have developed a three-dimensional (3D) fully nonlinear model for the driven mass transport on the surface of a thin film that is electrically conducting, coherently strained, epitaxially grown on an elastically deformable substrate, and subjected simultaneously to the electric and thermal fields. The model accounts for curvature-driven and stress-driven surface diffusion, electromigration and thermomigration fluxes driven by the external fields, as well as surface diffusional anisotropy. We have derived an evolution equation for the film's height and solved the elastostatic, electrostatic, and thermal boundary-value problems (BVPs) that are coupled with it. Combining the solutions to the BVPs with the evolution equation and linearizing, we have derived the corresponding amplitude evolution equation for the study of surface morphological stability and pursued it using linear stability analysis. We have examined in detail the synergy and competition between the applied electric and thermal fields and determined their proper alignments for optimal stabilization of the planar film surface morphology. For such alignments, we have predicted the critical electric-field strength required for film surface stabilization as a function of material properties, heteroepitaxial system parameters, and the strength of the applied temperature gradient.

We have also analyzed the surface morphological stabilization of heteroepitaxial films on substrates by combining the simultaneous application of the two external fields with well-known strain engineering techniques, including the use of thin compliant substrates; such substrates provide some elastic accommodation of the lattice-mismatch strain in the epitaxial film due to the substrate's ability to relax parallel to its interface with the epitaxial film. Employing substrate engineering techniques alone is not sufficient for stabilizing the planar morphology of the epitaxial films; for such stabilization, the simultaneous action of the externally applied electric field is required. Our analysis demonstrates that a properly oriented electric field, acting simultaneously with an external thermal field and in combination with the use of a compliant substrate, can reduce the electric-field strength requirement for planar film surface stabilization by several (over three) orders of magnitude compared to that required for a typical thick substrate and without any thermal aid. Our analysis generates experimentally testable hypotheses and motivates experimental measurements that can be compared directly with the theoretical predictions.

(272g) Molecular Simulations of Thermal Transport in Solid-Fluid and Solid-Solid Interfacial Layers

Sohail Murad, Chemical Engineering, University of Illinois at Chicago, Chicago, IL and Ishwar K. Puri, Engineering Science and Mechanics, Virginia Tech, Blacksburg, VA

Using molecular simulations, we have investigated heat transfer across the solid-fluid interface between water and silicon and silica wafers, and solid-solid interfaces in superlattices and thin solid films. The system set up has allowed us to focus on the resistance associated with both the fluid and solid interfaces. For instance, by maintaining the solid phase at a constant temperature we can focus solely on the fluid side resistance. Our results show that the thermal or Kapitza resistance at fluid side of the solid-fluid decreases significantly as the surface is made more hydrophilic. This is primarily due to increases in fluid adsorption and absorption at the surface, which enhance the intermolecular collision frequency at the interface. Increasing this frequency also reduces the dependence of thermal transport on variations in the interfacial temperature and pressure. Hence, decreasing the density diminishes the intermolecular collision frequency, which increases the thermal resistance. By maintaining the fluid at a constant temperature we have also examined the interface resistance on the solid side. Our results show that these interfacial resistances can diminish the wall heat flux by an order of magnitude in comparison with a hypothetical system for which the overall fluid–solid contact resistances are negligible. Finally, we consider the solid phase as a superlattice in which case the interfacial resistances produced between different solid layers can significantly lower the heat transfer. Our simulations show significant resistance to thermal transport between thin films of the solid phase that constitute the superlattice, providing insight into how a superinsulator can be designed.

Finally we have also investigatedthrough molecular dynamics simulations if the influence of an organized mass distribution in a fluid also leads to thermal rectification. OUr results show that by either using an external force to create a density gradient in the sytem between the hot and cold surfaces or by changing the surface morphology (by making the walls more hydrophilic for example) thermal rectification can be clearly observed. Thermal rectification has many applications in many industrial applications.

(272h) Modeling of Two-Phase Fluid Flow Over Superhydrophobic Surfaces

Margaret Freeman, Alberto Striolo and Dimitrios V. Papavassiliou, School of Chemical Biological and Materials Engineering, The University of Oklahoma, Norman, OK

Superhydrophobic surfaces, which combine the chemical composition of a hydrophobic surface with microscopic roughness elements to achieve an apparent contact angle between a drop of fluid and a surface of at least 150°, have attracted increasing attention in recent years for their potential use in a number of applications, including separation processes, bioadhesion, and microfluidic devices. These surfaces result in a reduced contact area between the liquid and solid by trapping air between the roughness elements, causing the flowing liquid to form a meniscus that may extend into the space between the roughness elements. One of the most potentially transformative uses of these surfaces is for sustainable drag reduction in laminar and in turbulent flow regimes. One common approach to modeling these surfaces with computational fluid dynamics assumes the meniscus formed between the roughness elements is flat and involves the single phase simulation of liquid flowing over a surface with alternating free shear and no slip boundary conditions [1, 2], or single phase flow with specified slip [3]. The exclusion of the vapor phase in these simulations, however, could prevent a robust theoretical understanding of drag reduction over superhydrophobic surfaces from being developed. This study evaluates the validity of the flat free surface assumption with emphasis in turbulent drag reduction and compares the results obtained using a volume of fluid multiphase model with previously published direct numerical simulation data. The dependence of the slip length of the liquid on the Reynolds number and shape, size, and configuration of the superhydrophobic roughness elements is also examined in order to better understand the mechanism by which slip results in flow over superhydrophobic surfaces.

ACKNOWLEDGEMENTS

The financial support of the National Science Foundation (CBET-0853657) is acknowledged.

REFERENCES

1. Daniello, R.J., N.E. Waterhouse, and J.P. Rothstein, "Drag Reduction in Turbulent Flows Over Superhydrophobic Surfaces," *Phys. Fluids*, **21**(8), 085103 (2009).
2. Rothstein, J.P., "Slip on Superhydrophobic Surfaces," *Annu.Rev. Fluid Mech.*, **42**, 89-109 (2010).
3. Spencer, N.B., L.L. Lee, R.N. Parthasarathy and D.V. Papavassiliou, "Turbulence Structure for Plane Poiseuille-Couette Flow and Implications for Drag Reduction Over Surfaces with Slip," *Can. J. Chem. Eng.*, **87**(1), 38-46 (2009).

(272i) Investigation of Falling Liquid Film Flow On Multi-Baffled Plate

Bo Sun, State Key Laboratory of Chemical Engineering , TianJin University, Tianjin, China, Ming Zhu, East China Engineering Science and Technology CO., LTD, hefei, China, Botan Liu, State Key Laboratory of Chemical Engineering , Tianjin University, Tianjin, China, Chunjiang Liu, State Key Laboratory of Chemical Engineering, Tianjin University and Xigang Yuan, State Key Laboratory of Chemical Engineering, Tianjin University, Tianjin, China

Falling liquid film flow accompanied by mass transfer is a common phenomenon in chemical processes, such as distillation and absorption. And it is found that different flow patterns of falling liquid film lead to different mass transfer efficiency. By changing the detailed configuration of corrugated sheet of structured packing, flow patterns of the falling liquid film flow on the packing can be changed. In this paper, a novel structured packing on which the liquid film flow with a new kind of flow pattern is proposed and mass transfer experiment demonstrate that the packing has a high mass transfer efficiency.

However, the complicated structure of the packing limits a deeply understanding of the mass transfer mechanism. So a simplified model called multi-baffled plate is proposed with the similar hydrodynamic characteristic. Based on the proposed model, experiment and numerical simulation are carried out with isopropanol desorption process on multi-baffled and ordinary plate respectively. Experimental results show that the mass transfer efficiency of multi-baffled configuration is about 50% higher than that of ordinary plate.

In order to investigate the detailed structure of the flow and mass transfer behavior, Computational Fluid Dynamics (CFD) model based on Volume of Fluid (VOF) method is developed, and marangoni effect induced by mass transfer is considered in the model. On the basis of the simulation results, it can be concluded that there are two important factors leading to the higher mass transfer efficiency on the multi-baffled plate. Firstly, the renewal rate is increased accompanied by vortices near the liquid film. When the liquid film flows across the junction of two baffles, the surface flow on the upstream plate is turned to be the bottom flow on the downstream plate, and the bottom flow on the upstream plate is turned to be the surface flow on the downstream plate. Secondly, flow instability is increased in the condition of the changing of flow direction at the junction of two baffles. Based on the reasons above, the mass transfer efficiency on the multi-baffle plates is increased.

Effect of perfluorocarbon vapor concentration on surface tension of water

Vasiliy Chernyshev and Mikhail Skliar
Chemical Engineering
University of Utah

Perfluorocarbons (PFCs) are highly hydrophobic, biocompatible compounds that have found a variety of medical applications. For example, stabilized perfluorocarbons are used as contrast agents during ultrasound imaging and in targeting drug delivery. Other applications include their use in aerosol propellants, lung ventilation, and anesthesia. Despite significant interest in these compounds, the study of their interfacial properties has received limited attention. Following our previous work, we investigated the influence of PFC vapor on the surface tension of a water droplet at room temperature. Our main focus was on perfluoropentane (PFP), one of the molecules in the PFC family that has a unique boiling temperature of about 29C that is below normal human body temperature and above the room temperature.

The droplet surface tension, volume and surface area measurements were carried out by means of real-time pendent-drop tensiometer developed by us using MATLAB software. The Laplace equation presented as nonlinear differential equations was used for the calculations. The study followed an experimental procedure summarized in Fig. 1 where the surface tension of water was followed in real time before and after the injection of PFP inside the sealed cuvette to produce PFP vapor. The measurement was stopped when the change of surface tension over time would become negligible. The result indicated a substantial decrease in surface tension of water with time (Fig. 2) as a result of PFP vaporization. A similar result was seen in the case where perfluorohexane (PFH) was used in the vapor phase; however a smaller influence was found. We further investigated such change in surface tension after the injection of PFC by determining the concentration change of PFC vapor over time. FT-IR experiments agreed well with a COMSOL model in identifying the change of PFP and PFH concentration over time in the area where the droplet was located during the experiments (Fig.3). Using this information, it was found that the surface tension decreases linearly with the increase of the PFP and PFH concentration in the gas phase, as shown in Figure 4 and 5. The results obtained in this work will be used in further investigation of liquid/vapor interfaces.

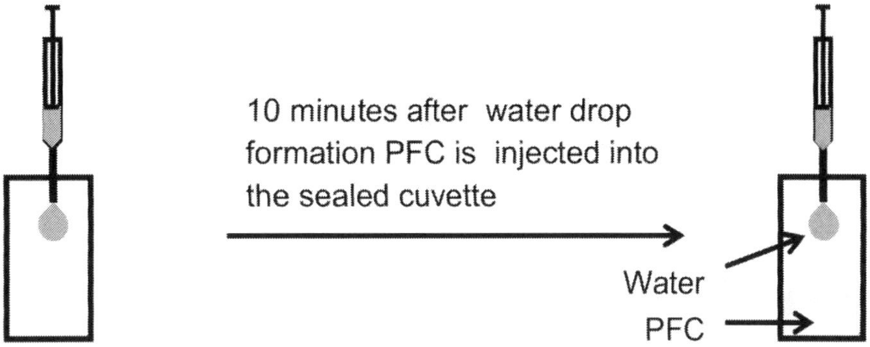

Fig. 1 Experimental set-up

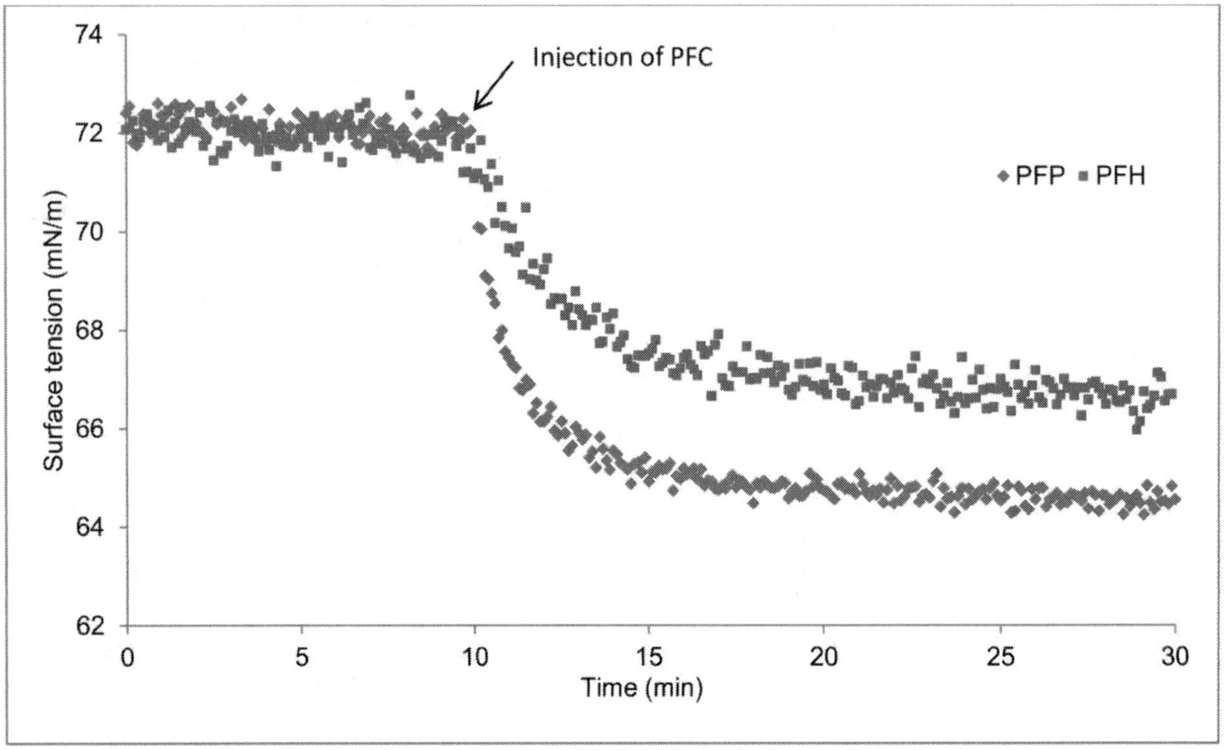

Fig. 2 Decrease in surface tension of water after the injection of PFP and PFH

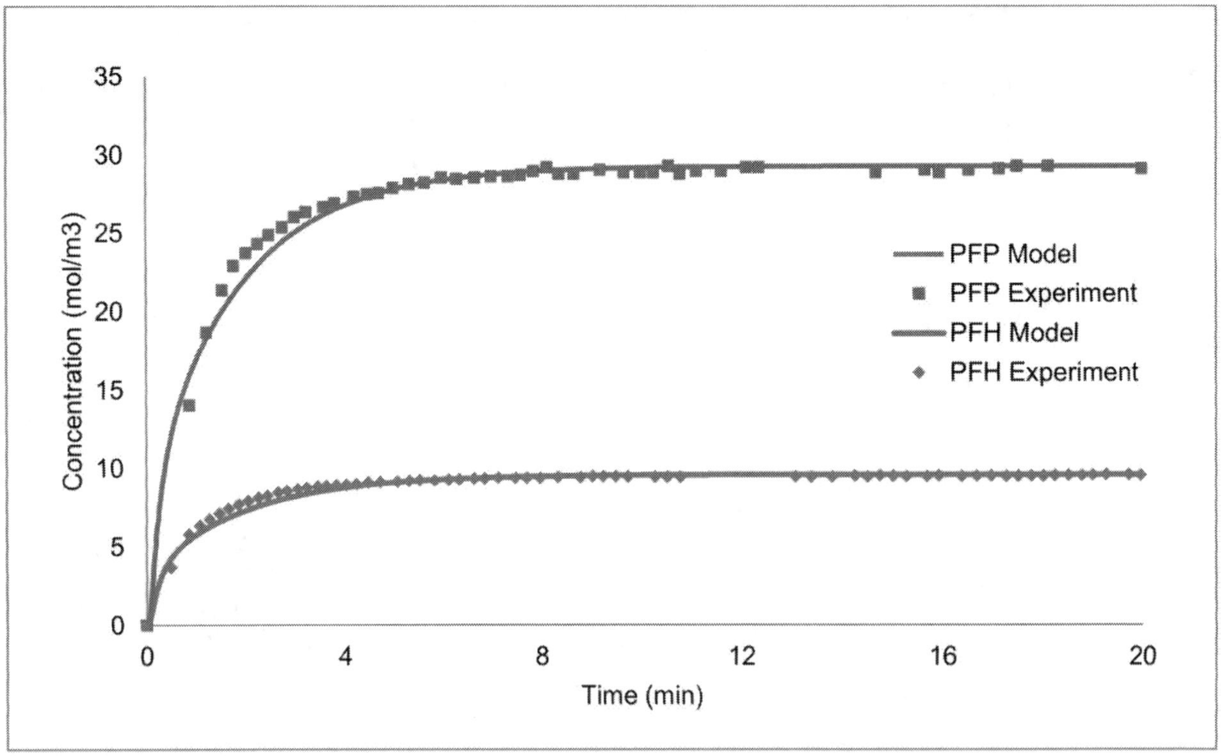

Fig. 3 Change of PFP and PFH vapor concentration after their injection into the cuvette

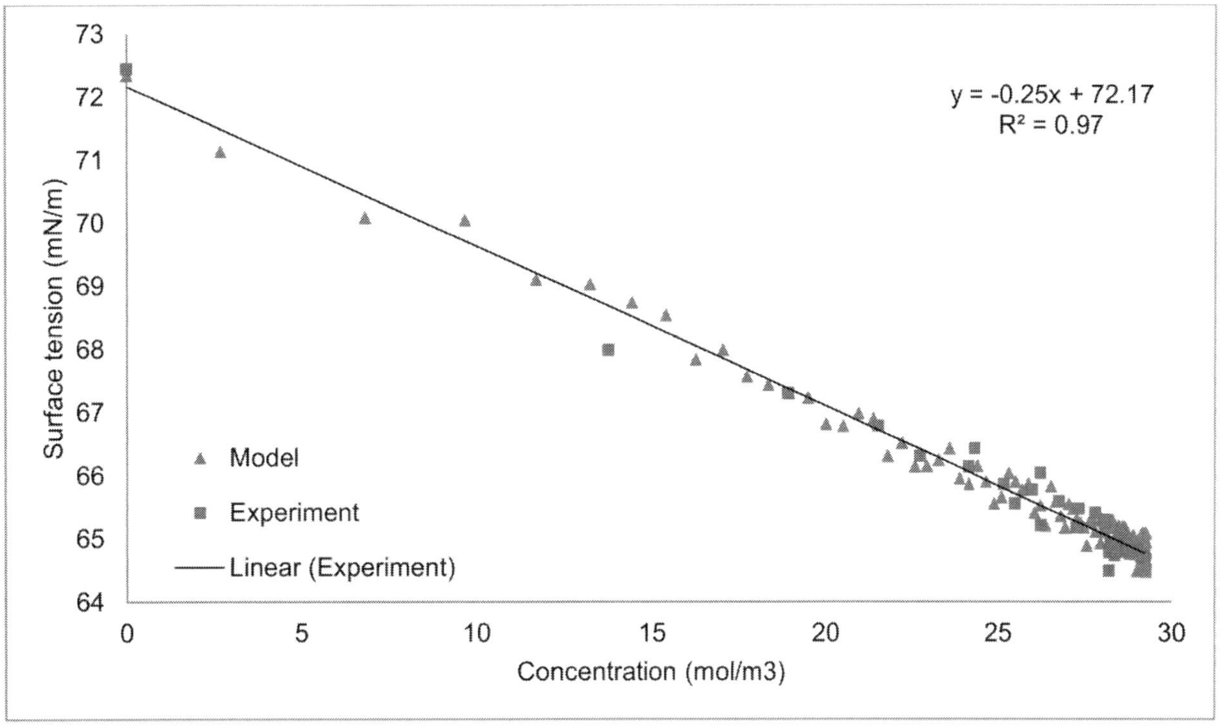

Fig. 4 Correlation between water surface tension and PFP concentration (FT-IR and model)

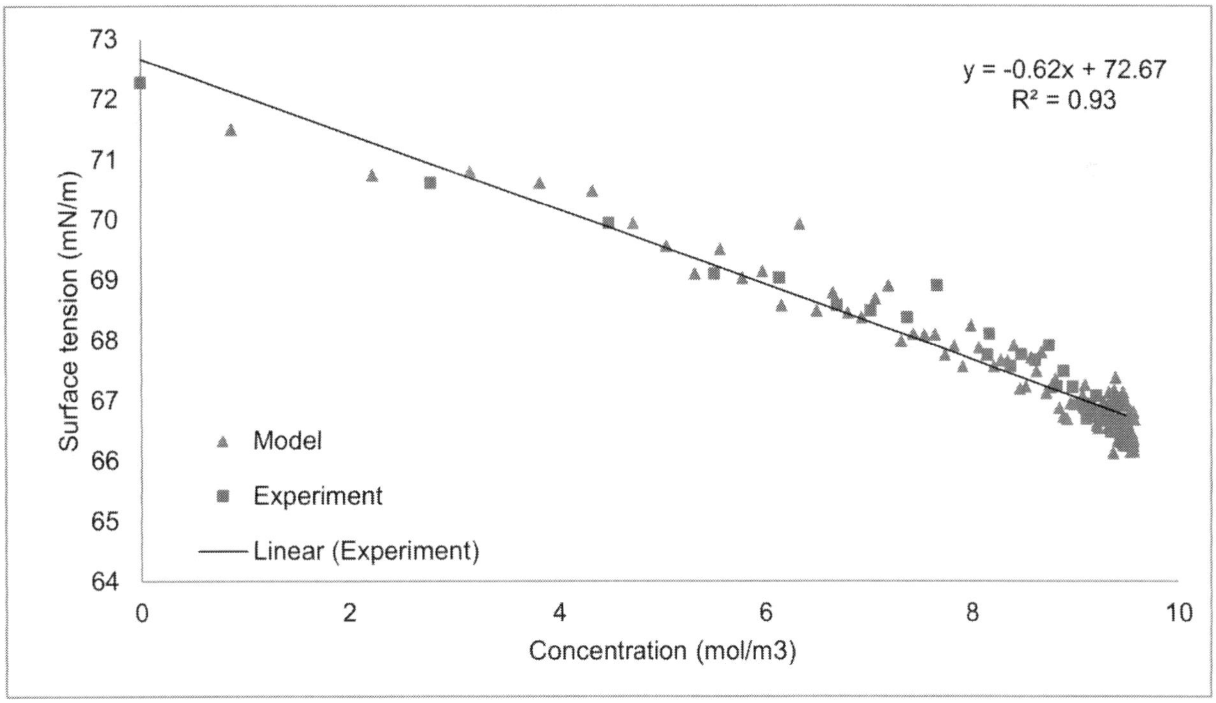

Fig. 5: Correlation between water surface tension and PFH concentration (FT-IR and model)

(274b) Modeling Interfaces with Step Potential for Equilibria and Dynamics (iSPEAD)

Ahmadreza F. Ghobadi, Chemical and Biomolecular Eng., The University of Akron, Akron, OH and J. Richard Elliott, Chemical and Biomolecular Engineering, The University of Akron, Akron, OH

Modeling Interfaces with Step Potential for Equilibria And Dynamics (iSPEAD)

Ahmadreza F. Ghobadi and J. Richard Elliott[*]

Department of Chemical and Biomolecular Engineering, The University of Akron,

Akron OH 44325

An effective combination of Discontinuous Molecular Dynamics (DMD) simulation and second order Thermodynamic Perturbation Theory (TPT) can be employed to simulate the entire phase diagram and also to offer a rigorous procedure for evaluating the perturbation contributions of any SAFT-like equation of state (EOS). In DMD/TPT methodology, the athermal reference fluid is fused hard sphere chain and is simulated by DMD at one single temperature and several densities. The size of each site type (σ) is adjusted to reproduce experimental liquid density. Then, after the simulation, the first and second order attractive contributions of Helmholtz free energy can be isolated and rigorously computed in accordance with the Barker–Henderson (BH) formalism. To do so, Elliott and co-workers optimized and evaluated a site-base transferable step potential with four wells for more than 500 compounds comprising 30 different families. Parameters of interaction potential (ε_i, $i = 1,4$) were adjusted to reproduce experimental saturated liquid density and vapor pressure. Following this procedure, Elliott and Gray {Journal of Chemical Physics, 123 (2005), p. 184902} proposed SAFT-like correlations to the results of simulation known as Step Potential for Equilibria And Dynamics (SPEAD). Ghobadi and Elliott {Fluid Phase Equilibria, v. 306 i.1 (2011) p. 57} recently improved the correlation for the critical point region.

The main difference between SPEAD and SAFT-family approaches is the nature of evaluating perturbation contributions. In SAFT formalism, a segment-base continuous interaction potential is considered and adjustable parameters are fitted to the experimental data while SPEAD benefits from site-base discontinuous interactions and raw simulation results. Although site-base approach offers transferability and physical consistency, a segment-base interaction potential becomes absolutely beneficial where Renormalization Group (RG) theory or Density Functional Theory (DFT) is applied on the classical Equation of State (EOS).

In this work, a segment-base Lennard-Jones (LJ) interaction potential is developed by re-evaluating the perturbation contributions of SPEAD. The Mean Field and local density approximations are implemented to calculate first and second order perturbation terms, respectively. The first order term is computed using an average radial distribution function at an effective packing fraction. The relation between density and effective packing fraction introduces two adjustable parameters for each component. These parameters along with the minimum of the LJ potential (ε) are fitted to the BH-generated perturbation contributions of SPEAD. The number of segments (m) is regressed to the results of DMD simulation of the reference fluid. Therefore, each component is characterized by 5 parameters: σ, m, ε, $c1$ and $c2$.

Using the segment-based potential described above and SPEAD+DFT EOS (named as iSPEAD), liquid-vapor density profile and surface tension of several pure components from n-alkanes, branched alkanes, alcohols, acids, ester and so forth were calculated. It should be noted that hydrogen bonding contribution is considered by Wertheim theory. The results show satisfactory agreement with available experimental data. Also, adsorption isotherms of various mixtures were studied by iSPEAD. It is demonstrated that adsorption at the liquid-vapor interface happens when there is a noticeable difference between parameters of constituent components of the mixtures. The final aim of this work is to investigate the potential of DFT to indicate the effect of different experimentally-examined surfactant molecules on the interfacial properties. When dealing with large molecules such as amphiphiles, individual segment densities contain valuable information about the microstructure of the fluid. In such cases, a three-fluid model will be used to obtain the density profiles for hydrophilic head, hydrophobic tail and tail-end segments. The three-fluid model isolates the influence of each part on interfacial properties and provides more in-depth information such as average end-to-end distance of surfactants, surface area per surfactant, aggregation number and so forth.

*Corresponding Author, please contact elliot1@uakron.edu, Office: (+1) 330 972 7253

Room Temperature Bubble Point Tests on Porous Screens: Implications for Cryogenic Liquid Acquisition Devices

Jason Hartwig, Research Aerospace Engineer, Zin Technologies, Brookpark, OH, 44142, USA
J. Adin Mann Jr., Professor of Chemical Engineering, Case Western Reserve University, Cleveland OH, 44106, USA

Gravity affects many fluidic processes, such as the positioning of the liquid and vapor phases within a propellant tank. In a standard 1-g environment the density of the individual fluid phases dictates the location of vapor and liquid phase; the heavier fluid (liquid) settles to the bottom while the lighter fluid (vapor) rises to the top. However, in the reduced gravity of a space environment, surface tension becomes the controlling mechanism for this liquid/vapor separation. When transferring cryogenic propellant from a storage tank to an engine or depot receiver tank, it is necessary to transfer only a homogeneous liquid phase. Multiple liquid acquisition devices (LAD) may be required to ensure liquid phase sufficiently covers the tank outlet in varying thermal and gravitational environments experienced during a space mission. A popular type of LAD is a screen-channel LAD. The channel side that faces the tank wall is covered by a fine mesh screen, which creates micron sized pores that act as a barrier to vapor ingestion. Liquid is wicked across the screen during tank outflow, preventing the pores from drying out when in contact with vapor. The total pressure loss across the screen and in the LAD channel must be less than the bubble point to prevent vapor ingestion into the channel. While certain screen-channel LAD meshes have proven flight heritage with storable propellants, finer meshes may be required to safely deliver low surface tension cryogenic propellants to an engine.

We present experimental results for room temperature bubble point tests conducted at the Cedar Creek Road Cryogenic Complex, Cell 7 (CCL-7) at the NASA Glenn Research Center. The purpose of these tests was to investigate the performance of three different fine mesh screens in room temperature liquids to provide pretest predictions in cryogenic liquid nitrogen (LN_2) and hydrogen (LH_2) as part of NASA's microgravity LAD technology development program. Bench type tests based on the maximum bubble point method were conducted for a 325x2300, 450x2750, and 510x3600 mesh sample in pure room temperature liquid methanol, acetone, isopropyl alcohol, water, and mixtures of methanol and water to cover the intermediate to upper surface tension range.

A theoretical model for the bubble point pressure is derived from the Young-LaPlace equation for the pressure drop across a curved interface. Governing equations are reduced in complexity through a set of simplifying assumptions to permit direct comparison with the experimental data. Screen pore sizes are estimated from scanning electron microscopy (SEM) to make pretest predictions. Pore sizes based on SEM analysis are compared with historical data available in the literature for the 325x2300 and 450x2750 screens as well with data obtained from bubble point tests conducted in this work.

Experimental results show that bubble point pressure is proportional to the surface tension of the liquid. We show that there is excellent agreement between data and model for pure fluids when the data is corrected for non-zero contact angle measured on the screens using a modified Sessile Drop technique. For mixtures, bubble point pressure scales with the methanol - water mole fraction composition for the range of 0% to 50% methanol. However, data obtained at methanol mole fractions greater than 50% show significant deviation from the predicted values. Deviation from the model is likely attributable to small differences in mixture compositions within the micron sized pores.

SEM image analysis of the three meshes indicated that bubble point pressure would be a maximum for the finest mesh screen. The pore diameters based on SEM analysis and experimental data obtained here are in excellent agreement for the 325x2300 and 450x2750 meshes, but not for the finest 510x3600 mesh. Therefore the simplified model can be used to interpolate predictions for low surface tension cryogenic liquids only when pore diameters are based on room temperature bubble point tests and not SEM analysis as presently implemented.

Predictions indicate that bubble point pressures in excess of 750 Pascals are achievable in LH_2 with the 450x2750 mesh, which represents a 25% margin over the baseline 325x2300 screen. Results here have implications for future LH_2 LADs technology development, as higher bubble points imply the LAD can deliver higher flow rates to an engine or depot receiver tank in micro-gravity.

(274d) Electrohydrostatics of Capillary Switches

Krishnaraj Sambath and Osman A. Basaran, School of Chemical Engineering, Purdue University, West Lafayette, IN

Two liquid droplets, one sitting on the top and the other hanging from the bottom of a plate, that are connected via a cylindrical hole of radius R that is filled with the same liquid as the drops is referred to as a capillary switch (CS). These coupled droplets could either maintain fixed contact lines or fixed contact angles with the plate. In situations where the contact line of the droplets are pinned, the CS is known to exhibit two stable equilibrium states when the combined volume of the two droplets is greater than that of a sphere of radius R. This fact is exploited in various applications, including optical lenses and adhesion, where the main challenge is to come up with ways to "toggle" the CS between its two stable states that are reliable, are energy efficient, and have fast response. The use of an electric field to achieve these goals is explored theoretically in this work in which the axisymmetric shapes and stability of a CS where the liquid is a perfect conductor and the fluid surrounding the two droplets is a passive dielectric gas are determined as a function of the applied field strength. The equilibrium shapes of the CS and the electric potential in the surrounding gas are governed by the augmented Young-Laplace equation and the Laplace equation, respectively, along with a volume constraint. These equations are solved numerically by using the Galerkin finite element method. Simulation results that give the variation of the dimensionless difference in volume between the two droplets as a function of the electrical Bond number (the ratio of electric to surface tension force) are used to infer whether an electric field represents an effective means of toggling a CS. The analysis is then extended to situations where the droplets make fixed contact angles with the plate and the results are compared to those obtained where the contact lines are pinned. When the contact angles are fixed, the electrohydrostatics of the CS is also amenable for study by domain perturbation analysis. Aside from providing additional insights into the behavior of electrified CSs, the domain perturbation results also provide a convenient framework for benchmarking the simulation results.

(274e) Interfacial Properties of Self-Assembled Polymer/Surfactant Mixed Nanoparticles

Yan Gao, Chemical and Petroleum Engineering, University of Kansas, Lawrence, KS, Jenn-Tai Liang, Chemical & Petroleum Engineering, University of Kansas, Lawrence, KS and **Prajnaparamita Dhar**, Chemical Engineering, University of Kansas, Lawrence, KS

Interfacial behavior of novel nanoparticle systems have been the focus of recent research efforts in colloid science due to their applications in consumer products, petroleum engineering and biotechnology. We will discuss the surface adsorption and surface rheological properties of self-assembled polyelectrolyte-surfactant aggregates (with radii 50-200 nm). Using the Wilhelmy plate method, we find that these nanoparticles are more surface active than the surfactant or the polyelectrolyte can lower the critical micelle concentration of the surfactant alone by a factor of 6. What is even more surprising is that these nanoparticles show a concentration dependent two-stage adsorption to the air/water interface. We propose that these nanoparticles undergo hydrophobic interface induced "disintegration", which is dominated by the concentration of nanoparticles, and the concentration of the entrapped surfactant at the interface. Further evidence of this interface-inducced release is provided by a sharp change in surface viscosity at times corresponding to the "disintegration" or release of entrapped surfactant. Three orders of magnitude increase in the sensitivity of our recently developed nanorod interfacial rheometer allows us to detect this change in the surface viscosity during the rearrangement of the nanoparticles at the interface.

(274f) Molecular Packing and Phase Behavior of Heterogeneous Lipid Monolayers in the Presence of n-Butanol

Yogi Kurniawan[1], Geoffrey D. Bothun[1], Keerthi P. Venkataramanan[2] and Carmen Scholz[3], (1)Department of Chemical Engineering, University of Rhode Island, Kingston, RI, (2)Biotechnology Science and Engineering Program, University of Alabama in Huntsville, Huntsville, AL, (3)Department of Chemistry, University of Alabama, Huntsville, Huntsville, AL

Interest in butanol as a biofuel is increasing due to its physicochemical properties (high energy density, low vapor pressure, hydrophobicity) and the ability to use it directly in gasoline engines. However, there are barriers to producing butanol cost-effectively at the industrial scale. Butanol production via fermentation is limited to low bacterial solvent tolerance due to cell membrane fluidization or destabilization. Bacteria respond to membrane destabilization by adjusting their membrane composition; typically modifying the ratio of saturated to unsaturated lipids (S/U). Nevertheless, mechanisms of lipid restructuring due to butanol partitioning are unclear.

In this research we are examining the effects of n-butanol on the molecular packing, phase behavior, and mechanical properties of monolayers composed of 1,2-dipalmitoyl-sn-glycero-3-phosphocholine (DPPC, a saturated lipid) and 1,2-dioleoyl-sn-glycero-3-phosphocholine (DOPC, an unsaturated lipid). Surface pressure-area per molecule isotherms were conducted at the air/water interface at 25 °C as a function of the DPPC:DOPC ratio (1:0, 3:1, 1:1, 1:3, and 0:1). DPPC undergoes a liquid expanded (LE) to liquid condensed (LC) phase transition upon compression, whereas DOPC remains in a LE phase.

In the absence of n-butanol, the presence of DOPC within the monolayer increased the surface pressure, relative to pure DPPC, and the position of the LE-LC transition of DPPC. DOPC also decreased the area compressibility modulus of DPPC/DOPC monolayers, denoting a more rigid structure. These results can be attributed lipid immiscibility that arises due to tail mismatch between saturated DPPC and unsaturated DOPC. When n-butanol was added to the sub-phase it partitioned into the monolayers and increased the surface pressure, the effective area per lipid molecule, and compressibility modulus. This effect was most pronounced when DPPC was the continuous phase. However, when DOPC was the continuous phase, n-butanol partitioning had little effect on the isotherm. Differential effects of n-butanol on DPPC and DOPC are attributed to lipid structure. DOPC can accommodate n-butanol within pockets created by tail kinking, while DPPC cannot.

(274g) Interfacial Dynamics of Oil Recovery Using Surfactant Floods At Nanoscale

I-Cheng Chen, Texas A&M, College Station, TX and **Mustafa Akbulut**, Chemical Engineering, Texas A&M University, College Station, TX

Using primary and secondary recovery, only 20-40% of oil can be extracted from an oil reservoir. The remaining 60-80% of oil stays underground where oil is adsorbed on rock surfaces, leading to oil entrapment within the rock pores. At a time of surging global energy demand, researchers are now actively investigating ways to squeeze extra barrels from oilfields via various enhanced oil recovery (EOR) techniques such as gas injection, chemical injection, microbial injection, or thermal recovery.

Surfactant flooding, which is one type of EOR, involves the addition of surfactants to water floods to reduce oil-water interfacial tension to ultra-low values so that mobility of oil droplets is increased. The technical feasibility of oil recovery via surfactant flooding were demonstrated in early *field tests* by various authors. Since then, numerous studies focusing on the bulk oil recovery properties of surfactant floods have been conducted: Majority of these studies have investigated the effects of phase behavior, wettability, interfacial tension, water mobility, and foaming performance on the behavior of oil recovery using surfactant floods. However, while there are a numerous studies on enhanced oil recovery via surfactant floods at the bulk scale, studies on the nanoscale dynamics of oil recovery via surfactant floods are very limited. In this work, we have investigated the kinetics and mechanism of heavy oil recovery using a model non-ionic surfactant solution, 4-nonylphenyl-polyethylene glycol at nanoscale. This study was conducted using four complementary techniques: quartz crystal microbalance with dissipation (QCM-D), atomic force microscopy (AFM), ellipsometry, and dynamics light scattering (DLS).

(279a) On the Chemistry of Deformation: Theory of Solute Strengthening in Aluminum and Magnesium Alloys

Louis Hector Jr., Chemical Sciences and Materials Systems, General Motors R&D, Warren, MI

Multiscale modeling techniques are currently being used to predict material properties that rely upon information that spans atomic to engineering length scales. Often absent in existing models is physics-based materials science that does not rely on data fitting. In this presentation, we will describe a quantitative, parameter-free theory of solute strengthening of metal alloys, with application to both aluminum and magnesium. An expression for the alloy yield strength as a function of temperature and strain rate is informed with solute-dislocation energetics computed with first principles density functional theory (DFT) and a lattice Green's function method. For Al alloys, strengthening due to Mg, Cr, Cu, and Si solutes is quantified. For Mg, focus will be on strengthening of basal flow stress in Mg-Al alloys. Key differences in strengthening due to the size of the equilibrium dislocation core structures will be emphasized. The model is readily extended to other solute chemistries and dislocation types and provides a basis for exploring the chemistry of deformation.

(279b) Semi-Empirical AM1, PM3 and PM6 Calculations On the Metakaolin Molecular Structure: Practical Application in Geopolymer Cement Production

Ojas Chaudhari[1], Joseph Biernacki[1] and Scott Northrup[2], (1)Chemical Engineering, Tennessee Technological University, Cookeville, TN, (2)Department of Chemistry, Tennessee Technological University, Cookeville, TN

Portland cement, the basic ingredient in concrete, is possibly one of the most common construction material used in the world. However, portland cement production results in the generation of green house gases from both energy consumption and chemical changes associated with the production process. In general, production of one ton of portland cement results in approximately one ton of the green house gas, CO_2, released into the atmosphere. In an effort to reduce our dependence on traditional portland cement-based concrete, ongoing research is concentrating on development of new low environmental impact alternative cements. Geopolymer cement concrete is one recently developed alternative. Geopolymer concrete uses different raw materials and different production processes than portland cement, which enable a five to six times reduction in green house gas emissions.

The process for geopolymer production includes mixing an anhydrous alumniosilicate material with alkaline solution such as sodium or potassium hydroxide. As such, large volumes of industrial by-products and waste such as fly ash or blast furnace slag can be used as potential sources of the alumniosilicate materials required in geopolymer cement production. However, metakaolin, a modified naturally occurring material, is a preferred raw material due to its high rate of dissolution in the alkaline solutions. The formation of geopolymer consists of three-stages: (1) dissolution; (2) reorientation; and (3) polycondensation. These, three stages take place almost simultaneously; therefore it is difficult or impossible at this time to conduct experimental studies on the isolated steps.

Understanding the dissolution of metakaolin is of great importance and represents a first step toward developing efficient geopolymeric cements. Dissolution of metakaolin in alkaline solution was investigated using quantum mechanical semiempirical methods. Calculations were performed using the AM1 (Austin model 1), PM3 (Parametric method 3) and PM6 (Parametric method 6) semiempirical methods on a metakaolin structure proposed by Yunsheng and Wei [1].

The assumed metakaolin structure consists of a SiO_4 and AlO_4 tetrahedron (single six-membered rings). The semi-empirical calculations were performed on six-membered ring clusters. Heat of reaction calculations were performed in both *local* (aqueous) and alkaline (NaOH or KOH) environments. Heats of formation were obtained for reactants and products using the mentioned semiempirical methods, and the difference between products and reactants was used to compute (estimate) the heat of reaction. The ion paring reactions between broken six-membered SiO_4 and AlO_4 rings, anion and alkaline cation, are considered to elucidate the dissolution processes. The recently developed PM6 semi-empirical method was successfully used to study dissolution of ring clusters in both environments.

Thus estimated heat of reactions in the local environment suggested that the AlO_4 ring is more easily dissolvable than the SiO_4 ring. Furthermore, the use of an alkaline environment is expected to accelerate the dissolution process with high exothermic energy values. Finally, SiO_4 and AlO_4 ring clusters demonstrate contraction after dissolution in the highly alkaline solution, which indicates that defect structure of metakaolin will contract during dissolution. Such observations can be used to understand the dissolution process of metakaolin and are expected to forward the development of geopolymeric cements made from waste and by-product materials in the near future.

[1]Yunsheng, Z. and S. Wei, "Semi-empirical AM1 calculations on 6- membered alumino-silicate rings model: implications for dissolution process of metakaolin in alkaline solutions " Journal Of Material Science, 42, 2007, pp 3015-3023.

(279c) Theoretical and Experimental Studies of Water Absorption Into Ionic Liquids

Wei Shi, National Energy Technology Laboratory, U.S. Department of Energy, Pittsburgh, PA, David Luebke, US DOE/NETL, Pittsburgh, PA, Hunaid Nulwala, NETL, Pittsburgh and Krishnan Damodaran, Deaprtemt of Chemistry, University of Pittsburgh, Pittsburgh

Thermodynamic and transport properties of water absorption in ionic liquids (ILs) are studied from both classical and ab initio simulation methods and experiments. Ab initio gas phase calculations show that anions are categorized into three groups in terms of interaction with water, i.e., the weak (< 45 kJ/mol) , intermediate (50-65 kJ/mol) and strong types (70-85 kJ/mol), which correspond to hydrophobic ILs and ILs with high and very high water solubilities, respectively. Replacing the -CH3 (electron dinating) with the -CF3 (elelectron withdrawing) group in the anion colud decrease water-anion interactions about 11-17 kJ/mol for two different types of anions. The electrostatic interactions between water and ions are significantly stronger than van der Waals interaction. Cation effects on water solubility will also be discussed. Both simulations and experimental data show that the dynamics for the IL studied is significantly improved at high water concentrations. Very imteresting, at about 40-50% water concentrations, the stuided IL and water exhibit a local minimum of self-diffusivity both from simulations and experiments. The self-diffusivity for H2, N2, O2, CO2, CH4, and H2O in two different ILs will also be compared.

(279d) Understanding Backbiting and Beta-Scission Reactions in Self-Initiated Polymerization of Methyl Acrylate: A Theoretical Study

Shi Liu[1], Sriraj Srinivasan[2], Michael C. Grady[3], Masoud Soroush[4] and Andrew M. Rappe[1], (1)Makineni Theoretical Laboratories, Department of Chemistry, University of Pennsylvania, Philadelphia, PA, (2)Arkema, Philadelphia, PA, (3)DuPont Experimental Station, Wilmington, DE, (4)Department of Chemical and Biological Engineering, Drexel University, Philadelphia, PA

The polymer industry uses free-radical polymerization of acrylates widely to manufacture variety of paints, coatings and adhesives[1,2]. High-temperature (> 100°C) polymerization of alkyl acrylates has been extensively studied and is known to allow for generating low molecular weight, low solvent resins[3,4]. Recent studies have shown that secondary reactions such as self-initiation[4], backbiting and beta-scission occur in high-temperature polymerization of alkyl acrylates[5]. A better quantitative understanding of these reactions can facilitate the rational design of chemically self-regulated polymerization processes.

Quantum chemical calculations can be applied to identify the rate constants, intermediates and mechanisms of individual reactions such as backbiting and beta-scission without a need for a macroscopic model. Both wave function-based quantum chemical methods and density-functional theory (DFT) have been successfully applied to study the mechanisms and kinetics of different types of reactions in free-radical polymerization of ethylene[6-8], styrene[9], methyl methacrylate[10,11], and methyl, ethyl and n-butyl acrylate[12,13].

In this work, we apply density functional theory (DFT) methods to investigate the mechanisms of backbiting and beta-scission reactions in thermal polymerization of methyl acrylate. This monomer was chosen to make the large size of the polymer system computationally manageable. We choose three different functionals, B3LYP, M06-2X and PBE0 with the 6-31G* basis set. Backbiting reactions from propagating radicals that have been initiated by two types of radicals (1) monomeric mono-radical (MMR) and (2) dimeric mono-radical (DMR) are investigated. Four backbiting mechanisms and two types of beta-scission reactions are explored. We have found that the 1:5 backbiting mechanism with a six-membered ring transition state and 1:7 backbiting with an eight-membered ring transition state are kinetically more favorable than 1:3 backbiting and 1:9 backbiting. Reactions involving longer polymer chains tend to have higher energy barriers to undergo intramolecular hydrogen transfer. The beta-scission reaction of a mid-chain radical (MCR) is found to be symmetric as the radical can form double bond with its closest left or right secondary carbon atom with nearly equal activation energy. The nuclear magnetic resonance (NMR) chemical shifts (^{1}H and ^{13}C) of species generated from these mechanisms are computed and compared. The chemical shifts of carbon nucleus in non-branched polymer chains and branched polymer chains have been estimated for poly(methyl acrylate).

References

[1] Adamsons, K.; Blackman, G.; Gregorovich, B.; Lin, L.; Matheson, R. *Prog. Org. Coat.* **1998**, *34*, 64.

[2] Sorensen, P. A.; Kiil, S.; Dam-Johansen, K.; Weinell, C. E. *J. Coat. Technol. Res.* **2009**, *6*, 135.

[3] Grady, M. C.; Simonsick, W. J.; Hutchinson, R. A. *Macromol. Symp.* **2002**, *182*, 149.

[4] Quan, C. L.; Soroush, M.; Grady, M. C.; Hansen, J. E.; Simonsick, W. J. *Macromolecules* **2005**, *38*, 7619.

[5] Zorn, A. M.; Junkers, T.; Barner-Kowollik, C. *Macromol. Rapid Commun.* **2009**, *30*, 2028.

[6] Heuts, J. P. A.; Gilbert, R. G.; Radom, L. *Macromolecules* **1995**, *28*, 8771.

[7] Wong, M. W.; Radom, L. *J. Phys. Chem.* **1995**, *99*, 8582.

[8] Wong, M. W.; Radom, L. *J. Phys. Chem. A* ***1998***, *102, 2237.*

[9] Khuong, K. S.; Jones, W. H.; Pryor, W. A.; Houk, K. N. *J. Am. Chem. Soc.* **2005**, *127*, 1265.

[10] Srinivasan, S.; Lee, M. W.; Grady, M. C.; Soroush, M.; Rappe, A. M. *J. Phys. Chem. A* ***2011***, *115, 1125.*

[11] Zhang, C. L.; Wang, X. Y.; Liu, L. M.; Wang, Y. L.; Peng, X. Y. *J. Mol. Model.* **2008**, *14*, 1053.

[12] Srinivasan, S.; Lee, M. W.; Grady, M. C.; Soroush, M.; Rappe, A. M. *J. Phys. Chem. A* ***2009***, *113, 10787.*

[13] Srinivasan, S.; Lee, M. W.; Grady, M. C.; Soroush, M.; Rappe, A. M. *J. Phys. Chem. A* ***2010***, *114, 7975.*

(279e) A Theoretical Study of Mechanisms for Chain Transfer to Monomer Reactions in Alkyl Acrylates

Nazanin Moghadam[1], Masoud Soroush[1], Andrew M. Rappe[2], Shi Liu[2], Sriraj Srinivasan[3] and Michael C. Grady[4], (1)Department of Chemical and Biological Engineering, Drexel University, Philadelphia, PA, (2)Makineni Theoretical Laboratories, Department of Chemistry, University of Pennsylvania, Philadelphia, PA, (3)Arkema, King of Prussia, PA, (4)DuPont Experimental Station, Wilmington, DE

Alkyl acrylates are widely used as primary binders in coatings formulations for the automobile industry [1-3]. The basic nature of acrylic resins and the plants producing the resins have changed considerably over the past decades, as a result of environmental limits on allowable volatile organic contents (VOCs) of resins [4-6]. High temperature (>100°C) polymerization of alkyl acrylates allows for the production of high-solids low-molecular-weight resins, but it also permits secondary reactions to occur at higher rates, including spontaneous initiation (in the absence of known added initiators), chain transfer to monomer (CTM) and polymer, back-biting, and β-scission [1, 2, 7-11]. The polydispersity index, which is a measure of the breadth of the polymer chain length distribution, was found to be 1.5 to 2.2 in high-temperature homo-polymerization of alkyl acrylates [10, 12]. This indicates that various types of chain transfer reactions occur abundantly at high temperatures.

Propagation rate constants of alkyl acrylates and methacrylates were determined using different density functional theory (DFT)-based hybrid functionals [13]. In our previous work, we used B3LYP/6-31G* to study several mechanisms of CTM reaction in methyl acrylate (MA) [14]. Hydrogen abstraction from methyl substituent group in the monomer by the live chain radical was identified as the most probable mechanism for CTM in MA. Before this work, there was no computational/theoretical comparison of CTM mechanisms in high-temperature free-radical polymerization of alkyl acrylates.

This paper presents newer results from a detailed computational investigation of four possible mechanisms of CTM in self-initiated homo-polymerization of methyl, ethyl and *n*-butyl acrylate. We have used B3LYP, X3LYP and M06-2X, and WB97X-D functionals, and 6-31G(*d*), 6-31G(*d,p*), 6-311G(*d*), and 6-311G(*d,p*) basis sets. Energy barriers and rate constants of the reactions involved in the four mechanisms have been estimated. The effects of live polymer chain length, the type of mono-radical that initiated the live polymer chain, and the type of live polymer chain radical (tertiary *vs.* secondary) on the kinetics of the reactions have been studied. The rigid rotor harmonic oscillator (RRHO) approximation has been applied to investigate the thermodynamics and kinetics of the reactions. Hydrogen abstraction from methylene substituent group in the monomer by the live chain radical has been found to be the most probable mechanism for CTM in ethyl acrylate and *n*-butyl acrylate. EA and *n*-BA live chains, initiated by two different types of monoradicals obtained via self initiation, have showed nearly similar rate constants of hydrogen abstraction from monomer. We have determined that increasing the chain length of the live polymer has negligible effect on the barriers and rate constants of chain transfer to monomer. The transition state geometries and activation barriers have been validated using different levels of theory.

References:

[1] Grady, M. C.; Simonsick, W. J.; Hutchinson, R. A.; Studies of Higher Temperature Polymerization of *n*-Butyl Methacrylate and *n*-Butyl Acrylate, *Macromol. Symp.*2002, 182, 149-168.

[2] Rantow, F. S.; Soroush, M.; Grady, M.; Kalfas, G.; Spontaneous Polymerization and Chain Microstructure Evolution in High Temperature Solution Polymerization of *n*-Butyl Acrylate, *Polymer*2006, 47, 1423-1435.

[3] Barth, J.; Buback, M.; Russell, G. T.; Smolne, S.; Chain-Length-Dependent Termination in Radical Polymerization of Acrylates, *Molecular. Chem. Phys.*2011, 212, 1366-1378.

[4] Superintendent of Documents, Title 1, US Government Printing Office, Washington, DC, 1990, P. 1.

[5] Superintendent of Documents, Clean Air Act Amendments of 1990, Title 111, US Government Printing Office, Washington, DC, 1990, P. 236.

[6] R. S. Reisch; Paints & Coatings, *Chem. Eng. News,* 1993, 71 (42), 34.

[7] Srinivasan, S.; Lee, M. W.; Grady, M. C.; Soroush, M.; Rappe, A. M.; Self-Initiation Mechanism in Spontaneous Thermal Polymerization of Ethyl and *n*-Butyl Acrylate: A Theoretical Study, *J. Phys. Chem. A*2010, 114, 7975-7983.

[8] Srinivasan, S.; Lee, M. W.; Grady, M. C.; Soroush, M.; Rappe, A. M.; Computational Study of the Self-Initiation Mechanism in Thermal Polymerization of Methyl Acrylate, *J. Phys. Chem. A*2009, 113, 10787-10794.

[9] Nikitin, A. N.; Hutchinson, R. A.; Wang, W.; Kalfas, G. A.; Richards, J. R.; Bruni, C.; Effect of Intramolecular Transfer to Polymer on Stationary Free-Radical Polymerization of Alkyl Acrylates, 5 – Consideration of Solution Polymerization up to High Temperatures, *Macromol. React. Eng.*2010, 4, 691-706.

[10] Quan, C.; Soroush, M.; Grady, M. C.; Hansen, J. E.; Simonsick, W. J.; High-Temperature Homopolymerization of Ethyl Acrylate and *n*-Butyl Acrylate: Polymer Characterization, *Macromolecules* 2005, 38, 7619-7628.

[11] Peck, A. N. F.; Hutchinson, R. A.; Secondary Reactions in the High Temperature Free Radical Polymerization of Butyl Acrylate, *Macromolecules*2004, 37, 5944-5951.

[12] Srinivasan, S.; Kalfas, G.; Petkovska, V. I.; Bruni, C.; Grady, M. C.; Soroush, M.; Experimental Study of Spontaneous Thermal Homopolymerization of Methyl and *n*-Butyl Acrylate, *Journal of Applied Polymer Science*2010, 118, 1898-1909.

[13] Yu, X.; Pfaendtner, J.; Broadbelt, L. J.; Ab Initio Study of Acrylate Polymerization Reactions: Methyl Methacrylate and Methyl Acrylate Propagation, *J. Phys. Chem. A*2008, 112, 6772-6782. [14] Moghadam, N.; Soroush, M.; Srinivasan, S.; Rappe, A. M.; Grady, M. C.; Computational Study of Chain Transfer to Monomer Reactions in Thermal Polymerization of Methyl Acrylate, Paper 749c, AIChE Annual Meeting 2011.

[14] Moghadam, N.; Soroush, M.; Srinivasan, S.; Rappe, A. M.; Grady, M. C.; Computational Study of Chain Transfer to Monomer Reactions in Thermal Polymerization of Methyl Acrylate, Paper 749c, AIChE Annual Meeting 2011.

(279f) From Quantum Chemistry to Diesel Injector Deposits: Revisiting Liquid Phase Oxidation

Amrit Jalan[1], Jorge Aguilera-Iparraguirre[1], Ionut M. Alecu[1], Joshua W. Allen[1], Caleb Class[1], Connie W. Gao[1], Gregory R. Magoon[1], Shamel S. Merchant[1], Richard H. West[2] and William H. Green Jr.[1], (1)Department of Chemical Engineering, Massachusetts Institute of Technology, Cambridge, MA, (2)Department of Chemical Engineering, Northeastern University, Boston, MA

Liquid phase oxidation of hydrocarbons has been the subject of experimental and theoretical investigations for over 75 years owing to its applications in many areas of practical importance.[1] Although the chemical identity of primary and secondary products of liquid phase oxidation are well established, elementary reactions leading to their formation and consumption are known for only a few. Understandably, kinetic modeling of these processes is challenging and requires tracking thousands of intermediates and reactions. The commonly accepted paradigm in the literature has been to build models manually, often lumping isomers and skipping reaction intermediates to keep them tractable. While these methods may help reproduce experimental observations, their predictive capability is unclear. Moreover, with this approach, mechanistic insight is not easily transferable from one system to another.

The experiments performed by Korcek and co-workers[2-6] over 30 years ago remain the most detailed analysis of primary and secondary oxidation products resulting from the liquid phase oxidation of long chain alkanes. Through the use of complex analytical techniques, Korcek *et al.* quantified the yields of primary products like mono-, keto- and di-hydroperoxides. Through thermal decomposition experiments they also hypothesized that ketohydroperoxides were precursors to the formation of carboxylic acids and methyl ketones which are the main secondary products. No detailed kinetic models have been reported to quantitatively explain these experiments. Furthermore, the new pathways for carboxylic acids and methyl ketones that were hypothesized have remained largely ignored in the literature.

In this work, we combine computational chemistry and automated kinetic modeling tools to model the chemistry of liquid phase oxidation. We provide theoretical evidence for pathways leading from the ketohydroperoxide to carboxylic acids and carbonyl compounds via cyclic peroxide intermediates, confirming Korcek's hypothesis. Next, we incorporate this and other relevant pathways in a detailed kinetic model built automatically using the Reaction Mechanism Generator (RMG) software[7] to explain product yields observed by Korcek *et al.* The RMG approach significantly improves the efficiency of kinetic modeling for such large molecules but may require user-guidance in certain cases. Recent efforts at incorporating solvent effects into RMG will also be briefly discussed.[8] Finally, we demonstrate the application of these chemistry and kinetic modeling tools to understand the origin of oxidative deposits formed in diesel injectors using a multi-scale approach incorporating kinetics, transport and phase separation.

References

(1) Denisov, E. T.; Afanas'ev, I. B. *Oxidation and Antioxidants in Organic Chemistry and Biology*; CRC Press, 2005.

(2) Hamilton, E. J.; Korcek, S.; Mahoney, L. R.; Zinbo, M. *Int J Chem Kinet* **1980**, *12*, 577.

(3) Jensen, R. K.; Korcek, S.; Mahoney, L. R.; Zinbo, M. *J Am Chem Soc* **1979**, *101*, 7574.

(4) Jensen, R. K.; Korcek, S.; Mahoney, L. R.; Zinbo, M. *J Am Chem Soc* **1981**, *103*, 1742.

(5) Jensen, R. K.; Korcek, S.; Zinbo, M.; Johnson, M. D. *Int J Chem Kinet* **1990**, *22*, 1095.

(6) Jensen, R. K.; Zinbo, M.; Korcek, S. *Journal of Chromatographic Science* **1983**, *21*, 394.

(7) William H. Green, Joshua W. Allen, Robert W. Ashcraft, Gregory J. Beran, Caleb A. Class, Connie Gao, C.

Franklin Goldsmith, Michael R. Harper, Amrit Jalan, Gregory R. Magoon, David M. Matheu, Shamel S. Merchant, Jeffrey D. Mo, Sarah Petway, Sumathy Raman, Sandeep Sharma, Jing Song, Kevin M. Van Geem, John Wen, Richard H. West, Andrew Wong, Hsi-Wu Wong, Paul E. Yelvington, Joanna Yu; "RMG - Reaction Mechanism Generator v3.3", 2011, http://rmg.sourceforge.net/

(8) Jalan, A.; West, R. H.; Green, W. H. In *AIChE Annual Meeting*Minneapolis, MN, USA, 2011.

(279g) Pathways of Liquefied Petroleum Gas Pyrolysis to Acetylene in Hydrogen Plasma: A Density Functional Theory Study

Xiaoyuan Huang, Dang-guo Cheng, Fengqiu Chen and Xiaoli Zhan, Department of Chemical and Biological Engineering, Zhejiang University, Hangzhou, China

Liquefied petroleum gas (LPG) is an important by-product from petroleum refining and can be converted to acetylene by plasma pyrolysis. In this work, we employed a density functional theory (DFT) to investigate the reaction pathways of propane and *n*-butane, the main components of LPG, for understanding the LPG pyrolysis in hydrogen plasma. Over sixty possible reactions were considered. The reaction enthalpies and activation energies of these reactions were calculated with a Gaussian method of B3LYP and basic set of 6-31G (d,p), and then analyzed. The results indicate that the main products of LPG pyrolysis are acetylene, ethylene, methane, ethane and hydrogen. Ethylene mainly comes from the decomposition of n-C_3H_7, n-C_4H_9 and C_2H_5 radicals. High temperature facilitates the formation of acetylene from propylene and ethylene. Methane and ethane are produced during the quenching process through the combination of methyl radicals with H radicals and themselves. There are many hydrogen abstraction reactions, and they are responsible for the production of hydrogen gas. Active H radicals in hydrogen plasma were found to play a very important role in reactions, because they can remarkably lower the energies needed for reactions. The main routes of the pyrolysis of propane and n-butane in plasma can be summarized as following:

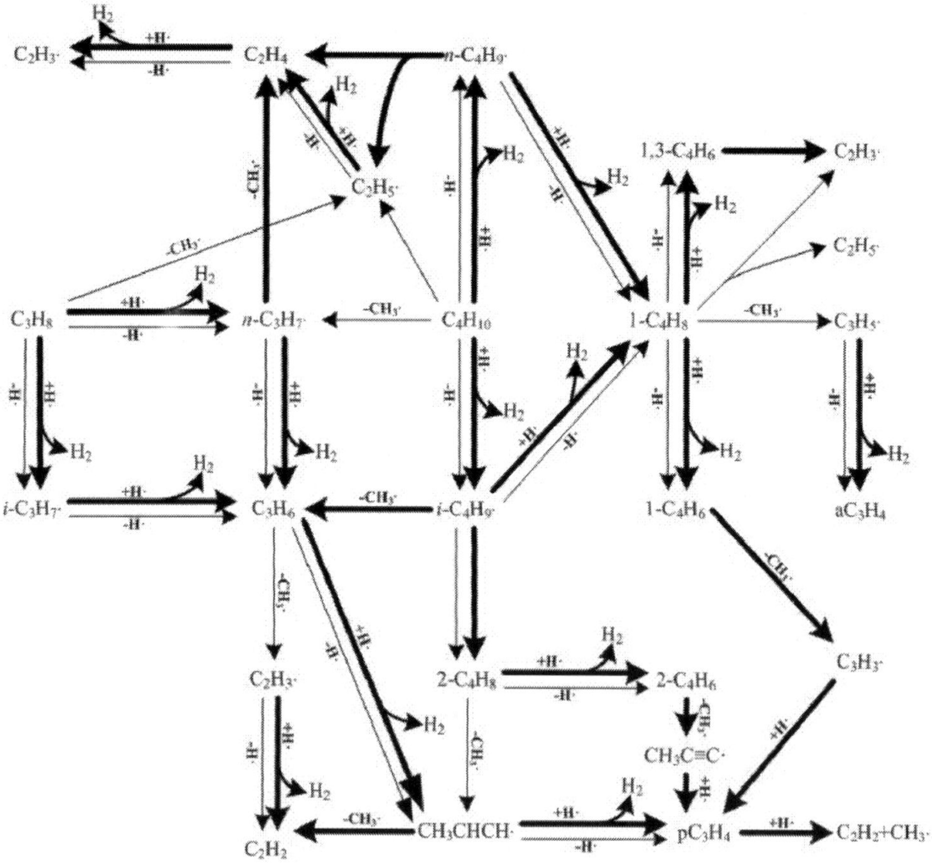

AUTHOR INDEX

AUTHOR INDEX

AUTHOR INDEX

AUTHOR INDEX

AUTHOR INDEX

AUTHOR INDEX

AUTHOR INDEX

AUTHOR INDEX

AUTHOR INDEX

AUTHOR INDEX

AUTHOR INDEX

AUTHOR INDEX

AUTHOR INDEX

AUTHOR INDEX

AUTHOR INDEX

AUTHOR INDEX

AUTHOR INDEX

AUTHOR INDEX

AUTHOR INDEX

Engineering Sciences and Fundamentals

Core Programming Topic at the 2012 AIChE Annual Meeting

Pittsburgh, Pennsylvania, USA
28 October - 2 November 2012

Volume 2 of 2

ISBN: 978-1-62276-726-7

Printed from e-media with permission by:

Curran Associates, Inc.
57 Morehouse Lane
Red Hook, NY 12571

Some format issues inherent in the e-media version may also appear in this print version.

Printed by Curran Associates, Inc. (2013)

For permission requests, please contact AIChE
at the address below.

AIChE
3 Park Avenue
New York, NY 10016-5991

Phone: (203) 702-7660
Fax: (203) 775-5177

www.aiche.org

Additional copies of this publication are available from:

Curran Associates, Inc.
57 Morehouse Lane
Red Hook, NY 12571 USA
Phone: 845-758-0400
Fax: 845-758-2634
Email: curran@proceedings.com
Web: www.proceedings.com

TABLE OF CONTENTS

VOLUME 1

VOLUME 2

(299a) Toward Tailored Elasticity of Particulate Soft Matter by Control of Structural Rigidity

Lilian C. Hsiao, Department of Chemical Engineering, University of Michigan, Ann Arbor, MI, Richmond Newman, University of Michigan, Ann Arbor, MI, Sharon C. Glotzer, Chemical Engineering , University of Michigan, Ann Arbor, MI and **Michael J. Solomon**, Chemical Engineering, University of Michigan, Ann Arbor, Ann Arbor, MI

The mechanism of yielding in colloidal gels, particularly for high strain-rate deformations that are typical of practical applications, is not fully understood. Recently, we have explored the concepts of isostaticity and structural rigidity as a guiding principle that can be used to understand and design the nonlinear elasticity in colloidal gels undergoing deformation. In our studies, we impose simple shear flows of variable strain on colloidal particulate gels formed through depletion interactions. We used in situ confocal laser scanning microscopy to characterize the resulting 3D microstructural changes. Contrary to methods based on ensemble-averaged parameters, we found that bond breakage in gels occurs mainly due to the erosion of rigid clusters that persist long after the gel has yielded. We develop a simple correlation between the elastic modulus and the stress-bearing volume fraction that is valid over a range of volume fractions and gelation conditions. We discuss implications of this successful correlation based on structural rigidity, including the effect of particle shape and roughness, as well as the indirect role of pair interactions on the non-linear elasticity of gelled particulate systems.

(299b) Aging in Colloidal Glasses: The Kovacs Signatures in Concentration Jump Conditions

Xiaojun Di, Xiaoguang Peng and Gregory B. McKenna, Chemical Engineering, Texas Tech University, Lubbock, TX

The analogy between colloidal dynamics and the dynamics of molecular glasses remains an important area of study. Of particular interest to our team is the aging responses of the two systems. We have been investigating the dynamics of colloidal systems composed of thermosensitive particles that change diameter upon change of temperature and comparing the behavior to what is expected in molecular glass-formers. In particular, we have found that concentration jumps in these systems mimic three important behaviors of molecular glasses that were catalogued by Kovacs[1] in the 1960s: the intrinsic isotherm (down-jump in temperature or increase in concentration), the asymmetry of approach (comparison of down-jump and up-jump experiments to demonstrate non-linearity of response), and memory (two-step temperature (concentration) history in which the second step shows non-monotonic response). In our early work [2] we were able to show, using multispeckle diffusing wave spectroscopy, that although the three signatures are observed in the concentration jump conditions, they are not identical to the observations in molecular glasses. In the present work we expand our investigations to the rheological domain and following the aging behavior of the dynamic moduli. Such experiments were developed by Struik[3] for molecular glasses and he found that the viscoelastic response was well described by free volume ideas, which are directly translatable to ideas of the fictive temperature. The concept of free volume is not applicable to experiments in colloids in which the response evolves at constant concentration. Hence, we are examining the idea of a fictive concentration (derived from the fictive temperature idea) and its applicability to complex concentration jump histories. The systems to are modifications of PNIPAAM as a corona grafted onto polystyrene cores. A series of different particle PNIPAAM fractions (different thermal sensitivity) is being investigated. Current results suggest that we see the aging suggested by the Kovacs signatures and quantitative comparisons between molecular and colloidal glasses will be made.

[1] A.J. Kovacs, Fortschritte der Hochpolymeren-Forschung, 3, 394(1963).
[2] X.J. Di, K.Z. Win, G.B. McKenna, T. Narita, F. Lequeux, S.R. Pullela, Z. Cheng, Phys. Rev. Lett., 106, 095701(2011).
[3] L.C.E. Struik, Physical Aging in Amorphous Polymers and Other Materials, Elsevier, Amsterdam, 1978.

This work is supported by NSF CBET-1133279

(299c) Multi-Scale Kinetics of a Field-Directed Colloidal Phase Transition

Eric M. Furst and James W. Swan, Department of Chemical and Biomolecular Engineering, University of Delaware, Newark, DE

Suspensions of polarizable colloids are expected to form crystalline equilibrium phases when exposed to a steady, uniform field. However, when colloids become localized this field-induced phase transition arrests and the suspension persists indefinitely as a kinetically trapped, percolated structure. The accomanying fluid-solid transition makes magneto-rheological (MR) fluids useful as shock absorbers, while electro-rheological (ER) fluids can be used as haptic controllers and tactile displays in micro-electronics devices. However, kinetic arrest ultimately prevents the formation of equilibrium phases encoded by the colloidal interactions or particle shape. In this work, we demonstrate that, by toggling the field strength at varied frequencies, such gels formed from MR fluids can be annealed. This allows the arrested structure to relax periodically to equilibrium. There is a stark boundary as a function of magnetic field strength and toggle frequency distinguishing percolation from phase separation. These results demonstrate how kinetic barriers to a colloidal phase transition are subverted through measured, periodic variation of driving forces. Such directed assembly may be harnessed to create novel materials from dispersions of colloids and nanoparticles.

(299d) Drift and Diffusion of Rigid and Flexible Fibers in Simple Shear Flow

Jianghui Wang[1], Emilio Tozzi[2], Michael D. Graham[3] and Daniel Klingenberg[1], (1)Chemical and Biological Engineering, University of Wisconsin Madison, Madison, WI, (2)Chemical Engineering and Materials Science, University of Calfiornia Davis, Davis, CA, (3)Department of Chemical and Biological Engineering, University of Wisconsin-Madison, Madison, WI

Economic biomass processing can be challenging because of difficulties associated with mixing and transporting fibrous biomass slurries. The dynamics of fibers in biomass slurries are studied by computer simulation in order to gain insight into the factors that govern their transport behavior. A new phenomenon was found for the motion of isolated, curved, rigid, neutrally-buoyant, non-Brownian, nonchiral fibers in simple shear flow of an incompressible Newtonian fluid at low Reynolds number. For some initial orientations, fibers with small curvature drift steadily in the gradient direction without external forces or torques. The average drift velocity and direction depend on the fiber aspect ratio, curvature and initial orientation. The drift results from the coupling of rotational and translational dynamics, and the combined effects of flipping, scooping, and spinning motions of the fiber. In suspensions of flexible fibers, the shear-induced diffusion caused by collisions between fibers depends on equilibrium shape, fiber aspect ratio, static friction between contacts, and concentration. While collisions between fibers increase the diffusivity, the formation of fiber networks decreases the mobility of individual fibers and decreases the diffusivity.

(299e) Shear Alignment of Fibers in Oscillatory Flows

Braden Snook and **Jason Butler**, Chemical Engineering, University of Florida, Gainesville, FL

Rigid fibers suspended in a viscous, Newtonian fluid can be aligned perpendicular to the flow-gradient plane by applying an oscillatory shear flow. Direct comparisons with published experiments demonstrate that a simple model, which considers only excluded volume and self-mobilities, can accurately predict the orientation distributions. Stresses calculated from the numerical simulations are also reported and compared to the experimentally measured rheology.

These simulations reveal that the alignment occurs over a narrow range of strain amplitudes and concentrations. Furthermore the fibers must be highly confined between the bounding walls. For example simulations of fibers of aspect ratio 11 sheared between planar walls separated by a distance of 1.5 times the fiber length align strongly in the vorticity direction at a volume fraction of 15% and strain amplitude of 3, whereas simulations with fully periodic boundary conditions under identical conditions indicate a preference to align with the direction of flow. For a volume fraction of 15%, the rods exhibit no change in orientation distribution at strain amplitudes less than one and the distribution is similar to that of steady shearing flows, with alignment in the flow direction, for strain amplitudes in excess of five. At a strain amplitude of 3, the extent of alignment of the orientations with the vorticity direction is a non-monotonic function of the concentration.

(299f) Hydrodynamically-Driven Assembly in Dip Coating From Particle Suspensions

Carlos E. Colosqui[1], Jeffrey F. Morris[1,2] and Howard A. Stone[3], (1)Levich Institute of Physico-Chemical Hydrodynamics, New York City, NY, (2)Chemical Engineering, The City College of City University of New York, New York City, NY, (3)Mechanical and Aerospace Engineering, Princeton University, Princeton, NJ

Our computational study of dip coating from a particle suspension reports striped assemblies as those observed experimentally when the thickness of the withdrawn film is comparable to or smaller than the particle size. In this so-called thin-film entrainment regime, particles can assemble within the entrained film due to capillary interactions that minimize the free surface energy. Below a critical withdrawal speed, however, we find that the assembly process mainly occurs within the dynamic meniscus before particles can actually enter the thin film. This type of assembly process is observed when large interfacial forces required to deform the dynamic meniscus prevent particles from entering the film. Interfacial forces are overcome by hydrodynamic drag when a certain number of particles assemble in a close-packed formation. Hence, the assembly process within the dynamic meniscus is driven by hydrodynamic forces imposed by the background flow configuration. At low particle concentrations, the hydrodynamically-driven assembly significantly enhances the order and periodicity of the formed particle arrays.

(299g) The Effect of Orthogonal Shear Flow On the Sedimentation of Particles in Viscoelastic Fluids

Sourav Padhy[1], Sreenath Krishnan[1], **Eric S.G. Shaqfeh**[2], Gianluca Iaccarino[1], Jeffrey F. Morris[3] and N. Tonmukayakul[4], (1)Mechanical Engineering, Stanford University, Stanford, CA, (2)Chemical Engineering and Mechanical Engineering, Stanford University, Stanford, CA, (3)Levich Institute of Physico-Chemical Hydrodynamics, New York City, NY, (4)Halliburton, Duncan, OK

During the fracture simulation of oil and gas wells, suspensions of solids in polymeric solutions are pumped to help prop open the fracture. These high-density solids in suspension sediment due to gravity and also experience an orthogonal shear flow. Experimental data (van den Brule & Gheissary, J. Non-Newton. Fluid Mech., vol. 49, pp. 123-132, 1993; Tonmukayakul et al., 82nd SoR Annual Meeting, 2010) has shown that both the shear thinning and the elasticity of the suspending polymeric solutions significantly affect the settling rate of the solids. The mechanism by which the elasticity of the carrying fluid affects the settling rate is not well understood. In the present work, we use simulations of viscoelastic flow past a single, torque-free sphere with a cross shear flow to study the effect of carrier fluid elasticity on the drag experienced by the sphere and thus on its settling rate. We also have completed experiments thus measuring the sedimentation rates of the solids in different viscoelastic fluids. In the numerical simulations, we use the FENE-P or the Giesekus constitutive model to represent the viscoelastic fluid with parameters obtained by fitting rheological data. For weakly shear thinning fluids, we obtain an increase in drag, i.e. a decrease in settling rate, as shear Weissenberg number is increased in both the numerical simulations and the experiments. The simulations are in quantitative agreement with the experiments at small Weissenberg number (Wi <2). At higher Weissenberg number, the numerical results are in qualitative agreement with settling experiments but, in general, the magnitude of the simulated decrease in settling rate is larger than that observed in the experiments. We present the detailed physical mechanism for the increase in the drag experienced by the sphere in the simulations and we show that polymer shear stress, T^{P}_{13}, (with 1, the sedimentation direction and 3, the shear direction) is the cause of the increase in sphere drag.

(299h) Bacteria-Swimming-Induced Hydrodynamic Diffusion of Tracer Particles

Kasyap Thottasserymana Vasudevan[1], Donald L. Koch[1] and Mingming Wu[2], (1)Chemical and Biomolecular Engineering, Cornell University, Ithaca, NY, (2)Biological and Environmental Engineering, Cornell University, Ithaca, NY

We present the first theoretical estimate of the hydrodynamic diffusion of tracer particles in a homogeneous, isotropic suspension of swimming bacteria producing Stokesian hydrodynamic disturbances. While there are many experimental and numerical investigations of enhanced diffusion due to swimming bacteria in the literature, no theoretical estimate for the enhancement in tracer diffusivity has been proposed so far. The theory assumes pair-wise interactions between a bacterium and a colloidal tracer particle with the hydrodynamic disturbance due to the bacteria estimated using slender-body theory. The method of averaged equations is then employed to determine the hydrodynamic diffusivity D_{hyd} of the beads at long length and time scales when a bead has undergone many interactions with bacteria. The hydrodynamic diffusivity D_{hyd} depends upon the Peclet number Pe=UL/D and the non-dimensional bacteria concentration nL^3 where U is the swimming speed of the cells, L is the total length of cells including the flagella bundle, D is the Brownian diffusivity of the tracer bead, and n is the number of cells per unit volume. At high Peclet numbers, the hydrodynamic diffusion has a pure convective origin and $D_{hyd} \sim UL(nL^3)$. At low Peclet numbers, the bead can sample the hydrodynamic disturbance of a bacterium diffusively throughout the period τ between tumbles leading to a characteristic radial distance for interactions of $r \sim (D\tau)^{1/2}$ and an enhancement to the diffusivity proportional to $UL(nL^3)U(\tau/D)^{1/2}$. We complement our theory with numerical simulations of bacteria-bead interactions and experiments on moderately dense suspensions of *E. coli* bacteria with concentrations ranging from $nL^3 = 0.6$ to 14. Numerical simulations of an interacting bead-bacteria pair allow us to assess the importance of excluded volume, which limits the ability of beads to sample near-field hydrodynamic disturbances. Experiments involve time resolved imaging of fluorescent beads in a bacterial suspension confined in a microchannel and tracking the images to obtain the bead diffusivity. Comparison of theory, simulation and experimental data reveals the remarkable fact that the pair-wise interactions model can adequately estimate hydrodynamic diffusion in suspensions with nL^3 as large as 14.

(299i) Novel Aspects of the Hydrodynamics of Particle-Laden Liquids: Example of a Successful Transient Flow Immunoassay (TrIA)

Anand B. Subramaniam[1], Olga Taran[1], E. Jane Maxwell[1] and George M. Whitesides[2], (1)Chemistry and Chemical Biology, Harvard University, Cambridge, MA, (2)Department of Chemistry and Chemical Biology, Harvard University, Cambridge, MA

In medicine, measuring the amount of solid precipitate formed when antigens react with antibodies is a foundational principle used to conduct a wide variety of diagnostic tests. Typically, the degree of precipitation of antigen-antibody complexes is determined by measuring the light scattered off the suspended solids. To obtain accurate measurements of the suspended solids, specialized light sources and detectors are required; thus dedicated and well-calibrated instruments are essential. Here we show that the amount of solids suspended in liquids can be determined using a novel aspect of the hydrodynamics of particle-laden liquids. Using our method, the amount of solids in a liquid can be determined with minute quantities of liquid (< 20 microliters) and short times (< 10 minutes) without the need for dedicated optics and instruments. As an example of a successful application, we conduct an immunoassay to determine the amount of total serum IgG, which is an antibody present in blood.

(303a) Modulus, Confinement and Temperature Effects On Surface Capillary Wave Dynamics in Bilayer Polymer Films near the Glass Transition

Christopher M. Evans[1], Suresh Narayanan[2], Zhang Jiang[2] and **John M. Torkelson**[3], (1)Chemical and Biological Engineering, Northwestern University, Evanston, IL, (2)X-ray Science Division, Argonne National Laboratory, Argonne, IL, (3)Departments of Chemical and Biological Engineering, Materials Science and Engineering, Northwestern University, Evanston, IL

For the first time, thin and ultrathin polymer layers [polystyrene] supported on substrates with modulus varying by 5 orders of magnitude [poly(isobutyl methacrylate), crosslinked polydimethylsiloxane, poly(4-vinyl pyridine), and silicon wafers] were characterized by synchrotron-based X-ray photon correlation spectroscopy (XPCS). At the polystyrene (PS) Tg + 9 K, the PS surface wave relaxation times characterized by XPCS were found to track with substrate modulus, with lower modulus leading to faster PS surface relaxations even when the PS top layers exceeded 100 nm in thickness. These results show that an immiscible polymer domain may significantly influence the dynamics of a second, neighboring immiscible polymer at a temperature near the Tg of the second polymer and over length scales that greatly exceed those of both cooperative segmental mobility near Tg (~1-4 nm) and the polymer radius of gyration. Results of this study also reveal the effect of confinement on surface wave relaxation time (t) in polymer films near Tg and at low qh (<1), where q is the scattering wavevector and h is the top layer thickness. In particular, at the PS Tg + 9 K, the values of t/h increased with decreasing PS layer thickness when measured on a given substrate. This is in direct contrast with prediction of simple capillary wave theory which does not consider confinement or substrate modulus effects. Instead, simple capillary wave theory indicates that values of t/h plotted as a function of qh should overlap for all film thicknesses tested. The fact that the data at Tg + 9 K differ dramatically from this prediction indicates the important role of confinement in altering the capillary wave response near Tg. In contrast, both substrate modulus and confinement effects are negligible when XPCS measurements are take at the PS Tg + 40 K. At this higher temperature, the XPCS response follows predictions of simple capillary wave theory. These results demonstrate the important role of temperature relative to Tg in defining the magnitude of confinement effects of polymer related dynamics, including surface capillary wave relaxations: confinement effects that may be substantial at temperatures within 10 K of Tg may be negligible at temperatures 40 K above Tg.

(303b) Measurement of Viscoelastic Properties of Supported Thin Polymer Films

Yunlong Guo and Rodney D. Priestley, Chemical and Biological Engineering, Princeton University, Princeton, NJ

For nearly two decades, research has been devoted to understanding the impact of nanoscale confinement and interfaces on the glass transition temperature (T_g) and mobility of thin polymer films. Prior work in supported thin polymer films has utilized spectroscopic methods to measure T_g and mobility, while wrinkling methods have been applied to measure the instantaneous elastic modulus. Here we present a non-contact method to measure the viscoelastic properties of supported thin polymer films. The approach is based on the blow-off method originally developed by Derjaguin [1]. The method entails flowing gas under laminar flow conditions through a narrow channel in order to generate a well-define shear stress on a properly placed sample. Our design is modernized in numerous ways, including, the ability to simultaneously control the gas and substrate temperature. This new feature will allow for the measurement of temperature-dependent viscoelastic properties.

[1] B.V. Derjaguin, V.V. Karasev, V.M. Starov, E.N. Khromova, Research into the Boundary Viscosity of Volatile Liquids by the Blow-off Method, J. Coll. Inter. Sci., 67 465 (1978)

(303c) Mechanics of Thin Polystyrene Films: Role of Additives and Chain Architecture

Bryan D. Vogt, Department of Polymer Enigneering, University of Akron, Akron, OH

For many applications, additives are included in polymer formulations to provide the desired processing characteristics or functionality. The use of additives is particularly important for the microelectronics industry where polymeric photoresists are used to transform UV exposure to physical patterns. These photoresists are highly engineered formulations with photoactive small molecules that catalyze the chemical modification of the polymer to generate solubility differences to create the features. Additionally, the casting solvent generally does not fully evaporate. The nanoscale features formed by these photoresists have commonly been utilized as practical motivation for examination of the properties of polymer thin films. Feature fidelity is predominately associated with the modulus of the polymer, so the mechanical properties of polymers as a function of film thickness provide some indication if confinement will adversely impact the feature quality.

However, very few studies have examined how additives impact the physical properties of polymer thin films. The work of Torkelson and co-workers has illustrated that small molecules that interact favorably with the polymer through relatively strong interactions (H-bonding, pi-pi overlap) can eliminate the nanoconfinement effect in the glass transition temperature. We have previously shown that one of these additives can also eliminate any thickness dependence in the elastic moduli as well. Here, we will discuss how a series of small molecules with vastly different solubility parameters impact the thin film modulus of polystyrene. Further, the polystyrene chain architecture is varied through controlled branching to demonstrate how available conformations of the polymer in solution can potentially impact how nanoconfinement impacts the mechanical properties of polystyrene.

(303d) Calorimetric Tg of Single Polystyrene Ultrathin Films

Siyang Gao, Texas Tech University, Lubbock, TX and **Sindee L. Simon**, Chemical Engineering, Texas Tech University, Lubbock, TX

Tg depressions are generally observed for supported and freely standing polystyrene ultrathin films. However, the field is controversial with nanocalorimetric studies made by other researchers over the past ten years showing no changes in Tg for polystyrene. Suggestions have even been made that the Tg depressions observed in polystyrene thin films are due to artifacts caused by sample preparation, plasticization, and/or degradation. In order to address this controversy, the calorimetric glass transition (T_g) is measured for single polystyrene ultrathin films using Flash Differential Scanning Calorimeter (DSC) as a function of cooling rate, molecular weight, film preparation method, and film thickness. The T_g values of films greater than approximately 100 nm in thickness are in good agreement with conventional DSC bulk data. For thinner films, a T_g depression occurs when films are spuncast and placed on the calorimetric chip with a thin layer of grease between the film and sensor; these "liquid-floating" samples show a Tg depression which decreases with increasing cooling rate and its magnitude is similar to values found in the literature for the most mobile solid-supported films, which might be expected since our films are "liquid floating" on a thin layer of grease. On the other hand, when polystyrene is spuncast directly on the chip sensor, no Tg depression is observed. These results explain the lack of a Tg depression observed in prior nanocalorimetric studies where films where spuncast directly on the chip sensor. The role of annealing above Tg is also examined. We find for all polystyrene ultrathin films that show a Tg depression, that T_g values revert to the bulk value after long-time annealing at 160 °C. Based on atomic force microscopy studies, the recovery process is due to dewetting and hole growth which results in a thickening of the film to above 100 nm. The annealing results confirm that the Tg depression is not due to degradation or artifacts involving plasticization by small molecules.

(303e) Surface Dynamics of Macrocyclic Polystyrene Films

Shih-fan Wang[1], **Mark D. Foster**[1], Zhang Jiang[2] and Suresh Narayanan[2], (1)Department of Polymer Science, The University of Akron, Akron, OH, (2)X-ray Science Division, Argonne National Laboratory, Argonne, IL

Thermally stimulated fluctuations of the surface of a melt of macrocyclic polymers have been studied for the first time. The surface fluctuations of macrocyclic polystyrene (cPS) of 2k, 7k, 17k, and 37k molecular weight (M) were probed using x-ray photon correlation spectroscopy (XPCS), a recently-developed technique that has already been applied to study surfaces of linear PS melts.

The surface fluctuations for the cPS films are slower than those of linear chain analogs for all M. However, the glass transition temperatures (Tg) of the cyclic chains are higher than those of the linear analogs, with the discrepancy decreasing with increasing M. A continuum hydrodynamic theory of thermally stimulated capillary waves with a nonslip boundary condition is adequate to rationalize the behavior of the cPS films. When results from cPS of different M are plotted as a function of T/Tg the data nearly collapse to a single curve, indicating that Tg is a key parameter for the surface dynamics of macrocyclics in the temperature range studied.

Acknowledgements: NSF support (CBET 0730692)

(303f) Construction of an Integrated Zwitterionic Device of Continuous Blood Glucose Monitoring, Real Time Self-Regulated Insulin-Releasing and Non-Fouling Capabilities

Tao Bai[1], **Lei Zhang**[2] and Shaoyi Jiang[2], (1)ChemE, University of Washington, Seattle, WA, (2)Chemical Engineering, University of Washington, Seattle, WA

In order to detect glucose molecules, natural proteins including lectins and glucose oxidase were once utilized. However, denaturation problems with these proteins must be taken into consideration in those systems. At the same time, Boric acid (BA) group was discovered in order to obviate protein instability problem because it can selectively form complex with diol group on glucose molecules. So far, researchers prefer to copolymerize BA contained monomers with amino contained monomers in order to enable BA group to work in physical PH value. However, it is foreseeable that, due to existence of abundant amino, that copolymer is easy to non-specifically absorb proteins, which can deactivate the device in short time. As a result, actual application of this device is rather limited. To our best knowledge, one multi-functionalized device which integrates continuous glucose-detection, real time overdose feedback response and non-fouling capabilities into together is not reported, but highly desirable. Therefore, in our paper, we prefer to synthesize BA group contained zwitterionic molecule in order to realize them at the same time.

(303g) Synthesis and Characterization of All-Conjugated Block Copolymers/Supramolecular Block Copolymers in Thin Films

Yen-Hao Lin[1], Rafael Verduzco[1] and Seth Darling[2], (1)Chemical and Biomolecular Engineering, Rice University, Houston, TX, (2)Center for Nanoscale Materials, Argonne National Laboratory, Argonne, IL

All-conjugated block copolymers with p- and n-type blocks represent a promising approach to improving the performance of all-polymer organic photovoltaics (OPVs). Large scale phase separation can be avoided and block copolymer self-assembly may lead to ideal structures for charge dissociation and transport. In this work, we present both covalent and supramolecular conjugated block copolymers as potential routes to self-assembled polymer OPVs. Covalent block copolymers are made using a bromine-terminated poly(3-hexyl thiophene) macroreagent and palladium catalyzed poly-condensation reaction to attach a second poly(fluorene) (PFO) or poly(fluorene-co-benzothiadiazole) (PFBT) polymer block. A combination of nuclear magnetic resonance spectroscopy (NMR) and size-exclusion chromatography (SEC) provides the absolute molecular weight, the molecular weight ratio, and an estimate of homopolymer impurities. Supramolecular block copolymers can be made using 2-ureido- 4[1H]-pyrimidinone (UPy) quadruple hydrogen bonding interactions between distinct polymer blocks. A practical advantage of UPy interactions compared with all-conjugated block copolymers is a greatly simplified synthetic route and straightforward preparation of a series of supramolecular block copolymers through solution blending. UPy terminated polymers are synthesized by coupling hydroxyl or primary amine terminated polymers to a reactive isocyanate-UPy group in a one-step reaction. All polymer films are subsequently prepared by solution casting onto a surface and thermal and/or solvent annealing. Thin film microstructure and polymer crystallinity is analyzed by optical microscopy, atomic force microscopy (AFM), and grazing-incidence wide-angle x-ray scattering (GIXS). Optical microscopy images show that UPy terminated polymer blends do not undergo large-scale phase separation after extended thermal annealing at 160 ^{o}C, which are conditions typically used to prepare bulk heterojunction polymer OPVs. AFM together with GIXS reveal the self-assembled nanostructures and crystallinity of block copolymers and polymer blends. The crystallinity of the resulting film depends on the relative molecular weights of each block. In the case of a large PFO block, crystallization of PFO dominates the morphology, while a larger P3HT results in exclusively P3HT crystallization. In UPy polymer blends, only nano fibers and P3HT crystallinity are observed under AFM and GIWAXS, respectively. These results show that covalent block copolymers can be used to control the resulting nanostructures and crystallinity by changing the ratios of each block, and UPy-mediated interactions can suppress large scale phase-separation in bulk heterojunction polymer blends at temperatures and processing conditions typically used to prepare bulk-heterojunction OPVs.

(319b) Development of Imidazoles As a Highly Tunable Platform for CO2 Capture and Materials Applications

Matthew S. Shannon[1], Michelle S. Hindman[1], A. Christopher Irvin[1], John W. Whitley[1], Jason E. Bara[1], C. Heath Turner[1] and Sergey P. Verevkin[2], (1)Chemical & Biological Engineering, University of Alabama, Tuscaloosa, AL, (2)Physical Chemistry, University of Rostock, Rostock, Germany

N-functionalized imidazoles are a class of heterocyclic organic molecules, and are well-known as components of pharmaceuticals and also serve as building blocks for imidazolium-based ionic liquids (ILs). While IL-based solvents have been the focus of many research efforts in areas such as CO2 capture, imidazoles have not been considered for these same applications. Imidazoles possess many of the same desirable characteristics as ILs such as tunable chemical structures, low vapor pressure and good chemical/thermal stability. Despite these promising features, it has only been very recently that reports on the the relationships between imidazole structure and physical properties were reported, and the study of imidazoles as tailored solvents is still dwarfed by the body of knowledge available for ILs. Our research in this area has focused on developing structure-property relationships for a wide range of *N*-functionalized imidazoles, with emphasis on density, viscosity, vapor pressure, pKa, heat capacity, solubility of CO2, CH4 and other gases, as well as the underlying thermodynamics. Because imidazoles possess such tunable structures, we also forsesee broad applications of this platform in polymers and self-assembling materials such as liquid crystals.

Furthermore, we have utilized molecular dynamics as well as COSMOtherm simulations to computationally model and predict properties as well as parametrically study the influences of various functional groups (e.g. alkyl, ethers, nitriles, etc.) on physical properties. While imidazoles have some similar properties to ILs, other properties such as viscosity can be an order of magnitude (or more) different. Furthermore, some unique chemistries are available due to the basicity/nucleophilicty of imidazoles that are unavailable in ILs. This presentation will detail our synthetic methods, property characterizations, COSMO modeling and provide comparisons to these properties in ILs.

(319c) Computationally Efficient Determination of Thermodynamic Stability of Five Metal Hydrides Using First Principles Including Hydrogen Isotope Effects

Kelly Nicholson and David Sholl, School of Chemical and Biomolecular Engineering, Georgia Institute of Technology, Atlanta, GA

In Next Generation Nuclear Plants an application of metal hydrides as tritium sequestration materials requires materials that are thermodynamically stable to temperatures in excess of 1000 K. While the thermodynamics of metal hydrides at low to moderate temperatures have been successfully described by T = 0 K total energies and simple harmonic models, it is unclear as to what level of theory is sufficient to describe high temperature materials in which anharmonic effects may be significant. Since it is advantageous to apply computationally efficient methods in large scale material screening, we explored the predicted stabilities of five materials – ZrH_2, HfH_2, TiH_2, LiH, and NaH – with four levels of theory using Density Functional Theory (DFT) to predict the dissociation temperatures, T_d, of very stable metal hydrides at one bar H_2 pressure. We also investigated the magnitude of the change in stability of hydrogen isotope-substituted metal hydrides with the aim of understanding if hydrogen isotope effects should be accounted for in material screening. Prediction of the Helmholtz free energy for the solid phases are based on 0 K total energy calculations, harmonic calculations at ground state volumes, quasiharmonic calculations, and quasiharmonic calculations with explicit anharmonic correction. It is determined that levels of theory which account for vibrational corrections to the crystal lattice (thermal expansion) are not necessary to get an accurate description of relative stabilities of metal hydrides. The simplest model with T_d described by DH/DS and DS=0.130 kJ (mol-K)$^{-1}$ predicts T_d to within 75 K of the much more computationally expensive quasiharmonic calculation for all materials. The shift in T_d upon substitution of tritium relative to protiated metal hydrides is within 50 K for all materials studied with larger shifts for lighter materials as expected. Accounting for vibrational effects due to isotope substitution in metal hydrides is unnecessary to accurately predict the relative stabilities of metal hydrides at high temperatures.

(319d) Exploring Retrograde Vaporization and Derivative Properties of near-Critical Fluids Using Monte Carlo Simulations and Equations of State

Angel D. Cortés-Morales[1], Nikolaos I. Diamantonis[2,3], Ioannis G. Economou[2,3], Cor J. Peters[2] and J. Ilja Siepmann[1], (1)Depts. of Chemistry and of Chemical Engineering & Materials Science, University of Minnesota, Minneapolis, MN, (2)Dept. of Chemical Engineering, The Petroleum Institute, Abu Dhabi, United Arab Emirates, (3)Molecular Thermodynamics and Modeling of Materials Laboratory, National Center for Scientific Research "Demokritos", Aghia Paraskevi Attikis, Greece

Configurational-bias Monte Carlo simulations in the Gibbs, isobaric-isothermal, and canonical ensembles and various equations of state are applied to investigate two important phenomena found for near-critical fluids: (i) double retrograde vaporization in vapor-liquid phase equilibria of binary mixtures near the critical point of the lower-boiling compound, and (ii) non-monotonic behavior of thermodynamic derivative properties as function of density or pressure at temperatures above, but close to the critical temperature. For the retrograde behavior, binary mixtures of methane/*n*-alkanes and carbon dioxide/(*n*-alkanes or alkylbenzenes) are discussed. For the thermodynamic derivative properties, various butane-like models are investigated at a temperature of $T/T_{crit} = 1.1$, where the critical temperature is determined from Gibbs ensemble simulations for each molecular model to allow for comparison of all models at the same reduced temperature. The molecular simulations allow one to provide molecular-level understanding of these near-critical phenomena, whereas the equations of state serve as supporting data and as guide for the set-up of Gibbs ensemble simulations (i.e., by providing information on appropriate overall composition and volume/pressure to be used).

(319e) Exploring the Low-Temperature Phase Behavior of ST2 Water with Well-Tempered Metadynamics

Jeremy C. Palmer, Department of Chemical and Biological Engineering, Princeton University, New Jersey , NJ, Pablo G. Debenedetti, Chemical and Biological Engineering, Princeton University, Princeton, NJ and Roberto Car, Princeton University, Princeton, NJ

Many of the well-known thermodynamic anomalies of water, such as its negative thermal expansion and increased compressibility upon cooling, become more pronounced when it is cooled below its freezing point into a metastable liquid. On the basis of numerical evidence from molecular simulation of the ST2 water model[1], Poole et al.[2] proposed that these anomalies may be explained by the presence of a second critical point associated with the coexistence of two distinct metastable liquid phases in the deeply supercooled regime, a low density liquid (LDL) and high density liquid (HDL). Since the hypothesized LDL-HDL coexistence region is well below the homogenous nucleation temperature of bulk water, obtaining direct experimental evidence to either support or falsify its existence remains a significant challenge. Subsequent investigations using molecular simulation techniques have reported conflicting results. Liu et al.[3] used histogram reweighting methods in the grand canonical ensemble to examine the low-temperature behavior of the ST2 model and found evidence for the existence of two metastable liquid phases. Sciortino et al.[4] also reported similar results for the ST2 model, using umbrella sampling grand canonical Monte Carlo simulations with a different treatment of long-ranged electrostatic contributions. However, Limmer and Chandler[5] recently used two-dimensional umbrella sampling and histogram reweighting to examine the underlying free energy surface of ST2 water in density and bond-orientational order space, and concluded that one of the observed phases in coexistence is not a liquid but in fact a crystal. We present findings from our recent study of the low temperature phase behavior of ST2 water using the well-tempered metadynamics technique of Parrinello and co-workers[6]. We discuss the method in detail and demonstrate that it is able to provide reliable and accurate estimates of the free energy surface of ST2 in density and bond-orientational order space in a fraction of the time required by other methods such as umbrella sampling. In agreement with Liu et al.[3] and Sciortino et al.[4], we show that the free energy surface is comprised of two distinct basins, both of which correspond to liquid-like phases with low bond-orientational order.

[1]F.H. Stillinger and A. Rahman, *J. Chem. Phys.*, 60, 1545 (1974).

[2]P. H. Poole, F. Sciortino, U. Essmann, and H. E. Stanley, *Nature*, 360, 324 (1992)

[3]Y. Liu, A.Z. Panagiotopoulos and P.G. Debenedetti, *J. Chem. Phys.*, 131, 104508 (2009).

[4]Sciortino, I. Saika-Voivod and P.H. Poole, *Phys. Chem. Chem. Phys.*, 13, 19759 (2011).

[5]D. Limmer and D. Chandler, *J. Chem. Phys.*, 135, 134503 (2011)

[6]A. Barducci, G. Bussi and M. Parrinello, *Phys. Rev. Lett.*, 100, 020603 (2008).

(319f) Characterization of Xenon Hydrates Formed At High Driving Forces and Advancements in Hydrate Equilibrium Predictions

Patrick G. Lafond, Chemical Engineering, Colorado School Of Mines, Golden, CO and Amadeu K. Sum, Chemical & Biological Engineering, Colorado School of Mines, Golden, CO

Clathrate hydrates are inclusion compounds composed of water and "guest" molecules. The water molecules form a crystalline packing of polyhedral cages with their hydrogen bond network. The guest molecules become trapped, or enclathrated, in the cage-like polyhedra giving hydrates their inclusion properties. Guest molecules are typically small and hydrophobic, including methane, ethane, and carbon dioxide. In the gas and oil industry, hydrates are often regarded as a nuisance as their formation can plug subsea pipelines; however, hydrates have recently gained interest as viable options for energy storage, water desalination, and carbon dioxide sequestration. While much is known about the thermophysical properties and crystal structure of hydrates at moderate conditions, less is known on the behavior of hydrates formed at high-driving forces conditions. In this study, we present our findings of xenon hydrates formed *in situ* at high driving forces in a differential scanning calorimeter (DSC), which reveals an ill-defined equilibrium temperature often displaying multiple dissociations, or endotherms characteristic of the melting of multiple solid structures. This apparent violation of the Gibbs phase-rule has been seldom observed with other guests, which raises the question of how well the xenon-hydrate crystal structure is actually defined. In this study, we probe the unusual dissociation behavior of xenon hydrates using DSC, powder x-ray diffraction, and nuclear magnetic resonance and revisit the van der Waals-Platteeuw model to incorporate guest-guest interactions for more realistic phase predictions.

(319g) Prediction of 1-Octanol/Water Partition Coefficients for Ionic Liquids by the Application of Adaptive Biasing Force Method

Ganesh Kamath, Chemistry, University of Missouri-Columbia, Columbia, MO, Navendu Bhatnagar, Chemical Engineering, Wayne State University, Detroit, MI, Gary A. Baker, Chemistry, University of Missouri-Columbia, Columbia and **Jeffrey J. Potoff**, Chemcial Engineering and Materials Science, Wayne State University, Detroit, MI

Ionic liquids are fast emerging as compounds representing a new class of solvents with a variety of applications in the chemical industry such as organic synthesis [1], transition metal catalysis [2], liquid phase exfoliation of graphene [3] and carbon capture [4] to name a few. Ionic liquids are tagged as environmental friendly substitutes for volatile organic solvents as they possess low vapor pressures and also have the ability to act as bio catalysts. Other advantageous properties include high chemical and thermal stability, high ionic conductivity and non-inflammability [1]. Also, with a plethora of cations and anions available for generation of ionic liquids, a large set of permutations is available depending on the problem at hand. However, ILs are also associated with certain eco-toxicological concerns as well marked by factors like sorption, hydrolysis and degradation. One prime parameter to quantify the environmental impact of any compound is the 1-octanol/water partition coefficient which provides a measure of solute hydrophobicity and also serves as a rough guideline for predicting "enzyme compatible" media for non-aqueous biocatalysis [5]. Experimental methods such as shake-flask followed by concentration determination using UV–vis spectroscopy or high-performance liquid chromatography (HPLC) have been used primarily for imidazolium-based ILs due to the ability to detect the imidazolium ring by UV–vis [6]. Theoretical methods such as linear free energy relation correlations (LFER) [7] and COSMO-RS [8] have been used to predict log K_{ow} for ILs albeit with a disadvantage that these require an extensive training set of experimentally-determined K_{ow} to parameterize the coefficients for solvation free energy correlation.

In this work, difference in Gibbs free energy of solute transfer between two solvent phases is calculated by the application of a recently developed method by Darve *et al.* [9-11] known as adaptive biasing force (ABF). This free energy is then used for estimation of log K_{ow} for imidazolium based ILs. ABF follows an unconstrained molecular dynamics based approach where the system is allowed to evolve freely enabling the solute to visit multiple states encountered between the initial and final state points. Molecular dynamics simulations were performed with NAMD [12] and trajectory visualization and analysis was done with VMD [13]. All atom force field developed by Lopes *et al.*[14] was employed to model imidazolium cations and anions. TraPPE-UA mode developed for alcohols [15] and SPC/E [16] model was used to describe the interactions of 1-octanol and water respectively. ILs made up of 1-butyl-3-methylimidazolium ($[bmim]^+$) cation and paired with six different anions were studied in this work. Free energy of hydration and solvation are calculated initially depending on whether the solute is transferred into vacuum from the water or 1-octanol phase respectively. Results were found to be in good agreement with the experimental data with a marked improvement when water saturated 1-octanol phase is used instead of dry 1-octanol.

References

1. Zhao, H and Malhotra, S. V. Aldrichimica Acta (2002) 35, 75-83
2. Wasserscheid, P. and Keim, W. Angew. Chem. Int. Ed. (2000) 39, 3772-3789
3. X. Wang, P. F. Fulvio, G. A. Baker, G. M. Veith, R. R. Unocic, S. M. Mahurin, M. Chi and S. Dai, *Chem. Commun.*, 2010, **46**, 4487-4489.
    1. E. D. Bates, R. D. Mayton, I. Ntai and J. H. Davis, *JACS Commun.*, 2001, **124**, 926-927.
    2. U. Kragl, M. Eckstein and N. Kaftzik, *Curr. Opin. Biotechnol.*, 2002, **13**, 565-571.
    3. L. Ropel, L. S. Belvèze, S. N. V. K. Aki, M. A. Stadtherr and J. F. Brennecke, *Green Chem.*, 2005, **7**, 83-90.
    4. M. H. Abraham, A. M. Zissimos, J. G. Huddleston, H. D. Willauer and R. D. Rogers, *Ind. Eng. Chem.*

Res, 2003, **42**, 413-418.
5. C.-W. Cho, U. Preiss, C. Jungnickel, S. Stolte, J. Arning, Ranke, A. J. Klamt, I. Krossing and J. Thoming, *J. Phys. Chem. B*, 2011, **115**, 6040-6050.
6. E. Darve and A. Pohorille, *J. Chem. Phys.*, 2001, **115**, 9169-9183.
7. E. Darve, M. A. Wilson and A. Pohorille, *Mol. Sim.*, 2002, **28**, 113-144.
8. E. Darve, David Rodríguez-Gómez and A. Pohorille, *J Chem Phys*, 2008, **128**, 144120.
9. J. C. Phillips, R. Braun, W. Wang, J. Gumbart, E. Tajkhorshid, E. Villa, C. Chipot, R. D. Skeel, L. Kale and K. Schulten, *J. Comp. Chem.*, 2005, **26**, 1781-1802.
10. Humphrey, W., Dalke, A. and Schulten, K., *J. Molec. Graphics*, 1996, vol. 14, pp. 33-38.
11. J. Lopes, J. Deschamps and A. Padua, *J. Phys. Chem. B*, 2004, **108**, 2038-2047.
12. B. Chen, J. J. Potoff and J. I. Siepmann, *J. Phys. Chem. B* 2001, **105**, 3093.
13. H. J. C. Berendsen, J. R. Grigera and T. P. Straatsma, *J. Phys. Chem.*, 1987, **91**, 6269-6271.

(320a) Studies of Realistic Roughness Effects in Wall Turbulence

Kenneth Christensen, Mechanical Science and Engineering Dept., University of Illinois at Urbana-Champaign, Urbana, IL

This keynote lecture will highlight on-going studies of highly-irregular surface roughness commonly encountered on practical flow surfaces, such as turbine blades, hulls of ships and within transport pipelines, for example. Such realistic roughness is typically due to cumulative damage to a flow surface that was likely aerodynamically or hydrodynamically smooth prior to deployment but has become significantly roughened via cumulative damage and/or deposition. Previous efforts have shown such realistic roughness patterns to generate bulk flow and heat-transfer characteristics quite different than that predicted from correlations developed from studies of much more idealized roughness topographies (often single scale arranged in an ordered manner). Thus, the present effort is devoted toward identifying the impact of highly-irregular roughness replicated from a turbine blade damaged by deposition of foreign materials on wall turbulence. To this end, several particle-image velocimetry (PIV) experiments were conducted for flow over the aforementioned surface, both within and outside the roughness sublayer, to explore its impact on the canonical behavior of smooth-wall turbulence. The impact of this roughness on various statistical and structural aspects of wall turbulence will be highlighted. In addition, the possibility of devising low-order topographical representations of this complex surface using singular value decomposition (SVD) is considered. Truncation of this decomposition so that only the largest topographical scales are included in a given model surface allows one to isolate the impact of these scales from the intermediate and finer-scale details of the roughness. The fidelity of two different low-order reconstructions in generating a similar impact on the flow as the original surface is evaluated using 2D and stereo PIV measurements within and outside the roughness sublayer.

(320b) Structure of Poiseuille-Couette Flow in an Infinite Channel with Direct Numerical Simulations

Quoc T. Nguyen and Dimitrios V. Papavassiliou, School of Chemical Biological and Materials Engineering, The University of Oklahoma, Norman, OK

Super hydrophobic surfaces (i.e., those with contact angle larger than 150 degrees) have been found to produce not only fluid slip in microchannel flows but also to induce considerable drag reduction for laminar flows [1]. Furthermore, applications of such surfaces for drag reduction in turbulent flows are also under consideration [2]. Previous works by Rothstein and his coworkers [3,4] have indicated that turbulent drag reduction may be obtained by using super hydrophobic surfaces appropriately. Another study by Spencer et al. [5] has pointed out that a specified fluid slip at the wall leads to a decrease of turbulent intensities and of turbulent kinetic energy production as well as a decrease in drag force at the channel wall. The present simulation study would give further information on how specified slip velocity at the wall affects near-wall flow structures.

A Poiseuille-Couette flow is simulated in an infinite channel in order to impose a streamwise slip boundary condition on one of the channel walls, based on the movement of the walls in this kind of flow. The pressure gradient is kept unchanged, and the top wall moves in the flow direction, while the bottom one is kept moving in the opposite direction compared to the top wall. A computational box size of $8\pi h$ x $2h$ x $2\pi h$ (with a resolution of 512 x 129 x 256) in x,y and z directions, respectively, with periodic boundary conditions is used to simulate this Poiseuille-Couette flow (h is half of the channel height, and is 300 in viscous wall units). The relative velocity between the two walls were set to be 1, 2 and 4 in viscous wall units, which are calculated based on the shear stress at the wall, and the density and viscosity of fluid under study. Due to the Galilean invariance of the Navier-Stokes equation, subtracting the mean velocity at the bottom wall would give us a corresponding case in which the bottom wall is stationary, while the top wall moves in the positive streamwise direction with a velocity equal to the relative velocity between the two walls. This is a way to simulate a super hydrophobic surface at the top wall of the channel. Our results show that the presence of streamwise wall slip modifies turbulence at the near wall region to get closer to the limiting state of the Lumley triangle (one component sate) and to satisfy both two-component limit and axisymmetry at large and small scales. Earlier analytical work has proved [6] that this would lead to a significant suppression of small scale turbulence and rapid laminarization close to the wall, leading to drag reduction. Correlation coefficients and analysis of the turbulence structure close to the wall will also be discussed.

References

[1] Ou, J., and J. P. Rothstein, Phys. Fluids **17**(10), Art. No. 103606 (2005).

[2] Fukagata, K., Kasagi, N., and P. Koumoutsakos, Phys. Fluids 18(5), Art. No. 051703 (2006).

[3] J. P. Rothstein, Annu. Rev. Fluid Mech. 42, 89-109 (2010).

[4] M. B. Martell, J. P. Rothstein, and J. B. Perot, Phys. Fluids 22(6), Art. No. 065102 (2010).

[5] N. B. Spencer, L. L. Lee, R. N. Parthasarathy, and D. V. Papavassiliou, The Canadian Journal of Chemical Engineering 87(1), 38-46,(2009).

[6] J. Jovanovic and R. Hillerbrand, Thermal Science 9, 3 (2005).

(320c) Large Eddy Simulation of a Turbulent Incompressible Wake Flow - Comparison with Particle Image Velocimetry Data

Katrine M. Nilsen[1], Bo Kong[2], Rodney O. Fox[2], James C. Hill[2] and Michael Olsen[1], (1)Mechanical Engineering, Iowa State University, Ames, IA, (2)Chemical & Biological Engineering, Iowa State University, Ames, IA

Large Eddy Simulation (LES) is becoming a more and more popular tool for prediction of turbulent flows both in industrial applications and for research purposes. In many situations the averaged flow fields from Reynolds Averaged Navier-Stokes (RANS) simulations do not provide sufficient information; in cases where unsteadiness and turbulent structures are important, RANS simulations fall short. While Direct Numerical Simulation (DNS) can give very detailed information about the flow, it is computationally expensive and thereby limited to low Reynolds number flows and simple geometries. LES, which can provide detailed information down to the smallest resolved case, is at the same time applicable to a much wider range of flows. However, LES involves modeling and validation of the method is therefore required.

In this work Large Eddy Simulation (LES) is employed to study turbulence characteristics in an incompressible wake flow. Experimental data from Particle Image Velocimetry (PIV) of the same flow is used to validate the simulation results.

The flow domain is a rectangular channel of cross section 60 mm by 100 mm. The channel features three inlet streams separated by a pair of splitter plates. Using equal flow rate for all three inlet streams, a wake is produced downstream of each splitter plate. The Reynolds number in this study is 37 500 based on the channels hydraulic diameter. This flow setup has the advantage that is resembles flow configurations found in industrial applications. A further advantage is that this configuration produces a complex flow well suited for evaluation of LES beyond the frame of simplified reference flows traditionally used for validation.

For the LES, the effect of time step, grid resolution, turbulent inlet conditions and subgrid model is investigated through a series of preliminary simulations. The inlet condition proved to be very influential to the results over most of the domain while the subgrid models that were tried gave only small differences in the simulation results.

For both simulation and experiment large ensembles (10 000) of instantaneous velocity fields from the channels center plane is collected. A thorough comparison between experiment and simulation data including mean and root-mean-square values of velocity, shear stress and energy dissipation rate as well as spatial velocity correlations were performed. Good agreement is found between simulation and experiment, showing that LES with the employed method to generate turbulent inflow conditions is successful in reproducing the wake flow.

(320d) Beyond Boussinesq: A Realizable Closure Model for the Reynolds Stress

Charles A. Petty, Chemical Engineering and Materials Science, Michigan State University, East Lansing, MI, Karuna S. Koppula, Chemical Engineering, Rochester Institute of Technology, Rochester, NY and Andre Benard, Mechanical Engineering, Michigan State University, East Lansing, MI

The Reynolds-averaged Navier-Stokes (RANS-) equation for constant property Newtonian fluids is unclosed due to the explicit appearance of the normalized Reynolds (NR-) stress and the turbulent kinetic energy. A closure model for the normalized Reynolds stress (either algebraic or differential) must produce non-negative operators for all turbulent flows, including flows in rotating frames-of-reference. In this presentation, a recently developed algebraic closure for the NR-stress, referred to as the URAPS-closure (see Koppula et al., 2009, 2011), is used to calculate the influence of frame rotation on the components of the NR-stress for homogeneous and for inhomogeneous flows. The URAPS closure is formulated as a mapping of the NR-stress into itself and is, thereby, a non-negative operator for all turbulent flows. Unlike closure models based on an eddy-viscosity theory, the new closure predicts the redistribution of the turbulent kinetic energy among the three components of the fluctuating velocity for statistically stationary rotating channel flows.

Koppula, K.S., A. Bénard, and C. A. Petty, 2009, "Realizable Algebraic Reynolds Stress Closure", 2009, *Chemical Engineering Science*, 64, 22, 16 November 2009, 4611-4624.

Koppula, K. S., A. Bénard, and C. A. Petty, 2011, "Turbulent Energy Redistribution in Spanwise RotatingChannel Flows", *Ind. Eng. Chem. Res.*, pubs.acs.org/IECR, dx.doi.org/ 10.1021/ie1020409.

(320e) Analysis of near-Contact Inertial Particle Pair Statistics in Turbulence

Lance R. Collins, Sibley School of Mechanical & Aerospace Engineering, Cornell University, Ithaca, NY

It is well known that turbulence can dramatically increase the collision frequency of a population of droplets or particles embedded in the flow. This has been used to partially explain the discrepancy between the observations of relatively rapidly evolving atmospheric clouds and the slower predictions of microphysical models that do not take turbulence into consideration (i.e., assume collisions are driven solely by gravitational settling). Recently there has been a lot of effort to better understand and predict this so-called "acceleration" of the drop size distribution due to turbulence using direct numerical simulations (DNS), high-precision laser based experiments and theory. What is known is that the enhancement, driven by the response of particles to the turbulence, is very sensitive to the droplet size, leading to a very complex cumulative effect on the size distribution. However, we can decompose this behavior into the weighted sum of pairwise interactions, which in turn can be described in terms of the average number density and relative velocity statistics of particle pairs at contact. For small drops this can be challenging to observe in DNS because it requires a huge population of particles. This is also challenging to measure in experiments due to experimental errors. In this talk, we describe an alternate simulation strategy we call satellite particle simulations (SPS). With SPS, the flow around a primary particle is linearized and the motion of a neighboring cloud of satellite particles is computed. In this way, the near-contact flow can be accurately represented. We show there is very good agreement between SPS and DNS for these statistics over a broad range of particle Stokes numbers (particle sizes). Moreover, these tools have proven extremely useful for analyzing two recent theories for the collision kernel in the literature.

(320f) Particle Dynamics in a Turbulent Particle-Gas Suspension At High Stokes Number

V Kumaran, Chemical Engineering, Indian Institute of science, Bangalore, India and Partha S. Goswami, CSIR Centre for Mathematical Modelling and Computer Simulation, Bangalore, India

The `fluctuating force' model has been developed to capture the effect of turbulent fluid velocity fluctuations on the particle phase in a turbulent gas-solid suspension in the limit of high Stokes number, where the particle relaxation time is large compared to the correlation time for the fluid velocity fluctuations.
The force exerted by the turbulent fluid velocity fluctuations, considered as anisotropic Gaussian white noise, is incorporated in the equation of motion of the particles. The noise amplitude is determined from the time correlations of the spatially varying and non-isotropic turbulent fluid velocity fluctuations obtained from Direct Numerical Simulations (DNS). The particle velocity distribution in the gravity driven flow in a vertical channel is analysed using a Langevin formulation for the force due to turbulent fluctuations, and the results are compared with DNS simulations incorporating one-way coupling (the force due to turbulent velocity fluctuations on the particle is incorporated, but not the reverse force due to the particle on the fluid turbulence). For monodisperse particles, there is quantitative agreement between the fluctuating force simulations (FFS) and DNS simulations, provided the time correlation of the fluid velocity fluctuations is calculated in a `moving Eulerian' reference frame moving with the fluid mean velocity. Comparisons are also made with experimental measurements of the particle velocity distributions using PIV and Particle Tracking Velocimetry. Good agreement is obtained only if the polydispersity in the particle size distribution is incorporated in the FFS. At lower mass loading (0.19), there is no significant modification of the fluid phase turbulent intensity, and the particle velocity distribution is well predicted by the FFS with one way coupling. At a higher particle mass loading of 1.7, there is a significant increase in the turbulent velocity fluctuations due to the particles. It is then necessary to incorporate the modified fluid turbulence intensity in the fluctuating force simulation.

PARAMETRIC SENSIBILITY ANALYSIS OF REYNOLDS STRESS MODEL IN A COAXIAL CONFINED JET FLOW

Lizoel Buss[1], Rodrigo K. Decker[1], Waldir P. Martignoni[2], Henry F. Meier[1]
[1] Regional University of Blumenau – FURB, Department of Chemical Engineering, São Paulo St., 89030-000, Blumenau – Brazil.
[2] PETROBRAS/AB-RE/TR/OT, República do Chile Av., 65, 20031-912, Rio de Janeiro – Brazil.

The objective of this study is to investigate the parametric sensibility of Reynolds Stress Model with version of Speziale, Sarkar and Gatski in relation to the C_μ, $C_{\varepsilon\text{-}1}$ and $C_{\varepsilon\text{-}2}$ parameters, in a confined coaxial jet flow for different velocity ratios. The numerical results were compared with experimental data, which was obtained by a two-dimension particle velocimetry system (2D-PIV), in the region of fully developed flow. The experimental and numerical results were evaluated in two different axial positions (L/D = 25 and L/D = 37.5) in terms of axial mean velocity and turbulence kinetic energy. The results indicated that changing the C_μ parameter in 30% above its default value, the RSM-SSG model represents more precisely the physical phenomena related to this kind of flow.

Introduction

Confined and coaxial jet flows are found in several industrial applications such as jet pumps, burners, ejectors, among others. In confined jets the entrainment of sufficient mass from surroundings is strictly related to the cortical structures of the flow which may vary according to the velocity ratio, $R_U = U_{out}/U_{in}$, where U_{in} is the jet velocity and U_{out} is the co-flow velocity [1]. In order to understand the flow behavior and the mixture process in coaxial jet flows several authors [1, 2, 3, 4, 5] developed experimental and numerical studies to investigate the influence of the inlet conditions for coaxial jets. Ahmed and Sharma [2] investigate the turbulent mixing of two coaxial and confined jets by means of LASER Doppler velocimetry system. The authors made experiments to investigate the velocities ratio effect between the two streams. Lin [3] studied a confined swirling coaxial jet modeling evaluating the influence of grid density and turbulence models. Kiwata *et al.*, [4] made experiments using hot film anemometry, particle image velocimetry (2D and stereoscopic) and LASER induced fluorescence to identify the interaction between axisymmetric and streamwise vortices in the unexcited and excited coaxial jet. Balarac *et al.*, [5] realized direct numerical simulations to investigate the flow dynamics and the mixing properties of coaxial jets at moderate Reynolds number. Mahmoud *et al.*, [6] proposed a numerical study of an axisymmetric turbulent jet discharging into co-flowing stream with R_U ranging between 0 and 4.5. The standard *k–ε* model and the Reynolds Stress Model were applied. The authors observed that both turbulence models are valid to predict the average and turbulent flow sizes. Decker *et al.* [1] presented numerical studies using four RANS models to investigate the behavior of a confined coaxial jet in the region of development flow.

Methodology

Experimental Setup

In the test facility showed in **Figure 1**, a Pitot tube (A) is connected to a differential pressure transmitter in order to measure the mass flow ratio in the know transversal section. As the mass flow ratio inside the chamber is the same in the measurement section, the inlet mean velocity of the gas phase in the chamber (U_{out}) can be determined. This velocity is controlled by a Programmable Logical Controller coupled to the radial ventilator located at the end of test facility (B), which is also responsible for maintaining the entire system under negative pressure. The jet velocity (U_{in}) is provided by a compressed air system, which is connected to a relief valve and a flow meter (C). The measurements are obtained in the Plexiglass chamber (D) for different axial positions from the nozzle by a particle image velocimetry (PIV) system (E). In order to measure the gas flow velocity, dioxide titanium (TiO_2) particles were used due to its ability to follow the gas flow (Stokes number << 1). The TiO_2 particles are loaded in the flow system by an ejector located in the inlet region (F).

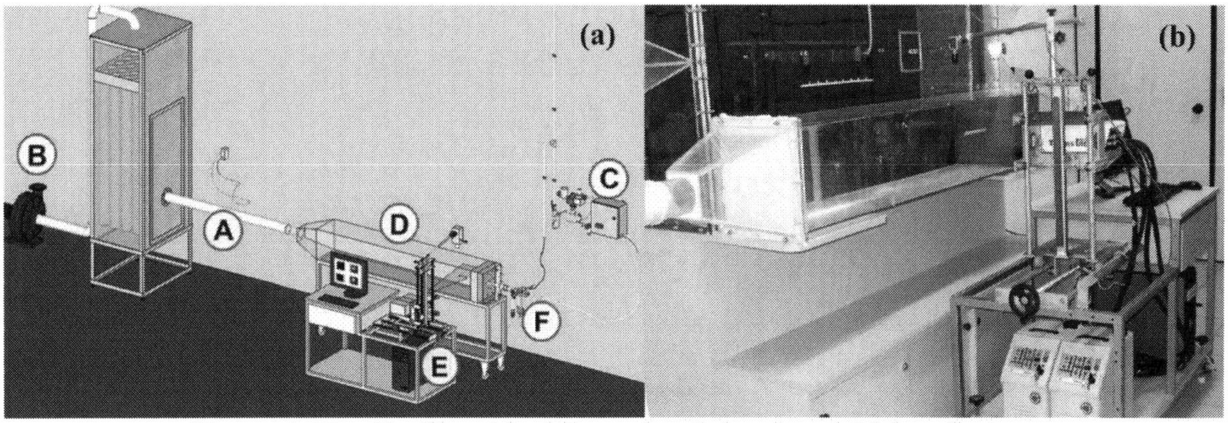

Figure 1. Test Facility: (a) tridimensional sketch and (b) installation.

Operational and Geometric Conditions

In the test facility, experiments were realized for two different initial velocities in the jet and in the chamber, which are presented in **Table 1**. This setup provides four different velocities ratios, which are presented in **Table 2**.

Table 1. Initial velocities in the jet and in the chamber.

Chamber Inlet		Jet Outlet	
Velocity (U_{out}) (m/s)	Reynolds Number	Velocity (U_{in}) (m/s)	Reynolds Number
0.8	15,576	9.64	6,962
1.0	19,345	11.78	8,507

Table 2. Velocities ratios.

Velocities Ratios R_U	Chamber Velocity U_{out} (m/s)	Jet Velocity U_{in} (m/s)
0.083	0.8	9.64
0.104	1.0	9.64
0.068	0.8	11.78
0.085	1.0	11.78

The experiments were developed at 25 °C. The PIV measurements were taken in a single horizontal plane from the middle of the jet nozzle. The measurement region is 510 mm length x 110 mm width, which is subdivided in six sections with 90 mm length x 110 mm width (top view in **Figure 2**). The initial and final positions of each of the six sections are presented in **Table 3**. After calibration of the 2D-PIV system, 1000 frames were taken in the measuring region with a time step of 150 μs. Then two different axial positions were selected to investigate the radial profiles in terms of mean velocity and turbulence kinetic energy. These positions are located in a chamber length per jet diameter relation (L/D) of 25 and 37.5. The geometrical conditions are presented in **Figure 2**, where D is the nozzle diameter.

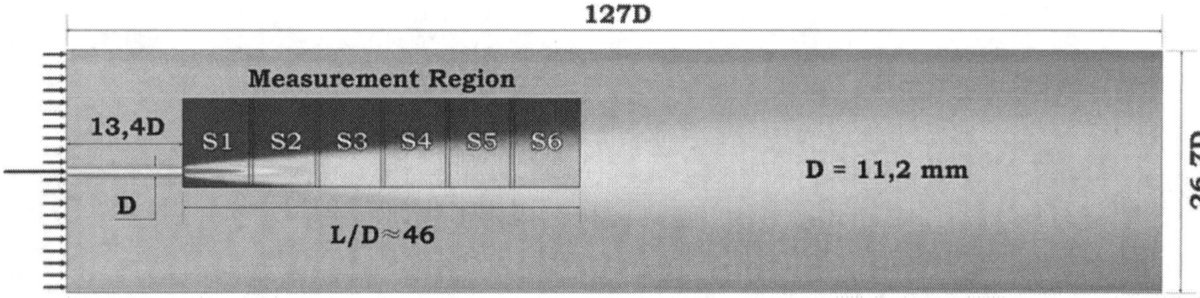

Figure 2. Geometrical conditions and measurement region.

Table 3. Sections for the experimental measurements.

Section	Initial position (mm)	(L/D) approx.	Final position (mm)	(L/D) approx.
S1	0	(0.0)	90	(8.0)
S2	85	(7.6)	175	(15.6)
S3	170	(15.2)	260	(23.2)
S4	255	(22.8)	345	(30.8)
S5	340	(30.4)	430	(38.4)
S6	425	(37.9)	515	(46.0)

Numerical Simulations

The objective of this study is to investigate the behavior of Reynolds Stress Model with version of Speziale, Sarkar and Gatski (RSM-SSG) [7] in a confined coaxial jet flow for different velocity ratios, changing the default values of some parameters of the model. This is a second order turbulence model, usually called a Second Order Closure. In this model, the eddy viscosity approach has been discarded and the Reynolds stresses are directly computed. For this investigation, the numerical simulations were performed varying the C_μ, $C_{\varepsilon\text{-}1}$ and $C_{\varepsilon\text{-}2}$ parameters in +30% and –30% of its default values, which are showed in **Table 4**. The C_μ parameter is related to the turbulent viscosity and the $C_{\varepsilon\text{-}1}$ and $C_{\varepsilon\text{-}2}$ parameters are related to the dissipation rate.

In order to analyze the influence of variation of these three parameters on the numerical simulations results, an experimental design of three factors at three levels was performed, which is presented in **Table 5**.

Table 4. Investigated parameters and its values to RSM-SSG model.

	C_μ	$C_{\varepsilon\text{-}1}$	$C_{\varepsilon\text{-}2}$
-30%	0.063	1.008	1.281
Default value	0.090	1.440	1.830
+30%	0.117	1.872	2.379

Table 5. Design of experiments.

Simulation	C_μ	$C_{\varepsilon\text{-}1}$	$C_{\varepsilon\text{-}2}$
1	0.063	1.008	1.281
2	0.063	1.008	1.830
3	0.063	1.008	2.379
4	0.063	1.440	1.281
5	0.063	1.440	1.830
6	0.063	1.440	2.379
7	0.063	1.872	1.281
8	0.063	1.872	1.830
9	0.063	1.872	2.379
10	0.090	1.008	1.281
11	0.090	1.008	1.830
12	0.090	1.008	2.379
13	0.090	1.440	1.281
14	0.090	1.440	1.830
15	0.090	1.440	2.379
16	0.090	1.872	1.281
17	0.090	1.872	1.830
18	0.090	1.872	2.379
19	0.117	1.008	1.281
20	0.117	1.008	1.830
21	0.117	1.008	2.379
22	0.117	1.440	1.281
23	0.117	1.440	1.830
24	0.117	1.440	2.379
25	0.117	1.872	1.281
26	0.117	1.872	1.830
27	0.117	1.872	2.379

The numerical simulations were conducted by a CFD commercial package, the FLUENT 13. The geometrical grid used in these simulations consists in approximately 291,640 hexahedral elements refined in direction to the nozzle. The advection scheme used in the simulations was High Resolution due to the strong convective component of the gas transport. The convergence criterion for all cases was 10^{-4} in the norm of the residues. The iterations were solved with a time step size of 10^{-4} s, in 5 s of real time simulation.

The boundary conditions for physical frontiers of the coaxial confined jet flow are:

- **Inlet (Chamber and Jet):** It is adopted an uniform and constant properties of the gas flow with an initial velocity condition.
- **Outlet:** Constant pressure with continuity conditions considered for all flow properties.
- **Wall:** No-slip conditions and wall logarithm function for the turbulent properties.

Results and Discussions

The numerical results of the axial mean velocities were compared with experimental data by means of the integrated absolute value of the error (IAE) criterion, where the set-point is the experimental data. The simulations results to the axial mean velocity are presented in **Table 6.**

Table 6. Value of the errors determined by IAE criterion for the axial mean velocity.

Simulation	Axial Position: 280 mm (L/D = 25) IAE	Axial Position : 420 mm (L/D = 37.5) IAE
1	56.90	61.66
2	371.91	310.90
3	429.92	359.04
4	722.83	779.65
5	98.84	113.02
6	381.16	333.84
7	725.64	776.38
8	687.09	778.01
9	152.83	171.00
10	72.45	98.80
11	337.09	283.65
12	376.14	323.50
13	734.91	790.29
14	73.42	72.92
15	327.58	283.47
16	739.67	791.41
17	718.39	800.91
18	93.72	128.06
19	101.49	125.93
20	294.26	249.73
21	364.61	335.27
22	735.83	786.78
23	55.80	50.35
24	275.87	239.68
25	747.39	800.45
26	730.34	807.13
27	67.06	85.16

In the **Table 6** it is possible to observe that the results to the simulations 1, 10, 23 and 27 are closer to the experimental data in relation to the default results (simulation 14) than the others simulations. These four best results are selected to be analyzed in graphical form in terms of mean axial velocity profiles and turbulence kinetic energy profiles, as can be seen in the Figure 3 and in the Figure 4, respectively, to two different axial positions.

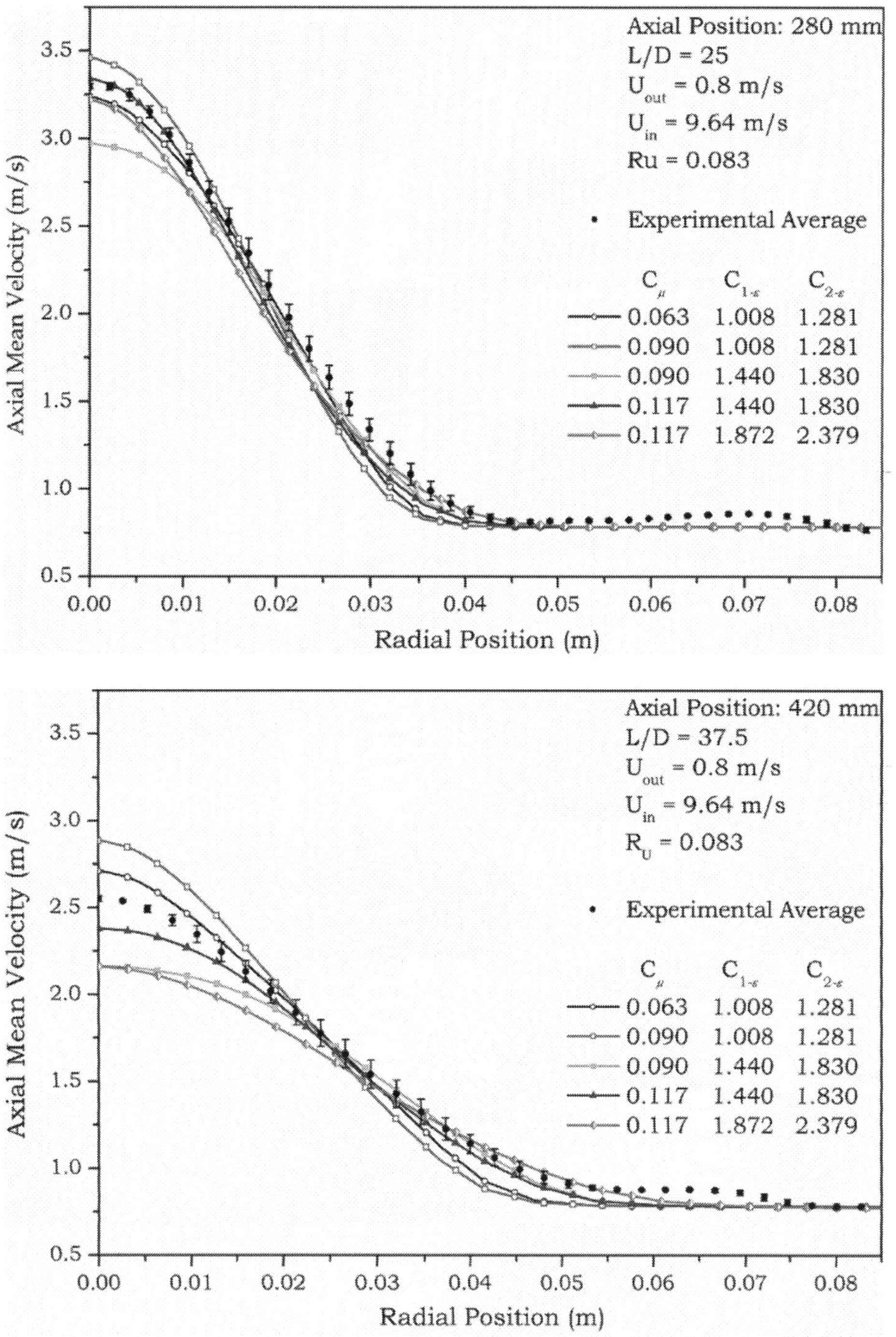

Figure 3. Axial mean velocity profiles for two different axial positions.

In **Figure 3** it is possible to observe that the best numerical result is obtained when the C_μ parameter is set to +30% to its default value and the two other parameters are fixed in default values (blue line). Extending this analysis, the **Figure 4** presents the turbulence kinetic energy profiles. To the both axial positions it is possible observe that the simulation with default values parameters presents the best result, however the simulation result which the C_μ parameter is adjusted in +30% to default value is also close to the experimental data, mainly in the L/D=37.5.

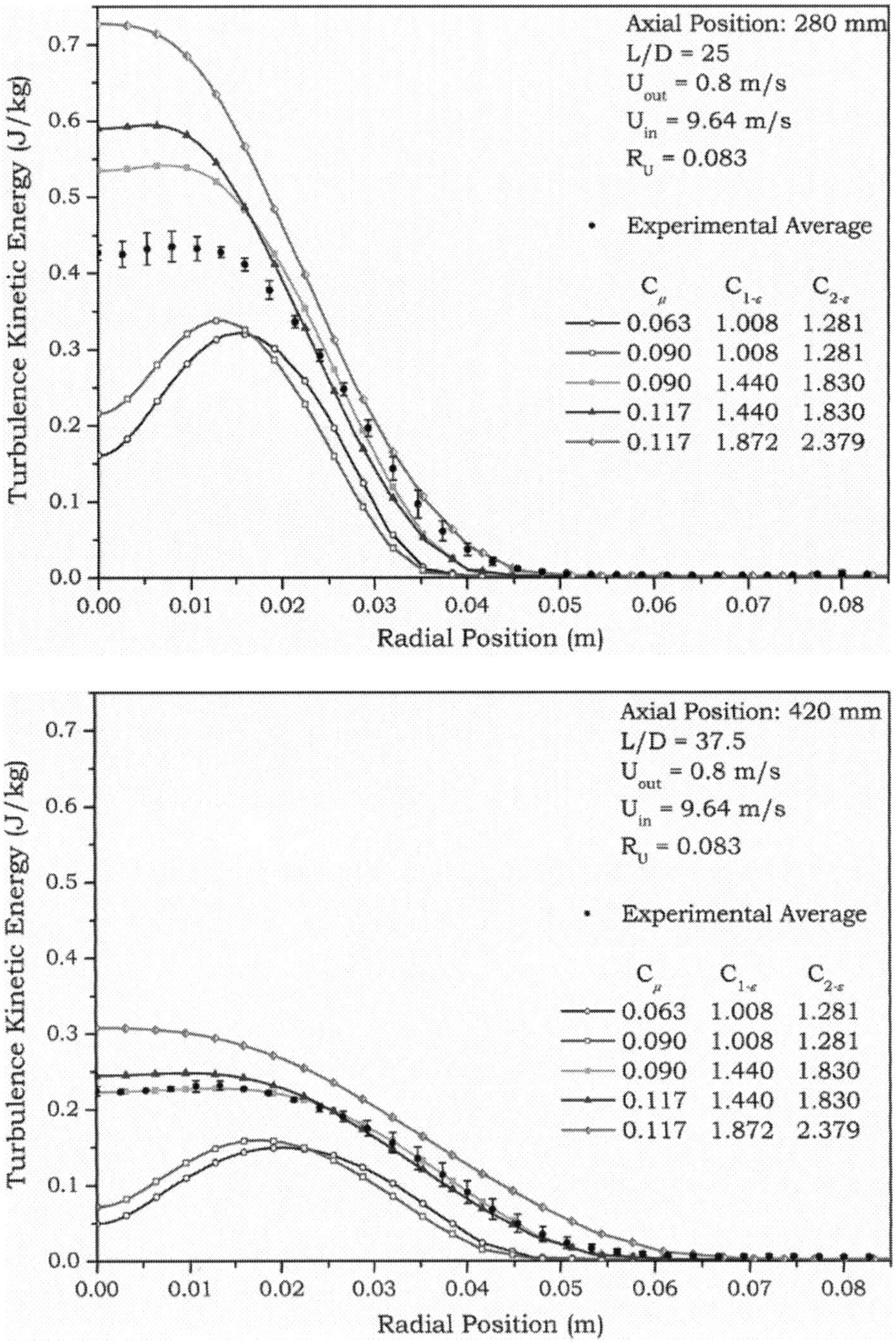

Figure 4. Turbulence kinetic energy profiles for two different axial positions.

These 27 numerical simulations were also performed to the others velocities ratios and the results obtained were similar to those presented previously.

Conclusions

The comparison between 27 simulations results with PIV experimental data showed that the best result is obtained when the C_μ parameter is modified from its default value (0.09) to a value 30% above the default value (0.117). These results indicate that the parametric sensibility analysis is a tool that can be used to determine the best fit of the model parameters. As future works, the investigation of the influence of variation of others parameters of the RSM-SSG model on the numerical results can be performed.

Acknowledgments

The authors are grateful to PETROBRAS, cooperation term 0050.0045549.08.4, for the financial support that guaranteed the development of this work.

References

1. Decker, R.K.; Buss, L.; Wiggers, V.R.; Noriler, D.; Reinehr, E.L.; Meier, H.F.; Martignoni, W.P.; Mori, M., (2011), Numerical Validation of a Coaxial and Confined Jet Flow, Chemical Engineering Transactions, 24, 1459-1464.
2. Ahmed, M. R.; Sharma, S. D., (2000) Effect of velocity ratio on the turbulent mixing of confined, co-axial jets. Experimental Thermal and Fluid Science, 22, 19-33.
3. Lin C.A., (1998), Modeling a confined swirling coaxial jet. Center for Turbulence Research, Annual Research Briefs.
4. Kiwata K., Ishii T., Kimura S., Okajima A. and Miyazaki K., (2006), Flow visualization and characteristics of a coaxial jet with a tabbed annular nozzle, JSME International Journal, Series B, 49, 4.
5. Balarac, G., Métais,O., Lesieur, M., (2007), Mixing enchancement in coaxial jets through inflow forcing: A numerical study, Physics of Fluids, 19, 75-102.
6. Mahmoud H., Kriaa W., Mhiri H., Le Palec G. and Bournot P., (2010), A numerical study of a turbulent axisymmetric jet emerging in a co-flowing stream, Energy Conversion and Management, 51, 2117-2126.
7. Speziale, C.G., Sarkar, S., Gatski, T.B., (1991), Modeling the Pressure-Strain Correlation of Turbulence: an Invariant Dynamical Systems Approach, Journal of Fluid Mechanics, Vol. 227, pp. 245-272.

(320h) Experimental and Computational Fluid Dynamics Study of a Confined Jet with Low Reynolds Number and High Jetting Velocity

Ning Chai, Hua Bai and Billy Smith, The Dow Chemical Company, Freeport, TX

Chemical vapor deposition (CVD) is a chemical process used to produce high-purity, high performance solid materials. CVD processes are typically conducted in a confined chamber. Gases are injected into the chamber using one single jet or multiple jets. The deposition (thickness) profile is affected by the flow pattern inside the deposition chamber, which depends primarily on the jetting behavior of the injection jets. For the particular CVD process of interest, the corresponding gas jets have unusually high velocity (100-1000 m/s) and high Mach number but relatively low Reynolds number (100-1000). Whether the jet is in the laminar, transitional, or turbulent flow regime can have significant impact on the jetting behavior thus the flow pattern. Generally speaking, the jetting energy (jet velocity) dissipates much faster in the turbulent flow regime than in the laminar flow regime. Study of such jetting condition is rare and hard to find in the public domain. Understanding the jetting behavior is critical for process improvement such as increasing throughput, improving deposition uniformity and reducing defects and failure rates. This paper aims to describe the flow behavior of confined gas jets under atmospheric and low pressure process conditions. Helium at the room temperature is used to mimic the low-Re and high-Ma jetting conditions in CVD reactor. In order to understand the jetting behavior in CVD reactors, a combined experimental and computational approach is adopted. This is a crucial step for model validation as well as providing insights to the process.

Laser light Mie scattering is utilized to study the overall flow transition behavior for the confined gas jet configuration. Gas phase particle image velocimetry (PIV) is then utilized for further illustration of the local flow behavior of the confined jet. The flow transition behavior observed in the experiment is then studied using computational fluid dynamics (CFD). Several popular RANS turbulence models are compared including the k-ε, k-ω SST Transitional models and the large eddy simulation (LES) model. The results suggest that with proper definition of the inlet flow boundary conditions, LES model is able to capture the flow transition behavior well while k-ε and k-ω SST model fail to do so. Compared with k-ε model, k-ω SST model predicts slower velocity decay in the initial stage and faster velocity decay in the later stage for the transitional flow.

(335a) Applications of the Reaxff Method to Metal and Metal Oxide Surface Chemistry

Adri van Duin, Department of Mechanical and Nuclear Engineering, Penn State University, State College, PA, Sung-Yup Kim, Mechanical and Nuclear Engineering, Pennsylvania State University, University Park, PA, Chenyu Zou, Mechanical Engineering, Pennsylvania State University, State College, PA, Thomas Patrick Senftle, Department of Chemical Engineering, Penn State University, State College, PA, Muralikrishna Raju, Dept. of Physics, The Pennsylvania State University, University Park, PA, Dan C. Sorescu, Computational Chemistry, National Energy Technology Laboratory, Pittsburgh, PA and Michael Janik, Department of Chemical Engineering, Pennsylvania State University-University Park, University Park, PA

The ReaxFF method provides a highly transferable simulation method for atomistic scale simulations on chemical reactions at the nanosecond and nanometer scale. It combines concepts of bond-order based potentials with a polarizable charge distribution.

Since it initial development for hydrocarbons in 2001 [1], we have found this concept to be highly transferable, leading to applications to elements all across the periodic table, including all first row elements, metals, ceramics and ionic materials (for example, [2-5]). For all these elements and associated materials we have demonstrated that ReaxFF can accurately reproduce quantum mechanics (QM)-based structures, reaction energies and reaction barriers, enabling the method to predict reaction kinetics in complicated, multi-material environments at a relatively modest computational expense.

In this presentation we will provide an overview of recent developments of the ReaxFF method, including its availability in parallel simulation environments, and recent application of this method to simulations on reactions at metal and metal oxide interfaces. In particular, we will discuss recent results of ReaxFF simulations on water/TiO_2 interfaces [6], protein/TiO_2 interfaces [7] and hydrocarbon catalytic conversion reactions at iron [8] and palladium surfaces. In all these cases we will describe how the ReaxFF parameters were developed by training against a predominantly QM based data set and how these parameters were subsequently used to perform a series of simulations that could be directly or indirectly validated against experiment.

[1] van Duin, A. C. T.; Dasgupta, S.; Lorant, F.; Goddard, W. A. Journal of Physical Chemistry A 2001, 105, 9396.

[2] Weismiller, M. R.; van Duin, A. C. T.; Lee, J.; Yetter, R. A. J. Phys. Chem. A 2010, 114, 5485.

[3] van Duin, A. C. T.; Bryantsev, V. S.; Diallo, M. S.; Goddard, W. A.; Rahaman, O.; Doren, D. J.; Raymand, D.; Hermansson, K. Journal of Physical Chemistry A 2010, 114, 9507.

[4] Joshi, K.; van Duin, A. C. T.; Jacob, T. Journal of Materials Chemistry 2010, 20, 10431.

[5] Fogarty, J. C.; Aktulga, H. M.; Grama, A. Y.; van Duin, A. C. T.; Pandit, S. A. J. Chem. Phys. 2010, 132, 174704/1.

[6] Kim, S.-Y.; Kumar, N.; Persson, P.; Sofo, J.; van Duin, A. C. T.; Kubicki, J. D. J. Comp. Chem. 2012, submitted.

[7] Monti, S.; van Duin, A. C. T.; Kim, S.-Y.; Barone, V. J.Phys.Chem. A 2012, 116, 5141.

[8] Zou, C.; van Duin, A. C. T.; Sorescu, D. Topics in Catalysis 2012, in print.

(335b) Liquid-Vapor and Liquid-Liquid Phase Equilibria in Binary Alloys with Modified Quantum-Corrected Sutton-Chen Potentials

Dongsheng Zhang, Materials Science and Engineering, University of Texas at Dallas, Richardson, TX and Lev Gelb, University of Texas at Dallas, Richardson, TX

The embedded atom method (EAM) has been extensively used to study metallic systems, and many different EAM potentials have been proposed, built from different functional forms and parameterized against different experimental and/or ab initio data, such as lattice constants, elastic constant, cohesive energies, and cluster structures. We consider here phase equilibria and thermophysical properties in binary alloys as modeled by the quantum-corrected Sutton-Chen potential (qSC) [1] and various mixing rules for unlike-element interactions [2]. Calculations of the enthalpy of mixing show that standard mixing rules do not generally agree well with experimental data [3-5]. We empirically determine modified cross-interactions which give satisfactory agreement, and consider how improved mixing rules may be developed for this class of systems. Using these improved potentials we then make predictions of liquid-vapor and liquid-liquid phase envelopes and other phase boundaries in selected systems.

1. S. N. Luo, et al., Phys. Rev. B 68 (2003) 134206
2. H. Rafii-Tabar, A.P. Sutton, Philos. Mag. Lett. 63 (1991) 217
3. K. Fitzner, Q. Guo, J. Wang, O. J. Kleppa, J. Alloys Compd. 291 (1999) 190
4. U. K. Stolz, I. Arpshofen, F. Sommer and B. Predel, J. Phase Eq. 14 (1993) 473
5. V. T. Witusiewicz, U. Hecht, S. G. Fries and S. Rex, J. Alloys Compd. 385 (2004) 133

(335c) Force Field for Describing the Colloidal Synthesis of Ag Nanostructures Via the Polyol Process

Ya Zhou and Kristen Fichthorn, Dept. of Chemical Engineering, Pennsylvania State University, University Park, PA

Polyvinylpyrrolidone (PVP) and polyethylene oxide (PEO) are structure-directing agents that exhibit different performance in the polyol synthesis of Ag nanostructures. The success of these structure-directing agents in selective nanostructure synthesis is often attributed to their selective binding to Ag(100) facets. We use first-principles, density-functional theory (DFT) calculations in a vacuum environment to show that PVP has a stronger preference to bind to Ag(100) than to Ag(111), whereas PEO exhibits much weaker selectivity. In an effort to understand the role of solvent in the surface-sensitive binding, we develop classical force fields to describe the interactions of the structure-directing (PVP and PEO) and solvent (ethylene glycol) molecules with various Ag substrates. We parameterize the force fields through force matching to DFT results using simulated annealing. We validate the force fields by comparisons to DFT and experimental binding energies. Our force fields reproduce the surface-sensitive binding predicted by DFT calculations. Molecular-dynamics simulations based on these force fields can be used to reveal the role of solvent, polymer chain length, and polymer concentration in the selective synthesis of Ag nanostructures.

(335d) Trappe-Zeo: Transferable Potentials for Phase Equilibria Force Field for All-Silica Zeolites

Peng Bai[1], Michael Tsapatsis[1] and J. Ilja Siepmann[1,2], (1)Chemical Engineering and Materials Science, University of Minnesota, Minneapolis, MN, (2)Chemistry, University of Minnesota, Minneapolis, MN

The transferable potentials for phase equilibria (TraPPE) force field is extended to allow for the investigation of adsorption behavior in all-silica zeolites where the zeolite host is treated as a rigid framework. The new TraPPE-zeo force field provides Lennard-Jones parameters and partial charges for the framework atoms. As for other parts of the TraPPE force field, Lorentz-Berthelot combining rules are employed to determine the parameters for all cross interactions, making the TraPPE-zeo force field applicable to any sorbate compound as long as suitable Lennard-Jones parameters and partial charges are available to describe sorbate-sorbate interactions; a significant advantage over force fields that explicitly tabulate the individual cross interaction parameters. The TraPPE-zeo force field is parameterized using a concurrent multi-dimensional fitting strategy to match a target set of experimental adsorption isotherms including n-alkanes, carbon dioxide, and ethanol as sorbates; that is, the TraPPE-zeo force field can treat the adsorption of non-polar, weakly polar, polar, and hydrogen-bonding compounds in a consistent manner. High accuracy of adsorption isotherms is also found for a diverse set of validation compounds over a wide range of pressures and temperatures.

(335e) Development of Force Field for Salts and Their Aqueous Solutions

Lukas Vlcek, Chemical Sciences Division, Oak Ridge National Laboratory, Oak Ridge, TN, Ariel A. Chialvo, Chemical Science Division, Oak Ridge National Laboratory, Oak Ridge, TN and John M. Simonson, Chemical and Engineering Materials Division, ORNL, Oak Ridge, TN

A balance between ion-ion and ion-water interactions plays a crucial role in the crystallization and dissolution of salts. Correct description of these interactions is therefore paramount for the understanding of natural processes leading to the precipitation of minerals, as well as for the design of new technologies, such as nuclear waste separation and storage. Our motivation is to investigate salt crystallization from mixed aqueous electrolyte solutions for the rational design of complexation agents used in selective crystallization of homologous series of oxoanions.

A reliable description of molecular and ionic interactions is needed to understand atomic-scale mechanisms underlying the thermodynamics and dynamics of electrolyte solutions. Thus, our goal is to develop a consistent force field parameterization for the study of oxoanions of general formula XO_4^{2-} (X=S, Se, Cr, Mo, W). While several force fields already exist for the description of simple alkali metal and alkali-halide series [1], and models for individual oxoanions have been also published [2], a consistent force field for oxoions suitable for comparative studies is not yet available.

The choice of an appropriate potential model form depends on the intended applications, and must balance accuracy and computational efficiency. Since salt nucleation and crystallization occur over large time and length scales, the simplicity of the model is of high importance. While most of the above ions are highly polarizable, their mineral environment is consistently polar, justifying the use of effective pair potentials to account for those polarizable contributions. Therefore we consider pair potentials represented by a combination of point charges and Lennard-Jones interactions that are compatible with the SPC/E water model [3]. We also test the limit of the effective description of the interactions and discuss ways to incorporate polarizability.

The resulting force field is optimized against experimental data including hydration free energy at infinite dilution [4], chemical potential at finite concentration based on the Kirkwood-Buff formalism [5], lattice constants and energies for selected crystals, and diffusion coefficient [4]. The potential parameters (at least 5 for each oxoion) were determined using global optimization based on the coupling parameter technique [6].

[1] Joung et al (2008) *J. Phys. Chem. B* **112**(30), 9020-9041.

[2] Cannon et al. (1994) *J. Phys. Chem.* **98**(24), 6225-6230.

[3] Berendsen et al. (1987) *J. Phys. Chem.* **91**(24), 6269-6271.

[4] Marcus, Y., *Ion properties* 1997, New York: Marcel Dekker.

[5] Gee et al. (2011) *J. Chem. Theory Comput.* **7**(5), 1369-1380.

[6] Vlcek et al. (2011) *J. Phys. Chem. B* **115**(27), 8775-8784.

Acknowledgements. This work was supported as part of the "Center for Nanoscale Control of Geologic CO_2", an Energy Frontier Research Center funded by the U.S. Department of Energy, Office of Science, Office of Basic Energy Sciences, and by the Division of Chemical Sciences, Geosciences, and Biosciences, Office of Basic Energy Sciences, U.S. Department of Energy.

(335f) Tetrahedrality At Hydrophobic Interfaces in a Coarse-Grained Water Model

Aviel Chaimovich and M. Scott Shell, Department of Chemical Engineering, University of California Santa Barbara, Santa Barbara, CA

The Lum-Chandler-Weeks theory highlights that hydrophobicity behaves differently for nonpolar spheres of different sizes: entropy dominates the hydration of small hydrophobes, while enthalpy dominates the hydration of large ones [1]. Similarly, Sarupria and Garde have found corresponding signatures for solvent density at hydrophobic interfaces, with low fluctuations near small solutes and high fluctuations near large ones [2]. One outstanding question is how this crossover with length scale can be interpreted in terms of the tendency of water to form a tetrahedral network. While tetrahedrality arguments have been frequently used to describe hydrophobicity, an exact picture of water's structure near nonpolar molecules remains unclear.

In this work, we address water's tetrahedral structure near hydrophobic colloids using coarse-grained models in molecular simulations. The water model is an isotropic ("core-softened") potential that is developed through a novel procedure which minimizes the information lost upon coarse-graining [3]. For the hydrophobes, we assume they are composed of Lennard-Jones points, and we use the Hamaker approximation to attain an isotropic potential for colloids whose size can vary. While the system might appear overly simplified, such a description has been shown to adequately capture several signatures of hydrophobic association [4]. Here, we show that the models, despite being isotropic in nature, manifest tetrahedrality through a careful balancing of driving-forces at different length scales. To enumerate the effects of tetrahedrality on hydrophobicity, we modify a well-studied tetrahedral order parameter so that it is suitable for interfaces, and we also propose several novel related metrics. We measure the disturbance of the hydrophobe on interfacial tetrahedrality, and we find that a select few of these order parameters show signatures of the corresponding Lum-Chandler-Weeks crossover. Finally, we present simple arguments for this crossover, emphasizing the manner in which this description manifests tetrahedrality, [1,2].

1. Lum, K., D. Chandler, and J.D. Weeks, *Hydrophobicity at Small and Large Length Scales.* The Journal of Physical Chemistry B, 1999. **103**(22): p. 4570-4577.

2. Sarupria, S. and S. Garde, *Quantifying Water Density Fluctuations and Compressibility of Hydration Shells of Hydrophobic Solutes and Proteins.* Physical Review Letters, 2009. **103**(3): p. 037803.

3. Chaimovich, A. and M.S. Shell, *Anomalous waterlike behavior in spherically-symmetric water models optimized with the relative entropy.* Physical Chemistry Chemical Physics, 2009. **11**(12): p. 1901-1915.

4. Hammer, M.U., et al., *The Search for the hydrophobic force law.* Faraday Discussions, 2010. **146**: p. 299-308.

(335g) The Importance of Compressed Liquid Density in Unambiguously Characterizing Molecular Interactions

J Richard Elliott[1], Ehsan Moharreri[1] and Ahmadreza F. Ghobadi[2], (1)Chemical and Biomolecular Engineering, The University of Akron, Akron, OH, (2)The University of Akron, Akron, OH

The compressed liquid density was studied using step potential models to improve understanding of what is necessary for a potential model to achieve an accurate description. In general, four isochores were for each compound. Properties for spherical molecules and n-alkanes through dodecane were available to high reduced pressures and temperatures from the NIST Webbook. Comparisons were based on compressibility factor (Z = PV/RT), temperature, and liquid density. Deviations are summarized in terms of the difference between temperatures needed to achieve the experimental Z at the specified density. The compressibility factor approaches zero near saturation, so the temperature intercept of the isochore precisely characterizes the saturation condition as well as compressed density. Accurate densities are an important preliminary to accurate prediction of transport properties. The Lennard-Jones equation of state (LJEOS) Kolafa and Nezbeda is applied as a basis for comparison for spherical molecules. LAMMPS simulations of the TraPPE-UA were compared for n-alkanes.

An investigation using the square-well model exhibited remarkable sensitivity to the square-well diameter when applied in this manner. An optimal well-width of 1.7 led to much smaller deviations than the commonly assumed value of 1.5. On the other hand, the LJEOS yielded accurate correlations for spherical molecules without modification. For n-alkanes, however, the TraPPE model (also based on the LJ model) was not as accurate as the step potential model. The greatest reduction in error derives from details of the attractive range of the potential model, rather than the repulsive shoulder. The accuracy and utility of thermodynamic perturbation theory are assessed in addressing this problem. In conclusion, accurate compressed liquid densities probe details of the potential model with surprising sensitivity.

(335h) A Predictive Transferable Force Field for Primary, Secondary and Tertiary Amines

Gustavo A. Orozco[1], Carlos Nieto-Draghi[1], Allan D. Mackie[2] and Veronique Lachet[1], (1)Thermodynamic and Molecular Simulation, IFP Energies nouvelles, Rueil- Malmaison, France, (2)Departament d'Enginyeria Química, Universitat Rovira i Virgili, Tarragona, Spain

Based on the Anisotropic United Atom approach AUA, we propose a transferable force field that includes primary, secondary and tertiary amines. A set of partial charges was calculated using Density Functional Theory (DFT) while for the Lennard-Jones parameters an optimization procedure was carried out fitting vaporization enthalpies, liquid densities and vapor pressures of a few reference molecules. For the hydrocarbonated groups involved in the studied molecules, we have transferred the parameters of the AUA4 force field for hydrocarbons. In order to include different molecular structures in our study, a set of torsional potentials has been proposed including linear, branched and multi-functional amines.

Monte Carlo simulations have been carried out in the Gibbs ensemble in order to calculate thermodynamic properties at different temperatures such as phase equilibrium diagrams, vaporization enthalpies, vapor pressures, normal boiling points, critical densities, critical temperatures and critical pressures.

The transferability of the force field has been evaluated for different amine molecules obtaining an excellent accuracy on the predictions of equilibrium properties as well as on critical coordinate estimations. Our force field is also able to reproduce with high accuracy the normal boiling points of the different studied amines with average deviations less than 2%.

We have also studied the capability of our force field to predict transport properties. Viscosity coefficients have been calculated using molecular dynamics in the isothermal-isobaric ensemble by means of the Einstein formalism. We have found a good agreement between the predictions of our force field and the experimental data.

(345a) Molecular Simulation Study of Wetting of Fluid Mixtures

Vaibhaw Kumar, Chemical and Biological Engineering, University at Buffalo, Buffalo, NY and Jeffrey R. Errington, Chemical and Biological Engineering, University at Buffalo, The State University of New York, Buffalo, NY

Fluid mixtures can exhibit a rich variety of wetting behavior in the vicinity of a solid surface. This diversity is due, in part, to the presence of multiple species, which allow for the possibility of liquid-vapor as well as liquid-liquid phase equilibrium. In this presentation we describe our recent efforts to study the wetting behavior of mixtures via molecular simulation. We focus on methodological developments that enable one to compute interfacial properties of fluid mixtures in an efficient manner. Our basic approach involves calculating the surface excess free energy (also known as the interface potential) as a function of the surface density of the fluid. The interface potential provides information regarding qualitative wetting features (such as wetting regime and the order of wetting transition) as well as access to quantitative information, such as the contact angle and interfacial tensions. We outline several strategies for acquiring the components of the interface potential that are necessary to evaluate interfacial properties of interest. We demonstrate the utility of these methods by computing interfacial properties of different Lennard-Jones mixtures which display i) simple, ii) azeotropy, iii) and liquid-liquid phase separation in the bulk phase. We also discuss how these techniques can be used to understand the wetting behavior of model oil-water systems.

(345b) Brownian Particle Adsorption At Fluid-Fluid Interfaces: The Effects of Geometry and Surface Roughness

Carlos E. Colosqui[1], Jeffrey F. Morris[1,2] and Joel Koplik[1,3], (1)Levich Institute of Physico-Chemical Hydrodynamics, New York City, NY, (2)Chemical Engineering, City College of City University of New York, New York City, NY, (3)Department of Physics, The City College of City University of New York, New York, NY

Via detailed molecular dynamics simulations we study the adsorption of Brownian particles at a fluid-fluid interface. In the present study, we focus on the technologically important case of Brownian particles that exhibit a discontinuous or irregular surface morphology. We observe that the position of the center of mass and the angular orientation of the studied particles can present a multiplicity of metastable states. The observed metastable states correspond to local minima in a nontrivial landscape of the system free energy. The presence of multiple metastable states can significantly retard, or even prevent, the relaxation to a state of minimum free energy that is predicted by continuum theory for ideally smooth particles. We find that the shape and size of the microscale roughness of the particle determine the energy barriers between observed metastable states and the particle mobility at the interface. These results shed important insights in the dynamics of colloidal assembly at fluid-fluid interfaces.

(345c) Measuring Hamaker Constants by Atomic Force Microscopy From the "Jump-Into-Contact" Distance: Quasi-Static Models and Dynamic Simulations

Jiannan Dong[1], Hou T. Ng[2], Elias I. Franses[1] and David S. Corti[1], (1)School of Chemical Engineering, Purdue University, West Lafayette, IN, (2)Hewlett-Packard Laboratories, Hewlett-Packard Co., Palo Alto, CA

The Hamaker constant is a quantitative measure of the fundamental attractive van der Waals interactions among particles, between particles and a flat surface, and other geometries. It plays an important role in numerous phenomena and behaviors of materials, such as colloidal stability, interfacial adhesion, and nanoparticle self-assembly.

Improved methods and procedures for determining the Hamaker constant from atomic force microscopy (AFM) measurements are presented. A quasi-static analysis is used to correct an earlier method based on the "jump-into-contact" distance represented by the delta d-value, accounting for the tip-surface interaction forces or energies. It is further extended to three tip-surface models in addition to the simplest sphere-plane model. The sensitivity of all four models to predicting the delta d-values and from them the Hamaker constants is investigated. A mechanical-energy-based analysis is also reported. It predicts the same results as the quasi-static force analysis, but offers additional insights on the mechanical stability of the tip.

In addition, a novel dynamic analysis of the tip motion is developed to simulate the tip motion for determining the time-dependent tip deflection as a function of two dimensionless groups, a and b. The parameter a is the ratio of the characteristic time of the tip to the time scale of the tip motion. The parameter b involves some characteristic properties of the tip, i.e. the radius of curvature and the spring constant, and an operational parameter, which is the initial tip height relative to the surface. Effective delta d-values depend on the number of data points sampled from the dynamic curve describing the motion of the tip. As the number of data points increases, the thus-determined dynamic delta d-value decreases. As the approaching speed of the cantilever decreases, the dynamic delta d-value approaches the quasi-static delta d-value. Therefore, the dynamic analysis provides insights on how to improve the experimental procedures for obtaining more accurate and reliable Hamaker constants.

Finally, the quasi-static delta d-values are compared with the dynamic AFM data for three model systems of glass (amorphous SiO_2), polystyrene, and crystalline copper phthalocyanine (CuPc). The measured data are in fair agreement with the predictions from four models using the literature values of the Hamaker constants. To further test the methods quantitatively, additional measurements with improved experimental procedures, AFM instrument designs, and tips and materials, are needed.

Prediction of Interfacial Tensions between Demixed Solvents

Tommy Grunert, Sabine Enders, Chair of Thermodynamics and Thermal Separation Processes, Technical University of Berlin, Berlin, Germany

Introduction

The study of interfacial properties is more and more in the focus of the industrial as well as the academic world. This is caused by the large impact of interfacial properties in chemical engineering, chemistry and pharmacy. The special challenge in this field consists in the experimental difficulties to measure the interfacial properties, like interfacial thickness, interfacial profiles in terms of density and concentration of each species in the interface. Only the interfacial tension itself can be measured with a high accuracy. From the thermodynamic point of view it will be desirable to develop a method able to predict the interfacial properties based on the bulk properties, for instance the involved phase equilibria, and interfacial tension of the corresponding subsystems. One possibility to reach to goal is the application of the density gradient theory (DGT) originally developed by van der Waals [1] and Rayleigh [2] and rediscovered by Cahn and Hilliard [3].

This theory is widely used for the calculation of surface tensions of pure components or mixtures (e.g. [4,5,6,7,8,9,10,11,12,13,14,15]). For this purpose the DGT must be combined with an equation of state (EOS). Seldom, the DGT was applied to the interface between demixed liquids[6,16,17,18]; however, these studies are limited to binary mixtures. For practical point of view multi-component mixtures are more interesting. Along this line, this contribution focuses on the prediction of the interfacial tension of ternary mixtures based on experimental interfacial tensions of binary subsystem and bulk data of the ternary system. Recently[19], a theoretical framework was provided in order to study a system having only one miscibility gap. For the study of the interface between to liquids exists two possibilities, the first is a combination with a suitable EOS[6,16,17] and the other one is the combination with a model for the Gibbs energy of mixing [6,18,19]. The optimization of extraction columns is mostly based on a model for the Gibbs energy of mixing. For this reason this method is chosen for this contribution.

For example, the ternary system composed of toluene + acetone + water has a large miscibility gap and relatively high values for the interfacial tension. This mixture was chosen as reference system for the investigation of extraction processes. The phase behavior[20] as well the interfacial properties[21,22,23,24] of this system were studies in the literature using different experimental methods.

Theory

Within the DGT, in its original version[1], the Helmholtz-energy of the inhomogeneous system is developed in a Taylor series. At constant temperature and pressure the densities and the composition will be changed during this interface[4]. In this case the DGT needs a suitable equation of state for the calculation of the Helmholtz-energy of the inhomogeneous system and the densities in the coexisting phases. However, if it is assumed that the pressure has only a negligible influence on equilibrium between two liquid phases (LLE) only the concentrations change during the interface must be considered[6]. In this case the density gradients will be replaced by the concentration gradients and the Helmholtz-energy density by the Gibbs- energy.

Gibbs energy of mixing

In thermodynamics of low-molecular weight mixtures, usually, the derivation of a real mixture from an ideal mixture is considered. This is in contrast to polymer solution thermodynamics, where the derivation of a real mixture from an ideal-athermic mixtures is handled[25,26], meaning that entropic effects arising from the difference in molecular size are automatically taken into account. In our theoretical framework[19] we take advantage from the ideal-athermic mixture rather than the ideal mixture also for mixtures composed from low-molecular weight substances, for instance for solvent mixtures. For these type of mixtures the concentration is measured by the segment mole fraction, Φ_i, of component i in the mixture. The introduction of the segment fraction requires the definition of the segment number, r_i. The segment number r_i is defined by $r_i = \frac{M_i}{M^+}$, where the reference molar mass M^+ is the molar mass of the smallest molecule in the mixture. Using this definition, it can be shown that the weight fraction is equal to the segment fraction.

Derivations of the real mixture from the ideal-athermic one can be modeled with the ansatz developed by Koningsveld-Kleintjens [27,28]. For a mixture made from k components the following expression for the Gibbs enthalpy of mixing is valid:

$$\frac{\Delta_{mix} g}{RT} = \sum_{i=1}^{k} \frac{\Phi_i \ln \Phi_i}{r_i} + \sum_{i=1}^{k-1} \sum_{j=i+1}^{k} \frac{\chi_{ij}(T)\Phi_i \Phi_j}{1-\gamma_{ij}\Phi_i} \tag{1}$$

where $\chi_{ij}(T)$ as well as γ_{ij} are the adjustable model parameter. The temperature-independent quantities γ_{ij} take the concentration dependency of the classical χ-parameter, introduced by Flory[25] and Huggins[26], into account. Applying these parameter allows to calculate also un-symmetrically miscibility gaps. The model parameter must be fitted to experimental bulk data. For binary systems having only one critical point (lower-critical solution temperature [LCST] or upper critical solution temperature [UCST]) the temperature dependency of the $\chi_{ij}(T)$ can modeled using:

$$\chi_{ij}(T) = \beta_{1,ij} + \frac{\beta_{1,ij}}{T} \tag{2}$$

For simplicity, a_{ij}-parameter occurring originally in Eq. (1)[27,28] is put to 0.

Density gradient theory (DGT)

In our theoretical framework[19] is assumed that the pressure has no impact on the LLE and the usually considered density gradients were replaced by the concentration gradients of each component between the two liquid bulk phases. For a binary mixture this situation is illustrated schematically in Fig. 1, where z is the direction perpendicular to the interface. Both profiles run from the constant equilibrium concentration of one liquid bulk phase continuously to the equilibrium concentration of the coexisting phase. Considering only the direction perpendicular to the interface mean that a flat interface is assumed. Often the role of capillary wave are discusses in the literature, controversially. Here is postulate that no capillary wave occurs.

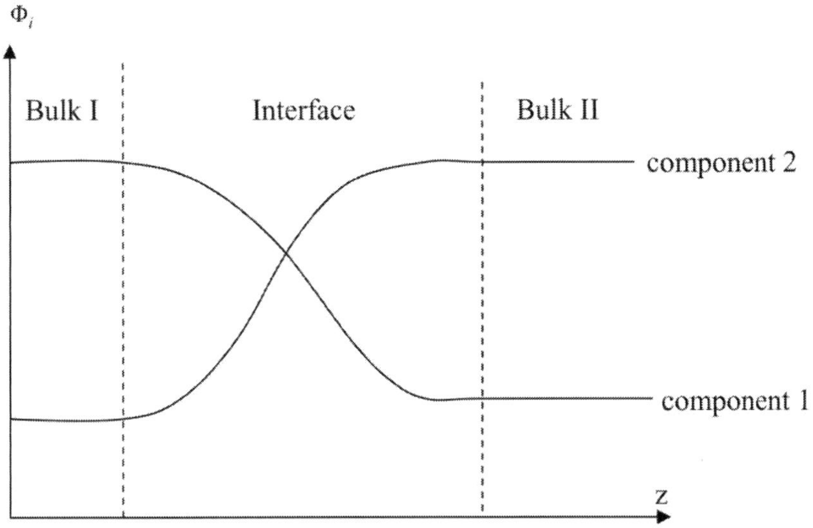

Fig. 1. Schematic drawing of the continuously concentration variation inside the interfacial region of a binary system.

According to Cahn and Hilliard theory[3] of inhomogeneous systems with an interface between the coexisting bulk phases, the interfacial tension is proportional to the grand thermodynamic potential, $\Delta\Omega$. If the pressure plays a role than the grand thermodynamic potential can be calculated via the Helmholtz energy density[3]. If the density of the whole system can be regarded as constant, the Helmholtz energy is equal to the Gibbs energy and hence, the grand thermodynamic potential, can be expressed with help of the Gibbs energy of mixing. The grand thermodynamic potential $\Delta\Omega$ for a binary system reads[6]:

$$\Delta\Omega(\Phi_1) = \Delta_{Mix}g(\Phi_1) - \Phi_1\mu_1^{Eq} - (1-\Phi_1)\mu_2^{Eq} \tag{3}$$

where μ_i^{Eq} is the chemical potential of the component i in the equilibrium which is obtained from the phase equilibrium conditions. At equilibrium the grand thermodynamic potential must be zero. According to the work of Enders and Quitzsch[6], the interfacial tension is proportional to the grand thermodynamic potential:

$$\sigma = \int_{\Phi_1^I}^{\Phi_1^{II}} \sqrt{2\kappa'\Delta\Omega(\Phi_1)}d\Phi_1 \tag{4}$$

The interfacial tension of a binary system depends on the integral over the grand thermodynamic potential $\Delta\Omega$, which is a function of the local composition, and on the so-called influence parameter κ', which has a physical meaning[3]. Let us consider the case of a system containing a pure component in its liquid-vapor equilibrium. In this situation, κ represents the second moment of the auto correlation function. Usually, κ is determined using one experimental surface tension at one temperature[5,6,7,12]. For mixtures, this value can be obtained by mixing rules[5,6,7,12]. However, replacing the Helmholtz energy density by the Gibbs energy means that parameter κ' lost his physical meaning and must be treated as adjustable parameter. In contrast, to the equation used for the calculation of the surface tension of a pure components[5,6,7] the density difference between the coexisting phases is missing and the grand thermodynamic potential is measured in $J/_{mol}$ instead of Pa for the Helmholtz energy density, if the DGT is combined with an EOS for the calculation of the surface tension. For this reason, the unit of κ' is not identical to the unit of κ used for the surface tension calculations. Usually, κ is measured in $J\,m^5\,mol^{-2}$ and κ' in $J\,mol\,m^{-4}$. The difference between both values corresponds to $\kappa = \kappa'/\rho^3$, where ρ is the density, which can be obtained experimentally.

In a LLE of a binary system there is just one free variable inside of the interfacial region, the segment mole fraction, for instance of component 1. For a ternary system two independent variables describing the concentration within the interface are needed. Using the ideal-athermic mixture as reference state, the concentration variables are the segment mole fractions of component 1 and 2. This is the same number of free variables in comparison to the vapor-liquid-equilibrium (VLE) of a binary system, where the density and the composition are changing during the interface. In order to deal with two variables, Poser and Sanchez[4] introduced the so-called partial density, which is the product of the mole fraction of component i and the local density. Hence, the partial densities for a VLE of a binary system equates to the segment mole fractions for a LLE of a ternary system. Therefore, a modified equation, derived by Poser and Sanchez[4], can by transfer to a ternary LLE. Applying these ideas leads to[19]:

$$\kappa' = \kappa_B + 2\sqrt{\kappa_A \kappa_B}\left(\frac{d\Phi_3}{d\Phi_1}\right) + \kappa_A \left(\frac{d\Phi_3}{d\Phi_1}\right)^2 \qquad (5)$$

In this notation, component 1 and 3 generate a binary LLE and component 2 is the transition component, which is mostly present at low concentration. In the regime of infinite low concentration of the transition component the concentration gradient in Eq. (5) has to converge to minus one, resulting in[19]:

$$\kappa' = \kappa_B - 2\sqrt{\kappa_A \kappa_B} + \kappa_A \qquad (6)$$

The parameter κ' in Eq. (6) can be expressed by two new parameter, namely κ_A and κ_B. But, κ_A and κ_B are not the influence parameter of pure components A and B inside of this theoretical framework[19]. More precisely, they are two adjustable parameters (κ_A and κ_B) in Eq. (5) for the calculation the interfacial tension. Eq. (6) connects the parameter κ', which can be obtained from the interfacial tension of the binary subsystem, and the two adjustable parameters κ_A and κ_B of the ternary system. To reduce the number of parameters to calculate the interfacial tension of a ternary LLE, the parameter κ_A is put equal zero. Insert this assumption in Eq. 6 it results in[19].

$$\kappa' = \kappa_B \qquad (7)$$

This assumption has the advantage, that the interfacial properties of the ternary system can be predict, if the influence parameter κ' is known from interfacial tension data of the binary subsystem forming the miscibility gap. Additionally, the composition profiles inside the interfacial region are determined by[19]:

$$z = z_0 + \int_{\Phi_1^I}^{\Phi_1^*} \sqrt{\frac{\kappa'}{\Delta\Omega}} d\Phi_1 \qquad (8)$$

The reference coordinate z_0 can be chosen to zero and the segment molar fraction Φ_1^* varies between the bulk concentrations. Thus, any value of the segment molar fraction inside the interfacial region can be assigned to one spatial coordinate. Caused by the missing density the calculated z has the unit mol/m^2. In order to obtain the physical spatial coordinate the calculated value has to be divided by the density difference of the coexisting phases measured in mol/m^3. Based on the concentration profile inside the interface the interfacial thickness and the relative adsorption of one component can be calculated.

Results

Recently[19], the theory was applied successfully to two systems, namely 1-butanol + ethanol + water and toluene + acetone + water. Both systems differ strongly in their demixing behaviour. Whereas the first system has only a relative small miscibility gap, the miscibility gap of the second system is quite large. Consequently, the interfacial tension of the second system is much higher then the interfacial tension of the first system. However, the temperature dependency was only studied for the first system. Therefore, in this contribution, we focus our attention to the influence of temperature on the interfacial tension of the second mixture made from toluene + acetone + water. The binary subsystem toluene + water creates the miscibility gap in the ternary mixture. Following our notation, introduced above, the transition component 2 is acetone. Toluene is the component 1 and water represents the component 3.

Table 1. Parameter set for the model (Eq (2)) for the LLE of the ternary system toluene (1) + acetone (2) + water (3).

i and j	$\beta_{1,ij}$	$\beta_{2,ij}$ $[K]$
1-2	-0.8769	-300
1-3	9.06	-1716
2-3	2.60	-559

The phase equilibria were measured by Hackl et al.[20]. Using these data the parameter of Eq. 1 can be adjusted. If we have in mind extraction prosesses, the most important region of the phase diagram is far away from the critical point. Therefore, the focus is located at low acetone concentrations. The experimental phase diagram including binodal line and tie-lines can be described with a good accurancy[19]. The temperature-independent $\gamma_{i,j}$-values were fitted to experimental data at 293.15 K[19]. For the estimation of the parameter occurring in Eq. (2) experimental data[20] at 283.15K, 293.15K and 303.15K were used. The obtained values are listed in Table 1. The large value, obtained for $\beta_{1,13}$ (Table 1) can be explained by the large miscibility gap occurring between water and toluene. The sign of the $\beta_{2,13}$-value is negative; with other words the binary subsystem exhibits a miscibility gap with an upper critical solution temperature.

For the interfacial tension calculation the parameter κ' in Eq. (4) is required. This value can be estimated using the experimental data of the binary subsystem water + toluene. The interfacial tension was measured by Prokop et al. [29] at 293.15 K resulting in $35.2\,mN \cdot m^{-1}$. Based on this value, the parameter κ' can be adjusted to $\kappa' = 7.5 \cdot 10^{-7}\, J \cdot mol \cdot m^{-4}$. Now, it is possible to predict the interfacial tension for the ternary system as function of the mass fraction of acetone in the aqueous phase. Fig. 2 demonstrates clearly that the theoretical prediction (black line in Fig. 2) is excellent in comparison with the experimental data (filled squares in Fig. 2) from Misek et al.[21].

If the DGT is applied for the calculation of the surface properties is often found that the influence parameter κ is temperature-independent and can be extrapolated over a large temperature range[6,12,13]. Now, the question arises, if the adjustable parameter κ' can also be handled as a temperature-independent value, although his physical meaning is lost. Based on the experimental data from Backes et al. [24] for the interfacial tension of the binary subsystem water + toluene at different temperatures, the parameter κ' was calculated at different temperatures. The adjusted parameter shows a small linear temperature dependency according:

$$\kappa' = 10^{-7}\left(3.824 - \frac{0.01 \cdot T}{K}\right)\frac{J \cdot mol}{m^4} \tag{9}$$

The reason for this finding can be the not perfect model for the Gibbs energy of mixing, which takes not into account the self-association of water or the polar properties of acetone. Nevertheless, Eq. (9) allows the prediction of the interfacial tension of the ternary system at different temperatures.

By using the set of parameter for the energy model (Eq. (1)) and the parameter κ' at different temperatures (Eq. (9)), the presented model of the DGT is able to predict the interfacial tensions at these different temperatures. In Fig. 2 can be seen, that the interfacial tension is decreasing by increasing the temperature. This behavior is connected to the UCST behavior of the binary subsystem water + toluene.

With the help of Eq. (8) it is possible to predict the concentration profiles of each component inside the interfacial region. It can bee seen in Fig. 3 that the concentrations varies continuously from bulk phase to bulk phase. Dependent on the mole fraction of acetone in the aqueous phase, the thickness of the interfacial region will increased as expected with approaching proximity to the critical point. The enrichment of acetone inside the interfacial region can have an influence on the mass transfer. A similar enrichment of acetone was also found in our previous study[19] at different acetone concentrations.

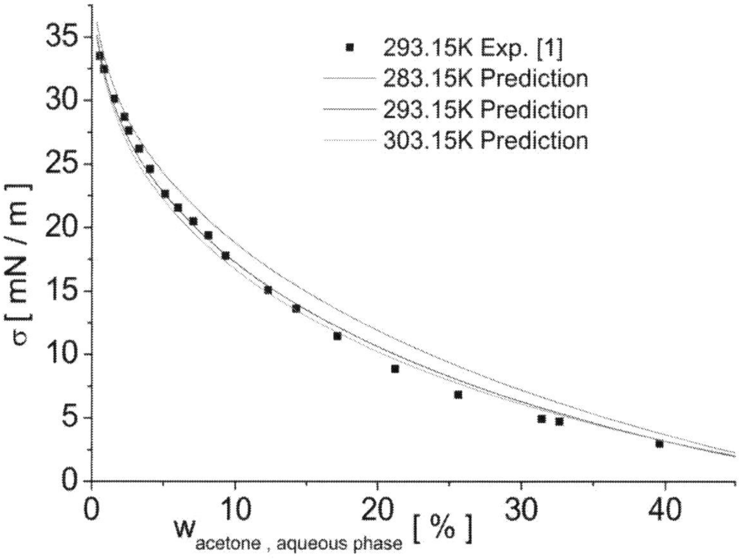

Fig. 2. Predicted interfacial tensions for the system toluene + acetone + water at different temperatures (black squares: experimental data at 293.15K[21]; blue line: 283.15 K; black line: 293.15 K[19]; red line: 303.15 K).

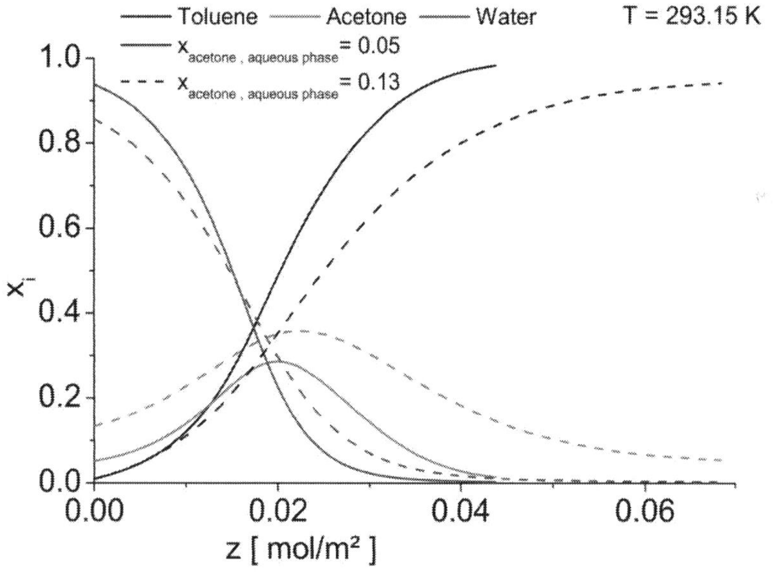

Fig. 3. Predicted mole fraction profiles (black line: toluene; red line: acetone; blue line: water; solid line: $x_{acetone,aqueous\,phase} = 0.05$; dashed line: $x_{acetone,aqueous\,phase} = 0.13$) inside the interfacial region for the system toluene + acetone + water at 293.15 K.

Conclusions

We extended our previously presented version of the DGT combined with a model for the Gibbs energy of mixing. The usage of this model allows the prediction of the interfacial tension of the ternary systems, especially toluene + acetone + water for different temperatures based on information of the interfacial tension of the binary subsystems having a miscibility gap and the phase equilibria data of the ternary system. The prediction of the interfacial properties, including interfacial tension, concentration profiles, interfacial thickness, and relative enrichment of one component, based on the assumption that the so-called influence parameter keeps constant for a special miscibility gap in a binary system, if a transition component is added to the binary system. For the studied system, this parameter depends linearly on temperature.

References

[1] J. D. van der Waals, Z. Phys. Chem. **13** (1894) 657.
[2] Lord Rayleigh, Philos. Magn. **33** (1892) 1.
[3] J. W. Cahn, J. E. Hilliard, J. Chem. Phys. **28** (1958) 258.
[4] C. I. Poser, I. C. Sanchez, Macromolecules **14** (1981) 361.
[5] P.M.W. Cornelisse, C.J. Peters, J. de Swaan Arons, J. Chem. Phys. **106** (1997) 9820.
[6] S. Enders, K. Quitzsch, Langmuir **14** (1998) 4606.
[7] C. Miqueu, B. Mendiboure, A. Graciaa, J. Lachaise, Ind. Eng. Chem. Res. **44** (2005) 3321.
[8] A.J. Queimada, C.Miqueu, I.M. Marrucho, G.M. Kontogeorgis, J.A.P. Coutinho, Fluid Phase Equilibria **228-229** (2005) 479.
[9] C. Miqueu, B. Mendiboure, A. Graciaa, L. Lachaise, Fuel **87** (2008) 612
[10] K. Glavatskiy, D. Bedeaux, Phys. Rev. E **77** (2008) 061101.
[11] M.B. Oliveira, I.M. Marrucho, J.A.P. Coutinho, A.J. Queimada, Fluid Phase Equilibria **267** (2008) 83.
[12] O. G. Nino-Amezquita, S. Enders, Ph. T. Jaeger, R. Eggers, Ind. Eng. Chem. Res. **49** (2010) 592.
[13] O. G. Niño Amézquita, S. Enders, P.T. Jaeger, R. Eggers, J. Supercrit. Fluids **55** (2010) 724-34.
[14] T. Lafitte, B. Mendiboure, M.M. Pineiro, D. Bessieres, C. Miqueu, J. Phys. Chem. B **114** (2010) 11110.
[15] C. Miqueu, J.M. Míguez, M.M. Piñeiro, T. Lafitte, B. Mendiboure, J. Phys. Chem. B **115** (2011) 9618.
[16] H. Kahl, S. Enders, Phys. Chem. Chem. Phys. **4** (2002) 931.
[17] S. Enders, H. Kahl, Fluid Phase Equilibria **263** (2008) 160-167.
[18] T. Zeiner, P. Schrader, S. Enders, D. Browarzik, Fluid Phase Equilibria **302** (2011) 321.
[19] T. Grunert, H. Rudolph, S. Enders, Zeitschrift für Physikalische Chemie, submitted, 2012.
[20] A. Hackl, W. Solar, G. Ziebland, Eur. Fed. Chem. Eng. London (1978).
[21] T. Misek, R. Berger, EFCE Publication Series **46** (1985).
[22] S. Enders, H. Kahl, J. Winkelmann, J. Chem. Eng. Data **52** (2007) 1072-1079.
[23] N.F. Murphy, J.E. Lastovica,J.G.Fallis, Ind. Eng. Chem. **49** (1957) 1035.
[24] H. M. Backes, "Measurement and Correlation of the Interfacial Tension Between two Liquid Phases.", Fortschr. Ber. VDI Z. Reihe **3** (1984) 97.
[25] P. J. Flory, J. Chem. Phys. **9** (1941) 660.
[26] M. L. Huggins, J. Chem. Phys. **9** (1941) 440.
[27] R. Koningsveld, L. A. Kleintjens, Macromolecules **4** (1971) 637.
[28] R. Koningsveld, L. A. Kleintjens, A. R. Schultz, J. Polym. Sci. Part A-2 **8** (1970) 1261.
[29] R. M. Prokop, M. L. Hair, A. W. Neumann, Macromolecules **29** (1996) 5902.

Jump Enhancement of Nucleation in Polymer and Carbon Dioxide Mixtures by Metastable Phase Transition

Xiaofei Xu,[1] Diego E. Cristancho,[2] Stéphane Costeux,[2] and Zhen-Gang Wang[1,*]

[1] *Division of Chemistry and Chemical Engineering, California Institute of Technology, Pasadena, CA 91125*

[2] *The Dow Chemical Company, Midland, MI 48674*

We combine a newly developed density-functional theory with string method to calculate minimum free energy path of bubble nucleation in compressible poly(methyl methacrylate)-CO_2 mixtures. As increasing initial pressure in the vicinity of a metstable phase transition, the terminal nucleated state jumps from the branch of CO_2-rich-vapor phase to the branch of CO_2-rich-liquid phase with a jump reduction for free ernergy barrier and radius of critical nucleus. This result suggests that the nucleation rate is enhanced as a jump discontinuous function of the initial pressure around the boundary of metastable phase transition.

PACS numbers: 82.60.Nh; 64.60.Q; 47.57.Bc

Nucleation taken place around a metastable phase transition has attracted considerable interests to enhance and control nucleation rate[1–3]. Close to metastable critical point, the free energy barrier for nucleation can be reduced by the large fluctuations[1]. Below the critical point, there can be a reduction for the barrier when physical conditions of nucleation pass through the boundary of metastble phase transition[2, 4]. These effects provide opportunities for enhancing the nucleation rate in polymer-carbon dioxide (CO_2) mixtures.

There are enormous practical interests to study the bubble nucleation of polymer-CO_2 mixtures in industry of making foam materials, particularly for its early stage which is crucial in determining cell size and pore density[5]. Despite many efforts to this problem, it is still far from satisfactorily understood owing to the lack of theoretical models that can accurately describe the mixtures at molecular level[6]. For compressible polymer-CO_2 mixtures, the solvent-rich phase can be CO_2-rich-liquid (CRL) or CO_2-rich-vapor (CRV) phase [7, 8]. Under unequilibrium condition, the CRL and CRV phase may coexists giving a metastable condensation point that could be utilized to enhance the nucleation rate. In this letter, we combine a newly developed density-functional theory (DFT) [9] with string method to investigate the nucleation of compressible poly(methyl methacrylate) (PMMA)-CO_2 mixtures. The DFT gives an accurate description for the mixtures and the string methods capture the early stage of bubble formation along minimum free energy path(MFEP). We will hold our attention on the enhancement of nucleation by the metastable condensation transition.

The molecular units of both components are coarse-grained as spherical particles with a hard core. The CO_2 molecule is modeled as one sphere. The molecular weight of PMMA is 89230 g/mol and is modeled as a freely jointed chain with 2855 identical spherical segments that are tangentially connected. The interaction between two arbitrary species (i.e. polymer segments or CO_2) is described by an effective Lennard-Jones potential. The DFT that is extended from the perturbed-chain statistical associating field theory equation of state[10], satisfactorily captures the phase and interfaical behavior of the mixtures[9]. The Helmholtz free energy functional for the system is expressed as a sum of an ideal-gas term and an excess term accounting for contributions from excluded volume effect, chain connectivity, association and dispersion effect. The details for the molecular model and free energy functional are given in our previous paper[9].

In a mean field picture, the nucleation proceeds along MFEP on the functional surface of grand potential $W[\rho_1(\boldsymbol{r}), \rho_2(\boldsymbol{r})]$, where ρ_1 and ρ_2 are the number density of CO_2 and PMMA segments, respectively. The MFEP $\{\rho_j(\boldsymbol{r}, s) | j = 1, 2, 0 \leq s \leq 1\}$(where s is the normalized reaction coordinate on the path) connects the initial state $\{\rho_j(\boldsymbol{r}, 0)\}$(the metastable bulk state of mixtures) and the terminal state $\{\rho_j(\boldsymbol{r}, 1)\}$(a well-developed bubble). It is defined such that the tangent along the path is parallel to the free energy gradient,

$$\begin{cases} \dfrac{\delta W}{\delta \rho_1} - \left(\tau_1 * \dfrac{\delta W}{\delta \rho_1}\right) \tau_1 = 0 \\ \dfrac{\delta W}{\delta \rho_2} - \left(\tau_2 * \dfrac{\delta W}{\delta \rho_2}\right) \tau_2 = 0 \end{cases} \tag{1}$$

with

$$\tau_j(\boldsymbol{r}, s) = \frac{\partial \rho_j(\boldsymbol{r}, s)}{\partial s} \Big/ \left(\frac{\partial \rho_j(\boldsymbol{r}, s)}{\partial s} * \frac{\partial \rho_j(\boldsymbol{r}, s)}{\partial s}\right)^{1/2}$$

is the normalized tangent along the path and $*$ denotes the inner product defined as $f * g = \int f(\boldsymbol{r}) g(\boldsymbol{r}) \mathrm{d}\boldsymbol{r}$. Equation (1) can be solved by using a modified steepest descent algorithm termed as string method[11],

$$\begin{cases} \dfrac{\partial \rho_1(\boldsymbol{r}, s; t)}{\partial t} = -\left[\dfrac{\delta W}{\delta \rho_1} - \left(\tau_1 * \dfrac{\delta W}{\delta \rho_1}\right) \tau_1\right] + k_1 \tau_1 \\ \dfrac{\partial \rho_2(\boldsymbol{r}, s; t)}{\partial t} = -\left[\dfrac{\delta W}{\delta \rho_2} - \left(\tau_2 * \dfrac{\delta W}{\delta \rho_2}\right) \tau_2\right] + k_2 \tau_2 \end{cases} \tag{2}$$

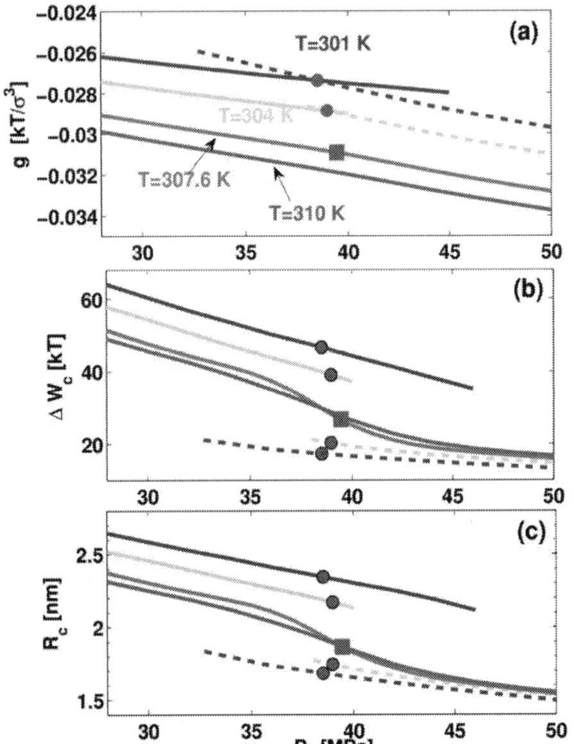

FIG. 1: (color online). (a) Grand potential density at the terminal nucleated state for T =301 K (blue lines), 304 K (green lines), 307.6 K (red lines) and 310 K (dark green lines). The solid lines at T =301K and 304 K are CRV phases and the dashed lines are CRL phases, respectively. σ = 0.25nm is the reduced length scale of volume. The solid circles are the condensation points, and the solid square is the critical point of condensation. (b) The free energy of critical nucleus formation. (c) The radius of critical nucleus.

where k_1 and k_2 are the Lagrange multipliers, and t is a fictitious time for evolving the equations on the free energy landscape. We adopt an explicit forward time splitting to solve Equation (2). The iteration starts at initial density profiles ($t = 0$) of a simple linear interpolation between the initial state ($s = 0$) and the terminal state ($s = 1$). After every iteration, we reparameterize the states of density profile equidistantly along the path. Iterations stops when the maximum free energy deviation along the path between two consecutive iterations is less than 10^{-5}.

The nucleation is initiated by a pressure drop from the coexistence point (initial pressure, P_0) to the ambient pressure 0.1 MPa. The density is allowed to relax, while maintaining the same composition(weight fraction of CO_2) to the new pressure value, resulting in a metastable bulk state from which nucleation takes place. The terminal nucleated state, i.e. the density at the center of a well-developed bubble, is determined by maintaining the equality of chemical potential with the bulk phase

$$\mu_j(\rho_1^{\text{center}}, \rho_2^{\text{center}}) = \mu_j(\rho_1^{\text{bulk}}, \rho_2^{\text{bulk}}), j = 1, 2 \qquad (3)$$

The superscript "center" and "bulk" denotes the value of the variable at the center of a well-deveoped bubble and in bulk phase, respectively.

At equilibrium, the CRL and CRV phases can co-exists with polymer-rich phase giving a triple point[7]. Previous research by Müller et al.[8] showed that the occurrence of triple point alters interface and nucleation behavior. Under unequilibrium condition, the CRL and CRV phase may coexists giving a metastable condensation point. Figure 1(a) shows the grand potential density as a function of initial pressue at the terminal nucleated state. At T =301 K, the CRV condenses into CRL phase at P_0 = 38.5 MPa. The CRV phase exists as $P_0 \leq$ 45.1MPa, and the CRL phase exists as $P_0 \geq$ 33.2MPa. The end points (P_0 = 33.2 and 45.1 MPa) of the two phases form the metastable spinodal for the condensation. The gap between the metastable spinodal and condensation points decreases as increasing temperture to T=304 K. It finally vanishes and there forms a critical point at T_c =307.6 K. As approaches T_c, the properties of CRL and CRV phase converge as one phase at the critical point. As $T > T_c$, the CO_2-rich phase becomes supercritical.

It should be mentioned that, the solubility in metastable bulk phase and equilibrium phase after nucleation both are continuous functions at these temperatures and initial pressures (results not shown), which suggest that there does not have phase transition. Thus, the phase transitions shown in Figure 1(a) do not origin in bulk phase behavior of both initial and terminal equilibrium phases, and they can only accessible inside a bubble by nucleation.

We begin our discussion with the classical nucleation theory (CNT). In CNT, the free energy to form a typical bubble of radius R is given by a balance between a surface and a volume contribution,

$$\Delta W_{\text{CNT}} = 4\pi R^2\gamma + \frac{4\pi R^3}{3}(g_{\text{center}} - g_{\text{bulk}}) \qquad (4)$$

where γ is the interfacial tension, and g is the grand potential density. At the condensation point in Figure 1(a), the two phases have the same grand potentials, but different interfacial tension between the polymer-rich phase and CO_2-rich phase. In general, the tension value for the interface of CRL phase is lower than that of CRV phase[9]. Therefore, CNT would predicts a jump for the free energy barrier around the condenstaion point. As is shown in Figure 1(b) and 1(c), string method gives the same predictions. At $T = 301K$ and $T = 304K$, the free energy barrier and critical radius at the two branches both end at the metastable spinodal points. As increasing the initial pressure, they jump to a lower

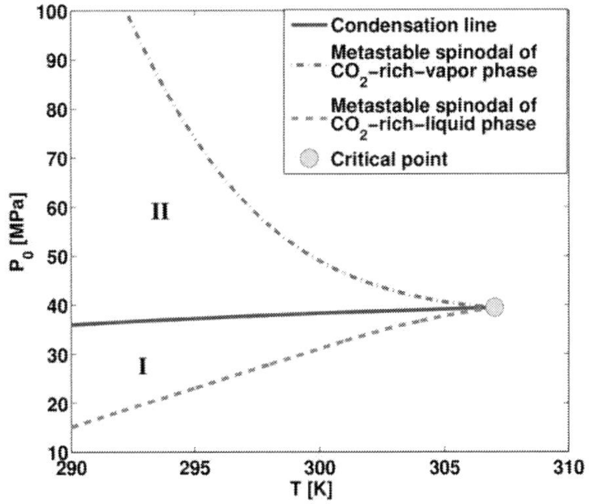

FIG. 2: (color online). Phase diagram for the terminal nucleated state in $T - P_0$ plane.

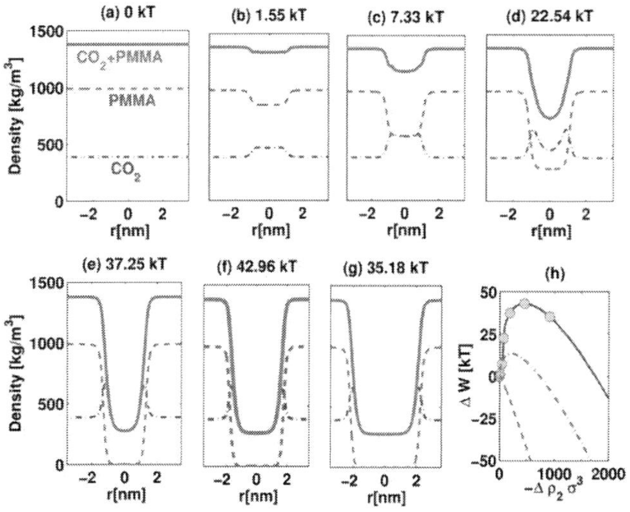

FIG. 3: (color online). (a) to (g): Evolution of the density profiles for a vapor-like bubble along MFEP at $P_0 = 38.5$ MPa and $T = 301$ K. The free energy of bubble formation is shown at the head of each sub-figure. Sub-figure (f) is the critical nucleus. (h): The free energy of bubble formation ($\Delta W = W - W_{\mathrm{bulk}}$) as a function of excess PMMA density($\Delta\rho_2 = 4\pi \int_0^\infty \mathrm{d}r r^2[\rho_2(r) - \rho_2^{\mathrm{bulk}}]$). The solid lines are the results of string method. The solid circles mark the position of the nucleus that shown in (a) to (g). The green dashed line is the CNT predictions that used the equlibrium tension value $\gamma = 1.91$ mN/m, and the purple dot-dashed line is the prediction of a modified CNT using tension value $\gamma = 6.86$ mN/m.

value from the branch of CRV phase to the branch of CRL phase within the region between the two spinodal points. It suggests that the nucleation rate is enhanced as a jump discontinuous function of initial pressure around the boundary of metastable phase transition. The height of this jump reduction is temperature-dependent, and decreases as approaching the critical point T_c. The two branches join at the critical point and emerges as one continuous curve as $T \geq T_c$.

It reveals that the nucleation of polymer-CO_2 mixtures is strongly related with initial pressure and temperature. Figure 2 shows the phase diagram at the terminal nucleated state. As $T < T_c$, in the region below the curve of metastable spinodal for CRL phase, the mixture can only nucleate to CRV phase to form a vapor-like bubble. In the region above the curve of metastable spinodal for CRV phase, the mixture can only nucleate to CRL phase to form a liquid-like bubble. In the intermediate region between the two spinodal curves (region I and II), CRL and CRV phases both exist. The mixtures may nucleate in either or both the CRL and CRV phase, depending on the kinetics in practice and how high barrier the nucleus could overcome. In general, the free enery barrier for CRL phase is lower than that for CRV phase; see Figure 1(b). Thus, the mixtrues should prefer nucleating to the liquid-like bubble. However, it shoud be noted that, CRV phase in region I has a lower grand potential than that of CRL phase; see Figure 1(a). Based on Ostwald's step rule[12, 13], the liquid-like bubble may further nucleate to CRV phase. In region II, CRL phase has a lower grand potential density, which means that the liquid-like bubble does not need to further nucleate. As $T \geq T_c$, the mixtures nucleate to the supercritical CO_2-rich phase.

The liquid-like and vapor-like bubble differs in their structure and ways of formation. Figure 3 and Figure 4 shows, respectively, the evolution of a vapor-like and liquid-like bubble along MFEP at the metastable condenstaion point ($P_0 = 38.5$ MPa) of T=301 K. We found some features that are incapable of being captured with crude theories based on incompressible assumption or a pseudo-one-component free energy function[14]. At the very early stage of vapor-like bubble formation (Figure 3(b)), the total density profiles present a slight oscillation. In Figure 3(b) and (c), the CO_2 density overshoots (i.e., exceeds the bulk coexistence value and the value in the well-developed bubble) on its way to the critical nucleus. When the bubble becomes large enough, the density in the core decrease to a value of slightly lower than the bulk density of CO_2. At the interface, there is a noticeable enhancement of CO_2. The evolutions agree well with the process of two step nucleation [1]: (i) the first step toward the critical nucleus is the formation of a small liquid-like nucleus. (ii) Then, beyond a certain size, a vapor-like nucleus forms inside the small liquid-like bubble, which gives a core-shell structure for a well-developed bubble. Compared with vapor-like bubble, the formation of a liquid-like bubble shown in Figure 4 is simple. The CO_2 density keep increasing to a value that is quite higher than the density in bulk phase. Then, in Figure 4(e), it slightly decrease which gives a slight surface

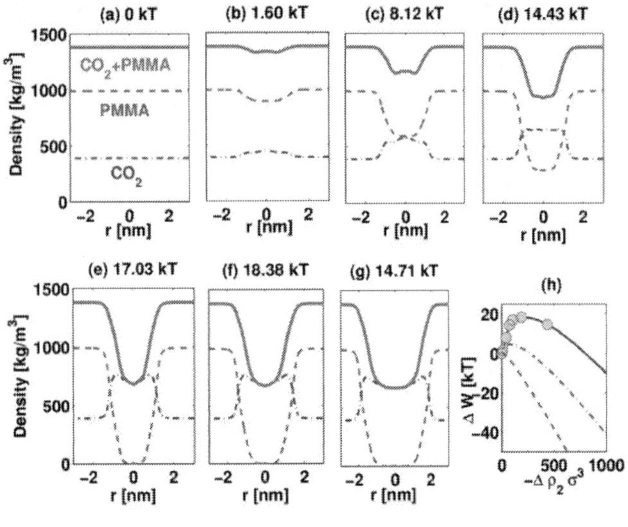

FIG. 4: (color online). (a) to (g): Evolution of the density profiles for a liquid-like bubble along MFEP at P_0 = 38.5 MPa and T = 301 K. Sub-figure (f) is the critical nucleus. (h): The free energy of bubble formation as a function of excess PMMA density. The tension value in modified CNT is $\gamma = 4.58$ mN/m.

enhancement of CO_2 at the interface.

To further show the adavantages of our methodology, we compare the results with predictions of CNT for free energy barrier in Figure 3(h) and 4(h). CNT gives an extremely low barrier and greatly underestimates the height of free energy barrier[15–17], which origins partially in the use of equilibrium interfacial tension. To examine this effect, we also give a modified version of CNT using the tension value fitted from the free energy of large nuclei. It gives a higher barrier value than CNT, but still lower than that of string method, which origins in the assumption of a sharp interface through the nucleated process in CNT. At the early stage of bubble formation, the grand potential difference from volume contribution in string method gradually changes its value from zero to $g_{\text{center}} - g_{\text{bulk}}$ (see the total density profiles in Figure 3(a) to (e)) , while in CNT it is fixed at $g_{\text{center}} - g_{\text{bulk}}$. The gradually increasing style reveals the fact that the volume contribution is minor and the surface contribution dominates the bubble nucleation at its early stage. Therefore, our predictions from string method are more reasonable.

In conclusion, we have investigated the nucleation of compressible PMMA-CO_2 mixtures along MFEP by combining a newly developed DFT with string method. As increasing initial pressure in the vicinity of the metstable condensation, the terminal nucleated state jump from the branch of CRV phase to the branch of CRL phase with a reduction for free ernergy barrier and radius of critical nucleus. This result suggests that the nucleation rate is enhanced as a jump discontinuous function of initial pressure around the boundary of metastable phase transition. In the region of below the metastable spinodal curve of CRL phase, the mixtures nucleate to vapor-like bubble, while they nucleate to liquid-like bubble in the region of above the metastable spinodal curve of CRV phase. In the intermediate region between the two spinodal curves, the mixtures may nucleate in either or both the CRL and CRV phase, depending on the kinetics in practice and how high barrier the nuclei could overcome.

Within the framework of mean field theory, the present work respresents the most advanced and accurate methodology in treating nucleation of polymer-CO_2 mxitures. It captures some features that are incapable by previous crude ones[14], and gives more reasonable predictions for free energy barrier than CNT and its modified version. The combination of the string method and DFT opens the way to studying nucleation in complex polymeric systems, such as polymer-CO_2 mixtures in the presence of nanoparticles and copolymer-CO_2 mixtures.

The Dow Chemical Company is acknowledged for funding and for permission to publish the results. The computing facility on which the calculations were performed is supported by an NSF-MRI grant, Award No. CHE-1040558.

---

* Electronic address: zgw@caltech.edu

[1] P. R. ten Wolde and D. Frenkel, Science **277**, 1975 (1997).
[2] R. P. Sear, J. Phys. Condens. Matt. **19**, 033101 (2007).
[3] O. Galkin and P. Vekilov, Proc. Natl. Acad. Sci. **97**, 6277 (2000).
[4] Z. Tavassoli and R. P. Sear, J. Chem. Phys. **116**, 5066 (2002).
[5] A. Nawaby *et al.*, Polym. Inter. **56**, 67 (2007).
[6] K. Binder *et al.*, *Polymer plus solvent systems: Phase diagrams, interface free energies, and nucleation* (Springer-Verlag Berlin, Berlin, 2005), vol. 173 of *Adv. Polym. Sci.*, pp. 1–104.
[7] X. F. Xu *et al.*, Ind. Eng. Chem. Res. **51**, 3832 (2012).
[8] M. Müller *et al.*, J. Chem. Phys. **117**, 5480 (2002).
[9] X. F. Xu *et al.*, J. Chem. Phys. **51**, 3832 (2012).
[10] J. Gross and G. Sadowski, Ind. Eng. Chem. Res. **40**, 1244 (2001).
[11] W. N. E and E. Vanden-Eijnden, Ann. Rev. Phys. Chem. **61**, 391 (2010).
[12] P. R. ten Wolde and D. Frenkel, Phys. Chem. Chem.Phys. **1**, 2191 (1999).
[13] W. Ostwald, Z. Phys. Chem. **22**, 289 (1897).
[14] Y. Kim *et al.*, Soft Matt. **7**, 7351 (2011).
[15] D. W. Oxtoby and R. Evans, J. Chem. Phys. **89**, 7521 (1988).
[16] M. Iwamatsu, J. Chem. Phys. **130**, 244507 (2009).
[17] J. F. Lutsko, Phys. Rev. E **74**, 021603 (2006).

Adsorption Behaviour of a Calcium Carbonate Particle to Solid Walls Having Different Hydrophilic Characteristics

Takafumi Toyoda, Hiroshi Suzuki, Yoshiyuki Komoda and Ruri Hidema, Kobe University, Kobe, Japan
Yutaka Shibata, Daikin Industries, Ltd, Osaka, Japan

Abstract

Adsorption characteristics of a calcium carbonate particle onto various walls having different hydrophilic characteristics have been investigated. The adsorption force of a calcium carbonate particle onto solid walls in a saturated solution of calcium carbonate was directly measured by use of an atomic force microscope. The crystallization characteristics of calcium carbonate onto the glass, copper without and with coatings were investigated by a fouling experiment. From the results, it was found that the adsorption force decreases with hydrophilic characteristics of walls. The mass of the $CaCO_3$ crystal adsorbing onto the solid surface increases with time on the bare copper wall, while both adsorption and desorption behaviors are observed in the cases of glass and the hydrophilic coated copper wall. It was also found from X-ray diffraction measurements that the structure of calcium carbonate adsorbing to the copper surface was an aragonite crystal. Thus, it was concluded that the fouling of calcium carbonate particles in heat exchangers occurs due to the adsorption of aragonites on a hydrophobic surface from the solution. Moreover, it is expected that fouling of the calcium carbonate can be controlled by the treatments of the hydrophilic processing to the surface wall.

Introduction

In recent years, saving energy has been required for the environment issues and for resource shortage problems at various fields. Heat exchangers were often used for saving thermal energy and for recover them in many processes. However, there exists a problem on using heat exchangers for water heat media. That is called fouling.

When hard water is heated, the scale (deposit) of calcium carbonate is built up in a heat exchanger or a evaporator due to the decrease of solubility. This is the basic mechanism of fouling. The scale of calcium carbonate forms a low heat transfer layer on walls of heat exchangers. It causes heat transfer reduction. Furthermore, the flow path reduction of the heat transfer media and the increase of pressure drop are caused, and then, the pumping power increases. At the final stage, it results a pipe blocking. Thus, it is important to control such fouling phenomena.

Calcium carbonate has two kinds of crystallographic structures of calcite and of aragonite in nature. Especially, the aragonite crystal is considered as a main scale material because an aragonite crystal has more adherability than a calcite crystal. In order to prevent the fouling of the calcium carbonate, some researches have been reported. Kim et al. (2002) observed the fouling process of $CaCO_3$ with a microscopic imaging technique. The validity of an electronic anti-fouling technology in a single tube heat exchanger was investigated by Cho et al., (1999). It was found that aragonite was predominantly precipitated in the presence of NaOH by Konno et al., (2002). Škapin et al. (2010) reported that the presence of ethylene glycol leads to the formation of aragonite, while the presence of diethylene glycol and tetraethylene glycol leads to the formation of calcite. Catalytic material was used for the mitigation of mineral fouling by Lee et al., (2006). However, the mechanism of the fouling has not been clarified sufficiently.

In this study, the adsorption characteristics of the calcium carbonate onto solid walls having different hydrophilic characteristics were investigated in order to investigate the fouling mechanism.

Experimental methods

Adsorption Force Measurements

The adsorption forces of the calcite and of the aragonite to several kinds of walls were directly measured by use of an atomic force microscope (AFM) as used by Wallqvist *et al.* (2006) and Yilmaz *et al.* (2007). Heavy calcium carbonate (calcite) and aragonite particles were tested in calcium carbonate saturation solution. The particle with 6.7μm of averaged diameter was bonded onto the tip of a cantilever as shown in Figure 1. The particle approaches and releases onto wall by use of the AFM. Figure 2 shows typical force curves on the distance between the particle and the wall. The minimum absolute value of the interaction force observed in the releasing curve as shown in the figure was defined as the adsorption force on the particle onto the solid wall. The contact area between the particle and the wall was calculated from a Hertz's elastic contact theory.

Figure 1. Tip of cantilever with a particle

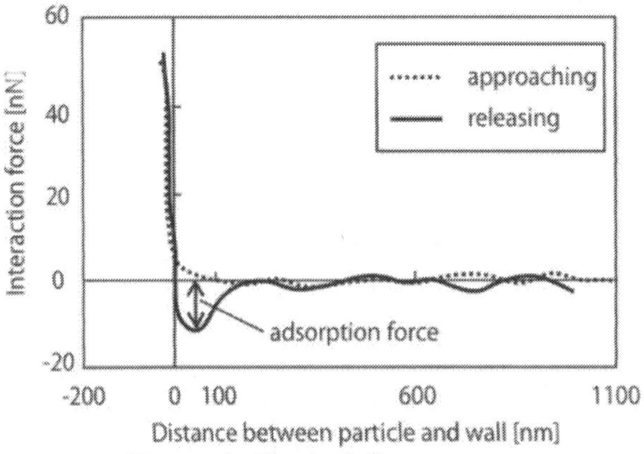

Figure 2. Typical force curve

$$b = \sqrt[3]{\frac{3\pi}{4} r_i \left(\delta_i + \delta_w\right) P} \tag{1}$$

Here, b [m] is the radius of the circle in the contact area between the particle and the wall, r_p [m] is the radius of the particle, and P [N] is the compressing force, respectively. Strain parameters, d_p [$m^2 \cdot N^{-1}$] and d_w [$m^2 \cdot N^{-1}$] of the particle and of the wall are calculated by the followings.

$$\delta_p = \frac{1 - \nu_p^2}{E_p \pi} \quad (2)$$

$$\delta_w = \frac{1 - \nu_w^2}{E_w \pi} \quad (3)$$

Here, ν_p and ν_w are Poisson's ratios of the particle and of the wall. E_p and E_w are Young's moduli of the particle and of the wall, respectively. The diameter of calcium carbonate particle was measured at five points from the photomicrograph of the non-spherical particle bonded to the cantilever. The mean value was used for the representative radius of the particle, r_p.

Hydrophilic characteristics of walls were evaluated from contact angle measurement. The liquid droplet contacted onto the wall is saturated solution of $CaCO_3$ at the room temperature of 20°C.
As the materials of walls, glass, aluminum, copper were selected. In order to develop a technique for preventing $CaCO_3$ fouling, three types coatings on the copper walls were also considered in this study: silane coupling, titanium dioxide (titania: TiO_2) coating and silicon dioxide (silica: SiO_2) coating.

Silane coupling treatment method is as follows. First, the silane solution was prepared by dropping triethoxyvinylsilane to acetic acid solution under stirring. Silane concentration is set at 1.0wt% versus filler, and the solvent of the acetic acid solution is water and ethanol, an acetic acid concentration is 2.0%. Then, the copper plate was soaked into the silane solution, and dried at 110°C for 5 min.

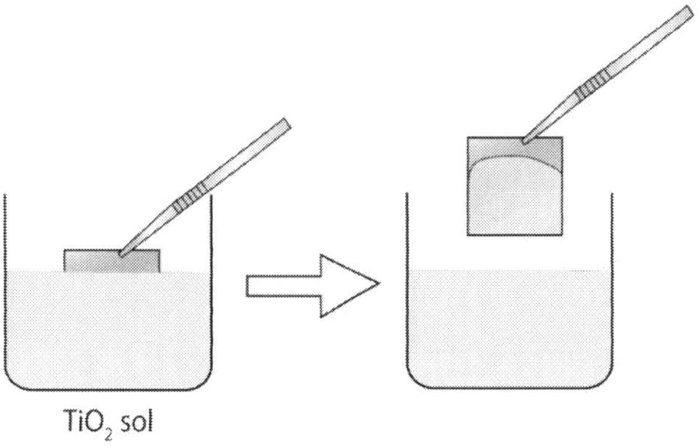

Figure 3. Dip coating system for TiO_2 coating.

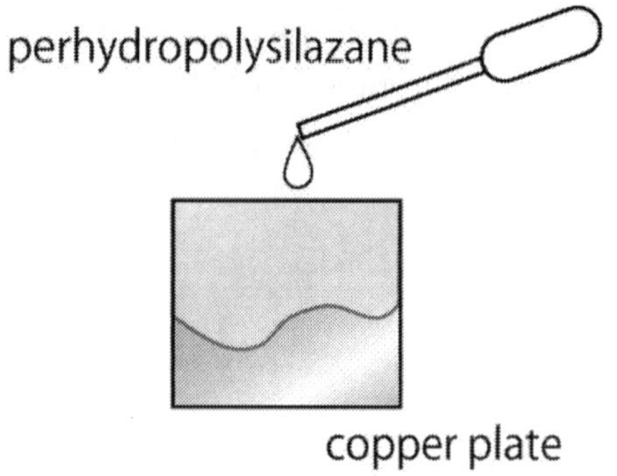

Figure 4. Flow coating system for SiO_2 coating

Titanium dioxide (TiO_2) coating method is as follows. First, the TiO_2 sol was prepared. The composition of this sol is 5wt% of the anatase-type titanium dioxide, 85wt% of ethanol + glycerin, 5wt% of acetylacetone, and 5wt% of polyethyleneglycol. Then, the copper plate was applied to this sol by a dip coating as Figure 2, dried at room temperature for 1 hour, and calcined at 500°C for 1 hour.

Silica (SiO_2) coating method is that the flow coating of a small amount of the perhydropolysilazane solution was applied to a copper plate as Figure 3, and the plate was dried at 500°C for 1 hour. The perhydropolysilazane react with the moisture in the air, and converts into silica glass at room temperature for two weeks. Eq. (1) shows the conversion of the perhydropolysilazane. In this study, the perhydropolysilazane was converted at 500°C in order to accelerate the reaction.

$$SiH_2NH + 2H_2O \rightarrow SiO_2 + NH_3 + 2H_2 \quad (1)$$

Fouling Experiments

The crystallization characteristics of calcium carbonate onto the glass, the copper and copper with coatings were also investigated by a fouling experiment. The substrate was soaked into the solution of 5mM $CaCl_2$ with 5mM $NaHCO_3$ in a bottle as shown in the figure 5. After the bottle was heated for 20 minutes to 9 hours at 80°C by a thermostat bath and these walls were picked up and dried, the weight of the calcium carbonate crystals adsorbing to these surfaces were measured. In addition, the crystals floating in the solution and adsorbing to walls were collected, and their structures were analyzed by an X-ray diffraction system (XRD).

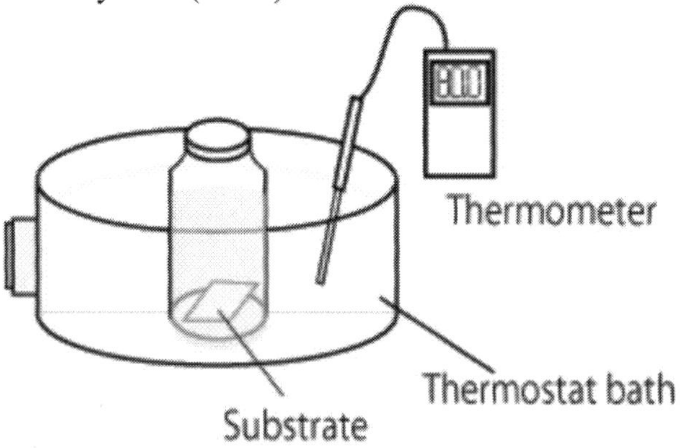

Figure 5. Fouling experimental device

Results and Discussions

Adsorption Force Between Particle and Wall Surface

Figure 6 shows the interaction force per unit area between particle and wall surface obtained by the adsorption force measurement for each material wall. In this figure, the abscissa is the contact angle of surface and the symbols show average values within the error bar designated in the figure.

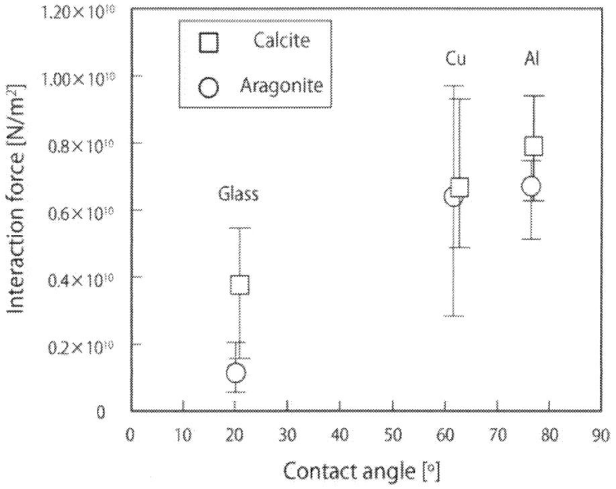

Figure 6. Interaction forces between particle and solid walls

From this figure, it is found that the adsorption force on the wall decreases with hydrophilic characteristics of wall surface. The absorption force of aragonite is also found to be slightly smaller than that of calcite for any case.

The results of the adsorption force measurement for the cases with coating are shown in Figure 7 with data shown in Figure 6.

From theses results, it was found that the adsorption force decreases with hydrophilic characteristics of walls including coated cases. Thus, it can be expected that the hydrophilic treatment on the wall surface is effective for preventing the absorption of $CaCO_3$ particles onto the walls

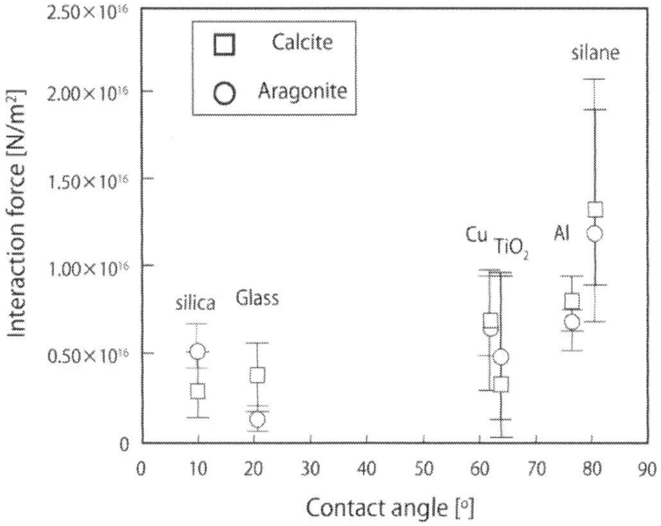

Figure 7. Effect of coatings on interaction forces between particle and walls

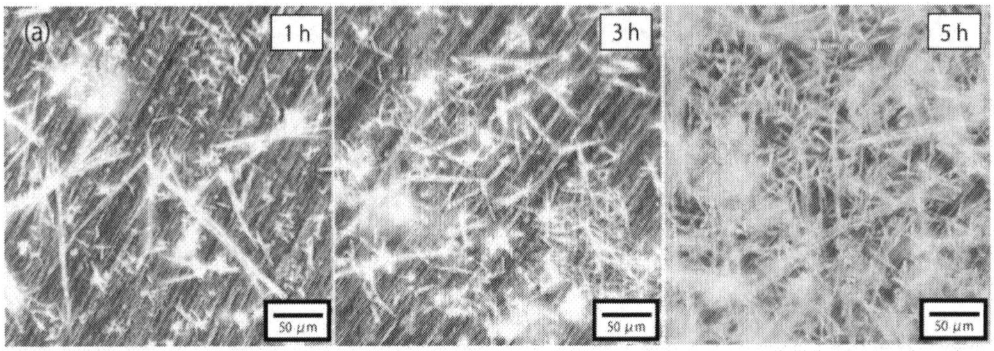

Figure 8. Photomicrograph on copper surface at 1 hour, 3 hours and 5 hours after heating

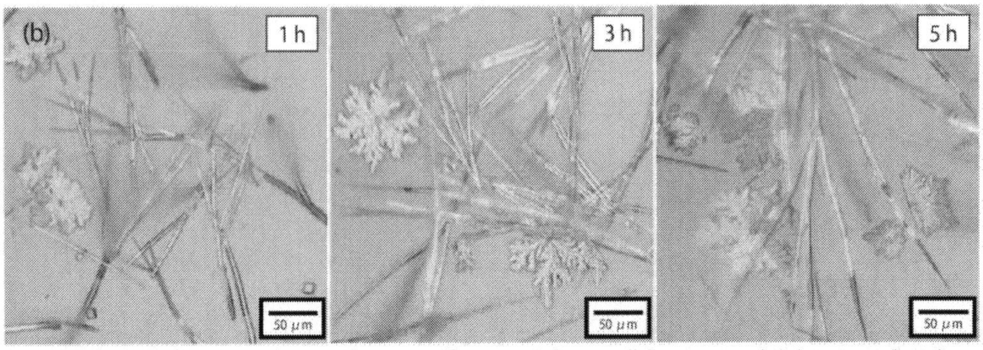

Figure 9. Photomicrograph on glass surface at 1 hour, 3 hours and 5 hours after heating

Fouling Characteristics

Figures 8 and 9 show the results of the fouling experiments on the bare copper and glass surfaces. The photomicrographs in each figure are obtained at the heating time of 1hour, 3 hours, and 5 hours, respectively.

As shown in Fig. 8 of the bare copper case, the needle-like crystals are deposited onto the wall and that the number of crystals increases with time. Such a needle-like crystal is typical aragonite. On the other hand, not only the needle-like but also the petaline-like crystals can be observed on the surface of glass. The number of crystals is not significantly increased with time, but the size grows.

Figures 10 and 11 show the crystal total mass deposited on the surface and the ratio of the crystal adsorbed onto the wall to the floating in solution for each case.

From these figures, it is found that the masses of the $CaCO_3$ crystal adsorbing onto the glass and copper with coating surface are a low value than the case of the untreated copper. Moreover, the ratio of the adsorbing crystal to the floating crystal is small in the glass, silane coupling copper and silica coating copper cases. In the glass and silica coating copper cases, the value of the crystal mass is unstable, because both adsorption and desorption behaviors are observed. This result is considered to correspond to the adsorption force characteristics.

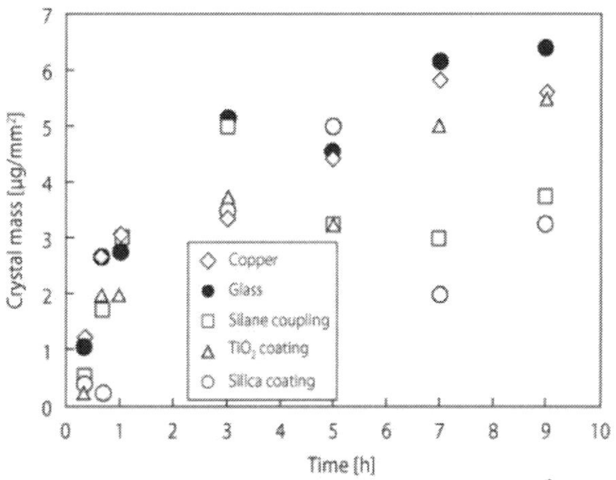

Figure 10. Crystal total mass of $CaCO_3$ per 1 mm^2 on the surface

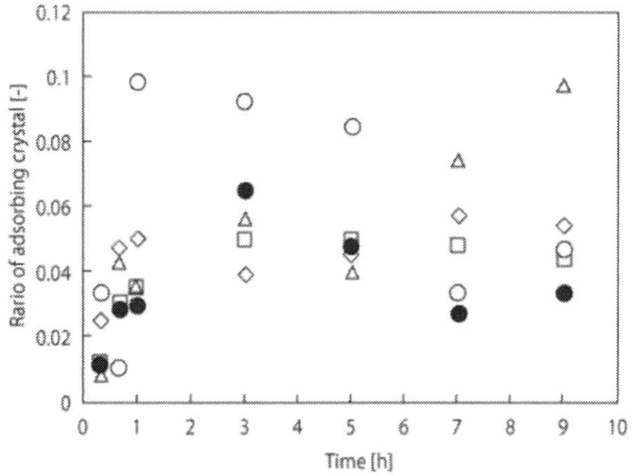

Figure 11. Ratio of adsorbing crystal to the floating crystal in the solution

Figure 11 shows that the XRD profile of the calcium carbonate crystal adsorbing to the surface. It was found from this figure that the structure of calcium carbonate adsorbing to the copper surface was an aragonite crystal. On the other hand, not only an aragonite crystal but also a vaterite crystal was observed on the glass and silica coating copper surface. Thus, it was concluded that the fouling of calcium carbonate particles in heat exchangers occurs due to the adsorption of aragonites on the hydrophobic substrate from the solution where the aragonites are formed.

In the case of the coating wall, the crystal structures are almost same be compared with copper without any coating. From this, it was found that the calcium carbonate crystal is hard to adhere by the weak adherability even where it is the same structure. Thus, it is expected that the fouling of the calcium carbonate can be controlled by use the hydrophilic treatment for the heat-transfer surface.

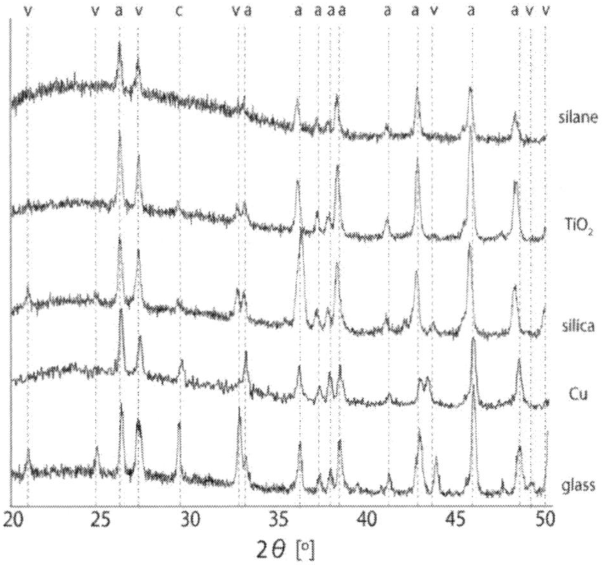

Figure 11. XRD profile of the calcium carbonate crystal adhering to each surface

Conclusions

Adsorption characteristics of a calcium carbonate particle onto various walls having different hydrophilic characteristics have been investigated. The adsorption force of a calcium carbonate particle onto solid walls in a saturated solution of calcium carbonate was directly measured by use of an atomic force microscope. The crystallization characteristics of calcium carbonate onto the glass, copper without and with coatings were investigated by a fouling experiment.

From the results, it was found that the adsorption force decreases with hydrophilic characteristics of walls. The mass of the $CaCO_3$ crystal adsorbing onto the solid surface increases with time on the bare copper wall, while both adsorption and desorption behaviors are observed in the cases of glass and the hydrophilic coated copper wall.

It was also found from X-ray diffraction measurements that the structure of calcium carbonate adsorbing to the copper surface was an aragonite crystal. Thus, it was concluded that the fouling of calcium carbonate particles in heat exchangers occurs due to the adsorption of aragonites on a hydrophobic surface from the solution.

From these results, it is expected that fouling of the calcium carbonate can be prevented by the treatments of the hydrophilic treatment to the surface wall.

Nomenclature

b = radius of circle in contact area between particle and wall [m]
E_p = Young's modulus of particle [Pa]
E_w = Young's modulus of wall [Pa]
P = compression force [N]
r_p = radius of a particle [m]
δ_p = strain parameter of a particle [$m^2 \cdot N^{-1}$]
δ_w = strain parameter of a solid wall [$m^2 \cdot N^{-1}$]
ν_p = Poisson's ratio of particle [-]

v_w = Poisson's ratio of wall [-]

References

1. Cho Y. I. and B.-G. Choi (1999), "Validation of an electronic anti-fouling technology in a single-tube heat exchanger," *International Journal of Heat andMass Transfer*, 42, pp.1491-1499
2. Kim W. T., C. Bai and Y. I. Cho (2002) "A study of CaCO3 fouling with a microscopic imaging technique," *International Journal of Heat and Mass Transfer*, 45, pp. 597-607
3. Konno H., Y. Nanri and M. Kitamura (2002) "Crystallization of aragonite in the causticizing reaction", *Powder Technology*, 123, pp. 33-39
4. Lee G. J., L. D. Tijing, B.C. Pak, B. J. Beak and Y. I. Cho (2006) "Use of catalytic materials for the mitigation of mineral fouling," *International Communications in Heat and Mass Transfer*, 33, pp. 14-23
5. Škapin S. D. and I. Sondi (2010) "Synthesis and characterization of calcite and aragonite in polyol liquids: Control over structure and morphology," *Journal of Colloid and Interface Science*, 347, pp. 221-226
6. Yilmaz H., K. Sato and K. Watari (2007) "AFM interaction study of a-alumina particle and c-sapphire surfaces at high-ionic-strength electrolyte solutions," *Journal of Colloid and Interface Science*, 307, pp. 116-123
7. Wallqvist, V., P. M. Cleasson, A. Swerin, J. Schoelkopf and P. A. C. Gane (2006) "Interaction forces between talc and hydrophobic particles probed by AFM," *Colloids and Surface A: Physicochem. Eng. Aspects*, 277, pp. 183-190

(345g) Dewatering Oil Sand Tailings Using Calcite

Anwesha Basu, Chemical Engineering, Pennsylvania State University, University Park, PA and Darrell Velegol, Chemical Engineering, Penn State University, University Park, PA

Bituminous sands contain petroleum that can be extracted by various means. As the petroleum is obtained from the sands, an additional side product consists of various types of clays and minerals, which compose the "tailings". Currently there are more than 50 square kilometers of tailings ponds in Canada. One component of the tailings is "mature fine tailings" (MFTs), which are clay particles that have settled, but still contain a large enough fraction of water that the material is not trafficable, and thus cannot be used for roads or buildings. There are various strategies that have been developed for consolidating MFTs, such as Consolidated Tailings; however, each technology has challenges, especially for already-existing tailings ponds. One type of tailing technology being pursued is the addition of chemical species that improves the consolidation process. For example, gypsum ($CaSO_4$) has been added, which improves the dewatering of the MFTs. In this talk we study the addition of calcite to tailings ($CaCO_3$), using various methods. We examine the effects on dewatering, showing that we obtain an enhanced removal of water. A further interesting aspect of $CaCO_3$ use is the resulting layering of the MFTs, including a final bitumen layer atop the others.This talk will discuss the approach used with $CaCO_3$, the results for dewatering and consolidation, improved trafficability of clay and potential transport, thermodynamic and surface chemistry mechanisms of action.

(348b) A Reassessment of the Precision and Accuracy of Atomistic Calculations of Elastic Constants of Amorphous Engineering Polymer Glasses

Paul W. Saxe, Materials Design, Inc, Angel Fire, NM, Clive M. Freeman, Materials Design, Inc., San Diego, CA and **David Rigby**, Materials Design, Inc, San Diego, CA

A Reassessment of the Precision and Accuracy of Atomistic Calculations of Elastic Constants of Amorphous Engineering Polymer Glasses

Efficient and cost effective development of new polymer-based materials for aerospace, automotive and related engineering applications relies on the availability of accurate values of material constitutive properties such as thermal and mechanical behavior. In the case of mechanical properties of amorphous polymers below their glass transition temperature, theoretical methods required to obtain predictions of elastic constants from atomistic simulations were developed early in the evolution of computational materials science [1, 2]. These methods were shown to give reasonable estimates of stiffness coefficients and the associated moduli [2, 3]. However, since the size of typical simulated systems is at most a few tens of Angstroms, and since amorphous polymers pack molecularly in a large number of configurations with comparable probabilities, the resulting structures tend to give rise to a broad distribution of calculated elastic constants, leading to considerable uncertainty in the elastic constants of macroscopic samples of material. This problem was addressed by Suter and Eichinger, who visualized the macroscopic structure of glassy polymers as collections of nanoscopic domains and applied theoretical methods developed by Hill and Walpole for composites to derive relatively narrow bounds estimates for the macroscopic material elastic constants [4]. Surprisingly however, since this work was first published, there have been few reported attempts to examine the effective precision achievable in prediction of elastic constants and to apply the methods to important engineering polymers.

In recent work, we have performed a detailed investigation of the precision and accuracy with which elastic constants can be predicted, and have begun to perform calculations on a variety of materials. In particular, in this presentation we discuss the scale of calculations required to obtain acceptable precision of elastic constants, and also illustrate the accuracy achievable by comparison of predicted results with experimental values for experimentally well-studied common amorphous polymers. In addition, we report on application of the methods developed to calculation of the elastic constants of known compatible engineering polymer blend systems containing polystyrene and poly(phenylene oxide), and further examine the results to determine the extent to which the precision is adequate to probe synergistic behavior in the blend systems.

References

1. Parrinello, M. and Rahman, A., J. Chem. Phys. **76**, 2662 (1982). 2. Theodorou, D.N. and Suter, U.W., Macromolecules **9**, 139 (1986). 3. *see*, Hutnik, M, Argon, A.S. and Suter, U.W., Macromolecules **24**, 5956 (1991); Fan, C.F., Cagin, T., Chen, Z.M. and Smith, K.A. Macromolecules **27**, 2383 (1994). 4. Suter, U.W. and Eichinger, B.E., Polymer **43**, 575 (2002). 5. MacKnight, W.J., Karasz, F.E. and Fried, J.R., in "Polymer Blends, Vol I", D.R. Paul and S. Newman (eds), Academic Press, New York (1978).

(348c) Understanding and Predicting Multimodal Ligand-Protein Interactions Using Molecular Simulations

Siddharth Parimal, Melissa A. Holstein, Shekhar Garde and Steven M. Cramer, Department of Chemical and Biological Engineering, Rensselaer Polytechnic Institute, Troy, NY

Recent developments in multimodal chromatography made by our group and by others have successfully shown that enhanced selectivities can be achieved for protein separations in multimodal systems as compared to traditional single-mode chromatography by the appropriate choice of ligand chemistry and mobile phase modifiers. The mechanisms underlying these unique binding behaviors are not well understood. While chromatographic retention times provide the overall results of a combination of kinetic and affinity phenomena, individual cases need to be studied in more detail to shed light on the principles that govern mulitmodal interactions. An integrated approach of experimental and theoretical techniques is therefore essential to probe different aspects of protein-ligand binding in these multimodal systems. The work presented here demonstrates how Molecular Dynamics (MD) simulations can be employed to gain fundamental insights into the protein-ligand interactions at an atomistic level. In particular, the interaction of proteins with multimodal ligands was investigated.

Protein-ligand binding in multimodal systems involves a combination of electrostatic and hydrophobic interactions. The accurate measurement of individual contributions of both these modes to the overall binding behavior has been broken down into two steps. The first step requires the characterization of the molecules involved in binding in terms of their electrostatic and hydrophobic potential. While electrostatic interactions between two molecules can be easily calculated, characterizing hydrophobicity at a molecular level is very difficult. The advancements made in the later part of the last decade towards identifying molecular signatures of hydrophobicity and the development of better sampling techniques in simulations have enabled the hydrophobic characterization of biological macromolecules. These methods were used to map regions on proteins which are involved in hydrophobic association to ligands. Multimodal ligand-protein binding is further complicated by the fact that the cooperative effects of the multiple modes present on a ligand, i.e., the synergy present in multimodal interactions, has not yet been characterized. This constitutes the second step in understanding multimodal interactions, where MD simulations were performed with a series of increasingly complex ligands to identify the key contributors to synergistic interactions in multimodal ligand-protein interactions. Simulations were also performed to quantify the effect of fluid phase modifiers in modulating the different types of interactions in multimodal systems.

The results were used to develop a new technique for mapping synergistic regions on protein surface, which have the potential to associate with multimodal ligands, and for understanding avidity effects on resins to make predictions of protein retention behavior in multimodal systems. The change in protein retention times as a response to changes in fluid phase conditions (e.g. changes in temperature, cosolutes like urea) has been explained. The knowledge base created using this study can have a profound impact on the way we think about multimodal systems, especially in the rational design of ligands and choice of appropriate fluid phase conditions for achieving novel bioseparations.

(348d) Coarse-Graining Protocol for Mesoscale Simulations of Nanostructured Polymers

Aleksey Vishnyakov, Ming-Tsung Lee and Alexander V. Neimark, Chemical and Biochemical Engineering, Rutgers, The State University of New Jersey, Piscataway, NJ

We suggest a straightforward technique for obtaining parameters for quasiparticles that can be used in coarse-grained molecular dynamics (CGMD) and dissipative particle dynamics (DPD) simulations of nanostructured polymeric materials.

The molecule of interest is first dissected into fragments of approximately similar size. Based on the chemistry of the fragments, we choose reference compounds that represent the corresponding sections of the molecule. We fit the parameters that describe the conservative interactions of these fragments to simple thermodynamic properties of the solutions of reference compounds, in particular, to the infinite dilution activity coefficients γ_{inf} of the corresponding binary solutions. This means that γ_{inf} for model solutions composed of quasiparticles impersonating different fragments of the molecules of interest have to match the activity coefficient of real solutions of representative compounds. The latter should be taken from experimental data when available or obtained with thermodynamic models, such as UNIFAC or COSMO-RS.

While fitting coarse-grained parameters to thermodynamic properties of reference systems is by no means new, our approach has several important advantages. Firstly, it is applicable to a variety of coarse-grained methods, including DPD, lattice MC and coarse-grained MD; (2) this scheme requires only standard and inexpensive calculations that do not require extensive experience in molecular simulations; (3) the reference correlations between the parameters of the coarse grained models and γ_{inf} are determined only by the form of the coarse-grained potentials and therefore have to be calculated only once and apply to parameterization of any system.

We verify our coarse-graining scheme against experimental data on aggregation numbers and critical micelle concentrations of several non-ionic, zwitterionic and ionic surfactants and demonstrate its advantages and limits of applicability. Then, we apply our scheme in DPD simulations of surfactant–polymer systems. SEM studies suggest that Carbopol forms a foam-like gel, which is destructed as the surfactant concentration increases. Our simulations shed light on this phenomenon. We demonstrate that surfactant molecules form micelles inside the polyacrylic acid structure without any appreciable interactions with the polymer. At the same time, surfactants interact intensively with hydrophobic linkers that may cause polymer network rapture. Using DPD simulations, we show how the gel nanostructured morphology and the disjoining pressure in the polymer foam films depend on the chemical composition and surfactant concentration.

(348e) Ionic Effects On the Stability and Morphology of AOT-WATER-Isooctane Reverse Micelles: Molecular Dynamics and Dynamic LIGHT Scattering Perspective

Victor R. Vasquez, Chemical and Materials Engineering, University of Nevada, Reno, Reno, NV

We use molecular dynamics (MD) and Dynamic Light Scattering measurements in the AOT-water-isooctane system to study the morphology and stability of the reverse micelle regime under the presence of various salts. These include zirconyl chloride, ammonium hydroxide, and aluminum nitrate. In this presentation, we discuss the results obtained with both methods. DLS measurements show that the average reverse micelle size decreases with the addition of the salts. Further increase in salt addition eventually destabilizes the reverse micelle structures, which is observed by the formation of cloudy solutions. An initial hypothesis suggested that the high valence free ions were causing the destabilization, but preliminary MD modeling results show that the presence of potential hydrolyzed structures might be playing a significant role on the stability and morphology of the system. MD simulations using only free ions in solution do not seem to destabilize the reverse micelles; however, the presence of structures such as zirconyl chloride hydrated dimers can cause instability and changes in the morphology. We discuss how we use MD combined with experimental data from DLS measurements to systematically analyze and gain physical insight into the behavior of reverse micelle systems for particle synthesis and other potential applications. In general, we find good qualitative agreement between MD and DLS in the size and morphology analysis of the system. The combination of MD with DLS allows a better interpretation of the experimental results, in particular for conditions where the structures are nonspherical, commonly observed at lower water-to-surfactant ratios or distorted by the presence of highly ionic solutions. Significant challenges still remain including the analysis of the effect of all possible hydrated states and the respective structures of the high valence ions in solution. We show preliminary progress with this, but better force fields are required to describe more accurately the complex chemistry and hydrated structures of these ions in solution. Other challenges include the typical modeling issues dealing with the long time and length scales characteristic of the dynamics in reverse micelle systems.

(348f) The Investigation of Green Tea Catechin Binding to Keratins by Molecular Dynamics and Experimental Validation

Jan Marzinek[1,2], Guoping Lian[1], Athanasios Mantalaris[2], Efstratios N. Pistikopoulos[2] and Yanyan Zhao[3], (1)Unilever Discover, Unilever R&D Colworth Science Park, Sharnbrook, United Kingdom, (2)Centre for Process Systems Engineering, Department of Chemical Engineering, Imperial College London, London, United Kingdom, (3)China Agricultural University, Beijing, China

Binding of small molecules to macromolecules of human bio-substrates regulates the sub-cellular disposition and the subject is fundamental to intestinal absorption of nutrients and drugs as well as transdermal permeation of cosmetics active [1, 2]. A number of studies have been reported on the binding properties of plant actives to bio-substrate proteins, using various experimental techniques [3-6]. The knowledge of how active molecules bind to proteins at the molecular level is still limited. Molecular Dynamics (MD) simulations provide an alternative approach to understand the interaction between active molecules and biological substrates at the atomic level. MD simulations allow researchers to extract the most detailed information about the interactions between proteins and ligands and extracted information can be validated by experiments.

This study reports recent progresses in using MD simulations to investigate the thermodynamic and equilibrium properties of small molecule binding to macromolecules of human substrates using keratin-polyphenol as an example. The small molecule is epigallocatechin-3-gallate (EGCG), which is the most active polyphenol found in green tea and reported to have antioxidant and antimutagenic properties as an active compound [7]. Keratin 1 and keratin 10, the most abundant SC keratins, have been chosen as the biosubstrate. The umbrella sampling technique together with the weighted histogram analysis method are used to calculate the binding free energy of EGCG to different keratin segments (α-helical rod, coiled coil of α-helixes and unstructured terminals). The influence of various conditions of temperature, salt concentration, EGCG/keratin molar ratio on the binding free energy and the number of hydrogen bonds between different protein segments and ligand has been assessed. Results have been also obtained on protein conformational shifts and the solvent accessible surface area changes in the presence of EGCG. It is shown that EGCG binding to keratin segments is driven by hydrophobic interaction involving the aromatic attraction and hydrogen bonding. EGCG molecule has 8 hydrogen donors and 11 hydrogen acceptors. It can form hydrogen bonds with both hydrophilic and hydrophobic residues of keratin proteins. The strongest keratin-EGCG interaction is found to be with the aromatic residues of keratin terminals and helical rod. The results from MD simulations are in a good agreement with fluorescence quenching and Isothermal Titration Calorimetry (ITC) experiments. Based on the results of both computer simulation and experiment, a new insight into EGCG-keratin interactions has been gained.

Keywords: EGCG, Stratum Corneum, Keratin, Binding free energy, Molecular Dynamics, Polyphenols

References

1. Hansen, S. et al. 2011. An Extended Database of Keratin binding. *J. Pharm. Sci.* **100**:1712-1726
2. Wang, L. et al. 2010. Determination of Partition and Binding Properties of Solutes to Stratum Corneum. *Int. J. Pharm.* **398**:114-122.
3. K. J. Siebert et al. 1996. Nature of Polyphenol-Protein Interactions. *J. Agric. Food Chem.* **44**: 80-85.
4. A. Bennick. 2002. Interaction of Plant Polyphenols with Salivary Proteins. *CROBM 2002* **13**: 184.
5. A. J. Charlton et al. 2002. Polyphenol/peptide Binding and Precipitation. *J. Agric. Food Chem.* **50**:1593-1601.
6. A. J. Charlton et al. 1995. Tannin Interactions with a full-length human salivary proline-rich protein display a stronger affinity than with single praline-rich repeats. *FEBS Letters*. **382:** 289-292.
7. B. N. Singh et al. 2011. Green Tea Catechin Epigallocatechin-3-gallate (EGCG): Mechanism, perspectives and clinical applications. *Biochemical Pharmacology*. **82**: 1807-1821.

(348g) Rational Platform for Biosurfactants Design Based On Molecular Dynamics and Experimental Validation Using Escherichia Coli Outer Membrane Protein A

Angie Paola Macías[1], Diana Marcela Carrero Pinto[1], Nathalia Garces Ferreira[2], Harold Enrique Castro Barrera[2], Oscar A. Álvarez[3] and Andrés Fernando González Barrios[1], (1)Chemical Engineering, Universidad de los Andes, Bogotá, Colombia, (2)Systems and Computing Engineering , Universidad de los Andes, Bogotá, Colombia, (3)Chemical engineering, Universidad de los Andes, Bogotá, Colombia

The industry of surfactants has played an important role in several industries such as cosmetic and cleaning. Nevertheless, mostly all compounds of this variety are derived from non-renewable sources such as hydrocarbons columned on the petrochemistry industry (Dewulf & Van Langenhove, 2006). Then the need for alternative strategies for obtaining such important products seems to be obvious. Polymers with amphiphilic capability such as peptides or proteins have been targeted as potential replacements due to its biodegradability and diversity from the synthesis approach perspective (Dexter & Middelberg, 2008). However, the selection of suitable sequences that could stabilize the emulsion is mostly based on a "trial and error" style (Dexter, Malcolm , & Middelberg, 2006). Here, we present a molecular dynamics (MD) approach to determine the stability of emulsions. We utilized MD to evaluate the viscosity, radial distribution function, and final configuration in n-dodecane/water emulsions for ten peptides sequences and transmembrane *E. coli* proteins. *In silico* results were experimentally validated and evaluating the stability of dodecane/water emulsions using purified *E. coli* OmpA. Protein. We obtained a positive effect on the emulsion stability as emulsions remain stable around [4-7.5] hours and droplet sizes around [365-6500] nm of diameter.

References

Dewulf, J., & Van Langenhove, H. (2006). Oleochemical and Petrochemical Surfactants. En J. Dewulf, H. Van Langenhove, & H. V. Jo Dewulf (Ed.), *Renewables-based technology: sustainability assessment* (Vol. II, págs. 265-279). Chichester, England: John Wiley & Sons.

Dexter, A., & Middelberg, A. (2008). Peptides As Functional Surfactants. *Industrial & Engineering Chemistry Research*, 6391–6398.

Dexter, A., Malcolm , A., & Middelberg, A. (2006). Reversible active switching of the mechanical properties of a peptide film at a fluid–fluid interface. *Nature Materials*, 502 - 506 .

(353a) Anisotropic Particle Migration and Assembly At Fluid Interfaces

Kathleen J. Stebe, Chemical & Biomolecular Engineering, University of Pennsylvania, Philadelphia, PA

As a community, our ability to create particles with well-defined shapes and surface energies is now well established, driving the need for assembly schemes to create new materials. In my group, we have focused on the role of shape anisotropy in capillary-driven assembly. On planar interfaces, far from contact, particles of arbitrary shape interact as capillary quadrupoles. Near contact, rod-like particles interact and assemble in a manner that depends strongly on particle shape. We demonstrate this dependence by comparing right circular cylinders, ellipsoids, and super-ellipsoids in terms of their preferred orientations and the mechanics of the microstructures formed when they assemble. On curved interfaces, rod-like particles migrate to highly curved sites to form hierarchical assemblies influenced by both particle-particle interactions and the energy owing to the interfacial curvature field. At the microscale, all of these capillary interactions are remarkably strong, and the structures that they form are often kinetically trapped. Strategies to impart near field repulsion between particles are described.

(353b) The Effects of Polydisperse Particle Populations On the Shear Rheology of Solid Stabilized Interfaces

Sourav Barman and **Gordon F. Christopher**, Department of Mechanical Engineering, Texas Tech University, Lubbock, TX

When particles stabilize drop interfaces in an emulsion, a "Pickering Emulsion" is formed. Pickering Emulsions provide stability through increased mechanical protection of an interface and enhanced surface viscoelasticity, They are used in food science, cosmetics, drug delivery, and energy applications. Many of these applications use polydisperse particle populations. However, there have been no systematic studies of the effects of polydispersity on the mechanical properties of solid stabilized interfaces. Changes in particle properties affect the interfacial rheology of a solid stabilized interface which in turn affects emulsion stability; therefore it is of great importance to elucidate the effects of polydispersity on surface rheology.

In order to examine the effects of particle polydispersity on packing and interfacial rheology, an AR-G2 rheometer is used with a double wall ring geometry. This tool is capable of measuring complex surface moduli down to 10^{-5} Pa m of both oil/water or air/water interfaces, which is sensitive enough to measure the shear rheology of solid stabilized interfaces. Furthermore, we have developed a system to allow real-time imaging of interfaces during testing. This unique setup allows the interfacial shear rheology to be directly tied to structural changes observed in solid stabilized interfaces. The result is an interfacial shear rheometer with the ability to visualize an interface and capture the dynamics of individual particles, of ~1 micron or larger, as they are sheared. We are able to examine of the effects of particle size, polydispersity, surface coverage and applied deformation on packing structure of an interface, and correlate changes in ordering to interfacial shear moduli. Using 2D analogs of standard rheological tests for suspensions, we examine the effects of polydispersity on polystyrene spheres at an air/oil interface at surface concentrations large enough to create hexagonal packing.

In suspensions of solid spheres, particle polydispersity affects particle ordering and maximum packing fraction, reducing the viscoelasticity of the system. Polydispersity has similar effects on 2D surfaces. As polydispersity increases the maximum packing fraction increases, long range order decreases, and particle mobility on the interface increases in comparison to monodisperse systems. As surface concentration increases, particle packing and structure becomes more ordered for all systems. However, polydisperse systems continue to show less order than monodisperse ones. We observe decreasing interfacial moduli and viscosity with increasing particle polydispersity over a range of surface concentrations. For all systems as maximum packing fractions is approached, the interfacial moduli and viscosity increase; however, the interfacial moduli and viscosity are lower for polydisperse systems in comparison to monodisperse systems at all surface concentrations. Polydispersity not only affects the magnitude of the measured interfacial moduli and viscosity, but also decreases surface relaxation times.

This work is the first to systematically study the shear rheology of interfaces stabilized with polydisperse particle populations. It shows the importance of particle population uniformity on the interfacial rheology of solid stabilized interfaces, which could affect a number of known applications.

(353c) The Dynamics and Break-up of Confined Bubbles Translating in Surface-Active Colloidal Dispersions

Anthony Kotula, Chemical Engineering, Carnegie Mellon University, Pittsburgh, PA and Shelley L. Anna, Chemical Engineering and Mechanical Engineering, Carnegie Mellon University, Pittsburgh, PA

The adsorption of particles to fluid-fluid interfaces is a key step in the generation of colloidosomes and particle-stabilized emulsions. Microfluidic channels are a promising tool for generating particle-stabilized drops and bubbles with independent control over the bubble size and the concentration of particles adsorbed at the fluid interface. In the first part of this presentation, we describe experimental observations of the adsorption of a nanoparticle-surfactant suspension to confined bubbles translating along a microchannel. Long bubbles exhibit a unique two-lobed shape that is linked to the adsorption of surface-active particles to the interface at a timescale comparable to the residence time in the channel. An accompanying decrease in the bubble velocity results from the added viscous drag at the bubble interface. We develop a transport model to describe the rate of particle adsorption to the interface and find good agreement between the model estimates of bubble shape changes and experimental observation. In the second part of this presentation, we focus on the bubble break-up that can occur in the microchannel during the particle adsorption process due to a significant difference in the front and rear bubble velocities. We describe the flow conditions necessary for bubble break-up and analyze the rate of neck thinning observed during the break-up process. The experimental thread break-up is compared to standard inviscid break-up models in order to determine the stabilizing effect of surface-active colloidal particles against rupture.

(353d) Thin Film Drainage Between Micro-Drops Under Applied Cyclical Drives to Mimic Micro-Fluidic Pumping Scenarios

Rico Tabor, Chu Wu, Franz Grieser, D.Y.C. Chan and **Raymond R. Dagastine**, the Particulate Fluids Processing Centre, The University of Melbourne, Parkville, Australia

Our research has focused on the development of innovative methods to quantitatively study the nano-scale thin films between micro-drops or micro-bubbles during their collision and coalescence. The forces that control these collisions are crucial for understanding applications as wide ranging as purification steps in mineral and pharmaceutical processing to emulsion and foam stability in food and personal care products as well as micro-fluidic devices. Previously, we have developed experimental and theoretical methods developed to study the dynamic interactions between two drops or bubbles using Atomic Force Microscopy (AFM) for single collisions. The agreement between experimental measurements and the quantitative modeling of the thin film drainage between deformable drops and bubbles has shown that the governing physics is dependent on a combination of equilibrium surface forces, hydrodynamic drainage forces and interfacial deformation at a wide range of collision speeds and varied solution conditions. Both experimental measurements and quantitative modeling have been able to shed light on behavior as diverse as the charging mechanism for drops and bubbles to counter-intuitive observations in micro-fluidic devices such as coalescence of drops on separation.

This talk will focus on what we can learn from the direct force measurement of sequential collisions of pairs of micro-drops in order to mimic pumping conditions in micro-fluidic devices. A number of diverse velocity drives profiles were used to drive to drops together with repeated collisions including waveforms for sinusoidal, peristaltic and diaphragm pumping. In most cases, it was shown that when equilibrium interactions of the drops were a meta-stable state, small perturbations in pumping conditions could cause drop coalescence. This was further probed through using simple constant velocity approach and retract cycles where the approach and retract velocities were varied independently. Agreement between quantitative modeling of these collisions and the experimental measurements allows for visualization of the thin film and pressures profiles during these processes. In addition, the validated model was then used to produce drop stability maps as a function of pumping condition parameters.

(353e) Multiphase Flow of Miscible Liquids: Jets and Drops

Travis W. Walker[1], Alison N. Logia[2] and Gerald G. Fuller[1], (1)Chemical Engineering, Stanford University, Stanford, CA, (2)Stanford University, Stanford, CA

Drops and jets of liquids that are miscible with the surrounding bulk liquid are present in many processes from cleaning surfaces with the aid of liquid soaps to the creation of biocompatible implants for drug delivery. Although the interactions of immiscible drops and jets show similarities to miscible systems, the small, transient interfacial tension associated with miscible systems create distinct outcomes such as intricate droplet shapes, break-up resistant jets, and spreading sessile drops.

Experiments and simulations have been conducted to understand several basic multiphase flow problems involving miscible liquids. Direct comparisons for the free-surface pendant drops resulting from drop impaction and for the dissolution of sessile drops in a miscible bath have been completed. Using high-speed imaging of the morphological evolution of the flows and finite element analysis to solve the coupled Navier-Stoke/Cahn-Hilliard model using diffuse-interface theory, we have been able to show that these processes are controlled by interfacial tensions. Since the measurements of the surface stresses in a miscible system are difficult to measure, the ability to match simulations to the experimental results allows a deeper understanding of the evolution of the apparent interfacial energy of the system.

Further multiphase flows include investigating miscible jets, which allow the creation of fibers and tubular shapes from inelastic materials that are otherwise difficult to process due to capillary breakup. This work shows that stabilization from the diminishing interfacial tensions of the miscible jets allow various elongated morphologies to be formed. When combined with a mechanism to freeze these fibers, highly oriented materials, such as anisotropic collagen fibers used as scaffolds for regeneration of anisotropic tissues, can be created.

(353f) Droplet Dynamics in Channel Flow with Sudden Expansion

Robert Carroll, Chemical Engineering, University of New Hampshire, Durham , NH and Nivedita Gupta, University of New Hampshire, Durham, NH

Drop Dynamics in a Sudden Expansion

Robert M. Carroll and Nivedita R. Gupta

Department of Chemical Engineering

University of New Hampshire

Durham, NH 03824

Abstract

Drops in channels of varying cross-section are encountered in a variety of applications. To understand the behavior of drops as they translate through a sudden expansion, we conduct a numerical study using a three-dimensional axisymmetric hybrid volume-of-fluid method combined with a front-tracking scheme. The simulations begin with a drop in the small channel at its expected steady shape. As it moves through the expansion, the drop elongates perpendicular to the direction of flow taking on an oblate shape. Further away from the expansion region, the drop reverts back to the steady shape predicted for the larger channel size. We study the dynamics of this process at a low but finite Reynolds number and a variety of drop sizes, capillary numbers, and channel expansion ratios. We specifically collect data on maximum deformation, drop speed relative to the fluid, and changes in drop surface area. For large drops, bullet- and parachute-shaped drops as well as drops with a negative curvature are observed. We validate the numerical simulations by comparing the results with experimental studies of glycerol-water drops in silicon oil flowing through a PDMS channel. We extend the numerical results to study the effect of soluble surfactants in the adsorption-desorption limit. The surfactants are modeled using a Langmuir adsorption framework. The non-uniform distribution of surfactants at the interface during the deformation process gives rise to Marangoni stresses that affect the maximum deformation and drop surface area.

(353g) Faraday Wave Dynamics of Immiscible Systems in Finite Cells

William R. Batson III[1], Farzam Zoueshtiagh[2] and Ranga Narayanan[1], (1)Chemical Engineering, University of Florida, Gainesville, FL, (2)Institut d'Electronique de Microélectronique et de Nanotechnologie (IEMN) CNRS 8520, Université Lille 1, Villeneuve d'Ascq, France

The motivation of this work is to produce an immiscible Faraday wave system with minimized sidewall stresses, permitting a comparison to available theory for parameter spaces where wavenumber selection is governed by sidewall boundary conditions. The invsicid model of Benjamin and Ursell[1] provides substantial insight into the fundamental physics of the instability, but is insufficient for making predictions in physical systems, in part due to its offering of instabilities arising from perfect resonance for infinitesimal forcing amplitudes. Linear damping has been incorporated into this model to aid matching with experiment, but this requires a phenomenological parameter and can't be used to make predictions a priori. The viscous model of Kumar and Tuckerman[2] treats the viscous effects of the system rigorously, and its predictions have been validated for conditions where sidewalls do not effect wavenumber selection. Prediction of threshold amplitudes where the excited wavelength is of the order of the lateral cell dimensions is much more challenging due to wetting and contact non-idealities at the sidewalls.

In these experiments the sidewall contact line is pinned, but the formation of a sidewall film of the upper fluid allows for an apparent free motion of the bulk phases and interface. Minimization of the upper fluid viscosity shows the thickness of this film decrease and in turn the no-stress limit is approached. Making the stress-free assumption, threshold amplitudes for a cell mode and its bounding co-dimension 2 points are well predicted by the Kumar and Tuckerman model, with slight deviation due to residual sidewall effects. Experiments near the threshold suggest a forcing frequency-dependent transition from subcritical to supercritical bifurcations. In the subcritical parameter space, unbounded growth and wavebreaking is common, while saturation to standing waves is more common in the supercritical space. In the saturation to standing waves, it is seen that the linear spatial form of the mode is well preserved for forcing amplitudes near the instability threshold, and sufficient increase promotes the development of secondary instabilities that serve as higher order system damping. This behavior helps explain the differences of what can be expected in a two liquid system versus a single liquid system, a central question to this work.

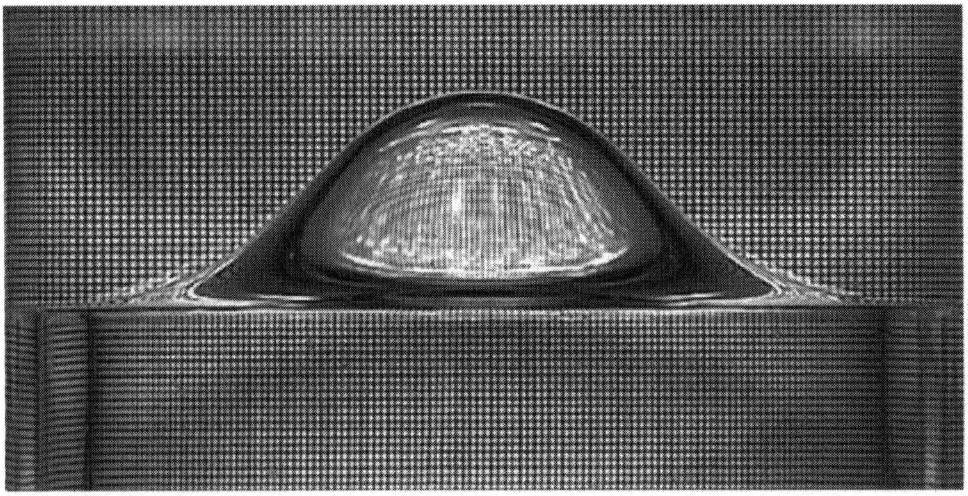

Fig. 1 Experimental excitation of (0,1) cylindrical mode in an immiscible liquid system

[1] Benjamin, T. and Ursell, F., *Proc. R. Soc. London*, Ser. A 225, pp. 505-515, 1954.

[2] Kumar, K. and Tuckerman, L., ***J****. Fluid Mech.*, 279, pp. 49-68, 1994.

(353h) Scaling in the Transition From Selective Withdrawal to Entrainment

Chris A. Pommer, Michael T. Harris and **Osman A. Basaran**, School of Chemical Engineering, Purdue University, West Lafayette, IN

In selective withdrawal (SW), fluid is withdrawn through a tube that has its tip suspended a distance S above a flat interface separating two fluids. When the withdrawal rate Q is low, the interface forms a steady-state hump and only the upper fluid is withdrawn. When Q is increased (or S decreased), the interface undergoes a topological transition so that the lower fluid is entrained with the upper one, forming a steady-state spout. Here, this discontinuous transition is analyzed computationally when both fluids are incompressible and Newtonian. The numerical method employed is an implicit method of lines ALE algorithm which uses finite elements with elliptic mesh generation. In particular, it is shown that the critical withdrawal rate at which the aforementioned transition occurs scales with the nozzle separation raised to some power n. Moreover, two distinct types of transition from SW to entrainment are identified. The first, which occurs at low flow rates, involves a transition from SW to viscous entrainment (VE) in which normal viscous stresses overcome capillary pressure to cause the transition. Here, the value of the scaling exponent is found to equal to three (n=3). This finding agrees with previous experiments and resolves a long standing controversy involving previous simulations. The second transition, which has heretofore not been known, involves one from SW to inertial entrainment (IE) in which viscous stresses play no role. Here, the value of the scaling exponent is found to equal to two (n=2). Scaling arguments are then used to rationalize the simulation results, the existence of these two distinct regimes, and the transition from the first regime, where n=3, to the other, where n=2.

(353i) Scaling of Dispersed Phase Flow Transitions in a Microfluidic Flow Focusing Device

Jihye Kim, Chenical Engineering, Texas Tech University, Lubbock, TX and Siva A. Vanapalli, Chemical Engineering, Texas Tech University, Lubbock, TX

Microfluidic production of drops using a flow-focusing device is an integral component of many lab-on-chip technologies. Several studies have documented the different behaviors of dispersed phase in flow-focusing devices. These behaviors include break-up modes such as dripping, jetting, and a non-break-up mode involving co-flowing laminar streams. Despite the identification of these behaviors, the control parameters that govern the transitions between these behaviors are not fully known.

Recently, Utada *et al.* [1] have shown that the transition between dripping and jetting in a cylindrical microcapillary device is dictated by the capillary number (*Ca*) of the outer fluid and Weber number (*We*) of the inner fluid. In this study, we investigate whether these dimensionless numbers also govern the different behaviors observed in the more complex geometry of a flow-focusing device. We fabricated flow-focusing devices in polydimethyl(siloxane) with a square orifice of size 50μm. We varied the capillary number based on the outer fluid from $\sim 10^{-4} - 0.1$, and the capillary number based on the inner fluid from $\sim 10^{-5} - 1$. The viscosity ratio of the inner to outer fluid was varied from $\mu_i/\mu_o = 0.03 - 100$.

In the microfluidic flow-focusing device, for viscosity ratio ~ O(1), we find that the transition from dripping to co-flowing laminar streams occurs at Reynolds number, Re ~ O(1), where Re is based on the inertial force of the inner fluid to the viscous force of the outer fluid. The dripping to jetting transition is found to roughly occur at $We_o \sim O(1)$, based on the outer fluid. For fluid viscosity ratios greater than or less than 1, initial results indicate that the transition between these different behaviors have a complex dependence on Re and We_o.

[1] A. S. Utada, A. Fernandez-Nieves, H. A. Stone, and D. A. Weitz, Physical Review Letters **99** (9), 4 (2007).

(354a) Bipolar Applied Field Experiments to Characterize Dielectric Relaxation and Describe Charge Transport in Low-k Materials

Joel L. Plawsky, The Isermann Department of Chemical and Biological Engineering, Rensselaer Polytechnic Institute, Troy, NY

Time dependent dielectric failure (TDDB) continues to be an important part of interconnect design as industry pursues integration of sub-45 nm process-technology nodes. The TDDB literature has provided key information about the role played by diffusing species such as electrons, holes, ions and neutral impurity atoms. However, no mechanism has been shown to describe how these species interact and affect dielectric failure. Charge relaxation in low-k dielectrics was studied using bipolar field experiments[2] coupled with a charge transport model. The goal is to tie changes in dielectric relaxation behavior to the injection and transport of charged species such as electrons, ions and traps to dielectric breakdown. We show that current relaxation upon inversion of the applied field can be described with classical dielectric relaxation models such the stretched exponential function, providing insight into charge transport in disordered materials. The kinetics of charge trapping events have been found to be consistent with a time-dependent reaction rate constant of the form $t^{\beta-1}$, $0<\beta<1$. Such dynamics have previously been observed in studies of charge trapping reactions in amorphous solids by Hamill and Funabashi. We explain the relaxation process in charge trapping events by developing a nonlinear charge transport model. Experiments for leakage relaxation at various temperatures and field strengths have been correlated to predictions from the model. These studies promise to help on designing robust low-κ and ultra-low-κ materials by providing insight into the dynamics of charge transport and charge relaxation during bias temperature stress.

1. Hamill, W. H. and Funabashi, K. "Kinetics of electron trapping reactions in amorphous solids; a non-Gaussian diffusion model," Phys. Rev. B,vol. 16, Dec. 1977. 2. J. Borja, J. L. Plawsky, T-M. Lu, W.N. Gill "Impact of frequency from bipolar applied field on dielectric breakdown for low-k materials," under review in IEEE Transactions on Electron Devices.

(354b) Modeling of Particle-Particle and Particle-Wall Contact Points in the 3D CFD Simulation of Transport in Fixed Beds of Spheres

Anthony G. Dixon[1], Michiel Nijemeisland[2] and E. Hugh Stitt[2], (1)Worcester Polytechnic Institute, Worcester, MA, (2)Johnson Matthey Catalysts, Billingham, United Kingdom

The use of computational fluid dynamics (CFD) as a tool for obtaining detailed flow and heat and mass transport information in packed beds is growing, despite the apparent intractability of the geometry of the packing. In particular, the simulation by CFD of fixed beds at higher particle Reynolds number (Re > 1000) and low tube-to-particle diameter ratio (2 < N < 8) is of interest as highly-exothermic partial oxidation reactors and highly-endothermic steam reforming reactors operate under these conditions. Recently, studies that presented simulations of beds of a large number of particles (usually between 100 – 1000 spheres) have appeared in the literature, for flow and heat transfer simulations.

In the development of meshes for CFD simulations of transport in fixed beds of spheres, particle-particle and wall-particle contact points often present difficulties. This paper presents a comparison between four methods of dealing with contact points in CFD simulations of both fluid flow and heat transfer in fixed beds of spheres, at commercially-relevant flow rates. These methods represent unselective (global) and selective (local) geometric manipulation of the spherical particles through either reducing or enlarging them. We give results for drag coefficient (C_D) and heat flow (Q) for flow past sphere-sphere and wall-sphere contact points, focusing on higher flow rates typical of industrial steam reformers (500 < Re < 10,000). Global methods, in which all particles in a bed are either shrunk or enlarged uniformly, change bed voidage giving erroneous results for C_D. Local methods, in which bridges are inserted or spherical caps are removed only at the points of contact, give much better results for C_D. The bridges approach is preferable for heat transfer, as fluid gaps reduce heat transfer too much, and particle overlaps increase it. A set of graphs is presented to allow estimation of the error introduced by the various methods of dealing with contact points. Examples are presented of the application of these methods to fixed bed transport simulations.

(354c) Non-Classical Reactive Transport of Solute Through a Porous Medium

Valentina Prigiobbe, Petroleum and Geosystems Engineering, University of Texas at Austin, Austin, TX, Marc Hesse, Jackson school of geosciences, University of Texas at Austin, Austin, TX and Steven Bryant, Petroleum and geosystems engineering, University of Texas at Austin, Austin, TX

Non-classical reactive transport of a solute through a porous medium

Valentina Prigiobbe[1], Marc A. Hesse[2,3], Steven L. Bryant[1,3]

[1]Department of Petroleum and Geosystems Engineering, University of Texas at Austin, USA.

[2]Department of Geosciences, University of Texas at Austin, USA.

[3]Institute for Computational Engineering and Sciences

Field observations [1] have shown that the transport of radionuclides may be much faster than predicted by standard models applied for the description of transport of dilute solutes in reactive porous media which are based on the theory of chromatography [2]. Numerical simulations motivated by these field observations showed that under certain conditions radionuclides can be strongly retarded as predicted by the theory but also travel without retardation (Figure 1). This transport without retardation is a notable exception from the theory and represent a non-classical reactive transport behavior [3,4,5] which is complement to the well-known fast transport pathways such as colloid-facilitated transport and flow in fractures.

Figure 1. Concentration of strontium (Sr^{2+}) as a function of the longitudinal coordinate. Sr^{2+}front: (a) The numerical simulation consists of a shock, a rarefaction, and the non-classical wave or pulse. (b) The analytical solution from the theory of chromatography consists of a spreading wave followed by a shock wave.

In this work, we analyze the non-classical reactive transport behavior using 1D model for incompressible flow through an iron-oxide porous medium of Sr^{2+}, H^{+}, Na^{+}, and Cl^{-}. We combine the theory of hyperbolic systems of conservation laws [6] with surface complexation and we derive the mathematical framework for the analysis of the system and the definition of the necessary conditions under which the non-classical reactive transport behavior arises [5]. Under the assumption of Na^{+} and Cl^{-} conservative, local chemical equilibrium, negligible hydrodynamic dispersion, and incompressible flow, the mathematical problem reduces to a strictly hyperbolic 3x3 system of conservation laws for effective anions, which are defined as the difference of the conservative species, the total protons, given by the difference of H^{+} and OH^{-}, and Sr^{2+}. The mass conservation laws have the nonlinearity in the accumulation term due to adsorption and they are coupled by the two Langmuir isotherms of H^{+} and Sr^{2+}. One characteristic field is linearly degenerate while the other two are not genuinely nonlinear due to one inflection point in the adsorption isotherms.

For this system, we solved the Riemann problem (constant initial and injected states) and the analytical solution consists of three waves separated by two intermediate points and comprises nine combinations of rarefactions, shocks, shock-rarefaction, and contact discontinuity. The slow and the intermediate waves are either a rarefaction, or a shock, or a shock–rarefaction while the fast wave is a contact discontinuity. Highly resolved numerical solutions at large Péclet numbers show excellent agreement with the analytic solutions in the hyperbolic limit (negligible hydrodynamic dispersion) except under certain conditions when a pulse of Sr^{2+} arises ahead of the retarded Sr^{2+} front which travels at the average fluid velocity. These conditions define the necessary conditions for the occurrence of the non-classical reactive transport of Sr^{2+}, which we verify in the laboratory performing column-flood experiments [7]. In the absence of colloids and fractures, the experiments confirmed this non-classical behavior with a strongly Sr^{2+} retarded front predicted by the theory and an isolated pulse of Sr^{2+} traveling at the average fluid velocity (Figure 2).

Figure 2. Measured concentration of Sr^{2+} (a, b) and of Na^{+} (b) towards the pore volume injected (PV). It is possible to see that the pulse of Sr^{2+} travels at the same speed of the conservative cation Na^{+}.

This non-classical reactive transport behavior is due to the interaction of hydrodynamic dispersion and of a strongly pH-dependent adsorption. The strong pH-dependence makes the retarded front unstable to small perturbations due to hydrodynamic dispersion which broadening the retarded front leads to the formation of the unretarded pulse.

These results raise important questions regarding the prediction of the migration of toxic compounds in the subsurface, e.g., Ba^{2+}, Ca^{2+}, Mg^{2+}, Co^{2+}, and Ni^{2+}, which are characterized by an adsorption isotherm similar to Sr^{2+}, and may have implications for the design of industrial chromatographic separation processes. Furthermore, it raises interesting questions for the theory of hyperbolic systems of equations, as it appears that the vanishing diffusion solution does not approach the diffusion free (hyperbolic) limit uniformly in all cases.

References

[1] Olsen et al. (1986) Geochim Cosmochim Acta 50, 593-607.

[2] Appelo (1996) Reviews in Mineralogy 34, 193-227.

[3] Rhee et al. (1989) First-order partial differential equations, theory and application of hyperbolic systems of quasilinear equations II Prentice-Hall.

[4] Bryant et al. (2000) Ind Eng Chem Res 39, 2682-2691.

[5] Prigiobbe, Hesse, Bryant (2012) Transport Porous Med 93, 127-145.

[6] Lax (1957) Comm Pure Appl Math 10, 537-566.

[7] Prigiobbe, Hesse, Bryant (2012) Geophys Res Lett, submitted.

(354d) Concepts Rendered False by Advances In Understanding and Obsolete by Advances In Computation

Stuart W. Churchill, Chemical and Biomolecular Engineering, University of Pennsylvania, Glen Mills, PA

The formulation and adoption of useful concepts and generalizations, and in many instances unique ones, have helped chemical engineering to flourish as an academic subject and as a profession. The broadest ones are illustrated by unit operations, unit processes, and transport phenomena. Narrower ones are illustrated by the equilibrium stage, the perfectly mixed continuous reactor, and the Colburn j-factor. These latter concepts have resulted from observations in plant operations and in the laboratory as well as from theoretical analyses and flashes of insight. All in all, they constitute an essential element of both education and practice in chemical engineering.

With the passage of time, most of the broad concepts have been replaced, modified, or supplemented. Many of the specific ones have been found to be false or unnecessary yet remain in textbooks and computer packages out of inertia and/or out of misdirected respect for those who originated them. The objective of this presentation is to identify some of the obsolete and/or false ones. Within this restricted space, identifications can only be illustrative, so an attempt has been made to choose examples that are diverse in subject and in the nature of the shortcomings. The results of the advances by virtue of computation will be shown graphically in the presentation. It is hoped that these examples will inspire each of you to question the continued viability of the concepts and idealizations in your personal professional portfolio, including the books you depend upon.

The formulation of each of the concepts discussed herein depends upon one or more ingenious idealizations and/or simplifications. With time some of these have become unnecessary because of advances in computer hardware and software, and some have been proven to be false by advances in understanding. Their identification is an essential part of the process of re-examination of a concept, and often the critical one.

Reaction engineering

The concept of reaction engineering was a great advance over chemical kinetics, which was originally taught as a subtopic in physical chemistry. It continues to have vitality but has accumulated a number of false concepts, a few examples of which are noted here.

Plug flow

This is a misleading, inaccurate, and totally unnecessary concept. It should be excised from our literature and replaced by realistic fluid-mechanical models.

Space velocity and space time

These concepts invoke a Lagrangian framework and thereby the shortcomings of plug flow.

They too should be excised.

Perfect radial mixing in tubular flow

This is a more accurate verbal description of the behavior resulting from the postulate of plug flow in that it can be considered as an asymptote for $Pr \to 0$ and $Sc \to 0$, but it does not ameliorate the errors in the model.

Fully developed flow with no radial mixing

This is the counterpart of perfect radial mixing, that is, it corresponds to $Pr \to \infty$ and $Sc \to \infty$, and as such is physically conceivable. It is useful as a lower-bound for the conversion and as an upper-bound for the length of the reactor, which are conservative predictions and thereby good engineering as contrasted with the predictions of perfect radial mixing which are non-conservative and thereby poor engineering.

Turbulent tubular flow

Chemical conversions rarely occur in the turbulent regime because the reactor is then almost certainly of excessive length. Plug flow is not approached; the velocity remains nearly parabolic.

Initiation

The initiation of a reaction over the entire cross-section at the inlet is implied in most models although it is impossible to achieve physically. This idealization is difficult to replace, because the alternatives involve developing flow and thereby an increase of an order of magnitude in complexity of the numerical modeling. Until such modeling is included in computer packages, the affects of this non-conservative idealization should not be ignored.

Rate mechanisms

The expression of reaction rates in terms of molecular concentrations rather than activities betrays the principles of chemical engineering thermodynamics. To brag about its rigor and generality as compared to that of mechanical engineering and chemistry and then cling to this erroneous concept is untenable. It is difficult to re-do rate constants in terms of concentrations because that requires a knowledge of the pressure, temperature, and composition at which the measurements were made. The error in the predictions is serious only if the conditions differ significantly from those of the measurements, but the time has come for the editors of chemical engineering publications to establish a new standard.

Over-correlation

As shown definitively in an article by Mayer and Stowe, models of the type proposed by Hougen and Watson for chemical conversions in flow through a bed of catalyst particles introduce more empirical coefficients than can be justified by the precision and extent of the experimental data. It follows that the resulting models have no physical significance. This approach should have been abandoned long ago. This is an example of the misuse of numerical computation. This listing could go on and on and on; the number of obsolete and false concepts and the unnecessary idealizations and simplifications in reaction engineering is almost endless, but some space must be reserved for other topics.

Combustion

Combustion has progressed from art to an engineering science by virtue of advances in computation and in the determination of free-radical mechanisms. It should be a sub-topic, such as Separations, within chemical engineering but because of the failure to encompass it in reaction engineering it has become a separate profession. The probable reason for such short-sightedness is that combustion occurs primarily in unconfined flow whereas chemical engineers are more comfortable with tubular flow. Two false concepts related to the modeling of combustion deserve mention. First, solids, liquids, and gases do not burn – only free radicals; therefore combustion must be preceded by pyrolysis. Second, the concentration of free radicals never attains a stationary state. Global modeling should be abandoned.

Separations

Molecular separations are, along with reaction engineering, unique to chemical engineering.

There is space to mention only one ingenious but now obsolete concept, namely, the ideal- stage. The many inherent idealizations, most of which are now avoidable owing to advances in computation, should not go unmentioned.

Heat Transfer

Chemical engineers have more or less ceded leadership in heat transfer to mechanical engineers although heat exchange remains essential in chemical processing and in petroleum refining. Space allows the mention of only two obsolete concepts and one overlooked effect.

Correlations in the form of products of powers of Re and Pr

A power-dependence actually occurs, if at all, only in an asymptotic sense. As an example, the Colburn analogy is functionally in error in every respect and, as a consequence, seriously in error numerically. The formulation and use of far more accurate correlating equations in the form of power-means of the asymptotes has been encouraged and abetted by advances in computation.

The hyperbolic equation of conduction

This model is totally false. In the instance of a pressure wave generated by impulsive heating, such as thunder by lightning, the behavior can be predicted exactly by taking the compressibility and the conservation of momentum into account. The computational investment is however high owing to stiffness.

The enhancement of convection by an energetic reaction

This effect, which is completely overlooked in all textbooks has been resolved by differential models and finite-difference computations.

Fluid Mechanics

This subject is shared with mechanical engineers, civil engineers, and others but chemical engineers remain the pace-setters in special topics such as non-Newtonian fluids, packed beds, and fluidized beds. Obsolete concepts are illustrated by the following examples.

The power-law fluid

Such behavior is simply a necessary artifact of the transition between pseudoplastic and dilatant behavior.

The orifice coefficient

The numerical expression derived for a sharp-edged orifice on the basis of a free streamline is a misleading approximation. The coefficient, including its asymptotic value, varies with D_o/D and *Re*, and should be computed numerically.

The Ergun equation

The attribution of the Re^2 term of the Ergun equation to turbulent rather than inertial flow is an example of a pervasive misinterpretation.

DNS (direct numerical simulation) and LES (large eddy simulation)

These methodologies exemplify self-defeating limitations. DNS has proven to be viable only for the hypothetical case of a parallel-plate channel and for Reynolds numbers barely above that for fully developed turbulence. LES has proven to have the same geometrical limitation and to require very empirical supplementation.

Mass Transfer

The analogy between heat and mass transfer

The substitution of *Sh* and *Sc* for *Nu* and *Pr* has been shown both experimentally and by Lagrangian *direct numerical simulation*to be invalid.

The Lewis number

In a session in his honor at the 57th *AIChE*Annual Meeting, Warren K. Lewis jokingly complained that the dimensionless group named in his honor commemorates one of his worst mistakes, namely the presumption that because that grouping happened to be nearly equal to unity in some of his experiments with water vapor in air it was unity for all conditions and systems.

(354e) On Boundary-Layer Flows Driven by Translating, Flexible Sheets

Talal T. Al-Housseiny, Chemical and Biological Engineering, Princeton University, Princeton, NJ and Howard A. Stone, Mechanical and Aerospace Engineering, Princeton University, Princeton, NJ

The interaction between fluid flow and flexible structures is important in a wide range of systems ranging from beating flagella to waving a flag. These coupled fluid-body problems are typically difficult to study theoretically and computationally. In this work, we study a process where a sheet of material is drawn through a fluid medium at a high velocity. Such scenario occurs in many industrial settings such as extrusion processes. The translating sheet deforms in the direction of drawing and induces a flow in the surrounding fluid. Unlike the classical problem in the literature where the mechanics of the sheet are neglected, and kinematics are prescribed, we herein couple the dynamics of both the sheet and the surrounding fluid. The characteristics of the induced flow are studied by means of boundary-layer theory. The coupled system we consider allows for analytical solutions that are valid under limiting conditions, which we also identify.

(354f) Transport in Stochastic Systems with Non-Adiabatic Coupling Between Different Degrees of Freedom

Dmitry I. Kopelevich, Department of Chemical Engineering, University of Florida, Gainesville, FL

Transport in complex molecular systems is frequently modeled by a reduced stochastic model, such as the Langevin equation. A common method of analysis of tramsport in stochastic systems with multiple degrees of freedom is based on the assumption that the system follows a minimum energy path (MEP), i.e. a path such that the free energy is minimized in all directions transversal to the path direction. This assumption effectively reduces the multi-dimensional stochastic process to a quasi-one-dimensional process confined to the MEP. However, the MEP assumption is not always valid. In particular, our earlier work has demonstrated that substantial deviations from the MEP may occur even in a relatively simple process, such as a molecular transport across a liquid-liquid interface or a lipid membrane. These deviations from the MEPs occur due to non-adiabatic coupling between translational motion of the solute molecule and thermal fluctuations of the interface.

In this talk, we present a theoretical analysis of such non-adiabatic processes. This analysis is based on applying the path integration formalism to solution of the Langevin equation. This formalism yields a probability distribution of paths and enables us to identify the most likely path (MLP) to be taken by the system. We demonstrate that the MLP may significantly deviate from the MEP, which effectively creates an additional energy barrier to transport. We investigate dependence of this dynamic barrier on several system parameters characterizing the strength of coupling between different degrees of freedom.

(367a) Collective Dynamics of Particle Clusters Flowing in a Quasi-Two-Dimensional Microchannel

William E. Uspal[1], H. Burak Eral[2] and **Patrick S. Doyle**[1], (1)Chemical Engineering, Massachusetts Institute of Technology, Cambridge, MA, (2)Chemical Engineering, MIT, Cambridge, MA

Control of flowing suspensions of particles is central to many emerging microfluidic applications, including cell sorting, information processing, and assembly of complex structures. Likewise, spatially ordered equilibrium states – crystals – and their excitations – phonons – are the mainstay of condensed matter physics. Flowing, non-equilibrium crystalline states of microparticles and droplets are desirable for microfluidic logic, assembly, and control, and have been achieved in recent work via exploitation of viscous hydrodynamic interactions in geometric confinement. For the most part, these studies considered large ensembles of particles and, accordingly, large scale collective modes arising from small displacements of individual particles. Via theoretical modeling and computational simulations, here we show that for small clusters of flowing particles tightly confined in a shallow, "quasi-two-dimensional" microchannel, new types of ordered behavior emerge, with complex and individualistic particle motion. We elucidate principles and techniques for the *a priori* construction or rapid numerical discovery of these states, which could be exploited for the orchestration of particle motion in lab-on-a-chip devices and other applications. Our theoretical studies are complimented by experiments in which we can "print" custom-shaped particles in predefined positions in a microchannel. The experiments allow for the unique ability to precisely set initial conditions for a cluster and then watch it evolve in flow.

(367b) A Macrotransport Equation for the Flow of a Concentrated Suspension of Non-Colloidal Particles Through a Capillary Tube

Arun Ramachandran, Department of Chemical Engineering and Applied Chemistry, University of Toronto, Toronto, ON, Canada

By implementing a two time-scale perturbation expansion of the suspension balance model[1], we formally derive a cross-section-averaged convection-dispersion equation for the flow of a concentrated suspension of neutrally buoyant, non-colloidal particles through a tube. The Taylor-dispersion coefficient D_{eff} in this macrotransport equation scales as $h(\Phi)UR^3/a^2$; here U is the velocity of the suspension in the tube, R is the tube radius, a is the particle radius and h is a monotonically increasing function of the particle volume fraction Φ. The linear dependence of D_{eff} on U implies that changes in the cross-section averaged axial concentration profile are dependent only on the total axial strain experienced by the suspension. This stipulates that the spatial evolution of the mixing length scale for a given fluctuation in the inlet concentration of the suspension, or the spatial evolution of the width of the mixing zone between two suspensions of different concentrations in the tube will be independent of the suspension velocity in the tube. The result contrasts from the quadratic dependence on the velocity that is expected from the classical problem of Taylor dispersion of a passive tracer in the same geometry, which results in faster dispersion at higher average velocities. A second major point of difference between the two problems is that the effective velocity of the particulate phase is concentration-dependent, and this can lead to either sharpening or relaxation of concentration gradients. Concentration pulses relax asymmetrically, and positive concentration gradients along the flow direction can lead to time-independent distributions in an appropriately chosen frame of reference; both effects are not observed in solutal dispersion. The results in this paper are relevant for microchannel flows, and can be used to design geometries that minimize or enhance mixing in suspension flows through microchannels.

References

1. P. R. Nott and J. F. Brady, Pressure-driven flow of suspensions: simulation and theory, *J. Fluid Mech*, **275**, 157-199 (1994).

2. D. Leighton and A. Acrivos, The shear-induced migration of particles in concentrated suspensions, *J. Fluid Mech*, **181**, 415-439 (1987).

(367c) Dispersion of Suspension Plugs in Microchannels

Glareh Azadi, Engineering, Brown University, Providence, RI, Kyongim Yeo, Lawrence Berkeley National Laboratory, Berkeley, CA, Martin Maxey, Applied Mathematics, Brown University, Providence, RI and **Anubhav Tripathi**, School of Engineering, Brown University, Providence, RI

Plugs of microbeads can be used as a model for understanding the behavior of suspensions in micofluidic devices. Despite a large volume of literature in the field of microfluidic suspension flows; the fundamental understanding of the dispersion of hard and soft particle suspension plugs of finite length has not be studied in detail. Here we focus on the dispersion characteristics of non-Brownian, low Reynolds number bead suspensions in microfluidic channels. The effect of initial plug lengths (10μm -15μm), bead size(1μm -3μm DI) and dilution(25%-75%) on the early stages of dispersion of these beads in a pressure driven flow will be presented. Plugs of magnetic beads are mixed with fluorescent tracer beads. Combined effect of the molecular diffusion and dispersion is evaluated and the Taylor dispersion coefficient is determined. Numerical simulations of suspension shear flow for test configurations relevant to plug dispersion were performed in order to present a new level of continuum models for particle stresses and particle fluxes in confined shear flows. These results can contribute to the design of high throughput microfluidic systems for cell screening and bio-separation.

(367d) Rheology and Complex Flow of a Concentrated Bimodal Suspension

Kun Yu, Bidya Prasad and **Nina C. Shapley**, Chemical and Biochemical Engineering, Rutgers, the State University of New Jersey, Piscataway, NJ

In this work, the rheology of a bimodal suspension of neutrally buoyant, noncolloidal spheres in a viscous, Newtonian liquid is investigated. The smaller particles are alginate gel microbeads, which are deformable and porous, and the larger particles are rigid PMMA spheres with a diameter approximately three times greater. Preliminary results at a total particle volume fraction of 0.5 show that the viscosity increases and becomes increasingly shear thinning as the fraction of larger, rigid particles increases. Comparisons to monomodal suspensions of either particle type will be presented. Rheological phenomena can also be related to nuclear magnetic resonance imaging (NMRI) observations of the suspensions flowing through an abrupt, axisymmetric 1:4 contraction-expansion. At high fractions of large particles, the small particle fraction appears to act as a lubricant near the wall of the contraction-expansion and promote plug flow of the large particle fraction, which suggests the potential of small particle additives for extrusion processing of composites.

(367e) Direct Measurement of Suspension Microstructure: Effects of Combined Hydrodynamic Electrostatic, and Viscoelastic Interactions

Bu Xu, M. Tharanga Perera and **James F. Gilchrist**, Department of Chemical Engineering, Lehigh University, Bethlehem, PA

Microstructure is key to understanding rheological behaviors of flowing particulate suspensions. Previous results obtained numerically via Stokesian Dynamics reveal an anisotropic structure forming under high Peclet number conditions. This work outlines an experimental technique based on confocal microscopy to study microstructure of a colloidal suspension in an index-matched fluid flowing in a microchannel. Results of simulation-like quality are produced and the effect of Pe and interparticle force is investigated. Microstructure is shown to be consequential from competitions between hydrodynamic, thermal, electrostatic, and elastic forces depending on the suspending medium. In highly screened systems, the stress of suspension calculated based on the microstructural information is compared with previous computations. In systems where screening is reduced and the Debye length is significiant, the particles interact both through soft repulsion and hydrodynamic interactions, relaxing the structure observed in systems having pure hydrodynamic interactions. We also explore non-Newtonian suspending fluids, such as Boger fluids, to gauge the effect of elasticity on suspension structure.

(367f) On the Existence of Flows Driven by Body Couples and Couple Stresses In Magnetic Nanoparticle Suspensions

Carlos Rinaldi, Chemical Engineering, University of Puerto Rico, Mayaguez, PR

Suspensions of magnetic nanoparticles, so-called ferrofluids, are a commercially and scientifically relevant example of fluids whose description requires consideration of the balance of internal angular momentum. For over 40 years, researchers in this field have sought to understand the mechanisms responsible for observations of flows of these fluids driven solely by the action of rotating uniform magnetic fields. One possible mechanism is that these flows are driven by the action of body couples, exerted by the magnetic field on the particle's magnetic dipoles, and so-called couple stresses, which account for the short-range transport of internal angular momentum. However, other possible mechanisms have been advanced in the literature, such as flow driven by traveling wave magnetic field gradients or by surface phenomena, and experiments have failed to shed light on the matter due to limitations in past experimental approaches. In this talk I will briefly summarize our past theoretical and experimental work investigating this phenomenon, as well as present new experimental results using model ferrofluids consisting of suspensions of magnetic nanoparticles with thermally-blocked magnetic dipoles, for which quantitative agreement can be obtained between experiment and theoretical predictions taking into account the action of body couples and couple stresses.

(367g) A Modified Kinetic Theory Model Bridging Dense and Dilute Regimes of Frictional Granular Flow

Sebastian Chialvo and Sankaran Sundaresan, Chemical Engineering, Princeton University, Princeton, NJ

We investigate the rheology of granular materials in both the dense and dilute inertial regimes via molecular dynamics simulations of homogeneous, simple shear flows of soft, frictional spheres. Though traditional kinetic theories are often used for continuum modeling of granular materials, they fail to describe flow behavior in dense systems near the jamming transition and do not account explicitly for interparticle friction. On the basis of our simulations and a recently-developed dense- regime model [1], we propose modifications to the commonly-used Garzó-Dufty kinetic theory [2], originally developed for frictionless particles at low-to-moderate solid fraction. These modifications include 1) a new expression for the radial distribution function at contact and 2) correction factors to the shear stress and energy dissipation expressions. These changes allow the new model to predict the increase in pressure in the dense regime as well as the decrease in pressure in the dilute regime that result when increasing the interparticle friction coefficient. Additionally, they enable the prediction of a yield stress in the close-packed limit, a feature not found in previous kinetic theories.

[1] S. Chialvo, J. Sun, and S. Sundaresan. Phys. Rev. E 85, 021305 (2012).
[2] V. Garzó and J.W. Dufty. Phys. Rev. E 59, 5895 (1999).

(367h) Intermediate Asymptotics of Axial Diffusion of Tumbled Granular Materials

Ivan C. Christov and Howard A. Stone, Mechanical and Aerospace Engineering, Princeton University, Princeton, NJ

Though granular materials consist of macroscopic particles that do not perform Brownian motion, "self-diffusion" can arise from agitation processes that cause non-equilibrium velocity fluctuations. Contrary to previous results [1], it has recently [2] been suggested that the axial diffusion process of granular matter in a long rotating cylinder might be "anomalous," i.e., the mean-squared-displacement of particles follows a power law in time with exponent less than unity. Further numerical [3] and experimental studies [4] were unable to settle this apparent paradox. We show two possible resolutions to the observations in [2] without the need to appeal to anomalous diffusion: (a) by considering the evolution of arbitrary initial data towards the self-similar intermediate asymptotics of diffusion, and (b) by accounting for the concentration-dependent diffusivity in bidisperse mixtures. Furthermore, since granular flow in the axial cross-section of a long rotating drum is composed of a thin surface shear layer over a fixed bed in static equilibrium, the cross-sectionally-averaged concentration can be separated into a diffusing and a non-diffusing component. We extend the standard axial diffusion model by using first-order kinetics to account for the continuous exchange of particles between the flowing layer and the fixed bed. The resulting transport equation is identical to the one governing the diffusion of momentum in a viscoelastic Jeffreys (also known as Oldroyd-B) fluid above an impulsively-moved plate. Results will be presented on the intermediate asymptotics of this new axial diffusion model equation with both constant and concentration-dependent diffusivities.

[1] R. Hogg, D. S. Cahn, T. W. Healy, and D. Fuerstenau, Chem. Eng. Sci. 21, 1025 (1966).
[2] Z. S. Khan and S. W. Morris, Phys. Rev. Lett. 94, 048002 (2005).
[3] N. Taberlet and P. Richard, Phys. Rev. E 73, 041301 (2006).
[4] D. Fischer, T. Finger, F. Angenstein, and R. Stannarius, Phys. Rev. E 80, 061302 (2009).

(367i) Predicting Clustering Instabilities in Granular Materials: Kinetic-Theory Simulations Vs. Molecular Dynamics Simulations

Peter P. Mitrano[1], John R. Zenk[1], Sofiane Benyahia[2] and **Christine M. Hrenya**[1], (1)Chemical and Biological Engineering, University of Colorado at Boulder, Boulder, CO, (2)U.S. Department of Energy, Morgantown, WV

Instabilities in freely cooling granular materials have been widely observed via kinetic-theory-based models and molecular dynamics (MD) simulations. Recently, the quantitative ability of kinetic-theory-based predictions for the *onset* of instabilities was previously assessed through comparison with MD simulations [Mitrano *et al.* 2011, Mitrano *et al.* 2012]. However, the ability of kinetic-theory descriptions to accurately predict the *evolution* of instabilities has not previously been addressed. In this work, we carry out such an assessment via a direct comparison between (i) transient simulation using kinetic theory, (ii) a linear stability analysis of the same theory, and (iii) MD simulation. The results reveal two important findings. First, an excellent agreement is obtained between the transient theoretical predictions and MD simulations for the critical clustering length scale. This agreement is somewhat surprising given the small-Knudsen (small-gradient) assumption of the Navier-Stokes-order theory coupled with the presence of large gradients in the unstable system. However, it is also encouraging as it implies that the range of validity of the theory is wider than expected from a strict interpretation of its assumptions. Second, the predictions of the same length scale from the *linear* stability analysis over-estimate those of MD and transient simulations. This result confirms earlier work indicating the important role of *nonlinear* mechanisms in the evolution of instabilities.

(384a) Phase Behavior of Gases in Ionic Liquids

Mark B. Shiflett, DuPont, Central Research and Development, Wilmington, DE and Akimichi Yokozeki, DuPont, Spencerport, NY

This presentation will emphasize the importance of characterizing the global phase behavior of gases in ionic liquids and how this can provide insight into new applications. Solubility measurements of several gases in ionic liquids are discussed and important experimental details are highlighted when measuring the vapor-liquid equilibria using a gravimetric microbalance and vapor-liquid-liquid equilibria using a mass-volume technique. The global phase behavior has been well correlated with a modified cubic equation of state and many of the systems described exhibit type-III and type-V phase behavior according to the van Konynenburg-Scott classification. We have found that gases can exhibit different solubility behaviors in ionic liquids (*i.e.* physical and chemical absorption) and that these behaviors can be analyzed with the present equation of state using a simple association model and excess thermodynamic functions. Gas separation using ionic liquids is discussed which includes examples of ternary phase behavior (experiments and model calculations) and what will be required to develop future commercial processes.

(384b) Pure and Mixture Thermophysical Properties and Phase Behavior of Lipidic Ionic Liquids

Kevin N. West[1], James H. Davis Jr.[1,2], John R. McCabe[1], Ali AL-Hashem[1], T. Kyle Zimlich[1], Arsalan Mirjafari[2] and Richard A. O'Brien[2], (1)Chemical & Biomolecular Engineering, University of South Alabama, Mobile, AL, (2)Chemistry, University of South Alabama, Mobile, AL

Recently, our group described the synthesis and solid/liquid equilibrium of lipidic ionic liquids: ionic liquids with long alkyl chains (C_{16}-C_{20}) that, like natural lipids, contain cis-unstaurations which dramatically lower their melting points relative to their saturated analogs. These compounds are of interest for a number of applications including reactions and separations involving non-polar compounds, models for biological membranes, separation of naturally occurring lipids from biological media and advanced lubricants. The first generation of these compounds utilized cis-double bonds to impart asymmetry in the alkyl chain. While effective at lowering melting points, the unsaturations create sites for oxidative degradation, much like rancidity in natural oils. Subsequently, we have developed other strategies to impart asymmetry in the alkyl chain including mid-chain branching and the inclusion of heteroatoms in the alkyl chain. Thioether groups, introduced through thiol-ene "click chemistry," have been a particularly valuable as they are synthetically very accessible and allows for a large degree of structural tunability as the position of the sulfur group in the chain is varied.

For these compounds to be useful in the potential applications mentioned above, the physical properties and phase behavior of these compounds must be explored. Additionally, property trends throughout the different subclasses must be described to identify where specific species may be employed and where expanded synthetic strategies would be beneficial. We are currently working to characterize these ionic liquids as pure species and as mixtures with other ionic liquids and with molecular species. This work will present our results relating to the pure and multicomponent thermophysical properties and phase equilibria of these systems.

(384c) Thermophysical and Electrochemical Properties of Self-Assembling Amphiphilic Ionic Liquids

Sitaraman Krishnan[1], Janice Losenge Lebga[1], Lin Wu[1], John B. McLaughlin[1], Simon Rock[2] and Dipankar Roy[2], (1)Department of Chemical and Biomolecular Engineering, Clarkson University, Potsdam, NY, (2)Department of Physics, Clarkson University, Potsdam, NY

Because of their low volatility, high thermal stability, and ability to conduct ions, ionic liquids are attractive as electrolytes for energy conversion and storage devices. Many of these ionic liquids are "fragile" liquids [1], that is, they undergo significant changes in molecular aggregation and fluid microstructure with relative small changes in temperature. We studied the influence of molecular self-assembly and temperature on the thermophysical and transport properties of ionic liquids with ammonium or imidazolium cations, and iodide, bistriflamide, mesylate, or triflate anions. Nonpolar alkyl groups, polar poly(ethylene glycol) (PEG) groups, and amphiphilic PEGylated fluoroalkyl groups were incorporated in the chemical structures of the cations. The selected cation and anion structures allowed us to systematically investigate the roles of charge delocalization and microstructure formation on ionic liquid properties.

The ionic liquids were characterized using techniques such as differential scanning calorimetry, thermogravimetry, cone-and-plate viscometry, pulsed-field gradient NMR spectroscopy, electrochemical impedance spectroscopy, and cyclic voltammetry. Molecular self-assembly was studied using self-consistent mean field simulations. The degree of dissociation of the cations and anions in the ionic liquid, that is, the ionicity of the ionic liquid, was characterized using measurements of ionic conductivities and ion diffusion coefficients.

PEGylation of the cations resulted in better solvation and cation–anion separation because of hydrogen bonding interactions of the oxygen atoms in PEG with the acidic protons of the cation [2]. The melting and crystallization phase transitions depended on the molecular weight of the PEG tails. Nonpolar alkyl groups, attached to cation, strongly influenced physical properties such as density, viscosity, and conductivity of the PEGylated ionic liquid. In ionic liquids with the fluorinated amphiphilic groups, the nonpolar fluoroalkyl groups exhibited microphase segregation from the polar PEG and ionic species in the liquid [3]. The aprotic ionic liquids with imidazolium cations and iodide anions exhibited greater ionicity than the protic ionic liquids with trialkylammonium or 3-alkylimidazol-1-ium cations and mesylate or triflate anions. In the case of protic ionic liquids, the mechanism of proton transport (vehicular vs. Grotthuss diffusion) was influenced by the fluid microstructure.

In this talk, the structure-property relationships developed in this study will discussed with the aim of optimal design electrolytes for devices such as dye-sensitized solar cells, lithium ion batteries, and high-temperature proton exchange membrane fuel cells.

REFERENCES

1. Angell, C. A. Formation of glasses from liquids and biopolymers. Science **1995**, *267*, 1924-1935.
2. Ganapatibhotla, L. V. N. R.; Zheng, J.; Roy, D.; Krishnan, S. PEGylated imidazolium iodide ionic liquid electrolytes: Thermophysical and electrochemical properties. *Chem. Mater.* **2010**, *22*, 6347-6360.
3. Ganapatibhotla, L. V. N. R.; Wu, L.; Zheng, J.; Jia, X.; Roy, D.; McLaughlin, J. B,; Krishnan, S. Ionic liquids with fluorinated block-oligomer tails: Impact of self-assembly on transport properties. *J. Mater. Chem.* **2011**, *21*, 19275-19285.

Determination of Infinite Dilution Activity Coefficients in Ionic Liquids by Open-Tubular Column Gas Chromatography

Qianqian Xu, *Baogen Su, Xinyi Luo, Huabin Xing, Zongbi Bao, Yiwen Yang, Qilong Ren, Zhejiang University, Hangzhou, China*

Infinite dilution activity coefficients (γ^{∞}) provide information regarding the solute-solvent interactions in the absence of solute-solute interactions and are used, in particular, for the selection of solvents for extraction and extractive distillation. Several methods developed for the measurement of γ^{∞} include ebulliometry method, static method, gas-stripping method (or dilutor method), liquid-liquid chromatography, gas-liquid chromatography (GLC) and headspace GLC, of which GLC is a rapid and exact method for experimental determination of γ^{∞}. Open-tubular column gas chromatography (OTCGC) has several advantages over PGC for γ^{∞} measurement such as high column efficiency and lower surface-to-volume ratio. Nonetheless, OTCGC has rarely been applied to the determination of γ^{∞}. This is probably due to the large errors raised by the low carrier gas flow rate and the small amount of stationary phase.

Ionic liquids (ILs) are considered to be a promising substitute for traditional solvents. The properties of extremely low vapor pressure even at high temperatures, high viscosity, high thermal stability and high polarity make ILs suitable as stationary phases for gas chromatography. Infinite dilution activity coefficients of organic solvents in a great deal of ILs have been determined by packed column gas chromatography (PGC) over the last decade.

In this work, we rationally revised the calculation equation of γ^{∞} for OTCGC, which, as a consequence, eliminated the corresponding errors mentioned above. The respective γ^{∞} values of organic compounds including alkanes, alkenes, alkynes, cycloalkanes, cycloalkenes, aromatic hydrocarbons, alcohols, aldehydes, ethers, ketones, esters, chloromethanes and nitroalkanes in 1-butyl-3-methylimidazolium hexafluorophosphate at 40-60 °C were successfully determined by OTCGC using the new equation. Our γ^{∞} values agreed well with literatures, the results further proved the feasibility of the revised equation for OTCGC to investigate infinite dilution activity coefficients. Using the γ^{∞} values determined by OTCGC with ionic liquid stationary phases, selectivities of selected compounds could be easily calculated.

(384e) Corresponding States Principle and Group Contribution Method for the Prediction of Physicochemical Properties of Ionic Liquids

Ying Huang, Haifeng Dong, Xiangping Zhang and Chunshan Li, Institute of Process Engineering, Chinese Academy of Sciences, Beijing, China

Ionic liquids (ILs) refer to liquids composed entirely of ions that are fluid around or below 100°C[1] and continue to attract burgeoning interest in both academic research and industrial field which often requires knowledge of their physicochemical properties. It is difficult to measure all the properties of different ILs at wide conditions due to the vast amount of possible combination of cations and anions. Alternatively, generalized reliable predictive models for physicochemical properties are of fundamental importance in exploiting new ionic liquids and process design or simulation. Group contribution method has been extended to predict the thermophysical properties of ionic liquids[2], but the research on the critical properties is scarce, which are essential for the calculation of physicochemical properties using the generalized equations or appropriate equation of state. A new method combined Group Contribution with Corresponding States Principle (GC-CSP) method is developed herein to predict the normal boiling temperatures, the acentric factors and the critical properties (namely, critical temperature, pressure and volume) of ILs. The reliability of the GC-CSP method developed here is verified by the prediction of density and surface tension which used the critical properties predicted using this method. With the calculated critical properties, new equations based on corresponding states principle for the estimation of density, surface tension, heat capacity and thermal conductivity are proposed via data regression. The results show that the GC-CSP method gives reliable critical properties of ILs which can be successfully applied to density and surface tension prediction. The calculated density, surface tension, heat capacity and thermal conductivity by the proposed equations display a good agreement with the corresponding experimental densities with the average absolute relative deviation all less than 4%.

References

1. Rogers RD, Seddon KR. Ionic liquids - Solvents of the future?. *Science*. Oct 31 2003;302(5646): 792-793.

2. Gardas RL, Coutinho JAP. Group Contribution Methods for the Prediction of Thermophysical and Transport Properties of Ionic Liquids. *Aiche Journal*. May 2009;55(5):1274-1290.

(384f) A Theory-Framed Quantitative Structure-Property Relationship (QSPR) Model for Liquid-Liquid Equilibrium

S. B. Gebreyohannes, K. M. Yerramsetty, B. J. Neely and K. A. M. Gasem, School of Chemical Engineering, Oklahoma State University, Stillwater, OK

A thorough understanding of phase behavior properties of chemicals is essential for designing and optimizing processes that involve separation of components from a mixture. The non-random two, liquid (NRTL) model is an activity coefficient model used widely in phase equilibria calculations, which has three adjustable parameters that are determined through regression of experimental data for a specific system. Generalizing a model can reduce time, money and effort required to carry out experiments. This work focuses on application of a theory-framed quantitative structure-property relationship (QSPR) modeling approach. A theoretical framework is employed to develop the behavior models, and the QSPR model is used to generalize the substance-specific parameters.

A database of 342 binary low temperature (10 – 40°C) LLE data was employed in this work. Data regression analysis was performed to determine the parameters of the NRTL model. The structural descriptors of the molecules were generated and used in developing a QSPR model to estimate the regressed NRTL parameters. The regression analysis yielded %AADs of 2, 13, 11, 17 for mole fractions of component 1 in phase 2 (component 1 rich) and phase 1 (component 2 rich) and partition coefficient of components 2 and 1, respectively.

The newly developed QSPR model yielded predictions with 8, 37, 40 and 44 %AAD for the mole fractions and partition coefficients, respectively. These errors are approximately 2 to 3 times the errors found from the regression analysis. There is also a failure rate associated with the property predictions due to convergence issues, which in the case of our model was 8%. The application of the popular and often employed UNIFAC model resulted in 3 to 7 times the errors obtained through regression analysis and a failure rate of 35%. The smaller prediction errors as well as the lower failure rate of the QSPR model demonstrates the superiority of the model in providing improved and reliable predictions, as well as an enhanced range of applicability when compared to the UNIFAC model for LLE systems.

(384g) Thermodynamics of Electrolyte-Containing Solutions and Aerosols

Cari S. Dutcher, Xinlei Ge, Anthony S. Wexler and Simon L. Clegg, Air Quality Research Center, University of California at Davis, Davis, CA

Solute and solvent activity coefficients of complex fluids over a wide range of concentrations govern numerous physical phenomena, from microstuctural transformations seen in plant cells as a function of intracellular water content, to phase partitions seen in atmospheric aerosols as a function of equilibrium relative humidity. Accurate predictions of water and solute activities in, for example, complex atmospheric aerosol mixtures to very low water content are essential for reliable climate predictions. A powerful method has been recently developed (Dutcher et al. JPC C, 2011, 2012) for capturing the thermodynamic properties of multicomponent solutions and aerosols at high solute concentrations by applying the principles of multilayer adsorption to ion hydration found in electrolyte solutions. In these works, statistical mechanics is used to model adsorption of a solvent on to *n* energetically distinct layers in the hydration shell surrounding the solute molecule in aqueous mixtures.

Here, we extend the statistical mechanical model of the adsorption free energy to include the dilute limit, mixed solute charge types, mixed solvents and surface tensions, allowing for a unified thermodynamic treatment of complex electrolyte containing solutions over wide ranges of solvent contents. Gibbs free energy, solvent and solute activity, solute concentration, as well as interfacial energies are derived, with 1-2 adjustable parameters per solute which can be estimated from the physical and electrostatic properties, if available, of the solution constituents or fit directly to osmotic and activity coefficient data. The resultant thermodynamic model is compared to existing activity models that have been extended to electrolyte containing solutions, such as statistically associating fluid theory based models, local composition based models, and aerosol organic/inorganic based models, and found to yield unprecedented agreement of the solute concentration and osmotic coefficients for solutions over the entire solute concentration range. Finally, the effect of the long- and short-range electrostatics on the mixing relationship is discussed.

(384h) Thermodynamics of Pharmaceutical Systems

Gabriele Sadowski[1], Anke Prudic[2] and Jan Cassens[2], (1)Department of Biochemical and Chemical Engineering, Laboratory of Thermodynamics, Technische Universität Dortmund, Dortmund, Germany, (2)TU Dortmund, Dortmund, Germany

Pharmaceuticals are complex compounds often exhibiting very low aqueous solubility. Because of this, they only slowly dissolve in the body when administered as crystalline solids, which leads to a low bioavailability. Several approaches to increase the bioavailability of pharmaceutical compounds have been presented in literature.

One of them is salt formation by ionization of functional groups which can be achieved by adding an acid or a base and therewith changing the pH in the solution. The observable increase in solubility depends on the acid and base constants of the pharmaceutical itself as well as on the amount and nature of the acid or base added.

Another possibility is to form a so-called solid solution in which the pharmaceutical compound is integrated into an amorphous hydrophilic polymer which acts as carrier matrix. In this matrix, the pharmaceutical is integrated in its amorphous state rather than in the crystalline state and therewith faster accessible upon dissolution of the polymer. However, if the solid solution is not thermodynamically stable, phase separation and even recrystallization of this solid solution might occur during storage. Therefore, thermodynamic phase behaviour of the solid solution has to be known and the appropriate polymer has to be chosen for a given pharmaceutical.

However, approaches as described above are so far usually found by trial and error procedures. A thermodynamic modeling which enables for the prediction of the phase behavior is the key step in choosing the appropriate formulation of a drug.

The talk gives an overview about the liquid-liquid as well as solid-liquid phase equilibria in systems containing pharmaceuticals. The influence of temperature, solvents, co-solvents, polymers as well as of pH will be discussed. Based on numerous examples, general conclusions will be drawn.

Moreover, it will be shown, that despite of the fact that pharmaceuticals are complex compounds, thermodynamic modelling allows for reliable correlations and even predictions of the phase behaviour of aqueous and organic solutions (in pure and mixed solvents) as well as for solid solutions.

(407a) Interaction of Tau Protein with Model Lipid Membranes Induces Membrane Disruption

Emmalee M. Jones[1], Manish Dubey[2], Saurabh Singh[3], Ann Junghans[3], Phillip J. Camp[4], Briana C. Vernon[1], Jacek Biernat[5], Eckhard Mandelkow[5], Jaroslaw Majewski[6] and Eva Y. Chi[7], (1)University of New Mexico, Albuquerque, NM, (2)Chemical Engineering, University of Washington, Seattle, WA, (3)Los Alamos Neutron Science Center, Los Alamos National Laboratory, (4)Department of Chemical and Nuclear Engineering, University of New Mexico, Albuquerque, NM, (5)Max Planck Unit for Structural Biology, Hamburg, Germany, (6)Los Alamos Neutron Science Center, Los Alamos National Laboratory, Los Alamos, NM, (7)Chemical and Nuclear Engineering, University of New Mexico, Albuquerque, NM

The misfolding and aggregation of the intrinsically disordered, microtubule-associated tau protein into neurofibrillary tangles is implicated in the pathogenesis of Alzheimer's disease. However, the mechanism by which tau aggregation causes neuronal dysfunction remains unclear. Recent work has shown that lipid membrane permeabilization may serve as a pathway by which protein aggregates exert toxicity through cell membrane disruption, resulting in the alteration of ion homeostasis and disregulation of neuronal signal transduction.We investigated tau's propensity to interact with membranes and elucidated the disruptive structural perturbations these interactions induce in the membrane. We show that tau selectively inserts into anionic DMPG lipid monolayers at the air/water interface over neutrally charged membranes. Upon insertion into DMPG monolayer, tau disrupts the morphology of lipid condensed domains in the membrane. On the molecular scale, grazing incidence X-ray diffraction data show that tau insertion disrupts the packing of lipid tails. Neutron reflectivity data show that tau completely disrupts supported DMPG bilayers while leaving the neutral DPPC bilayer intact. To further investigate whether tau induced membrane structural changes lead to permeabilization, we also perform vesicle dye leakage experiments, which show that tau can induce dye leakage from vesicles containing anionic lipids. Our results indicate that tau's strong interaction with anionic lipids induces disruptions to membrane lipid packing, morphology, and structural integrity, suggesting possible membrane-based mechanisms of tau toxicity in neurodegenerative diseases.

(407b) Kinetics of Tau Protein Adsorption to a Model Hydrophobic Interface: The Role of Phosphorylation and Other Point Mutations

Alexandra Hyler, University of Kansas, Lawrence , KS, Benjamin Combs, University of Kansas, Lawrence, T. Christopher Gamblin, Molecular Biosciences, University of Kansas, Lawrence and **Prajnaparamita Dhar**, Chemical Engineering, University of Kansas, Lawrence, KS

Abnormal aggregation of microtubule associating protein, tau, into neurofibrillar aggregates , due to protein mutations, is a defining hallmark of several neurological diseases. Recent research indicates that the polymerization of soluble tau proteins into paired helical filaments may be influenced by the hydrophobic properties of its monomers, the presence of inducers and the local environment. In this talk, we will discuss our recent results showing the similarities and differences in the adsorption and aggregation kinetics of tau proteins, in the presence and absence of mutations. In particular, we will discuss the role of hyperphosphorylation, as well as point mutations at the N and C terminus of the protein on the surface activity. We find that even though tau is a soluble protein, it is highly surface active at nanomolar concentrations. We also find that these proteins demonstrate a two-step adsorption to the hydrophobic interface, and that the adsorption kinetics is dependent both on the concentration and isoform of the protein. We will also discuss our surface rheological data which suggests surface induced aggregation in these proteins.

(407c) Interfacial Protein Dynamics Resolved Using Single-Molecule Tracking

Mark J. Kastantin[1], Blake B. Langdon[1] and Daniel K. Schwartz[2], (1)Chemical and Biological Engineering, University of Colorado, Boulder, CO, (2)Chemical and Biological Engineering, University of Colorado at Boulder, Boulder, CO

Protein adsorption at the solid-liquid interface is fundamental to many applications including biocompatible materials, biofilm fouling, biosensing, and protein separations. Despite abundant research into this topic, a complete mechanistic understanding of dynamic surface behaviors is lacking. The potential for multiple soft (i.e. noncovalent) interactions between protein and surface (e.g. hydrophobic, hydrogen bonding, ionic interactions) often leads to composite behaviors of the average population in which it is difficult to identify the fundamental contributing factors. In contrast to methods that measure average properties (e.g. net adsorption or mean diffusion coefficient), single-molecule methods allow separation of different modes of protein-surface interaction for separate analysis. In this work, total internal reflection fluorescence microscopy is used to attain single-molecule resolution of protein adsorption onto model hydrophilic and hydrophobic solid surfaces using the proteins bovine serum albumin and fibrinogen. Our results demonstrate that heterogeneity is a common theme for all proteins studied on all surfaces.

Proteins were found to adsorb in different aggregation states (i.e. monomers, dimers, etc.) as evidenced by fluorescence intensities that were roughly integer multiples of the lowest observed intensity. Aggregation number was found to subsequently affect their surface affinity and mobility, with larger aggregates exhibiting longer surface residence times and slower diffusion. All protein species were found to be capable of multiple modes of surface interaction as evidenced by multiple observed characteristic residence times and diffusion coefficients, even for monomers. Temperature studies of desorption and diffusion allowed us to determine the apparent Arrhenius activation energy of each mode of surface interaction.

Notably, activation energies of all modes of desorption and diffusion were relatively weak ($< 5\ k_BT$). Coupled with the observation that the mean residence time of a single protein is relatively short (~1s) for all proteins on all surfaces, it is apparent that direct protein-surface interactions are insufficient to explain the formation of stable protein layers. Future work will focus on protein-protein interactions and the indirect role of the surface in mediating these interactions.

(407d) Adsorption of Helices and Coils On Negatively Charged Surfaces

Daniel Forciniti, Cbe, Missouri University of Science and Technology, Rolla, MO

Protein adsorption is a complex phenomenon partially because of the variety of amino acids present in a protein. The interaction between the protein and the surface is the result of the individual contributions of patches or segments found on the protein surface. The properties of these patches are defined by the individual amino acids that constitute them. There is a growing interest in adsorption studies using short peptide instead of the whole molecule because one can mimic the behavior of those patches. By using tailored peptides consisting of domains with specific structure and functionality, much is to be gained in the understanding the adsorption mechanism. We have experimentally explored the adsorption of a variety of diblock peptides on different substrates in the past. In this paper we present Monte Carlo simulations of the adsorption of one of such peptides onto negatively charged surfaces with the objective of elucidating the adsorption mechanism.

Monte Carlo simulations were performed to study the adsorption of a positively charged diblock peptide, consisting of five residues each of lysine (Lys) and alanine (Ala), onto negatively charged surfaces in the presence and absence of water. Initial conformations of helix and random coil were used. The procedure developed by Mungikar and Forciniti (Biomacromolecules, 7:239 (2006)) was adopted to perform the MC simulations. The peptide was placed at the center of the simulation box allowed to move towards a surface.

RasMol was used in this work to get a snapshot of the whole system at the end of the simulation and to determine the peptide's secondary structure (helices, turns and random coils) after adsorption. These structures were confirmed/rejected by generating Ramachandran plots, which help monitoring the extent to which the peptide reorganizes at the surface. These plots were constructed using the torsion angles (Φ, Ψ) of the peptide generated during the production run of the simulation. In addition to mapping the conformational space sampled by the peptide at the surface, these plots were used to corroborate that RasMol definitions of structures (such as helix, turn and random coil) were reasonable. The Peptide End-to-End Distance was defined as the distance separating the N atom of the N-terminal Ala^1 residue and the C^α of the C-terminus Lys^{10}. A change in this value is an indication of a significant reorganization of the peptide's global structure. Monitoring of changes in intra-molecular and peptide-solvent hydrogen bonding network of the peptide allows studying the structural rearrangement of the peptide at the surface. RasMol was used to study the changes in the intra-molecular hydrogen-bonding network of the peptide. Hydrogen bonding involving water was determined using the pair correlation functions between selected peptide atoms and water.

For both helix and random coil orientations, in water as well as vacuum, the Lys block rests near the surface whereas its Ala block is relatively away from the surface. Electrostatic force is a driving force for the adsorption of the peptide at the solid surface. In water, the helical peptide adsorbed in a side-on orientation whereas the random coil peptide had its Ala block extended into the bulk. This apparent 'repulsion' for the surface is actually due to the effect of solvent and is not seen in vacuum. The high-density water patch at the surface is probably the main reason. Transients in the adsorption process were observed. The peptide reaches the solid surface almost instantaneously and the slowly relax to reach a stable, and more compact, conformation.

(407e) Investigation of the Impact of Interfacial Adsorption On Droplet PCR Using a Novel Tablet Platform

Stephanie Angione[1], Anuj Chauhan[2] and Anubhav Tripathi[1], (1)School of Engineering, Brown University, Providence, RI, (2)Dept. of Chemical Engineering, University of Florida, Gainesville, FL

We present a novel droplet based tablet platform for temporal polymerase chain reaction in microliter droplets. We utilized this platform to successfully amplify DNA at concentrations as low as 2.0 copies/ul. We further utilized the platform to investigate the effect of performing PCR in small volumes on reaction performance by specifically examining adsorption of reagents at the oil/water interface. We determined that adsorption of *Taq* polymerase at the biphasic interface reduces yield and impairs reaction performance at standard concentrations. We investigated the effect of polymerase adsorption at the interface by examining various polymerase concentrations for three different droplet sizes, 1ul, 2ul and 4ul compared to full volume batch PCR. Our data shows the well known phenomenon of protein adsorption at the water/oil interface. The only volume which displayed amplification for all the polymerase concentrations examined, 0.01-0.05 units/ul was the full volume 50ul PCR controls. The effect of interfacial adsorption was pronounced for all the droplets investigated although significant differences between the droplet volumes were not observed. The reason not much difference is observed between 1ul, 2ul and 4ul is that the relevant controlling parameter is the surface to volume ratio, which is A/V ~ 3/R for a spherical interface. We examined the data points by developing a mass balance equation to calculate , the interfacial enzyme concentration. To evaluate monolayer formation at the interface the thermodynamic variable, for polymerase was calculated to be 0.0046 units/mm^2, or 3.65×10^8 molecules/mm^2. Here, we have used our experimental data to evaluate the bulk and loss concentrations for the 1ul, 2ul and 4ul droplets. The results indicate that for droplet applications involving enzymatic reactions like PCR, higher concentrations of enzyme are required. The typical full volume PCR reaction utilizes .02-.025 units/ul of polymerase, whereas our results indicate that for a 1ul droplet, to get maximum amplification .04-.045 units/ul are necessary. We compared calculations for monolayer formation of polymerase using the radius of gyration of native polymerase with that required by our data. We then determined that the enzyme is partially unfolded at the oil/water interface using a random coil approximation for further investigation of monolayer coverage. Thus, enzyme adsorption and unfolding at the interface leads to a lack of activity and lower PCR yield. Thus, microdroplet PCR reactions require additional polymerase to achieve sufficient amplification and we project that for applications utilizing nanodroplets or picodroplets like digital applications, even greater concentrations of polymerase are required to achieve desired results.

(407f) Protein Stability At the Silicone Oil-Water Interface Using Single Molecule TIRF and Interfacial Tensiometry

Aaron C. McUmber[1], Indira Sriram[1], Robert Walder[1], Theodore W. Randolph[2] and Daniel K. Schwartz[1], (1)Chemical and Biological Engineering, University of Colorado at Boulder, Boulder, CO, (2)Department of Chemical and Biological Engineering, Center for Pharmaceutical Biotechnology, University of Colorado, Boulder, CO

Therapeutic protein solutions have become a popular strategy in disease treatment. However, protein aggregation is a significant problem that must be understood and minimized in order to develop safer and more effective treatments. Aggregation at the silicone oil-water interface is of particular interest due to increased aggregation seen within prefilled glass syringes coated with silicone oil. In this presentation, the authors investigate the aggregation behavior of three model proteins (lysozyme, insulin, and IgG) at the silicone oil-water interface. Single molecule total internal reflection fluorescence microscopy (SMTIRFM) and interfacial tensiometry were used to identify the aggregation mechanics of the three proteins on the molecular level as well as demonstrate a connection between the microscopic data and bulk solutions. SMTIRFM captures molecular interactions by directly observing fluorescently-labeled proteins within a developing non-fluorescent protein layer allowing the collection of molecular interfacial diffusion rates as well as absolute adsorption kinetics. By directly capturing molecular interactions, SMTIRFM delves into within the regime where interfacial tensiometry does not measure any change in interfacial tension, known as the lag phase of protein adsorption.

(407g) The Adhesion Mechanism of Mussel Adhesive Protein-3 to TiO2 Surface

Jing Yu[1], Wei Wei[2], J.Herbert Waite[3] and Jacob N. Israelachvili[1], (1)Chemical Engineering, University of California, Santa Barbara, Santa Barbara, CA, (2)Materials Research Laboratory, University of California, Santa Barbara, Santa Barbara, CA, (3)Department of Molecular, Cell & Developmental Biology, University of California, Santa Barbara, Santa Barbara, CA

The underwater adhesion of marine mussels relies on mussel foot proteins (mfps) rich in the catecholic amino acid 3, 4-dihydroxyphenylalanine (dopa). As a side-chain, dopa is capable of strong bidentate interactions with a variety of surfaces, including many minerals and metal oxides. TiO2 is one of the most common used medical implants materials. Studying the binding mechanism of dopa to TiO2 surface is therefore of theoretical and practical interests. Using a surface forces apparatus, we explored the force distance profile and the adhesion energy of mussel foot protein 3 (mfp-3) to TiO2 surface in three different pHs (pH3, 5.5, and 7.5). At pH3, mfp-3 showed the strongest adhesion force on TiO2, with an adhesion energy of ~ -7.0 mJ/m^2. Increasing the pH gives rise to two opposing effects: it causes the oxidation of dopa, which eliminates the dopa adhesion; meanwhile, higher pH increases the binding affinity of dopa-Ti coordination bond, leading to the further stabilization of the dopa group and an increase of adhesion force. The two competing effects give rise to a higher adhesion force of mfp-3 on TiO2 surface at pH 7.5 than the adhesion force measured at pH 5.5. Our results suggest that dopa containing proteins and synthetic polymers have a great potential to be used as coating materials for medical implants.

(413a) Computer Simulation of Vesicles and Vesicle Suspensions in Flow

Eric S.G. Shaqfeh, Chemical Engineering and Mechanical Engineering, Stanford University, Stanford, CA

It is well known that individual vesicles or liposomes (i.e. fluid enclosed by a lipid bilayer membrane suspended in a second fluid) are characterized by a remarkable dynamics in flow. For vesicles that are "near spheres" this dynamics includes at least 5 different types of orbits in shear flow that are functions of the viscosity ratio between the inner and outer fluid as well as the Capillary number based on the bending modulus. However, this dynamics becomes even more rich as the reduced volume falls below about 0.65 where now there are atleast three equilibrium shapes (prolates, discocytes, and stomatocytes) which are linearly stable. It is therefore not surprising that a suspension of vesicles is characterized by fascinating collective behavior as well. I will discuss our recent development of a numerical code (based on Loop subdivision) which allows the Stokes flow simulation of non-dilute suspensions of vesicles and capsules at essentially any value of the reduced volume. We will then use these numerical simulations to examine a number of interesting phenomena including: 1) The lift of a vesicle away from a wall and the resulting "Fahraeus-Lindqvist" layer for the flow of a wall-bound suspension of vesicles, 2) The effective rheology and dynamics of a non-dilute vesicle suspension under shear, and 3) The effect of Brownian motion on the dynamics of vesicles in flow.

(413b) Microstructural Origins of Two-Step Yielding and Strain-Induced Strengthening in Dilute Colloidal Gels

Hubert K. Chan and **Ali Mohraz**, Department of Chemical Engineering & Materials Science, University of California, Irvine, Irvine, CA

Colloidal gels are particulate soft materials traditionally utilized in the food, ceramics, and cosmetics industries as well as a number of emerging technologies including direct write assembly. In such applications, colloidal gels are often subjected to transient flow during their processing and handling, and their end-use properties are governed by the coupling between their microstructure and transient nonlinear rheology. In this work we investigate the nonlinear rheology of dilute depletion-induced colloidal gels as model systems, and report that they can undergo yielding via a two-step process, with mechanisms fundamentally different from dense suspensions. We attribute the two-step yielding in dilute gels to interparticle bond rotation and bond breakage at the respective yield points. Implications of this behavior for transient deformation in step-stress experiments are examined. These include remarkable creep profiles at intermediate applied stress values, highlighted by strain-induced anisotropic hardening and flow arrest at very large accumulated strains. Quantitative time-resolved confocal microscopy in conjunction with a custom-built shear cell is used to investigate the microstructural origins of this behavior. The results are relevant to the processing or final consumer use of particulate suspensions with attractive interparticle interactions, and will be discussed in light of a balance between the globally imposed stresses and the thermodynamic interparticle forces.

(413c) Structure, Rheology and Optical Properties of Self-Assembled Micelle-Nanoparticle Plasmonic Nanogels: Experiments and Molecular Dynamics Simulations

Tao Cong, Georo Zhou, Abhinanden Sambasivam and **Radhakrishna Sureshkumar**, Department of Biomedical and Chemical Engineering, Syracuse University, Syracuse, NY

Cationic surfactants such as cetyl-trimethylammonium bromide (CTAB) have the ability to self assemble with metallic nanoparticles to form a corona or a double-layer vesicular structure [1]. Such structures, upon further interaction with wormlike micelle fragments, are hypothesized to form micelle-nanoparticle elastic networks [2, 3]. Recently, we have shown that self-assembly of metallic nanoparticles with wormlike micelles in surfactant solutions is a robust route for producing stable multicomponent nanogels with remarkable color uniformity [4, 5]. The optical properties of the gels can be easily tuned by varying the type, shape and/or concentration of nanoparticles. In particular, multicomponent gels capable of broadband absorption of visible light can be robustly manufactured via the self-assembly route. Small angle X-ray scattering and rheological studies suggest that the nanoparticles materially participate in the wormlike micelle network to form a more compact double network. Such plasmonic gels exhibit rich rheological behavior depending on the nanoparticle concentration and the salt/surfactant ratio. Specifically, non-monotonic dependence of zero shear viscosity on nanoparticle concentration, rheopexy, shear thickening, shear banding and shear thinning are observed. The nanogels exhibit enhanced viscoelasticity upon the addition of more nanoparticles and are thermoreversible. The mechanisms contributing to the rheological and structural properties of the plasmonic nanogels are studied by coarse-grained Molecular Dynamics simulations based on the MARTINI force field framework [6, 7]. The results suggest that vesicular nanoparticle-surfactant structure can form stable bridges with cylindrical micelles through two mechanisms: end-cap opening and lateral attachment. The energetics of bridge formation, the effect of flow shear on the stability of the end-cap and lateral bridges and potential applications to efficient light trapping in photovoltaic cells and optofluidic devices will be discussed.

Acknowledgements: We acknowledge National Science Foundation grant CBET-1049454 for partial support of this research. Syracuse University has filed US and PCT patent applications based on the findings of tis work.

References:

1. B. Nikoobakht and M.A. El-Sayed, Langmuir, 17, 6368 (2001)
2. F. Nettesheim, M.W. Liberatore, T.K. Hodgdon, N.J. Wagner, E.W. Kaler, and M. Vethamuthu, Langmuir, 24, 7718 (2008).
3. M.E. Helgeson, T.K. Hodgdon, E.W. Kaler, and N.J. Wagner, Langmuir, 26, 8049 (2010).
4. T. Cong, S.N. Wani, P.A. Paynter, and R. Sureshkumar, Applied Physics Letters, 99, 043112 (2011).
5. T. Cong, S.N. Wani, G. Zhou, E. Baszczuk, and R. Sureshkumar, Proc. SPIE, 8097, 80970L (2011).
6. A.V. Sangwai and R. Sureshkumar, Langmuir, 27, 6628 (2011).
7. A.V. Sangwai and R. Sureshkumar, Langmuir, 28, 1127, (2012).

(413d) Extracting Fundamental Materials Properties From Rheological Measurements

Folarin Latinwo and Charles M. Schroeder, Chemical and Biomolecular Engineering, University of Illinois at Urbana-Champaign, Urbana, IL

Extracting Fundamental Materials properties from Rheological Measurements

Folarin Latinwo[1] and Charles M. Schroeder[1]

[1]Department of Chemical & Biomolecular Engineering, University of Illinois at Urbana-Champaign, Urbana, IL 61801

A long standing problem in complex fluids has been the determination of fundamental materials properties (e.g. elasticity) from dynamic far-from-equilibrium measurements (e.g. transient stresses). We recently developed a framework to extract information about polymer elasticity from trajectories of conformational changes of single molecules induced by fluid flow [1]. In our approach, we applied tools from non-equilibrium statistical mechanics namely the Jarzynski relation [2] to complex fluids via Brownian dynamics (BD) simulations of bead-spring models of polymer molecules. The relation allows for the calculation of free energy changes from work distributions of arbitrarily far-from-equilibrium processes and has been applied as originally conceived only to single molecule studies. In our current work, we demonstrate the further application of this new tool to complex fluids by determination of free energy changes from transient bulk stresses. Using BD simulations of bead-spring models of dsDNA and polystyrene molecules in various flow types (shear and extensional) and standard expressions for the stress tensor (e.g. Kramers-Kirkwood), we make a connection between stresses, work distributions of bulk measurements and the Jarzynski relation. In doing this, we developed a formalism to extract new fundamental information such as free energies (which intrinsically defines a system) from previously existing measurements.

[1] Latinwo, F. and Schroeder, C. M. Physical Review Letters, submitted.

[2] Jarzynski, C. Phys. Rev. Lett. 78, 2690--2693 (1997).

(413e) Rinsing Flows: Exploiting Viscoelastic Liquids

Travis W. Walker, Tienyi Theresa Hsu, Curtis W. Frank and Gerald G. Fuller, Chemical Engineering, Stanford University, Stanford, CA

Rinsing flows refer to processes where a jet of one fluid (the jet fluid) is used to ablate a second liquid that coats a planar substrate (the coating liquid). Using an impinging jet of water to rinse coating fluids of varying rheology has been considered to understand the flow structures of the resulting hydraulic jumps qualitatively and quantitatively. We investigated the interactions of the two-fluid system during the transient growth of the flow profile, observing rheological dependencies on the magnitude, velocity, and topography. Various instabilities were also observed, including Saffman-Taylor fingering and a flow profile dependency on the local viscosity ratio of the rinsing fluid to the coating fluid.

Four classes of test fluids, each having approximately equal viscosities at low shear, have been chosen for this study: a Newtonian solution, a viscoelastic polymer solution, a Boger fluid, and a shear-thinning, inelastic colloid suspension. The presence of shear-thinning of the coating liquid influenced the overall velocity of the radial growth, while determining the geometry of the driving front. The dependence of these results on the local rheology was supported by experiments on Newtonian fluids of various shear viscosities. The presence of elasticity was seen to dampen the disturbances of the hydraulic jump, influence the overall jump height, and vary the radial growth of the jump. These observations were supported by experiments of Boger fluids with varying elasticity.

(413f) X-Ray Scattering Investigation of Structural Relaxation in an Ordered Block Copolymer Melt in Uniaxial Extensional Flow

Erica McCready and Wesley Burghardt, Department of Chemical and Biological Engineering, Northwestern University, Evanston, IL

The structural dynamics of an ordered styrene-ethylene butylene-styrened triblock copolymer have been studied in uniaxial extensional flow using in situ x-ray scattering. Experiments were performed in a custom instrument consisting of an SER extensional flow fixture housed in a convection oven designed to facilitate x-ray access. Use of synchrotron radiation provided sufficient time resolution to study the structural response during inception of uniaxial flow, and as a function of time following flow cessation. The sample studied here exhibits hexagonally packed cylindrical microdomains of polystyrene embedded in a poly(ethylene butylene) matrix. Application of extensional flow produces multiple structural effects, including deformation of the microphase-separated morphology, and a complex reorientation process in which elongated PS microdomains progressively orient along the stretching axis. A series of experiments was run in which samples were stretching to varying Hencky strains, allowing investigation of the nature of structural relaxation from a variety of flow-induced structural states induced during extensional flow. Significant differences in structural relaxation are found depending on the total applied extensional strain.

(413g) Capillary Break-up, Gelation and Extensional Rheology of Hydrophobically Modified Cellulose Ether Solutions

Vivek Sharma[1], Simon Haward[1], Asa Soderlund[2], Phil Threlfall-Holmes[3] and Gareth H. McKinley[1], (1)Mechanical Engineering, Massachusetts Institute of Technology, Cambridge, MA, (2)AkzoNobel, Stenungsund, Sweden, (3)AkzoNobel, Gateshead, United Kingdom

Associative polymer solutions are used extensively in the formulations of functional soft materials like inks, paints, food, nasal sprays, cosmetics, insecticides, fertilizers and bio-assays to control the rheology and processing behavior of multi-component complex fluids. These complex dispersions are processed and used over a broad range of shear rates ($1\text{-}10^6\ s^{-1}$) and extensional rates ($1\text{-}10^5\ s^{-1}$), and their non-Newtonian behavior displayed under processing conditions is often not measurable in the conventional rheometric measurements. In this talk, we use the example of cellulose ether – ethyl-hydroxyethyl-cellulose (EHEC) – and its hydrophobically modified analog (HMEHEC) to illustrate the influence of hydrophobic stickers on the linear and nonlinear rheology of their aqueous solutions. We show that a constitutive model based on fractional calculus can be used to characterize the physical gelation in HMEHEC solutions, and we use microfluidic rheometry to extend viscosity measurements to high shear rates ($10^3\text{-}10^6\ s^{-1}$). The transient extensional viscosity response of the associative polymer solutions dictates the complex pinch-off dynamics during jetting and spraying, where inertial, viscous and elastic effects moderate capillary break-up of cylindrical fluid elements into droplets. We compare the responses of EHEC and HMEHEC solutions to extensional deformations during capillary break-up using (a) capillary break-up extensional rheometer (CABER) and (b) Rayleigh Ohnesorge Jetting Extensional Rheometer (ROJER), an imaging-based technique capable of attaining high extension rates that has been developed as a part of this study. Furthermore, we measure extensional viscosity using excess pressure drop and flow induced birefringence in the vicinity of a stagnation point in a cross-slot extensional rheometer. This study highlights a number of complimentary techniques that are critical for the design and characterization of functional soft materials used in complex processing operations such as jetting and spraying.

(413h) Enhancement of Polymer Globule Formation by Shear Flow

Rangarajan Radhakrishnan and Patrick T. Underhill, The Isermann Department of Chemical and Biological Engineering, Rensselaer Polytechnic Institute, Troy, NY

Polyelectrolytes, polymers in poor solvents, and other systems with attractions and repulsions show formation of globules/structures in equilibrium. Typically these transitions are driven by changes in temperature, salt concentration, pH, etc. In some systems, shear flow can act to accelerate or cause collapse, instead of the common case of shear acting to stretch the polymer. Since all atom simulations of such systems are computationally expensive, we have developed a new coarse-grained model. Brownian Dynamics simulations of this model with short ranged attractions and repulsions have been used to examine the long-time dynamics in shear flow and globule formation. We analyze the effect of shear flow on such systems as a barrier hopping process in a free energy landscape. For small shear rates, the flow acts to alter the first passage process across the barrier. At higher shear rates, the flow causes a hopping back and forth between the two states.

(413i) Modeling Dilute and Semi-Dilute Flexible Polymer Solutions in Extensionally-Dominated Flows

Antony N. Beris, Chemical and Biomolecular Engineering, University of Delaware, Newark, DE and Kostas D. Housiadas, Mathematics, University of the Aegean, Samos, Greece

Extensionally-dominated flows are of importance in several applications involving dilute and semi-dilute polymer solutions, such as those involving the use of polymer solutions in liquid jets, in flow through porous media in tertiary oil recovery as well as in polymer-induced drag reduction. Traditionally, dilute polymer solutions are modeled at the continuum level by an entropic spring-type model, such as the Finitely Extensible of Non-linear Elastic dumbbells with the Peterlin approximation (FENE-P) model. On the other hand, semi-dilute solutions are often modeled by an anisotropic mobility model, such as the Giesekus model. It is straightforward to also imagine several refinements based on these ideas, such as the use of the inverse Langevin function instead of the Peterlin function to represent finite extensibility effects, or, especially aided by the non-equilibrium thermodynamics formulation [1], by a combination of that with the anisotropic mobility, defining therefore a modified Giesekus model. This is indeed the more appropriate model for semi-dilute solutions of finite extensibility flexible macromolecules. In fact, such a model has been proposed and compared favorably against atomistic simulations for unentangled polymer melts [2].

In the present work we develop such a model for semidilute polymer solutions and we utilize it for the modeling of the highly extensional flow associated with fast filament stretching in an effort to test the model against the very recent experimental results obtained for such a flow by Tembely et al. [3]. We show there that, if restricted in the modeling (as is typically the case) by the use of one relaxation mode, in addition to the above-mentioned effects, one needs to consider a significant increase to the effective relaxation time taking place during the flow in order to achieve a good fit of the results. This is in agreement with one of the suggestions offered by that work which, however, only considered constant relaxation time models, resulting in a partial fit of the results. The implication is that when a significantly truncated model is used, such as a on relaxation mode continuous viscoelastic model, the parameters entering that description are not to be seen as constants, simply connected to the underlying molecular structure, but as effective, renormalized, ones, with a dependence on the flow and/or flow history. The nature and further modeling of that dependence remains one area of active investigation.

References

[1] Beris A.N. and Edwards B.J., Thermodynamics of Flowing Systems with Internal Microstructure, Oxford U. Press, 1994.

[2] Stephanou P.S., Baig C. and Mavrantzas V.G., J. Rheology, 53:309-337 (2009).

[3] Tembelya M., Vadillo D., Mackley M.R. and Soucemarianadin, A.. J. Rheology, 56:159-183 (2012).

(435a) Bioactive Surfaces in Microfluidic Devices

Shashi Murthy, Chemical Engineering, Northeastern University, Boston, MA

Microfluidic channels with immobilized proteins and antibodies have been utilized for a wide range of applications, from clinical diagnostics to tissue engineering. This presentation will describe our work in the isolation of stem and progenitor cells for diagnostics and therapeutics from complex samples such as blood and digested tissue. A key element in our approach is the design of alginate hydrogel-based surface coatings that can be applied onto the surfaces of microfluidic channels. These hydrogels are functionalized with antibodies, enabling specific capture of target cells, but more importantly they also have the ability to release captured cells without loss of viability or function. Specific examples to be covered in the presentation will include isolation of endothelial progenitor cells for cardiovascular diagnostics and vascular tissue engineering and isolation of intestinal stem cells for intestinal organoid unit regeneration.

UNSTEADY FLOW FEATURES IN A T-SHAPED MICRO-MIXER

Chiara Galletti, Elisabetta Brunazzi, Roberto Mauri

Dipartimento di Ingegneria Chimica, Chimica Industriale e Scienza dei Materiali, Università di Pisa, I-56126 Pisa, Italy.

Abstract. A Direct Numerical Simulation of the mixing process between two water streams in a T-shaped micro-device indicates that for a certain range of Reynolds numbers the flow regime might be unsteady, with time-periodic structures. Pulsating flows strongly enhance the degree of mixing, that can be as large as twice the one that is observed in the steady engulfment regime. Then, when we increase the Reynolds numbers beyond the pulsating range, the flow is no more unsteady, with a consequent reduction of the mixing efficiency. In addition, we also observe that for our investigated geometry the unsteady, periodic behavior does not occur when the inlet flows are blunt and not fully developed.

Keywords: Flow instability, laminar flow, hydrodynamics, Direct Numerical Simulation.

1. INTRODUCTION

Effective mixing in small volumes is a crucial step in many processes, such as miniature fuel cells, molecular diagnostics and, in general, micro-reactors, where a fast homogenization of reactants is required. Typical uses of these microfluidic devices require these systems to be inexpensive and simple to operate, such as passive micro-mixers, where mixing is promoted by stretching and recombining of the flow fields without the help of any external power source [1].

One of the simplest designs of passive micro-mixers is a T shape, in which the inlets join the main channel with T-shaped branches. This type of mixer is also suitable for carrying out fundamental studies, as the T-shaped micro-mixer is often encountered as a junction element in more complex micro-systems. The efficiency of T-shaped micro-mixers for liquid mixing has been largely proved in literature [2].

So far, different flow regimes have been identified in the T-shaped micro-mixer depending on the Reynolds number, *Re*, namely: the stratified, the vortex and the engulfment flow regimes [3][4][5][6]. In the stratified flow regime, the inlet streams come into contact in the mixing zone and then flow side by side through the mixing channel, resulting in a completely segregated flow. In the vortex regime, above a critical *Re*, a secondary flow in the form of a double vortex pair occurs, due to the instabilities induced by the centrifugal forces at the confluence, leaving however the two incoming streams well segregated. In the engulfment regime, occurring at higher *Re*, fluid elements reach the opposite side of the mixing channel, thus largely increasing the degree of mixing. Moreover, for specific *Re* ranges above the engulfment, but still in the laminar regime, some kinds of instabilities have been recently observed. For example, based on numerical simulations on a T-mixer with a 2:1 aspect ratio (i.e. the ratio between width, *W*, and height, *H*, of the mixing channel), Bothe et al. [5] reported that when $Re > 240$, thereby above the onset of engulfment occurring at $Re = 146$, the flow exhibits a periodic behavior. For T-shaped micro-mixers having the same *2:1* aspect ratio, Dreher et al. [7] focused their attention on the flow behavior above the engulfment point, confirming that when $Re > 240$ the steady state simulations do not converge and transient ones should be adopted instead. According to the authors, this was due to a periodic behavior of the flow; in particular, they identified a periodic pulsating flow for $240 < Re < 400$, characterized by a Strouhal number $0.15 < St < 0.29$. Such periodic pulsations were found to disappear for larger *Re*, i.e. when $Re > 500$, where a chaotic, pseudo-turbulent motion was observed. Importantly, such instabilities do enhance the mixing process: indeed, optimum mixing is reached for $240 < Re < 700$. An experimental evidence on the occurrence of such flow instabilities was provided also by Thomas et al. [8][9], who investigated through planar and discrete-point Laser Induced Fluorescence the water flow in a T-shaped channel with $W = 2W' = 40\ mm$ and $H = 20\ mm$ for $56 < Re < 422$.

The present work shows results from the Direct Numerical Simulation of a T-shaped micro-mixer, characterized by a different aspect ratio than the one used by previous investigators [5][7][8][9], also for Reynolds numbers above the engulfment regime, discussing the features of flow instabilities and their relevance for mixing. Moreover, the effect of inlet flow conditions on the instability region is discussed, as understanding factors affecting the instabilities may help mixing optimization and control.

2. TEST CASE AND NUMERICAL MODEL

In the micro-mixer that we have considered (see Figure 1) the mixing channel has a rectangular cross section with a *3:2* aspect ratio (i.e. width $W = 3H/2$ and depth *H*), with length $L = 15\ H$, while the inlet channels are identical, with a *3:4* aspect ratio (i.e. width $W' = 3H/4$ and depth *H*) and length $L' = 7H/4$. Consequently, as we assume that the two inlet streams have equal flow rates, the mean velocity *V* at the inlets is equal to that in the mixing channel; in addition, flow rates were varied so that *Re* in the mixing channel ranges from *100* to *530*, with *Re* defined in terms of the hydraulic diameter, $D = 6H/5$.

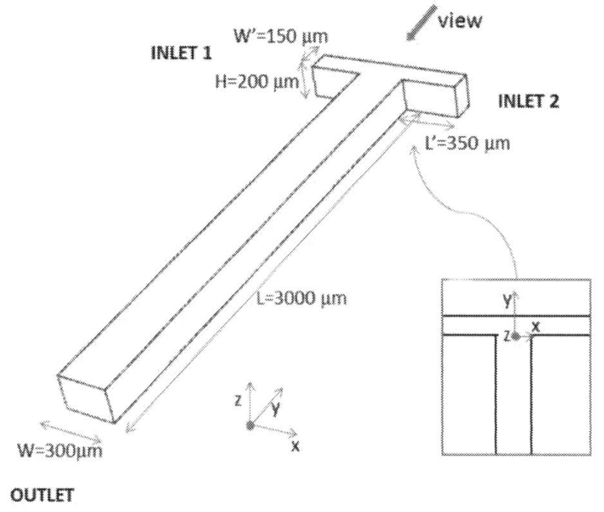

Figure 1 - Sketch of the T micro-mixer.

Simulations were performed using the commercial software Ansys 13, using the FLUENT fluid dynamics package, where length and time were scaled in terms of *D* and *D/V* [10].

After a careful dependency study on the velocity fields (Galletti *et al.* [11]), a structured grid with regular cubical elements of *H/50* size was chosen, corresponding to *3.3 M* elements. In particular, there are *76x50* elements in each cross section within the mixing channel, thus in agreement with the recommendation by Hussong *et al.* [12].

Mixing fluids are assumed to consist of the same Newtonian liquid, i.e. incompressible and with constant viscosity, with c denoting the normalized concentration (i.e. ranging from 0 to 1) of one of the two inlet streams. Constant, e.g., ambient, pressure is set at the outlet of the mixing channel, while no-slip boundary conditions are applied to the walls. Fixed mean velocity is set at both inlets, assuming that the velocity profile is fully developed. Therefore, solving the Navier-Stokes equations with no-slip boundary conditions at the walls and with constant axial pressure gradient G [13][14], we have,

$$v(y,z) = -\frac{G}{2\mu}y(Y-y) - \frac{4GY^2}{\mu\pi^3}\sum_{k\ odd}\frac{1}{k^3}\sin\left(k\pi\frac{y}{Y}\right)\left[Cosh\left(k\pi\frac{z}{Y}\right) - Tanh\left(\frac{k\pi}{2\eta}\right)Sinh\left(k\pi\frac{z}{Y}\right)\right] \quad (1)$$

where Y and Z are, respectively, the width and depth of the inlet conduit cross sections (W' and H in our case), while η is the conduit aspect ratio. The constant pressure gradient G can be derived as a function of the mean velocity V as:

$$G = -\frac{12\mu V}{Y^2}\left[1 - \frac{192}{\pi^5}\eta\sum_{k\ odd}\frac{1}{k^5}Tanh\left(\frac{k\pi}{2\eta}\right)\right]^{-1}. \quad (2)$$

A second order discretization scheme was used for all equations, as higher order solvers were found to provide negligible corrections. Some simulations were carried out with a steady solver. A sufficient number of iterations were performed in order to obtain residuals of less than 10^{-13} for all equations. The transient simulations were performed with an explicit solver, choosing a discretization time step Δt ensuring a Courant number lower than 1, i.e.

$$\frac{\Delta t}{D/V} < \frac{1}{60}\frac{V}{V_{mx}}, \quad (3)$$

where V_{mx} is the maximum fluid velocity (approximately, $V_{mx} \cong 2\text{-}3\ V$.

3. RESULTS

Figure 2 shows the tracer distribution in the mixing channel cross-section at $y = -L/3$ for different Re. Figure 2a refers to $Re = 144$ in the vortex flow regime, and indicates a poor mixing. Figure 2b and Figure 2c show the engulfment regime at $Re = 196$ and 216, respectively, characterized by non-symmetrical vortical structures resulting in enhanced mixing. Finally, Figure 2d refers to $Re = 480$: in this case, it can also be observed the presence of vortical structures, although the pattern is different from those at lower Re. For all such conditions, a solution could be found with a steady solver. In particular, the $Re = 480$ run was also performed with a transient solver, indicating steady flow. However, for intermediate Re (ranging, approximately, from 230 to 410), an unsteady behavior was observed. Figure 3 shows an example of velocity time series evaluated through a DNS simulation for $Re = 360$ at different locations in the mixing channel.

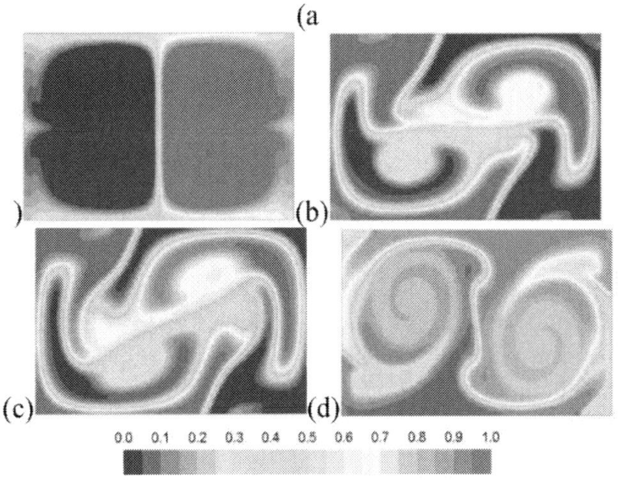

Figure 2 – Tracer distribution in the mixing channel cross section at y = $-L/3$ with (a) $Re = 144$; (b) $Re = 192$; (c) $Re = 216$ (d) $Re = 480$.

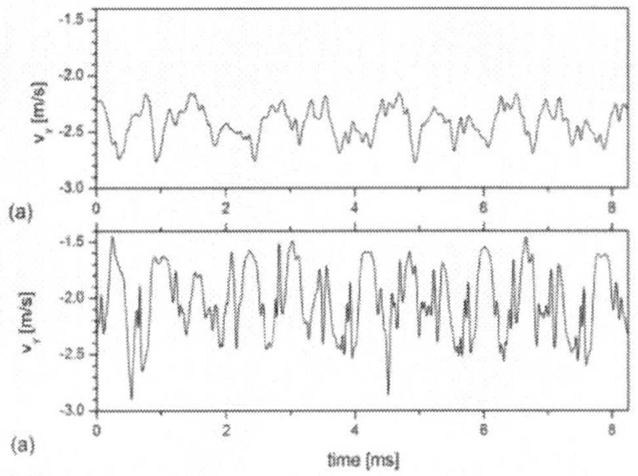

Figure 3– Time series of the y component of velocity in the center of the mixing channel for (a) $y = -2L/3$ and (b) $y = -L/3$.

In this case, it can be observed the presence of unsteady phenomena. Such time series seem to indicate a rather periodic behavior, with at least two characteristic frequencies, $f_1 \cong 0.085\ V/D$ and $f_2 \cong 0.25\ V/D$. Therefore, considering that $St = fD/V$ is the Strouhal number, we see that f_2 would lead to a $St = 0.25$, which is in excellent agreement with the result by Dreher *et al.* [7], who measured a *470 Hz* frequency using a T-mixer having $W = 2W' = 2H = 600\ \mu m$ and for $Re = 300$.

Figure 4 shows the region of the T-mixer where the local velocity is above its mean value V, evaluated every $1/f$ time interval in order to describe the unsteady flow behavior. Pulsating structures originating from the interaction between the inlet streams can be observed. Consequently, the tracer distribution along the mixing channel varies greatly with time, as depicted in Figure 5 for the cross-section at $y = -L/3$. At first glance it can also be noticed that the degree of mixing is larger than that observed for all the *Re* of Figure 2, being the extension of the green region larger.

Quantitatively, the mixing efficiency is evaluated through the bulk concentration (cup mixing concentration), that is the flow-weighted mean concentration, i.e.:

$$\bar{c} = \frac{\langle cv \rangle}{\langle v \rangle}; \qquad \langle cv \rangle = \frac{1}{A}\int_A cv\, dA; \qquad \langle v \rangle = \frac{1}{A}\int_A v\, dA, \tag{3}$$

where the overbar indicates the cup mixing average, while the brackets denote the volume (i.e. un-weighted) average on a cross section of area A. The cup mixing average is preferred over the more usual volume average concentration, because it describes the concentration that "one would measure if the tube were chopped off and if the fluid issuing forth were collected in a container and thoroughly mixed" [15].

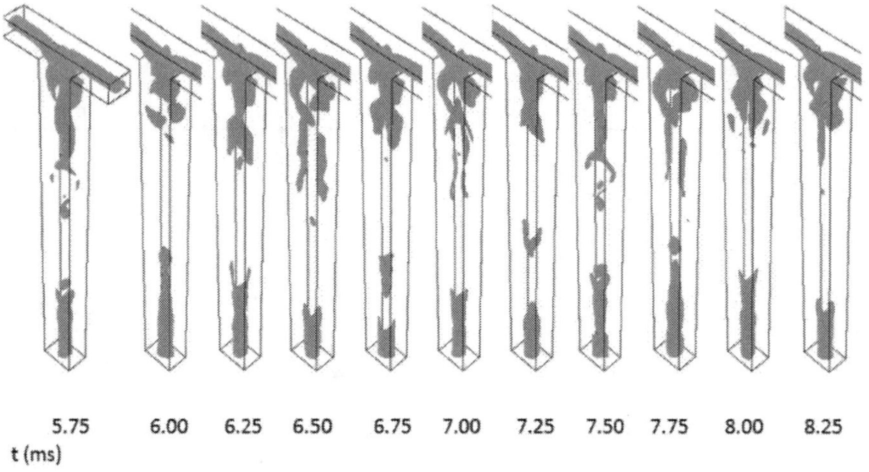

Figure 4 – Regions of the micro-mixer with velocity above *2.5 m/s*. *Re =360.*

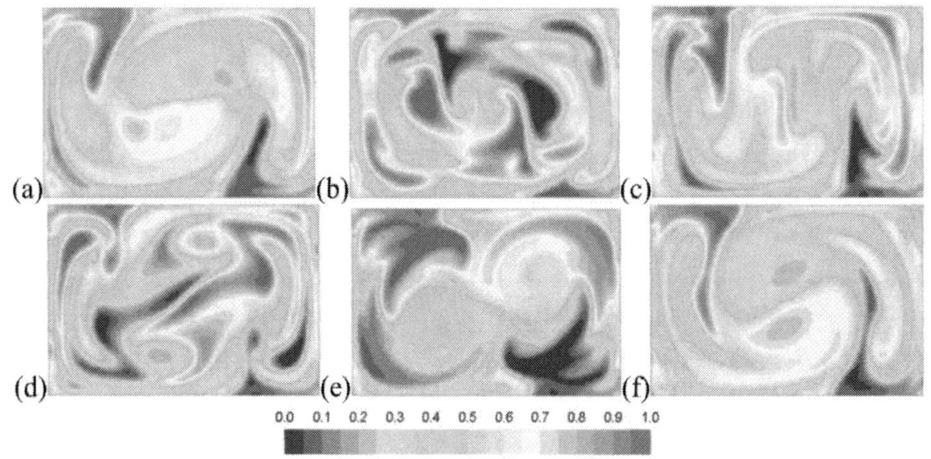

Figure 5 – Tracer distribution in the mixing channel cross-section when *Re = 360* at *y = -L/3 μm* and at different times (scaled in *D/V* unit): (a) *42.5*; (b) 44; (c) *45.5*; (d) *47 ms*; (e) *48.5 ms*; (f) *50.*

In the same way, a bulk, or cup mixing, mean square concentration difference is defined as:

$$\overline{(\Delta c)^2} = \frac{\left\langle \left(c-\bar{c}\right)^2 v \right\rangle}{\langle v \rangle}; \qquad \sigma_{cm}^2 = \overline{(\Delta c)^2} / \overline{(\Delta c)^2}_{\max}, \tag{4}$$

where its maximum value is used as a normalizing factor. Clearly, as the maximum mean square concentration difference occurs when the two streams are completely unmixed (i.e., $c = 0$ in half of the cross section and $c = 1$ in the other half) then $\overline{(\Delta c)^2}_{\max} = \frac{1}{4}$. For sake of convenience, here the following definition of degree of mixing will be used:

$$\delta_m = 1 - \sigma_{cm} \tag{5}$$

as it varies from *0*, when the two streams are perfectly unmixed, to *1*, when they are completely mixed.

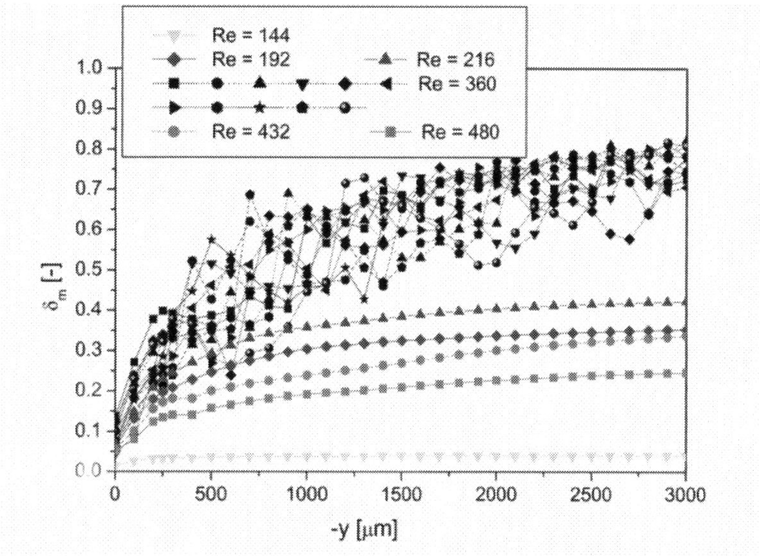

Figure 6 – Degree of mixing along the mixing channel for different regimes: vortex flow regime (green symbols); engulfment flow regime (blue symbols); unstable flow regime (black symbols); steady flow regime for high *Re* (blue symbols).

Figure 6 shows the degree of mixing along the mixing channel evaluated for different flow regimes. In the vortex regime, the degree of mixing is very low (*4%* maximum) indicating a segregated flow. In the engulfment regime, mixing is enhanced with degree of mixing at the outlet equal to *35%* when *Re =192*, and *42%* when *Re = 216.* The unstable regime is indicated by the black symbols, each of them calculated for different time steps; the unsteady structures originate values of the degree of mixing that are well above those observed for the steady flow conditions. For instance, in the outlet section, the mixing degree varied from *70* to *82%* when *Re = 360.* Similarly also pressure drops in the T-mixer were found to vary significantly (i.e. of about *12%*) in the unstable regime for *Re = 360.* Then, as

Re is increased further, the flow field returns to be steady and the degree of mixing decreases to values that are comparable with those observed in the engulfment regime. For instance, when *Re = 480*, we find that the in outlet section the degree of mixing is $\delta_m \cong 25\%$, which is lower than the value (i.e. $\delta_m \cong 40\%$,) at *Re = 200*, due to the lower residence time in the mixing channel caused by the larger mean velocity.

Finally, it should be stressed that the appearing of an unsteady, pulsating regime depends not only on the geometric characteristics of the device and the operating Reynolds number, but also on the type of flow conditions at the inlets. In fact, Galletti *et al.* [11] showed that flow instabilities occur only when the velocity profile at the inlet is peaked and therefore do not appear when, for example, the inlet flow is not fully developed (i.e. the inlet channels are not long enough to ensure that fully developed flow conditions are reached).

5. CONCLUSIONS

An unsteady flow regime is observed for *Re* ranging from *228* to *408* in a T-shaped micro-mixer having a 3:2 aspect ratio, thus confirming the finding by Dreher at al. [7], who nevertheless used a 1:1 aspect ratio. The unsteady flow is characterized by pulsating structures, which strongly enhance mixing, leading to values of the degree of mixing that are almost twice as large as those achieved in the engulfment flow regime. For larger *Re*, above the unsteady flow regime, a steady flow was observed, with a strong reduction of the mixing efficiency. Therefore, the precise knowledge of the operating conditions for which the unstable flow occurs are of great importance for micro-mixer optimization. Moreover, the instabilities may be enhanced or dumped by specific conditions: for instance, Galletti *et al.* [11] showed that non-fully developed velocity profiles at the inlets may suppress all unstable flows.

REFERENCES

[1] V. Kumar, M. Paraschivoiu, K.D.P. Nigam, Single-phase fluid flow and mixing in microchannels, Chem. Eng. Sci. 66 (2011) 1329-1373.

[2] D. Bökenkamp, A. Desai, X. Yang, Y. Tai, E. Marzluff, S. Mayo, Microfabricated silicon mixers for submillisecond quench-flow analysis, Anal. Chem. 70 (1998) 232–236.

[3] M. Hoffmann, M. Schlüter, N. Räbiger, N., Experimental investigation of liquid–liquid mixing in T-shaped micro-mixers using μ-LIF and μ-PIV, Chem. Eng. Sci. 61 (2006) 2968-2976.

[4] M. Engler, N. Kockmann, T. Kiefer, P. Woias, Numerical and experimental investigations on liquid mixing in static micromixers, Chem. Eng. J. (2004) 315-322.

[5] D. Bothe, C. Stemich, H.-J. Warnecke, Fluid mixing in a T-shaped micro-mixer, Chem. Eng. Sci. 61 (2006) 2950-2958.

[6] N. Kockmann, D.M. Roberge, Transitional flow and related transport phenomena in curved microchannels, Heat Transfer Engineering 32 (2011) 595-608.

[7] S. Dreher, N. Kockmann, P. Woias, Characterization of laminar transient flow regimes and mixing in T-shaped micromixers, Heat Transfer Eng. 30 (2009) 91-100.

[8] S. Thomas, T. Ameel, J. Guilkey, Mixing kinematics of moderate Reynolds number flows in a T-channel, Phis. Fluids 22 (2009) 013601 (10 pp).

[9] S. Thomas, T. Ameel, J. Guilkey, Mixing kinematics of moderate Reynolds number flows in a T-channel, Phys. Fluids 22 (2010) 013601 (10 pp).

[10] J.M. Ottino, S. Wiggins, Introduction: mixing in microfluidics, Philos. Trans. R. Soc. London, Ser. A 362 (2004) 923-935.

[11] C. Galletti, M. Roudgar, E. Brunazzi, R. Mauri, Effect of inlet conditions on the engulfment pattern in a T-shaped micro-mixer, Chem. Eng. J. 185–186 (2012) 300-313.
[12] J. Hussong, R. Lindken, M. Pourquie, J. Westerweel, Numerical study on the flow physics of a T-shaped micro mixer, in: IUTAM Symposium on Advances in Micro- and Nanofluidics (B.V. M. Ellero et al., eds), IUTAM Bookseries 15, Springer Science, New York, 2009.
[13] P.C. Chatwin, P.J. Sullivan, The effect of aspect ratio on the longitudinal diffusivity in rectangular channels, J. Fluid Mech. 120 (1982) 347-358.
[14] J. Happel, H. Brenner, Low Reynolds Number Hydrodynamics, Prentice Hall, New York, 1965 (Eq. 2-5.24).
[15] R.B. Bird, W.E. Stewart, E.N. Lightfoot, Transport Phenomena, Wiley, New York, 1960 (pp. 297 and 641).

(435c) Stop Flow Lithography: Beyond PDMS Microfluidic Devices

Ki Wan Bong[1], Jingjing Xu[2], Jong-Ho Kim[3], Stephen C. Chapin[4], Michael S. Strano[2], Karen K. Gleason[5] and Patrick S. Doyle[4], (1)Chemical Engineering, MIT, Cambridge, MA, (2)Department of Chemical Engineering, Massachusetts Institute of Technology, Cambridge, MA, (3)Chemical Engineering, MIT, Cambridge, (4)Chemical Engineering, Massachusetts Institute of Technology, Cambridge, MA, (5)Chemical Engineering, Massachusetts Insitute of Technology, Cambridge, MA

Stop flow lithography has become a powerful technique to functionalize complex functional microparticles [1,2]. Currently, flow lithography relies on the use of polydimethylsiloxane microchannels, because the process requires local inhibition of polymerization, near channel interfaces, via oxygen permeation. The dependence on polydimethylsiloxane devices greatly limits the range of precursor materials that can be processed in flow lithography. Here we present oxygen-free flow lithography via inert fluid-lubrication layers for the synthesis of new classes of complex microparticles [3]. We use an initiated chemical vapour deposition nano-adhesive bonding technique to create non-polydimethylsiloxane-based devices. We successfully synthesize microparticles with a sub-second residence time and demonstrate on-the-fly alteration of particle height. This technique greatly expands the synthesis capabilities of flow lithography, enabling particle synthesis, using water-insoluble monomers, organic solvents, and hydrophobic functional entities such as quantum dots and single-walled carbon nanotubes. As one demonstrative application, we created near-infrared barcoded particles for real-time, label-free detection of target analytes.

1. Dendukuri, D., Gu, S.S., Pregibon, D.C., Hatton, T.A., and Doyle, P.S., "Stop flow lithography in a microfluidic device", *Lab on Chip*, 7, 818-828, 2007.

2. Dendukuri, D., Pregibon, D.C., Collins, J., Hatton, T.A. and Doyle, P.S. "Continuous Flow Lithography for High-Throughput Microparticle Synthesis", *Nature Materials*, **5**, 365-369, 2006

3. Bong, K.W., Xu, J., Kim, J.-H., Chapin, S.C., Strano, M.S., Gleason, K.K., and Doyle, P.S., "Non-polydimethylsiloxane devices for Oxygen-Free Flow Lithography", *Nature Communications*, **3**, 805 2012.

(435d) Sculpting and Atomizing Pinned Drops with Localized Acoustic Pressures of Surface Acoustic Waves: Exponentially Small Contact Angles

Daniel Taller[1], David Go[1] and Hsueh-Chia Chang[2], (1)Aerospace and Mechanical Engineering, University of Notre Dame, Notre Dame, IN, (2)Chemical & Biomolecular Engineering, University of Notre Dame, Notre Dame, IN

Surface acoustic wave microelectromechanical (MEMS) devices are piezoelectric crystals with an interdigitated transducer on the surface of the crystal. Upon the application of a high frequency electrical signal, a surface acoustic wave (SAW) propagates along the surface of the crystal as a Rayleigh wave. Recently, SAW devices have been coupled with small amounts of liquid (~ 0.1 mm^3) in microfluidic devices. When a SAW wave interacts with a liquid film, it generates acoustic streaming enabling the manipulation of the liquid film, and, at sufficient amplitude, aerosolizes the film. Due to viscous dissipation, the SAW wave diffracting from the solid substrate into the liquid drop produces an exponentially decaying time-averaged pressure in the drop. In this work, we show that if the drop is pinned against a bounding wall such as the filter paper used in this study, the localized acoustic pressure generates a sequence of surface droplets at the contact line, whose dimensions decay in the same manner as the acoustic pressure. The undulating interfacial profile near the contact line also inherits this exponential decay, such that the averaged contact angle is exponentially small. The size distribution of surface drops is collapsed under the exponential scaling that depends only on the SAW decay rate and amplitude. This paper also presents a numerical and experimental study of the streaming and aerosolization of liquids via surface acoustic waves. A high-speed imaging system is used to characterize the fluid's free surface and aerosolization as a function of power and liquid properties. Several regimes, including a stable liquid film, initial droplet formation, thin film aerosolization, and cavitation are identified and explored. A two-dimensional numerical model based on the Laplace-Young equation is developed to model the profile of the liquid film. It is further shown that the threshold power for atomization scales as 5th power of the SAW decay rate for very localized acoustic pressures.

(435e) Microfluidic Static Droplet Arrays with Tunable Concentration Gradients

Meng Sun, Swastika S. Bithi and **Siva A. Vanapalli**, Chemical Engineering, Texas Tech University, Lubbock, TX

Recent advances in drop-based microfluidics are creating new opportunities for conducting high throughput analysis with significantly reduced sample and reagent volumes. In particular, microfluidic production of arrays of nanoliter-scale drops at prescribed coordinates could emerge as a powerful and inexpensive alternative to conventional multi-well-plate-based screening. However, the challenge is to develop effective strategies to vary the reagent concentration from drop-to-drop in the static droplet arrays (SDA). Here we describe a one-step passive strategy to create such concentration gradients in immobilized droplet arrays. The technique exploits controlled coalescence, and mixing of materials between moving plugs and stationary drops in a microfluidic parking network (MPN). We show that the concentration of one or more reagents can increase or decrease from drop-to-drop enabling a wide range of composition variation. This unique approach also generates concentration gradients of soluble reagents in drops containing insoluble materials such as particles and cells at well-defined time points.

(435f) Modeling, Predicting, and Controlling Microscale Tipstreaming

Todd Moyle[1], Lynn Walker[1] and Shelley L. Anna[2], (1)Chemical Engineering, Carnegie Mellon University, Pittsburgh, PA, (2)Departments of Mechanical Engineering and Chemical Engineering, Carnegie Mellon University, Pittsburgh, PA

Microscale tipstreaming is one method to overcome the limiting length scale in microfluidics allowing for production of submicron sized droplets. Tipstreaming is the ejection of small drops from a liquid thread formed by interfacial tension gradients and convective transport of surfactant. Controlling and understanding this process is essential for successful application in areas such as synthesis of nano-scale particles, manipulation of biomolecules, enzyme activity studies and others. However, models that predict operating conditions for tipstreaming in microfluidics do not currently exist and the periodic nature of thread formation has hindered its use in applications. In this work, we develop an analytical model aimed at capturing the essential physics of the tipstreaming mechanism. The model relies on observations of a fixed interfacial shape indicative of thread formation to simplify the fluid flow and surfactant transport equations. The result is an interfacial mass balance of surfactant. Conditions where the mass balance can be satisfied define the operating conditions for microscale tipstreaming. Good agreement is found between model predictions and experiments. Lastly, modeling the interfacial shape in this manner evokes the idea of fixing the interfacial shape experimentally. We implement an active feedback control algorithm using a custom LabVIEW code for image analysis to hold the interfacial shape fixed during tipstreaming. Controlling the interfacial shape experimentally increases thread production, potentially expanding the number of applications where tipstreaming can be used.

(435g) Block-and-Break Generation of Microdroplets with Fixed Volume

Volkert van Steijn[1], Piotr M. Korczyk[2], Adam R. Abate[3], Ladislav Derzsi[2], David A. Weitz[4] and Piotr Garstecki[2], (1)Department of Chemical Engineering, Delft University of Technology, Delft, Netherlands, (2)Institute of Physical Chemistry, Polish Academy of Sciences, Warsaw, Poland, (3)Bioengineering and Therapeutic Sciences, alifornia Institute for Quantitative Biosciences, University of California, San Fransisco, CA, (4)Department of Physics and School of Engineering and Applied Science, Harvard University, Cambridge, MA

Microfluidic devices produce monodisperse drops with a high-level of control over size and contents, making them useful for applications in biotechnology, material synthesis, optofluidics, and medical diagnostics. Drop volume depends on the flow rates, which makes it easy to tune. It, however, also sets strict requirements on the stability of flows for the generation of monodisperse drops, as variations in flow lead to corresponding variations in volume.

In this work, we introduce a novel type of microfluidic droplet generator that produces droplets of a volume set by the geometry of the droplet generator, and not by the rates of flow of the liquids. The generator consists of a classic T-junction with a bypass channel. This bypass directs the continuous fluid around the forming droplets, so that they can fill the space between the inlet of the dispersed phase and the exit of the bypass without breaking. Once filled, the droplets block the exit of the bypass and are squeezed by the continuous fluid, which leads to their pinch-off. We illustrate the potential of this fixed-volume droplet generator for i) forming monodisperse droplets from a source of varying rates of flow, ii) forming monodisperse droplets containing a gradation of solute concentration, and iii) producing monodisperse droplets in parallel.

(435h) Microbead Arrays Formed by Flow Entrapment in a Microfluidic Cell and the Effect of the Trap Geometry On the Mass Transfer of Targets to Probe Molecules On the Particle Surfaces

Xiaoxiao Chen, Chemical Engineering, The City College of New York, The School of Engineering, New York, NY and Charles Maldarelli, Levich Institute and Chemical Engineering, The City College of New York, New York, NY

Platforms which display biomolecules in an array for the multiplexed screening of the binding interactions of these molecules with molecular targets are important tools in biomedical research (drug discovery), clinical diagnostics (biomarker identification) and sensor development. Platforms which integrate the array and the binding detection in a single, microfluidic format have the advantage of reduced volumes of reagents and targets, high sensitivity, and increased throughput. Microbead-based arrays are biomolecular displays in which the probe molecules are first linked to micron-sized beads which are arranged and fixed to a surface. These arrays can easily be constructed in a microfluidic platform by using a concourse of traps designed to hydrodynamically capture the beads as they stream through the course. The arrangement of the traps serves as the template for the microbead array. By trapping the beads in a sequence of stages, in which at each stage, beads displaying only one probe are trapped, with their entrapment locations mapped, a fully indexed array can be formed. The array can then be screened against probes which are passed through the concourse.

The traps used to capture the microbeads have to properly designed to not only efficiently capture the microbeads, but also to minimize, during the assaying step, diffusion limitations in the transport of the target molecules to the surface of the captured particle. Eliminating diffusion barriers increases the dynamic response of the array, which is then only limited by the kinetics of the surface binding of the target to the probe. This presentation will examine experimentally and theoretically how trap geometries can be designed to reduce diffusion barriers. We study a general trap design consisting of an open cavity or chamber, facing the flow, in which the particle is captured within the chamber. The traps are placed in a staggered arrangement. To illustrate the effect of the trap shape on the mass transfer of the target, we examine the binding of the ligand biotin attached as the probe to the surface of the microbead, Neutravidan, which binds selectively to the biotin, is streamed through the cell. By fluorescently labeling the Neutravidin, and recording the accumulating fluorescence at each bead location by using confocal laser scanning fluorescence microscopy and epifluorescent microscopy, the rate of conjugation of the protein to the displayed ligand is measured as a function of time. Traps with large cavities sequestering beads deeply in the enclosing chamber are found to have significant diffusion barriers relative to traps in which the microbeads are held in smaller cavities. The capture efficiency is reduced by the smaller capture area of these traps, but we demonstrate that hydrodynamic arraying is still effective. Numerical solutions of the convective diffusion equations are presented for the diffusion and binding of the protein to the biotin surface of the microbeads for different trap geometries. These simulations confirm the experimental results, and provide an accurate measurement of the kinetic constants for the attachment of the protein to the biotin.

(435i) Microfluidic Hydrogel Structures for Controlled Concentration Gradients and Fast Solution Switching

Joel S. Paustian, Rodrigo Nery-Azevedo and Todd M. Squires, Chemical Engineering, University of California, Santa Barbara, Santa Barbara, CA

Hydrogels are aqueous polymer networks that are solid under stress, but are permeable to water and solutes. We present a simple microfluidic technique to synthesize in-channel hydrogel structures with precise, high resolution microscale geometries. We create create hydrogel membrane walls, with feature sizes under 10 microns, that remain sturdy under high shear from adjacent flows and withstand pressure gradients. At the same time, the hydrogel membranes allow solute, solvent, and electric fields without passing an appreciable fluid flow, thus enabling chemical conditions on one side of the membrane to be changed without mixing, stirring, or diluting.

These patterned, permeable membranes allow us to set up defined concentration or pH gradients, rapidly modify solute composition in a microchamber, or pass electric currents between channels in specified regions. Microscale dynamics, transport, and transient rheology can be probed in intriguing new ways. We provide several examples of the novel experiments enabled by these micropatterned hydrogels.

(443a) Evaporation Rate of Water in Hydrophobic Confinement

Pablo G. Debenedetti and Sumit Sharma, Chemical and Biological Engineering, Princeton University, Princeton, NJ

The drying of hydrophobic cavities is believed to play an important role in biophysical phenomena such as the folding of globular proteins, the opening and closing of ligand-gated ion channels, and ligand binding to hydrophobic pockets. We use forward flux sampling to compute the rate of capillary evaporation of water confined between two hydrophobic surfaces separated by nanoscopic gaps, as a function of gap, surface size and temperature. Over the range of conditions investigated (gaps between 9 and 14 Å and surface areas between 1 and 9 nm^2) the free energy barrier to evaporation scales linearly with the gap between hydrophobic surfaces, suggesting that line tension makes the predominant contribution to the free energy barrier. The exponential dependence of the evaporation rate on the gap between confining surfaces causes a ten order-of-magnitude decrease in the rate when the gap increases from 9 to 14 Å. The computed activation energies are of the order of $50kT$, and we present umbrella-sampling calculations of the free energy barriers. Evaporation rates per unit area are found to be two orders of magnitude faster in confinement by the larger (9 nm^2) than by the smaller (1nm^2) surfaces considered here, at otherwise identical conditions. We show that this is a direct consequence of the dependence of hydrophobic hydration on the size of solvated objects. For sufficiently large surfaces, the critical nucleus for the evaporation process is a gap-spanning cylindrical vapor tube.

(443b) Towards a Molecular Theory for Homogeneous Bubble Nucleation in Metastable Liquids

Korosh Torabi[1,2] and David S. Corti[1], (1)School of Chemical Engineering, Purdue University, West Lafayette, IN, (2)Department of Chemistry, Northwestern University, Evanston, IL

Homogeneous bubble nucleation refers to the process by which a first-order phase transition occurs within a bulk metastable liquid, in the absence of impurities or external perturbations. We report an equilibrium embryo definition for homogeneous bubble nucleation based on the following two order parameters: the number of particles n contained within a spherical volume v, which are in constrained equilibrium with the surrounding superheated liquid. The reversible work of formation, or the underlying free energy surface, of these (n,v) embryos within the superheated Lennard-Jones liquid is generated via several novel Monte Carlo simulation procedures. By doing so, we also emphasize several key aspects that any well-formulated equilibrium based theory of homogeneous nucleation must address. Namely, within the suggested embryo definition any configuration of the particles at a fixed location in space must be capable of being mapped onto one and only one point in the order parameter space. Moreover, the suggested equilibrium-based model must be able to identify the transitional configurations that, based on a committor probability analysis, are dynamically relevant to the nucleation process. Another novel characteristic of the developed theory is that we define a limit for the configuration space of the metastable phase. Hence, those configurations that have a zero committor probability are not included in the free energy surface. Finally, we develop a simulation-based algorithm that non-redundantly relates the number density of the critical nucleation sites to their free energies of formation (The critical nuclei are located close to the boundary of the metastable phase configuration space where the committor probability is about one half). In turn, this allows us to determine the forward rate coefficient of the critical nuclei based on their average life times calculated via MD simulation of a large number trajectories initiated from the transitional region of the configuration space. We compare our nucleation rate predictions to those obtained from other methods, and discuss the limitations of all of these approaches.

(443c) Crystal Nucleation and Growth in Molecular Systems

Jerome Delhommelle and Caroline Desgranges, Department of Chemistry, University of North Dakota, Grand Forks, ND

Controlling the crystallization of molecules from supersaturated solutions is a long-standing issue in solid state chemistry. Since polymorphs may have very different physical properties, it is crucial to control in which polymorph a molecule crystallizes in many applications, such as e.g. in the pharmaceutical industry. We discuss in this presentation our recent work on the molecular simulation of the crystallization of molecules from the liquid phase. We simulate the nucleation step using a combination of molecular dynamics simulation with the so-called umbrella sampling technique. This sampling method, when used together a reaction coordinate (or order parameter) characteristic of the crystalline order for the chosen system, is particularly well suited to study activated processes (or rare events) such as e.g. the nucleation event, which requires the system to overcome a large free energy barrier of nucleation. Throughout the nucleation and growth processes, we focus on examining how the selection of a specific polymorph takes place.

Loop Heat Pipes Exploiting Nano-porous Wicks
- Development and Application of Materials Sustaining Fluid under Tension

Among many different technologies for transferring heat, the heat pipe (HP) – closed-circuit systems in which a working fluid transfers its latent heat as it cycles between an evaporator and a condenser by capillary pumping force – is one of the most efficient systems known today and has offered an attractive alternative in an array of applications[1]. However, two main limitations have constrained the application of conventional HPs: 1) the maximum capillary head developed by the micro-porous wicks of ~1 bar that limits the maximum heat loads, the length, and their orientation with respect to gravity or accelerations.; and 2) the saturation state in the condenser region creates an unbounded liquid-vapor interface that causes undesired transients and additional resistance to heat transfer.

The transpiration process in vascular plants operates by a mechanism similar to the recirculation of liquid through the wick of HPs[2]. Unlike in HPs, though, the water pressure in plants is able to reach large negative values (down to -100 bars) to overcome gravity, viscous drag in the liquid path, and the reduced chemical potential of water in sub-saturated soils. The secret behind the amazing abilities of plants is a the structure of plants cell walls, which serve as nano-porous membranes, allowing the water pressure to be less than zero (under tension). Inspired by plants, our research is focused on the development of a nano-porous loop heat pipe (NPLHP) exploiting nano-porous wicks, including a theoretical model of the coupled transport and thermodynamic phenomena and an experimental platform with which to test design concepts for NPLHPs.

After presenting the design considerations of a NPLHP, we will focus on a central experimental challenge in its development: nano-porous membranes that allow for the pure liquids to come to metastable equilibrium at negative pressures with sub-saturated vapors. We will discuss several membrane structures developed in our laboratory and our experimental methods for characterizing them. We will then turn to an intriguing open question about water under tension: several experimental studies focusing on cavitation thresholds (the pressure at which the thermodynamically metastable liquid turns into vapor by nucleation) of water under tension have been reported, yet there is no consensus[3]. Many experiments using various methods have found the stability limit of water at room temperature is between -20 ~ -30MPa[4-6], while experiments in quartz inclusions appear to reach a stability limit of -140 MPa (with the use of an extrapolated equation of state). We will describe how measurements based our distinct experimental approach compare with those in the literature and discuss the implications for our understanding of the phase diagram of water and on the practical limitations on the use of water at negative pressure in technologies such as NPLHPs.

1 Faghri, A. *Heat Pipe Science and Technology*. (Taylor and Francis, 1995).

2 Nobel, P. S. *Physicochemical and Environmental Plant Physiology*. 2nd edn, (Academic Press, 1999).

3 Caupin, F. & Herbert, E. Cavitation in water: a review. *Comptes Rendus Physique* **7**, 1000-1017 (2006).

4 Briggs, L. J. Limiting Negative Pressure of Water. *Journal of Applied Physics* **21**, 721-722 (1950).

5 Wheeler, T. D. & Stroock, A. D. Stability Limit of Liquid Water in Metastable Equilibrium with Subsaturated Vapors. *Langmuir* **25**, 7609-7622, doi:10.1021/la9002725 (2009).

6 Davitt, K., Arvengas, A. & Caupin, F. Water at the cavitation limit: Density of the metastable liquid and size of the critical bubble. *EPL* **90**, 16002 (2010).

(443e) Energetic Origins of Metastable Zone, Induction Supersaturation, and Implications towards Nucleation

Venkateswarlu Bhamidi, Paul J.A. Kenis and Charles F. Zukoski, Department of Chemical & Biomolecular Engineering, University of Illinois at Urbana Champaign, Urbana, IL

For decades, the metastable zone in solution crystallization – a supersaturated region of the phase diagram in which no significant nucleation of crystals occurs – has been viewed as a kinetic phenomenon that originates from the strong dependency of the rate of nucleation on solute concentration. Here we show that, for several different compounds, when the system is slowly supersaturated, the observed metastability limit (MZL) becomes independent of both the rate of change of supersaturation and the time required to reach MZL, a result that suggests an MZL that in fact has energetic origins.

Our experiments are based on the crystallization of a variety of compounds from aqueous solutions. In these studies we employed an evaporation-based microfluidic crystallization platform, in which several small (5 μL) droplets of aqueous solutions were subjected to slow, regulated evaporation. By exploring the metastability limit exhibited by a large number of droplets, we show that the observed probability distribution of nucleation is well described by the classical descriptions of the work required to produce a critical cluster. From this probability distribution we extract an *induction supersaturation* that occurs when the energy barrier to nucleation has a magnitude of $1kT$, the product of Boltzmann's constant and the absolute temperature. This induction supersaturation is determined by the energetics of cluster formation and can be regarded as the limiting lower boundary of metastability. This surprising result raises important concerns regarding the current interpretation of the processes responsible for the nucleation of crystals from solution.

(443f) Ostwald-Like Ripening of the Anomalous Mesoscopic Clusters in Protein Solutions

Peter G. Vekilov[1], Ye Li[2] and Vassiliy Lubchenko[2], (1)Chemical Engineering, University of Houston, Houston, TX, (2)University of Houston, Houston, TX

Anomalous clusters of mesoscopic dimensions composed of protein-rich liquid exist in protein solutions, both in the homogenous region of the solution phase diagram and in the region supersaturated with respect to an ordered solid phase, such as crystals. They are of interest because in supersaturated solutions they are crucial nucleation sites for the ordered phase. We monitor over 71 hours the evolution of the anomalous clusters in solutions of the protein lysozyme employing dynamic and static light scattering and Brownian microscopy. For the sake of experimental clarity, we choose conditions, under which no condensed phases, liquid or solid, are stable or present as long-lived metastable domains. We show that cluster formation is reversible and the clusters are in near equilibrium with the solution, up to a capillary correction. We observe that en route to full equilibration, the average cluster size grows with time following a universal law $t^{0.26\pm0.03}$, independent of the cluster volume fraction. We show that the cluster size distribution is scale-invariant at all stages of the cluster evolution. These observations indicate that, similar to classical Ostwald ripening of new phases, cluster evolution is driven by minimization of the surface free energy and occurs by the exchange of single molecules. The observed cluster evolution may help to understand nucleation mysteries, such as variable nucleation rates under steady conditions, and others.

(452a) Direct Force Measurement Between Chemically Patterned Heterogeneous Surfaces

Christian Pick, German Drazer and Joelle Frechette, Chemical and Biomolecular Engineering, Johns Hopkins University, Baltimore, MD

Artificially patterned surfaces provide a means of precisely tailoring the interactions between particles and surfaces. As a consequence, such surfaces have a variety of potential uses in the fabrication of sensors and separation devices. They can also provide a template to direct assembly at the micro and nanoscale. However, to utilize artificially patterned surfaces for such applications, it is important to fully understand how patterning or surface heterogeneities alters surface forces. One way patterning can alter surface forces is through control over the distribution and size of charge heterogeneities. In this work, charge heterogeneities consisting of alternating stripes of a cationic polyelectrolyte were patterned onto mica via microcontact printing. The force profile between the patterned surface and bare mica were directly measured in the surface force apparatus. This force profile was compared to the force profiles between two bare mica surfaces and two surfaces uniformly coated in polyelectrolyte. Comparisons were made between the measured force profiles and DLVO theory. The effects of charge heterogeneity on surface forces have been explored experimentally through direct force measurement.

(452b) Understanding the Mechanism of Simultaneously Oleophobic/Hydrophilic Surface: When a Nanometer-Thick Polymer Coating Meets An Attractive Solid Substrate

Lei Li[1], Yongjin Wang[1] and Jianing Sun[2], (1)University of Pittsburgh, Pittsburgh, PA, (2)J.A. Woollam Co., Inc., Lincoln, NE

Understanding the mechanism of simultaneously oleophobic/hydrophilic surface: When a nanometer-thick polymer coating meets an attractive solid substrate

Lei Li[1], Yongjin Wang[1] and Jianing Sun[2]

1. Department of Chemical and Petroleum Engineering, University of Pittsburgh, 3700 O'Hara Street, Pittsburgh, PA 15261

2. J.A. Woollam Co., Inc., 645 M Street, Suite 102, Lincoln, NE 68508

Simultaneously hydrophilic/oleophobic surfaces are highly desirable in many important applications. However, such wetting behavior are not expected for traditional surfaces since previous experimental results showed that, in general, a solid surface is more wettable to oil than to water. Interestingly, a couple of recent studies showed that it is possible to achieve simultaneous "oleophobicity/hydrophilicity" by applying some nanometer-thick polymer on some solid substrate. The underlying mechanism, which is critical to fabricating "oleophobic/hydrophilic" surfaces, remains unclear to date. In the current presentation, we report our recent experimental work in uncovering the mechanism of "hydrophilic/oleophobic" surfaces. First, another "hydrophilic/oleophobic" surface, which is fabricated by dip-coating a nanometer-thick perfluoropolyether on a silicon wafer and shows higher water contact angle (WCA) than hexadecane contact angle (HCA), has been identified. Second, it was found that the HCA on this surface changes significantly with time, which indicates that the observed "oleophobicity/hydrophilicity" is kinetic in nature. Third, the effect of the chemicals structure of both polymer and substrate on the peculiar wetting behavior has been studied. The experimental results suggest that the interaction between the nanometer-thick polymer and the substrate is the key to the appropriate size of the intermolecular "hole" in the polymer layer, which in turn determines the kinetics of the penetration of water and hexadecane through the polymer nanofilm. Only when the "hole" size is appropriate so that the penetration rate of hexadecane is much slower than that of water, the peculiar "hydrophilic/oleophobic' wetting performance is observed. On the basis of these findings, the design criteria to fabricating hydrophilic/oleophobic surfaces have also been proposed.

(452c) Single Molecule Studies of the Hydrophobic Interaction At the Solid-Liquid Interface

Nathaniel Nelson, Robert Walder and Daniel K. Schwartz, Chemical and Biological Engineering, University of Colorado at Boulder, Boulder, CO

The interactions between adsorbate molecules and hydrophobic surfaces are of significant interest due to their importance in a variety biological and separations processes. While macroscopic observations of hydrophobic interactions are often reported (e.g. using AFM or SFA), it is challenging to extrapolate ensemble-averaged measurements to molecular-level phenomena. Using total internal reflection fluorescence microscopy to image individual molecules at the solid-liquid interface, we directly observe dynamic behavior, including surface diffusion and adsorption/desorption kinetics, that are due to non-specific interactions. Since these phenomena are examined at the single molecule level, we are able to identify unique diffusive modes/mechanisms and characteristic residence times of multiple molecular populations on chemically homogeneous surfaces without relying on models to quantify these behaviors. The data presented here show a trend towards a stronger probe-surface interaction with increasing alkyl chain length monolayers despite these monolayers having the same water contact angle. On longer (16-18 carbon) alkylsilane monolayers our probe molecule, fluorescently labeled dodecanoic acid, has significantly slower diffusion and exhibits far fewer desorption mediated "flights" than on short (4-6 carbon) monolayers. Additionally, probes on the longer chain surfaces are much more likely to exhibit extended surface residence times as opposed to short transient surface visits. We hypothesize that this increase in interaction is due to an increase in the monolayer's ability to entangle the fatty acid probe.

(452d) Formation of Carboxylic Acid Nanorods On Oleylamide-Capped Au Nanoparticles

Sunxi Wang, Li Li and Guangzhao Mao, Chemical Engineering and Materials Science, Wayne State University, Detroit, MI

2012 AIChE Annual Meeting: Pittsburgh, PA, USA, Oct 28-Nov 2

This paper describes nanorod formation induced by nanoparticle seeds. Carboxylic acids self-assemble on graphite into an epitaxial pattern. However, this persistent pattern is changed into nanorods by nanoparticles. Two control parameters have been identified including nucleus/substrate contact angle and seed to critical nucleus size ratio. Effective carboxylic chain length and nanoparticle size range are identified, which can be related to the seed size effect. Some nanorods are detached when methyl-terminated nanoparticles are used and the percentage of detached nanorods increases with increasing carbon chain length. In contrast, the nanorods do not detach from carboxyl-terminated nanoparticles. This study offers direct experimental evidence to molecular simulation prediction that nuclei display a confined morphology on a highly curved seed surface due to the difficulty of maintaining an unstrained structure and they readily detach from the surface during heterogeneous crystal nucleation. This study contributes a solution-based method for the making of nanorods.

(452e) Quantification of Nucleic Acid Hybridization near Solid Liquid Interface

Jingjing Wang, Center for Biomedical Engineering, Brown University, Providence, RI, Kenneth Morabito, Division of Engineering and Warren Alpert School of Medicine, Brown University, Providence, RI and Anubhav Tripathi, School of Engineering, Brown University, Providence, RI

Given low concentration of target RNA in biological samples, RNA isolation methods needs fundamental understanding of hybridization process near interfaces. In this study, we present a novel microfludic method for sDNA isolation near solid liquid interface. We have examined the key factors that have effects on the hybridization efficiency, which include incubation time (15min, 25min, and 35min), temperature (25C, 40C and 70C) and location of target sequence (5' end, middle, and 3' end) in sDNA. We present extensive results on the capacity of solid liquid interfaces in capturing RNA targets that incorporates the advantages of microfluidics.

We found that there is minimal incubation time for optimal hybridization to happen and beyond that there is no significant change. We also found that heating the solid liquid system will enhance the hybridization efficiency. There is 3-fold increase in hybridization capacity with 40C treatment and 2-fold increase with 70C treatment after 15min incubation compares to that in room temperature. This study is useful for examining target sDNA hybridization capacity with biotinylated capture near solid liquid surface. This investigation presents the parameters of this quick method on a microchip platform with many possible applications. The technique can be used to improve a number of assays, including DNA/RNA isolation and amplification, blood plasma separation and peptide isolation. Our study provides a quantitative guidance to clinically relevant RNA detection methods.

Acknowledgements: We acknowledge support from the National Institute of Health (US) and the China Scholarship Council.

(452f) Using Monte Carlo Simulations to Study Wetting Behavior of Ionic Liquids

Kaustubh S. Rane and Jeffrey R. Errington, Chemical and Biological Engineering, University at Buffalo, The State University of New York, Buffalo, NY

Room temperature ionic liquids (RTILs) are finding applications in many emerging technologies due to their low vapor pressures, high conductivities and excellent solvent capabilities. Understanding the wetting behavior of ionic liquids is crucial for developing these technologies. This presentation focuses on our efforts to use Monte Carlo simulations to study the effects of chemical nature of ionic liquids on their wetting properties. To understand the role of dispersion and coulombic interactions we study simple models of ionic liquids comprising of spherical cations and anions with variable strengths of electrostatic interactions relative to dispersion interactions. The range of relative ionic strengths that we investigate includes those observed in RTILs and molten alkali halides. Results are presented to show the dependence of contact angle on temperature and substrate strength for fluids with different relative ionic strengths. We also present results for realistic models of imidazolium based RTILs on graphite surface to show the effect of temperature and chemical nature of cation on the wetting properties. We briefly explain the challenges encountered in Monte Carlo simulations of realistic models of RTILs and the strategies we use to overcome them.

(452g) Investigation of Free-Energy Barriers for Droplet Wetting-Mode Transitions On Structured Surfaces Using Forward Flux Sampling

Azar Shahraz, Chemical Engineering, Pennsylvania State University, University Park, PA, Ali Borhan, Department of Chemical Engineering, The Pennsylvania State University, University Park, PA and Kristen Fichthorn, Dept. of Chemical Engineering, Pennsylvania State University, University Park, PA

Liquid droplets exhibit multiple wetting modes on textured surfaces, depending on the surface topography, intrinsic contact angle, and droplet size. In addition, it has been observed experimentally that the wetting state of a droplet can depend on its initial position on the substrate. In this case, the wetting state with the lowest free energy may not be achieved due to the existence of a free-energy barrier between two wetting modes. In this work, we use molecular dynamics simulations to study the wetting states of droplets on grooved hydrophobic solid surfaces. We observe both Cassie and Wenzel wetting modes, in which the droplet resides on top of the grooves or penetrates within the grooves, respectively. As in the experiments, the wetting state we observe can depend on the initial configuration of the droplet, indicating the presence of long-lived metastable states. We use the forward flux sampling method to study transitions between the Cassie and Wenzel states. Our studies show that metastable wetting states can exist over human time scales, and that the free-energy barrier for transitions between these states depends on topological parameters characterizing the grooved surface.

(455a) Crossover From a Hydrophobic to Hydrophilic Force Law in a Coarse-Grained Water Model

Aviel Chaimovich, Rahul R. Lele, Jacob Israelachvili and M. Scott Shell, Department of Chemical Engineering, University of California Santa Barbara, Santa Barbara, CA

Water-mediated forces, namely the so-called hydrophobic and hydrophilic interactions, govern many biological mechanisms, such as the folding of proteins, the forming of membranes, as well as the assembly of cellular machines. Despite numerous efforts on these fundamental forces, there remains significant work for a complete understanding of them [1]. In particular, recent experiments for surfactant-coated interfaces suggest that both hydrophobic and hydrophilic interactions can be cast under a single force law [2]. Nevertheless, no definitive theory yet exists for such a surprising combined water-mediated interaction.

In this work, we address the crossover from hydrophobic to hydrophilic interactions using coarse-grained models in molecular simulations, addressing in turn the possibility of a collective force law for the water-mediated interaction. We immerse a pair of colloids in a water bath, computing the free energy as a function of their respective separation distance [1]. To resolve the free energies, we use a novel Monte Carlo technique based on the flat-histogram and transition-matrix methods. Importantly, we can control the colloid's binding energy with water, going from very hydrophobic to very hydrophilic character. We also tune the colloid's size, going from very small to very large diameter. In the latter case, we retrieve the empirical observation, namely, an attraction for hydrophobic surfaces and a repulsion for hydrophilic surfaces, both exhibiting an exponential decay with colloidal separation distance [2]. We find the emergence of specific scaling laws and transitions in driving-forces (entropic and enthalpic) for both hydrophobic and hydrophilic interactions. Overall, these findings provide a unified picture of hydrophobic and hydrophilic interactions, and we suggest a simple interpretation of the corresponding water-mediated force law.

1. Hammer, M.U., et al., *The Search for the hydrophobic force law.* Faraday Discussions, 2010. **146**: p. 299-308.

2. Donaldson, S.H., et al., *General hydrophobic interaction potential for surfactant/lipid bilayers from direct force measurements between light-modulated bilayers.* Proceedings of the National Academy of Sciences, 2011. **108**(38): p. 15699-15704.

(455b) Extended Flory-Huggins Model for Poly-(ethylene) Oxide Solutions

Mangesh Chaudhari, Department of Chemical and Biomolecular Engineering, Tulane University, New Orleans, LA and Lawrence R. Pratt, Chemical and Biomolecular Engineering, Tulane University, New Orleans, LA

Extended Flory-Huggins model for poly-(ethylene) oxide solutions

M. I. Chaudhari, L. R. Pratt

Tulane University, New Orleans, LA 70118

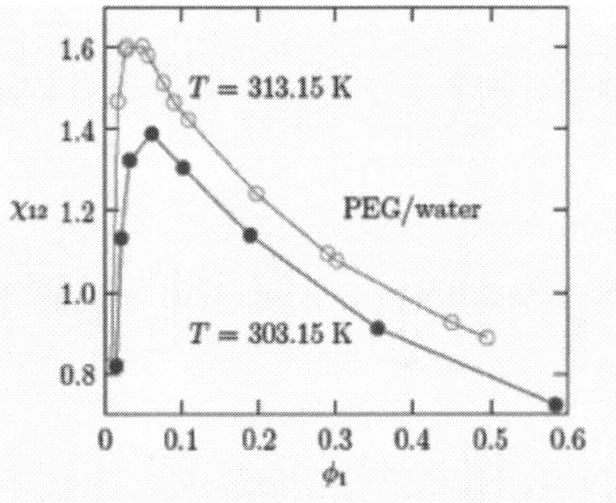

Classic Flory-Huggins model for polymer phase equilibria is based on a lattice model and widely used to understand thermodynamics of polymer solutions. Nevertheless, this model is only applicable to polymer solutions that are sufficiently similar components. Bae et.al. (1993) showed that this model is unsatisfactory for aqueous poly-(ethylene) oxide solutions (FIG 1). Composition independence of interaction parameter and assumption of zero volume of mixing is often inadequate for such systems. Here, we present extended Flory-Huggins model based on evaluation of the activity of the small molecular solvent (water). Molecular simulations of poly-(ethylene) oxide chains molecules capped with small hydrophobic groups solvated with water are analyzed for generating solution structure information required for extended model. We will survey the structural information that is a basis for a recent proposal for high-angle neutron scattering for these solutions. Then we present the computed solvent activities, and the implied Flory-Huggins interaction parameter (X) are presented. Our aim is to validate this extended model by molecular dynamic simulation and understand thermodynamics of these PEO polymers in solutions.

FIG. 1: Composition dependence of the Flory-Huggins interaction parameter interred from the data of reference below for the PEG/water system for two temperatures. $^{-}{}_{1}$ is the volume fraction of the small-molecule solvent. Assumption of composition independence of X_{12} is less satisfactory for PEG/water system. The temperature dependence is is stronger, and thus clearer, for this system, and the interaction strength increases with increasing temperature, reflecting significant entropic contribution.

1. Bae, Y.C., Shim, J.J., Soane, D.S., and Prausnitz, J.M., J. App. Poly, Sci., 47, 1193-1206(1993)

(455c) Investigating Polymer Relaxation in Nanochannels and Nanoslits

Yeng-Long Chen[1,2,3], Jen-Fang Chang[3] and Pokeng Lin[3], (1)Department of Chemical Engineering, National Tsing-Hua University, Hsinchu, Taiwan, (2)Department of Physics, National Taiwan University, Taipei, Taiwan, (3)Institute of Physics, Academia Sinica, Taipei, Taiwan

We investigate polymer relaxation in nanoslits and nanochannels with the lattice Boltzmann-Brownian dynamics (LB-BD) method. Classical theories of deGennes and Odijk predict a transition of polymer properties as the confinement height H becomes smaller than the polymer persistence length, at which the confinement strongly affects the stiffness and restricts the conformational entropy of a semi-flexible polymer such as DNA. With LB-BD, the stretch relaxation time and fluctuation correlation decay time (tr) are measured in nano-confinements of height H, from H larger than the polymer radius of gyration Rg to smaller than the polymer persistence length P. A clear transition in tr is observed for chains in nanochannels. tr is found to have a non-monotonic dependence on H, as found in previous experimental observations [1]. In contrast, a clear transition is not observed for chains in nanoslits [2,3]. The difference between nanoslits and nanochannels may be attributed to the confinement-induced changes on chain relaxation mechanisms. Hydrodynamic interactions are found to influence the chain relaxation time, even in confinement smaller than P.

[1] W. Reisner et al., Physical Review Letters 94, 196101 (2005).

[2] PK Lin et al., Macromolecules 45, 2920 (2012).

[3] J. Tang et al., Macromolecules 43, 4368 (2010).

(455d) Fast Monte Carlo Simulation of Liquid Crystal Nanodroplets by Theoretically-Informed Metropolis Sampling of Alignment-Tensor Fields

Juan Pablo Hernandez-Ortiz, Department of Materials and Bioengineering, Universidad Nacional de Colombia, Sede Medellin, Medellin, Colombia, Orlando Guzman, Departamento de Fisica, Universidad Autonoma Metropolitana-Iztapalapa, Mexico D. F., Mexico, Tyler Roberts, Department of Chemical and Biological Engineering, University of Wisconsin Madison, Madison, WI and Juan J. de Pablo, Department of Chemical and Biological Engineering, University of Wisconsin-Madison, Madison, WI

We have developed a fast theoretically informed Monte Carlo simulation method for liquid crystals, which works on multiple scales of coarse-graining, and have applied it to describe the morphology of nematic droplets of nanoscopic size. The method relies on stochastic changes to the alignment-tensor field $\mathbf{Q}(\mathbf{r})$ that describes the state of the liquid crystal inside the droplet, which are accepted on the basis of sampling criteria dictated by the desired free energy functional $F[\mathbf{Q}(\mathbf{r})]$ of the system. We take F to consist of the usual sum of bulk and surface terms and represent the liquid crystal configuration $\mathbf{Q}(\mathbf{r})$ in terms of a radial-basis-function interpolation.

By realizing that different values of parameter β in the Boltzmann factor $\exp(-\beta\ \Delta F)$ correspond to different levels of the coarse-graining length-scale L, Monte Carlo methods capable of sampling simultaneously multiple values of β, such as parallel-tempering or replica exchange, become particularly useful for multi-scale descriptions of the system, while also accelerating sampling of the system's phase space.

We apply this method to predict the phase diagram of liquid-crystal morphologies inside nematic droplets over a range of nanoscopic sizes and anchoring conditions. A comparison to previous results, from particle-based molecular dynamics and Ginzburg-Landau continuum methods, reveals excellent agreement with past work and serves to highlight the improved performance of our proposed approach.

(455e) How Can Talc Surfaces Be Both Hydrophilic and Hydrophobic?

Benjamin Rotenberg, Laboratoire PECSA, UMR 7195, CNRS et UPMC-Paris, **Amish J. Patel**, Chemical and Biological Engineering, Rensselaer Polytechnic Institute, Troy, NY and David Chandler, Chemistry, University of California, Berkeley, CA

Experiments have shown that individual water molecules adsorb strongly on a talc surface (hydrophilic behavior). However, a droplet of water beads up on the same surface (hydrophobic behavior). To rationalize this dichotomy, we investigate the influence of surface microstructure and the strength of adhesive (surface-water) interactions on surface hydrophobicity. We show that at low relative humidity, the competition between adhesion and the favorable entropy of being in the vapor phase determines the surface coverage. However, at saturation, it is the competition between adhesion and cohesion (water-water interactions) that determines surface hydrophobicity. Because adhesive interactions in talc are strong enough to overcome the unfavorable entropy, water adsorbs strongly on talc surfaces. However, they are too weak to overcome the cohesive interactions, and water thus beads up on talc surfaces. Surprisingly, we find that even talc-like surfaces that are highly adhesive do not fully wet at saturation. Instead, a water droplet forms on top of a strongly adsorbed monolayer of water. Our results imply that the interior of hydrophobic zeolites suspended in water may contain adsorbed water molecules at pressures much lower than the intrusion pressure.

(455f) Structure and Properties of Sticky Colloids in Confined Systems

Deepti Ballal, Chemical and Biomolecular Engineering, Rice University, Houston, TX and Walter G. Chapman, Rice University, Houston, TX

Recently, experimental methods have been developed to synthesize sticky colloidal particles which have sticky/patchy sites on their surface through which they form directional bonds[1]. Interest in these systems is due to the fact that specific target structures can be formed by controlling the intermolecular potential[2]. Research has focused on understanding the phase diagram and the different parameters affecting the phase behavior of the system[3].

Although the bulk properties of sticky colloids are fairly well understood, little is known about the structure of sticky colloids and mixtures in confined systems. Here we study the thermodynamics and structure of sticky colloids in bulk and confined systems using Wertheim's thermodynamic perturbation theory[4] in a density functional theory formalism[5]. The effect of number of sticky sites, temperature and concentration on the structure including the potential of mean force is analyzed.

References

1. Bianchi, E., R. Blaak, and C.N. Likos, 2011. **13**(14): p. 6397-6410.

2. Glotzer, S.C. and M.J. Solomon, Nat Mater, 2007. **6**(7): p. 557-562.

3. Sciortino, F., European Physical Journal B -- Condensed Matter, 2008. **64**(3/4): p. 505-509.

4. Wertheim, M.S., J. Stat. Phys., 1986. **42**(3): p. 459.

5. Segura, C.J., W.G. Chapman, and K.S. Shukla, Mol. Phys., 1997. **90**: p. 759.

(455g) Scaled-Particle Theory and the Thermodynamics of Solvation Over a Range of States

Hank Ashbaugh, Chemical and Biomolecular Engineering, Tulane University, New Orleans, LA

The hydrophobic effect plays a predominant thermodynamic impetus for self-assembly. Theories of the hydrophobic driving force typically parse solvation into the introduction of a hard excluded volume solute into solution followed by the addition of realistic attractive van der Waals interactions. Scaled particle theory describes the process of introducing hard excluded volume cavities into solution over a range of sizes, from the molecular scale to macroscopic scale solutes, relevant to self-assembly processes. In this presentation, we describe ongoing work in our group in applying scaled particle theory to dissolution in aqueous, organic, Lennard-Jones, and Hard Sphere solvents. We also consider dissolution in the Lennard-Jones solvent from state points along the saturation curve well into the compressed liquid, as well as the extension of scaled particle theory to cylindrical solutes. Results from simulations of model solvents are applied to water as a solvent as well in an effort to interpret hydrophobic contributions to solvation over a range of conditions.

(455h) An Eyring-Based Model to Estimate the Dynamic Viscosity of Nanofluids

Ricardo Macias-Salinas, ESIQIE-Departamento de Ingeniería Química, Instituto Politécnico Nacional, Mexico, D.F., Mexico

Nanofluids are a new class of heat transfer fluids obtained by suspending nano-sized particles into a base liquid. It has been shown[1] that an uniform dispersion of low concentrations of nanoparticles (e.g. TiO_2, CuO, Al_2O_3) into a conventional liquid such as water may yield a significant enhancement in the thermal performance of the resulting suspension. Another important transport property that also undergoes a substantial change is the dynamic viscosity of the nanofluid. In fact, the increase in viscosity of nanofluids may occur thus leading to serious operating limitations in terms of increased pressure drops and therefore costly pumping requirements. Consequently, the importance of accurately calculating the effective dynamic viscosity of a nanofluid seems paramount in establishing if its use is indeed advantageous with respect to its pure base liquid. Well-known colloid dispersion theories by Einstein[2,3], Brinkman[4] and Lundgren[5] have been recently used to predict the dynamic viscosity of nanofluids; however, all these theories significantly under-estimate this property. The aim of this work is to present the development of a viscosity model under the basis of the Eyring theory to properly estimate the effective dynamic viscosity of nanoparticle suspensions (aqueous and non-aqueous) within sufficiently wide ranges of temperature, particle volume fraction and particle diameter.

[1] Choi SUS. Developments and applications of non-Newtonian flows, vols. 31/MD–66. New York, NY, FED: ASME Publ.; 99–105, 1995.

[2] Einstein A. Ann Phys; 19:289–306, 1906.

[3] Einstein A. Ann Phys; 34:591–92, 1911.

[4] Brinkman HC. J Chem Phys; 20:571, 1952.

[5] Lundgren T. J Fluid Mech; 51:273–99, 1972.

(456a) Role of Bond Flexibility On the Diffusion and Shear Viscosity of Water

Gabriele Raabe, Institut für Thermodynamik, Technische Universität Braunschweig, Braunschweig, Germany and Richard J. Sadus, Centre for Molecular Simulation, Swinburne University of Technology, Hawthorn, Victoria 3122, Australia

Molecular dynamics simulations for the shear viscosity and self diffusion coefficient of pure water were performed to investigate the effect of including intramolecular degrees of freedom in simple point charge (SPC) models over a wide range of state points. Results are reported for the flexible SPC/Fw model, its rigid SPC counterpart, and the widely used SPC/E model. The simulations covered the liquid phase at 277.15 K and 353.15 K and the supercritical phase at 673.15 K and pressures up to 200 MPa. The flexibility exhibited by the SPC/Fw model results in slowing down of the dynamics. That is, it results in higher shear viscosities and lower diffusion coefficients than can be obtained from the SPC model, and with this in better agreement with experimental data. Significantly, the SPC/Fw model can be used to adequately predict the diffusion coefficients at ambient and supercritical temperatures over a wide range of pressures. In the ambient temperature range, the simulation results for the shear viscosity by SPC/Fw are superior to those of the conventional rigid bond models.

(456b) Cold Water As a Novel Supercritical-Fluid Solvent: A Two-State Model

Daphne A. Fuentevilla, Vincent Holten and Mikhail A. Anisimov, University of Maryland, College Park, MD

Water above its vapor-liquid critical point is commonly used as a supercritical-fluid solvent, in particular in the destructive oxidation of organic waste. However, many scientists believe that deep in the supercooled region, not directly accessible to bulk-water experiments, there exists a critical point of liquid-liquid separation. We have applied the concepts of critical phenomena and thermodynamics of supercritical fluids to explore the possibility of viewing supercooled water as a novel supercritical solvent. Using a phenomenological mean-field "two-state" model, we clarify the nature of the phase separation in a polyamorphic single-component liquid. Considering supercooled water as a mixture of two structures, controlled by thermodynamic equilibrium, we describe water as an athermal "solution" with phase separation driven by the non-ideal entropy of mixing. We then elaborate on this model with respect to the addition of solutes. The addition of a solute generates critical lines emanating from the pure-water liquid-liquid critical point. Because of the anomalously significant change of pressure with respect to temperature along the liquid-liquid transition line, the value of the Krichevskii parameter, which determines the selectivity of supercritical-fluid solvent to fine variations of temperature and pressure, may be very large even if the critical temperature weakly depends on the solute concentration. Physically, it means that even a very small addition of the solute may significantly affect the properties of cold water and aqueous solutions.

(456c) Thermophysical and Structural Properties of Electrolyte Solutions Confined in a Carbon Nanotube

Daniel W. Siderius[1], Jeffrey A. Fagan[2] and Vincent K. Shen[1], (1)Chemical and Biochemical Reference Data Division, National Institute of Standards and Technology, Gaithersburg, MD, (2)Polymers Division, National Institute of Standards and Technology, Gaithersburg, MD

Recently, single-wall carbon nanotubes (SWCNT) have found use as sensing materials where, for example, a particular nanotube chirality will bond to certain strands of DNA and, through further electronic measurements, can identify particular base sequences in a DNA strand [1]. When this type of technology has been sufficiently developed, solutions of SWCNTs could, in principle, be used for lab-on-chip biometric applications for gene identification. A key barrier at present is the satisfactory identification of SWCNTs in solution. It is difficult to separate SWCNTs by conventional analytical techniques and most significant progress has only occurred recently [2]. Furthermore, SWCNTs do not readily solubilize in water and a surfactant is typically used to disperse nanotube clusters.

Recent efforts to develop centrifugation techniques that efficiently and reliably separate SWCNTs by chirality have yielded a puzzling result: preliminary measurements have suggested that the density of the surfactant solution in a SWCNT exceeds the bulk density of water by nontrivial amounts for some SWCNT species [3]. If this effect is genuine, it must be a consequence of cation solvation, since the anionic surfactant itself is unable to enter the SWCNT due to size constraints. It is speculated that the alteration of the fluid density is a response to cation-induced restructuring of water in the nanotube.

We address this issue via grand-canonical transition-matrix Monte Carlo simulation, in an effort to eliminate effects that may obscure a clear measurement of the fluid density in experiment. Using the venerable SPC/E model of water [3], we compute solution density as a function of pressure and electrolyte content and compare our results to existing simulations of pure water in SWCNTs [4]. Our results present a clearer view of the phase behavior of electrolyte solutions in SWCNTs which will aid future efforts in reliably separating solutions of SWCNTs.

[1] Tang et al., *Nano. Lett.*, 6:1632 (2006), Tu et al., *J. Am. Chem. Soc.*, 133:12998 (2011)

[2] Fagan et al., *Langmuir*, 24:13880 (2008), Fagan et al., *ACS Nano*, 5:3943 (2011)

[3] Berendsen et al., *J. Phys. Chem.*, 91:6269 (1987)

[4] Kyakuno et al., *J. Chem. Phys.*, 134:244501 (2011), Pascal et al., *Proc. Natl. Acad. Sci. USA*, 108:11794 (2011)

(456d) Bridge Functionals Representing Higher-Order Terms in the Perturbative Classical Density Functional Theory of Solid-Fluid Transitions

Anurag Verma, Chemical Engineeering, University of Massachusetts, Amherst., Amherst, MA and **David Ford**, Chemical Engineering, University of Massachusetts Amherst, Amherst, MA

Classical density functional theory (DFT) is a useful tool for predicting the solid-fluid phase transition based on a pairwise interaction potential model; it is more accurate than perturbation theory and less expensive than simulation. The key element in DFT is an accurate and computationally feasible expression for the excess Helmholtz free energy as a functional of the density. Weighted DFTs model this excess free energy as a volume integral over an appropriately weighted free energy density; these DFTs are highly accurate for hard spheres but rather cumbersome to apply to more complex potentials. Perturbative DFTs use a formally exact functional expansion of the solid free energy about the liquid reference state, but in practice are typically limited to second order due to difficulty in computing the higher order terms. We have been exploring a closure-based DFT that represents the sum of these higher order terms as a bridge function B, which may be represented as a function in real space B(r) or as a functional of the second order term B[gamma(r)] or the density B[rho(r)]. We present what we have learned about this bridge function for a variety of interaction potentials (hard sphere, power law repulsion, Lennard-Jones) using an inverse DFT approach. There is a remarkable universality in the shape of the bridge function across different potentials, and only a portion of the bridge function seems to be relevant to the thermodynamic properties at coexistence. We also present the formal graph theory expression for B[rho(r)] and comment on the prospects for calculating the values of the graphs.

(456e) Liquid-Vapor Coexistence Properties of Complex Molecules Using Grand-Canonical and Isothermal-Isobaric Temperature Expanded Ensemble Simulations

Kaustubh S. Rane, Sabharish Murali and Jeffrey R. Errington, Chemical and Biological Engineering, University at Buffalo, The State University of New York, Buffalo, NY

The chemical industry is rapidly evolving with the discovery of new tunable molecules and novel applications of conventional compounds. The knowledge of liquid-vapor coexistence of such compounds is necessary for the design of numerous technological applications. In this presentation we describe a Monte Carlo simulation based approach to evaluate the liquid-vapor co-existence properties of complex molecules over a broad range of temperatures. We address the general difficulties encountered in performing Monte Carlo simulations of large, flexible and associating molecules. To improve sampling of configuration space we use a combination of reservoir GCMC and expanded ensemble addition and deletion moves. Hybrid Monte Carlo moves are used to sample the intramolecular degrees of freedom of large molecules and distance-biased moves are used to sample strongly associating species like ionic liquids. We first discuss the direct determination of coexistence properties via grand canonical and isothermal-isobaric simulations. We then discuss the use of temperature expanded ensemble simulations for tracing coexistence curves over a wide range of temperatures. This strategy reduces computational expenses and avoids the need to overcome large free energy barriers between liquid and vapor phases at low temperatures. Results are presented for the liquid-vapor phase coexistence properties of neutral molecules like octane, squalene and pyrene as well as polar molecules like propanol and water and ionic liquids like 1,3-dimethylimidazolium tetrafluoroborate ([C_1mim][BF_4]).

(456f) Molecular Modeling of Vapor-Deposited Ultra-Stable Glasses

Sadanand Singh, Chemical and Biological Engineering, University of Wisconsin - Madison, Madison, WI, Devin B. Averett, Chemical & Biological Engineering, University of Wisconsin - Madison, Madison, WI and Juan De Pablo, Department of Chemical and Biological Engineering, University of Wisconsin-Madison, Madison, WI

Glasses are generally prepared by cooling from the liquid phase, and their properties depend on the cooling rate. Computer simulations of glass forming liquids have shed important insights into the structure and properties of glasses, but the cooling rates employed in past simulations have been many orders of magnitude faster than those used in laboratory experiments. In this work we present the formation of model glasses whose properties correspond to those of a material prepared at a cooling rate that would be more than nineteen orders of magnitude slower than previously possible, thereby creating highly stable glasses whose energy is the lowest ever achieved on a computer. Such glasses conceivably represent the state that would be reached if one could age the materials for thousands of years. Their properties are shown to closely follow those of experimental vapor-deposited stable glasses. Our findings will be discussed in the context of results for binary Lennard-Jones glasses, trehalose glasses, and glycerol glasses. In particular, in the case of vapor-deposited ultra-stable glasses, we find enhanced mechanical properties and slow-down of alpha relaxation processes as compared to ordinary glasses prepared by slow cooling of the liquid.

(456g) Topological Thermodynamic Tools for a Consistent Representation of the Acetic Acid-Isoamyl Alcohol-Isoamyl Acetate-Water Reactive System

Wilmar Osorio Viana[1], Miguel Duque Bernal[1], Jesús David Quintero Arias[1], Izabela Dobrosz Gómez[2], Javier Fontalvo[1] and Miguel Angel Gómez-García[1], (1)Departamento de Ingeniería quimica, Laboratorio de Intensificación de Procesos y Sistemas Híbridos – G.I.A.N.T.: Grupo de Investigación en Aplicación de Nuevas Tecnologías, Universidad Nacional de Colombia Sede Manizales, Manizales, Colombia, (2)Departamento de Física y Química, Laboratorio de Intensificación de Procesos y Sistemas Híbridos – G.I.A.N.T.: Grupo de Investigación en Aplicación de Nuevas Tecnologías, Universidad Nacional de Colombia Sede Manizales, Manizales, Colombia

Fusel oil, an agroindustrial residue from the bioethanol production, has isoamyl alcohol as a major component. The esterification of acetic acid with that alcohol is a promising option to produce isoamyl acetate. However, as required for process analysis, design and simulation, available activity coefficient models for this system are not reliable. In this work, a new set of parameters for NRTL activity coefficient model for the quaternary mixture acetic acid/isoamyl alcohol/isoamyl acetate/water were correlated from experimental vapor-liquid and liquid-liquid equilibrium data. A complete and consistent thermodynamic topological characterization is presented. It includes the Residual Curve Maps, of the non-reactive and the reactive mixtures, and the Phase Stability Test (Wasylkiewicz & Ung, Fluid Phase Equilibria, 175 (2000) 253). The obtained new set of parameters has better predictive capabilities than current available models. A successful liquid-liquid equilibrium phase stability test using the adjusted model was also performed. Those results can be used to perform more reliable conceptual design and simulation for conventional and intensified chemical process for the isoamyl acetate production.

(456h) Molecular Dynamics Study of the Phase Behaviour of H2O-CO2-NaCl and H2O+NaCl Systems

Thomas Lafitte, Yang Liu, Pablo G. Debenedetti and Athanassios Z. Panagiotopoulos, Chemical and Biological Engineering, Princeton University, Princeton, NJ

The vapor-liquid phase behavior for the systems H_2O-CO_2-NaCl and H_2O+NaCl is investigated through interfacial molecular dynamics simulations with GPU acceleration.

While these systems have been studied previously by means of analytical equations of state and molecular simulations, a complete understanding of the phase behaviour near the critical region is still lacking. Molecular simulation represents the most powerful approach to understand the structure of the fluid mixtures as function of temperature and pressure, and provide the basis for developing thermodynamic models. However its application to the phase behaviour of H_2O-CO_2-NaCl and H_2O-NaCl systems at conditions near the critical points remains limited. In the case of Gibbs ensemble simulations, the molecule-transfer move of NaCl turns out to be the bottleneck of the approach, and the traditional assumption is to neglect the NaCl content in the gas phase. This issue can be overcome by using direct interface simulation since no approximation is made regarding the location of the ions in each phase. Nevertheless, its application remains challenging due to the system size that needs to be considered in order to obtain good statistics for the solubility of NaCl in both vapour and liquid phases. We show that the use of molecular dynamics with GPU acceleration allows the study of the systems while maintaining reasonable computing time.

(457a) Effect of Particle Size On Hydrogen Production by Reaction of Liquid Water and Aluminum Powders

Hongqi Nie[1], Mirko Schoenitz[1] and Edward L. Dreizin[2], (1)NJIT, Newark, NJ, (2)Otto H. York Department of Chemical, Biological, and Pharmaceutical Engineering, New Jersey Institute of Technology, Newark, NJ

Abstract

Reaction of aluminum with water is of interest for hydrogen generation for a wide range of applications, from in-situ operated fuel cells to propulsion of underwater vehicles. This reaction is extensively studied and various aluminum-based composite materials and alloys are under development with the aims to accelerate the rate of hydrogen production and enable controlled operation of the hydrogen-generating systems. It would be particularly interesting to achieve the desired rate of hydrogen production at low operating temperatures, using liquid water as opposed to water vapor. However, the understanding of the rate-limiting processes is lacking for the reaction of pure aluminum and liquid water. Recent research by different authors showed, somewhat unexpectedly, that reaction dynamics are different for micron sized spherical aluminum powders with different size distributions. This result can be interpreted considering evolution of aluminum hydroxide layers and possible changes in their integrity and diffusion resistance when the layer thickness becomes comparable to its radius of curvature. In addition, porosity in such layers can develop when they are sufficiently thick; however, such thick layers may not necessarily grow on finer Al particles. This work attempts to understand the effects of particle size on its reaction dynamics in liquid water using experimental data from time-resolved micro-calorimetry. The experimental data tracking energy release caused by the Al/H_2O reaction are processed using the actual measured particle size distributions for three different Al powders. The reaction rate is assumed to be proportional to the particle surface; thus the entire heat flow measured by the calorimeter is partitioned among Al particles based on their size distribution. Evolution of particle sizes and thickness of the grown $Al(OH)_3$ layers are tracked during the micro-calorimetry experiment. The results are processed using a simplified reaction model and diffusion coefficients are found for particles of different sizes and at different reaction stages. Ideally, such coefficients should only depend on temperature, so that their observed dependencies on particle size and reaction progress serve to indicate where the modifications of the simplified reaction model are necessary. Furthermore, reaction rates are compared for individual particles of the same size which are present in aluminum powders with different size distributions. Once the correct reaction model is implemented, the reaction rates for such identical particles must coincide. Additional comparisons and measurements will be presented and discussed, aimed at establishing a corrected reaction model for aluminum and liquid water. Based on the developed reaction mechanism, a composition and morphology of an optimized, Al-based material for hydrogen generation will be proposed.

(457b) Aqueous Solubility, Octanol-Water Partition Coefficients and Carbon Adsorption of 1, 3, 3-Trinitroazetidine (TNAZ)

Veera Boddu, Environmental Processes Branch, US Army Engineer Research and Development Center, Champaign, IL, **Carmen Costales-Nieves**, Environmental Processes Branch, US ARMY ERDC-CERL, Champaign, IL, Rayan Zilner, Environmental Processes Branch, US Army ERDC-CERL, Champaign, IL and Reddy Damavarapu, Energetics and Warheads Division, U.S. Army - Armament Research, Development and Engineering Center, Picatinny Arsenal, NJ

A high energy explosive component, 1, 3, 3-Trinitroazetine (TNAZ), has the potential to replace TNT and other conventional explosives in various formulations. With a melting point of about 373 K, it is steam-castable (melt-castable) and is also a thermally stable material for up to 500 K. Its physical properties are important to estimate the environmental and toxicological consequences of residuals released from manufacturing, packaging, testing and training operations. Accurate values of these properties at ambient temperatures are very essential for estimating its environmental distribution, biotransformation, and the development of potential water treatment processes. In this study, the aqueous solubility (Sw) and octanol-water partition coefficient (Kow) were experimentally measured in the temperature range of 288 – 318 K. The experimental results were compared with literature data and predictions using SPARC model. Other physical properties including vapor pressure and Henry's law constant were also predicted using the SPARC. Preliminary isothermal adsorption of TNAZ on granular activated carbon (GAC) was also conducted in the above temperature range. The results show that the aqueous solubility increases with temperature and varies ~400-1000 mg/L in the temperature range.. Experimental solubility and octanol-water partition coefficients, SPARC predictions, and the results of the GAC adsorption studies will be presented.

(457c) Combustion Dynamics of Aluminum in Turbulent Flows

Amy Corcoran[1], Vern K. Hoffmann[2] and Edward L. Dreizin[1], (1)Otto H. York Department of Chemical, Biological, and Pharmaceutical Engineering, New Jersey Institute of Technology, Newark, NJ, (2)Mechanical Engineering, New Jersey Institute of Technology, Newark, NJ

Combustion dynamics of aluminum in turbulent flows

A.L. Corcoran, V.K. Hoffmann, E.L. Dreizin

New Jersey Institute of Technology

Predictive mechanisms for particle ignition and combustion rates are required in order to develop optimized propellant and energetic formulations using micron-sized metal powders, such as aluminum. Most current descriptions are based on laboratory experiments performed in stationary or laminar combustion configurations. However, turbulent environments exist in most applications and validity of the present descriptions for such environments has not been established. This experimental study is aimed to measure burn times versus size of fine aluminum powders burning in environments with different levels of turbulence. An initial laminar oxygen-acetylene flame is produced, and auxiliary tangential jets of air with adjustable flow rates are used to achieve different controlled levels of turbulence. Aluminum powder is forced into the center of the flame using a flow of nitrogen. A spherical powder with nominal particle size of 6.5 μm is used. Metal mass feed rate is adjusted using a custom-made powder feeder. The flame with burning aluminum is photographed using a Sony DSC-H50 camera. A spinning disk with multiple holes is used to interrupt the photo-exposure with a pre-set frequency. In the obtained images, the streaks of luminous burning particles appear dashed, and the particle burn times are determined by counting the dashes for each streak. Image processing produced statistical information on particle burn times. The particle burn times are correlated with the particle size distributions to obtain the burn times as a function of the particle size for flows with different turbulence levels. The data are compared with similar correlations available in the literature. Additional spectroscopic flame measurements will be performed to further characterize aluminum combustion at different flow conditions.

(457d) Evaluation of Predictive Capabilities for Pyrotechnics Using Different Thermodynamic Software Packages(INVITED)

Jan Puszynski[1], Paul Redner[2], Karl Oyler[3], Christopher Csernica[4], Jesse Sabatini[4], Jay Poret[4], Anthony Shaw[4], Neha Mehta[3], Gartung Cheng[3] and Deepak Kapoor[5], (1)Chemical and Biolgocial Engineering Department, South Dakota School of Mines and Technology, Rapid City, SD, (2)Advanced Materials Branch, US Army, RDECOM-ARDEC, Picatinny Arsenal, NJ, (3)Energetics and Warheads, US Army RDECOM-ARDEC, Picatinny Arsenal, NJ, (4)Pyrotechnics, US Army, RDECOM-ARDEC, Picatinny Arsenal, NJ, (5)Advanced Materials, US Army, RDECOM-ARDEC, Picatinny Arsenal, NJ

Thermodynamic codes play a valuable role in the downselection of the constituents of energetic systems. This is particularly true for thermite and other systems in which the ratio of fuel to oxidizer can be greatly varied and are not contained within the same molecule. If the correct code is used properly, it becomes a very important tool in identifying material systems with key thermophysical properties. Having knowledge of these properties will, in turn, reduce the time and number of experiments necessary to reach the desired endpoint because it will enable one to focus on higher-value materials.

This study explores the use of PEP2100, CHEETAH 6.0 and HSC Chemistry software packages to predict the adiabatic reaction temperature, enthalpy, and product species of various illuminant as well as nanothermite reacting systems. The predicted thermochemical data are compared with experimental observations and correlated to other characteristics that cannot be predicted using computational software, such as light output and color ratio.

(457e) Direct Calculation of 1-Octanol–Water Partition Coefficients From Adaptive Biasing Force Molecular Dynamics Simulations

Navendu Bhatnagar, Chemical Engineering, Wayne State University, Detroit, MI, Ganesh Kamath, Chemistry, University of Missouri-Columbia, Columbia, MO, Isaac Chelst, Chemical Engineering and Materials Science, Wayne State University, Detroit, MI and **Jeffrey J. Potoff**, Chemcial Engineering and Materials Science, Wayne State University, Detroit, MI

The 1–octanol–water partition coefficient log K_{ow} of a solute is a key parameter used in the prediction of a wide variety of complex phenomena such as drug availability and bioaccumulation potential of trace contaminants. In this work, the application of adaptive biasing force molecular dynamics simulations (ABF–MD)[1-4] to the calculation of 1-octanol-water partition coefficients is presented. Unlike traditional methods, such as thermodynamic integration and free energy perturbation, which produce relative free energy values, the ABF-MD method predicts absolute free energies of transfer and eliminates the need for reference solutes. Two approaches are evaluated; the direct transfer of the solute from 1–octanol to water phase, and separate transfers of the solute from the water or 1–octanol phase to vacuum, with both methods yielding statistically indistinguishable results. Simulations performed on a variety of system sizes are used to show that the calculated potential of mean force is remarkably insensitive to the microstructure of the solvent, a finding that suggests relatively small system sizes may be used to determine accurate free energies of transfer. The effect of various simulation parameters on the sampling efficiency is discussed.

Calculations are performed on a variety of systems, including n-alkanes, alcohols, chemical warfare agents, energetic materials and ionic liquids. The effect of wet vs. dry octanol and choice of water model is assessed for each system. The predictions of ABF-MD are found to be in excellent agreement with prior calculations performed with thermodynamic integration[5] or Gibbs ensemble Monte Carlo[6].

1. E. Darve and A. Pohorille, J Chem Phys **115** (20), 9169-9183 (2001).

2. E. Darve, M. A. Wilson and A. Pohorille, Mol Simulat **28** (1-2), 113-144 (2002).

3. J. Henin and C. Chipot, J Chem Phys **121** (7), 2904-2914 (2004).

4. D. Rodriguez-Gomez, E. Darve and A. Pohorille, J Chem Phys **120** (8), 3563-3578 (2004).

5. N. M. Garrido, A. J. Queimada, M. Jorge, E. A. Macedo and I. G. Economou, J Chem Theory Comput **5** (9), 2436-2446 (2009).

6. J. I. Siepmann and B. Chen, Journal of Physical Chemistry B **110** (8), 3555-3563 (2006).

(470a) The Motility of Ameboid Cells of the Immune System: Traction Stresses and Directional Motion

Daniel A. Hammer, Chemical and Biomolecular Engineering, University of Pennsylvania, Philadelphia, PA

Cells of the immune system must traffic throughout the body to identify and eliminate foreign objects, transmit molecular information to other effectors cells, and to maintain immunological homeostasis. The trafficking of immune cells depends on directed cell motility in response to chemokines, where differential occupancy of receptors leads to directed cell motion. Because these cells crawl quickly (several microns/minute), they exert small forces, and measuring traction stresses is a challenge. In this talk, we describe methods to measure the traction stresses during directed cell motility of two critical immune effector cells that are derived from the same progenitor - neutrophils and dendritic cells - under imposition of a gradient of chemokine. For neutrophils, traction measurements were performed using beads embedded in gels, and the traction field was deconvolved from the optical flow (strain field) using two dimensional elasticity theory worked out by Micah Dembo (Boston University). In dendritic cells, we used molecularly stamped polymer micropost arrays, in collaboration with Chris Chen's laboratory at Penn, in which the known elasticity and deflection of the microposts could be used to quantify forces. We found that neutrophils exert squeezing contractile forces in the rear, whereas dendritic cells pull through filopodial extension and contraction in the front. In maximal chemotaxis, the average force exerted by dendritic cells is about a factor of five smaller than that of a neutrophil, and we estimate that a single filopod uses 400 myosin-IIB motors to pull a cell forward. There was a direct correlation between directional motiuon and the magnitude of the root mean squared force required exrted on the substrate. Our spatio-temoral maps of mechanochemical stresses during dendritic cell motility paves the way for detailed modeling of directed cell motion using physiochemical and hydrodynamic models that incorporate contractile and adhesive stresses.

(470b) A Nonlocal Theory to Predict the Concentration Distribution of Red Blood Cells Flowing in a Microchannel

Vivek Narsimhan, Chemical Engineering, Stanford University, Stanford, CA, Hong Zhao, Mechanical Engineering, Stanford University, Stanford, CA and Eric S.G. Shaqfeh, Chemical Engineering and Mechanical Engineering, Stanford University, Stanford, CA

When red blood cells move throughout our microcirculation, they do not flow with a uniform distribution across the blood vessel, but instead leave a significant layer of clarified fluid near the vessel wall, a phenomenon famously coined the "Fahraeus-Lindqvist effect." The size and shape of this clarified layer plays an important role in reducing the blood's effective viscosity in the microcirculation, reducing bleeding times by concentrating platelets near vessel walls, and controlling adsorption of nanoparticles in the blood stream. We develop a kinetic model to predict the concentration distribution of deformable particles, and hence the Fahraeus-Lindqvist layer, in two different wall-bounded flows: Couette flow, and Poiseuelle flow. This theory balances the hydrodynamic lift on deformable particles (in this case, blood cells or vesicles) with the flux due to particle collisions, which we represent non-locally using a Boltzmann-type approximation. We determine the scaling of the Fahraeus-Lindqvist layer with wall shear rate, channel height, and volume fraction (aka, Hematocrit). We also predict a significant layering of particles near the channel wall, consistent with many recent simulations of suspensions of deformable particles. Our theory is not self-sufficient, as it requires data (simulation or experimental) on the details of a binary collision process and the hydrodynamic lift of a single particle in a wall-bounded shear flow. Nevertheless, it represents a significant improvement in terms of time savings and predictive power over current large-scale numerical simulations of suspension flows. Our theory agrees reasonably well with recent boundary integral simulations of suspensions of deformable particles (red blood cells and vesicles).

(470c) Intra-Thrombus Fluid and Molecular Transport Based On in-Vivo Imaging with Single Platelet Resolution

Roman S. Voronov[1], Timothy J. Stalker[2], Lawrence F. Brass[2] and Scott L. Diamond[1], (1)Chemical and Biomolecular Engineering, University of Pennsylvania, Philadelphia, PA, (2)School of Medicine, University of Pennsylvania, Philadelphia, PA

The assembly of platelet deposits and fibrin polymerization results in over 1 million heart attacks and strokes each year in the US. Conversely, deficiencies in these processes result in bleeding risks that confront surgeons on a regular basis. Motivated by recent ability of intravital microscopy experiments to image thrombus structure with a single platelet resolution, we set out to study intra-thrombus microenvironment via an approach that combines in-vivo laser injury experiments in mouse cremaster muscle arterioles, fluorescent confocal imaging with single platelet resolution, Optical Doppler velocimetry and supercomputer Lattice Boltzmann method fluid dynamics simulations.

In this presentation we report fluid velocities, surface stresses and Darcy permeabilities that were calculated locally to the thrombus structure. Furthermore, in order to explore effects of molecular size on transport through the thrombus, movement of coagulation molecules corresponding to small (Ca), medium (ADP) and large (Factor X) molecular weights was studied using passive scalar tracking coupled with Brownian motion. In one mode of release the molecules were released at the injury site below the thrombus. Their motion was tracked under the in-vivo blood flow conditions and without flow, for comparison. In the second mode of release we studied how molecules from the bulk lumen diffused to the injury site. The methodology and findings reported herein enhance the understanding of fundamental transport processes that occur within the thrombus, and could have implications for therapeutic strategies that target drug delivery to local areas within the thrombus structure.

(470d) Exploring the Role of Flagella in Bacterial Swarming

Douglas B. Weibel, Hannah H. Tuson and Matthew F. Copeland, University of Wisconsin-Madison, Madison, WI

Proteus mirabilis is an opportunistic pathogen that causes urinary tract infections and the biofouling of urinary catheters. A fundamental unanswered question has centered on the mechanism by which this bacterium invades and colonizes the urinary epithelium. Swarming may provide a clue to these infections.

During swarming on nutrient agar in the lab, *P. mirabilis* becomes morphologically differentiated and is characterized by long (20-100 um long), multinucleate cells that have a remarkably high dense coating of flagella. Back-of-the-envelope calculations of the energy balance in these cells suggests that the assembly and actuation of flagella may account for as much as ~50% of the total energy of the cell. This phenotype is clearly important or it would have been lost long ago.

We explored the physiological and behavior importance of the density of flagella on swarming *P. mirabilis* cells. By investigating the swimming velocity of different *P. mirabilis* cell morphologies—each with a different combination of cell length and flagellar density—we discovered that swarmer cells have a larger swimming velocity in solutions with a microviscosity ranging from 10-1000 cP, compared to the other morphologies that we tested. Swarmer cells were motile in fluids with a viscosity that approached ~9000 cP; the motility of all other cells types we studied was inhibited at this viscosity. Overexpression of $FlhD_4C_2$—the transcription factor controlling the flagellar gene cascade—increased the swimming velocity of vegetative cells in all of the viscous fluids we tested. Furthermore, $FlhD_4C_2$ overexpression rescued vegetative cell motility at ~9000 cP. Both immunofluorescence microscopy and quantitative Western blots demonstrate a correlation between the density of flagella and cell velocity and suggests a threshold density for increasing the velocity of cells in high viscosity fluids. From these data and the reported values for the number of flagella associated with other swarming species, we speculate that an increase in flagellar density may be a universal requirement for swarming motility. Our hypothesis is that the high density of flagella on *P. mirabilis* may be responsible for cell motility in high viscosity environments, including the mucus secreted by the urinary epithelium, thereby providing a route to infection.

Vertical Migration of Small Organisms in a Stratified Fluid

Amin Doostmohammadi and Arezoo M. Ardekani
University of Notre Dame, Notre Dame, IN, USA

The appearance of vertical density gradients due to variation in temperature or salinity is ubiquitously observed in aquatic environments. Important environmental processes are correlated with stratification and the presence of stratified layers can have a significant impact on the gravitational settling of particles and vertical migration of swimming organisms [1-3]. We have performed direct numerical simulation of the migration of small organisms and settling of particles at pycnoclines. Our recent results reveal the full nonlinear effects of stratification on the bio-locomotion and settling dynamics, a pertinent step in explaining accumulation of organisms and particles and intense biological activity associated with pycnoclines.

We show that for an archetypal swimmer model, squirmer, stratification leads to significant change in the flow signature and markedly affects the nutrient uptake rates and energy expenditure at pycnoclines [3-4]. The buoyancy effects are more pronounced for a salt stratified fluid compared to a temperature stratified fluid due to the larger deflection of isopycnals. In addition, we unveil the effects of stratification on the time-dependent behavior of settling particles. Our results show that stratification suppresses the vertical descent of a non-neutrally buoyant particle settling in a linearly stratified fluid as shown in figure 1.

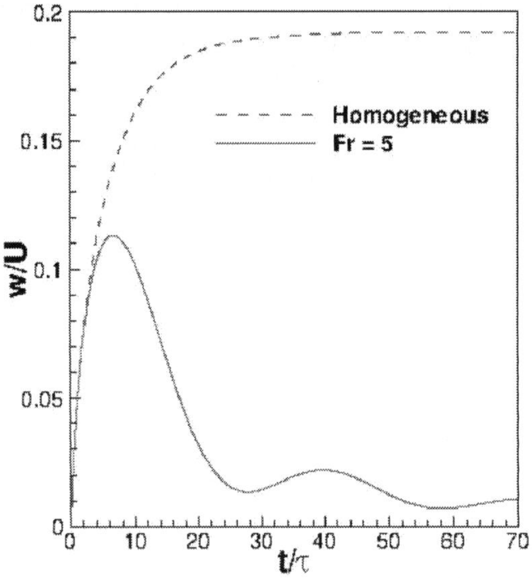

Figure 1. Time evolution of settling velocity of a particle of size *d*, settling due to the gravitational acceleration, *g*. $U=\sqrt{gd}$ and $\tau=\sqrt{d/g}$ are the characteristic velocity and the characteristic time, respectively. The ratio of the particle density to the reference density of the fluid is *1.14*, the Reynolds number is *40*, the Prandtl number is *700* and the Froude number is defined as $Fr=U/Nd$, where *N* is the Brunt-Väisälä frequency.

References

[1] C.R. Torres, H. Hanazaki, J. Ochoa, J. Castillo, M. Van Woert "Flow past a sphere moving vertically in a stratified diffusive fluid" Journal of Fluid Mechanics, 417:211-236, 2000.

[2] A.M. Ardekani, R. Stocker "Stratlets: Low Reynolds number point-force solutions in a stratified

fluid," Physical Review Letters, 105:084502, 2010.

[3] A. Doostmohammadi, R. Stocker, A.M. Ardekani "Low-Reynolds number swimming at pycnoclines," *Proceedings of National Academy of Sciences*, 109:3856-3861, 2012.

[4] A.Doostmohammadi, A.M. Ardekani "Vertical migration of the small organisms in a stratified fluid," *Proceedings of ASME 2012 Fluids Engineering Summer Meeting*, July 8-12, 2012, Rio Grande, Puerto Rico.

(470f) A Micro-Swimmer Model with Dipolar, Quadrupolar and Rotlet Dipolar Flows

Ronald G. Larson, University of Michigan, Ann Arbor, MI and Nobuhiko Watari, Macromolecular Science and Engineering, University of Michigan, Ann Arbor, MI

To study the collective swimming of a peritrichous bacterium such as *Escherichia coli*, we model a micro-swimmer using a Low-Order Multipole Swimming (LOMS) model. This model consists of an ellipsoidal particle that migrates at a constant velocity along its long axis and generates a flow of a superposition of a Stokeslet dipole, a Stokeslet quadrupole and a Rotlet dipole. The magnitude of these three flow components are found by modeling a single *E.coli* cell using a bead-spring model containing 120 beads. The bead-spring model accounts for: 1), the hydrodynamic and the mechanical interactions among the cell body and multiple flagella; 2), the reversal of the rotation of a flagellum in a tumble; and 3), the associated polymorphic transformations of the flagellum. Because a flexible hook connects the cell body and each flagellum, the flagella can take independent orientations with respect to the cell body. This simulation reproduces the experimentally observed behaviors of E. coli, including the steady clockwise swimming near a wall. The reduction of this 120-bead model to a simple LOMS is therefore the first multi-scale simulation study on micro- swimmers. On the contrary to the previous study using dipolar flow swimmer, it is found that the hydrodynamic interaction between micro-swimmers gives a significant contribution to the cell-cell scattering, or rotational diffusion of swimmers, mainly because of the higher-order flows, i.e. the Stokeslet quadrupole and Rotlet dipole.

(470g) RNA/Protein Droplet Assembly by Intracellular Phase Transitions

Clifford Brangwynne, Chemical and Biological Engineering, Princeton University, Princeton, NJ

Ribonucleoprotein (RNP) bodies are non membrane bound macromolecular assemblies that form from the dynamic interactions of RNA and protein. They are crucial for cell growth, embryonic development, and many other fundamental biological processes. Among dozens of examples include germ granules and processing bodies in the cytoplasm, and Cajal bodies and nucleoli in the nucleus. We show that these structures behave as liquid droplets of RNA and protein. RNP droplets assemble from components in the surrounding cytoplasm by a de-mixing phase transition, whose set-point is biologically controlled in space and time. For several types of RNP bodies, this phase transition appears to be controlled during different developmental stages to accommodate growth-dependent changes in RNA metabolism. To understand the biophysical properties of these structures, we use a micro-rheological approach to characterize their viscoelastic moduli, which appear to depend on their intrinsic biological activity. These findings shed light on how RNP bodies can self-assemble from soluble RNA/protein components, and simultaneously function as micro-reactors for these components, speeding up rate-limiting steps in RNA metabolism by localizing high concentrations of reactants.

(470h) Bridging Length Scales: Hierarchical Coarse-Graining of Elastic Biopolymer Models

Elena F. Koslover, Biophysics Program, Stanford University, Stanford, CA and Andrew J. Spakowitz, Chemical Engineering, Biophysics, Stanford University, Stanford, CA

One of the key challenges in modeling biomolecular systems lies in bridging the gap between length scales at which we have a concrete understanding of the fundamental molecular mechanics and the much larger length scales relevant for biological function. A classic example is the DNA polymer, whose molecular structure gives rise to an elastic rigidity at the nanometer length scale
whereas contour lengths up to several meters must be packaged, processed, and accessed by the cell. A meaningful approach to coarse-graining polymer models is essential for modeling systems of biological relevance. We present here a highly general, straight-forward procedure for generating coarse-grained polymer models. We demonstrate how a polymer may be modeled as a stretchable, shearable wormlike chain on increasingly large length scales while keeping long chain statistics intact.

As an example application for this multiscale modeling approach, we consider the effect of periodically bound DNA-bending proteins on the shape and behavior of the DNA chain. Specifically we study the physical properties of DNA-nucleosome arrays, which form the initial step in the packaging of the eukaryotic genome into a hierarchical chromatin structure. We employ our coarse-graining procedure to determine looping probabilities for nucleosome arrays on different length scales, examining the effects of linker length and local geometry on the distribution of loops. These looping calculations are then leveraged to investigate the thermodynamics of chromosome condensation via the binding of bridging proteins. In addition, we look at the role of DNA elasticity and nucleosome geometry in the structure of compact fibers. Our multiscale models aim to address how the complex hierarchy of chromatin packing arises from the basic polymer properties of DNA and interacting proteins.

The novel polymer coarse-graining procedure presented here is generally applicable and can be used to systematically zoom out from a detailed polymer model based on localized mechanics to a simple effective model that can be used to calculate large length-scale behavior.

(470i) Mobility of a DNA Molecule Confined in a Nanochannel

Douglas R. Tree[1], Yanwei Wang[2] and **Kevin D. Dorfman**[1], (1)Chemical Engineering and Materials Science, University of Minnesota, Minneapolis, MN, (2)Polymer Science and Engineering, Soochow University, Suzhou, China

Nanochannels are an ideal platform for studying the basic physics of confined polymers, using DNA as the model polymer. In the case of weak confinement, de Gennes predicted that the hydrodynamic mobility of a confined flexible chain will scale inversely with its extension due to the hydrodynamic interactions inside the blobs of the confined chain. In this presentation, we will show that the de Gennes theory breaks down for DNA in a high ionic strength buffer due to the large ratio between the persistence length (53 nm) and the effective width (4.6 nm) of DNA. We have computed the hydrodynamic mobility using Monte Carlo sampling of the Kirkwood mobility and a numerical solution for the Green's function of the confined chain. We show that there is a broad plateau in which the hydrodynamic mobility is independent of the fractional extension of the chain. Moreover, we directly connect the existence of this plateau to a transition regime for the extension of a semiflexible chain where the fractional extension scales inversely with the channel width. The width of the transition regime depends on the ratio of the persistence length to the effective width, and thus vanishes for flexible chains. For DNA, the latter ratio is so large that this Rouse-like regime is predicted to persist for the entire range of experimentally observed extensions of DNA in a nanochannel.

(473a) Incorporation of Amphiphilic Peptides to Lipid/Polymer Nanobubbles

Joseph V. Badami, Chemical Engineering, The City College of New York, New York, NY and Raymond Tu, Department of Chemical Engineering, The City College of City University of New York, New York, NY

Microbubble systems have become an attractive avenue for theranostic drug delivery, as they provide a means to image, target, and administer therapeutics. While successful in applications such as echocardiography, extending the potential usage of these materials to treat a greater diversity of conditions will require a reduction in bubble size to allow for delivery and permeation through tissue. Simultaneously, the inclusion of additional therapeutic/targeting domains requires a greater understanding of the molecular interactions that take place amongst shell constituents to move towards a more robust and multifunctional interface. In this work, we investigate the introduction of rationally designed, amphiphilic, alpha-helical peptides of differing charge distribution to polymerically stabilized, lipid nanobubbles as a means to control bubble size and provide site-specific targeting capabilities. We explore the impact that these combinations of molecules have on both the curvature and stability of such bubble systems. Our experiments aim to quantify the fundamental effects that the peptides have on interfacial phase behavior, in-plane structure and elasticity.

(473b) Membrane-Mediated Interactions Between Carbon Nanotubes Embedded in a Lipid Bilayer

Young-Min Ban, ACI Unit Process Development Group, Samsung Electro-Mechanics, Gyeonggi-Do, South Korea and **Dmitry I. Kopelevich**, Department of Chemical Engineering, University of Florida, Gainesville, FL

Understanding behavior of nanoparticles embedded in lipid membranes is important for development of novel biomedical devices and assessment of potentially harmful effects of nanomaterials on living cells. The current work is focused on theoretical and computational investigation of interactions between nanoparticles embedded into a lipid bilayer. In this talk we present results of our coarse-grained molecular dynamics (MD) simulations of carbon nanotubes (CNTs) in a dipalmitoylphosphatidylcholine (DPPC) lipid bilayer. There are no long-range interactions between the considered nanoparticles in solution. However, embedding them into a bilayer may give rise to long-range membrane-mediated interactions. Indeed, our simulations indicate presence of long-range interactions between CNTs embedded in a DPPC bilayer.

One possible cause of these interactions is associated with a stress induced by the nanoparticles in the membrane interior. In order to minimize this stress, the nanoparticles would either repel or attract each other, depending on the specifics of their interactions with the surrounding lipids. However, our simulations rule this mechanism out. We observe that perturbations of the membrane structure and stress profile induced by a single CNT residing in the hydrophobic core of the membrane are very short-ranged.

Therefore, we investigate an alternative mechanism of the membrane-mediated interactions between CNTs. This alternative mechanism is associated with effect of the CNTs on the membrane fluctuations. CNTs embedded in a membrane impose a significant constraint on the membrane undulations. This, in turn, creates a force acting between the nanotubes. This effect was predicted theoretically almost two decades ago [Golestanian et al., Phys. Rev. E, **54**,6725 (1996)]. However, to the best of our knowledge, it has not been observed in MD simulations or experiments for nanoparticles embedded into a lipid membrane. In the current talk, we present computational evidence for this effect and perform a quantitative comparison between the theory and the MD simulations.

(473c) Effects of Myelin Basic Protein (MBP) On the Interaction Forces and the Rheology Between Myelin Bilayers Using Surface Forces Apparatus (SFA)

Dong woog Lee[1], Xavier Banquy[1], Kai Kristiansen[1], Joan Boggs[2] and Jacob Israelachvili[1], (1)Chemical Engineering, University of California, Santa Barbara, Santa Barbara, CA, (2)The Hospital for Sick Children, Molecular Structure and Function, Toronto, Canada

Myelin Basic Protein (MBP) plays an important role in myelin structure, strongly holding the cytoplasmic membrane sides together mainly by electrostatic and hydrophobic interactions [1]. MBP is also known to affect lipid domains, increasing their size by lowering the line tension [2]. In this work, the rheological properties of MBPs between model myelin lipid bilayers and the effect of MBP on interaction forces are studied using the Surface Force Apparatus (SFA) technique. Results show that above a certain critical bulk concentration, C_{crit}, MBP adsorbs abruptly on lipid membranes, forming a 3 nm - thick layer in which the viscosity is 2 to 3 times higher than the bulk MBP solution. Also, above this C_{crit}, normal myelin membranes show higher adhesion and less repulsion compared to EAE (a model for multiple sclerosis) myelin lipid bilayers while below C_{crit}, there was no evidence of MBP adsorption and no significant difference between the force profiles of normal (healthy) and EAE (diseased) myelin bilayers. These results suggest that MBP adsorbs "cooperatively" in clusters/domains to myelin bilayers, and provide new insights in to how lipid-protein domains affect inter-membrane interactions.

References:

[1] Min, Y., et al., Interaction forces and adhesion of supported myelin lipid bilayers modulated by myelin basic protein. *PNAS*, 2009. **106**(9): p. 3154-3159.

[2] Y. Min, et al., Critical and Off-Critical Mixing-Demixing Transitions in Model Extracellular and Cytoplasmic Myelin Lipid Monolayers. *Biophys. J.* (2011) **100** 1490-1498.

(473d) Stochastic Studies of Influenza Virus Fusion to Biomimetic Membranes

Deirdre A. Costello, Chemical Engineering, Cornell University, Ithaca, NY and Susan Daniel, Chemical and Biomolecular Engineering, Cornell University, Ithaca, NY

Influenza viruses are membrane-enveloped, negative-strand RNA viruses that employ membrane fusion to release its RNA into host cells and initiate replication. Influenza enters the cell via clathrin-mediated endocytosis. Fusion of the viral and endosomal membrane is facilitated by the conformational change of the viral protein hemagglutinin (HA) at low pH. Traditional bulk fusion assays rely on the fusion fluorescently labeled viruses to synthetic lipid vesicles to obtain kinetic data about the fusion pathway. However, fusion is a stochastic event and only ensemble averages of fusion kinetics are obtained from bulk measurements. To obtain more detail, we use fluorescence dequenching and total internal reflection microscopy (TIRFM) to track and quantify fusion of individual viruses to supported lipid membranes. Imaging individual virus fusion events enables hemifusion kinetics to be differentiated from pore formation kinetics. Our assays are carried out in high-throughput microfluidic devices where fusion is initiated by reducing the pH in the device. Hemifusion lag times are determined for each individual fusing virus and from this data we can determine hemifusion rate constants (k_H) and the number of steps in the hemifusion pathway (N). By using a distinct two-fluorophore labeling approach, we can also measure the time to pore formation of individual viruses following the hemifusion event. In this study we compare the fusion characteristics of strains of influenza H3N2 that have undergone varying levels of laboratory adaptations. Adaptations occur when strains of influenza, such as the commonly studied X:31, undergo multiple passages in eggs to improve virus yield. Multiple passages in eggs result in strains that exhibit significant morphological changes and altered receptor-specificity compared to circulating strains. Our result shows that some less adapted strains exhibit a significant shift in the optimal pH for endosomal fusion compared to the highly adapted X:31.

(473e) Polysaccharides Pattern the Phase Behavior of Lipid Membranes

Anand B. Subramaniam, Chemistry and Chemical Biology, Harvard University, Cambridge, MA; School of Engineering and Applied Sciences, Harvard University, Cambridge, MA, Guido Guidotti, Department of Molecular and Cellular Biology, Harvard University, Cambridge, MA, Vinothan N. Manoharan, School of Engineering and Applied Sciences and Department of Physics, Harvard University, Cambridge, MA and Howard A. Stone, Mechanical and Aerospace Engineering, Princeton University, Princeton, NJ

Hydrated networks of glycans (polysaccharides) in the form of cell walls, periplasms or gel-like matrices are ubiquitously present adjacent to cellular plasma membranes. Yet despite their abundance, the function of glycans in the extracellular milieu is largely unknown. Here we show that glycans as part of hydrated networks can interact directly with lipids in model membranes. Because of this interaction, the spatial configuration of the glycans controls the phase behaviour of multiphase lipid membranes: inhomogeneous glycan networks stabilize large lipid domains at the characteristic length scale of the network, while homogeneous glycan networks suppress macroscopic lipid phase separation. We find that glycans pattern lipid phase separation due to a preference for ordered lipid phases, and that the patterning is thermally reversible, which is an indication that the effect is not kinetic but rather is thermodynamic. The discovery here that glycan networks can control the length scale of domains in adjacent lipid membranes necessarily has implications for plasma membrane mediated transport processes and potentially rationalizes some long-standing puzzling observations that differentiate the behaviour of native plasma membranes and model membranes. Our results may also indicate a more intimate coupling between cellular lipidomes and glycomes than realized currently.

(473f) Polyoxometalate (POM) Nanocluster-Induced Phase Transition in Phospholipid Biomembranes

Benxin Jing and **Y. Elaine Zhu**, Department of Chemical and Biomolecular Engineering, University of Notre Dame, Notre Dame, IN

Polyoxometalate (POM) nanoclusters that are transition metal oxygen clusters with well defined atomic coordination structures have recently emerged as new and functional nanocolloidal materials used as catalysts, anti-cancer medicines, and building blocks for novel functional materials. However, their implications to human health and environment remain poorly investigated. In this work, we examine the interaction of highly charged anionic and neutral POM nanocluters with phospholipid bilayers as a model cell membrane system. It is observed that upon the adsorption of highly anionic POMs of 3-5 nm in diameter, lipid dynamics is significantly suppressed and lipid bilayers are disrupted with resultant pore and budding-like structural formation. Direct calorimetric experiment of anionic POM interaction with lipid bilayers of varied lipid compositions confirms the POM-induced fluid-to-gel phase transition in lipid bilayers, due to strong electrostatic interaction between POM nanocluster and lipid head groups. In contrast, for neutral or divalent Zn-S nanocluster of 1-2 nm in diameter, it is revealed that the uptake and insertion of Zn-S nanoclusters into lipid bilayer can induce dynamic supramolecular self-assembly. Additionally, the assembly and disassembly process on lipid bilayer in distinct gel and fluid phases at varied temperature is investigated, suggesting the interfacial assembly critically relies on the phase state of lipid bilayers.

(484a) Accurate and Robust Classical Density Functional Theory for the Adsorption of Fluids Into Nanoporous Materials

Stepan Hlushak[1], Clare McCabe[1] and Peter T. Cummings[1,2], (1)Chemical and Biomolecular Engineering, Vanderbilt University, Nashville, TN, (2)Center of Nanophase Materials Sciences, Oak Ridge National Laboratory, Oak Ridge, TN

Classical density functional theory for square-well and multi-yukawa simple and chain fluids was implemented in two dimensions and used to describe supercritical adsorption of the model fluids into cylindrical pores of various size and wall-fluid interaction. The implementation uses Fast Fourier Transform (FFT) to integrate the convolution integrals appearing during the calculation of the nonlocal terms, which significantly increases computing performance. Pair attractive interactions are taken into account using the Fourier transform of the direct correlation function (DCF), calculated from the first-order mean spherical approximation (FMSA) theory [1,2].

The theory was tested against Monte-Carlo simulation data for adsorption of the model fluids into narrow cylindrical pores. Because of the confinement effects in the pore, the density of the absorbed fluid is significantly different from the bulk fluid, which constitutes a challenge for the usual theories based on the expansion of the thermodynamics around the bulk fluid density. The overcome the difficulty we proposed and tested reference fluid density functional approach of Gillespie et al [3] and an approach with a functional that mimics energy route thermodynamics. The approaches tend to improve predictions of the density profiles and adsorption isotherms for almost all the pores studied.

The case of the adsorption of chain fluids is more interesting. Confinement limits number of possible conformations of the chains, which are consequently depleted near the walls and repelled out of the narrow pores. Developed density functional theory manages to capture these depletion effects, but because of accumulations of different approximations adopted in the theory, the agreement with simulation is more qualitative than quantitative. This seems to be mainly due to inaccurate inhomogeneous contact value, used to account for chain connectivity. Inclusion of the contact value greatly improves predictions for the adsorption isotherms, while the density profiles are improved only at low densities.

(484b) Aqueous Carbon Dioxide Environments Under Silica Nano Confinement

Ariel A. Chialvo, Chemical Science Division, Oak Ridge National Laboratory, Oak Ridge, TN, Lukas Vlcek, Chemical Sciences Division, Oak Ridge National Laboratory, Oak Ridge, TN and David R. Cole, School of Earth Science , Ohio State University, Columbus, OH

The CO_2 geological capture and sequestration have been considered potential approaches toward the mitigation of the greenhouse effects of CO_2 release into the atmosphere, a process that relies on the low hydraulic permeability of caprocks resulting from fluid-substrate interfacial and confinement behavior. Fluids at interfaces and under confinement exhibit microstructural, dynamical, and thermophysical behavior quite different from their bulk counterparts, a feature that also highlights the inherent inability of current modeling approaches to describe properly the fluid-caprock interfacial mechanisms underlying the geological CO_2 sequestration.

Experimental and modeling studies most often focus on the interactions of either water-rich CO_2 or pure CO_2 environments with representative caprock substrates, and the ensuing modeling is based on that assumption, *i.e.*, the impact of the CO_2-contaminants are rarely addressed, despite the fact that CO_2 always carries small quantities of 'contaminants' including SO_2, H_2S, NO_x whose impact in terms of rock interactions cannot be ignored.

The overarching goal of this investigation is to address key fundamental issues regarding the behavior at aqueous-CO_2 solutions at the fluid-silica interface, including (a) how the degree of surface polarity and/or surface mismatch affect/s the interfacial structure, (b) how the overlapping of interfacial structures affects the confined fluid composition, and (c) how contaminants might affect the preferential adsorption and composition of the interfacial layers. For that purpose we carry out a molecular-based study of the microstructural and dynamical behavior of CO_2-aqueous solutions at silica surfaces under extreme confinement at conditions relevant to geologic capture and sequestration of carbon dioxide, involving either water-rich CO_2 or CO_2-rich phases. This effort comprises (globally) isobaric-isothermal and (locally) grand-canonical molecular dynamics simulations of aqueous-CO_2 systems whose unlike-pair interactions are described by a recently optimized force-field parameterization that provides an accurate and simultaneous representation of the water-rich CO_2 and CO_2-rich phases at LLE conditions [1]. Moreover, we employ three types of slit-pore configurations to represent two extreme cases of surface polarity, and a mismatched pair of plates, to interrogate the fluid behavior at and confined between heterogeneous surfaces. Based on this study we illustrate how the interplay between these types of fluid-surface interactions and extreme fluid confinement, *i.e.*, strong overlapping of interfacial structures, can induce a drying out of the pore environment whose immediate consequence is a significant enhancement of the pore CO_2 concentration relative to that of the corresponding bulk environment [2]. This behavior highlights some often overlooked implications including (a) the pore fluid environments cannot be represented by that of a bulk counterpart at the prevailing pore state conditions when either interpreting or modeling the process at a macroscopic level, (b) the chemical processes occurring in nano-pore aqueous environments strongly depend on the nature of the confining mineral surfaces, (c) the fluid environment in contact with the mineral surface behaves considerably different from either that a few molecular diameters away from the mineral surface or that representing the corresponding bulk, and (d) the competitive co-adsorption of CO_2 contaminants at the mineral surfaces can radically modify the interfacial equilibrium and dynamical behavior, and consequently, the mineral surface chemical stability.

[1] Vlcek, L.; Chialvo, A. A.; Cole, D. R. *Journal of Physical Chemistry B* **2011**, *115*, 8775.

[2] Chialvo, A. A.; Vlcek, L.; Cole, D. R. *Journal of Physical Chemistry C* **2012**, *Submitted.*

(484c) Molecular Simulation Study of the Dynamics of Ionic Liquid (IL) 1-n-Butyl-3-Methyl-Imidazolium-Bis(trifluoromethanesulfonyl)Imide [Bmim][NTf2] Confined in Graphite Slit Pore

Samir Budhathoki, Jindal Shah and Edward J. Maginn, Chemical & Biomolecular Engineering, University of Notre Dame, Notre Dame, IN

Ionic liquids (ILs) are salts with melting temperatures below 100 °C. They are usually made up of unsymmetrical organic cation and charge delocalized anion. ILs exhibit properties such as negligible vapor pressure, excellent solvation ability, high thermal and electrochemical stability, and good ionic conductivity, which make them viable for broad range of applications as in solar cells, gas separations, and catalysis. The viscosity of bulk ILs are, however, usually high which presents challenges when ILs are used in such applications. The high viscosity is generally attributed to the strong intermolecular forces in ILs. We explore, in this study, if nanoconfinement of ILs leads to decrease in viscosity. To this end, a molecular simulation study is performed to examine the effect confinement has on the dynamics of the ionic liquid (IL) 1-n-butyl-3-methylimidazolium bis(trifluoromethanesulfonyl)imide [bmim][NTf2] at different temperatures. Self-diffusion coefficients and activation energies are reported, and compared with results obtained for the bulk IL. We also analyze the dynamics of the IL as a function of the distance of ions from the pore wall. Additionally, the velocity auto correlation function, rotational correlation function and van Hove correlation functions are presented in order to provide a detailed perspective on the dynamics of the system.

(484d) Statistical Mechanics of Fluids At Interfaces and Under Confinement Via Mayer-Sampling Monte Carlo Simulation

Jung Ho Yang, Andrew J. Schultz, Jeffrey R. Errington and David A. Kofke, Chemical and Biological Engineering, University at Buffalo, The State University of New York, Buffalo, NY

The physical properties of small systems are fundamental importance in many industrial fields, such as membrane separation, oil recovery and colloidal stability. Engineering such processes requires molecular understanding of interfacial or confined systems. The study of surface phenomena is far less well developed compared to a vast knowledge of the phase behavior of infinitely large systems. When a system volume is reduced to micro/nano-size level, the physical properties are no longer size independent. Experiments and molecular simulations show that a confined fluid behaves differently from the corresponding bulk fluid. A finite system of interacting molecules itself makes the interaction on a given molecule different. In the present work, we study the effect of surface fluid interaction on the thermodynamic properties of simple fluids in slit pores using cluster expansion methods. The relevant surface variables for simple fluids confined in slit pores are directly expressed in terms of molecular interactions. The Mayer-sampling Monte Carlo (MSMC) method is applied to calculate the necessary cluster integrals which contains surface contributions. MSMC is a general technique for evaluation of the integrals appearing in the expression for the virial coefficients, using ideas from free-energy calculations. We report the results of calculation of surface cluster integrals evaluated by using MSMC. These results are used to obtain critical data under confinement, even close to molecular diameters . Also we discuss the behavior of the surface virial coefficients as varying slit width and surface fluid interaction strength. Finally we analyze the influence of finite size on the phase transition, shift of the critical point, and critical temperatures predicted by surface virial equation of state for confined fluids through molecular clustering.

(484e) Molecular Dynamics Simulation of the Ionlic Liquid [EMIM+][TFMSI-] Confined Inside a Slit Rutile (110) Nanopore

Ramesh Singh[1], Nav Nidhi Rajput[2], Joshua D. Monk[1] and Francisco R. Hung[1], (1)Cain Department of Chemical Engineering, Louisiana State University, Baton Rouge, LA, (2)Cain Department of Chemical Enginering, Louisiana State University, Baton Rouge, LA

The unique properties of ionic liquids (ILs) make them a viable replacement for conventional electrolytes in dye-sensitized solar cells (DSSCs.) Therefore it is desirable to understand the atomic level interactions between the ILs and possible electrode materials such as titanium oxide, zinc oxide and copper oxide used in DSSCs. In the present work, structural and dynamical properties of the IL [EMIM$^+$][TFMSI$^-$] confined inside a rutile (110) slit nanopore of size H = 5.2 nm are investigated by means of classical molecular dynamics simulations. The simulations are carried out at 333 K varying amounts of IL inside the pore. Various equilibrium and time dependent properties such as density profiles, radial density profiles (RDF), orientation of the ions, auto-correlation functions and mean square displacements (MSDs) are calculated and the effect of the confinement on the above properties are studied. It is found that confinement has profound effects on the properties of the ions; densities of the ions are inhomogeneous with higher densities near the rutile (110) surface. Variation in the amount of IL inside the pore has only slight effects on the densities of the ions in the first layers, whereas significant changes are observed in the center of the pore. RDF results indicate a completely different structure of the ions in the first layers at different loadings and in different regions of the pores. The imidazolium ring of the cations prefers to align tilted near the wall with the ring's hydrogen atoms nearest to the surface, whereas no preferential orientation is observed in the center region of the pore. The dynamics of the IL decreased by an order of magnitude near the surface as compared to the ions in the center of the pore. Reductions in the amount if IL inside the pore can lead to the formation of small vacuum regions in the center of the pore. The dynamics of the ions close to the boundary with these vacuum regions are found to be very slow as compared to those observed in the bulk IL..

(484f) Molecular Modeling of Gases Absorption in 1-n-Hexyl-3-Methylimidazolium Bis(trifluoromethylsulfonyl)Imide ([hmim][Tf2N]) Confined in Silica Slit Pores

Wei Shi, National Energy Technology Laboratory, U.S. Department of Energy, Pittsburgh, PA

An experimental group [1] recently in 2012 for the first time verify our theoretical predictions in 2010 [2] that ionic liquid under confinement could exhibit about 100 times faster self-diffusivity relative to the bulk IL. The encouraging experimental data strongly suggest that ionic liquid properties could be modified through confinement. To further understand gases separation in IL confined in the nanoporous materials, we systematically study CO_2, H_2, N_2, and H_2O absorption in the [hmim][Tf_2N] confined in silica slit pores from molecular modeling.

Two-dimensional $NP_{xy}T$ and isostress-osmotic Monte Carlo simulations were performed to compute the [hmim][Tf_2N] IL confined in silica slit pores and gases absorption in the silica confined IL at 313-373 K. A 9-3 analytical potential was used to describe the sorbate-sorbent interactions in most of the calculations. Simulations show that the apparent density of the confined [hmim][Tf_2N] in the 25 Å- 45 Å silica pores are smaller than pure bulk IL density. These results suggest that the silica slit pore confined [hmim][Tf_2N] IL is less packed relative to the pure IL. This implies that more free volume is available when the IL is confined in the silica pore. The CO_2 amounts of absorption in the 25 Å-silica confined IL is about 1.1-1.7 times larger than the CO_2 absorption in the bulk IL. This is consistent with the literature experimental data. Additionally, H_2 and N_2 amounts absorption in the IL are increased 1.6-3 times when the IL is confined. The increased CO_2, H_2, and N_2 amounts of absorption in the confined IL are partly due to the less packed IL structures in the silica pores. In contrast, the amounts of water absorption in the silica-confined IL are similar to that in the bulk IL. This is partly due to the strong hydrogen bonding like interactions between water and the [Tf_2N]- anion.

At low solute concentrations, CO_2 and H_2 molecules are absorbed close to the silica pore regions which are not accessible to the large [hmim][Tf_2N] IL molecules. This in turn leads to 2-8 times faster self-diffusivity for CO_2 and H_2 molecules in the confined IL than in the unconfined bulk IL. CO_2 Self-diffusion in the silica-confined IL using an atomistic model for the silica also exhibits similar improvement as using the 9-3 analytical potential. In contrast, water molecules are not absorbed close to the silica wall regions and the water self-diffusivity in the confined IL is only increase about 1.1-1.8 times. Water diffuses faster in the confine IL partly because the specific water-[Tf_2N]- pair interactions are broken under the confinement. Very interesting, the confined [hmim][Tf_2N] IL exhibits decreased dynamics at high temperature of 373 K, but improved dynamics at low temperature of 313 K.

[1] Iocab, C et al. *Soft Matter*, **2012**, *8*, 289

[2.] Shi, W. et al. *J. Phys. Chem. B* **2010**, *114*, 15029

(484g) Melting Transition of Confined Lj Solid

Chandan K. Das, Chemical Engineering, IIT Kanpur, Kanpur, India and Jayant K. Singh, Chemical Engineering, Indian Institute of Technology, Kanpur, Kanpur, India

Melting transition of confined LJ solid

Chandan K. Das, Jayant K. Singh*

Department of Chemical Enggineering

Indian Institute of Technology, Kanpur, India

**Corresponding author E-mail: jayantks@iitk.ac.in*

In recent year greater attentions have given to study the melting/freezing phenomena in confinement due to its importance in different field of modern technology, like microfluidics, lubrication, adhesion and nanotechnology. Lowering of freezing temperature has been reported for oxygen in sol-gel glasses[1]. Large effects on freezing and melting temperature have been reported for Indium in porous silica glasses [2]. In contrast of these results, some other experiments investigated an increase in the melting temperature[3]. Depression or elevation of freezing temperature of LJ solid have been observed under repulsive or attractive slit confinement[4]. Both the experiments and the simulations of fluids for confined cylindrical pore suggest that, for pore diameters smaller than 20σ the confined phase does not crystallize into a homogeneous solid phase, while for diameters smaller than 12σ the confined solid phase was amorphous throughout the pore[5]. An abnormality in the freezing/melting point corresponding to the difference in the crystal structure has been reported[6].

We study the melting transition of crystalline solid under confined attractive and repulsive slit pores, with the help of molecular dynamics simulation. In particular, we investigate in details, density-temperature hysteresis for various pore sizes ranging from 8 to 3 molecular diameters. In addition we examine the amplitude of atomic vibration at a certain fraction of nearest neighbor distance, also widely known as Lindemann parameter. We also compare the hysteresis loop, Lindemann criteria along with the correlation of the second and fourth moments of a 3D distribution of Drwith time t; for melting transition and generalize the criteria for locating the melting transition under confinement. We obtain shift in the melting temperature with respect to the bulk for various pore surfaces and predict the correlation of $\Delta Tm \sim H^{-n}$. Using the correlation, crossover behavior is observed from 3D to 2D with n ranging from 2.1 to 0.6.

[1] J. Warnock, D. D. Awschalom, and M. W. Shafer, Phys. Rev. Lett. 57 (14), 1753 (1986).

[2] K. M. Unruh, T. E. Huber, and C. A. Huber, Phys. Rev. B 48 (12), 9021 (1993).

[3] J. Klein and E. Kumacheva, Science 269, 816 (1995).

[4] M. Miyahara and K. E. Gubbins, J. Chem. Phys. 106, 2865 (1997).

[5] M. Sliwinska-Bartkowiak, G. Dudziak, R. Sikorski, R. Gras, R. Radhakrishnan, and K. E. Gubbins, J. Chem. Phys. 114 (2), 950 (2001).

[6] T. Kaneko, T. Mima, and K. Yasuoka, Chem. Phys. Lett. 490, 165 (2010).

(490a) In-Situ PEFC Oxygen Sensing: An Ultra-Microelectrode Sensor with Oxygen Flux Interrupt Measurements

William K. Epting, Dept of Mechanical Engineering, Carnegie Mellon University, Pittsburgh, PA and Shawn E. Litster, Mechanical Engineering, Carnegie Mellon University, Pittsburgh, PA

Oxygen transport losses continue to plague polymer electrolyte fuel cells (PEFCs). These losses are due to gas-phase oxygen diffusion through the diffusion medium (DM) and electrode, as well as the dissolution into and diffusion through the electrode's ionomer binder. Direct measurements of the partial pressure of oxygen at the DM/electrode interface would help resolve the relative contributions of these losses. Here we present a new sensor and method enabling direct, *in-situ* measurements of oxygen partial pressure at the DM/electrode interface using an ultra-microelectrode sensor embedded in a thin-film polymer housing and a novel oxygen flux interruption measurement technique. The oxygen flux interrupt method resolves the competing effects of oxygen depletion and sensor hydration that confound partial pressure quantification with calibration curves. Using this method, an instantaneous calibration point is obtained with each measurement. Here, the level of oxygen depletion is determined and analyzed for various PEFC operating current densities and humidification levels.

(490b) Achieving Ultra-High Platinum Utilization Via Optimization of PEM Fuel Cell Cathode Catalyst Layer Microstructure

Jeffrey Marquis and Marc-Olivier Coppens, Chemical & Biological Engineering, Rensselaer Polytechnic Institute, Troy, NY

Abstract: Inefficient usage of expensive platinum catalyst has plagued the design of PEM fuel cells and contributed to the limited production and use of fuel cell systems. This work optimizes the cathode catalyst layer microstructure in two ways: 1) with respect to platinum utilization (measured as kilowatts of electricity produced per gram of platinum) and 2) with respect to current density. A one-dimensional agglomerate model that accounts for liquid water saturation, as initially developed by Nguyen *et al.* [1, 2] is modified for use in this study. The cathode catalyst layer microstructure is optimized by manipulating the platinum loading (m_{Pt}), platinum-to-carbon ratio (Pt|C), catalyst layer thickness (t_{cl}), and catalyst layer void fraction (). Results are compared to those of Secanell *et al.* [3] in which current density was maximized. It is shown that a twenty-fold increase in platinum utilization can be achieved over existing catalyst layer designs while maintaining power densities above 0.20 W/cm^2.

References:

[1] G. Lin, W. He, T. V. Nguyen, *Modeling liquid water effects in the gas diffusion and catalyst layers of the cathode of a PEM fuel cell.* J. Elec. Soc., 151, A1999-A2006 (2004)

[2] X. Wang, T. V. Nguyen, *Modeling the effects of capillary property of porous media on the performance of the cathode of a PEMFC.* J. Elec. Soc., 155, B1085-1092 (2008)

[3] M. Secanell, K. Karan, A. Suleman, N. Djilali, *Multi-variable optimization of PEMFC cathodes using an agglomerate model.* Electrochimica Acta, 52, 6318-6337 (2007)

(490c) Impact of Temperature On PEMFC Gas-Diffusion Layer Properties

Anthony Kwong[1,2], Haluna Gunterman[1,2] and Adam Weber[2], (1)University of California Berkeley, Berkeley, CA, (2)Lawrence Berkeley National Laboratory, Berkeley, CA

Impact of Temperature on PEMFC Gas-Diffusion Layer Properties

Anthony Kwong[1,2], Haluna P. Gunterman[1,2], Adam Z. Weber[2]

[1]University of California, Berkeley, CA 94720-1462

[2]Lawrence Berkeley National Laboratory

1 Cyclotron Rd, MS 70-108B

Berkeley, CA 94720

The characterization of gas-diffusion layer (GDL) properties has been critical in understanding two-phase flow of water and gas within proton-exchange-membrane fuel cells (PEMFCs). In recent years, many studies of the capillary pressure versus saturation relationship for GDLs have been performed at room temperature. However, PEMFC operating temperatures are typically between 50C to 80C. Thus, the water-uptake studies to date may not be accurate representations of the water transport phenomenon within the PEMFC during operation. In addition, most other transport properties including permeability, thermal conductivity, breakthrough pressure, and surface adhesion forces are also only measured at room temperature. In this talk, the impact of temperature (from ambient up to 80C) on these properties and system response will be discussed.

The GDL samples used were SGL 24AA and SGL 24EA. These samples have a PTFE loading of 0% and 30% respectively and a thickness of 0.19 mm. Water-uptake measurements were conducted to measure the capillary pressure – saturation curve in a setup and methodology developed by Gostick et al.[1] The experiments were conducted at 25C, 60C and 70C in an environmental chamber. Humidity was also varied from 0% to 90% but showed no measurable effect; thus, humidity was maintained at 50% for all temperature studies. Figure 1 and figure 2 show the capillary pressure- saturation curves for SGL 24AA and SGL 24EA, respectively. As temperature increases, the residual saturation also increases. This is to be expected since the surface tension of water decreases as temperature increases. Thus, at higher temperatures, water does not adhere to itself as well as it does in lower temperatures. This allows for the formation of more disconnected water pathways within the sample and hence, the residual saturation increases as temperature increases. The contact angle of water on the sample also varies as a function of temperature, and the surface adhesion force at higher temperatures was measured using a sliding-angle technique.[2] At higher temperatures, the sample becomes more wetting (or more hydrophilic). This is evident in our data by noticing the shift towards the negative capillary pressure as temperature is increased. From the Young- Laplace equation, the pressure difference resulting from the temperature changes can be calculated. For our temperature range of 25C to 70C, the corresponding pressure differences are -14 Pa and -12 Pa respectively. Comparing the SGL 24AA and SGL 24EA, it is evident that the overall net effect of temperature change (increase in residual saturation and increase in wettability) is greater for SGL 24AA. This is explained by the presence of more exposed carbon within the 24AA sample (no polytetrafluoroethylene content) which oxidizes at higher temperatures, in agreement with ex-situ-aged sample results.[2]

In addition to adhesions force and capillary properties, the permeability and thermal conductivity of the samples were measured at various temperatures. The determined relationships were fit and used in a sample two-phase flow model to predict the changes due to temperature on liquid versus vapor transport through the GDL.

Acknowledgements

This work was funded by the Assistant Secretary for Energy Efficiency and Renewable Energy, Fuel Cell Technologies Program, of the U. S. Department of Energy under contract number DE-AC02-05CH11231.

References

1. Gostick JT, Ioannidis MA, Fowler MW, Pritzker MD., *Electrochem. Comm.*, **10**, 1520 (2008).

2. P. K. Das, A. Grippin, A. Kwong, A. Z. Weber, *J. Electrochem. Soc.*, **159**, B489 (2012).

Figure 1. Capillary-pressure – saturation curve of SGL 24AA. Temperature is increased from low to high.

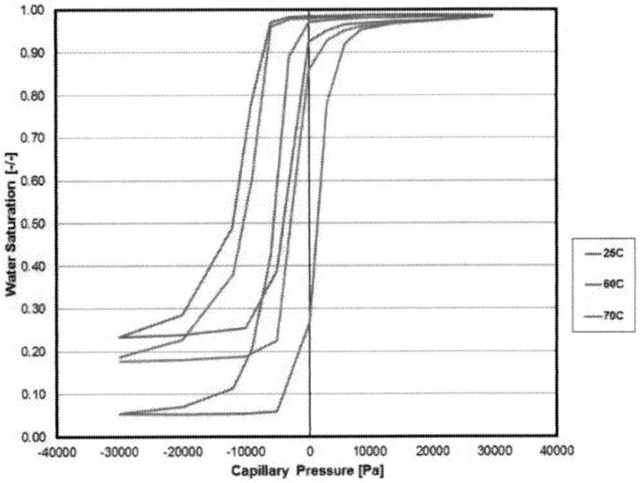

Figure 2. Capillary-pressure – saturation curve of SGL 24EA. Temperature is increased from low to high.

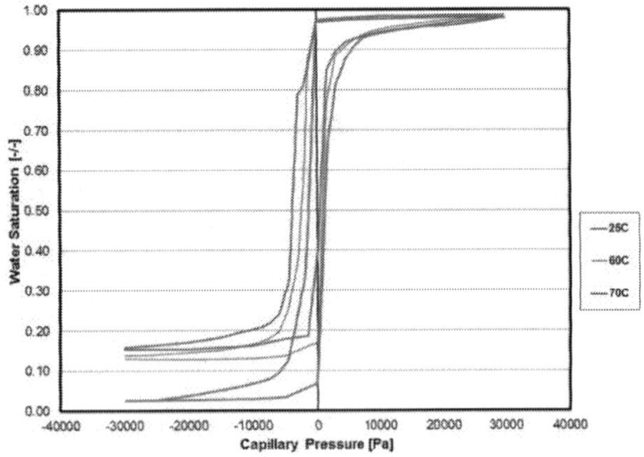

Water Management in a Seven-Layer Model of Proton Exchange Membrane Fuel cell

S.Rakhshanpouri[a], *S. Rowshanzamir*[b-i]
a: Chemical Engineering Department, Iran University of Science and Technology (IUST), Tehran, Iran
b: Chemical Engineering Department & Green Research Canter, Iran University of Science and Technology (IUST), Tehran, Iran

Abstract - The most critical problems to overcome in the proton exchange membrane (PEM) fuel cell technology are the water management. In former years, Springer and colleagues have presented a five-layer model that neglected the anode and cathode catalyst layer. On the other hand, Chen and Chang have presented another the five-layer model that was ignored the input anode and cathode channel. In this model, a steady-state, one-dimensional, isothermal and isobar model in a single PEM fuel cell is presented with a 117 Nafion membrane. A seven-layer theoretical model is proposed which includes anode and cathode inlet channels, anode and cathode gas diffusion layers (GDLs), catalyst layers (CLs), and the layer of proton exchange membrane. Also, the transport phenomena are considered by changing several crucial system parameters such as the relative humidity of reactant gas, stoichiometric ratio of reactant gases, the porosity of GDL, and the membrane thickness. The results show that with sufficient levels of humid reactant gases, the water management would improve for larger porosities of GDLs or a thinner membrane, and the resistance and overvoltage of the membrane can be reduced significantly as well. Furthermore, it is found that if the relative humidity is reduced to about 80%, membrane can be kept in the suitable moisture and flooding of water can be controlled. In addition, using the relative humidity 60% and 80%, stoichiometry coefficients 6 and 4, respectively for the cathode and anode, the cell performance will be optimized and water management is controllable. This model will help to select the amount of humidity, stoichiometric ratio, porosity of gas diffusion layer and membrane thickness so that the fuel cell would not suffer from dehydration and flooding.

Keywords: Fuel Cell, Water Management, Modelling, Seven-Layer Model, Flooding.

1. Introduction

One of the most promising alternatives to the burning of fossil fuel in the production of electric power is the Proton Exchange Membrane (PEM) Fuel Cell. The PEM fuel cell has many advantages to the current internal combustion engine. Since it is not limited by the Carnot cycle, it can achieve much higher efficiencies than a combustion cycle. The PEM fuel cell is also environmentally friendly. When operated on hydrogen, the PEM fuel cell is a zero emission power source. Even when operated on reformate hydrogen (hydrogen evolved from hydrocarbon fuel sources); the emissions are lower than in a standard combustion engine. Other advantages of this fuel cell system include the simplicity of design, low noise operation and economic

[i] -Corresponding author. Fuel Cell Laboratory, Green Research Center, Iran University of Science and Technology, Narmak, Tehran, Iran. Tel./fax:+98 21 77491242.
E-mail address: rowshanzamir@iust.ac.ir (S. Rowshanzamir).

independence. It will help developing countries that lack the power grid infrastructure by using fuel cells for localized electric power stations. Fuel cells can eliminate the need for large, noisy power plant hundreds of miles from the eventually usage location. Furthermore, by localizing power generation the power loss due to transmission through long distance power lines is eliminated (Nguyen et al., 2003).

PEMFC is an electrochemical energy conversion device that converts hydrogen and oxygen into water and, in the process, it produces heat and electricity. The oxygen required for a fuel cell comes from the air pumped into the cathode. However, hydrogen has some limitations that make it impractical for use in most applications. One of the most critical aspects of a PEMFC is the water management. There is a delicate balance between the membrane hydration and avoiding cathode from flooding. Cathode flooding occurs when the water produced exceeds the water removal rate. Flooding in the cathode reduces the oxygen transport to the reaction sites and decreases the effective catalyst area. In modern fuel cells, the cathode (or the anode) consists of: a catalyst layer, gas diffusion layer (GDL) and bipolar plates. The GDL is a porous medium electrically conductor. Liquid water that builds up at a fuel cell cathode decreases the performances and inhibits robust operation (Najjari et al., 2008).

Water is one of the most important factors in the PEMFC operation because the performance of a PEMFC depends strongly on the water balance within the cell. As it is well known, electro-osmotic drag and back diffusion in the membrane affect the water distribution in the PEMFC competitively (Hao et al., 2008).

A study from 1991 to 2008 to develop a theoretical model to describe the phenomenon of water in the PEMFC is performed most reliable conditions for this model have been expressed (Li et al., 2008).

2. Mathematical Model

In former years, Springer et al. (1991) presented a five-layer model that considered the anode and cathode catalyst layers very thin. On the other hand, Chen et al. (2007) presented another five-layer model that ignored the anode and cathode channels input. While the proposed mathematical model is a one dimensional, steady state, isothermal and isobar model to describe the water transport phenomena in a PEMFC with a seven-layer structure, in which consists of anode and cathode inlet channels, anode and cathode gas diffusion layer (GDL), catalyst layers (CLs) and the proton exchange membrane. The seven layer polymer fuel cell in the form (2-1) is shown.

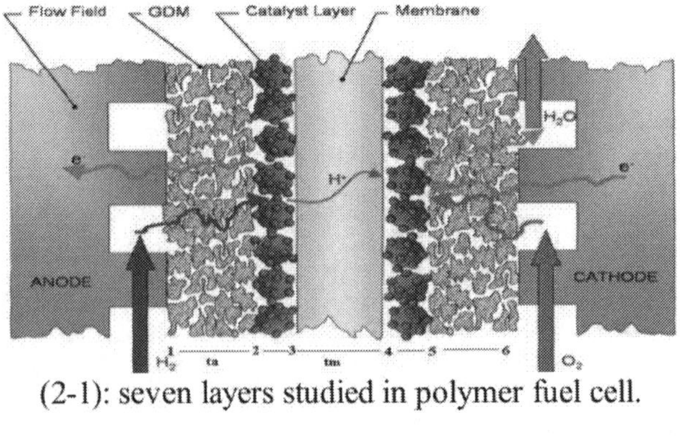

(2-1): seven layers studied in polymer fuel cell.

2.1. Anode and cathode inlet channels (Springer et al., 1991)

Consumption hydrogen molar flux through the anode (or production water molar flux through the cathode) in steady state condition is:

$$J = {}^{I}/_{2F} \qquad \text{(mole/cm}^2\text{s)} \qquad (1)$$

That ***I*** is Current density (A/cm²) and ***F*** is Faraday constant (96484 C/mole). Also, the fluid composition in the anode and cathode flow channel is assumed to be uniform. The anode and cathode feed streams are each assumes to be saturated with vapour at the humidifier temperatures, as:

$$x_{WA}^{I} = \frac{P_A^{sat}}{P_A} \quad , \quad x_{WC}^{I} = \frac{P_C^{sat}}{P_C} \qquad (2)$$

From flow balance, the anode inlet water flow less the exit water flow is the flow crossing interface 1:

$$N_{WA}^{I} - N_{WA}^{L} = N_{WA} \qquad (3)$$

$$\frac{\upsilon_H x_{WA}^{I}}{1-x_{WA}^{I}} J - \frac{(\upsilon_H - 1)x_{W1}}{1-x_{W1}} J = \alpha J \qquad (4)$$

That υ_H is stoichiometric coefficient and is equal to $\frac{N_{H2}^{I}}{N_{H2,1}}$. On the cathode side there are three species. The inlet dry gas mole fraction of oxygen X_{ON} is known. The inlet O_2 mole fraction of the saturated gas is $X_O^I = (1 - X_{WC}^I)X_{ON}$. The three species inlet flow rates and exit flow rates expressed proportional to *I* are:

$$N_{WC}^{I} = \frac{\upsilon_O x_{WC}^{I}}{2(1-x_{WC}^{I})x_{ON}} J \quad , \quad N_O^{I} = \frac{\upsilon_O}{2} J \quad , N_N^{I} = \frac{\upsilon_O(1-x_{ON})}{2x_{ON}} J \qquad (5)$$

$$N_{WC}^{L} = N_{WC}^{I} + (1+\alpha)J \ , \ N_O^{L} = \frac{(\upsilon_O - 1)}{2} J \ , N_N^{L} = \frac{\upsilon_O(1-x_{ON})}{2x_{ON}} J \qquad (6)$$

By summing the three flows in Eq. (6), the total cathode exhaust flow is:

$$N_{total,C}^{L} = \left(\frac{\upsilon_O}{2(1-x_{WC}^{I})x_{ON}} + \alpha + \frac{1}{2}\right) J \qquad (7)$$

The mole fraction of water and oxygen at interface 4 and throughout the cathode flow channel is the ratio of the water or oxygen flow to the total flow through interface 4. Thus:

$$x_{W6} = \frac{x_{WC}^{I}\upsilon_O + 2(1+\alpha)(1-x_{WC}^{I})x_{ON}}{\upsilon_O + (2\alpha+1)(1-x_{WC}^{I})x_{ON}} \qquad (8)$$

$$x_{O6} = \frac{(\upsilon_O-1)(1-x^I_{WC})x_{ON}}{\upsilon_O+(2\alpha+1)(1-x^I_{WC})x_{ON}} \tag{9}$$

Thus, the mole fraction of water at interface 1:

$$x_{W1} = \frac{x^I_{WA}\upsilon_H-\alpha(1-x^I_{WA})}{x^I_{WA}-\alpha(1-x^I_{WA})+\upsilon_H-1} \tag{10}$$

2.2. Gas diffusion in the electrodes (Springer et al., 1991)

Anode and cathode gas mixtures with water vapour act as ideal gases. Let z increase in the direction of reactant flow. The Stefan-Maxwell equation for multi-component diffusion defines the gradient in mole fraction of the components:

$$\frac{dx_i}{dz} = RT\sum_j \frac{x_iN_j-x_jN_i}{PD_{ij}} \tag{11}$$

The binary diffusion coefficients were calculated for mass diffusivity from the Slattery and Bird estimation at low pressures:

$$PD_{ij} = a\left(\frac{T}{\sqrt{T_{CA}T_{CB}}}\right)^{b}(p_{CA}p_{CB})^{1/3}(T_{CA}T_{CB})^{5/12}\left(\frac{1}{M_A}+\frac{1}{M_B}\right)^{1/2}\varepsilon^{3/2} \tag{12}$$

The unit of pressure-binary diffusion is *atm-cm²/s*. Where a = 0.0002745, b = 1.832 forH_2, O_2, andN_2, and a = 0.000364 and b = 2.334 for water vapour. The $\varepsilon^{3/2}$ term is a Bruggemann correction of the diffusion coefficients to account for the porosity of the electrode.

From Eq. (11), for mole fraction of water in anode (x_{WA}) and mole fraction of oxygen and water in cathode (x_{OC} , x_{WC}), it is:

$$\frac{dx_{WA}}{d_z} = \frac{RTJ}{P_AD_{WH}}\left(x_{WA}(1+\alpha)-\alpha\right) \tag{13}$$

$$\frac{dx_O}{d_z} = \frac{RTJ}{P_C}\left(\frac{x_O(1+\alpha)+0.5\,x_{WC}}{D_{WO}}+\frac{1-x_{WC}-x_{ON}}{D_{ON}}\right) \tag{14}$$

$$\frac{dx_{WC}}{d_z} = -\frac{RTJ}{P_C}\left(\frac{(1-x_{WC}-x_O)(1+\alpha)}{D_{WN}}+\frac{0.5\,x_{WC}+x_O(1+\alpha)}{D_{WO}}\right) \tag{15}$$

Equation (13) can be integrated across the anode thickness, so:

$$x_{W2} = \left(x_{W1}-\frac{\alpha}{1+\alpha}\right)exp\left(\frac{RTJt_a}{P_AD_{WH}}\right)+\frac{\alpha}{1+\alpha} \tag{16}$$

Eq. (14) and (15) can be integrated during at interface 5 to obtainx_{W5}.

2.3. Catalyst layers (Chen et al., 2007)

The catalyst layer contains voids and electrolyte phase. Hence, three mechanisms may contribute to water transport in CLs. In voids the water vapour is transported by Knudsen diffusion, while in the electrolyte phase the water transport is mainly due to diffusion and drag. The flux in the electrolyte phase is related to a correction factor that accounts for the amount of electrolyte in the CL and can be approximated by the volume fraction of the electrolyte phase. Since the correction factor is generally quite small which makes the water transport due to the electrolyte phase be negligible, we can assume the water transport is primarily in vapour form through the voids due to Knudsen diffusion. Accordingly, at the anode *CL*. It is:

$$J_2 = -D_{cl}\,(\frac{\partial C_2}{\partial z}) \tag{17}$$

D_{cl} Is the Knudsen diffusion coefficient of water vapour in the voids. The Knudsen diffusion coefficient is assumed to 0/0012 (cm^2/s). From the mass conservation equation, we have:

$$\mathcal{E}_c \frac{\partial C_2}{\partial t} = -\frac{\partial J_2}{\partial z} \tag{18}$$

Because of steady state, we have:

$$-\frac{\partial J_2}{\partial z} = 0 \tag{19}$$

The boundary conditions at interface 2 are:

$$C_1 = C_2 \tag{20}$$

$$D_{gl}\left(\frac{\partial C_1}{\partial z}\right) = D_{cl}\left(\frac{\partial C_2}{\partial z}\right) \tag{21}$$

Above equations can be solved during at interface 2 to obtainx_{W3}. Also for anode we have:

$$\mathcal{E}_c \frac{\partial C_2}{\partial t} = -\frac{\partial J_4}{\partial z} + \frac{S_W Q}{n F} \tag{22}$$

$$0 = -\frac{\partial J_4}{\partial z} + \frac{S_W Q}{n F} \tag{23}$$

$$J_4 = -D_{cl}\,(\frac{\partial C_4}{\partial z}) \tag{24}$$

The boundary conditions at interface 5 are:

$$C_4 = C_5 \tag{25}$$

$$D_{gl}\left(\frac{\partial C_4}{\partial z}\right) = D_{cl}\left(\frac{\partial C_5}{\partial z}\right) \tag{26}$$

Above equations can be solved during at interface 5 to obtainx_{W4}.

2.4. Water transport in membrane (Springer et al., 1991)

λ, is given as the ratio of the number of water molecules to the number of charge (SO3H +) sites. A fit of the experimental relationship of λ *vs.* water vapour activity, a, used in the model is:

$$\lambda = 0.043 + 17.81\,a - 39.85\,a^2 + 36\,a^3 \qquad \text{for } 0 < a \leq 1 \tag{27}$$

That$\alpha = \frac{x_W P}{P_{sat}}$, Also:

$$\lambda = 14 + 1.4\,(a - 1) \qquad \text{for } 1 \leq a \leq 3 \tag{28}$$

Water transport in membrane has two dominant mechanisms. One of that is electro-osmotic drag and another is back diffusion transport. So, water flux through the membrane in effect of electro-osmotic drag (drag (mole/cm2s)) is:

$$N_{W,drag} = 2J.n_{drag},\ n_{drag} = n_{drag}{}^{sat}\,\lambda/22 \tag{29}$$

$$n_{drag}{}^{sat} = 0.2 \pm 2.5 \tag{30}$$

Eq. (32) shows the water molar flux through the membrane in effect of back diffusion. In order to eliminate tracking the membrane swelling in the model, we convert to the equivalent water concentration in a dry membrane ($\lambda\rho_{dry}/M_m$). To account for membrane swelling, dry membrane thickness dimensions are expanded by the factor ($1+s\,\lambda$) (Zhang et al., 2008). So:

$$N_{w,diff} = -D' c_w \frac{d(\ln a)}{d\lambda}\frac{d\lambda}{dz} = -D'\left(\frac{\frac{\lambda\rho_{dry}}{M_m}}{(1+s\,\lambda)^2}\frac{\lambda}{a}\frac{da}{d\lambda}\right)\frac{d\lambda}{dz} = -\frac{\rho_{dry}}{M_m} D_\lambda \frac{d\lambda}{dz} \tag{31}$$

$$D_\lambda = 10^{-6} exp\left[2416\left(\frac{1}{303} - \frac{1}{273+T_{cell}}\right)\right](2.563 - 0.33\lambda + 0.0264\lambda^2 - 0.000671\lambda^3) \tag{32}$$

So, the net water flux through the membrane is:

$$N_w = n_{drag}{}^{sat}\frac{\lambda}{22}(2J) - \frac{\rho_{dry}}{M_m} D_\lambda \frac{d\lambda}{dz} = \alpha J \tag{33}$$

$$\frac{d\lambda}{dz} = [2n_{drag}\frac{\lambda}{22} - \alpha]\frac{JM_m}{\rho_{dry}D_\lambda} \tag{34}$$

Now we can calculate a new α and check this value to old value. If new α is closer to old α, our calculation is correct. The key to solving the water transport through the PEMFC, we must apply in iteration to determine the proper value of α. We guess a value for α and calculate all mole fractions, and then we distinguish new α which is obtained following the integration of Eq. (35). These two values will be equal when the proper value of is used.

3. Result and Discussion

The base-case input parameters were listed in Table 1. The computed variables of this model and the Springer model were shown in Table 2.

Table 1: The base-case input parameters for a cell with Nation 117 membrane.

Input Parameters		
I	A/cm^2	0.5
j_0	A/cm^2	0.01
υ_H		4
υ_O		6
ε		0.4
P_A	atm	3
P_C	atm	3
t_{mem}	cm	0.0175
t_a, t_c	cm	0.0365
t_{cat}	cm	0.0020
T_{cell}	K	353.115
V_{oc}	V	1.1

Table 2: The computed variables of this model and the Springer model.

computed variables	**this model**	**Springer model**
V_{cell} (Volt)	0.6795	0.6680
R_{mem} (Ω-cm^2)	0.2699	0.2850
α	0.3836	0.4
x_{W1}	0.1024	0.1015
x_{W2}	0.1010	0.1013
x_{W3}	0.1	0.2327
x_{W4}	0.2347	0.2264
x_{W5}	0.2242	
x_{W6}	0.2239	
x_{O5} x_{O3}	0.1405	0.1329
x_{O6} x_{O4}	0.1408	0.1403

x_{liq}	0.0811	0.077
$Rel\ Hum\ 1$	0.6668	0.6515
$Rel\ Hum\ 6$ $Rel\ Hum\ 4$	1.4574	 1.4535

It can be seen in Table 2, the membrane resistance of this model is lower than Springer model, which voltage is also more. Also, the calculated relative humidity of this model at interface 6 and the Springer model at interface 4 is more than 1. These values indicate that there is some liquid water.

Model results indicate that high porosity of gas diffusion layer and reduce relative humidity in both anode and cathode reactant gases (about 80%) will reduce flooding. Also with thin cell membrane thickness (about 50 μm), the performance will increase, but increases the risk of flooding in the cathode. Therefore not suitable for water management and is not recommended. Also high hydrogen stoichiometry has best performance, which also is not suitable for water management. Therefore, the reduce relative humidity to 60% in cathode gas input, to 80% in anode gas input, and using hydrogen stoichiometry to 4, oxygen stoichiometry to 6, for control of water management and increase fuel cell efficiency is recommended.

References

1- Chen F., Chang M., Fang C. (2007). Analysis of Water Transport in a Five-layer Model of PEMFC. Journal of Power Sources, 164, 649-658.

2- Hao L., et al., (2008). Transient Behaviour of Water Generation in a Proton Exchange Membrane Fuel Cell. Journal of Power Sources, 177, 404-411.

3- 3- Li H., et al., (2008). A Review of Water Flooding Issues in the Proton Exchange Member Fuel Cell. Journal of Power Sources, 178, 103-117.

4- Najjari M., Nasrallah S., Khemili F. (2008). The Effects of the Cathode Flooding On the Transient Responses of A PEM Fuel Cell. Renewable Energy, 33, 1824-1831.

5- Nguyen T., et al. (2003). A Liquid Water Management Strategy for PEM Fuel Cell Stacks. Journal of Power Sources, 114, 70–9.

6- Springer T. E., et al., (1991). Polymer Electrolyte Fuel Cell Model, Electrochemical society, 138, 2334-2342.

7- Zhang L., Pan M., Quan Sh. (2008). Model predictive control of water management in PEMFC. Journal of Power Sources, 180, 322-329.

(490e) Investigation of Cathode Contaminants within Microfluidic and Laminar-Flow Fuel Cells

Matthew S. Naughton, Department of Chemical & Biomolecular Engineering, University of Illinois at Urbana-Champaign, Urbana, IL and Paul J.A. Kenis, Department of Chemical & Biomolecular Engineering, University of Illinois at Urbana Champaign, Urbana, IL

Fuel cells hold promise as alternative power sources due to their ability to bypass Carnot efficiency limitations by directly converting chemical energy into electrical energy. However, the use of hydrogen fuel is not compatible with existing distribution infrastructure [1]. Liquid fuels such as methanol, ethanol, and sodium borohydride offer an alternative energy-dense means to produce power, but losses from fuel crossover have limited the usable fuel concentrations at the anode [2]. Quantification of these losses and/or mitigation of crossover losses is necessary for optimal fuel cell design.

Previously, we developed pH-flexible flowing electrolyte microfluidic fuel cells and laminar-flow fuel cells, which use an external reference electrode to individually analyze cathode and anode performance [3,4]. This microfluidic configuration combines the versatility of a traditional three electrode cell with the conditions found in an operating fuel cell, allowing for *in-situ* studies of catalyst and electrode performance. External control over the flowing electrolyte stream allows for controlled introduction of contaminants and maintains their concentrations over the course of experimentation [5].

Here, we present our work on the effects of fuels such as methanol, ethanol, and sodium borohydride as cathode contaminants in acidic and alkaline fuel cells. The effects of these contaminants on various cathode catalysts such as Pt, Ag, and Cu triazole are quantified. We also investigate the effect of cathode oxidizers such as hydrogen peroxide at the anode. Understanding of crossover losses in an operating fuel cell has the potential to reduce the cost of fuel cell systems and enable contaminant-resistant catalyst design.

[1] L. Carrette, K. A. Friedrich, U. Stimming, ChemPhysChem, 2000, 1, 162-193

[2] J .S. Spendelow, A. Wieckowski, Physical Chemistry Chemical Physics, 9, 2007, 2654

[3] F. R. Brushett, W. P. Zhou, R. S. Jayashree, P. J. A. Kenis, Journal of the Electrochemical Society, 2009, 156, B565

[4] E. R. Choban, L. J. Markoski, A. Wieckowski, P. J. A. Kenis, Journal of Power Sources, 2004, 128, 54-60

[5] M. S. Naughton, F. R. Brushett, P. J. A. Kenis, Journal of Power Sources, 2011, 196, 1762-1768

(490f) Molybdenum Dioxide (MoO2)-Based Anode Fabrication and Its Performance Analysis for SOFC Applications

Byeong Wan Kwon[1], Oscar Marin-Flores[2], Shuozhen Hu[2], M. Grant Norton[3] and Su Ha[2], (1)Chemical engineering, Washington State University, Pullman, WA, (2)Washington State University, Pullman, WA, (3)School of Mechanical and Materials Engineering, Washington State University, Pullman, WA

Molybdenum Dioxide (MoO_2)-Based Anode Fabrication and Its Performance Analysis for SOFC Applications

Byeong Wan Kwon[a], Oscar Marin-Flores[a], Shuozhen Hu[a], M. Grant Norton[b], Su Ha[a]

[a] *The Gene and Linda Voiland School of Chemical Engineering and Bioengineering, Washington State University, Pullman, WA 99164*

[b] *School of Mechanical and Materials Engineering, Washington State University Pullman, WA, 99164-2920*

Solid oxide fuel cells (SOFCs) are a promising fuel flexible technology for clean and efficient conversion of chemical energy into electrical energy. In SOFCs using hydrocarbons as fuel, hydrocarbons are oxidized on the surface of the Ni-based anode through a pathway that produces carbon deposits known as coking. These carbon deposits deactivate the anode and lead to a poor long-term stability of SOFC. Therefore, a new anode material is required for direct liquid SOFCs having a good coking resistance for a long-term stability.

Our group conducted the fabrication and performance of a porous molybdenum dioxide (MoO_2)-based anode for direct utilization of liquid fuel SOFCs instead of Ni-based anode, which can directly convert liquid fuels into electrical energy without external fuel processors. The MoO_2-based anode was manufactured onto yttria-stabilized zirconia (YSZ) electrolyte with Sr-doped $LaMnO_3$ (LSM) cathode via combined electrostatic spray deposition (ESD) and direct painting methods. The cell performance was measured by directly feeding *n*-dodecane, which is a model compound of jet A fuel, to the MoO_2-based anode at 750°C. The stabilized power densities from our MoO_2-based SOFC were 2000 mW/cm^2 at 0.6V using *n*-dodecane as fuels. To test the long-term stability of MoO_2-based SOFC against coking, *n*-dodecane was continuously fed into the cell for 24 h at its open cell potential of ~0.86V. During this long-term testing, voltage-current density plots were periodically obtained and they showed no significant changes over 24 hr. The X-ray diffraction patterns obtained from the spent MoO_2-based anode after 24 hr of operation indicated that the bulk structure of the MoO_2-based anode exhibit no phase transition. Based on our visual observations, there was no coking. Therefore, the catalytic activity toward the electrochemical oxidation of fuel remains stable even after 24hr operation. On the other hand, SOFCs with conventional nickel (Ni)-based anodes under the same operating conditions showed a significant amount of coke formation on the Ni surface, which led to a rapid drop in cell performance. Hence, the present work demonstrates that MoO_2-based anodes exhibit an outstanding tolerance to coke formation and the power density of the MoO_2 anode is much higher than that obtained with the Ni based anode.

(490g) Predictive Modeling of Effective Conductivity in Infiltrated Solid-Oxide Fuel Cell Anodes

Ryan C. Snyder, Michael D. Gross and Micheal J. Synodis, Chemical Engineering, Bucknell University, Lewisburg, PA

The Solid Oxide Fuel Cell (SOFC) is an electrochemical device that converts chemical energy into electrical energy by directly and efficiently oxidizing hydrocarbon fuels. The SOFC anode, which is where the fuel oxidation reaction occurs, must have the following four properties to be functional: 1) porosity for mass transport, 2) catalytic activity for the oxidation reaction, 3) oxygen ion conductivity to provide an oxidizing agent to the fuel, and 4) electronic conductivity to transport the electrons released during the oxidation reaction. Recent work has shown that anode fabrication and modification by infiltration of an electronic conductor and oxidation catalyst into an oxygen ion conducting porous scaffold can result in outstanding electrochemical performance. In addition, infiltration allows one to separate the calcination temperature of the various active components, requires less electronic conducting material to form an interconnected conductive pathway, relieves dimensional instabilities within the fuel cell, and allows the porosity and pore morphology to be easily controlled. While the infiltration method provides great flexibility in composite composition and structure, the influence of the controllable parameters are not well understood. Furthermore, fabrication and testing of anode composites representing the full range and number of working parameters would be time consuming. Thus, a model to guide experiments is desired.

In this talk, we will highlight our recent work on using a model to predict properties of the fuel cell anode as a function of the important processing variables. The model is based upon the underlying physics of the process including the mixing of the anode materials, evaporation of the solvent, burning off of the pore formers, sintering of the remaining YSZ backbone and infiltrating with conducting particles. The model then is used to predict the effective conductivity of the fuel cell anode as a function of conductor particle fraction. These results are favorably compared to existing experimental results and demonstrate effective prediction with no adjustable parameters both in regard to the conductor fraction where conduction first takes place and the quantitative effective conductivity at realistic loadings. We will also highlight other important characteristics such as the conductor particle coordination number. Beyond the validation of the model, we also will demonstrate the models capability of predicting the effective conductivity as a function of important model parameters such as pore former and YSZ particle size as well as pre-infiltrate porosity. Finally, we will discuss the opportunities to advance this type of modeling to incorporate other properties of the anode such as thermal expansion to provide design optimization.

(490h) Spatially-Resolved Modeling of Oxygen Reduction Reaction Inside Polymer Electrolyte Fuel Cell Cathode's Water-Filled Mesopore

Iryna V. Zenyuk and Shawn E. Litster, Mechanical Engineering, Carnegie Mellon University, Pittsburgh, PA

In this work we study the effects of water-filled mesopores on the oxygen reduction reaction (ORR) in the cathode catalyst layer of a polymer electrolyte fuel cell. We model reactions in water-filled pores with a multi-scale, multi-dimensional numerical framework, where oxygen and proton transport, kinetics and surface chemistry are spatially resolved. Poisson-Nernst-Planck theory describes charge transport and potential distributions in the water-filled pore, adjacent Nafion polymer electrolyte film and carbon wall. The model incorporates spatially resolved, coupled electric double layers (EDLs) at water|Nafion and water|electrode interfaces. We also use the Stern modification for the Guoy-Chapman EDL theory to account for dielectric saturation and ion specific adsorption at the inner Helmholtz plane (IHP), which is modeled as a finite-thickness domain. The carbon wall domain includes a surface complexation model to account for carbon black catalyst support's surface chemistry and a space-charge layer to account for the electron spillover effect. The ORR is modeled with multi-step, double-trap model where the driving potential is the difference between the potential at the wall and that at the outer Helmholtz plane. Our results show that water-filled pore is kinetically limited due to proton depletion at high interfacial potentials. The pore surface area next to the Nafion electrolyte produces the most current, whereas the rest of the pore contributes very little to the overall ORR reaction. We also compare the results of water-filled pore to Nafion-filled pore and present the values for ORR effectiveness factor.

(490i) Incorporating a Size Distribution and Interfacial Oxygen Resistance Into a PEFC Agglomerate Model

William K. Epting, Dept of Mechanical Engineering, Carnegie Mellon University, Pittsburgh, PA and Shawn E. Litster, Mechanical Engineering, Carnegie Mellon University, Pittsburgh, PA

In the past, the popular agglomerate model for polymer electrolyte fuel cells (PEFCs) has typically simplified the electrode by assuming a single, representative agglomerate diameter. In reality, the electrode contains a distribution of solid agglomerate sizes. In this work, we examine the effect of incorporating an agglomerate size distribution (ASD) into the agglomerate model. The ASD is obtained from our own prior nanoscale X-ray CT measurements. To further study oxygen transport through the ionomer binder that surrounds agglomerates, we additionally incorporate an interfacial oxygen mass transport resistance at the gas/ionomer interface. This resistance accounts for the rate-dependent disparity between the equilibrium concentration just inside the ionomer (as expected from Henry's law) and the actual concentration there when oxygen flux is underway.

We find a substantial difference (or "error") between the agglomerate model with an ASD and the model with a single agglomerate size. This error is a strong function of overpotential, and of the choice of agglomerate size employed in the single-size model. We find the single agglomerate diameter that gives the lowest maximum error across all overpotentials for our particular electrode geometry, and when considering the interfacial mass transport resistance). This diameter is 105 nm, in which case the single-agglomerate version of the model still errs by as much as +/- 20% versus the ASD case. We also present the ideal (no error) values of single-size agglomerate choices, or the "effective diameter." This value is a function of overpotential and of whether or not the interfacial resistance is used. The effective agglomerate diameter becomes lower at higher overpotentials. This is because at higher current, transport is a greater issue, forcing the smaller (and less transport-hindered) agglomerates to contribute disproportionately more to the reaction rate.

(499a) Consequences of Interfacial Viscoelasticity On Thin Film Stability

Liat Rosenfeld and Gerald G. Fuller, Chemical Engineering, Stanford University, Stanford, CA

The phenomenon of dewetting is frequently observed in our everyday life. It is of central importance in many technological applications as well as in a variety of physical and biological systems. The presence of non-soluble surfactants at an air/liquid interface may affect the dewetting properties of the aqueous layer.

The tear film is comprised of an aqueous layer covered with a ~100nm thick blanket of lipids, known as the meibomian lipids. Interfacial rheological measurements of meibomian lipids reveal that these films are remarkably viscoelastic. Tear film dewetting is an extremely important parameter that influences tear film stability.

In this research, in order to better understand the role of the meibomian lipids in the tear film stability, we have developed a methodology to characterize dewetting of thin aqueous layers covered with insoluble surfactants. Meibum was used to form a viscoelastic monolayer. As comparison to meibum we have studied DPPC, a phospholipid which is also viscoelastic, and the Newtonian arachidyl alcohol (AA). The apparatus allows control of the surface pressure of the monolayer, which is a key feature since this variable controls the surface viscoelasticity.

We have found, both experimentally and theoretically, that monolayers of viscoelastic surfactants are able to stabilize thin films against spontaneous dewetting. As the surface pressure of these layers is increased, their effectiveness is enhanced. Moreover, these surfactants are able to reduce the critical thickness for dewetting. Meibum is particularly effective in stabilizing thin films. Our results suggest that the meibomian lipids play a vital role in maintaining tear film stability in addition to suppressing evaporation. Furthermore, changes in meibomian lipid composition that alter the lipid melt temperature and viscoelastic properties, may ultimately influence tear film stability, which in clinical cases may result in dry eye disease.

(499b) On the Surface Shear Viscosity of Soluble Surfactants

Zachary A. Zell[1], Suraj Deshmukh[2], Christopher Tucker[2] and Todd M. Squires[1], (1)Chemical Engineering, University of California, Santa Barbara, Santa Barbara, CA, (2)The Dow Chemical Company, Midland, MI

The stability and flow behavior of multiphase materials like foams and emulsions are closely linked to their interfacial properties, which can be greatly affected by the presence of surfactants. One such important property is the surface shear viscosity, as it is generally thought to enhance stability via reduced film drainage. The understanding of surfactant effects on surface shear viscosity is thus crucial to the study of such materials. However, for a common anionic surfactant used to make foams, sodium dodecyl sulfate(SDS), the reported values of measured surface viscosities vary by several orders of magnitude (0.01-100μNás/m). In this study, we use ferromagnetic microdisks deposited at the SDS solution - air surface, as active, microrheological probes of the interface. Our oscillatory shear measurements reveal that the values of surface shear viscosity are on the order of 0.01μNás/m or lower, at which point the probe becomes Boussinesq limited. We therefore question the interpretation of previous measurements, and thus of the reported values of the surface viscosity of SDS solutions. Instead, we suggest that the non-viscometric flow (including compression and dilation) gives rise to enhanced drag forces that do not reflect shear viscosity. Rather, we propose that the enhanced drag coefficients in the literature are likely due to Marangoni forces and not the surface shear viscosity. Our results suggest that the generally-believed correlation between the surface shear viscosity of surfactants, and the stability of foams and emulsions made using them, is not correct, but instead reflects other physico-chemical processes.

(499c) Visualization of Foam Sweep and Oil Displacement in Fractured Porous Media Micromodels

Charles Conn, Kun Ma and Sibani Lisa Biswal, Chemical and Biomolecular Engineering, Rice University, Houston, TX

Enhanced oil recovery (EOR) applications are considering foam for gas mobility control to ensure pore-trapped oil can be effectively displaced. In fractured reservoirs gas tends to channel only through the high-permeable regions, bypassing the low-permeable porous matrix, where most of the residual oil remains. Because of the unique transport problems presented by the large permeability contrast between fractures and adjacent porous media, we aim to understand the mechanism by which foam transitions from the fracture to the matrix, and how initially-trapped oil can be displaced and ultimately recovered.

This research presents direct visual observation of foam in fractured porous media micromodels with dimensions on the order of tens of micrometers. High-speed microscopy videos highlight tunable bubble generation via a flow-focusing microchannel geometry, bubble stability at the foam-oil interface, and dynamic foam behavior at the pore scale (including both fractures and model porous media). Foam sweep and oil displacement is studied as a function of foam quality, bubble size, surfactant type, and fracture-matrix permeability. Comparisons are made with pure gas and surfactant-free flooding, showing improved sweep and oil mobilization (up to 98% oil-in-place displaced) for foam systems.

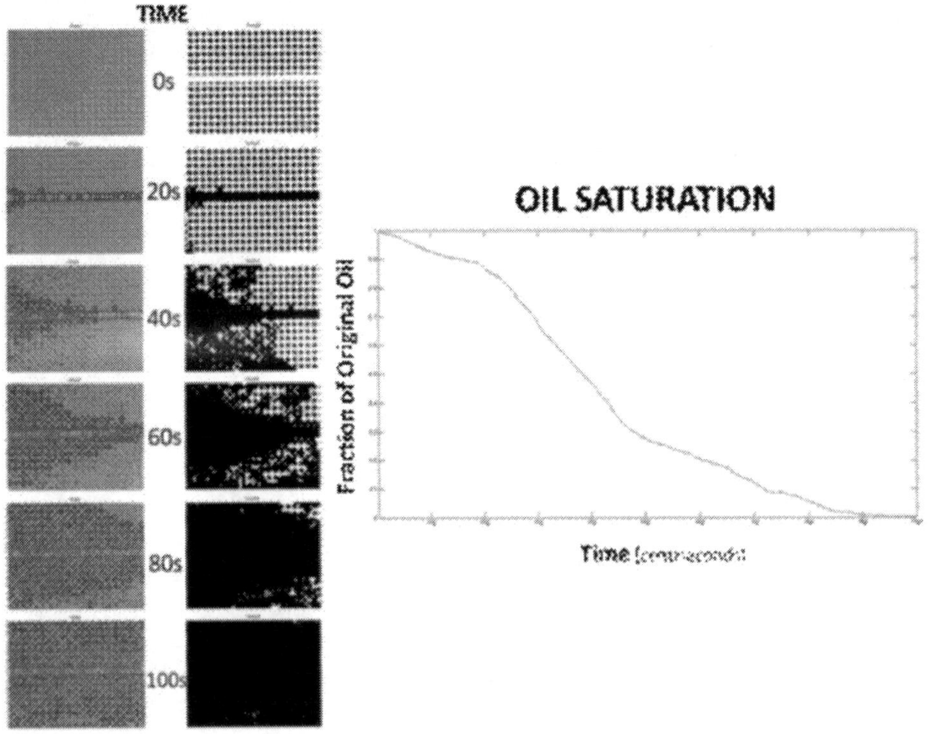

Figure: Foam flooding experiment of a fractured porous media micromodel. Oil is dyed red (white in threshold); surfactant solution and bubbles are grey, with bubbles having dark black lamellae. The plot shows fraction of white pixels (representing regions of oil) relative to the initial white pixel count as a function of time. After 100 seconds, <2% of the original oil-in-place remains.

(499d) Electrohydrodynamic Instabilities of Bilayer Membranes

Jacopo Seiwert[1], Paul F. Salipante[1], Michael Miksis[2] and Petia M. Vlahovska[1], (1)School of Engineering, Brown University, Providence, RI, (2)ESAM, Northwestern University, Chicago, IL

The interface defining a biological cell is a thin membrane, which acts as a leaky capacitor. We investigate the influence of capacitance and conductivity on the stability of a planar membrane subjected to DC or AC electric fields, typically used in cell electroporation. We develop a zero-thickness model of the membrane, in which the bilayer finite thickness is effectively accounted for by membrane electro-mechanical properties such as bending modulus, capacitance and conductance. The linear stability analysis shows that membrane conductance and asymmetry in the embedding electrolyte solutions destabilize the interface. However, the capacitive charging acts to stabilize the system under conditions where an ordinary fluid- fluid interface is unstable. We also find that while a purely capacitive membrane is linearly stable in DC fields, it can become unstable in AC fields. We identify the frequency correspoding to the most unstable mode and correlate it with the membrane charging time. Finally, the instability can lead to vesicle burst, and the timescale of collapse shows a t~1/E^2 dependence.

(499e) Linear Stability of the Belt-Constrained Rayleigh Drop

Joshua B. Bostwick, Mathematics, North Carolina State University, Rayleigh, NC; Theoretical and Applied Mechanics, Cornell University, Ithaca, NY and **Paul H. Steen**, Chemical and Biomolecular Engineering, Cornell University, Ithaca, NY

The eigenfrequencies and eigenmodes of inviscid vibrations of a drop held by surface tension and constrained by a spherical belt are solved for. The belt support extends between two latitudes. The full problem is solved with focus on eigenvalue multiplicities over the 2-parameter plane of geometrical constraints. Rayleigh (1879) and Strani and Sabetta (1984) results are recovered. An approximation as two coupled harmonic oscillators is derived and shown to be effective. Constrained oscillations are of importance in a number of emerging applications.

Surface tension resists deformation of a liquid mass from rest configurations in much the same way as the linear-elastic spring resists excursions of a mass from a rest position. Indeed, under broad conditions, the stability of a liquid mass held by surface tension is determined by an operator equation on small disturbances to the interface of the same form as that of the damped harmonic oscillator. Motions are assumed inviscid and irrotational. The governing equations for the velocity potential within the drop are solved for general boundary conditions, following the boundary-integral approach. The resulting operator equation maps one function space to another, where the geometrical constraints of the belt are incorporated into the function space. The linear operator equation is an integro-differential boundary-value problem and has the form of an eigenvalue problem. The eigenvalue problem is reduced to a set of algebraic equations using a Rayleigh-Ritz procedure, in the standard way. Solutions are obtained by using basis functions expressed as series of Legendre polynomials, suitably orthonormalized.

The two spherical-cap interfaces are coupled through the underlying fluid and act as oscillators. Depending on the position and extent of the constraint, qualitatively different couplings can occur. In particular, there are regions of multiplicity and near-multiplicity where different mode shapes have the same or nearly the same frequencies. In a neighborhood of these geometries, the spectrum is particularly sensitive to geometry. The regions of sensitivity may be of importance to application since there any coupling through nonlinearity (not considered in this paper) will be most dramatic. Finally, for much of the plane of geometrical constraint, the spectral dependence can be thought of as arising from a superposition of single-interface constrained drops.

(499f) Leveraging Flow Geometry to Control Interfacial Instabilities

Talal T. Al-Housseiny[1], Peichun A. Tsai[2] and Howard A. Stone[2], (1)Chemical and Biological Engineering, Princeton University, Princeton, NJ, (2)Mechanical and Aerospace Engineering, Princeton University, Princeton, NJ

In most cases, interfacial instabilities hinder the operation of industrial processes. For instance, fluid-fluid instabilities remain a major challenge for enhanced oil recovery processes such as water flooding. On the other hand, these instabilities can be beneficial to chromatographic separation and can improve mixing in small-scale devices. The fact that, depending on the application, either a stable or an unstable interface is desirable makes the ability to control interfacial instabilities essential in design and technology.

It is universally accepted that, in a uniform medium, flow displacement is unstable when a low viscosity fluid invades a fluid of higher viscosity: the classical viscous fingering instability. Consequently, once fluid properties are specified, opportunities for system control become very limited. However, real systems where displacement instabilities occur, such as porous structures, lung airways and printing devices, are rarely uniform. We find that the simplest heterogeneity can lead to fundamentally different interfacial behaviors. We leverage our findings to devise a strategy to manipulate instabilities in fluid-fluid systems.

(499g) To Bounce or Not to Bounce: Impact of Viscoelastic Drops On Dry, Textured Surfaces

Vivek Sharma[1], Kyoo-Chul (Kenneth) Park[1], Robert E. Cohen[2] and Gareth H. McKinley[1], (1)Mechanical Engineering, Massachusetts Institute of Technology, Cambridge, MA, (2)Chemical Engineering, Massachusetts Institute of Technology, Cambridge, MA

The deposition of aqueous drops on non-wetting surfaces is an important canonical problem for many applications, including suppressing splash or rebound of sprayed herbicides on intrinsically hydrophobic plant leaves. The addition of a small amount of high molecular weight polymer has been demonstrated to suppress drop rebound on smooth hydrophobic surfaces. The high extensional viscosity of polymer solutions and the increased viscous dissipation near the receding contact line are cited as two distinct anti-rebound mechanisms. Using drop impact experiments on micro- and nano-textured surfaces in addition to smooth hydrophobic surfaces with controlled wetting characteristics, we examine the roles of viscosity, elasticity and inertia on expansion, retraction, and rebound of well-characterized viscoelastic fluids. We show that when a viscoelastic drop impacts a textured hydrophobic surface, the answer to the question "to bounce or not to bounce" depends critically on both the fluid viscoelasticity and the surface characteristics, including details of texture and its wetting characteristics. By adopting a stick-slip flow model on textured surfaces with various topographic length scales and solid area fractions, we rationalize the dynamics leading to complete rebound following drop impact on nanotextured surfaces even for viscoelastic fluids.

(499h) On the Diffusiophoretic Motion of Biocolloids with Saturable Kinetics

Elizabeth A. Knapp, Chemical Engineering, City College of New York, New York, NY and Charles Maldarelli, Levich Institute and Chemical Engineering, The City College of New York, New York, NY

Phoretic transport is the process by which colloidal particles migrate due to a gradient of a thermodynamic variable, such as electric potential, temperature, or concentration, which is applied across the particle. Due to the difficulties associated with imposing an external gradient to move a particle over small length scales, self-propulsion is a promising alternative. One mechanism for autonomous motion is self-diffusiophoresis, in which a concentration gradient of a solute across the particle gives rise to unbalanced Van der Waals forces on the particle which move the particle forward. These self-propelled colloidal particles ("swimmers") can function as motors to transport cargo over small scale landscapes.

A self-diffusiophoresic motor has been designed previously in which hydrogen peroxide reacts on a platinum catalyst on one side of the colloid. Consumption of the fuel and production of oxygen generates concentration gradients that drive the diffusiophoretic motion. As a result of the caustic reaction involved, it is imperative to find alternatives to the peroxide system in order to use colloidal motors in a broader range of applications.

In this presentation we develop a diffusiophoretically swimming colloid fueled by bio-conjugation. These are a class of new, biologically compatible engines which find application in emerging technologies requiring motors to transport cargo as in cellular drug delivery, microfluidic "lab-on-a-chip" platforms for point of care diagnostics and the bottom-up fabrication of nanoparticle assemblies. Colloids with one functionalized surface (Janus particles) displaying protein receptors are immersed in a solution containing a ligand which conjugates to the receptor. Conjugation on the active side of the colloid creates a concentration gradient of the ligand that drives the diffusiophoretic motion. As the receptors become saturated the motion decreases to zero and, as a result, this scheme can be used to propel colloids over a prescribed time interval. Using optical microscopy and particle tracking, we compare the randomized Brownian motion of colloids to the directed motion of Janus motors to verify the concept of self-diffusiophoretic swimming driven by asymmetric bioconjugation and the timed nature of the motion.

Experimental measurements of the self-propulsion velocity compare favorably to an asymptotic theory in which the length scale of the Van der Waals interaction is much smaller than the particle radius. The concentration field driving the motion is determined by the unsteady binding of the ligand to the surface receptor and hence captures the fact that the velocity can only be sustained for a finite time.

(499i) Oscillatory Responses of Mechanically-Excited Sessile Drops and Polymer Balloons

Chun-Ti Chang[1], Joshua Bostwick[1], Susan Daniel[2] and Paul Steen[2], (1)Theoretical and Applied Mechanics, Cornell University, Ithaca, NY, (2)Chemical and Biomolecular Engineering, Cornell University, Ithaca, NY

Sessile drops and rubber balloons possess contrasting mechanical responses. Whereas the former exhibits the one-fold static behavior predicted by Young-Laplace equation, up to two folds exist on the same characteristic curve of the latter. With the coverage of soft polymer film on its free surface, a drop changes its static behavior and stability as a result of the re-enforced surface tension. The oscillatory modal responses vary accordingly, as the fluid inertia force and solid elasticity compete. Similar behaviors between resonating drops and balloons, such as their modal wave patterns and the corresponding harmonic and sub-harmonic responses, have been repeatedly observed in various experiments in spite of the apparent difference between the two media: a liquid versus a liquid-solid composite continuum. Such similarity in their dynamic responses essentially qualifies balloons as drops with a much higher effective surface tension. On the theoretical side we study mechanisms of symmetry-breaking and change of stability. On the application side, we manipulate geometric arrangements and topological configurations of rubber balloons for implementing active surfaces with adjustable binary properties.

(499j) Computational Investigations Into the Effects of Viscoelasticity On Liquid Transfer During a Gravure Process

J. Alex Lee and Matteo Pasquali, Chemical and Biomolecular Engineering, Rice University, Houston, TX

Liquid coating and printing processes are attractive for bringing large-scale manufacturing capabilities to potentially important advanced functional materials such as structured films for which fabrication is typically limited to the lab-scale. For example, gravure coating of dilute block copolymer solutions has been shown as facile fabrication method for thin nano-pillared films. Accordingly, there is an emerging interest in modeling the transfer of liquids during gravure-like processes, both experimentally and computationally.

Two salient features of the action of liquid transfer during gravure are the extensional nature of the flow (c.f. axisymmetric filament stretching) and de-wetting of the liquid from the gravure cell wall associated with emptying. Coating liquids often exhibit complex rheologies due to the presence of polymers or particles; in the block copolymer example above, viscoelasticity of the liquid can be expected to play an important role. While filament stretching studies abound in the literature, the consideration of dynamic wetting in such flows is lacking, and the problem of computing viscoelastic flows with dynamic (de)-wetting in general is largely an open one.

We extend the method of computation for viscoelastic filament stretching flows [1] with a model for dynamic de-wetting based on the commonly used Navier-slip law at the solid-liquid boundary and a dynamic contact angle. For ALE methods such as ours, the slip law causes fatal problems at the dynamic contact line when computing the evolution of the conformation tensor due to unresolvable stress there. Allowing the contact angle to be dynamic can mitigate these problems.

Recent experiments [2] demonstrate the effects of viscoelasticity in a model process whereby liquid is pulled out of a cavity (truncated cone) onto a moving flat piston. Fraction of liquid transferred is examined as a function of piston speed for Newtonian and viscoelastic liquids. Viscoelasticity is shown to affect liquid transfer either adversely or favorably depending on the interplay of gravitational and inertial effects. We discuss computational results in relation to the experiments, and show qualitative success in predicting the enhancement or reduction in liquid transferred out of the cavity when the liquid is viscoelastic.

[1] Bhat, P., Basaran, O., & Pasquali, M. (2008). Dynamics of viscoelastic liquid filaments: Low capillary number flows. *J. Non Newtonian Fluid Mech.*, *150*(2-3), 211–225.

[2] Sankaran, A. K., & Rothstein, J. P. (2012). Effect of viscoelasticity on liquid transfer during gravure printing. *J. Non Newtonian Fluid Mech.*, 175-176(C), 64–75.

(509a) Using SAFT-FMT-DFT to Model the Adsorption of Light Gases and Hydrocarbons in Activated Carbon

Lucas A. Mitchell[1], Bryan J. Schindler[1], Carolina dos Ramos[1], Clare McCabe[1], Peter T. Cummings[1,2] and M. Douglas LeVan[1], (1)Chemical and Biomolecular Engineering, Vanderbilt University, Nashville, TN, (2)Center of Nanophase Materials Sciences, Oak Ridge National Laboratory, Oak Ridge, TN

The use of molecular simulation, the development of accurate models, and the comparison of theoretical predictions with experimental results help to advance our understanding of adsorption and its application. One prominent molecular model is density functional theory (DFT), which is commonly used to treat the adsorption of spherical molecules in slit-shaped carbon pores. In this paper, we couple DFT with an accurate molecular-based equation of state to calculate the fluid thermodynamic properties using fundamental measure theory (FMT), which is a rigorous theory for homogeneous and non-homogeneous hard-sphere chain fluids. An accurate model is obtained with adsorbing molecules treated as hard-sphere chains with square well interactions. The Mansoori-Carnahan-Starling-Leland and Carnahan-Starling-Boublik equations of state are used for the hard sphere interactions, and the statistical associating fluid theory for potentials of variable range (SAFT-VR) is used to describe the square-well fluid. For adsorption in pores, various potentials can be used for the interaction of the chain molecules with the wall. After comparing model predictions with results from Monte Carlo simulations, the model is used to obtain a pore size distribution for activated carbon from experimental data for adsorption of nitrogen, which is treated as a non-spherical molecule. The adsorption of n-pentane in slit-shaped carbon pores of various widths is then predicted from the model, and together with the pore size distribution, adsorption isotherms are obtained for comparison with experimental data.

(509b) Removal of Oxygenated Species From Liquid Toluene by Adsorption

Cristian Brunchi, Andrzej I. Stankiewicz, Herman J. M. Kramer and Thijs J. H. Vlugt, Process & Energy, Delft University of Technology, Delft, Netherlands

Removal of oxygenated species from liquid toluene by adsorption

Cristian C. Brunchi, Andrzej I. Stankiewicz, Herman J.M. Kramer and Thijs J.H. Vlugt

Process& Energy Laboratory, Delft University of Technology, NL-2628 CA Delft, the Netherlands

Corresponding author e-mail: c.c.brunchi@tudelft.nl

The goal of this study is to develop a general methodology that can allow one to select a good zeolite adsorbent for any specific case study that requires removal of traces of mixtures from bulk liquid systems. As a case study, we are investigating the removal of mixtures of oxygenated odor compounds from liquid toluene using adsorption as a separation technology. The typical odor compounds present in the toluene stream are: butanal, ethyl-hexenal, dimethylcyclohexanone, trimethylphenol and trimethylanisole. Pure component adsorption isotherms in Na-Y zeolite are computed using Monte Carlo simulations. Binary, ternary and 6-component mixture isotherms are generated using the Adsorbed Solution Theory (AST). The fugacity of the liquid bulk mixture is calculated with two approaches: using the Peng-Robinson EOS and by calculating the activity coefficients in the bulk liquid mixture (gama-phi approach). Finally the generated adsorption isotherms are compared with experimentally obtained adsorption isotherms. Preliminary results show that the system behaves strongly non-ideal in the adsorbed phase.

(509c) Multicomponent Adsorption of Alcohols Onto Silicalite-1 From Aqueous Solution: Assessment of Ideal Adsorbed Solution Theory

Peng Bai[1], Michael Tsapatsis[1] and J. Ilja Siepmann[1,2], (1)Chemical Engineering and Materials Science, University of Minnesota, Minneapolis, MN, (2)Chemistry, University of Minnesota, Minneapolis, MN

Configurational-bias Monte Carlo (CBMC) simulations in the isobaric-isothermal version of the Gibbs ensemble (GE) were carried out to probe the adsorption from aqueous solutions of methanol and/or ethanol on silicalite-1. This methodology requires neither specification of the chemical potential nor any reference to activity models based on experimental data. The CBMC-GE methodology can be applied to the complete range of mixture compositions from pure water to pure alcohols, and can also be used when multiple solute types are present at high concentrations. The simulations demonstrate high selectivities for the alcohols (alpha (ethanol) > alpha (methanol)) almost over the entire composition range. The ideal adsorbed solution theory is found to substantially under predict the amount of adsorption for water and leads to very large errors in selectivity at low alcohol concentrations in the solution phase. The simulations indicate that, at lower loadings, the adsorbed alcohol molecules can serve as seeds for water adsorption but, at higher loadings, alcohols displace water molecules from their preferred region.

(509d) Molecular Simulations for Adsorption, Diffusion and Separation of Noble Gas Mixtures In Metal Organic Frameworks

Yeliz Gurdal, Koc University, Istanbul, Turkey and Seda Keskin, Chemical and Biological Engineering, Koç University, Istanbul, Turkey

Since the number of different types of metal organic frameworks (MOFs) that has been synthesized is enormous, molecular modeling of these materials has played an important role in providing information about the performance of MOFs for specific applications. MOFs have garnered much attention for gas separation and gas storage due to their tunable pore sizes. A large number of molecular simulation studies investigated adsorption of CO_2, CH_4, N_2 and H_2 in addition to separation of binary mixtures of these gases. On the other hand, only very few publications addressed capture and separation of noble gases, although noble gas separations play significant role in industry. For example, Xe has several usage are including lighting, imaging, anesthesia, nuclear magnetic resonance, ion propulsion engines. The capture and separation of Xe from Kr is crucial to obtain high purity Xe. Identifying and developing materials which show higher capture and separation capacity for noble gas mixtures would contribute to noble gas separation industry by reducing energy consumption.

In this work, we used atomically detailed modeling methods to examine both single component and mixture adsorption of Xe, Kr and Ar in ten different MOFs. We used both GCMC simulations and Ideal Adsorbed Solution Theory to compute adsorption of mixtures in MOFs. In addition to calculating adsorption-based selectivities, we also computed self diffusivities of each component in Xe/Kr and Xe/Ar mixtures using molecular dynamics and predicted permeation selectivity of MOFs for these mixtures for the first time in the literature. The working capacity and gas permeability of each MOF were calculated using the data obtained from molecular simulations and the performances of MOFs in noble gas separations were compared with well known porous materials. Based on this comparison, we addressed which material properties are desirable to develop more selective MOF materials for Xe/Kr and Xe/Ar separations.

(509e) CO2 /CH4 Separation in Zeolite-Like Metal Organic Frameworks

Baris Demir, Department of Chemical Engineering , Istanbul Technical University, Istanbul, Turkey and M. Göktug Ahunbay, Department of Chemical Engineering, Istanbul Technical University, Istanbul, Turkey

Zeolite-like metal organic frameworks with sodalite-type framework (sod-ZMOFs) are promising candidates for natural gas (CO_2/CH_4) separation and purification applications. They have anionic frameworks and extra-framework ions in the pores of molecular dimensions increasing the interactions with guest molecules and improving their separation, storage, or ion-exchange capability. The indium-based sod-ZMOF, with imidazoledicarboxylic acid as the organic linker, has higher CO2 adsorption capacity than ZIF-8, which has an analogous structure based on sodalite-topology. The charge-compensating extra-framework ions inside the cavities of these MOFs can be ion-exchanged with alkali metals to improve CO_2/CH_4 separation performance. Furthermore they can be used in the preparation of polyimide-based mixed-matrix membranes due their good compatibility and dispersion properties.

In this study, adsorption and diffusion of CO_2 and CH_4 gases in the sod-ZMOFs were studied using Grand Canonical Monte Carlo and Molecular Dynamics simulations. The simulations were carried out with "as-synthesized" as well as Li^+, Na^+ and K^+ exchanged sod-ZMOFs. The selectivity CO_2 /CH_4 selectivity of the sod-ZMOFs were compared based the type of the cation and the ion-exchange ratio over a wide range of pressure. Adsorption and diffusion simulations showed that the selectivity of these MOFs increases two orders of magnitude when it is fully exchanged with Na^+ ions. Furthermore, the selectivity depends not only on the cation type and ion-exchange ratio but also on the location of the cations in the framework.

(509f) Carbon Dioxide Adsorption On Mesoporous Amorphous Silica Surfaces with Tethered Propylamine Groups

Aleksey Vishnyakov, Chemical and Biochemical Engineering, Rutgers, The State University of New Jersey, Piscataway, NJ, Aatto Laaksonen, Department of Materials and Environmental Chemistry, Stockholm Univerisity, Stockholm, Sweden and LiJun Chen, Chemistry, Stockholm University, Stockholm, Sweden

Using a combination of classical and ab initio modeling methods, we study adsorption of carbon dioxide and nitrogen on amorphous silica grafted with propylamine chains at moist conditions that may be found in post-combustion gases. We model physisorption and chemisorption aspect separately, using classical and ab initio tools, respectively, and consider different chemisorption mechanisms leading to formation either carbamates or bicarbonates. The carbamate mechanism requires two amine groups per each CO2 molecule chemisorbed, while bicarbonate formation requires a single amine. Thus, bicarbonate formation is desired from the practical point of view, but favorable only when reactants are surrounded by water molecules. Earlier experimental results indicated bicarbonate mechanisms; however, a few recent studies insist on the carbamate mechanism of CO2 chemisorption.

Ab initio modeling shows that activation barriers are rather low for both chemisorption mechanisms. Therefore, in these systems, physisorption mechanisms strongly influence chemisorption mechanisms, because physisorption precedes chemisorption. Since published experimental results on water sorption at amine-grafted silica vary dramatically, and we modeled CO2 physisorption at various levels of hydration using classical molecular dynamics and grand canonical Monte Carlo methods. Generally, we conclude that the physisorption scenarios are unfavorable to bicarbonate formation at all but highest hydration levels, which are not typically observed in experiments. In particular, water resides closer to the silica surface than CO2. CO2 molecule in the process of chemisorption reaction act is typically surrounded by alkylamines that are known to favor carbamate mechanism.

(509g) First-Principles Study of H2O Adsorption On the NaBH4(100) Surface

Ping Li, University of Pittsburgh, Pittsburgh, PA, J. Karl Johnson, Dept. of Chemical and Petroleum Engineering, University of Pittsburgh, Pittsburgh, PA and Michael A Matthews, Chemical Engineering, University of South Carolina, Columbia, SC

Sodium borohydride ($NaBH_4$) is a promising hydrogen storage material for niche applications due to its high gravimetric (10.7 wt%) and large volumetric (115 kg of H_2 m^{-3}) hydrogen content. The hydrolysis of $NaBH_4$ releases hydrogen with both fast kinetics and high extent of reaction under technical conditions by using steam deliquescence of $NaBH_4$. The $NaBH_4$ crystal is strongly hygroscopic, rapidly absorbing water vapor from the atmosphere and then dissolving into solution, even at fairly low relative humilities. The hydrolysis reaction takes place in solution at sufficiently high temperatures to produce H_2 and hydrated $NaBO_2$. To the best of our knowledge, there are no experimental data on the adsorption energy or structure of adsorbed water on $NaBH_4$ surfaces. It is difficult to characterize the initial adsorption processes on $NaBH_4$ surfaces from experiments because the adsorption and dissolution of the crystal occurs rapidly. We have therefore used first-principles density functional theory (DFT) to probe the details of the initial adsorption process at the atomic level. Standard DFT functionals do not include an adequate description of van der Waals interactions, which are expected to be important for the energetics and structure of H_2O adsorption on $NaBH_4$. We have therefore used two different approaches for including van der Waals forces within DFT calculations and have compared our calculations with standard generalized gradient approximation DFT calculations. We have computed the adsorption energies and geometries for various coverages of H_2O on the $NaBH_4(100)$ surface. We characterize the H_2O-H_2O and H_2O-surface interaction energies as the number of H_2O molecules on the surface increases. Diffusion of water on the surface is also an important process for cluster growth. We have used the Climbing Image-Nudged Elastic Band method to compute the diffusion pathways for an isolated water molecule on $NaBH_4(100)$.

(514a) Crystals Are Like People: Growth and Defects Are What Make Them Interesting

Michael F. Doherty, Chemical Engineering, UC Santa Barbara, Santa Barbara, CA

"Crystals Are Like People: Growth and Defects Are What Make Them Interesting"

There is a growing interest in predicting and controlling the size and shape of crystalline particles. Multidimensional population balances have been developed to accomplish this task but they suffer from the drawback of needing the absolute growth rate for every family of faces that may appear on the crystal surface. Such growth rates are known for only a handful of crystalline materials and prospects are bleak for extending the library of growth rate data. This raises the question of where the surface growth rates for all the families of faces will come from to drive multidimensional population balance engineering technology. One answer is "from first principles." In this lecture it will be shown how to reformulate multidimensional population balances in terms of relative growth rates and how to create first principles mechanistic models to calculate these quantities for real molecular crystals as a function of supersaturation.

(514b) Determination of Crystal Growth Rates for Solids Growing From Multiple-Conformers Solutions and Evidence of Growth Inhibition by the Wrong Conformer

Lotfi Derdour, Drug Product Science and Technology, Bristol-Myers Squibb co., New Brunswick, NJ and Chandrasekar Sivakumar, Chemical Engineering, Carnegie Mellon University, Pittsburg, PA

The validity of a model derived for crystal growth rate for solids crystallized from solutions containing multiple conformers is evaluated. A simple and straightforward method for determining crystal growth rates using FBRM and FTIR/ATR is utilized to maximize growth rate data collected from a single experiment. Crystal growth rate data obtained experimentally were in good agreement with model predictions. In particular, the data indicated that when the approach of the right conformer (Derdour et al[[1]]) applies, crystal growth rate is inversely proportional to crystal size. Further, the results indicate that growth rates determined from supersaturation measurements can be over-estimated because of surface nucleation and/or agglomeration that can result from supersaturation spikes during anti-solvent addition. Lastly, crystal growth data obtained at different temperatures indicated that the slow growth rate observed at low temperatures may be caused in part by surface poisoning by the wrong conformer.

---

1 Derdour et al., Crystallization from solutions containing multiple conformers: A new modeling approach for solubility and supersaturation, Chem Eng Sci., 66(1), pp. 88-102, 2011.

(514c) Molecular Driving Forces of Crystallization: A Study of the Crystal Growth Rate From the Fluctuation-Dissipation Theorem

Jia Fu, Chemical and Environmental Engineering, University of California, Riverside, Riverside, CA and Jianzhong Wu, Chemical and Environmental Engineering, UC Riverside, Riverside, CA

A perennial challenge for crystal engineering is identification of thermodynamic conditions favoring crystal growth with controlled size and morphology. While experimental investigations are instrumental to understanding the effects of macroscopic variables on the kinetic processes, analysis and utilization of the experimental data toward a rational design require a fundamental knowledge linking molecular driving forces to thermodynamic variables dictating crystallization. In this work, the crystal growth rates are calculated from NVT molecular dynamics simulations based on the fluctuation-dissipation theorem. Special attention is given to the effect of intermolecular forces on the interfacial behavior and the kinetics of crystallization. By calculating of crystal growth rates at the (001) direction of model fluid-solid systems over a broad range of the parameter space, we find that crystal growth in protein solutions or in colloids can be qualitatively different from that in simple fluids and that the kinetics of colloidal crystallization can be achieved by an effective control of the potential of mean force in addition to thermodynamic conditions.

(514d) Simulation of the Crystal Shape Distribution in the Presence of Growth Modifiers

Aniruddha Majumder[1], Zoltan K. Nagy[1], Sean K. Bermingham[2] and Hassan Mumtaz[2], (1)Department of Chemical Engineering, Loughborough University, Loughborough, United Kingdom, (2)Process Systems Enterprise, London, United Kingdom

Crystal shape and its distribution are of great interest due to the crucial role it plays in determining the quality of the final products and the efficiency of downstream processes. Selected additives as growth modifiers (GMs) are often used in the industries to achieve the desired crystal shape distribution. Thus, prediction and control of crystal shape distribution are necessary for product design, quality control and economical operation point of view. Morphological population balance modelling (PBM) is widely used as a tool to describe the evolution of crystal shape distribution in a crystallizer [1,2,3]. Despite the fact that the effect of impurities (often called additives or growth/nucleation modifiers) on the crystal shape has been demonstrated in many experimental studies, most of the previous works related to the modelling of crystal shape distributions consider only the effect of operating conditions such as supersaturation, cooling profile and seed distribution on the crystal shape distribution. Recently the effect of GMs has been incorporated in single dimensional PBM model, to explain the effect of GMs on the overall growth of particles and on the crystal size distribution (CSD) [4]. However, to the best of our knowledge, no work has been presented in the literature, which incorporates the effect of growth modifiers on the evolution of crystal shape distribution.

In this work, we present a modelling framework and simulation results that can be used to predict the evolution of crystal shape distribution in a crystallizer in the presence of GM. The growth rate of the crystals depends not only on the concentration of the GM, but also the time during which the crystals are exposed to the GM in the mother liquor [5]. These two aspects have recently been incorporated in a mono-dimensional PBM proposed by Fevotte and Fevotte [4]. We further extend this model to a multi-dimensional PBM so that morphological PBM describing the evolution of crystal shape distribution can also be incorporated.

Following the model by Kubota and Mullin [6], the growth rate dependence on the GM concentration is described by surface coverage of the adsorbed species on the growing crystal and a proportionality constant denoting the effectiveness of the adsorbed species. On the other hand, in order to track the contact time of the crystals with GM, the density function is redefined with an additional property that takes into account the exposure time of the crystals to impurity. It is assumed that the transient coverage process can be described by a first order dynamics. The resulting PBMs are simulated using commercially available software gCRYSTAL (Process Systems Enterprise, UK) for 1D problems. On the other hand, well-established high resolution finite volume technique is used for simulation of all the 1D and multidimensional problems [7]. This study would be very useful in understanding the effect of GM as well as in achieving tailored crystal shape distribution.

References

[1] C. Borchert, N. Nere, D. Ramkrishna, A. Voigt and K. Sundmacher. On the prediction of crystal shape distributions in a steady-state continuous crystallizer. *Chem. Eng. Sci.*, 64(4): 686-696, 2009.

[2] G. Yang, N. Kubota, Z. Sha, M. Louhi-Kultanen and J. Wang. Crystal shape control by manipulating supersaturation in batch cooling crystallization. *Cryst. Growth Des.*, 6(12):2799-2803, 2006.

[3] C. Y. Ma and X. Z. Wang. Model identification of crystal facet growth kinetics in morphological population balance modeling of l-glutamic acid crystallization and experimental validation. *Chem. Eng. Sci.*, 70(0):22-30, 2012.

[4] F. Fevotte and G. Fevotte. A method of characteristics for solving population balance equations (PBE) describing the adsorption of impurities during crystallization processes. *Chem. Eng. Sci.*, 65(10):3191-3198, 2010.

[5] N. Kubota. Effect of impurities on growth kinetics. *Cryst. Res. Technol.*, 36(8-10):749-769, 2001.

[6] N. Kubota and J. W. Mullin. A kinetic model for crystal growth from aqueous solution in the presence of impurity. *J. Crys. Grow.*, 152(3):203-208, 1995.

[7] R. Gunawn, I. Fusman and R. D. Braatz. High resolution algorithm for multidimensional population balance equations. *AIChE J*, 50:2738-2749, 2006.

(514e) On Predicting Nuclei Shape Distribution

Meenesh R. Singh and Doraiswami Ramkrishna, School of Chemical Engineering, Purdue University, West Lafayette, IN

Modeling crystallization processes must account for the rate of nucleation as well as the shape distributions of nuclei. Most models that are based on the classical nucleation theory have limited considerations for non-uniform shapes and shape distributions of nuclei, anisotropic surface energies, and non-ideal solution behavior. We present a formalism to predict nuclei shape distribution accounting for the foregoing factors. The methodology includes formulation of Gibbs free energy function for all possible morphologies of a nucleus followed by multi-dimensional maximization to obtain critical shape and size of the nucleus at some supersaturation. Furthermore, the local fluctuations in the supersaturation due to Brownian motion of solute molecules can cause variations in the size and shape of nucleus. Such variations can be modeled using the Fokker-Planck equation provided their transition probabilities are known. The transition probabilities for supersaturation fluctuations are obtained from the framework of density fluctuations.[1, 2] The corresponding Ito's equation is solved following multidimensional optimization in morphology domains to obtain nuclei shape distribution of Potassium Acid Phthalate.

References:

1. Smoluchowski, M., Drei vortrage uber diffusion, brownsche bewegung und koagulation von kolloidteilchen. *Zeitschrift fur Physik* 1916, 17, 557-585.

2. Chandrasekhar, S., Stochastic problems in physics and astronomy. *Reviews of modern physics* 1943, 15, (1), 1-89.

(514f) Molecular Dynamics Simulations of Vapor Bubble Growth and Detachment On A Heating Surface

Jinyong Bao, Department of Chemical Engineering, City College of New York ,City University of New York, New York, NY, Joel Koplik, Department of Physics, Levich Institute, City College of New York and Graduate Center,City University of New York, New York, NY and David S. Rumschitzki, Department of Chemical Engineering, The City College of The City University of New York, New York, NY

Nucleate boiling is an attractive alternative for systems that require large heat transfer from a hot solid due to the large latent heats that are accessible at relatively low wall superheat above the fluid's saturation temperature. Based on our previous simulations of conductively-driven quasi-static vapor bubble growth in an axisymmetric, cylindrical cell comprised of solid and liquid phases of finite thicknesses with fluid motion and heat transfer obeying small Reynolds, Peclet, Capillary and Bond numbers, we found that the appearance and the motion of three-phase contact line (CL) is a critical element, central to physics needed to explain the large heat enhancement in nucleate boiling. However, our simulations also suggested that different ad hoc models for the poorly-understood motion of the CL as the vapor bubble grows do have a significant effect on vapor bubble deformation and detachment when the Bond number is no longer small. Therefore, we have used molecular dynamics (MD) to simulate a nanoscale version of our three-phase system. Molecular interaction physics is the only input, to this calculation and we then observe the resulting CL motion. Because MD not only includes heat transfer, but also fluid flow, this simulation removes many of the restrictions of our earlier continuum calculation for this nano-sized system. Since terrestrial gravity is utterly negligible at the nanoscale, we introduce a fictitious body force to mimic relevant Bond numbers. In our MD simulations, we study not only the heterogeneous birth of a vapor nucleus on the solid-fluid interface at constant pressure by virtue of heat transfer through the solid and its achievement of stability, but also its subsequent growth, deformation and detachment.

Our MD simulation domain of interest is a cuboid, which is composed of the fluid region sandwiched between two parallel solid walls. The fluid phase is made of argon atoms interacting via a 12-6 Lennard-Jones (LJ) potential. The atoms of two solid walls are tethered via springs to their lattice positions and also interact with both the fluid and each other via LJ potentials. By considering the effects such as the wettability of the solid surface, we find the appropriate interaction parameters between the solid and fluid atoms in LJ form, which are critical parameters for being able to observe nucleation and heat transfer. Periodic boundary conditions are applied in 4 vertical boundaries of the domain. After achieving thermal equilibrium of all the phases in the simulation domain (solid walls and liquid argon) at the uniform reduced temperature of T=0.75 (the fluid saturation temperature), we start to expend the top wall gradually at constant reduced temperature T=0.75 to some specified pressure at which no vapor bubble appears. Next, we instantaneously increase the temperature at the bottom layer of the bottom wall while keep the original lower temperature at the top wall, both maintained by thermostating. This maintains a temperature gradient through the fluid phase. Meanwhile, the pressure at the top wall is kept fixed by allowing the top surface to freely slide up and down. Our simulations show that after some vapor patches appear and disappear on the solid surface randomly in space and time, one of these patches successfully grows into a stable vapor bubble, whose growth and CL motion we then trace.

As noted, by artificially applying a uniform body force that creates a relevant Bond number, we have successfully nucleated and grown a vapor bubble to a moderate size and observed bubble necking and detachment. We found that, despite the existence of temperature slip between solid and liquid, the vapor bubble volume still grows at $t^{3/2}$, t being time, as in our earlier conduction-only continuum calculations. This scaling is also in agreement with numerous experimental observations. Furthermore, our temperature profiles show that vapor bubble surface and the three-phase CL appear to be at the liquid saturation temperature, which confirms the assumptions made in our continuum calculations' boundary condition at liquid-vapor interface. By further increasing this artificial uniform body force (i.e., Bond number) and keeping it constant, we find the body force indeed deforms the bubble and causes it to detach from the surface. During this process, we reset/raise the temperature at the top wall to avoid vapor condensation at the top of vapor bubble as the bubble grows, detaches and rise into regions of lower temperature, without changing the magnitude of the uniform body force. These simulations result in, among other things, the time evolution of the CL motion. The radius of CL initially expands with the growth of the bubble at low Bond number. Then, as the bubble grows to a size where the body force becomes significant, the CL contracts sharply but continuously as the bubble deforms until detachment. In addition, we also consider the effects of different parameters, such as those that influence the temperature slip between solid and liquid, the magnitude of the body force, the temperature of the bottom wall and the thickness of the bottom wall on vapor bubble growth and detachment.

(520a) Tuning Protein Transport in Nanostructured Block Copolymer Hydrogel Systems

Vicki Cheng and Lynn Walker, Chemical Engineering, Carnegie Mellon University, Pittsburgh, PA

Water soluble PEO-PPO-PEO triblock copolymers are capable of forming thermoreversible gels that consist of packed micellar aggregates with nanoscale periodicity. These micellar crystals have interstitial spacings on the order of several nanometers, which can be used to template comparably sized nanoparticles or globular proteins with hydrodynamic diameters (D_h) ranging from 4-7 nm. The resulting structure of such a nanoparticle-hydrogel composite system is that of spatially ordered nanoparticle arrays which are embedded within a nanostructured block copolymer matrix. This work characterizes the transport behavior of nanoparticles (proteins) in these micellar crystals using fluorescence recovery after photobleaching (FRAP). Relevant parameters for controlling particle transport in these systems such as nanoparticle loading, particle size, and block copolymer phase are identified and quantified. The diffusivity of three fluorescently tagged proteins (Lactalbumin, D_h=3.8 nm; BSA, D_h=7 nm, Gamma globulin, D_h=9.4nm) is measured in a series of different block copolymer hydrogels with different unit cell dimensions. Protein diffusivity is measured in a wide range of block copolymer concentrations, which covers block copolymer phases spanning from disorded micellar fluids to packed micellar structures.The measured protein diffusivities in the gels are reduced by several orders of magnitude when compared to diffusion in free solution, and demonstrate the versatility of these self-assembled systems for controlling nanoparticle transport.

(520b) Block Copolymer - Nanoparticle Co-Assembly Facilitated by Solvents

Biswajit Sarkar and **Paschalis Alexandridis**, Department of Chemical and Biological Engineering, University at Buffalo - The State University of New York (SUNY), Buffalo, NY

The spatial arrangement of nanoparticles in polymer matrices can result in interesting hybrid nanocomposite materials. Block copolymers provide self-assembled systems with great versatility as to the types of nanostructures formed and the conditions under which they are stable. We are exploring the interplay between block copolymer organization and nanoparticle surface properties as affected by the solvents present. We discuss here systems of silica nanoparticles dispersed in micellar solutions and lyotropic liquid crystalline phases formed by poly(ethylene oxide)-poly(propylene oxide) block copolymers (Pluronics) in water and its mixtures with polar organic solvents. We report on the effects of nanoparticle surface chemistry, size, and loading on the block copolymer organization.

The spatial arrangement of nanoparticles in polymer matrices can result in interesting hybrid nanocomposite materials. Block copolymers provide self-assembled systems with great versatility as to the types of nanostructures formed and the conditions under which they are stable. We are exploring the interplay between block copolymer organization and nanoparticle surface properties as affected by the solvents present. We discuss here systems of silica nanoparticles dispersed in micellar solutions and lyotropic liquid crystalline phases formed by poly(ethylene oxide)-poly(propylene oxide) block copolymers (Pluronics) in water and its mixtures with polar organic solvents. We report on the effects of nanoparticle surface chemistry, size, and loading on the block copolymer organization.

(520c) NMR Investigation of Surfactant Structure Effects On Mixing in Cationic/Carbohydrate Surfactants

Saikat Das, Chemical and Materials Engineering , University of Kentucky, Lexington, KY and Stephen E. Rankin, Chemical and Materials Engineering, University of Kentucky, Lexington, KY

Mixtures of cationic and non-ionic surfactants have recently been the subject of a number of investigations due to the synergistic effects that arise in their mixing and properties. In these mixed micelles, the molecular interactions between the surfactants plays a determining role in micellization thermodynamics and therefore characteristics such as critical micelle concentration (cmc), micelle size, aggregation number, etc. Although several investigations have reported global thermodynamics of mixed surfactants with the help of a regular solution theory-based interaction parameter (β) (see for instance the review of Ruiz in *Sugar-Based Surfactants: Fundamentals and Applications*, Surfactant Science Series vol. 143, CRC Press: 2009, p. 413), this parameter is only an indicator of net enthalpy changes associated with mixing in micelles. The relative positions of the surfactants in the mixed micelle and the interaction sites responsible for observed enthalpy changes in cationic-non-ionic surfactants are not as thoroughly understood.

Here, a variety of NMR (nuclear magnetic resonance) techniques are applied to gain insight into the relative positions and the interaction sites between alkyltrimethylammonium-based cationic and carbohydrate-based (non-ionic) alkyl ether surfactants in mixed micelles assembled in D_2O (deuterated water). Our main objective is to determine combinations of surfactants in which the sugar head group is positioned slightly outside of the micelle so that there is a good chance that this surfactant combination can be used in future experiments to provide good interactions with a metal oxide matrix for molecular imprinting. We use spin-spin relaxation measurements (T_2 analysis) and nuclear Overhauser effect spectroscopy (1D NOE spectra and 2D NOESY) for this purpose. Effects of carbohydrate head group hydrophilicity (gluopyranoside vs. xylopyranoside) and alkyl tail length are investigated. The NOESY evidence suggests that in mixed micelles of n-octyl-b-D-glucopyranoside (C8G1) and hexadecyltrimethylammonium bromide (C_{16}TAB), the glucose headgroup is located just inside of the corona region of the micelle, and therefore that it is not exposed to the external solution. Trends in the T_2 values and NOE spectra indicate that changing the headgroup to the less hydrophilic xylopyranoside exacerbates the problem by driving the headgroup more deeply into the core of the mixed micelle. We also study the effects of the relative lengths of the alkyl tails in the cationic and carbohydrate surfactants. As the chain length of the cationic surfactant decreases, the favorable interaction with the sugar headgroup found for the C8G1 / C_nTAB mixture is known to decrease [Ruiz]. Our NMR results are consistent with the carbohydrate being "pushed" outwards towards the corona of the micelle as this occurs, although in extreme cases (such as when the length of the cationic surfactant tail is less than that of the carbohydrate surfactant) this may have the undesirable consequence of reducing the level of mixing between the two surfactants in micelles.

(520d) Molecular Simulation of Trisiloxane Surfactants At the Air/Water Interface

Rolf E. Isele-Holder and **Ahmed E. Ismail**, Mechanical Engineering, RWTH Aachen University, Aachen, Germany

Trisiloxane surfactants are widely used in various products such as paints, inks, and herbicides, because of their capacity to enable "superspreading"—the greatly enhanced spreading of aqueous solutions on hydrophobic surfaces. Despite numerous studies on trisiloxane surfactants and superspreading, the mechanisms that explain superspreading have not yet been identified. Proposed explanations include the extreme reduction of the air/water surface tension in the presence of trisiloxane surfactants, the self-association of the surfactants, the transport of the surfactants within water droplets to the interfaces, and the interaction of the surfactants with the substrate.

Molecular dynamics simulations have the potential to provide deeper insight into the superspreading mechanism. However, before the superspreading problem can be addressed, improved algorithms and molecular models for trisiloxane surfactants are required. For efficient and accurate calculation of long-range dispersion forces, which are known to be especially relevant in simulations with surfaces, we have developed and optimized a PPPM method for dispersion interactions. Furthermore, we optimized the interactions of an existing all-atom force field for the polyethylene oxide chain of the surfactant to the TIP4P/2005 model, the best non-polarizable water model currently available for describing the liquid-vapor interface, to obtain an improved behavior of the trisiloxane surfactants in aqueous solutions. We use this model to study the adsorption of the trisiloxane surfactants at the air/water interface in various concentrations. The reduction of the surface tension, as well as the structural behavior of the trisiloxane surfactants and water near the interface that explain the simulated surface tensions, are reported. We also compare our findings to experimental results from sum frequency generation.

(520e) Low-Density Self-Assembled Monolayers for Functional Host-Guest Thin Films

Duhyun Shin and Stephen M. Martin, Chemical Engineering, Virginia Tech, Blacksburg, VA

We have investigated the exchange of functional guest molecules for tunable surfaces with host-guest thin film systems. Self-assembled host-monolayers containing pre-existing cavities have been prepared on a quartz-crystal or gold substrate via a stepwise reaction. The host-monolayers consist of an integrated spacer, usually aromatic group such as anthracene-derivatives, at the base of the cavity. Guest molecules consisting of a functional moiety and a long alkane tail can be then inserted into the host-cavities in order to form functionalized surfaces. The insertion and replacement processes of guest molecules from an organic solution into the host-cavities have been measured using quartz crystal microbalance with dissipation monitoring, QCM-D. In addition to the frequency (mass) change of inserted quest molecules, changes in the film structure upon guest insertion was also obtained via X-ray reflectivity. The resultant host-guest thin films were also characterized by contact angle measurements, Fourier Transform infrared spectroscopy, and X-ray photoelectron spectroscopy.

The Micellization Coupled Polymerization Behavior of Sodium Conjugated Linoleate

*Ma Lin, Fan Ye, Liu Yuhang, Li Yandi, Liu Zebang, Chen Zeyang, Fang Yun**
(School of Chemical & Material Engineering, Jiangnan University, Wuxi, Jiangsu 214122, China)

Abstract

Polymerized micelles synthesized from self-assembled aggregates of the precursor polymerizable surfactants have attracted more and more attention because they are more stable than the precursor self-assembled micelles or ordinary micelles from corresponding conventional surfactants. As its sp[1]ecial structure, polymerized micelles have important applications in many fields, such as drug delivery carrier, microreactor and templating functional nanomaterials. In this paper, sodium conjugated linoleate (CLA-Na) was used as the precursor polymerizable surfactant, which self-assembled to form micelles in aqueous solution and followed the stable polymerized micelles of CLA-Na were obtained by initiating thermal polymerization of the micellized CLA-Na molecules. At certain concentration and pH value, the polymerization of CLA-Na polymerizable micelles was investigated in details by changing initiator, polymerization time and temperature. The pro- or post-polymerized micelles of CLA-Na were characterized and compared with each other by UV-Vis and TEM. The optimum polymerization was reaction at 60 °C for 6 h with APS as initiator. At pH 8.5 and when the total concentration of CLA/CLA-Na was 5×10^{-3} mol/L, the polymerized micelles possessed spherical morphology and uniform diameter (D=60-80 nm).

Key words: conjugated linoleic acid, polymerized micelle, self-assembly

Introduction

As small ordinary self-assembled micelles of surfactants relied on weak non-covalent interaction to maintain its morphology, they were easily influenced by environmental changes. Such dynamic characteristic and concentration-dependent of small self-assembly of surfactants impeded its structure characterization to large extent. Besides, the instability, which was lack of enough strength, limited its applications in drug delivery carrier, microreactor and templating functional nanomaterials.

In recent years, the self-assembly of polymerizable surfactants, as well as the research in fastening the ordered aggregation via polymerization, have been reported at home and abroad. Although most of the researches were concentrated on the synthesis of polymerizable surfactants[1-7], there were rarely study on natural polymerizable surfactants of unsaturated fatty acids. Actually, the synthesis and applications of polymerizable surfactants have been early

*Corresponding author: yunfang@126.com. Supported by the National Natural Science Foundation of China (21276113).

reported by Freedman[8] in 1958, but it were obtained widely attentions until the last two decades. Introducing double bonds into surfactants to form polymerizable surfactants, made it self-assemble to ordered self-assembly, which was fixed via polymerization. It was not only having theory significances, but also broadened its applications, which has more prospects.

In this paper, sodium conjugated linoleate (CLA-Na) was used as the precursor polymerizable surfactant to obtain polymerized micelles. At certain concentration and pH value, the polymerization of CLA-Na polymerizable micelles was investigated in details by changing initiator, polymerization time and temperature. The pro- or post-polymerized micelles of CLA-Na were characterized and compared with each other by UV-Vis and TEM.

Experimental Section

Materials

Sodium hydroxide, borax, boric acid, azodiisobutyronitrile(AIBN), and ammonium persulfate(APS) were all purchased from Sinopharm Chemical Reagent Co., Ltd. (SCRC). Ultrapure water, with the resistivity of 18.2MΩ·cm, was twice distilled and filtered.

Polymerization

Samples for polymerization were prepared with ultrapure water and adjusting the pH of the solution. And then the samples were depleted of oxygen by bubbling with nitrogen for 25 min. Adding initiator into the solution with the flu of nitrogen for another 10 min. The CLA-Na micelles were polymerized by heating the solution to certain temperature for certain hours.

UV-Vis Analysis

When the total concentration of CLA/CLA-Na was 3×10^{-5} mol/L, pro- or post-polymerized solution of CLA/CLA-Na was analysis by UV-Vis spectrophotometer (Beijing Purkinje General Instrument Co., Ltd.).

TEM Observation

At pH 8.5, samples containing pro- or post-polymerized micelles of CLA/CLA-Na were observed by transmission electron microscopy (Japanese Electronics Co., Ltd.).

Results and Discussion

Factors

As conjugated double bonds had obvious ultraviolet absorption peak at 233 nm, the peak would be decreased or disappeared after polymerization. Changing initiator, polymerization time and temperature, the polymerization of CLA-Na polymerizable micelles was investigated in details by UV-Vis. Under the same conditions, the water-soluble APS as initiator had better effect than oil-soluble AIBN, and the optimum polymerization was reaction at 60 °C for 6 h.

Self-assembly morphology

Fig. 1 shows the TEM micrographs of pro- and post-polymerized micelles of CLA-Na with pH 8.5 and the total concentration of CLA/CLA-Na was 5×10^{-3} mol/L. The solution of CLA/CLA-Na self-assembled on this condition (Figure 1 B). The molecules of carboxylic acid and carboxylate anions were coexisted and co-assembled, which aggregated spherical morphology and uniform diameter (D=30-60 nm). On this basic, with APS as initiator, the polymerized micelles of CLA-Na (Figure 1 A) possessed spherical morphology and uniform diameter (D=60-80 nm), which were bigger than the micelles before polymerization. It could be explained that during the polymerization, the neighboring CLA/CLA-Na micelles occurred polymerization, which assembled bigger polymerized micelles.

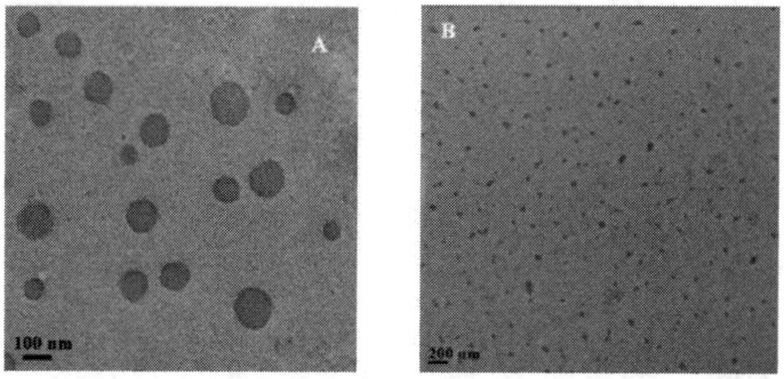

Figure 1. The TEM micrographs of pro- and post-polymerized micelles of CLA-Na (A: pH=8.5, c=5 mmol/L, APS, 10 h, 80 °C; B: pH=8.5, c=5 mmol/L)

Conclusions

At certain concentration and pH value, the polymerization of CLA-Na polymerizable micelles was investigated in details by changing initiator, polymerization time and temperature. In order to obtain stable polymerized micelles of CLA-Na, the pro- or post-polymerized micelles of CLA-Na were characterized and compared with each other by UV-Vis and TEM. The conclusions were as follows:

(1) The optimum polymerization was reaction at 60 °C for 6 h with APS as initiator.

(2) At pH 8.5 and when the total concentration of CLA/CLA-Na was 5×10^{-3} mol/L, the self-assembled aggregates spherical morphology and uniform diameter (D=30-60 nm). And the polymerized micelles of CLA-Na possessed spherical morphology and uniform diameter (D=60-80 nm).

References

1. Abe M, Tsubone K, Koike T, et al (2006), "Polymerizable cationic gemini surfactant," *Langmuir*, 22(20), pp. 8293-8297.
2. Abe M, Koike T, Nishiyama H, et al (2009), "Polymerized assemblies of cationic gemini surfactants in aqueous solution," *Journal of Colloid and Interface Science*, 330(1), pp. 250-253.
3. Alkilany A M, Nagaria P K, Wyatt M D, et al (2010), "Cation exchange on the surface of gold nanorods with a polymerizable surfactant: polymerization, stability, and toxicity evaluation," *Langmuir*, 26(12), pp. 9328-9333.
4. Zhu Z Y, González Y I, Xu H X, et al (2006), "Polymerization of anionic wormlike micelles," *Langmuir*, 22(3), pp. 949-955.
5. Lijima M, Kobayakawa M, Yamazaki M, et al (2009), "Anionic surfactant with hydrophobic

and hydrophilic chains for nanoparticle dispersion and shape memory polymer nanocomposites," *Journal of the American Chemical Society*, 131(45), pp. 16342-16343.

6. Liu S Y, González Y I and Kaler E W (2003), "Structural fixation of spontaneous vesicles in aqueous mixture of polymerizable anionic and cationic surfactants," *Langmuir*, 19(26), pp. 10732-10738.
7. Zhu Z Y, Xu H X, Liu H, et al (2006), "Stabilization of catanionic vesicles via polymerization," *Journal of Physical Chemistry B*, 110(33), pp. 16309-16317.
8. Freedman H H, Mason J P and Medalia A I (1958), "Polysoaps: II. Preparation of vinyl soaps," *Journal of the American Chemical Society*, 23, pp. 76-82.

(522a) Dynamics of Magnetic Field Alignment and Field Effects On Phase Behavior in Block Copolymers by in-Situ SAXS

Chinedum Osuji, Chemical Engineering, Yale University, New Haven, CT

The use of external fields to direct block copolymer self-assembly is well documented. Magnetic fields offer particular promise due to their space-pervasive nature and the ability to produce arbitrary alignments over truly macroscopic length scales in appropriate systems. We present here the results of in-situ SAXS studies performed using a custom superconducting magnet integrated with lab-scale x-ray scattering instruments. We consider the case of side-chain liquid crystalline diblock copolymers ordering under high magnetic fields. Despite the coincidence of the block copolymer order-disorder transition (ODT) and the LC clearing temperature in these weakly segregated materials, there is no measurable effect of the field on the ODT of the system, up to 6 T. This is in line with estimates based simply on the magnitudes of the relevant energy scales - the free energy of field interaction and the enthalpy of the isotropic-LC transition. We show that the alignment of the system is critically limited by the viscosity of the mesophase such that alignment can only be advanced by residence in a small temperature window near the ODT. This residence produces a weakly aligned system which thereafter transitions to a strongly aligned state on cooling even in the absence of the field.

This work was conducted with support from NSF under DMR-0847534.

(522b) Melt Memory of Prior Crystallization of Random Ethylene Copolymers

Benjamin Reid, Florida State University, Tallahassee, FL and Rufina Alamo, Chemical and Biomedical Engineering, Florida A&M University and Florida State University, Tallahassee, FL

We report the effect of molecular weight and comonomer content on the crystallization of model random ethylene 1-butene copolymers. A large set of narrow un-branched polyethylenes (PE) was used as a control. The samples were crystallized from a melt state above the equilibrium melting temperature and cooled at a constant rate. The exothermic heats of the melt-solid transition were reported as the crystallization temperatures (T_c). Following expectations, the T_c of un-branched PE samples was constant and independent of the initial melt temperature. The same independence was observed for copolymers with low molar mass. However, above a molar mass of 4500 g/mol, the T_c of copolymers depends on the temperature of the initial melt, increasing as the temperature of the melt decreases. We attribute the increase in T_c to a melt memory effect from the prior crystallization and correlate this phenomenon with the presence of partial order in the copolymer's initial melt that acts as a nucleant. We attribute the residual order in the melt to increased restrictions to reach the initial random topology of the copolymer upon melting. Erasing memory of the prior sequence selection in copolymer melts requires much higher temperatures than the theoretical equilibrium value.

(522c) Self -Assembly of Conjugated Diblock Copolymers within Nanofibers for Solar Cell Applications

Alda Kapllani, Qinsu Niu, Chau Tran and Vibha Kalra, Chemical and Biological Engineering, Drexel University, Philadelphia, PA

We report the fabrication and self-assembly characterization of nanofibers of oligothiophene-*block*-oligophenylenevinylene (OT-*b*-OPV), a new class of fully conjugated diblock copolymers, using a technique called electrospinning. Electrospinning is a fiber formation technique that uses strong electric field to accelerate and thin a polymer melt/solution jet forming nanoscale fibers with diameters in the range of 50-500 nm. The unique self assembly between electron conducting (OPV) and hole conducting (OT) blocks will minimize recombination of electrons and holes created on photo-excitation of the active layer in photovoltaics, which can potentially increase the efficiency of organic solar cells. The formation of nanofibers of these unique materials will facilitate the development of ultra-light weight breathable smart textiles with integrated devices.

Prior to electrospinning, solution-cast films of OT-*b*-OPV were fabricated and characterized to better understand the effects of electrospinning on block copolymer self assembly. Within the thin film, covalently bonded OT and OPV blocks micro-phase separate to form self-assembly structures with approximately 5 nm domain spacing. In addition, the oligothiophene (OT) blocks exist as highly ordered crystalline domains with spacing of less than 2 nm through intermolecular packing with their alkyl side chains aligned normal to the substrate and the thiophene rings aligned parallel to the substrate through π-π stacking, thereby resulting in hierarchically-ordered self assembly.

Oligomers are not electrospinnable by themselves due to the lack of chain entanglement in solution, which is a prerequisite for the electrospinning process. To overcome this limitation, in this study, nanofibers were fabricated by coaxially electrospinning OT-OPV (oligothiophene-co-oligophenylenevinylene) as the core with poly vinyl pyrrolidone (PVP) as the shell. PVP serves as a spinning aid and allows the formation of a continuous OT-OPV jet in the core. Pure OT-OPV nanofibers were successfully obtained by selective removal of PVP shell post-electrospinning. The external nanofiber morphology was characterized using scanning electron microscopy (SEM). The block copolymer undergoes micro phase separation and exhibits hierarchical self-assembly inside the fibers as seen via both transmission electron microscopy (TEM) and wide and small angle x-ray scattering (SAXS/WAXS). Prior to TEM, nanofibers were first microtomed into 60 nm thin sections to study the self assembly inside nanofibers both along and perpendicular to the fiber axis.

Owing to the fast solvent evaporation and strong extensional deformation during electrospinning, the OT-OPV exhibit a disordered, but percolating self assembly within nanofibers WAXS data reveals the reduction in the p-p distance in nanofibers compared to films. Similar effect has been seen in nanofibers of pure polythiophene (P3HT) compared to films of the same materials. Annealing studies of both films and nanofibers show an enhancement in the materials crystallinity, which is known to enhance conductivity.

(522d) Dynamic Mechanical Strengthening of Polycarbonate

Robert H. Lambeth and Alex Hsieh, U.S. Army Research Laboratory, Aberdeen Proving Ground, MD

Polymers are viscoelastic materials and exhibit a wide range of mechanical deformation behavior. Polycarbonate (PC) has excellent impact strength against small fragments, thus is the choice of material for use in safety lenses as well as in spall layers for plastic/plastic or glass/plastic composites. However, PC not only has very poor resistance to abrasion but also is very notch sensitive. PC becomes prone to crack initiation and propagation, which occurs often in a target with thickness greater than about 5-6 mm, particularly under extreme service (i.e. low temperature) conditions. The geometry dependence has been attributed to a change of stress state from the plane stress to plane strain, thus suppressing yielding and favoring craze initiation and growth and subsequently leading to a catastrophic failure. This characteristic notch sensitivity limits the useful parts thickness of PC components. A common solution is to laminate thin PC sheets together to achieve thicker parts but leads to other issues such as delamination. To improve the performance of polycarbonate under extreme service conditions, we have prepared various co-polycarbonates with varying compositions of bisphenol A (BPA), 4,4'-dihydroxydiphenol (DOD), and triptycene hydroquinone (THQ). Inclusion of DOD and THQ in the polymer composition is expected to reduce the cross-sectional area of the polymer chains and enhance chain packing to improve crazing thresholds. Initial thermal and mechanical testing has shown an enhancement in the glass transition temperature, yield stress, and flow stress of the copolymers over commercial PC when DOD and THQ are incorporated in the feed stock.

(522e) Synthesis and Characterization of Poly(styrene)-b-Poly(4-fluorostyrene)-b-Poly(isobutylene) Using Atom Transfer and Cationic Polymerization

Edward M.A. Guerrero-Gutierrez[1], Michelle Gonzalez[2] and David Suleiman[2], (1)Chemical Engineering, University of Puerto Rico, Mayaguez, Mayaguez, PR, (2)Chemical Engineering, University of Puerto Rico, Mayaguez Campus, Mayaguez, PR

This investigation studied the synthesis and characterization of a novel multiblock copolymer using atom transfer polymerization (ATRP) and cationic polymerization. The block copolymer proposed was made of poly(styrene), poly(4-fluorostyrene) and poly(isobutylene). The diblock copolymer of poly(styrene)-b-poly(4-fluorostyrene) was synthesized using ATRP. The third block, poly(isobutylene), was incorporated using cationic polymerization. This methodology was proposed as an alternative to obtain an advanced fluorinated, elastomeric and glassy polymer capable of being functionalized for several energy applications such as direct methanol fuel cells (DMFC).The multi block copolymer properties were measured using different characterization techniques such as gel permeation chromatography (GPC), thermogravimetric analysis (TGA) and differential scanning calorimetry (DSC). The polymer composition was monitored with Fourier Transform Infrared spectroscopy (FTIR) and Nuclear Magnetic Resonance (NMR) spectroscopy. Significant differences were observed in the thermal and physical properties after the block copolymer was synthesized. Finally, the resulting polymers could be sulfonated to create ionic fluoropolymers that can be used as proton exchange membranes for fuel cell applications.

(522f) Structure and Properties of Polyimide-Poly (ethylene glycol) Films for Fuel Cell Applications

Elyse Coletta, Chemical Engineering, Stanford University, Stanford, CA, Michael Toney, SLAC National Accelerator Laboratory, Menlo Park, CA and Curtis W. Frank, Department of Chemical Engineering, Stanford University, Stanford, CA

Polymer electrolyte fuel cells have the potential to provide cleaner and more efficient energy. Because of the conduction mechanism for current polymer electrolyte membranes, their performance is limited to relatively low temperatures and high humidity. By using different polymers than have been traditionally used, more versatile operating conditions and better efficiencies may be achieved. Because many different properties are needed for an effective polymer electrolyte membrane, multiple polymers often need to be used. Aromatic polyimides are known to be very stable, both thermally and chemically. Poly (ethylene glycol) has been known to provide proton conduction properties under certain conditions. In the current work, aromatic polyimide and poly (ethylene glycol) (PEG) composite membranes have been synthesized for higher temperature fuel cell polymer electrolyte membrane applications in order to exploit the drastically different properties of the polymers. The polyamic acid precursors were synthesized by a classical one pot poly-condensation method that has been shown to work for many different types of polyamic acids. The poly (ethylene glycol) (Mn 1500) was used in pre-polymerized form and incorporated with the polyamic acids in various ways including both blending and chemical bonding to achieve structural features on a small length scale. These polymer systems were converted to polyimide-PEG systems through thermal imidization. This thermal processing can achieve a variety of film properties easily if varied. The polyamic acid precursors were characterized using Fourier transform infrared spectroscopy. The general structure of the films and their thermal properties were analyzed using thermal gravimetric analysis. Small angle x-ray scattering was used to determine the polymer structure on a length scale between 0.5 and 80 nanometers, because it has been shown that features on this length scale can often lend themselves to achieving good conduction properties. Since acids are often needed to provide any significant conduction properties to polymers, phosphoric acid was incorporated into the films by soaking. The films were then analyzed for protonic conductivity properties using a cyclic voltammetry method. Effects of temperature and stability were studied with regards to the conductivity. These composite systems were analyzed in order to determine composition-structure-property relationships for fuel cell membrane applications as many films are studied for structure or properties but rarely the relationship.

(522g) Thin Film Processing and Characterization of Sulfonated Poly(Styrene-Isobutylene-Styrene) Triblock Copolymers

Omar Movil[1] and Agnes Padovani[1,2], (1)Department of Chemical Engineering, University of Puerto Rico, Mayaguez, PR, (2)Department of Engineering Science & Materials, University of Puerto Rico, Mayaguez

Thin films of sulfonated block copolymers are of particular interest for applications in microelectrochemical devices such as sensors, biosensors, fuel cells, and actuators. This study focuses on the fabrication and materials characterization of sulfonated poly(styrene-isobutylene-styrene) (S-SIBS) triblock copolymer films. More specifically, the morphological and nanomechanical properties of the films were targeted such as to understand their integration capability. For these studies, S-SIBS thin films were deposited onto silicon wafers via spin-coating and critical parameters such as sulfonation percent and polymer concentration were evaluated. The effects of a thermal annealing treatment on film morphology were also investigated. A variety of techniques were used for the materials characterization including profilometry, atomic force microscopy (AFM), and nanoindentation. Profilometry results showed that at high polymer concentrations, the films were approximately 10 times thicker as compared to the films fabricated at low polymer concentration. The latter had thickness values on the order of $< 1\mu m$. The film thickness also increased with the sulfonation percent in the film. AFM results revealed a morphological transition from spherical domains to a network-like structure as the sulfonation percent increases from 0% to 45%. In addition, the polystyrene (PS) domain size also increases as a function of the sulfonation percent. Significant effects on the phase-separated morphology were also observed as a result of the thermal annealing treatment, but only in the films fabricated from membranes with a T_g lower than the annealing temperature. Nanoindentation results indicate that both, the elastic modulus and hardness, increase with the sulfonation percent in the film and the differences observed correlate with changes in morphology.

(525a) Thermodynamic Model for Phase Boundary of Gas Hydrates with and without Additives

Min-Kang Hsieh[1], Yen-Ting Yeh[1], Wan-Yi Ting[1], Po-Chun Chen[2], Yan-Ping Chen[1], Shiang-Tai Lin[1] and Li-Jen Chen[1], (1)Department of Chemical Engineering, National Taiwan University, Taipei, Taiwan, (2)Central Geological Survey, P. O. Box 968, New Taipei City, Taiwan

In this work, we propose a new Langmuir adsorption constant for the modeling of the phase boundary of clathrate hydrates over a wide range of conditions. The proposed pressure- and temperature- dependent Langmuir adsorption constant is designed to produce a reduced free volume available to the encapsulated gas molecules as the pressure increases. The Peng-Robinson-Stryjek-Vera (PRSV) equation of state, combined with the COSMO-SAC activity coefficient liquid model through the modified Huron–Vidal (MHV1) mixing rule and with this new Langmuir model can be used to describe various types of three phase coexisting conditions of gas hydrates, from vapor-ice-hydrate equilibrium (VIHE) at low temperatures, to vapor-liquid-hydrate equilibrium (VLHE) at higher temperatures, and to liquid-liquid-hydrate equilibrium (LLHE) at high pressures, using a single set of parameters. The average relative deviations in the equilibrium pressure are found to be 4.57 % for 12 pure gas hydrates over a large range of temperatures (148.8 K to 323.9 K) and pressures (5.35×10^{2} Pa to 8.27×10^{8} Pa). Furthermore, the retrograde behavior observed in CH_4, CO_2, C_3H_8, and i-C_4H_{10}pure gas hydrate systems can all be successfully modeled by the change of free volume at high pressures.

By using this method, the prediction of phase boundary change in gas hydrates with the addition of organic inhibitors and electrolytes is achieved. We examined the accuracy of this method using 5 organic inhibitors and 9 electrolytes, and over a range of temperature (259.0 K to 303.6 K) and pressure (1.37×10^{5} Pa to 2.08×10^{8} Pa). The average relative deviations in the predicted equilibrium temperatures are found to be 0.72 % with organic inhibitors and 0.18 % with electrolytes, respectively. We believe that this method is useful for many gas hydrate related engineering problems such as the screening of inhibitors for gas hydrates in flow assurance.

(525b) Theory of Gas Hydrates: Effect of the Approximation of Rigid Water Lattice

Hrushikesh Pimpalgaonkar, Shivanand Vessam, Srikanth Ravipati and **Sudeep Punnathanam**, Chemical Engineering, Indian Institute of Science, Bangalore, India

One of the assumptions of the van der Waals and Platteeuw theory for gas hydrates is that the host water lattice is rigid and not distorted by the presence of guest molecules. In this work, we study the effect of this approximation on the triple-point lines of the gas hydrates. We calculate the triple-point lines of methane and ethane hydrates via Monte Carlo molecular simulations and compare the simulation results with the predictions of van der Waals and Platteeuw theory. Our study shows that even if the exact intermolecular potential between the guest molecules and water is known, the dissociation temperatures predicted by the theory are significantly higher. This has serious implications to the modeling of gas hydrate thermodynamics, and in spite of the several impressive efforts made toward obtaining an accurate description of intermolecular interactions in gas hydrates, the theory will suffer from the problem of robustness if the issue of movement of water molecules is not adequately addressed.

(525c) Thermodynamic Properties of 2,5-Dimethylfuran + Furfuryl Alcohol, 2,5-Dimethylfuran + Methylisobutylketone, or Furfuryl Alcohol + Methyl Isobutylketone Binary Mixtures At Several Temperatures: Measurements and Modeling

Latifa Negadi[1], Lamia Kara Zaitri[1], Ilham Mokbel[2] and Jacques Jose[2], (1)Laboratoire de Thermodynamique Appliquée et Modélisation Moléculaire, Université de Tlemcen, Tlemcen, Algeria, (2)Laboratoire des Sciences Analytiques-UMR 5280, UCB-Lyon1, Villeurbanne, France

Diminishing fossil fuel reserves and growing concerns about global warming indicate that sustainable sources of energy are needed in the near future. For fuels to be useful in the transportation sector, they must have specific physical properties that allow for efficient distribution, storage and combustion; these properties are currently fulfilled by non-renewable petroleum-derived liquid fuels. Ethanol, the only renewable liquid fuel currently produced in large quantities, suffers from several limitations, including low energy density, high volatility, and contamination by the absorption of water from the atmosphere. Recently, a team from the University of Wisconsin-Madison1 has developed a two-step catalytic process that can convert fructose into a potentially better liquid biofuel than ethanol, 2,5-dimethylfuran. 2,5-Dimethylfuran has 40%-higher energy density and a higher boiling point than ethanol, and is not water soluble. Fructose can be made directly from biomass or from glucose and although there's some work needed before 2,5-dimethylfuran production can be made commercially viable, this new catalytic process looks promising.

The present paper reports the vapor pressures of (2,5-dimethyl furan + furfuryl alcohol), (2,5-dimethylfuran + methylisobutylketone) and (Furfurylalcohol + methylisobutylketone) binary mixtures measured using a static device at temperatures between 313 and 393 K.

The data were correlated with the Antoine equation. From these data excess Gibbs functions (G^E) were calculated for several constant temperatures and fitted to a third-order Redlich-Kister equation using the Barker's method. The three binary systems exhibit positive deviations in G^E for all investigated temperatures over the whole composition range.

The NRTL and UNIQUAC models have also been used and good results were obtained in the prediction of the total pressure.

Acknowledgements

This work has been done in the framework of the bilateral project TASSILI (Ref. 09MDU761).

(525d) Phase Behavior of Binary Systems of Liquid Crystals with CO2

Mariette De Groen, Process and Energy, Delft University of Technology, Delft, Netherlands, Thijs JH Vlugt, Process & Energy, Delft University of Technology, Delft, Netherlands and Theo W. de Loos, Engineering Thermodynamics, Delft University of Technology, Delft, Netherlands

Rod-like liquid crystals (LCs) are compounds having a structured liquid phase, called either nematic (N) or smectic phase (Sm) depending on the degree of ordering. In the phase diagram, this structured liquid phase is located between the solid (S) and the usual isotropic liquid phase (I). The first order phase transition between the structured and isotropic liquid can be controlled by varying the temperature. CO_2 is less soluble in the structured liquid than in the isotropic liquid. Therefore, a liquid crystal may be used as a solubility switch: CO_2 can be dissolved in the isotropic liquid phase and by cooling the isotropic mixture a few degrees, the structured phase is formed and CO_2 is released again. The pressure is kept constant during this process. This CO_2 capture process has the potential to consume less energy than current processes. Nematic liquid crystals are most interesting for capturing CO_2 because the viscosity of nematic LCs is much lower than that of the smectic LCs.

Two parameters are of particular importance for the functioning of this solubility switch: the solubility of CO_2 in the isotropic liquid phase and the solubility difference between the structured- and isotropic liquid phase. The solubility of CO_2 in liquid crystals is determined by the chemical structure of the liquid crystal. Solubility predictions with the Predictive Soave-Redlich-Kwong (PSRK) equation of state indicate that the presence of ether or ester groups leads to a higher CO_2 solubility than for carboxylic acid and alcohol groups. The solubility difference between the structured and isotropic liquid phases is determined by the phase transition enthalpy. A larger transition enthalpy leads to a larger solubility difference.

Phase diagrams (p,T) were measured for samples of pure liquid crystals and for mixtures of liquid crystals with up to 5% CO_2 in a closed environment using a Cailletet setup. The liquid crystals tested can be divided in three different classes: apolar liquid crystals, polar liquid crystals and weakly polar liquid crystals. Preliminary results confirm part of the predictions concerning the solubility in the isotropic liquid: the solubility of CO_2 is influenced by the polarity of the liquid crystal. The apolar liquid crystals have the lowest mass based solubility and weakly poloar molecules the highest solubility. However, the presence of an ether group next to a benzene ring lead to a lower instead of higher solubility. A possible explanation for this behavior is a distorted quadrupolar moment of the benzene ring, leading to decreased affinity for CO_2.

The relative solubility difference depends on the enthalpy: the liquid crystal with the highest enthalpy shows a larger change in the pressure needed to keep the CO_2 dissolved in the two phases. Further research is needed to optimize these parameters.

(525e) Vapor-Liquid Phase Behavior of Reactive System of Isoamylene with Methanol to TAME

Yonghong Li, School of Chemical Engineering and Technology, Tianjin University, Tianjin, China

Vapor-Liquid Phase Behavior of Reactive System of Isoamylene with Methanol to TAME

Yonghong Li[1,2] , Chuanzhuang Feng[1] and Jilian Dong[1]

(1) Key Laboratory for Green Chemical Technology of State Education Ministry, School of Chemical Engineering, Tianjin University, Tianjin 300072, P. R. China

(2) National Engineering Research Center for Distillation Technology, Tianjin University, Tianjin 300072, P. R. China

Abstract

Etherification of isoamylene with methanol over cation-exchange resin catalysts to produce tert-amyl methyl ether (TAME) is the major chemical reaction for etherification of light gasoline which is expected to upgrade light gasoline by raising octane number. The etherification is reversible and exothermic so that it is beneficial to carry out the etherification in a catalytic distillation column. In this process, vapor¨Cliquid phase equilibrium (VLPE) is coupled with reaction which results in complex phase behavior.

In this paper, experiments of VLPE of quaternary mixtures, including methanol, TAME and two isoamylenes (2M1B and 2M2B), were performed in a novel circulation equilibrium cell packed with resin catalyst of D005II. The vapor-liquid phase behaviors coupled with the reaction were investigated at temperature of 338.15K, 348.15K and 358.15K. The reaction equilibrium constants at different temperatures were calculated using experimental data of phase equilibrium and reaction equilibrium equations, The activity-coefficient-models, including Wilson, NRTL and UNIQUAC, as well as the Soave-Redlich-Kwong equation of state were used to correlate the experimental data respectively. The correlation results showed that a good consistency between the experimental data and the Wilson model. This study provides with important thermodynamic information for the process design of etherification catalytic distillation of light gasoline.

Acknowledgments

We acknowledge gratefully the financial support of the Program of China Universities' Innovative Research Terms (No. IRT0936) and the financial support of CNPC (No.09020109).

APPLICATION OF LATTICE CLUSTER THEORY TO POLYMER PHASE BEHAVIOR

Sabine Enders, TU Berlin, Berlin, Germany
C. Browarzik, TU Berlin, Berlin, Germany
M. Fischlschweiger, TU Berlin, Berlin, Germany
D. Browarzik, University of Halle-Wittenberg, Germany

Introduction

The phase behavior of polymer containing systems, like polymer solutions or polymer blends; depend strongly on the molecular weight and the architecture of the polymer in terms of linear, branched or hyperbranched polymers. In order to model the influence of architecture on the phase behavior the Lattice Cluster Theory (LCT), developed by Freed and coworker [1], can be utilized. This approach takes the architecture of the polymer by the short range correlations directly into account. The theoretical framework provides an analytical expression for the Helmholtz energy, where the architecture is incorporated by the architecture coefficients. These coefficients can be estimate by the united-atom approach using graph theory [2]. Even for linear polymers, the LCT offers several advantages in comparison with the classical Flory-Huggins theory, without the application of additional adjustable parameters. The influence of pressure on the thermodynamic properties can by handled by incorporation of empty lattice sites in the theoretical framework resulting in a compressible version of the theory.

In this contribution the LCT is applied to the liquid-liquid equilibria (LLE) of hyperbranched polymer solutions in a single solvent or a solvent mixture [3,4,5,6,7,8], to vapor-liquid equilibria (VLE) of hyperbranched polymer in a supercritical fluid [4] and finally to the solid-liquid equilibrium (SLE) of linear and branched polymers in a solvent [9].

Theoretical Background

The basic idea of the LCT [1] is the extension of the well-known Flory Huggins theory [10], often applied for polymer solutions. In the framework of FH theory, individual monomers are treated as single entities, devoid of any chemical structure. Flory and Huggins [10] employed a very simple mean-field approximation that essentially ignores the details of the polymer chain connectivity and, therefore, cannot distinguish between linear, star, branched, and comb polymer architecture. In 1991, and Freed and Dudowicz [e.g. 1] have developed a systematic expansion of the partition function of a lattice polymer using the LCT. This model takes into account the effect of branching on the thermodynamic properties of polymer blends and emerges in the form of a cluster expansion. As the LCT is an extension of the FH theory, both theories have the same fundamental idea of a lattice, which is occupied by a polymer and so the polymer can be divided in different segments. Using this idea, the exact partition function of a polymer blend, where two segments i and j interacts with the energy ε_{ij} could be found [1]. In the framework of LCT, the corrections to the Helmholtz free energy are derived in form of a cluster series expansion in the inverse coordination number $1/z$ and in the reduced interaction energy $\varepsilon_{\mu\nu}/k_BT$, taking into account the growing correlations between near segments on the same molecule. By introducing these cluster expansions and fundamental statistical thermodynamics, the Helmholtz free energy of a k component mixture is obtained. I detailed description can be found in the literature [1,2,6].

The number and the kind of functional groups can be handled via the Wertheim perturbation theory, applicable for directed forces between the functional groups and the surrounding solvent molecules [2,3,6]. An alternative is a quasi-chemical model, where association equilibria were assumed [7,8].

Results and Discussion

Liquid-Liquid-Equilibria

The best studied hyperbranched polymers are the commercially available polyester from the Boltorn family. They are based on 2,2-bis(hydroxymethyl)propionic acid (bis-MPA) as AB_2-monomer and 2,2-bis(hydroxymethyl)1,3-propanediol as core molecule and were synthesized in a pseudo-one-step reaction, where the core to monomer ratio determines the size and OH-functionality and thus the generation number [11,12]. Polymers prepared by polycondensation usually exhibit a broad polymerization distribution and the properties depend on the polymerization conditions, especially the LLE in water [13]. The experimental cloud-point curves for two production charge of Boltorn H20, named by Boltorn H20a [3] and Boltorn H20b [13], in water are depicted in Figure 1. The dependence of the thermodynamic properties on the production condition can clearly be seen in this figure. In the case of hyperbranched polymers the influence of the polar functional groups has to be considered. This can be realize by two methods, namely by assumption of strong physical interaction using the Wertheim theory [2,3,13] or by assumption of a chemical equilibrium, the so-called ECALM-model [7]. Both approaches allow two kinds of self-association, namely of the polar solvent, mostly water, and of the polymer, and the cross-association of the solvent and the polar end-groups of the polymer. In Figure 1 can be recognized that the LCT in combination with the ECALM model is able to describe the observed phase behavior. The calculations performed with the LCT in combination with the Wertheim theory [2,13] lead to very similar results. The occurring association forces lead to a broadening of the cloud-point curve. For the binary system Boltorn H40 + water the opposite effect was figured out [7]. The reason for this finding is the different molecular size. In the case of Boltorn H20 less water molecules than in the case of Boltorn H40 can form hydrogen bonds with the inner functional groups. Altogether the water content of the cross associates of Boltorn H20 is lower than that of Boltorn H40.

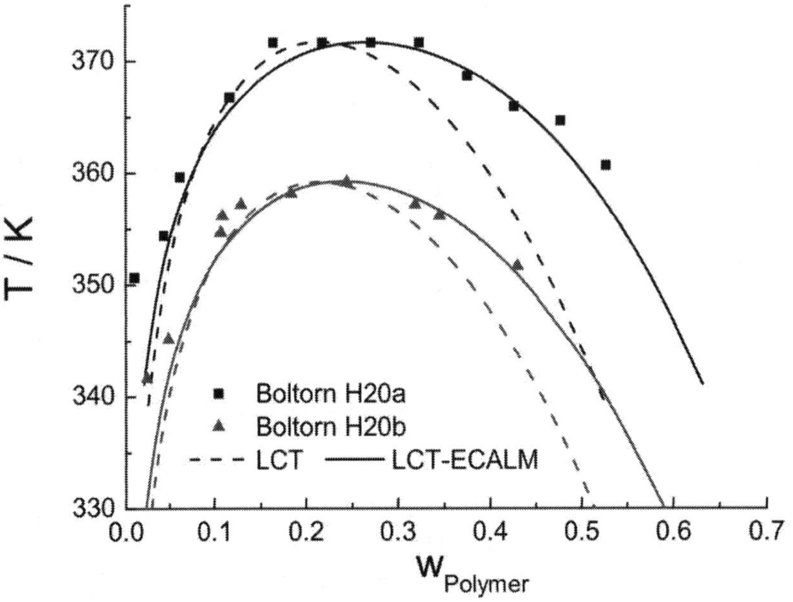

Figure 1: Experimental (black squares: Boltorn H20a + water [3], blue triangles: Boltorn H20b + water [13]) and modeled LLE (solid line: LCT + ECALM, broken line: LCT) [7].

The LCT can also be applied to ternary systems [2,3,5]. For ternary systems, where all binary subsystems exhibit a miscibility gap, the prediction of the ternary phase behavior based on binary experimental data is possible. One example is the ternary mixture Boltorn H20 + water + butan-1-ol. A detailed description of the calculation procedure is given in the literature [2]. In brief, all model parameter were fitted to independent experimental data of the corresponding binary subsystems. The experimental phase diagram, together with the prediction by LCT in combination with the Wertheim theory, is shown in Figure 2. If not all binary subsystems show a demixing behavior the required interaction parameter can also estimated using the VLE of the corresponding binary subsystem. This was demonstrated for the system H20 + water + propan-1-ol, where the binary subsystem water + propan-1-ol is completely miscible [3]. This approach leads to a qualitative agreement with experimental findings [3]. A small readjustment of the interaction parameter permits also a satisfactory quantitative agreement [3].

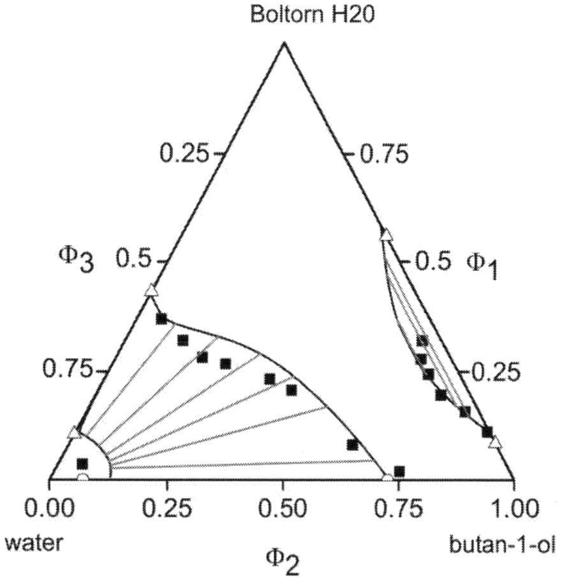

Figure 2: Experimental (squares: experimental cloud point data of the ternary system; open triangles: experimental tie-lines of the corresponding binary systems) and predicted phase behavior (black line: binodal line, red line: tie lines) of the water + butan-1-ol + Boltorn H20 system[2] at 353.15 K.

Vapor-Liquid-Equilibria

Usage of the compressible version of the LCT [2,4,14] allows the calculation of VLE. For systems composed of polymers and near or supercritical fluids often an interesting phase behavior is observed. For example, the binary system propane + Boltorn H3200 shows a VLE and a LLE [15] as function of temperature, pressure and composition. This situation leads to a three-phase equilibrium, where two liquid phases and one vapor phase are in equilibrium. The calculation of this type of phase behavior is quite challenging. An attempt is demonstrated in Figure 3 for the system propane + Boltron H3200, which is a fatty acid modified Boltorn H30 molecules with generation number g=3. In Figure 3, it can be seen a satisfactory agreement between the experimental data and the calculated phase diagram is achieved. Details of the calculation procedure and the parameter estimation method are given elsewhere [4].

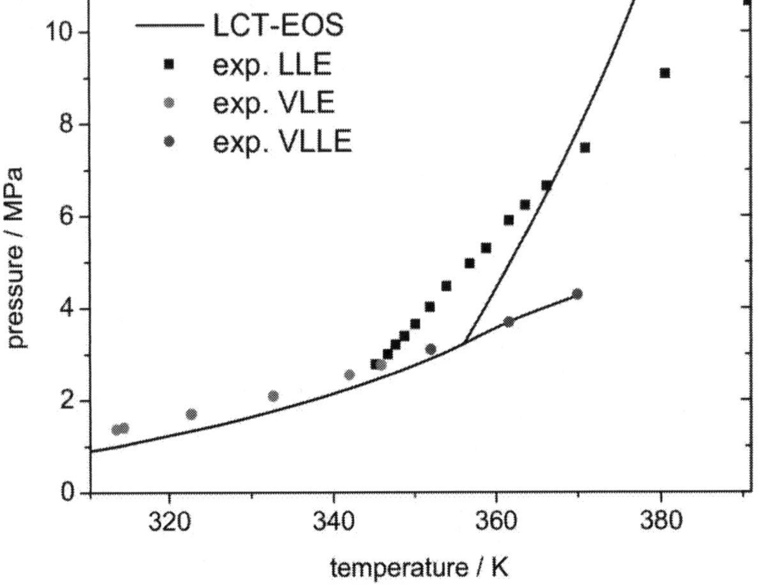

Figure 3: Experimental [15] (black squares: LLE; red circles: VLE; blue circles: VLLE) and calculated [4] phase behavior of propane + Boltorn H3200 with a polymer mass fraction of $w_B = 0.701$.

Solid-Liquid-Equilibria

The SLE forms the thermodynamic background of temperature rising elution fractionation (TREF) and crystallization analysis fractionation (CHYRSTAF). In order to obtain quantitative information about the distribution of lengths of crystallizable polymer sequences mathematical models relating the crystallization conditions inside the TREF or CRYSTAF columns with the microstructural details of the polymer are needed. The LCT, in its in-compressible version, is applied for the calculation of the activity coefficients in the liquid phase and the melting enthalpy and the melting temperature were taken from DSC-measurements.

In order to prove the basic idea the SLE of a simple binary mixture, made from hexane + eicosane, was calculated and the results were compared with experimental data measured by Domanska et al. [17] (Figure 4). In these calculations was assumed that only eicosane forms the solid phase. The pure-component properties of eicosane ($\Delta_{SL}H = 75.4\,kJ \cdot mol^{-1}; T_M = 310K$) were measured by Palczewska-Tulinska et al. [18] using differential scanning calorimetry. Applying LCT for this purpose only one parameter, namely ε / k_b, must be estimated. The experimental data can be modeled with a high accuracy using $\varepsilon / k_b = 5K$ (Figure 4). This very low value can be explained by the chemical nature of both substances. Marsh et al. [16] measured the enthalpy of mixing of a hexane + hexadecane mixture at $T = 313.15K$. These experimental values can be modeled with help of LCT using $\varepsilon / k_b = 5.7K$ [9. This situation allows the conclusion that the ε / k_b-value can be used for all mixtures containing alkanes. For comparison, the calculation results assuming ideal behavior are also added in Figure 4. The large derivation between these calculations and the experimental values shoes the large effect of the entropic contribution modeled with LCT on the SLE.

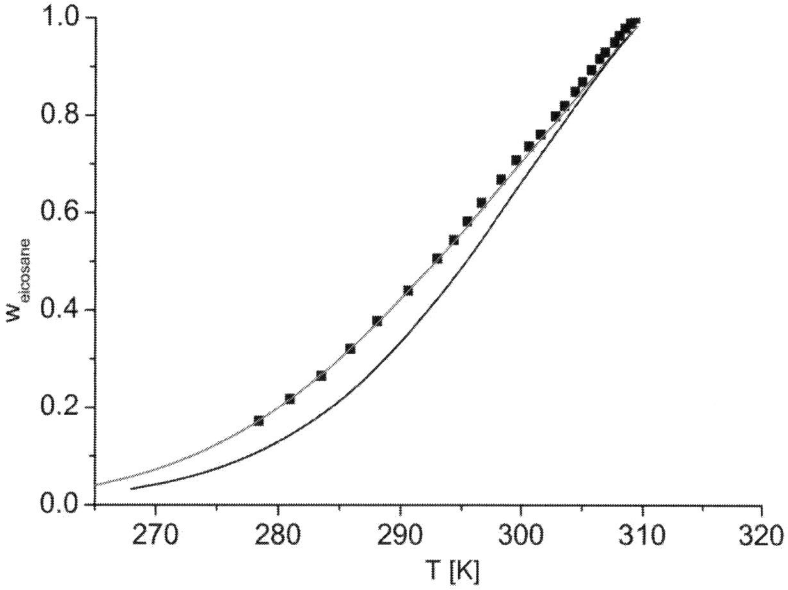

Figure 4: Experimental [17] (solid squares) and calculated [9] (red line: LCT with $\varepsilon / k_B = 5K$, black line: ideal solution approach) SLE of the system hexane + eicosane. The pure-component properties of eicosane ($\Delta_{SL}H = 75.4 kJ \cdot mol^{-1}; T_M = 310K$) were taken from the literature [18].

Going back to our original goal to deliver the thermodynamic background for TREF or CHRYSTAF the influence of the polymer structure is investigated by choosing a model system. The model system consists of a liner polymer or a short-chain branched polymer (maybe polyethylene) and a solvent occupying only on lattice place. The phase diagram is plotted in Figure 5. From experimental studies [19] it is known that the branching leads to lower melting temperatures. Exactly, this situation is predicted by the LCT, without any adjustable parameter. The difference between both melting temperatures increases with decreasing polymer concentration. Especially, in the low concentrated regime (insert in Figure 5) the influence of short-chain branching is high, which agrees qualitatively with CRYSTAF.

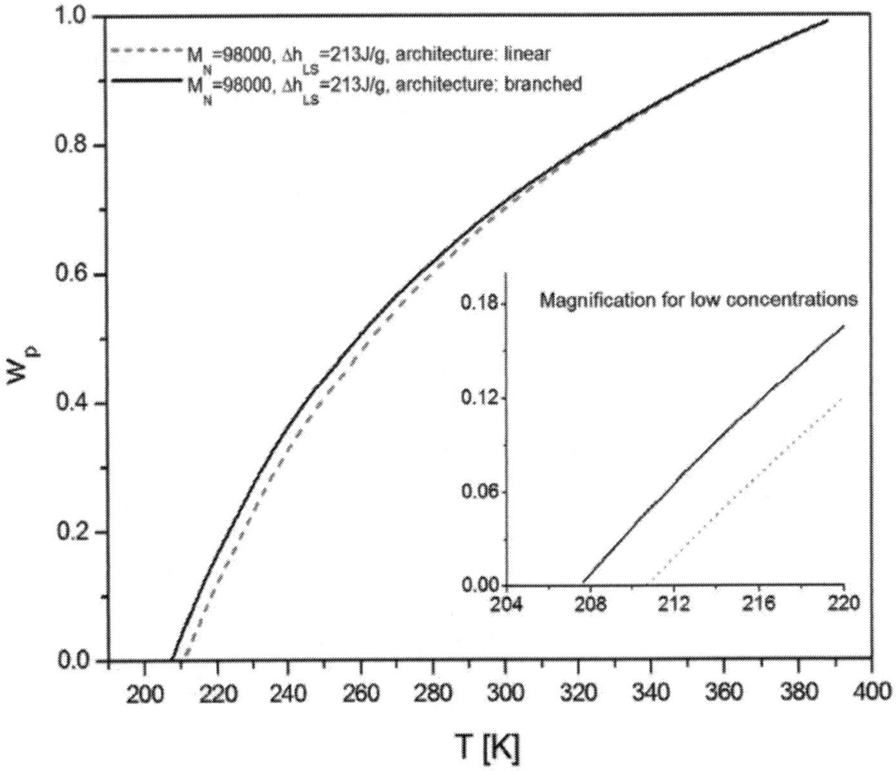

Figure 5: Influence of the architecture on solid-liquid equilibria [9].

Summary and Outlook

The LCT in its compressible version or in its in-compressible version was applied successfully to different kind of phase diagrams. The common feature of these phase diagrams is the influence of the structure of the hyperbranched polymers. For binary systems composed of hyperbranched polymers and a solvent the liquid-liquid equilibria, the vapor-liquid equilibria, as well as the vapor-liquid-liquid equilibria could by calculated in satisfactory agreement with experimental data. The proposed thermodynamic theory is able to describe the most important experimental findings [19], especially the influence of the short-chain branching on the SLE. For this reason, the new approach can be used for the extraction of information of the polymer based on TREF or CHYRSTAF data. The next step will be the quantitative comparison between the experiments and the theory.

References

[1] K.F. Freed, J. Dudowicz, Adv. Polym. Sci. 183 (2005) 63-126.
[2] S. Enders, K. Langenbach, P. Schrader, T. Zeiner, Polymers 4 (2012) 72–115.

[3] T. Zeiner, S. Enders, Chem. Eng. Sci. 66 (2011) 5244–5252.
[4] K. Langenbach, S. Enders, C. Browarzik, D. Browarzik, J. Chem. Thermodynamics, submitted September 2012.
[5] D. Browarzik, S. Enders, C. Browarzik, Fluid Phase Equilibria, in preparation.
[6] S. Enders, T. Zeiner, "Application of Lattice Cluster Theory to the Calculation of Miscibility - and Interfacial Behavior of Polymer Containing Systems" in „*Polymer Phase Behavior*", Eds.: T.P. Ehlers, J.K. Wilhelm, Nova Science Publishers, Inc., Hauppauge, USA. 2011.
[7] C. Browarzik, D. Browarzik, S. Enders, Fluid Phase Equilibria 328 (2012) 49-60.
[8] T. Zeiner, P. Schrader, S. Enders, D. Browarzik, Fluid Phase Equilibria 302 (2011) 321-330.
[9] M. Fischlschweiger, S. Enders, Fluid Phase Equilibria, in preparation.
[10] Flory, P.J. *Principles of polymer chemistry*; Cornell University Press, 1953.
[11] E. Malmström, M. Johansson, A. Hult, Macromolecules 28 (1995) 1698-1703.
[12] E. Malmström, M. Johansson, A. Hult, Macromolecules 29 (1996) 1222-1228.
[13] T. Zeiner, M.J. Puyan, P. Schrader, C. Browarzik, S. Enders, Molecular Physics 110 (2012) 1359-1373.
[14] K.O. Langenbach, S. Enders, Fluid Phase Equilibria 331 (2012) 58–79.
[15] V.M. Portela, E.J.M. Straver, Th.W. de Loos, J. Chem. Eng. Data 54 (2009) 2593-2598.
[16] K.N. Marsh, P.P. Organ, J. Chem. Thermodyn. 17 (1985) 835-841.
[17] U. Domanska, K. Kniaz, Int. Data Ser. Sel. Data Mixtures Ser. A, 2 (1990) 83-93.
[18] M. Palczewska-Tulinska, D. Wyrzykowska-Stankiewicz, A.M. Szafranski, J. Cholinski, Solid and Liquid Heat Capacity Data Collection, Part 2: C6-C33, DECHEMA e. V. 4 Germany, 1997.
[19] B. Monrabal, P. Del Hierro, Anal. Bioanal. Chem. 399 (2011) 1557-1561.

(525g) Prediction of Solid – Liquid – Gas Equilibria for Carbon Dioxide + Organic Compound System Using Peng-Robinson Equation of State and GE Type Mixing Rule

Yusuke Shimoyama, Department of Chemical Engineering, Tokyo Institute of Technology, Tokyo, Japan and Ng Brendan, Tokyo Institute of Technology, Tokyo, Japan

Particle micronization using supercritical fluids have been drawing increasing interest due to the relatively mild operating temperatures and non-toxicity of product formed. To design and operate such processes, accurate modeling of the system's phase equilibrium is required. In this work, the solid – liquid – gas equilibria for supercritical carbon dioxide + solid organic compound systems were predicted by Peng-Robinson equation of state (PR EOS) and excess Gibbs free energy (GE) type mixing rule. The GE type mixing rules tested were the Modified Huron – Vidal (MHV1) and the Linear Combination of Vidal and Michelsen (LCVM) mixing rules. Activity coefficient models using molecular surface charge density were applied for the MHV1 and LCVM mixing rules. Different model combinations of activity coefficient model, combinatorial term and mixing rule were applied to several carbon dioxide + solid organic compound systems for solid – liquid – gas equilibrium calculation. The organic compounds investigated were classified into three classes: aromatic hydrocarbons, organic alcohols and organic acids. The effects of mixing rule and activity coefficient model on the prediction performance of the solid – liquid – gas equilibria were investigated.

(525h) Experimental Solubility Data of Boldine in Supercritical Carbon Dioxide Using Ethanol As a Co Solvent

Moisés S. Durán-Zenteno, Laboratorio de Termodinámica, SEPI-ESIQIE, Instituto Politécnico Nacional, Mexico D.F., Mexico, Luis A. Galicia-Luna, INSTITUTO POLITECNICO NACIONAL-ESIQIE, Mexico, Mexico and Octavio Elizalde-Solis, Departamento de Ingeniería Química Petrolera, INSTITUTO POLITECNICO NACIONAL-ESIQIE, México, D. F., Mexico

Abstract

New solubility data of Boldine in supercritical carbon dioxide using ethanol as a co-solvent at 333 K and pressures over the critical point of the mixture (solvent + co-solvent) were determined using an equipment based on static-analytic method. Amount of solids dissolved in supercritical solvent were determined on-line by coupling a high performance liquid chromatograph. The sampler uses a six-way two port valve which holds 20 microliters of fluid phase sample. Two calibrated platinum thermometers were used to determinate the average temperature of the mixture inside a cell. A calibrated pressure sensor was used to monitor the pressure in the system. The HPLC detector was calibrated using Boldine (certified purity of 0.99 mole fraction), the standard solution was prepared by dissolving 1000 ppm of Boldine in HPLC-grade methanol. Each data was obtained after getting the chemical, mechanical and thermal equilibrium. These data were compared with data obtained using supercritical carbon dioxide as a solvent.

(537a) Receptor-Ligand Interactions of Thrombin with Glycosaminoglycans On Self-Assembled Monolayer Platforms

Yingzi Jin[1], Umesh Desai[1] and **Vamsi K. Yadavalli**[2], (1)Department of Medicinal Chemistry, Virginia Commonwealth University, Richmond, VA, (2)Chemical and Life Science Engineering, Virginia Commonwealth University, Richmond, VA

Heparan sulfate and heparin are sulfated polysaccharides that interact with a number of proteins and mediate a host of biological responses. Heparin is known to regulate several steps of the coagulation cascade. Its interaction with thrombin, factor Xa and factor XIa, key coagulation enzymes, is the primary reason for its use as an anticoagulant for the past 7 decades. Yet, the fundamental molecular bases for these interactions remain unclear. Understanding the specificity of these interactions may result in better anticoagulants. Atomic force microscopy (AFM) is a valuable tool for uncovering bio-recognition at a molecular level. Here we present a novel platform that allows us to elucidate covalently immobilized glycosaminoglycans (GAGs) to a mixed self-assembled monolayer (SAM) platform of *N*-hydroxysuccinimide (NHS) and oligoethylene glycol (OEG)-terminated thiols on ultraflat gold surfaces. Directional immobilization of chemical modified heparin as 2,6-diaminopyridinyl heparin *via* the reducing end allowed us to probe the interactions of thrombin in real time and under varying conditions (interactions in the presence of NaCl, pH and varying chain lengths). The interaction profiles and rupture forces for these binding under physiological conditions can be used to quantitatively understand the role of GAGs in blood coagulation. This experimental platform provides an attractive strategy to study protein interactions at interfaces as well as to screen for novel anticoagulants.

(537b) Enantiomeric Interactions Between Liquid Crystals and Organized Monolayers of Tyrosine-Containing Dipeptides

Yiqun Bai, Chemical and Biological Engineering, University of Wisconsin- Madison, Madison, WI and **Nicholas L. Abbott**, Department of Chemical and Biological Engineering, University of Wisconsin-Madison, Madison, WI

We have examined the orientational ordering of nematic liquid crystals (LCs) supported on organized monolayers of dipeptides with the goal of understanding how peptide-based interfaces encode intermolecular interactions that are amplified into supramolecular ordering. By characterizing the orientations of nematic LCs (4-cyano-4'-pentylbiphenyl (5CB) and TL205 (a mixture of mesogens containing cyclohexane-fluorinated biphenyls and fluorinated terphenyls)) on monolayers of either L-cysteine-L-tyrosine, L-cysteine-L-phenylalanine or L-cysteine-L-phosphotyrosine formed on crystallographically textured films of gold, we conclude that patterns of hydrogen bonds generated by the organized monolayers of dipeptides are transduced via macroscopic orientational ordering of the LCs. This conclusion is supported by the observation that the ordering exhibited by the achiral LCs is specific to the enantiomers used to form the dipeptide-based monolayers. The dominate role of the –OH group of tyrosine in dictating the patterns of hydrogen bonds that orient the LCs was also evidenced by the effects of phosphorylation of the tyrosine on the ordering of the LCs. Overall, these results reveal that crystallographic texturing of gold films can direct the formation of monolayers of dipeptides with long-range order, thus unmasking the influence of hydrogen bonding, chirality and phosphorylation on the macroscopic orientational ordering of LCs supported on these surfaces. These results suggest new approaches based on supramolecular assembly for reporting the chemical functionality and stereochemistry of synthetic and biological peptide-based molecules displayed at surfaces.

(537c) Immobilized Oligopeptide Microarrays for Detecting Proteases

Kun-Lin Yang, Department of Chemical and Biomolecular Engineering, National University of Singapore, Singapore, Singapore and Chih-Hsin Chen, Chemical and Biomolecular Engineering, National University of Singapore, Singapore, Singapore

We describe a highly sensitive, substrate-specific, label-free, and multiplexed protease assay which reports proteases activities as an optical bar chart, allowing test results to be easily assessed by laymen with the naked eye. First, an oligopeptide microarray having six rows of immobilized oligopeptides, with well-controlled orientations and concentration gradients, is immersed in a buffer solution containing proteases. Then, a thin layer liquid crystal is supported on the microarray to transduce the oligopeptide cleavage event into an optical bar chart of different colors and lengths. This type of optical bar chart provides very rich information such as protease concentration, incubation time, surface densities of oligopeptides, etc. Both trypsin and chymotrypsin can be detected by using this assay within 3 h. The capability of the multiplexed protease assay opens up possibilities for detecting toxins such as botulinum neurotoxins which are known to cleave proteins and affect the docking and fusing synaptic vesicles. Some preliminary experimental results on botulinum neurotoxin A and B detection will also be presented.

(537d) Insights Into the Lytic Mechanism of Antimicrobial Piscidin 1 and 3 Using QCM-D

Mirco Sorci[1], Keith Bradley[1], Georges Belfort[2] and Myriam L Cotten[3], (1)Department of Chemical and Biological Engineering and Center for Biotechnology and Interdisciplinary Studies, Rensselaer Polytechnic Institute, Troy, NY, (2)Chemical and Biological Engineering and Center for Biotechnology and Interdisciplinary Studies, Rensselaer Polytechnic Institute, Troy, NY, (3)Department of Chemistry, Hamilton College, Clinton, NY

Antimicrobial peptides (AMPs) are naturally occurring antibiotics which induce cell death in bacteria by destabilizing their membranes. Unlike traditional antibiotics, AMPs do not induce antibiotic resistance in pathogens and have additional immunomodulatory and metabolic functions, making them good candidates as novel agents in antibacterial therapies.

In this work, we focused on piscidin 1 (P1) and 3 (P3), which are twenty-two-residue-long amphipathic, cationic AMPs isolated from hybrid striped bass. Extensive work has been done to characterize the atomic-level *structure* of both P1 and P3 in magnetically and mechanically aligned bilayers using solid-state NMR. Here we investigated the *mechanism* of action of both P1 and P3 in supported lipid bilayers (SLB), using a quartz crystal microbalance with dissipation monitoring (QCM-D). In order to mimic bacterial cell membranes, we selected a 3:1 ratio of 1-palmitoyl-2-oleoyl-sn-glycero-3-phosphocholine (POPC) : 1-palmitoyl-2-oleoyl-sn-glycero-3-phosphoglycerol (POPG). Liposomes were deposited onto a QCM-D silicon oxide crystal until SLB formation, then P1 and P3 were introduced into the system and the interaction between the peptide and the lipid system was monitored by changes in frequency (Δf) and energy dissipation (ΔD). Different concentrations (2-20 μM) of the peptide and buffer conditions (pH 6.0 and 7.4) were tested. In particular we observed a Δf increase (~ mass removal) as we increased the concentration of P1 at pH 7.4, suggesting lipid removal, membrane thinning, or pore formation; while at pH 6.0 the Δf decrease (~ mass addition) was probably due to the electrostatic interaction between the more cationic peptide and the anionic POPG. P1 and P3 actions were also compared. In all cases, ΔD values were a clear indication of the destabilization of the SLB. Overall, this study, which shows how P1 and P3 disrupt a bacterial membrane mimic, helps us better understand the mode of action of piscidin and related AMPs.

(537e) The Interaction of Bacteriophages with Materials Depends On Their Surface Chemistry

Rebecca Cademartiri, Chemical and Biological Engineering and Materials Science and Engineering, Iowa State University, Ames, IA

Bacteriophages (phages) are viruses that attack and often destroy bacteria. They can be used as an alternative to standard antibiotics in a variety of systems (e.g., drinking water, food, animals, humans). When compared to standard antibiotics, phages have the advantages that (i) they are specific to their host bacteria (i.e., they leave other bacteria intact), and that (ii) they spontaneously evolve (i.e., they are better suited to deal with mutating bacteria). Bacteriophages are considered safe for use in humans and animals and have been approved by the FDA for use in food.

While the behavior of bacteriophages towards bacteria is well studied, there is no detailed understanding of their interaction with the surface of materials. Recent progress (e.g., in drug delivery, tissue engineering), has shown that materials are becoming increasingly important as supports for devices in medicine, biodetection, filtration, and sensing. Unraveling the interaction of bacteriophages with surfaces will be necessary to take full advantage of them in medical or environmental applications.

Here I will present results that illustrate how the binding of bacteriophages to materials (e.g., metals, ceramics, polymers) depends on the chemistry of their surface. Most bacteriophages are anisotropic and display a specific site with which they bind to the bacteria. This anisotropy and directionality of interaction suggests that the total number of bacteriophages adsorbed to a surface and their orientation are important. It will be crucial to understand how the surface properties (e.g., hydrophilic vs. hydrophobic or cationic vs. anionic) of a material influence the adsorption and orientation of phages to make them more easily applicable. Understanding the rules – either mechanistic or heuristic – underlying the adsorption of phages to surfaces will allow us to engineer materials that display the desired behavior towards phages (e.g., adsorption and release).

(537f) Porous Nanofilm Biomaterials: Biomolecular Adsorption and Cellular Adhesion

Connie Wu[1], Seyma Aslan[1], Adeline Gand[2], Joseph Wolenski[1], Emmanuel Pauthe[2] and **Paul Van Tassel**[1], (1)Yale University, New Haven, CT, (2)Universite de Cergy-Pontoise, Cergy-Pontoise, France

Hydrogel-like biomaterials are often too soft to support robust cell adhesion, yet methods to increase mechanical rigidity (e.g. covalent cross-linking of the gel matrix) can compromise bioactivity by suppressing the accessibility / activity of embedded biomolecules. Nanoparticle templating is reported here as a strategy toward porous, layer-by-layer assembled, thin (ca. 100 nm) polyelectrolyte films i) of sufficient mechanical rigidity to promote initial cell adhesion, and ii) exhibiting significant adsorption of bioactive species within the pore space. Latex nanoparticles are incorporated during layer-by-layer assembly, and following EDC-NHS cross-linking of the polyelectrolyte film, are removed *via* exposure to tetrahydrofuran (THF). THF exposure results in only a partial reduction in film thickness (as observed by ellipsometry), suggesting the presence of internal pore space. The attachment, spreading, and metabolic activity of pre-osteoblastic MC3T3-E1 cells cultured on templated, cross-linked films are statistically similar to those on non-templated films, and much greater than those on non-cross-linked films. Laser scanning confocal microscopy and quartz crystal microgravimetry indicate a high capacity for bioactive species adsorption (ca. 10% of film mass) wthin nanoparticle templated films. Porous nanofilm biomaterials, formed *via* layer-by-layer assembly with nanoparticle templating, promote robust cell adhesion and exhibit high bioactive species loading – and thus appear to be excellent candidates for cell-contacting applications.

(537g) Interfacial Binding of Anionic Liposome and Anionic SPIO Nanoparticles

Aihong Xi, Yanjing Chen and Geoffrey D. Bothun, Chemical Engineering, University of Rhode Island, Kingston, RI

Interactions between biological membranes and inorganic nanoparticles play an important role in nanomedicine and nanotoxicology. Adhering nanoparticles can affect lipid phase behavior and transmembrane permeability, and, when strongly adhering, can lead to pore formation and nanoparticle invagination. This work examines the effects of temperature and electrolyte concentration on the interaction between partially anionic lipid bilayer vesicles and anionic superparamagnetic iron oxide nanoparticles. Although the liposomes and nanoparticles are like-charged, they exhibit different charge densities. Extruded liposome (~200 nm hydrodynamic diameter) were prepared in water using dipalmitoylphosphatidylcholine (DPPC) and dipalmitoylphosphatidylglycerol (DPPG) at a molar ratio of 3:1. The nanoparticles had a core diameter of 5 nm and were coated with a carboxylated polymer to yield an average hydrodynamic diameter of 16 nm. Despite being like-charged, preliminary dynamic light scattering data at 25°C shows nanoparticle binding to gel-phase liposomes and causes liposome aggregation. Isothermal titration calorimetry results suggest that this process was driven by entropic (hydrophobic) interactions. Increasing temperature to 45°C or adding salt (NaCl or phosphate buffered saline, PBS) reversed liposome aggregation. Current work is focused on elucidating the effects of temperature and electrolyte concentration on nanoparticle binding, liposome aggregation, and liposome fusion.

(556a) Electrochemistry and Transport Limitations of Non-Aqueous Li-Air Batteries From First-Principles

Venkatasubramanian Viswanathan[1], Jens Hummelshoj[2], Alan Luntz[3] and Jens K. Norskov[1], (1)Chemical Engineering, Stanford University, Stanford, CA, (2)SLAC National Laboratory, Menlo Park, CA, (3)IBM Almaden, San Jose, CA

Li-air batteries have much higher gravimetric energy storage density compared to all other battery chemistries, and this has led to a strong interest in seeing if such batteries could be developed for powering EVs, enabling driving ranges comparable to gasoline powered automobiles. However, many fundamental challenges exist in these batteries that need to be solved before these batteries can become practical.

In this work, we will discuss a detailed electrochemical model describing the discharge process of the growth of Li_2O_2 on terraces, steps and kinks. The different kinds of growth mode along the different kinds of sites have vastly different thermodynamic overpotential. This leads to an explanation of the high Tafel slope observed in Li-air batteries. We will also discuss the effect of deep discharge and the origin of 'sudden death' in these lithium-air batteries and we will show that the origin of this 'sudden death' in flat electrode cells is due to the limitation imposed by electron transport to support the required electrochemistry. Using a simple metal-insulator-metal model, we show that discharge products beyond > 10 nm, have serious transport limitations and this leads to a bias rise across the film. In addition, we also discuss the effects of the presence of a carbonate layer which has been recently identified experimentally. We will also discuss the development of a model incorporating details from some of the recent experiments that show the presence of a carbonate layer. We conclusively show that the origin of potential rise during recharge of the lithium-air batteries is due to a rising concentration of carbonate impurities at the Li_2O_2-electrolyte interface. This rise in the overpotential leads ultimately to the decomposition of the electrolyte and limits the rechargability of lithium-air batteries.

(556b) Nanofiber-Based Novel Electrode Architectures for Batteries and Supercapacitors

Chau Tran, Nataliia Mozhzhukhina and Vibha Kalra, Chemical and Biological Engineering, Drexel University, Philadelphia, PA

The ability to control the nano-architecture within electrode materials is critical to the development of high performing energy storage devices. We will present our work on development of nanofiber-based hierarchically structured materials for application in lithium-air batteries and electric double-layer capacitors. For supercapacitors, we report a facile method for obtaining extremely high surface area (>1500 m^2/g) and uniformly porous carbon nanofibers. As a first step, blends of polyacrylontritrle (PAN) and a sacrificial polymer in dimethyl formamide (DMF) were electrospun into non-woven nanofiber mats with diameters in the range of 200-300 nm. Fast evaporation of solvent (~200 nl/s) and high elongational flow rate (~10^5 s^{-1}) during electrospinning allowed us to prevent phase separation and develop a co-continuous morphology of PAN and the sacrificial polymer in the nanofibers. As a second step, electrospun nanofiber mats were subjected to stabilization and carbonization processes to obtain porous carbon nanofibers (CNFs) as PAN converted to carbon and the sacrificial polymer decomposed out to create intra-fiber pores. Unlike other studies so far, we chose the sacrificial polymer possessing a high decomposition temperature and chain rigidity which prevented fibers from shrinking and collapsing during the PAN stabilization process at ~280 °C. Here we demonstrated that using Nafion as a sacrificial polymer allowed us to obtain CNFs with specific surface area of up to 1600 m^2/g without any activation process. We exhibit the tunability of the pore sizes within CNFs by varing material composition. These materials exhibit a large specific capacitance (>200 F/g and >30 F/cm^3) possibly due to the presence of a large fraction of meso-pores (2-4 nm) compared to activated carbons (pores <2 nm), which leads to an increase in the accesible carbon surface, thereby improving specific capacitance.

For lithium-air batteries, we incorporate (via in-situ synthesis) manganese dioxide as the catalyst in porous carbon nanofibers to fabricate multi-functional air cathodes. The aim is to establish well-defined multi-phase boundaries within the cathodes such that all reactants; oxygen (via electrode pores), lithium ions (via electrolyte in pores) and electrons (via carbon) can simulatneously reach the catalyst nanoparticles for the oxygen reduction reaction to take place efficiently. Strutural characterization of all nanofiber samples was conducted using scanning electron microscopy (for external nanofiber morphology), transmission electron microscopy of microtomed fiber sections (for internal structure), and nitrogen sorption isotherms (for pore size distribution, pore volume and surface area). Electrochemical performance was studied using cyclic voltammetry at different scan rates, galvanostatic charge-discharge measurements at different current densities, and electrochemical impedance spectroscopy (EIS).

(556c) Hybrid Li-Air Battery with Sulfuric Acid Electrolyte and Buckypaper Air Cathode

Yunfeng Li, Chemical and Biological Engineering, Missouri University of Science and Technology, Rolla, MO, Kan Huang, Chemical and Biological Engineering, Missouri University of Science and Technology, rolla, MO and Yangchuan Xing, Dept. of Chemical and Biological Engineering, Missouri University of Science and Technology, Rolla, MO

Li-air battery (LAB) with aqueous-based catholyte and a water-stable lithium electrode using a LISICON-type lithium ion conducting glass ceramic membrane can eliminate the problems of Li-air battery with organic electrolytes, such as air-breathing cathode choking and degradation of Li anode. We present a rechargeable dual-electrolyte hybrid Li-air battery (HyLAB) with outstanding discharging performances has been demonstrated. The HyLAB has sulfuric acid as the catholyte and has a large active electrode area of 10 cm^2. Two types of air-breathing cathode with active layer made from a buckypaper with 5 wt.% Pt supported on carbon nanotubes (Pt/CNTs) and 5 wt.% Pt on carbon black (Pt/CB) as electrocatalyst for oxygen reduction were prepared, and their cell discharge performances were also compared. The cell with Pt/CNTs showed a much better performance than the cell with Pt/CB at high current densities. A better performance of Pt/CNTs was attributed to the more open porous structures formed in the fibrous CNTs. The HyLAB with Pt/CNTs can achieve a discharging capacity of 306 mAh/g at a cut-off voltage of 3.15 V and current density of 0.2 mA/cm^2. The corresponding specific energy is 1067 Wh/kg based on the total weight of the sulfuric acid. Detailed results will be presented in this presentation.

(556d) Modeling Phase Separating Materials in Porous Electrodes Using Non-Equilibrium Thermodynamics

Todd R. Ferguson, Chemical Engineering, Massachusetts Institute of Technology, Cambridge, MA and Martin Z. Bazant, Chemical engineering and Mathematics, MIT, Cambridge, MA

Phase separating materials are prevalent in lithium-ion batteries. Commercial lithium-ion battery anodes typically consist of graphite, a material known to go through multiple "stages" as the filling fraction (i.e. percentage) of lithium in the material changes. Other phase separating materials, such as lithium iron phosphate (LFP), have gained a lot of interest due to rate and capacity improvements.

Despite their widespread use, the behavior of phase separating materials inside batteries isn't very well known. The disappearance of voltage plateaus with increased discharge rates and the existence of a non-vanishing voltage gap in the limit of zero current are two phenomena which are not very well understood.

We present a thermodynamically consistent form of porous electrode theory. This set of equations, derived from concentrated solution theory using non-equilibrium thermodynamics and a general reaction rate equation, is used to model phase separating materials in lithium-ion batteries. These models demonstrate phenomena often observed in phase separating battery materials, such as a non-vanishing voltage gap, the mosaic effect, current dependent suppression of phase transformation, and porous electrode effects in phase separating materials (i.e. discrete filling).

This model can be expanded to address a wide range of materials, including LFP and graphite, using thermodynamic models for the materials. The kinetics, equilibrium potential, and particle dynamics are all coupled to the thermodynamics of the material. These dynamics are then coupled to transport in the electrolyte during simulation of the cell, producing a wide variety of behavior typically observed in phase separating materials.

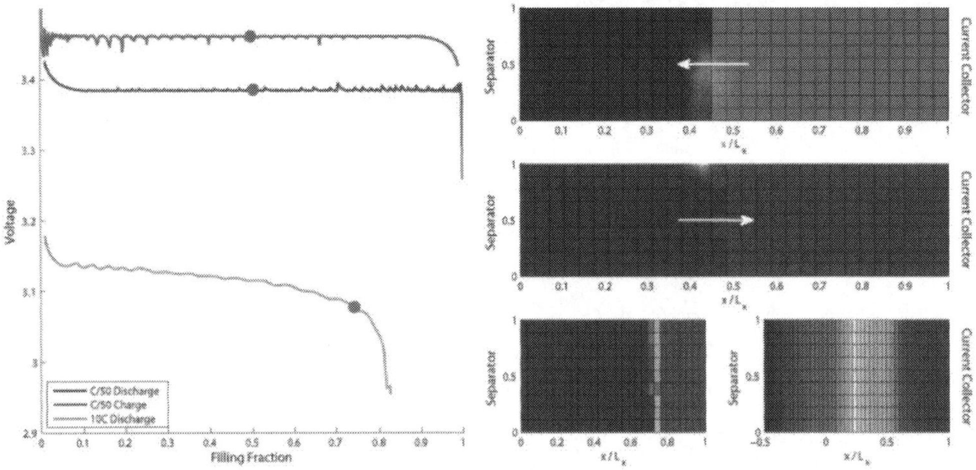

(556e) Improve the Mass Transfer to Graphite Electrode Through Cell Design in Vanadium Redox Flow Batteries

Feng Chen, Pacific Northwest National Laboratory, Richland, WA; Materials Science, Washington State University, Richland, WA

Improve the mass transfer to graphite electrode through cell design in vanadium redox flow batteries

Feng Chen, Yuyan Shao, Qingtao Luo, Wei Wang, Liyu Li, Zimin Nie, Gary Yang

Pacific Northwest National Lab, Richland, WA99354

Abstract

Electrolyte mass transfer is a key factor that affects the performance of vanadium redox flow battery (VRFB). Flow batteries with graphite felts inserted with graphite compounded plates were designed to investigate the influence of mass transfer between electrolyte and graphite fiber on the cell performance. Different amounts of graphite plates were applied in the cell design to explore the extent of influence of plate-like barriers on the mass transfer and diffusion rate of electro-active ions. Vanadium redox couples were used as the electro-active species. The flow batteries with new design were also tested under different flow rates to examine the combinational effect of flow rate and graphite plate barriers. Another set of PVC plate barriers were also employed in the new design to study the effect of barrier material on the cell performance. It was shown that the coulombic, energy, and voltage efficiencies were greatly improved by applying the new electrode design in the flow battery system. The testing of cells with PVC insulated plates is still on-going. However, the testing of cell with new design under different flow rates needs to be repeated and verified.

In-Situ Investigation of Vanadium Ions Transport in Redox Flow Battery

Qingtao Luo,[1] Liyu Li,[2] Zimin Nie,[1] Wei Wang,[1] Xiaoliang Wei,[1] Feng Chen,[1] Bin Li,[1] Guan-guang Xia,[1] Baowei Chen,[1] Zhenguo Yang,[2] Sprenkle Vincent[1]

[1] Pacific Northwest National Laboratory, P. O. Box 999, Richland, WA 99354, USA

[2] UniEnergy Technologies, LLC, 4333 Harbour Pointe Blvd SW, Unit A, Mukilteo, WA 98275, USA

The all-vanadium redox flow battery (VRB) is probably the most promising and extensively researched flow battery system [1-3]. The ion exchange membrane (IEM) is one of the most important components in a RFB system; it prevents the crossover of the active materials in the positive and negative electrolytes while supporting proton transport to complete the circuit. Perfluorinated polymer membranes (e.g., NAFION_ membranes) have become the choice of IEMs for VRBs due to their high conductivity and excellent chemical stability in highly oxidizing positive electrolytes [4, 5]. However, severe capacity decay of VRBs during cycling were found and reported by several groups [6-8]. Although the capacity fading mechanism is not entirely understood by the VRB research community, the inevitable crossover of vanadium ions across the membrane may play a significant role in causing the capacity fading during cycling. In order to shed light on the VRB capacity fading mechanism, it is necessary to investigate the transport behaviors of vanadium ions across the membrane.

In this work, redox flow batteries with vanadium and iron redox couples were constructed to investigate the transport behaviors of vanadium ions. It was shown that the transport of vanadium ions was affected by a combination of different factors, such as the charge/discharge process, current density, and SOC. The electric field accelerated the positive-to-negative and reduced the negative-to-positive transport of vanadium ions in the charging process and affected the vanadium ion transport in the opposite way during discharge. With the increase of current density, the vanadium ion transport rate increases during the charge process but decreases during discharge. At each given current density, the vanadium ion transport rate, in general, demonstrated a trend of decreasing with the increase of SOC. Furthermore, A simplified Nernster-Planck equation was established to simulate the vanadium ion transport, and the concentration-gradient diffusion coefficients and electric-migration coefficients were calculated on the basis of experimental data.

Reference

[1] B. Dunn, H. Kamath, J.-M. Tarascon, Science 334 (2011) 928-935.
[2] Z. Yang, J. Zhang, M.C.W. Kintner-Meyer, X. Lu, D. Choi, J.P. Lemmon, J. Liu, Chem. Rev. 111 (2011) 3577-3613.
[3] M. Skyllas-Kazacos, M.H. Chakrabarti, S.A. Hajimolana, F.S. Mjalli, M. Saleem, J. Electrochem. Soc. 158 (2011) R55-R79.
[4] Q. Luo, H. Zhang, J. Chen, D. You, C. Sun, Y. Zhang, J. Membr. Sci. 325 (2008) 553-558.
[5] X. Li, H. Zhang, Z. Mai, H. Zhang, I. Vankelecom, Energy Environ. Sci. 4 (2011).
[6]S. Kim, J. Yan, B. Schwenzer, J. Zhang, L. Li, J. Liu, Z. Yang, M.A. Hickner, Electrochem. Commun. 12 (2010) 1650-1653.
[7] C. Jia, J. Liu, C. Yan, J. Power Sources 195 (2010) 4380-4383.
[8] T. Sukkar, M. Skyllas-Kazacos, J. Appl. Electrochem. 34 (2004) 137-145

(556g) Energy Storage in Redox-Active Organic Molecules

Fikile R. Brushett, Lu Zhang, John T. Vaughey and Andrew N. Jansen, Chemical Sciences & Engineering Division, Argonne National Laboratory, Argonne, IL

Low-cost, scalable energy storage systems are required to improve the energy efficiency of the electrical grid (*e.g.*, loading-level, frequency regulation) and to enable the widespread penetration of intermittent renewable energy sources (*e.g.*, wind, solar) [1]. Redox flow batteries may provide the best combination of cost, efficiency, and scalability to enable these applications. Present flow battery technologies have low energy densities (< 40 Wh/L), due to the use of aqueous electrolytes with low redox species solubilities (~2 M) and low operating voltages (< 1.5 V, limited by water electrolysis), which, in turn, lead to high system-level costs ($280-450/kWh [2]). Employing non-aqueous electrolytes offers a wider window of electrochemical stability which enables cell operation at dramatically higher potentials (2-4 V) [3-5]. If appropriate redox couples can be identified, operating at higher cell voltages leads to greater system energy (and power) densities and higher roundtrip efficiencies. Moreover, to achieve the same output as an aqueous system, fewer stack layers, lower flow velocities, smaller tanks, and fewer ancillaries would be required, significantly reducing hardware costs and enhancing system reliability.

Redox-active organic molecules may be good candidate materials for non-aqueous flow batteries. This is an attractive approach to energy storage as it enables flexibility to design molecules with favorable properties (*i.e.*, redox potential, activity, solubility) by tailoring either the redox-active moiety or the surrounding molecular structure. Furthermore, research in overcharge protection materials for lithium (Li)-ion batteries has led to the identification of a number of promising materials and to the establishment of some key design principles. Here, we will present our research efforts on a non-aqueous Li-ion redox flow battery employing 2,5-di-tert-butyl-1,4-dialkoxybenzene and quinoxaline as the high- and low-potential active species.

[1] Z. Yang, J. Zhang, M.C.W. Kintner-Meyer, X. Lu, D. Choi, J.P. Lemmon, J. Liu, *Chem. Rev.*, **2011**, *111*, 3577.

[2] D. Rastler, *Electric Perspectives*, **2008**, *Sept/Oct,* 30.

[3] Y. Matsuda, K. Tanaka, M. Okada, Y. Takasu, M. Morita, T. Matsumura-Inoue, *J. Appl. Electrochem.*, **1988**, *18*, 909.

[4] Q. Liu, A.A. Shinkle, Y. Li, C.W. Monroe, L.T. Thompson, A.E.S. Sleightholme, *Electrochem. Commun.*, **2010**, *12*, 1634.

[5] J.Y. Mun, M.-J. Lee, J.-W. Park, D.-J. Oh, D.-Y. Lee, S.-G. Doo, *Electrochem. Solid-State Lett.*, **2012**, *15*, A80.

(561a) Determination of a Biocarbon Fuel Cell Electrolyte Chemistry

Maider Legarra Arizaleta[1], Ashley S. Blitz[1], Zsuzsanna Czégény[2] and Michael J. Antal[1], (1)Hawaii Natural Energy Institute, University of Hawaii, Honolulu, HI, (2)Hungarian Academy of Sciences

The electrochemical combustion of carbon converts carbon particles directly into electricity with a theoretical efficiency of 100%. As a result of this surprising aspect, different attempts have been undertaken to develop a reliable carbon fuel cell for more than a century. In our laboratory, an aqueous-alkaline carbon fuel cell that works at moderate temperatures (over 200 °C) and high pressures (≈48 bar) has been under development since 2000. Our findings indicated that under these conditions, biocarbon can indeed undergo vigorous oxidation and thereby generate power in a fuel cell at efficiencies exceeding 80%. However, serious engineering challenges confront the development of a reliable carbon fuel cell. Due to the precipitation of bicarbonate crystals that were found when disassembling the fuel cell, we became concerned about the electrolyte instability that results from the reaction of CO_2 with carbonate ions to produce bicarbonate. Therefore, we decided to study the bicarbonate/carbonate chemistry at the fuel cell conditions. To understand the $KHCO_3/K_2CO_3$ equilibria at the fuel cell conditions, we built a "tubing bomb" reactor that was suitable for use at pressures of 2000 psi and temperatures of 300 °C. Moreover, it could be quickly heated and cooled in a fluidized sand bath. The liquid phase chemistry was studied by exposing solutions of potassium bicarbonate in the pressure vessel to temperatures near 300 °C at their saturation pressure, different time frames and different cooling treatments. Three different analyses of the final solutions were utilized: (1) pH measurement, (2) TGMS of the dry crystals in the Hungarian Academy of Science and (3) Titration.

The significant pH increase of the final solutions suggested the potassium bicarbonate decomposed via the reaction $2HCO_3^- \leftrightarrow CO_3^{2-} + CO_2 + H_2O$. We learned that the TG-MS results were not reliable because of the high sensitivity of the carbonate/bicarbonate solutions to the presence of CO_2. The drying of the sample under atmospheric conditions shifts the above reaction to the left due to the atmospheric CO_2 and acidifies the sample; on the other hand, we also observed that the drying by exposing the solution to vacuum shifts the reaction to the right as the CO_2 escapes to the gas phase and the sample becomes more alkaline. Therefore, the bicarbonate/carbonate solutions drying process modifies the composition of the sample. To preserve the original solution, the bicarbonate/carbonate solutions storage should be done by capping them in bottles with little dead air space to minimize its exposure to air. The solution can be determined reliably by titration. This titration analysis confirmed the bicarbonate decomposition into carbonate and CO_2 and gave more accurate values of the extents of reaction.

Based on the titration analysis of the final solutions, we determined the dependence of the equilibrium constant with the temperature in this range and extracted the thermodynamic properties - Gibbs free energy, enthalpy and entropy- from a Van't Hoff plot. We found that bicarbonate ions are not stable at fuel cell conditions. Our findings indicate that at these conditions, the bicarbonate decomposition reaction into CO_2 and carbonate is spontaneous and endothermic, with a positive change in entropy. The troublesome crystals observed in the fuel cell disassembly were formed while the vessel was cooled under pressure and precipitated thereafter. Our carbon fuel cell should experience no problems with the formation of $KHCO_3$, providing the cell is quickly cooled and pressure is quickly released. These findings have been a cause of optimism for the carbonate fuel cell. In this presentation, we describe the results of both the electrolyte study and the ongoing carbon fuel cell study.

*** Formatting Issue - Best Available Paper/Graphic ***

(561b) Control of a Proton Exchange Membrane Fuel Cell, DC-DC Power Converter, an Ultracapacitor System Using Sliding Mode Controls

James E. Smith, Univ. of AL in Huntsville, Huntsville, AL

Control of a Proton Exchange Membrane Fuel Cell, DC-DC Power Converter, an Ultracapacitor System Using Sliding Mode Controls

Roshini.S. Ashok and Yuri B. Shtessel, Department of Electrical and Computer Engineering and James. E. Smith, Jr, Department of Chemical and Materials Engineering,

The University of Alabama in Huntsville, Huntsville, AL 35899

In this paper we consider control of a power system that comprises three controlled components: Proton Exchange Membrane fuel cell (PEMFC), ultracapacitor, and a boost DC-DC power converter, using sliding mode control techniques. Control of fuel cells using sliding mode control (SMC) technique (Levant, A., (1993), Fridman, L. and Levant, A., (2002) and Shtessel, Y., et. al., (2011)) that is robust to the fuel cell model uncertainties and the external disturbances is studied in Sabanovic, A., et. al., (2004). Sliding mode control design for power systems, which consists of fuel cell and boosts DC-DC power converter, is presented in (Spurgeon S. and Davies, R., (1993), Shtessel, Y. and Ashok, R.S., (2011)). The use of ultracapacitor was addressed in the work (T. Azib, et. al., 2010). All three subsystems are controlled to provide a required power system's performance.

I. PHYSICAL DESCRIPTION OF PEMFC

The schematic of a Proton Exchange Membrane (PEM) fuel cell has two electrodes (anode and cathode) isolated by a membrane acting as an electrolyte as shown in Figure 1. (Matraji, I. 2010).

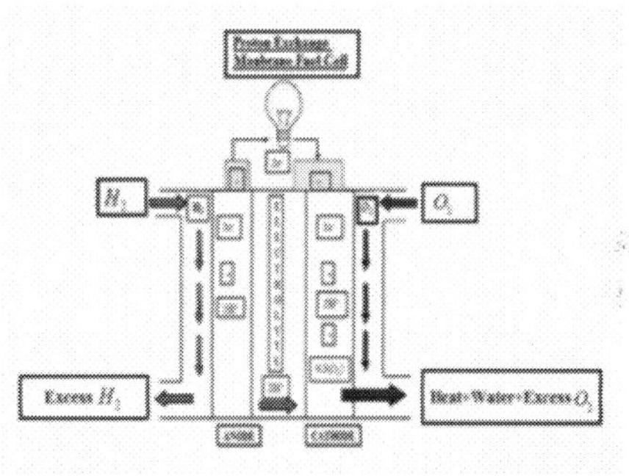

Figure 1. Schematic of a Fuel Cell Operation

The performance of a PEM fuel cell increases, when its operating temperature is around 70-80°C at a partial reactant pressure of 3-5atm. The chemical reactions in PEMFC can be described by the following equations (Larminie J. and Dicks, A. (2003), Kunusch, C. (2006)).

$$\left.\begin{array}{l} H_2 \rightarrow 2H^+ + 2e^- \\ 2H^+ + \frac{1}{2}O_2 + 2e^- \rightarrow H_2O \\ H_2 + \frac{1}{2}O_2 \rightarrow H_2O \end{array}\right\} \quad (1)$$

QUOTE

II. MATHEMATICAL MODELS

The mathematical model of the fuel cell electric power unit is derived dynamically, by combining the thermodynamic flow rate, boost DC-DC power converter and the ultra-capacitor. A typical PEMFC has the following V-I characteristic or polarization curve as shown in Figure 2.

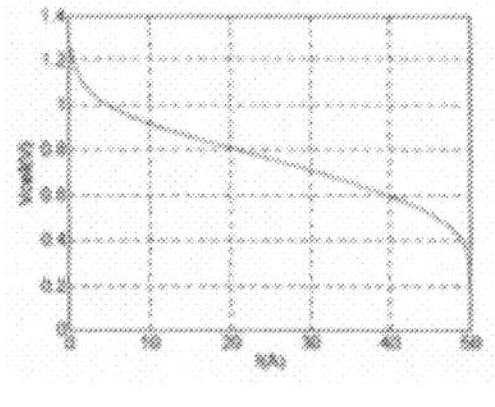

*** Formatting Issue - Best Available Paper/Graphic ***

Figure 2. V-I Characteristics of the Fuel Cell

The voltage across the fuel cell (V_{fc}) is a function of the stack current, partial pressure of reactants, temperature and humidity of the membrane.

[illegible]

[illegible] (2)

where n the number of cells in the stack, V_{st} the total voltage across the stack, E_0 the thermodynamic cell potential, V_{act} activation overvoltage, V_{ohm} ohmic voltage drop, and V_{conc} the mass transport voltage drop or the concentration.

A. Reversible Cell Voltage

The open circuit voltage E_0 , (T. Azib, et. al., 2010.), is calculated using

[illegible] (3)

where [illegible] is a change in entropy, [illegible] Faraday's constant, [illegible] universal gas constant, [illegible] , P_{H_2} partial pressure of hydrogen in the cell, P_{O_2} partial pressure f oxygen in the cell, P_{H_2O} partial pressure of water and T_0 the standard temperature ($278^\circ K$). The operating temperature T is in $^\circ K$. The sum of the partial pressures equals the total pressure [illegible] .

B. Activation Over Potential

Slow transfer of charge at the electrodes causes this loss. A part of the electrode potential is used for electron transfer to match the current demand. Hence the voltage at the fuel cell drops by;

[illegible] (4)

where C_{dl} is a double layer capacitance, and R_{act} is the equivalent resistor to activation.

The double layer of charge (*dl*) is needed to understand the dynamics of the fuel cell. Whenever two materials of opposite charge come in contact, there is an accumulation of the charge on the surface. The charge layer that is deposited on the interface of the electrode or electrolyte acts as storage of the electrical charges and acts as an electrical capacitor C_{dl} (Correa, J., et. al., (2004)).

C. Ohmic Over Potential

The ohmic over potential V_{ohm} results from the resistance to electron transfer in the electrolyte and is modeled by;

[illegible] (5)

R_{ohm} is the equivalent resistor to the ohmic over potential, and i_{fc} is the current through the fuel cell.

D. Concentration over potential

The electrical current and the physical characteristics of the system are directly proportional to the pressures of the oxygen and hydrogen. Maximum current density is used to determine the voltage drop, under which the fuel is being used at the same rate. Thus, the drop in voltage is caused due to the mass transport,

[illegible] exp( ) (6)

where coefficients vary with the temperature and are given to by;

[illegible] ,

[illegible] .

A stack of 300 fuel cell units are used, along with an Ultracapacitor, are the primary source of electrical energy for the boost DC-DC power converter. The equivalent circuit (Larminie J. and Dicks, A. (2003), Kunusch, C. (2006)) is presented in Figure 3 below;

Vtfc

*** Formatting Issue - Best Available Paper/Graphic ***

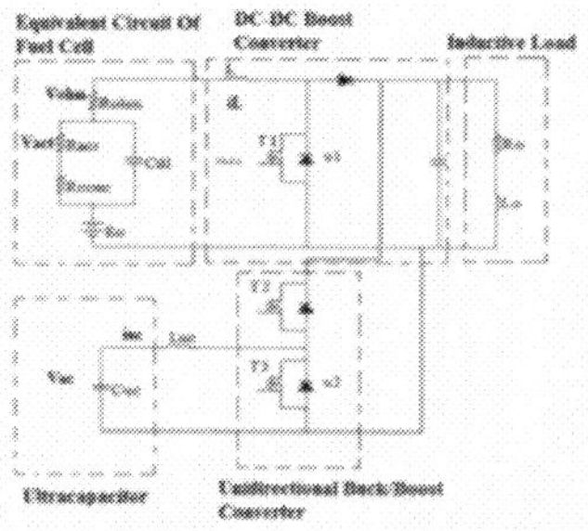

Figure 3. Circuit Diagram of Fuel Cell System

i_{fc} → current of the fuel cell, same as i_L , the inductance current,

R_o → resistance of the load,

L_o → inductance of the load,

C_{dl} → charge build up due to diffusion and reactions between the electrons in the electrode and protons in the electrolyte,

R_{act} → resistance that causes activation loss (V_{act}) due to slow rate of the reaction,

R_{ohm} → resistance that causes the ohmic over potential (V_{ohm})

R_{conc} → resistance that causes the drop in concentration of reactants at the reaction sites

V_{fc} → total voltage across the fuel cell,

V_o → output load voltage of the DC/DC convertor,

QUOTE R_{conc} i_o → current through the load inductance. <>**III. SIMULATIONS**

The system of electrical energy supply consists of the boost DC-DC power converter, which receives electrical energy from a stack of 300 fuel cell elements and is controlled by SMC and ASTW control is simulated. The parameters of the model are presented in Table 1.

Table 1. Model Parameters

Parameters	Values
LL_o	3.5×10 [illegible] H
CC_{dl}	1×10 [illegible] F, 68.5×10 [illegible] F
$i_L(0)$	20 A
$V_o(0)$	10 V
R_o	20 [illegible]
R_{uc}	0.08/
V_{uc}	25 V
C_{uc}	125F
L_{uc}	140×10 [illegible] H
T_c	353 K
R_{ohm}	0.06/

The load voltage command profiles were selected in accordance with (Sabanovic, A., et. al., (2004), Spurgeon S. and Davies, R., (1993) and Utkin, V., et. al., (1999)):

***** Formatting Issue - Best Available Paper/Graphic *****

$$v_{ref} = \begin{cases} 400 & 0.75 \\ 751.1 \\ 400 \end{cases} \quad (7)$$

The load resistor was changed at $t = 0.85$:

$$R_L = \begin{cases} 20, & 0.85 \; t \\ 30, & 0.85 \; t \end{cases} \quad (8)$$

The fuel cell varying resistor changed its value at $t = 1.05$

$$R_{ohm} = \begin{cases} 0.75, & 1.05 \; t \\ 1.00, & 1.05 \; t \end{cases} \quad (9)$$

The simulation plots are presented in Figures 4-10. The plots on Figure 4 illustrate high accuracy direct tracking of the load voltage command profile via classical sliding mode control (Figure 8) in the presence of unknown load resistor while the voltage command profile is changing twice during simulation time. High accuracy tracking of the fuel cell current command profile that is generated on line is confirmed via plots in Figure 5. The ultracapacitor current i_{uc} is shown in Figure 6. The time history of adaptive control gains is demonstrated in Figures 7 and 8. The control functions of the PEMFC current and DC-DC boost converter are shown in Figures 9 and 10.

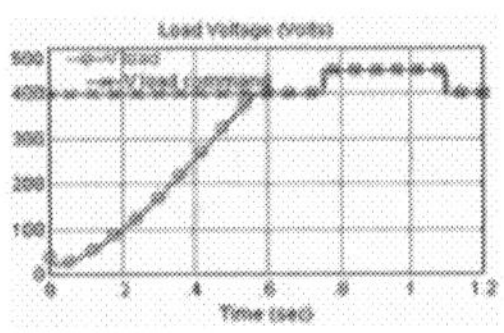

Figure 4. Load Voltage

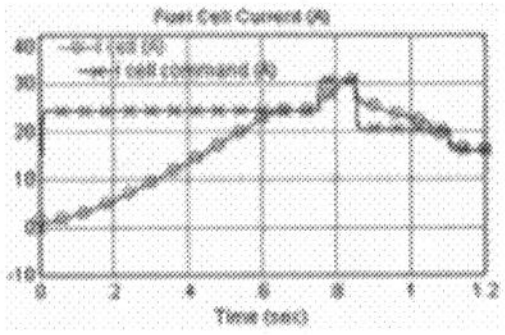

Figure 5. Fuel Cell Current

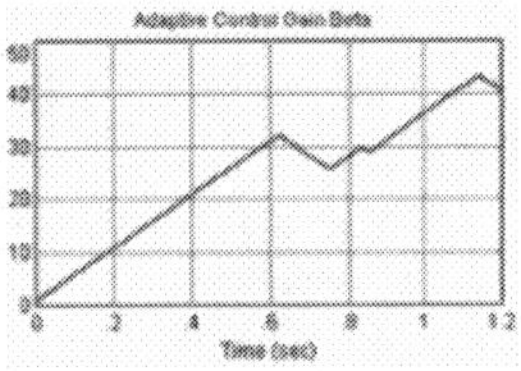

Figure 8. Adaptive Control Gain β

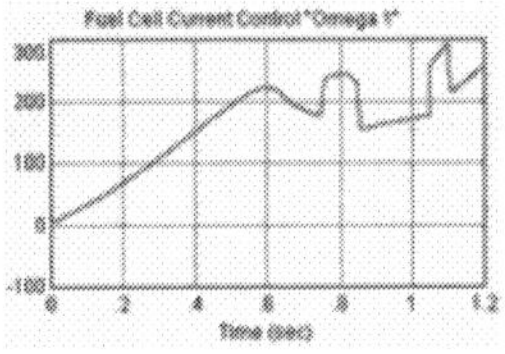

Figure 9. PEMFC Current Control Γ_1

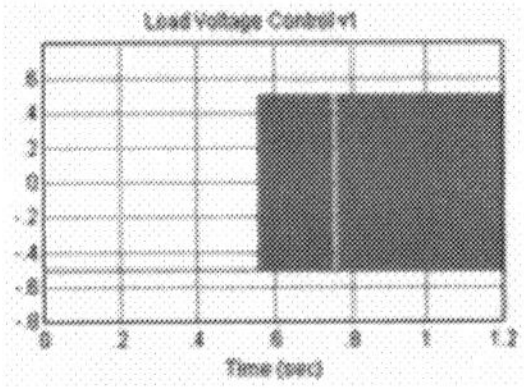

Figure 10. DC-DC Boost Power Converter Control v_1

<>IV CONCLUSIONS

Two types of sliding mode control (traditional SMC and adaptive-gain 2-SMC, continuous super-twisting control) are used at the same time for controlling PEMFC power system. It is observed that a three-fold SMC feedback control structure (adaptive 2-SMC PEMFC current and traditional SMC voltage and ultracapacitor current controllers) avoids non-minimum phase property of boost DC-DC power converter, which output voltage is controlled directly and simplifies the controller design. The ultracapacitor not only acts as an auxiliary power source when there is an interruption in power supply but also responds rapidly to a fast load demands. The energy produced from a fuel cell is managed efficiently using appropriate Sliding mode control techniques. The efficacy and robustness of the SMC and 2-SMC controllers are confirmed via computer simulations.

V REFERENCES

Azib, T., Talj,, R Bethoux, O and Marchand, C. Member,IEEE.*"Sliding Mode Control and Simulation of a Hybrid Fuel-Cell Ultracapacitor Power System"*, 3425-3430, 2010.

Correa, J., Farret, F., Canha L., and Simoes, M., *"An Electrochemical-Based Fuel- Cell Model Suitable for Electrical Engineering Automation Approach"*, Vol.51, NO.5, October 2004

Fridman, L. and Levant, A., "Higher Order Sliding Modes," in *Sliding Mode Control in Engineering*. New York:

*** Formatting Issue - Best Available Paper/Graphic ***

Marcel Dekker, Inc., 2002, ch. 3, pp. 53–101.

Kunusch, C.,, "Second order sliding mode control of a fuel cell stack using a twisting algorithm," M.S. thesis, Electr. Dept., Nat. Univ. La Plata, Argentina, Mar. 2006.

Larminie J. and Dicks A., *Fuel Cell Systems Explained*, 2nd ed. New York: Wiley, 2003.

Matraji, I., Laghrouche, S. and Wack, M., *"Second order Sliding Mode Control for PEM Fuel Cells"*49th IEEE Conference on Decision and Controls, Atlanta, GA, USA, 2010.

Sabanovic, A., Fridman, L., and Spurgeon, S., *Variable Structure Systems: From Principles to Implementation*. London, U.K.: IEE, 2004.

Shtessel, Y and Ashok, R.S. "PEM Fuel Cell/ DC-DC Boost Power Converter System Control Via Traditional and Higher Order Sliding Modes," *Proceedings of the Conference on Decision and Control*, 2011.

Shtessel, Y., Plestan, F. and Taleb, M., "Lyapunov design of adaptive super-twisting controller applied to a pneumatic actuator," *Proceedings of IFAC World Congress*, Milan, Italy, August 2011.

Spurgeon S. and Davies, R., "A nonlinear control strategy for robust sliding mode performance in the presence of unmatched uncertainty," *Int. J. Control*, vol. 57, no. 5, pp. 1107–1123, 1993.

Utkin, V., Gulder, J., and Shi, J., *Sliding Mode Control in Electro-Mechanical Systems*, London, U. K.: Taylor and Francis, 1999.

(561c) Nonlinear Analysis of Polymer Electrolyte Fuel Cell Voltage Instability Under Flooding Conditions

Michael Burkholder[1], Nick Siefert[2] and Shawn E. Litster[1], (1)Mechanical Engineering, Carnegie Mellon University, Pittsburgh, PA, (2)U.S. Department of Energy, National Energy Technology Laboratory, Pittsburgh, PA

In this work, we apply a nonlinear chaos analysis to the electrochemical instability induced by water accumulation in the cathode of an operating polymer electrolyte fuel cell. The fuel cell voltage signal is analyzed using techniques designed to estimate invariants typical of deterministic systems with high sensitivity to initial conditions, such as chaotic cathode water accumulation and removal. Voltage data are taken under varying fuel cell operating conditions, and noise in the data is reduced using a nonlinear noise reduction algorithm. The chaotic strange attractor of the system is reconstructed in phase space using time-delay embedding. Correlation sums over the strange attractor are calculated to estimate the fractal correlation dimension of the system. Estimations of the Kolmogorov entropy provide an additional measure of the complexity of the strange attractor. The values of the chaotic invariants are compared across varying degrees of cathode flooding to discern how they change with two phase flow regimes and fuel cell operating conditions. Local recurrence times within the attractor are calculated to characterize the stability of underlying quasi-stable periodic orbits, illuminating potential control mechanisms. Future work will involve leveraging the chaotic understanding of two phase flow with chaos control methods to increase the power stability.

(561d) Temperature Dynamics and Control of a Fuel Cell Stack System

Daniel O'keefe[1], Mohamed El-Sharkh[2] and **Dr. Srinivas Palanki**[1], (1)Department of Chemical and Biomolecular Engineering, University of South Alabama, Mobile, AL, (2)Department of Electrical and Computer Engineering, University of South Alabama, Mobile, AL

Hydrogen-powered fuel cells are of current interest as an alternative source of energy for both mobile as well as stationary applications. While there are multiple theoretical models available for a fuel cell stack, they only consider the dynamics of the reaction itself inside of the cell and assume that the stack temperature can be maintained at the desired set-point via natural air convection. However, when the power requirements are high, it may be necessary to use a water-cooled system to maintain a desired stack temperature. In this research, a time-varying proportional-integral (PI) controller is designed for controlling the temperature of a water-cooled 5 kW hydrogen fuel cell stack. This controller is designed using a mathematical model for a Ballard Mark V PEM fuel cell stack, which is derived using basic chemical engineering material and energy balances. The controller affects the stack temperature by changing the flow rate of cooling water that passes across the stack. The model is then analyzed using a number of realiastic power demand profiles to determine the effectiveness of the controller. In particular a power profile from experimental data from a residential house is used to test the efficacy of the controller. The results show that a time-varying PI controller is adequate for maintaining the stack temperature within 5 K of the target point.

(561e) The Effect of Electron Donor On Microbial Fuel Cell Performance

Hao Wang[1], Donald Comfort[1], Piyush Shah[2], Andrew Sarangan[2] and John Rowe[3], (1)Dept. of Chemical and Materials Engineering, University of Dayton, Dayton, OH, (2)Electro-Optics Program, University of Dayton, Dayton, OH, (3)Biology, University of Dayton, Dayton, OH

Microbial fuel cells (MFCs) are a fuel cell technology that utilizes microorganisms as catalysts in the conversion of electron donors into electrical power. Similar to traditional fuel cells, MFCs can be set up in two-chamber systems separating the anode and cathode chambers by a proton exchange membrane (PEM). Unlike traditional hydrogen or methanol fuel cells which efficiently utilize a single energy source, MFCs are capable of utilizing a variety of electron donors. The spectrum of electron donors is dependent upon the metabolic capabilities of the microorganism that is selected as the biocatalyst for the process. The ability of MFCs to utilize the wide variety of electron donors, brings up the question what the power production is under these different substrates, which will be important for applications using a variety of energy sources.

In these studies, the γ-proteobacterium *Pseudomonas aeruginosa* (PAO1) was utilized as the anode biocatalyst to process a variety of electron donors. An H-style MFC utilizing 1 mil Nafion® PEM with 175 mL anode and cathode chambers was utilized to assess the electrical production performance of a variety of electron donors (glucose, formate, succinate, acetate, and cellobiose). The anode was carbon cloth and the cathode was platinum foil. Mineral salt medium supplemented with 5 g/L of each individual substrate was as the anolyte. Succinate and glucose were the most effective catalysts in terms of power production at 46 and 41 mW/m^2. No observable power was produced from formate, acetate, or cellobiose. This electricity production by *P. aeruginosa* from glucose and succinate, but not formate, acetate, or cellobiose indicates that the composition of the electron donor feed sources needs to be considered when implementing microbial fuel cells utilizing complex feed sources.

(561f) Dynamic Studies of a Polymer Electrolyte Membrane Fuel Cell (PEMFC)

Srinivasarao Modekurti, Department of Chemical Engineering, West Virginia University, Morgantown, WV, Debangsu Bhattacharyya, Department of Chemical Engg., West Virginia University, Morgantown, WV, Srinivasan Raman, Department of Chemical Engineering, Texas Tech University, Lubboock, TX and Raghunathan Rengasamy, Department of Chemical Engineering, Texas Tech University, Lubbock, TX

Polymer electrolyte membrane fuel cells (PEMFCs) used in the commercial applications should operate efficiently in the face of disturbances such as change in the load, feed composition, etc. and malfunctions such as flooding, drying, starvation of reactants, etc. A dynamic model can be used for developing effective control strategies to maintain the efficiency of healthy as well as faulted cells under changing operating conditions .

Though many steady state models are available in the open literature, detailed dynamic models with experimental validation are few in number. In this work, a detailed two-dimensional two-phase dynamic model of a PEMFC is developed. Both anode and cathode catalyst layers are modeled using spherical agglomerate characterization. The model includes mass conservation equations for the chemical species in the flow channel, gas diffusion layer (GDL), microporous layer (MPL), and the catalyst layer (CL). In addition, conservation of charged species is modeled in the GDL, MPL, and CL. As the liquid water plays a crucial role in the PEMFCs, the effect of liquid water is modeled in all the layers. The transport of liquid water through polymer membrane is modeled considering electro-osmotic drag and back-diffusion. The dynamic model is validated with the transient data generated from an in-house cell by changing various operating conditions. The validated model is utilized to study dynamics of various process variables under normal and faulted operation of the cell. An in-situ measurement of these variables is very difficult, if not impossible, with current state-of-the-art technology. The presentation will include a number of key observations that can help to improve the operational efficiency of the PEMFCs under changing operating conditions.

(561g) Electrochemical Equipment Design for Hybrid Vehicles

Syed Kaschif Ahmed, Chemical and Biological Engineering, Illinois Institute of Technology, Chicago, IL and Donald J. Chmielewski, Illinois Institute of Technology, Chicago, IL

The notion of a fuel cell powered vehicle has captured the imagination of many as a clean/efficient alternative to the internal combustion engine. However, the capital cost of a fuel cell power unit is significant, and threatens to be a show-stopping hurtle. As such, many have looked to hybridize the fuel cell with lower cost energy storage devices; a rechargeable battery and / or super-capacitor. These low energy / high power density devices will provide supplemental power during periods of large demand and thus allow for the selection of a smaller / lower cost fuel cell unit. The basic design challenge is to size of the fuel cell and supplemental energy storage units so as to minimize overall cost.

Due to the drive cycle expected during vehicle operation, the proposed design scheme should include power demand variability as an underlying element. It is further recognized that a high level controller is needed to coordinate device interactions as well as ensure that demanded power is sent to the electric motor driving the vehicle. Clearly, the characteristics of this high level controller will influence the required size of each device. As such, controller characteristics have been included as design variables. Additional design criteria include; max power and energy density for each type of device and losses associated with energy conversion.

Concerning computational issues, we will show that nearly all of the nonlinear constraints discussed above can be converted to a convex form. The exception is a limited set of scalar reverse-convex inequalities, used to define the relation between variance and standard deviation for key process variables. Given this convex / reverse-convex form a computationally efficient branch-and-bound algorithm has been developed to obtain globally optimal solutions to the proposed hybrid vehicle design problem.

(561h) Nanostructured Photosystem Composites As Biological Fuel Cell Catalysts

Ramaraja P. Ramasamy, The University of Georgia, Athens, GA

Manipulating photosynthesis for electricity generation is an interesting concept. Here we report for the first time, the demonstration of light to electrical power conversion in photosynthetic biological fuel cell using plant photosystems complexes and enzyme biocatalysts. The system works by combining the photo-induced water oxidation reaction on nanostructured photosystem protein based anode and oxygen reduction reaction onbio-cathode. Both the photosystem complex and cathode biocatalyst molecules were immobilized onto multiwalled carbon nanotubes through molecular tethers on gold electrodes. The nanostrcutured electrodes possess high surface area for high biomolecule loading and the carbon nanotubes served as excellent immobilization support for the biocatalysts. The photosystem-carbon nanotube composite exhibited high photo-electrochemical activity upon light illumination. For the first time ever, direct electrochemical communication was observed between plant proteins and the electrode. The photo-induced water splitting reaction has been confirmed to be the main source of electrons for photo-current generation. The favorable electron transfer pathways in the photosynthetic process for photo-electrochemical activity have been identified. The direct redox electrochemistry observed has significant impact in artificial photosynthesis research.

(561i) Characterizing Anode Potential Losses in Microbial Fuel Cells Using Electrochemical Techniques

César I. Torres, Chemical Engineering, Arizona State University, Tempe, AZ and Sudeep C. Popat, Arizona State University, Tempe, AZ

Characterizing Anode Potential Losses in Microbial Fuel Cells Using Electrochemical Techniques

Sudeep C. Popat and César I. Torres

Anode-respiring bacteria (ARB) are unique in their capacity to transfer electrons from organic substrates to a solid anode. The result of anode respiration is an electrical current that can be used for various bioenergy applications including power generation in microbial fuel cells (MFCs). An efficient MFC must have minimal potential losses at the anode, the cathode, or the electrolyte. The anode of an MFC presents an interesting challenge as it is composed of anode-respiring bacteria (ARB) catalyzing the oxidation of organic matter into CO_2. As a living catalyst, ARB form a porous biofilm that extends tens of micrometers. Inside the ARB biofilm, concentration and potential gradients determine the current production by ARB. Anode resistances can be categorized as those associated with (1) mass transport, (2) ARB intracellular metabolism, and (3) extracellular electron transfer (EET). Although some efforts have been made to characterize these anodic resistances, their contributions to MFC potential losses are largely unknown.

Electrochemical measurements performed recently have begun to provide quantitative measurements of EET resistances (Torres et al., 2008; Torres et al., 2010; Strycharz et al., 2011; Manklavar et al., 2011). Measurements performed using cyclic voltammetry (CV) in inter-digitated microelectrode arrays (IDAs) provide biofilm conductivity measurements that allow us to estimate EET resistances. In addition, the studies performed by Manklavar et al. (2011) complement these measurements and provide specific details on the nature of the EET conductive matrix. Our research group used CVs to interpret anode resistances in MXCs using the Nernst-Monod equation (Torres et al., 2007; Torres et al., 2008). All these efforts have one commonality: the conclusion that EET resistances are low, indicative of an efficient electron-transport mechanism through the conductive biofilm.

Although anode potential losses can be measured with CV, they cannot be characterized individually. Theoretical assumptions, such as those proposed by the Nernst-Monod equation, are the only tools available to further characterize resistances in CVs. In order to address this problem of separating individual polarization losses, fuel-cell researchers rely on Electrochemical Impedance Spectroscopy (EIS). EIS has been used extensively in measuring the polarization losses at the anodes and cathodes of solid oxide fuel cells. A key feature of EIS is that it can help distinguish multiple processes contributing to polarization loss on a single electrode, if the characteristic frequency at which each process controls is different. The use of EIS in MFCs has only begun to be explored, with the most notable studies being on characterizing total anodic, ohmic, and cathodic polarization losses during long-term operation (Borole et al., 2010) and in detecting shuttles produced by ARB (Ramasamy et al., 2009).

Our preliminary EIS analyses using *G. sulfurreducens* as model ARB seem to indicate that at least two anodic resistances can be resolved using EIS within typical frequency ranges (Figure 1). The resistances we observed reveal very interesting properties of ARB's electron transfer. The first resistance (exhibited at high frequencies) appears to be constant as a function of anode potential (not shown). More explicitly, it acts as a linear resistance. Enzymatic processes always have a sigmoidal response to potential; therefore, this resistance must be attributed to EET (e.g., nanowires acting as a linear resistor). The second resistance is affected by anode potential (not shown) and follows a sigmoidal response, which is typical of enzymatic processes. However, the response obtained from this second resistance is that of a Nernst-Monod sigmoid with $n = 2$, not $n = 1$. Most researchers have assumed an $n=1$ for the Nernst-Monod fitting based on the fact that cytochromes carry out a one-electron transfer and are known to be key proteins in the EET pathway. These results underscore that we need to revisit the $n = 1$ assumption for the Nernst-Monod equation. If $n = 2$ for the Nernst-Monod curve, a two-electron transfer ultimately limits the flow of electrons to the anode. This process most likely occurs within or before the electron-transport chain, in which a few reactions are known to have n = 2. Thus, we hypothesize that the second resistance observed in Figure 1 is attributable to an intracellular resistance, basically an energy loss *within* the cell. If this is correct, EIS becomes a method to differentiate between EET resistances and intracellular resistances.

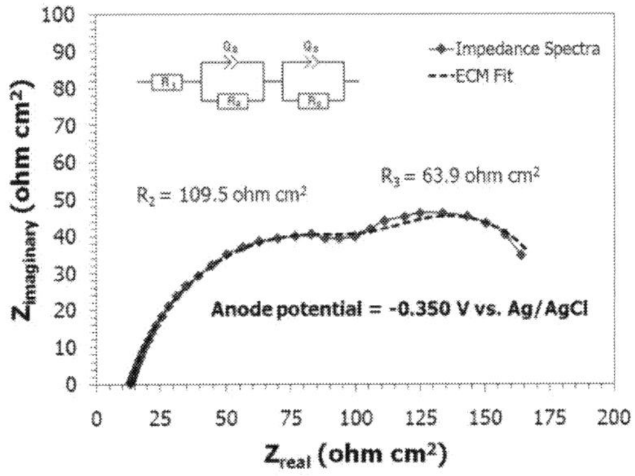

Figure 1. Nyquist plots example of an ARB biofilm tested at -0.35 vs Ag/AgCl.

References:

Borole, A. P., D. Aaron, et al. (2010). Environ Sci Technol 44(7): 2740-2744.

Malvankar, N. S., M. Vargas, et al. (2011). Nat Nanotechnology 6(9): 573-579.

Ramasamy, R. P., V. Gadhamshetty, et al. (2009). Biotechnol Bioeng 104(5): 882-891.

Strycharz, S. M., A. P. Malanoski, et al. (2011). Energ Environ Sci 4(3): 896-913.

Torres, C. I., A. Kato Marcus, et al. (2007). Appl Microbiol Biotechnol 77(3): 689-697.

Torres, C. I., A. K. Marcus, et al. (2008). Environ Sci Technol 42(17): 6593-6597.

Torres, C. I., A. K. Marcus, et al. (2010). FEMS Microbiol Rev 34(1): 3-17.

(582a) The Growth of a Vapor Bubble On a Heated Surface Using the Lattice Boltzmann Calculation: Flow, Heat Transfer and Phase Change

Lin Zheng, Chemical Engineering, The City College of City University of New York, New York, NY, Taehun Lee, Mechanical Engineering, The City College of City University of New York, New York, NY and David S. Rumschitzki, Department of Chemical Engineering, The City College of The City University of New York, New York, NY

Nucleate boiling is a very attractive method for cooling down a hot surface because, due to typically large latent heats, evaporation of a small amount of liquid can absorb a large amount of heat from the solid. The danger, however, is that the solid surface can even temporarily dry out, and the consequent hiatus in cooling could be catastrophic. One would therefore like to understand the physics of nucleate boiling well enough to prevent dryout.

Previously we reported conduction-only calculations for this problem that, surprisingly, yielded results that agreed with a large amount of phenomenology and experimental observations for these systems. These calculations showed, however, that nearly all of the evaporation takes place at the three-phase contact line (CL). Moreover the poorly-understood motion of the CL is critical to the evolution of vapor bubbles as gravity deforms them and drives them toward detachment from the solid surface, processes that determine which boiling regimes the system is in and that are clearly central to system's deciding whether dryout will occur. Since CL motion with bubble growth is an input to a traditional continuum calculation, and since, even in the absence of phase change, is a subject of much disagreement in the literature, we seek to employ a method that obviates the need for ad hoc assumption about CL motion.

As such, we employ and extend the lattice Boltzmann equation (LBE) method, one of a class of methods known as diffuse interface methods, to include heat transfer, fluid flow and phase transition phenomena and implement it in two dimensions. The advantage of this method is that it considers the liquid and the vapor as a single phase of variable density. As such, the liquid-vapor interface is in the fluid domain and, in particular, the CL is part of its solid-fluid boundary, and their positions are outputs of the calculation, rather than input to it. The price that one pays for this approach is that the interface no longer remains nanoscopic, but rather is resolved over 3-5 gridpoints. The way that we employ this method is by using phase field theory to determine the transport properties of the multiphase flow such as pressure tensor and chemical potential from the Helmholtz free energy. This uses a known LBE formulation for the flow field. For the temperature field, we derive a LBE that, via Chapman-Enskog expansion techniques, can recover the correct energy equation (in an entropy formulation that allows a straightforward incorporation of the latent heat) from thermodynamic theory by ignoring the viscous dissipation term. We verify the accuracy of these calculations by comparison with problems having known analytic (e.g., similarity) solutions, such as Laplace's law, Stephan problem and Scriven's growth of a single bubble in an infinite fluid.

We look at the growth of a vapor bubble on an infinitely thin heated solid surface. Since we require the solid surface to be isothermal, it fixes the CL temperature as the surface temperature, unlike our earlier conduction-only calculation that included the solid and set the CL temperature equal to the saturation temperature. We calculate the temperature and flow fields and the motion of the CL at various time points up to and beyond bubble detachment and compare with (3D) data in the literature. We consider the size of the escaping bubble and of the vapor residue and its fate, including the descent and impingement of colder fluid that replaces the escaping detached bubble. Time permitting, we consider the case with a finite thickness solid as well and how issues of finite interface thickness play out in the solutions.

(582b) Rare Events Calculations of Methane Clathrate Hydrate Nucleation From Solution At Moderate Supersaturations

Brandon C. Knott[1], Valeria Molinero[2], Michael F. Doherty[1] and Baron Peters[1,3], (1)Chemical Engineering, UC Santa Barbara, Santa Barbara, CA, (2)Chemistry, University of Utah, Salt Lake City, UT, (3)Chemistry and Biochemistry, UC Santa Barbara, Santa Barbara, CA

Clathrate hydrates are crystalline ice-like inclusion compounds in which water molecules form cages around hydrophobic guest molecules. Methane clathrate hydrates form at high pressures and low temperatures and are found in abundance on the ocean floor and under permafrost. Additionally, their formation in oil and natural gas pipelines can severely affect pipeline economics and safety. Past computational studies aimed at elucidating the mechanism of hydrate nucleation used "brute force" molecular dynamics (MD) trajectories. These trajectories are run at extremely high driving force (i.e. supersaturation) compared to driving forces that exist in nature. We used a coarse-grained model of methane hydrates[1] to perform phase coexistence calculations and to compute supersaturations. Then we used rare events methods to examine methane hydrate nucleation at moderate supersaturation. These calculations enable us to estimate nucleation rates and draw conclusions regarding the reaction mechanism.

References

[1] L.C. Jacobson and V. Molinero, "A Methane-Water Model for Coarse-Grained Simulations of Solutions and Clathrate Hydrates", J. Phys. Chem. B, **2010**, 114, 7302-7311.

(582c) Quantifying Clathrate Hydrate Nucleation and Growth From Molecular Simulations

Amadeu K. Sum[1], David T. Wu[2], Somendra Nath Chakraborty[1] and Eric Grzelak[3], (1)Chemical & Biological Engineering, Colorado School of Mines, Golden, CO, (2)Chemical Engineering and Chemistry Departments, Colorado School Of Mines, Golden, CO, (3)Chemical & Biological Engineering Department, Colorado School of Mines, Golden, CO

Despite their industrial significance and ubiquity, the formation mechanism of hydrates remains poorly understood. The challenges associated with describing hydrate nucleation and growth mechanisms remain at the forefront of fields such as hydrogen storage, carbon dioxide sequestration, and hydrocarbon transportation and recovery. Various mechanisms of hydrate formation have been postulated, most involving the annealing of an amorphous or partially-ordered aggregation of cages into the thermodynamically-preferred crystal structures. In this talk, we will present an overview of our efforts in understanding clathrate hydrate nucleation and growth using molecular simulations, including defining the thermodynamics phase space and direct simulations capturing the different stages through the nucleation and growth processes. The molecular events during nucleation and growth are quantified by applying different order parameters to represent the structuring of water and gas molecules. One of the approaches is based on a Voronoi tessellation analysis, which provides information on the cooperative ordering of gas and water molecules. In applying the tessellation to the nucleating trajectories, we observe that several of the Voronoi properties (polyhedra volume, total surface area, asphericity parameter) show strong signatures of change when hydrates nucleate. The results suggest that the Voronoi tessellation analysis is a powerful and efficient tool that can be applied to obtain further insight into the mechanisms associated with the ordering of methane and water molecules in the nucleation and growth processes of hydrates.

(582d) Insights Into Hydrate Nucleation Using Molecular Simulations

Sapna Sarupria, Chemical and Biomolecular Engineering, Clemson University, Clemson , SC and Pablo G. Debenedetti, Chemical and Biological Engineering, Princeton University, Princeton, NJ

Gas hydrates represent an important class of crystalline solids that have applications in wide ranging fields such as petrochemical industry, environment and energy. While the thermodynamics of hydrate formation is well understood, we have very little insights into the kinetics of hydrate nucleation. Nucleation studies in experiments are challenging due to the difficulty in accessing the nanometer length and nanosecond timescales that are relevant to nucleation processes. Conventional molecular dynamics simulations, which sample nanometer length scales and nano-to-microsecond time scales are ideally suited to probe this phenomenon. However, the long waiting times involved in observing nucleation events in simulations make it computationally challenging to sample statistically relevant number of trajectories. Therefore, we approach this problem from two perspectives - first we generated several microsecond long trajectories of methane hydrate nucleation, and performed extensive analysis of the events leading to nucleation. This provides valuable insights into various factors that feature in hydrate nucleation. Secondly, based on the insights from the nucleation trajectories, we carried out detailed studies of these factors. Together, our studies provide insights into aspects governing hydrate nucleation, and will help design further studies focused on the kinetics of hydrate nucleation, and growth.

(582e) A Microscopic View of Nucleation in the Anatase-to-Rutile Transformation

Ya Zhou and Kristen Fichthorn, Dept. of Chemical Engineering, Pennsylvania State University, University Park, PA

We use molecular simulation techniques to investigate the anatase-to-rutile tansformation in titania nanocrystals. A thermodynamic analysis indicates that edge and corner atoms significantly influence the critical size at which rutile nanocrystals become energetically preferred over anatase. Using molecular dynamics simulations, we probe kinetics of the transformation in individual anatase nanocrystals and in nanocrystal aggregates. We follow structural evolution using simulated X-ray diffraction. Additionally, we develop a local order parameter to distinguish individual Ti ions as anatase, rutile, or anatase {112} twin-like. We apply our local order parameter to track the formation and growth of rutile nuclei. Anatase {112} twins form easily at surfaces and interfaces of nanocrystal aggregates and rutile nucleates among the twins. Stable rutile nuclei maintain {101} facets during growth, as a result of nucleation from layers of alternating anatase {112} twins, until the growth is inhibited at surfaces and grain boundaries. Our results are in agreement with experiment and indicate the central role of {112} twin-like anatase in the transformation.

(582f) Gel-Induced Selective Crystallization of Polymorphs

Ying Diao[1], Allan S. Myerson[2], T. Alan Hatton[1] and Bernhardt L. Trout[3], (1)Chemical Engineering, Massachusetts Institute of Technology, Cambridge, MA, (2)Department of Chemical Engineering, Massachusetts Institute of Technology, Cambridge, MA, (3)Massachusetts Institute of Technology, Cambridge, MA

Although nanoporous materials have been explored for controlling crystallization of polymorphs in recent years, polymorphism in confined environments is still poorly understood, particularly from a kinetic perspective, and the role of the local structure of the substrate has largely been neglected. Herein, we report the use of a novel material, polymer microgels with tunable microstructure, for controlling polymorph crystallization from solution and for investigating systematically the effects of nanoconfinement and interfacial interactions on polymorphic outcomes. We show that the polymer microgels can improve polymorph selectivity significantly. The polymorphic outcomes correlate strongly with the gel-induced nucleation kinetics and are very sensitive to both the polymer microstructure and the chemical composition. Further mechanistic investigations suggest that the nucleation-templating effect and the spatial confinement imposed by the polymer network may be central to achieving polymorph selectivity. We demonstrate polymer microgels as promising materials for controlling crystal polymorphism. Moreover, our results help advance the fundamental understanding of polymorph crystallization at complex interfaces, particularly in confined environments.

(582g) Model Based Modified Brust-Schiffrin Method for Synthesis of Monodispersed Gold Nanoparticles

Sivaramakrishna Perala, Department of Chemical Engineering, Indian Institute of Science, Bangalore, India and Sanjeev Kumar, Department of chemical engineering, Indian Institute of Science, Bangalore, India

The synthesis of thiol capped gold nanoparticles using Brust-Schiffrin Method[1] (BSM) produces sub 5 nm gold nanoparticles on gram scale. The high yield of extremely stable particles and varied options for surface functionalization have made BSM a widely used synthesis protocol in a number of fields. One of its key shortcomings, high polydispersity of particles, is overcome by size fractionation, a tedious procedure which also decreases the yield of particles in the desired size range. In this work, we have developed a comprehensive model for BSM of synthesis and used it to develop modified synthesis protocols to produce particles with improved monodispersity. The new protocols have been verified experimentally.

The two-phase BSM involves complex reduction reactions facilitated by phase transfer agents, nucleation, and simultaneous growth and capping of nanoparticles. The existing body of experimental findings, including the most recent one implicating micelles of phase transfer agent as the reaction sites[2], was analyzed and subjected to additional experimentation to develop a mechanistic picture of the synthesis process. Since populations of particles with different sizes and different free surface area available for growth need to be tracked, the framework of bi-variate population balance equations was used to capture the particle synthesis process. The model successfully captures the experimental findings available in the literature. It unravels competition between capping and continued nucleation as key to size and polydispersity control. The model is next used to simulate particle synthesis for alternative, non-obvious mixing protocols. Two improved strategies are proposed through insight gained from the model developed. Experiments conducted by using modified protocols exhibited better control over the monodispersity than the original BSM protocols.

References:

1.   Brust, M.; Walker, M.; Bethell, D.; Schiffrin, D. J.; Whyman, R., *J. Chem. Soc., Chem. Commun.*, 1994, **7**, 801-802. 2.   Li, Y.; Zaluzhna, O.; Tong, Y., *Langmuir*, 2011, **27**, 7366-7370.

(588a) Selective Colloidal Self-Assembly As a Means of Single-Walled Carbon Nanotube Purification

Constantine Khripin[1], Nicholas A Arnold-Medabalimi[1], Xiangyun Qiu[2] and Ming Zheng[1], (1)Polymer Division, National Institute of Standards and Technology, Gaithersburg, MD, (2)Physics, George Washington University

Colloidal dispersion of single-walled carbon nanotubes (SWCNTs) with DNA, forming the DNA-SWCNT hybrid colloid, has opened up many possibilities in the field of SWCNT purification. DNA-SWCNT can be chromatographically fractionated both by length and chirality, achieving samples of un-paralleled homogeneity. However, available methods require a high level of equipment and expertise. Here, we report a series of simple methods based on the selective self-assembly of colloidal DNA-SWCNT. Under molecular crowding conditions, DNA-SWCNT self-assemble into nematic rod-like clusters. When the crowding polymer is non-interacting, e. g. polyethylene glycol (PEG), self-assembly is length dependent. We use this effect to fractionate DNA-SWCNT by length. When the polymer is interacting, e. g. polyvinyl pyrrolidone (PVP), self-assembly becomes SWCNT chirality dependent. This selectivity varies with the identity of the salt ions present, as per the Hofmeister effect, and with the sequence of the DNA dispersant. We identify two sequences, $(GT)_{20}$ and $(AC)_{20}$, able to produce useful samples with low SWCNT chirality variability. Finally, we report that DNA-SWCNT self-assembly can be induced simply by a high concentration of salt. This effect is length dependent, and can also be used for length fractionation. The solubility behavior suggests a high degree of cooperatively in self-assembly, especially for long SWCNTs. Furthermore, solubility shows the Hofmeister effect with a partial inversion of the series, reportedly associated with comparatively hydrophobic colloids. The many colloidal self-assembly phenomena observed in DNA-SWCNT suggest this versatile particle may be useful for the testing of theoretical formulations on self-assembly.

(588b) Floret-Shaped Solid Domains On Giant Fluid Lipid Vesicles Induced by pH

Amey Bandekar, Biomedical Engineering and Chemical and Biochemical Engineering, Rutgers University, Piscataway, NJ and **Stavroula Sofou**, Biomedical Engineering, and Chemical and Biochemical Engineering, Rutgers University, Piscataway, NJ

Lateral lipid phase separation of titratable PS or PA lipids and their assembly in domains induced by changes in pH are significant in liposome-based drug delivery: environ- mentally responsive lipid heterogeneities can be tuned to alter collective membrane properties such as permeability (altering drug release) and surface topography (altering drug carrier reactivity) impacting, therefore, the therapeutic outcomes. At the micrometer scale fluorescence microscopy on giant unilamellar fluid vesicles (GUVs) shows that lowering pH (from 7.0 to 5.0) promotes condensation of titratable PS or PA lipids into beautiful floret-shaped domains in which lipids are tightly packed via hydrogen-bonding and van der Waals interactions. The order of lipid packing within domains increases radially toward the domain center. Lowering pH enhances the lipid packing order, and at pH 5.0 domains appear to be entirely in the solid (gel) phase. Domains phenomenologically comprise a circular "core" cap beyond which interfacial instabilities emerge resembling leaf-like stripes. At pH 5.0 stripes are of almost vanishing Gaussian curvature independent of GUVs' preparation path and in agreement with a general condensation mechanism. Increasing incompressibility of domains is strongly correlated with a larger number of thinner stripes per domain and increasing relative rigidity of domains with smaller core cap areas. Line tension drives domain ripening; however, the final domain shape is a result of enhanced incompressibility and rigidity maximized by domain coupling across the bilayer. Introduction of a transmembrane osmotic gradient (hyperosmotic on the outer lipid leaflet) allows the domain condensation process to reach its maximum extent which, however, is limited by the minimal expansivity of the continuous fluid membrane.

(588c) Self-Assembly of Clathrin Protein 3D Structures

Manickam Adhimoolam Arunagirinathan[1,2], Brian J Gibbons[3], Alia P. Schoen[1,2], Andrew J. Spakowitz[1,4] and Sarah C. Heilshorn[1,2], (1)SLAC National Accelerator Laboratory, Stanford Institute for Materials and Energy Sciences, Menlo Park, CA, (2)Materials Science and Engineering, Stanford University, Stanford, CA, (3)Structural Biology, Stanford School of Medicine, Stanford, CA, (4)Chemical Engineering, Stanford University, Stanford, CA

Clathrin is a ~630 kDa protein that resembles a three-legged pinwheel, *i.e.*, a triskelion, with each leg having an extended length of ~50 nm and width of ~ 3 nm. Due to molecular recognition interactions between legs and the non-planar geometry of the basic building block, clathrin self-assembles to form 3D assemblies resembling fullerene structures. *In vivo*, clathrin helps in transportation of extracellular components across the cell membrane with the help of adaptor proteins through receptor mediated endocytosis by forming enclosed polyhedral cages. *In vitro*, clathrin self-assembles in the absence of lipid membranes or adaptor proteins to form a variety of three-dimensional shapes like tetrahedra, cubes and spheres. The underlying mechanisms that enable clathrin to self-assemble into such a diversity of nanostructures is not well understood. Here we use Cryo Transmission Electron Microscopy (cryo-TEM) and Dynamic Light Scattering (DLS) to explore the influence of solvent conditions (pH and ionic strength) on the kinetics and morphology of clathrin self-assembly and to discover potential new nanostructures that clathrin may form.

Our results demonstrate that clathrin has strikingly different assembly kinetics at pH 5 (below the isoelectric point) versus pH 6 (above the isoelectric point). Within 30 minutes of inducing self-assembly at pH 6, discrete spherical cages of diameter ~60-100 nm were observed under cryo-TEM. In contrast, inducing assembly at pH 5 causes the formation of large (~200-600 nm) disordered aggregates. Interestingly, upon aging for 4 days at room temperature, the large aggregates undergo substantial remodeling to form discrete 60-100 nm spherical cages, while the samples assembled at pH 6 remained unchanged. This corroborates DLS data that indicate clathrin self-assembly is a function of both pH and time, with large length-scale structural rearrangements occuring over long time-scales (*i.e.*, days). The stability of clathrin cages at pH 6 and the transformation of cages at pH 5 lend new insight into the role of kosmotropic and chaotropic interactions in mediating protein self-assembly via electrostatic interactions. We hypothesize that clathrin initially forms large, irregular aggregates at pH 5 due to the sudden decrease in electrostatic repulsion between building blocks that is experienced below the isoelectric point. Over time, the mechanical strain within the aggregate is resolved by reorganization of clathrin triskelion into discrete cage structures. Current efforts are underway to construct a phase diagram of clathrin nanostructures and to compare our experimental results with computational predictions that consider the strength of leg-leg interactions and the elasticity of individual legs.

(588d) Molecular Dynamics Simulations of Peptide Self-Assembly Under Confinement

Cade B. Markegard[1], Iris W. Fu[1] and Hung D. Nguyen[2], (1)Chemical Engineering and Materials Science, University of California, Irvine, Irvine, CA, (2)Department of Chemical Engineering and Materials Science, University of California, Irvine, Irvine, CA

Molecular dynamics simulations of self-assembled peptide aggregates have commonly been investigated under bulk conditions, however to more accurately represent the crowded, heterogeneous environment of a cell, it is necessary to perform studies of protein folding and aggregation under confinement. In this work, we look at three peptide sequences (KA14K, WQVQVEVQVEVQVQVQV, & WQVEVQVQVQVQVEVQV), to study their individual aggregation processes starting from a random configuration when confined between two surfaces. In general, we investigate how confinement influences the kinetics of aggregation compared to bulk. Since long-range interactions especially those between charged groups are most affected by confinement, we examine how the location of charged amino acid residues modify the kinetics and final self-assembled structures. To achieve this we use an extended version of a coarse-gained amino acid model, PRIME (Protein Intermediate Resolution Model), and utilize Discontinuous Molecular Dynamics in order to obtain the long time scales needed to study aggregation. Our results indicate that confinement increases the propensity to form beta-structures. Moreover, the position of the charged residues determines the location and length of a beta-strand. In addition, we propose general guidelines for how the placement of electrostatic residues in a peptide affects the assembly of ordered structures.

(588e) A Computational Study of the Self-Assembly of Diblock Peptides

Daniel Forciniti, Cbe, Missouri University of Science and Technology, Rolla, MO and Fazeem Rehman, Chem. Eng., Missouri S&T, Rolla, MO

Model compounds have been used to study the formation of amyloid deposits in a controlled manner and to address the question: Are all proteins capable of forming amyloid deposits? .

We have identified a family of diblock peptides that forms amyloid fibrils at moderate conditions (pH 4 and 60 °C). The kinetics of fibril formation seems to be affected by the overall length of the peptide but not by their composition (beyond some critical block concentrations). Longer peptides form fibrils faster than shorter ones. Fibrillation depends on minute differences in amino acid composition. For example, diblock peptides of lysine and isoleucine form fibrils after a critical length is reached whereas diblock peptides of lysine and leucine do not form fibrils irrespective of the length of the blocks. The fibrillation process escapes a simple explanation.

To advance our knowledge about the formation of fibrils by this type of peptides we have started a computationally study. A Monte Carlo code developed by our group years ago (Mungikar and Forciniti, Biomacromolecules, 7:239 (2006)) was modified to to handle the simulation of multiple peptide molecules (the old code was written for an infinite dilution situation). Most simplified peptide models will not be able to capture the differences between isoleucine and leucine that seems to be critical in our case. Unfortunately and all-atom simulation in the presence of explicit water will allow us only to scratch the surface of the nucleation process (fewer than 10 peptides) but it will be prohibitive to study the peptide concentrations found during fibril formation. To overcome this problem, we used an all atoms model for the peptides but an implicit model to account for solvent effects. The solvent is treated implicitly using a peptide density dependent switch function that accounts for intra and intermolecular hydrogen bonding and hydrophobic interactions. Initial simulations suggest the initial formation of amorphous aggregates (fibril nuclei) formed by multiple contacts of the non polar amino acids. Slow conformational changes are then induced by repulsions between the charged lysine groups.

(588f) Polyacrylamide Induced Thinning and Increase in Head Group Area of Supported Cationic Bilayers

Saurabh Das, Chemical Engineering, University of California, Santa Barbara, Santa Barbara, CA, Stephen H. Donaldson Jr., Chemical Engineering, UC Santa Barbara, Santa Barbara, CA and Jacob Israelachvili, Department of Chemical Engineering, University of California at Santa Barbara, Santa Barbara, CA

The interaction between supported cationic bilayers and the effect of uncharged polymer on theses interactions was studied using a surface forces apparatus (SFA) with a radius of curvature of the surfaces of ~2 cm. Bilayers of the cationic surfactant di(tallow ethyl ester) dimethyl ammonium chloride (DEEDMAC) were deposited on molecularly smooth mica surfaces using the Langmuir-Blodgett technique and the forces between the bilayers were measured in a 4.5 mM $CaCl_2$ solution at pH=4. The forces between the bilayers were repulsive and followed DLVO theory with a modified Debye length to bilayer-bilayer separation of ~1nm. The addition of polyacryamide, an uncharged polymer, decreased the repulsive force of interaction between the bilayer surfaces. The polyacrylamide chain tends to lay on the surface of the bilayer due to hydrophobic force between the backbone of the polymer chain and the hydrocarbon chain of the DEEDMAC bilayer and caused thinning of the hydrocarbon chains and increase in the headgroup area.

(588g) Oppositely Charged Polyelectrolyte Assembly in Solution and in Multilayer Films

Biswa P. Das, Ankitkumar I. Fajalia and Marina Tsianou, Department of Chemical and Biological Engineering, University at Buffalo - The State University of New York (SUNY), Buffalo, NY

Multilayered thin films have been the focus of intense research because of their vast array of applications. We investigate here the relationship between polyelectrolyte multilayer growth and the phase behavior of the corresponding polyelectrolyte complexes in solution. We examine the growth and disassembly under different pH conditions of layer-by-layer (LbL) films composed of (strong or weak) oppositely charged polyelectrolytes (including poly(diallyldimethylammonium chloride) (PDDA), poly(styrene sulfonate) (PSS), and poly(acrylic acid) (PAA)) and we compare these to the formation and dissolution of the complexes in solution. We monitor the assembly growth by UV-vis spectroscopy, estimate the assembly thickness by scanning electron microscopy (SEM) cross-section images and the surface roughness by atomic force microscopy (AFM). We further examine the charge density of films by Fourier transform infrared (FTIR) spectroscopy. Our results demonstrate that in films different assembly regimes are attained by manipulating the deposition pH of the weak polyelectrolyte solution. The results are correlated with the phase behavior of the polyelectrolyte complexes in solution. Our findings provide insights into the sensitivity and response of films or complexes to external stimuli during various applications.

(591a) Development of a Group Contribution Method to Predict Sublimation Enthalpies of Solids Constituted by Aromatic and/or Polycyclic Aliphatic Rings. Application for Solid-Vapour Equilibria

Antonio Tabernero, Eva M. Martín del Valle and Miguel A. Galán, Department of Chemical Engineering, University of Salamanca, Salamanca, Spain

Over the last few years, significant research has been conducted with regard to particle formation using supercritical fluid technologies. Non-solvent particles with controlled particle size distribution can be produced by using these environmentally friendly processes. As a consequence, different industries, such as the pharmaceutical, can take advantage of these techniques for different purposes.

Solubility of the solid in the supercritical fluid is crucial to determine which supercritical fluid technique must be used. Isofugacity criterion, together with cubic equations of state, is the most used approach to fit and estimate these equilibrium data. Nevertheless, different solid properties should be known before starting this estimation.

One of those properties is the sublimation pressure. As a matter of fact, different works indicate that a bad choice of the method to calculate this pressure might yield high deviations in the outcome. Clausius-Clapeyron equation can be used for instance to calculate that pressure. However, it is required the knowledge of the sublimation enthalpy. Therefore, a consistent group contribution method to estimate this enthalpy, based only on the molecular structure of the compound, can speed up this calculation process.

A group contribution method to estimate the sublimation enthalpy of any solid constituted by aromatic and/or polycyclic aliphatic rings is proposed in this work. Pharmaceuticals are usually constituted by this type of "chemical" rings. A training set of sublimation enthalpies was built by using literature (90% of the values) or calculated (estimated from experimental solubility) data.

Sublimation pressure of the naphthalene will be calculated with our method, and subsequently the solid-vapour equilibrium of the system CO_2-naphthalene at high pressure will be determined.

Sublimation enthalpies for this type of solids were predicted with our group contribution method with an average absolute relative deviation of less than 9%. Moreover, equilibria data CO_2-naphthalene at high pressure were fitted with Peng-Robinson equation of state with an average absolute relative deviation of less than 15%.

These results highlight that this group contribution method can be applied to calculate sublimation enthalpies of any pharmaceutical, with the aim of using the obtaining value for fitting or predicting solid-vapour equilibria at high pressure.

(591b) Modeling of Alkane Hydrate Dissociation Conditions in the Presence of Electrolyte Solutions and Alcohols Using Ion-Based Statistical Associating Fluid Theory

Hao Jiang, Chemical and Petroleum Engineering, University of Wyoming, Laramie, WY and Hertanto Adidharma, Department of Chemical & Petroelum Engineering, University of Wyoming, Laramie, WY

Modeling of Alkane Hydrate Dissociation Conditions in the Presence of Electrolyte Solutions and Alcohols Using Ion-Based Statistical Associating Fluid Theory

Ion-based statistical associating fluid theory (SAFT2), coupled with van der Waals and Platteeuw model for hydrate phase, is applied to describe the hydrate dissociation conditions for methane, ethane, and propane in the presence of electrolyte solutions, alcohols, and mixed electrolyte - ethylene glycol solutions.

The ions are treated as nonassociating spherical segments, the parameters of which are obtained by fitting experimental activity coefficient and liquid density data of single aqueous salt solution. The binary interaction parameters for correcting the short-range interactions between ions and water, and between ions and alkanes are not needed. The model can be conveniently extended to mixed electrolytes solutions without introducing additional binary interaction parameters. The inhibition effects of NaCl, KCl, and $CaCl_2$ solutions, including their mixtures, on alkane hydrate are found to be well predicted.

The alcohol parameters are obtained by fitting experimental saturation pressure and liquid density data of pure alcohol. The binary interaction parameters between alcohols and water are regressed from the vapor-liquid equilibrium data of binary water – alcohol mixtures. The binary interaction parameters between alkanes and alcohols are not needed. With this approach, the alkane hydrate dissociation conditions in the presence of ethanol, ethylene glycol (MEG), and glycerol are also well predicted.

Without additional fitted parameters, except for the constant binary interaction parameter between Ca^{2+} and ethylene glycol, the model is successfully applied to the prediction of alkane hydrate dissociation conditions in the presence of solutions containing both electrolytes (NaCl, KCl, $CaCl_2$) and ethylene glycol (MEG). In contrast to the commonly used thermodynamic models for mixed electrolyte solutions, which in many cases require concentration-dependent parameters, we find that the proposed model, with a simple extension to mixtures, accurately predicts the alkane hydrate dissociation conditions and well-captures the effect of different compositions of mixed inhibitors.

(591c) Prediction of Phase Behaviors of Strong Association Systems by PR+Cosmosac Equation of State

Chan-Chia Hsu, Wei-Lin Chen and Shiang-Tai Lin, Chemical Engineering, National Taiwan University, Taipei, Taiwan

Organic acids (e.g., formic acid and acetic acid) are widely used in the industry especially in the esterification process. Most of the thermodynamic models fails describe the phase behaviors of mixtures containing acids because of the strong association between acid molecules that may lead to the formation of acid dimers (or higher multi-mers). Currently, the Hayden-O'Connell equation of state is one of most successful model for mixtures containing association species. However, rooted from virial expansion, the application of this method is limited to the vapor phase. In this work, we extend the previously developed PR+COSMOSAC equation of state to model systems containing strongly associating species. The association of acid molecules is considered as a chemical reaction, and the association constant can be determined from partial molar Gibbs free energy (chemical potential) of acid monomer and dimmer. We show that a comparable accuracy compared with the Hayden-O'Connell equation of state is achieved for the compressibility factor of formic, acetic, and proponic acids in the vapor phase. The cross association of formic acid and acetic-acid mixture is also well-described by this new method. Furthermore, the PR+COSMOSAC equation of state is readily applicable to the liquid phases. We will illustrate is application the VLE and LLE behaviors of some acid containing systems.

(591d) Computational Cluster Methods for Development of Molecular-Based Equations of State: Beyond the Virial Equation

Hye Min Kim, Shu Yang, Andrew J. Schultz and David A. Kofke, Chemical and Biological Engineering, University at Buffalo, The State University of New York, Buffalo, NY

Thermodynamic models for bulk behavior are most effective and have greatest predictive power when they are developed with consideration of molecular phenomena. Methods based on computation of cluster integrals are attractive in this context because they can accommodate full molecular detail while providing simple formulas for the bulk behavior. The best-known such method is the virial equation of state (VEOS), which has a long history. It has been the subject of a number of studies that emerged after the development of the Mayer-sampling Monte Carlo method[1] for calculation of cluster integrals. The VEOS has well known limits, and it is worthwhile to consider other cluster-based formulations as a means to develop descriptions of the bulk via computation of cluster integrals.

Here we examine two such approaches. First we continue our previous work for associating systems. VEOS is not particularly effective at capturing the strong directional interactions of associating fluids. In our work we have considered instead Wertheim's multi-density formalism for associating fluid systems. A Wertheim equation of state (WEOS) is derived based on Wertheim's association theory[2, 3]. In previous work, the effectiveness of Wertheim's multi-density formalism has been verified with a simple Lennard-Jones model having one-, two-, and four-sites formed from conical square-well potential[4]. In current work, we examine WEOS for two realistic models, the Gaussian charge polarizable model (GCPM) of water[5] and the improved TraPPE-UA acetic acid[6]. GCPM water is a polarizable model, so we need to extend the Wertheim treatment to handle such systems. We used a three-association-site model to represent 1 oxygen and 2 hydrogen sites. To describe the behavior of acetic acid molecules, one-site and two-site models are examined and the flexible contributions are considered. We compare the pressures determined from various truncated WEOS to values from VEOS and from Monte Carlo simulation. The thermodynamic properties of GCPM water obtained from WEOS converge much better to NPT simulation results compared to the conventional virial formalism at both sub- and super-critical temperatures. We obtain similar improvement in WEOS over VEOS for acetic acid.

Second, we investigate the effectiveness of cluster-integral approaches for ionic solution. The long-range electrostatic force among ionic particles make the traditionally-defined virial coefficients diverge. To overcome this difficulty, Mayer[7] proposed a theory to generate cluster integrals via virial expansion of the Coulomb potential with the addition of an exponential factor, and then resummed the resulting integrals to attain convergence. These cluster integrals depend on ionic concentrations and can be conveniently used to estimate the thermodynamic properties, such as osmotic pressure and activity coefficients. We applied 1-1 restricted primitive model (RPM) to examine our approach. The 2nd order cluster integrals are reduced to single integrals because of spherical symmetry. While some researchers employed Fourier Transforms to calculate some 3rd order cluster integrals, we use Mayer-sampling Monte Carlo simulation together with Fast Fourier Transform to calculate all cluster integrals that have significant contributions to the properties up to the 4th order and thereby extend the range of concentration described by the theory. Comparison with simulation data for the same model is presented.

References

1. J.K. Singh, and D.A. Kofke, *Mayer Sampling: Calculation of Cluster Integrals using Free-Energy Perturbation Methods*. Physical Review Letters, 2004. **92**(22): p. 220601/1-220601/4.

2. M.S. Wertheim, *Fluids with highly directional attractive forces. I. Statistical Thermodynamics*. Journal of Statistical Physics, 1984. **35**(1): p. 19-34.

3. M.S. Wertheim, *Fluids with highly directional attractive forces. III. Multiple Attraction Sites.* Journal of Statistical Physics, 1986. **42**(3): p. 459-492.

4. H.M. Kim, A.J. Schultz, and D.A. Kofke, *Molecular Based Modeling of Associating Fluids via Calculation of Wertheim Cluster Integrals.* Journal of Physical Chemistry B, 2010. **114**: p. 11515-11524.

5. P. Paricaud, M. Predota, A.A. Chialvo, P.T. Cummings, *From dimer to condensed phases at extreme conditions: Accurate predictions of the properties of water by a Gaussian charge polarizable model.* Journal of Chemical Physics, 2005. **122**: p. 244511/1-244511/14.

6. G. Kamath, F. Cao, and J.J. Potoff, *An Improved Force Field for the Prediction of the Vapor-Liquid Equilibria for Carboxylic Acids.* Journal of Physical Chemistry B, 2004. **108**: p. 14130-14136.

7. J.E. Mayer, *The theory of ionic solutions.* Journal of Chemical Physics, 1950. **18**: p. 1426-36.

(591e) Phase, Interfacial, Derivative and Transport Properties of Aqueous Systems with the Soft-SAFT EoS

Felix Llovell, MATGAS Research Center, Bellaterra, Barcelona, Spain, Oriol Vilaseca, MATGAS Research Center and ICMAB-CSIC, Bellaterra, Barcelona, Spain and Lourdes F. Vega, MATGAS Research Center and Carburos Metálicos, Air Products Group, Barcelona, Spain

Water is a fascinating molecule that exhibits a high complex non-ideal behaviour both as a pure fluid and in mixtures from a thermophysical point of view. The modelling of water and their mixtures is still an issue for scientists, as there is still not any single model able to accurately reproduce all its thermophysical properties. However, aqueous systems are present everywhere, and reliable tools are needed.

The goal of this contribution is to present a simple but quite efficient model, within the framework of the soft-SAFT molecular-based equation of state, for representing an important amount of thermophysical properties of water and several aqueous systems, including water + CO_2, water + 1-alkanols and water + ionic liquids. The vapour-liquid and liquid-liquid equilibria of these systems are described in a wide range of temperatures and pressures in good agreement with the experimental data. The calculation of the interfacial tension of some of these mixtures is adressed using the Density Gradient Theory approach into soft-SAFT, while the viscosity of these mixtures is also evaluated using the Free-Volume-Theory (FVT). A further discussion about the ability of the model to calculate some derivative and excess properties of interest is included.

***Acknowledgements**: F. Llovell acknowledges a TALENT fellowship from the Generalitat de Catalunya. This work was partially financed by the Spanish Government under projects CTQ2008-05370/PPQ and CEN-2008-1027. Additional support from Carburos Metálicos and the Catalan Government was also provided (2009SGR-666).*

(591f) Predicting the Phase Behavior of Fluids with Multiple Polar Groups Using the GC-SAFT-VR Equation of State

Gaurav Das, Jessica D. Haley and Clare McCabe, Chemical and Biomolecular Engineering, Vanderbilt University, Nashville, TN

Predicting the Phase Behavior of Fluids with Multiple Polar Groups Using the GC-SAFT-VR Equation of State

Gaurav Das,[1] Jessica D. Haley[1], and Clare McCabe[1,2]

[1]*Department of Chemical and Biomolecular Engineering, Vanderbilt University, Nashville, TN*

[2]*Department of Chemistry, Vanderbilt University, Nashville, TN*

The statistical associating fluid theory[1] is a popular molecular-based equation of state (EOS) that has been successfully applied to predict the thermodynamics and phase behavior of a wide range of fluid systems. In recent developments, group contribution approaches have been combined with the SAFT equation in order to develop more predictive models. The GC-SAFT-VR[2,3] equation is one such approach, based on the SAFT-VR equation in which attractive interactions are described through a potential of variable range (VR). The GC-SAFT-VR EOS models chains composed of segments of different size and/or energy of interaction at both the monomer and chain levels of theory, thus molecules are described by heterosegmented chains in which each type of segment represents a functional group present in the molecule. Parameters for key functional groups (such as CH_3, CH_2, CH, CH_2=CH, C=O, C_6H_5, ether and ester, OH, NH_2, CH=O, COOH) have been obtained by fitting to experimental vapor pressure and saturated liquid density data and their transferability tested by comparing the theoretical predictions with experimental data for pure fluids and binary mixtures not included in the fitting process, as well as by studying the VLE and LLE of small molecules in polymer systems.[4,5-7] In general the GC-SAFT-VR approach is found to predict the phase behavior of the systems studied in excellent agreement with experimental data without adjusting the group parameters to binary mixture data; however deviations were seen for some of the polymer systems studied which contained multiple polar groups. Here we incorporate an explicit dipolar term into the GC-SAFT-VR EOS following the work of Zhao and McCabe. The new equation is tested through the study of fluorinated ether (HFE's) molecules and their mixtures with alcohols, ethers, and ketones.

1. A. Gil-Villegas, A. Galindo, P. J. Whitehead, S. J. Mills, G. Jackson and A. N. Burgess, "Statistical associating fluid theory for chain molecules with attractive potentials of variable range," J. Chem. Phys., 106, 4168-4186 (1997).

2. C. McCabe, A. Gil-Villegas, G. Jackson and F. del Rio, "The thermodynamics of heteronuclear molecules formed from bonded square-well (BSW) segments using the SAFT-VR approach," Mol. Phys., 97, 551-558 (1999).

3. Y. Peng, H. G. Zhao and C. McCabe, "On the thermodynamics of diblock chain fluids from simulation and heteronuclear statistical associating fluid theory for potentials of variable range," Mol. Phys., 104, 571-586 (2006).

4. Y. Peng, K. D. Goff, M. C. dos Ramos and C. McCabe, "Developing a predictive group-contribution-based SAFT-VR equation of state," Fluid Phase Equilib., 277, 131-144 (2009).

5. Y. Peng, K. D. Goff, M. C. dos Ramos and C. McCabe, "Predicting the Phase Behavior of Polymer Systems with the GC-SAFT-VR Approach," Industrial & Engineering Chemistry, 49, 1378-1394 (2010).

6. M. C. dos Ramos and C. McCabe, "On the prediction of ternary mixture phase behavior from the GC-SAFT-VR approach: 1-Pentanol + dibutyl ether plus n-nonane," Fluid Phase Equilib., 302, 161-168 (2011).

7. M. C. dos Ramos, J. D. Haley, J. R. Westwood and C. McCabe, "Extending the GC-SAFT-VR approach to associating functional groups: alcohols, aldehydes, amines and carboxylic acids," Fluid Phase Equilib., 306, 97-111 (2011).

(591g) A Molecularly-Constructed Cubic Equation of State for Nonassociating Fluids

Arthur S. Gow, Dept of Chemistry & Chemical Engineering, University of New Haven, West Haven, CT and Shenali Parikh, Dept. of Chemical and Biomolecular Engineering, University of Pennsylvania, Philadelphia, PA

Classical engineering cubic equations of state (e.g., Soave-Redlich-Kwong and Peng-Robinson EOS) present difficulties when applied to systems at extreme conditions or which contain polar components. Conversely, molecularly-based equations of state (i.e., statistical associating fluid theory EOS and its many variants) are much more accurate when successfully applied to such systems; however, their application is computationally complex due to their large number of molar volume roots and nested iterations. An optimum solution to this dilemma is to develop a formalism for design of equations of state which retain the computationally attractive cubic in molar volume form while incorporating the rich molecular level behavior of the species involved.

This talk will investigate one such route to obtaining a molecularly-based cubic EOS. Molecules are considered (as in the SAFT approach) to be made of bonded segments (chains), which undergo repulsive, dispersive, and structure-dependent multi-polar interactions with each other. A simple molecular theory is used for each contribution to the Helmholtz free energy, which remarkably results in an equation of state which is cubic in molar volume. The repulsive contribution to the compressibility factor is obtained from an exact fit of hard-sphere molecular simulation data. Moreover, this simple expression produces a self-consistent analytical form of the reference hard-sphere fluid radial distribution function (RDF), which is used with the square-well intermolecular potential to obtain analytical solutions to the perturbation integrals for dispersion and dipole-dipole interactions. Furthermore, Wertheim's (first thermodynamic perturbation theory) TPT1 is applied to the reference fluid RDF to obtain an expression for the Helmholtz free energy of chain formation. The EOS contains two universal and four substance specific parameters. Finally, some typical results for pure fluids are presented, and recommendations for future work are discussed including a variation of scaled-particle theory, which is proposed for association.

(591h) Gsaft: Application of the SAFT-γ Mie Group Contribution EoS in the Oil/Gas Industry - From Academic Research to Industrial Deployment

Laurent Avaullee[1], Claire S. Adjiman[2], Filipe Calado[3], Pierre Duchet-Suchaux[1], Javier Fuentes[3], Amparo Galindo[2], George Jackson[2], Thomas Lafitte[2], Costas C. Pantelides[2,3], **Vasileios Papaioannou**[2] and Thomas H. Williams[3], (1)Total S.A., Pau, France, (2)Centre for Process Systems Engineering, Department of Chemical Engineering, Imperial College London, London, United Kingdom, (3)Process Systems Enterprise Ltd., London, United Kingdom

SAFT-γ Mie is a new equation of state recently developed by the Molecular Systems Engineering group at Imperial College London. It is an advanced group-contribution form of the SAFT equation of state making use of the Mie potential for a more accurate and flexible description of the dispersive/repulsive interactions between segments. One of its key characteristics is the accurate description of vapour/liquid phase equilibria, including the region of the critical point, as well as the second-derivative thermodynamic properties such as the thermal expansivity, isothermal compressibility, heat capacity, Joule-Thomson coefficient, and speed of sound.

In 2009, Process Systems Enterprise (PSE) acquired the exclusive intellectual property rights associated with SAFT-γ Mie and related work, for the purpose of incorporating these developments within its gSAFT advanced thermodynamics technology for process modelling. In late 2010, TOTAL, PSE and Imperial College embarked on a joint project aimed at exploring in detail the applicability, benefits and limitations of this technology on a wide range of mixtures of interest to the oil & gas industry. The current phase of the project is primarily focused on mixtures of hydrocarbons (alkanes and aromatics), carbon dioxide, water and methanol. The main output is a single, consistent set of group parameters capable of accurately describing the behaviour of these generic mixtures within the SAFT-γ Mie framework.

Starting with a brief overview of the SAFT-γ Mie equation of state, this paper primarily focuses on the systematic methodology employed in developing the corresponding like and unlike group parameters. This comprises a sequence of steps including the choice of representative components and mixtures, the definition of an appropriate set of groups required to describe them, the collection of the necessary experimental data, a streamlined set of software tools and workflows employed for the accurate, efficient and error-free estimation of the relevant parameters, and ultimately the deployment of the gSAFT software within process modelling tools. Results illustrating the predictive accuracy of gSAFT for the above mixtures are presented, including representative examples of the simultaneous description of vapour-liquid and liquid-liquid equilibria, the densities of the coexisting phases, and excess properties of mixing.

Interaction Forces Between Colloidal Particles within a Nanoparticle Suspension

Introduction

Exploring the interaction forces for a hard sphere-plate system has significant fundamental importance in understanding the interparticle interaction for controlling the stability of a colloidal system. The surface interaction between a spherical particle and a plate is a special case of the particle-particle interaction with one particle having an infinitely large diameter. The direct interaction between a single hard sphere and a hard wall in vacuum is rather straightforward in that the attractive van der Waals interaction which has no value at large separation distances, drops to a minimum as the sphere approaches contact with the wall. This interaction grows to infinity as the sphere begins to contact the wall, reflecting the mutual impenetrability of the hard surfaces.[1] However, when this sphere-plate system is submerged in a fluid medium, the interaction becomes mediated by the suspending fluid.

In addition to the theoretical modeling of the sphere-plate interaction force,[2-7] the colloidal probe technique[8] has become a popular method for direct surface force measurement. There are several representative scientific review articles detailing the introduction and application of the colloidal probe technique published in the past two decades.[9-11] The interaction forces between identical surfaces [12-16] and between dissimilar surfaces[17-18] have been extensively investigated. Most of the measurements focus primarily on sphere-plate interaction regulated by different background fluids in terms of DLVO forces.[12-18] However, with specifically chosen surfaces and fluids, direct force measurement can be extended to investigate depletion forces,[19-21] oscillatory structural forces,[22,23] hydrodynamic forces,[24,25] and the hydrophobic attraction.[26,27]

The direct force measurements of the sphere-plate interaction have been conducted in a number of the background fluids. Besides regular aqueous electrolyte solutions containing a variety of salts,[12, 28-32] mediums with various concentrations of water and zirconium octanoate[33] and different 1,2-propanediol and water mixtures[34] were used to investigate the stabilization mechanism of copper phthalocyanine dispersions and nano-zirconia dispersions, respectively. Recently, nanoparticle or dispersant-containing suspensions have been employed to study the interactions of sphere-plate systems. These force measurements have demonstrated significant effects on the interaction between silica surfaces through the addition of sodium dodecyl sulfate dispersions[35] and either silica or alumina nanoparticle suspensions.[36] The influence of the dispersant (sodium polyacrylate) molecular weight on the force curves between titanium dioxide sphere-plate surfaces has also been investigated by a colloidal probe microscope.[37]

A previous paper by our group was the first to document the direct measurement the interaction force in a nanoparticle haloing system using the colloidal probe technique.[38] In nanoparticle haloing, a suspension of weakly charged silica spheres can be stabilized by negatively charged nanoparticles due to the formation of a dynamic halo around the colloidal particles.[39-40] While the silica-zirconia is the most investigated system, other systems such as polystyrene-silica,[41] alumina-silica,[42] and silica-polystyrene[43] have been identified as being stabilized by the nanoparticle haloing mechanism. In this paper, it was shown that a transition occurs in the interaction between a silica sphere and plate in a series of zirconia nanoparticle suspensions ranging in volume fraction from 10^{-6} to 10^{-3} at a pH of 1.5. The interaction, which is dominated by an attractive van der Waals interaction at low volume fractions and a repulsive electrostatic interaction a high volume fractions, undergoes a transition through a metastable state at a nanoparticle volume fraction of 10^{-5} where the nanoparticle haloing mechanism is thought to develop.[38-40] However, even though this is the only time thus far that the interaction forces have been directly measured at the initiation of nanoparticle haloing, the contribution of the various fundamental forces to the haloing mechanism still remained ambiguous. Despite the number of papers discussing the stabilization mechanism from a theoretical point of view, few studies have attempted to describe the interaction force from a force measurement perspective primarily due to the limited amount of data available for such an analysis. In this paper, we demonstrate the feasibility of using an effective zeta potential model for fitting the experimental force curves and provide a potentially promising fitting model for other nanoparticle haloing systems.

Experimental Method

The force measurements between silica surfaces were taken with an XE-100 Atomic Force Microscope (Park Systems, Santa Clara, CA) utilizing contact mode in zirconia nanoparticle suspensions being contained by a Petri dish. A very smooth silica circular plate (height of 1/16 in., diameter of 1/2 in., and root mean square surface roughness of <2 nm, Quartz Scientific, Fairport Harbor, OH) was used as the flat substrate for the experiment. After ten minutes of sonication, it was cleaned alternatively by 18 MΩ RO water and anhydrous ethanol three times and dried in a laminar flow hood prior to each experiment. Force measurements were made using a V-shaped silicon nitride cantilever with attached 600 nm and 1μm silica colloidal probe (NOVASCAN, Ames, IA). The average diameter of the hydrous zirconia nanoparticles we used is 8nm, 50nm, and 100nm.

Theory

The direct force measurements were conducted at four different microsphere to nanoparticle size ratios: 100 (1μ/ 8nm), 70 (600nm/8nm), 20 (1μ/ 50nm) and 10 (1μ/ 100nm), respectively. Since the zeta potential of silica in a HNO_3 solution of pH 1.5 has a negligible value of +1 mV, the electrostatic double layer interaction between silica surfaces can be ignored. However, with the addition of the zirconia nanoparticles, a nanoparticle halo begins to form around the silica microsphere which leads to an effective zeta potential of the binary composite sphere (BCS: the entity of the silica sphere and non-adsorbing halo of zirconia particles around the silica colloid).

For the effective zeta potential model, it is assumed that the effective zeta potential is the same for both the nanoparticle regulated silica plate and BCS. In addition, a continuum assumption is made for the overlapping of the effective double layers when a sphere approaches the plate where the nanoparticle halo exists. Therefore, an effective electrostatic force between the nanoparticle-regulated sphere (BCS) and the plate is considered for this model. Hogg et al. derived the general Hogg-Healy-

Fuerstenau (HHF) equation for the repulsive potential (V_R) of unequal spheres having surface potentials Ψ_1 and Ψ_2, which is given by[44]

$$V_R(D) = (\frac{1}{4})\varepsilon_0\varepsilon_r\pi\frac{a_1a_2}{a_1+a_2}(\Psi_1^2+\Psi_2^2) \times[\frac{2\Psi_1\Psi_2}{\Psi_1^2+\Psi_2^2}\ln(\frac{1+\exp(-\kappa D)}{1-\exp(-\kappa D)}) + \ln[1-\exp(-2\kappa D)] \tag{1}$$

where ε_0 is the vacuum permittivity, ε_r is the dielectric constant, a_1, a_2 are the radii of the particles, κ is reciprocal of the Debye length and D is the separation distance of the closest approach between the sphere and the plate. For the sphere-plate interaction system in which the experimental conditions are $\psi_1=\psi_2=\psi_{effective}$, $a_1=a_{sphere}$, and $a_2=a_{plate}=\infty$, the HHF equation (1) simplifies to

$$V_{effective}(D) = \varepsilon_0\varepsilon_r\pi a_{sphere}\Psi_{effective}^2\exp(-\kappa D) \tag{2}$$

Considering the nature of the deferential relationship between the potential and the interaction force, the force equation can be derived as

$$F_{effective}(D) = \kappa\varepsilon_0\varepsilon_r\pi a_{sphere}\Psi_{effective}^2\exp(-\kappa D) \tag{3}$$

Among various surface forces, the attractive van der Waals force ubiquitously occurs between two surfaces at a given distance. For simplicity, we have ignored the effect of the retardation on the van der Waals force (F_v) between silica surfaces.[12] Therefore the physics of the interaction is expressed mathematically as a function of separation distance D, such that

$$F_V(D) = -A_DR/6D^2 \tag{4}$$

where A_D is the Hamaker constant of 0.8×10^{-20} J[45] and R is the radius of the microsphere (a_{sphere}).

The stability of polydisperse systems of colloids, natural rubbers, micelles, or polymer coils can potentially be influenced by depletion forces. For our case when a sphere approaches a plate in a nanoparticle suspension at low volume fraction, an attractive depletion force may begin to evolve. The nanoparticles cannot penetrate into the gap between the two surfaces, creating a negative osmotic pressure leading to an outflux of solvent which leads to a further attraction between the surfaces. A simple yet qualitatively correct model of nonadsorbing polymer-colloid mixtures was originally proposed for the depletion force by Asakura and Oosawa[46] in 1954. In this model, hard spherical colloids are dispersed in a mutually noninteracting rigid macromoleculear solution where the macromolecules cannot penetrate the hard colloidal particles. More recently, Walz et al. have developed an equation for the depletion force between a single spherical particle and a flat plate in a solution of nonadsorbing hard spherical macromolecules based on the Asakura and Oosawa (AO) model.[47]

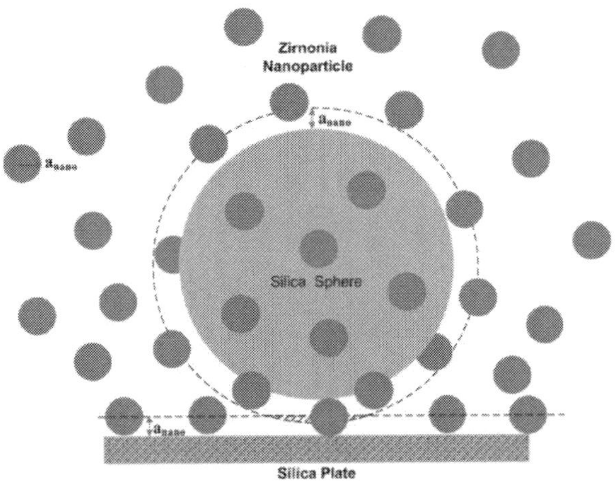

Figure 1. Schematic description of the experimental setup and the origin of the depletion interaction between a spherical particle and flat plate. The overlap of excluded volume regions is indicated by the shaded area.

Consider the case of a microsphere and a plate separated by a width D immersed in a suspension of hard spherical nanoparticles of radius a_{nano} as shown in Figure 1. The microsphere excludes a spherical volume of radius $(R+a_{nano})$ from the nanoparticles. The plate excludes a layer with the width of a_{nano}. When the microsphere approaches sufficiently close to the plate, these two excluded volumes overlap by the volume of the shaded region in Figure 1, and the total excluded volume of this system actually decreases by this amount. The depletion force can now be written as[47]

$$F_{sphere-plate} = \begin{cases} -(4+2D/R-2D/r-D^2/rR)(\rho_\infty kT\pi Rr) & for \quad D<2r \\ 0 & for \quad D\geq 2r \end{cases} \tag{5}$$

where ρ_∞ is the bulk number density of the nanoparticles, kT is 1.411×10^{-19}J and $r = a_{nano}$ is the radius of the nanoparticle.

Results and Discussion

In this model, the total interaction force between nanoparticle-regulated silica surfaces in different zirconia nanoparticle suspensions was computed by a combination of the van der Waals force between silica surfaces, the effective electrostatic force of the BCS, and the attractive depletion force. Under the condition of nanoparticle haloing, highly charged zirconia nanoparticles self-organize around a negligibly charged silica microsphere to form a halo-like structure that stabilizes the suspension. By considering the effect of the surrounding zirconia nanoparticles which form a loosely organized layer at a small distance (~2.15 nm) from the surface of the microspheres,[48] it is impossible to directly compute the theoretical Debye length using existing theory. Due to the obvious effect of the nanoparticles on the electrostatic double layer of the silica surface, a variable value of the effective Debye length was defined as a nanoparticle-regulated Debye length and employed for the current model. Additionally, it was found that the effect of the depletion force (F_{dep}) could be ignored in this particular system due to the fact that F_{dep} has a value of 5.8×10^{-5} nN at nanoparticle volume fraction of 10^{-3} which is much smaller than the scale of the as-measured interaction force ($\sim10^{-1}$ nN) seen in Figure 2. Therefore, due to the negligible depletion force, the model equation of the total interaction force for this system, (including the effective electrostatic force, van der Waals force, and the depletion force) can be determined according to:

$$F_t = F_v + F_{effective} \tag{6}$$

and by substituting Equation (3) and (4) into Equation (6), we obtain

$$F_t = -A_D R / 6D^2 + \kappa \varepsilon_0 \varepsilon_r \pi a_{sphere} \Psi^2_{effective} \exp(-\kappa D) \tag{7}$$

Having established the effective zeta potential fitting model, a fit of the experimental force curves was conducted. Fitting results are summarized in Figure 2. At $\alpha \approx 70$, the effective zeta potentials of the zirconia nanoparticle halo and the silica sphere entity in nanoparticle suspensions at volume fractions of 10^{-6}, 10^{-5}, 10^{-4} and 10^{-3} have values of 30 mV, 41 mV, 53 mV, and 77 mV,[38] respectively. Similar trends are found at volume fractions of 10^{-6}, 10^{-5}, 10^{-4} with α=100 and 50. As shown in Figure 2 (a), as the volume fraction of the nanoparticles increases from 10^{-6} to 10^{-5}, the interaction between silica surfaces shows a dramatic change from a purely attractive interaction to a repulsive interaction which agrees with the increase of the effective zeta potential values. At high nanoparticle concentration (10^{-5}, 10^{-4}, and 10^{-3}), the repulsive barriers continue to increase with the volume fraction of nanoparticles.

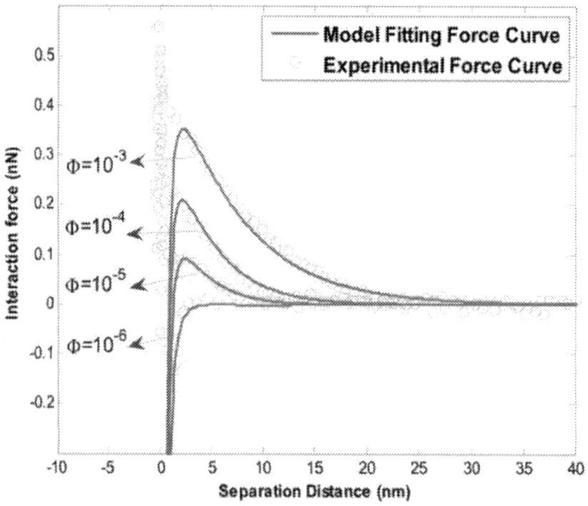

a

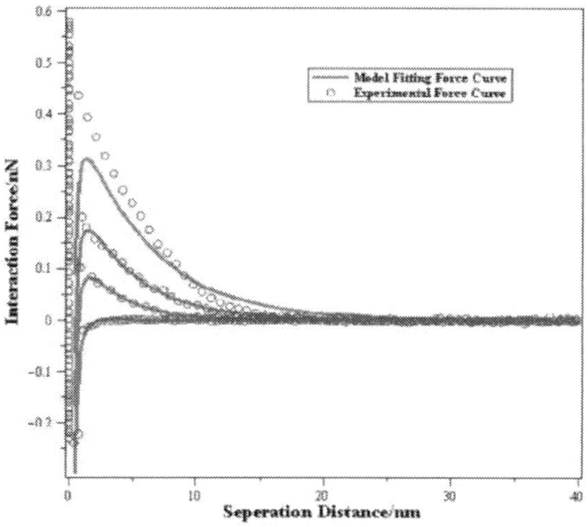

b

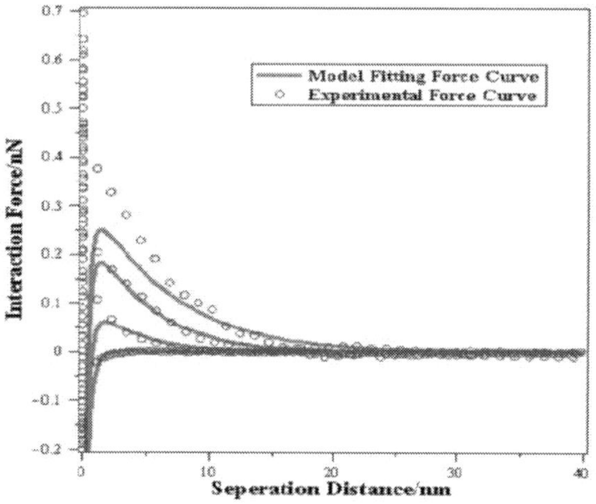

c

Figure 2. Effective zeta potential fitting for the experimental force curves between a silica sphere and a silica plate in zirconia nanoparticle suspensions with different volume fractions at a pH of 1.5 with size ratios of a) 70 (600nm/8nm), b) 100 (1μ/ 8nm), and c) 20 (1μ/ 50nm), respectively.

Under the continuum assumption, we substitute the values of the effective zeta potential of the BCS into the total force equation and adjust the values of the effective Debye length until the theoretical and experimental force curves match. Final fitting values were found to be 2.5 nm, 3 nm, 4 nm and 6.5 nm, which shows a successive increase in the effective Debye length with a corresponding increase in the nanoparticle concentration. The values of the effective Debye lengths of the zirconia nanoparticle-regulated double layer are larger than the Debye length of silica in a pure nitric acid solution of pH 1.5 (1.7 nm) which is to be expected. This change in the Debye length may be attributed to a slight increase in the size of the BCS. As the nanoparticle volume fraction is increased, the nanoparticle number density in the halo leads to a rise in the gap between the silica microsphere and haloing nanoparticles (seen in Figure 3) in order to minimize the interaction between nanoparticles in the halo for a stabilized system. Additionally, for the effective Debye length one would have to count the contribution from the surface charge over the entirety of the nanoparticles which means that the effective plane of charge of the halo would have to be slightly larger than the distance of separation between the nanoparticles and the silica sphere. This is indeed what is observed at a nanoparticle volume fraction of 10^{-5} where the ~3 nm effective Debye length is greater than the 2.3 nm separation distance reported in our previous work.[38]

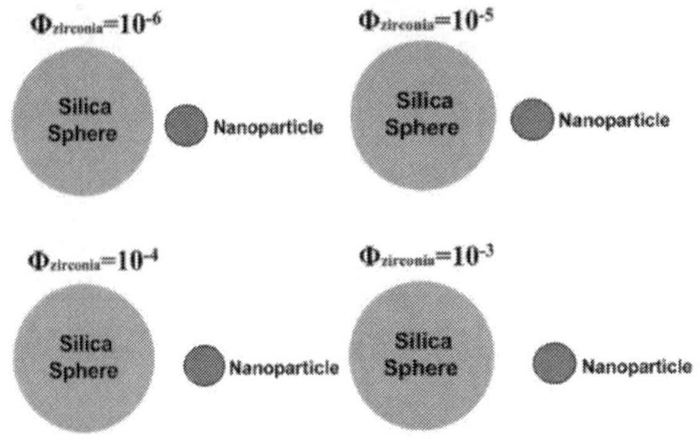

Figure 3. Schematic description of the effect of the nanoparticle volume fraction on the gap between the silica sphere and a zirconia nanoparticle.

At the volume fractions of 10^{-5} through 10^{-3}, the observed interaction force continues to increase until the two silica surfaces approach contact, which does not follow the model prediction of the dominance of the van der Waals force at smaller distances (<2 nm). The lack of an observed van der Waals force may be a combination of the negligible depletion force and the effect of trapped nanoparticles between the silica sphere and plate to keep the surfaces apart at the distance of < 2 nm. The discrepancy between the experimental curve and the theoretical curve at small separations is most likely caused by the continuum assumption used to develop this model which does not exist in this system at very small separation distances. However, this model predicts the different fundamental interaction forces at larger separation distances which account for more than 90% of the total effective interaction distance.

After the zeta potential model successfully fit the experimetal data at α=70, we extend this model to systems with α=100 and 20. As seen is Figure 2 (b&c), the model fit the experimental curve well at nanoparticle concentrations of 10^{-6}, 10^{-5} and 10^{-4}, which demonstrates the zeta potential model is applicable in a general nanoparticle-halo systems.

Conclusions

In this paper, an effective zeta-potential fitting model is proposed to explain the interaction forces prevalent in a nanoparticle haloing system. As the nanoparticle volume fraction increases, there is an increase in the effective zeta potential of the nanoparticle regulated silica sphere in zirconia nanoparticle suspensions at a pH of 1.5. The interaction forces between silica surfaces remain repulsive after a dramatic transition from a purely attractive to a purely repulsive force. Based on a continuum assumption, the effective double layers of nanoparticle regulated silica surfaces begin to overlap when the sphere approaches the plate, which causes an effective electrostatic interaction force between the surfaces. Total surface forces from this model (which combines the van der Waals attraction and effective electrostatic force) are shown to fit the experimental force curves very well when using the effective Debye length as a fitting parameter. The effective Debye length is found to be dependent upon the concentration of the nanoparticle suspensions. As the nanoparticle concentration is increased, the nanoparticle-regulated effective double layer is shifted to larger values. Based on this model, it is found that the nanoparticle haloing stabilization mechanism can be explained almost entirely through van der Waals and electrostatic interactions. Additionally, attractive depletion forces do not influence the overall interaction force, though they may become important at very small separation distances (<2 nm) where the continuum assumption begins to fail.

(624b) Direct Measurement of Conservative and Dissipative Colloidal Forces in the Presence of Adsorbing and Nonadsorbing Polymer

Daniel J. Beltran-Villegas, Tara D. Edwards and Michael A. Bevan, Chemical and Biomolecular Engineering, Johns Hopkins University, Baltimore, MD

Colloidal and macromolecular dispersions are important in many industrial applications and in biological and environmental systems. The macromolecular induced interactions have been shown to be important mediators in biological processes and drug delivery applications. A good understanding of these interactions is key to practical and scientific approaches in soft matter and biophysics. In this talk we report the implementation of a novel dynamic analysis to measure colloidal conservative and dissipative forces from ensemble colloidal particle probe excursions normal to an underlying substrate. The analysis is based on Bayesian Inference analysis of particle trajectories to find a set of potential energy and particle diffusivity profiles. This analysis enables the direct estimation of both Thermodynamic and Hydrodynamic interactions between single particles and a wall with non-intrusive probes. We test this analysis by measuring particle-wall interactions for particles stabilized by adsorbed polymer brushes in the presence of non-adsorbing polymer in solution that induce a depletion attraction between particles and the underlying wall. Our results show agreement with standard Boltzmann probability analysis of long trajectories for potential energy profiles and theory. The resulting particle diffusivity profiles are shown to follow known hydrodynamic contributions that include the surface separation dependence and the effect of both adsorbed polymer brush and non-adsorbing polymer in solution. This is the first time that these theories have been validated by means of a non intrusive experimental methodology. Further work on this topic includes the measurement of biomacromolecule forces (both conservative and dissipative) to allow a deeper understanding of the fundamental mechanisms of life.

(624c) Out-of-Equilibrium Forces Between Colloids

Indira Sriram, Chemical and Biological Engineering, University of Colorado at Boulder, Boulder, CO and Eric M. Furst, Department of Chemical and Biomolecular Engineering, University of Delaware, Newark, DE

Two colloidal probe particles are held with optical traps orthogonal to a uniformly flowing suspension of colloidal bath particles. Using confocal microscopy, we characterize the local bath suspension microstructure as a function of the probe separation and flow velocity. We find that at sufficiently close separations, bath particles are excluded from passing between the probes, resulting in an asymmetric, non-equilibrium microstructure in which the major features are a depleted region between the probes and dense boundary layers along the surfaces that face away from the neighboring probe. Consequently, the drag force acting on the probes is lower than that acting on a single probe and a net force pushes the probes together along their line of centers. The strength of the latter mutual force increases with increasing flow velocity. These experiments demonstrate that depletion-like forces can be induced between two particles by a non-equilibrium microstructure in a strongly driven suspension.

(624d) Localized Electro-Osmosis (LEO) Induced by Colloidal Motors

Tso-Yi Chiang and Darrell Velegol, Chemical Engineering, Penn State University, University Park, PA

Experiments have shown that speed of colloidal motors decreases as the size of the motors increases. However, previous electrokinetic models have shown that the colloidal motor speed is independent of size, at least for the case of thin double layers for reaction-limited catalysis, which is the usual case observed in experiments. Why do we get conflicting results from experiments and modeling? We hypothesize that it is due to localized electro-osmotic flows at the wall, caused by the colloidal motors. Here, we propose an electrokinetic model that includes wall effects to describe the colloidal motor speed. With localized electro-osmosis, we are able to show that motor speed is dependent upon size, and the result from our model is compared to data in the literature. Also, we compare the electro-osmotic flow induced by an applied electric field with that by a self-generated electric field to distinguish situations where each is important, such as confined spaces and pores.

(624e) Interaction of Two Particles in a Chemically Reactive Medium

Sergey Shklyaev, Department of Chemistry, The Pennsylvania State University, University Park, PA; Department of Chemical Engineering, University of Puerto Rico–Mayagüez, Mayagüez, PR and Ubaldo Cordova, Chemical Engineering, University of Puerto Rico at Mayagüez, Mayaguez, PR

An increasing number of experiments on catalytically-driven (active) colloidal particles have shown that the interaction of chemically active particles is more complicated than usual interaction of two nonreactive (passive) particles. Indeed, each chemically active particle changes the distribution of reactants which, in turn, generates an overall force on other particles. First, we consider a pair of spherically symmetric catalytic particles, which are far from each other, in a colloidal dispersion of reactants. In this case there appears an attractive force decaying inversely proportional to the squared distance between the particles. This effect was first demonstrated by Wei & Jan [1]. We generalize their calculations to a more realistic case, when both reactant and product species are present in the dispersion; here the force can be either attractive or repulsive depending on the stoichiometry factor introduced in [2]. In fact, the interaction force can be thought of as a force between two charged particles which can bear charges of either the same or opposite signs depending on the stoichiometry factor. Another analogy is so-called secondary Bjerkness force between two bubbles oscillating in fluid. The leading term in interaction of two osmotic motors (Janus particles) is the same as for two particles uniformly covered with catalyst. The force does not depend on the mutual orientation of self-propulsion. Next, we deal with interaction between catalytic and passive (cargo) particles. It is demonstrated that the force on a cargo is exactly the same as the force imposed by a catalytic particle on another one. On the other hand, the force on a catalytic particle imposed by the cargo is much smaller; it is inversely proportional to quintic degree of the interparticle distance. Note that if the cargo and catalytic particles are connected by a rigid thin rod, the pair starts moving; thus it operates as a catalytic motor [3]. Within the above-mentioned electrostatic analogy, the cargo particle is equivalent to a particle of vanishing permittivity. We also generalize the model to finite ratio of particle radii and the interparticle distance and to include hydrodynamic interactions.

1. H.-H. Wei & J.-S. Jan, J. Fluid Mech. 657, 64--88 (2010).

2. U. M. Cordova-Figueroa, & J. F. Brady, Phys. Rev. Lett. 100, 158303 (2008).

3. S. Thakura & R. Kapral, J. Chem. Phys. 133, 204505 (2010).

(624f) On the Brownian Coagulation of Colloidal Dispersions

Aniruddha V. Kelkar, Jiannan Dong, Elias I. Franses and David S. Corti, School of Chemical Engineering, Purdue University, West Lafayette, IN

The classical theories of "perikinetic" or Brownian coagulation of colloidal systems proposed by Smoluchowski and Fuchs are applicable to the early stages of aggregation of hard or interacting spheres in dilute dispersions. The analysis can be extended to the full aggregation process by using population balance equations. Herein, it is often assumed that the rate constant of aggregation for dissimilar associations is independent of particle size. These classical analyses also assume that aggregation occurs due to the transport of particles towards each other via a steady-state flux.

A critical evaluation of the assumptions inherent to the classical approaches is carried out. The coagulation dynamics are reassessed using a non-steady-state flux. Brownian Dynamics Simulation (BDS), which is a stochastic-dynamics approach, is employed to provide an accurate description of the phenomenon. BDS results are used as benchmarks for comparisons of the predictions of the non-steady-state flux approach to the classical approach. These comparisons highlight the importance of accounting for the initial transient dynamics, particularly for large particle sizes. Population balance schemes utilizing both the Brownian kernel and a size-independent kernel are employed to study the full aggregation dynamics. BDS results are used to test the differences in the predictions of these two approaches. The size-independent kernel was found to work reasonably well for most cases. The BDS results also establish the concentration range over which the classical theories apply (particle volume fractions ≤ 0.05).

A new model has also been developed that utilizes a chemical potential gradient as the driving force for coagulation. In addition, the diffusion coefficient is chosen to be a function of the particle volume fraction in the dispersion. This accounts for both 'thermodynamic' and 'hydrodynamic' concentration effects. Thus, this analysis is not restricted to dilute colloidal systems (*i.e.*, systems with particle volume fraction less than 0.05). By directly including non-steady-state dynamics and hydrodynamic interactions, this new approach provides an accurate description of Brownian coagulation for particle volume fractions up to at least 0.40.

(624g) Aggregation Kinetics At High Concentrations: Beyond Smoluchowski's Diffusion-Limited Kernel

Marco Lattuada, Department of Chemistry and Applied Biosciences , Institute for Chemical and Bioengineering, ETH Zurich, Zurich, Switzerland; Adolphe Merkle Institute, University of Fribourg, Marly, Switzerland

Smoluchowski's equation for the rate of aggregation of colloidal particles under diffusion-limited conditions has set the basis for the interpretation of kinetics of aggregation phenomena. Nevertheless, its use is limited to sufficiently dilute conditions. In this work we propose a correction to Smoluchowski's equation by using a result derived by Richards (J. Phys. Chem. 1986, 85, 3520) within the framework of trapping theory. This corrected aggregation kernel, which accounts for concentration dependence effects, has been implemented in a population-balance equations scheme and used to model the aggregation kinetics of colloidal particles undergoing diffusion-limited aggregation under concentrated conditions (up to a particle volume fraction of 30%). The predictions of population balance calculations have been validated by means of Brownian dynamic simulations. It was found that the corrected kernel can very well reproduce the results from Brownian dynamic simulations for all concentration values investigated, and is also able to accurately predict the time required by a suspension to reach the gel point. On the other hand, classical Smoluchowski's theory substantially underpredicts the rate of aggregation as well as the onset of gelation, with deviations becoming progressively more severe as the particle volume fraction increases.

References:

Lattuada M., J Phys. Chem. B 2012, 116, 120–129

(628a) Interfacial Rheology of an Oscillating Spherical Cap

Anthony Kotula, Chemical Engineering, Carnegie Mellon University, Pittsburgh, PA and Shelley L. Anna, Chemical Engineering and Mechanical Engineering, Carnegie Mellon University, Pittsburgh, PA

Interfacial viscoelasticity is an important contributor to the stability of emulsions and foams with adsorbed macromolecules and particles. A common method of measuring interfacial mechanics involves blowing a bubble at a capillary tip and measuring the pressure jump across the interface of the expanding or oscillating interface. We have implemented a similar method of measuring apparent interfacial elasticity in a microtensiometer capable of measuring interfacial properties of small (radius ~10^{-5} m) droplets. In this device, the interface forms a spherical cap at the end of a drawn capillary tube that is tens of micrometers in diameter. The imposed pressure inside the spherical cap oscillates sinusoidally with amplitude of the order of 100 Pa and frequencies from 0.05 to 0.5 Hz. The microtensiometer has several advantages for rapid measurement of novel surface active species, including small sample volumes, faster transport timescales, and simple interfacial shape analysis. The common assumption that the interface is a sphere expanding radially in all directions is not valid since the radius of the capillary and the bubble are comparable and the interface is pinned at the edges of the capillary tube. In the present study, we use toroidal coordinates to derive the surface deformation tensor, and show that the deformation of an expanding or oscillating spherical cap pinned at a capillary tip is purely dilatational with a non-uniform dilatation rate along the interface. We discuss the tangential velocity at the interface of the growing bubble in a Newtonian fluid via a solution of the stream function and show the conditions when this flow contributes significantly to the interfacial deformation. The interfacial dilatation rate is greatest at the tip of the spherical cap and smallest at the contact line, leading to the equivalent of a non-viscometric flow for an interface. We use existing interfacial rheological models to derive the expected pressure jump as a function of oscillation frequency and amplitude, the capillary and interfacial radii, and the interfacial mechanics for several model interfaces. We use these model systems to demonstrate the appropriate interpretation of oscillatory interfacial measurements for deforming spherical caps, and apply our model to experiments measuring the interfacial elasticity of particle and surfactant adsorbed interfaces. Finally, we discuss experimental protocols that can allow for the separation of various contributions to the interfacial elasticity.

(628b) Unexpected Dynamics of Drop Coalescence

Santosh Appathurai[1], Michael T. Harris[1], Osman A. Basaran[1], Joseph Paulsen[2], Justin Burton[2] and Sidney Nagel[2], (1)Purdue University, West Lafayette, IN, (2)University of Chicago, Chicago, IL

Drop coalescence is central to diverse processes involving dispersions of drops of one fluid in a second fluid. Common practical occurrences of coalescence include demulsification of crude oil, phase separation in emulsion-based products including salad dressing, and sintering of ceramic powders. In nature, drop coalescence is central to raindrop growth and, in biology, remarkable parallels exist between coalescence and fusion of nucleoli. During coalescence, two drops first touch and then merge as the liquid neck connecting them grows from initially microscopic scales to a size comparable to the drop diameters. The neck's expansion is controlled by the Laplace pressure, which diverges when the curvature of the interface is infinite at the point where the drops first touch. Thus the change in topology, as two drops become one, is inextricably linked to a singularity in the dynamics. Conventionally, this process has been thought to have just two dynamical regimes: a viscous and an inertial regime. According to conventional wisdom, the initial regime is always the viscous one regardless of the value of the liquid's viscosity, which is plausible given neck's microscopic size. Conventional wisdom further dictates that high viscosity liquids remain in the viscous regime but that low viscosity liquids transition from the viscous to the inertial regime as coalescence proceeds. We use experiments and simulations to reveal that the regime that describes the initial dynamics of coalescence has been missed. A new phase diagram of coalescence is presented. The results have profound implications to diverse situations where hydrodynamic singularities arise.

(628c) Wall Effects in Stokes Experiment with a Liquid Foam

Haijing Gao[1], Hariprasad J. Subramani[2], Michael T. Harris[1] and **Osman A. Basaran**[1], (1)School of Chemical Engineering, Purdue University, West Lafayette, IN, (2)Flow Assurance, Chevron Energy Technology Company, Houston, TX

Liquid foams are widely used in numerous applications ranging from the oil and gas industry to beauty, healthcare, and household products industries. A fundamental understanding of the relationships between the properties of liquid foams and their flow responses is, however, still in its infancy compared to that involving the fluid dynamics of simple fluids. In this talk, the flow of a dry liquid foam around a spherical bead, i.e. the Stokes problem for liquid foams, is studied experimentally. In contrast to previous work (cf. Cantat and Pitois, Phys. Fluids 2006), the focus of the present research is to probe the effect of a solid wall that is located a few bubble radii from the bead. The new experimental results show that the elastic modulus of dry liquid foams is directly proportional to the surface tension of the foaming agents and inversely proportional to the average bubble size in the foams, in agreement with previous theoretical and experimental studies. The experiments further show that the close proximity of the solid wall causes profound structural changes to the gas bubbles as the foam flows past the bead. A good understanding of these structural changes and how they can affect the elastic modulus of foams can be indispensable in formulating improved models for accurately describing the dynamical response of foams within the realm of continuum mechanics.

(628d) Cosurfactant Leads to Successful Dilute Emulsion Formation Via Microemulsion

Hanseung Lee[1], Eric Morrison[2] and Alon V. McCormick[1], (1)Chemical Engineering and Materials Science, University of Minnesota, Minneapolis, MN, (2)Ecolab, Eagan, MN

Cosurfactants play an important role in self-emulsification in three-component systems (water, oil, and non-ionic surfactant). The addition of a very small amount of cosurfactant can abruptly change the interfacial tension and curvature. It can determine the region of composition/temperature producing stable microemulsions. Such one-phase microemulsions have been known to serve as useful precursors to nanoemulsions – dilute long-lived emulsions of nanoscale droplete.

Here, we create very stable vesicular structures from the microemulsions; we found that a sufficient amount of cosurfactant determines the success of microemulsion processing. We have investigated the phase behavior of water/n-hexadecane/nonylphenol ethoxylates system and created kinetically stable nanoparticles.

To study this, we use three complementary methods: cryogenic transmission electron microscopy (Cryo-TEM), cryogenic scanning electron microscopy (Cryo-SEM), and freeze-fracture electron-beam evaporated replicate TEM (FF-TEM). The complementary use of these techniques shows the behavior of these complex fluid microstructure transitions at multiple scales. Cryo-TEM alone cannot show crucial, large-scale structures, but cryo-SEM can. Cryo-SEM alone cannot show where the oil is, but FF-TEM can.

(628e) Pickering Foams Stabilized by α-Zirconium Phosphate-Propylamine Nano-Sheets

Juan S. Guevara[1], **Andres F. Mejia**[1], Ya-Wen Chang[1], Min Shuai[1], M. Sam Mannan[2] and Zhengdong Cheng[3], (1)Chemical Engineering, Texas A&M University, College Station, TX, (2)Mary Kay O'Connor Process Safety Center, Artie McFerrin Department of Chemical Engineering, Texas A&M University, College Station, TX, (3)Artie McFerrin Department of Chemical Engineering, Texas A&M University, College Station, TX

We demonstrate the stabilization of Pickering foams utilizing partially hydrophobic inorganic nano-sheets with a thickness at atomic scale. The nano-sheets are obtained by intercalation and exfoliation of ZrP crystals using propylamine as a guest molecule. Subsequently, the hydrophobicity of the nano-sheets is adjusted by adsorbing different quantities of propylamine using the ion-exhange properties of ZrP. The adsorption of propylamine on the nano-sheet surfaces grants an inherent hydrophobicity that enables the nano-sheets to adsorb on the air-water interface. The foam stability was studied by calculating the rate of water drainage as a function of time. The foams exhibit excellent stability even at low nano-sheet concentrations. In addition, the foam-to-water ratio was evaluated by varying the nano-sheet concentration and the degree of hydrophobicity. Cross-polarizing micrographs and scanning electron microscopy observations revealed the adsorption of the ZrP-propylamine nano-sheets on the air-water interface. We observed that the foams are stabilized by several layers of nano-sheets that prevent the diffusion of gas between the bubbles.

(628f) Development an Evaluation of a Perfluorocarbon-Based Oxygen Carrier Mixed with HES 6% and Blood

Diana M. Vasquez Gutierrez[1], Camila I. Castro Paez[2], Oscar A. Alvarez Solano[2] and Juan C. Briceño Triana[3], (1)Chemical Engineering Department, Universidad de los Andes, Bogotá, Colombia, (2)Chemical Engineering Department, Universidad de los Andes, (3)Biomedical Engineering Department, Universidad de los Andes

In recent decades the field that presents the greatest attention from the scientific community is the research on the circulatory system and particularly the tissue oxygenation. The reason for this is that the main causes of deaths worldwide are related to cardiovascular diseases and hemorrhage; the basic principle of treatment is the control of the bleeding site and replace of the lost volume with blood transfusions. Although this is a routine procedure, there are risks associated with its performance as the transmission of pathogens and changes in blood properties such as viscosity, pH and affinity of hemoglobin for oxygen. An alternative for the multiple disadvantages of the blood transfusions are the perfluorocarbon-based oxygen carrier (PFCOCs) which are direct emulsions O/W that can replace blood in its function of transport oxygen to tissues.

The objective of this study is to develop a stable O / W perfluorocarbon emulsion, with perfluoro-octyl bromide (PFOB) (30% v/v) as disperse phase, an electrolyte solution as the continuous phase and soybean lecithin as surfactant; able to transport oxygen intravascular for cases of acute hemorrhage.

Perfluorocarbon emulsions have been widely studied and characterized, leading to the understanding of the system and its properties, but their influence on the real intravascular scenario has not been objectively analyzed in terms of the interaction between the emulsion and blood properties. For this reason, an in vitro model of normovolemic hemodilution was implemented to analyze the effects of PFCOCs on blood rheology and predict PFCOCs performance when infused.

For the development of the emulsion, the most favorable composition of surfactant is determined by analyzing the stability of the emulsion evaluating the particle size distribution and rheological behavior over time, by varying the concentration of surfactant 3.5% w/v, 4.5% w/v and 5.5% w/v. The behavior at the microscopic level is set using the technique of dynamic light scattering DLS with the Zetasizer Nano ZS from Malvern® and the rheological behavior is measured from steady-state testing using the AR-G2 Magnetic Bearing Rheometer from TA Instruments ® with a concentric cylinder geometry at shear rates of 0.1 to 1000 s-1 and 1000 to 0.1 s-1.

For the normovolemic hemodilution, fresh porcine blood was collected and stored in blood transfusion bags using CPDA-1 as anticoagulant, and mixed with hydroxyethyl starch (HES Tetraspan 6%, B. Braun). The in vitro procedure consisted of 4 exchanges, each one replacing 40% of volemia. In the first two exchanges, volume was replaced with HES and in the last two with a mixture of 20% HES and 80% PFCOCs. After each exchange a sample was taken, characterized microscopically at 100X, and macroscopically measuring viscosity with the rheometer.

As a result of the normovolemic hemodilution, viscosity of the mixture increased with the increasing perfluorocarbon emulsion-to-blood volume ratio, up to 9.4 cP after the last exchange. There is particle aggregation in the blood-emulsion mixture with sizes of 17 μm and they may impede the flow of erythrocytes. In the in vivo intravascular application these aggregates may induce occlusion of the micro vascular system.

(628g) Break-up Patterns and Conditions of Water Droplet in High-Voltage High-Frequency Pulse Electric Field

Zhiqian Sun, College of Chemical Engineering, China University of Petroleum (East China), Qingdao, China

Break-up behavior of water droplets in emulsion under high electric field is one of the most important phenomena during electrocoalescence process. Many processes, such as demulsification of W/O emulsion and electrical dispersion, are accompanied with tear-up of water droplets and rupture of oil-water interfacial film. However, electrostatic break-up mechanism of droplets still lacks of comprehensive and in-depth analysis up to the present. Consequently, the microscopic experimental study was carried out to investigated the break-up patterns and conditions of water droplet in high-voltage high-frequency pulse electric field comprehensively in this paper. Microscopic experimental results indicated that, there are mainly three break-up patterns of droplet in high-voltage high-frequency pulse electric field: break-up at one pole, break-up at both poles and break-up in middle part. When electrostatic tensile stress exceeds additional compressive stress formed by interfacial tension at the pole of polarizational droplet, top or bottom of the droplet broken up, spraying out a series of tiny droplets with particle size of 1 to 3 microns, which called "break-up at one pole". With electric field strength increasing, both poles of the water droplet break up, which called "break-up at both poles". When electric field strength increases further, droplet is quickly stretched and deformed, the middle of the droplet is stretched to filamentous shape and then forms a string of small water chain with particle size ranging from several microns to tens of microns, which called "break-up in middle part". Subsequently, combined with microscopic experimental results, the conditions of droplet polarization, break-up at pole(s) and break-up in middle part under high-voltage high-frequency pulse electric field were obtained on the basis of theoretical analysis. The research results laid good foundation for further investigation of high-voltage high-frequency electrostatic emulsion-breaking mechanism.

(635a) Stretched Perfluorosulfonic Acid Membranes for Hydrogen/Air and Direct Methanol Fuel Cells

Wenjing Zhang, Remington E. Fischer and **Peter N. Pintauro**, Department of Chemical and Biomolecular Engineering, Vanderbilt University, Nashville, TN

To develop a durable and high performance proton exchange membrane for hydrogen/air and direct liquid methanol fuel cell applications, we have systemically examined uniaxial stretching effects on perfluorosulfonic acid (PFSA) polymers. PFSAs represent a family of commercially available polymers with a perfluorinated backbone and side chains that terminate in hydrophilic sulfonic acid groups. Three PFSA materials were studied: DuPont's Nafion® (with a long and branched side chain of $-O-CF_2CFCF_3-O-(CF_2)_2-SO_3H$), 3M Company PFSA (a short-side-chain version of Nafion with $-O-(CF_2)_4-SO_3H$ pendant groups), and Aquivion® from Solvay Solexis (with a very short side chain of $-O-(CF_2)_2-SO_3H$). The effect of uniaxial stretching on the crystallinity and fuel cell properties of these polymers was probed, where solution-cast membranes were stretched at a temperature above the polymer's α-transition temperature, followed by a polymer annealing step.

Nafion 1100EW, 3M 1000EW and 825EW, Aquivion 830EW PFSA show an increase in crystallinity upon uniaxial stretching, with a decrease in methanol permeability and no change in proton conductivity. A direct methanol fuel cell employing a stretched 3M 825 EW PFSA membrane with a draw ratio (DR) of 4, for example, exhibits a 31% improvement in power density at 0.4V, as compared to a fuel cell with a Nafion 117 (1100 EW) membrane. The higher power output is associated with lower methanol crossover.

Interestingly, 3M's 733 EW PFSA polymer has no crystallinity before or after stretching. The nanostructure of ionic domains, however, varied with draw ratio. From wide-angle x-ray diffraction studies, the ionomer peaks of a stretched membrane shift to higher values of 2θ and a new ionomer peak appears when the draw ratio is greater than 4. For a stretched 733 EW membrane (DR=4), there is a 120% increase in the storage modulus with a proton conductivity equal to that of an unstretched film. This combination of properties makes stretched 3M 733 EW PFSA a promising PEM material for a hydrogen/air fuel cell. From an open circuit voltage humidity cycling fuel cell test at 80°C, for example, it was found that the lifetime of a stretched 3M 733 EW PFSA membrane (DR=4) was extended by a factor of 3.5, as compared to that of an unstretched membrane.

(635b) Influences of Confinement and Annealing On the Behavior of Polymer Electrolyte Materials

Brandon W. Rowe, Kirt Page and Christopher Soles, Polymers Division, National Institute of Standards and Technology, Gaithersburg, MD

Polymer electrolyte materials are of great interest in the continuing development of fuel cell and battery technologies, which offer cleaner, more efficient alternatives to power generated through tradition technologies. While substantial progress has been made in improving the performance of polymer electrolyte membrane (PEM) fuel cells, challenges in their development remain. Due to the complex structure present in these PEM materials, including ionic clusters and crystallites, their performance is intimately linked to the specific morphology of the system. As film thickness is decreased to improve performance and lower costs, the morphology of the system (*i.e.*, size and connectivity ionic domains and crystallites) will become more strongly influenced by interactions at the material interfaces, impacting the performance of the membrane layer and the polymer electrolyte in the catalyst layer. Nafion, the most prevalent PEM material, has been studied in detail in its bulk form; however, relatively little information has been published on the behavior of thin films of Nafion. The focus of this work is to quantify how the morphology and material properties of PEM materials change under confinement at interfaces, determine how these interfacial effects are influenced by processing, and to establish a correlation between these effects and fuel cell performance.

Using a variety of experimental techniques, including: quartz crystal microbalance (QCM), grazing incidence small angle x-ray scattering (GISAXS), ellipsometry, and x-ray reflectivity, we have investigated the influences of confinement and thermal annealing on the behavior of thin Nafion films. Consistent with previous studies, the sorption of water in thin (40 nm) Nafion films is lower than in bulk membranes. Additionally, the previous history of these films, i.e., thermal annealing, has a significant impact on water uptake behavior. The nanostructure within the Nafion films, as studied using GISAXS, was also influenced by confinement and annealing. In addition to changes in the ionic domain morphology in this material, which can be connected to the water uptake behavior, changes in the crystalline regions were also observed that relate to changes in the film strength and stability. To complement the studies on structure and transport properties, we have also investigated the mechanical properties and stress change upon hydration of these films using a cantilever bending technique. The observed deviations from bulk behavior in these thin Nafion films highlight the importance of understanding the influences of confinement in detailed modeling efforts, which typically use bulk-film parameters to describe transport in thin PEM layers and at catalyst interfaces.

(635c) Computation Study On Water and Charge Transport in the Cross-Linked and Sulfonated Poly (1,3-Cyclohexdiene) (xsPCHD) Based Proton Exchange Membranes

Qifei Wang, Chemical and Biomolecular Engineering, University of Tennessee Knoxville, Knoxville, TN and David J. Keffer, Chemical and Biomolecular Engineering, University of Tennessee, Knoxville, Knoxville, TN

Abstract

Using different computational approaches (Molecular Dynamics Simulation, Confined random walk (CRW) simulation and an analytical theory), this work studied the cross-linked and sulfonated Poly (1,3-Cyclohexdiene) (xsPCHD) hompolymer, xsPCHD-polyethylene glycol (PEG) block copolymer and xsPCHD/PEG blend proton exchange membrane (PEM) molecular level structures and water and charge transport in the membranes. The membrane structure was first studied through pair correlation functions (PCFs) from MD simulation. Interesting results are found for the copolymer and blend membranes. The PEG-water PCF shows PEG has greater affinity to water in the blend than that in the copolymer membrane, suggesting PEG is well mixed with water and serves as a cosolvent in the blend membrane. The PEG-H_3O^+ PCFs show that the end OH group of PEG has much stronger affinity to H_3O^+ while the backbone of PEG has weaker affinity to H_3O^+ in the copolymer membrane than that in the blend membrane. The features of the PCFs suggest very different proton transport mechanisms in the copolymer and blend membranes. The CRW simulation results show that bigger water clusters formed in the copolymer membrane. The existence of PEG in the blend membrane slows down the water diffusion but helps the structural diffusion of charge. Using a recently developed analytical theory, the proton conductivity is calculated and compared with that of experimental measurements. Good agreement is obtained for homopolymer membrane but the accuracy of the theory is affected when applied to copolymer and blend membrane due to the existence of PEG in these membranes.

(635d) Multi-Scale First-Principles Approach to Three-Phase Model of Polymer Electrolyte Membrane Fuel Cell

Giuseppe F. Brunello, Ji Il Choi and **Seung Soon Jang**, School of Materials Science and Engineering, Georgia Institute of Technology, Atlanta, GA

Multi-scale First-Principles Approach to Three-Phase Model of Polymer Electrolyte Membrane Fuel Cell

Giuseppe F. Brunello, Ji Il Choi and Seung Soon Jang

Computational NanoBio Technology Laboratory, School of Materials Science and Engineering, Georgia Institute of Technology, 771 Ferst Drive NW, Atlanta, GA 30332-0245

It is desirable to achieve fundamental understanding of the three-phase system consisting of carbon support, polymeric ionomer, and Pt catalyst nanoparticle, where the most essential electrochemical events for fuel cell operation take place in the presence of other molecular species such as water, proton, and oxygen. Although variety of cutting-edge in-situ and time-resolved probing techniques would be available to investigate nanometer-scale systems, the structure and dynamic behavior of the three-phase system of fuel cell has not been thoroughly analyzed at molecular level, which is mainly due to such complicated multi-component feature of the three-phase system in several tens of nanometers. In this context, we simulate the three-phase system using multi-scale first-principles modeling approach consisting of quantum mechanical density functional theory (DFT) and molecular dynamics (MD) simulations in order to investigate the nanophase-segregation of polymeric ionomers and water molecules, surrounding Pt nanoparticle on graphitic carbon support. For this, we develop a force field based on DFT computations and ran large-scale MD simulations. The nanophase-segregated structures and transport properties of water, proton, and oxygen are analyzed, and especially the spatial arrangement of molecular species around Pt particle is characterized in detail. All the simulation results are compared to those of bulk Nafion phase with the same water contents.

(635e) Hybrid Membranes for Proton Exchange Membrane Fuel Cell

Zhongyi Jiang, Hong Wu, Jingtao Wang, Lingli Nie, Weikang Yuan, Bin Zheng, Xiaohong Zheng, Yabo Wang, Weiqiang Hou, Lulu Xiao, Yuning Zhao, Tao Xu and Xiaohui Shen, Key Laboratory for Green Chemical Technology of Ministry of Education, Sch Chem Engn & Technol, Tianjin Univ, Tianjin, China

Hybrid membranes prepared by incorporating inorganic fillers into the polymer matrix are promising candidates for proton exchange membranes (PEMs) and have spurred increasing attention. The work in our group mainly focuses on the regulation of membrane microstructure (e.g. free volume) and the intensification of membrane performances (e.g. methanol barrier property, proton conductivity). Inorganic fillers can regulate the free volume of the resultant hybrid membrane by interfering the mobility of polymer chains. It is found that the polymer-inorganic interfacial interaction governed the membrane microstructure: i) attractive force would generate strong interfacial stress, which inhibits the mobility of polymer chains and leads to dense chain packing, in turn reducing free volume cavity size; and ii) repulsive force enhances the mobility of polymer chains, which allows for a loosened chain packing and increased free volume cavity size. The methanol transport of through PEM can be described by solution-diffusion mechanism. Inorganic fillers can change the hydrophilic/hydrophobic nature of membrane surface, which in turn tunes the methanol dissolution. Meanwhile, inorganic fillers can tune the methanol diffusion by adjusting the free volume and methanol affinity properties of membrane bulk. Both the filler structure (e.g. shape, particle size, pore size) and filler composition (e.g. hydrophilic/hydrophobic nature, acid/base property) can affect methanol dissolution and diffusion properties of hybrid membranes. Accordingly, the methanol barrier property of hybrid membrane can be intensified. The less-conducting inorganic fillers generally dilute the concentration of proton conducting group and reduce water uptake, resulting in reduced proton conductivity of hybrid membranes. Therefore, the filler surface is modified to introduce proton-conducting groups, which render additional pathways for proton transfer. Furthermore, the proton transfer barrier can be reduced by optimizing proton donor/acceptor ability of the conducting groups. In addition to polymer-inorganic hybrid membrane, inspired by the water storage mechanisms in plant cells, polymeric microcapsules are synthesized and incorporated into polymer matrix to prepare polymer-organic "hybrid membrane". The microcapsules serve as water reservoirs and enhance water retention properties in particular under low humidity, endowing hybrid membranes with high proton conductivity.

(636a) Fundamentals of Electrochemical Energy Storage

Sri Narayan, Chemistry, University of Southern California, Los Angeles, CA

Electrochemical energy storage systems are used to efficiently store and transport electrical energy. The applications for such systems range from portable electronics to vehicles and large-scale power plants. This tutorial will describe the operating principles, materials requirements, rate processes and the design approaches relating it to the requirements of the applications. The tutorial will discuss the features and performance characteristics of metal-hydride batteries, lithium-ion batteries, metal-air and metal-sulfur batteries, and redox flow type batteries. The technical challenges to designing inexpensive, eco-friendly, and robust electrochemical energy storage systems and the recent technological advances will be discussed.

(636b) Materials Challenges of Electrochemical Energy Storage Technologies

Arumugam Manthiram, Materials Science and Engineering, University of Texas at Austin, Austin, TX

Electrical energy storage (EES) is critical for the efficient utilization of renewable energies like solar and wind as well as for electrifying the transportation sector. Rechargeable batteries and supercapacitors are the primary candidates for EES. Cost, safety, cycle life, energy, power, and environmental impact are the major issues hampering the adoption of these technologies. This presentation will first provide an overview of the basic principles involved in electrochemical energy storage, followed by an account and status of the existing electrode materials for lithium-ion batteries. After pointing out the materials challenges and opportunities, the presentation will then focus on (i) next generation of cathode and anode materials for lithium-ion batteries, (ii) high energy, high power lithium-sulfur batteries, and (iii) high-energy density dual-electrolyte lithium-air cells employing a lithium-ion conducting solid electrolyte. Particularly, the importance of novel synthesis approaches, surface modification, electrode architecture, and optimum cell configurations to realize the necessary performance parameters will be emphasized.

(636c) Promise and Challenges of Materials for Electrochemical Capacitors

Prashant Kumta, Bioengineering, University of Pittsburgh, Pittsburgh, PA

Nanotechnology is a revolutionary area that has impacted several areas of materials science and technology including the area of electrochemically active materials for lithium-ion batteries, electro-active catalysts for fuel cell systems, and electrochemical capacitors. The growing worldwide demand for energy makes the need for sustainable energy and power storage systems highly critical. In this regard electrochemical supercapacitors are emerging as devices of prime importance owing to their superior characteristics unparalleled by any other charge storage device. These include: high power densities at relatively high energy densities and long cycle life. Electrochemical capacitors comprise electrochemical double layer (EDLC), pseudo-capacitors, and supercapacitors. All of these systems are differentiated by their intrinsic distinct fundamental mechanisms and are all capable of delivering large bursts of charge over a very short period of time. It is due to this unique capability that electrochemical capacitors find a very important role in the growing alternative energy field. High surface area carbons, hydrated ruthenium oxide, nanostructured VO_x and MnO_2, high surface area and highly electronically conducting graphene, carbon nanotubes (CNTs) are among the primary materials being considered for supercapacitor applications. In recent years there has also been significant activity in transition metal non-oxide based electrochemical capacitors. This presentation will cover the fundamental principles of electrochemical capacitors and discuss the current materials of choice. The presentation will culminate by providing insights into the challenges and promise of novel next generation materials for use in electrochemical capacitors for high power applications.

(636d) Fundamentals for Designing Flow Batteries

J.W. Van Zee, Karson Leperi and Victor Caldwell, Chemical Engineering, University of South Carolina, Columbia, SC

Selected flow battery systems will be used to present the fundamentals of and advances in electrochemical engineering design. Applications of these design principles will allow assessment of performance, degradation, and opportunities for various load cycles. The interaction of economics with the principals of thermodynamics, kinetics, mass transfer, and potential distributions will be discussed. In the spirit of a tutorial, analogies will be drawn between common chemical engineering concepts and electrochemical engineering principles. The goal is to provide the basis for an understanding of the limitations and the advantages of flow batteries. Examples of the use of experimental designs to isolate limiting behavior and opportunities for improvement will be discussed. Opportunities for improvements using commercial computational fluid dynamics packages will also be discussed.

The fundamentals are introduced with a simple algebraic model for a hybrid flow battery and stack that uses zinc-bromine chemistry. Performance is shown to be lost though subtractions from the thermodynamic voltage by ohmic losses which can be a function of geometry, by columbic efficiency losses which can be a function of separator design, and by voltage efficiency losses which are related to kinetic, mass transfer in the cell, and shunt currents in the stack. A review of more sophisticated models and the opportunities for cell and stack improvement will be discussed.

Then the paper will discuss the necessary concepts for extension to a complete flow battery with an example of replacing the zinc negative electrode with hydrogen. Novel designs and limitations are discussed for this system with a brief review of porous electrode theory. The economics of the system are discussed components as well as the interactions of the effects of geometry, component selection, and auxiliary power and energy requirements.

(636e) Quantifying Solute-Solute Interactions in Flow-Battery-Membrane Transport

Lucas Griffith, Chemical Engineering, University of Michigan, Ann Arbor, MI and Charles W. Monroe, Department of Chemical Engineering, University of Michigan, Ann Arbor, MI

Redox Flow Batteries (RFBs) have the potential to provide the high-capacity energy storage needed to improve power grid efficiency through peak shaving, as well as aid in the integration of renewable energy sources, such as wind energy and solar energy. The high cost of traditional ion-selective membranes has motivated the exploration of the many different membranes available for use in RFBs[1]. In order to facilitate this exploration it is necessary to characterize mass transport in various membranes, so that the effect of using different membranes on RFB performance can be modeled.

Although treatment of binary diffusion in all-vanadium RFBs is ubiquitous in the literature[2], there is a surprising lack of effort to describe the effect of ion-ion interactions during multicomponent transport across ion-selective membranes. Heintz *et al*[3] have investigated these effects for sodium and hydrogen diffusing across Neosepta membranes and have seen that the magnitude of the ion-ion interactions is comparable with the magnitude of the ion-solvent interactions. This shows that the importance of including ion-ion interactions in models of RFBs should be tested, so we have measured ion-ion interactions for $VOSO_4$ and H_2SO_4in two different membranes.

This talk will briefly discuss how Barnes's transient diffusion model of dialysis cells[4] can be solved and used to measure binary diffusion coefficients and sorption equilibria simultaneously. Results of binary experiments can be used as a basis for describing ternary interdiffusion measurements, which can be modeled by applying the Stefan-Maxwell transport formalism. For n diffusing species, this system of n force-explicit equations can be reduced to a system of n-1 flux explicit equations by taking the membrane velocity as the reference and applying the Gibbs-Duhem equation[5]. Taylor and Krishna[6] have shown that a batch dialysis cell will decay exponentially toward equilibrium at long times, with decay constants that are the eigenvalues of the system of transport equations. Therefore, if the binary diffusion coefficients are known, solute-solute interactions can be measured with an interdiffusion experiment in a batch dialysis cell. The values of the binary diffusion coefficients as well as the solute-solute interaction parameters will be given for $VOSO_4$ and H_2SO_4 (species relevant to aqueous all-vanadium RFBs) in two exemplary hydrated membranes.

1. Skyllas-Kazacos, M., M. H. Chakrabarti, et al. (2011). J. Eletrochem Soc **158**(8): R55-R79

2. Wiedemann, E., A. Heintz, et al. (1998). J. Membrane Sci. **141**(2): 215-221.

3. Heintz, A., E. Wiedemann, et al. (1997). J. Membrane Sci. **137**(1-2): 121-132.

4. Barnes. (1934). Physics **5**(1): 5.

5. Meyers, J. P. and J. Newman (2002). J. Electrochem. Soc. **149**(6): A718-A728.

6. Taylor, R. and R. Krishna (1993). Multicomponent mass transfer, Wiley.

(642a) Density Functional Theory for the Structure and Dynamics of Electric Double Layers of Ionic Liquids

Ke Wang, Department of Chemical and Environmental Engineering, University of California, Riverside, Riverside, CA, De-en Jiang, Chemical Sciences Division, Oak Ridge National Laboratory, Oak Ridge, TN and Jianzhong Wu, Chemical and Environmental Engineering, UC Riverside, Riverside, CA

Engineering Sciences and Fundamentals

01C11 Interfacial Aspects of Electrochemical Energy Storage Systems
Dan Steingart and Holly J. Martin

Density functional theory for the structure and dynamics of electric double layers of ionic liquids

Ke Wang[1], De-En Jiang[2] and Jianzhong Wu[1]

[1]Department of Chemical and Environmental Engineering, University of California, Riverside, California 92521 and [2]Chemical Sciences Division, Oak Ridge National Laboratory, Oak Ridge, Tennessee, 37831

Abstract

Ionic liquids have been used in diverse electrochemical systems including energy storage devices such as supercapacitors, batteries and fuel cells. Because of the intrinsic complexity of organic ions and strong electrostatic correlations, the electrochemical properties of ionic liquids often defy the descriptions of conventional mean-field methods such as the Gouy-Chapman-Stern (GCS) theory for equilibrium properties and the Poisson-Nernst-Plank (PNP) equation for the dynamics. In this work, we introduce the classical density functional theory (DFT) as an alternative to conventional methods to describe equilibrium and time-dependent properties of electric double layers unique to ionic liquids. By considering the molecular size, topology, and electrostatic correlations, we have examined the equilibrium structures of ionic liquids near electrodes and the responses of ionic profiles, electrostatic potential, surface charge density and surface power density during electrode charging. We have also examined major factors responsible for the unique features of electric-double layers in ionic-liquids including capacitance oscillation in porous electrodes and formation of long-range alternating structures of cations and anions at charged surfaces.

(642b) Modelling of Contact Resistance for Electrode-Interconnect Interface in Solid Oxide Fuel Cell

Tapobrata Dey[1,2], Debanand SingDeo[1], Prakash C. Ghosh[1], R N Basu[2] and **Manaswita Bose**[1], (1)Energy Science and Engineering, Indian Institute of Technology Bombay, Mumbai, India, (2)Fuel Cell and Battery Division, Central Glass and Ceramic Research Institute, CSIR, Kolkata, India

Solid oxide fuel cells (SOFCs) are electrochemical energy conversion devices which operate at temperatures between 650 to 1000 ^{O}C. It consists of an electrolyte having good ionic conductivity such as YSZ, with porous Ni based anode and LSM as cathode material. Steel interconnects are used on both the anode and cathode sides as current collectors. The high operating temperatures provide additional benefits such as electricity generation using multiple fuel choices and Combined Heat and Power (CHP) with efficiencies as high as 80%. Scale up of performance with increase in active area is one of the major challenges for SOFC. One of the important factors which lead to poor performance is the contact resistance of the cell. It is reflected in the form of resistive voltage loss across the electrode-electrolyte interface. Contact is achieved between the SOFC cell components by applying loads across the cell area. In the event of improper contact between interconnect and electrodes; the performance of the cell is affected. On the other hand, if the load is very high the cell components may get damaged and micro cracks may appear during operation. The objective of this work is to investigate the behaviour of contact resistance using experiments and theoretical modelling at different loading conditions and temperatures. The contact resistances are measured using four probe method. Samples are prepared for both the electrodes and interconnect in dimension of 1 cm X 1 cm. The anode (Ni-YSZ) is fabricated by the tape casting technique. Cathode material is prepared using the LSM powder through the auto ignition technique. The material used for interconnect is Crofer 22 APU. The samples are arranged facing each other. On both the sides, a very Pt fine mesh is used for the electrical contact of the four probes. Finally, the assembly is placed between a pair of alumina plates for insulation as well as mechanical support. Constant current is passed using current source and the potential difference across the sample is measured using high precession multimeter at various applied loads. The total resistance can be expressed as the sum of the electrode-interconnect bulk materials resistance and the contact resistance across the interface. The contact resistance values are obtained by measuring electrode and interconnect separately and subtracting from the total resistance. The experimental results indicate that contact resistance values are strongly influenced by the load applied. The load range used in the study varies from 180 to 1200 gm cm^{-2}. At low load, such as 180 gm cm^{-2} and temperature of 800 ^{O}C the contact resistance is 90 mΩ cm^{-2} for the anode-electrode interface. Under the same conditions the cathode-interconnect interface shows contact resistance of 359 mΩ cm^{-2}. The contact resistance decreases sharply with increase in load and reaches a constant value when load is 800 gm cm^{-2}. Reduction in the contact resistance with increase in the applied load is expected as the actual interfacial area increases with the compression force.

In the present work, we have developed a model to investigate the effect of the compressive pressure on the contact resistance. Any surfaces possess micro-inhomogenity, also called asperities. The asperities of adjacent surfaces come into contact when a certain load is applied. During contact the materials deform locally. The deformation may be of two types, elastic or plastic depending on the load applied. The elastic deformation of the asperities are described with the help of the Hertzian contact theory [1]. We have also included the effect of the plastic deformation for the applied loads greater than a certain critical value [2]. We have assumed that the summits are spherical in shape and the heights are normally distributed. We have also assumed the deformations are local. The contact resistance for a deformed asperity is obtained using Holm theory for electrical contacts [3]. The necessary surface parameters are obtained by finding the moments asperity distribution [4]. In the present work, we have observed that the anode-interconnect interface has high density of asperity summits and a low band-width parameter compared to the cathode-interconnect interface. These factors are essential in achieving low contact resistance values. Figure 1 shows that the model prediction agrees very well with the experimental observation. The minimum load required to obtain proper contact for an SOFC single cell is determined as the contact resistance reaches a plateau. It is observed that the minimum load required for the cathode-interconnect interface is 650 gm cm^{-2} at operating temperatures of 800 ^{O}C. The proposed model can be employed to predict the contact resitance between two surfaces in a solid oxide fuel cell which will in turn enable to estimate the performace of a cell.

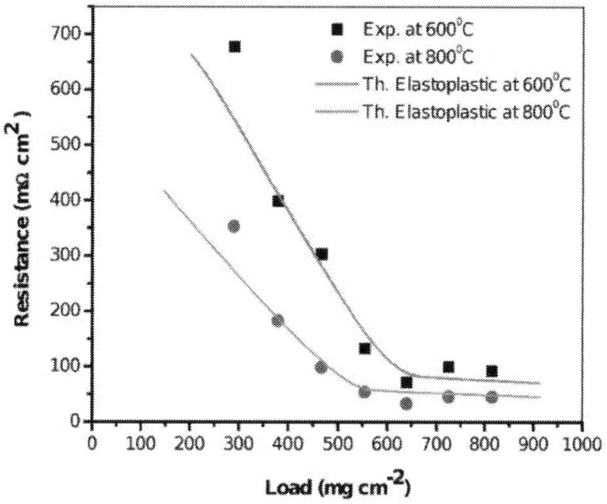

Figure 1 Experimental and theoretical results for contact resistance between cathode and interconnect at 600 and 800 °C

References:

1. J.A. Greenwood, J.B.P Williamson, Contact of nominally flat surface, Proceedings of the Royal Society of London. Series A, Mathematical and Physical Sciences, (1966), 295(1442), 300-319.

2. R.L. Jackson, I. Green, A statistical model of elasto-plastic asperity contact between rough surfaces, Tribology International, (2006), 39, 906-14.

3. R. Holm, Electric Contacts: Theory and Application, Springer-Verlag, Berlin (1967).

4. P.R. Nayak, Random Process Model of Rough Surfaces, Journal of Lubrication Technology, Trans. ASME, (1971), 93 (3), 398-407.

(642c) Phase Changes in Secondary Manganese Dioxide Electrodes for Grid-Scale Batteries

Joshua Gallaway[1], Can Erdonmez[2], Lev Sviridov[1], Abhinav Gaikwad[1], Tal Sholklapper[1], Nilesh Ingale[1], Michael Nyce[1], Benjamin Hertzberg[1], Sanjoy Banerjee[1] and Daniel Steingart[1], (1)Energy Institute and Chemical Engineering, City College of New York, New York, NY, (2)Energy Storage Group, Brookhaven National Laboratory, Upton, NY

The low cost, high availability, and safety of the alkaline electrolytic manganese dioxide (EMD) electrode make its use in large, rechargeable, stationary grid-scale batteries desirable. However, manganese dioxide is widely used only as a primary battery electrode. This is due to its propensity to form electrochemically irreversible phases while cycling, limiting its cycle life.

One strategy to extend the cycle life of manganese dioxide electrodes is to discharge in a limited window, corresponding to some fraction of one electron per manganese atom. Commercial EMD is composed of particles on the order of ten microns in diameter, and the diffusion of both protons and electrons through this crystal are required for cycling. Literature has shown that the effective diffusion coefficient for this transport is on the order of 1×10^{-15} cm^2/s.[1] It is believed that approaching one electron of reaction per manganese atom causes an irreversible phase change to Mn_2O_3 or Mn_3O_4.[2,3] Quantifying and avoiding this change is our goal.

An added complication in studying the manganese dioxide active material is that for practical use it is always compressed in a porous electrode with graphite, which is electrically conductive, and electrolyte, which is ionically conductive. A poor distribution of current across a porous electrode can lead to disparate states of charge across the active material, presenting problems with interpreting experimental results. Electrode modeling and microscopy must be used to assure that reaction rates in the electrode are understood and accounted for.[4,5]

E-TEM was used to observe phase changes in the MnO_2 active material. These phase changes were provoked in one of two ways: ex situ electrochemical cycling, or in situ heating. In situ X-ray diffraction (XRD) was performed on a modified manganese dioxide-zinc coin cell, which was fitted with a transparent window and a beam path through the zinc anode. These experiments were performed at the National Synchrotron Light Source (NSLS) on beamline X-14. After three shallow cycles of the cell in the range from 100% to 75% state of charge, the lattice parameters of ramsdellite were found to decrease from the charged to discharged state, while the lattice parameters for ε-MnO2 increased. This is only in partial agreement with previous literature, in which lattice parameters only increased.[6] We believe a unified approach involving both spectroscopy and microscopy is important for understanding the phase changes provoked by shallow cycling.

Recent battery efforts at the CUNY Energy Institute have incorporated parallel experimentation across many cell sizes, with the target being large cell stacks.[7]

ACKNOWLEDGEMENTS

The authors would like to thank ARPA-E under award number DE-AR0000150 for generous support.

REFERENCES

1. A.P. Malloy and S.W. Donne, *J Power Sources*, **179**, 371-380 (2008).

2. P. Ruetschi, R. Giovanoli, *J Electrochem Soc*, 135, 2663-2669 (1988).

3. A. Kozawa, R.A. Powers, *J Electrochem Soc*, 113, 870-878 (1966).

4. J. S. Newman and C. W. Tobias, *J Electrochem Soc*, **109** (12), 1183-1191 (1962).

5. J. S. Chen and H. Y. Cheh, *J Electrochem Soc*, **140** (5), 1213-1218 (1993).

6. C. Mondoloni, M. Laborde, J. Rioux, E. Andoni, C. Levy-Clement, *J Electrochem Soc*, **139**, 954-959 (1992)

7. Y. Ito, M. Nyce, R. Plivelich, M. Klein, D. Steingart, S. Banerjee, *J Power Sources*, **196**, 2340-2345 (2011).

(642d) Adsorption Behaviors of Sodium Dodecyl Sulfate Onto Clathrate Hydrates in Electrolyte Solutions

Oluwaseun Salako[1], Chi Lo[2], Junshe Zhang[3], Ponisseril Somasundaran[4], Alexander Couzis[5] and Jae W. Lee[3], (1)Chemical Engineering, CCNY, New York, NY, (2)Chem Eng Dept, The City College of New York, New York, NY, (3)Department of Chemical Engineering, The City College of New York, New York, NY, (4)Columbia University, New York, NY, (5)Chemical Engineering, City College of New York, New York, NY

Clathrate Hydrates are ice-like non-stoichiometric crystalline compounds where water molecules (host molecules) form cages that trap low molecular weight molecules (guest molecules). Clathrate hydrates have attracted lots of attention since the discovery of large reserves of natural gas hydrates in oceanic deposits and permafrost areas because of the potential of methane hydrate as an energy resource. On the other hand, synthetic methane hydrates provide a promising gas storage capacity with 170 vol./vol. of hydrates. Both methane recovery from the hydrate deposits and methane storage in hydrate forms require solid understanding of hydrate formation kinetics. Gas hydrate formation is slow due to the diffusion limitation of guest molecules into a liquid-solid growth front but this can be subdued by mechanical agitation and by addition of a small amount of surfactant. Sodium dodecyl sulfate (SDS) has been shown to be a potent surfactant for improving hydrate formation kinetics of low molecular weight hydrocarbon guest molecules like methane but the mechanism by which SDS improves the kinetics of hydrate formation is not well understood. The adsorption of SDS on cyclopentane hydrate-water interface was reported to be a main factor for the enhanced kinetics and possible orientations of SDS molecules at the hydrate-water interface are proposed [1]. Another study observed a marked increase in hydrate formation when 20 ppm SDS in 1 mM NaCl solution was used for a methane-cyclopentane (CP)-water hydrate system. However, the effect of salt on SDS adsorption was not clearly understood. It was not also understood as to why the highest enclathration rate was observed for the case of 1 mM NaCl and how this is related to the SDS adsorption behavior. In this presentation, we try to understand the adsorption behaviors of SDS at CP hydrate-water interface in NaCl solutions by determining SDS adsorption isotherms and SDS solubility. Zeta potentials of CP hydrate particles will be shown to serve as backup data for the adsorption isotherm.

Literature Cited:

[1] C. Lo, J.S. Zhang. A. Couzis, P. Somasundaran, J.W. Lee. J. Phys. Chem. C 114 (2010) 13385–13389.

[2] S. Zhang, J. A. Salera, J. W. Lee. Ind. Eng. Chem. Res. 49 (2010) 8267-8270.

(642e) Electrochemical Performance of Binary Mixture of Ionic Liquids Electrolytes in Supercapacitors

Song Li[1], Guang Feng[1], Pasquale F. Fulvio[2], Patrick Hillesheim[2], Chen Liao[2], Sheng Dai[2] and Peter T. Cummings[1,3], (1)Department of Chemical and Biomolecular Engineering, Vanderbilt University, Nashville, TN, (2)Chemical Sciences Division, Oak Ridge National Laboratory, Oak Ridge, TN, (3)Center of Nanophase Materials Sciences, Oak Ridge National Laboratory, Oak Ridge, TN

Ionic liquids (ILs), as promising electrolyte candidates for batteries and supercapacitors have been attracting increasing research interest. Recently, it was reported that a binary mixture of ionic liquids extends the operation temperature range for supercapacitors to -50 - 100 °C [1], which makes the mixture of ionic liquids potential electrolytes for applications in severe weather conditions. In this work, a equimolar mixture of 1-methyl-1-propylpyrrolidinium bis(trifluoromethylsulfonyl)imide ([C_3mpy][Tf_2N]), 1-methyl-1-butylpiperidinium bis(trifluoromethylsulfonyl)imide ([C_4mpip][Tf_2N]) was investigated by classic molecular dynamics (MD) simulation. Differential scanning calorimetry (DSC) measurement verified that the binary mixture exhibited lower glass transition temperature than either of the pure ILs. Moreover, the binary mixture gave rise to the higher conductivity than the neat ILs at lower temperature range. In order to study its capacitive performance in supercapacitors, the mixture and the neat ILs were used as electrolytes near onion-like carbon (OLC) electrode at varying temperatures. Both the differential capacitance and integral capacitance exhibited independence of the electrical potential applied for three electrolytes, which is in agreement with previous report on a different IL [2]. Anomalous temperature dependence of the capacitance was observed and it was dominated by the electrical double layer (EDL) thickness, which is for the first time substantiated in MD simulation. In summary, this study demonstrated that the binary mixture of ILs outperformed neat ILs as supercapacitor electrolyte with widened operation temperature, higher conductivity and a near-flat capacitance as a function of electric potential. The anomalous temperature dependence of the capacitance dominated by the EDL thickness provides a theoretical model for interpreting the anomalous temperature dependence observed in experiments.

References

[1] R. Y. Lin, P. L. Taberna, S. Fantini, V. Presser, C. R. Perez, F. Malbosc, N. L. Rupesinghe, K. B. K. Teo, Y. Gogotsi, P. Simon, Capacitive Energy Storage from -50 to 100 °C Using an Ionic Liquid Electrolyte,Journal of Physical Chemistry Letters,2011. 2(19):2396-2401

[2] G. Feng, D. Jiang, and P. T. Cummings, Curvature Effect on the Capacitance of Electric Double Layers at Ionic Liquid/Onion-like Carbon Interfaces,Journal of Chemical Theory and Computation, 2012, 8(3):1058–1063

(642f) Mass Transport At Internal Interfaces of Inorganic Materials and Implications to Electrochemical Energy Systems

Kedarnath Kolluri, Materials Science in Radiation and Dynamic Extremes, Los Alamos National Lab, Los Alamos, NM

Interfaces of inorganic solids have gained prominence in electrochemical energy systems because of enhanced mass transport at interfaces as well as their ability to store a lot and retrieve quickly guest species. Despite several experimental observations, quantitive relations tying mass transport to interface structure are lacking. Since electrochemical energy systems are much complicated, we study simpler metal-metal interfaces as the first step. We use atomistic simulations to study the formation, migration, and clustering of delocalized vacancies and interstitials at several model metal-metal semicoherent interfaces. We find that point defects migrate between interfacial trapping sites through a multi-step mechanism that may be described analytically. Similar mechanisms operate in the formation, migration, and dissociation of interfacial point defect clusters. We show that effective mass transport rates may be computed using the harmonic approximation of transition state theory with a temperature dependent prefactor. Our results demonstrate that vacancies and interstitials at some interfaces may be governed by mechanisms of higher complexity than conventional point defects in crystalline solids. Implications of our studies to mass transport at grain boundaries and heterointerfaces of ceramics are discussed.

(650a) Lipid Diversity: Is It Important in Modeling Organism and Organelle Membranes?

Jeffery Klauda, Department of Chemical and Biomolecular Engineering, University of Maryland, College Park, MD

Biological membranes form a barrier to protect the cell from its environment and selectively control the entrance/exit of small molecules. These membranes consist of lipids, such as phospholipids and sterols, and proteins that act as gates to control molecule transport across the amphiphillic surface. Results from both experimental and simulation methods are often required to elucidate structural, mechanical, and dynamical properties of membranes at the molecular level. The importance of lipid composition and lipid types will be presented with focus on model singled-celled organisms (bacteria and yeast). Methods for developing accurate force field parameters from quantum mechanics will be briefly discussed. For bacterial membranes, chain branching and replacement of carbon-carbon double bonds with a cyclopropane-moiety influence structural membrane properties. In addition to lipid diversity on membrane properties, the influence of membrane composition on membrane-protein interaction will also be presented. Although the focus has been on developing model membranes for computational purposes, our results also have implications to experimental *in vitro* studies.

(650b) Multiscale Modeling of Membrane Remodeling by the Protein Epsin

Ryan P. Bradley, Chemical and Biomolecular Engineering, University of Pennsylvania, Philadelphia, PA and Ravi Radhakrishnan, Department of Bioengineering, University of Pennsylvania, Philadelphia, PA

The protein epsin plays a key role in both sensing and inducing curvature during clathrin-mediated endocytosis (CME), affecting a multitude of downstream cellular signals. Both the constitutive activation of cell surface receptors regulated by CME, including epidermal growth factor receptor (EGFR), and defects in the endocytic machinery itself may be oncogenic. Therefore, treatment of the resulting cancers requires molecular resolution of endocytosis. To this end, we employ multiscale modeling techniques to study the process by which the epsin N-terminal homology domain (ENTH) remodels the cell membrane. Coarse-grained molecular dynamics (CGMD) simulations under the MARTINI force field [1] capture the mechanism by which a single ENTH domain induces curvature in a large patch of the lipid bilayer. Calculation of the three-dimensional stress tensor [2] reveals both the distribution of stresses in the bilayer, and characterizes the curvature field induced by ENTH. Simulations with interacting pairs of ENTH in various configurations characterize the additivity of their curvature fields, and inform continuum mechanics models for vesicle budding [3]. To parameterize the coarse-grained model for epsin-membrane interactions, we must turn to more detailed all-atom simulations of both the protein-membrane dynamics and the action of a key lipid binding partner, phosphatidylinositol (4,5)- bisphosphate (PIP2), which associates with epsin's embedded helix. All-atom simulations of mixed lipid bilayers containing PIP2 under the CHARMM36 force field [4] provide structural and dynamic properties of PIP2 for use in CGMD simulations. Likewise, all-atom simulations of ENTH parameterize a heterogeneous elastic network model which reproduces the dynamics of ENTH embedded in the membrane. By imposing a variety of restraints on our model system, we probe the roles of membrane surface tension, interactions of multiple epsins, and association with PIP2. These methods apply equally well to other membrane-remodeling proteins, including the proteins N-BAR, I-BAR, and exo70, which are thought to generate either positive or negative curvature with distinct mechanisms. In each case, the flow of information between CGMD and AAMD levels explains how molecular interactions are translated to larger length scales.

[1] Marrink, S. J.; Risselada, H. J.; Yefimov, S.; Tieleman, D. P. & de Vries, A. H. The MARTINI Force Field: Coarse Grained Model for Biomolecular Simulations, *Journal of Physical Chemistry B*, **2007**, 111, 7812-7824.

[2] Ollila, O. H. S.; Risselada, H. J.; Louhivuori, M.; Lindahl, E.; Vattulainen, I. & Marrink, S. J. 3D Pressure Field in Lipid Membranes and Membrane-Protein Complexes Physical Review Letters, *American Physical Society*, **2009**, 102, 078101.

[3] Liu, J.; Tourdot, R.; Ramanan, V.; Agrawal, N. J. & Radhakrishanan, R. Mesoscale simulations of curvature-inducing protein partitioning on lipid bilayer membranes in the presence of mean curvature fields, *Molecular Physics*, **2012**, In Press.

[4] Bjelkmar, P.; Larsson, P.; Cuendet, M. A.; Hess, B. & Lindahl, E. Implementation of the CHARMM Force Field in GROMACS: Analysis of Protein Stability Effects from Correction Maps, Virtual Interaction Sites, and Water Models, *Journal of Chemical Theory and Computation*, **2010**, 6, 459-466.

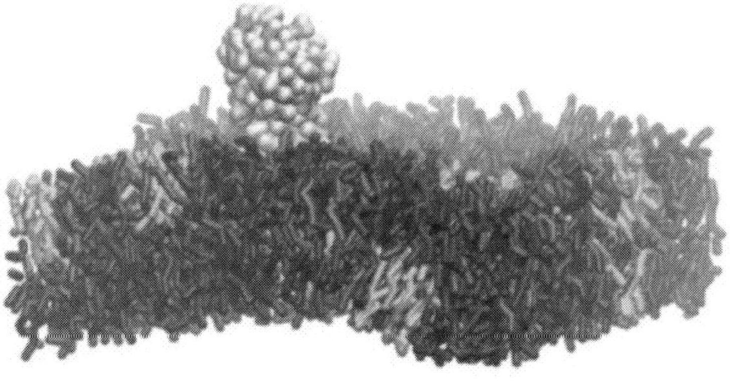

Figure. Snapshot of CGMD simulation of ENTH domain embedded in a mixed DOPC(blue)-DOPS(purple) bilayer with PIP2 (green) clusters (water not pictured).

(650c) Twisting Coarse-Grained DNA

Margaret C. Linak and Kevin D. Dorfman, Chemical Engineering and Materials Science, University of Minnesota, Minneapolis, MN

DNA can exhibit a wealth of complex shapes and functions in both *in vitro* and *in vivo* systems depending on its sequence, environmental conditions, and the forces acting upon the molecule. These complex configurations are thought to act as beacons and recognition sites for proteins involved in transcription, regulation, and repair along the length of DNA. However, in order to be able to capture such diverse features, a suitable model must be developed. Using an experimentally validated, coarse grained model of DNA, we examined the effect of twisting on nucleic acid systems that cannot be captured with other coarse grained models either because they lack the non-canonical interactions needed for advanced structures or because they are parameterized, and thus biased, for double-stranded B-form DNA. Including features such as Hoogsteen bonding and a smooth transition between single-, double-, and triple-stranded systems allows P-DNA (duplex with bases on the outside) and H-DNA (triplex formed from cruciform structure) systems to be captured.

Predicting anti-HIV-1 activity of TIBO Compounds by QSAR Approach Using a New Topological Index

Qiang WANG [a]*, *Qingzhu JIA* [a], *Peisheng MA,* [b] *Shuqian Xia,* [b] *Hongmei Tang* [a], *Yu Deng* [a]

a. School of Material Science and Chemical Engineering, Tianjin University of Science and Technology, 13St. TEDA, Tianjin, 300457, People's Republic of China

b. School of Chemical Engineering and Technology, Tianjin University, Tianjin 300072, People's Republic of China

* To whom correspondence should be addressed. E-mail: *wang_q@tust.edu.cn*

Abstract

The search and development of anti-HIV drugs is currently one of the most urgent tasks of pharmacological studies. In this work, a new topological index, a extended distance matrix, was proposed. This new topological index is composed of the distance matrix and the atomic character matrix, and further applied to build up a positional distributive structure-activity relationship (PDSAR) model for predicting anti-HIV-1 activity ($\log(1/IC_{50})$) of 89 tetrahydroimidazo[4,5,1-jk][1,4]benzodiazepinone (TIBO) compounds. Results indicate that the new model based on this topological index provides very satisfactory result. The overall average absolute difference for $\log(1/IC_{50})$ predictions of 89 TIBO compounds is 0.411693 and R^2 value is of 0.849292. Comparing with R. Garg et al.' method, our PDSAR method performed better both in accuracy and stability.

Keywords: Anti-HIV-1 activity; TIBO compounds; Structure-activity relationship; Topological index; Prediction

Introduction

Acquired immunodeficiency syndrome (AIDS) is a collection of symptoms and infections resulting from the specific damage to the immune system caused by the human immunodeficiency virus (HIV).[1-3] In the HIV life cycle, three enzymes are essential for replication of this virus inside host, reverse transcriptase (RT), protease (PR) and integrase (IN). Theoretically, an anti-HIV agent may exert its activity by inhibiting a variety of steps in the life cycle of the virus. Therefore, all of them are considered to be promising targets for the development of anti-HIV drugs.[4-6]

In order to search for anti-HIV drugs with fewer side effects and high efficacy, modeling the biological activity to propose new candidate molecules is an important approach. So, over the last few years, the quantitative structure-activity relationship (QSAR) studies have been widely carried out for different series of HIV-1 inhibitors, such as HIV-1 RT inhibitors[1, 7-10], HIV-1 IN inhibitors [2, 11-13] and HIV-1 protease inhibitors.[14-15]

Nowadays, QSAR models using topological indexes (TIs) have been more frequently used to obtain physicochemical and biological properties. Based on the distance and adjacency matrices, various TIs have been proposed and used in QSAR approaches.[16-20] However, until now no single topological index could be used universally in optimal correlations. Therefore, it is necessary for the researcher to see if a single set of descriptor, or a single topological index can be used to build a universal model in order to predict good values for all properties.

Recently, authors proposed a universal positional distributive group contribution (PDGC) theory for the prediction of various properties of a diverse set of organics compound.[21] Our previous works suggests that it is possible to use a totally same universal framework to predict various properties of organic compounds containing various functionalities. Very recently, a topological index was proposed based on atom characters (e.g. atom radius and atom electronegativity et al.) by our research group. And it has been successfully used for predicting the decomposition temperature and toxicity of Ionic Liquids. [22,23]

Therefore, the major objectives of this study are: (i) to propose a new topological index, (ii) to develop a more accurate and stable positional distributive structure-activity relationship (PDSAR) model for biological activity prediction of HIV-1 inhibitors involving various drugs.

Biological data

In this study, a series of 89 tetrahydroimidazo[4,5,1-jk][1,4]benzodiazepinone (TIBO) compounds were taken under consideration in this study.[1,3] And the anti-HIV activity IC_{50} is the minimum inhibitory concentration of the compound required to achieve 50% protection of MT-4 cells against the cytopathic effect of virus. The logarithm of the inverse of this parameter was used as biological end points (log $1/IC_{50}$) in this work.

Optimization of molecules structure

Structure building and the entire modeling works were carried out using the software of heperchem.7.0 . Energy minimization of the molecules was effected using Allinger's MM force field followed by semi empirical PM3 method. Most stable structure for each compound was generated and used for calculating various physicochemical descriptors like thermodynamic, steric and electronic values of descriptors.

Method proposed in this work

Based on chemical graphs, a new topological index calculated from a molecular graph was introduced. This newly proposed topological index is an extended distance matrix (MD), from which the extended adjacency matrix(Ma), the extended interval matrix(Mb) and the extended interval jump matrix(Mc) are deduced. In deed, the extended distance matrix(M_D) is composed of the distance matrix (D) and the atomic character matrix (Me). Our topological index quantitatively describes the structural information of molecules, taking into account parameters such as eletronegativity fraction, Van der Waals radius, minimum bond length with adjacent atom except hydrogen, number of adjacent hydrogen atom, and the number of adjacent atom except hydrogen. All the above information are considered for developing the atomic character matrix (Me). Then, the extended adjacency matrix M_A, the extended interval matrix M_B and the extended interval jump matrix M_C are deduced, respectively.

Here, matrix considered in this presentation are as follows.

$Md = (a_{ij})$ distance matrix

$$Ma = (a_{ij}) \quad a_{ij} = \begin{cases} 1 & \text{if } i \text{ and } j \text{ are adjacent} \\ 0 & \text{otherwise} \end{cases}$$

$$Mb = (a_{ij}) \quad a_{ij} = \begin{cases} 1 & \text{if the path length between atoms } i \text{ and } j \text{ is 2} \\ 0 & \text{otherwise} \end{cases}$$

$$Mc = (a_{ij}) \quad a_{ij} = \begin{cases} 1 & \text{if the path length between atoms } i \text{ and } j \text{ is 3} \\ 0 & \text{otherwise} \end{cases}$$

And composition of the extended matrix Me is of the following.

$$Me = \begin{bmatrix} \text{eletronegativity fraction} \\ \text{Van der Waals radius} \\ \text{the minimum bond length with adjacent atom except hydrogen} \\ \text{the number of adjacent atom except hydrogen} \\ \text{number of adjacent hydrogen atom} \end{bmatrix};$$

Thus, an extended distance matrix, which is the new topological index of our work, is defined as below.

$$M_D = \begin{bmatrix} Md \\ Me \end{bmatrix};$$

Then, the extended adjacency matrix MA, the extended interval matrix MB and the extended interval jump matrix MC are deduced, respectively.

$$M_A = \begin{bmatrix} Ma \\ Me \end{bmatrix}; M_B = \begin{bmatrix} Mb \\ Me \end{bmatrix}; M_C = \begin{bmatrix} Mc \\ Me \end{bmatrix};$$

Based on the new topological index, our approach is performed with ordinary least squares (OLS) regression. Also, in order to optimization the molecular structure, Gaussian software has been used for optimization the minimum bond length.

In this work, using the new topological index, a positional distributive structure-activity relationship (PDSAR) model for log ($1/IC_{50}$) prediction is expressed as Eq. (1).

$$Log(1/IC_{50})=MD+MA+MB+MC+b_1 exp(1/N)+b_2 exp(1/MW)+M_0 \qquad (1)$$

Here,

$$MD=a_{d1} \cdot norm(M_D,1)+ a_{d2} \cdot norm(M_D,2)+ a_{d3} \cdot norm(M_D, fro)$$

$$MA=a_{a1} \cdot norm(M_A,1)+ a_{a2} \cdot norm(M_A,2)+ a_{a3} \cdot norm(M_A, fro)$$

$$MB=a_{b1} \cdot norm(M_B,1)+ a_{b2} \cdot norm(M_B,2)+ a_{b3} \cdot norm(M_B, fro)$$

$$MC=a_{c1} \cdot norm(M_C,1)+ a_{c2} \cdot norm(M_C,2)+ a_{c3} \cdot norm(M_C, fro)$$

Where, norm(M_D, 1) means the largest column sum of matrix M_D, norm(M_D, 2) means the largest singular value of matrix M_D, norm(M_D, fro) is the frobenius-norm of matrix M_D, N for total number of atoms, MW is molecular weight, M_0 is the constant added, and a, b_1, b_2 are regression parameters.

Results and discussion

Prediction results of anti-HIV-1 activity for 89 TIBO derivatives with our PDSAR model are shown as in Table 1. Also, in order to evaluate the performance of our PDSAR model, the log ($1/IC_{50}$) prediction results with R. Garg et al.' QSAR method [1] had been used for comparison and the results are shown in Table 1. The plot of predicted versus experimental values about log ($1/IC_{50}$) activity for the TIBO derivatives is shown in Fig.1. And the statistical parameters of various models for log ($1/IC_{50}$) activity of the TIBO derivatives are listed in Table 2.

Results shown in Table 1 and Fig.1 indicate that the predicted log ($1/IC_{50}$) agree well with the "experimental results", which demonstrates that the new topological index for predicting log ($1/IC_{50}$) has good overall accuracy. The average absolute difference (AAD) for log ($1/IC_{50}$) prediction of 89 TIBO compounds is 0.411693. Also, our high-quality prediction model is evidenced by a R^2 value of 0.849292 and a PRESS value of 27.32721. On the other hand, based on R. Garg et al.' QSAR method, the *AAD* for log ($1/IC_{50}$) prediction is 0.523933. The R^2 and PRESS values are 0.743363 and 46.5349, respectively. Comparing with R. Garg et al.' method, our PDSAR method performed better both in accuracy and stability.

Conclusion

A new topological index, calculated directly from molecular structure, was proposed and applied to build up a positional distributive structure-activity relationship (PDSAR) model for predicting anti-HIV-1 activity of 89 TIBO compounds in this work. Results indicate that log ($1/IC_{50}$) was successfully predicted with our PDSAR model. It is evident that the proposed topological index can be used to predict log ($1/IC_{50}$) with a significant degree of confidence. The overall average absolute difference for log ($1/IC_{50}$)predictions of 89 TIBO compounds is found to be 0.411693. Comparing with R. Garg et al.' method,[1] our method performed better both in accuracy and generality.

Acknowledgements. Research reported in this work was supported by the National Natural Science Foundation of China (No. 20976131, and No. u1162104). Also, we would give much thanks to Fangyou Yan who has contributed for valuable advice and discussion.

Literature Cited

[1] R. Garg, S.P. Gupta, H. Gao, M.S. Babu, A.K. Debnath, Chem. Rev. 1999, 99, 3525–3601.

[2] Horrick Sharma, Shivaputra Patil, Tino W. Sanchez, Nouri Neamati, Raymond F. Schinazi, John K. Buolamwini, Bioorganic & Medicinal Chemistry 2011, 19, 2030–2045

[3] Kukla, M. J.; Breslin, H. J.; Pauwels, R.; Fedde, C. L.; Miranda, M.; Scott, M. K.; Sherrill, R. G.; Raeymackers, A.; Van Gelder, J.; Andries, K.; Janssen, M. A. C.; De Clercq, E.; Janssen, P. A. J. J. Med. Chem. 1991, 34, 746.

[4] Rongjing Hu, Jean-Pierre Doucet, Michel Delamar, Ruisheng Zhang, Eur. J. Med. Chem. 2009, 44, 2158–2171.

[5] Sabet R, Fassihi A, Moeinifard B. Journal of Molecular Graphics and Modelling 2009, 28, 146–155.

[6] J. Huuskonen, J. Chem. Inf. Comput. Sci. 2001, 4, 425–429.

[7] Darnag, R.; Mostapha Mazouz E.L.; Schmitzer, A.; Villemin, D.; bdellah Jarid, A.; Cherqaoui, D. Eur. J. Med. Chem. 2010, 45, 1590–1597.

[8] Ravichandran, V.; Prashantha Kumar, B.R.; Sankar, S.; Agrawal, R. K. Eur. J. Med. Chem. 2009, 44, 1180-1187.

[9] M.A. Sattwa, R. Kunal, Eur. J. Med. Chem. 2009, 44 (4), 1509–1524.

[10] B. Hemmateenejad, S.M. Tabaei, F. Namvaran, J. Mol. Struct. Theochem. 2005, 732 (1–3), 39–45.

[11] Zhengjun Cheng, Yuntao Zhang, Weizhong Fu, Eur. J. Med. Chem. 2010, 45, 3970-3980.

[12] Ravichandran,V.; Shalini, S.; Sundram, K .; Sokkalingam, A. D. Eur. J. Med. Chem. 2010, 45, 2791-2797.

[13] Peng Lu, XiaWei, RuishengZhang, European Journal of Medicinal Chemistry 2010, 45, 3413-3419.

[14] Noslen Hernández, Rudolf Kiralj, Márcia M.C. Ferreira, Isneri Talavera, Chemometrics and Intelligent Laboratory Systems. 2009, 98, 65–77.

[15] A. Srinivas Reddy, Sunil Kumar, Rajni Garg. Journal of Molecular Graphics and Modelling 2010, 28, 852–862.

[16] Khadikar, P. V.; Kale, P. P.; Deshpande, N. V.; Karmarkar, S.; Agrawal, V. K., J. Math. Chem. 2001, 29, 143-150.

[17] Alexandru T, B., Chem. Phys. Lett. 1982, 89, 399-404.

[18] Hosoya, H., Bull Chem. Soc. Jpn. 1971, 44, 2332-2339.

[19] Biye, R., Comp. Chem. 2002, 26, 357-369.

[20] Estrada, E., J. Chem. Inf. Comp. Sci. 1995, 35, 701-707.

[21] Wang Qiang, Jia Qingzhu, Ma Peisheng, J Chem Eng Data 2012, 57, 169-189.

[22] Yan, F.; Xia, S.; Wang, Q.; Ma, P. Sh. J Chem Eng Data 2012, 57 (3), 805–810, DOI: 10.1021/je201023a.

[23] Yan, F.; Xia, S.; Wang, Q.; Ma, P. Sh. J Chem Eng Data 2012, *57* (8), 2252–2257, DOI: 10.1021/je3002046

Table 1 Prediction results of anti-HIV-1 activity for 89 TIBO derivatives with R. Garg et al.' method and our PDSAR model

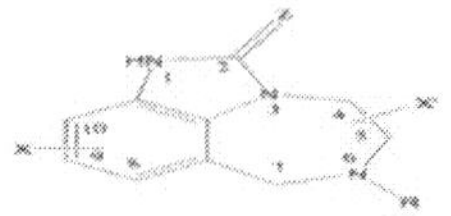

Substituents				log (1/IC$_{50}$)	R. Garg et al.' method[1]		This work	
No.	X	Z	R	X'	Expt.	Cal.	Diff	Cal.	Diff
1	H	S	DMA[a]	5-Me(S)	7.36	6.72	0.64	6.88	0.4762
2	9-Cl	S	DMA	5-Me(S)	7.47	6.99	0.48	7.89	-0.4210
3	8-Cl	S	DMA	5-Me(S)	8.37	7.92	0.45	7.74	0.6325
4	8-F	S	DMA	5-Me(S)	8.24	7.18	1.06	7.60	0.6378
5	8-SMe	S	DMA	5-Me(S)	8.30	7.75	0.55	7.90	0.4038
6	8-OMe	S	DMA	5-Me(S)	7.47	7.09	0.38	7.21	0.2603
7	8-OC$_2$H$_5$	S	DMA	5-Me(S)	7.02	7.30	-0.28	7.40	-0.3779
8	8-CN	O	DMA	5-Me(S)	5.94	6.01	-0.07	6.15	-0.2053
9	8-CN	S	DMA	5-Me(S)	7.25	7.19	0.06	7.23	0.0219
10	8-CHO	S	DMA	5-Me(S)	6.73	7.16	-0.43	6.99	-0.2559
11	8-CONH$_2$	O	DMA	5-Me(S)	5.20	5.44	-0.24	5.03	0.1706
12	8-Br	O	DMA	5-Me(S)	7.33	6.87	0.46	6.25	1.0788
13	8-Br	S	DMA	5-Me(S)	8.52	8.16	0.36	7.75	0.7732
14	8-I	O	DMA	5-Me(S)	7.06	7.21	-0.15	6.79	0.2698
15	8-I	S	DMA	5-Me(S)	7.32	8.49	-1.17	8.24	-0.9192
16	8-C=−CH	O	DMA	5-Me(S)	6.36	6.15	0.21	6.09	0.2662
17	8-C=−CH	S	DMA	5-Me(S)	7.53	7.52	0.01	7.49	0.0370
18	8-Me	O	DMA	5-Me(S)	6.00	6.15	-0.15	6.04	-0.0354

19	8-Me	S	DMA	5-Me(S)	7.87	7.52	0.35	7.56	0.3142
20	9-NO_2	O	CPM[b]	5-Me(S)	4.48	4.25	0.23	4.47	0.0118
21	8-NH_2	O	CPM	5-Me(S)	3.07	4.11	-1.04	4.21	-1.1410
22	8-NMe_2	O	CPM	5-Me(S)	5.18	4.65	0.53	5.40	-0.2241
23	9-NH_2	O	CPM	5-Me(S)	4.22	3.70	0.52	3.38	0.8359
24	9-NMe_2	O	CPM	5-Me(S)	5.18	4.24	0.94	5.34	-0.1605
25	9-NHCOMe	O	CPM	5-Me(S)	3.80	3.52	0.28	4.04	-0.2356
26	9-NO_2	S	CPM	5-Me(S)	5.61	5.45	0.16	5.27	0.3382
27	9-F	S	DMA	5-Me(S)	7.60	6.77	0.83	7.42	0.1841
28	9-CF_3	O	DMA	5-Me(S)	5.23	5.84	-0.61	5.76	-0.5322
29	9-CF_3	S	DMA	5-Me(S)	6.31	7.06	-0.75	5.97	0.3448
30	9-Me	O	DEA[c]	5-Me(S)[e]	6.50	5.01	1.49	5.36	1.1440
31	10-OMe	O	DMA	5-Me(S)	5.18	5.30	-0.12	5.49	-0.3116
32	10-OMe	S	DMA	5-Me(S)[e]	5.33	6.62	-1.29	6.39	-1.0622
33	9,10-di-Cl	S	DMA	5-Me(S)	7.60	7.23	0.37	7.60	0.0002
34	10-Br	S	DMA	5-Me(S)	5.97	7.05	-1.08	7.22	-1.2523
35	H	O	$CH_2CH{=}CH_2$	5-Me(S)	4.15	4.10	0.05	3.91	0.2377
36	H	O	2-MA[d]	5-Me(S)	4.33	4.26	0.07	4.57	-0.2434
37	H	O	CH_2CO_2Me	5-Me(S)	3.07	3.78	-0.71	3.16	-0.0873
38	H	O	CH_2C=−CH	5-Me(S)	3.24	3.84	-0.6	4.20	-0.9615
39	H	O	CH_2-2-furanyl	5-Me(S)	3.97	4.03	-0.06	4.07	-0.1029
40	H	O	$CH_2CH{=}CH_2$[S(+)]	5-Me(S)	4.18	4.10	0.08	3.91	0.2687
41	H	O	$CH_2CH_2CH{=}CH_2$	5-Me(S)	4.30	4.23	0.07	4.12	0.1840
42	H	O	$CH_2CH_2CH_3$	5-ME(S)	4.05	4.15	-0.1	4.45	-0.4029
43	H	O	2-MA[S(+)]	5-ME(S)	4.72	4.26	0.46	4.57	0.1478
44	H	O	CPM	5-Me(S)	4.36	4.18	0.18	4.62	-0.2645
45	H	O	$CH_2CH{=}CHMe$(E)	5-Me(S)	4.24	4.24	0	4.10	0.1391
46	H	O	$CH_2CH{=}CHMe$(Z)	5-Me(S)	4.46	4.24	0.22	4.10	0.3592
47	H	O	$CH_2CH_2CH_2Me$	5-Me(S)	4.00	4.35	-0.35	4.46	-0.4584
48	H	O	DMA	5-Me(S)	4.90	5.35	-0.45	5.36	-0.4568
49	H	O	$CH_2C(Br){=}CH_2$	5-Me(S)	4.21	4.37	-0.16	4.17	0.0353
50	H	O	$CH_2C(Me){=}CHMe$(E)	5-Me(S)	4.54	4.40	0.14	4.62	-0.0754
51	H	O	DMA[R(+)]	5-Me(S)	4.66	5.35	-0.69	5.36	-0.6982
52	H	O	DMA[S(+)]	5-Me(S)	5.40	5.35	0.05	5.36	0.0430
53	H	O	$CH_2C(C_2H_5){=}CH_2$	5-Me(S)	4.43	4.40	0.03	4.42	0.0140
54	H	O	$CH_2CH{=}CHC_6H_5$(Z)	5-Me(S)	3.91	4.73	-0.82	3.91	-0.0018
55	H	O	$CH_2C(CH{=}CH_2){=}CH_2$	5-Me(S)	4.15	4.30	-0.15	4.11	0.0396
56	8-Cl	S	DMA	H	7.34	7.66	-0.32	7.70	-0.3591
57	9-Cl	S	DMA	H	6.80	6.73	0.07	7.86	-1.0627
58	H	O	2-MA	5,5-di-Me	4.64	4.40	0.24	5.03	-0.3864
59	H	O	2-MA	4-Me	4.50	4.14	0.36	3.66	0.8396
60	9-Cl	S	2-MA	4-Me(S)	6.17	5.91	0.26	6.09	0.0751
61	9-Cl	S	CPM	4-Me(R)	5.66	5.83	-0.17	6.31	-0.6459
62	H	O	C_3H_7	4-$CHMe_2$	4.13	4.51	-0.38	4.40	-0.2690
63	H	O	2-MA	4-$CHMe_2$	4.90	4.62	0.28	4.59	0.3059
64	H	O	2-MA	4-C_3H_7	4.32	4.67	-0.35	4.76	-0.4396
65	H	O	DMA	7-Me	4.92	5.35	-0.43	4.87	0.0487
66	8-Cl	O	DMA	7-Me	6.84	6.64	0.2	5.85	0.9938
67	9-Cl	O	DMA	7-Me	6.80	5.71	1.09	6.25	0.5502
68	H	S	C_3H_7	7-Me	5.61	5.52	0.09	5.58	0.0275
69	H	S	DMA	7-Me	7.11	6.72	0.39	7.08	0.0279
70	8-Cl	S	DMA	7-Me	7.92	7.93	-0.01	7.21	0.7052
71	9-Cl	S	DMA	7-Me	7.64	6.99	0.65	7.65	-0.0092
72	H	O	DMA	4,5-di-Me(cis)	4.25	5.55	-1.3	4.15	0.0992

73	H	S	DMA	4,5-di-Me(cis)	5.65	6.92	-1.27	5.75	-0.1031
74	H	S	CPM	4,5-di-Me(trans)	4.87	5.75	-0.88	4.89	-0.0204
75	H	S	DMA	4,5-di-Me(trans)[e]	4.84	6.92	-2.08	5.75	-0.9141
76	H	S	DMA	5,7-di-Me(trans)	7.38	6.92	0.46	6.16	1.2242
77	H	S	DMA	5,7-di-Me(cis)	5.94	6.92	-0.98	6.16	-0.2197
78	9-Cl	O	DMA	5,7-di-Me(R,R-trans)	6.64	5.91	0.73	6.23	0.4127
79	9-Cl	S	DMA	5,7-di-Me(R,R-trans)	6.32	7.20	-0.88	7.56	-1.2392
80	H	S	DMA	4,7-di-Me(trans)[e]	4.59	6.92	-2.33	5.73	-1.1430
81	9-Cl	O	DMA	5-Me(S)	6.74	5.71	1.03	6.57	0.1704
82	9-Cl	S	CPM	5-Me(S)[e]	7.47	5.83	1.64	6.95	0.5220
83	H	S	CPM	5-ME(S)[e]	7.22	5.55	1.67	6.14	1.0773
84	H	O	C_3H_7	5-Me	4.22	4.15	0.07	4.45	-0.2348
85	H	S	C_3H_7	5-Me	5.78	5.52	0.26	6.05	-0.2745
86	H	O	2-MA	5-Me	4.46	4.26	0.2	4.57	-0.1124
87	H	S	DMA	5-Me	7.01	6.72	0.29	6.88	0.1262
88	H	O	DMA	5-Me(S)	5.48	5.35	0.13	5.36	0.1230
89	H	S	2-MA	5-Me(S)[e]	7.59	5.63	1.96	6.27	1.3217

[a] 3,3-Dimethylallyl. [b] Cyclopropylmethyl. [c] 3,3-Diethylallyl. [d] 2-Methylallyl. [e] Data points not included in deriving equation.

Table 2 Statistical parameters of the models about the TIBO derivatives

	R. Garg et al.' method[1]	This work
R^2	0.743363	0.849292
PRESS	46.5349	27.32721
MEAN	0.523933	0.411693

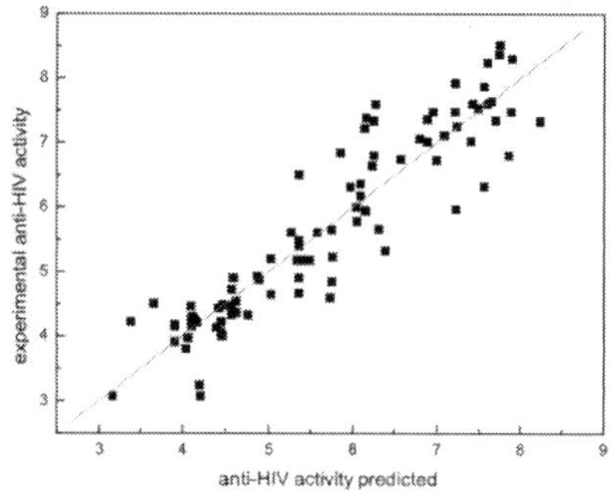

Fig.1 log (1/IC_{50}) activity of TIBO derivatives observed experimentally versus predicted with our PDSAR model

(650e) Understanding the Activation Mechanism of the Insulin Receptor Kinase Domain Using Enhanced Conformational Sampling and Free-Energy Calculations

Harish Vashisth, Chemistry and Biophysics, University of Michigan, Ann Arbor, MI and Cameron F. Abrams, Department of Chemical and Biological Engineering, Drexel University, Philadelphia, PA

Receptor tyrosine kinases (RTKs) are tightly regulated ligand-activated transmembrane glycoproteins that catalyze the phosphorylation of specific tyrosines on protein substrates. Activation of the insulin receptor (IR), an RTK, is dependent upon trans-autophosphorylation of three activation loop (the A-loop) tyrosines located in each cytoplasmic kinase domain of IR (IRKD). Crystal structures of the inactive and active states of bi-lobal IRKD have revealed critical differences in the conformations of the A-loop and the alpha-C-helix in the N-lobe, both of which are significantly displaced on activation. The highly conserved residues Asp1150, Phe1151, and Gly1152 at the N-terminus of the A-loop (the "DFG" motif) collectively "flip" to bury the Phe1151 underneath alpha-C-helix, and simultaneously present Asp1150 for ATP binding. However, the exact mechanism of the conformational change in the A-loop which leads to the "DFG-flip" as well as assembly/disassembly of the regulatory (R) and catalytic (C) spines remains elusive, chiefly due to the unavailability of structural data on intermediate conformations of the A-loop. In this work, we have characterized the inactive-to-active conformational change in the A-loop via a judicious combination of untargeted temperature-accelerated molecular dynamics (TAMD) and free-energy calculations using the string method. TAMD simulations consistently show folding of the A-loop into a helical conformation followed by unfolding to an active conformation, causing the highly conserved DFG-motif to switch from the inactive “D-out/F-in” to the active “D-in/F-out” conformation. The minimum free-energy path (MFEP) computed from the string method confirms existence of these helical intermediates along the inactive-to-active path, and the thermodynamic free-energy differences are consistent with previous work on various kinases. The conformational change in the A-loop also suggests that the R-spine can be dynamically assembled/disassembled both via DFG-flip or the movement of the alpha-C-helix.

(650g) Free Energy Landscape of C Rugosa Lid Closing by Well-Tempered Metadynamics

Patrick R. Burney and Jim Pfaendtner, Chemical Engineering, University of Washington, Seattle, WA

Lipase family enzymes are common components of most organisms' lipid metabolism pathway. In humans and many yeasts and bacteria, lipases hydrolyze triglycerides. Their importance in this role explains why human lipases have been implicated in a variety of health conditions, including heart disease and diabetes. Industrial utilization of lipases, which primarily come from yeasts, has predominantly been for producing food additives and in detergents, yet in the past decade for the production of biodiesel and other sustainable fuels from plant-derived oils, including waste oils from the food industry.

Despite their importance in both technological application and human health, little is known about the dynamics of lipases at the atomic level or the atomic scale features of key molecular events in the action of lipases. Previous experimental investigations have shown a common structural motif among lipases is a lid-like domain that protects the active site from the solvent until in the enzyme contacts a solvent-lipid interface. The C. rugosa lipase (CRL) structure has been resolved for both the lid-open and lid-closed states. The two states have nearly identical structure, except for the 26 amino-acid lid, which differs by 17 Å between the open and closed states. Given the importance of this conformational change in regulating lipase activity, the molecular details of the lid mechanism are essential knowledge for engineering improved lipases. However, direct observation of this process is difficult or impossible via experiment alone. Therefore, we have used simulation to investigate the behavior of the CRL lid in explicit water and a number of different solvents that show varying levels of lipase activity. The major challenge of using classical MD in investigating conformational rearrangement and rare events (e.g., lid opening/closing) lies in sufficient sampling. As we have previously demonstrated, even microsecond-long simulations of the all-atom (AA) system cannot provide enough sampling to calculate the equilibrium probability distribution of this process. Building off of our previous AA simulation of CRL behavior in water, ionic liquid and organic solvents, we have applied coarse-graining and other multiscale techniques toward investigating the lid behavior for CRL. We use the AA simulation trajectories in addition to coarse-grained (CG) models to develop a set of collective variables that describe the lid's low energy states. Finally, we use the metadynamics [1] algorithm to calculate the free-energy landscape along these collective variables and identify key structural transitions in the opening and closing process.

[1] A. Laio and M. Parrinello (2002). "Escaping free-energy minima." PNAS 99(20): 12562-12566.

(650h) Molecular Dynamics Study of Polymer Separation Using a Nanofluidic Staircase

Frederick R. Phelan Jr. and Christopher Forrey, Polymers Division, NIST, Gaithersburg, MD

The diffusive behavior of isolated polymer chains in a nanofluidic staircase has recently been studied experimentally [1–5] and by simulation [6]. Chains are observed to exhibit spontaneous 1-D biased diffusion from regions of high to low confinement, without the use of external forces, under conditions where the local confinement lies in either the Odijk or de Gennes regimes. The nanostaircase has potential for a number of applications in polymer measurement science and transport, an important one of which could be separations. In this work, we examine polymer separation in the nanofluidic staircase using the molecular dynamics simulation software LAMMPS [7], [8]. Length based separations of linear polymers as applicable to DNA separations are the main topic of the study, but the effect of more complex architectures such as branching and star-like polymers more common in commodity and specialty polymers are also examined.

References

[1] E. A. Strychalski, S. M. Stavis, M. Gaitan, and L. E. Locascio, "Nanoslinky: DNA Entropophoresis Down A Nanofluidic Staircase," presented at the The 14th International Conference on Miniaturized Systems for Chemistry and Life Sciences, Groningen, The Netherlands, 2010, pp. 2071–2073.

[2] S. M. Stavis, J. Geist, M. Gaitan, L. E. Locascio, and E. A. Strychalski, "DNA molecules descending a nanofluidic staircase by entropophoresis," *Lab on a Chip*, vol. 12, no. 6, pp. 1174–1182, 2012.

[3] E. A. Strychalski, J. Geist, M. Gaitan, L. E. Locascio, and S. M. Stavis, "Quantitative measurements of the size scaling of linear and circular DNA in nanofluidic slitlike confinement," *Macromolecules*, vol. 45, pp. 1602–1611, 2012.

[4] S. M. Stavis, E. A. Strychalski, and M. Gaitan, "Nanofluidic structures with complex three-dimensional surfaces," *Nanotechnology*, vol. 20, no. 16, p. 165302, Apr. 2009.

[5] S. M. Stavis, J. Geist, and M. Gaitan, "Separation and metrology of nanoparticles by nanofluidic size exclusion," *Lab Chip*, 2010.

[6] Frederick R. Phelan Jr., Christopher Forrey, Lilian Johnson, Samuel M. Stavis, Jon Geist, and Elizabeth A. Strychalski, "The Nanofluidic Staircase: A Brownian Motor for Polymer Characterization and Transport," *in preparation*, p. 2012.

[7] S. Plimpton, "Fast Parallel Algorithms for Short-Range Molecular Dynamics," *Journal of Computational Physics*, vol. 117, no. 1, pp. 1–19, Mar. 1995.

[8] "LAMMPS Molecular Dynamics Simulator." [Online]. Available: http://lammps.sandia.gov/. [Accessed: 02-May-2011].

(656a) Synthesis of CZTS Nanomaterials Via a Continuous Flow Supercritical Carbon Dioxide Process

Michael J. Casciato, Galit Levitin, Dennis W. Hess and Martha Grover, School of Chemical & Biomolecular Engineering, Georgia Institute of Technology, Atlanta, GA

Synthesis of CZTS Nanomaterials Via a Continuous Flow Supercritical Carbon Dioxide Process

Michael J. Casciato*, Galit Levition, Dennis W. Hess, Martha A. Grover

School of Chemical & Biomolecular Engineering, Georgia Institute of Technology, Atlanta, GA 30332 USA

*michael.casciato@chbe.gatech.edu

Direct band gap thin film solar cells have attracted attention over the last 20 years as a potential alternative to indirect band gap silicon-based solar cells[1]. In particular, copper indium gallium selenide ($CuIn_xGa_{1-x}Se_2$, CIGS) thin film solar cells have achieved solar conversion efficiency nearing 20%[2]. However, CIGS thin films require the use of expensive, scarce, and toxic compounds which may inhibit wide scale implementation of CIGS devices[3]. A related chalcogenide material, copper zinc tin sulfide (Cu_2ZnSnS_4, CZTS), has been explored recently because it is composed of relatively less expensive, more abundant, and less toxic materials compared to CIGS. The highest photoconversion efficiency reported using the CZTS material is 10.1%[4], although this device was based on CZT(S,Se), which possesses selenium in addition to CZTS and therefore is less green than CZTS alone.

Sputtering[5], electrodeposition[6], and coevaporation[7] of metals followed by sulfurization and annealing are three widely used techniques to fabricate CZTS. Metal dithiocarbamate precursors have been used in liquid[8] and chemical vapor deposition[9] (CVD) processes to form CZTS nanocrystals and thin films directly, without sulfurization. However, none of these techniques is well-suited for depositing CZTS particles or thin films onto structures that possess high aspect ratios or high tortuosity due to solubility and/or transport limitations. Nonetheless, it may still be desirable to decorate high aspect ratio/tortuosity nanostructures with CZTS particles or thin films. For example, CZTS particles could be anchored on Si nanowires or carbon nanotubes to improve photoelectrochemical cell performance[10]. Thus, a technique that can efficiently cover high aspect ratio/tortuosity nanostructures with CZTS particles or films would be advantageous.

Much work has been performed detailing the deposition of metal thin films and nanoparticles into high aspect ratio/tortuosity structures using supercritical carbon dioxide (sc-CO_2) because sc-CO_2 possesses liquid-like solubility and gas-like diffusivity[11]. Furthermore, sc-CO_2 is a green, sustainable processing technique, employing recycled CO_2 as a solvent for precursors.

In this work, a sc-CO_2 continuous flow reactor (CFR) is employed to deposit CZTS micro- and nanoparticles on a silicon wafer from metal dithiocarbamate precursors. This work demonstrates that the sustainable sc-CO_2 CFR process is a viable technique for fabricating CZTS particles, establishing the potential for future studies to deposit CZTS in high aspect ratio/tortuosity nanostructures. Moreover, no postprocessing was required to form the kesterite CZTS phase, although annealing at high temperature and selenization may still be necessary to yield an efficient solar device, as mentioned above[4].

References:

[1] A. Shah, P. Torres, R. Tscharner, N. Wyrsch, H. Keppner, *Science* **1999**, *285*, 692.

[2] K. Ramanathan, M. A. Contreras, C. L. Perkins, S. Asher, F. S. Hasoon, J. Keane, D. Young, M. Romero, W. Metzger, R. Noufi, J. Ward, A. Duda, *Progress in Photovoltaics: Research and Applications* **2003**, *11*, 225.

[3] F. Kessler, D. Rudmann, *Solar Energy* **2004**, *77*, 685.

[4] D. A. R. Barkhouse, O. Gunawan, T. Gokmen, T. K. Todorov, D. B. Mitzi, *Progress in Photovoltaics: Research and Applications* **2012**, *20*, 6.

[5] J.-S. Seol, S.-Y. Lee, J.-C. Lee, H.-D. Nam, K.-H. Kim, *Solar Energy Materials and Solar Cells* **2003**, *75*, 155.

[6] J. J. Scragg, P. J. Dale, L. M. Peter, *Thin Solid Films* **2009**, *517*, 2481.

[7] T. Tanaka, D. Kawasaki, M. Nishio, Q. Guo, H. Ogawa, *physica status solidi (c)* **2006**, *3*, 2844.

[8] A. Khare, A. W. Wills, L. M. Ammerman, D. J. Norris, E. S. Aydil, *Chemical Communications* **2011**, *47*.

[9] K. Ramasamy, M. A. Malik, P. O'Brien, *Chemical Science* **2011**, *2*.

[10] a)P. K. Sarswat, M. L. Free, *physica status solidi (a)* **2011**, *208*, 2861; b)K.-Q. Peng, X. Wang, X.-L. Wu, S.-T. Lee, *Nano Letters* **2009**, *9*, 3704; c)L. Sheeney-Haj-Ichia, B. Basnar, I. Willner, *Angewandte Chemie International Edition* **2005**, *44*, 78.

(656b) Coupling Homogeneous Reactions with Heterogeneous Separations Toward Sustainable Production of Pharmaceuticals

Wilmarie Medina-Ramos[1], Wesley H. Woodham[1], William Heaner[2], Ali Z. Fadhel[1], Steven R. Saunders[1], Elizabeth J. Biddinger[1], Aman A. Desai[3], Pamela Pollet[2], Charles A. Eckert[1] and Charles L. Liotta[2], (1)School of Chemical and Biomolecular Engineering, Georgia Institute of Technology, Atlanta, GA, (2)School of Chemistry and Biochemistry, Georgia Institute of Technology, Atlanta, GA, (3)Process Chemistry and Development, The Dow Chemical Company, Midland, MI

Designing changes in phase equilibria enables the facile and efficient recovery of homogeneous catalysts and their subsequent recycle. Organic-Aqueous Tunable Solvents (OATS) can be used to couple homogeneous reactions (fast rates and high selectivity) with heterogeneous separations. Upon reaction completion, the separation is simply conducted by applying CO_2 pressure to manipulate the phase behavior of the initial monophasic organic−water mixtures. As a result, a phase split can be induced resulting in two phases: one organic-rich and one aqueous-rich phase. The products are separated at this stage (organic-rich phase) from the catalysts-containing phase which can now be recycled. Several examples will be presented and discussed, including the Rhodium-catalyzed hydroformylation of model compounds applicable to the synthesis of ibuprofen. We will demonstrate that very modest CO_2 pressures (around 3 MPa) enable the recovery of up to 99% of the product and 99.9 % of the relatively hydrophobic catalyst. The presentation will be aimed at demonstrating the breath of OATS technology to develop alternative strategies toward the sustainable production of pharmaceuticals.

(656c) Designing Hydrofluoroalkane- and Hydrofluoroolefine-Philic Moieties for Medical Aerosols

Lin Yang, Department of Chemical Engineering and Materials Science, Wayne State University, Detroit, MI and Sandro R.P. da Rocha, Wayne State University, Detroit, MI

Hydrofluoroalkanes (HFAs) and Hydrofluoroolefins (HFOs) are widely used propellants in a variety of applications, including as medical aerosols in pressurized metered-dose inhalers, nasal sprays, topical sprays and medical foams. A common problem associated with formulating therapeutic molecules in HFAs and HFOs medical aerosols is the fact that many therapeutics have low solubility in these propellants. The need then arises to design moieties that are capable of imparting stability to formulations, which is also a challenge given the somewhat hydrophobic and lipophobic nature of these semi-fluorinated solvents.

In this work *ab initio* calculations are used to guide the design of HFA- and HFO-philic chemistries. We investigate three propellants, namely HFA134a and HFA227, which are approved for use in metered-dose inhalers, and HFO-1234ze, whose potential in medical sprays in being currently evaluated – currently in use as refrigerants. We determine the binding energy between the propellants and fragments of relevant chemistries including alkanes, ethers and esters. The relative binding strength, that between propellant-fragment to that of fragment-fragment, which may serve as a better predictor of solubility and solvation of those chemistries is also discussed. The effect of degree of branching of those chemistries are also studied, by varying pendant methyl group within each category. The results show that ether and ester groups interact more favorably with HFAs/HFOs than their non-oxygenated analogs by participating in stronger bonds with the dipole of the propellant molecules. Adding a pendant methyl group can decrease the self-interaction energy between fragments, thus indirectly enhancing HFA/HFO-philicity.

(656d) Essential Oil with Therapeutic Potential: Solubility in Compressed CO2

Odell L. Glenn Jr. and Michael A Matthews, Chemical Engineering, University of South Carolina, Columbia, SC

Previous work have shown that essential oils such as tea tree oil, lavender oil and lemon oil may act as acaricides, antimicrobial agents or antiseptic agents. Liquid and supercritical CO_2 may be a useful vehicle for delivery of these agents in a variety of indoor environments. Tea tree oil (TTO) is the essential oil steam-distilled from *Metaleuca alternifolia,* an Australian native plant. Several *in vitro* studies have investigated its antimicrobial properties and in recent years, TTO has become popular as a naturally occurring antimicrobial and antiseptic agent. Lavender oil is an essential oil obtained by distillation from the flower spikes of certain species of lavender. Like all essential oils, it is not a pure compound but a mixture of naturally occurring phytochemicals, including linalool and linyl acetate. Lemon oil is an essential oil extracted from lemon peels. Lemon oil is used in many personal care products, home cleaning products and furniture polishes. Constituents of lemon oil include limonene, which is responsible for the aroma and citral, for what we recognize as the lemon note in the aroma profile. We report the solubility of tea tree oil, lavender oil and lemon oil in compressed CO_2 at several temperatures and pressures. The Peng-Robinson equation has been used to model the phase behavior of each binary system.

(656e) Supercritical Fluid Extraction of Acai Berries

Rosanna Ayala, University of South Florida, Tampa, FL and Aydin K Sunol, Chemical & Biomedical Engineering, University of South Florida, Tampa, FL

Isolation of nutrient and anti-oxidants from acai berries were successfully performed using supercritical carbon dioxide and carbon dioxide ethanol mixtures. Two different semi batch pilot plants were utilized; one for moderate pressures (up to 200bar), while the other one was for ultra-high pressures (up to 600 bar). The extract yields were as high as 40% and compared well with those of soxhlet extraction. The products were analyzed using a portfolio of techniques including HPLC and FTIR. Furthermore, solubility of model components of the berries, nutrients and anti-oxidants, in supercritical carbon dioxide and carbon dioxide ethanol mixtures were studied using a variable volume synthetic phase analysis experimental set-up. The modeling of phase behavior of the system was performed using Peng Robinson equation of state with Wong Sandler mixing rules to regress the binary interaction parameters from the experimental cloud point data. The utilization of the extraction residue as biomass feedstock and fermentation to wines were also explored.

(663a) Molecular Simulation Studies Relating Thermodynamics of Polycation-DNA Binding to Non-Viral DNA Delivery

Arthi Jayaraman, Department of Chemical and Biological Engineering, University of Colorado, Boulder, Boulder, CO and Robert Elder, Chemical and Biological Engineering, University of Colorado at Boulder, Boulder, CO

In this talk I will present our recent work involving computational design of polymers as carriers for gene delivery. Viruses are extremely effective as gene delivery vectors but elicit dangerous immunogenic responses. Non-viral gene delivery agents, like polycations, while not as effective as viral vectors have a significant advantage of being non-immunogenic. Polycations have emerged as promising non-viral delivery agents due to their propensity to bind the polyanionic DNA backbone, neutralizing the negative charge of DNA backbone, and facilitating gene delivery. Recent experiments by our collaborators T. Emrick and coworkers [1] has shown that the architecture of poly-lysine affects, in a non-trivial manner , the efficiency of these polycations as transfection agents. Our work using molecular simulations, with both atomistic [2] and coarse-grained models, connects poly-L-lysine architecture to DNA binding thermodynamics, and explains the molecular reasons behind the experimentally observed trends in DNA delivery.

1. S.S. Parelkar, D. Chan-Seng, T. Emrick Reconfiguring polylysine architectures for controlling polyplex binding and non-viral transfection *Biomaterials* 32 (9) pp 2432-2444 (2011)

2. R. Elder, T. Emrick, and A. Jayaraman,'Understanding the effect of polylysine architecture on DNA binding using molecular dynamics simulations' *Biomacromolecules* 12 (11), pp 3870-3879 (2011)

(663b) Order-Disorder Transitions in Strongly-Segregated Block Copolymers

Sangwoo Lee, Timothy M. Gillard and Frank S. Bates, Department of Chemical Engineering and Materials Science, University of Minnesota, Minneapolis, MN

Leibler's seminal work, the mean field treatment on the block copolymer thermodynamics near the order-disorder transition has given valuable understandings on this subject, but it assumes an infinitely large degree of polymerization (N) with an infinitesimally small Flory-Huggins interaction parameter (χ). All experimentally accessible block copolymer compounds have N and χ far from these limits. The correction of mean field theory with the fluctuation effect from finite N near ODT has been attempted, but it is still not successful. We have investigated the thermodynamics of poly(1,4-isoprene-*b*-DL-lactide) diblock copolymers near the ODT with small N and relatively large χ using small angle X-ray scattering (SAXS), dynamic mechanical spectroscopy (DMS), and differential scanning calorimetry (DSC) techniques. These studies provide fresh insights into the role of fluctuations on the state of segregation and ordering kinetics in symmetric and asymmetric diblock copolymers, prompting new questions regarding the application of mean-field and fluctuation corrected theories to block copolymer melts with finite molecular weights.

Molecular Simulation of Thermal Responsive Polymers PNIPPAM (poly (N-isopropylacrylamide))

Introduction During the last decades, the developing focus of smart polymers mainly relied on achieving the polymer controllable response to the external stimuli such as PH, temperature, irradiation. Lower critical solution temperature (LCST) polymer is typical one of thermal responsive polymers which undergo swelling to maintain its topological structure. Because LCST polymer had been exploited in many areas like drug delivery systems, tissue engineering, and epicardial restraint therapies, Significance in understanding, describing and predicting the attractive stimuli-response properties of LCST polymers to engineer the conformational trigger should been shed a light on. PNIPPAM (poly-isopropylacrylamide) is one of the most extensively studied LCST polymers with a negative thermal response and has drawn great attention as a drug-delivery polymer. PNIPPAM exist as coil structure at room temperature, but undergoes a discontinuous, dramatic phase transition to a collapsed state above LCST (~305K)[1]. Substantial experimental work using tools like FT-IR spectroscopy [2] and differential scanning calorimeter [3] has been done to examine PNIPPAM's counter-intuitive coil-to globule transition and underlying thermodynamic mechanisms. This transition is generally considered to be the competitive results of hydrophobic interaction of pendent isopropyl group as well as hydrogen bonding association between amide groups with water. However specific mechanisms leading PNIPPAM's collapse is still under debated. One of the promising paths to obtain adequate molecular details is employing Molecular Dynamic (MD) simulation. Previous MD simulation studies [4, 5] had demonstrated that structure change of PNIPPM from stretched chain to a collapsed one is able to be observed by using OPLS+SPC/E force fields. Nevertheless, molecular level mechanism involved in this structure transition especially hydrophobic effect of solute and its thermodynamic explanation is still poorly understood. We performed a more in-depth investigation to probe the nature of PNIPPAM's counter-intuitive collapse behavior by employing replica exchange simulation[6] to ultimately determine the global polymer conformation and multiple data analysis approaches to evaluate solvent intermediated interaction between constituent hydrophobic groups.

Simulation Methodology A Series of REMD simulations related with PNIPPAM homopolymer were conducted under isothermal-isobaric ensemble using GROMACS 4.0.7[7]. Various solutes included propane, Polyacrylamides and PNIPPAMs were solvated in water box consisting varying number (1000 to 3113) water molecules. SPC/E[8] water model and OPLS-AA[9] force field had been adopted to describe solvent and solute respectively. The Parrinello-Raham barostat was used to maintain pressure at 1 bar and Nose-hoover thermostat was used to maintain temperature for each replicas. A 1fs time step was used to integrate equation of motion. Particle mesh Ewald method was used to calculate long-range electrostatic interaction and L-J interaction was evaluated out of 1.4nm. LINCS algorithm was used to constrain the geometry of water and fix carbon bond length. Temperatures of 15 replicas were assigned using an exponential distribution over a range from 280K to 357K. Replica Exchange interval was set 0.25ps (250 steps). Before REMD simulation, each replica was equilibrated for 5ns. An implemented REMD program in GROMACS package was used to perform 50ns REMD simulation for various solutes.

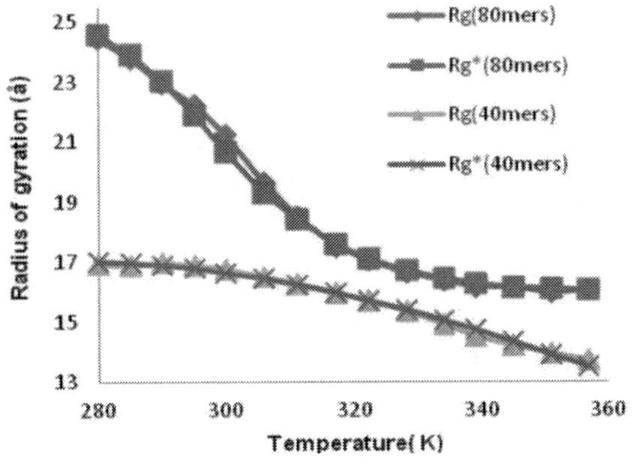

Figure 1. R_g of PNIPPAMs with various lengths as function of temperature. R_g^* is the fitted value according the two-state model's prediction

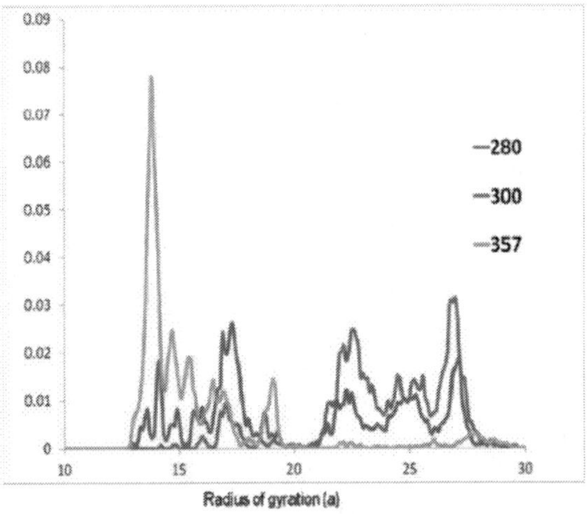

Figure 2. Probability distribution of radius of gyration for 80mers PNIPPAM under 280K,300K and 357K

Results and Discussion Using simulation technique mentioned above, radius of gyration R_g was calculated from simulation results to characterize the structure of PNIPPAMs. Fig 1 presents 80 mers and 40 mers PNIPPAM's results which are concluded as the average R_g for each replica. R_g of 80mers PNIPPAM displayed a first-order phase transition which had reached a good agreement with previous experimental observation. For 40mers PNIPPAM, though the tendency of first-order phase transition is not as obvious as longer chain temperature induced collapse still undergoes. The probability distribution of R_g (Fig 2) was also analyzed in terms of various selected temperatures and indicated that PNIPPAM exhibited a two-state behavior during our simulation which means its conformation preferred to be either at extended state or collapse state but unlikely existed in a transitional state. A two-state model which usually described protein folding hereby was adopted to extract thermodynamic driving forces for chain collapse. In the derivation of model, the equilibrium constant of transition between globule and coil state K_{eq} is defined as ratio of P_C the probability of coil state and P_G the probability of globule state and the free energy involved in the transition process can be calculated based on following equation: $\Delta G = RT \times \ln K_{eq}$. In

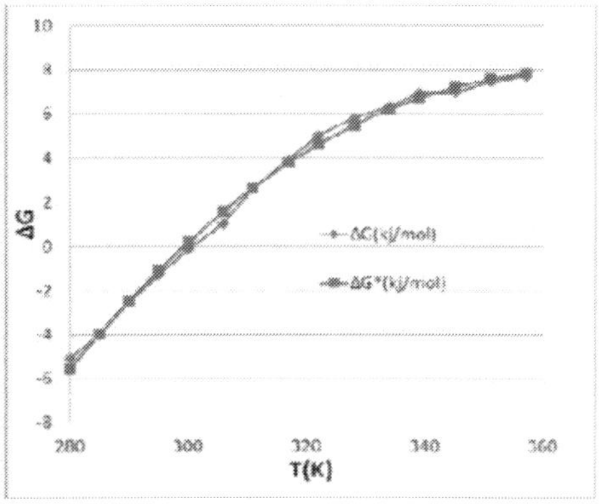

Figure 3. Free energy (80mers PNIPPAM) as function of temperature, blue line is summarized from the simulation data while the red line is from thermodynamic empirical equation fitting.

order to describe the relationship between free energy and temperature, we can introduce a thermodynamic empirical equation to fit results derived from simulation by using least square fitting method,

$$\Delta G^* = a + b(T - T_0) + cT\ln\frac{T}{T_0}$$

As results showing in figure 3, values obtained from equation fitted well with simulation data with T_0 set as 300K. Increasing temperature lead free energy change from negative to positive which means a conversion between extended states to collapse state. By using fitted data, A predicted R_g value (R_g^* in Figure 1) which reflected transition straightforwardly can also be obtained through reversely derivation of two-state model under the assumption that R_g^2 is in a linear relation with P_C.

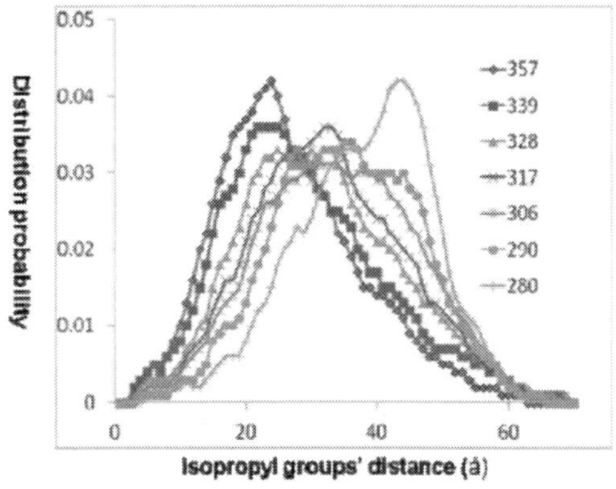

Figure 4. Probability distribution of isopropyl groups' distance within 40 monomers length under various temperatures' conditions

One of the known factors promoting free energy is entropy increase due to increasing temperature resulting in a reduction of the total number of water molecule structured around the hydrophobic solutes. Therefore the hydrophobic interaction arisen by isopropyl pending groups of PNIPPAM take an important role in understanding theoretical background of its collapse which inspire our multiple comparative simulations and their related analysis. First analysis we conducted is to study the isopropyl groups' distance within various length of monomers. In figure 4, Statistical results of the distance's distribution of isopropyl group within a 40mers' length manifested the hydrophobic groups became approach within closer distance with temperature increasing, which can also be interpreted as a phenomenon of stronger interaction. To evaluate collapse behavior's dependence of hydrophobic groups, REMD simulation was performed for polyacrylamide which is the similar polymer as PNIPPAM but substituting isopropyl group with hydrogen. Analysis of average R_g of 40mers and 80mers polymers indicated there was no occurrence of obvious thermal-responsive collapse as PNIPPAM's. To take a quantitatively evaluation of hydrophobic groups' response with temperature, another REMD simulation of pure propane solvating with water as long as its analysis of radial distribution function to compare with PNIPPAM's results was undertook. As shown in Figure 7, even though the scale of g(r) of PNIPPAM's alpha carbon on isopropyl group is higher than alpha carbon of propane due to larger amount of PNIPPAM's isopropyl group and stronger restriction of their spatial coordination, the quantitative ratios of first g(r)'s peaking value at 357K and one at 280K are 1.6 for both two simulations. Results elucidated that PNIPPAM's isopropyl group behave more interacted with each other in a similar way as known propane aggregation phenomenon due to temperature strengthening hydrophobic interaction.

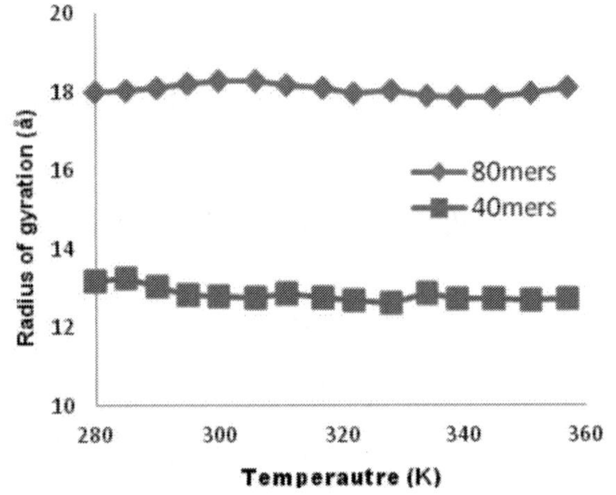

Figure 5. Average R_g of 80mers and 40mers polyacrylamide as a function of temperature

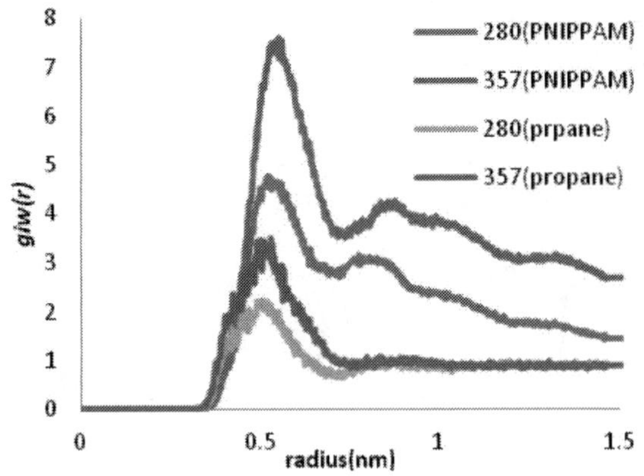

Figure 6. Radius distribution function for alpha-carbons of isopropyl group of PNIPPAM (red & blue) and propane (purple & green) respectively at highest and lowest temperature

Conclusion Our simulation work for PNIPPAM polymers has reproduced the thermal induced first-order phase transition over a broad range of temperature. A two-state model was applied to fit our simulation data and extract the thermodynamic driving forces for PNIPPAM's collapse behavior. Analysis of distance distribution of isopropyl groups exhibited that the hydrophobic isopropyl groups' thermal-responsive interaction grew stronger as temperature increase. On the other hand R_g results from REMD simulation of polyacrylamide doesn't show obvious collapse behavior as PNIPPAM's simulation. Further comparative analysis of radial distribution function for propane groups' carbon atoms displayed consistent thermal responsive change in magnitude between propane solution and PNIPPAM. In all, we suggest hydrophobic group of PNIPPAM make the main contribution for PNIPPAM's thermal-responsive collapse behavior.

Reference

1. Wu, C.; Wang, X. H., *Physical Review Letters* **1998,** *80* (18), 4092-4094.
2. Sun, S. T.; Hu, J.; Tang, H.; Wu, P. Y., *Journal of Physical Chemistry B* **2010,** *114* (30), 9761-9770.
3. Otake, K.; Inomata, H.; Konno, M.; Saito, S., *Macromolecules* **1990,** *23* (1), 283-289.
4. Walter, J.; Ermatchkov, V.; Vrabec, J.; Hasse, H., *Fluid Phase Equilibria* **2010,** *296* (2), 164-172.
5. Algaer, E. A.; van der Vegt, N. F. A., *Journal of Physical Chemistry B* **2011,** *115* (46), 13781-13787.
6. Earl, D. J.; Deem, M. W., *Physical Chemistry Chemical Physics* **2005,** *7* (23), 3910-3916.
7. Hess, B.; Kutzner, C.; van der Spoel, D.; Lindahl, E., *Journal of Chemical Theory and Computation* **2008,** *4* (3), 435-447.
8. Berendsen, H. J. C.; Grigera, J. R.; Straatsma, T. P., *Journal of Physical Chemistry* **1987,** *91* (24), 6269-6271.
9. Jorgensen, W. L.; Maxwell, D. S.; TiradoRives, J., *Journal of the American Chemical Society* **1996,** *118* (45), 11225-11236.

(663d) A Thermodynamic Model for Hydrogels

Markus C. Arndt, Department of Biochemical and Chemical Engineering, Laboratory of Thermodynamics, Technische Universität Dortmund, Dortmund, Germany and Gabriele Sadowski, Department of Biochemical and Chemical Engineering, Laboratory of Thermodynamics, Technische Universität Dortmund, Dortmund, Germany

Hydrogels are strongly hydrophilic polymers which are capable of storing immense amounts of water and other hydrophilic solvents. Next to established applications such as drying agents, recent research focuses on unique properties of so-called smart polymers for novel utilisations. Prominent representative of these are hydrogel networks based on poly(N-isopropylacrylamide) PNIPAAm, a biocompatible polymer with high sensitivity towards external stimuli. Based on the thermodynamic phenomenon of a liquid-liquid miscibility gap in the aqueous polymer solution, PNIPAAm hydrogel networks exhibit a characteristic and reversible temperature response: passing a critical temperature of approximately 35 °C (95 °F) the gel drastically switches its degree of swelling from highly swollen to collapsed. Similar swelling transitions are known in response to changes in the surrounding medium composition: the addition of co-solvents or solutes triggers a dramatic change in the degree of swelling. This sensitivity towards physiological substances as well as the thermal instability point near the body temperature make PNIPAAm highly attractive for applications in biological environments; prominent examples are pharmaceutical drug depots, wound moisture regulators or biomechanical sensors and actuators.

It is clear, that for such applications the understanding and modelling of the phase behaviour are essential factors, and a physical explanation of the swelling transition phenomena is required from the point of view of the molecular level. In our approach of a thermodynamic description of such hydrogel systems, we use the Perturbed-Chain Statistical Associating Fluid Theory PC-SAFT for modelling the physical attractive forces between the substances of the system to obtain the Helmholtz energy as significant thermodynamic quantity, and combine the benefits of this general equation of state with an additional elastic contribution for the Helmholtz energy. The distinct consideration of the elastic energy in the polymer network attributes to the stretching of the cross-linked polymer chains, when invading solvent molecules enforce a deformation of the network. Secondary to the elastic energy, an elastic excess pressure is induced in the gel phase depending on the degree of swelling. With an explicit calculation of this pressure difference, we are able to precisely solve modified isofugacity criteria for the equilibrium calculations between the phases of solution and gel. This way considering molecular interactions and elastic forces within the PC-SAFT equation of state, not only fundamental thermodynamic properties of the system can be modelled but also the degree of swelling of hydrogels and the exact gel composition, which is the crucial factor aiming at the applicability of hydrogels. In the present work hydrogel equilibria have been calculated successfully in good accordance with experimental data and a significant predictive capability of our model is proven, while varying both the temperature over a wide range and as well the composition of the surrounding aqueous solution phase, taking binary and ternary systems with PNIPAAm, water, several alcohols (methanol, ethanol, n-propanol, i-propanol), salts ($NaCl$, $NaNO_3$, $(NH_4)_2SO_4$), other organic compounds (acetone) and even oligomers into account (poly(ethylene glycol)), approaching biologically relevant conditions.

Solution Behavior at High Pressures: Linear to Star Polymers in SCF and Liquid Solvents

Yue Wu[1,], Matthew S. Newkirk[1], Mark A. McHugh[1], Val Krukonis[2], Kara Williams[2]*

[1] *Virginia Commonwealth University, Department of Chemical and Life Science Engineering, Richmond, Virginia, USA 23220*
[2] *Phasex Corporation, Lawrence, Massachusetts, USA 01843*
[*] *Corresponding author: wuy@vcu.edu; Phone: (+1) 804 827-7000, ext: 460*

Abstract

We describe the techniques used to determine the phase behavior and density behavior of linear and star polymers in supercritical fluid (SCF) solvents to 150°C and 25,000 psia. Attention is given to the impact of the amount and length of the arms, the functional groups on the arm, and the polymer molecular weight on the solution behavior at high pressures. The polymers are fractionated using SCF technology to obtain samples used to elucidate the impact of polymer architecture relative to molecular weight and molecular weight distribution. The resultant phase behavior and density data are modeled using the perturbed chain statistical associating fluid theory (PC-SAFT), as described in the talk.

Introduction

Star polymers are essentially globular, they do not exhibit chain entanglements, they possess a large number of functional groups within a single molecule, and the functional groups can be readily modified. They can be synthesized via one-step or two-step reactions, making them promising for industrial applications. Star polymers are used as fuel additives for lubrication in the oil industry [1, 2], as drug encapsulates for controlled release [3, 4], as selective solvents for extraction, in extractive distillation, and in absorption processes [5], etc. Many fundamental properties of star polymers are still lacking, such as their solubility behavior in different solvents and even pure component PVT properties, especially at high pressures. The high-pressure solution behavior is affected by the amount and length of the arms in the star polymer, the type of functional groups on an arm, and, to a lesser degree, the molecular weight of the star polymer [6-8]. Further, high-pressure density data are not available for most star polymers. Fundamental PVT data are needed to estimate equation of state parameters [9]. This study reports star polymer-propane phase behavior and density data for star polymers with a divinylbenzene (DVB) core and acrylate copolymer arms. The star polymers are fractionated using supercritical fluid propane and CO_2 to purify and narrow the molecular weight distribution of the polymer samples. The phase behavior of the fractionated star polymers in propane is measured and is compared with that of the parent star polymer and the acrylate copolymer arms in the same solvent. The resultant star polymer-solvent solution behavior is modeled using the PC-SAFT equation of state (EOS).

Materials and Methods

Materials

The star polymer is synthesized with reversible addition-fragmentation-transfer (RAFT) process, shown in Figure 1. Figure 1 also shows the structure of star polymers with a divinylbenzene (DVB) core and arms, which are statistically random methacrylate copolymers consisting of 50 mol% methyl (C1) and 50 mol% lauryl (C12) groups, as shown in Figure 2.

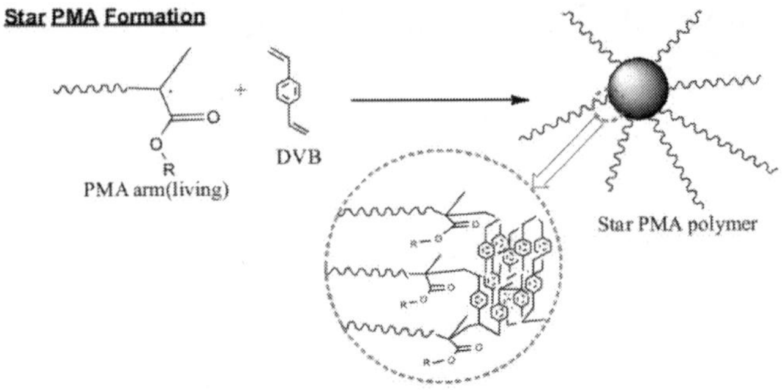

Figure 1. The synthesis process and structure of the star polymer.

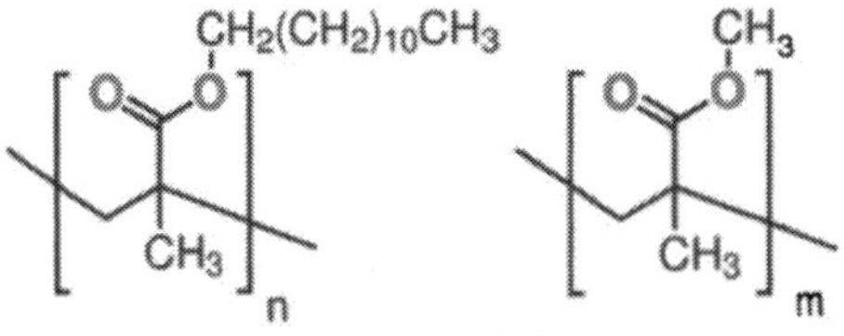

(A) (B)

Figure 2. The structure of (A) poly (lauryl methacrylate) and (B) poly (methyl methacrylate) repeat groups copolymerized to create an "arm".

Fractionation

The parent star polymer is initially synthesized in an "oil" comprised of 95+ vol% saturated hydrocarbons and ~ 5 vol% saturated cyclic hydrocarbons. Figure 3 shows a schematic diagram of the apparatus used in this study. Typically 10 to 20 grams of the polymer-oil mixture are loaded into a column packed with stainless steel Goodloe packing to ensure distributed flow of the SCF solvent through the column [10]. The fluid of interest is delivered to a diaphragm compressor where it is pressurized and charged to a surge tank maintained at 10,000 psia. The fluid is then throttled through an in-line pressure reducing regulator, delivered to a preheat column, and then to the column packed with the star polymer. The pressure in the packed column is controlled to within ± 60 psia using the regulator and the temperature of the packed column is maintained to within ± 1.0°C. At each pressure level the fluid extracts a portion of the parent material. The loaded SCF fluid flows through a valve causing the polymer to precipitate into a preweighed beaker or U-tube submerged in an ice-water bath. The gas is routed to a dry-test meter to monitor the total volume passed through the extractor. Table 1 shows the results of one of the fractionation experiments. This study utilizes various fractions from these fractionation studies to determine phase and density behavior in supercritical propane.

Phase Behavior Determination

The high-pressure phase behavior and density measurements are performed using a variable-volume cell shown in Figure 4. The cell is constructed of a high nickel content steel (Nitronic 50) with a working volume of ~24 ml. A sapphire window is fitted to one end of the cell and is sealed with an elastomeric O-ring and a backup ring. The system pressure is measured on the water side of the piston with a pressure transducer calibrated against a Heise pressure gauge. The type-k thermocouple used to measure the temperature of the fluid in the view cell is calibrated against a precision immersion thermometer (Fisher Scientific, 35 to 200°C, precise to 0.10°C, and recalibrated by ThermoFisher Scientific Company at four different temperatures using methods traceable to NIST standards). Polymer, to within ±0.0001 grams, is loaded into the cell followed by the gas of interest to obtain a ~5 wt% polymer-SCF solution. The cell contents are compressed to the desired operating pressure by an internal piston displaced with water pressurized by a pressure generator. The mixture in the cell is first isothermally compressed to a single phase and then the pressure is slowly decreased until the solution becomes cloudy as recorded by a light sensor that measures the amount of visible light traveling from one side port to the sensor in an opposing side port. The phase-transition pressure is the average of the pressure when light intensity just begins to decrease and when the light intensity reaches a minimum.

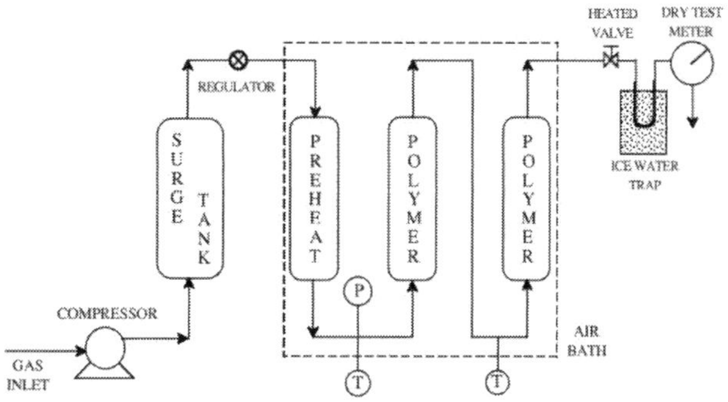

Figure 3. Schematic diagram of the high-pressure fractionation system.

Table 1. Typical results from the SCF fraction of star polymer.

Fraction	Weight collected, g	Observations
Charge	12.51	Operating at ~130°C
F-1	2.68	Oil extracted with CO_2 at ~5,000 psig
F-2	0.55	More oil extracted with Propane at 800 psig
F-3	0.89	3,000 psig Propane
F-4	1.65	4,400 psig Propane
F-5	3.03	5,300 psig Propane
F-6	2.34	6,700 psig Propane

Density Measurements

High-pressure density measurements are obtained in the single-phase region for the polymer-SCF solvent solutions. The internal volume of the cell is determined using a linear variable differential transformer (LVDT, Schaevitz Corp., Model 2000 HR) that tracks the location of the internal floating piston as shown in Figure 4B. Connected to the piston is a rod with a magnetic end piece, called a core, and as the piston moves the magnetic core travels through the LVDT located outside the high-temperature air bath. The view cell volume is calibrated using n-decane at 150°C [11]. With the known total mass and the cell volume, the density can be calculated at a specific pressure and temperature. The expanded uncertainties in the reported densities (k = 2) are estimated to be 0.60% of the reported density.

Results and Discussion

Figure 5A shows the gel permeation chromatography (GPC) results of the parent polymer. The parent polymer has a peak M_w ~19,000, a "free arm" peak at M_w ~34,800, and a low M_w oil peak. "Free arm" refers to the arms not attached to the polymer. High-pressure fractionation removes the free arms and the oil from the parent star polymer. Figures 5B through 5D show the GPC traces for fractions F-1, F-2, F-5, and F-6. F-1 and F-2 show the oil extracted by CO_2 and propane. F-5 and F-6 show star polymers extracted at 5,300 and 6,700 psig, respectively. Compared with Figure 5A, the GPC trace of the parent polymer, F-5 and F-6 are star polymers without oil and low molecular weight unreacted arms.

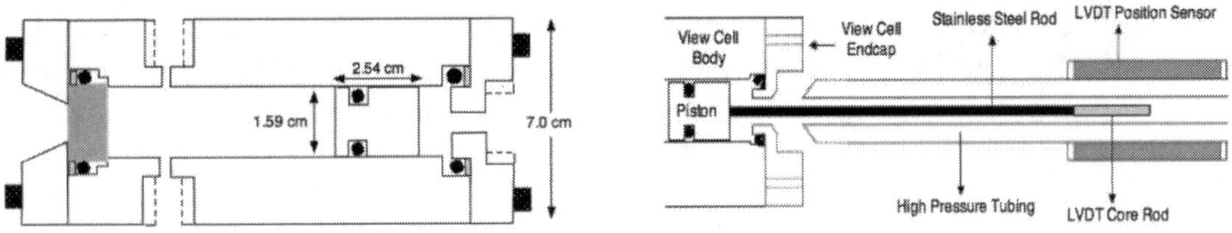

(A) (B)

Figure 4. A) Schematic diagram of the high-pressure, high-temperature variable-volume view cell used in this study. B) Schematic diagram of the rod connecting the piston to the LVDT.

The temperatures and pressures for effective fractionation are based on the conditions needed to dissolve the neat oil solvent and the star polymer in the fractionation solvent. Initial experiments show that the 5 wt% star polymer - oil solvent mixture does not dissolve in CO_2 even at 150°C and 25,000 psia, although the mixture can be dissolved in propane at ~5,000 psia within the above temperature range. Figure 6 shows the phase behavior of ~5 wt% oil in CO_2 and in propane, respectively. The pressure-temperature (P-T) traces of the Fluid → Liquid + Liquid boundaries have a shape observed for other CO_2-saturated hydrocarbon and propane-saturated hydrocarbon mixtures [12]. The oil easily dissolves in propane and the dissolution pressure increase with an increase in temperature. In contrast, the oil-CO_2 phase transition pressure increases dramatically at temperatures below 50°C. Since CO_2 dissolves the oil at ~4,000 psig and temperatures above 50°C, the fractionation is performed with CO_2 at ~5,000 psig and 130°C to extract the oil from the polymer-oil mixture and then propane is used at pressures from 800 to 6,700 psia to remove any residual oil, extract any free unattached arms, and fractionate the parent star polymer at each pressure level as already shown in Table 1.

Figure 7 shows the phase equilibrium data of ~5 wt% homogeneous poly (lauryl methacrylate) (PLMA), and PLMA-PMMA copolymer with 35 wt% PMMA (LMA-MMA_{35}) in propane to investigate the phase behavior of just the free arms in propane. PLMA can be dissolved in propane below 3000 psia while the phase transition pressure of the copolymer is above 15,000 psia. Given the molecular weight of homogeneous PLMA and the PLMA-PMMA copolymer are almost the same, ~90,000, the phase transition pressure increase from PLMA to the copolymer should be due to the addition of more methyl methacrylate (MMA) group. The results are not surprising, because MMA is more polar than lauryl methacrylate (LMA) and it is harder to dissolve a more polar group in the nonpolar propane. This argument can be also verified by the fact that poly (methyl methacrylate) (PMMA) cannot be dissolved in propane at the temperatures up to 200°C and pressures up to 30,000 psia.

Figure 8 shows the phase behavior of the two fractionated star polymers, F-5 and F-6, in propane. It is not surprising that the polymer, F-6, with higher molecular weight is dissolved in the supercritical propane at higher pressures. A crystallinity region is found below 70°C for both fractions, which is also verified by DSC measurements, is likely due to the self-polymerization of the DVB core. The most interesting observation is about the architectural effects of star polymers with copolymer arms in comparison of just the copolymer arms. Considering that the star polymers consist of LMA-MMA copolymer arms with 50 mol% MMA, and the Mw of star polymer is larger than the LMA-MMA_{35} copolymer with 35 mol% MMA, it was expected that the phase transition pressure of star polymers should be higher than the copolymer. However, the phase transition pressure of the star polymer is lower, which is likely due to architectural and/or free volume effects.

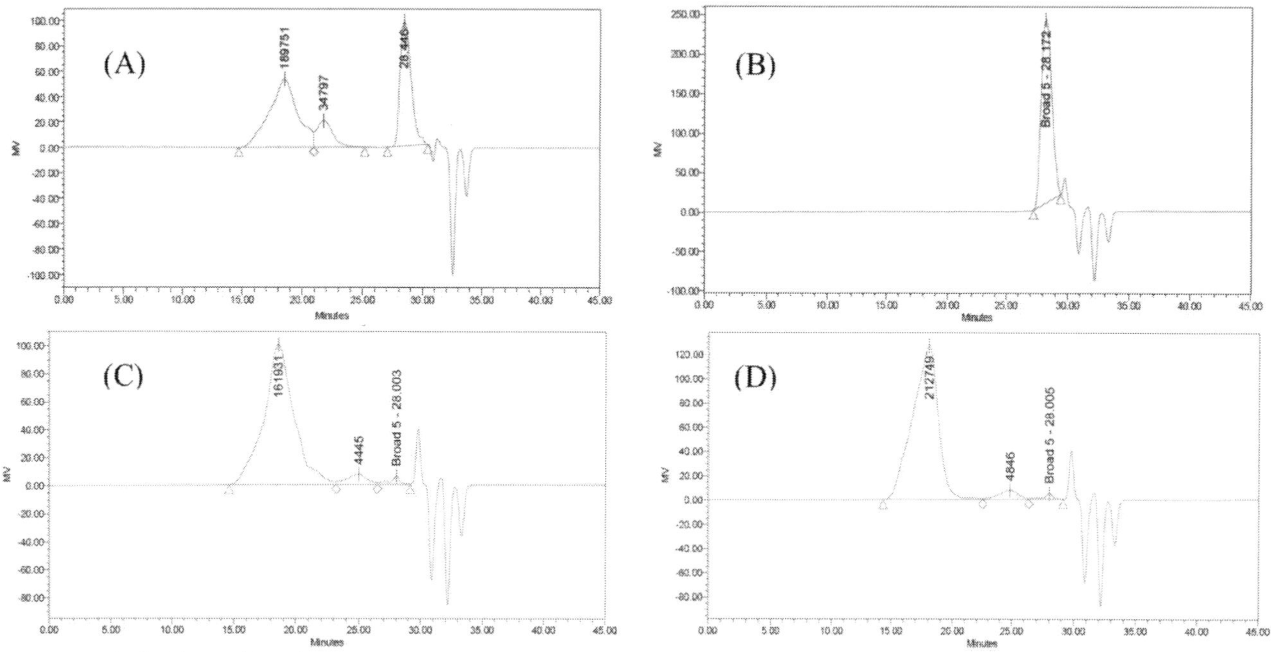

Figure 5. GPC of (A) the parent star polymer, (B) Fraction 1 and 2, (C) Fraction 5, and (D) Fraction 6.

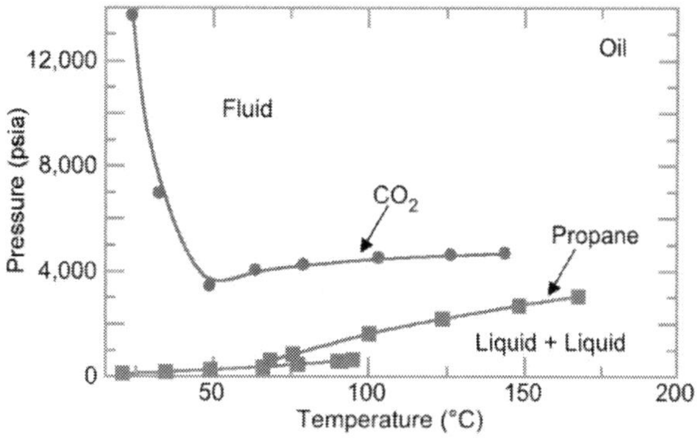

Figure 6. Phase behavior of ~5 wt% oil in CO_2 and propane. The solid lines are used to guide the eye.

Density data for the star polymer solutions provides fundamental PVT data needed for modeling studies. Figure 9 shows the density data of the single-phase region for ~5 wt% F-6 in propane.

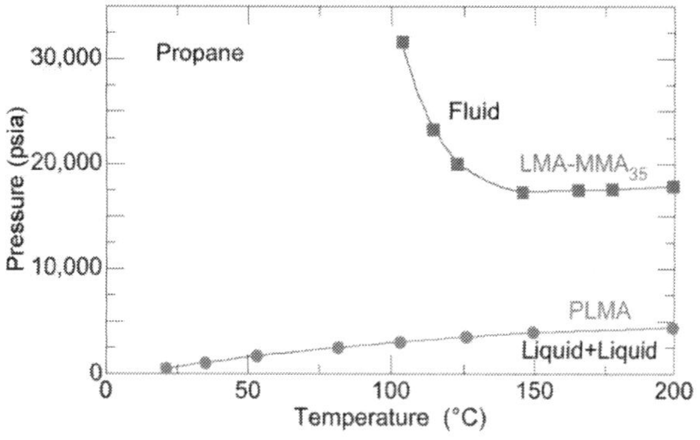

Figure 7. Phase behavior of ~5 wt% poly (lauryl methacrylate) (PLMA), and PLMA-PMMA copolymer with 35 wt% PMMA ($LMA\text{-}MMA_{35}$) in propane

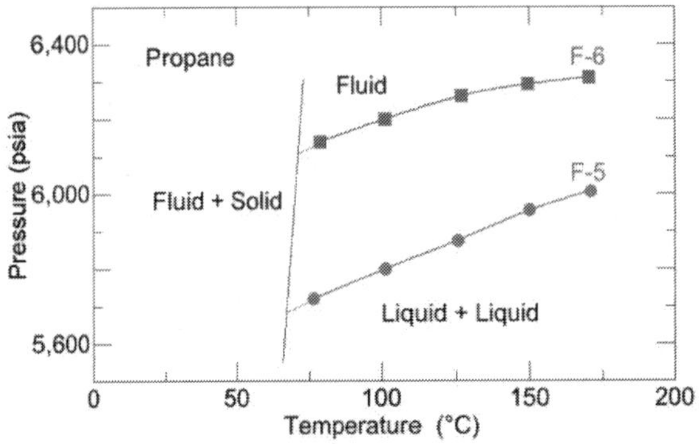

Figure 8. Phase behavior of the two fractionated star polymers, F-5 and F-6, in propane.

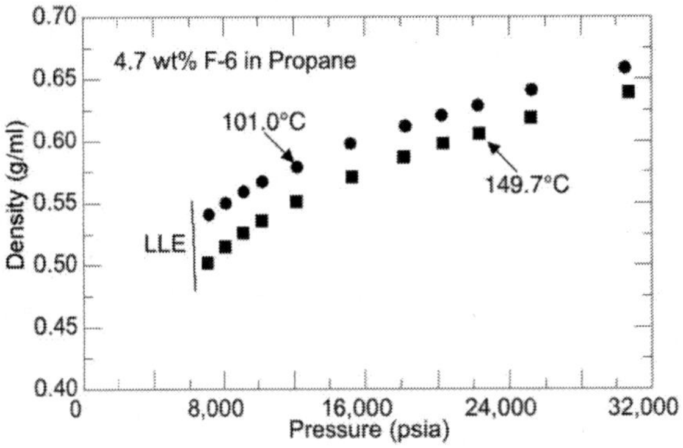

Figure 9. Density of the single-phase region for ~5 wt% F-6 in propane

PC-SAFT Modeling

The PC-SAFT EOS [13] is used to model the phase behavior of each polymer solution in propane. For the non-associating system, the equation is given as a sum of Helmholtz free energies.

$$\tilde{a}^{res} = \tilde{a}^{hc} + \tilde{a}^{disp} \tag{1}$$

where $\tilde{a}^{hc}$ and $\tilde{a}^{disp}$ are the reduced hard chain reference and dispersion terms, respectively. For a multi-component fluid,

$$\tilde{a}^{hc} = m\tilde{a}^{hs} + \tilde{a}^{chain} = m\tilde{a}^{hs} + \sum_i x_i (1 - m_i) \ln\left(g_{ii}^{hs}\right) \tag{2}$$

$$m = \sum_i x_i m_i \tag{3}$$

$$\tilde{a}^{hs} = \frac{1}{\xi_0}\left[\frac{(\xi_2)^3 + 3\xi_1\xi_2\xi_3 - 3\xi_1\xi_2(\xi_3)^2}{\xi_3(1-\xi_3)^2} - \left[\xi_0 - \frac{(\xi_2)^3}{(\xi_3)^2}\right]\ln(1-\xi_3)\right] \tag{4}$$

$$\xi_k = \left(\frac{\pi\rho_{num}}{6}\right)\sum_i x_i m_i (d_i)^k;\ k = 0,1,2,3 \tag{5}$$

$$d_i = \sigma_i\left[1 - 0.12\exp(-3\varepsilon_i / k_B T)\right] \tag{6}$$

$$g_{ii}^{hs} = \frac{1}{1-\xi_3} + \frac{3d_i}{2}\frac{\xi_2}{(1-\xi_3)^2} + 2\left[\frac{d_i}{2}\right]^2 \frac{\xi_2^2}{(1-\xi_3)^3} \tag{7}$$

In these equations g_{ii}^{hs} represents the hard-sphere fluid radial distribution function. The terms m_i, σ_i, and ε_i/k_B represent, for component i, the pure-component PC-SAFT parameters, which are the segment number, temperature-independent segment diameter and the energy parameter respectively, while d_i represents the hard-sphere diameter at a given temperature. The variable ρ_{num} is the total number density of molecules (Å^3), while ξ_0, ξ_1, ξ_2, and ξ_3 are functions of the average pseudo-molecular size. For PC-SAFT theory, the dispersion energy is calculated for a reference fluid comprised of hard chains and it is given in equation 13, where PC-SAFT applies the second-order perturbation theory of Barker and Henderson [14, 15] to a hard chain reference fluid.

$$\tilde{a}^{disp} = -\pi\rho_{num}\left[2I_1(\eta,m)\sum_i\sum_j x_i x_j m_i m_j \sigma_{ij}^3 \frac{\varepsilon_{ij}}{k_B T} + mI_2(\eta,m)C\sum_i\sum_j x_i x_j m_i m_j \sigma_{ij}^3\left(\frac{\varepsilon_{ij}}{k_B T}\right)^2\right] \tag{8}$$

$$\sigma_{ij} = 0.5(\sigma_i + \sigma_j) \quad \& \quad \varepsilon_{ij} = (1-k_{ij})\sqrt{\varepsilon_i \varepsilon_j} \tag{9}$$

The terms I_1 and I_2 in equation 13 represent complicated integrals that can be approximated as power series in density [13], while the combining rules given in equations 14 and 15 are used to calculate the size and energy parameters σ_{ij} and ε_{ij} for a pair of unlike segments. The binary interaction parameter k_{ij} is used to correct the segment-segment interactions for these unlike segments. C is given as

$$C = \left(1 + m\frac{8\eta - 2\eta^2}{(1-\eta)^4} + (1 - m)\frac{20\eta - 27\eta^2 + 12\eta^3 \quad 2\eta^4}{[(1-\eta)(2-\eta)]^2}\right)^{-1} \tag{10}$$

$$\eta = \frac{\pi}{6}\rho_{num}\sum_i x_i m_i d_i^3 \tag{11}$$

Three parameters, m, σ, and ε/k, and the binary interaction parameter, k_{ij}, are needed to model the phase behavior. Table 2 lists the parameters determined by fitting the equation to phase and density data when the binary interaction parameter is set to zero. More detailed results will be presented in this talk.

Table 2. PC-SAFT parameters obtained by fitting to the experimental phase behavior and density data.

	M_w	M_n	m	σ (Å)	ε/k (K)
F-5	201 k	123 k	2451.7	3.4381	199.03
F-6	316 k	226 k	5969.7	3.4553	201.69

Conclusion

High-pressure fractionation is used to purify and reduce the molecular weight distribution of the star polymer. The phase behavior of the fractionated star polymers in propane is determined and compared with the phase behavior of PLMA and of PLMA-PMMA copolymer in propane. The phase behavior and density properties are modeled using the PC-SAFT equation of state. The three parameters, m, σ, and ε/k, are calculated by fitting the equation to the experimental data obtained in this study.

References

[1] Ye, Z.B., Zhu, S.P., Macromolecules, Vol. 36, **2003**, p. 2194.
[2] Ye, Z.B., AlObaidi, F., Zhu, S.P., Macromol. Chem. Phys., Vol. 205, **2004**, p. 897.
[3] Kim, T.H., Cook, S.E., Arote, R.B., Cho, M.H., Nah, J.W., Choi, Y.J., Cho, C.S., Macromol. Biosci., Vol. 7, **2007**, p. 611.
[4] Liu, H.J., Chen, Y., Zhu, D.D., Shen, Z., Stiriba, S.E., React. Funct. Polym., Vol. 67, **2007**, p. 383.
[5] Seiler, M., Fluid Phase Equilib., Vol. 241, **2006**, p. 155.
[6] Seiler, M., Rolker, J., Arlt, W., Macromolecules, Vol. 36, **2003**, p. 2085.
[7] Seiler, M., Rolker, J., Mokrushina, L., Kautz, H., Frey, H., Arlt, W., Fluid Phase Equilib., Vol. 221, **2004**, p. 83.
[8] Schacht, C.S., Bahrarnali, S., Wilms, D., Frey, H., Gross, J., de Loos, T.W., Fluid Phase Equilib., Vol. 299, **2010**, p. 252.
[9] Allal, A., Boned, C., Baylaucq, A., Phys. Rev. E, Vol. 64, **2001**, p.
[10] Meilchen, M.A., Hasch, B.M., McHugh, M.A., Macromolecules, Vol. 24, **1991**, p. 4874.
[11] Caudwell, D.R., Treulser, J.P.M., Vesovic, V., Wakeham, W.A., J. Chemical and Engineering Data, Vol. 54, **2009**, p. 359.
[12] McHugh, M.A., Krukonis, V.J., Supercritical Fluid Extraction: Principles and Practice, 2nd Edition ed., Butterworth-Heinemann, Boston, 1994.
[13] Gross, J., Sadowski, G., Ind. Eng. Chem. Res., Vol. 40, **2001**, p. 1244.
[14] Barker, J.A., Henderson.D., J. Chem. Phys., Vol. 47, **1967**, p. 2856.
[15] Barker, J.A., Henderson.D., J. Chem. Phys., Vol. 47, **1967**, p. 4714.

(663f) Pressure-Induced Melting and Crystallization of Poly(epsilon caprolactone) in Dense Carbon Dioxide

Shinya Takahashi and Erdogan Kiran, Chemical Engineering, Virginia Tech, Blacksburg, VA

A new experimental approach has been developed to investigate the melting and crystallization dynamics in semi-crystalline polymers that are exposed to dense carbon dioxide. The approach involves placing a melt pressed film of the polymer with thicknesses in the range from 100 to 300 μm in between two microscope glass plates, and then putting the sandwich inside a mold which serves as a holding plate. The mold is machined to have a large circular opening that allows full observation of the film placed in the mold. The assembly with the sample with its hold plate is then placed inside a variable- volume high pressure view- cell equipped with sapphire observation windows. The cell is charged with carbon dioxide and the pressure is adjusted using a pressure generator. The system incorporates capabilities to monitor the transmitted light intensity in real time with a dedicated computer. At a given temperature, pressure scans are carried out in both the increasing and then decreasing directions of pressure while recording the transmitted light intensities. The transmitted light intensity *vs* pressure plots provide the details of the crystallization process during pressure reduction, and the melting process during pressure increase. Carrying out these at a number of different temperatures permits the generation of thermodynamic P-T phase diagrams that provide the crystallization and melting boundaries. The results will be discussed for selected polyesters, focusing mostly on Poly (-caprolactone) (Mw = 87 K and PDI = 1.67) over a temperature range from 25 – 45 °C and over a pressure range up to about 50 MPa.

(663g) Inferring Phase Behavior of Polymer Solutions From Molecular Simulations and Extrapolation to the Long Chain Limit

J Richard Elliott, Chemical and Biomolecular Engineering, The University of Akron, Akron, OH and Amir Vahid, Chemical and Biomolecular Engineering, University of Akron, Akron, OH

Molecular simulation of polymeric systems is impeded by the slow equilibration and large simulations inherent in directly treating these systems. Work by Ghobadi and Elliott (1) and Elliott and Gray (2) showed how molecular simulations of the terms contributing to perturbation theory exhibit asymptotic trends in the long chain limit. This observation forms the basis for a general approach to treating polymers of any molecular weight of structure. For polymer solutions, detailed simulations of mixtures over broad ranges of molecular weight, branching, and ring content form the basis for mixing rules to interpolate between state points of various compositions, densities, and molecular weights.

This methodology is applied to predict the compatibility of several plasticizers in polyVinylChloride and polyLactide with an emphasis on naturally occurring and sustainable plasticizers. Works Cited

1. *Evaluating perturbation contributions in SAFT models by comparing to molecular.* **Ghobadi, Ahmadreza F. and Elliott, J.R.** 2011, Fluid Phase Eq., Vol. 306, pp. 56-67.

2. *Asymptotic Trends in Thermodynamic Perturbation Theory.* **Elliott, J.R. and Gray, N.H.** 2005, J. Chem. Phys. , p. 123:184902.

(670a) A Future Energy Supply Based On Liquid Organic Hydrogen Carriers (LOHC)

Wolfgang Arlt[1], Peter Wasserscheid[2], Daniel Teichmann[1], Karsten Müller[1] and Katharina Stark[1], (1)Chair of Separation Science and Technology, University of Erlangen-Nuremberg, Erlangen, Germany, (2)Chemical Reaction Engineering, Friedrich-Alexander University Erlangen-Nuremberg, Erlangen, Germany

This contribution describes a concept for the storage of renewable energy using Liquid Organic Hydrogen Carrier (LOHC) compounds. These compounds are characterized by the fact that they can be loaded and un-loaded with considerable amounts of hydrogen in a cyclic process (energy carrying compounds, ETS). The system N-Ethylcarbazole / Perhydro-N-Ethylcarbazole is the most explored and consequently best understood LOHC system so far. It can store six moles of hydrogen per one mole of carrier substance and therefore allows for a hydrogen storage capacity of 5.8 wt-%.

In contrast to other hydrogen storage methods (like compressed gaseous or cryogenic hydrogen) LOHCs can be safely stored under ambient conditions in big quantities. As they have many physico-chemical similarities to Diesel, LOHCs can make use of the existing energy infrastructure (e.g. storage tanks or road tankers).

The focus of this contribution lies on the implementation of an energy storage system for residential and commercial buildings or groups of buildings based on LOHC. The transition of the mainly fossil-fuelled energy system towards regenerative sources (in Germany called 'Energiewende' – 'energy sea change') has already shifted the role of buildings from mere energy consumers to energy producers (e.g. via rooftop photovoltaics). In a future energy system buildings could in addition support the integration of intermittent renewable energies by acting as decentralized energy storage facilities.

Furthermore, options for mobility are shortly reported.

Preparation of Biobased Jet Fuel Surrogates

Presented at AIChE Annual Meeting, Pittsburgh, PA

1 November 2012

HN Conkle, GM Marcum, EW Griesenbrock, SM Howe, Battelle;
RW Morris, Jr. , AFRL/RZQF;
HJ Robota and DK Thomas, UDRI

Battelle
The Business of Innovation

Integrity ★ Service ★ Excellence

Public Release Case No. : 88ABW-2012-88ABW-2012-4763

 AFRL 1

Abstract

- The Air Force has certified its entire fleet using a 50:50 blend of petroleum : first- and second-generation alternative fuels
- The Air Force Fuels Directorate seeks to understand the characteristics of next generation fuels based on pretreated seed oils, biomass pyrolysis, and fermentation
- Assured Aerospace Fuels Research Facility (AAFRF) located at Wright-Patterson Air Force Base is being operated to produce next generation biobased fuel surrogates needed to assess emissions, fuel compatibility, and combustion performance of these future aviation fuels

 AFRL 2

Presentation Outline

- Air Force Alternative Aviation-fuel Initiative
- Sample Preparation Unit (SPU) at the AAFRF
- Isomerized Tetradecane Results
- Isomerized Decane Results
- Future Plans
- Collaboration Opportunities
- Acknowledgements

 AFRL 3

Air Force Alternative Aviation-fuel Initiative

Certify entire AF Fleet to use a 50/50 SynFuel blend

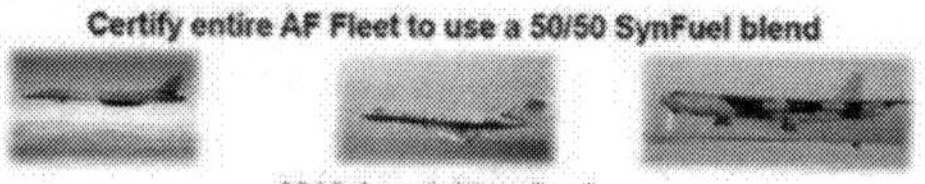

<u>2016 Acquisition Goal</u>

Cost competitively acquire 50% of Air Force's domestic aviation fuel requirement via alternative fuel blends in which the alternative component is derived from domestic sources produced in a manner that is greener than fuels produced from conventional petroleum

 AFRL 4

Alternative Fuel Generations

- 1st: Fischer Tropsch (from coal and natural gas (F-T)*
- 2nd:Hydrotreated esters and fatty acids to make Hydrotreated renewable jet (HRJ)*
- 3rd: Alcohols dehydration and oligomerization to make Alcohol to Jet (ATJ)
- 4th and 5th: Seed oil de-oxygenation, biomass pyrolysis, direct fermentation

* AF fleet certified

 AFRL 5

New Generation of Fuels Do not Look like SPK or HRJ

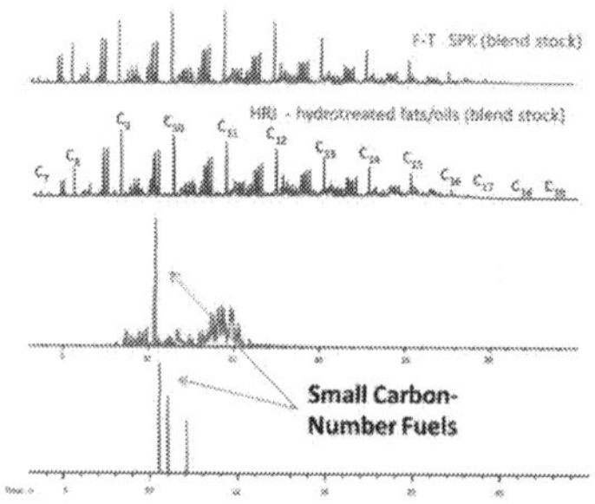

Surrogates can be Produced in the AAFRF Sample Preparation Unit

 AFRL 6

AAFRF Building 496

Approved for public release; distribution unlimited AFRL 7

AAFRF Facility

Approved for public release; distribution unlimited AFRL 8

AAFRF and Fuel Farm

Approved for public release; distribution unlimited AFRL 9

SPU Features

- Dual packed-bed reactors can be operated in series, parallel, or independently
- Three distillation columns for intermediate and product separation; ability to distill final product from 100 to 2 psia
- Mass flow measurement on all streams
- Recycle of H_2 and vacuum column underflow
- iFIX® Supervisory Control and Data Acquisition (SCADA) system
- Safety PLC

Approved for public release; distribution unlimited AFRL 10

SPU Parameters

SPU Parameters	
System Pressure Design Limits	
Reaction System	600-2,000 psig
Naphtha Splitter	10-25 psig
Naphtha Fractionator	20-85 psig
Vacuum Column	2-10 psia
Compressors	up to 2,500 psig
System Temperature Design Limits	
Reaction System	350 – 850 °F
Naphtha Splitter	Reboiler : 650 °F Condenser: <100 °F (end)
Naphtha Fractionator	Reboiler : 650 °F Condenser: <100 °F (end)
Vacuum Column	Reboiler : 650 °F Condenser: <100 °F (end)
Wax Containing Equip	150-175 °F
H_2/Light Gas Recycle	70-150 °F

Approved for public release; distribution unlimited AFRL 11

Bldg 496 Layout

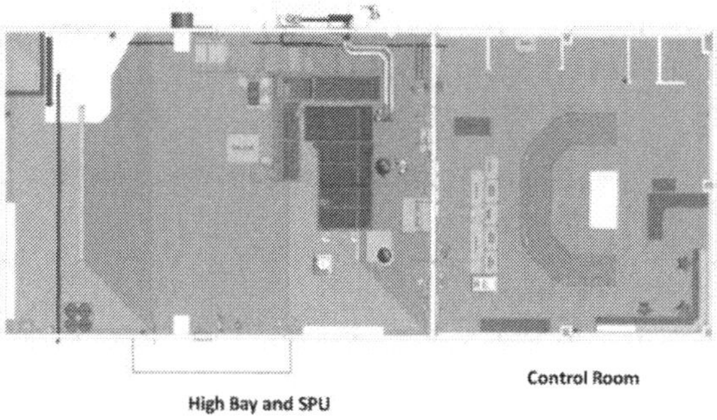

Approved for public release; distribution unlimited AFRL 12

Sample Preparation Unit Control Room

Typical Process Screen (Reactors)

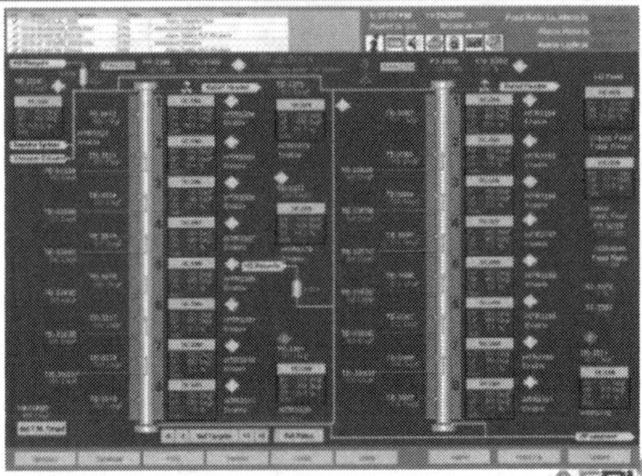

SPU in High Bay

Equipment Layout and Relative Dimensions

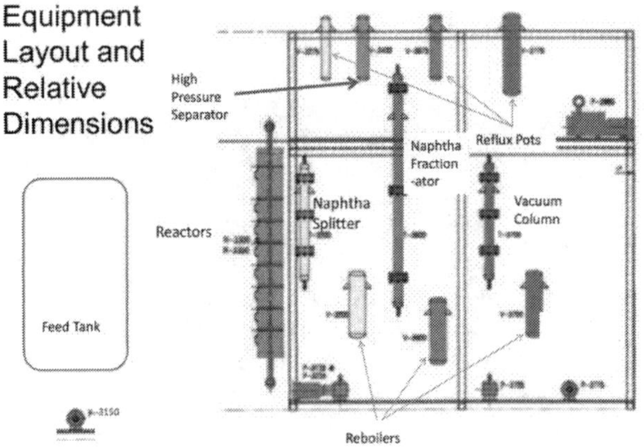

SPU Process Safety Features

- Class I Div 2 Group B design
- Dual PLC controls (Honeywell)
  - Safety shut-down separated from process controls
  - Both systems report to iFIX® SCADA system
- H_2 detection systems
  - 6 detectors on equipment over most likely leak points
  - Integrated with ventilation and facility alarms

SPU Process Safety Features, CONT

- Continuous N_2 purge for every vessel/process element not containing hydrogen
- Pressure relief on every vessel
- Dual Header System
  - Process vent header
  - Separate pressure relief header
  - Flame arrestor on each header
- Remote H_2 storage to minimize quantity in cell
- Safety permit required before operations

New SynFuels Processes are in the Early Stages of Development

- **Need**
  - A way to produce large quantities of 4th and 5th generation fuels now for early evaluation and testing
- **Solution**
  - Produce multi-hundred gallons of fuel surrogates using the AAFRF SPU
    - Run 6: Isomerized tetradecane (C_{14})
    - Run 7: Isomerized decane (C_{10})

Approved for public release; distribution unlimited AFRL 19

Run 6: Produce Isomerized C_{14} with Low Freezing Point

- **Run 6A (in SPU)**
  - Hydrogenate and isomerize tetradecane
  - Separate out cracking products
- Run 6B (in Solvent Dewaxing facility)
  - Purify Run 6A C_{14} by mixing with solvent and chilling to -65°C to solidify the n-C_{14} constituent
  - Filter and evaporate to recover low n-C_{14} product and solvent (for reuse)
- Run 6C (in SPU)
  - Distill and recover low n-C_{14} product in the overhead stream of the Fractionator Column

Approved for public release; distribution unlimited AFRL 20

SPU Process Diagram
(For C_{14} and C_{10} Production)

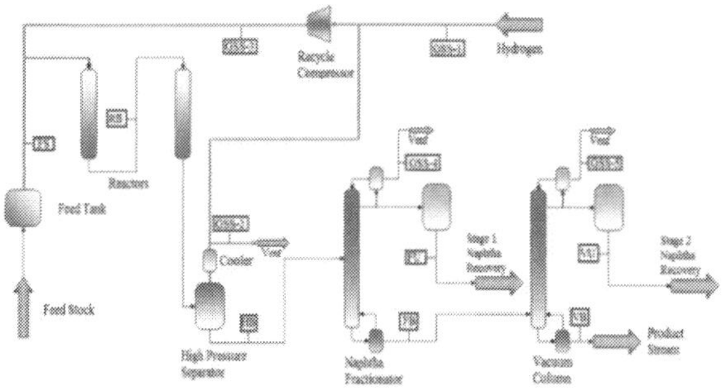

Approved for public release; distribution unlimited AFRL 21

Consistent Selectivity Demonstrated

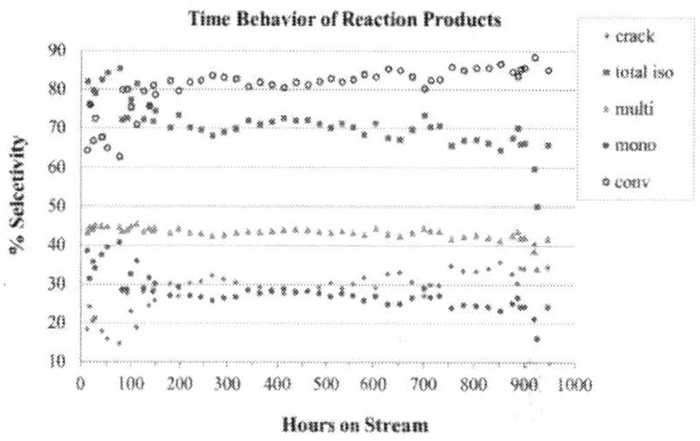

Approved for public release; distribution unlimited AFRL 22

Run 6A Results

Hydrogenation, hydro-isomerization and Separation Results

Parameter	Objective	Test Results
Quantity recovered	> 1,700 gallons	1,768 gallons
Multi-C_{14}/(Mono-C_{14} + Multi-C_{14})	> 50%	61 %
n-C_{14}	< 60%	26 %
Non-C_{14} compounds	<5%	2%

Approved for public release; distribution unlimited AFRL 23

Run 6: Produce Isomerized C_{14} with Low Freezing Point

- Run 6A (in SPU)
  - Hydrogenate and isomerize tetradecane
  - Separate out cracking products
- **Run 6B (in Solvent Dewaxing facility)**
  - Purify Run 6A C_{14} by mixing with solvent and chilling to -65°C to solidify the n-C_{14} constituent
  - Filter and evaporate to recover low n-C_{14} product and solvent (for reuse)
- Run 6C (in SPU)
  - Distill and recover low n-C_{14} product in the overhead stream of the Fractionator Column

Approved for public release; distribution unlimited AFRL 24

10-Gallon Solvent Dewaxing Unit

Solvent Dewaxing System

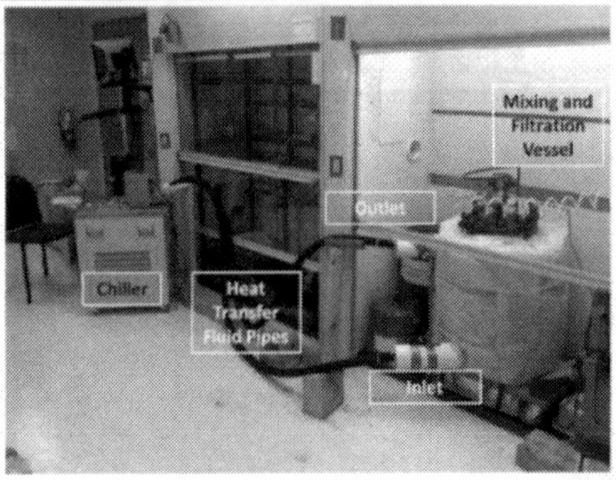

Run 6B Results

10-Gallon Scale-Up Solvent Dewaxing Unit Results

Parameter	Objective	Initial Test Results: With 60% n-C_{14} feed	Initial Test Results: With 24% n-C_{14} feed
Residual n-C_{14}	< 1.5%	1.4%	0.6%
Freeze point	< -47°C	-49°C	-62°C

Run 6: Produce Isomerized C_{14} with Low Freezing Point

- Run 6B (in Solvent Dewaxing facility)
  - Purify Run 6A C_{14} by mixing with solvent and chilling to -65°C to solidify the n-C_{14} constituent
- Run 6A (in SPU)
  - Hydrogenate and isomerize tetradecane
  - Separate out cracking products
  - Filter and evaporate to recover low n-C_{14} product and solvent (for reuse)
- **Run 6C (in SPU)**
  - **Distill and recover low n-C_{14} product in the overhead stream of the Fractionator Column**

Naphtha Fractionator

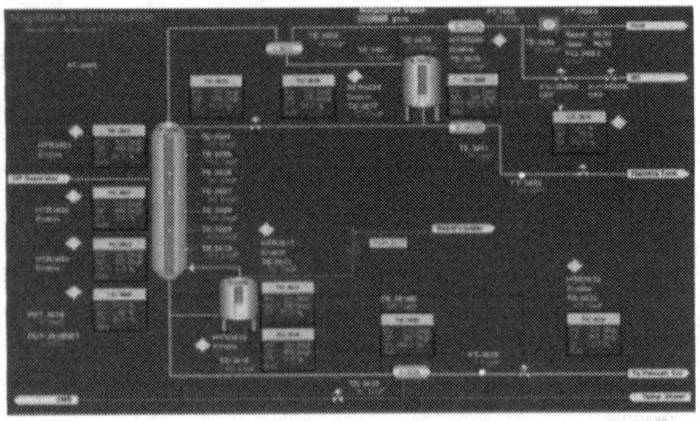

Run 6C Results

9-Stage Fractionator Column Results

Parameter	Objective	Test Results*
Quantity recovered	≥ 400 gallons	400+
n-C_{14}	< 2.6%	2.3%
Freezing point	< -40°C	-42 °C

* Projected as of 22 September 2012

Conclusions from Run 6 Used to Make C_{14} Single-Carbon Surrogate

- Run 6A:
  - 1,768 gallons of C_{14} fuel surrogate produced from 96% tetradecane
- Run 6B:
  - Demonstrated the capability to produce purified tetradecane with freeze point less than -47°C
- Run 6C
  - 400+ gallons of C_{14} fuel surrogate produced
  - Freeze point less than -40°C

Conclusions from Run 7 Used to Make C_{10} Single-Carbon Surrogate

- Run 7: For the Production of Isomerized Decane
  - > 844 gallons of C_{10} fuel surrogate produced
  - 98% decane
  - 44 °C flash point (the minimum allowable JP-8 flash point is 38 °C)

Future Plans

•Hydrogenation and de-oxygenation of seed-oil based bio crude to produce synthetic jet and diesel fuels

•Hydro de-oxygenation and hydrotreating of biobased feedstock

Facility is Available for Contract Research

•Feedstock upgrading evaluations

•Optimized recovery of products

•Catalyst evaluations

•Production of specialized fuels

Acknowledgement

This research was supported, in part, by the U. S. Air Force Cooperative Grant Numbers F33615-03-2-2347 and FA8650-10-2-2934 with Mr. Robert W. Morris Jr. serving as the Air Force Grant Monitor. The research was also sponsored by the State of Ohio Subrecipient Award No. COEUS # 005909 to the University of Dayton (Dr. Dilip Ballal as the Grant Monitor) under the "Center for Intelligent Propulsion and Advanced Life Management," program with the University of Cincinnati (Prime Award NO. TECH 09-022). The authors gratefully acknowledge this grant support.

Questions?

For More Information:

Go to: http://wbi-icc.com/aafrf

Or Contact

Nick Conkle
Fuels and Energy Branch
Wright-Patterson Air Force Base
Harold.Conkle@WPAFB.AF.MIL
Phone (office): 937-255-1408

(670c) Production of Hydrocarbons by Catalytic Dehydration of Bio-Butanol (mixture of 1-butanol, acetone, and ethanol)

Shaima Nahreen and Ram B. Gupta, Chemical Engineering, Auburn University, Auburn, AL

Drop-in transportation fuel can be produced by dehydration of bio-butanol to produce alkenes followed by oligomerization to produce liquids. Bio-butanol, which contains n-butanol (63 wt%), acetone (29 wt%) and ethanol (8 wt%), can be produced by using *Clostridia* bacteria which can effectively utilize both C_5 and C_6 sugars from multiple lignocellulosic feedstocks. In this work the catalytic dehydration reactions of bio-butanol mixture is studied in a continuous packed bed tubular reactor. The product streams (gas, aqueous liquid, and organic liquid) are analyzed and compared with those from the individual components. Two catalysts, γ-Al_2O_3 and ZSM-5, are tested for their effectiveness at different temperatures. The products from the bio-butanol mixture feed are mostly unsaturated hydrocarbons in the range of C_2-C_{15}. The difference in the higher heating values (HHV) of the liquid products and infrared spectra of the gas products indicate that the products from the bio-butanol feed are different from the products from the individual feed components. This difference is attributed to the synergistic inter-component reactivity during the dehydration reactions. HHV of the liquid product increases with the residence time. A better conversion of bio-butanol in terms of liquid product quality and HHV was found in dehydration on γ-Al_2O_3 catalyst than on ZSM-5, when all other parameters (i.e., temperature and residence time) are kept constant. Composition analyses and yield data are obtained in this study for rationale understanding of the dehydration step.

(670d) Fischer-Tropsch Synthesis Using A PILOT Scale Fixed BED Compact HEAT Exchange Reactor

Steven J. Vallee, Zhijun Jia and Hani Gadalla, Chart Energy & Chemicals, La Crosse, WI

The Fischer-Tropsch reaction is a highly exothermic reaction (157 kJ/mol CO converted) used to convert syngas (a mixture of CO and H_2) to liquid hydrocarbons. Conventional fixed bed reactor technology suffers from heat transfer limitations. Operating conditions must be limited to prevent reactor runaway, resulting in lower productivity per unit reactor volume. Non-isothermal operating conditions make the process more difficult to start up and control and can lead to increased catalyst deactivation and unwanted methane side-product formation.

A Compact Heat Exchange Reactor (CHER) combines a reactor and compact heat exchanger into a single device, and provides a solution to the heat transfer limitations commonly observed for conventional fixed bed reactors. A pilot scale CHER was designed and constructed by Chart Energy & Chemicals in La Crosse WI, for the purpose of synthesis of diesel fuel via the Fischer-Tropsch (FT) reaction from syngas. Results are presented for a pilot scale CHER reactor for Fischer-Tropsch synthesis for >1 bpd liquid fuel production that is larger than our previously presented CHER results.

Syngas production by the SCR of methane over Ca-Ni/MgAl catalyst

Yun Ju Lee[a], Eun-Hyeok Yang[a,b], Jin Hee Lee[a,b], Arunabha Datta[a], Young Chul Kim[c] and Dong Ju Moon[a,b] *

a :Clean Energy Research Center, KIST, Seoul , South Korea
b: Clean Energy & Chemical Engineering, UST, Daejeon, South Korea
c: Department of Advanced Chemicals and Engineering, Chonnam National University, Gwangju Chonnam, South Korea

Summary : Steam CO_2 reforming(SCR) reaction of methane for the production of syngas was investigated over Ni based catalysts. The Ni based catalysts were prepared by the solid phase crystalization and impregnation methods. The catalysts before and after the reaction were characterized by N_2 physisorption, CO chemisorption, TPR, XRD, SEM and TEM techniques. The H_2/CO ratio produced in the SCR of methane showed a strong dependence on the feed composition. The conversion of CH_4 was increased with increasing the concentration of steam and CO_2 in the feed. For Fischer-Tropsch synthesis reaction, the feed ratio of $CH_4/H_2O/CO_2$ for the production of $H_2/CO=2$ was estimated by the modeling of commercial simulation package, and identified in the fixed bed reactor system. It was found that the Ca modified Ni based hydrotalcite catalyst showed higher catalytic stability with the restriction of the carbon formation in the SCR of methane under the tested condition.

Keywords : Methane; Steam CO_2 reforming; PRO-II Simulation.

Introduction

Recently, a gas to liquid(GTL) process for the production of clean alternative petroleum from natural gas has been one of the most promising projects in the energy industry. In these processes, there are some key issues to produce syngas with adequate H_2/CO ratio required for Fischer–Tropsch synthesis and to develop the catalyst with high activity as well as stability. SCR has attracted much attention from industry for syngas production [1].

In general, reforming processes such as steam reforming(SR) and carbon dioxide reforming(DR) of methane have some disadvantages because both reforming reactions require an additional process to adjust the H_2/CO ratio [2]. In addition, partial oxidation(POX) of methane which meets the H_2/CO ratio of 2 creates difficulties in controlling the process because of the hot spots and explosion dangers [3]. On the contrary, SCR is an available process for the direct control of the H_2/CO ratio which would be suitable for the Fischer–Tropsch(FT) process by adjusting the feed ratio of steam and carbon dioxide [4].

$$CH_4 + H_2O \rightarrow 3H_2 + CO \quad (1)$$

$$CO + H_2O \rightarrow CO_2 + H_2 \quad (2)$$

$$CH_4 + CO_2 \rightarrow 2H_2 + 2CO \quad (3)$$

Generally, the steam reforming of methane is used to produce hydrogen in commercial process (reaction (1)). It is usually followed by the water gas shift (WGS) reaction(reaction (2)), so as to produce additional hydrogen. However, the presence of carbon dioxide in the feed would decrease the hydrogen production activity. The carbon dioxide reforming of methane is to produce synthesis gas (reaction (3)) [5].

The supported Ni based catalysts have been used for the industrial reforming processes because they are economical compared to noble metal-based catalysts. Also the anionic clays based hydrotalcite-like compounds have been researched as the catalysts in various reactions such as hydrogenation, polymerization and reforming. It is known that the hydrotalcites or hydrotalcite-like compounds after calcination have high surface area, basic properties, formation of homogeneous mixtures of oxides with very small crystal size, memory effect, which allows the reconstruction of the original hydrotalcite structure when contacting the product of the thermal treatment with water solutions containing various anions under mild conditions. K. Takehira et al.[6] reported that Ni-loaded catalyst prepared from Mg-Al hydrotalcite-like anionic clay have high and stable activity for CO_2 reforming, steam reforming and auto thermal reforming (ATR) of CH_4 [7].

In this work, steam CO_2 reforming of CH_4 was estimated by PRO-II simulation and investigated over Ni-based catalyst for application in GTL.

Experimental

Catalyst Preparation

Nickel nitrate [$Ni(NO_3)_2 \cdot 6H_2O$], Aluminum nitrate [$Al(NO_3)_3 \cdot 9H_2O$], and Magnesium nitrate [$Mg(NO_3)_2 \cdot 6H_2O$] with the purity of about 98 ~ 99% (Aldrich and Alfa Aesar)were used without additional purification. The distilled and deionized water was used throughout whole the experiment.

* Corrresponding author
E-mail : djmoon@kist.re.kr
Phone/fax : +82-2-958-5867/+82-2-958-5809

The Ni/MgAl oxide catalysts using modified hydrotalcite were prepared by the co-precipitation method reported by Takehira et al. with minor modification [6]. The obtained Mg(Ni)-Al hydrotalcite-like precursors were calcined in a furnace in an air atmosphere by increasing the temperature from ambient temperature to 850 °C at a rate of 5 °C/min followed by keeping at 850 °C for 5 h to form the precursor of Ni/MgAl [8]. The Ca modified Ni-based hydrotalcite catalyst(Ca-Ni/MgAl) was prepared by co-precipitation and dipping methods, and characterized by N_2 physisorption, CO chemisorption, XRD, SEM, TEM and ICP.

Steam CO_2 Reforming Reaction

The steam CO_2 reforming of methane was performed in a fixed bed reactor system. The preheater (0.008 m O.D. and 0.25 m length) and steam reforming reactor (0.008 m O.D. and 0.25 m length) were made of an Inconel 600 tube, respectively. Before the reaction, the Ni-based catalysts were reduced at 800 °C for 3 h in the hydrogen atmosphere. The unreacted H_2O was removed by a cold trap and then a gas effluent was analyzed by an on-line gas chromatograph (HP 7890A Series, TCD) and GC/MS (HP 7890A Series, MS detector). All runs were carried out at a temperature ranges of 650 ~ 850 °C at atmospheric pressure with a space velocity of 10,000 h^{-1} and feed molar ratio of $CH_4 : H_2O : CO_2 = 1.0 : 2.0 : 1.0$.

Results and Discussion

The metal loading and textural properties of the prepared catalysts are summarized in Table 1. BET surface area and crystallite size of Ni/MgAl, Ca-Ni/MgAl, Ni/Al_2O_3, Ni/MgO and Ru/TiO_2 catalysts were measured by N_2 physisorption and XRD, respectively. It was found that Ni/MgAl, Ca-Ni/MgAl hydrotalcite catalyst has higher BET surface area than conventional catalysts.

Table 1. Physical and chemical properaties of prepared Ni-based catalysts

Catalyst	Metal loading (wt%)		BET S.A. (m^2/g)[a]	Crystallite size (nm)[b]
	Ca/Ru	Ni		
Ni/MgAl		20.0	197.7	14.0
Ca-Ni/MgAl	0.1	20.0	160.0	20.3
Ni/Al_2O_3		20.0	81.2	13.3
Ni/MgO		20.0	42.9	29.4
Ru/TiO_2	2.0		34.3	27.1

a : BET suface area was measured by N_2 physisorption
b : Particle size was calculated by XRD line broadening method

Figure 1 shows the (a) CH_4 conversion and (b) CO_2 conversion with time on stream in the SCR of methane over Ni/MgAl, Ca-Ni/MgAl, Ni/Al_2O_3, Ni/MgO and Ru/TiO_2 catalysts. The SCR of methane was carried out at 850 °C, feed molar ratio of $CH_4 : H_2O : CO_2 = 1 : 2 : 1$ and GHSV of 10,000 h^{-1}. There was almost 100 % CH_4 conversion while the CO_2 conversion was about 60 % under the tested conditions. It was found that Ca-Ni/MgAl and Ni/MgO catalysts showed higher catalytic activity and higher stability with time on stream than other catalysts under the tested condition.

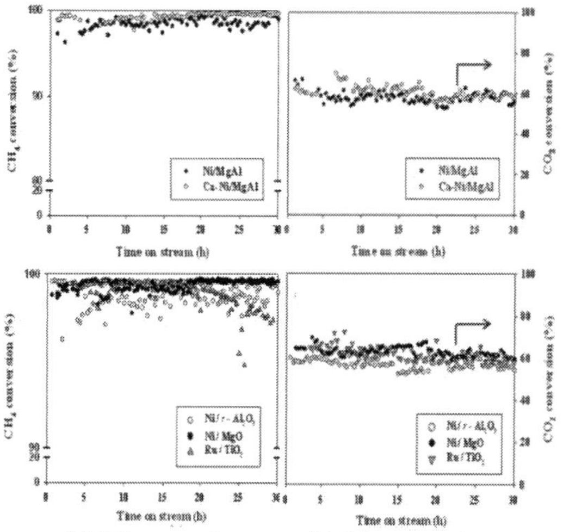

(a) CH_4 conversion (b) CO_2 conversion

Fig. 1. Catalytic activity and stability for SCR of methane over Ni/MgAl, Ca-Ni/MgAl, Ni/Al_2O_3, Ni/MgO and Ru/TiO_2 catalysts for 30 h. Reaction condition: $CH_4 : H_2O : CO_2 = 1 : 2 : 1$, GHSV=10,000 h^{-1}, 850℃ and 1 bar.

The results suggest that the prepared catalyst can be applied to the production of synfuel in GTL-FPSO system [9].

Conclusions

The optimized feed molar ratio for the production of synthesis gas with molar ratio of $H_2/CO = 2$ via steam CO_2 reforming of methane for FT synthesis application is $CH_4 : H_2O : CO_2 = 1.0 : 2.0 : 1.0$, which was estimated by PRO-II simulation, and identified by experimental in a fixed bed reactor system. It was found that the Ca modified Ni based hydrotalcite catalyst showd higher catalytic stability with respect to carbon formation in the SCR under the tested conditions.

References

[1] J. R. Rostrup-Nielsen, *Catal. Today*, 18 (1993) 305-324.
[2] Y. J. Lee, S. I. Hong and D. J. Moon, *Catal. Today*, 174 (2011) 31-36.
[3] S. Rabe, T. B. Truong and F. Vogel, *Appl. Catal. A*, 292 (2005) 177-188.
[4] V. R. Choudhary and K. C. Mondal, *Appl. Energ.*, 83 (2006) 1024-1032.
[5] J. S. Kang, D. H. Kim, S. D. Lee, S. I. Hong and D. J. Moon, *Applied catalysis A*, 332 (2007) 153-158.
[6] K. Takihira, T. Shishido, P. Wang, T. Kosaka and K. Takaki, *J. Catal.*, 221 (2004) 43.
[7] F. Cavani, F. Trifirò and A. Vaccari, *Catal. Today*, 11 (1991) 173-301.
[8] D. H. Kim, J. S. Kang, Y. J. Lee, N. K. Park, Y. C. Kim, S. I. Hong, D. J. Moon, *Catal. Today*, 136 (2008) 228-234.
[9] D.J. Moon, Y.J. Lee, J.S. Kang, M.J. Park, S.J. Kim, S.D Lee, H.J. Lee, Korea Patent Application, 10-2009-0116000 (2009) and U.S. Patent Application, 2011/0130474A (2011).

(670f) Effect of Hydrogen On Catalytic Decarboxylation of BROWN Grease to Green Diesel

Elvan Sari[1], Manhoe Kim[1], Steven O. Salley[1] and K. Y. Simon Ng[2], (1)Chemical Engineering and Materials Science, Wayne State University, Detroit, MI, (2)Department of Chemical Engineering and Materials Science, Wayne State University, Detroit, MI

Brown grease (BG) is a mixture of waste fats and oils which may contain 50-100 % free fatty acid (FFA). BG is a potential low cost and non-food competing feedstock for biofuels. Due to its high FFA content, BG is a good candidate for decarboxylation reaction where the oxygen is removed as carbon dioxide, producing "green diesel." This process is a low cost alternative to hydrodeoxygenation process. Approximately 75 % of BG consists of unsaturated FFAs with oleic acid having the largest portion (1). Decarboxylation of oleic acid studies showed that the hydrocarbon selectivity is low because the unsaturated C=C bond in the oleic acid tends to be saturated under the reaction conditions in the presence of H_2 and decarboxylation of oleic acid yields to stearic acid and n-heptadecane as products (2). In the case of absence of H_2 the key challenge is fast catalyst deactivation. We explored the conversion of brown grease to green diesel via a pre-hydrogenation/decarboxylation reaction route with minimum hydrogen consumption. Results demonstrated that C=C bonds in FFAs can be easily saturated under 15 bar and 100 ^oC in the presence of Pd/C under continuous flow of 10 vol% H_2 - 90 vol% Ar in a semi- batch reactor. The saturated FFA content in the BG was then converted to n-paraffins via decarboxylation by increasing the temperature to 300 ^oC. The main n-paraffin products consist of tridecane, pentadecane and heptadecane as a result of decarboxylation of BG FFAs. Due to the high unsaturation level of BG feedstock, formation of heavier products with carbon number more than 25 was observed. Pre-hydrogenation step helps to decrease heavier compounds yield and to increase olefinic C13-C18 compounds yield. H_2- pretreatment also avoids C=C bond cleavage. Increasing H_2 amount from 0.4/1 ratio of H_2/BG to excess amount (4/1 ratio of H_2/BG) increases the FFA conversion more than 2 times. While decarboxylation and decarbonylation reactions are the major pathways for pre-hydrotreated BG conversion, decarboxylation and dimerization are the major pathways for conversion of non-pretreated BG under stoichiometric H_2 amount (H_2/BG ratio of 2/1 mol/mol). Increasing BG initial concentration resulted n-paraffins selectivity to decrease and heavy compounds selectivity to increase. This is because the higher BG concentration in reaction atmosphere means the higher unsaturated FFAs concentration; therefore, more heavy compounds formation via dimerization. With a dilute solution of BG, side reactions are minimized and primary n-paraffins production increases. A continuous decarboxylation of BG over 5%Pd/C catalyst conducted in a fixed bed tubular reactor showed pretty stable behavior of BG decarboxylation for 4 days. However, decarboxylation reaction sites of Pd/C catalyst was partially poisoned by impurities in BG.

(670g) Conversion of Fischer-Tropsch Light Hydrocarbons Over Ga-ZSM-5 to Assist in the Production of Fuels

Benjamin D. Harris, Chemical Engineering, University of Michigan, Transportation Energy Center, Ann Arbor, MI

Currently, synthetic fuel used for military purposes is created from a mixture of 50% petroleum-derived and 50% synthetically-created hydrocarbons. The mixing of these two blends of hydrocarbons is required to meet the many fuel specifications required for military use. The reformed, synthetic hydrocarbons produced by the Fischer-Tropsch (FT) process do not contain aromatic compounds. In JP-8, aromatic compounds (typically C8-C16) are required to meet such properties as energy density and specific gravity. Additionally, prior use of aromatic-containing JP-8 has made elastomers in the current fleet susceptible to leaking when introduced to a non-aromatic-containing fuel blend. For these reasons, synthetic fuel produced currently requires the addition of single-ring aromatic compounds.

This work examines the synthesis of aromatic compounds from a model light gas stream available in a synthetic fuel plant. The light gas stream in this plant would often be reformed to produce hydrogen or flared for heat duty. However, using the catalyst Ga-ZSM-5, C6-C10 aromatics can be produced from light gases (C2-C6) in the presence of hydrogen, carbon monoxide, and carbon dioxide. This work examines the potential aromatic yields that can be obtained from such a stream, and it examines the potential effects of carbon monoxide and carbon dioxide on the catalyst. Finally, this work explores the additional step of converting from unreacted alkenes and C6-C10 aromatics to an aromatic spectrum that is appropriate for blending with FT synthetic fuel.

(670h) Coal and Biomass to Liquids and Power Generation Via Fischer-Tropsch Synthesis and Fuel Cells

Julia Valla, Shoucheng Du and George M. Bollas, Department of Chemical, Materials & Biomolecular Engineering, University of Connecticut, Storrs, CT

Introduction

Security of energy supply, economic sustainability, and concerns over global climate change are strategic energy objectives of oil-importing countries [1,2]. Thus, considerable interest is focused on the so-called coal to liquid (CTL) processes, based on Fischer-Tropsch (F-T) synthesis. The perspective of coal as energy source depends on the success of emerging "clean coal technologies" (CCT) [2]. However, synthetic fuels made from coal, even with CO_2 capture and storage (CCS), have greenhouse gas (GHG) emission rates equal or higher than the petroleum products they replace. One approach to address this issue is to identify options that will offset the GHG emissions of coal processing. Co-feeding of biomass in CTL processes (called coal and biomass to liquids - CBTL) is one of the options considered. Biomass is among the most promising renewable energy sources, due to its neutral CO_2 life cycle. The option of biomass as a stand-alone source of fuel production is not yet feasible, because sustainably-recovered biomass is expensive. However, co-feeding of biomass in coal-based plants provide a viable solution for controlling coal GHG emissions. Moreover, coal power plants are based on combined cycle (CC) technologies. Of scientific interest is the replacement of the CC by fuel cells (FC), to improve process efficiency and reduce carbon footprint. The objective of this work is to evaluate the potential of emerging and efficient technologies for the production of clean energy (liquid fuels and electricity) based on coal and biomass. The technology options, evaluated include: (1) *co-feeding of biomass in CTL plants; and (2) partial or complete replacement of CCs with FCs.*

Simulation Design Strategy

Coal and biomass to liquids through F-T synthesis is a multi-step and energy intensive process. Briefly, carbonaceous material (coal and/or biomass) is gasified and the gas is processed to make purified synthesis gas. The F-T process polymerizes syngas into valuable gasoline and diesel range hydrocarbons. In this work, the feasibility of biomass addition to coal-based F-T synthesis was analyzed using Aspen Plus. Southern pine wood and Illinois coal #6 were used as feedstocks in this study [3]. As shown by the respective ultimate analysis (dry basis) of Figure 1(b), Illinois #6 coal has approximately 20wt% more carbon and 30wt% less oxygen than Southern Pine, corresponding to a more valuable feedstock (in terms of heating value), but also higher CO_2 emissions. The high oxygen content in biomass diminishes its heating value. On the other hand, coal has 12wt% ash and high sulfur content, which need to be removed during the process.

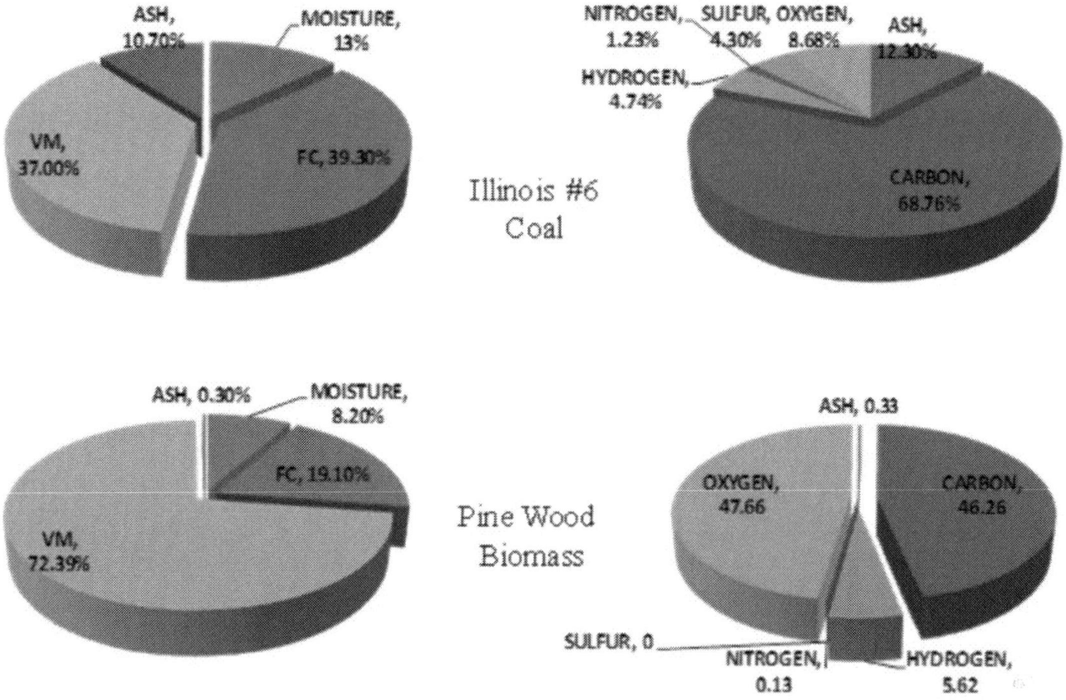

Figure 1: Illinois #6 coal and Southern Pine wood proximate (a) and ultimate (b) analysis

The process flowsheet developed is briefly shown in Figure 2 and consists of the following subsystems: (1) preparation unit where coal and/or biomass are milled, dried and grinded; (2) air separation unit (ASU) where high purity oxygen (95vol%) is produced for the gasification using conventional cryogenic technology; (3) entrained-flow oxygen-blown gasification unit, operating at *ca.* 2500°F/600psia, where coal and/or biomass are gasified to syngas; (4) water gas shift (WGS) unit for the adjustment of H_2:CO ratio at 2.0 (as required for Co-based F-T synthesis); (5) acid gas removal unit based on the Rectisol process for the recovery of CO_2 and H_2S; (6) Claus unit for the conversion of H_2S to elemental sulfur; (7) low temperature Co-based F-T synthesis unit (operating at 480°F/450psia), where the clean syngas is converted to valuable liquid fuels; and (8) power production unit. For the power island two cases were considered: a) gas and steam CC, and b) solid oxide fuel cell (SOFC). In the CC, the gas turbine utilizes tail gas from the F-T synthesis and steam turbine utilizes steam generated in the Heat Recovery Unit (HRU). In the case of the SOFC system the tail gas from the F-T unit first enters a pre-reforming step, where all the light hydrocarbons are reformed and then enters the FC Anode, where any remaining CH_4 is reformed, CO is shifted and H_2 is oxidized with oxygen from the Cathode. The SOFC operates at 1650°F and 1atm, at 85% fuel utilization factor. Part of the unconverted fuel is recycled to the pre-reformer, and the remaining is combusted. Aggressive CO_2 capture (after F-T synthesis) was considered, but was not included in this study due to its high energy penalty.

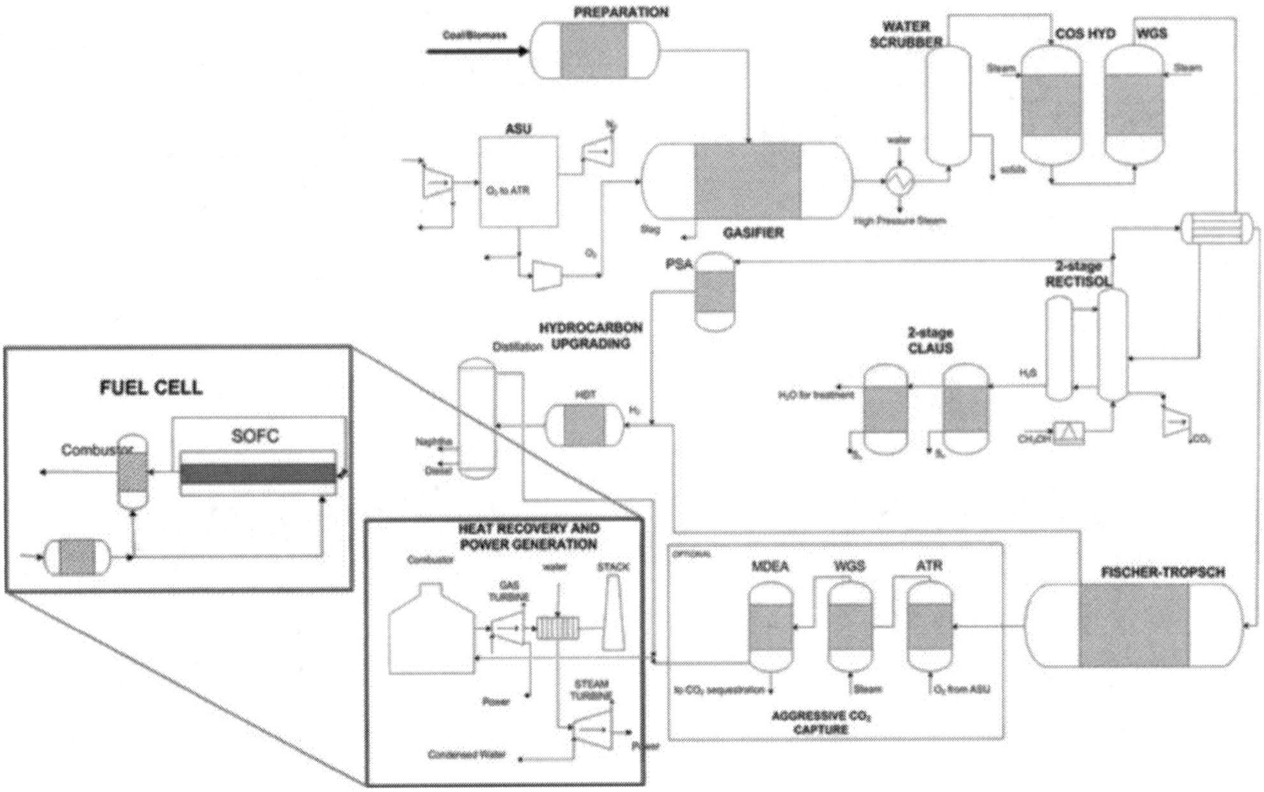

Figure 2: Flowsheet of CBTL-CC and CBTL-FC

Results and Discussion

Five scenarios of biomass to coal ratio were evaluated (0, 15, 50, 75 and 100 wt%) with respect to their overall thermodynamic efficiency (to liquids and power). The Co-based F-T synthesis was simulated on the basis of a semi-empirical mathematical model [4]. In Figure 3 the effect of biomass addition on the major plant inputs is presented for a base case of processing 1.50E+06 lb/hr of coal and/or biomass. The oxygen demand for the gasification process is gradually reduced as the addition of biomass in the CBTL process is increased. The oxygen content of biomass can serve as an oxidant source, reducing the requirements in oxygen supply. Consequently, the power demands in the ASU unit (the most energy intensive unit in the plant) are significantly reduced as more biomass is added in the plant. Moreover, the steam requirements in the WGS unit are significantly reduced by increasing biomass addition. Biomass has a higher H:C ratio than coal, resulting in higher H_2:CO ratios after gasification; hence, smaller steam requirements. Note that the shift steam is taken from an intermediate extraction point in the HRU. Thus, any steam extracted for the shift is not expanded in the steam turbine to produce power; therefore the WGS step is typically viewed as a burden on the steam cycle.

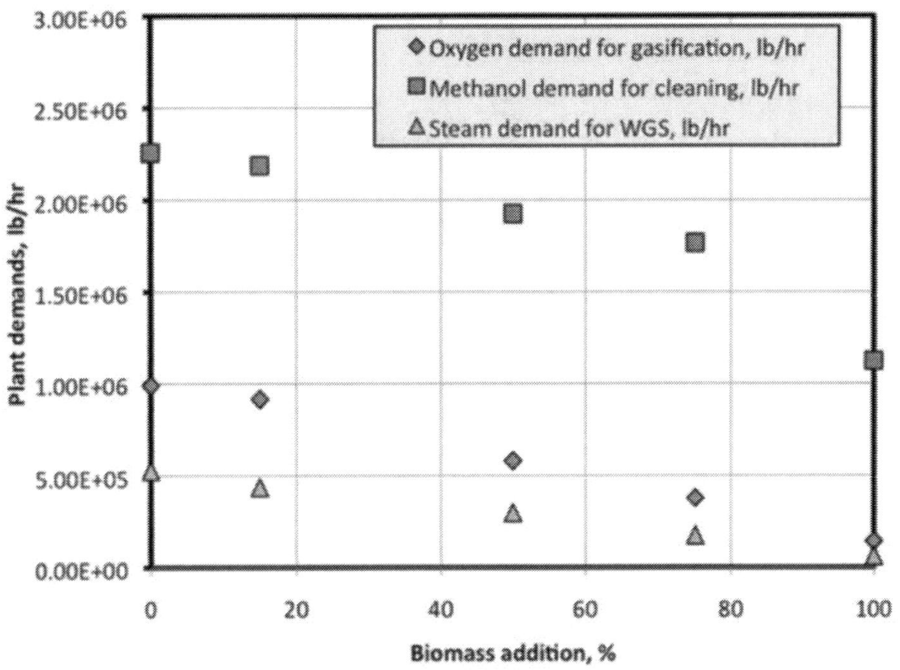

Figure 3: Effect of biomass addition on the plant demands

Finally, the methanol required in the Rectisol unit is lower with increasing biomass addition. The reason for this is the very small sulfur content of biomass and the lower CO_2 of the shifted syngas. It should also be noted that the CO_2 formed by biomass gasification is not considered as emission (assuming a close-to-zero carbon life cycle for biomass). Thus, the energy demand is significantly reduced by co-feeding biomass to the plant. However, less energy is produced (both liquids and electricity), due to its lower energy density. In Figure 4(a) the energy input (based on HHV of feed), the energy production (based on HHV of liquids and electricity via the CC) and the plant energy consumption are plotted as a function of biomass addition. In Figure 4(b) the efficiency of the CBTL plant to liquids and CC electricity, and the overall efficiency are plotted as a function of the biomass addition. It is obvious that, albeit the lower power production, the overall efficiency of the CBTL plant increases by adding biomass, due to the lower plant energy requirements.

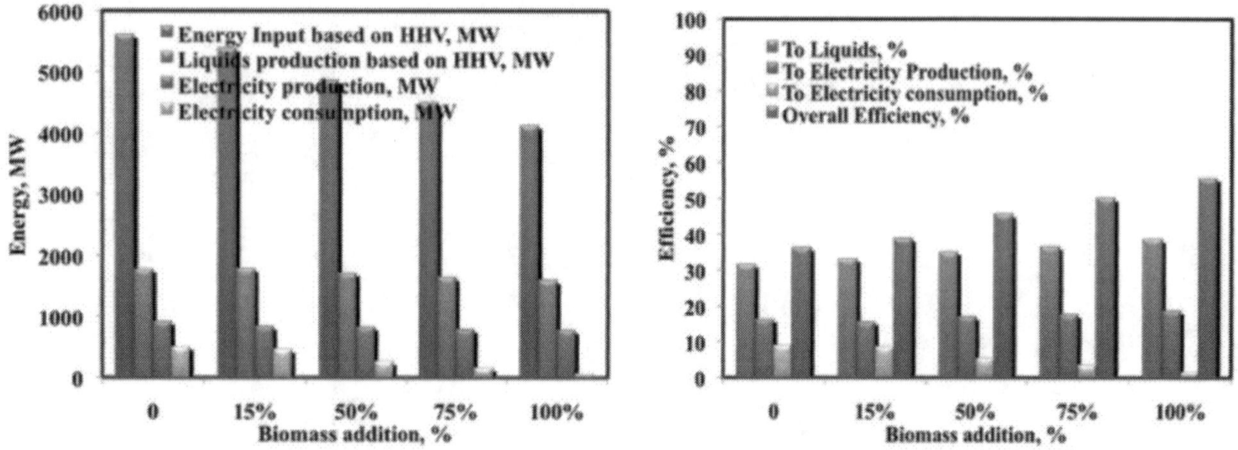

Figure 4: a) Energy input, production and consumption, and b) Energy efficiencies as a function of biomass addition in the CBTL plant with parallel CC

To increase the overall electricity production, the replacement of the CC with SOFCs was studied. A tubular SOFC system, developed by Siemens Power Generation Inc. (SPGI) [5], was considered. The current density was set to $2000 A/m^2$. The results are presented in Table 1. The SOFC system significantly improves the electricity production compared to the CC.

Table 1: Net electricity production based on CC and SOFC as a function of biomass addition

Biomass addition %/Net electricity	0	50	100
CC, MW	907	820	770
SOFC, MW	2010	1935	1880

Conclusions

Biomass co-feeding in existing coal-fired power plants was considered as an alternative to control GHG emissions in F-T liquid fuels synthesis. The results of this study show that as the biomass addition is increased liquid production and power generation are reduced. However, the overall efficiency of the plant is increased, due to the lower power consumption. The replacement of the conventional CC for power generation by SOFCs was also studied showing enhancements in the efficiency of the power generation island.

Acknowledgment

This work was supported in part by the Connecticut Center for Advanced Technologies and the Faculty Large Grant of the University of Connecticut.

References

[1] T. J. Tarka, "Affordable, Low-Carbon Diesel Fuel from Domestic Coal and Biomass," DOE/NETL-2009/1349

[2] T. G. Kreutz, E. D. Larson, G. Liu, R. H. Williams, "Fischer-Tropsch Fuels from Coal and Biomass," 25th Annual International Pittsburg Coal Conference, 29 September – 2 October 2008, Pittsburg, Pennsylvania, USA

[3] G. Vaughan, S. Luz-Acosta, C. Roberts and M. R. Eden "Systems Analysis of Coal and Biomass Based Fuel Production Strategies" Consortium for Fossil Fuel Science Meeting, Pittsburg, PA, August 3-4, 2010

[4] C. N. Hamelinck, A. P.C. Faaij, H. den Uil, H. Boerrigter, "Production of FT transportation fuels from biomass: technical options, process analysis and optimization, and development potential," Energy 29 (2004) 1743-1771

[5] W. Doherty, A. Reynolds, D. Kennedy, "Computer simulation of a biomass gasification solid oxide fuel cell power system using Aspen Plus," Energy 35 (2010) 4545-4555

(677a) Liquid and Supercritical CO_2 Extraction of Fat From Rendered Materials

José Luis Orellana and Christopher L. Kitchens, Department of Chemical and Biomolecular Engineering, Clemson University, Clemson, SC

The processing of rendered materials (RM) allows the separation of fat which results in value-added co-products for energy and up-graded feed sources for animals. The current industry utilizes steam and continuous mechanical press to extract the fat from the RM. The main objective of this work is to investigate the use of CO_2 as a solvent for the extraction of the remaining fat from RM, which will result in ecological and economic benefits for the rendering industry.

Supercritical ($SCCO_2$) and liquid CO_2 (LCO_2) were employed for the extraction of the fat in RM. Lipids were characterized by gas chromatography and by alkaline titration to determine the fatty acids (FA) composition and the amount of free FA, respectively. A continuous-flow extraction unit was used for the extraction. The impact of pressure, temperature, and mass of CO_2 on the extraction yield and solubility of the fat on CO_2 was studied. The solubility of the fat at the different conditions was calculated from the slopes of the extraction curves. The experimental data was fitted to the Chrastil and Modified Chrastil equations to correlate the solubility as a function of density and temperature.

A maximum extraction yield of approximately of 93% was obtained for all the conditions tested. This means that no significant difference was observed in the maximum total amount of fat extracted as an effect of temperature and pressure only. However, the mass of CO_2 used for the extraction was greatly changed due to the solubility of the fat. The highest solubility (6.47 $g.L^{-1}$) was reached at 25°C and 5,000psi – being this the highest density used in this work. The solubility of fat significantly increased when using higher densities of CO2. This behavior was very well correlated by both models used to predict the solubility, with values of average absolute relative deviation (AARD) around 5.5%. It can be concluded that high fat extraction yields can be obtained using CO_2 in RM; and that at the same pressures, LCO_2 is more effective than $SCCO_2$.

(677b) Microwave-Mediated Transesterification of Algal Biomass to Biodiesel Under Supercritical Ethanol Conditions

Prafulla Dinkarrao Patil[1], Harvind Reddy[1], Tapaswy Muppaneni[1], Peter Cooke[2], Tanner Schuab[3], Nagamany Nirmalakhandan[4], Peter Lammers[5] and Shuguang Deng[1], (1)Chemical Engineering, New Mexico State University, Las Cruces, NM, (2)Core University Research Resource Laboratory, New Mexico State University, Las Cruces, NM, (3)College of Agricultural, Consumer and Environmental Sciences, New Mexico State University, Las Cruces, NM, (4)Civil and Environmental Engineering, New Mexico State University, Las Cruces, NM, (5)Energy Research Laboratory, New Mexico State University, Las Cruces, NM

We demonstrated an integrated novel approach to convert lipid-rich, wet/dry algae (*Nannochloropsis Salina*) into fatty acid ethyl esters (FAEE) under microwave-mediated supercritical ethanol conditions. This process enables simultaneous extraction of lipids from algal biomass and convert/transesterify them into algal biodiesel in a shorter reaction of time and may reduce the energy consumption due to simplified separation and purification steps. High conversion rates can be possible when the extractive-transesterification of algal biomass was performed near-critical or supercritical conditions resulted due to ionic conduction and dipolar polarization microwave effect. Experimental runs were designed to optimize the process parameters to evaluate the effect on the algal biodiesel under controlled microwave power conditions. We described the use of passive heating elements made of silicon carbide (SiC) to aid the microwave-mediated heating process at higher temperatures. The algal biomass characterization and algal biodiesel analysis were performed using various analytical instruments such as FTIR, SEM-EDS, TEM, TLC and GC-MS. The confocal laser scanning microscopy and fluorescence spectroscopy has been implemented for the detection of algal biodiesel fuel obtained from dry and wet algal biomass. The thermogravimetric analysis (TGA) under nitrogen and oxygen environment was performed to check the thermal and oxidative stability of ethyl esters. This simple in situ transesterification process using greener solvent and catalyst-free approach can be potential route for algal biodiesel production.

(677c) Production of Biodiesel From Triglycerides in Binary Supercritical Mixtures of Carbon Dioxide and Methanol: Co-Solvent Effects

Jared S. Bouquet[1], Aaron C. Gonzales[2], Ryan E. Tschannen[3] and Sunggyu Lee[2], (1)Department of Chemical & Biological Engineering, Missouri University of Science and Technology, Rolla, MO, (2)Chemical and Biomolecular Engineering, Ohio University, (3)Department of Chemical and Biomolecular Engineering, Ohio University, Athens, OH

Biodiesel is a promising source of renewable energy because of its ability to be used in existing automobile engines and due to the large number renewable feedstocks that can be used in its production. The most widely used method for the creation of biodiesel involves the transesterification of triglycerides with methanol, and the process is catalyzed by a sodium hydroxide or similar base. This process has six moles of methanol fed per mole of triglyceride, and is a slow process which requires hours of reaction time and the separation of the salt catalyst afterwards. Another emerging process utilizes supercritical methanol without the catalyst to perform the transesterification reaction. This process is much quicker, only taking three to five minutes, but requires a much higher excess of methanol (approximately 42 moles per mole of triglyceride) to be fed. One possible solution to this is to partially supplement of the excess supercritical methanol with supercritical carbon dioxide. The conditions for carbon dioxide to become supercritical are not as severe as those for methanol, which can reduce some of the operating costs of the supercritical methanol production method. An experimental study was conducted using a 0.1 L Haynes® 282 reactor system with 42:1 to 13:1 methanol-to-triglyceride molar ratios, and 0 to 3 SLPM of carbon monoxide. The experiments were performed at 350 ˚C and at 1200 psi.

KINETICS OF BIODIESEL SYNTHESIS VIA TRANSESTERIFICATION IN SUPERCRITICAL METHANOL

Tao Cong, Jiuxu Liu, George Anitescu, Ronghong Lin, and Lawrence L. Tavlarides, Syracuse University, Syracuse, NY, USA

Introduction

Biodiesel is one of the most attractive renewable fuels. It can be made from triglycerides found in various biomass sources such as animal fats[1], plant oils[2-4], and algae[5]. Biodiesel is currently produced through transesterification reactions using acid or alkali as catalysts. These conventional processes require a high purity of feedstock with very low free fatty acid and water content. Furthermore, the complexity of the separation step to remove catalyst and glycerol from biodiesel drives up the biodiesel cost.[6]

A catalyst-free method for the transesterification of plant oils and animal fats using supercritical alcohol has been proposed recently.[3,7-10] At high temperatures and pressures, methanol and triglycerides mix and form a homogeneous phase which will accelerate the transesterification reactions due to removal of the oil-alcohol interphase that limits mass transfer and hence reaction rates.[11] Furthermore, separating biodiesel product from glycerol byproduct is simple, because the products are immiscible at ambient temperature and no soap or other side products need to be removed.[5-10]

In this study, we conducted continuous flow reactions of methanol and a model triglyceride compound, triolein ($C_{57}H_{104}O_6$), under various conditions to study the consecutive reactions of this compound and to better understand the reaction kinetics. A three-step kinetic model was proposed based on experimental data. These results should provide a better understanding of high temperature transesterification reactions to produce biodiesel fuels.

Materials and experimental set-up

Materials

Triolein (99.9%), N-Methyl-N-(trimethylsilyl) trifluoroacetamide (MSTFA), ASTM D6584 individual standard solution were purchased from Sigma Aldrich. HPLC-grade methanol and n-heptane were purchased from Thermo Fisher Scientific and used without further purification.

Experimental Setup

A schematic diagram of the experimental setup is illustrated in **Figure 1**. The experimental setup consists of two high pressure syringe pumps (Teledyne ISCO) for triolein and methanol delivery, two preheaters (red dot rectangle), a coiled tubular reactor (316 SS, ID1.753mm×6m) placed in a furnace (Briskheat)(red rectangle), a micrometering valve (Autoclave Engineers) for pressure control, a water bath to cool the products, several thermocouples (T1-T5), a pressure transducer, and a data acquisition system (National Instruments) for recording and monitoring temperature and pressure.

Experimental Conditions

Supercritical transesterification experiments of triolein were conducted at 355-400 °C and 150 bar. The residence time was 0.5-10 min, and a constant methanol-to-triolein molar ratio of 9:1 was used.

As soon as steady-state conditions were achieved (after at least three residence time), effluent stream samples were collected and then analyzed without further treatment.

Analytical Methods

Free and total glycerol was analyzed by GC-FID (HP 5890 series II) equipped with a Restek Rtx-Biodiesel triglyceride column (10 m × 0.32 mm I.D. × 0.1 μm film thickness). The ASTM standard method D6584 was followed to prepare calibration curves and perform sample analysis.

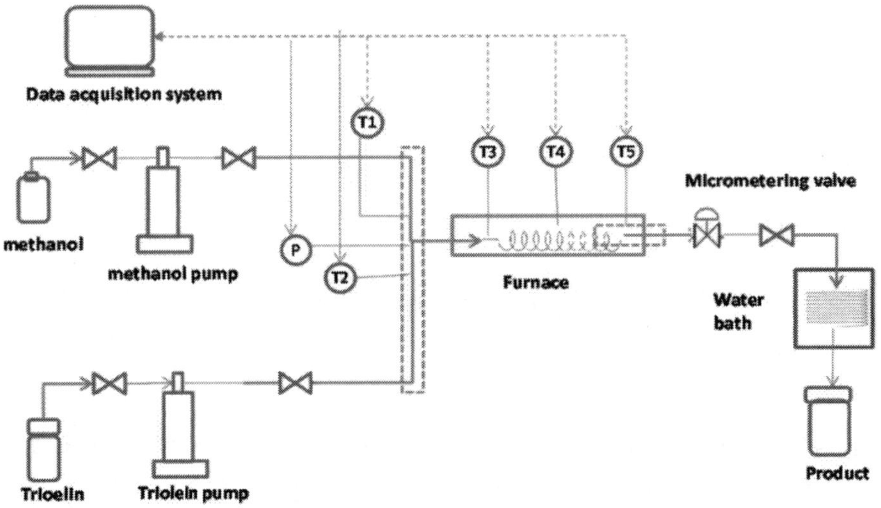

Figure 1. Schematic diagram of experimental setup for supercritical transesterification reactions.

Results and Discussion

Component Profiles

Triglycerides are converted to diglycerides which are further converted to monoglycerides and finally to glycerol. If the reactions were incomplete, there would be triglycerides, diglycerides, and monoglycerides left in the mixtures. The glycerol portion of these compounds is referred to as bound glycerol, which affects the quality of biodiesel. **Table 1** summarizes the reaction conditions and content profiles of each compound in collected samples.

The transesterification conversion (X_{TE}) and BG percentage were calculated by the following equations based on comparison of GC-FID results of reaction products with that of 99% pure triolein:

$$X_{TE}(\%) = 100 * \left(1 - \frac{C_{TO}+C_{DO}+C_{MO}}{C_0}\right) \quad (1)$$

$$\text{bound glycerol(wt\%)} = 0.2591{*}C_{MO} + 0.1488{*}C_{DO} + 0.1044{*}C_{TO} \quad (2)$$

Here C_{TO}, C_{DO}, C_{MO} are contents of triolein, diolein and monoolein after reaction, respectively, in weight %. C_0 is the initial content of triolein before reaction (99.9 wt%). The conversion increased from 91.32% to 97.66% as the residence time increased from 2 to 10 min at 385 °C. Near complete conversion was achieved at 400 °C for a residence time of 6 min. As shown in **Table 1** a methanol-to-triolein molar ratio of 9:1 and residence time greater than 4 minutes yields a high triolein conversion and high FAMEs yield at temperatures greater than 355°C where the reaction system is in the homogeneous state.

Table 1. Species content in collected samples determined by GC-FID.

Run #	T (°C)	P (bar)	τ[a] (min)	GL[b] (wt%)	MO[c] (wt%)	DO[d] (wt%)	TO[e] (wt%)	conversion (%)	BG[f] (wt%)	TG[g] (wt%)
1	385	150	0.5	0.007	1.12	16.15	31.55	50.9	5.98	5.98
2	385	150	1.0	0.026	2.85	14.23	12.17	70.6	4.12	4.20
3	385	150	1.5	0.265	3.20	6.55	8.05	82.2	2.64	2.90
4	385	150	2.0	0.54	3.45	4.07	3.46	89.0	1.25	1.79
5	385	150	4.0	0.50	2.69	0.98	0.15	96.2	0.86	1.36
6	385	150	6.0	0.44	2.30	0.90	0.07	96.7	0.74	1.18
7	385	150	8.0	0.38	2.12	0.69	0.06	97.1	0.66	1.04
8	385	150	10.0	0.36	1.76	0.54	0.04	97.7	0.54	0.90
9	355	150	6.0	0.44	3.58	0.87	1.17	94.4	1.18	1.62
10	370	150	6.0	0.51	3.83	0.89	0.57	94.7	1.19	1.70
11	400	150	6.0	0.18	1.18	0.36	0.03	98.4	0.36	0.54

a- the residence time; b- free glycerol; c- monoolein; d- diolein; e- triolein; f- bound glycerol; and g- total glycerol (GL+BG). The molar ratio of methanol to triolein is 9:1.

It is noted that the product is composed of two phases, the biodiesel or ester phase consisting of triolein, diolein, monoolein, free glycerol and methyl oleate, and the second phase of unreacted methanol and glycerol. As only the mass composition of the product in the biodiesel phase was measured for each set of flow conditions in the reactor, after steady state and stabilized reaction conditions were achieved, a mass balance over the reactor was employed to obtain the weight fraction of each compound in the product. The result is shown in **Table 2**.

Table 2. Species content in the reaction system.

Run #	T (°C)	P (bar)	τ[a] (min)	MO[c] (wt%)	DO[d] (wt%)	TO[e] (wt%)	ME[f] (wt%)	MeOH[g] (wt%)
1	385	150	0.5	0.78	11.28	22.04	35.55	20.72
2	385	150	1.0	1.96	9.78	8.36	48.51	19.32
3	385	150	1.5	2.25	4.61	5.67	57.87	18.3
4	385	150	2.0	2.44	2.88	2.45	63.1	17.74
5	385	150	4.0	1.95	0.71	0.11	69.8	17.02
6	385	150	6.0	1.68	0.66	0.05	70.61	16.93
7	385	150	8.0	1.55	0.51	0.04	71.15	16.87

8	385	150	10.0	1.3	0.4	0.03	71.96	16.78
9	355	150	6.0	2.56	0.62	0.84	67.66	17.25
10	370	150	6.0	2.74	0.64	0.41	67.65	17.25
11	400	150	6.0	0.88	0.27	0.02	73.2	16.65

a- the residence time; b- free glycerol and decomposition products; c- monoolein; d- diolein; e- triolein; f- methyl oleate; and g- methanol.

Critical Properties of Triolein-Methanol and Reaction Mixtures

To determine whether the reaction mixtures are in a supercritical state the critical temperatures and pressures of triolein-methanol mixtures were estimated by Peng-Robinson Equation of State.[12] The critical properties and other parameters of the pure components are listed in **Table 3**. The critical temperatures of triolein-methanol mixtures as a function of methanol-to-triolein molar ratio are plotted in **Figure 2** which shows that triolein-methanol feed mixtures (385 °C, 150 bar, molar ratio 9:1) were not in a supercritical state.

Table 3. Physical properties of triolein and methanol[13]

Compound	M (kg/kmol)	Tc (°C)	Pc (bar)	ω
Methanol	32	239	80	0.565
Triolein	884	705	3.34	1.987

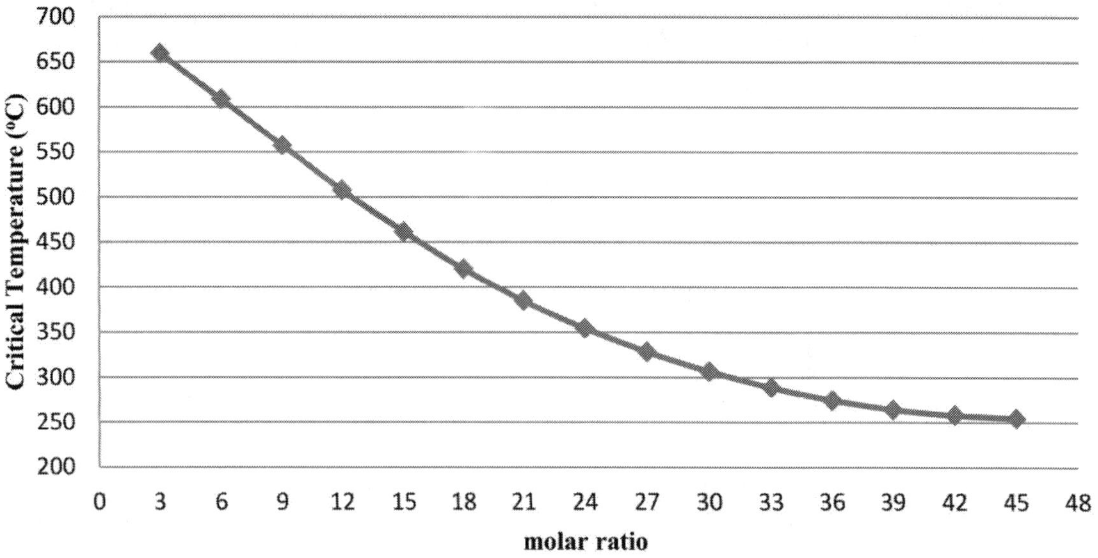

Figure 2. Critical temperature of triolein-methanol mixtures at different molar ratios.

Further, to demonstrate that the evolving reaction mixtures during the course of the reaction did not reach a supercritical state, the phase envelope of the reaction mixtures at different residence times (**Table 2** , Runs 1-8) in the reactor were calculated and shown in **Figure 3.** It is seen that the reaction system is in a homogeneous liquid state (385 °C, 150 bar) from the start of the reaction assuming complete mixing. With the consumption of triolein and methanol and the appearance of methyl oleate, the enclosed area under P-T phase envelope shrinks quickly until four minutes when reaction reaches near equilibrium. Thus the mixtures in the reactor should be in a homogeneous liquid state. The critical temperature drops from 557°C to 467°C in 0.5 min with the appearance of methyl oleate and other reaction products, and is 425 °C at 10 minutes. Thus the supercritical state of the reaction mixture was not reached.

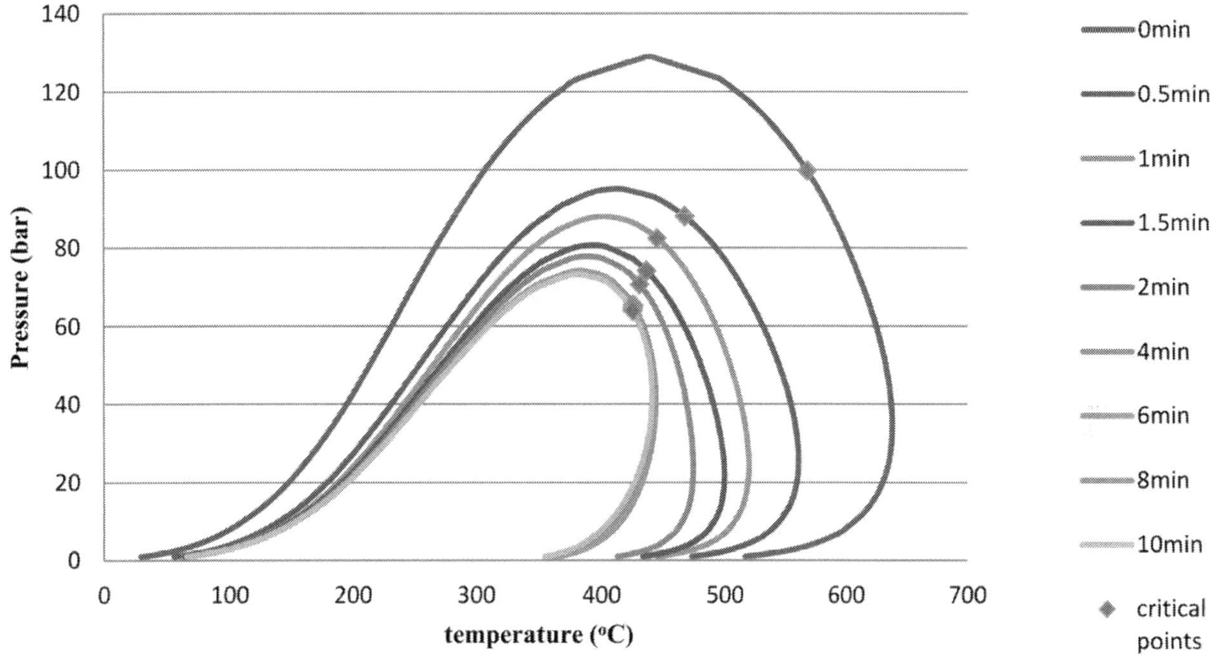

Figure 3. Pressure- temperature phase envelope of reaction mixtures at different residence time in the reactor.

Kinetic model

Following the consecutive thermal transesterification reactions proposed by Noureddini and Zhu,[14] a three-step kinetic model was constructed below to evaluate the experimental data:

$$TO + MeOH \xrightarrow{k_1} DO + ME \tag{3}$$

$$DO + MeOH \underset{k_{-2}}{\overset{k_2}{\rightleftarrows}} MO + ME \tag{4}$$

$$MO + MeOH \underset{k_{-3}}{\overset{k_3}{\rightleftarrows}} GL + ME \tag{5}$$

where TO, DO, MO and ME are triolein, diolein, monoolein and methyl ester.

The kinetics of the transesterification process can be described by the following differential equations [15], wherein, it is assumed that the first reaction is irreversible, the last two are reversible and all steps are second order. Concentrations units are mass fraction. Data in **Table 2** were analyzed by this model and comparisons are shown in **Figure 4.** The Averaged Absolute Relative Deviation is 0.13.

$$\frac{dX_{TO}}{d\tau} = -\frac{1}{Mw_{MeOH}} k_1 X_{TO} X_{MeOH} \tag{6}$$

$$\frac{dX_{DO}}{d\tau} = \frac{Mw_{DO}}{Mw_{TO} \times Mw_{MeOH}} k_1 X_{TO} X_{MeOH} - \frac{1}{Mw_{MeOH}} k_2 X_{DO} X_{MeOH} + \frac{Mw_{DO}}{Mw_{MO} \times Mw_{ME}} k_{-2} X_{MO} \times X_{ME} \tag{7}$$

$$\frac{dX_{MO}}{d\tau} = \frac{Mw_{MO}}{Mw_{DO} \times Mw_{MeOH}} k_2 X_{DO} X_{MeOH} - \frac{1}{Mw_{ME}} k_{-2} X_{MO} X_{ME} - \frac{1}{MeOH} k_3 X_{MO} X_{MeOH} + \frac{Mw_{MO}}{Mw_{GL} \times Mw_{ME}} k_{-3} X_{GL} X_{ME} \tag{8}$$

$$\frac{dX_{GL}}{d\tau} = \frac{Mw_{GL}}{Mw_{MO} \times Mw_{MeOH}} k_3 X_{MO} X_{MeOH} - \frac{1}{Mw_{ME}} k_{-3} X_{GL} X_{ME} \tag{9}$$

$$\frac{dX_{MeOH}}{dt} = -\frac{1}{Mw_{TO}} k_1 X_{TO} X_{MeOH} - \frac{1}{Mw_{DO}} k_2 X_{DO} X_{MeOH} - \frac{1}{Mw_{MO}} k_3 X_{MO} X_{MeOH} + \frac{Mw_{MeOH}}{Mw_{MO} \times Mw_{ME}} k_{-2} X_{MO} X_{ME} + \frac{Mw_{MeOH}}{Mw_{GL} \times Mw_{ME}} k_{-3} X_{GL} X_{ME} \tag{10}$$

$$\frac{dX_{ME}}{d\tau} = \frac{Mw_{ME}}{Mw_{TO} \times Mw_{MeOH}} k_1 X_{TO} X_{MeOH} + \frac{Mw_{ME}}{Mw_{DO} \times Mw_{MeOH}} k_2 X_{DO} X_{MeOH} + \frac{Mw_{ME}}{Mw_{MO} \times Mw_{MeOH}} k_3 X_{MO} X_{MeOH} - \frac{1}{Mw_{MO}} k_{-2} X_{MO} X_{ME} - \frac{1}{Mw_{GL}} k_{-3} X_{GL} X_{M} \tag{11}$$

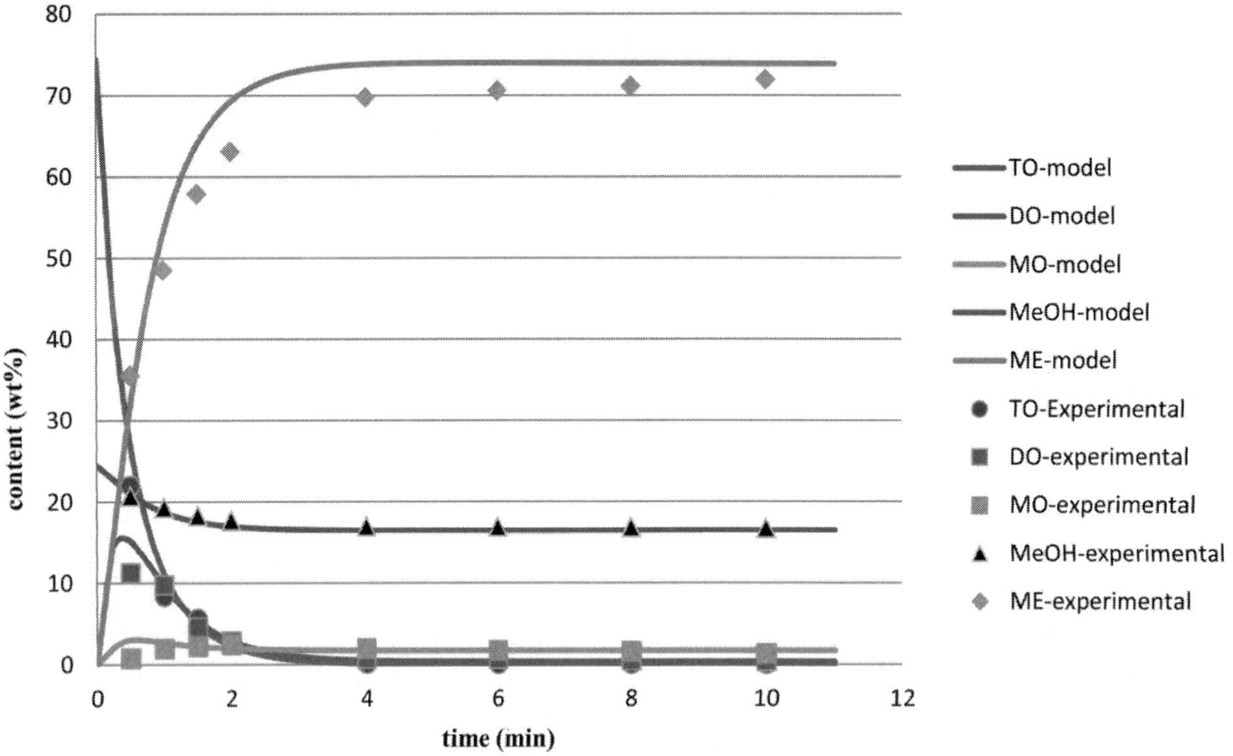

Figure 4. Comparison of the three-step kinetic model simulation with experimental results in Table 2 for 385 °C and 150 bar.

Conclusions

Transesterification of triolein with methanol was carried out at 355 – 400 °C, 150 bar, methanol-to-triolein molar ratio of nine, and residence times of 0.5 – 10 min. At the reaction conditions of 385 ^{0}C and 150 bar, the reaction system is shown to be in the liquid state throughout the reaction. A three step, consecutive second order reaction mechanism was employed to model the system. The evolution of concentration of each component in the homogenous reaction system can be well predicted by the kinetic model.

References

1. Marulanda, V.; Anitescu, G.; Tavlarides, L. *J. Supercrit. Fluids* 2010, *54,* 53.
2. Rathore, V.; Madras, G. *Fuel* 2007, *86,* 2650.
3. Saka, S.; Kusdiana, D. *Fuel* 2001, *80,* 225.
4. He, H.; Wang, T.; Zhu, S. *Fuel* 2007, *86,* 442.
5. Pinnarat, T.; Savage, P. *Ind. Eng. Chem. Res* 2008, *47,* 6801.
6. Warabi, Y.; Kusdiana, D.; Saka, S. *Bioresource Technol.* 2004, *91,* 283.
7. Kusdiana, D.; Saka, S. *Fuel* 2001, *80,* 693.
8. Demirba, A. *Energ. Convers. and Manage.*2002, *43,* 2349.
9. Demirba, A. *Energ. Convers. and Manage.* 2003, *44,* 2093.
10. Marulanda, V.; Anitescu, G.; Tavlarides, L. *Energ. Fuel.* 2010, *24*, 253.
11. Srivastava, A.; Prasad, R. *Renew. and sust. energ. rev.* 2000, *4,* 111.
12. Tang, Z., Do, Z. X., Min, E. Z., Gao, L., Jiang, T., & Han, B. X. (). *Fluid Phase Equilibria.*2006, *239*(1), 8-11.
13. Demirbas, A. *Bioresource Technol.* 2008. *99,* 1125.
14. Noureddini, H.; Zhu, D. *J. Am. Oil Chem. Soc.* 1997, *74,* 1457.
15. Diasakou, M.; Louloudi, A.; Papayannakos, N. *Fuel* 1998, *77,* 1297.

(677f) Reaction Kinetics for the Production of Hydrogen From Bioethanol in an Inconel 625 Reactor From 600 to 700°C

Jonathan E. Wenzel, Chemical Engineering, Kettering University, Flint, MI and Lisandro Hernandez de la Pena, Chemistry and Biochemistry, Kettering University, Flint, MI

Ethanol produced from renewable resources may be reformed in supercritical water to produce hydrogen. For this reaction, supercritical water acts both as a highly energetic reactant and solvent. The reformation of ethanol was studied in a 926 mL tubular Inconel 625 Grade 1 reactor without the aid of a heterogeneous catalyst. The reformation reactions were evaluated between 600 and 700 C at 24.2 MPa with space times between 160 and 500 seconds and ethanol to water mass feed ratios between 1 and 17%. Furthermore, ethanol and water formed a binary supercritical mixture at the inlet conditions of the reactor. The products in order of descending order of prevalence were hydrogen, carbon dioxide, methane, ethane, and carbon monoxide. All of the ethanol fed into the reactor was reacted. The global reactions and revised reaction kinetic rate information will be discussed. In addition, the proposed rate laws and regressed activation energies will be presented.

(686a) Exploring and Manipulating Dense Packing Properties of Particles in Different Continuous Phase

Tianying Jiang, Chemical and Biomolecular Engineering, University of Illinois Urbana-Champaign, Urbana, IL and Charles F. Zukoski, Department of Chemical & Biomolecular Engineering, University of Illinois at Urbana Champaign, Urbana, IL

When introducing size polydispersity or concentrating colloid dispersion quickly, dynamic arrested state has been entered with particles packing randomly without formation of long range order to create crystal phase. In this region, the long range diffusivity of the particles is greatly hindered by cooperative motions. Developed knowledge in this dense packing state is important for practical application considering the potential in designing materials with novel optical and mechanical properties. Efforts have been made in the past to understand the relaxation behavior of colloidal dispersions in this dynamic arrested state both theoretically and experimentally. However, no conclusion has been drawn to explain the dynamics of colloids when approaching the condition where the dispersion cannot be processed. Here we use rheology study to explore the dynamics of colloid in different continuous phases when approaching the maximum processable volume fraction. Results of linear elasticity, yielding stress and nonlinear rheology are compared to understand the short-range diffusion and long-range relaxation to ultimately explain the macroscopic change tendency near the volume fraction where dispersions cannot flow. Different continuous phases are applied to control the particle interaction which is deformed when bringing particles into close proximity. Of particular interest is our ability to alter the particle maximum dense packing volume fraction by working with a binary mixture of particles where we vary the fraction of the total volume fraction occupied by a certain species, and different behavior of this binary mixture is also observed in flow properties compared to homogeneous colloidal dispersions.

(686b) Re-Entrant Behavior in Colloidal Disks with Adsorbing Polymer

Anand Atmuri, Chemical Engineering, University of Massachusetts Amherst, Amherst, MA and **Surita Bhatia**, Chemical Engineering, University of Massachusetts, Amherst, Amherst, MA

We report a re-entrant colloidal glass transition in dispersions of model disk shaped colloid, laponite®, with the addition of an adsorbing low molecular weight polymer, poly(ethylene oxide) (PEO). In basic solutions laponite® forms a repulsive colloidal glass. With the addition of low molecular weight PEO, we observe a retardation in formation of the glass, and the system behaves as a viscous fluid. Upon aging, a glassy phase re-forms, with the time scale depending on the concentration of polymer. For samples with polymer concentration high enough to saturate the laponite surface, we believe this arrested state to be an attractive glass where excess PEO chains in the suspension lead to a weak interparticle attraction. For the polymer molecular weight we considered, all the samples eventually age to form a disordered state. We combine our results with previous work on laponite®-PEO dispersions to propose a qualitative state diagram with attractive glass, repulsive glass, liquid, and gel regimes as a function of polymer molecular weight, polymer concentration, and aging time. To our knowledge, the progression of repulsive glass-liquid-attractive glass states has not been previously observed in colloids with adsorbing polymers.

(686c) Re-Entrant Viscoelastic Behavior in Thermoresponsive Nanoemulsion Organohydrogels

Juntae Kim, Chemical Engineering, UC Santa Barbara, Santa Barbara, CA and Matthew E. Helgeson, Chemical Engineering, University of California, Santa Barbara, Santa Barbara, CA

Recently, we reported the discovery of novel gelling suspensions based on oil-in-water (O/W) nanoemulsions that thermoreversibly form "organohydrogels" – solid-like gels comprised of up to 90% oil and water – at high temperature due to the presence of a thermoresponsive polymer [1]. Here, we report on similar O/W nanoemulsions which exhibit re-entrant viscoelastic behavior, whereby gel-like phases are observed at both low and high temperatures, with an intermediate excursion into a low-viscosity liquid phase. Using both linear and non-linear rheological probes to fingerprint the mechanical properties at both temperature extremes reveals significantly different viscoelasticity driven by relatively mild changes in polymer solution behavior. At low temperatures the material forms a transient gel where the droplets act as a physical cross-linker for the polymer, whereas at high temperature it forms a percolated network of droplets. Coupling rheological techniques with *in situ* scattering measurements allows for detailed characterization of the microstructure in the respective phases, and in particular how the structure responds to large-scale deformation to give rise to a number of nonlinear rheological phenomena, including yielding and shear banding. Ultimately, we find that the nature and behavior of both the low and high-temperature phases is critically linked to the strength and the range of inter-droplet attractions. This key result serves as a general guideline for the design of viscoelastic phases in colloid-polymer mixtures, and using it we show how the re-entrant viscoelastic and non-linear behavior can be programmed based on several key parameters, including the nanoemulsion droplet size as well as polymer and surfactant chemistry.

[1] M.E. Helgeson, S.E. Moran, and P.S. Doyle, "Thermogelling, photocrosslinkable nanoemulsions for the formation of tunable mesoporous organohydrogels", Nature Materials, 2012, 11(4): 344-352.

FUNDAMENTAL INTERFACIAL BEHAVIOR OF CONFINED COLLOID-POLYMER SYSTEMS

Mariam Nouri[*], *Rice University, Houston, TX, Ryan McGorty, Harvard University, Cambridge, MA, Vinothan N. Manoharan, Harvard University, Cambridge, MA, Marc Robert, Rice University, Houston, TX*

Abstract

The phase equilibria of confined aqueous colloid-polymer systems are studied experimentally using fluorescence microscopy. The colloidal particles are fluorescent polystyrene spheres and the polymer is polyacrylamide, for which the ratio of the radius of gyration to the colloidal radius equals 0.73. The mixture is confined between a glass slide and a coverslip, such that the spacing between the glass plates varies between 3 - 6 microns.

The phase diagram is determined as a function of the colloidal and polymer concentrations. Liquid–liquid phase coexistence between a colloid-rich phase and a polymer-rich phase occurs at intermediate polymer concentrations, while liquid–solid phase coexistence between a polymer-rich liquid and a colloid-rich solid is observed at high polymer concentrations. The thickness of the interface between these coexisting phases is measured using image analysis. For liquid-liquid coexistence, the interfacial thickness is found to diverge as the critical point is approached. It is also observed that the colloid-rich solid and liquid domains coarsen mainly by Ostwald ripening.

Introduction

Colloid-polymer mixtures and their phase behavior in three dimensions have been subject of intensive research. In contrast to three-dimensional colloid-polymer systems [1], little is known about their two-dimensional counterparts. Qualitatively, one would expect two-dimensional systems to behave similarly to three-dimensional ones, exhibiting in particular similar phase transitions. However, many fundamental questions are yet to be answered, in particular concerning interfacial properties and their critical behavior, which are predicted to be strongly abnormal [2]. Consequently, one of the main objectives of this study is to investigate the properties of the interface between coexisting liquid phases in two dimensions.

Many features of phase equilibria are shared between mixtures of colloids and polymers and mixtures of simpler molecules. An important advantage of using colloidal systems is that for sufficiently large colloids, they are directly visible under the microscope, which allows one to identify the nature of the phases. It also enables one to obtain information about the interface between coexisting phases in two-dimensional systems, which is not visible to the naked eye in contrast to the case of three dimensions.

Phase separation in colloidal suspensions can be caused by addition of non-absorbing polymer. This is due to an effective attraction between colloidal particles, which is induced by an imbalance in osmotic pressure, as two colloidal particles approach each other and exclude the polymer. This effective attractive force is called the depletion force, a theoretical model of which was first proposed by Asakura and Oosawa [3]. This phase separation caused by depletion has been described theoretically [4] and observed experimentally [5]. For a colloid of diameter σ and a polymer with a radius of gyration r_g, it is well known that below a polymer-to-colloid size ratio $s \sim .31$, which is defined as

[*] corresponding author: mn9@rice.edu

$2r_g/\sigma$, liquid-liquid coexistence is predicted not to occur according to mean-field theory, but it does according to computer simulations [6].

Early considerations of fluids considered the interface between two fluid phases to be of essentially zero thickness, a reasonable assumption far from criticality. However, van der Waals introduced the idea of an interface of nonzero thickness, which arises from a continuous density gradient between the phases, as one crosses the interface. The free energy of the system is assumed to be a function of this density gradient, and so by minimizing the free energy, it is possible to directly compute that density profile, and consequently, determine the interfacial thickness. His ideas were later expanded by widom [7].

Another view of the interface comes from capillary-wave theory, which describes the interface as some hypothetical sharp boundary, upon which are superposed waves of varying amplitudes and wavelengths [8]. These waves are assumed to be thermally driven fluctuations, and the amplitudes of each mode are determined by the surface tension – larger amplitudes increase the effective surface area, and so are less probable at high surface tension. If one averages over all these waves, a continuous density profile obtains in the form of an error function, which is different from that predicted by van der Waals theory, which gives a hyperbolic tangent in its classical formulation and a modified exponential in its non-classical formulation [7].

Furthermore, if one looks at the critical divergence of the interfacial thickness in each theory, they give the same result in three or more dimensions. However, remarkably, their predictions disagree in two dimensions, where they give different values for the critical exponent for the divergence of the interfacial thickness [2]. Additionally, capillary-wave theory predicts that this divergence exponent depends on the presence and nature of an external field, while the van der Waals theory assumes that external fields are irrelevant, only serving to render the interface planar. It is left to experiments to resolve the discrepancy between the competing theories.

Experimental methods

Materials

Studies of colloid-polymer systems are generally carried out in aqueous media, as this is a convenient and innocuous solvent. Incidentally, there are a variety of colloidal particles and polymers which are used frequently for their solubility in water. It is necessary, however, to consider the size ratio s, and to select a combination of colloid and polymer which will maintain this ratio at a value greater than ~.31 to ensure the presence of a liquid-liquid coexistence region [4]. At the same time, care must be taken to ensure that the colloid particles are resolvable under microscopy. If the particles are not easily seen under the microscope, the measurement of the interfacial thickness may be difficult to perform accurately. On the other hand, the polymer must not be too large, as high molecular weight polymers will increase the viscosity of the medium significantly, making it practically difficult to work experimentally. In some cases, they may even form gels. Thus an upper bound on the molecular weight of the polymer is called for.

A polymer with one of the largest radii of gyration which does not immediately form gels in aqueous solution is polyacrylamide (PA), which is available commercially at high molecular weight. PA at an average molecular weight of ~18,000,000 corresponds to a radius of gyration of 73nm. In order to maintain our minimum size ratio s, we must pair this polymer with a colloid no larger than 470nm. We chose particles significantly smaller than these upper bounds, in order to drive up the ratio s, which broadens the two-phase region, making the task of finding the critical point easier. Silica and polystyrene are both available as colloids in a very wide range of sizes, spanning these bounds and well below. Particles of either of these materials are also available with fluorescent tags.

In our studies, samples are made by mixing solutions of fluorescent polystyrene beads (PS) and polyacrylamide (PA) in water. PA has a molecular weight of 18 million and the PS beads have a diameter of 200 nm. The s value is equal to 0.73, and hence this system is theoretically expected, according to mean-field theory, to exhibit fluid-fluid coexistence for appropriate values of the concentrations. It may be noted that another colloid-polymer system consisting of poly-N-Isopropylacrylamide (PNIPAM) and polyacrylamide has been studied and also been found to exhibit fluid-fluid coexistence. PNIPAM has a radius of 110 nm and the polyacrylamide with a molecular weight of 5 to 6 million has a radius of gyration of 36 nm [9].

Sample preparation

Achieving a truly two-dimensional system is difficult. Previous attempts have been made to construct such systems at interfaces between two bulk three-dimensional phases, i.e. at air-water interfaces. One of the biggest challenges facing this kind of system is the distortions introduced to the surface by the colloid particles, whose presence causes a curvature of the interface, and can introduce an unwanted effective surface tension-driven potential.

On the other hand, more success has been achieved with confined systems, where a three-dimensional system is constructed in such a fashion that one dimension is restricted to a scale on the order of the size of the particles. This is frequently achieved in practice through the use of thin spacers to separate parallel walls of the system, where the dimensions of the walls are orders of magnitude larger than the thickness of the spacers. Spacers may be thin sheets, such as Mylar films, placed around the edges of the constructed cell, or even a small concentration of colloidal particles larger than those under study.

Generally, concentrated stocks of both colloid and polymer are kept on hand, and these are diluted to the final concentrations for each sample in tubes at volumes around 0.5ml. Each of these samples is large enough for several hundred sample cells – their large volume, however, is needed to guarantee good mixing. Such samples may be allowed to rest for as much as a day to allow the system to achieve homogeneity. From these bulk samples, a small drop is prepared in a cell, between clean glass slides and coverslips sealed with epoxy, as described below. To clean the coverslips and glass slides, they are rinsed with deionized water, left to dry, and rinsed with alcohol (200 proof), and left to dry again.

A day after making such a sample, it is sonicated again for 10-15 min and then a small amount of it, usually 2-5 μL in form of a drop, is placed on a glass slide coverslip, allowing the system thickness to be defined by the volume of the drop and the area of the coverslip. A 22X40 mm coverslip is carefully placed on top of the drop and the sample is sealed with regular (5-minute) epoxy glue. Assuming that the drop fills the gap uniformly, 2-5μL theoretically results in a gap of 2.3-5.7μm. For 22X22 mm coverslips, the drops are 2-3μL in volume, and the theoretical thickness of the sample is 4.1-6.2μm. In order for the sample to equilibrate it is left to rest for at least 3 hours. The sample is then examined under the microscope as often as possible. Sometimes it takes days to see phase-separated regions.

Microscopy and image analysis

Observation of the equilibrated samples takes place under fluorescent microscopy with an oil-immersion 100x/1.40 objective, to visualize the fluorescently-tagged colloid particles. It is possible to ascertain the qualitative phases visually – that is, whether the system lies in the solid-liquid, liquid-

liquid, or single phase regions of the phase diagram. In the solid-liquid region, the boundaries between the two coexisting phases are sharp, and the solid regions tend to have hexagonal edges. On the other hand, in the liquid-liquid region, the interface between the two coexisting phases is more diffuse and undulations are observed to propagate on the interface. Videos are taken from different areas of the samples which lie in the region of interest, i.e., in the liquid-liquid region near the critical point.

Fluorescent tags guarantee high contrast between the colloidal particles and the aqueous polymer medium, allowing image intensity to serve as a proxy for colloid concentration. Thus, droplets of colloid liquid will appear as contiguous bright patches in these images. To make this kind of measurement more precise, the background intensity is measured independently, and calibration curves correlating colloid concentration to intensity are determined.

From images taken of single droplets, we then measure how the intensity changes from the center of a droplet to the bulk medium. In this way, the colloid density gradient across the interface can be directly measured, and the interfacial thickness determined as per van der Waals theory. Likewise, with video, the wave-like fluctuations of the interface are also measured by Fourier analysis, and the interfacial thickness determined also as per capillary wave theory. It is anticipated that both of these measurements should yield diverging interfacial thickness as the critical point is approached. Note that these measurements are static and should not depend on time; however, the longer a droplet is observed the more configurations there are to be sampled. If one only looks at a droplet for a short amount of time, the interface structure does not have the time to evolve and provide sufficient information on its equilibrium value.

Determination of interfacial tension by Fourier analysis of capillary waves

The interface between two coexisting phases is not flat. Undulations on this rough interface are known as capillary waves. The statics of thermally driven capillary waves for two-dimensional droplets was studied theoretically by Tüzel at al [10]. The following figure (Fig 1) illustrates a fluctuating drop in two dimensions, whose coordinates are mapped from Cartesian (x, y) to polar coordinates, i.e., (r, θ). Here it is also assumed that the droplet is incompressible, and as a result the enclosed area of the droplet will be fixed.

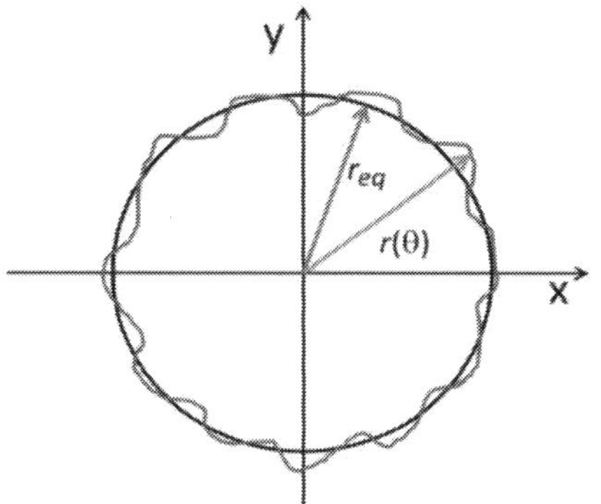

Figure 1- A schematic of a fluctuating drop

Fourier analysis of the capillary waves will result in an expression for the interfacial tension in terms of temperature and average values of the Fourier coefficients, given by

$$\langle {a_k}^2 \rangle + \langle {b_k}^2 \rangle = \frac{2k_B T}{\pi r_{eq} \sigma} \left(\frac{1}{k^2 - 1} \right) \tag{1}$$

As stated above, in the capillary wave model, the finite thickness of an interface results from the wave-like thermal fluctuations on a flat sharp interface. In order to find this flat sharp interface for a droplet, it first needs to become isolated from the rest of the image (steps A and B in Fig. 2). Then the gray-scale droplet image is thresholded and converted to a black and white image (step C). Edge detection routines in MATLAB allow for the isolation of the perimeter (step D). The boundary of the droplet is then parameterized in distance from the drop centroid r and the polar angle θ (step E). The centroid is defined as the geometric center of a noncircular object. $r = 0$ corresponds to the average droplet radius. The Fourier coefficients are then computed for 100-2000 configurations (frames) and averaged. The relation between the averaged Fourier coefficients and the mode number is then used according to Eqn. 1 to determine the interfacial tension.

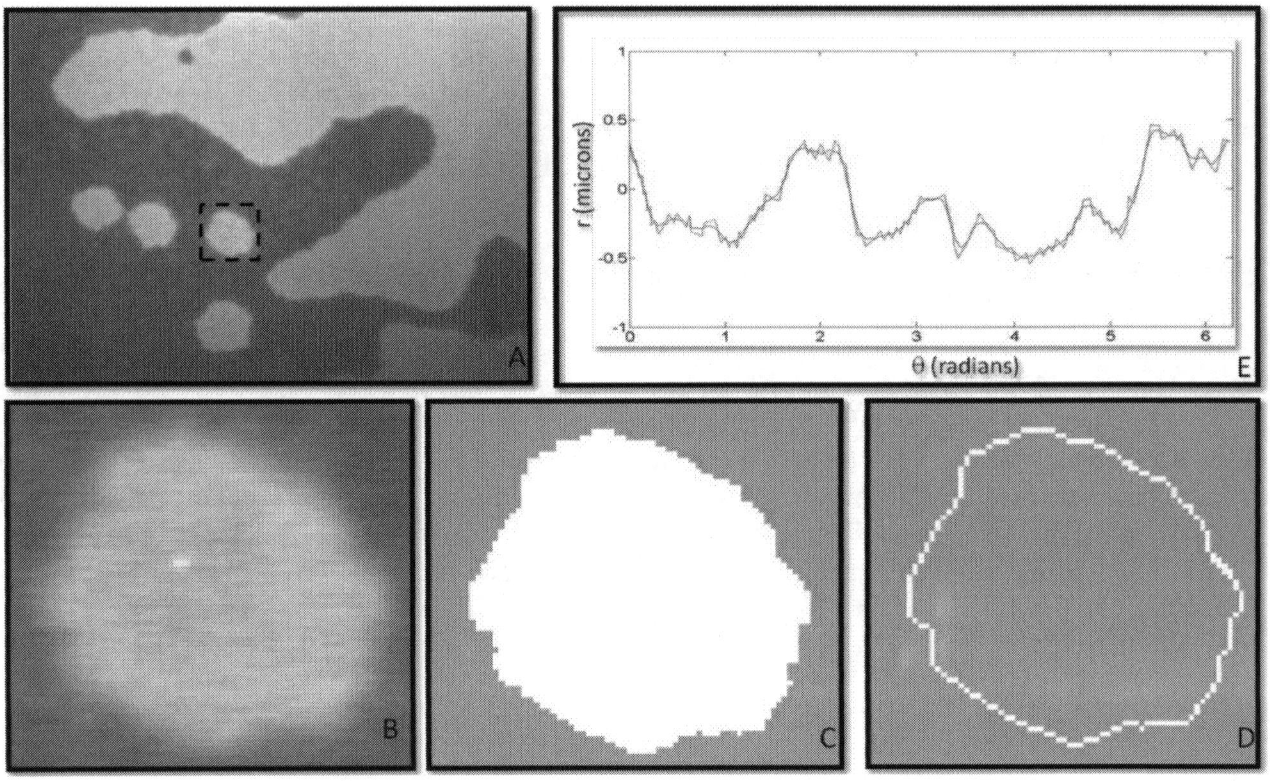

Figure 2 - Steps of image processing for determination of the interfacial tension

Determination of interfacial thickness

Drops are isolated from the rest of the image and their centroids are found. Light intensity is measured as a function of distance from the centroid. The intensity profile is then fitted to an error function and the interfacial thickness is computed from the fitting parameters. The interfacial thickness is then normalized by the droplet radius.

Results

The experimental phase diagram is presented in Fig. 3 below. Sample images of the solid-liquid, liquid-liquid, and the one-phase region are shown on the right. As described above, one can readily identify each phase by direct observation. For four samples, labeled a, b, c and d, the interfacial thickness and tension are determined as explained above.

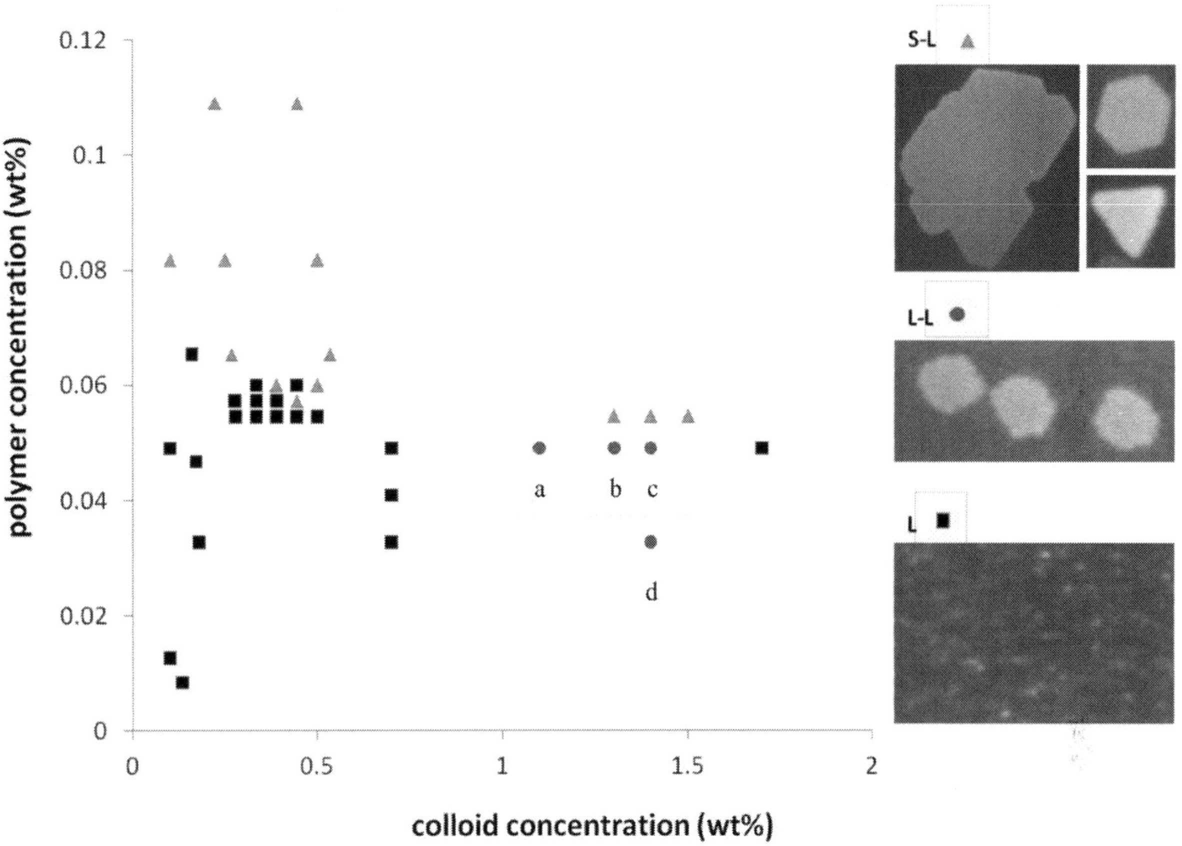

Figure 3- Left: Phase diagram of PA and PS. Solid-liquid, liquid-liquid and one-phase liquid are shown by green triangle, red circle, and black square symbols respectively. Right: Sample images of each phase

The critical point has not yet been located precisely. In contrast to thermal systems, where the temperature can be readily be tuned and controlled, in athermal systems, such as the present one, concentrations can only be changed by preparation of a completely new sample, making it difficult to find the exact location of the critical point. However, there is a strong evidence for the approximate location of the critical point from the analyses of the interfacial thickness. Keeping the polymer concentration constant and moving from point *a*, to points *b* and *c*, the average interfacial thickness increases, implying that the critical colloid concentration must be larger than the colloid concentration at point *c*. Keeping the polymer concentration constant and moving from points *c* to *d*, an increase in the interfacial thickness is observed. Therefore, it can be inferred that the critical polymer concentration must be smaller that the polymer concentration at point d. Unfortunately, until the critical point is exactly located, the values of the critical exponents cannot be accurately determined.

However, using image analysis techniques and Fourier analysis of the capillary waves, we can qualitatively demonstrate that, as the critical point is approached, the interfacial tension σ vanishes and the interfacial thickness L diverges. The data presented in table 1 are obtained from droplets with an average radius of about 5 microns. Only droplets which are roughly circular are considered in the analysis. For each sample, over 10 droplets are analyzed except for point *d*, where the droplets are not well-defined, and it is difficult to find droplets which maintain a constant area and do not coalesce instantly with neighboring droplets. In the determination of the interfacial tension, it is assumed that the interfacial fluctuations are relatively small. But as one approaches the critical point, this assumption becomes inappropriate because the interfacial fluctuations become large and can no longer be decomposed into single-values sinusoidal waves. An example of these large fluctuations is presented below in Fig. 4. The weak contrast between the phases is a manifestation of the fact that the concentrations of the coexisting phases become increasingly close to each other, as they do upon approach to criticality.

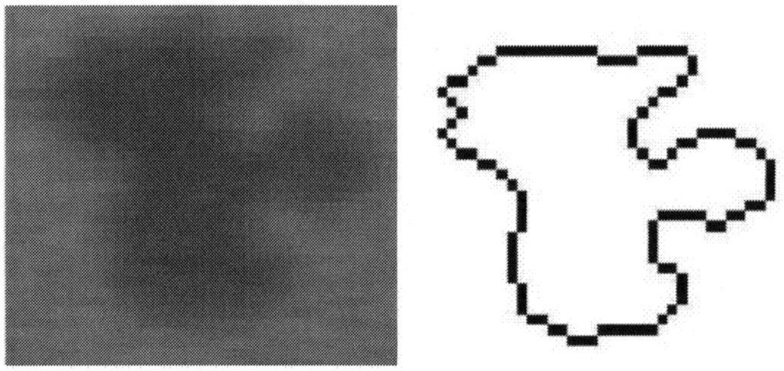

Figure 4 - Large interfacial fluctuations close to the critical point. Left: original image. Right: Droplet boundary

Table 1. Average interfacial thickness and tension

sample name	a	b	c	d
normalized interfacial thickness	0.249	0.328	0.441	1.271
interfacial tension (10^{-20} N)	3.73	1.71	0.762	0.627

Conclusion

The phase diagram of PA (18 million molecular weight) and PS (200 nm diameter) is mapped out in composition space. Two-phase solid-liquid and liquid-liquid coexistences and the one-phase liquid are observed experimentally. It is also shown experimentally, in accord with theory, that the interfacial thickness and the interfacial tension diverges and vanishes, respectively, upon approaching the critical point. This work reports the experimental observation of depletion induced liquid-liquid coexistence, as well as the structure of interface separating coexisting liquid phases in two-dimensional colloid-polymer systems. The liquid-liquid coexistence region in the phase diagram in Fig. 3 is rather small, and the proximity of the critical point difficult to avoid, making it difficult to explore coexisting phases that do not exhibit large interfacial fluctuations. Consequently, we are planning to use smaller colloids to increase the s-value and broaden the liquid-liquid coexistence region, which will enable us to sample points that are far enough from the critical point, in accord with the small fluctuation assumption of our analysis.

Acknowledgement

We thank Dominique Langevin for helpful discussions, and Matteo Pasquali for access to his laboratory. Support from the National Science Foundation under Grant No. 1034421 is gratefully acknowledged.

References

1. D. Aarts, M. Schmidt, and H. Lekkerkerker, Science **304**, 847 (2004)
2. M. Robert, Physical Review A, **30** 2785-2787 (1984).
3. S. Asakura, F. Oosawa, Journal of Chemical Physics, 22 1255-1256 (1954).
4. J.T. Lee, M. Robert, Physical Review E, **60** 7198-7202 (1999).
5. T.C. Lee, J.T. Lee, D.R. Pilaski, M. Robert, Physica A, **329** 411-430 (2003)
6. C.Y. Chou, T.T.M. Vo, A.Z. Panagiotopoulos, M. Robert, Physica A, **369** 275-290 (2006).
7. S. Fisk and B. Widom, J. Chem. Phys. **50**, 3219- 3228 (1969)
8. F.P. Buff, R.A. Lovett, F.H. Stillinger, Physical Review Letters, **15** 621-623 (1965).
9. R. McGorty, PhD thesis, Harvard University, 2011
10. E. Tüzel, G. Pan, and D. Kroll, J. Chem. Phys. **132**, 174701 (2010)

(686e) Confinement-Induced Solidification of Colloid-Polymer Mixtures

Melissa Spannuth[1] and **Jacinta C. Conrad**[1,2], (1)Department of Chemical and Biomolecular Engineering, University of Houston, Houston, TX, (2)Petroleum Engineering Program, University of Houston, Houston, TX

Using a model colloid-polymer suspension, we show that confinement induces solidification in attractive colloidal suspensions via a fundamentally different route from that active in hard sphere colloidal suspensions. For a range of polymer concentrations, the suspensions undergo a phase transition from a colloidal fluid of clusters to a colloidal gel as confinement increases while polymer and particle concentration are held constant. In both fluid- and solid-like attractive suspensions, effects of confinement on the structure and dynamics appear at much larger thicknesses than for hard-sphere suspensions. The solidification does not originate from structuring of the colloids by the walls. Instead, by analyzing cluster size distributions in the fluid phase and particle dynamics in the gel phase as a function of confinement, we find that the strength of the effective interparticle attraction increases as the samples are confined. We show that the increase in the effective attraction can be understood as a consequence of the increasing importance of excluded volume due to the walls to the free energy of the polymer as confinement is increased.

(686f) Gelation Versus Phase Separation: Gravitational Effects On Adhesive Hard Sphere Colloidal and Nanoparticle Dispersions

Jung Min Kim, Department of Chemical and Biomolecular Engineering , University of Delaware, Newark, DE, Aaron P. R. Eberle, Advanced Characterization, ExxonMobil Research and Engineering Company, Annandale, NJ, Jun Fang, Analytical and Systems Research , Arkema Inc. , King of Prussia, PA, Ramón Ramón, Physical Engineering Department, University of Guanajuato, Leon, Mexico and Norman J. Wagner, Department of Chemical and Biomolecular Engineering, University of Delaware, Newark, DE

Tuning either the volume fraction or the interparticle potential of colloidal dispersions can lead to a diverse spectrum of dynamically arrested states ranging from repulsive driven- and attractive driven- glasses [1], nonergodic gels [2, 3], and to self-similar fractals [4]. The most popular model systems for studying the microstructure, properties, and formation kinetics of colloidal gels are colloid-polymer mixtures [2, 5] and thermoreversibly gelling octadecyl silica particles [3, 6]. Differences in the gel strength, microstructure, and phase behavior are reported in the literature between these systems, which can be related in part to the differences in the nature of the attractive interactions. In this work, we report experimental results on thermoreversible gelling systems with varying particle size and fixed coating, such that the range of attraction is systematically varied. Rheology, small angle neutron scattering, and fiber optic quasi-elastic light scattering are used to determine the gel point, microstructure and strength of the attractive interaction and thereby develop a state diagram. The gel boundary is found to depend systematically on particle size, thus violating the proposed principle of corresponding states by Noro and Frenkel [7]. A density-mismatch is suspected to be the possible reason for this unexpected behavior based on prior experimental and theoretical studies [8, 9] on the effect of gravity on the competition between gelation and phase separation of colloidal dispersions. To test our hypothesis, we studied gelation in the presence of increased gravitational acceleration by centrifugation. To understand this behavior we calculate the gravitational Péclet number (Pe_g), a comparison of a relative magnitude between Brownian diffusion and gravitational settling of the aggregates. We find that Pe_g can predict the interplay between gravity, particle size, and volume fraction and the occurrence of the gravitational settling. Our experimental findings are corroborated by Monte Carlo simulations, which display the same quantitative separation boundary and densification of the particle-rich phase when the attraction is turned on at a given Pe_g. Thus, the density-mismatch is responsible for the particle size-dependence of the gel transition of octadecyl silica particles. These results provide guidance for the rational design of materials based on colloidal gels and show that gelation is not necessarily preceded by phase separation at low to intermediate particle concentrations.

1. Pham, K.N., et al., *Multiple glassy states in a simple model system.* Science, 2002. **296**(5565): p. 104-106.

2. Lu, P.J., et al., *Gelation of particles with short-range attraction.* Nature, 2008. **453**(7194): p. 499-504.

3. Eberle, A.P.R., et al., *Dynamical Arrest, Percolation, Gelation, and Glass Formation in Model Nanoparticle Dispersions with Thermoreversible Adhesive Interactions.* Langmuir, 2012. **28**(3): p. 1866-1878.

4. Krall, A.H. and D.A. Weitz, *Internal dynamics and elasticity of fractal colloidal gels.* Physical Review Letters, 1998. **80**(4): p. 778-781.

5. Conrad, J.C., et al., *Arrested fluid-fluid phase separation in depletion systems: Implications of the characteristic length on gel formation and rheology.* Journal of Rheology, 2010. **54**(2): p. 421-438.

6. Grant, M.C. and W.B. Russel, *Volume-fraction dependence of elastic-moduli and transition-temperatures for colloidal silica-gels.* Physical Review E, 1993. **47**(4): p. 2606-2614.

7. Noro, M.G. and D. Frenkel, *Extended corresponding-states behavior for particles with variable range attractions.* Journal of Chemical Physics, 2000. **113**(8): p. 2941-2944.

8. Poon, W.C.K. and M.D. Haw, *Mesoscopic structure formation in colloidal aggregation and gelation.* Advances in Colloid and Interface Science, 1997. **73**: p. 71-126.

9. Rogers, R.B., et al., *Compact laser light-scattering instrument for microgravity research.* Applied Optics, 1997. **36**(30): p. 7493-7500.

(686g) Producing Reversible Gels to Control the State of Aggregation in Nanofluids

Saba Lotfizadeh and Themis Matsoukas, Department of Chemical Engineering, The Pennsylvania State University, University Park, PA

The presence of nanoparticles in thermal fluids (nanofluids) has been shown to improve the heat transfer properties of the fluid but the magnitude of the enhancement and the physical mechanism by which this enhancement occurs are open to debate. The standard classical model is Maxwell's mean field theory, which is borrowed from electricity and offers a prediction for the conductivity of the fluid as a function of the conductivity of particles and fluid. The standard theory produces two limiting bounds depending on whether the dispersed phase is the solid or the fluid. Since colloidal particles invariably suffer from aggregation, aggregate networks are thought to increase conductivity somewhere between the two bounds of the Maxwell theory. Direct experimental tests of this hypothesis have not been made because controlling the degree of aggregation in typical colloids in a systematic way is not possible. In this work we develop a colloidal system with the ability to undergo \textit{reversible} aggregation/deaggregation and produce stable aggregates of desired size, all the way from fully dispersed nanoparticles (30 nm in diameter) to fully gelled systems (infinite size). We accomplish this by treating Ludox silica with N-[3-(Trimethoxysilyl)propyl]ethylenediamine to produce a monolayer of the amine that is chemically bonded onto the particle. At basic pH the amino groups are positively charged and allow the full dispersion of the nanoparticles. At lower pH the amino groups are in neutral form such that the state of the suspension is dominated by the interaction between the hydrophobic chains and the surrounding fluid (water) causing the colloid to aggregate. The degree of aggregation depends on the degree of ionization of the amino groups and this in turn is tuned via pH. Since the surface of silica is covered, these aggregates are not permanent clusters but can be either grown to larger size, by decreasing the pH, or dissolved, by increasing the pH. This allows us to produce stable nanofluids ranging from 30 to 500 nm. We provide colloidal characterizations of this system and report the thermal conductivity of the nanofluids.

(690a) A Tutorial On Electrocatalysis: Fundamental Concepts, Methods, and Applications

Thomas F. Jaramillo, Chemical Engineering, Stanford University, Stanford, CA

In recent years, there has been a significant increase in research and development efforts for technologies that involve electrochemical transformations. Examples include solar fuels, flow batteries, and fuel cells. Electrocatalysis plays a crucial role in these (and related) technologies, as the efficiency of the device is closely correlated to catalyst efficiency. This talk will serve as a tutorial in the area of electrocatalysis, geared for non-experts with limited experience in electrochemistry.

The talk will cover fundamental concepts in the field, including electrochemical thermodynamics and kinetics. Examples of topics include: how to read and understand cell potentials, how to interpret current-voltage data, and a primer on how to rank catalysts. The focus will primarily be on chemical transformations on surfaces. Experimental methods will be briefly described, along with electrocatalyst figures of merit that are commonly employed in electrocatalysis studies. The talk will conclude with application notes relevant to selected electrochemical energy conversion reactions. The ultimate goal of this tutorial is to introduce non-experts to the fundamental knowledge and conceptual framework needed to follow research presentations and published literature in the area of electrocatalysis.

(690b) Using Electronic Structure Calculations to Understand Electrocatalysis: Applications to Carbon Dioxide Reduction

Andrew A. Peterson, School of Engineering, Brown University, Providence, RI

In this tutorial, the basic electronic structure tools as applied to electrocatalysis will be presented. Included will be discussions of modeling the computational hydrogen electrode, understanding the liquid-solid interface, exploring field effects, utilizing adsorption energy scaling, and the generation of activity volcanoes. The application of these methods to the study of the electoreduction of carbon dioxide will be highlighted.

(690c) Experimental Methods to Benchmark Electrocatalysts for Hydrogen Evolution and Oxygen Evolution

Charles C. L. McCrory, Joint Center for Artificial Photosynthesis, California Institute of Technology, Pasadena, CA, Jonas C. Peters, Chemistry, California Institute of Technology, Pasadena, CA and Thomas F. Jaramillo, Chemical Engineering, Stanford University, Stanford, CA

Objective evaluation of electrocatalytic activity for oxygen evolving and hydrogen evolving catalysts is of fundamental importance for the development of solar water-splitting devices. However, current testing methods employed in catalysis research are far from standardized, making it difficult to cross-compare the activity and stability of these systems. Even the definition of activity can vary significantly from study to study; for instance, when comparing catalytic activity some authors choose to report the exchange current density while others choose to report either the overpotential required to achieve a fixed current density or the measured current density at a fixed operating potential. Moreover, current density itself can be defined either per geometric area, BET-active surface area, electrochemically active surface area, and so on. These differences in testing protocols and reporting results can complicate comparisons of catalyst activity. In this tutorial session, an overview of the standard methods used within the Benchmarking facility of the Joint Center for Artificial Photosynthesis (JCAP) will be discussed. In particular, the presentation will focus on standard techniques used for determining electrochemically-active surface areas, measuring electrocatalytic activity and stability, and analyzing electrocatalytic products. The tutorial will also highlight some of the challenges and limitations regarding these measurements. Figures of merit for electrocatalytic hydrogen evolving and oxygen evolving reactions and ways of presenting relevant data to broad scientific audiences will also be discussed.

(690d) Atomistic Modeling of Electrochemical and Electrocatalytic Processes

Michael J. Janik, Chemical Engineering, Pennsylvania State University, University Park, PA

First principles, or quantum mechanical (QM) methods are widely used to investigate the energetics of adsorption and surface reactions. These methods may be applied to investigate electrode redox processes, including ion adsorption, surface oxidation/reduction, and electrocatalytic reaction energetics. As an example, QM methods have been used to determine key reaction energetics and recommend improved compositions of electrocatalysts for fuel cells. In this tutorial session, the development and application of QM modeling approaches for electrochemical processes will be reviewed. An atomistic picture of the electrode-electrolyte interface will be developed, and the challenges to applying QM approaches this interface will be discussed. A succession of models will be detailed that build in complexities of this interface, including thermochemical approaches to representing the electron and ion chemical potentials, inclusion of electrode charging and interfacial electric fields, and representation of electrolyte distribution at the interface. Brief discussion will be given to the use of force-field models to further represent the longer length and time scales associated with the electrochemical interface. Numerous applications will be discussed to illustrate the application of atomistic modeling in electrochemistry, including surface oxidation and electrocatalytic reactions in fuel cells and electrolysis. Examples systems chosen from research in our group will include borohydride oxidation, oxygen reduction and hydrogen oxidation in various electrolytes, and carbon dioxide reduction.

(691a) Understanding Droplet Bridging in Ionic Liquid-Based Pickering Emulsions

Denzil S. Frost, Jared J. Schoepf, Elizabeth M. Nofen and Lenore L. Dai, School for Engineering of Matter, Transport and Energy, Arizona State University, Tempe, AZ

Understanding Droplet Bridging in Ionic Liquid-Based Pickering Emulsions

Denzil S. Frost, Jared J. Schoepf, Elizabeth M. Nofen, and Lenore L. Dai

We have studied the unique bridging behavior of solid-stabilized oil-in-ionic liquid (IL) and water-in-ionic liquid emulsions with respect to particle concentration, particle size, and droplet phase using a confocal laser scanning microscope. Surprisingly, in contrast to the consensus originating from oil/water Pickering emulsions in which the solid particles equilibrate at the oil-water droplet interfaces and provide effective stabilization, here the polystyrene microparticles treated with sulfate, aldehyde sulfate, or carboxylate dissociable groups mostly formed monolayer bridges among the droplets rather than residing at the droplet-ionic liquid interfaces. The bridge formation inhibited individual droplet-droplet coalescence; however, due to low density and large volume (thus the buoyant effect), the aggregated droplets actually promoted phase separation and distressed emulsion stability The emulsions exhibited three morphology regimes: (1) single, sparingly covered droplets, (2) bridged clusters of droplets, and (3) fully covered droplets. The degree of bridging was directly proportional to the total potential bridging area, which can be determined from the particle size and concentration. A "phase" diagram of this behavior is proposed and shown in Figure 1. This type of emulsion diverges from much of the conventional wisdom of oil-water Pickering emulsions regarding the particle self-assembly onto droplet interfaces and liquid film stability. While the focus here is the bridging regime, we also report interesting observations, specifically, the deformed oil droplets and the transport of excess solid particles into the water droplets, in the fully covered droplet regime. The work identified new self-assembled particle structure and morphology in solid-stabilized emulsions.

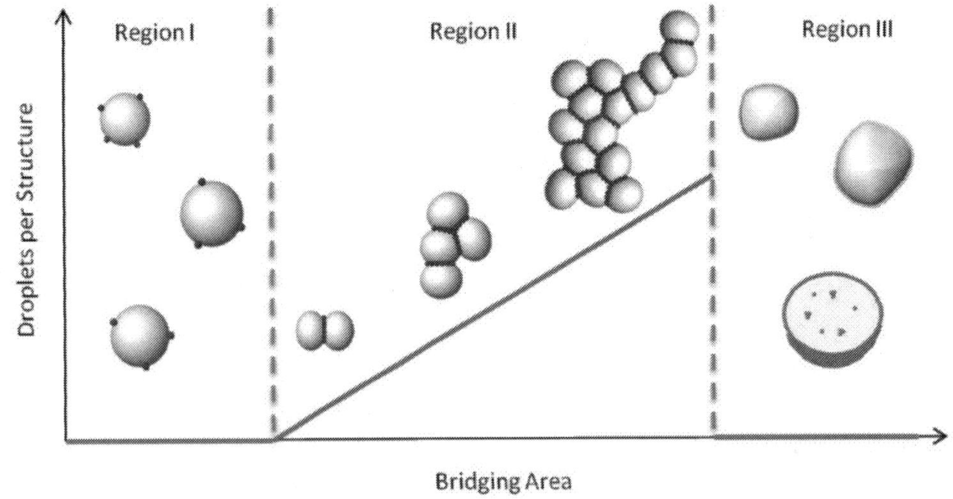

Figure 1 "Phase" diagram of the oil-in-water and water-in-oil Pickering emulsions discussed in this paper. The regimes are denoted as: Region I, single sparingly covered droplets; Region II, bridged clusters of droplets; Region III, fully covered droplets. In the third phase, sulfate-treated and carboxylate treated particles are represented in green and red, respectively.

(691b) Rheology and Structure in Particle-Containing Polymer Blends

Shailesh Nagarkar, University of Pittsburgh and **Sachin Velankar**, University of Pittsburgh, Pittsburgh, PA

Various types of particles readily adsorb at the interface between two immiscible liquids or at the surface of a liquid. Such particles bear some similarity to conventional molecular surfactants. However, unlike surfactants, particles adsorb almost irreversibly at interfaces – a fact that can lead to interesting phenomena such as the stability of non-spherical jammed drops and particle-bridged emulsions. This talk will describe the effects of such interfacially-adsorbed particles on the structure and flow properties of two-phase polymer blends.

Experiments were conducted on blends of polyethylene oxide (PEO) drops dispersed in a polyisobutylene (PIB) matrix with less than 2 vol% of monodisperse silica particles. The blends were examined by rheological experiments and Scanning Electron Microscopy (SEM). Particles with two different surface chemistries, which correspond to two different contact angles at the PEO/PIB interface, were considered. We find that particles that are preferentially-wetted by the drop phase are able to glue together two drops by bridging across them. The morphology therefore consists of particle-bridged clusters in which the individual drops are non-spherical and resemble polyhedral foam bubbles. Such blends show strong gel-like rheological behavior including a low-frequency plateau in modulus in linear viscoelastic experiments and a yield stress in steady flow experiments. In contrast, particles that are equally-wetted by both phases are not capable of bridging across drops; in this case, the morphology consists of nearly-spherical drops that are not clustered, and a yield stress is not evident rheologically. In summary, in polymer blends with added particles, the particle wettability has a major effect on the morphology and flow characteristics.

(691c) Nano-Mechanical Properties of Clay Armoured Emulsion Drops As a Function of Solution Conditions

Sin Ying Tan, Rico Tabor, Lydia Ong, G. W. Stevens and **Raymond R. Dagastine**, the Particulate Fluids Processing Centre, The University of Melbourne, Parkville, Australia

Solid-stabilized emulsions have received increasing attention over the last two decades due to their superior stability when compared to traditional surfactant stabilized emulsions. These emulsions can be found in a wide range of industries such as food, pharmaceutical, paint and petrochemical. Non-spherical particles such as plate-like or needle-like have had more recent attention and it has been shown that they can stabilize emulsions at a much lower solids volume fraction than spherical particles. Clay platelets, have been used as stabilizers in emulsions and bubbles and are common to many dispersions. Clay-stabilized bubbles, or armoured bubbles, have been shown to be very robust behaviour in water and other liquids, and show evidence of size selective permeability, *i.e.* they are semi-porous[1].

Here, the mechanical properties of clay-armoured emulsion droplets were investigated using laser scanning confocal microscopy with an in situ atomic force microscopy measurement[2]. This combination allows the visualization of droplet shape as a function of applied force. The emulsion droplets were found to be mechanically robust, stable against coalescence during drop collisions and able to recover from large deformations without disintegration. A Hookean constitutive law was used to extract the surface Young's modulus of the clay-armoured droplets as a function of a range of solution conditions. The clay-armoured droplets were relatively insensitive to changes in solution ionic strength, pH and the addition of surfactant. The elasticity measurements in this study should help illuminate the impact of the clay-armoured droplets on macroscopic properties of emulsions including rheological properties and emulsion stability.

1. A. B. Subramanian, J. Wan, A. Gopinath and H. A. Stone, *Soft Matter*, 2011, **7**, 2600-2612.
2. S.-Y. Tan, R. F. Tabor,· L. Ong, G. W. Stevens, and R.R. Dagastine, *Soft Matter*, 2012, **8**, 3112-3121.

(691d) Towards Stimulus-Responsive Control of Foam and Emulsion Stability Using Nanoparticles

Filippo Gambinossi, Lafayette College, Easton, PA and **James K. Ferri**, Department of Chemical and Biomolecular Engineering, Lafayette College, Easton, PA

The physics and engineering of switchable stability in foams and emulsion systems depends of active control of the fluid/fluid interface. We utilize nanoparticles with designed surface chemistry to control the contact angle of the nanoparticle at the fluid/fluid interface, the free energy of attachment, ultimately the interfacial elasticity. Interrogation of the aggregation kinetics and rheology of the fluid/fluid interface enables elucidation into fundamental mechanisms of foam and emulsion stabilization.

Here, we discuss a new class of thermosensitive polymeric materials based on random copolymers which display a lower critical solution temperature (LCST) in water that is dependant on co-polymer ratio and ionic strength. Gold nanoparticles grafted with thermosensitive polymer brushes of these copolymers have switchable aggregation kinetics (k11) and interfacial elasticity (E*). We show the dependence of k11 and the complex interfacial modulus (E*) on environmental conditions such as salt concentration and temperature; it is demonstrated that switchable stabilization is controlled by bulk aggregation rate, rather than interfacial rheology.

(691e) Molecular Dynamics Simulations of Colloidal Particles Approaching Liquid-Liquid Interfaces

Sepideh Razavi, Chemical Engineering, City College of New York, New York, NY, Joel Koplik, Department of Physics, Levich Institute, City College of New York and Graduate Center,City University of New York, New York, NY and Ilona Kretzschmar, Chemical Engineering, The City College of New York, The City University of New York, New York, NY

The tendency of colloidal particles to reside at a fluid interface at equilibrium position as predicted by Young's equation is the basic concept underlying the behavior of Pickering emulsions. Recently colloids of tunable wettability have been produced by partial modification of their surface, which yields so-called Janus particles with asymmetric surface properties. Despite the vast literature on colloidal particles at interfaces, little information is available with respect to the behavior of Janus particles and the dynamics of colloidal particles moving towards and onto an interface. We have investigated the movement of colloids toward the interface between two immiscible fluids via Molecular Dynamics (MD) Simulations. Although micron-sized particles are used for most practical applications, we focus on nanometer-sized particles in our simulations due to computational constraints. However, our initial results show that the behavior of larger particles is captured reasonably well by the MD calculations. We present simulation results for both ordinary symmetric particles and Janus particles, and compare the results to experimental observations.

(691f) Electrostatics At the Oil-Water Interface: Effect of Image Force and Dipole Field

Hongzhi Wang and Sven Behrens*, Chemical & Biomolecular Engineering, Georgia Institute of Technology, Atlanta, GA

Our work proposes two widely neglected electrostatic effects for the current understanding of Pickering emulsions via investigating the emulsification with charged latex particles. The emulsification results firstly suggest the effect of repulsive image force across the interface on the formation of Pickering emulsions via hindering the adsorption of particle to the oil-water interface, and secondly suggest the effect of self energy of particle- counterion on the contact angle of particle at the oil-water interface and concomitant emulsion type. Centering on these two arguments, the subsequent study attempts to support our claims with exploring the behavior of particle near and at the oil-water interface. Experimental observations of the deformation of oil-water interface as approaching to the particle indicate preliminarily the effect of repulsive image force. Also, the real contact angle of particle at the oil-water interface (counterpart to the macroscopic measured contact angle) measured by gel trapping method combined with AFM provides some evidences for effects of the particle-dipole. All these improved fundamental understandings will hopefully benefit the industrial application of Pickering emulsions and the development of novel emulsion –template materials.

(691g) SANS and SAXS Study of Nanoparticle Adsorption At Oil-Water Interfaces

Kjersta Larson and **Danilo Pozzo**, Chemical Engineering, University of Washington, Seattle, WA

Nanoparticle-stabilized emulsions show various potential applications in the development of nanomaterials such as colloidosomes and composite particles. In this talk we will examine the adsorption of both charged nanoparticles and amphiphilic nanoparticle surfactants at dispersed oil-in-water emulsion interfaces. The adsorption of nanoparticles at emulsion interfaces has been typically difficult to study because of the wide range of size scales that are involved. Here, we show how Small Angle Neutron Scattering (SANS) and Small Angle X-ray Scattering (SAXS) can be used to quantitatively examine nanoparticle adsorption at dispersed oil-water interfaces directly and without any sample manipulation (e.g. drying). We have developed a detailed scattering model that makes it possible to quantitatively determine the fraction of an oil-water interface that is covered with adsorbed particles. From this model, we are able to systematically probe the influence of colloidal interactions on the adsorption of nanoparticles at these interfaces. Finally, we will also examine the emergence of new optical properties that result from collective plasmonic interactions in emulsion systems that are stabilized by metallic nanoparticles.

(697a) Hydroxide Conducting Polymerized Ionic Liquid Block Copolymers for Alkaline Fuel Cells

Yuesheng Ye, Department of Chemical and Biological Engineering, Drexel University, Philadelphia, PA, Tsen-Shan Wang, University of Pennsylvania, Philadelphia, PA, Karen I. Winey, Materials Science and Engineering, University of Pennsylvania, Philadelphia, PA and **Yossef A. Elabd**, Chemical and Biological Engineering, Drexel University, Philadelphia, PA

In previous work in our laboratory, we reported on the chemical stability of an imidazolium-based hydroxide conducting polymerized ionic liquid (PIL), poly(1-[(2-methacryloyloxy)ethyl]-3-butylimidazolium hydroxide) (poly(MEBIm-OH), over a broad range of humidities, temperatures, and alkaline concentrations using the combined techniques of electrochemical impedance spectroscopy and nuclear magnetic resonance spectroscopy [1]. This polymer provides a good building block for future investigations since chemical stability of hydroxide conducting polymers is a significant issue as it pertains to long-term alkaline fuel cell (AFC) operation. In this talk, we will present the hydroxide conductivity and morphology of a series of PIL diblock copolymers, poly(MMA-*b*-MEBIm-OH), composed of the IL monomer, MEBIm-OH, and the non-ionic monomer, methyl methacrylate (MMA), at various PIL compositions. PIL block copolymers conjoin the advantages of block copolymers and PILs: self-assembled nanostructured morphologies and unique electrochemical properties. To date, there are few reports of PIL block copolymers in the literature. Therefore, a fundamental understanding of morphology and ion transport in these materials will be highly beneficial, specifically for the optimization of AFC performance.

1. Ye, Y.; Elabd, Y.A. Relative Chemical Stability of Alkaline Exchange Polymerized Ionic Liquids. *Macromolecules* **2011**, *44* (21), 8494-8503.

(697b) Hydroxyl Functionalized Graphene: A Promising Anhydrous Proton Exchange Membrane

Pabitra Choudhury and J. Karl Johnson, Chemical and Petroleum Engineering, University of Pittsburgh, Pittsburgh, PA

We predict that graphene functionalized with OH groups can be used as an efficient proton exchange membrane capable of functioning in the complete absence of water. Fast anhydrous proton transport takes place on the surface of functionalized graphene as a result of the nanostructured spacing of the functional groups. We have considered two model membranes to test our hypothesis: (1) a membrane consisting of graphane (fully hydrogenated graphene) where a one-dimensional (1-D) chain of the hydrogen atoms has been replaced by OH groups (hydroxyl graphane), as shown in Figure 1, and (2) a membrane consisting of graphane completely saturated with OH groups on one side (i.e., every other C atom) of the graphane. This membrane forms a two dimensional (2-D) hydrogen bonded network, as shown in Figure 2. The fully hydroxylated structure has the advantage of providing redundant pathways for proton diffusion. However, the saturated OH coverage on the 2-D surface could reduce the proton diffusivity. We have used first principles density functional theory to study the structural, energetic, and transport properties of model membranes. Our calculations predict that the proton mobility on the surface of a 1-D hydrogen bonded graphane membrane will be close to the diffusivity of H^+ in bulk water and about 10 times faster than through Nafion at room temperature. In contrast, the proton diffusivity along the surface of a 2-D hydrogen bonded graphane membrane is about two orders of magnitude slower than the estimated diffusivity of proton across the 1-D hydrogen bonded chain of graphane membrane. We predict that functionalized graphene proton exchange membranes could operate over a wide temperature range, making them more efficient than membranes requiring hydration.

Acknowledgement: We gratefully acknowledge the support by DTRA under Contract No. HDTRA1-09-1-0008. We also gratefully acknowledge NSF XSEDE (TeraGrid) resources under allocation numbers TG-DMR100097, TG-DMR110091 and TG-SEE090006. We thank the Center for Simulation and Modeling at the University of Pittsburgh for providing computational support.

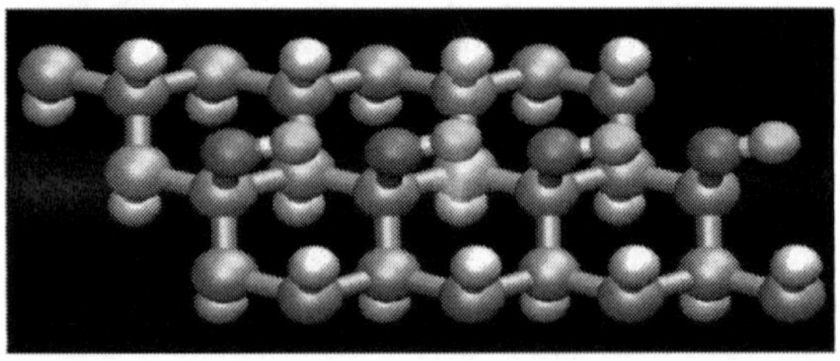

Figure 1. Functionalized 1-D hydrogen bonded hydroxyl graphene (sp^3 like structure). The cyan color represent C, white color represent hydrogen atom, and red-green combination represent O and H in OH groups, respectively.

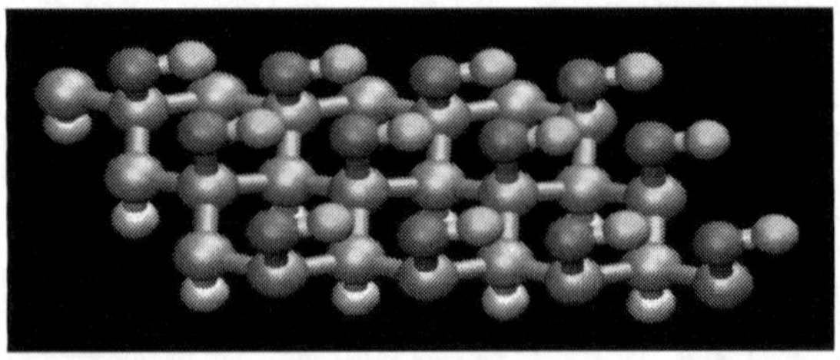

Figure 2. OH saturated 2-D hydrogen bonded graphane membrane. The cyan color represent C, white color represent hydrogen atom, and red-green combination represent O and H in OH groups, respectively.

(697c) Gas Permeation in Perflurosulphonated Membranes: Influence of Temperature and Relative Humidity

Marco Giacinti Baschetti[1], Matteo Minelli[2,3], Jacopo Catalano[4] and Giulio Sarti[1], (1)Chemical Engineering, Mining and Environmental Technology, University of Bologna, Bologna, Italy, (2)University of Bologna, Bologna, Italy, (3)Advanced Applications in Mechanical Engineering and Materials Technology Interdepartmental Center for Industrial Research, CIRI-MAM, Bologna, Italy, (4)Worcester Polytechnic Institute

Gas permeation in Perflurosulfonated membranes: influence of temperature and relative humidity

Marco Giacinti Baschetti, Matteo Minelli, Jacopo Catalano[1], Giulio C. Sarti

Dipartimento di Ing. Chimica, Mineraria e delle Tecnologie Ambientali (DICMA)

Alma Mater Studiorum - Università di Bologna, via Terracini 28, 40131, Bologna, Italy.

Perfluorosulfonated ionomer (PFSI) are of high interest in energy application due to their ability to be used as polyelectrolyte membranes (PEM) inside hydrogen or methanol based fuel cells (FC).[1,2] In fuel cells (FC) applications, the polyelectrolyte membranes (PEM) are exposed to different gases such as hydrogen, oxygen, nitrogen and carbon dioxide, depending on the type of FC considered, at varying operative conditions, both in terms of temperature and relative humidity. For these reasons, the knowledge of gas permeability of different gases in such materials as a function of these parameters is of interest for a proper description and modeling of the system. Different studies appeared in the literature addressing these issues [3-10], however systematic investigation of these phenomena are relative scarce and a thorough characterization is still missing for many of the PSFI now available.

In the present study, gas permeability in two different PFSI, namely Nafion® and Aquivion®, have been investigated as a function of temperature and relative humidity. In particular, permeation of gases such as O_2, N_2 and CO_2 and also Helium, used to mimic H_2, have been studied from 25 to 50°C and at relative humidity up to 80%.

The results showed that both materials posses very similar permeability values and a common behavior with respect to temperature and relative humidity. The permeability of the different gases in the two ionomers increases as the temperature or the humidity increase, the latter parameter having a really dramatic effect on the gas flow rate across the membrane. The measured permeability indeed showed increases of up to two orders of magnitude when humidified gas were flown in the system.

The effect of water on the membrane transport properties is rather complex since the permeability has a sharp increase already at very low humidity, and then shows a more gentle, exponential rise with RH that seems to lead towards values comparable with those of gas permeability in pure water.

The comparison between the two materials does not show a clear trend; in general, however, Aquivion seems to have higher permeability than Nafion at low relative humidity and at the lower temperatures inspected, while Nafion shows higher permeation rates at larger R.H. and high temperatures.

These findings can be related to the fact that at a given activity Nafion in general sorbs less water than Aquivion, due to the higher equivalent weight. Such difference seems to be significant at low activity, where the water content is still low. It becomes then less important at higher activity where interconnected hydrophilic domains are formed in both membranes.

The temperature dependence follows an Arrhenius type of behavior for all gases in the two polymers, as expected for permeation processes substantially related to a solution diffusion mechanism. The activation energies were calculated for both materials as a function of R.H, and showed higher values in Nafion than in Aquivion, in agreement with the permeability behavior described above.

In all cases, for both materials the activation energy ranges between the value of the dry polymer and that obtained for the gas permeation in pure polymer. In the case of Nafion, a well defined trend of permeability with activation energy can be found with water activity, at least in the case of nitrogen and oxygen. In the case of Aquivion, on the other hand, no clear trend was detected.

All the experimental evidences suggest that gas permeation in humid PFSI is driven by gas sorption and diffusion in the hydrophilic domains of the materials and ultimately in the water channels which are usually considered to form in the hydrated matrix. This hydrated phase indeed has a largely higher gas permeability than that of the hydrophobic phase and controls the overall permeation process.

References:

1. Fuel Cell Handbook (7th Edition), EG&G Technical Services, Inc. (2004) Contract No. DE-AM26-99FT40575

2. Marcinkoski, J., Kopasz, J., & Benjamin, T. Progress in the US DOE fuel cell subprogram efforts in polymer electrolyte fuel cells. Int. J. Hydrogen Energy, 2008;33: 3894.

3. Sakai T, Takenako H, Wakabayashi N, Kawami Y, Torikai E. Gas permeation properties of solid polymer electrolyte (SPE) membranes. J Electrochem Soc 1985;132:1328

4. Ogumi Z, Takehara Z, Yoshizawa SI. Oxygen permeation through Nafion and NEOSEPTA. J Electrochem Soc 1984;131:769.

5. Pellegrino J, Kang KS. CO2/CH4 transport in polyperfluorosulfonate ionomers: effects of polar solvents on permeation and solubility. J Membr Sci 1995;99:163.

6. Buchi FN, Wakizoe M, Srinivasan S. Microlectrode investigation of Oxygen permeation in perfluorinated proton exchange membranes with different equivalent weights. J Electrochem Soc 1996;143:927

7. Ma S, Odgaard M, Skou E. Carbon dioxide permeability of proton exchange membranes for fuel cells. Solid State Ionics 2005;176:2923.

8. Mohamed HFM, Ito K, Kobayashi Y, Takimoto N, Takeoka Y, Ohira A. Free volume and permeabilities of O2 and H2 in Nafion membranes for polymer electrolyte fuel cells. Polymer 2008;49:3091.

9. James Cw A, RoyMcGrath JE, Marand E. Determination of the effect of temperature and humidity on the O2 sorption in sulfonated poly(arylene ether sulfone) membranes. J Membr. Sci. 2008;309:141.

10. Catalano J., Myezwa T., De Angelis M.G., Giacinti Baschetti M., Sarti G.C. The effect of relative humidity on the gas permeability and swelling in PFSI membranes Int. J. Hydrogen Energy 2012;37:6308

---

[1] Present address: Center for Inorganic Membrane Studies, Department of Chemical Engineering, Worcester Polytechnic Institute, 100 Institute Rd, Worcester, MA 01609, USA.

(697d) Pore Filling Anion Exchange Membranes and Membrane Electrode Assemblies for Solid-State Alkaline Fuel Cells

Takeo Yamaguchi[1], Han Zhang[2], Hyangmi Jung[1], Hidenori Ohashi[2] and Takanori Tamaki[2], (1)Chemical Resources Laboratory, Tokyo Institute of Technology, Kanagawa, Japan, (2)Chemical Resources Laboratory, Tokyo Institute of Technology, Yokohama, Japan

In solid-state alkaline fuel cells (SAFCs), stability improvement of the membranes and water management of the MEAs are essential to achieve good performances. To improve the stability, suppressing water mobility by porous substrate matrix leads to improve the membrane durability. The pore-filling membranes exhibited higher stability compared with normal cast films made of the same polymer. Next, water management is important for SAFCs because water reacts at the cathode and is generated at the anode, in contrast to proton-exchange membrane fuel cells (PEMFCs). Thus, optimization is required to establish design methodology for SAFCs with pore-filling anion exchange membranes. Water movement in a SAFC system has been observed accurately using a water collection method. We confirm for the first time the movement of water from the anode to the cathode during an actual fuel cell performance test. Compared with a PEMFC, water movement occurs in the opposite direction leading to difficult water management in the SAFC. With the knowledge obtained, controlling membrane performances and careful control of humidity have achieved water management in the SAFCs.

(697e) Hybrid Proton-Conducting Membranes Doping with Various Functionalized Titania for Direct Methanol Fuel Cells

Hong Wu, Zhongyi Jiang, Tao Xu, Weiqiang Hou and Xiaohui Shen, Key Laboratory for Green Chemical Technology of Ministry of Education, Sch Chem Engn & Technol, Tianjin Univ, Tianjin, China

Organic-inorganic hybrid proton-conducting membranes have gained tremendous attention in the field of direct methanol fuel cells (DMFCs). The hybrid membranes combine the intrinsic physical and chemical properties of both inorganic and organic components. A variety of polymers have been doped with various functionalized inorganic particles or inorganic phase to achieve high proton conductivity, low methanol crossover, good mechanical and chemical properties. The tuning of the fuctionalized groups on the inorganic fillers, which could provide additional pathways for proton transfer, is a key issue in fabricating hybrid membranes. Chelation grafting of functional groups onto titania allows facile and fine modifications of the inorganic surface. Titania submicrospheres grafted with phosphonic acid groups (-PO_3H_2), sulfonic acid groups (-SO_3H), carboxylic acid groups (-CO_2H), amino groups (-NH_2) and amino acid groups (-CO_2H, -NH_2) were respectively incorporated into chitosan (CS) or sulfonated poly(ether ether ketone) (SPEEK) membrane matrix to fabricate hybrid membranes with improved proton conductivity. The conductivities at room temperature and 100% RH of the chitosan/phosphonic acid-modified titania membrane, the SPEEK/sulfonic acid-modified titania membrane, the SPEEK/carboxylic acid-modified titania membrane and the SPEEK/amino acid-modified titania membrane were 0.011, 0.053, 0.071 and 0.066 S cm^{-1}, respectively, with 15 wt% of fillers. Besides, due to the different proton acceptor/donor capability of functional groups, the hybrid membrane embedded with amino-functionalized titania exhibited higher values at elevated temperature of 60 ^oC compared with carboxylic-functionalized titania. And the hybrid membrane embedded with amino acid-functionalized titania displayed a superior proton-conducting ability with the proton conductivity of 0.258 S cm^{-1} at 60 ^oC. In addition, SPEEK/carboxylic acid-functionalized titania and SPEEK/amino acid-functionalized titania membrane hybrid membranes were also fabricated via in situ sol-gel method and their proton conductivities were 0.063 and 0.047 S cm^{-1}, respectively, nearly three times higher than the control membranes. In conclusion, all the hybrid membranes exhibited enhanced proton conductivity, low methanol crossover, better mechanical and chemical properties, and are promising candidates for DMFCs.

(700a) Fundamental Concepts in Catalytic Oxidative-Coupling Reactions On Metallic Gold

Bingjun Xu, Chemical Engineering, Caltech, Pasadena, CA, Robert J. Madix, School of Engineering and Applied Sciences, Harvard University, Cambridge, MA and Cynthia M. Friend, Chemistry and Chemical Biology, Harvard University, Cambridge, MA

Despite the chemical inertness of metallic gold under ambient conditions, once activated by atomic oxygen, gold surface is extremely active and selective towards many key oxidative chemical transformations. A molecular-level understanding of the mechanism is critical in the discovery and design of "green" catalytic processes mediated by gold-based materials. Our work establishes of the mechanistic framework for oxidative-coupling reactions on the gold surface through fundamental surface science studies on the model system, atomic oxygen activated Au(111). The self-coupling mechanism for alcohols is unequivocally established by accurately predicting the cross-coupling of alcohols and aldehydes, which also expands the scope of the mechanism to other oxidative-coupling reactions. The mechanistic framework further demonstrates its generality by predicting new oxidative-coupling reactions between amines and aldehydes, amines and alcohols, as well as carbonylations. Strong parallels between reactions taking place on the model surface at low pressure and reactions carried out on nanoporous gold or supported gold catalysts under working catalytic conditions demonstrates the practical relevance of the fundamental mechanistic studies.

(700b) Catalytic Reformation of Glycolaldehyde, Glyceraldehyde and Glucose On Pd(111): A Mechanistic Study

Jesse R. McManus and John Vohs, Chemical and Biomolecular Engineering, University of Pennsylvania, Philadelphia, PA

The environmentally motivated shift towards a carbon neutral, sustainable hydrogen economy has spotlighted the need for advances in biomass reforming science. Utilization of biomass and biomass derivatives for hydrogen production can provide hydrogen fuel with a considerably smaller carbon footprint than the ubiquitous natural gas reforming process. However, the catalytic conversion of complex biomass feed molecules to hydrogen is not well understood, therefore there is a distinct need to elucidate the fundamental science for larger oxygenate bond scission. With a more comprehensive understanding of the catalytic bond scission of biomass derived oxygenates, extensions can be made to functionally similar molecules and catalytic chemistry predicted. An approach for bond scission characterization of the cellulosic monomer D-glucose and two simpler functional analogues on a Pd(111) catalyst is presented.

Characterization difficulties arising from the cellulosic sugars' low vapor pressures have so far precluded their use in ultra high vacuum (UHV) surface science studies that have proven so successful in mapping out reaction mechanisms. Demonstration of the successful implementation of glucose in a UHV surface science study shows that this is no longer the case. A reproducible molecular glucose flux is verified using mass spectrometry and its adsorption on the catalyst surface is corroborated using temperature programmed desorption (TPD). Adsorption of glucose, glyceraldehyde and glycolaldehyde on the Pd(111) surface is characterized using high resolution electron energy loss spectroscopy (HREELS) in conjunction with TPD. Density functional theory calculations provide additional insights into the favored adsorption configurations. Results indicate an initial open ring, η_1-adsorption through the aldehyde oxygen of glucose followed by transition into the energetically favored η_2 di-sigma bonding configuration. This is observed similarly with glyceraldehyde and glycolaldehyde molecules pointing to initial chemistry being dominated by the aldehyde functionality.

Evolution of the energy losses in the HREELS data with increasing temperature outlines the sequence of bond scission events. Following transition into the η_2 bonding configuration, the smaller molecules see O-H cleavage and subsequent adsorption through the alcohol oxygen. Continuing the sequence, C-C scission ultimately results in the production of CO and H_2 products. These and additional similarities in adsorption mechanics and reaction chemistry between D-glucose and the simpler glyceraldehyde and glycolaldehyde molecules provide support for the use of functionally similar model compounds to predict the chemistry of complex biomass derived oxygenates in future studies.

(700c) Optically Excited Plasmonic Metal Nanostructures As Selective Direct Propylene and Ethylene Epoxidation Catalysts

Marimuthu Andiappan[1], Hongliang Xin[1], Phillip Christopher[2] and **Suljo Linic**[1], (1)Chemical Engineering, University of Michigan, Ann Arbor, MI, (2)Department of Chemical and Environmental Engineering, University of California, Riverside

In this contribution, we demonstrate that plasmonic silver and copper catalysts, optically excited with low intensity (order of Solar intensity) visible light, can achieve significantly higher selectivity compared to other conventional heterogeneous catalysts in the direct epoxidation (using O_2 as an oxidant) of propylene and ethylene to form propylene and ethylene oxide respectively.

We will show that the underlying mechanisms governing bond making (breaking) on optically excited plasmonic metals are fundamentally different than on semiconductors and those related to classical thermo-catalytic processes. Consequently, plasmonic metals exhibit profoundly different behavior. For example, unlike semiconductors, these materials exhibit a positive relationship between reaction rates (and quantum efficiencies) and light intensity. Also, plasmonic nanostructures exhibit an exponential increase of photo-catalytic rate (and quantum efficiency) on operating temperature.

We have developed a molecular model based on first-principles calculations that describes photo-chemical transformations on plasmonic metals. We show that the model captures the unique, experimentally observed features of the photo-catalytic processes on plasmonic metals.

1. S. Linic, P. Christopher and D. B. Ingram, Plasmonic-metal nanostructures for efficient conversion of solar to chemical energy, **Nature Materials, 10**, 911, 2011.

2. P. Christopher, H. Xin and S. Linic, Visible light enhanced catalytic oxidation reactions on plasmonic silver nanostructures, **Nature Chemistry,** 3, 467, 2011.

(700d) Metal-Modified Carbides for Conversion of Alcohols to Syngas and Aldehydes

Thomas G. Kelly, Department of Chemical Engineering, University of Delaware, Newark, DE and Jingguang G. Chen, Chemical Engineering, University of Delaware, Newark, DE

Metal-Modified Carbides for Conversion of Alcohols to Syngas and Aldehydes

Thomas G. Kelly and Jingguang G. Chen[1]*

[1]Center for Catalytic Science and Technology, Department of Chemical and Biomolecular Engineering, University of Delaware, Newark, Delaware 19716 (USA)

tgkelly@udel.edu, *jgchen@udel.edu

In the search for energy sources to replace fossil fuels, biorenewable fuels have become a distinct possibility. One class of biorenewable molecules are small-chain alcohols. These alcohols can be reformed to produce syngas or thermochemically converted to useful products. Recently, 3d/Pt(111) bimetallic surfaces have been shown to be active for methanol and ethanol reforming. The Ni/Pt(111) surface displayed higher reforming yield than both of the parent metals [1,2]. However, Pt-based bimetallics face the problems of Pt scarcity and poor overlayer stability. Metal carbides such as tungsten carbide (WC) and molybdenum carbide (Mo_2C) have been shown to exhibit Pt-like properties [3]. Also, WC can be used as a less expensive and more stable support in place of Pt, as was shown in using Ni/WC to replace Ni/Pt for ethanol decomposition [4]. In the current study, we have extended our previous surface science studies for methanol on Pt-modified WC [5] to C1-C3 alcohols on carbides modified with Ni, Rh, Cu, and Au. Furthermore, an understanding of bond scission sequence should also provide insights into the utilization of carbides for catalytic conversion of alcohols.

Density Functional Theory (DFT) was used to calculate the binding energies of alcohols and relevant intermediates on model surfaces of WC and Mo_2C, which showed the possibility to correlate reactivity with binding energy. Temperature-programmed desorption (TPD) was used on model WC and Mo_2C surfaces modified with different metals to quantitatively determine the activity of the surfaces and their selectivity towards C1-C3 alcohol decomposition. High resolution electron energy loss spectroscopy (HREELS) identified different reaction intermediates on carbide and metal-modified carbide surfaces. The first step in alcohol decomposition on these surfaces was O-H bond cleavage to form an adsorbed alkoxy species. In the case of methanol, the predominant reaction product on clean WC was methane; adding small amounts of Ni or Rh promoted the C-H bond scission and shifted the selectivity towards CO and H_2. Au simply acted as a site blocker since activity decreased with increasing Au coverage [6]. Similar experiments have been conducted for ethanol and propanol decomposition on Rh/WC to investigate the pathway for C-C bond scission. The selectivity on clean WC was for C-O bond scission to produce ethylene. Adding Rh led to C-H and C-C bond cleavage; HREELS showed that the C-C bond in ethanol is broken by 200 K. Lastly, the clean Mo_2C surface was shown to completely decompose ethanol to its constituent elements with a small amount of ethylene and methane production. Ni-modified Mo_2C produced CO as the major product, indicating C-C bond scission. Similarly to methanol on Au-modified WC, ethanol desorbed intact on Au-modified Mo_2C. Interestingly, a pathway divergent from reforming was observed on Cu-modified Mo_2C, which produced acetaldehyde. Apparently Cu was efficient at breaking the O-H and α C-H bonds but further dehydrogenation was not favored over desorption.

Of further interest is the bond scission sequence of additional C2 and C3 oxygenates on metal-modified WC. HREELS experiments on WC and Rh/WC were conducted for ethanol, acetaldehyde, propanol, and propanal to investigate the bond scission sequence in these oxygenates. On WC, ethanol followed a decomposition pathway similar to acetaldehyde; the spectra suggested that both molecules formed a di-σ species that facilitates C-O bond cleavage. This behavior was mirrored by propanol and propanal on WC. On Rh/WC, ethanol and acetaldehyde again followed similar pathways, except that the pathway in this case is through C-C bond scission instead of C-O scission.

The results of decomposition for C1-C3 alcohols on carbides hold promise for applications in heterogeneous reforming. On native carbides, C-O scission occurred to produce hydrocarbons. Ni and Rh supported on carbides have been shown to be efficient at cleaving C-H and C-C bonds of alcohols while leaving the C-O bond intact. Alternatively, modifying Mo_2C with Cu resulted in acetaldehyde production from ethanol. The ability to selectively decompose alcohols by using various admetals demonstrates the versatility of metal-modified carbides. Future experiments will involve synthesis of supported metal-modified carbide catalysts and reactor testing for alcohol decomposition.

References

1. Skoplyak, O., Barteau, M.A., Chen, J.G. *J. Phys. Chem. B.* **110** (2006) 1686.

2. Skoplyak, O., Menning, C.A., Barteau, M.A., Chen, J.G. *J. Chem. Phys.* **127** (2007) 114707.

3. Chen, J.G. *Chem. Rev.* **96** (1996) 1477.

4. Ren, H., Hansgen, D.A., Stottlemyer A.L., Kelly, T.G., Chen, J.G. *ACS Catal.* **1** (2011) 390.

5. Stottlemyer, A.L., Liu, P., Chen, J.G. *J. Chem. Phys.* **133** (2010) 104702.

6. Kelly, T.G., Stottlemyer, A.L., Ren, H., Chen J.G. *J. Phys. Chem. C.* **115** (2011) 6644.

(700e) Enantioselective Separation of Aspartic Acid On Naturally Chiral Cu (3 1 17)$^{R\&S}$ Surfaces

Yongju Yun, Bharat S. Mhatre and Andrew J. Gellman, Chemical Engineering, Carnegie Mellon University, Pittsburgh, PA

Production of enantiomerically pure chemicals is a major concern in pharmaceutical industries because left- and right-handed enantiomers of chemicals have significantly different physiological impact when ingested. One strategy for producing enantiomerically pure chemicals is to employ chiral surfaces for enantioselective chemical processes. In this work, enantioselective adsorption of aspartic acid on naturally chiral Cu (3 1 17)$^{R\&S}$ surfaces was investigated in UHV using temperature programmed desorption. We have developed a novel technique in which the two enantiomers can be differentiated using mass spectroscopy by incorporation of ^{13}C labels. Exposure of a racemic mixture of D- and *L-aspartic acid to the naturally chiral surfaces results in enantioselective equilibrium adsorption with an enantiomeric excess of ~40%. This is the first categorical demonstration of enantiospecific adsorption and separation on naturally chiral metal surfaces. This result implies that the enantiospecific interactions between chiral molecules and naturally chiral surfaces can serve as the basis for enantioselective separations or enantioselective catalysis.

(700f) Chemical Reaction of •CH_3 with the Basal Plane of Graphite

Pabitra Choudhury[1], J. Karl Johnson[1], Lynn Mandeltort[2] and John T. Yates Jr.[2], (1)Chemical and Petroleum Engineering, University of Pittsburgh, Pittsburgh, PA, (2)Department of Chemistry, University of Virginia, Charlottesville, VA

Pristine graphite is known to be resistant to chemical attack. In this work, we present both experimental and theoretical work showing facile reaction between methyl radical species and the basal plane of graphite. We have used van der Waals corrected spin-polarized periodic density functional theory (DFT) to identify the reaction mechanism and energy barriers for catalytic dissociation of CH_3Cl on Li-doped graphite, followed by reaction of CH_3 radicals on the graphite surface. The energy barrier for CH_3 binding to graphite is found to be less than ~0.3 eV. Experiments have been performed on high quality HOPG (highly oriented pyrolytic graphite) with low defect density (~10^9 cm^{-2}) to mitigate the effects of step edges and defects on the graphite surface chemistry. Theoretical calculations show that CH_3 radicals become mobile over an energy barrier of ~ 0.7 eV, consistent with experimental observations of CH_4 desorption at and above 330 K. The calculations show that mobile CH_3 radicals abstract hydrogen from neighboring CH_3 radicals with an energy barrier of 0.56 eV in CH_4 production. Experimental observations indicate that ~ ¾ of the methyl radicals remain on the graphite surface up to 700 K at puckered sp^3 carbon sites, while ¼ of the CH_3 radicals participate in CH_4 formation with small amounts of C_2 and C_3 hydrocarbons formation through the onset of surface mobility of CH_3 radicals bonded to the graphite surface.

Acknowledgement: We gratefully acknowledge the support by DTRA under Contract No. HDTRA1-09-1-0008. We also gratefully acknowledge NSF XSEDE (TeraGrid) resources under allocation numbers TG-DMR100097, TG-DMR110091 and TG-SEE090006. We thank the Center for Simulation and Modeling at the University of Pittsburgh for providing computational support.

(700g) Autocatalytic Decomposition Mechanism of Aspartic Acid On Cu(110) Surfaces

Bharat S. Mhatre, Department of Chemical Engineering, Carnegie Mellon University, Pittsburgh, PA and Andrew J. Gellman, Chemical Engineering, Carnegie Mellon University, Pittsburgh, PA

Surface explosions are surface reaction mechanisms wherein an autocatalytic increase in the reaction rate results in the reaction proceeding to completion over a very narrow temperature range during heating. While several substrate-adsorbate systems exhibiting decomposition via surface explosion mechanism have been identified, a detailed understanding of the mechanism remains unclear. We have successfully identified aspartic acid as a suitable probe for studying surface explosion mechanism on Cu(110) surface. Because a wide range of isotopically labeled varieties of aspartic acid are commercially available, we have been able to conduct a detailed investigation of its autocatalytic reaction mechanism. Using specifically labeled aspartic acid molecules, we have identified the reaction products and identified the origins of atoms in the decomposing aspartic acid molecules. The explosive nature of the reaction mechanism has enabled us to study its kinetics using isothermal methods by heating the Cu(110) surface to a constant temperature and then monitoring the product desorption as a function of time. We observe a significant lag time during which nucleation of the reaction is occurring without observable desorption of products. Once the reaction begins, it the proceeds to completion over a relatively short time period. Our preliminary studies on chiral $Cu(643)^{R\&S}$ surfaces indicate that aspartic acid exhibits enantiospecific surface decomposition kinetics. The ultimate objective of this work is to be able to identify the mechanism and in particular, the steps in the mechanism which are responsible for the enantiospecificity.

(702a) Investigation of the Oxidation of Metal/Metal Oxide in the Corrosion Environment Using the Reaxff Reactive Force Field

Chenyu Zou, Osvalds Verners, Yun-Kyun Shin and Adri van Duin, Mechanical Engineering, Pennsylvania State University, State College, PA

A ReaxFF reactive force field description for Fe, Ni and Al metal and metal oxide and their interaction with oxygen has been developed. The force field parameters were trained against an extensive set of ab initial data including heats of formation and equation of state of varieties of bare metal, alloy and metal oxide. The potential developed here were validated using molecular dynamics simulations on relatively small systems (~1000atoms) addressing the process of Ni self diffusivity, Al/Fe impurity diffusivity in fcc Ni and O/H diffusivity in Al-Fe-Ni. Results from these studies agree with those from first principle calculations. Based on the force field parameters, we studied the surface segregation of different types of $Al_xFe_yNi_{(1-x-y)}$ alloys and the adsorption of O_2 and subsequent oxidation process. Results of these studies will contribute to revealing the corrosion and aging of the material.

This project aims at closing the gap between atomistic-scale studies (e.g. Quantum Mechanics calculation) and experimental observations. The method will contribute to the development of reliable computational tools to predict the performance of novel Ni based superalloys in conditions relevance to advanced energy systems. Specifically, the large system sizes (>>1000 atoms) and large simulations times (>1 ns) will greatly enhance our capability to design and test materials in environment and time scales relevant to these systems.

(702b) On the Mechanism of Surface Convex and Concave Reduction Via Outward Solid-Phase Ionic Diffusion In Gas-Solid Reactions

Zhenchao Sun, Qiang Zhou, Pengpeng Qi, Hao Chi and Liang-Shih Fan, William G. Lowrie Department of Chemical and Biomolecular Engineering, The Ohio State University, Columbus, OH

Gas solid reaction is a typical chemical reaction, which is ubiquitous in process industry. One of the properties that characterize the gas-solid reactions is the surface structure, which determines the solid exposure to the gas reactant in the reaction process. Generally, a solid surface can be regarded as combinations of convexes and concaves. Many gas-solid reactions (*e.g.* M + O_2 -> MO) induce decreases of surface area and pore volume. Such surface structure change is widely attributed to sintering and molar volume increase. Sintering effect tends to agglomerate smaller particles into big ones and fuse surface pores. In most gas-solid reactions, solid reactants capture partial or entire gas component, which usually induces molar volume increase. Thus, the increased molar volume of solid particle narrows the inter-particle(grain) distance and plugs pores (concave). In this study, we found that there exists a third mechanism for surface convex and concave reduction via outward ionic diffusion. An iron oxidation reaction is used to represent all the gas-solid reactions involving outward ionic diffusion of solid species. It was found that surface concave tends to be buried and surface convex tends to flatten as the reaction proceeds. A 2-D continuum diffusion model is also established to simulate this solid-phase diffusion process, which shows consistent results as the experiments.

(702c) Study of Sulfidation Behavior On the Ordered Metallic Alloy Surface Using the Reaxff Force Field

Yun Kyung Shin[1], Hyunwook Kwak[2], Alex Vasenkov[2] and Adri C.T. van Duin[1], (1)Department of Mechanical and Nuclear Engineering, Pennsylvania State University, University Park, PA, (2)CFD Research Corporation, Huntsville, AL

Although sulfur as an impurity on metallic alloys plays an important role in reducing the stability and corrosion resistance of the protective oxide layer and degrading the mechanical properties of the material, the underlying mechanism has not been clearly described yet. To investigate this mechanism, we have expanded our ReaxFF force fields for Fe/Al/Ni alloy[1] and the alloy oxide to Al/Ni//O/S interactions. This ReaxFF force field for Al/Ni//O/S interactions is fitted against quantum mechanical (QM) calculations. The QM data includes equation of state for crystal phases, heat of formation, adsorption on various metal surface sites and diffusion for the development of the force fields capable of describing the surface reactions as well as bulk properties. Although QM methods have served as useful tools for predicting molecular properties in small chemical systems, their computationally expensive cost has been a major practical obstacle to the application to large systems. On the other hand, ReaxFF performs the reactive molecular dynamics simulations at much cheaper computational cost. ReaxFF implemented in parallel molecular dynamics (MD) simulation packages is suitable for large scale systems larger than 10^6 atoms over time periods in the nanosecond regime.

In order to understand the sulfide formation process, we investigated the formation of sulfide layer on an Al_2O_3 layer and a metal surface at high temperatures (1000-2000 K) by performing ReaxFF MD simulations. We examined the sulfide structure (e.g., connectivity and angle distribution) and the charge distribution across the sulfide structure at the interface as well as the sulfide formation behavior on the surface. Our simulations indicate that this formation process is dominated by the outward diffusion of aluminum in the alloy/metal. Understanding the mechanism of the sulfide formation at high temperature provides a theoretical basis for corrosion protection of the metallic alloy in sulfidizing environments.

[1] Y. Shin, S. Kwak, A. Vasenkov and A.C.T. van Duin, Development of ReaxFF force field for Fe/Al/Ni alloys and the atomistic study of segregation, manuscript in preparation.

(702d) The Effect of the SO2 :HCl Ratio On Active Oxidation of WTE Superheater Materials

Timothy Sharobem, Department of Earth and Environmental Engineering, Henry Krumb School of Mines (HKSM), Columbia University, New York, NY, Paul F. Duby, Earth and Environmental Engineering , Columbia University, New York , NY and Marco J. Castaldi, Department of Earth and Environmental Engineering (HKSM), Columbia University in the City of New York, New York City, NY

Boiler corrosion in waste-to-energy facilities, municipal solid waste (MSW) combustion, is much greater than in comparison to other power plants. This is largely attributed to the chloride concentration in MSW and the subsequent reactions that occur with generated gaseous HCl and volatilized metal chlorides under elevated metal temperatures. In an effort to engineer new methods to mitigate corrosion, the addition of SO2 to WTE flue gas streams has been shown in previous works to reduce corrosion. The following presentation is part of ongoing work to understand the effect of the SO2:HCl ratio on the corrosion of alloys exposed under superheater temperatures and flue gas compositions. Previous results by the author have shown that in the absence of metal chlorides, the addition of sulfur to increase the SO2 : HCl ratio had no consistent trend on the corrosion rate of SA178A, a commerical low carbon steel, under various typical flue gas water vapor concentrations. This work will further discuss the influence of water vapor on the system on the effect of the SO2 : HCl ratio as well as the influence of metal chlorides.

(702e) Corrosion-Stress Relaxation Effects On Surface and Tensile Properties of an AZ31 Magnesium Alloy

Chris Walton, Holly J. Martin, M.F. Horstemeyer, Wilburn Whittington and P.T. Wang, Center for Advanced Vehicular Systems, Mississippi State University, Mississippi State, MS

Magnesium alloys exhibit the attractive combination of low densities and high strength per weight ratios, along with good damping capacity, castability, weldability, and machinability [1-4]. Because of these characteristics, Mg alloys are being increasingly used in electronics, automobile, and aerospace industries [2-5]. Of the various commercial magnesium alloys, those developed from the Al-Zn ternary system (i.e. the as-named AZ alloys) have found the largest number of industrial applications [3]. Although Mg alloys are some of the best candidates of lightweight structural materials with a relatively high strength-to-weight ratio, they have not been used more widely in vehicles because of their low resistance to corrosion [4, 6-9] and creep [10]. While the creep/stress relaxation properties can be improved considerably through alloy chemistry design, cost-effective methods for mitigation of corrosion are still lacking. When combining both corrosion and creep/stress relaxation mechanisms, AZXX Mg alloys are highly susceptible to stress corrosion cracking [11]. This is primarily due to the $Mg_{17}Al_{12}$secondary phase located at the grain boundaries (intergranular) causing microgalvanic cells and the microstructure interaction (transgranular) with a water environment causing hydrogen embrittlement [11]. This synergetic interaction between mechanical and chemical processes then becomes most important. Localized corrosion taking place on slip planes due to mechanically enhanced anodic dissolution causes crack initiation and propagation then the plasticization of the metal assisted by the chemical reaction causes additional dislocation flux and enhanced plasticity [10].

Researchers at the Center for Advanced Vehicular Systems (CAVS) at Mississippi State University have begun to investigate this phenomenon as it relates to tensile and surface properties. An extruded AZ31 alloy was exposed to two environments, an immersion bath and a cyclic salt spray chamber, while under a constant strain ($\varepsilon_a \approx 0.0036$ mm/mm) over 60 hours. Tensile and surface properties were examined at 0, 1, 4, 12, 36, and 60 hours to examine the effect that the coupled chemical and mechanical mechanisms had on the specimens and to determine if or when stress corrosion cracking developed.

[1] Ghali, E. Magnesium and magnesium alloys, (2000) *Uhlig's Corrosion Handbook*, p. 793.

[2] Y. Chen, Q. Wang, J. Peng, C. Zhai, W. Ding, J Mater. Process. Tech. 182 (2007) 281-285.

[3] M. Marya, L. Hector, R. Verma, W. Tong, Mat. Sci. Eng. A418 (2006) 341-356.

[4] Pardo, A., Merino, M.C., Coy, A.E., Arrabal, R., Viejo, F., Matykina, E. (2008) *Corros. Sci.*, 50 (3), pp. 823-834.

[5] Feliu, S., Maffiotte, C., Galván, J.C., Barranco, V., (2011) Corros. Sci. 53 (5), pp. 1865-1872.

[6] Alvarez, R.B., Martin, H.J., Horstemeyer, M.F., Chandler, Mei.Q., Williams, N., Wang, P.T., Ruiz, A., (2010) *Corros. Sci.* 52 (5), pp. 1635-1648

[7] Chamos, A.N., Pantelakis, S., Spiliadis, V., (2010) *Mater. Des.* 31 (9), pp. 4130-4137

[8] Song, G.L., Atrens, A., (1999) *Adv. Eng. Mater.*, 1 (1), pp. 11-33.

[9] Renlong, X., Li, B., Li, L., Liu, Q., Mater. Des., 32 (8-9), pp. 4548-4552.

[10] Y. Unigovski, Z. Keren, A. Eliezer, E.M. Gutman, Mats. Sci. and Eng. A398, 2005, 188-197

[11] A. Atrens, N. Winzer, W. Dietzel, Adv. Eng. Mater. 13, 2011, 11-18

(702f) Real Time Corrosion Monitoring of Microbially Induced Corrosion in Water-Cooling Systems

Alvaro A. Rodriguez, Chemical and Biomolecular Engineering, The University of Akron, Akron, OH, Chelsea N. Monty, Department of Chemical and Biomolecular Engineering, The University of Akron, Akron, OH and Mariano Iannuzzi, The University of Akron, Akron, OH

Microbially induced corrosion (MIC) is an insidious form of corrosion that is difficult to detect. Moreover, it is extremely challenging to quantify the degree to which certain structures and components may be affected by MIC. There is, therefore, a significant risk associated with microbial activities related to corrosion, especially for critical assets operating in remote areas for long periods of time. Water-cooling systems and heat exchangers are known to suffer microbial contamination to various degrees. Moreover, for certain corrosion resistant alloys (CRAs) used in pressure-retaining components there is always a possibility that localized corrosion will eventually transition into stress corrosion cracking (SCC). Therefore, assessing whether microbial contamination of water-cooling systems leads to MIC is of great scientific and technological significance. The ability to monitor the kinetics of MIC in real time is also paramount.

Shewanella putrefaciens strain MR-1 is an iron-reducing bacterium used in this study to understand the mechanism of MIC in aerobic and anaerobic conditions. This system produces changes in the form of pH, oxygen and ion concentrations, bilayer formation, metabolic pathway, among others. The objective of this work is to assess the impact of MIC in carbon steel (UNS C1018) by using electrochemical techniques such as linear polarization resistance (LPR), impedance and Tafel plots using the split cell method. The outcome of this investigation will help elucidating one of the most challenging questions in MIC research of water-cooling systems, i.e. under which conditions microbial contamination of a particular system by a given community of microorganisms leads to microbially induced corrosion.

(702g) Welded Aluminum 6061: The Effect of Corrosion On Yield Strength

Holly J. Martin, Esteban Marin and Roger King, Center for Advanced Vehicular Systems, Mississippi State University, Mississippi State, MS

Transportation vehicles contain five thousand or more spot welds, with more welds present for larger or heavier vehicles [1]. The strength and dependability of these vehicles are greatly improved with the use of spot welds [1]. Vehicles are also exposed to environments containing salt, such as de-icing salt on roads or salt in the air near saltwater bodies. The presence of salt can then affect the undercarriage or body of these vehicles, including the welds. The effects of corrosion on metals include mass loss and hydrogen embrittlement, both of which reduce the strength of the metal. Hydrogen embrittlement can also cause a reduction in yield strength and the development of cracks that can result in catastrophic failure. Unfortunately, the effects of salt on the yield strength of these spot welds are unknown.

Corrosion is a large problem that has led to the development of multiple ASTM standards, including immersion (G-31), salt spray (B-117), and accelerated corrosion (G-85, B-368) [2-5]. Unfortunately, the ASTM standards do not translate well to the automotive industry, as they do not contain a pollution phase, a wet phase, and a dry phase [6-8]. These specific phases are needed to replicate various conditions experienced by vehicles, including de-icing salt, mud, and condensation [6-8]. In order for any corrosion testing to be useful for the automotive industry and to adequately quantify the effects of a corrosive environment on the yield strength of welded aluminum joints, a pollution phase (acidified salt spray), a wet phase (100% humidity), and a drying phase are all necessary [7].

Developing a corrosion model that can be coupled with a damage model to describe the interaction of corrosion and mechanical stresses is the goal of our research. Determining how the yield strengths of three different weld production types, low, nominal, and high, are affected by a corrosive environment is a necessary part of the development of a coupled corrosion-damage model for welds. At the Center for Advanced Vehicular Systems (CAV) at Mississippi State University (MSU), samples of aluminum alloy 6061 are being exposed to two environments, an immersion environment and a cyclical salt spray environment, using 5% sodium chloride aqueous solution and acetic acid for 240 hours. During the 240 hours, the experiment is interrupted at various time intervals to determine how the environment is affecting the surface and the yield strength of the weld. The surface information gathered will consist of data measuring various aspects of pit nucleation, pit growth, and pit coalescence that will be used in the development and calibration of an internal state variable model [9]. This model relates total corrosion to general corrosion, pitting corrosion, and intergranular corrosion [9] and can be coupled with a fully established damage model. Combining these two models will allow us to accurately predict the failure of a material, in this case aluminum, due to the embrittlement effects on the mechanical properties caused by a corrosive environment.

[1] T. Snow, "Digital Spot Welding and Car Safety", *Dassault Systemes*, Published March 11, 2011; Viewed April 13, 2012. http://perspectives.3ds.com/manufacturing/digital-spot-welding-and-car-safety/

[2] "G-31, Standard Practice for Laboratory Immersion Corrosion Testing of Metals" (2008) *ASTM International*.

[3] "B-117, Standard Practice for Operating Salt Spray (Fog) Apparatus" (2008) *ASTM International*.

[4] "G-85, Standard Practice for Modified Salt Spray (Fog) Testing" (2008) *ASTM International*.

[5] "B-368, Standard Test Method for Copper-Accelerated Acetic Acid-Salt Spray (Fog) Testing (CASS Test)" (2008) *ASTM International*.

[6] K.R. Baldwin, C.J.E. Smith, (1999) "Accelerated corrosion tests for aerospace materials: current limitations and future trends", *Aircr. Eng. Aerosp. Technol.* 71, 239–244.

[7] N. LeBozec, N. Blandin, D. Thierry, (2008) "Accelerated corrosion tests in the automotive industry: a comparison of the performance towards cosmetic corrosion", *Mater. Corros*. 59, 889–894.

[8] G. Song, D. St. John, C. Bettles, G. Dunlop, (2005) "The corrosion performance of magnesium alloy AM-SC1 in automotive engine block applications", *JOM* 57, 54–56.

[9] H.J. Martin, R.B. Alvarez, J. Danzy, M.F. Horstemeyer, P.T. Wang, (2012) "Quantification of Corrosion Pitting Under Immersion and Salt-Spray Environments on an As-Cast AM60 Magnesium Alloy", *Corrosion*, In Press.

(710a) Understanding the Mechanisms of Cellulose Dissolution in Ionic Liquids

Brooks D. Rabideau[1], Animesh Agarwal[2] and Ahmed E. Ismail[1], (1)Mechanical Engineering, RWTH Aachen University, Aachen, Germany, (2)RWTH Aachen University, Aachen, Germany

Heightened interest in energy independence and environmental awareness has accelerated the search for combustible fuel sources which are both renewable and have a low environmental impact. One promising route is the conversion of waste biomass into tailor-made fuels. An important aspect of this process is the low-energy separation of cellulose from the biomass. Ionic liquids (ILs) have been demonstrated capable of dissolving cellulose while being essentially non-volatile, making them strong candidates for "green" processing. However, full exploitation of ILs as a commercially viable processing medium requires better understanding of the dissolution mechanism. However, in spite of extensive research, many parts of this process remain uncertain. Using molecular dynamics simulations, we examine the behavior of cellulose in the ionic liquids [BMIM]Cl, [EMIM][Ac] and [DMIM][DMP] as well as water. All three of these ionic liquids have been observed to dissolve cellulose quite well yet have differently sized anions. We explore these differences and the impacts they have on their interactions with cellulose. First, we examine the dynamics of a single cellulose strand in these ionic liquids, and study the structural and dynamic properties, such as the radius of gyration and the number and lifetime of hydrogen bonds that are formed between the anions and cellulose. Next, we probe the dissolution mechanism of multiple, bound cellulose oligomers of varying lengths, examining interactions at the IL/cellulose interface and the breakup of inter-cellulose hydrogen bonds. We also study the relationship between the anion strength and the hydrogens in hydroxyl groups in determining the relative aggregation of cellulose chains in these different solvent media.

(710b) Modeling Biomolecular Structure and Excitation Energy Transfer in Photosynthetic Pigment-Protein Complexes

William P. Bricker, Energy, Environmental & Chemical Engineering, Washington University in St. Louis, St. Louis, MO, Linda N. Holyoke, Physics, Saint Louis University, St. Louis, MO and Cynthia S. Lo, Energy, Environmental and Chemical Engineering, Washington University in St. Louis, St. Louis, MO

In photosynthetic organisms, the mechanisms by which photosynthetic pigment-protein complexes transfer harvested energy efficiently to their photosystems must be better understood at the molecular scale, so that these mechanisms can be applied to building inexpensive and efficient bio-hybrid photovoltaics. Using the peridinin-chlorophyll a-protein (PCP) from marine algae as a model photosynthetic light-harvesting complex (LHC), we model excitation energy transfer in the complex using both a Förster resonance energy transfer (FRET) model and a semi-classical propagation scheme. To model the rates and efficiencies of excitation energy transfer in PCP using FRET, we first calculate absorption/fluorescence spectra and transition dipole moments from the electronic excited states using time-dependent density functional theory (TD-DFT). We use a QM/MM approach to determine the biomolecular structure of PCP and the effect of the protein scaffold on the rates and efficiencies of excitation energy transfer. We also present our development of a semi-classical propagation scheme to observe the dynamical evolution of the excited state wavefunctions, by constructing an initial model Hamiltonian of the PCP complex and calculating the time correlation of the wavefunctions. Both models are compared to experimental spectroscopic measurements, and implications for quantum coherence in bio-hybrid photovoltaics are discussed.

(710c) Polyelectrolyte Interactions At a Conducting Interface: A Simulation Study

Maria Sammalkorpi, Aalto University, Helsinki, Finland and **Paul Van Tassel**, Yale University, New Haven, CT

Recent experiments demonstrate polyelectrolyte adsorption to a conducting surface to become continuous, i.e. to scale linearly with time over hours, upon application of a modest electric potential. Continuous polyelectrolyte adsorption offers the possibility of polymer films of tailored nanoscale thickness realized in a single step, but raises fundamental mechanistic questions. A key feature of one hypothesized mechanism involves a net attraction among like-charged polymers at the electrified interface, caused by a highly correlated counter-ion distribution and/or the presence of image charges. In this talk, we present a Monte Carlo simulation study of polyelectrolyte-polyelectrolyte interactions, with a goal of developing mechanistic insight into like-charge attraction at the electrified interface, and hence also into the continuous adsorption process. Our system consists of two parallel polymer chains, composed of charged tangent spheres, above a conducting surface, in the presence of spherical counter-ions and salt ions. We observe conditions where counter-ions preferentially locate between the polymers, and act to promote a net polymer-polymer attraction. We discuss the sensitivity of the attractive regime to a Coulombic coupling parameter, to counter-ion size, and to the nature of the underlying substrate. We also compare simulation results to those of a strong Coulomb coupling statistical mechanical model, and to our ongoing experiments.

(710d) Molecular Dynamics Simulations of Thermoresponsive Poly (N-isopropylacrylamide) and Its Copolymer

Hongbo Du, Chemical Engineering, University of Arkansas, Fayetteville, AR, S. Ranil Wickramasinghe, Department of Chemical Engineering, University of Arkansas, Fayetteville, AR and Xianghong Qian, Ralph E Martin Department fo Chemical Engineering, University of Arkansas, Fayetteville, AR

Salt ions play a critical role in many biological processes including protein solubility, stability, denaturation and aggregation. Understanding ion-protein interaction is crucial for elucidating the fundamental underlying mechanisms. The Hofmeister series refers to the ability for salt ions to precipitate proteins from an aqueous solution based on empirical observations. However, to date, the fundamental mechanisms remain elusive. Previously it was thought that ions in solution can affect the bulk water structure thus affecting the hydrophobic-hydrophilic transition of proteins in aqueous solution. Ions are considered chaotropic (structure breaking) or kosmotropic (structure making). Recently spectroscopic and calorimetry studies show that ions only affect the closest hydration shells and the bulk water structure is not altered by the presence of the salt ions. In addition, attempts have been made to correlate the Hofmeister series with surface tension, the solvation entropy of the ions and other physical and thermodynamics properties of the ions. However, all these correlations have failed to explain the entire series and the inverse Hofmeister series sometimes observed. A fundamental molecular level understanding is also lacking.

PNIPAM has been widely studied as a prototype temperature responsive polymer which exhibits hydrophobic-hydrophilic phase transition at its lower critical solution temperature (LCST). It has been found that salt ions have a significant impact on the LCST. Most salts are found to reduce the transition temperature and the effect of ions on LCST follows the Hofmeister series. Classical molecular dynamics simulations were carried out to investigate the ion-PNIPAM interactions in aqueous solutions. The salt investigated include MCl_n (M=Li, Na, K, Rb, Cs, Mg, and Ca) and NaX (X=Cl, Br and I). The effects of salt on LCST transition as well as ion specific interactions with PNIPAM will be elucidated.

(710e) Molecular Simulation of Hydration of Zwitterion

Qing Shao and Shaoyi Jiang, Chemical Engineering, University of Washington, Seattle, WA

Zwitterionic materials are an important new type of biomaterials for biological and chemical applications such as drug delivery, marine coating and molecular sensing. One important research topic of zwitterions is to understand their structure-property relationship. In this work, we carried out a series of quantum calculations and molecular simulations to study the zwitterionic strength and hydration of zwitterion molecules with various negatively and positively charged groups, distance between charged groups and additional groups. We investigated the electrostatic potential surface of the molecules and the coordination number, orientation distributions and residence dynamics of water molecules around the zwitterion molecules. We also investigated the preferential association of various zwitterions with ions. The simulation results illustrate the effect of distance between charged groups and additional groups on the zwitterionic strength of molecules with various charged groups, as well as their hydration. The correlation we observed between the charged group distance/addtional groups and the hydration of zwitterion offers a strategy for the rational design of zwitterionic materials.

STRUCTURE OF RIGID HARD-RING FLUIDS

Mariam Nouri[*] and Marc Robert, Rice University, Houston, TX

Abstract

The structure of hard-ring fluids is studied in two and three dimensions, using Monte Carlo computer simulations in the canonical ensemble. These fluids are composed of molecules consisting of hard-spheres linked together to form planar rigid rings. For rings of various sizes and for a wide range of densities, results are reported for the pair distribution function of the ring centers and for the pair distribution of the ring orientations. In a dense monolayer, where the rings are completely confined to a plane, a "shoulder", precursor of the freezing transition, is observed in the second peak of the pair distribution function of the ring centers, as previously seen in the simple hard-sphere fluid. In two dimensions, where the ring centers are confined to a plane but the rings themselves are free to rotate around all axes, nematic and tetratic ordering is observed at sufficiently high densities. Results are also presented for three dimensions.

Introduction

Ring-shaped molecules are ubiquitous, and found in particular among organic, polymeric, and biological molecules. Recently, the synthesis and characterization of shape-persistent macrocycles have been drawing increasing attention [1]. These molecules are fairly rigid and possess non-collapsible backbones and nanoscale tunable cavities [2]. They exhibit fascinating structural properties such as organization of remarkable patterns at interfaces [3]. Also, the stacking of rings enables formation of tubular structures similar to the cyclicoligopeptide channels reported by Ghadiri [5]. Furthermore, because of their nonspherical shape, they are likely to give rise to a wide variety of liquid crystalline phases [4].

Hard molecules, i.e., molecules with purely repulsive interactions, have played a central role in the current understanding of real molecular fluids. Indeed, spherical or nonspherical hard-core models play the role of the reference fluid in thermodynamic perturbation theories, which predict the properties of a real fluid where molecules also exert attractive forces: the latter are so weak compared to the infinite repulsions, that they can be treated perturbationally. From a computational point of view, hard-core interactions are favored over more complicated potentials for their simplicity and for the computational efficiency they offer.

The rigid hard-ring model is also of interest because, despite its simplicity, it displays nematic, tetratic, and possibly smectic phases, in addition to the isotropic and crystalline phases. Previous computer simulations of hard-core bodies, as reviewed by Frenkel [6], Allen [7], and Care and Cleaver [8], have revealed that nonspherical hard molecules can exhibit a rich phase behavior, and that attractive interactions are not necessary for the occurrence of liquid crystalline order, as previously observed in the hard-rod fluid.

[*] corresponding author: mn9@rice.edu

Molecular model

Our model to describe ring-shaped molecules consists of hard-spheres linked together to form a planar rigid ring (shown in Fig. 1). In this pearl-necklace model, it is assumed that there are no intermolecular attractions between the molecules, the only interactions present being the hard-core repulsive forces. The diameter of the ring can be altered by changing the number of hard spheres on the ring.

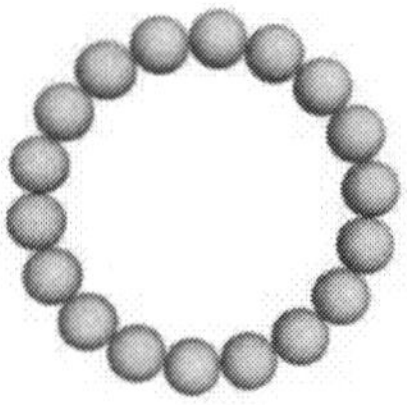

Fig.1 Model of rigid hard-ring molecule

The potential between any two hard spheres on two different rings is given by:

$$U(r) = \begin{cases} \infty & r \leq \sigma \\ 0 & r > \sigma \end{cases}, \qquad (1)$$

where r is the distance between two spheres and σ is the diameter of a sphere.

Our rigid hard-ring model fluid is studied in three different cases in two dimensions, and in one case in three dimensions. These cases are defined as follows, denoting by d the dimension of space, and by n that of the order parameter:

Two dimensions ($\mathbf{d = 2}$)

The systems studied in two dimensions differ from each other by the degree of orientational freedom of the molecules, while the space in which the centers of the molecules are allowed to move remains two-dimensional in all three cases. As has become well-known, the absence of true long-range positional or orientational order is one of the most fascinating properties of a large class of two-dimensional systems.

$\boldsymbol{n = 1}$

Here, the centers of the molecules are confined to a plane, and the molecules themselves are only allowed to rotate within that same plane, as illustrated in Fig. 2 (left). Clearly, the hole of a molecule cannot be penetrated by other molecules.

$\boldsymbol{n = 2}$

In this case, while the centers of the molecules are only allowed to move in a plane, the molecules' normals lie and rotate in the same plane, as shown in Fig. 2 (middle). In other words, the molecules are perfectly standing on their edges.

$\boldsymbol{n = 3}$

Again, the centers of the molecules are restricted to move in a two-dimensional plane, but the molecules themselves are allowed to wobble out of that plane. In other words, molecules possess fully three-dimensional rotational freedom, as illustrated in Fig. 2 (right).

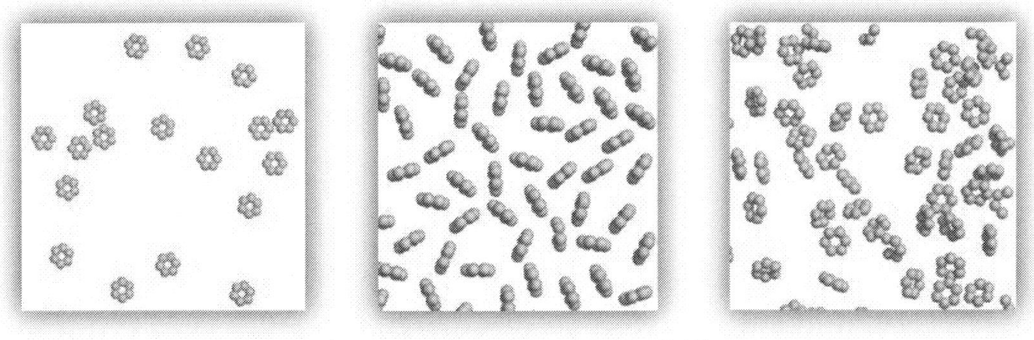

Fig.2 Two dimensional hard-ring fluid. Left: $n = 1$. Middle: $n = 2$. Right: $n = 3$.

Three dimensions $(\boldsymbol{d = 3, n = 3})$

In this case, the molecules are free to fully move and rotate arbitrarily in a three-dimensional space, as illustrated in Fig. 3.

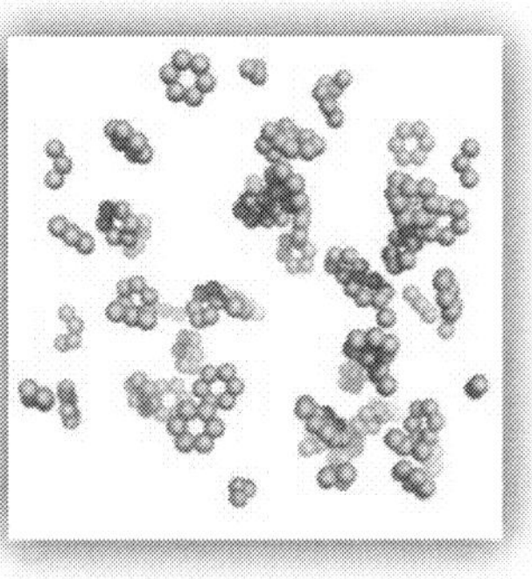

Fig.3 Three-dimensional hard-ring fluid $(d = 3, n = 3)$

Analogy with lattice models of a ferromagnet

Translating the above cases with different degrees of orientational order into the ferromagnetic lattice model language, one obtains in two dimensions systems analogous to the two-dimensional Ising $(d = 2, n = 1)$, XY $(d = 2, n = 2)$ and Heisenberg $(d = 2, n = 3)$ models, and in three dimensions analogous to the three-dimensional Heisenberg model $(d = 3, n = 3)$.

Not all two-dimensional systems exhibit phase transitions; for example, the two-dimensional Heisenberg model of a ferromagnet, in which the spins can point in any direction and for which the order parameter is thus three-dimensional, does not exhibit any phase transition at all [9]. The same holds for the spherical model [10]. However, two-dimensional solids, while lacking crystalline order, do exhibit a remarkable phase transition, the famous BKT (Berezinskii [11]-Kosterlitz-Thouless [12]) transition, at which the orientational pair correlation function becomes long-ranged, like the positional pair correlation function of an ordinary fluid at its critical point. This fascinating property is unique to certain two-dimensional systems, and is not shared by any systems in one or three dimensions. Theoretically, for our models, both the isotropic-nematic and the isotropic-tetratic phase transitions, if they occur, are expected to be BKT transitions.

It should be observed that the analogy between a lattice model of a ferromagnet and a continuum model of rigid hard-rings is not a perfect one. In a ferromagnetic lattice model, two next-nearest-neighbor spins being aligned in one direction will not necessarily force the spin between them to point in the same direction. This is illustrated in Fig. 4 (left), where the fact that the two blue spins are parallel does not impose the orientation of the spin between them.

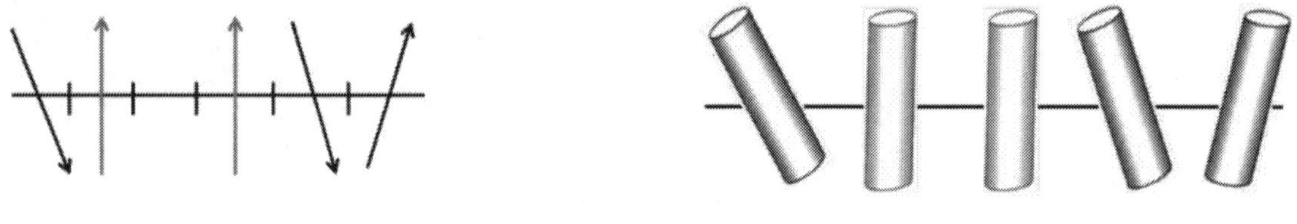

Fig.4 Left: Spins on a lattice. Right: Hard rods in the continuum

In contrast, in the continuum, for two hard anisotropic molecules with cylindrical symmetry which point in one direction, and are located such that only one molecule can fit between them, if there is a molecule between the two other molecules, the repulsive interactions will force the former to align with its two neighbors. If there were a hard rod present between the two blue rods shown in Fig. 4 (right), it would necessarily be orientated in the same direction as that of its neighbors.

However, the analogy between these models is meaningful, because spins at low temperatures and hard rods at high densities, behave similarly and their alignment is favored. Indeed, it has also been shown that charged colloidal disks in an applied electrical field that are forced to stand on their edges $(d = 2, n = 2)$ in an applied field exhibit a BKT transition [13], the same as that which occurs in the XY lattice model.

Computational methods

The following Fig. 5 shows the two characteristic lengths in the hard-ring model: σ, the diameter of the hard spheres, is shown in black, and r, the radius of the ring on which the centers of the hard spheres are located, is illustrated with a red arrow.

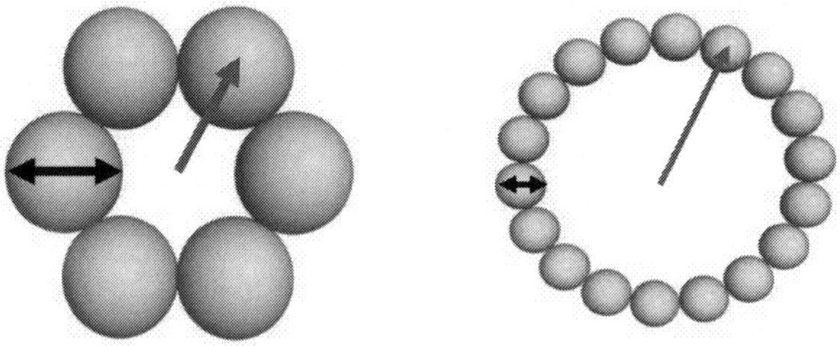

Fig.5 Diameter of hard spheres σ shown in black. Ring radius shown in red
Left: A 6-membered ring. Right: An 18-membered ring.

Ring orientations are represented by a normal vector perpendicular to the surface of the ring, shown in Fig. 6.

Fig.6 Ring orientation represented by a normal vector.

Monte Carlo simulations in the canonical ensemble are performed in a box of volume V, at a prescribed value of the density. Conventional periodic boundary conditions are imposed in d dimensions. Molecule displacements and acceptance criteria are performed according to the Metropolis [14] procedure.

Rotational moves are based on generating random vectors on the surface of a sphere, as proposed by Marsaglia [15]. At higher densities, to increase computational efficiency, the random vectors are chosen within a restricted region on the surface of a sphere. Positional and rotational moves are performed simultaneously as a combined random trial move, which is then accepted or rejected.

As the molecular interactions are purely hard, avoiding overlaps is the only criterion to reject generated configurations. The conventional hard-spheres overlap detection procedure [16] is employed to detect overlaps between hard rings. If one assumed for a moment that the fluid only consisted of hard spheres that are subject to the positional constraint to be linked on a ring, then if any two hard spheres are overlapping, the two corresponding hard rings must overlap too. To speed up this process, a nearest-neighbors list is generated for each molecule. This list contains all other molecules whose centers are located within a distance from the center of the reference molecule, shorter than the molecular diameter. If no overlaps are detected between the hard spheres, the configuration requires further investigation to reject the interlocking but non-overlapping configurations.

An interlocking algorithm is proposed to reject the moves that result in two interlocking rings. In this algorithm a virtual disk of radius equal to that of the ring, and of thickness equal to the hard-sphere diameter, is assumed to be located inside the ring. To detect whether ring A and ring B are interlocked, the following procedure is carried out. If an overlap is detected between the spheres on ring A and the virtual disk inside ring B, *and* if the spheres on ring B also overlap with the virtual disk inside ring A, ring A and ring B are interlocked; if not, ring A is only penetrating the hole inside ring B, which is an acceptable move. Configurations of interlocking and penetrating molecules are shown in the following Fig 7.

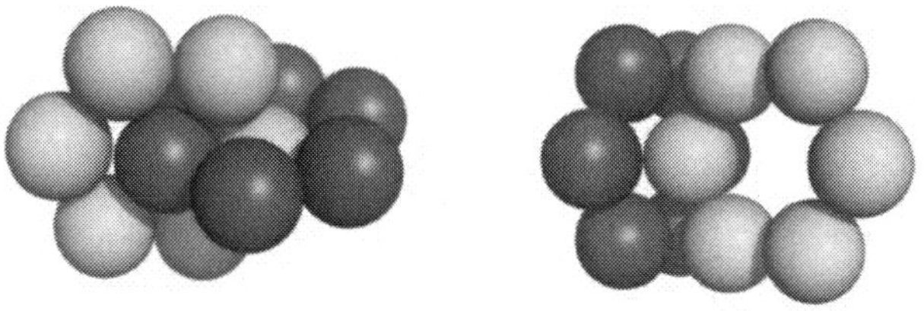

Fig.7 Left: Two inerlocked rings. Right: Penetrating, but not interlocking rings

The positional pair distribution function and the orientational pair correlation functions are computed for each case and the nematic and tetratic order parameters are computer only for those cases with $n > 1$. The order parameters S are defined as:

$$S^2 = \lim_{r\to\infty} < f(\cos(\theta_{ij})) > \tag{2}$$

where r is the distance between the centers of rings i and j, θ_{ij} is the angle between the normals of these rings and f is chosen as T_2 and T_4 for $n = 2$, and as P_2 and P_4 for $n = 3$, regardless of the dimensionality of space. T and P represent Chebychef and Legendre polynomials, respectively. In the isotropic phase $S = 0$, and in the nematic and tetratic phases $S \neq 0$.

Results

Two dimensions

For dense fluids in two dimensions and $n = 1$, a "shoulder", precursor of the freezing transition, is observed in the second peak of the pair distribution function of the ring centers, as previously seen in the simple hard-sphere fluid in two and three dimensions [17]. The split second peak is absent in the pair distribution function of the low-density fluid. It was previously incorrectly believed that the shoulder is a signature of amorphous solid, but our results support those of Truskett *et al.* [17], and indicate that the "shoulder" is actually a precursor to the development of long-range order. In the close-packed configuration shown in Fig. 8 (right), the centers of the molecules in the first layer (shown in green) about a center molecule (shown in red) lie on a circle. The centers of the molecules in the next layer lie on circles with different radii, as a result of the hexagonal symmetry. In a dense, but not close-packed phase, when forming the second layer, the center of the molecules will slightly shift from these two circles. Consequently, a "shoulder" appears in the second peak of the positional pair distribution function, as shown in Fig. 8 (left).

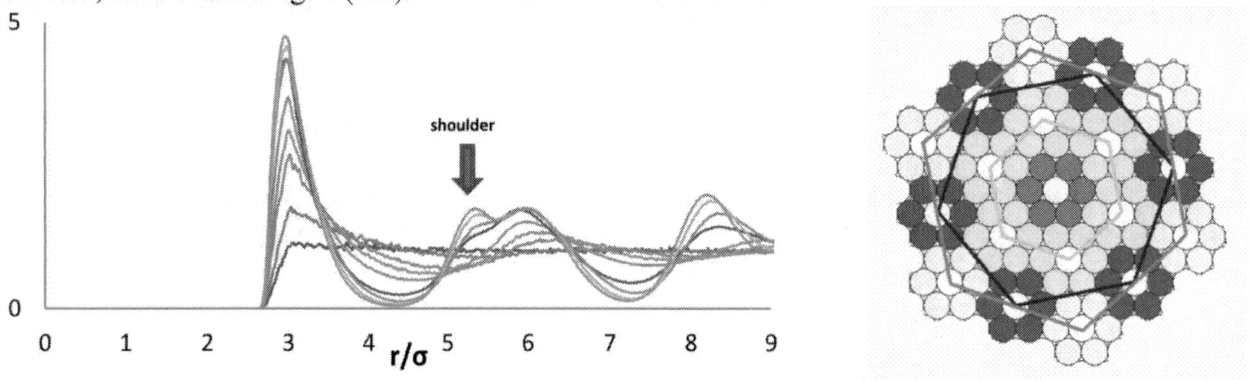

Fig.8 Left: Positional pair distribution function. Right: Close packing arrangement $(\boldsymbol{d} = \mathbf{2}, \boldsymbol{n} = \mathbf{1})$

For the $n = 2$ case, for 6-membered rings, although local nematic and tetratic ordering occurs, there is no evidence of orientational long-range order. Observation of nematic and weakly tetratic phases in colloidal platelets [13], however, suggests that a higher aspect ratio is required for the appearance of such orderings.

In contrast to the cases above, the value of the first peak of the positional pair distribution function does not increase upon increasing the system's density. Instead, different types of ordering, illustrated in Fig. 9, appear in the system.

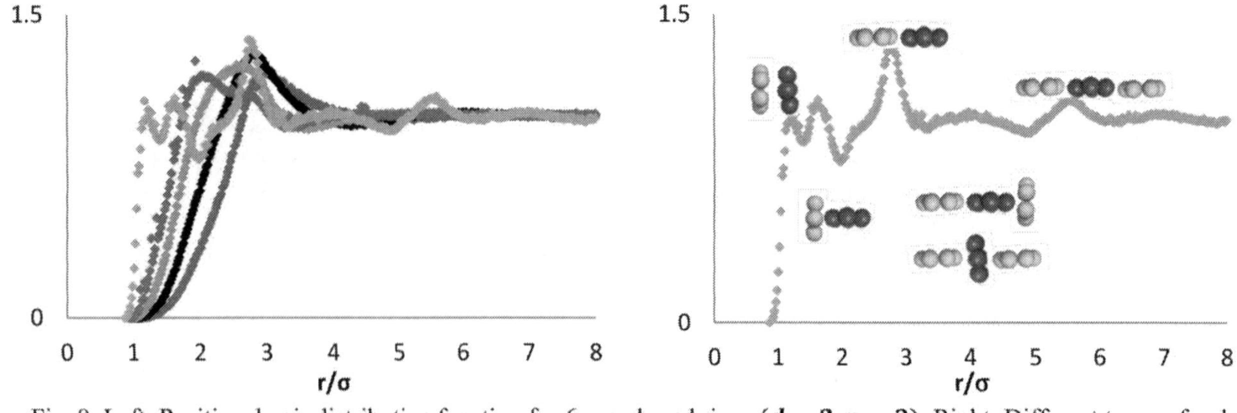

Fig. 9 Left: Positional pair distribution function for 6-membered rings $(\boldsymbol{d} = \mathbf{2}, \boldsymbol{n} = \mathbf{3})$. Right: Different types of order

The nematic order parameter exhibits an interesting behavior. When increasing the density, the value of S^2 first smoothly increases to a value slightly below 0.25, then remains roughly constant, and thereafter increases again. This can be explained by the fact that increasing the density limits the orientational freedom of the rings, so that most rings are forced to stand roughly on their edges. The normal to the plane of the rings is no longer free to rotate in all three dimensions and it is mainly limited to rotate in a two-dimensional plane. There will be a point where, while the three-dimensional order parameter is nonzero, it's two-dimensional counterpart vanishes. This only means that the orientational freedom of the rings is restricted, and is not necessarily an indication of nematic order (see Fig. 10). However, an increase of the order parameter beyond this point can only be attributed to nematic ordering. The same bahavior is observed for the tetratic order parameter.

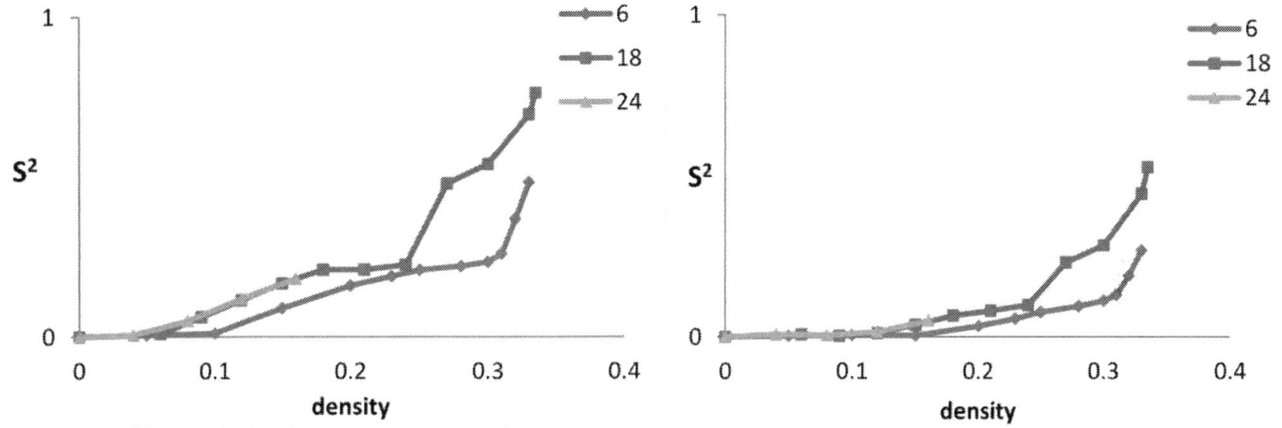

Fig.10 Nematic (left) and tetratic (right) order parameters for 6-, 18-, and 24- membered rings ($d = 2, n = 3$)

Three dimensions

The pair distribution function of the rings centers is shown below in Fig 11. At low densities, it is very similar to the that obtained by Wojcik and Gubbins [18] for oblate spherocylinders with an aspect ratio of 3, which is equal to that of a 6-membered ring. The only difference is that the fist peak in the pair distribution function occurs at a value less than 3 due to the lack of complete cylindrical symmetry in the ring molecules. At higher densities, however, more differences are observed in the structure of these fluids, which can only be attributed to the increased possibility of penetration of one molecule into the hole of the others at higher densities.

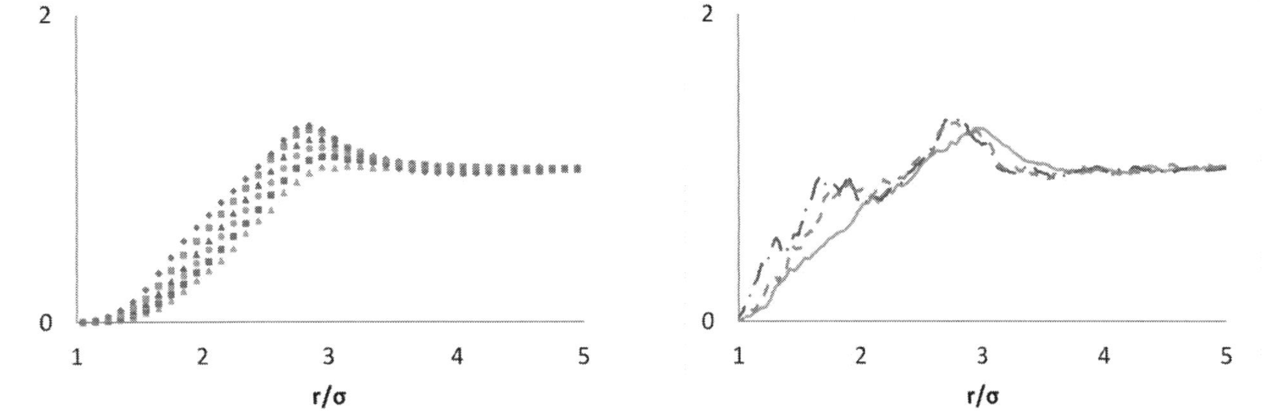

Fig.11 Positional pair distribution function for 6-membered rings of low (left) and high density (right) ($\boldsymbol{d} = \mathbf{3}, \boldsymbol{n} = 3$)

Acknowledgement

Support from the National Science Foundation under Grant No. 1034421 is gratefully acknowledged. This work was also supported by the Welch Foundation for fundamental research in chemistry under Grant No. C-1071.

References

1. W. Zhang and J. S. Moore, Angew. Chem. Int. Ed. 45, 4416-4493, 2006
2. M. Iyoda, Pure Appl. Chem 82, 831-841, 2010.
3. J. Kroemer, I. Rios-Carreras, G. Fuhrmann, C. Musch, M. Wunderlin, T. Debaerdemaeker, E. Mena-Osterlitz, and P. Baeurle, Angew. Chem. Int. Ed. 39, 3481, 2000.
4. B. Neumann, D. Joachimi, and C. Tschierske, Adv. Mater. 9, 241, 1997.
5. J. R. Granja and M. R. Ghadiri, J. Am. Chem. Soc. 116, 10785, 1994.
6. D. Frenkel, Comp. Phys. Comm. 44, 243, 1987.
7. M. P. Allen, Phil. Trans. R. Soc. Lond. A 344, 323, 1993.
8. C. M. Care and D. J. Cleaver, Rep. Prog. Phys 68, 2665, 2005.
9. J. Fröhlich, B. Simon, and T. Spencer, Phys, Rev. Lett 36, 804, 1976.
10. T. H. Berlin and M. Kac. Phys. Rev. 86, 821-835, 1952
11. V. L. Berezinskii and A. Ya. Blank, Sov. Phys. – JETP 37, 369, 1973.
12. J. M. Kosterlitz and D. J. Thouless, J. Phys. C: Solid State Phys. 6, 1181, , 1973.
13. K. Zhao, C. Harrison, D. Huse, W. B. Russel, and M. P. Chaikin, Phys. Rev. E 76, 040401, 2007
14. N. Metropolis, A. W. Rosenbluth, M. N. Rosenbluth, A. H. Teller, and E. Teller, Chem. Phys. 21, 1085, 1953.
15. G. Marsaglia, Ann. math. Stat 43, 645, 1972
16. M. R. Wilson and M. P. Allen, Mol. Phys. 80, 277-295, 1993.
17. T. M. Truskett, S. Torquato, S. Sastry, P. G Debenedetti, and F. H. Stillinger, Phys. Rev. E 58, 3083, 1998
18. M. Wojcik and K. E. Gubbins, Mol. Phys. 53, 397-420, 1984

(710g) Effects of Asphalt Compositions On Nanoscale Properties by Molecular Simulation

Mohammad Masoori, Chemical Engineering, University of Rhode Island, Kingston, RI and Michael L. Greenfield, University of Rhode Island, Kingston, RI

Having a deep understanding of all components in a mixture is the main key to understand and improve an industrial material. Asphalt is widely used material since ancient eras. Despite the fact that we used this material for over thousands years, we do not have a good understanding of its components and molecular structure of its components. Asphalt nano-scale properties have enormous effects on its properties. Another challenge of asphalt is the composition of its components, different compositions of same components may have different properties. Deep understanding of asphalt leads to pavement performance and durability. Because different asphalts are from different sources which have different components, they have different properties. Chemical composition and its effect on its properties has been investigated. To analyze storage and loss modulus, the inverse Fourier transform has been applied to stress correlation function results from molecular simulations of different model asphalts, including those of Canadian model asphalt system (AAA-1 in the Strategic Highway Research Program). Calculations for multiple temperatures investigate the effect on frequency-dependent storage and loss modulus.

(710h) Toward a Molecular Model of SiC Nanoporous Membrane: Application of Reactive Molecular Dynamics Simulation to the Study of the Pyrolysis of Hpcs Polymeric Precursor

Saber Naserifar, Mork Family Department of Chemical Engineering and Materials Science, University of Southern California, Los angeles, CA, Lianchi Liu, Shanghai Jiao Tong University, Shanghai, China, Theodore T. Tsotsis, Mork Family Department of Chemical Engineering and Material Science, University of Southern California, Los Angeles, CA, **Muhammad Sahimi**, Mork Family Department of Chemical Engineering and Materials Science, University of Southern California, Los Angeles, CA and William A. Goddard III, Chemistry and Chemical Engineering Division, California Institute of Technology, Pasadena, CA

Molecular dynamics (MD) simulations were carried out using a reactive force field, ReaxFF, that we have developed for pyrolysis of liquid HPCS polymer, in order to develop a new molecular-scale model of solid amorphous silicon carbide that is used in the fabrication of SiC nanoporous membranes.

The amorphous structure of SiC was generated by carrying out extensive MD simulations in the NVT ensemble at 2000 K, using ReaxFF. During the simulations the hydrogen radicals, which bond together to generate H_2 over the time, were removed from the simulation box. Long NVT-MD simulations at 2000 K were utilized afterward to equilibrate the structure and make the crosslinking of Si and C atoms more likely. Then, MD simulations in the NPT ensemble and at 2000 K were carried out to attain the density of the ceramic. The system was then gradually cooled down back to ambient temperature. It was then annealed between 300 K and 3000 K to rearrange the atoms to their equilibrium state. Eventually, long NPT-MD simulations were carried to arrive at the amorphous structure of SiC, which is essential to the development of a molecular model of SiC nanoporous membranes. To test the validity of the model, in addition to calculating the radial distribution function, we also computed X-ray diffraction pattern and the mechanical properties of the material. The results are compared with the experimental data.

Design of Multiple Brine Producer/Injector Configurations to Increase Carbon Dioxide Injectivity in Saline Formations

Q. Sun, T. Ertekin

Extended Abstract

Carbon dioxide injection into subsurface geological horizons is one of the more promising carbon storage technologies. Mt. Simon sandstone formation appears to be an excellent candidate of carbon dioxide sequestration for the following reasons:

1. Mt. Simon formation is an extensively high capacity storage formation due to its lateral extent and thickness. Accordingly, the injection potential of the reservoir is high.
2. There exists an un-permeable formation, Eau Clair shale, above Mt. Simon sandstone formation which serves as a strong seal to prevent the injected carbon dioxide from migrating into upper horizons.
3. Mt. Simon formation was once used to dispose industry waste water. The existing brine inside the formation can be produced to create more space for carbon dioxide to be injected.

The goal of the study discussed here is to examine the relationship between the carbon dioxide injectivity and the injector type and brine production by using numerical reservoir simulation techniques, and finally generate a solution for optimal injector type and arrangement of injector and producer to inject as much CO_2 as possible while producing minimum amount of brine. An injection efficiency (IE) term is defined , which is calculated as the ratio of cumulative carbon dioxide injection to the cumulative brine production to evaluate the effectiveness of an injector/producer pattern for a carbon dioxide injection scenario.

A significant parameter in evaluating the efficacy of CO_2 injection into brine formations is the value of the stabilized reservoir pressure of the CO_2- brine system. In designing the injection operation one should make sure that the reservoir pressure after stabilization is lower than the initial reservoir pressure to minimize the risk of CO_2 leakage and upwards migration. A dimensionless pressure depletion ratio term is defined, which is calculated as the ratio of stabilized injector block pressure to the initial reservoir pressure to evaluate the pressure in the CO_2-brine system after stabilization. The design should not just chase for high cumulative injection efficiency or high cumulative injection efficiency values. The pressure depletion ratio has to be less than one to ensure that the injection process is conducted under consideration of a safety factor.

In parallel to the aforementioned goals and challenges, a mathematical model is built allowing the numerical simulation of the study area. Instead of simulating the entire reservoir, the model focuses on a small unit of the reservoir. The small unit has the same reservoir properties range to represent one small part of the entire reservoir. In order to reduce the uncertainties from the various reservoir properties, the analysis is achieved by implementing the Monte Carlo simulation protocol. Reservoir property arrays are generated within the property ranges applicable to the study area. The model does multiple simulation runs with the reservoir property arrays. The model outputs the value of injection efficiency and the pressure depletion ratio after multiple Monte Carlo simulations runs and describes the results for the P90, P50 and P10 analysis of injection efficiency and pressure depletion ratio confidence intervals.

The commercial numerical model used in the study uses a compositional formulation to consider the thermodynamic behavior of CO_2-brine system under operating reservoir conditions.

The simulation study shows that the regular 4- spot injection pattern with horizontal injector has best injection performance compared with other types of injection patterns. For a given formation, the implicit expressions of injection efficiency and pressure depletion ratio can be written as:

$$Injection\ Efficiency\ (IE, Metric\ ton/bbl) = f(Q, A, L_w, P_{wf})$$

$$Pressure\ Depletion\ Ratio = f(Q, A, L_w, P_{wf})$$

Here Q represents for the injection rate;

A represents for the pattern area;

L_w represents for the horizontal extension of the horizontal well;

P_{wf} represents for the bottom hole pressure for the brine producers.

Since Monte Carlo simulation protocol is applied in this study, the model needs to call the high-fidelity model (the numerical simulator) for a number of times for each run which can be prohibitively large. In order to address this issue, the study includes an artificial neural network (ANN) based proxy model to minimize the excessively time consuming simulation runs. With the help of the proxy model a very large number of runs can be generated more efficiently and the resolution of the Monte Carlo analysis can be enhanced significantly.

Two artificial neural networks have been developed in this study. One is forwards solution network. The inputs of forward solution network are reservoir properties and engineering design of the injection pattern, which includes the injection rate, size of injection pattern, horizontal extension of horizontal injector and bottom hole pressure of brine producer. The outputs are cumulative injection CO_2, cumulative brine production and injector block pressure after stabilization. The other one is inversed solution network. The input of inversed solution network is the reservoir properties and the project target, which includes the goal of CO_2 injection, brine production and injector block pressure after stabilization. The outputs of the inversed solution are the required engineering design of the injection pattern to fulfill the given project target.

The two artificial neural networks are trained by large number of inputs and corresponding outputs. The capabilities of the networks are tested by comparing the result predicted from the network to the results generated by high-fidelity model. Table-1 shows the error between the results predicted by the forwards solution network and the result generated by commercial simulation model.

Table 1. Testing Results of Forwards Solution Network

Error	Mean error (%)	P90 error (%)
Cumulative CO_2 Injection	2.78	5.53
Cumulative Brine Production	2.65	4.88
Well Block Pressure after Stabilization	0.19	0.59

Table-2 shows the error between the results predicted by the inversed solution network and the result generated by commercial simulation model.

Table 2. Testing Results of Inversed Solution Network

Error	Mean error (%)	P90 error (%)
Injection Rate	5.90	13.58
Pattern Area	8.43	18.69
L_W	0.04	0.05
P_{wf}	0.50	1.67

Both of the forwards solution network and inversed solution network shows reasonable error values so that they could be treated as expert neural network system in this problem.

With the help of these two expert neural network system, one could generate results from large number of input to study the relationship between the injection efficiency, pressure depletion ration and injection rate, pattern size, horizontal extension of horizontal well, bottom hole pressure of brine producer.

(733b) Separation Work and the Thermodynamics of CO2 Capture From Flue Gas

Hugo S. Caram[1,2], Hans Thomann[2], Ramesh Gupta[2] and Walter Weissman[2], (1)Chemical Engineering, Lehigh University, Bethlehem, PA, (2)ExxonMobil Research and Engineering Company, Annandale, NJ

The economics and process selection for CO2 capture from flue gases is influenced to a significant extent by the energy requirements for the process. This paper presents a thermodynamic framework for estimating the energy requirements of CO_2 capture processes. The presented thermodynamic framework is general in nature and applies to CO_2 capture using absorption in amines and can be applied to any separation involving a mixture of gases.

Of specific interest are the processes using a first step of selectively capturing the gas to be separated by a solvent followed by stripping of the absorbed /adsorbed gas in a second step. Among the many possibilities these will include liquid absorption cycles, mostly amine based and chemical looping using materials such as calcium oxide or sodium carbonate-bicarbonate. While the capture process occurs spontaneously, stripping and regeneration of the solvent that releases the purified gas requires energy. Given the very large commercial scale of these processes it is of utmost importance to minimize the energy requirements. For example, the heat provided to the stripper reboiler may constitute more than 80% of the total energy requirements of the gas capture operations. It is then important to develop a conceptual understanding of the factors affecting the energy requirements. These include, rich and lean loadings, stripper pressure, energy losses in cross flow heat exchangers and the type of solvent deployed. The model being presented provides the lower bound for the energy consumption. The analysis will address the minimum separation work from a solvent and its relation to the minimum separation work from an ideal gas mixture. The model provides a criterion for solvent selection and for improving the process configuration . The results will be compared with reported experimental and industrial results.

(733c) Rate-Based Modeling of CO2 Absorption and Desorption Into Aqueous Monoethanolamine Solutions

Ying Zhang[1], Brian Hanley[2] and **Chau-Chyun Chen**[2], (1)AspenTech Limited, Shanghai, China, (2)Aspen Technology, Inc., Burlington, MA

Nineteen CO_2/monoethanolamine (MEA) absorption and desorption data sets from recent pilot plant studies at the University of Kaiserslautern[i] have been simulated with Aspen Plus. Simulations were performed using both rigorous rate - based calculations and the traditional equilibrium – stage approach. Rate - based results have been generated for several mass transfer/interfacial area correlations available in the open literature. Key performance measurements, including CO_2 removal efficiency in the absorber, temperature and CO_2 concentration profiles in the absorber and the desorber, and the desorber reboiler duty, are then compared to experiment. We have found that the quality of the rate based results is strongly tied to the choice of the underlying mass transfer/interfacial area correlations selected for use in the simulation. In contrast, the equilibrium-stage model fails to reliably predict these key performance variables.

---

[i] Mangalapally, H. P.; Hasse, H. Pilot Plant Study of Post-Combustion Carbon Dioxide Capture by Reactive Absorption: Methodology, Comparison of Different Structured Packings, and Comprehensive Results for Monoethanolamine, *Chemical Engineering Research and Design*, **2011(89)**, 1216-1228.

(733d) The Clathrate Process for Pre-Combustion Capture of Carbon Dioxide Employing a Novel Fixed Bed Reactor

Ponni Valavan Babu and Praveen Linga, Chemical and Biomolecular Engineering, National University of Singapore, Singapore, Singapore

Carbon capture and storage (CCS) from large point sources such as power plants is one of the important options for reducing anthropogenic CO_2 emissions. One such approach to reduce CO_2 emissions is to capture CO_2 prior to combustion. In a pre-combustion process, the fuel gas produced contains 40% CO_2 and 60% H_2 and comes out at a total pressure of 2.5-5 MPa. Some of the processes to capture CO_2 from a pre-combustion fuel gas stream are absorption, adsorption, membrane separation, clathrate process and cryogenic distillation. Hydrate Based Separation (HBS) process is one of the relatively new approaches for separation of CO_2 from the fuel gas mixture. The basis for the separation is the selective partition of CO_2 between the hydrate phase and the gas phase. The clathrate hydrate process is environmental friendly and economically attractive. In this study a novel fixed bed reactor approach for the hydrate based separation process was employed to separate CO_2 from a fuel gas mixture. Silica gel and silica sand were used as the medium in the fixed bed reactor and their performance on the separation was evaluated. Experiments were carried out at 9.0, 8.5 and 7.5 MPa and 274.15 K. In silica sand bed, water conversion up to 80% was achieved whereas in silica gel bed water conversion was low irrespective of the driving force. The initial rate of hydrate formation was faster in silica gel bed when compared to that of in silica sand bed. The hydrate crystals formed were decomposed by heating the system from 274.15K to 285.15K and the decomposed gas was collected separately in the reservoir. Almost 65% of the gas was recovered from both silica gel and silica sand bed within 3hr.

(733e) CO2 Capture Via Gas Hydrate Crystallization

A Basdeo[1] and **A Basdeo**[1,2], (1)Thermodynamics Research Unit, University of KwaZulu Natal, Durban, South Africa, (2)Centre Énergétique et Procédés, MINES ParisTech, CEP/TEP - Centre Énergétique et Procédés, Fontainebleau, France

CO_2 Capture via Gas Hydrate Crystallization

A. Basdeo [1], P. Naidoo[1], A.H Mohammadi[1, 2], D. Ramjugernath[1]

[1] Thermodynamics Research Unit, School of Chemical Engineering, University of KwaZulu-Natal, Howard College Campus, King George V Avenue, Glenwood, Durban, 4041, South Africa.

[2] MINES ParisTech, CEP/TEP - Centre Énergétique et Procédés, 35 Rue Saint Honoré, 77305 Fontainebleau, France.

Corresponding Author: *Email: annelinebasdeo@gmail.com; Tel:+27 31 260 2969. Fax : +27 31 260 1118*

Carbon dioxide (CO_2), in addition to other gases, allows for the transfer of solar radiation through the earth's atmosphere but prohibits the escape of heat. As a result the global surface temperature is increased. Since the Industrial Revolution, the combustion of fossil fuels has led to release of billions of tons of CO_2 into the atmosphere. According to the International Energy Agency (IEA), CO_2 emissions in 2050 may rise to twice those of 2007, if no mitigation options are implemented (IEA, 2010). The concerns of climate change have led to interest in mitigation strategies such as carbon dioxide capture and storage (CCS). Reports by the IEA proclaim that in order to stabilize CO_2 concentrations, carbon dioxide capture and storage (CCS) will need to contribute to one-fifth of the total reductions by 2050 (IEA, 2010).

CCS involves the capturing of CO_2, emitted from large industrial sources, before release into the atmosphere and storing it in secure geological formations deep underground. Gas hydrate crystallization is an attractive technology for capturing CO_2.Gas hydrates are non-stiochiometric crystalline compounds that consist of a lattice of water molecules that physically encage molecules of another component i.e. CO_2 (Lee et al., 2010).

Tetra-n-butyl ammonium chloride (TBAC) and tetra-n-butyl ammonium fluoride (TBAF) are additives that can reduce hydrate formation pressures to feasible industrial conditions however there is insufficient phase data available on (CO_2+TBAC) and (CO_2+TBAF) systems (Li et al., 2010).

Due to the current interest in CO_2 capture and storage by gas hydrate crystallization; there is a demand in experimental phase equilibrium data for the relevant CO_2 hydrate systems. In order to design efficient CCS processes, reliable and accurate phase equilibrium data is required. The present study involves the examination of CO_2 capture from a CO_2+N_2 mixture via gas hydrate crystallization and the measurement of hydrate phase equilibrium data for CO_2+N_2+TBAC+H_2O and CO_2+N_2+TBAF+H_2O systems by employing an isochoric pressure search method using static high pressure equipment. The effect of TBAF and TBAC concentrations on the CO_2 hydrate formation pressures and the CO_2+N_2 separation efficiency will be investigated.

A MATLAB programme will be developed to fit the measured hydrate phase equilibrium data. An ASPEN Tech simulation will be performed to simulate and predict results of the overall carbon capture process. In order to determine the efficiency of CO_2 capture, compositions of the gas phase will be measured using a ROLSI^{TM} capillary sampling device and analysed with a gas chromatograph. A material balance will be employed to determine the compositions of the hydrate and aqueous phases. The material balance will be solved using a mathematical algorithm which employs Newton's numerical method and differential evolution optimization strategy.

(733f) Hydrate Phase Equilibria of Clathrate Hydrates of Carbon Dioxide + Propane

Juan R. Macias-Pérez, Laboratorio de Termodinámica SEPI, Instituto Politécnico Nacional E.S.I.Q.I.E., Mexico, D. F, Mexico, Luis A. Galicia-Luna, INSTITUTO POLITECNICO NACIONAL-ESIQIE, Mexico, Mexico and Amir H. Mohammadi, CEP/TEP - Centre Énergétique et Procédés, Mines Paris-Tech, Fontainebleau, France

The gas hydrate research has particular relevance in the petroleum industry: however, it has evolved into investigations on the promotion of hydrate formation as a potential novel separation technology. In this work we describe the details of a new experimental equipment for measuring the conditions of dissociation of hydrates. The equipment is suitable for measurements at temperatures ranging from 243 to 423 K and pressures up to 40 MPa. Experimental dissociation conditions are reported for region clathrate hydrates of carbon dioxide + propane in gas–liquid water–hydrate (G–Lw–H) equilibrium. Experimental data were measured at temperatures between (272 and 282) K, pressures up to 5 MPa. The mixture was prepared according to the synthetic method. Composition of carbon dioxide in the feed are also varied. The experimental results are obtained using the isochoric method (search for equilibrium pressure). The dissociation data generated in this study together with literature data are compared with predictions of a thermodynamic model.

(741a) Temperature-Responsive Nematic Hydrogels

Andres F. Mejia[1], Ratna Ng[1], Peter Nguyen[1], Hugo Y. Acosta[1], M. Sam Mannan[2] and Zhengdong Cheng[1], (1)Chemical Engineering, Texas A&M University, College Station, TX, (2)Mary Kay O'Connor Process Safety Center, Artie McFerrin Department of Chemical Engineering, Texas A&M University, College Station, TX

We report the fabrication of temperature-responsive nematic hydrogels using large aspect ratio (ξ = *diameter/thickness*) zirconium phosphate (ZrP) monolayers embedded into a polymer network. To achieve an ordered nanostructure, these monolayers were immersed into a monomer solution. The monolayer suspension was macroscopically oriented by shear flow and polymerized *in situ* to obtain a nematic hydrogel. The alignment of the large aspect ratio nano-sheets by a weak external stimulus enables the formation of nematic textures during hydrogel formation. Cross-polarizing micrographs revealed the liquid crystallinity of the nematic hydrogels. We analyzed the influence of the initiator and the catalyst on the nematic textures. Furthermore, by setting the temperature above 32°C, we found that the nematic hydrogel experienced volume shrinking evolving the variation of the nematic textures. Finally, at low monolayer concentrations the hydrogels presented an isotropic-nematic (I-N) order transition on the edge. The nematic hydrogels synthesized in this work open new perspectives for technological applications of these systems to optical devices.

(741b) Feedback Controlled Colloidal Crystal Assembly Via Electric Field and Depletion Mediated Interactions

Tara D. Edwards and Michael A. Bevan, Chemical and Biomolecular Engineering, Johns Hopkins University, Baltimore, MD

The ability to assemble nano- and micro-sized components into ordered structures can serve as a basis for the scalable manufacture of meta-materials. However, a robust approach is not available to control the assembly of such structures due to the complexity of such processes and lack of available models. In this talk, we demonstrate the use of depletion and electric field mediated interactions in conjunction with free energy landscape (FEL) models to manipulate colloidal microstructures from fluid to crystal configurations. Externally applied electric fields and depletion induced via poly(N-isopropylacrylamide) (PNIPAM) hydrogel nanoparticles are used to tune kT-scale colloidal attraction via open- and closed-loop feedback control. The degree of colloidal crystallization is monitored using real-time particle tracking and order parameter computation. FELs are constructed using equilibrium (i.e. Boltzmann inversion) and non-equilibrium (i.e. fitting Smoluchowski equation coefficients) analyses of colloidal assembly trajectories. Our ability to directly measure and tune colloidal interactions, as well as model how these interactions affect dynamic colloidal microstructures, allows us to perform feedback control to manipulate the assembly, disassembly, and repair of colloidal crystals. These results provide a basis to produce perfect single colloidal crystals that can be translated to scalable manufacturing approaches.

(741c) Tuning the Isotropic Interaction Potential Between Paramagnetic Particles Using a Rotational Magnetic Field

Daniel Du[1], Dichuan Li[1] and Sibani L. Biswal[2], (1)Chemical and Biomolecular Engineering, Rice University, Houston, TX, (2)Chemical & Biomolecular Engineering, Rice University, Houston, TX

We present a novel method to control the attractive interaction potential between colloidal particles, from 5kT to 40kT using an AC magnetic field. This system consists of negatively charged superparamagnetic colloidal particle suspension which are confined between two coverslips with certain separation. A highly tunable long-range attractive interaction is induced by a high frequency rotating magnetic field generated by two pairs of coaxial air-core solenoids connecting with multi-frequency AC power supply. The solenoids are calibrated to be orthogonal to each other..A local field theory is put forward to characterize the attractive pair interaction and account for the effect of different parameters. The repulsive component to the interaction potential is the short-range electrostatic interactions, which can be described using DLVO theory.

We first study the frequency effect on the pair interaction by increasing the frequency up to 100Hz. At low frequencies, an attractive potential between particles is generated causing them to form chain-like structures and the magnetic torque causes the chain to rotate with the magnetic field. The torque will vanish at frequency above a critical frequency which is a function of field strength. There is a maximum threshold frequency to also consider which is related to relaxation time of the paramagnetic particles. At frequencies above this threshold frequency, the particles are not able to relax via either Néel relaxation or Brownian relaxation. Both frequencies are determined by simulation or experiment and only the critical frequency is shown to be dependent on field strength. The frequency used in our experiment is fixed on 20Hz due the trade-off between torque elimination and particle relaxation.

We measure the pair potential and three-body potential experimentally. Quantitative aggrement is found between experiments and local field pair potential theory at various field strengths. The effective three-body potential of a dilute trimer system is measured and compared with the superposition of exact pair potential to study the many-body effect on the system. The same deviation is observed for various field strengths. We believe this system provides an ideal platform to study phase behavior and dynamics of atomic system in two dimensions.

(741d) Multiblock Magnetic Nanowires

Sangyeul Hwang, Chemical Engineering Department, University of Michigan, Ann Arbor, MI, Tae-Hong Park, Chemical Engineering, U of Michigan, Ann Arbor, MI and Joerg Lahann, Department of Chemical Engineering, Macromolecular Science and Engineering, Biomedical Engineering, and Materials Science and Engineering, University of Michigan, Ann Arbor, MI

Magnetic particles (MPs) are known to align to form a chain structure under a parallel magnetic field. It is expected that different MPs are randomly arranged in these one-dimensional (1-D) chains. We introduce a novel method to direct a highly ordered dipole chains of different MPs. A magnetic field gradient is utilized to create a potential well, where a magnetic force decreases far away from a boundary line that has the highest magnetic force and thus the lowest potential energy. Two distinct types of MPs can be assembled at the boundary line to form diblock-like 1-D structures. In principle, many other sets of different MPs can be added into the assembled diblock MP chains, producing mutilblock particle organizations. To conserve the high ordered 1-D patterns in the absence of a magnetic field, we also report a new strategy that interconnects the aligned particles in-situ. By adding a dispersant (solvent such as tetrahydrofuran) into the aligned particles in anti-solvent (water), the adjacent particles that are contact in each other start to deform and are fused together. Quenching the deformation before complete polymer dissolution of the particles readily produces mutiblock magnetic nano/microwires.

(741e) Fabrication of Au Network Structure by Convective Self-Assembly for Transparent Conductive Films

Satoshi Watanabe, Tatsuya Hanafusa, Yasushi Mino and Minoru Miyahara, Department of Chemical Engineering, Kyoto University, Kyoto, Japan

Au network patterns are promising for transparent conductive materials because void spaces in the structures give transparency and Au particulate lines produce conductivity. Bottom-up approaches can offer a route to establishing a fabrication technique that is robust and cost-effective, and what is especially required is a template-free technique that can control the colloidal pattern. We have been studying the convective self-assembly (CSA) method, and successfully fabricated stripe, grid, and dot array patterns without any template. To further control the periodicity of the pattern, we recently modified the technique by incorporating a liquid-level manipulation (LLM) scheme in which the suspension is periodically pumped in or out. By using the modified technique, which we call CSA-LLM, we prepared stripe and grid networks of silica particles with controlled line width and spacing. In the present study, we fabricated grid patterns of Au nanoparticles by the CSA-LLM technique, and measured the transparency and conductivity of the patterns to clarify the relation between the periodicity of the pattern and the property as the transparent conductive films. We demonstrated superior performance of our technique to control the Au particulate line width and spacing. Furthermore we found out that the wettability of substrates varies the line thickness. Our technique can thereby produce transparent conductive films with a desired transparency and conductivity.

(741f) Supramolecular Assembly and Surfactant Behavior of Triblock Rod – Coil Amphiphiles At Liquid Interfaces Using Classical Density Functional Theory

Bennett D. Marshall II[1], Kenneth R. Cox[1] and Walter G. Chapman[2], (1)Chemical Engineering, Rice University, Houston, TX, (2)Rice University, Houston, TX

In the framework of Wertheim's perturbation theory we present a classical density functional theory for rod coil molecules with one rigid block and one or two flexible blocks. The theory is validated against molecular simulation for the case of an amphiphilic molecule in a selective pore. The theory is then applied to study the interfacial properties of the oil / water interface in the presence of triblock rod – coil amphiphiles (TBRCA's) consisting of a center rigid hydrophobic rod portion with flexible hydrophilic tails attached to each end. It is shown that TBRCA's preferentially adsorb to the interface in either a parallel or perpendicular orientation. The fraction of TBRCA's in each orientation can be manipulated by varying the length of the shortest flexible hydrophilic tail, length of the hydrophobic rod or the total number of TBRCA's at the interface. Also it is shown that the structural properties of TBRCA's can be adjusted to yield a broad range of surfactant behavior.

(741g) A Hierarchical Sequence of Colloidal Assembly

Adriana San Miguel and Sven H. Behrens, School of Chemical & Biomolecular Engineering, Georgia Institute of Technology, Atlanta, GA

In order to study the influence of nanoscale particle roughness on particle-stabilized emulsions, we recently developed a protocol for preparing such emulsions that involves (i) the bulk assembly of dissolved copolymers into functional nanoparticles, (ii) the subsequent assembly of these nanoparticles at the solid-liquid interface of a microparticle dispersion, followed by (iii) the assembly of the resulting composite microparticles at the liquid-liquid interface of an emulsion. The resulting particle-decorated droplets, in turn, offer great opportunities for further interfacial assembly into functional coatings. We will demonstrate the benefits of our particular protocol for its original purpose, the study of roughness effects on Pickering emulsions and briefly explain the outcomes of that study. The primary goal of this talk, however, is to go beyond this rather narrow context, and to highlight the extreme versatility of this type of hierarchical assembly as well as the control of the final material's properties offered at every assembly step.

(742a) Predicting Disorder-Order Transitions of Solvent-Free Nanoparticle-Organic Hybrid Materials

Hsiu-Yu Yu and Donald L. Koch, School of Chemical and Biomolecular Engineering, Cornell University, Ithaca, NY

Nanoparticle–organic hybrid materials (NOHMs) are a new type of complex fluid composed of hard, inorganic nanocores with oligomeric chains covalently grafted to the surface of the core in the absence of other solvent molecules. These materials provide us with a framework to study disorder-order transitions of "self-suspended" particles. In this talk, we present a classical density-functional theory to address the liquid to face-centered-cubic solid transition of NOHMs. In the absence of intervening solvent, the nanoparticle cores in both disordered and ordered phases are distributed in homogeneous structures governed by the requirement that the tethered oligomers must fill the interstitial space. The phase boundary is determined by comparing the total free energies of the two phases, which consist of the hard-sphere translational entropy as well as the oligomer configurational entropy. Since there are no macroscopic variations in the core volume fraction, the phase that exhibits the lower total free energy is the thermodynamically stable state. When the particles are localized near the lattice positions, the oligomers have more difficulty filling the space and have more stretched configurations. Therefore the theory predicts that NOHMs with stiffer tethered oligomers have stronger entropic penalties in ordered solid and favor the disordered phase. This is consistent with the experimental observation that NOHMs can be disordered soft glasses even when the core volume fraction is close to or higher than the hard-sphere freezing volume fraction [1]. When van der Waals interactions among the cores are included, anisotropic structures can form with the interparticle spacing in one direction different from the other two. Such structures have been observed in experimental studies and Monte Carlo simulations of polymer-grafted nanoparticles with solvent [2] and were recently observed in molecular dynamics of a solvent-free system [3]. Using the same theoretical approach, the transitions of isotropic fluid to string-like or sheet-like structures of NOHMs will be discussed.

[1] J. L. Nugent, S. S. Moganty, and L. A. Archer, *Adv. Mater.* **22**, 3677 (2010).

[2] P. Akcora, H. Liu, S. K. Kumar, J. Moll, Y. Li, B. C. Benicewicz, L. S. Schadler, D. Acehan, A. Z. Panagiotopoulos, V. Pryamitsyn, V. Ganesan, J. Ilavsky, P. Thiyagarajan, R. H. Colby, and J. F. Douglas, *Nature Mater.* **8**, 354 (2009).

[3] A. Chremos and A. Z. Panagiotopoulos, *Phys. Rev. Lett.* **107**, 105503 (2011).

(742b) Interfacial Routes to Colloidal Gelation

Matthew N. Lee, Jessica A. Witt, Hubert K. Chan and **Ali Mohraz**, Department of Chemical Engineering & Materials Science, University of California, Irvine, Irvine, CA

Colloidal gels are percolating particle networks with arrested dynamics and remarkable rheology, typically stemming from attractive interparticle interactions. These properties have been scientifically investigated and exploited in a variety of technological applications including food processing, cosmetics, oil industries, and direct-write assembly. In this talk, two new routes to colloidal gelation will be discussed, which exploit colloidal self-assembly at liquid-liquid interfaces. The first system contains a two-phase liquid mixture with particles that have a three-phase contact angle $\theta > 90°$. The obtuse contact angle leads to bridging of dispersed droplets by a monolayer of interfacial particles, and gelation results from a percolating three-dimensional particle/droplet network. The second system involves a suspension of neutrally wetting colloidal particles in a binary liquid mixture undergoing spinodal decomposition. Here, particles are swept up by the newly formed interface upon phase separation and become jammed when the interfacial area just accommodates all the particles. Gelation then results from two-dimensional particle jamming at the interface of a three-dimensional bicontinuous structure. Using quantitative confocal microscopy and bulk rheometry, the microstructural origins of gel-like rheology in these systems are examined and compared to more traditional colloidal gels in which attractive interparticle interactions are required for gelation. Finally, promising applications of these novel materials in a variety of technological areas will be discussed.

(742c) Electrostatic Interactions to Modulate the Reflective Assembly of Nanoparticles At the Oil-Water Interface

Mingxiang Luo and **Joelle Frechette**, Chemical and Biomolecular Engineering, Johns Hopkins University, Baltimore, MD

Assembly of nanoparticle at fluid interfaces could enable the design of nanoparticles-based plasmonic materials that exhibit promising opto-electronic and chemical properties. The driving force for the adsorption of small nanoparticles (5 - 10 nm) at the oil-water interface can be small and is particularly sensitive to the surface chemistry of the particles. Here we show that the interfacial assembly of 5 and 10 nm diameter gold nanoparticles functionalized with stoichiometric ion-pairs is reversible and can be tuned with the solution pH. We also show that at the interface the nanoparticles form a reflective layer with a mirror-like reflectance. Using titration we demonstrate that the mechanism for particle desorption from the interface is the electrostatic repulsion between the nanoparticles due to pH-dependent adsorption of hydroxide ions. By controlling electrostatic repulsion, we demonstrate two avenues to reversibly control the optical properties of the interface: 1) increase the pH of the aqueous solution to desorb particles from the interface, and 2) decrease the ionic strength in the aqueous phase to increase the spacing of the nanoparticles at the oil-water interface.

(742d) Synthesis and Self-Assembly of Nanoparticle Surfactants

Kjersta Larson and **Danilo Pozzo**, Chemical Engineering, University of Washington, Seattle, WA

A simple method is presented for the synthesis of amphiphilic gold nanoparticle surfactants that self-assemble into clusters of controllable structure. The synthesis technique is based on the sequential functionalization of the surface with thiol terminated polyethylene glycol (PEG) to sterically stabilize particles in water and short alkane-thiols rendering the particles amphiphilic. The nanoparticle surfactants are surface active and form rafts at the air-water interface and stable nanoparticle clusters of controllable size and shape in dispersion. The clusters are also reminiscent of micelles formed from molecular surfactants with aggregation numbers that can be controlled via the modification of the grafting density of the polymer on the nanoparticle surface. The nanoparticle surfactants are also shown to be highly effective emulsifying agents due to their high adsorption energy at oil-water and air-water interfaces and they form stable colloidosomes that are stable against coalescence. Both the clusters and the colloidosomes that are formed from these particles exhibit tunable shifts in plasmon resonance and show enhanced near-infrared optical absorption. This makes them useful for a wide range of applications in medicine and energy.

(742e) Hydrodynamic Boundary Condition for Particle-Laden Interfaces

Lorenzo Botto, Department of Chemical Engineering, Imperial College, London, United Kingdom

Colloidal particles can irreversibly adsorb to fluid interfaces. The presence of the particles can lead to a substantial resistance to the motion of the fluid adjacent to the interface, when the area fraction occupied by the particles is significant. We perform a fully-resolved numerical simulation study of a doubly-periodic monolayer of spherical particles adsorbed at a gas-liquid interface and subject to simple shear flow, in the small capillary number regime. The numerical results are used to discuss the suitability of a mixed slip boundary condition for modeling the flow resistance opposed by the particle layer. Such boundary condition could be applied, for instance, to model the formation of particle-laden bubbles in microfluidic devices and the hydrodynamics of particle-laden foams. Among other information, the simulations provide, for the microstructure considered, the dependence of the "slip length" on the area fraction. This parameter is shown to be related to the drag force on each particles.

(742f) Heterogeneous Nanoparticles Simulated At Water-Oil Interfaces: Implications for the Stability of Pickering Emulsions

Xuan-Cuong Luu[1], Heng Fan[1] and Alberto Striolo[2], (1)School of Chemical, Biological and Materials Engineering, The University of Oklahoma, Norman, OK, (2)School of Chemical Biological and Materials Engineering, The University of Oklahoma, Norman, OK

Pickering emulsions find many applications, including some in food processing, personal care products, and drug delivery devices. Because Pickering emulsions are stabilized by nanoparticles adsorbed at oil-water interfaces, the emulsions stability is naturally related to the structural and dynamical properties of the interfacial nanoparticles. Such properties are investigated here by means of dissipative particle dynamics simulations. The force field parameters have been derived to reproduce atomistic simulation results for silica-based nanoparticles at water-decane interfaces (*Langmuir* **2011,** 27, (9), 5264-5274). Several nanoparticles are considered, including Janus or homogeneous ones with different amounts of hydrophilic/hydrophobic surface groups (e.g., 25% hydrophilic – 75% hydrophobic surface groups). We focus on systems of different composition. For example, systems containing both homogeneous and Janus nanoparticles are investigated. For a given composition, simulations are conducted at increasing nanoparticle density. The results are quantified in terms of radial distribution function, hexagonal order parameter, contact angle, and self-diffusion coefficient. We sometimes find unexpected behavior. For example, at high surface density 'caging' phenomena are observed, where the diffusion of a nanoparticle is hindered by the rigid structure (i.e., cage) formed by surrounding nanoparticles. Cage relaxation and rearrangement are discussed. Radial distribution functions, and even the contact angle of a given nanoparticle are found to strongly depend on the system composition. Implications of such observations on macroscopic observables are discussed.

(742g) Experimental Examination of Irreversibility of Electrostatically Driven Adsorption of Silica Nanoparticles At Solid-Liquid Interface

Xue Li, Olivia Niitsoo and Alexander Couzis, Chemical Engineering, The City College of New York, New York, NY

Adsorption of colloidal nanoparticles (NPs) at solid-liquid interface is a scientifically interesting and technologically important phenomenon due to the fundamental importance in many industrial, environmental, and biological processes, such as wastewater treatment, chromatography, papermaking, or biocompatibility. The process is well understood theoretically by the random sequential adsorption (RSA) model, based on the assumption of irreversible adsorption. Irreversible adsorption is defined as a process in which, once adsorbed, a particle can neither desorb, nor to move laterally on the surface. A great number of reports have confirmed the accuracy of the RSA model. However, experiment data in literatures to verify the irreversibility of particle adsorption is very limited. In this study, we demonstrated the irreversibility of electrostatically driven nanoparticle adsorption utilizing a carefully selected set of experiments.

SiO_2 NPs with a diameter of 50-80 nm prepared by Stöber method (a sol-gel approach) was firstly electrostatically adsorbed onto amino-functionalized silicon wafer substrates. Then the substrates were rinsed but no drying procedure in order to get rid of the aging effect. Subsequently, Ag@SiO_2 core/shell NPs of the same size as well as surface chemistry with SiO_2 NPs were performed the second adsorption as exchange agent. Adsorption time for the second step was varied from 1min to one week. Surface coverage quantification has shown that the surface coverage of the first adsorbed SiO_2 NPs stays the same no matter how long the subsequent adsorption of Ag@SiO_2 core/shell NPs was performed. Furthermore, the surface coverage for one kind of NPs adsorption is equal to the total surface coverage for two kinds of NPs adsorption when the treated time is the same. All the results confirmed the irreversibility of electrostatic adsorption of NPs.

(750a) Morphological Transitions of Capillary Bridges in a Slit Pore Geometry

David Broesch and Joelle Frechette, Chemical and Biomolecular Engineering, Johns Hopkins University, Baltimore, MD

Morphological transitions of capillary bridges in a slit pore geometry

David Broesch and Joelle Frechette

Chemical and Biomolecular Engineering Department, Johns Hopkins University

Abstract

Investigations of capillary bridges are traditionally limited to the idealized geometry of an axisymmetric bridge between spheres or between a sphere and a plate. As capillary forces are employed to drive micro- and nanoscale assembly between complex objects, such as in flip chip technology, there is a need to better understand the shape (and associated forces) of capillary bridges in complex geometries. Here we investigate the morphological evolution of non-axysymetric capillary bridges in a slit-pore geometry. In our experimental system the liquid bridges are formed between two hydrophobic surfaces patterned with hydrophilic strips. The aspect ratio of the capillary bridges (length/width) is varied from 2.5 to more than 100 by changing the pore geometry or the fluid volume. As the separation between the surfaces increases (aspect ratio decreases), we observe a large increase in the mean curvature of the bridge. More specifically, we see that the mean curvature of the bridges change sign (goes from negative to positive for larger pore height). In addition we found a general scaling for which all the curves mapping the morphological evolution collapse. We also simulated our experiments using Surface Evolver and observed quantitative agreements with our experiments. Finally, we show that the morphology diagrams obtained from our 3D analysis are considerably different from those in the simpler 2D model presented in the literature.

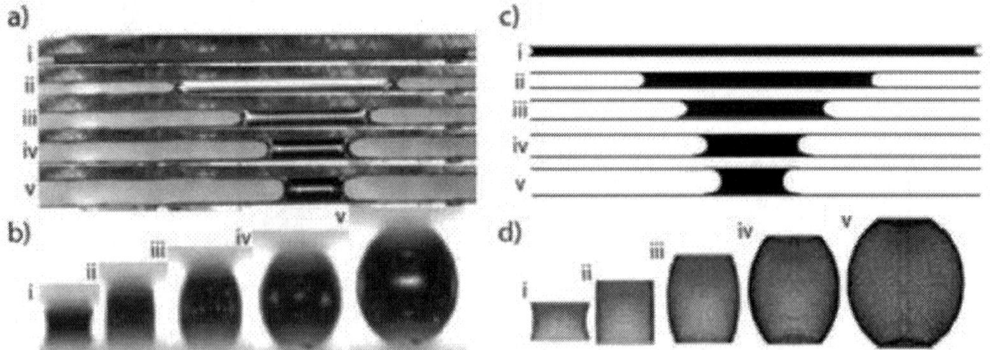

Images of liquid bridges with volume = 2.5μL, channel width = 500μm and separation i) 300, ii) 500, iii) 700, iv) 900 and v) 1100μm. a) and b) are experimental length and pinning angle data respectfully, while c) and d) are obtained from an energy minimization of the interface model.

(750b) Surfactant Transport, Interfacial Rheology, and Film Drainage of a Model Oil-Dispersant-Aqueous System

Matthew D. Reichert, Chemical Engineering, Carnegie Mellon University, Pittsburg, PA, Shelley L. Anna, Chemical Engineering and Mechanical Engineering, Carnegie Mellon University, Pittsburgh, PA and Lynn M. Walker, Chemical Engineering, Carnegie Mellon University, Pittsburgh, PA

Oil spills in marine environments continue to demonstrate the lack of basic information available for effective disaster response. Specifically, surfactant dispersants are deployed at both the sea surface and at leaking well heads; these are two situations that represent different, and often untested, conditions for dispersant use. Variables that have not been characterized include high temperature gradients, large pressures, convection, and rapid dispersant concentration changes. Understanding surfactant transport phenomena to the oil-aqueous interface is imperative to the decision-making process of when and where surfactant dispersants should be applied. In this work, we develop a model surfactant-oil-aqueous system using Tween-80 (a primary component in Corexit® dispersant), squalane, and both simulated sea water (SSW) and deionized water (DI) to investigate transport phenomena and interfacial mechanics, which impact system stability and the lifetime of an emulsion. We use a microtensiometer[1] to study the importance of convection, rapid bulk concentration changes, and interfacial mechanics of an oil-aqueous interface. We are able to measure the dynamic interfacial tension as surfactant adsorbs to an initially clean oil-aqueous interface, and we are able to remove the bulk surfactant at different points in the dynamics by rinsing the interface and continuously replacing the bulk fluid with surfactant-free aqueous phase. A critical interfacial tension arises that links the transport dynamics to the onset of partial reversibility in the system. By measuring the mechanical properties of pre-rinsed and rinsed interfaces, we also find a critical interfacial tension that leads to changes in the elasticity of the interfaces. This interfacial tension correlates well with the critical interfacial tension that indicates the onset of partial reversibility. To determine the effects of interfacial elasticity and concentration changes on the stability of an emulsion, we bring two surfactant-coated interfaces into contact and monitor coalescence times and initial droplet shape profiles. We discuss these coalescence results in the context of film drainage, and as drainage time relates to the elasticity of the contacted interfaces.

1. Alvarez, N.J., L.M. Walker, and S.L. Anna, *A microtensiometer to probe the effect of radius of curvature on surfactant transport to a spherical interface.* Langmuir, 2010. **26**(16).

(750c) Oil Emulsification by Surface-Tunable Carbon Black Particles

Amitesh Saha[1], Ani Nikova[2], Pradeep Venkataraman[3], Vijay T. John[3] and Arijit Bose[1], (1)Department of Chemical Engineering, University of Rhode Island, Kingston, RI, (2)Cabot Corporation, Billerica, MA, (3)Department of Chemical and Biomolecular Engineering, Tulane University, New Orleans, LA

Dispersants were used extensively to emulsify crude oil during the Deepwater Horizon oil spill. In contrast to surfactant-stabilized emulsions, ones that are stabilized by interfacially active particles can remain stable at large background dilutions and can also be used to absorb toxic polycyclic aromatic hydrocarbons and thus lower their concentrations in the aqueous phase. We examine particle stabilized octane-in-water emulsions prepared by modifying the pH or varying the salt concentration of an amino-benzoic acid terminated carbon black suspension in water, adding octane to this suspension and mixing. Varying the conditions in the suspension modifies the hydrophilic/hydrophobic balance of the particles, making them partially wettable in both water and octane. These particles then locate themselves at the oil-water interfaces after mixing, and stabilize the emulsion. The fractal nature of these particles provides increased specific surface area that helps promote adsorption of low molecular weight polycyclic aromatic hydrocarbons from the octane. At 0.015% w/w carbon black in the aqueous phase, cryo-SEM images reveal that the particles arrange into multiple layers at the octane water interfaces. We propose that this is a consequence of some carbon-black aggregation as the pH is lowered or salt concentration increased. At 0.0075% w/w carbon black, stable emulsions are created with incomplete coverage at the oil-water interfaces. We monitor the zeta potential of these carbon black particles at various pH and salt concentrations, show specific ion binding, and relate the hydrophobicity of the particles to this potential. When naphthalene, a model polycyclic aromatic hydrocarbon, is added to the octane, its partitioning into the aqueous phase is reduced dramatically from a control case with no carbon black. These particles have potential use for emulsifying crude oil subsequent to an oil spill.

(750d) Bacterial Biofilm Susceptibility to Silver Nanoparticle Dispersions: Impacts of Colloidal Stability, Humic Acid, and Dissolved Silver

Stacy M. Wirth, Department of Chemical Engineering, Carnegie Mellon University, Pittsburgh, PA, Gregory V. Lowry, Department of Civil and Environmental Engineering, Carnegie Mellon University, Pittsburgh, PA and Robert D. Tilton, Department of Chemical Engineering and Department of Biomedical Engineering, Carnegie Mellon University, Pittsburgh, PA

This work is motivated by the need to understand environmental risks posed by potentially biocidal engineered nanoparticles. Silver nanoparticles (AgNPs) are of particular concern since they are increasingly used in commercial and consumer products and their release to waste streams and the environment is inevitable. The broad spectrum biocidal properties that motivate AgNP use in consumer and medical applications could have detrimental consequences for environmental microorganisms that are crucial to normal ecosystem function.

In order to understand the environmental risks of engineered AgNPs, an understanding of how they interact with environmental bacteria, which typically exist as surface attached biofilms rather than the more frequently studied planktonic, or suspended, cultures, is necessary. This study therefore addressed AgNP toxicity towards single species *Pseudomonas fluorescens* biofilms using dye staining methods. Both silver nanoparticles and dissolved silver negatively impacted biofilm viability in a dose-dependent manner. However, no silver treatments (up to the very high concentration of 100 ppm AgNPs) resulted in 100% biofilm viability loss, even though these same concentrations caused complete viability loss in planktonic cultures, suggesting some biofilm tolerance to AgNP toxicity.

Exposure of AgNPs to environmental conditions will result in various processes, such as aggregation, dissolution and adsorption of naturally occurring macromolecules, which could impact nanoparticle toxicity towards biofilms. In this study, addition of the macromolecule humic acid (HA), which comprises a major fraction of the natural organic matter that is ubiquitous in most environments, to AgNP dispersions was found to decrease the dispersion toxicity. The decreased toxicity was found not to result from HA adsorption to particle surfaces but rather from HA complexation of dissolved silver introduced as part of the AgNP dispersions. Therefore, dissolved silver plays at least a partial role in the overall AgNP dispersion toxicity towards biofilms.

The effects of aggregation were assessed by comparing the toxicities of two types of AgNPs with different colloidal stabilities. The more stable dispersions exhibited a distinct nanoparticle-specific toxicity that was not observed for less stable, highly aggregated particles. This lack of particle-specific toxicity from highly aggregated particles suggests that the biofilm is somehow protected from nanoparticle aggregate toxicity. One potential explanation is that the biofilm extracellular polymeric substance presented a barrier to aggregate transport into the biofilm while the small, stable AgNPs were able to diffuse through the biofilm matrix and gain closer access to cells.

Changes in the amount of biofilm during silver exposure were measured using the crystal violet assay. High concentration silver treatments resulted in loss of adherent biomass, though a significant amount of biofilm remained in all cases. However, low concentration silver treatments actually stimulated biofilm growth, with greater amounts of adherent biomass resulting from these low concentration silver treatments than from silver-free controls. The enhanced growth was specific to adherent cells.

Overall, this study indicates that both silver nanoparticles themselves and dissolved silver originating from the nanoparticles contribute to overall AgNP dispersion toxicity towards bacterial biofilms. Environmental interactions which impact the colloidal stability of AgNPs or the bioavailability of dissolved silver will greatly impact toxicity towards biofilms.

(750e) Molecular Dynamic Simulations of Tween 80 Surfactant with Squalane/Water Mixtures, As a Model for the Deepwater Horizon Oil Spill

Xueming Tang, Chemical Engineering, University of Michigan, Ann Arbor, MI and Ronald G. Larson, University of Michigan, Ann Arbor, MI

In 2010, the explosion of Deepwater Horizon resulted in the release of about 4.9 million barrels of crude oil into the Gulf of Mexico[1]. To protect the coastline and marine environments, 1.84 million barrels of dispersant (Corexit) were sprayed into the gulf, either on the sea surface or deep under the water into the oil plume. Chemical dispersants accelerate the dispersal processes by breaking up oil slick and forming water-soluble micelles into the water column. One of the main active ingredients of Corexit is Tween-80[2], which is a nonionic long chain surfactant (contour length about 6 nm). The micellization properties of Tween-80 have never been studied at atomistic level. In this study, we carried out molecular dynamic simulations on systems composed of Tween-80, squalane (as a model oil), and sea salts at atomic united atom levels[3]. Some of the partial charges of Tween 80, as available, were adopted from existing force field parameters[4,5]. The rest of the partial charges were estimated using Gaussian density function method Bly3p at 6-31g(d,p) basis[6] followed by Boltzmann energy average method over the conformers. We then carried out molecular dynamics simulations of systems composed of Tween 80, squalane in the presence or absence of sea salt. In particular, we studied the stabilities of Tween 80 micelles by molecular dynamics simulations at varying aggregation numbers. 60 randomly distributed Tween 80 formed a spherical micelle within 10 ns. This is consistent with Anwar's study from a Coarse grained model[7]. Radii of gyration, solvent accessible surface area, probability distributions of different atoms of Tween 80 around the micelle center of mass, and radial densities of atoms around the micelle center of mass were investigated. We observed the formation of a surfactant-laden oil-water interface occurred within 20 ns after exposing squalane oil to a surfactant-laden aqueous phase in a simulation box of size $12\times12\times15$ nm^3. Initially randomly distributed Tween 80 and oil in water formed a suspended oil-filled micellar aggregate coated with hydrophilic head groups of Tween 80 within 20 ns. Such results might be used to help determine mass transport coefficients required for engineering dispersant systems for control of oil releases.

References:

1. http://en.wikipedia.org/wiki/Deepwater_Horizon_oil_spill
2. http://en.wikipedia.org/wiki/Corexit
3. D. R. Kester, I. W. Duedall, D. N. Connors, and R. M. Pytkowicz, Limnol. and Oceano., 1967, 12(1), 176-179
4. W. F. V. Gunsteren *et. al* . J. Peptide Sci., 2005, 11, 74–84.
5. W. F. V. Gunsteren *et. al.* J. Chem. Theory Comput., 2011, 7 (5), 1237–1243
6. Gaussian 09, Revision A.1, M. J. Frisch, *et. al*. Gaussian, Inc., Wallingford CT, 2009.
7. A. Amani, P. York, H. de Waard, and J Anwar, Soft Matter, 2011, **7**, 2900-2908.

(750f) Multi-Scale Approach to the Design of Emulsions: Water-in-Oil and Oil-in-Water with Dispersed Phase Between 10% w/w and 30% w/w

Natalia Murcia[1], Andrea Nieto[1], Diego Pradilla[1] and Oscar Alberto Alvarez[2], (1)Universidad de Los Andes, Bogotá, Colombia, (2)Chemical Engineering, Universidad De Los Andes, Bogotá, Colombia

A wide selection of products in food, pharmaceutical, cosmetic and oil industries are emulsions; therefore, they are constantly investigated in order to improve their properties and applications. Emulsion preparation process involves different factors such as process conditions, formulation and composition parameters and as a result of their combination, certain properties are achieved according to the level in which these factors are varied.

The multi-scale approach to the design of emulsions leads to a better comprehension of the effect of each factor and its relationship with the others through the analysis of the resulting measured properties in three different scales: macroscopic, microscopic and molecular.

The purpose of this investigation is to determine the relationship existing between the process conditions , the emulsion formulation and composition in the preparation of water-in-oil and oil-in-water emulsions containing low dispersed phase (between 10% w/w and 30% w/w) and non-ionic surfactants. The oil phase to be used is mineral oil, dodecane or decaline in order to study in each case the influence of the oil chemical structure and the non-ionic surfactant is a blend of Span 80 and Tween 20. The aqueous phase is deionized water.

The properties of the emulsions to be measured are: the rheological behavior as the macroscopic response, in both the steady state and oscillatory tests using a controlled stress rheometer (TA Instruments AR-G2), the particle size and distribution as the microscopic response, using a Malvern Instrument Model 3000 Mastersizer particle size analyzer and the near infrared spectroscopy as the molecular response, using a FOSS 5000 Smartprobe Analyzer.

The main process conditions studied are the impeller tip speed, tank and impeller geometry and fluid height which are linked by the measured energy required to prepare the emulsions. The analysis of the possible interactions is made by observing the effect of this added energy and the emulsion's formulation and composition on the obtained properties evaluated in the macroscopic, microscopic and molecular responses. One possible interaction for example, is that it may exist a relationship between the impeller's pumping capacity and the energy consumption for the preparation of the emulsions.

Paper number: 251970

IMPACT OF DIFFERENT FACTORS ON PHASE BEHAVIOR OF MICROEMULSION SYSTEM

Huie Liu, Xue Xia, State Key Laboratory of Heavy Oil Processing, China University of Petroleum, Qingdao, China

Introduction

Microemulsions are thermodynamically stable, optically transparent and isotropic dispersions of aqueous and organic liquids, stabilized by an interfacial film of surfactant molecules[1]. Microemulsion can exist in one of the three forms of equilibrium state, Winsor type I (two phase, O/W microemulsion coexisting with excess oil phase), Winsor type II (two phase, W/O microemulsion coexisting with excess water phase) and Winsor type III (three phase, middle phase microemulsion coexisting with excess water and oil phases) [2].

Microemulsion systems can be used in the disposal of organic contaminants[3~8]. Microemulsion solutions can increase the apparent solubility of the organics by several orders of magnitude via micellar solubilization[8]. Although microemulsion can easily achieve ultra-low interfacial tension (generally about 10^{-2} mN·m^{-1}) to solubilize organics, organics release and surfactant (cosurfactant) reuse are still desirable for such remediation schemes because of the surfactant (cosurfantant) cost and waste stream disposal. Winsor type II and III microemulsion can solubilize more organics and are more effective for solubilizing organics than Winsor type I microemulsion[9], and can treat organic contaminants under room temperature so as to reduce the energy consumption of the recovery. If the Winsor type II or III microemulsion convert into type I, the surfactant and cosurfactant will mostly migrate into the type I microemulsion, which has a lower solubility of organics, resulting in a large portion of organics released from the microemulsion as free phase[10, 11]. This may provide a way for the organics recovery and surfactants reuse.

Organic contaminants solubilization in microemulsion depends on many factors, including surfactant type, electrolyte and the molecular weight of organics. In order to use microemulsion more effectively, the influence factors of microemulsion should be studied firstly[12]. In this work the influence of different factors, including the types of surfactant, electrolyte and organics on the microemulsion system will be investigated, in order to provide conditions for the maximum solubilization of different types of organics and achieve the recovery of organics and reuse of surfactants.

Materials and methods

Surfactants used in this work include sodium dodecyl sulfate (SDS) and cetyltrimethyl ammonium bromide (CTAB), both CP, from Sinopharm Chemical Reagent Company. Cosurfactant is *n*-butanol, AR, from Sinopharm Chemical Reagent Company. NaCl, NaOH, KOH, KCl, $CaCl_2$ and Na_2CO_3 (AR, from Sinopharm Chemical Reagent Company) are chosen as electrolytes; Organics used includes *n*-heptane, cyclohexane (both AR, from Sinopharm Chemical Reagent Company), gasoline, and kerosene (both industrial grade). Another important reagent is laboratory prepared deionized water.

For determined organics, with equal volume of organics and water and fixed surfactant and cosurfactant concentration, if the concentration of salt in the system increases, the microemulsion will change from Winsor type I→III→II. This is called salinity scan method. This method was used in this work. During the scanning process, the volume of every phase was recorded after phase equilibrium achieved at different salinity, and then microemulsion phase diagram was drawn according to the data.

Results and discussion

Effect of cationic and anionic surfactant association

Sodium dodecyl sulfate (SDS) is a widely used anionic surfactant because of its halotolerancy and low price. Its Hydrophile-Lipophile Balance (HLB) value is about 40, having strong water-solubility and relatively weak oil-solubility. Cetyltrimethyl ammonium bromide (CTAB) is a cationic surfactant, with HLB value being about 14. The oil-solubility of CTAB is stronger than SDS. The association effect for SDS and CTAB was investigated with these two kinds of anionic and cationic surfactants. The content of CTAB increased gradually during the experiment, with the total content of surfactants (SDS and CTAB) held constant.

As shown in Fig. 1, Under the conditions of $V_{kerosene}$:V_{water} being 1:1, total surfactants content 4.44%, *n*-butanol content 7.95% and NaCl contetn 1.67%, a Winsor type I microemlsion was found for the single SDS surfactant. With the increase of the mass ratio between CTAB and SDS, the HLB of the mixed surfactants decreases. It can be seen that more organics and less water solubilized into the microemulsion system during this process. However, the solubility for oil is not the highest for the single CTAB surfactant. This means that under proper condition, the electrostatic attraction between the cations and anions promotes the association of the two kinds of surfactants, leading to higher solubility than the single ones.

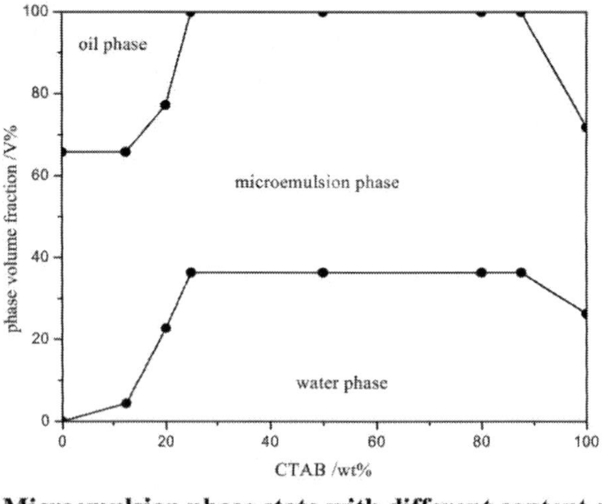

Figure 1 Microemulsion phase state with different content of CTAB
(Conditions: $V_{kerosene}$:V_{water} = 1:1, *w*(surfactants)= 4.44%, *w*(NaCl)= 1.67%, *w*(*n*-butanol)= 7.95%)

Microemulsion system for different organics

Cyclohexane, *n*-heptane, gasoline and kerosene are used to investigate the microemulsion formation for different organics. Fig.2 shows the phase behaviors of the microemulsion systems for different kinds of organics during the salinity scan using NaCl. The results show that the smaller the molecular weight of organics is, the more easily they are solubilized and the larger the volume of Winsor type III microemulsion is formed. The one for forming Winsor type III microemulsion most easily is cyclohexane, and then is gasoline, *n*-heptane and kerosene. The molecular weight of *n*-heptane is greater than cyclohexane, which makes the resistance much greater for the *n*-heptane molecules to pass through the interfacial membrane into the micelles. More NaCl is needed for *n*-heptane to reduce the interfacial tension and increase the interfacial flexibility and mechanical strength of the composite membrane formed by surfactant and co-surfactant. Gasoline is made up of hydrocarbons from about C_4 to C_{10}, containing more small-molecule hydrocarbons, which makes it be solubilized more easily and less NaCl is needed than for *n*-heptane. Kerosene is made up of many bigger molecules, so more NaCl is needed to form microemulsion.

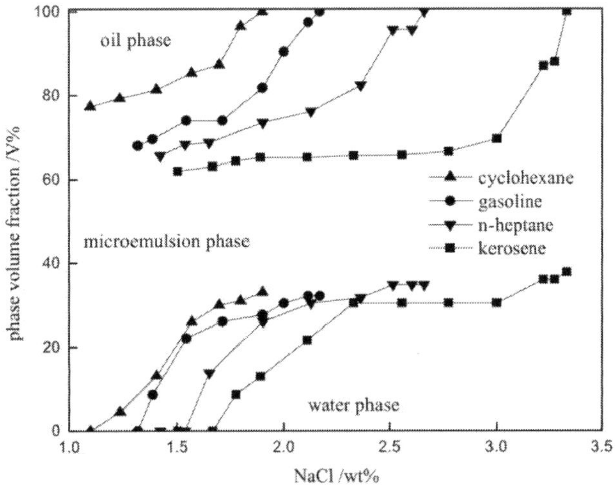

Figure 2 Phase diagram of salinity scan using different type of organics
(Conditions: $V_{kerosene}$:V_{water} = 1:1, *w*(SDS)= 4.44%, *w*(*n*-butanol)= 11.7%)

Influence of salts on microemulsion system

$CaCl_2$, KCl, KOH, NaOH, NaCl, and Na_2CO_3 are used in the salinity scan of the SDS/ *n*-butanol/ kerosene/ water microemulsion system, as shown in Fig.3. It was found that each type of salt had its typical length of salinity for forming Winsor type III microemulsions. The high surface charge density of Ca^{2+} makes it more effective than K^+ and Na^+ in decreasing the HLB of surfactant system, and Winsor type III and II microemulsion can be formed under much less salt concentration. The microemulsion phase behavior when using KCl (or NaCl) was similar to KOH (or NaOH). The results show that for the same cation (K^+ or Na^+), the type of monovalent anion (Cl^- or OH^-) has little influence on the SDS/ *n*-butanol/ kerosene/ water microemulsion state, even though KOH and NaOH are alkalis. It was observed that the effect of K^+ is stronger than Na^+. As to the effect of NaCl and Na_2CO_3, NaCl is more effective than Na_2CO_3, although they have the same monovalent cation. The reason is that the bivalent anion, CO_3^{2-} leads to a lower activity of Na^+ than the monovalent Cl^-.

It is attractive that when Winsor type II or III microemulsion was obtained using $CaCl_2$, it might be Winsor type I if $CaCl_2$ was replaced by NaCl (See Fig.3). So, if Ca^{2+} is replaced by Na^+, Winsor type II or III microemulsion will convert into Winsor type I, and most of the organics solubilized in the microemulsion will be released. Tests were carried out to approve this supposition. $CaCl_2$ was used to prepare Winsor type III microemulsion, as shown Fig.4, in which the volume fraction of solubilized kerosene is 48%, and then Na_2CO_3 was added into the system, causing the Ca^{2+} precipitate into $CaCO_3$, and the Ca^{2+} in the system was substituted by Na^+. When proper quantity of Na_2CO_3 was added, the Winsor type III microemulsion disappeared and type I formed. The volume fraction of kerosene solubilized in the Winsor type I microemulsion is only 17%. Thus, after exchanging Ca^{2+} with Na^+ through precipitation, 50% of the organics was released from the microemulsion. With the further addition of Na_2CO_3, it can be seen that the microemulsion system changes from Winsor type I→III→II, and the volume fraction of kerosene solubilized in the microemulsion gradually changes from 17% to 67%.

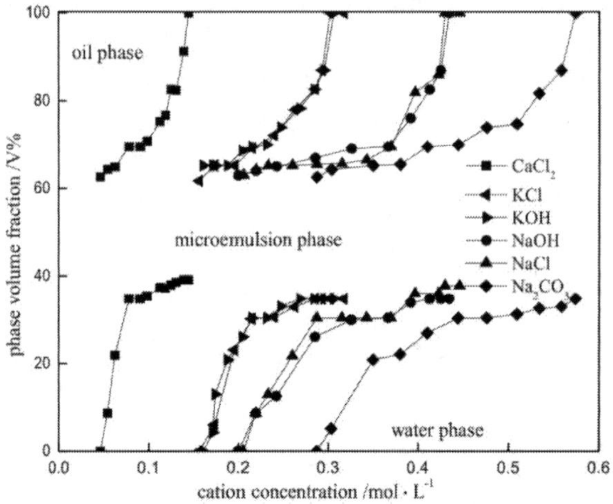

Figure 3 Phase diagram of salinity scan using different types of salts
(Conditions: $V_{kerosene}$:V_{water} = 1:1, w(SDS)= 4.44%, w(n-butanol)= 11.7%)

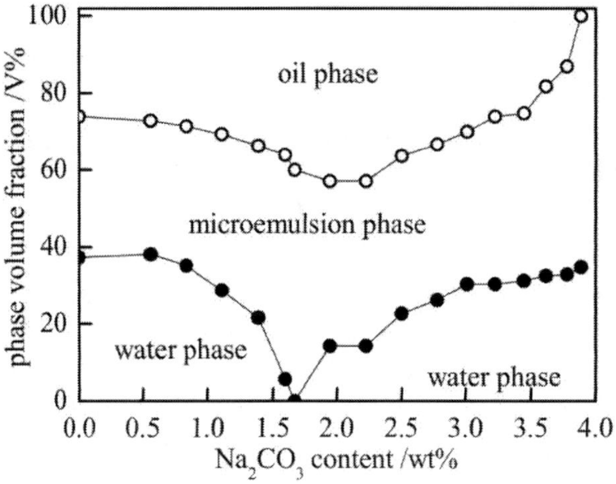

Figure 4 Phase inversion through salt substitution
Initial conditions: $V_{kerosene}$:V_{water} = 1:1, w(SDS)= 4.44%, w(n-butanol)= 11.25%, w($CaCl_2$)= 1.67%;

For the system using $CaCl_2$ as electrolyte, after the replacement of Ca^{2+} by Na^+ in aqueous solution, the Winsor type III (or II) can go back to type I, whose organics solubility is relatively small, resulting in the release of most of the initially solubilized organics. And with the addition of Na^+, the Winsor type I microemulsion converts to type III (or II) again and more organics can be solubilized into the microemulsion gradually. These results indicate that it is promising to use microemulsion phase behavior in the recovery of organic contaminants and the reuse of surfactant (cosurfactant). Further work is continued to realize the repeating use of the surfactant (cosurfactant).

Influence of water/oil ratio on microemulsion system

The influence of water/oil ratio was also investigated. Winsor type I microemulsion was formed with 10ml kerosene and 10ml water ($V_{kerosene}$:V_{water} = 1:1), initially. As shown in Fig. 5, water was then added gradually into the system so as to change the water/oil ratio in the system. What was attractive is that the change from microemulsion Winsor type II→III→I was observed and more and more oil was released gradually. When the volume of water increased from 10ml to 18ml, 9ml or 90% (v) oil was released from the microemulsion. These phenomena mean that when organic contaminants were microemulsified into Winsor type II or III microemulsion for the purpose of organic contaminants disposal, water may be added to release the organics from the microemulsion and the surfactants can be reused.

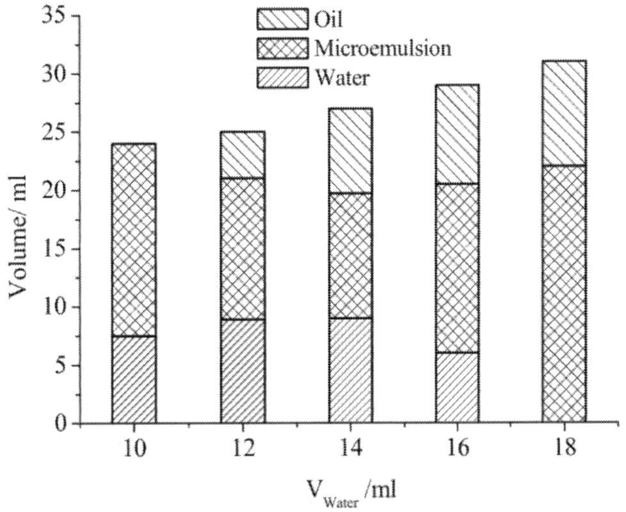

Fig. 5 Influence of water volume on phase behavior of SDS- n-butanol- kerosene- water microemulsion system
Initial conditions: Vkerosene:Vwater = 1:1 (10ml:10ml), *w*(n-butanol)= 17.1%, w(SDS)=4.4%, *w*(NaCl)=2.06%

Conclusions

The influences of surfactant, electrolyte, organics and water/oil ratio on microemulsion system were inverstigated. The effectiveness of surfactant mixture (anionic surfactant and cationic surfactant) was better than the single ones. For different types of organics, it was easier for smaller molecular-weight organics to form microemulsion than larger molecular-weight ones. The microemulsion system changes from Winsor type I→III→II with increasing salt concentration, but each type of salt has its typical length of salinity for forming Winsor type III microemulsions. Alkalinity has no distinct influence on SDS/ *n*-butanol/ kerosene/ water microemulsion system.

$CaCl_2$ was used to prepare Winsor type III microemulsion. Through chemical reaction with Na_2CO_3, Ca^{2+} was replaced by Na^+, and the Winsor type III microemulsion system can convert into Winsor type I microemulsion. The solubilized organics are released, and the surfactant and co-surfactants which shift into Winsor type I microemulsion can be reused. Through further addition of salt, Winsor type III or II microemulsion system can be formed and organics can be solubilized again. Winsor type III (or II) microemulsion changed into type I with the addition of water into the system. Most of the organics entered into the free phase during this process. These are promising ways for the release of organics from the type III (or II) microemulsion and the recycle of surfactants.

Acknowledgement

This research was financially supported by Chinese National Natural Science Foundation (No. 21106187), Promotive Research Funds for Excellent Young and Middle-aged Scientists of Shandong

Province (No. BS2011NJ021) and the Fundamental Research Funds for the Central Universities (No. 10CX04015A).

References

1. Rosano H. L., Cavallo J. L., Chang D. L., Whittam J. H. (1988), "Microemulsions: A commentary on their preparation," *J Soc Cosmet Chem,* 39, pp201-209.
2. Salager J. L., Antón R. E., Sabatini D. A., Harwell J. H., Acosta E. J., Tolosa L. I. (2005), "Enhancing Solubilization in Microemulsions--state of the Art and Current Trends," *Journal of Surfactants and Detergents*, 8(1), pp3-21.
3. Wu, B.; Shiau, B.-J.; Sabatini, D. A.; Harwell, J. H.; Vu, D. Q. (2000), "Formulating Microemulsion Systems for a Weathered Jet fuel Waste Using Surfactant/ Cosurfactant Mixtures," *Separation Science and Technology*, 35(12), pp1917-1937.
4. Childs, J.; costa, E.; Annable, M. D.; Brooks, M. C.; Enfield, C. G.; Harwell, J. H.; Hasegawa, M.; Knox, R. C.; Rao, P. S. C.; Sabatini, D. A.; Shiau, B.; Szekeres, E.; Wood, A. L. (2006), "Field demonstration of surfactant-enhanced solubilization of DNAPL at Dover Air Force Base, Delaware," *Journal of Contaminant Hydrology* , 2, pp1– 22
5. Tongcumpou, C.; Acosta, E. J.; Quencer, L. B.; Joseph, A. F.; Scamehorn, J. F.; Sabatini, D.A.; Chavadej, S.; Yanumet, N. (2003), "Microemulsion Formation and Detergency with Oily Soils: I. Phase Behavior and Interfacial Tension," *Journal of Surfactants and Detergents*, 6(3), 191-203
6. Tongcumpoua, C.; Acosta, E. J.; Quencer, L. B.; Josephc, A. F.; Scamehorn, J. F.; Sabatini, D. A.; Chavadej, S.; Yanumet, N. (2003), "Microemulsion Formation and Detergency with Oily Soils: II. Detergency Formulation and Performance", *Journal of Surfactants and Detergents*, 6(3), pp205-214
7. Tongcumpou, C.; Acosta, E. J.; Quencer, L. B.; Joseph, A. F.; Scamehorn, J. F.; Sabatini, D. A.; Yanumet, N.; Chavadej, S. (2005), "Microemulsion Formation and Detergency with Oily Soils: III. Performance and Mechanisms," *Journal of Surfactants and Detergents*, 8(2), 147-156
8. Shiau, B.; Sabatini, D. A.; Harwell, J. H. (1994), "Solubilization and microemulsification of chlorinated solvents using direct food additive (edible) surfactants," Ground Water, 32 (4), pp561–569.
9. Shiau, B.; Sabatini, D. A.; Harwell, J. H. (1994), "Solubilization and microemulsification of chlorinated solvents using direct food additive (edible) surfactants," *Ground Water*, 1994, 32 (4), pp561–569.
10. Cheng, H. F., Sabatini, D. A., (2002), "Phase-behavior-based surfactant-contaminant separation of middle phase microemulsions," *Separation Science and Technology*, 37(1), pp127-146.
11. Cheng H, Sabatini D A, (2007), "Separation of Organics compounds from surfactant solutions: A Review," *Separation Science and Technology,* 42, pp453-475
12. Liu, H.Z., (2005), "Microemulsion extraction technology and application," Science Publications, Beijing. (In Chinese)

(751a) Simulation of Lithium-Ion Cells and Batteries

Ralph E. White, Chemical Engineering, University of South Carolina, Columbia, SC

Simulation of lithium-ion cells and batteries includes predicting the performance and life of these devices. The beginning of life performance of lithium-ion cells can be predicted by using John Newman's porous electrode theory, which will be reviewed in this presentation. Predicting the rate of heat generation in lithium-ion cells is a critical aspect of designing heat removal components of a lithium-ion battery pack for electrical vehicle applications. The rate of heat generation of the components in a lithium ion cell can be predicted by using multiscale, multidimensional modeling technique, which will be presented in this talk. In order the reduce the computation burden for solving these model equations, reduced order models have been developed and will be outlined in this tutorial. The life of lithium-ion cells and batteries can be predicted by adding equations that describe the degradation processes to the model equations, or by using parameter estimation techniques to determine how the important physical parameters in the beginning of life model equations change with time. An overview of these life predicting techniques will be presented.

(751b) Design Aspects of Flow Batteries for Large Scale Energy Storage

Robert Savinell, Chemical Engineering, Case Western Reserve University, Cleveland, OH

Large scale energy storage technology is needed to facilitate integrating renewable energy generation technologies, such as wind and solar, into the electric grid because these energy generation technologies are inherently intermittent and thus matching supply to load demand is a challenge. Also, energy storage is needed for modernizing the grid into a "smart" electricity distribution system. There is a large variety of potential energy storage solutions available for this broad range of applications and understanding the various operating and economic constraints and requirements is necessary to choose a technology. Flow batteries of various chemistries are gaining attention again in recent years because of these markets and opportunities and of the inherent advantage of flow batteries. The advantages and disadvantages, and their technical and economic challenges will be addressed for a number of approaches. The presentation will focus its attention of electrode materials, electrode design, and cell design for redox flow batteries.

(751c) Electrode and Cell Design Principles in Batteries and Fuel Cells

Trung V. Nguyen, University of Kansas, Lawrence, KS

In a battery or fuel cell, the electrodes and other components in the cell must be carefully selected and designed to address the electrode potentials, reactions, and transport (electronic, ionic, molecular) processes, especially the limiting ones, of the system to meet the performance (discharge and charge rate and cycle life), shelf-life or storage life, and safety requirements. While high active surface area is needed for high power, its effect on the self-discharge rate and corrosion should also be considered. In addition to the active materials that participate in the electrochemical reactions, the role of the supporting materials such as the current collectors, catalyst supports, electrolytes, and binders, both conductive and non-conductive, must also be considered because they can have a significant effect on the performance and/or shelf-life of the system. This presentation will discuss the materials selection and components and cell design requirements for a few battery and fuel cell systems.

(751d) Electrowinning Principles and Practice: A Brief Introduction to the Electrochemistry and Engineering Behind Modern Electrowinning Plants

Dan Steingart, Energy Institute and Chemical Engineering, City College of New York, New York, NY

Electrowinning, or electroextraction, is the oldest industrially applied electrochemical process, and the backbone of the production of many metals, both precious and otherwise (aluminum, copper, zinc and lead, among others), as well as industrial gases (e.g. the chlor akali process). In the US alone electrowinning processes consume on the order of 1 to 2% of total electricity produced, and many of the most energy intensive processes are have much room for improvement of efficiency. In this tutorial we will broadly examine the theory and practice of both the electrochemical processes as well as the supporting mining and extraction operations. Beyond a discussion of the basic operating principles, we will examine issues of feed purity, flow control, and rate sensitivity. Finally, experimental and R&D scale approaches to efficiency improvements will be covered. While the aluminum, copper, and chlor-alkali systems will be systems of focus in this talk, the principles discussed will be applicable to any modern electrowinning process.

(751e) Design of Enzyme Electrodes for Sensing and Power Applications

Joshua Gallaway, Chemical Engineering, Energy Institute, City College of New York, New York, NY

The prospect of harnessing the catalytic functionality of enzymes for practical applications holds tremendous scientific and technological promise. Composed of folded proteins built around an active site, enzymes are biological catalysts that enable thousands of highly specific reactions to occur under mild conditions. In biological systems, electron exchange to and from enzymes happens diffusionally from small redox molecules. In contrast, practical systems designed by humankind often rely on electron exchange at a heterogeneous interface (an electrode). By using mediator molecules to transport electrons to enzymes they can be used on electrodes with relatively high currents, but the resulting systems can be complex. Electron hopping, Michaelis-Menton kinetics, materials aspects of hydrogels, and other complications make enzyme electrode design less than intuitive. This talk aims to demonstrate the connection between electrode modeling, synthesis techniques, and fabrication, which when used together can greatly improve the practical performance of enzyme electrodes. An example case will be presented in which current response of an oxygen reducing enzyme electrode was doubled by molecular engineering.

(759a) From Normal to Single-File Diffusion of Gas Molecules in Nanotube Systems by NMR

Muslim Dvoyashkin[1,2], Ipek Yucelen[3], Sankar Nair[4], Ji Zang[4], David Sholl[4], Aakanksha Katihar[2], Aiping Wang[2], Clifford R. Bowers[1] and **Sergey Vasenkov**[2], (1)Department of Chemistry, University of Florida, Gainesville, FL, (2)Department of Chemical Engineering, University of Florida, Gainesville, FL, (3)School of Materials Science and Engineering, Georgia Institute of Technology, Atlanta, GA, (4)School of Chemical & Biomolecular Engineering, Georgia Institute of Technology, Atlanta, GA

Studies of sorbate diffusion in systems of one-dimensional nanochannels are important due to significance of such studies for a number of applications including catalysis, separations, and nanofluidics. Confinement to one-dimensional channels can give rise to anomalous transport. One of the most fascinating and best known types of anomalous transport in channels is single-file diffusion, viz. diffusion under conditions when sorbate molecules cannot pass one another in narrow channels. Here we report results of microscopic studies of self-diffusion and tracer exchange of gas molecules in aluminosilicate nanotubes and in self-assembled L-alanyl-L-valine (AV) nanotubes. Spherical gas molecules with similar sizes (tetrafluoromethane and xenon) were chosen as sorbates for these studies. Diffusion studies were performed by C-13 and Xe-129 pulsed field gradient (PFG) NMR method under conditions of application of large (up to 30 T/m) gradients and high (17.6 T) magnetic field. Investigations of tracer exchange between nanotubes and the surrounding gas phase were performed using continuous flow hyperpolarized Xe-129 NMR spectroscopy. Depending on the relationship between the nanotube diameter and sorbate size, application of the two unique NMR techniques allowed observation of different types of dynamics corresponding to normal and single-file diffusion. The distinct signatures of both types of diffusion in the obtained experimental data will be discussed in detail. The experimental results for tetrafluoromethane diffusion in aluminosilicate nanotubes will be compared with the corresponding results of MD simulations.

(759b) Water Transport Through Carbon Nanotubes with Modified Water Models

Minseok Song, Mark A. Snyder and Jeetain Mittal, Department of Chemical Engineering, Lehigh University, Bethlehem, PA

Remarkably fast water flow through carbon nanotube (CNT) based membranes opens exciting possibilities for realizing both high-flux and high-selectivity molecular transport, manipulation, and separation at various scales. Broadening of the scope of nanotube research beyond CNT-water systems is daunted by the complex parameter space comprised of a wide range of possible NT materials (e.g., carbon, alumina, silica, aluminosilicate, etc.), vast potential nanotube modifications, and innumerable fluid compositions. Yet, it is critical for elucidating and tapping the unrealized potential of nanotube membranes for more generalized chemical transport. We will discuss the transport behavior of geometrically modified water models inside various diameters of CNTs using molecular dynamics simulations of the extended simple point charge (SPC/E) model of water. Here, we use the bond angle of water as a representative surrogate of non-aqueous molecular fluids. The geometric modification consists of altering the H-O-H angle of SPC/E water. We find that water transport inside CNTs varies in a nonmonotonic way as a function of bond angle, reaching a maximum for an bond angle of 90 degree. This trend could not be predicted based on water density inside and outside CNTs as a function of bond angle. We will present detailed analysis and theoretical justification for the observed behavior.

(759c) Confined Electrochemical Crystallization of Nanorods On Gold Nanoparticles

Li Li, Sunxi Wang and Guangzhao Mao, Chemical Engineering and Materials Science, Wayne State University, Detroit, MI

Integration of nano-components with spatial control and multi-functionalities remains a challenge in nano-manufacturing. Our previous work has demonstrated that thiolate-protected CdSe, CdS, and Au nanoparticles are capable of disrupting the 2-D crystalline structure of *n*-carboxylic acids and promoting confined nanorod crystallization on highly oriented pyrolytic graphite (HOPG). The confined growth of *n*-carboxylic acids nanorods is imposed by the high curvature of the nanoparticles as nucleation seeds. In this work, we use electrochemically deposited Au nanoparticles as seeds to nucleate and confine the electrochemical crystallization of organic semiconductors or conductors such as tetrathiafulvalene charge transfer salt (TTF) and partially oxidized potassium tetracyanoplatinate sesquihydrate (TCP) on HOPG. The particle/rod structure on HOPG is characterized by cyclic voltammetry (CV), AFM, and SEM. The size and density of deposited Au nanoparticles on HOPG are controlled by the precursor ion concentration, applied potential, and deposition time. The formation of particle/rod structure is found to be dependent on Au nanoparticle morphology and crystallization conditions. The results contribute to the fundamental understanding of seed-mediated nucleation and a novel strategy to manufacture organic nanorods at ambient conditions.

(759d) Chemically Modified Nanopores in an n-Doped Silicon Membrane: A Brownian Dynamics Study

Christopher McKinney, Clarkson University, Potsdam, NY

Solid-state nanopores have recently become an important area of study for the investigation of single biomolecule dynamics. Of particular interest is the use of nanopores in the detection and analysis of proteins and DNA. Several studies have considered nanopores of various sizes and compositions in their efficiency at the characterization of DNA and polymers. A novel approach to altering the functionality of a nanopore is to tether biomolecules (primarily DNA and polymers) to its inner surface in order to change the interaction of the pore with a passing species. We consider a single nanopore in an *n*-doped semiconductor membrane with polymer or DNA functionalization of its surface. This system is modeled using Brownian dynamics simulations with a Poisson modification to account for surrounding electrolyte solution. We systematically vary key parameters of the system, such as grafting density, graft size, graft composition, and applied voltage across the pore to detail and characterize the behavior of the functionalized nanopore. This study lays the groundwork for simulating tethered nanopores in more complex systems, such as those with a translocating biomolecule. Ultimately, results of this characterization will (1) determine if the functionalized nanopore system will be a useful tool in the investigation of single biomolecule dynamics, and (2) will define how key parameters should be altered to "tune" a nanopore to have a desired selectivity.

(759e) The Chemical Dynamics of Nanosensors Capable of Single-Molecule Detection

Ardemis A. Boghossian[1], Jingqing Zhang[1], Francois Le floch-Yin[2], Zachary Ulissi[1], Peter Bojo[3], Jae-Hee Han[4], Jong-Ho Kim[5], Jyoti Arkalgud[1], Nigel F. Reuel[1], Richard D. Braatz[1] and Michael S. Strano[1], (1)Department of Chemical Engineering, Massachusetts Institute of Technology, Cambridge, MA, (2)Department of Aeronautics and Astronautics, Massachusetts Institute of Technology, Cambridge, MA, (3)Department of Biological Engineering, Massachusetts Institute of Technology, Cambridge, MA, (4)Department of Energy IT, Kyungwon University, Seongnam, South Korea, (5)Department of Chemical Engineering, Hanyang University, Ansan, South Korea

Recent advances in nanotechnology have produced the first sensor transducers capable of resolving the adsorption and desorption of single molecules. Examples include near infrared fluorescent single-walled carbon nanotubes that report single-molecule binding via stochastic quenching. A central question for the theory of such sensors is how to analyze stochastic adsorption events and extract the local concentration or flux of the analyte near the sensor. In this work, we compare algorithms of varying complexity for accomplishing this by first constructing a kinetic Monte Carlo model of molecular binding and unbinding to the sensor substrate and simulating the dynamics over wide ranges of forward and reverse rate constants. Methods involving single-site probability calculations, first and second moment analysis, and birth-and-death population modeling are compared for their accuracy in reconstructing model parameters in the presence and absence of noise over a large dynamic range. Overall, birth-and-death population modeling was the most robust in recovering the forward rate constants, with the first and second order moment analysis very efficient when the forward rate is large ($>10^{-3}$ s^{-1}). The precision decreases with increasing noise, which we show masks the existence of underlying states. Precision is also diminished with very large forward rate constants, since the sensor surface quickly and persistently saturates.

(759f) A Structure-Function Relationship for the Optical Modulation of Phenyl Boronic Acid-Grafted, Polyethylene Glycol Complexed to Single Walled Carbon Nanotubes

Bin Mu, Jingqing Zhang and Michael S. Strano, Department of Chemical Engineering, Massachusetts Institute of Technology, Cambridge, MA

Phenyl boronic acids are important binding ligands to pendant diols useful for saccharide recognition. The aromatic ring can also function to anchor an otherwise hydrophilic polymer backbone to the surface of hydrophobic graphene or carbon nanotube. In this work, we synthesize a homologous series of seven phenyl boronic acid conjugated poly ethylene glycol 8 membered branched polymers using different phenyl boronic acids. We evaluate the resulting dispersion of single walled carbon nanotubes (SWNTs) in aqueous suspension, and the nIR fluorescence of these polymer-SWNT complexes. FTIR, UV-visible, NMR, and visible fluorescence spectroscopies were applied to characterize these polymers. The fluorescence of SWNTs wrapped by these polymers presents apparent variation in the intensity and the wavelength shift. An examination of the fluorescent excitation-emission profile demonstrated that the varied boronic acid structures have an ability to quench the fluorescence of the different SWNT species to the different level. Finally, the screening results of a library of sugar molecules demonstrated the success of the new design concept toward saccharide biosensor.

(777a) Comparative Study of the Vanadium Redox Flow Battery Efficient Electrolyte Electro-Catalytic Experiments and Simulation

Chun Yang[1], Chengming Zhao[1], Xiao-feng Xie[2] and Zongqiang Mao[1], (1)Institute of Nuclear and New Energy Technology, Tsinghua University, Beijing, China, (2)Institute of Nuclear and New Energy, Tsinghua University, Beijing, China

Comparative study of the vanadium redox flow battery efficient electrolyte electro-catalytic experiments and simulation

Chun YANG, Chengming ZHAO, Xiaofeng XIE *, Zongqiang MAO

Institute of Nuclear and New Energy Technology, Tsinghua University, Beijing 100084, China;

All vanadium redox flow battery (VRB) has been widely used in the large scale wind and photovoltaic power system for storage of electrical energy [1, 2]. There has an excellent solution for back-up rechargeable battery of nuclear energy plants, which might prevent the disaster like Fukushima nuclear accident. However, the VRB has a fatal flaw, namely the low energy density. If a catalyst is added to accelerate the valence change of vanadium ions, to improve the electrochemical speed, so as to achieve the purpose of improve the efficiency of the VRB.

In this study, the different metal catalyses are used for positive vanadium electrolyte for improving the VRB performance. The electrochemical tests show the Mn (IV) has particular catalytic properties, which increase the peak current both anodic and cathodic. The charge/discharge capacity of single VRFB with Mn (IV) added in positive electrolyte are shown in Fig1. When Mn (IV) is added in positive electrolyte, both charge and discharge capacity increase obviously.

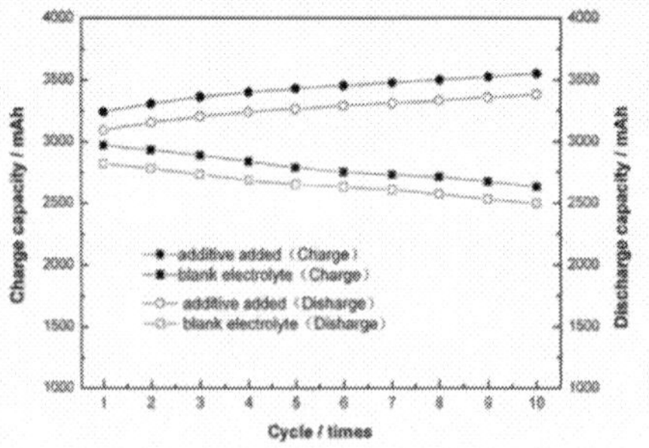

Fig1 charge and discharge capacity of positive electrolyte added in catalyst

Furthermore, the catalysis reaction of positive vanadium electrolyte is investigated by density functional theory (DFT). Geometric structure optimization and spectrum parameters are calculated by basis set B3LYP with considering the solvent effect. The optimized geometric structure and thermodynamic parameters have been obtained, and experimental data and simulation results are in good agreement.

Keywords: All vanadium redox flow battery; Positive electrolyte; Density functional theory; Electro-catalytic

Acknowledgements: The work is funded by the National High Technology R&D Program of China (2012AA051201), Tsinghua University Initiative Scientific Research Program (20101080623), and the National Natural Science Foundation of China (21176140).

References:

[1] M.H. Chakrabarti, R.A.W. Dryfe and E.P.L. Roberts, Electrochim. Acta, 52 (2007) 2189.

[2] M. Vijayakumar, Sarah D. Burton, Cheng Huang, et al. J. Power Sources, 195 (2010) 7709.

(777b) A Mathematical Model of a Regenerative H2-Br2 Fuel Cell

Venkata R. Yarlagadda, Chemical & Petroleum Engineering, University of Kansas, Lawrence, KS and Trung V. Nguyen, University of Kansas, Lawrence, KS

The regenerative hydrogen-bromine (H_2-Br_2) fuel cell is an attractive system for electrical energy storage because it has high energy conversion efficiency and high-power density capability and uses low cost active materials. Due to the fast reaction kinetics of the hydrogen and bromine reactions, the system is often limited by transport. To identify the transport processes that limit the performance of this system and component and cell designs that will minimize the effect of transport, a mathematical model of a H_2-Br_2 cell is developed. The model region consists of a porous bromine electrode filled with HBr-Br2 solution, a proton-conducting membrane and a porous hydrogen electrode. This presentation will discuss the model equations, solution technique, and results obtained from this theoretical study. Also discussed is how the model results have been used to guide the development of electrode components and cell design for this system.

Acknowledgement

This work was funded by NSF through grant number EFRI-1038234.

(777c) Performance and Modeling of Convection Battery

Galen Suppes and Michael J. Gordon, Chemical Engineering, University of Missouri – Columbia, Columbia, MO

The convection battery forces flow of electrolyte through the cathode, anode, and the separator between them. The goal is to increase ion fluxes (A/cm2) to realize the benefit of thicker electrodes, lower cost batteries, and reduced charge times. A pump that circulates electrolyte was turned off to create a diffusion cell that was used as a control to which the performance of the convection battery was compared.

Based on performance at <1.1 V overpotential (based on a 3.1 V open circuit) and similar capacity utilization, the convection battery provided a 5.6 fold increase in ion flux for these initial studies, increasing flux from 1.6 A/cm2 to 8-10 A/cm2. Little capacity fade was observed on the measured discharge cycles (10 cycles). These studies provided an important milestone in the research, development, and validation of a new battery design including cycling studies with lithium iron phosphate chemistry.

A model for the convection cell is presented based on porous electrode theory. The model uses a concentrated binary electrolyte. The model was applied to a convection cell without electrolyte flow electrolyte, and a convection cell with flow of electrolyte. The different models are compared and parametric studies were performed on the convection cell to determine the effect of flow rate on the cell.

(777d) Improved Cycle Life Performance of MnO2 Cathode and Zn Anode for Secondary Rechargeable Alkaline Batteries

Nilesh D. Ingale[1], Joshua Gallaway[1], Michael Nyce[1], Tal Sholklapper[1], Daniel Steingart[1], Alexander Couzis[2] and Sanjoy Banerjee[1], (1)Energy Institute and Chemical Engineering, City college of New York, New York, NY, (2)Chemical Engineering, City College and the Graduate Center of the City University of New York, New York, NY

Primary alkaline manganese dioxide batteries have been widely used in a variety of consumer applications and devices. Alkaline manganese dioxide batteries use manganese dioxide (MnO_2) as cathode and zinc (Zn) as anode material. As both zinc and manganese dioxide are abundant in nature and inexpensive, this battery has much potential in the rechargeable battery market, and may be able to compete with lead-acid and nickel-cadmium cells due to its low cost.

The cycle life of alkaline manganese dioxide rechargeable batteries has been limited to 20-30 cycles due to irreversibility associated with MnO_2 upon deep discharge and formation of electrochemically inactive phases. Thus the commercialization of this battery has been limited. Also, formation of a haeterolite ($ZnO:Mn_2O_3$) phase during discharge, when Zn penetrates into the lattice structure of MnO_2, has made battery cycling irreversible. The Zn anode also has limitations on cycle life due to the redistribution of Zn active material and formation of dendrites during recharge, causing short-circuits.

In order to design a predictable, inexpensive Zn-MnO_2 battery, we access a fraction of one electron capacity of the MnO_2 during each cycle. However, MnO_2 being inexpensive and abundant, low cost alkaline batteries based on MnO_2 cathodes can be developed by accessing only this fraction of the 1-eletron capacity of MnO_2. In the present work, we have demonstrated more than 1500 cycles of a MnO_2 cathode coupled with various anodes such as Zn and Cd with more than 80% energy efficiency. This demonstrates the possibility of making inexpensive secondary rechargeable batteries with long cycle life based on MnO_2. The problems associated with Zn anode has been taken care of with the help of strategy used in the present work.

(777e) Method for Measuring Gas Evolution of Lithium-Ion Battery Under Automotive Operation Conditions

Paul T. Yu, Fuel Cell Activities, General Motors Company, Honeoye Falls, NY and Rohit Makharia, Fuel Cell Activities, General Motor Company, Honeoye Falls, NY

Method for Measuring Gas Evolution of Lithium-Ion Battery under Automotive Operation Conditions

Paul T. Yu, Rohit Makharia

Electrochemical Energy Research Laboratory, General Motors Company, Honeoye Falls, NY 14472, USA

Abstract

Gas evolution due to electrolyte-solvent/electrode interaction in lithium-ion (Li-ion) battery is widely recognized. It is found that the gas generated during battery storage and cycling could result in capacity fade. Furthermore, the gas generated abruptly under an abuse condition and the time-dependent inflation of pouch-type battery cells strongly impacts the design of battery packs and modules in automotive applications. Gas evolution rate and its mechanism are not thoroughly understood due to the complicated chemistry of electrolyte-solvent/electrode interactions under various operation conditions. In this work, we focus on developing a test method to evaluate gas evolution rate and gas composition in an operating Li-ion cell. The test method includes an 8 cm^2 single cell connected with an on-line modified gas chromatograph (GC) to quantify the amount of gas evolution and to analyze the gas composition inside of a battery. Preliminary test results show that this test method can quantify gas evolution rate and its composition. Besides providing insight into solid-electrolyte interface (SEI) chemistry, the proposed method could be used for screening electrolyte and its additives, and optimization of operation window to minimize gas evolution while maximizing battery life.

Key words: Li-ion Battery, Gas Evolution, Solid-Electrolyte Interface, Gas Chromatograph

(777f) Computational Study On Stability of Hybrid Nanostructures of Silicon and Multiwalled Carbon Nanotubes As Lithium-Ion Battery Anode

Sameer Damle[1], Siladitya Pal[2], Rigved Epur[3], Prashant Kumta[2] and Spandan Maiti[2], (1)Chemical Engineering, University of Pittsburgh, Pittsburgh, PA, (2)Bioengineering, University of Pittsburgh, Pittsburgh, PA, (3)Mechanical Engineering and Materials Science, University of Pittsburgh, Pittsburgh, PA

Lithium-ion batteries are forerunners in energy storage systems and have found use in various applications ranging from mobile phones to electric vehicles. Traditional Li-ion batteries are comprised of graphite based anodes which exhibit Li intercalation capacity of 372 mAh/g. Continuous efforts are being made to explore alternate materials with higher Li intercalation capacity to achieve better battery performance. Research in silicon based anodes has gained importance as it can provide ~4200 mAh/g Li alloying capacity corresponding to the highly lithiated phase of $Li_{4.4}Si$. However, the huge volumetric expansion (~280%) associated with Li alloying in Si causes microcracking and loss of electrical contact leading to failure of electrode.

Due to the ability of nanostructures to accommodate large strains, various anode nanostructured configurations have been studied in the recent past. These include thin films, core-shell nanotubes, nanowires, nanospheres, etc. Carbon nanotube based configurations are favorable for use due to the associated short diffusion lengths, better electrical conductivity (1-D electronic transport), mechanical strength, and an increased ability to alleviate stress. Si deposited by chemical vapor deposition (CVD) on multi-walled carbon nanotubes (MWCNT's) has been explored earlier as a high capacity anode electrode material for Li ion batteries. However, during electrochemical cycling, there is a gradual loss of electrical contact between the MWCNTs and Si leading to capacity fade. Since the loss of electrical contact occurs at the interface between the MWCNT and the Si coating, a properly engineered interface between these two components can maintain the integrity of the nanostructured configuration. Hence, it is paramount to understand the mechanism and properties of the bulk materials and interface that affects the disintegration of the nanostructured configuration.

In this work, we utilize a thermodynamically consistent theoretical framework to model the Li alloying induced deformation as well as failure of the anode configuration. A novel cohesive law is used to model the debonding. Transport of Li is coupled with the mechanical equilibrium in a finite deformation setting. Finite element based computational procedure is utilized to simulate the coupled equations. We also investigate the effect of critical parameters that govern the mechanical stability of the interface between the MWCNT and Si coating. Based on the present investigation, critical parameters are identified to design better anode configurations which can provide prolonged electrochemical cycling of the anode without capacity fade.

(777g) Impact of Carbon BET Pore Volume On Sulfur Cathode Capacity

Galen Suppes, Chemical Engineering, University of Missouri – Columbia, Columbia, MO and Donald Dornbusch, University of Missouri, Columbia, MO

The effect of carbon surface area on capacity is investigated in cathodes for lithium sulfur batteries. Lithium sulfur batteries have high specific capacity (a theoretical capacity of 1672 mAh/g-S); however, additives like carbon are needed to overcome sulfur's low electrical conductivity. Cathodes were prepared at 12.5wt% to 68wt% sulfur on activated carbon having BET surface areas of 1220 to 3200 m^2/g. The utilization of the sulfur ranged from 33% to 83% of the theoretical loading with a good correlation to the accessible pore volumes. The data substantiate that sulfur can be successfully accessed on carbon supports through both intra-particle and inter-particle adsorption.

Nitrogen gas adsorption was performed to obtain specific surface area and pore volume for carbons based on BET adsorption theory; this method is commonly used for characterizing activated carbons. After the characterization, the activated carbons underwent a heat treatment in the presence of excess sulfur at 200°C for 6 hours and followed by 300°C for 3 hours. The goal was to allow sufficient time for sulfur pore penetration over the initial 6 hour period, then remove excess sulfur during the final 3 hours.

(777h) Electrodeposition As a Route Toward Fabrication of Three-Dimensional Electrochemical Energy Storage Devices

Derek C. Johnson[1], Matthew T. Rawls[1] and Amy L. Prieto[2], (1)Prieto Battery, Inc., Fort Collins, CO, (2)Department of Chemistry, Colorado State University, Fort Collins, CO

Electrodeposition as a Route Toward Fabrication of Three-Dimensional Electrochemical Energy Storage Devices

Derek C. Johnson, Matthew T. Rawls, and Amy L. Prieto

Chemistry Department, Colorado State University

Prieto Battery, Inc., Fort Collins, CO

Lithium-ion batteries have become the primary energy choice for low power applications, and are now being sought after for high power applications in hybrid electric vehicles (HEV), plug-in hybrid electric vehicles (PHEV), and electric vehicles (EV). To meet the performance requirements for high power density applications, the short comings that have confined the use of conventional lithium-ion batteries to low-power applications needs to be addressed. To address these problems, we utilize electrodeposition as the main synthetic tool for the fabrication of a three-dimensional (3D) solid-state lithium-ion battery. The ultimate goal is to use this synthetic technique to fabricate a solid-state battery with a significantly higher power density and longer cycle life when compared to traditional lithium-ion batteries containing a liquid electrolyte or a polymer gel electrolyte. Electrodeposition is used to synthesize high surface area anodes that do not require additional conducting materials such as graphite or polymer binders to retain structural integrity. Since the anode is highly conducting, it is subsequently used as the substrate to which a solid-state electrolyte is directly electrochemically deposited. An aqueous based cathode slurry is then applied to the solid-state electrolyte to complete the cell. Battery performance of these composite structures will be presented.

(783a) Simulation of Propylene Carbonate As Solvent for Electrochemical Double-Layer Capacitors

Xinli You, Lawrence R. Pratt and Noshir Pesika, Chemical and Biomolecular Engineering, Tulane University, New Orleans, LA

Simulation of Propylene Carbonate as Solvent for

Electrochemical Double-layer Capacitors

X. You, L. R. Pratt, and N. Pesika

Abstract Propylene carbonate (PC) is a popular choice of solvent for electric double-layer capacitors (EDLCs).[1,2] Despite recent encouraging simulations of EDLCs[3] based on carbon nanotube forests, thorough testing of the required force-field features has not been available. Recent work[3,4] shows that the modeled liquid PC completely wets a planar graphite surface (contact angle, =0), in disagreement with experimental droplet formation (=37°). Empirical adjustment of PC:graphite attractive interactions brings those results into agreement. Molecular dynamics simulations of PC droplets, bulk PC liquid, and a saturated solution of tetraethylammonium tetrafluoroborate in PC, using the GROMACS package, further characterize the assumed force-field. We present and discuss a range of interfacial properties including surface tensions, and their temperature and concentration dependences. We consider the applicability of the Kelvin equation, describing the vapor pressure of small droplets, for evaluating the curvature dependence of surface tension. [5]

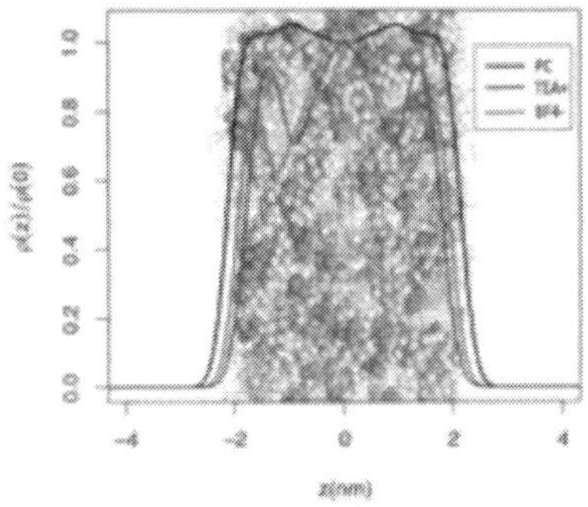

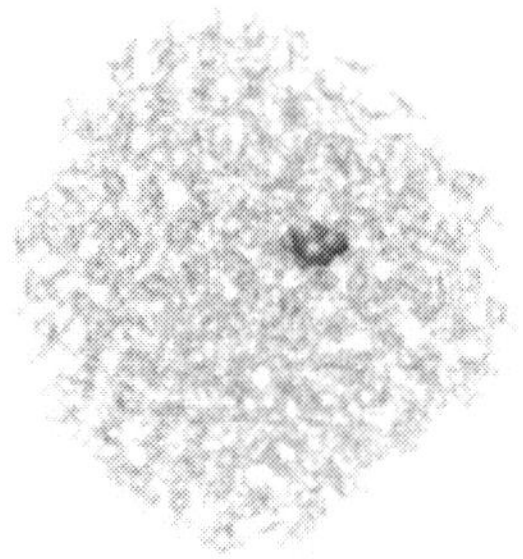

FIG: (left) Interfacial density profiles and simulation geometry for slab calculation of 1000 PC with 85 tetraethylammonium and 85 tetrafluoroborate at 300K. Compared with pure PC (surface tension = 40.8erg/cm2), the ionic material increases the surface tension to 41.5erg/cm^2. (right) Simulation configuration of a droplet with 600 PC molecules.

1. Yang, L., Fishbine, B. H., Migliori, A., and Pratt, L. R. *J. Am. Chem. Soc*, **131**, 12373 (2009).

2. Yang, L., Fishbine, B. H., Migliori, A., and Pratt, L. R. *J. Chem. Phys*. **132**, 044701 (2010).

3. W. Zhang, G. G. Hoffman[1], L. R. Pratt[1], N. Pesika[1], K. M. Aritakula[2] and S. W. Rick[2]([1]Tulane University, [2]University of New Orleans 2012)

4. "Simulation of Propylene Carbonate as Solvent for Electrochemical Double-layer Capacitors", Hoffman, G. G., Mariano, R., Pesika, N., Pratt, L. R., Rick, S. W., and X. You (Proc. LA EPSCoR RII LA-SiGMA 2011 Symp.), p 101-104.

5. Powles, J. G., Fowler, R. F., and Evans, W. A. B. *Chem. Phys. Letts*. **76**, 289 (1983).

(783b) Capacitance Behaviour of Supercapacitors Based-On Carbon Nanotube and Ionic Liquids

Guang Feng, Department of Chemical & Biomolecular Engineering, Vanderbitl University, Nashville, TN and Peter T. Cummings, Center of Nanophase Materials Sciences, Oak Ridge National Laboratory, Oak Ridge, TN

As a novel class of energy storage devices, electrical double layer (EDL) capacitors, also called supercapacitors, have attracted great attention, owing to their advantageous properties, such as high power density, high capacitance, and excellent durability. With high specific surface area and electrical conductivity, recently, carbon-based materials are the most widely used electrodes for supercapacitors, including activated carbons, templated and carbide-derived carbons (CDC), carbon nanotubes (CNTs), and onion-like carbons (OLCs), among others. Based upon our previous work on the ionic liquids near OLCs, supercapacitors composed of CNTs with diameters ranging in size from 0.67 nm to 2.02 nm in an ionic liquid were studied to investigate the dependence of capacitance on the pore size using molecular dynamics simulations. We find that the surface charge density increases almost linearly with the potential applied on EDLs near CNTs. This leads to a nearly flat shape of the differential capacitance versus the potential, which is similar with that near the OLCs rather than planar electrodes. Furthermore, our simulations reveal that the capacitance of EDLs on CNTs increases with the curvature of CNT surface increasing, which is in agreement with theoretical prediction. Both interesting capacitive behaviours of CNT-based supercapacitors invite experimental exploration to take advantage of these phenomena.

(783c) Mechanistic Insight Into Cellulose Dissolution In Ionic Liquids

Krishna Mohan Gupta, Chemical and Biomolecular Engineering , NUS , Singapore, Singapore, Zhongqiao Hu, Chemical and Biological Engineering, National University of Singapore, Singapore, Singapore and Jianwen Jiang, Chemical and Biomolecular Engineering, National University of Singapore, Singapore, Singapore

Currently, rapid growth in energy demand worldwide is deemed as promising issue in science and research community. To meet this demand, there has been considerable interest in environmentally benign energy sources. As the most abundant, biodegradable, natural material on the earth, cellulose is considered to be a viable energy source to produce biofuels. Because of the highly ordered structure and complex hydrogen-bonding network, however, cellulose is not readily soluble in water and common solvents. In the past decade, ionic liquids (ILs) have emerged as a unique class of green solvents due to relatively low melting point, negligible vapor pressure, and high thermal stability etc. Experimental studies suggest that ILs are promising solvents for cellulose, but the mechanism of cellulose dissolution in ILs remains largely elusive.

In the present article, molecular simulations have been performed to elucidate the cellulose dissolution mechanism in ILs. A crystalline Iβ structure is adopted for cellulose and two ILs are considered, namely 1-*n*-butyl-3-methylimidazolium hexafluorophosphate [BMIM][PF_6] and 1-*n*-butyl-3-methylimidazolium acetate [BMIM][Ac]. The AMBER force field is used to mimic the cellulose and ILs, and atomic charges were estimated from quantum chemical calculations. The simulated densities, crystalline lattice constants, thermal expansion coefficients and Young's modulus of cellulose crystal match fairly well with experimental data. The Young's modulus along the chain direction is high due to the strong glycosidic bonds in oligosaccharide chains. The hydroxyl groups of cellulose are found to be responsible for hydrogen-bonding within cellulose, including the intra-chain O2H2···O6 and O3H3···O5 and the inter-chain O6H6···O3.

Upon contact with solvents ([BMIM][PF_6], [BMIM][Ac] and water), the numbers of intra-chain and inter-chain H-bonds at the bulk layer and the intra-chain O3H3···O5 H-bonds at the surface layer of cellulose remain nearly a constant. However, the number of hydrogen-bonds at the cellulose surface, particularly for the inter-chain O6H6···O3 in [BMIM][Ac] deceases sharply. This is attributed to the orientational change of -OH groups at the surface by cellulose/solvent interactions. The simulation results reveal that solvation leads to the breaking of hydrogen-bonds at the cellulose surface. Among the three solvents, [BMIM][Ac] appears to have the strongest capability to break the hydrogen-bonds in cellulose. The present simulation study provides microscopic insight into the interactions of cellulose with ILs.

(783d) Ionic Liquid Media for Self-Assembly of Amphiphiles

Aikaterini Tsoutsoura and **Paschalis Alexandridis**, Department of Chemical and Biological Engineering, University at Buffalo - The State University of New York (SUNY), Buffalo, NY

In the context of developing self-assembly methodologies for the synthesis and manufacturing of complex formulations and materials, our group has a long-standing interest on the utilization of solvents for the modulation of the organization of amphiphiles such as surfactants and block copolymers. Having low vapor pressure, high thermal stability, and novel solvation properties, ionic liquids can expand the operating range and applications of self-assembled amphiphiles. We discuss here the structuring of poly(ethylene oxide)-poly(propylene oxide)-poly(ethylene oxide) (Pluronic) amphiphilic block copolymers in the ionic liquids ethylammonium nitrate (EAN), 1-butyl-3-methylimidazolium tetrafluoroborate (bmimBF4), and 1-butyl-3-methylimidazolium hexafluorophosphate (bmimPF6), and their mixtures with water or polar organic solvents.

In the context of developing self-assembly methodologies for the synthesis and manufacturing of complex formulations and materials, our group has a long-standing interest on the utilization of solvents for the modulation of the organization of amphiphiles such as surfactants and block copolymers. Having low vapor pressure, high thermal stability, and novel solvation properties, ionic liquids can expand the operating range and applications of self-assembled amphiphiles. We discuss here the structuring of poly(ethylene oxide)-poly(propylene oxide)-poly(ethylene oxide) (Pluronic) amphiphilic block copolymers in the ionic liquids ethylammonium nitrate (EAN), 1-butyl-3-methylimidazolium tetrafluoroborate (bmimBF4), and 1-butyl-3-methylimidazolium hexafluorophosphate (bmimPF6), and their mixtures with water or polar organic solvents.

(783e) Study of Interfacial Phenomena, Viscosity and Derivative Properties of Common Ionic Liquids with the Soft-SAFT EoS

Oriol Vilaseca, MATGAS Research Center and ICMAB-CSIC, Bellaterra, Barcelona, Spain, Felix Llovell, MATGAS Research Center, Bellaterra, Barcelona, Spain, Mariana Belo, University of Aveiro, Aveiro, Portugal, Joao A.P. Coutinho, Department of Chemistry, CICECO, Universidade de Aveiro, Aveiro, Portugal and Lourdes F. Vega, MATGAS Research Center and Carburos Metálicos, Air Products Group, Barcelona, Spain

Ionic Liquids (ILs) are a family of compounds with a very wide variety of properties due to their particular physicochemical characteristics. In particular, their extremely low volatility marks them as environmentally benign alternatives to volatile organic solvents for separation processes. Thermodynamic models tend to focus only on particular properties, such as the density or the solubility. However, it is difficult to find theoretical tools able to reproduce several features of ionic liquids and, in particular, their interfacial, derivative and transport properties.

In this contribution, within the framework of the soft-SAFT EoS [1] coupled with the Density Gradient Theory (DGT) [2] and the Free-Volume-Theory, the single-phase equilibrium, surface tensions, viscosities, as well as the critical temperature, pressure and density have been estimated for different ionic liquid families and compared with those reported in the literature from experimental or simulation data. In particular, two different cations (imidazolium and pyridinium) and several anions (BF4, PF6, Tf2N…) have been revised and compared. A correlation for the influence parameter as a function of the molecular weight was obtained, empowering the predictive capabilities of the equation for interfacial tensions of compounds of the family for which experimental data is scarce or unavailable [3]. In addition, surface thermodynamic properties were also derived from the dependence of the surface tension values, and compared with those obtained within the same framework. The viscosity of those compounds is also provided in good agreement with the experimental data. Finally, several second-order thermodynamic properties such as the heat capacities, the isothermal compressibility or the Joule-Thomson coefficient are evaluated. The effect of the molecular parameters on those properties is further discussed.

F. Llovell acknowledges a TALENT grant from AGAUR, Generalitat de Catalunya. M. B. Oliveira acknowledges the financial support from Fundação para a Ciência e a Tecnologia for her Post-Doctoral grant (SFRH/BPD/71200/2010). This work is partially financed by the Spanish Government under project CTQ2008-05370 and CEN2008-01027. Additional support from the Catalan Government is also acknowledged (SGR- 2009SGR-666).

[1] F.J. Blas, L.F Vega, Mol. Phys. **92** (1997) 135-150

[2] J.D. van der Waals, Z. Phys. Chem. **13** (1894) 657-725

[3] O.Vilaseca and L.F. Vega, Fluid Phase Equilib. **306** (2011) 4-14

(783f) Structure and Dynamics of an Ionic Liquid Confined Inside Model CMK-3 and CMK5 Ordered Mesoporous Carbons

Joshua D. Monk and **Francisco R. Hung**, Cain Department of Chemical Engineering, Louisiana State University, Baton Rouge, LA

Ionic liquids (ILs) have been considered as alternative electrolytes for energy storage in electrochemical double-layer capacitors, where the electrolyte is in contact with the nanoporous electrodes. Relevant macroscopic properties of these systems (e.g., electrical capacitance and resistance) are determined by the molecular-level properties of the confined ILs. However, many aspects of how the properties of nanoconfined ILs differ from their bulk counterparts remain to be fundamentally understood.

We report classical molecular dynamics simulation results of a typical IL, [dmim$^+$][Cl$^-$], confined inside model CMK-3 and CMK5 ordered mesoporous carbons. These materials consist of hexagonally-arranged nanorods (CMK-3) or nanopipes (CMK-5), where the walls are made of amorphous carbon. Our results suggest that variables such as pore size and the amount of IL inside the porous materials (pore loading) affect the structural properties of the confined IL (local density, ion orientation and radial distribution functions). We also discuss how pore size and pore loading affect the dynamical properties of the confined IL (mean squared displacements, diffusivities). Our results indicate that the dynamics of the confined ions are heterogeneous and depend strongly on the distance of the ions with the pore walls. The ions in the center regions of the pore have dynamics and relaxation times that are similar to those observed in the bulk IL at the same temperature. In contrast, the dynamics slow down appreciably as the ions get closer to the pore walls. Our results also suggest that these dynamical heterogeneities are linked to variations in local structural properties. The structure of the ions in the center of the pore is similar to that observed in the bulk IL; however, significant structural differences are observed between the ions near the pore walls and those in the center of the pore.

(783g) An Approach to Low Friction by Combination of Mesoporous Titanium Dioxide Film and Ionic Liquid

Xing Feng[1], Xiaohua Lu[1], Rong An[1], Longyun Qiu[1] and Yijun Shi[2], (1)chemical engineering, Nanjing university of technology, Nanjing, China, (2)science, Nanjing forestry university, Nanjing, China

Titanium (Ti) coated with native dense titanium dioxide (TiO_2) layers, as a robust substrate, provides advantages relative to silicon and the other metals in microelectromechanical systems (MEMS)/nanoelectromechanical systems (NEMS) due to its superior fracture toughness, better electrical conductivity, and excellent corrosion resistance. However, poor friction on TiO_2 surface affects MEMS/NEMS performance and in some cases, can even prevent devices from working, limits the durability and reliability of Ti on MEMS/NEMS.

Ionic liquid (IL), is called green solvents for their non-volatile, non-flammable and high thermal and chemical stabilities, having attracted increasing attentions on MEMS/NEMS. Its characteristics are also just what high-performance lubricants demand in MEMS/NEMS. However, the IL films form on solid surfaces by the way of weak physisorption rather than strong chemical bonding. Some approaches have been tried with chemical functional groups to fix IL film as a fluidic nature for reducing friction. However, the functional groups are susceptible to oxidation, hydrolysis or thermal degradation. Consequently, there exists an ongoing need for geometrical roughness-induced nano-patterned structures to hold IL films as lubricants to reduce friction. Atomic force microscopy (AFM) with a nanoscale sensitivity, is utilized to explore the molecular-level factors that govern in nanofrictional fields.

In this work, mesoporous TiO_2 film was fabricated on Ti based on titanate to provide nano-patterned pores due to the existence of small amount of TiO_2(B). 1,2-propanediol/choline chloride deep eutectic solvent, as an IL, is held in the nano-patterned pores to from stable IL films. The nano-patterned pores induced by TiO_2(B) could provide a fishing net-like support to contain IL films and ensure the maintenance of the IL within the pores by capillary force. In comparison, dense TiO_2 film was also prepared to support IL film, but it is not IL-philic enough, and IL can not from a stable film on the TiO_2 surface. Obviously, only IL weak physisorption on support could not form a stable lubricating film. By AFM, the friction coefficients of mesoporous TiO_2 film and IL filled in mesoporous TiO_2 film are measured respectively as 0.0026 and 0.0011. IL addition is in a large part responsible for the friction coefficient of mesoporous TiO_2 film decreasing to more than 1/2. With the support of nano-patterned mesoporous TiO_2 film, IL could form a stable film to act as a molecular-level lubricant for reducing the friction. It will have a strong impact on a broad range of applications of Ti on including but not limited to MEMS/NEMS. It is also valuable in fundamental understanding of interfacial phenomena in nanostructures to provide a bridge between science and engineering.

AUTHOR INDEX

AUTHOR INDEX

AUTHOR INDEX

AUTHOR INDEX

AUTHOR INDEX

AUTHOR INDEX

AUTHOR INDEX

AUTHOR INDEX

AUTHOR INDEX

AUTHOR INDEX

AUTHOR INDEX

AUTHOR INDEX

AUTHOR INDEX

AUTHOR INDEX

AUTHOR INDEX

AUTHOR INDEX

AUTHOR INDEX

AUTHOR INDEX

AUTHOR INDEX